THE INTERNATIONAL SERIES OF MONOGRAPHS ON PHYSICS

GENERAL EDITORS

W. MARSHALL   D. H. WILKINSON

ELECTRON PARAMAGNETIC RESONANCE OF TRANSITION IONS

BY

A. ABRAGAM

Directeur de la Physique au Commissariat à l'Energie Atomique
Professeur au Collège de France

AND

B. BLEANEY

Dr. Lee's Professor of Experimental Philosophy
Clarendon Laboratory, Oxford

CLARENDON PRESS · OXFORD
1970

Oxford University Press, Ely House, London W. 1

GLASGOW NEW YORK TORONTO MELBOURNE WELLINGTON
CAPE TOWN SALISBURY IBADAN NAIROBI DAR ES SALAAM
LUSAKA ADDIS ABABA BOMBAY CALCUTTA MADRAS KARACHI
LAHORE DACCA KUALA LUMPUR SINGAPORE HONG KONG TOKYO

© OXFORD UNIVERSITY PRESS 1970

Printed in Northern Ireland at The Universities Press, Belfast

TO
J. H. VAN VLECK

PREFACE

ELECTRON paramagnetic resonance was discovered in 1944 by E. Zavoisky in the U.S.S.R. and has developed into a major scientific technique. This book is designed to provide a comprehensive introduction to the subject, necessarily omitting a number of topics more suitable for specialist monographs. The presentation is on three levels. An introductory survey outlines the main features of the subject for the beginner in the field. This is followed by a general survey, covering the resonance phenomenon, the spin Hamiltonian, Endor, spin–spin and spin–lattice interactions, together with an outline of the known behaviour of the transition groups, at a level intended for the experimentalist. Finally a theoretical survey makes considerable use of group theory and symmetry properties in discussing the fundamentals of paramagnetism with special reference to the resonance spectrum.

A feature of the book is the inclusion of much hitherto unpublished material. The emphasis is on basic principles, with numerous references to publications containing further experimental results and more detailed developments of the theory.

The authors have pleasure in expressing their gratitude to numerous colleagues including Drs. G. M. Copland, E. R. Davies, E. A. Harris, J. Owen, and F. I. B. Williams, to name only a few, who read and suggested improvements to parts of the manuscript. We are particularly indebted to Dr. F. S. Ham who generously made available in advance of publication his article 'Jahn-Teller Effects in Electron Paramagnetic Resonance Spectra' (*Electron Paramagnetic Resonance*, edited by S. Geschwind, Plenum Press, New York) and to Dr. E. Beloritzky who read and made valuable comments on Chapters 11 to 20.

The authors are equally indebted to Madame Porneuf for her invaluable assistance both in proof-reading and in preparing much of the typescript.

A. A.
B. B.

ACKNOWLEDGEMENTS

The authors are indebted to the following for permission to use published diagrams as a basis for figures in the text—J. H. Anderson D. M. S. Bagguley J. M. Baker N. Edelstein R. Englman G. Feher S. Geschwind F. S. Ham I. H. Harrop W. Hayes K. H. Hellwege Chao-Yuan Huang C. D. Jeffries K. C. Krikorian D. Mergerian M. C. M. O'Brien R. Orbach J. Owen M. H. L. Pryce G. Seidel M. D. Sturge Satoru Sugano D. H. Templeton P. E. Wagner J. E. Wertz F. I. B. Williams W. P. Wolf American Institute of Physics American Physical Society Institute of Physics and Physical Society (London) Macmillan (Journals) Ltd. Physical Society of Japan Royal Society of London Springer-Verlag Taylor and Francis Ltd.

The following have kindly allowed us to include unpublished data—Hin Lew, P. Unsworth.

CONTENTS

PART I

PRELIMINARY SURVEY

                                                                   page

1. **INTRODUCTION TO ELECTRON PARAMAGNETIC RESONANCE**
   - 1.1. Electronic and nuclear magnetic dipole moments   1
   - 1.2. Hyperfine structure in a free atom or ion   5
   - 1.3. Magnetic resonance   9
   - 1.4. Effective spin and anisotropy   10
   - 1.5. 'Initial splittings' or 'fine structure'   16
   - 1.6. Magnetic hyperfine structure   25
   - 1.7. Hyperfine structure including nuclear electric quadrupole interaction   31
   - 1.8. A simple example   33
   - 1.9. Transition group ions and ligand fields   39
   - 1.10. Spin-spin interaction   52
   - 1.11. Spin-lattice interaction   60
   - 1.12. Dynamic nuclear orientation   74
   - 1.13. Endor   87
   - 1.14. Experimental aspects   92

PART II

GENERAL SURVEY

2. **THE RESONANCE PHENOMENON**
   - 2.1. Use of rotating coordinates   95
   - 2.2. Magnetic resonance   96
   - 2.3. Quantum-mechanical analysis   98
   - 2.4. Magnetic resonance in aggregated systems   102
   - 2.5. Adiabatic rapid passage   104
   - 2.6. Relaxation effects   108
   - 2.7. Radio-frequency pulses and spin-echoes   113
   - 2.8. Solution of the macroscopic equations for slow passage   115
   - 2.9. Intensity and line width   119
   - 2.10. Spectrometer sensitivity   125

3. **THE SPIN HAMILTONIAN AND THE SPECTRUM**
   - 3.1. The spin Hamiltonian   133
   - 3.2. The effect of anisotropy in the g-factor   135
   - 3.3. Multipole fine structure   139
   - 3.4. Fine structure in cubic fields ($S = \tfrac{5}{2}, \tfrac{7}{2}$)   142
   - 3.5. Electronic 'quadrupole' fine structure ($S = 1, \tfrac{3}{2}$)   151
   - 3.6. Electronic 'quadrupole' fine structure in a strong magnetic field   156
   - 3.7. Hyperfine structure I—introductory remarks   163

3.8.	Hyperfine structure II—strong external field	167
3.9.	Hyperfine structure III—nuclear electric quadrupole interaction	178
3.10.	'Forbidden' hyperfine transitions	186
3.11.	Ligand hyperfine structure	192
3.12.	The spectrum of a powder	200
3.13.	Effects of crystal imperfections	205
3.14.	Weak-field Zeeman interaction for non-Kramers ions	209

4. ELECTRON-NUCLEAR DOUBLE RESONANCE (ENDOR)

4.1.	Introduction	217
4.2.	The Endor spectrum	223
4.3.	Enhancement of the nuclear transition probability	228
4.4.	Endor on donors in silicon	234
4.5.	Endor on donors in silicon—relaxation effects	239
4.6.	Relaxation effects in Endor—general	243
4.7.	The hyperfine structure of europium	251
4.8.	The Endor spectrum of Nd^{3+} in $LaCl_3$	255
4.9.	Endor measurements of ligand hyperfine structure	259
4.10.	Endor line widths	264
4.11.	'Indirect' observation of Endor transitions	272
4.12.	Summary	274

5. THE LANTHANIDE (4f) GROUP

5.1.	Lanthanide compounds	277
5.2.	The free ions	282
5.3.	Crystalline field theory—C_{3h} symmetry	285
5.4.	Magnetic hyperfine structure	296
5.5.	Nuclear electric quadrupole interaction	301
5.6.	Experimental results for ethylsulphates and anhydrous chlorides	303
5.7.	Experimental results for the double nitrates, $Ln_2Mg_3(NO_3)_{12}, 24H_2O$	320
5.8.	Lanthanide ions in cubic symmetry	325
5.9.	Ions with a half-filled 4f-shell, $4f^7$, $^8S_{\frac{7}{2}}$. Eu^{2+}, Gd^{3+}, Tb^{4+}	335
5.10.	Higher-order terms in the spin Hamiltonian	341

6. THE ACTINIDE (5f) GROUP

6.1.	Ions and compounds of the actinide group	346
6.2.	Tripositive actinide ions	348
6.3.	Actinide ions in CaF_2	350
6.4.	Actinide ions in octahedral symmetry	354
6.5.	Neptunyl and plutonyl ions	359

7. IONS OF THE 3d GROUP IN INTERMEDIATE LIGAND FIELDS

7.1.	Introduction	365
7.2.	The intermediate crystal field approach	372
7.3.	The strong crystal field approach	377
7.4.	The effects of bonding	392
7.5.	The electronic spin Hamiltonian	398

7.6.	Magnetic hyperfine interaction	406
7.7.	Nuclear electric quadrupole and nuclear Zeeman interaction	414
7.8.	$3d^1$. Ti^{3+} in an octahedral field. ^2D, $L=2$, $S=\frac{1}{2}$	417
7.9.	$3d^2$. V^{3+}, Cr^{4+} in an octahedral field. ^3F, $L=3$, $S=1$	426
7.10.	$3d^3$. V^{2+}, Cr^{3+}, Mn^{4+} in an octahedral field. ^4F, $L=3$, $S=\frac{3}{2}$	430
7.11.	$3d^4$. Cr^{2+} in an octahedral field. ^5D, $L=2$, $S=2$	434
7.12.	$3d^5$. Cr^+, Mn^{2+}, Fe^{3+} in an octahedral field. ^6S, $L=0$, $S=\frac{5}{2}$	436
7.13.	$3d^6$. Fe^{2+} in an octahedral field. ^5D, $L=2$, $S=2$	443
7.14.	$3d^7$. Fe^+, Co^{2+}, Ni^{3+} in an octahedral field. ^4F, $L=3$, $S=\frac{3}{2}$	446
7.15.	$3d^8$. Co^+, Ni^{2+}, Cu^{3+} in an octahedral field. ^3F, $L=3$, $S=1$	449
7.16.	$3d^9$. Ni^+, Cu^{2+} in an octahedral field. ^2D, $L=2$, $S=\frac{1}{2}$	455
7.17.	$3d$ ions in tetrahedral symmetry	467

8. IONS OF THE d-GROUPS IN STRONG LIGAND FIELDS

8.1.	The ions and their compounds	472
8.2.	The strong ligand field octahedral complex	476
8.3.	Hyperfine interaction	478
8.4.	d^1 in strong octahedral field; $(d\epsilon)^1$, $(t_2)^1$, $S=\frac{1}{2}$	478
8.5.	d^2 in strong octahedral field; $(d\epsilon)^2$, $(t_2)^2$, $S=1$	479
8.6.	d^3 in strong octahedral field; $(d\epsilon)^3$, $(t_2)^3$, $S=\frac{3}{2}$	479
8.7.	d^4 in strong octahedral field; $(d\epsilon)^4$, $(t_2)^4$, $S=1$	480
8.8.	d^5 in strong octahedral field; $(d\epsilon)^5$, $(t_2)^5$, $S=\frac{1}{2}$	481
8.9.	d^6 in strong octahedral field; $(d\epsilon)^6$, $(t_2)^6$, $S=0$	486
8.10.	d^7 in strong octahedral field; $(d\epsilon)^6(d\gamma)$, $(t_2)^6e$, $S=\frac{1}{2}$	486
8.11.	d^8 in strong octahedral field; $(d\epsilon)^6(d\gamma)^2$, $(t_2)^6e^2$, $S=1$	487
8.12.	d^9 in strong octahedral field; $(d\epsilon)^6(d\gamma)^3$, $(t_2)^6e^3$, $S=\frac{1}{2}$	487
8.13.	d^1 in cubic (eightfold) coordination	490

9. SPIN-SPIN INTERACTION

9.1.	Introduction	491
9.2.	Magnetic dipole-dipole interaction	492
9.3.	Exchange interaction	495
9.4.	Multipole interactions	499
9.5.	Interaction between a pair of similar ions	502
9.6.	Interaction between a pair of dissimilar ions	509
9.7.	Line broadening by spin-spin interaction	514
9.8.	Line shape due to dipolar spin-spin interaction	521
9.9.	Effect of exchange interaction on line shape	527
9.10.	Magnetic dilution, and the spectra of pairs	529
9.11.	Temperature-dependent effects	535

10. SPIN-PHONON INTERACTION

10.1.	The attainment of thermal equilibrium	541
10.2.	The phonon radiation bath	547
10.3.	Spin-lattice relaxation by phonons—Waller processes	551
10.4.	Spin-lattice relaxation by modulation of the ligand field	557
10.5.	Summary and comparison with experiment	565
10.6.	The phonon 'bottle-neck' and phonon 'avalanche'	574

PART III

THEORETICAL SURVEY

11. THE ELECTRONIC ZEEMAN INTERACTION
- 11.1. The interaction between electrons and a magnetic field — 585
- 11.2. The Zeeman effect in a free atom (or ion) — 586
- 11.3. LS-coupling and the Landé formula — 587
- 11.4. Self-consistent field configurations — 589
- 11.5. Spin-orbit coupling — 592
- 11.6. Matrix elements between Slater determinants — 593
- 11.7. Introduction of the crystal field — 595

12. GROUP THEORY—AN OUTLINE
- 12.1. Invariance and degeneracy — 601
- 12.2. Linear representations, equivalence, and irreducibility — 602
- 12.3. Orthogonality relations, characters, and classes — 603
- 12.4. Reduction of a representation and calculation of the characters — 605
- 12.5. Splitting of a degenerate level by a perturbation of lower symmetry — 607
- 12.6. The direct product of two representations — 611

13. GROUP THEORY—THE ROTATION GROUP
- 13.1. Angular momentum — 615
- 13.2. The irreducible representations — 617
- 13.3. The coupling of angular momenta — 620
- 13.4. Multiple vector coupling and Racah symbols — 622
- 13.5. Irreducible tensor operators, the Wigner–Eckart theorem, and equivalent operators — 624

14. THE CUBIC GROUP AND SOME OTHER GROUPS
- 14.1. The cubic group — 629
- 14.2. The fictitious angular momentum — 632
- 14.3. The multiplets Γ_4 and Γ_5 in trigonal axes — 633
- 14.4. The double cubic group — 634
- 14.5. Groups of lower symmetry — 638
- 14.6. Improper rotations — 640

15. TIME REVERSAL AND KRAMERS DEGENERACY
- 15.1. Operations involving the time — 643
- 15.2. Complex conjugation — 644
- 15.3. Determination of the time reversal operator — 646
- 15.4. Kramers degeneracy — 647
- 15.5. Time-reversal operator in the $|J, M\rangle$ representation — 649
- 15.6. The 'Spin Hamiltonian' for a Kramers doublet — 650
- 15.7. The rhombic group — 654
- 15.8. Threefold symmetry — 654
- 15.9. Selection rules related to time-reversal — 656
- 15.10. The effect of an applied electric field on a paramagnetic ion — 659

16. ELEMENTARY THEORY OF THE CRYSTAL FIELD
16.1. The crystal field (or crystal potential) 665
16.2. Equivalent operators 670
16.3. Off-diagonal matrix elements of the crystal field 676
16.4. The electronic Zeeman interaction 677
16.5. Electron spin-spin interactions 678

17. HYPERFINE STRUCTURE
17.1. Electrostatic hyperfine interactions 680
17.2. Magnetic hyperfine interactions 687
17.3. Alternative derivation of the magnetic hyperfine interaction 690
17.4. Equivalent operators for the magnetic hyperfine interaction 692
17.5. The effect of s-electrons: configuration interaction 695
17.6. The effect of s-electrons: core polarization 702
17.7. Finer effects in the theory of hyperfine structure 706

18. IONS IN A WEAK CRYSTAL FIELD (f ELECTRONS)
18.1. Kramers ions in a weak crystal field 713
18.2. Rare-earth ions in cubic symmetry 719
18.3. The quadruplet Γ_8 721
18.4. Representation of an irreducible tensor within the quadruplet Γ_8—quadrupole coupling 731
18.5. Non-Kramers ions in the rare-earth group 732
18.6. Non-Kramers rare-earth ions in cubic surroundings 739

19. INTERMEDIATE CRYSTAL FIELDS (THE IRON GROUP)
19.1. Effect of the cubic crystal potential 742
19.2. 'Singlet' orbital ground state (ions of type A) 745
19.3. Triplet orbital ground state (ions of type B) 751
19.4. Departures from cubic symmetry 755
19.5. The influence of excited terms 758

20. THE EFFECTS OF COVALENT BONDING
20.1. Summary of the foregoing theory 761
20.2. The molecular orbitals model for covalent bonding 762
20.3. Bonding and anti-bonding orbitals, overlap, and covalency 764
20.4. The ground states in weakly covalent compounds 767
20.5. Orbital momentum and spin-orbit coupling in the presence of covalent bonding 773
20.6. Ligand hyperfine structure for ions of type A 777
20.7. Orbital singlets: correction terms for the ligand hyperfine structure 781
20.8. Ligand hyperfine structure for ions of type B 784
20.9. Ligand quadrupole hyperfine structure 788

21. THE JAHN–TELLER EFFECT IN PARAMAGNETIC RESONANCE

21.1.	Introduction	790
21.2.	The Born–Oppenheimer approximation and the Jahn–Teller theorem	792
21.3.	The magnetic properties of a ^2E level	797
21.4.	The static Jahn–Teller effect in a ^2E state	804
21.5.	Dynamic features of the static Jahn–Teller effect	807
21.6.	The dynamic Jahn–Teller effect in a ^2E state	808
21.7.	Motional narrowing of the Jahn–Teller spectrum	826
21.8.	Comparison with experiment	830
21.9.	The Jahn–Teller effect in a triplet state	832
21.10.	The Jahn–Teller effect in an orbital triplet with Γ_3 coupling	835
21.11.	The Jahn–Teller effect in an orbital triplet with Γ_5 coupling	841
21.12.	Comparison with experiment	846

APPENDIX A. Thermal and magnetic properties of a paramagnetic substance   848

APPENDIX B. Tables 1 to 26   856

BIBLIOGRAPHY   879

AUTHOR INDEX   893

SUBJECT INDEX   899

PART I.

PRELIMINARY SURVEY

1

INTRODUCTION TO ELECTRON PARAMAGNETIC RESONANCE

1.1. Electronic and nuclear magnetic dipole moments

A PARAMAGNETIC substance may be defined as one that possesses no resultant magnetic moment in the absence of an external magnetic field but acquires a magnetic moment in the direction of an applied field whose size is a function of the field. Such a resultant moment of the substance may be partly due to induced dipoles, which appear only through the action of the applied field and result from a change in the motion of the constituent electrons of each atom or ion. In all cases this gives rise to a negative induced moment (i.e. a moment anti-parallel to the applied field), and hence to the diamagnetism that is a basic property of all substances. Superimposed on this in certain cases there may be a positive induced moment, resulting in paramagnetism (this phenomenon is often referred to as 'Van Vleck paramagnetism'). Both of these effects are independent of temperature except in temperature regions where excited states are becoming appreciably populated.

A more important class of paramagnetic substances, with which we shall be mainly concerned, consists of those where the constituent atoms (or ions) have permanent magnetic moments of atomic or nuclear magnitude. In the absence of an external field such dipoles are randomly oriented, but the application of a field results in a redistribution over the various orientations in such a way that the substance acquires a net magnetic moment. Such permanent magnetic dipoles occur only when the atom or nucleus possesses a resultant angular momentum, and the two are related by the formula

$$\boldsymbol{\mu} = \gamma \mathbf{G}, \tag{1.1}$$

where $\boldsymbol{\mu}$ is the magnetic dipole moment, $\mathbf{G}$ the angular momentum (an integral or half-integral multiple of $h/2\pi = \hbar$, where h is Planck's constant), and γ is the magnetogyric ratio, which is of order (e/mc) for electrons and (e/Mc) for nuclei. When such a dipole is subjected to a magnetic field $\mathbf{H}$, it experiences a couple $\boldsymbol{\mu} \wedge \mathbf{H}$, and the equation of motion is

$$d\mathbf{G}/dt = \boldsymbol{\mu} \wedge \mathbf{H}, \tag{1.2}$$

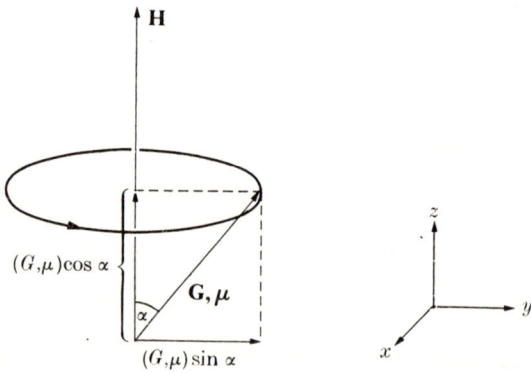

Fig. 1.1. Precession of a magnetic dipole $\boldsymbol{\mu}$ with angular momentum $\mathbf{G}$ about an applied magnetic field $\mathbf{H}$. The angular velocity $\boldsymbol{\omega}_L = -\gamma\mathbf{H}$ is positive if γ is negative, as for an electron; the direction of precession is then that of a right-handed screw advancing along $\mathbf{H}$, as shown.

which may be combined with eqn (1.1) to give

$$d\mathbf{G}/dt = \gamma \mathbf{G} \wedge \mathbf{H} \tag{1.3}$$

or

$$d\boldsymbol{\mu}/dt = \gamma \boldsymbol{\mu} \wedge \mathbf{H}. \tag{1.4}$$

If the magnetic field $\mathbf{H}$ is assumed to lie along the z-axis of a system of Cartesian coordinates, the solution of this equation can be written in the form:

$$\begin{aligned} G_x &= G \sin \alpha \cos(\omega_L t + \epsilon), \\ G_y &= G \sin \alpha \sin(\omega_L t + \epsilon), \\ G_z &= G \cos \alpha, \end{aligned} \tag{1.5}$$

with analogous equations for $\boldsymbol{\mu}$. The motion of the vectors $\mathbf{G}$, $\boldsymbol{\mu}$ consists of a uniform precession about $\mathbf{H}$ with angular velocity (cf. Fig. 1.1)

$$\boldsymbol{\omega}_L = -\gamma\mathbf{H}, \tag{1.6}$$

where the negative sign means that the precession is in the direction of a left-handed screw advancing along **H** if γ is positive, and vice versa. The component of **G**, **μ** along **H** remains fixed in magnitude, so that the energy of the dipole in the field **H** (the 'Zeeman energy')

$$W = -\boldsymbol{\mu} \cdot \mathbf{H} \tag{1.7}$$

is a constant of the motion.

When a free atom or ion has a resultant angular momentum in its electron system, it will possess a permanent magnetic dipole moment. The magnetogyric ratio is then

$$\gamma = -g(e/2mc), \tag{1.8}$$

where the negative sign occurs because of the negative charge on the electron; e, m are the charge and mass of the electron (both taken as positive numbers). The quantity g is a pure number of order unity, whose value depends on the relative contributions of orbit and spin to the total angular momentum. If only orbital momentum is present we can write

$$\mathbf{G} = \hbar \mathbf{L}, \qquad g = g_L, \tag{1.9}$$

where **L** is the quantum number of the total orbital momentum, and the value of g_L is unity apart from some small corrections due to diamagnetic and relativistic effects; these are usually less than 10^{-4} and we shall neglect them as lying outside the precision of most solid state experiments, which form our main concern. If only electron spin momentum is present we write similarly

$$\mathbf{G} = \hbar \mathbf{S}, \qquad g = g_S, \tag{1.10}$$

where **S** is the quantum number of the total spin momentum. Apart from diamagnetic and relativistic corrections which are of the same order as in the orbital case, we have the important quantum electrodynamical correction

$$g_S = 2(1 + \alpha/2\pi - \ldots) - 2 \cdot 0023, \tag{1.11}$$

where α is the fine-structure constant. For many (but not all) solid-state experiments it is sufficient to take $g_S = 2$.

When both orbital and spin momentum are present, the value of g depends on the nature of the coupling between them. In LS-coupling, the resultant angular momentum is associated with a quantum number **J**, where $\mathbf{J} = \mathbf{L} + \mathbf{S}$, and the appropriate value of g is

$$g_J = \frac{J(J+1)(g_L + g_S) + \{L(L+1) - S(S+1)\}(g_L - g_S)}{2J(J+1)}, \tag{1.12}$$

which reduces to the usual Landé formula

$$g_J = \frac{3}{2} - \frac{L(L+1)-S(S+1)}{2J(J+1)} \qquad (1.13)$$

if we set $g_L = 1$, $g_S = 2$ exactly. The resultant electronic magnetic dipole moment is

$$\boldsymbol{\mu}_J = -g_J \beta \mathbf{J}, \qquad (1.14)$$

where $\beta = (e\hbar/2mc)$ is the Bohr magneton (again a positive number). Equation (1.14) is valid so long as we can neglect any interactions that would admix states of different J; this means that the energy associated with such interactions (such as the Zeeman interaction with an external field) must be small compared with the energy differences between levels of different J. These energy differences are predominantly due to spin-orbit coupling, which in its simplest form can be represented by a term

$$\mathcal{H}_{\mathrm{SO}} = \lambda(\mathbf{L}\cdot\mathbf{S}). \qquad (1.15)$$

The energy of a level J is then given by the Landé formula

$$W_J = \tfrac{1}{2}\lambda\{J(J+1)-L(L+1)-S(S+1)\} \qquad (1.16)$$

so that the separation between successive levels is

$$W_J - W_{J-1} = \lambda J, \qquad (1.17)$$

a result known as the Landé interval rule. For transition group ions λ is of order 10^2 to 10^3 cm^{-1}, so that at temperatures well below room temperature (for which kT is 200 cm^{-1}) only the lowest level is occupied.

For a nucleus with spin quantum number I, the value of γ is

$$\gamma_\mathrm{n} = g_\mathrm{n}(e/2Mc), \qquad (1.18)$$

where M is now the mass of the proton and g_n is a number that is positive for many (but not all) isotopes. The nuclear magnetic moment is then

$$\boldsymbol{\mu}_I = g_\mathrm{n}\beta_\mathrm{n}\mathbf{I}, \qquad (1.19)$$

where $\beta_\mathrm{n} = (e\hbar/2Mc)$ is the nuclear magneton. It is often convenient to express the nuclear magnetic moment in terms of the Bohr magneton

$$\boldsymbol{\mu}_I = g_I \beta \mathbf{I}, \qquad (1.20)$$

where $g_I = g_\mathrm{n}(m/M) = g_\mathrm{n}/1836$ and is therefore of order 10^{-3}. (The nomenclature g_n, g_I is used rather indiscriminately in the literature, but the distinction we have made here is a convenient one and will be retained in later chapters.) As different sign conventions are used by different authors, we note here that we take both g_n and g_I to have the same sign as $(\boldsymbol{\mu}_I/\mathbf{I})$.

1.2. Hyperfine structure in a free atom or ion

In a free atom or ion which has a resultant electronic angular momentum **J**, and a nuclear spin **I**, the two are generally coupled together through the magnetic interaction between the electronic and

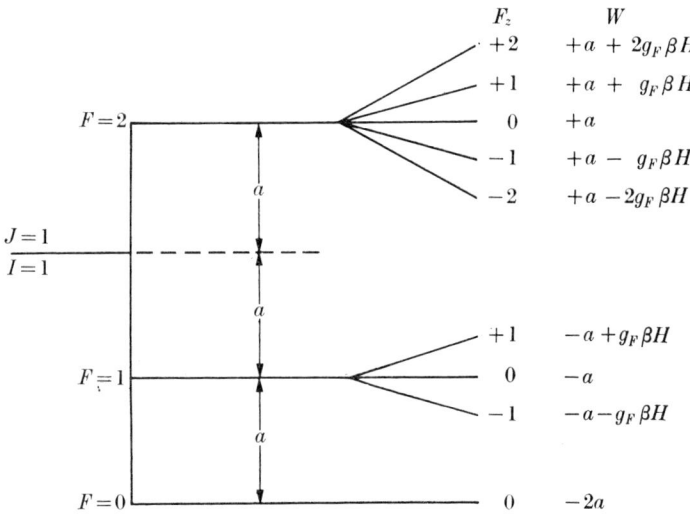

FIG. 1.2. Magnetic hyperfine multiplet for the system $J = 1$, $I = 1$ corresponding to the Hamilton (1.21), with weak field Zeeman splitting. Note that when $J = I$ the value of $g_F = \frac{1}{2}g_J$ for all values of F, see eqn (1.26); or more accurately, including the nuclear contribution, $g_F = \frac{1}{2}(g_J - g_I)$ for all values of F when $J = I$, from eqn (1.25).

nuclear dipole moments. Within a manifold of given J this usually takes the form
$$\mathscr{H} = a(\mathbf{J} \cdot \mathbf{I}), \tag{1.21}$$
which couples **J**, **I** together to form a set of levels with quantum numbers $F = (J+I), (J+I-1), \ldots, |J-I|$. The energy of a level with quantum number F is
$$W_F = \tfrac{1}{2}a\{F(F+1) - J(J+1) - I(I+1)\} \tag{1.22}$$
so that the separation between successive levels is
$$W_F - W_{F-1} = aF. \tag{1.23}$$
The splitting of the hyperfine multiplet (Fig. 1.2) given by eqn (1.23)

is similar to the Landé interval rule, eqn (1.17), and follows from the fact that the magnetic hyperfine interaction (1.21) is similar in form to the spin-orbit coupling (1.15). If both J and I are equal to 1 or more the electronic system and the nucleus may possess electric quadrupole moments, and interaction between these gives a coupling different in form from (1.21) so that the splitting of the hyperfine multiplet no longer obeys the interval rule.

In a weak external magnetic field, sufficiently small that the Zeeman energy is small compared with the hyperfine energy, each manifold of $(2F+1)$ states belonging to a given value of F is split by an amount proportional to H (see Fig. 1.2). The manifold behaves as though it had a magnetic dipole moment

$$\boldsymbol{\mu}_F = -g_F \beta \mathbf{F}, \tag{1.24}$$

where

$$g_F = \frac{F(F+1)(g_J-g_I)+\{J(J+1)-I(I+1)\}(g_J+g_I)}{2F(F+1)}, \tag{1.25}$$

a result that can be obtained directly from (1.12) by replacing J, L, S by F, J, I respectively, and g_L by g_J, g_S by $-g_I$. Since g_I is about $10^{-3} g_J$ it is often possible to use the approximation

$$g_F = g_J \left\{ \frac{F(F+1)+J(J+1)-I(I+1)}{2F(F+1)} \right\}, \tag{1.26}$$

which corresponds just to the projection of the electronic moment $\boldsymbol{\mu}_J$ onto $\mathbf{F}$, and neglects the corresponding projection of the nuclear moment $\boldsymbol{\mu}_I$.

The hyperfine interaction energy is seldom larger than about 10^{-1} cm^{-1}, and is often much smaller. Thus until we reach temperatures below 1°K (for 1°K, $kT = 0.7$ cm^{-1}) all levels in the hyperfine multiplet are nearly equally populated. More important for our purposes is the fact that the hyperfine energy is generally smaller than the Zeeman energy

$$W_Z = -\boldsymbol{\mu}_J \cdot \mathbf{H} - \boldsymbol{\mu}_I \cdot \mathbf{H} \tag{1.27}$$

(where for completeness we have included the nuclear Zeeman energy, though the important term is the electronic Zeeman energy), if H is of order 10^4 G. The scheme in which $\mathbf{J}$, $\mathbf{I}$ couple to form a resultant $\mathbf{F}$ is then not applicable, since the electronic Zeeman energy (~ 1 cm^{-1}) is larger than the hyperfine energy. In the strong field limit (the Back–Goudsmit region) the electronic moment precesses about the external field independently of the nuclear moment. The magnetic hyperfine interaction (1.21) then results in a magnetic field at the nucleus, set up by the precessing electronic magnetic moment, whose steady component $\mathbf{H_e}$ is parallel to the external field $\mathbf{H}$. In this approximation the nuclear magnetic moment precesses about the combined field $(\mathbf{H}+\mathbf{H_e})$. Conversely this sets up a nuclear magnetic field whose steady component $\mathbf{H_n}$ is again parallel to $\mathbf{H}$, and the electronic moment thus precesses about a combined field $(\mathbf{H}+\mathbf{H_n})$.

This approximation means that, if $\mathbf{H}$ is along the z-axis, we are retaining from the magnetic hyperfine energy (1.21) only the term

$$aJ_zI_z \equiv g_J\beta J_z H_n \equiv -g_I\beta I_z H_e. \tag{1.28}$$

Similarly we retain only the z-components of the Zeeman interactions, so that the total energy is just

$$W = g_J\beta J_z H + aJ_zI_z - g_I\beta I_z H. \tag{1.29}$$

For the electrons we have

$$W_e = g_J\beta J_z(H+H_n) \tag{1.30}$$

where, since H_n is proportional to I_z, it has $(2I+1)$ values, equally spaced. Similarly for the nucleus we have

$$W_n = -g_I\beta I_z(H+H_e) \tag{1.31}$$

where, since H_e is proportional to J_z, it has $(2J+1)$ values, equally spaced. A typical set of energy levels is shown in Fig. 1.3.

We have spelt out the position in the Back–Goudsmit region rather carefully, because it is the region we shall be most concerned with in electron paramagnetic resonance, and to emphasize the orders of magnitude involved. If we take as typical,

electronic Zeeman interaction ~ 1 cm^{-1},
magnetic hyperfine interaction $\sim 10^{-1}$ cm^{-1},
nuclear Zeeman interaction $\sim 10^{-3}$ cm^{-1},

then the magnetic fields involved are

$$H \sim 10^4 \text{ G}, \; H_n \sim 10^3 \text{ G}, \; H_e \sim 10^6 \text{ G},$$

showing that the precession frequency of the electronic moment is determined primarily by the external field H, with the nuclear field H_n adding a subsidiary amount, while the precession frequency of the nuclear dipole moment is determined mainly by the electronic field H_e, to which the external field H adds a small contribution.

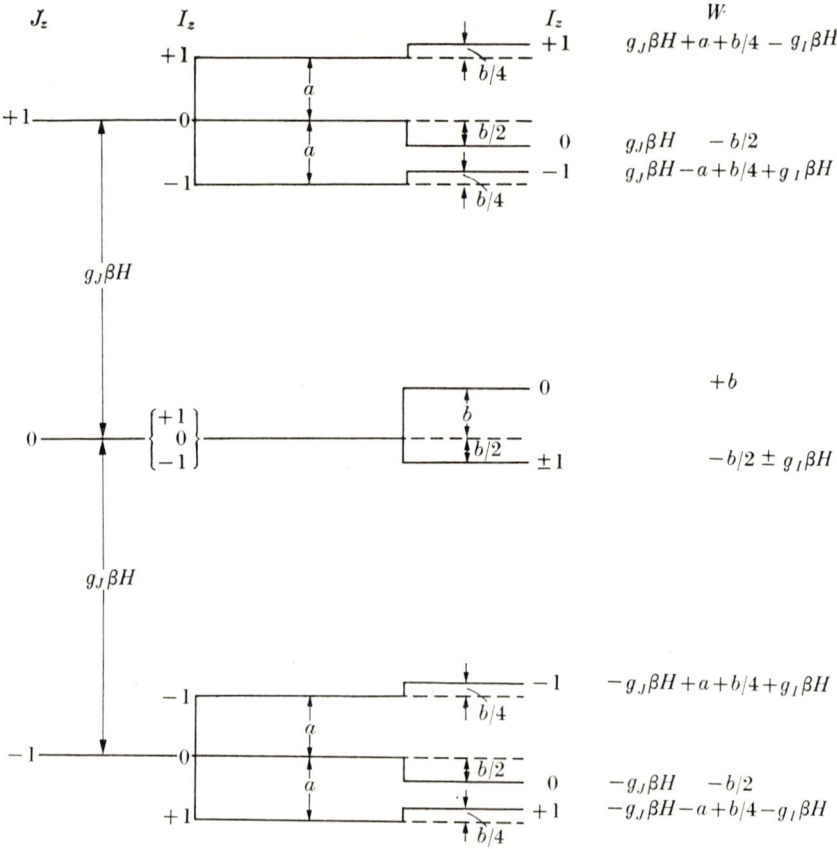

Fig. 1.3. Hyperfine energy levels in a strong external magnetic field H along the z-axis, for a system with $J = 1$, $I = 1$. The diagram corresponds to the strong field limit, eqns (1.29) and (1.32), except that displacements due to the nuclear Zeeman term $-g_I \beta I_z H$ are omitted. Note that the order of the nuclear levels due to the magnetic hyperfine interaction (proportional to J_z and hence to $-H_e$) is reversed top and bottom, unlike the quadrupole contributions.

For completeness we include here the contribution to the nuclear electric quadrupole interaction in the strong external magnetic field limit. This adds an energy

$$W = \frac{b}{4J(2J-1)I(2I-1)} \{3J_z^2 - J(J+1)\}\{3I_z^2 - I(I+1)\}, \quad (1.32)$$

where the coefficient $(b/4)$ is the value of W in the extreme case $J_z = J$, $I_z = I$. As before, the quadrupole term occurs only if both J, I are 1 or more. The effect of the quadrupole energy is illustrated in Fig. 1.3.

1.3. Magnetic resonance

We have seen that a magnetic dipole $\boldsymbol{\mu} = \gamma \mathbf{G}$ (see eqn (1.1)), when placed in a field $\mathbf{H}$, precesses about the field with angular velocity $\boldsymbol{\omega}_L = -\gamma \mathbf{H}$ (eqn (1.6)). This precession produces an oscillatory magnetic moment in any direction normal to the field $\mathbf{H}$, which can interact with an oscillatory magnetic field $H_1 \cos \omega t$, which is also normal to $\mathbf{H}$. The interaction has a marked effect on the motion of the dipole only when ω is close to the natural precession frequency ω_L, so that we are concerned with a resonance phenomenon. When the resonance condition $\omega = \omega_L$ is fulfilled, the component $\mu \cos \alpha$ of the dipole along the steady field $\mathbf{H}$ can be altered materially even by oscillatory fields whose amplitude $H_1 \ll H$. This effect is known as 'magnetic resonance'.

From eqn (1.7), a change in the component $\mu \cos \alpha$ means a change in the energy of the dipole, which is $W = -\mu H \cos \alpha$, which, from (1.1), we can write as

$$W = -\gamma G H \cos \alpha. \tag{1.33}$$

The stationary values of the component $G \cos \alpha$ of the angular momentum are restricted in quantum mechanics to $\hbar M$, $\hbar(M-1)$, etc., where the magnetic quantum number M has a series of integral or half-integral values differing by unity between successive values. The allowed transitions for this simple system, when a suitably oriented oscillatory magnetic field is applied, are given by the selection rule $\Delta M = \pm 1$, and hence require a quantum of energy

$$\hbar \omega = W_M - W_{M-1} = -\gamma \hbar H$$

or

$$\omega = -\gamma H = \omega_L. \tag{1.34}$$

The minus sign is significant only if circularly polarized radiation is used, in which case the applied magnetic field must rotate in the same sense as the dipole precesses about H in order to achieve resonance. In practice linearly polarized radiation is almost always used (unless an experimental determination of the sign of γ is required); such radiation can be decomposed into two circularly polarized components, one of which will then be in the correct sense for resonance.

The simple quantum approach gives the same resonance condition as we expected from classical considerations. We shall show in Chapter 2 that the equation of motion (1.2), (1.3), or (1.4) is quantum mechanically correct, and discuss both the classical and quantum approaches in detail. The numerical value of the resonance frequency is easily found. For an electronic dipole $\gamma = -g(e/2m)$, with e in e.m.u., so that, using the value $e/m = 1\cdot 758796(17) \times 10^7$ e.m.u./g,

$$\begin{aligned}\nu = \nu_L &= g(e/2m)H/(2\pi) \\ &= 1\cdot 3996 \times 10^6 \, (gH) \quad (\text{Hz, G}). \end{aligned} \quad (1.35)$$

The value of g is thus easily determined from a measurement of field and frequency, since

$$g = 0\cdot 71557 \times 10^{-6} \, (\nu/H) \quad (\text{Hz, G}). \quad (1.36)$$

Most measurements are made at centimetre wavelengths, so that in terms of wave-number $\bar{\nu}$ (in cm^{-1}) or wavelength λ (in cm) (1.36) becomes

$$\begin{aligned} g &= 21\cdot 4198 \, (\bar{\nu}/H) \quad (\text{cm}^{-1}, \text{kG}) \\ &= 21\cdot 4198 \, (\lambda H)^{-1} \quad (\text{cm, kG}). \end{aligned} \quad (1.37)$$

This shows that if $g = 2$, a field of $10\cdot 7$ kG is required for resonance at a wavelength of 1 cm. In practice the most commonly used wavelengths are about 3 cm ($\nu \approx 10$ GHz), 0·8 cm ($\nu \approx 40$ GHz), and harmonics of the latter.

Since nuclear magnetic moments are about $10^{-3}\beta$, their resonance frequencies are correspondingly smaller in the same field. For the proton the resonance frequency is

$$\nu_\text{P} = 4\cdot 2577 \times 10^3 H \quad (\text{Hz, G}), \quad (1.38)$$

so that the resonance frequency is 42·577 MHz in a field of 10 kG. In an electron resonance experiment, the proton magnetic resonance frequency is often determined in order to find the value of the magnetic field H at the sample, using (1.38). The value of g for the electronic dipoles can be found directly from the ratio of the electron and proton resonance frequencies, using the relation

$$g = 3\cdot 04208 \times 10^{-3} \, (\nu/\nu_\text{P}). \quad (1.39)$$

If the electron sample and proton sample are not located at the same point, it may be necessary to apply a small correction for the difference in magnetic field at the two samples.

1.4. Effective spin and anisotropy

For a free atom the electronic magnetic dipole moment can generally be written in the form $\boldsymbol{\mu}_J = -g_J\beta \mathbf{J}$ (eqn (1.14)) and the Hamiltonian for its Zeeman interaction with a field **H** is

$$\mathscr{H} = -(\mathbf{\mu}_J \cdot \mathbf{H}) = g_J \beta (\mathbf{H} \cdot \mathbf{J}). \tag{1.40}$$

For a stationary state whose magnetic quantum number is M, corresponding to a component $\hbar M$ of angular momentum in the direction of H, the energy is

$$W_M = g_J \beta H M \tag{1.41}$$

and allowed transitions of the type $\Delta M = \pm 1$ require an energy quantum

$$h\nu = g_J \beta H. \tag{1.42}$$

This simple form of the Zeeman interaction is valid when we are dealing with a group of $(2J+1)$ levels, degenerate when $H = 0$, and well separated from other levels, i.e. the Zeeman energy $g_J \beta H \ll \Delta$, where Δ is the energy difference from the next level or group of levels.

In the solid state we are concerned mostly with ions rather than with atoms. Many such ions have closed electron shells which have no resultant angular momentum and hence no permanent electronic magnetic dipole moment. Partly filled shells, with permanent dipole moments due to the orbital motion of the electrons, or to their intrinsic spin, or to both, occur in the 'transition groups'. These comprise the $3d$ ('iron' group), the $4d$ ('palladium' group), the $5d$ ('platinum' group), the $4f$ ('lanthanide' or 'rare earth' group), and the $5f$ ('actinide' group). Other paramagnetic ions may occur as defects or be created by irradiation. We shall mainly be concerned with localized states of transition group ions in non-conducting crystals. Furthermore, we are primarily concerned with 'paramagnetic' substances in which each permanent dipole is substantially unaffected by the presence of the other permanent dipoles. The theory then treats each dipole separately, and the interaction with neighboring permanent magnetic dipoles enters only in a subsidiary manner.

Nevertheless, in the solid state a paramagnetic ion is by no means 'free'. It is surrounded by a cage of diamagnetic ions, the nearest at distances of order 0·2 to 0·3 nm, the complex forming part of an extended lattice. These charged 'ligand' ions have a strong interaction with the paramagnetic ion, producing a strong electrostatic field (the 'ligand' field), through which the paramagnetic electrons must move. The energy associated with the interaction between the paramagnetic electrons and this ligand field varies roughly from 10^2 to 10^4 cm^{-1}; it may thus exceed the spin-orbit interaction, and in some cases it may exceed the electrostatic interaction with other electrons on the central ion responsible for LS-coupling. This ligand interaction is an additional complication in the already complex problem of a many-electron atom. In the 'crystal field' approach, the ligand ions are regarded as setting up an additional electrostatic potential (the 'crystal potential'), which reflects the symmetry of the complex and its immediate surroundings. The magnetic electrons, localized on the central ion and moving in this potential, experience a 'Stark splitting' of their orbital levels. The lowest levels are just those in which these negatively charged electrons are most successful in avoiding negatively charged ligand ions, thus reducing the energy due to their mutual electrostatic repulsion. In a more sophisticated approach the electrons are no longer regarded as localized on the central ion but are shared with the ligand ions and are spread out over the complex. They occupy molecular orbitals rather than atomic orbitals and may take part in both σ- and π-bonding, different orbitals having different energies. This approach allows for a certain amount of covalent bonding, in contrast to the purely ionic approach of the crystal-field method.

Whichever theoretical approach is adopted the result is a splitting of levels, leaving groups of rather small degeneracy. The degeneracy within each group (and, to some extent, the relative splittings between different groups) depends on the symmetry within the complex, and can thus be predicted by group theory. In the case of local cubic symmetry, groups of 1, 2, 3, or 4 degenerate levels are found, but in cases of lower symmetry levels may often be only single, or degenerate in pairs. An important over-riding theorem concerning the residual degeneracy is due to Kramers; in a system containing an odd number of electrons, at least twofold degeneracy must remain in the absence of a magnetic field. The pairs of states ('Kramers doublets') involved are time conjugate, one being obtained from the other by use of a time-reversal operator whose square is -1; they can be split by a magnetic perturbation (a time-odd operator), but not by an electrostatic perturbation, which is even under time reversal.

In electron paramagnetic resonance we are concerned with transitions between levels split at most by a few cm^{-1}. Hence we are immediately interested only in groups of levels that are degenerate (or nearly so) in zero magnetic field, and a convenient method is needed to represent the behaviour of such a group of levels when a magnetic field is applied to the system. A suitable method has been evolved which uses the concept of an 'effective spin' $\tilde{\mathbf{S}}$, which is a fictitious angular momentum such that the degeneracy of the group of levels involved is set equal to $(2\tilde{S}+1)$. For example, an isolated Kramers doublet with just two levels is assigned an effective spin $\tilde{S} = \frac{1}{2}$. This concept of an 'effective spin' is useful because it is possible to set up an 'effective spin Hamiltonian' that gives a correct description of the behaviour of the group of levels in terms as concise as those for a free atom or ion. In some cases a theoretical justification of the effective spin Hamiltonian can be given; if the local symmetry of the complex is known, the effective spin Hamiltonian is expected to reflect this symmetry, thus imposing restrictions on its form. If the observed spectrum does not conform to the spin Hamiltonian that has been assumed, other forms can be tried empirically until a fit is obtained.

The first term that we can expect to meet in the effective spin Hamiltonian is that representing the electronic Zeeman interaction. By analogy with eqn (1.40) we might expect this to be of the form

$$\mathscr{H} = g\beta(\mathbf{H} \cdot \tilde{\mathbf{S}}). \tag{1.43}$$

This has $(2\tilde{S}+1)$ states, which, again by analogy, we label by means of a magnetic quantum number M representing the value of $\tilde{S}_z$. Then, taking $\mathbf{H}$ to be along the z-axis, the energy is

$$W_M = g\beta H M \tag{1.44}$$

and the quantum of energy required for an allowed transition of the type $\Delta M = \pm 1$ is

$$h\nu = g\beta H, \tag{1.45}$$

giving a resonance condition similar to (1.42). For a doublet state which we label $\tilde{S} = \frac{1}{2}$, the energy levels are as shown in Fig. 1.4: two levels, diverging linearly with H, with slopes $\pm\frac{1}{2}g\beta H$.

The form of eqn (1.43) presupposes that the Zeeman interaction depends only on the angle between the effective spin vector $\tilde{\mathbf{S}}$ and the magnetic field. In practice this is commonly found not to be the case; the Zeeman interaction depends also on the angle that $\mathbf{H}$ makes with certain axes defined by the local symmetry of the magnetic complex. A more general form, which takes into account 'anisotropy of this kind, is

$$\mathscr{H} = \beta(\mathbf{H} \cdot \mathbf{g} \cdot \tilde{\mathbf{S}}), \tag{1.46}$$

which is a shorthand notation for

$$\mathscr{H} = \beta \begin{Bmatrix} g_{xx}H_x\tilde{S}_x + g_{yy}H_y\tilde{S}_y + g_{zz}H_z\tilde{S}_z + g_{xy}H_x\tilde{S}_y + g_{yx}H_y\tilde{S}_x + \\ + g_{yz}H_y\tilde{S}_z + g_{zy}H_z\tilde{S}_y + g_{zx}H_z\tilde{S}_x + g_{xz}H_x\tilde{S}_z \end{Bmatrix}. \tag{1.47}$$

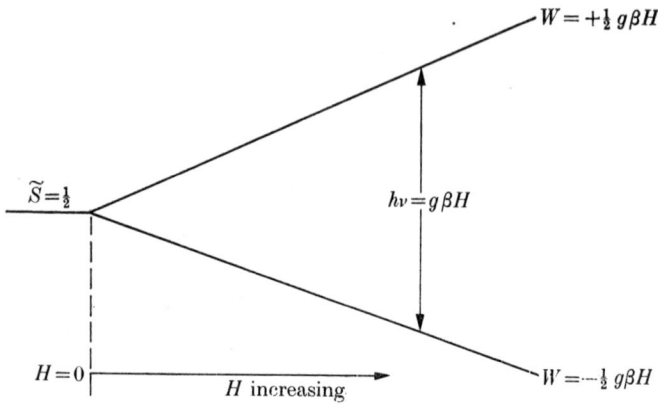

Fig. 1.4. Energy levels for $\tilde{S} = \frac{1}{2}$; Zeeman effect in an applied field H, with allowed transition at $h\nu = g\beta H$.

In the majority of cases (the exceptions being complexes with rather low symmetry) the quantities $g_{xy} = g_{yx}$, etc., and the cross terms can then be eliminated by a suitable choice for the x, y, z axes (known as the 'principal axes'), yielding the simpler form

$$\mathcal{H} = \beta(g_{xx}H_x\tilde{S}_x + g_{yy}H_y\tilde{S}_y + g_{zz}H_z\tilde{S}_z). \tag{1.48}$$

If the magnetic field **H** is applied in a direction with cosines (l, m, n) with respect to these principal axes, the energy levels are given by an equation of the form (1.44), with a value of g given by the relation

$$g^2 = l^2 g_{xx}^2 + m^2 g_{yy}^2 + n^2 g_{zz}^2. \tag{1.49}$$

With this value of g, the resonance condition is again given by eqn (1.45). Equations (1.44), (1.45), and (1.49) can be verified by a straightforward diagonalization of the Hamiltonian (1.48), or by the equivalent method (see § 3.2) of changing to a set of axes (x_e, y_e, z_e) for $\tilde{S}$ where the direction cosines of z_e are $(lg_{xx}/g, mg_{yy}/g, ng_{zz}/g)$ with respect to the principal axes. The direction of z_e is not the same as that of **H** unless **H** is along one of the principal axes, or there is no anisotropy. For example, if the complex has cubic symmetry, it follows necessarily that $g_{xx} = g_{yy} = g_{zz}$, so that the Zeeman interaction has the same Hamiltonian as for a free magnetic dipole. In axial symmetry

$$(g_{xx} = g_{yy} = g_\perp; g_{zz} = g_\parallel)$$

eqn (1.49) reduces to

$$g^2 = g_\parallel^2 \cos^2\theta + g_\perp^2 \sin^2\theta \qquad (1.49a)$$

where θ is the angle between $\mathbf{H}$ and the z-axis.

We have seen in § 1.1 that the precession of the magnetic dipole moment about the applied field $\mathbf{H}$ is in the right-handed sense if the magnetogyric ratio γ is negative; for an electronic magnetic moment, from eqn (1.8), this occurs when g is positive. If a circularly polarized oscillatory magnetic field is applied, it will be effective in interacting with the precessing magnetic moment only if it rotates in synchronism with it, and hence observation of which sense of rotation is effective in causing magnetic resonance transitions will determine experimentally the sign of g. For electrons in a free atom, this sign is always positive (since we have already allowed for the negative charge of the electron in eqn (1.8)), but this is not necessarily the case when we are dealing with a magnetic ion in a solid. A simple case where the effective value of g is negative is given in § 3.4: the weak-field Zeeman effect of the Γ_7 doublet belonging to a manifold with $\tilde{S} = \frac{5}{2}$ in a cubic field.

When anisotropy is present the situation is more complicated, since it is possible for g_{xx}, g_{yy}, g_{zz} to have different signs. (We remark here that experience shows that in this case it is necessary to take the value of g in eqn (1.49) as always positive, see § 3.2.) It is shown in § 3.2 that the use of circularly polarized radiation still gives magnetic resonance transitions of different intensity according to the sense of rotation, and that observation of which sense gives the greater intensity can be used to determine the sign of the product $(g_{xx} g_{yy} g_{zz})$. If we have axial symmetry $(g_{xx} = g_{yy} = g_\perp)$, this means that the sign of $g_z = g_\parallel$ can thus be established experimentally, but that of $g_\perp$ is indeterminate.

An experimental determination of the sign of g for an isotropic Kramers doublet (the ground state of NpF_6) has been carried out by Hutchison and Weinstock (1960). The general question of the sign and other properties of the g-'tensor' is considered further in §§ 15.6–15.8. We mention here also that for values of $\tilde{S} = \frac{3}{2}$ and greater, other types of Zeeman term linear in H may occur (see § 18.3 for a general discussion; a simple example is given in § 3.4).

We shall not consider here the question of the intensity of magnetic resonance transitions (for which see Chapter 2 and § 3.2), but mention that in axial symmetry, when $g_\perp = 0$, the intensity is zero (i.e. the transition is not allowed). For Kramers doublets this rule holds rigidly apart from small higher-order Zeeman terms in H^2 or H^3 for which further terms in the spin Hamiltonian are required (see § 18.3). For non-Kramers doublets distortions from axial symmetry due to crystal defects or strains may produce a situation in which transitions are allowed with the oscillatory magnetic field along the supposedly unique symmetry axis (see § 1.5).

It is important to note that $\tilde{S}$ is not necessarily the true angular momentum of the system, in which case the quantity g does not give the true magnetogyric ratio that would be measured in a classical experiment of the Barnett or Einstein–de Haas type. For this reason g is better called the 'spectroscopic splitting factor'. As already remarked, although the individual quantities g_{xx}, g_{yy}, g_{zz}, etc., are not necessarily all positive, we shall see in Chapter 3 that it is convenient always to regard g as a positive quantity, since then the energy levels always ascend in value with increasing values of M as far as the electronic Zeeman interaction is concerned. There is no loss of generality in this; it means only that along a principal axis p, for which g_{pp} is negative, the positive sense of z_e is in the opposite sense to the applied field H. In fact this is necessary in order to preserve a continuous rotation of the axis z_e as $\mathbf{H}$ is rotated through a principal axis for which g_{pp} is negative (see Fig. 3.1).

In this introductory chapter we shall seek primarily to present the main features of the spin Hamiltonian in a plausible way, as extensions of the spin Hamiltonian for a free atom. A simple example that illustrates this empirical approach is given in § 1.8. The relation between the spin Hamiltonian and the spectrum is discussed in detail in Chapter 3; a full theoretical discussion of the spin Hamiltonian, and its relation to the fundamental theory of the paramagnetic complex, is given in Chapters 14–21.

1.5. 'Initial splittings' or 'fine structure'

For a free atom or ion, with no hyperfine structure, the $2J+1$ levels are degenerate in zero magnetic field, and the states transform into one another under all spatial rotations as the system has full rotational symmetry. For an ion in a solid (with no hyperfine structure) the symmetry conditions may be more restricted. The concept of the effective spin $\tilde{S}$ is then valid (in its narrowest terms) when we are dealing with a set of $2\tilde{S}+1$ states that transform into one another under the symmetry operations appropriate to the crystal lattice in which the ion is placed. The $2\tilde{S}+1$ levels are then degenerate in zero magnetic field, and, except in the particular cases mentioned at the end of the last paragraph, the Zeeman effect in an applied field is represented by eqn (1.46),

$$\mathscr{H} = \beta(\mathbf{H} \cdot \mathbf{g} \cdot \tilde{\mathbf{S}}).$$

This gives a set of equally spaced levels, and the allowed transitions are between successive levels ($\Delta M = \pm 1$), which differ in energy by

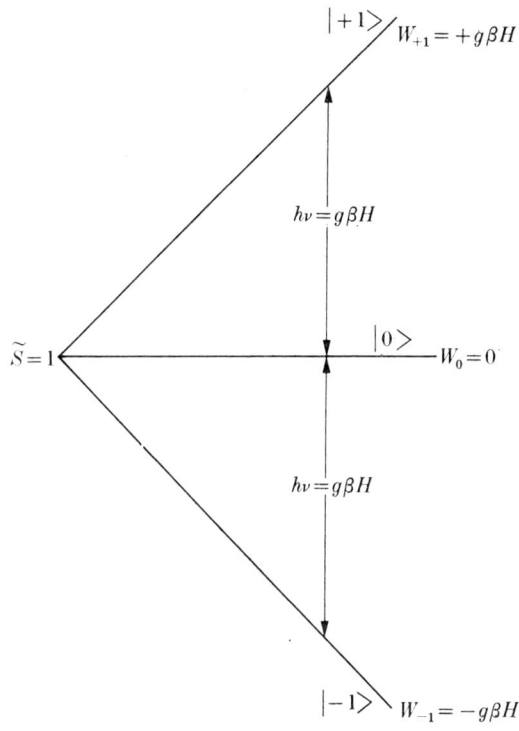

FIG. 1.5. Zeeman effect for triplet state $\tilde{S} = 1$, in the absence of any initial splitting of the levels. The two allowed magnetic resonance transitions coincide at $h\nu = g\beta H$.

$g\beta H$, so that the $2\tilde{S}$ transitions coincide at

$$h\nu = g\beta H. \tag{1.45}$$

This holds true whatever the direction of $\mathbf{H}$, which affects only the value of g. The situation is illustrated in Fig. 1.5 for $\tilde{S} = 1$; an example is the Ni^{2+} ion, $3d^8$, in a ligand field of exact octahedral symmetry (for which the g-factor is necessarily isotropic, since $g_{xx} = g_{yy} = g_{zz}$).

If we do not have exact octahedral (cubic) symmetry, the three levels may not be degenerate in zero magnetic field. This may easily be understood by reference to a special case of $\tilde{S} = 1$: a mythical ion with one electron in an atomic p-state, and no electron spin. The orbital level $l = 1$ then embraces three states, which may be written in the form

$$|(x/r)\rangle, \qquad |(y/r)\rangle, \qquad |(z/r)\rangle$$

as far as angular variation of the wave-functions is concerned. In a ligand field of cubic symmetry the directions x, y, z are equivalent, and all three states have the same energy in the absence of any external field. However, if the ligand field has only axial (or threefold, or fourfold) symmetry about the z-axis, only the x, y axes are equivalent. It follows that the $|(x/r)\rangle$, $|(y/r)\rangle$ states have the same energy, but this is not necessarily the same as that of the $|(z/r)\rangle$ state. In a ligand field of still lower (e.g. twofold) symmetry, none of the three axes is equivalent, and all states have different energies.

In the case of a spin or effective spin of $\tilde{S} = 1$, the difference of energy between the $\tilde{S}_z = \pm 1$ states and the $\tilde{S}_z = 0$ state in the case of axial, threefold, or fourfold symmetry about the z-axis may be represented by the addition of a term $D\tilde{S}_z^2$ to the effective spin Hamiltonian. It is generally preferable to write this in a form that includes an additive constant so as to leave the centre of gravity of the levels unchanged. On including the Zeeman interaction, we have then

$$\mathscr{H} = \beta(\mathbf{H} \cdot \mathbf{g} \cdot \tilde{\mathbf{S}}) + D\{\tilde{S}_z^2 - \tfrac{1}{3}\tilde{S}(\tilde{S}+1)\}. \tag{1.50}$$

The energy levels of this Hamiltonian for $\tilde{S} = 1$, when the magnetic field is applied along the z-axis, are

$$W_{\pm 1} = +\tfrac{1}{3}D \pm g_z \beta H_z, \qquad W_0 = -\tfrac{2}{3}D. \tag{1.51}$$

These are shown in Fig. 1.6, together with the allowed transitions ($\Delta M = \pm 1$), which occur at $h\nu = |(D \pm g_z \beta H_z)|$.

When $\mathbf{H}$ is not along the z-axis, the energy levels do not in general vary linearly with H and the position is more complicated. The reason for this is that in zero field the magnetic moment precesses about the unique axis (the z-axis), but in a very strong magnetic field ($g\beta H \gg D$) it changes to a precession about $\mathbf{H}$. In a first approximation the energy levels then diverge linearly with H, and the states may be approximately characterized as $|+1\rangle$, $|0\rangle$, $|-1\rangle$ corresponding to their components of angular momentum about an axis that is identical with $\mathbf{H}$ if there is no anisotropy in the g-factor. In this approximation the levels

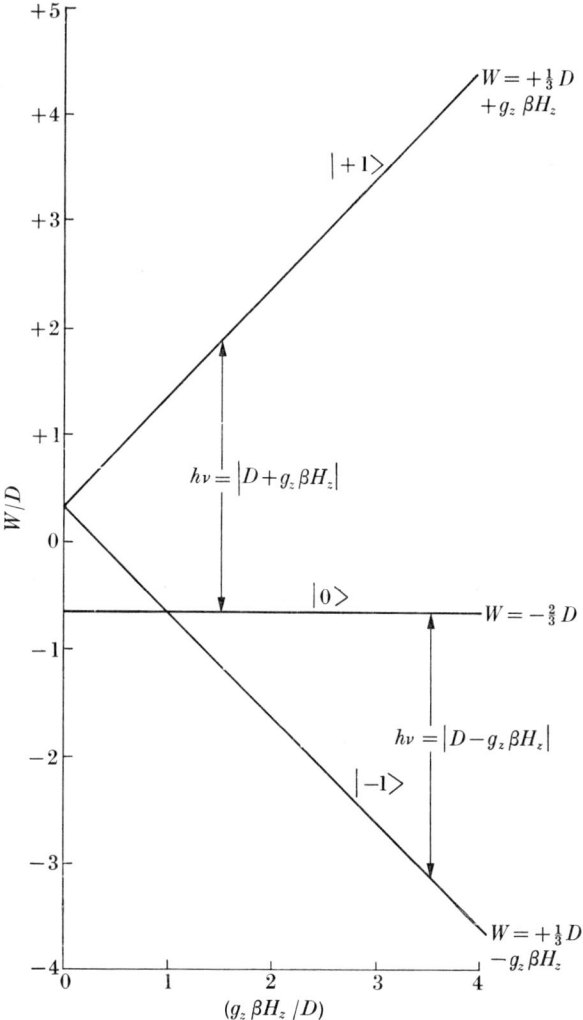

Fig. 1.6. Energy levels and allowed transitions for the Hamiltonian (1.50) with $S = 1$ and H along the z-axis, in reduced energy coordinates. The transitions are drawn for a constant frequency and variable H, such that $h\nu = \frac{5}{2}D$.

are (if g is isotropic, and **H** is at an angle θ with the z-axis)

$$W_{\pm 1} = +\tfrac{1}{6}D(3\cos^2\theta - 1) \pm g\beta H + \ldots \\ W_0 = -\tfrac{1}{3}D(3\cos^2\theta - 1) + \ldots, \qquad (1.52)$$

where the next terms are of order $D^2/(g\beta H)$. The allowed transitions are (again in first approximation)

$$|\pm 1\rangle \leftrightarrow |0\rangle, \qquad h\nu = |\tfrac{1}{2}D(3\cos^2\theta - 1) \pm g\beta H| \qquad (1.53)$$

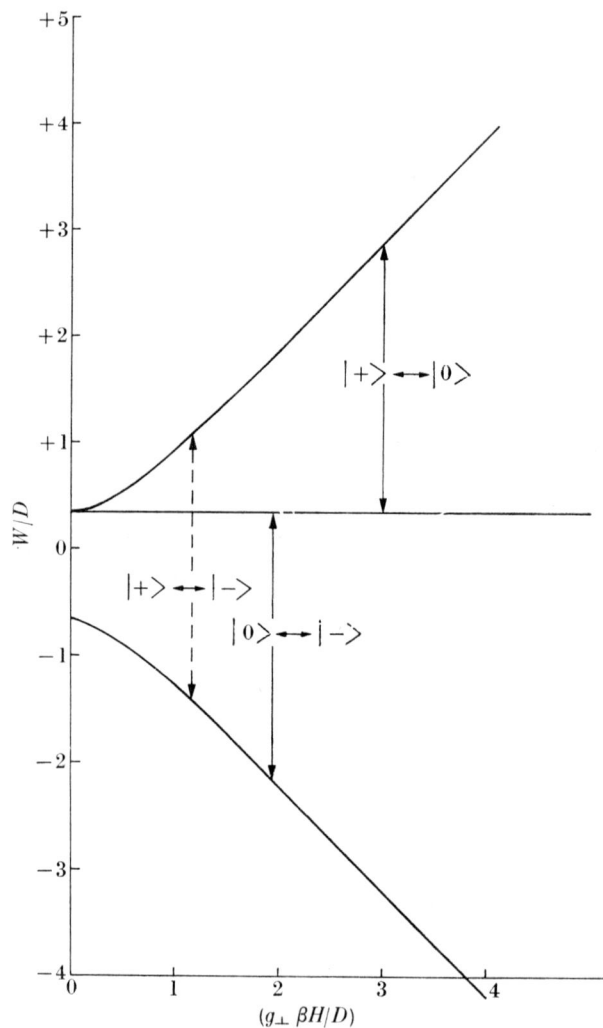

Fig. 1.7. Energy levels and allowed transitions for the Hamiltonian (1.50), with $\tilde{S} = 1$ and H perpendicular to the z-axis, in reduced energy coordinates. As in Fig. 1.6, the transitions are drawn for a constant frequency and variable H, such that $h\nu = \frac{5}{2}D$. The states are characterized by the nomenclature $|+\rangle$, $|0\rangle$, $|-\rangle$; in strong magnetic fields $(g_\perp \beta H \gg D)$ they become approximately $|+1\rangle$, $|0\rangle$, $|-1\rangle$. The transition $|+\rangle \leftrightarrow |-\rangle$ is then weaker than the other two by a factor of order $(D/g_\perp \beta H)^2$.

together with a weakly allowed transition, whose intensity is smaller by a factor of order $(D/g\beta H)^2$,

$$|+1\rangle \leftrightarrow |-1\rangle, \qquad h\nu = 2g\beta H. \tag{1.54}$$

For obvious reasons this is often referred to as a $\Delta M = 2$ transition, but this is a misleading characterization, since the levels really correspond to states that are linear combinations of the three basic states $|+1\rangle, |0\rangle, |-1\rangle$. In intermediate field strengths ($g\beta H \sim D$) the states are mixed in significant proportions and W varies non-linearly with H,

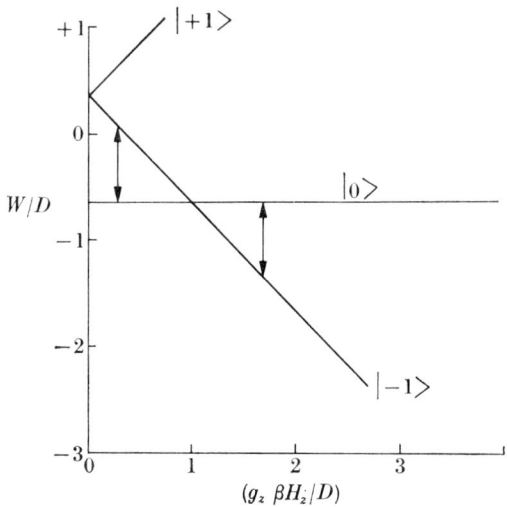

FIG. 1.8. Energy levels and allowed transitions for the Hamiltonian (1.50) with $\tilde{S} = 1$ and H along the z-axis, in reduced energy coordinates. The figure differs from Fig. 1.6 in that the transitions are drawn for $h\nu = 0.7D$; both correspond to $h\nu = |D - g_z\beta H_z|$, with $g_z\beta H_z$ respectively smaller and larger than D in the two cases. Each transition corresponds to $|0\rangle \leftrightarrow |-1\rangle$; the transition $|+1\rangle \leftrightarrow |0\rangle$ is not energetically possible when $h\nu < D$, and H is along the z-axis.
If D is negative the figure is similar but inverted, with the labels $|+1\rangle, |-1\rangle$ interchanged.

corresponding to a more complicated motion of the magnetic moment under the conflicting influences of the two terms in the spin Hamiltonian (1.50). The variation of the levels when H is normal to the z-axis is shown in Fig. 1.7.

Figures 1.6 and 1.7 have each been drawn showing the allowed transitions for $h\nu > D$. The corresponding transitions observed when $h\nu < D$ are shown (for **H** along the z-axis) in Fig. 1.8; when **H** is perpendicular to this axis it is clear from Fig. 1.7 that only one transition, $|+\rangle \leftrightarrow |0\rangle$, will be observed when $h\nu < D$.

The diagrams, Figs. 1.6–1.8, have been drawn to correspond to working at constant frequency and variable magnetic field, which is generally the most convenient experimental arrangement. It is obviously then appropriate to make H the dependent variable, and the strong field equations (1.53), where g is isotropic, can be written

$$g\beta H = h\nu \pm \tfrac{1}{2}D(3\cos^2\theta - 1) + \ldots \tag{1.55}$$

or

$$H = H_0 \pm \tfrac{1}{2}(D/g\beta)(3\cos^2\theta - 1) + \ldots, \tag{1.56}$$

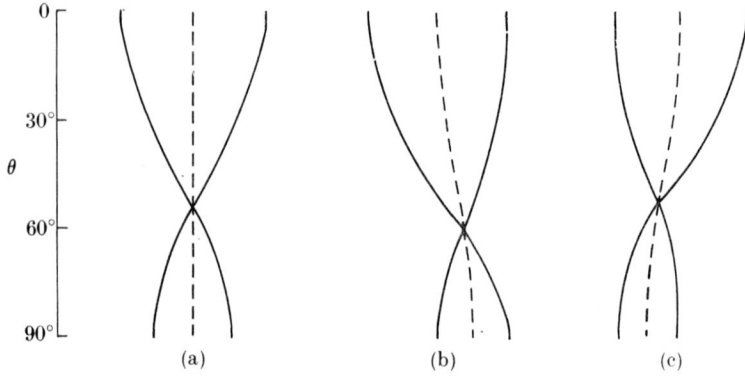

Fig. 1.9. Angular variation at constant frequency and variable H of the two strongly allowed transitions for $\tilde{S} = 1$ when H makes an angle θ with the z-axis and the spin Hamiltonian is

$$\mathscr{H} = \beta(\mathbf{H}\cdot\mathbf{g}\cdot\tilde{\mathbf{S}}) + D\{\tilde{S}_z^2 - \tfrac{1}{3}\tilde{S}(\tilde{S}+1)\}.$$

(a) g isotropic; the splitting varies as $(3\cos^2\theta - 1)$ in first approximation $(D \ll g\beta H)$.
(b) g axially anisotropic $(|g_\parallel| > g_\perp)$.
(c) g axially anisotropic $(|g_\parallel| < g_\perp)$.
Increasing field is towards the right; the broken lines show the variation in the position of the (coincident) transitions if D were zero.

where the omitted terms are of order $(D^2/h\nu)$ or $(D/g\beta)^2/H_0$, and $H_0 = h\nu/g\beta$, while $(D/g\beta)$ is a measure of D expressed in terms of field strength. Clearly the value of H_0 gives the value of g, while the separation between the two transitions (see Fig. 1.9(a)) gives that of $(D/g\beta)$. A check that the angular variation is $3\cos^2\theta - 1$ can be used to confirm that we have the correct form of the Hamiltonian, while the fact that H_0 is independent of θ shows that g is isotropic; the two transitions cross at $\theta = 54°\,44'$. These conclusions are of course exactly true only in the limit $(h\nu/D) \to \infty$, but the displacements in the transitions of order $D^2/h\nu$ are small so long as $h\nu \gg D$.

If g is anisotropic ($g_{zz} = g_\parallel$; $g_{xx} = g_{yy} = g_\perp$), then when **H** is at an angle θ to the z-axis, the value of g is given by

$$g^2 = g_\parallel^2 \cos^2\theta + g_\perp^2 \sin^2\theta \tag{1.49a}$$

and the transitions in strong fields come at

$$H = H_0 \pm \tfrac{1}{2}(D/g\beta)\left(3\frac{g_\parallel^2}{g^2}\cos^2\theta - 1\right) + \dots, \tag{1.56a}$$

so that the cross-over point is displaced to a different value of θ, and the mid-point of the two transitions (H_0 in this approximation) is now also a function of the angle θ. The two cases of $g_\parallel > g_\perp$ and $g_\parallel < g_\perp$ are sketched in Figs. 1.9(b) and (c).

Although it is generally experimentally convenient to work at a constant frequency, the strong field measurements also give much more information than a frequency-sweep system in zero magnetic field. In the present example, the latter would enable the value of $|D|$ to be found from the frequency at which maximum absorption is observed (though comparison of intensities at different frequencies needs careful instrumentation). In contrast, measurements at high fields make it possible in addition

(a) to determine the orientation of the unique axis (the z-axis), if not known;

(b) to determine the principal values of the g-tensor;

(c) to check the assumption of axial symmetry.

Similar advantages obtain through working in high fields in more complicated cases, as outlined below; the sign of D can also be found from observations of relative intensity at temperatures where $kT \sim h\nu$, as described in §§ 3.3, 3.6.

If $D \gg g\beta H$, the singlet and doublet are so well separated that resonance transitions between them cannot be made unless $h\nu \sim D$. When $h\nu \ll D$, we have only a 'weak field' Zeeman effect, which gives a first-order splitting of the doublet levels when **H** is along the z-axis but only a second-order splitting when **H** is normal to it. The isolated doublet can then be regarded as a 'non-Kramers' doublet for which $\tilde{S} = \tfrac{1}{2}$, whose first-order Zeeman effect is represented by the new spin Hamiltonian (with $\tilde{S} = \tfrac{1}{2}$)

$$\mathcal{H} = g_\parallel \beta H_z \tilde{S}_z, \tag{1.57}$$

where the effective value of $g_\parallel$ is now twice that appropriate to the spin Hamiltonian with $\tilde{S} = 1$, while $g_\perp = 0$. No transition within the doublet is allowed by this Hamiltonian.

If a small rhombic distortion is present the two states in (1.57) may be admixed and a slight initial splitting of the two levels is caused (see § 3.5). For states of the form $|+\rangle$, $|-\rangle$ given by eqn (3.27a), between which $\tilde{S}_z$ has a finite matrix element, resonance transitions are allowed for an oscillatory magnetic field along the z-axis. Thus a small distortion from axial symmetry will split the doublet slightly and give allowed transitions with this unusual orientation of the oscillatory field. A modified spin Hamiltonian with $\tilde{S} = \frac{1}{2}$ which allows for this is

$$\mathcal{H} = g_\parallel \beta H_z \tilde{S}_z + \Delta_x \tilde{S}_x + \Delta_y \tilde{S}_y. \tag{1.57a}$$

If the distortion from axial symmetry is due to local defects or strains the parameters Δ_x, Δ_y will have a random distribution of values that is often assumed to be Gaussian (for a fuller discussion, see § 18.5).

We shall not at this stage discuss in detail the many types of 'initial splittings' that can occur in practice. Their presence may be detected through the observation of transitions occurring at different values of the magnetic field, instead of all being superimposed at $h\nu = g\beta H$. With sufficient precautions to ensure that no transitions escape detection, the number of resonance lines ($= 2\tilde{S}$ in strong fields) may be used to check the value of $\tilde{S}$, or to determine it if not previously known. For values of $\tilde{S} = \frac{3}{2}$ or more, the transitions even in strong fields are unequal in intensity, that for $M \Leftrightarrow M-1$ being proportional to

$$S(S+1) - M(M-1). \tag{1.58}$$

This corresponds to the projection of the magnetic moment on the direction of the oscillatory magnetic field being larger for small values of $|M|$ than for higher values of $|M|$ (see § 3.3). This variation of intensity makes it readily possible to distinguish 'fine structure' due to such initial splittings from hyperfine structure (see below), where the lines corresponding to different nuclear orientations have the same intensity at temperatures where $kT \gg$ hyperfine energy.

In conclusion, we remark that the fine structure terms in the effective spin Hamiltonian must reflect the symmetry of the lattice point on which the paramagnetic ion is placed. In cubic symmetry, for example, there may be terms for $\tilde{S} = 2$ or more of the type

$$\tfrac{1}{6}a\{\tilde{S}_x^4 + \tilde{S}_y^4 + \tilde{S}_z^4 - \tfrac{1}{5}\tilde{S}(\tilde{S}+1)(3\tilde{S}^2 + 3\tilde{S} - 1)\}. \tag{1.59}$$

This will give a fine structure in the spectrum of the type shown in Fig. 3.3(a) for $\tilde{S} = \frac{5}{2}$ in strong magnetic fields; the structure is similar at all orientations of the magnetic field, but the line separations vary as $(l^4+m^4+n^4-\tfrac{3}{5})$, which is typical of cubic symmetry, (l, m, n) being the direction cosines of H with respect to the fourfold cubic axes. This variation corresponds to the cubic harmonic

$$x^4+y^4+z^4-\tfrac{3}{5}r^4 = r^4(l^4+m^4+n^4-\tfrac{3}{5});$$

the apparently more complicated constant term in the spin Hamiltonian (1.59) is due to the fact that allowance must be made for the non-commuting properties of the components of $\tilde{\mathbf{S}}$, a complication that does not arise with the components of $\mathbf{r}$.

Similarly the fine structure term discussed earlier in this section arises from the spherical harmonic

$$r^2(3\cos^2\theta-1) = 3z^2-r^2 = 3z^2-(x^2+y^2+z^2)$$

whose analogue in the spin Hamiltonian is

$$3\tilde{S}_z^2-(\tilde{S}_x^2+\tilde{S}_y^2+\tilde{S}_z^2) = 3\tilde{S}_z^2-\tilde{S}(\tilde{S}+1).$$

It corresponds to a quadrupolar distortion in the magnetization density on the ion, while terms of the fourth degree (such as the cubic term above) correspond to hexa-decapole distortions.

1.6. Magnetic hyperfine structure

For a free atom or ion in LS-coupling, the magnetic hyperfine structure term has the simple form (1.21)

$$\mathscr{H} = a(\mathbf{J}\cdot\mathbf{I}).$$

The interaction that this represents may be regarded either as that between the nuclear magnetic dipole moment and the magnetic field at the nucleus due to the electronic magnetization, or conversely, that of the electronic magnetization with the magnetic field due to the nuclear magnetic moment. In the solid state we have seen that the interaction between the electronic magnetization and an external magnetic field $\mathbf{H}$ may vary with the orientation of $\mathbf{H}$ with respect to the principal axes of the ligand field, and we may expect the same to be true for its interaction with the nuclear magnetic field. Thus instead of a simple direct analogue to (1.21)

$$\mathscr{H} = A(\tilde{\mathbf{S}}\cdot\mathbf{I}) \qquad (1.60)$$

we may anticipate a more complicated type of interaction analogous to that for the Zeeman interaction (1.46), (1.47). Both experimentally and theoretically we find that in the vast majority of cases the correct

term is of the form
$$\mathscr{H} = (\tilde{\mathbf{S}} \cdot \mathbf{A} \cdot \mathbf{I}), \tag{1.61}$$
which is shorthand for
$$\left.\begin{aligned}\mathscr{H} = A_{xx}S_xI_x + A_{yy}S_yI_y + A_{zz}S_zI_z + A_{xy}S_xI_y + A_{yx}S_yI_x + \\ + A_{yz}S_yI_z + A_{zy}S_zI_y + A_{zx}S_zI_x + A_{xz}S_xI_z.\end{aligned}\right\} \tag{1.62}$$

By a suitable choice of axes that, when the hyperfine structure is due to the nucleus of the paramagnetic ion itself and not to the nucleus of a ligand ion, nearly always means the principal axes of the g-tensor, this can be reduced to the form
$$\mathscr{H} = A_{xx}S_xI_x + A_{yy}S_yI_y + A_{zz}S_zI_z. \tag{1.63}$$
In a first approximation in a strong external magnetic field ($A \ll g\beta H$) the magnetic hyperfine energy is
$$W = AMm, \tag{1.64}$$
where M is the electronic magnetic quantum number and m the nuclear magnetic quantum number, and A is given by the relation
$$g^2A^2 = l^2g_{xx}^2A_{xx}^2 + m^2g_{yy}^2A_{yy}^2 + n^2g_{zz}^2A_{zz}^2, \tag{1.65}$$
where (l, m, n) are the direction cosines of H with respect to the principal axes (x, y, z). In the case of axial symmetry ($g_{xx} = g_{yy} = g_\perp$, $g_{zz} = g_\parallel$; $A_{xx} = A_{yy} = A_\perp$; $A_{zz} = A_\parallel$) this takes the form
$$g^2A^2 = g_\parallel^2 A_\parallel^2 \cos^2\theta + g_\perp^2 A_\perp^2 \sin^2\theta, \tag{1.66}$$
where θ is the angle between H and the unique axis (the z-axis). These equations can be verified by perturbation theory, or by the equivalent method (see § 3.8) in which we change to a set of 'electronic' axes (x_e, y_e, z_e) as mentioned in § 1.4 and also to a set of 'nuclear' axes (x_n, y_n, z_n) in which the direction cosines of the z_n-axis are
$$(lg_{xx}A_{xx}/gA,\; mg_{yy}A_{yy}/gA,\; ng_{zz}A_{zz}/gA),$$
which corresponds to the direction of the steady component of the electronic magnetic field at the nucleus. This direction is not the same as that of $\mathbf{H}$ nor that of z_e unless $\mathbf{H}$ coincides with one of the principal axes of the g-tensor and hence also (on our assumption) of the A-tensor, except when there is no anisotropy. As before, if the complex has cubic symmetry so that $g_{xx} = g_{yy} = g_{zz}$ and also $A_{xx} = A_{yy} = A_{zz}$, we have no need for the complexities of a tensor type interaction and the magnetic hyperfine interaction (apart from occasional extra terms) is

the same (eqn (1.60)) as we would have expected by simple analogy with a free ion.

Although we shall primarily discuss in this section the simple case of strong external magnetic field ($A \ll g\beta H$) it may be helpful at this point to draw an energy level diagram without this restriction. For simplicity we take the case of $\tilde{S} = \frac{1}{2}$, $I = 1$ with no anisotropy, for

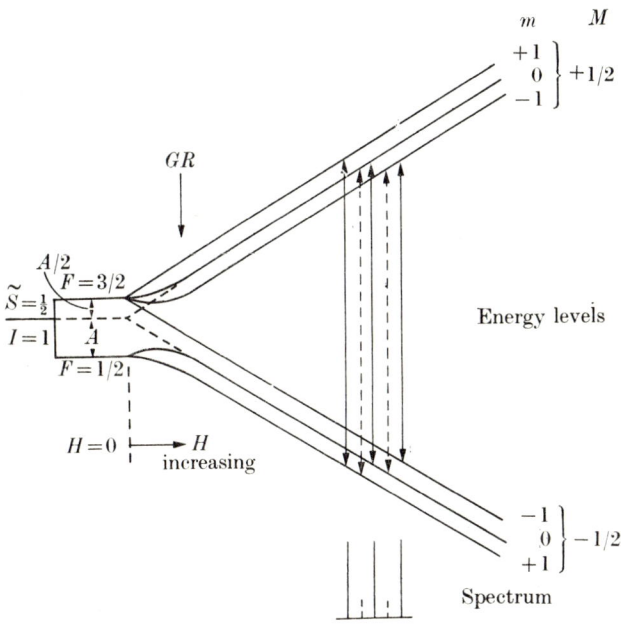

FIG. 1.10. Energy levels and allowed transitions for $\tilde{S} = \frac{1}{2}$, $I = 1$ for the Hamiltonian (with no anisotropy)
$$\mathscr{H} = g\beta(\mathbf{H}.\tilde{\mathbf{S}}) + A(\tilde{\mathbf{S}}.\mathbf{I}).$$
The strong $\Delta M = \pm 1$, $\Delta m = 0$ transitions are indicated by continuous vertical lines; the weak $\Delta M = \pm 1$, $\Delta m = \mp 1$ transitions by broken vertical lines. The diagram is drawn for constant frequency and variable magnetic field. The arrow marked GR indicates the working region for the Gorter–Rose method of nuclear polarization (see §1.12).

which the spin Hamiltonian is similar to that for a free ion

$$\mathscr{H} = g\beta(\mathbf{H}.\tilde{\mathbf{S}}) + A(\tilde{\mathbf{S}}.\mathbf{I}) - g_I\beta(\mathbf{H}.\mathbf{I}). \tag{1.67}$$

The last term is the nuclear Zeeman interaction, which is usually small and will be neglected for the time being. The levels are then shown as a function of H in Fig. 1.10. In zero magnetic field the six levels split into a quadruplet corresponding to a total angular momentum $F = \frac{3}{2}$

and a doublet $F = \frac{1}{2}$. In strong fields the levels diverge linearly and in first approximation are given by

$$W = g\beta HM + AMm - g_I\beta Hm, \qquad (1.68)$$

where the electronic magnetic quantum number $M = +\frac{1}{2}$ or $-\frac{1}{2}$ and the nuclear magnetic quantum number $m = +1$, 0 or -1. In intermediate fields ($A \sim g\beta H$) the behaviour is more complicated. If the magnetic hyperfine structure term is anisotropic, F is no longer a good quantum number and the levels in zero field are more split up (in the case under consideration, they split into three doublets). However, in the special case of axial symmetry with H along the unique axis, the quantity $(M+m)$, which is equivalent to the weak-field quantum number m_F for a free ion, is still a good quantum number (see § 3.9). When anisotropy is present (including anisotropy in the g-factor), we may expect the general behaviour of the energy levels to be yet more complicated; but in strong magnetic fields ($A \ll g\beta H$) eqn (1.68) is still valid in a first approximation, with g given by (1.49) or (1.49a) and A by (1.65) or (1.66). In this approximation, the energy levels diverge linearly with H but with separations that depend on the orientation of the magnetic field.

We now consider the possible transitions, which are of three types.

(a) *Transitions in which* $\Delta M = \pm 1$, $\Delta m = 0$

In a strong external field $H(A \ll g\beta H)$ these are the main allowed transitions, corresponding to a simple magnetic resonance of the electronic magnetization without change in the orientation of the nuclear magnet. They are indicated in Fig. 1.10 by continuous vertical lines for a system where H is varied and the frequency of resonance is constant. In the approximation of eqn (1.68) these transitions are equally spaced at

$$h\nu = g\beta H + Am. \qquad (1.69)$$

The three transitions corresponding to $m = +1$, 0, and -1 are equally intense at all ordinary temperatures ($kT \gg A$), since the three (or, more generally, $2I+1$) nuclear orientations are equally probable. As pointed out in § 1.2, this approximation is equivalent to the electronic magnetic resonance taking place in a field $(H+H_n)$, where $H_n = (Am/g\beta)$ is the steady component of the nuclear field in the direction of H; when anisotropy is present, the size of H_n depends on the orientation of H, as shown, for example, in Fig. 1.11.

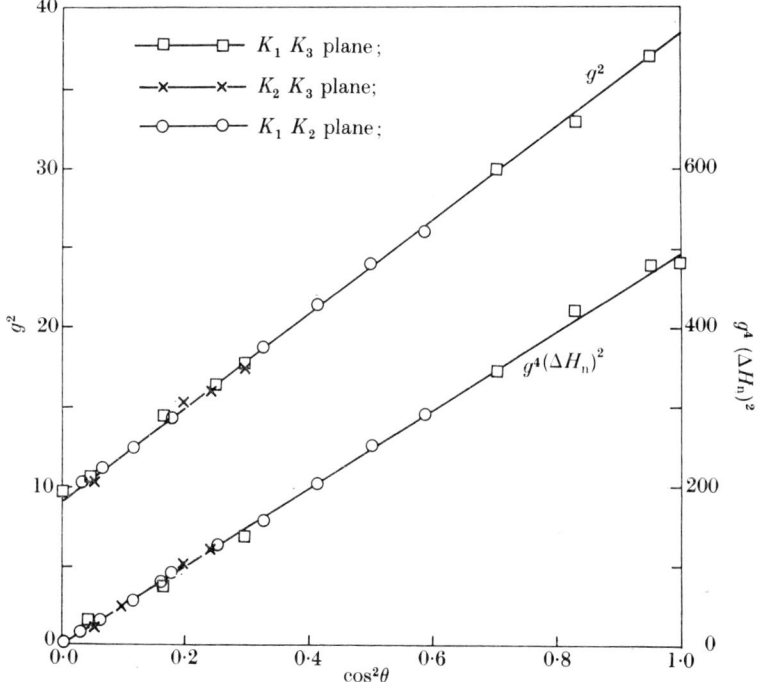

Fig. 1.11. Plot of the quantities g^2 and $g^4(\Delta H_\mathrm{n})^2$ against $\cos^2\theta$ for a highly anisotropic Co^{2+} ion with nearly axial symmetry, confirming the expected linear variation with $\cos^2\theta$. For g^2 this follows from eqn (1.49a). The quantity ΔH_n is the interval in field between successive hyperfine lines; i.e. from (1·69) $A = g\beta\Delta H_\mathrm{n}$ and hence (1·66) gives

$$\beta^2 g^4 (\Delta H_\mathrm{n})^2 = g_\parallel^2 A_\parallel^2 \cos^2\theta + g_\perp^2 A_\perp^2 \sin^2\theta.$$

In a second approximation equation (1.69) would contain terms in m^2 of order $(A^2/h\nu)$, corresponding to a slight curvature of the energy levels which is more marked in intermediate fields. These terms give rise to a slightly unequal spacing in the hyperfine structure (cf. Fig. 1.12); in nearly all cases (see § 3.8, and Figs. 3.9, 3.11) the spacing is slightly larger towards the high-field end of a group of hyperfine lines, when observed at constant frequency. When $\tilde{S} = 1$ or more, the average interval between hyperfine lines differs between different electronic transitions (see Fig. 3.9) because of similar second-order effects; this can provide some information about the sign of the hyperfine interaction if that of the initial splitting terms is known (see § 3.8). Although the quantities A_{xx}, A_{yy}, A_{zz} may not all have the same sign, it is simplest to assume that the quantity A in eqn (1.65) is always positive in sign, like that of g. The order of the levels is then known,

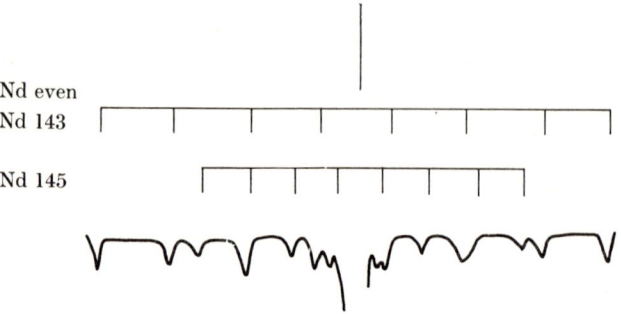

Fig. 1.12. Magnetic hyperfine structure in the magnetic resonance spectrum of $Nd^{3+}(4f^3)$ in lanthanum ethylsulphate. The two groups of 8 ($= 2I+1$) lines establish that the nuclear spins of the odd isotopes ^{143}Nd, ^{145}Nd are each $I = \frac{7}{2}$. The relative abundances of the two isotopes are 12 and 8 per cent respectively, so that the stronger hfs lines, (intensity $\sim \frac{1}{8}$(12 per cent) $= 1\frac{1}{2}$ per cent of the main line ('off scale') due to the even isotopes), belong to ^{143}Nd. The slight asymmetry and unequal separation are due to second-order effects (Bleaney and Scovil 1950).

those with positive values of M lying above those with negative values; and those with positive values of m lie above those with negative values for positive values of M, and vice versa for negative values of m (as in Fig. 1.10), provided that $|A| > |g_I \beta H|$.

(b) *Transitions in which* $\Delta M = \pm 1$, $\Delta m = \pm 1$

These transitions, which involve a change in the orientation of the nuclear magnetic moment as well as the electronic magnetic moment, are weakly allowed, their intensity being smaller than that of the $\Delta m = 0$ transitions considered above by a factor of order $(A/h\nu)^2$. When **g**, **A** are both isotropic, or if both are axially symmetric and **H** is along the unique axis, such transitions are allowed only when $\Delta M = \pm 1$, $\Delta m = \mp 1$ (that is, when $\Delta(M+m) = 0$). The transitions shown by broken vertical lines in Fig. 1.10 are of this type. For $\tilde{S} = \frac{1}{2}$, all $\Delta m = \pm 1$ transitions lie nearly midway between the $\Delta m = 0$ transitions, being slightly displaced by second-order effects of order $(A^2/h\nu)$ and also by the nuclear Zeeman energy $\sim g_I \beta H$. When the only weak transitions allowed are of the type $\Delta(M+m) = 0$, they require an oscillatory magnetic field parallel to the steady magnetic field, whereas $\Delta M = \pm 1$, $\Delta m = 0$ and $\Delta M = 0$, $\Delta m = \pm 1$ transitions require an oscillatory magnetic field normal to the steady field. When anisotropy is present, such polarization rules are not exact except in special cases (see § 3.9).

(c) *Transitions of the type* $\Delta M = 0$, $\Delta m = \pm 1$

In very strong external fields ($A \ll g\beta H$) these are purely nuclear transitions, but in intermediate fields they may have some weak electronic intensity of order $(A/g\beta H)^2$ relative to the $\Delta m = 0$ transitions. In the approximation of eqn (1.68) they require a quantum

$$h\nu = |AM - g_I \beta H| \qquad (1.70)$$

but they will be displaced by higher-order effects of order $(A^2/g\beta H)$. In general, the frequencies required are much less than those for transitions of the type $\Delta M = \pm 1$, thus making the intensity so low that these transitions are virtually unobservable directly. They are important because they give much fuller and more accurate information about the hyperfine structure than the previous transitions (a) and (b), and can be observed indirectly by means of the Electron-Nuclear Double Resonance technique (Endor), which is discussed below in § 1.13 and more fully in Chapter 4.

This discussion of magnetic hyperfine structure has been confined to the simplest terms. As in the case of the electronic Zeeman interaction, when $\tilde{S} = \frac{3}{2}$ or more, extra terms linear in I may occasionally be required (see § 18.3). The extent to which the quantity A can be regarded as a simple tensor is discussed in §§ 15.6–15.8. It should be noted also that a 'pseudo-nuclear Zeeman interaction', comparable with or larger than the true nuclear Zeeman interaction, may occur (see below, §§ 1.8 and 3.7), and the last term in (1.70) is not exactly $g_I \beta H$ when anisotropy is present.

1.7. Hyperfine structure including nuclear electric quadrupole interaction

In § 1.2 we considered briefly the form of the nuclear electric quadrupole interaction for a free atom or ion with $J = 1$, $I = 1$, and we now use this example to introduce this interaction for an ion in the solid state.

Suppose such an ion is subjected to a ligand field of axial symmetry which gives rise to an 'initial splitting' of the form $D\{J_z^2 - \frac{1}{3}J(J+1)\}$, with a negative value of D. Then the state $J_z = 0$ (where the z-axis is now that associated with D) will lie above the $J_z = \pm 1$ states by an amount $|D|$ in energy, which we assume to be much larger than the electronic Zeeman or nuclear hyperfine interactions. The latter two states comprise an isolated ground doublet (though not a Kramers doublet) of the type we meet in electron paramagnetic resonance. For each state of this ground doublet the quantity $3J_z^2 - J(J+1)$ has the value $+1$, so that in first approximation the nuclear electric quadrupole interaction (1.32) will have the form

$$W = P_\|\{I_z^2 - \tfrac{1}{3}I(I+1)\}, \tag{1.71}$$

where

$$P_\| = \frac{3b}{4J(2J-1)I(2I-1)}. \tag{1.72}$$

We note also that for the singlet state $J_z = 0$ the value of $3J_z^2 - J(J+1)$ is -2, so that it will have a nuclear electric quadrupole interaction of the form (1.71), but with a value of $P_\|$ of opposite sign and twice as large as that in (1.72). If D were positive, this would be the ground state, but as no electron magnetic resonance signal can be observed from an electronic singlet state we shall not consider it further, beyond remarking that for it the value of $P_\|$ can be found by nuclear magnetic resonance or nuclear electric quadrupole resonance.

The example we have just used to illustrate the nature of the nuclear electric quadrupole interaction was a non-Kramers doublet; a similar example for a Kramers doublet is given in § 1.8. In each case the interaction is of the form $P_\|\{I_z^2 - \tfrac{1}{3}I(I+1)\}$, which is appropriate to the case of axial symmetry. With this term, the spin Hamiltonian becomes

$$\mathscr{H} = \beta(\mathbf{H} \cdot \mathbf{g} \cdot \tilde{\mathbf{S}}) + (\tilde{\mathbf{S}} \cdot \mathbf{A} \cdot \mathbf{I}) + P_\|\{I_z^2 - \tfrac{1}{3}I(I+1)\} - g_I\beta(\mathbf{H} \cdot \mathbf{I}). \tag{1.73}$$

When $\mathbf{H}$ is at an angle θ to the z-axis, and $g\beta H$ is much greater than any other term in (1.73), the energy levels in first approximation are given by

$$W = g\beta H M + AMm + P\{m^2 - \tfrac{1}{3}I(I+1)\} - g_I\beta Hm, \tag{1.74}$$

where g, A are given by eqns (1.49a), (1.66) and

$$P = P_\| \left(3\frac{g_\|^2 A_\|^2}{g^2 A^2} \cos^2\theta - 1\right). \tag{1.74a}$$

The strongly allowed transitions are given by the rule $\Delta M = \pm 1$, $\Delta m = 0$, for which

$$h\nu = |g\beta H + Am|,$$

the same as eqn (1.69). Thus in first approximation, the strongly allowed transitions are not changed by the presence of a term such as (1.71). However, except when $\mathbf{H}$ is along the symmetry axis, the nuclear electric quadrupole interaction gives rise to terms off-diagonal in I, whose effects cannot be neglected unless $(P/A) \ll 1$. They result in transitions of the type $\Delta M = \pm 1$, $\Delta m = \pm 1$, ± 2 becoming weakly allowed; their relative intensity is $\sim (P/A)^2$ when $P < A$, comparable with the $\Delta m = 0$ transitions when $P \sim A$ (see §3.9). Transitions of this type make it possible to determine both $P_\|$ and g_I from the electron magnetic resonance spectrum, though much greater accuracy can be obtained using Endor (see § 1.13 and Chapter 4).

For brevity we shall not discuss here a number of small but important second-order effects in hyperfine structure. These are illustrated in the following section, together with other features of the spin Hamiltonian already mentioned, by means of a simple example. To retain simplicity in this example we shall again assume axial symmetry, so that the formulae derived will not be in their most general form. We shall also neglect any hyperfine interactions involving higher nuclear moments than quadrupole, whether these arise from genuine nuclear moments of such order, or through residual effects of lower order interactions.

1.8. A simple example

We have now presented the main features of the effective spin Hamiltonian, with some simple illustrative examples. At this point it may be helpful to take one single example, and use it to illustrate these features in a way that may serve as a summary of the points already covered, as well as introducing some of the second-order effects mentioned at the end of the last section.

For this example we take a $4f^1$ ion such as Ce^{3+}, whose ground state is $^2F_{\frac{5}{2}}$. We shall assume that it is split by a ligand field of axial symmetry, but that this splitting is small compared with the energy of the first excited state $^2F_{\frac{7}{2}}$, which for Ce^{3+} lies at about 2200 cm^{-1}. This means that we work entirely within the ground manifold $J = \frac{5}{2}$, and neglect admixtures of $J = \frac{7}{2}$ and of higher excited states. A ligand field of axial symmetry can be represented by a spin Hamiltonian (cf. Fig. 1.13)

$$\mathscr{H} = B_2^0\{3J_z^2 - J(J+1)\} + \text{higher degree terms in } J_z. \quad (1.75)$$

Since this contains only J_z and powers of J_z (restricted to even powers; see Chapter 16), and no terms such as $J_\pm$, the sixfold $J = \frac{5}{2}$ manifold splits into three doublets, characterized by

$$X^\pm, J_z = \pm \tfrac{5}{2}; \quad Y^\pm, J_z = \pm \tfrac{3}{2}; \quad Z^\pm, J_z = \pm \tfrac{1}{2}. \quad (1.76)$$

We shall not assume that any given doublet is the ground state, but take the energies as W_X, W_Y, W_Z respectively. The differences between

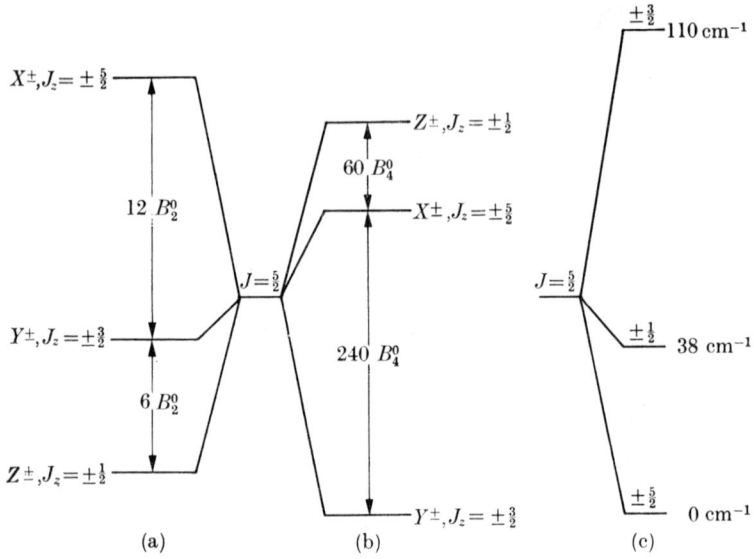

FIG. 1.13. Energy levels for $J = \frac{5}{2}$ in a ligand field of axial symmetry. Diagrams (a), (b) correspond respectively to the ligand field operators

(a) $B_2^0\{3J_z^2 - J(J+1)\}$
(b) $B_4^0\{35J_z^4 - 30J(J+1)J_z^2 + 25J_z^2 - 6J(J+1) + 3J^2(J+1)^2\}$

which are analogous to axial potential terms of the second, $r^2Y_2^0$, and fourth degree, $r^4Y_4^0$, respectively. Any linear combination of (a), (b), irrespective of the signs and magnitudes of B_2^0, B_4^0 would give the results quoted in §1.8, provided that the separations of the three doublets are large compared with the Zeeman and hyperfine interactions.

(c) gives the observed splittings for the ground state $^2F_{\frac{5}{2}}$ of $4f^1$, Ce^{3+} in $CeCl_3$.

these energies are small compared with that of the $J = \frac{7}{2}$ state, but large compared with any of the interactions considered below.

(a) *The electronic Zeeman interaction*

The Zeeman energy in an external field H is given by the matrix elements of the operator

$$\mathscr{H} = g_J\beta(\mathbf{H} \cdot \mathbf{J}). \tag{1.40}$$

Within each doublet, these are

$$\begin{array}{cccc}
 & \langle\pm| g_J\beta HJ_z |\pm\rangle & \langle\pm| \tfrac{1}{2}g_J\beta HJ_\pm |\mp\rangle & \\
X & \pm\tfrac{5}{2}g_J\beta H & 0 & \\
Y & \pm\tfrac{3}{2}g_J\beta H & 0 & \\
Z & \pm\tfrac{1}{2}g_J\beta H & \tfrac{3}{2}g_J\beta H.
\end{array} \tag{1.77}$$

These can be compared with those of an effective spin Hamiltonian with $\tilde{S} = \frac{1}{2}$ for each doublet

$$\mathcal{H} = \beta\{g_\parallel H_z \tilde{S}_z + g_\perp(H_x \tilde{S}_x + H_y \tilde{S}_y)\} \tag{1.78}$$

for which we have

$$\begin{matrix} \langle \pm | g_\parallel \beta H \tilde{S}_z | \pm \rangle & \langle \pm | \tfrac{1}{2} g_\perp \beta H \tilde{S}_\pm | \mp \rangle \\ \pm \tfrac{1}{2} g_\parallel \beta H & \tfrac{1}{2} g_\perp \beta H \end{matrix} \tag{1.79}$$

from which we see that the values of $g_\parallel$, $g_\perp$ for each doublet are

$$\begin{matrix} & g_\parallel & g_\perp \\ X & 5g_J & 0 \\ Y & 3g_J & 0 \\ Z & g_J & 3g_J. \end{matrix} \tag{1.80}$$

The last doublet gives a resonance with a rather anisotropic g-factor, but no transitions are allowed within the first two doublets, unless higher-order Zeeman effects are included. For instance, if **H** is along the x-axis, (1.40) will have the off-diagonal terms

$$g_J \beta H_x J_x = \tfrac{1}{2} g_J \beta H_x (J_+ + J_-), \tag{1.81}$$

which admix some of the $J_z = \pm \tfrac{1}{2}$ states into the $J_z = \pm \tfrac{3}{2}$ states. The doublet states Y then become approximately

$$\begin{aligned} |Y^+\rangle &= |+\tfrac{3}{2}\rangle + \frac{\sqrt{(2)} g_J \beta H_x}{(W_Y - W_Z)} |+\tfrac{1}{2}\rangle \\ |Y^-\rangle &= |-\tfrac{3}{2}\rangle + \frac{\sqrt{(2)} g_J \beta H_x}{(W_Y - W_Z)} |-\tfrac{1}{2}\rangle \end{aligned} \tag{1.82}$$

and further application of the same operator between these admixed states gives a matrix element between Y^+ and Y^- of magnitude

$$3(g_J \beta H_x)^3 / (W_Y - W_Z)^2, \tag{1.82a}$$

showing that there is a third-order Zeeman effect when H is normal to the z-axis. This gives a weakly allowed transition within the Y doublet, with an effective value of $g_\perp$ of

$$(g_\perp)_Y = 6 g_J \left(\frac{g_J \beta H_x}{W_Y - W_Z} \right)^2. \tag{1.83a}$$

The value of $g_\perp$ for the Z doublet is similarly modified, becoming

$$(g_\perp)_Z = 3g_J\left\{1 - 2\left(\frac{g_J\beta H_x}{W_Y - W_Z}\right)^2\right\}. \tag{1.83b}$$

We have here neglected matrix elements between the X and the Y states, but these elements together with those just considered give a fifth-order Zeeman splitting for the doublet X.

(b) *The magnetic hyperfine interaction*

In the same approximation where we can neglect matrix elements between states belonging to $J = \frac{5}{2}$ and $J = \frac{7}{2}$ the magnetic hyperfine operator is

$$\mathcal{H} = A_J(\mathbf{J} \cdot \mathbf{I}) \tag{1.84}$$

whose matrix elements within each doublet can be related to those of a term in the effective spin Hamiltonian with $\tilde{S} = \frac{1}{2}$,

$$\mathcal{H} = A_\parallel \tilde{S}_z I_z + A_\perp(\tilde{S}_x I_x + \tilde{S}_y I_y). \tag{1.85}$$

It is clear that the matrix elements needed are exactly the same as those for the Zeeman interaction, so that we have the following table:

$$\left.\begin{array}{ccc} & A_\parallel & A_\perp \\ X & 5A_J & 0 \\ Y & 3A_J & 0 \\ Z & A_J & 3A_J \end{array}\right\} \tag{1.86}$$

The fact that

$$A_\parallel/g_\parallel = A_\perp/g_\perp = A_J/g_J$$

or, for rhombic symmetry,

$$A_x/g_x = A_y/g_y = A_z/g_z = A_J/g_J, \tag{1.87}$$

holds generally for systems where we can neglect matrix elements between manifolds of different J.

(c) *The nuclear electric quadrupole interaction*

We have not so far quoted in full the nuclear electric quadrupole operator (given in eqn (1.108a)), and for our present example with axial symmetry we need only, in first approximation, the part given by eqn (1.32). The reason for this is that all matrix elements of the full operator vanish within each doublet X, Y, Z except those of $3J_z^2 - J(J+1)$.

These are

$$\begin{array}{ll} & \langle \pm \| 3J_z^2 - J(J+1) \| \pm \rangle \\ X & 3(\tfrac{5}{2})^2 - (\tfrac{3\cdot 5}{4}) = +10 \\ Y & 3(\tfrac{3}{2})^2 - (\tfrac{3\cdot 5}{4}) = -2 \\ Z & 3(\tfrac{1}{2})^2 - (\tfrac{3\cdot 5}{4}) = -8. \end{array} \quad (1.88)$$

Hence for each doublet the nuclear electric quadrupole interaction has the form

$$\mathscr{H} = P_{\|}\{I_z^2 - \tfrac{1}{3}I(I+1)\}, \quad (1.89)$$

where

$$P_{\|} = \frac{3b}{4J(2J-1)I(2I-1)} \langle \pm \| 3J_z^2 - J(J+1) \| \pm \rangle. \quad (1.90)$$

(d) *The pseudo-nuclear electric quadrupole interaction*

In some cases off-diagonal elements of the hyperfine operator may produce extra terms that cannot be neglected. From the magnetic hyperfine operator we have, for example, a term

$$\frac{\langle z| A_J(\mathbf{J}\cdot\mathbf{I}) |y\rangle\langle y| A_J(\mathbf{J}\cdot\mathbf{I}) |z\rangle}{W_z - W_y}, \quad (1.91)$$

where y, z are states belonging to different electronic doublets. We evaluate this term for the doublet $Z^\pm (J_z = \pm\tfrac{1}{2})$, for which the only matrix elements are clearly those with the doublet $Y^\pm (J_z = \pm\tfrac{3}{2})$. We have then

$$\frac{\langle \pm\tfrac{1}{2}| \tfrac{1}{2}A_J J_\mp I_\pm |\pm\tfrac{3}{2}\rangle \langle \pm\tfrac{3}{2}| \tfrac{1}{2}A_J J_\pm I_\mp |\pm\tfrac{1}{2}\rangle}{W_Z - W_Y}, \quad (1.92)$$

which gives

and

$$\begin{array}{l} \{2A_J^2/(W_Z-W_Y)\}I_+I_- \quad \text{for the } Z^+ \text{ state} \\ \{2A_J^2/(W_Z-W_Y)\}I_-I_+ \quad \text{for the } Z^- \text{ state.} \end{array} \quad (1.93)$$

Since

$$I_+I_- = I_x^2 + I_y^2 - i(I_x I_y - I_y I_x) = I_x^2 + I_y^2 + \hbar I_z$$

while

$$I_-I_+ = I_x^2 + I_y^2 - \hbar I_z$$

we have contributions of two types, one of which has the same sign for each component of the doublet while the other has opposite signs. For the first of these we can write

$$I_x^2 + I_y^2 = I(I+1) - I_z^2 = \tfrac{2}{3}I(I+1) - \{I_z^2 - \tfrac{1}{3}I(I+1)\}, \quad (1.94)$$

which, apart from a term in $I(I+1)$ that moves all levels up or down together, represents an additional term of the same form as the nuclear electric quadrupole interaction of magnitude

$$\Delta P_\parallel = -2A_J^2/(W_Z-W_Y). \tag{1.95}$$

The other contribution, linear in I_z and of opposite sign for the two components of the electronic doublet, represents an addition of the same form as the magnetic hyperfine interaction of magnitude

$$\Delta A_\parallel = 4A_J^2/(W_Z-W_Y), \tag{1.96}$$

where the extra factor 2 appears because we relate it to an operator $\Delta A_\parallel \tilde{S}_z I_z$ whose values are $\pm\frac{1}{2}\Delta A_\parallel I_z$ for the two components of the electronic doublet.

The magnitude of these additional terms needs to be evaluated in interpreting precise hyperfine measurements to find nuclear moments; note also in comparing values for two isotopes that such terms complicate the analysis, in that the ratio of the magnetic hyperfine interactions is not equal to the ratio of the nuclear magnetic moments (apart from any real hyperfine anomaly), and similarly the ratio of the nuclear electric quadrupole interactions will not equal the ratio of the nuclear electric quadrupole moments, because of the additional terms which vary as A_J^2.

(e) *Pseudo-nuclear Zeeman effect*

In addition to terms quadratic in the hyperfine structure, which are usually rather small compared with the first-order hyperfine interactions, there are cross terms involving products of electronic and hyperfine terms. One such term involves the electronic Zeeman interaction and the magnetic hyperfine interaction, and has the general form

$$\frac{\langle z| A_J(\mathbf{J}\cdot\mathbf{I}) |y\rangle\langle y| g_J\beta(\mathbf{H}\cdot\mathbf{J}) |z\rangle}{W_z-W_y} + \frac{\langle z| g_J\beta(\mathbf{H}\cdot\mathbf{J}) |y\rangle\langle y| A_J(\mathbf{J}\cdot\mathbf{I}) |z\rangle}{W_z-W_y}. \tag{1.97}$$

For the doublet $Z^\pm (J_z=\pm\frac{1}{2})$ the matrix elements are again those with the doublet $Y^\pm$, and can be evaluated as before. We obtain

$$\{2g_J\beta A_J/(W_Z-W_Y)\}\{I_\mp(H_x\pm iH_y)+(H_x\mp iH_y)I_\pm\} =$$
$$= \{4g_J\beta A_J/(W_Z-W_Y)\}(H_xI_x+H_yI_y). \tag{1.98}$$

This represents a term whose form is that of an anisotropic nuclear Zeeman effect

$$-\beta(\Delta g_x^{(I)} H_x I_x + \Delta g_y^{(I)} H_y I_y), \tag{1.99}$$

which is present in addition to the true nuclear Zeeman interaction $-g_I\beta(\mathbf{H}\cdot\mathbf{I})$. If we incorporate the latter into the more general expression $-\beta(\mathbf{H}\cdot\mathbf{g}^{(I)}\cdot\mathbf{I})$ we have in the present case

$$\left.\begin{array}{l}g_x^{(I)} = g_y^{(I)} = g_I - 4g_J A_J/(W_Z - W_Y),\\ g_z^{(I)} = g_I.\end{array}\right\} \quad (1.100)$$

Since g_I is about $\frac{1}{1000}$ of g_J, the 'pseudo-nuclear Zeeman effect' may easily outweigh the true nuclear Zeeman effect if the separation of the electronic doublets Z and Y is not too large.

1.9. Transition group ions and ligand fields

We have so far considered electron paramagnetic resonance in non-conducting solids in an almost purely phenomenological way. In the later sections of our introductory survey we attempt to clothe the bones of our skeleton by outlining the scientific areas in which the subject has made a significant contribution.

Magnetic resonance is observable only in a substance that contains a sufficient number of permanent magnetic dipoles; if these are electronic in origin then resonance is detectable with as few as 10^{11} dipoles, and in special cases, with even fewer (see § 2.10). Apart from transition group ions in paramagnetic salts, electronic dipoles in these low concentrations may arise in a number of ways, of which we quote a few examples.

(a) Point defects in natural crystals (such as nitrogen atoms in diamond) or in synthetic crystals (aluminium in quartz, SiO_2).

(b) Localized donor and acceptor states produced by doping of semiconductors (e.g. phosphorus in silicon).

(c) Point defects produced by electromagnetic or particle irradiation, or chemical treatment such as additive colouring; these may be neutral atoms (e.g. H in CaF_2; Ag in KCl), an electron occupying a negative ion vacancy (F-centres), a positive ion vacancy with an electron missing from an adjacent ion (V_1-centres), and similar defects more complicated in nature.

(d) Stable free radicals, such as diphenyl-trinitro phenyl hydrazyl ('g-marker'), a few gaseous stable molecules such as O_2, NO, NO_2, and many unstable free radicals ranging from short transient species in chemical reactions to long-lived varieties such as the excited spin-triplet state in naphthalene (Hutchison and Mangum 1958).

(e) Conduction electrons in metals and semiconductors.

A prominent feature of nearly all these cases is that we are dealing with single electrons with just one unpaired spin and almost no orbital momentum. The observed g-factor is very close to the free-spin value of 2·0023, but in precise measurements some information can be obtained from a small departure from the free-spin value, which may also reveal some anisotropy. In a few cases considerable molecular information is obtained from the spin-spin splitting of the $S = 1$ triplet, such as in O_2 or anthracene. In general, much more important information is obtained from hyperfine structure, if present. For atoms in $^2S_{\frac{1}{2}}$ states (H in CaF_2, Ag in KCl) the magnetic hyperfine constant may be quite close to the free atom values (Hall and Schumacher 1962), with a small hyperfine structure due to interaction with nuclear moments on adjacent nuclei (Delbecq, Hayes, O'Brien, and Yuster 1963; Bessent, Hayes, and Hodby 1967). On the other hand, interaction with nuclear moments of relatively distant atoms may reveal considerable spreading of the wave-function of the magnetic electron, as for phosphorus in silicon (Feher 1959). In some cases the hyperfine structure has provided essential information on the nature of the impurity (N in diamond, Al in quartz; see Bleaney and Owen 1965) or the location of the magnetic electron.

The position is very different with ions of the 'transition groups' that form the principal classes of normally paramagnetic compounds. Instead of single electron spins, we are now concerned with a partly-filled electron shell, containing in most cases a number of electrons with both orbital and spin angular momentum. Associated with these large amounts of angular momentum are large permanent magnetic dipole moments, making their compounds strongly magnetic. At high temperatures such compounds are paramagnetic, but when the temperature is lowered sufficiently they may enter an ordered magnetic state, either ferromagnetic, ferrimagnetic, or antiferromagnetic, with simple or complicated ordered spin arrangements. Classically they have been extensively investigated by bulk magnetic and thermal measurements, and more recently (particularly in the ordered state) by means of neutron diffraction. Judiciously chosen compounds have made it possible to attain temperatures as low as a few millidegrees absolute by the method of magnetic cooling; in this region their hyperfine and nuclear properties have been analyzed by the techniques of nuclear alignment and orientation.

Following the theoretical derivation of Curie's law (and the modified Curie–Weiss law), and the explanation on quantum theory of the origin of permanent atomic magnetic dipole moments, interest in paramagnetism has been sustained by the fascinating variety of behaviour observed in transition group salts. The primary cause of this is the interaction between the magnetic electrons and the surrounding (diamagnetic) ligand ions, which gives rise to complex splittings of the energy levels of the free ion. In some salts this results in departures from Curie's law owing to the presence of low-lying excited states, usually accompanied by high anisotropy in the magnetic susceptibility and by very short spin-lattice relaxation times. In other salts (particularly those with half-filled electron shells), all excited states lie high in energy (well above room temperature); the susceptibility may then follow Curie's law quite accurately down to below 1°K, there is little anisotropy, and spin-lattice relaxation times are comparatively long.

A further complication is the presence of spin-spin interaction. This may be purely magnetic in origin, arising from the interaction of one magnetic dipole with the local magnetic field of its neighbouring dipoles. In the paramagnetic state this local field varies from site to site, shifting the resonance in a random way and thus broadening it, with an additional effect if the ions are similar so that the precessing components of their dipole moments are in resonance. Purely magnetic broadening of this type may result in line widths of 10^2 to 10^3 G; in some salts this may be augmented by interactions due to electric moments associated with the highly anisotropic electric charge distributions of the magnetic electrons. In more concentrated salts exchange interactions may dominate other forms of spin-spin interaction. These complications (in themselves a field of study) obscure the behaviour of the single magnetic ion in its ligand field. To observe this, 'dilute' salts are grown, in which a small concentration of the magnetic ion under investigation is incorporated in an isomorphous diamagnetic compound. Such dilute salts (e.g. MgO containing Mn^{2+}, Fe^{2+}, Co^{2+}, etc.; lanthanum or yttrium ethylsulphate containing traces of other lanthanide ions) have been widely studied by magnetic resonance, so that the results may be compared with ligand field theory. A brief sketch of this is given below, with much more elaborate discussions in later chapters.

The ligand field is an additional interaction in the already complicated problem of a free many-electron ion. A perturbation approach is necessary in which terms must be considered in order of diminishing importance (i.e. diminishing interaction energy), and the ligand interaction must be introduced at the appropriate point relative to the sequence of interactions internal to the paramagnetic ion. In order of diminishing interaction energy, these are, briefly:

(a) Interaction of an electron with the Coulomb field of the nucleus, modified by the repulsive field of the other electrons,

$$\mathscr{H} = \sum_i \frac{p_i^2}{2m} - \sum_i \frac{Ze^2}{r_i} + \sum_{i>k} \frac{e^2}{r_{ik}}. \qquad (1.101)$$

With a suitably averaged electronic field with central symmetry, this Hamiltonian results in the electronic levels being grouped into configurations; an example is $3d^3$, a (ground) configuration lying some 10^5 cm^{-1} below the first excited configuration, $3d^2 4s$. The residual mutual electrostatic repulsion of the electrons, not represented by a central field, gives rise to LS-coupling with energy splittings of order 10^4 cm^{-1} between terms of different L, S built from the same configuration; for example the ground term of $3d^3$ is (by Hund's rules) ^4F, with $S = \frac{3}{2}$ and $L = 3$, some 10^4 cm^{-1} below the ^4P ($S = \frac{3}{2}$, $L = 1$) term belonging to the same ($3d^3$) configuration (see Fig. 1.14).

(b) The spin-orbit coupling

$$\mathscr{H} = \frac{\hbar^2}{2m^2c^2}\left(\frac{1}{r}\frac{\partial V}{\partial r}\right)(\mathbf{l} \cdot \mathbf{s}) = \zeta(\mathbf{l} \cdot \mathbf{s}), \qquad (1.102)$$

which in LS-coupling becomes

$$\mathscr{H} = \lambda(\mathbf{L} \cdot \mathbf{S}). \qquad (1.103)$$

For ground terms obeying Hund's rules (i.e. terms with maximum spin S), $\lambda = \pm(\zeta/2S)$ with the upper and lower signs appropriate to electron shells less or more than half-filled respectively. This splits a given term into a multiplet of levels with different values of J. The components of the multiplet are split by $\sim 10^2$ cm^{-1} for $3d$ electrons (see Fig. 1.14), and by larger amounts for ions with larger atomic numbers.

(c) The spin-spin interaction, magnetic in origin, for which one possible form is

$$\mathscr{H} = \sum_{i>j} 4\beta^2 \left\{\frac{\mathbf{s}_i \cdot \mathbf{s}_j}{r_{ij}^3} - \frac{3(\mathbf{r}_{ij} \cdot \mathbf{s}_i)(\mathbf{r}_{ij} \cdot \mathbf{s}_j)}{r_{ij}^5}\right\}. \qquad (1.104)$$

This plays a small but not altogether insignificant role in some magnetic problems (its interaction energy is usually <1 cm^{-1}); in LS-coupling it may be written in the form

$$\mathscr{H} = -\rho\{(\mathbf{L} \cdot \mathbf{S})^2 + \tfrac{1}{2}(\mathbf{L} \cdot \mathbf{S}) - \tfrac{1}{3}L(L+1)S(S+1)\} \qquad (1.105)$$

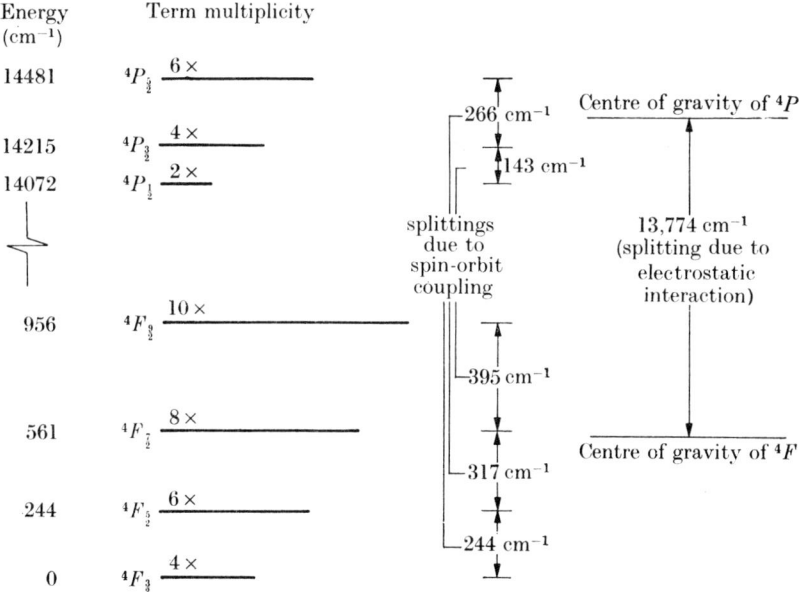

FIG. 1.14. Quartet energy levels of the free triply charged chromium ion, Cr^{3+}, $3d^3$. These are formed from the three electrons in the $3d$ shell: the splitting between the ^4F and ^4P terms is due to electrostatic repulsion between the electrons; the splittings between levels of different J within each term are due to spin-orbit coupling.

Doublet terms formed from $3d^3$ lie in the energy range 14000–37000 cm^{-1}. The next lowest levels are those belonging to the configuration $3d^2 4s$, and lie above 100000 cm^{-1}.

where ρ is a constant (see § 16.5). The spin-spin interaction represents the variation in the mutual magnetic dipole interaction energy with the direction of magnetization that may occur when the magnetization density is not spherically symmetric.

We turn now to consideration of the ligand field interaction. Its importance relative to the interactions just enumerated is different for the various transition groups, but they may conveniently be grouped as follows.

(i) $4f$ and, to a rather lesser extent, $5f$ electrons are, crudely speaking, not 'valence' electrons but 'inner' electrons, and their interaction with the ligand field is $\sim 10^2$ cm^{-1}, which is smaller than the spin-orbit coupling. In first approximation the effect of this 'weak ligand field' is partly to raise the $(2J+1)$ degeneracy of each level of the spin-orbit multiplet of given J (cf. Fig. 1.13). Exceptions arise for complex ions of the $5f$ group such as $[UO_2]^{2+}$, etc., where the $5f$ electrons are involved in bond formation in the linear O-U-O complex and the remaining (magnetic) electrons must be considered individually.

(ii) $3d$ electrons are valence electrons and the ligand interaction produces splittings of order 10^4 cm^{-1}, comparable with the LS-coupling energy. In the 'intermediate ligand field' approximation the ligand field is treated as interacting with the orbital momentum only, lifting the $(2L+1)$ degeneracy of the lowest L, S state (cf. Fig. 1.15). Since the ligand interaction is comparable with the LS-coupling energy, admixtures of excited L, S states may be needed in the ground state to give a better approximation. Interaction with the spin states occurs through the mechanism of the spin-orbit coupling.

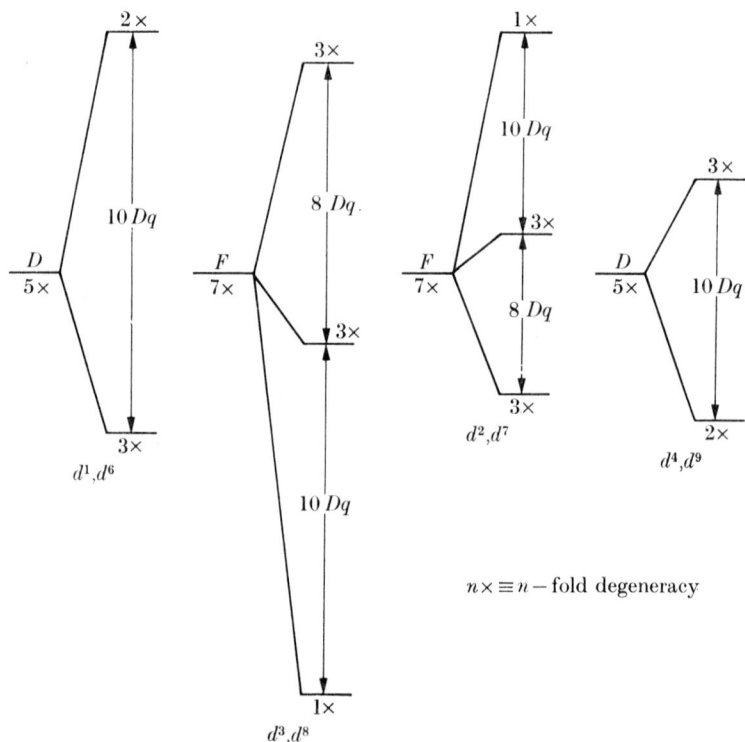

Fig. 1.15. Cubic crystal potential splittings of the orbital ground states in LS-coupling of partly filled d-shells. Octahedral coordination is assumed; the splittings are inverted for tetrahedral or 8-fold coordination. Typical values of the cubic field parameter Dq are given in Table 7.3, and examples of the overall splittings (in aqueous solution) are for the $3d$ group:

d^1, Ti^{3+}	20000 cm^{-1}	d^9, Cu^{2+}	12000 cm^{-1}
d^3, Cr^{3+}	32000 cm^{-1}	d^7, Co^{2+}	18000 cm^{-1}

The diagrams are drawn roughly to this scale. As the nuclear charge increases the d-electrons are drawn closer to the nucleus of the central ion and the influence of the ligand ions decreases.

(iii) $4d$, $5d$ electrons (and $3d$ electrons in some compounds such as the complex cyanides) are valence electrons for which the ligand interaction is much stronger than the LS-coupling. In this 'strong ligand field' limit, single electron states must be used, strongly split by the ligand field, the electrostatic and spin-orbit couplings within the atom being included subsequently or simultaneously.

In both groups (ii) and (iii) the ligand complexes have for the most part predominantly cubic symmetry, and it is the ligand interaction of cubic symmetry that is strong ($\sim 10^4$ cm^{-1} or more). The residual parts of the ligand interaction, representing departures from exact cubic symmetry, are smaller in magnitude ($\sim 10^2$ to 10^3 cm^{-1}), and their effects must often be considered at the same time as the spin-orbit interaction. In group (i) many salts have predominantly axial symmetry (three, four or sixfold), to which the 'simple example' of § 1.8 might be regarded as a zeroth-order approximation.

The simplest method of treating the ligand field is to consider it as a purely electrostatic interaction. The ligands are regarded as charged ions located on given lattice points, which give rise to a 'crystalline potential' V that reflects the local symmetry of the ionic environment. This results in level splittings for the magnetic electrons, corresponding to a 'Stark' effect with a rather complex electric potential. In this formalism three approaches arise (I, II, III, below), corresponding to the three extremes set out in the previous paragraph; a more sophisticated approach is outlined in (IV).

(I) *Weak ligand field (crystal potential < spin-orbit coupling)*

If the splittings are small compared with the spin-orbit coupling we can work in terms of J with, if necessary, corrections for coupling to states of different J (usually the first excited state, $J \pm 1$, will suffice). If then an external field $\mathbf{H}$ is applied, the diagonal electronic Zeeman interaction is of the form

$$\mathcal{H} = g_J \beta (\mathbf{H} \cdot \mathbf{J}) \tag{1.40}$$

and the corresponding magnetic hyperfine operator is

$$\mathcal{H} = A_J (\mathbf{J} \cdot \mathbf{I}). \tag{1.84}$$

These are the approximations used in the treatment of the 'simple example' of § 1.8.

An expression for g_J has already been given (eqn (1.12)) but we have not so far related the magnetic hyperfine constant A_J to the basic properties of the electron-nuclear system. Its value is

$$A_J = 2\beta(\mu_I/I)\langle r^{-3}\rangle\langle J\|\ N\ \|J\rangle, \qquad (1.106)$$

where we have used the nomenclature (μ_I/I) so that the nuclear moment can be expressed either in terms of the nuclear magneton (eqn (1.19)) or the Bohr magneton (eqn (1.20)). The quantity $\langle J\|\ N\ \|J\rangle$ is listed for the ground states of $4f$, $5f$ ions in Table 20 and $\langle r^{-3}\rangle$ is the mean inverse third power of the distance of the electron from the nucleus.

In many cases we must add to (1.106) a 'contact' or 'core polarization' term (see eqn (1.114) below) which, in terms of $\mathbf{J}$, may be written

$$\mathscr{H} = 2\beta(\mu_I/I)\langle r_c^{-3}\rangle(g_J-1)(\mathbf{J}\cdot\mathbf{I}), \qquad (1.107)$$

where (g_J-1) is the factor that allows for the projection of $\mathbf{S}$ upon $\mathbf{J}$ (see § 5.4).

For convenience we give here also the full form of the nuclear electric quadrupole operator. This is

$$\mathscr{H} = \frac{b}{2J(2J-1)I(2I-1)}\{3(\mathbf{I}\cdot\mathbf{J})^2 + \tfrac{3}{2}(\mathbf{I}\cdot\mathbf{J}) - I(I+1)J(J+1)\}, \qquad (1.108)$$

but this form (commonly used in work on free atoms) is often less convenient for our purposes than the alternative

$$\mathscr{H} = \frac{b}{4J(2J-1)I(2I-1)} \times$$

$$\times \begin{bmatrix} \{3J_z^2 - J(J+1)\}\{3I_z^2 - I(I+1)\} + \\ +\tfrac{3}{2}\{(J_zJ_+ + J_+J_z)(I_zI_- + I_-I_z) + (J_zJ_- + J_-J_z)(I_zI_+ + I_+I_z)\} + \\ +\tfrac{3}{2}\{(J_+^2I_-^2 + J_-^2I_+^2)\} \end{bmatrix}$$

$$(1.108a)$$

of which the first term in square brackets is just the diagonal part quoted in eqn (1.32). The empirical constant b is related to the electron-nuclear properties by the equation

$$b = -J(2J-1)e^2Q\langle r_q^{-3}\rangle\langle J\|\ \alpha\ \|J\rangle, \qquad (1.109)$$

where Q is the electric quadrupole moment of the nucleus, $\langle r_q^{-3}\rangle$ is an electronic average similar to $\langle r^{-3}\rangle$ above (see the discussion in § 17.7), and $\langle J\|\ \alpha\ \|J\rangle$ is a number listed in Table 20.

(II) *Intermediate ligand field* (*spin-orbit coupling* < *crystal potential splitting* < *LS-coupling*)

In this situation we may work in terms of L, S with, if necessary, corrections for coupling to excited states of different L, S. Within a state of given L, S the diagonal Zeeman interaction is

$$\mathscr{H} = \beta \mathbf{H} \cdot (\mathbf{L} + g_s \mathbf{S}). \tag{1.110}$$

The magnetic hyperfine operator is more complicated, and must be broken down into its component parts. The orbital motion of the electrons gives rise to a magnetic interaction with the nuclear dipole moment

$$\mathscr{H} = 2\beta(\mu_I/I)\langle r_l^{-3}\rangle(\mathbf{L} \cdot \mathbf{I}). \tag{1.111}$$

The interaction between the nuclear dipole moment and the electronic spin separates into two parts; for the spin magnetization 'outside' the nucleus we have

$$\mathscr{H} = 2\beta(\mu_I/I)\langle r_{sC}^{-3}\rangle \xi \{L(L+1)(\mathbf{I} \cdot \mathbf{S}) - \tfrac{3}{2}(\mathbf{L} \cdot \mathbf{I})(\mathbf{L} \cdot \mathbf{S}) - \tfrac{3}{2}(\mathbf{L} \cdot \mathbf{S})(\mathbf{L} \cdot \mathbf{I})\}, \tag{1.112}$$

where, for electrons of orbital momentum l, ξ (see § 17.4), is given by

$$\xi = \frac{2l+1-4S}{S(2l-1)(2l+3)(2L-1)}. \tag{1.113}$$

The quantity (1.112) is zero if the spin density outside the nucleus has spherical symmetry (so also is (1.111), since such symmetry is found only for $L = 0$). If there is a finite spherically symmetric spin density at the nucleus we have also a term

$$\mathscr{H} = 2\beta(\mu_I/I)\langle r_c^{-3}\rangle(\mathbf{I} \cdot \mathbf{S}), \tag{1.114}$$

where $\langle r_c^{-3}\rangle$ is a measure of the spin density at (in contact with) the nucleus. This 'contact' interaction is due to unpaired s-electron spin density, which arises from polarization of the filled electron shells through exchange interaction with the partly filled shells (the so-called 'core polarization' effect). There may also be a relativistic contribution (for further discussion see §§ 17.5–17.7).

For a state of given L, S the nuclear electric quadrupole operator is given by expressions analogous to (1.108), (1.108a), provided that $\mathbf{J}$ is everywhere replaced by $\mathbf{L}$. The parameter b is then given by the relation

$$b = -L(2L-1)e^2Q\langle r_q^{-3}\rangle \langle L\| \alpha \|L\rangle, \tag{1.115}$$

where $\langle L\| \alpha \|L\rangle$ is a number listed in Table 19, and which is also simply related to ξ (see § 17.4).

The quantities $\langle r_l^{-3}\rangle$, $\langle r_{sC}^{-3}\rangle$ are identical for hydrogenic wave-functions, but in real atoms may differ by small amounts (see § 17.7). The differences are usually ignored in solid-state electron paramagnetic resonance experiments, if only for lack of experimental evidence as to the magnitude, if any, of the difference. The same is true for the quantity $\langle r_q^{-3}\rangle$, though this may differ considerably from the two previous quantities because of electric shielding effects (see § 17.7). There may also be a contribution to the nuclear electric quadrupole interaction from the electric field gradient produced by the ligand ions. In most cases the overall effect can be expressed in the operator form

$$\mathcal{H} = P_{xx}I_x^2 + P_{yy}I_y^2 + P_{zz}I_z^2, \qquad (1.116)$$

of which (1.89) is a special form appropriate to axial symmetry.

In cubic symmetry the ligand ions make no contribution; for an electronic doublet ($\tilde{S} = \frac{1}{2}$) eqn (1.116) reduces to a constant (or to zero if $P_{xx}+P_{yy}+P_{zz} = 0$), since $P_{xx} = P_{yy} = P_{zz}$; but if $\tilde{S} = 1$ or more there may be a nuclear electric quadrupole interaction similar to (but not necessarily identical with) that for a free ion (see §§ 17.1, 18.4, 18.6).

(III) *Strong ligand field (LS-coupling < crystal potential)*

In the limit of very strong ligand field interaction, giving splittings large compared with those due to LS-coupling, our formulae must be expressed in terms of single electron states. For example, the Zeeman interaction becomes

$$\mathcal{H} = \beta \mathbf{H} \cdot \sum_i (\mathbf{l}_i + g_s \mathbf{s}_i) \qquad (1.117)$$

where the sum is over all electrons in partly-filled shells.

We shall not repeat the formulae for the hyperfine operators since these are analogous to those already given. As in the Zeeman interaction, instead of $\mathbf{L}$, $\mathbf{S}$ we have $\mathbf{1}_i$, $\mathbf{s}_i$ with a sum over all electrons. The value of ξ in the formula corresponding to (1.112) is simply

$$\xi = \frac{2}{(2l-1)(2l+3)}, \qquad (1.118)$$

as can be found from (1.113) by setting $L = l$, $S = \frac{1}{2}$. A similar replacement of $\mathbf{J}$ by $\mathbf{1}$ in (1.108a), with a sum over all electrons, gives the nuclear electric quadrupole operator, with

$$b = -l(2l-1)e^2Q\langle r_q^{-3}\rangle \langle l\| \alpha \|l\rangle, \qquad (1.119)$$

where $\langle l\| \alpha \|l\rangle$ is given in Table 18.

(IV) *The paramagnetic complex as a whole*

The crystalline potential approach to the ligand field problem outlined in the last three sections is fairly successful in interpreting the observed level splittings. However these are often dictated by the local symmetry at the paramagnetic ion, and empirical values of the parameters are deduced to fit the splittings. Attempts to calculate the crystalline parameters from electrostatic models have met with poor success. Furthermore, in such a model we would expect to use the free ion values of parameters such as the spin-orbit coupling and $\langle r^{-3} \rangle$, etc., but it has long been known (Owen 1955) that reduced values are required to fit the experimental results.

Instead of an approach in which the magnetic electrons are regarded as localized on the central ion, and move in an electrostatic potential field due to the ligand ions, it is necessary to allow for electron transfer between the central ion and the ligand ions. In other words, a purely ionic approach is replaced by one in which the complex consisting of the central ion and the ligand ions is treated as a whole. In the 'configuration interaction' approach the electron transfer is included by adding to the original ionic configuration an admixture of another configuration with different charge states. For example, in the octahedral complex $[NiF_6]^{4-}$ the situation may be represented symbolically by the linear combination

$$[Ni^{2+}(F_6)^{6-}] + \alpha[Ni^+(F_6)^{5-}]$$

where α is a small admixture parameter. In the alternative 'molecular orbital' approach the wave-function of each electron is taken to be a linear combination of atomic orbitals belonging to the central ion and the ligand ions. For example, in $[NiF_6]^{4-}$ one might represent a $2p$ electron belonging originally to a fluorine ion by the wave-function

$$\psi_{2p}(F^-) + \alpha' \psi_{3d}(Ni^{2+}),$$

where α' is another admixture parameter such that α'^2 is roughly the probability of finding the fluorine $2p$ electron transferred to a $3d$ orbit on the nickel ion.

The 'configuration interaction' and 'molecular orbital' approaches are equivalent if (as is seldom the case) no approximations are made in either treatment. The molecular orbital approach is most often adopted (for a review, see Owen and Thornley, 1966). The calculation starts from single electron states, making it rather complex for ions with several magnetic electrons. The Zeeman and other interactions must be evaluated for the molecular orbitals; the main effects may be summarized as follows.

(a) The effective orbital magnetic moment is reduced, so that the Zeeman interaction becomes

$$\mathscr{H} = \beta \mathbf{H} \cdot \sum_i (k\mathbf{l}_i + g_s \mathbf{s}_i), \tag{1.120}$$

where the parameter k is less than unity and may be anisotropic except for complexes with cubic symmetry.

(b) The spin-orbit coupling constant ζ is effectively reduced.

(c) Values of $\langle r^{-3} \rangle$ for hyperfine interaction with the nucleus of the central ion are effectively reduced.

(d) A 'ligand hyperfine structure', often rather complex if a number of ligand nuclei are involved, appears in addition to the central ion hyperfine structure (if any). The hyperfine splittings in the ligand hyperfine structure may be compared with analogous splittings for free ligand ions (measured, or estimated theoretically) to find the size of the admixtures of ligand wave-functions.

Effects (a), (b), (c) are generally revealed only by a comparison of theory and experiment, frequently involving a knowledge of the positions of the excited states gained from optical spectroscopy (Owen 1955, Owen and Stevens 1953, Stevens 1953a). Even so, such effects

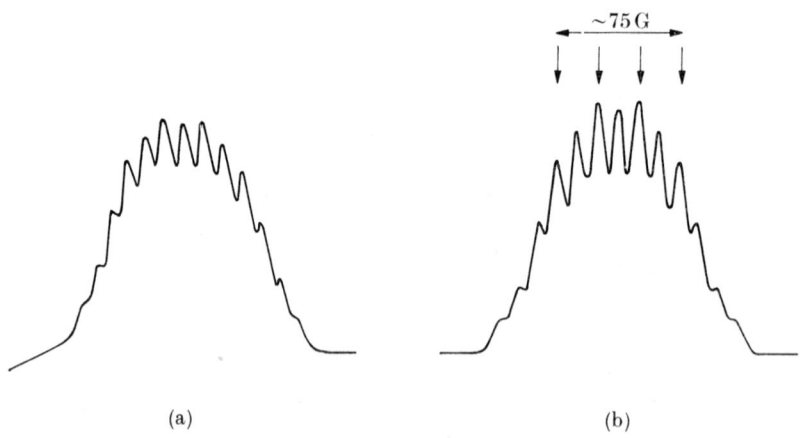

FIG. 1.16. Hyperfine structure in Na_2 (Ir, Pt)Cl_6, $6H_2O$ with H parallel to the x-axis. The arrows indicate the expected positions of the four hfs lines due to the nuclei ($I = \frac{3}{2}$) of the central Ir ions. Each of these four lines is further split into a hfs due to the nuclei of the six ligand chlorine ions (which also have nuclear spin $= \frac{3}{2}$). For the (IrCl$_6$) complex one would expect a maximum of $7 \times 4 = 28$ hyperfine lines, but many of these coincide, simplifying the observed structure. (a) observed (b) calculated best fit. (After Griffiths and Owen 1954.)

cannot always be ascribed completely to covalent binding, since there may be contributions from the Jahn–Teller effect (Chapter 21), and to a much lesser extent from the general spectrum of lattice vibrations through interactions between the magnetic ion and the phonons (an example ascribed to this effect is Ce^{3+}; see § 5.6). The effect (d), on the other hand, is directly visible in the paramagnetic resonance spectrum, as shown in Fig. 1.16.

Summary

We may sum up the position as follows. In the d-groups the strong cubic ligand field 'quenches' the orbital momentum very effectively when it leaves an orbital singlet as the ground state, and the magnetic effects are due almost entirely to the spin degeneracy remaining in this state. When an orbital triplet is lowest, the orbital momentum is only partly quenched, and plays a considerable (though reduced) part in determining the magnetic properties, but the orbital doublet states that are lowest for d^4, d^9 in Fig. 1.15 have rather unusual 'non-magnetic' properties. In the $4f$, $5f$ groups quenching of the orbital momentum is only a minor effect, and the resultant ground states often carry large amounts of both orbital and spin momentum, giving rise to abnormally large effective g-values and magnetic hyperfine structures. In the very strong ligand field compounds (mostly $4d$, $5d$, and a few $3d$) of octahedral symmetry the single electron d-states split as for d^1 in Fig. 1.15, but the splitting is so large that in the majority of chemically stable compounds magnetic electrons enter only the lower threefold orbital states, which can accommodate up to six electrons. Thus these form a magnetic sub-shell, which is closed at d^6, as in $K_3Co(CN)_6$, which has only a weak temperature-independent paramagnetism; (in an alternative approach, the remaining two orbital states have been used in forming σ-bonds with the ligands and are thus not available for magnetic electrons).

At this stage we shall not elaborate further the complex subject of paramagnetic ions in ligand fields. The theory is discussed much more fully in Chapters 16–21; the position in the various transition groups is surveyed in Chapters 5–8, where it is shown how the levels split for each ion, and details are given of the observed paramagnetic resonance spectra. At the same time it will be seen how very successful the effective spin Hamiltonian has been in describing the behaviour of each set of a rather few levels, isolated from other sets by the ligand field and the spin-orbit coupling, as far as magnetic resonance is concerned at centimetre and millimetre wavelengths. When a number of such groups must be treated together, as in infrared and optical spectroscopy of the solid-state paramagnetic compounds, the effective spin Hamiltonian can be used only in a rather limited way, for example, to describe the spin resonance spectrum of a small group of excited states detected by optical methods.

1.10. Spin-spin interaction

So far we have discussed the magnetic resonance behaviour of a paramagnetic ion either isolated or subjected to a ligand field due to the matrix of (diamagnetic) ions in which it is embedded in a crystal lattice. We have not considered interactions between neighbouring paramagnetic ions; these are known generically as 'spin-spin' interactions, and arise in a variety of ways.

Conceptually, the simplest of these interactions (and the only one readily amenable to calculation) is magnetic dipole interaction, arising from the influence of the magnetic field of one paramagnetic ion on the dipole moments of neighbouring paramagnetic ions. The average distance between such ions in a normally paramagnetic salt such as the double sulphates of the iron group (the alums or the tutton salts) is about 0·6 to 0·7 nm. If each ion carries a moment of one Bohr magneton (corresponding to $S = \frac{1}{2}$ and $g = 2$), the magnetic field that it produces at a neighbouring ion is roughly $\beta/r^3 \approx 50$ G. If there are a number of paramagnetic neighbours, whose fields at a given site add randomly and whose moments may be as much as 5β or 7β, the net local field may be 10^2 to 10^3 G. The actual local field at any given site will depend on the arrangement of the neighbours and the directions of their dipole moments (cf. Fig. 1.17); the latter gives a randomness to the local field in the paramagnetic state in both size and direction, a randomness that vanishes if the magnetic system enters an ordered state.

If an external magnetic field acts on the paramagnetic compound, the local field at each ion must be added vectorially to it. If the local field is small compared with the external field (which might be some kilogauss), only the component of the former parallel to the latter is important. The size of this component varies from site to site, giving a random displacement to the resonance frequency of each ion which is similar in its effect to that due to inhomogeneity in the external field. For this reason this effect is known as 'inhomogeneous broadening'; the resonance frequency of each ion is displaced, but the lifetime of the ion in a given quantum state is not reduced (cf. the Doppler effect for

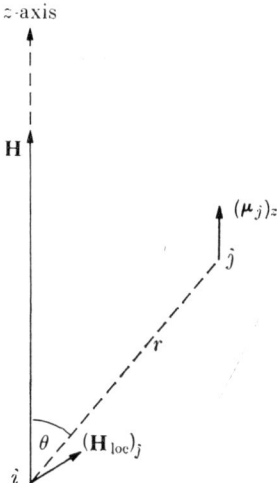

FIG. 1.17. The steady component of the local field $(\mathbf{H}_{\text{loc}})_j$ set up at site i by the steady component $(\boldsymbol{\mu}_j)_z$ of the magnetic dipole moment $\boldsymbol{\mu}_j$ at site j. The z-component of $(\mathbf{H}_{\text{loc}})_j$ at site i varies as $(3\cos^2\theta - 1)$ if the paramagnetic properties of ion j are isotropic. The presence of a strong external field $\mathbf{H}$ along the z-axis is assumed.

the molecules of a gas). If the paramagnetic ions are identical, so that they precess at the same frequency in the external magnetic field, there is an additional resonance interaction. The precessing components of one magnetic dipole set up an oscillatory field at another dipole which is just at the right frequency to cause magnetic resonance transitions, and vice versa. The mutual interaction produces resonance transitions that are equivalent to the exchange of quanta between neighbouring ions (in a strong external field this keeps the energy of the system constant). This extra interaction for identical spins gives an additional broadening (which may be of order 50 per cent in magnitude); it also shortens the lifetime of the individual ion in a given quantum state, and the broadening is no longer completely 'inhomogeneous'.

This finite lifetime in a given quantum state is frequently related to a 'spin-spin relaxation time' τ_2; in classical terms τ_2 is a measure of the average duration in time of a wave train emitted or absorbed by an ion in the process of paramagnetic resonance. Clearly τ_2 is infinite as far as a purely inhomogeneous broadening process is concerned, and it bears no simple relation to the observed line width. If we denote by $2\Delta H$ the width of the line at half the maximum intensity (cf. Fig. 1.18), we have from eqn (1.45),

$$h\,\Delta\nu = g\beta(\Delta H), \tag{1.121}$$

which we can relate to a parameter

$$\tau_2' = 1/(2\pi\,\Delta\nu). \tag{1.122}$$

If we define a homogeneously broadened line as one whose width is entirely due to processes that shorten and control the lifetime of the quantum states between which transitions are taking place, then for it $\tau_2' = \tau_2$. If the line width is due to spin–spin interaction, this relation does not generally hold, and in many cases τ_2 may be very much longer than τ_2'. For a line whose width is mainly due to inhomogeneous broadening the concept of a 'spin packet' is often used (cf. Fig. 1.18);

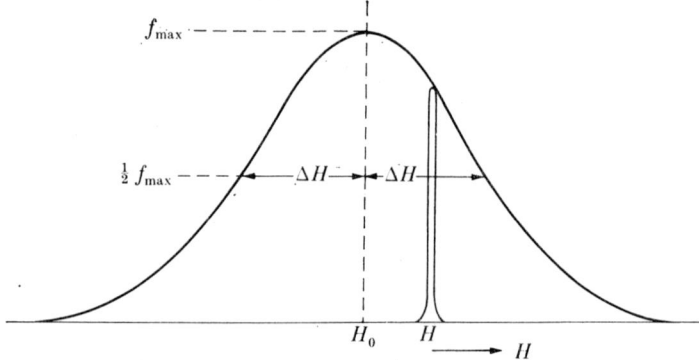

FIG. 1.18. A 'spin packet' in an inhomogeneously broadened line. The overall line shape is simply the envelope of the multitude of spin packets, with a line width parameter ΔH (width at half maximum intensity $= 2\Delta H$) which is not related to any relaxation time. An individual 'spin packet' with resonant frequency $h\nu = g\beta H$, where H is the net field acting on all spins in the packet, has a line width δH such that $g\beta\delta H \sim (\hbar/\tau_2)$, where τ_2 is the true relaxation time associated with spin-spin interaction, assuming this is shorter than the spin-lattice relaxation time τ_1.

the width of a spin packet is not quantitatively defined, but is presumably given by eqns (1.121), (1.122) provided that, instead of τ_2', we use the real value of τ_2 or of any other time constant associated with a genuine relaxation process, so that the 'spin packet' is homogeneously broadened.

A line width of 10^2 to 10^3 G in an electron magnetic resonance is a serious handicap to the resolution, since many fine or hyperfine splittings are much smaller. When isomorphous diamagnetic salts exist, these can be used to 'dilute' the paramagnetic salt by growing a single crystal of the diamagnetic salt 'doped' with a low concentration of paramagnetic ions. With sufficiently low concentrations, broadening due to spin-spin interactions with other paramagnetic ions can be reduced to a point at which it is less than that due to nuclear magnetic dipole moments that are not removed in the process of dilution. For example, in hydrated salts of the 3d or 4f group, where the paramagnetic ion is immediately surrounded by six or nine water molecules at a distance of about 0·2 nm, the residual width due to the nuclear moments of the twelve or eighteen protons in the waters of crystallization is in the region of 10 to 15 G. This can be further reduced by deuteration (growing the crystals from heavy-water solution) by a factor of about 3, corresponding to the smaller nuclear magnetic moment of the deuteron. In crystals where the abundance of nuclear moments is low or zero, such as MgO or diamond, line widths of a fraction of a gauss can be obtained.

In addition to magnetic dipole interaction between paramagnetic ions, there may also be interactions associated with electrical moments. In ions that have orbital momentum, the distribution of electric charge in the magnetic sub-states will be anisotropic, and sometimes very anisotropic. The charge distribution may possess an electric quadrupole or higher moment; in rare cases, at sites that do not have full inversion symmetry, the ion may possess an electric dipole moment. Such moments will interact with the electrostatic potential components associated with similar moments on adjacent ions, but the magnitude of the interaction is difficult to estimate because of uncertainties in the size of the electric moments, which depend on the radial wave-functions of the magnetic electrons, and in shielding (or anti-shielding) effects due to intervening ions. If the ground states of the ions are Kramers doublets, and no excited states are populated, the electrostatic interaction gives no first order spin-spin interaction because the two states of the doublet are related by time reversal and possess the same charge distribution in each state, so that no change in the electrostatic interaction arises from reversal of the magnetic dipole moment. In some other cases the electrostatic interaction may outweigh the magnetic interaction (see § 9.4).

The shape of a line broadened by spin-spin interaction cannot be readily computed; an assumption frequently made is that the shape is that of a Gaussian distribution of the type

$$f(H) = \frac{1}{(2\pi \langle H_i^2 \rangle)^{\frac{1}{2}}} \exp\left\{-\frac{(H-H_0)^2}{2\langle H_i \rangle^2}\right\}, \qquad (1.123)$$

where $f(H)$ is the intensity at a point H, $\dot{H}_0$ the centre of the resonance line being scanned in field, and $\langle H_i^2 \rangle$ is the mean 'second moment' of the line (see § 9.7). In some cases the assumption of a Gaussian shape gives a good fit (cf. Fig. 9.8). In others, when the number of nearest-neighbour ions is small (e.g. two collinear equidistant neighbours, as in the tutton salts of the iron group or the rare-earth ethylsulphates), the line width can vary sharply with the direction of the external field and the shape may be quite different from Gaussian (Figs. 1.19, 1.20); it may even show resolved structure (Fig. 9.9). This is a reminder that a line inhomogeneously broadened by spin-spin interaction is really an overlapping conglomerate of many relatively narrow lines (or 'spin packets') displaced from the central frequency by the interaction. If semi-dilute crystals are used, in which the concentration of

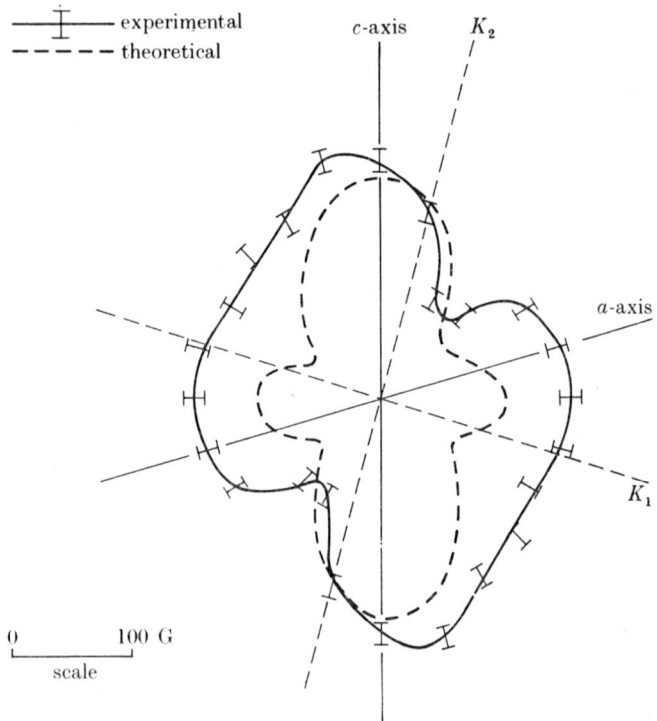

Fig. 1.19. Polar diagram of $\{\langle H_i^2 \rangle\}^{\frac{1}{2}}$ in the ac-plane for a copper tutton salt (copper ammonium selenate). The two nearest neighbours of a Cu^{2+} ion at the centre are two equidistant Cu^{2+} ions at a distance 0·61 nm along the c-axis. The broken curve shows a line width varying mainly as $|(3\cos^2\theta - 1)|$, where θ is the angle which the external magnetic field makes with the c-axis; this is the expected variation, if the interaction is magnetic dipolar and primarily with the two nearest Cu^{2+} ions along the c-axis. The continuous curve is drawn through the experimental points; the observed value of $\{\langle H_i^2 \rangle\}^{\frac{1}{2}}$ is larger near the K_1 axis (a principal susceptibility axis), partly because of unresolved hyperfine structure whose splitting is larger near K_1 than near K_2, and, to a lesser extent, because the electronic g-values vary in a similar way (Bleaney, Penrose, and Plumpton 1949).

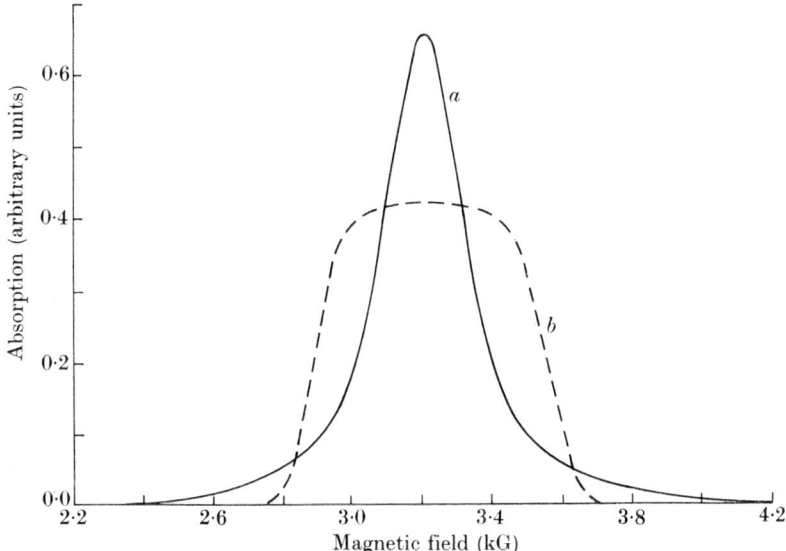

Fig. 1.20. Observed line shape along the crystallographic c-axis for (a) copper ammonium sulphate, (b) copper caesium sulphate. These are both tutton salts of similar structure. The flat-topped shape of (b) is probably due mainly to magnetic dipolar interaction; in (a) there is a considerable narrowing effect attributed to exchange interaction. Evidence from specific heat measurements confirms that exchange interaction is small for copper caesium sulphate, but considerably larger for copper ammonium sulphate (Benzie, Cooke, and Whitley 1955). (After Bleaney, Penrose, and Plumpton 1949.)

paramagnetic ions is sufficient to give a fair probability (say, a few per cent) of pairs of ions occupying adjacent sites, the spectra of such pairs (and 'triples') can be analyzed and information extracted concerning the magnitude and nature of the spin-spin interaction (see §§ 9.5, 9.6).

In many paramagnetic compounds in which the separation of the magnetic ions is less than about 0·5 nm, exchange interaction between neighbours exceeds the purely dipole interaction. The effects produced by exchange interaction are complex, but a brief summary of the principal features is as follows.

(a) Isotropic exchange interaction, represented by a term in the spin Hamiltonian of the form

$$\mathcal{H} = \mathcal{J}_{ij}(\tilde{\mathbf{S}}_i \cdot \tilde{\mathbf{S}}_j), \tag{1.124}$$

modifies the line shape in the following way. If the spins are identical, the value of the second moment $\langle H_i^2 \rangle$ is unchanged, but the line is narrowed in the centre and extended in the wings (cf. Fig. 1.21). The width at half maximum intensity is reduced (the phenomenon is

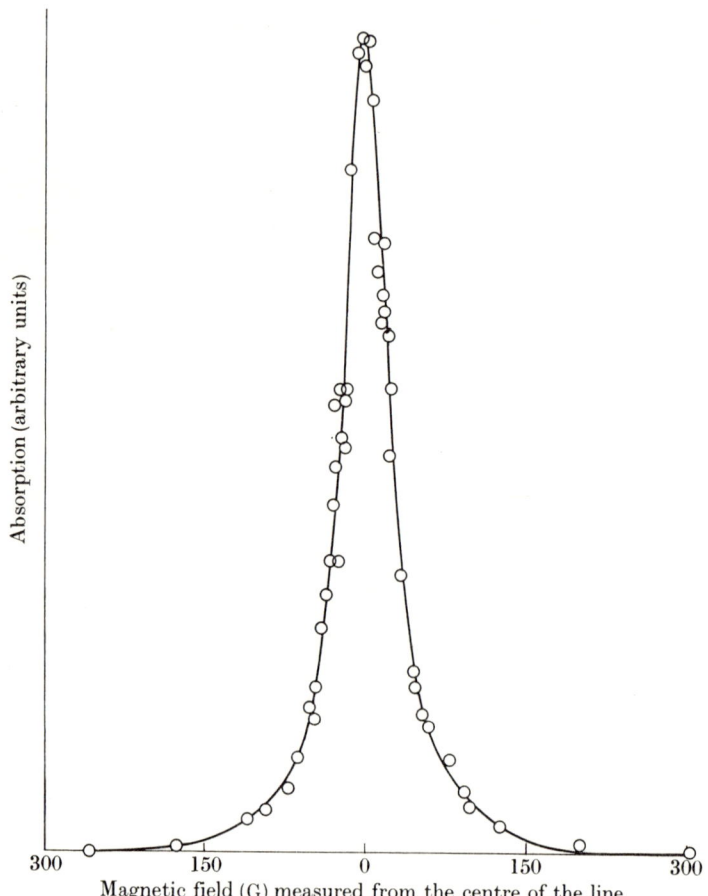

Fig. 1.21. Absorption curve for $CuSO_4, 5H_2O$ at a wavelength of 1·23 cm (about 24000 MHz), in a direction where the two Cu^{2+} ions are magnetically equivalent. The line width parameter ΔH at half intensity is only about 25 G, roughly one-tenth of the value expected from magnetic dipolar spin-spin interaction. This 'exchange narrowing' also gives a line close in shape to Lorentzian (see eqn (1.125)). In other directions, where the two Cu^{2+} ions are not magnetically equivalent, the line width is much greater, and two separate resonances can be resolved at millimetre wavelengths (Bagguley and Griffiths 1950).

therefore known as 'exchange narrowing'), and the shape approximates to a Lorentzian

$$f(H) = \frac{1}{\pi} \frac{\Delta H}{(\Delta H)^2 + (H - H_0)^2}, \qquad (1.125)$$

where ΔH is defined as in Fig. 1.18, rather than to a gaussian shape. The narrowing process resembles that found in nuclear magnetic resonance in a liquid: through exchange interaction with other neighbours, any given neighbouring dipole is changing its orientation at a rate of order ($\mathscr{J}/h$), so that its local field fluctuates at a similar rate. This local field is then less effective in broadening the resonance line, because (owing to the gyroscopic qualities of the spins) only local fields that remain substantially constant over a period, not just long compared with the Larmor precession period, but long compared with the duration τ_2 of the wave train, are effective in shifting the resonance frequency of any given spin.

If the spectrum of an isolated ion shows structure due to 'initial splittings' or hyperfine interaction, it cannot be counted as fully identical even with a similar neighbour, and exchange interaction comparable in size with the splitting terms producing the structure may give rise to broadening rather than to narrowing. If the exchange interaction dominates the splitting terms, its tendency is to give a single line at the average Zeeman frequency in which the structure is not resolved. In the limit of very high exchange interaction, appreciable narrowing of the line in the centre (and extension in the wings) may occur. Similarly, if exchange interaction links two ions with different g-values g_i, g_j, only a single resonance line at the mean g-value $\frac{1}{2}(g_i+g_j)$ may be observed if the exchange energy $\mathscr{J} \gg |(g_i-g_j)\beta H|$, the difference in the Zeeman energies; in this situation two lines can still be resolved if H is increased (by working at higher frequencies) so as to reverse the inequality.

(b) Anisotropic exchange, represented by a spin Hamiltonian of the form

$$\mathscr{H} = (\tilde{\mathbf{S}}_i \cdot \mathscr{J}_{ij} \cdot \tilde{\mathbf{S}}_j), \tag{1.126}$$

which is shorthand for a tensor type interaction similar to that in eqns (1.47) or (1.62), generally arises if the magnetic ions have anisotropic g-tensors, even if the exchange interaction between the true spins (not the effective spins) of the ions is itself isotropic (cf. § 9.3). Since the form of the interaction is then basically similar to that for magnetic dipole interaction, we may expect anisotropic exchange interaction also to give rise to broadening of the resonance line.

In concluding this section on spin-spin interaction we make two comments.

(i) It is a temperature independent interaction, giving a line-broadening that is in itself independent of temperature when $(g\beta H/kT) \ll 1$. There is a small shift in the centre of the line which increases as $g\beta H/kT$ increases, because orientation of neighbouring dipoles parallel to the external field is statistically more probable than anti-parallel orientation. When $(g\beta H/kT) \gg 1$, all dipoles are oriented parallel to the external field and the resonance lines become sharp, and appreciably shifted (see § 9.11). Broadening due to spin-spin interaction can also be markedly reduced at high temperatures if it is due predominantly to interaction with a different paramagnetic species that has a very short spin-lattice relaxation time (cf. Fig. 9.11).

(ii) The electromagnetic interactions between spins may be regarded as due to the exchange of 'virtual photons'. The spins are also coupled to the phonon bath (see following section), and forms of spin-spin interaction may arise through the exchange of 'virtual phonons'.

1.11. Spin-lattice interaction

Application of an oscillatory magnetic field of the correct frequency produces transitions between the levels shown in Fig. 1.4 both in the up and the down directions. The up transition corresponds to the absorption of a quantum of energy from the radiation field; in a down transition, energy is radiated back into the radiation field. Since spontaneous emission is negligible at radio-frequencies, we are concerned almost entirely with induced emission and absorption, which are coherent in phase with the radiation field. The net absorption is due to the excess in the number of up transitions over the down transitions; since each process is equally probable, the excess is proportional to the difference between the numbers of ions in the lower and upper states. If w_e is the rate at which electronic transitions are induced per ion by the applied oscillatory field, the power absorbed by the spin system will be

$$dW/dt = w_e(h\nu)(n_a - n_b), \quad (1.127)$$

where $h\nu$ is the quantum involved and n_a, n_b are the instantaneous populations of the lower and upper states. Since transitions are induced in the up direction at a rate $w_e n_a$, and in the down direction at a rate $w_e n_b$, more up transitions are induced than down transitions if $n_a > n_b$, and the population difference $(n_a - n_b)$ will decline continuously. A similar process would occur if we started from the unusual position of $n_a < n_b$, except that power would then be emitted by the spin system instead of being absorbed. In either case the populations would ultimately become equal, the spin temperature T_s, defined by

$$\frac{n_b}{n_a} = \exp\left(-\frac{h\nu}{kT_s}\right), \quad (1.128)$$

becoming infinite. The definition of the spin temperature for a two-level system follows naturally from Boltzmann statistics, since $h\nu$ is the difference in energy between the two states.

In a state of thermodynamic equilibrium we would expect that the two states would have populations, denoted by N_a, N_b, such that

$$\frac{N_b}{N_a} = \exp\left\{-\frac{h\nu}{kT_0}\right\}, \qquad (1.129)$$

where T_0 is the ambient temperature; the condition of equilibrium is just $T_s = T_0$. In order to attain thermal equilibrium there must be some interaction between the spin system and the thermal fluctuations that define the temperature of the matrix in which the spin system is embedded. As the populations n_a, n_b can only be altered by transitions between the two levels, such transitions must be induced by the thermal fluctuations; that is, by the molecular motion within the matrix, gas, liquid, or solid. In the last case the molecular motion consists of the lattice vibrations, which may or may not be localized, together with (in a conducting solid) the kinematic effects associated with movement of the conduction electrons (and, more rarely, of charged ions).

For a simple two-level spin system the appropriate rate equation for the processes producing thermal equilibrium is

$$\frac{d(n_a - n_b)}{dt} = \frac{1}{\tau_1} \{(N_a - N_b) - (n_a - n_b)\}, \qquad (1.130)$$

where τ_1 is a parameter known as the 'spin-lattice' relaxation time. From the form of the equation it is clear that the equilibrium situation is one in which $(n_a - n_b) = (N_a - N_b)$, and the rate of approach to this situation is proportional to the departure of the instantaneous population difference from the equilibrium value. The solution is

$$(n_a - n_b) = \{(n_a - n_b)_0 - (N_a - N_b)\}\exp(-t/\tau_1) + (N_a - N_b), \qquad (1.131)$$

showing that we have an exponential return to the equilibrium situation from an initial value of $(n_a - n_b)_0$ at $t = 0$, with time constant τ_1.

If a man-made oscillatory field producing resonance transitions is also present, we have $-(dn_a/dt) = (dn_b/dt) = w_e(n_a - n_b)$ so that, for it,

$$\frac{d(n_a - n_b)}{dt} = -2w_e(n_a - n_b), \qquad (1.132)$$

and on combining this with (1.130) we have

$$\frac{d(n_a - n_b)}{dt} = \frac{1}{\tau_1}(N_a - N_b) - \left(\frac{1}{\tau_1} + 2w_e\right)(n_a - n_b). \qquad (1.133)$$

Hence the steady-state solution, putting $d(n_a-n_b)/dt = 0$, is

$$\frac{(n_a-n_b)}{(N_a-N_b)} = \frac{1}{1+2w_e\tau_1}, \tag{1.134}$$

showing that the population difference is either slightly or greatly disturbed from the thermal equilibrium value according as $(2w_e\tau_1)$ is small or large compared with unity. With the powerful electromagnetic fields generated by electronic oscillators the quantity $(2w_e\tau_1)$ may easily be quite large, in which case (n_a-n_b) falls markedly below (N_a-N_b). The net magnetization is proportional to (n_a-n_b), and falls correspondingly; this phenomenon is known as 'power saturation' of the magnetic resonance signal. Under conditions such that $h\nu \ll kT_0$, kT_s we may approximate the exponentials in (1.128), (1.129) to give

$$\frac{n_a-n_b}{n_a+n_b} = \frac{1-\exp(-h\nu/kT_s)}{1+\exp(-h\nu/kT_s)} \approx \frac{(h\nu/kT_s)}{2}, \tag{1.135a}$$

$$\frac{N_a-N_b}{N_a+N_b} = \frac{1-\exp(-h\nu/kT_0)}{1+\exp(-h\nu/kT_0)} \approx \frac{(h\nu/kT_0)}{2}, \tag{1.135b}$$

so that (since $n_a+n_b = N_a+N_b$)

$$\frac{n_a-n_b}{N_a-N_b} = \frac{\tanh(h\nu/2kT_s)}{\tanh(h\nu/2kT_0)} \approx \frac{T_0}{T_s}. \tag{1.136}$$

The phenomenon of power saturation corresponds to a rise in spin temperature to the point where the power absorbed from the radiation field by the spin system is just balanced by the power fed out to the thermal bath via the spin-lattice relaxation mechanism. From (1.127), (1.134) the actual power absorbed is

$$\frac{dW}{dt} = \frac{(N_a-N_b)(h\nu)w_e}{1+2w_e\tau_1}, \tag{1.137}$$

which tends to an upper limit dependent only on τ_1 as w_e increases.

We discuss now the nature of the interaction that determines the value of τ_1. The only direct interaction of the spin system is through magnetic resonance transitions induced by an oscillatory magnetic field, and for relaxation effects this means with the thermal electromagnetic radiation bath. The energy density in this bath is

$$\rho\, d\omega = \frac{\hbar\omega^3}{\pi^2 c^3} \frac{d\omega}{\exp(\hbar\omega/kT_0)-1} \tag{1.138}$$

and at the frequencies and temperatures involved this is much too small to produce relaxation times of the right order. On the other hand, the energy density in the lattice vibrations (the 'phonon radiation bath') is greater by a factor of roughly $(c/v)^3$, and since the velocity of sound v in a solid is in the region of 3×10^3 ms^{-1}, the phonon energy density is greater by a factor of about 10^{15}. This may easily outweigh the fact that the interaction between spins and phonons involves mechanisms that are inherently weaker than the simple magnetic resonance interaction between spins and photons. The first discussion of such mechanisms by Waller (1932) was based on modulation of the spin-spin interaction by the phonons, which induces oscillatory components in the distances between paramagnetic ions (which we have hitherto regarded as embedded in a fixed lattice). Waller distinguished two processes in such interactions.

(a) *Direct process*. A phonon of the same energy ($\hbar\omega$) as the spin quantum required for a resonance transition is absorbed by the spin system, resulting in an 'up' transition in Fig. 1.4; or a phonon is emitted, accompanied by a 'down' transition within the spin system.

(b) *Raman process*. A phonon of any frequency ($\omega_p/2\pi$) may interact with a spin, causing a transition (up or down) within the spin system, the phonon being scattered with a different frequency $(\omega_p/2\pi) \mp \nu$ respectively, where ν is the magnetic resonance frequency. From the analogy with the corresponding phenomenon for electromagnetic radiation, this is known as a 'Raman process'.

The direct and Raman processes are illustrated schematically in Fig. 1.22. If we regard the direct process as a first order (one phonon) process, then in comparison the Raman process is a second order (two phonon) process with a correspondingly smaller coupling of the phonons to the spin system. Nevertheless it may play the major role in causing spin-lattice relaxation, as can be understood from the following discussion. As an example we take a copper tutton salt in a magnetic resonance experiment at 10^{10} Hz (3 cm wavelength), for which the line width is about 10^2 G ($\Delta\omega = 2 \times 10^9$ s^{-1}), maintained at 4°K. The number of phonons at this frequency within this band-width (given approximately by (1.138) with v instead of c in the denominator) per cm^3 of material is about 10^{15}. (By way of comparison we note that the number of spins is 3×10^{21}, while the number of photons is less than 1.) Only these phonons, which are at the same frequency as the resonance frequency, are on 'speaking terms' with the spins and can take part in the direct process for relaxation. On the other hand, the peak in the energy density

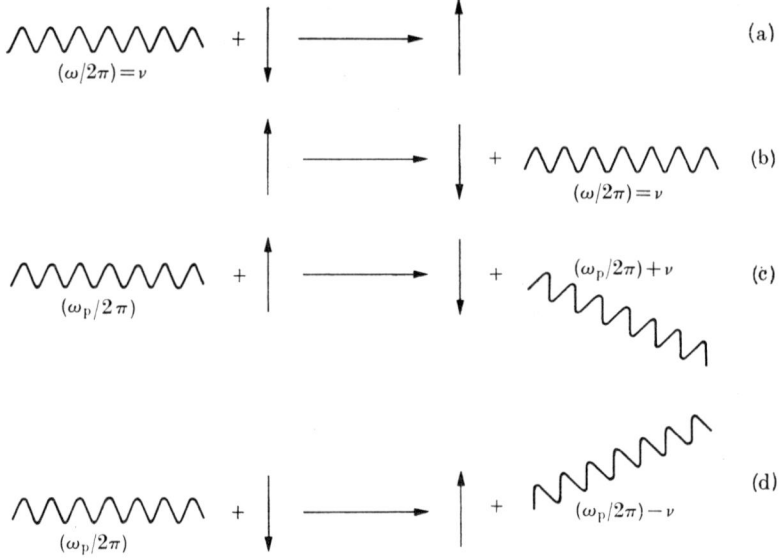

Fig. 1.22. Schematic illustration of the direct process (a), (b) and the Raman process (c), (d) in spin-lattice relaxation. (a) a phonon of frequency $(\omega/2\pi) = \nu$, the magnetic resonance frequency, is absorbed by a spin which makes a transition to the upper state. (b) a spin makes a transition to the lower state, emitting a phonon at the resonant frequency. (c) a phonon of frequency $(\omega_p/2\pi)$ is scattered as a phonon of frequency $(\omega_p/2\pi)+\nu$, accompanied by a 'down' transition of the spin. (d) a phonon of frequency $(\omega_p/2\pi)$ is scattered as a phonon of frequency $(\omega_p/2\pi)-\nu$, accompanied by an 'up' transition of the spin.

curve for the phonons (cf. Fig. 10.3) occurs at about $3kT_0$, which for the above example is nearly thirty times greater than the quantum $h\nu$ for magnetic resonance. Thus the number of phonons that can take part in the direct process is only an extremely small fraction of the total number. Since any phonon can participate in the Raman process, the whole phonon spectrum can contribute to relaxation within the spin system by this process, thus greatly enhancing its importance. Obviously this is true only when $(h\nu/kT_0) \ll 1$ (it is about 0·1 in the above example), and at lower temperatures where the overall number of phonons has fallen sufficiently to remove the overwhelming effect of their preponderance in number, the direct process will become the more important of the two.

The theory of Waller, based on modulation of the magnetic dipolar spin-spin interaction, led to relaxation times appreciably longer than those observed experimentally. A more potent mechanism and one which, unlike that of Waller, is independent of the degree of concentration of the magnetic ions, is modulation of the ligand field by the lattice vibrations. This produces primarily a fluctuating electric field, which modulates the orbital motion of the magnetic electrons. There is no direct interaction with the electron spins, but they feel the effect of the modulation of the orbital motion through the spin-orbit coupling, in the same way as they feel the effects of the static ligand field. The resultant spin-lattice relaxation times are therefore greatly dependent on the extent to which the orbital moment is absent in the free ion, or quenched by the static ligand field. As a rough rule long spin-lattice relaxation times are to be expected when the g-value is close to the free spin value 2·0023, and short relaxation times when g departs markedly from this value. Examples of the former type are isolated spins attached to defects, and transition group ions that have half-filled shells, whose ground states are S-states; on the other hand, short relaxation times are found for all ions of the $4f$, $5f$ groups (except those with the half-filled shell f^7), and those ions of the d-groups for which the ligand field leaves orbitally degenerate ground states (cf. Fig. 1.15).

The foundations of the quantitative formulation of the theory of spin-lattice relaxation through modulation of the ligand field were laid by Kronig (1939) and Van Vleck (1940), and have been extended by many others, notably Orbach (1961). There are many complications to be considered, but here we confine ourselves to the temperature dependence. In the majority of cases this can be written in the form

$$\frac{1}{\tau_1} = a \coth\left(\frac{h\nu}{2kT_0}\right) + bT_0^n + \frac{c}{\exp(\Delta/kT_0) - 1}, \qquad (1.139)$$

where the various terms arise from different processes that contribute simultaneously (but in widely varying degree) to the relaxation *rate*, that is, to $(1/\tau_1)$. Briefly these processes are as follows.

(a) Direct process, involving phonons of the same energy as the magnetic resonance quantum $h\nu$, which gives the first term in (1.139). The two limiting cases are

(i) $(h\nu/kT_0) \ll 1$, for which $\frac{1}{2} \coth(h\nu/2kT_0) \to (kT_0/h\nu)$, which is the average number of quanta per phonon mode in the 'classical' high temperature limit;

(ii) $(h\nu/kT_0) \gg 1$, for which $\coth(h\nu/2kT_0) \to 1$, giving a constant spin-lattice relaxation time determined by the rate of spontaneous emission of phonons from the upper state when phonons of energy $h\nu$ are no longer thermally excited.

(b) Raman process, a two-phonon process in which all phonons can take part, giving the strongly temperature dependent second term.

Typical values of the exponent n are:

>non-Kramers doublet, $\quad n = 7$;
>
>Kramers doublet, $\quad n = 9$;
>
>multiplet with small splitting, $\quad n = 5$.

(c) Orbach process, giving the third term. This involves absorption of a phonon by a direct process to excite the spin system to a much higher level (see Fig. 1.23), at an energy Δ above the ground doublet,

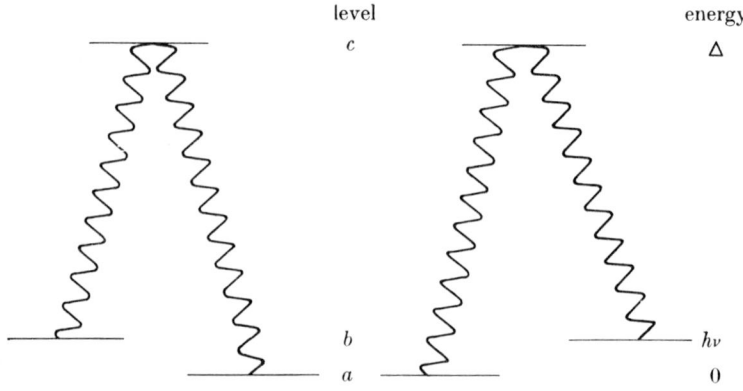

FIG. 1.23. A schematic illustration of the Orbach process. The paramagnetic ion has a ground doublet, with levels (a), (b) at energies 0, $h\nu$, together with an excited state (c) at a much greater energy Δ. Phonons at frequencies Δ, $\Delta-h\nu$ are absorbed or emitted by direct processes, causing transitions $(a) \leftrightarrow (c)$ and $(b) \leftrightarrow (c)$ respectively. The net result is to establish thermal equilibrium between the ground levels (a), (b). The rate is controlled by the number of phonons available at frequencies near Δ/h, which must therefore be an allowed phonon frequency.

followed by the emission of another phonon of slightly different energy so that the magnetic ion is indirectly transferred from one level to the other of the ground doublet. This process is also strongly temperature-dependent, being determined by the number of phonons of energy Δ available to excite the ion to the upper state at Δ. When $\Delta \gg kT_0$, the last term in eqn (1.139) approximates to $c \exp(-\Delta/kT_0)$.

Because of the high temperature dependence, values of τ_1 for the same salt vary rapidly and over a wide range, making experimental confirmation of the theory difficult. Figure 1.24 shows an elegant set of experimental values, fitted by a combination of direct and Raman terms, determined by observation of the transient recovery of the magnetic resonance signal after saturation by a high-power pulse; Fig. 1.25

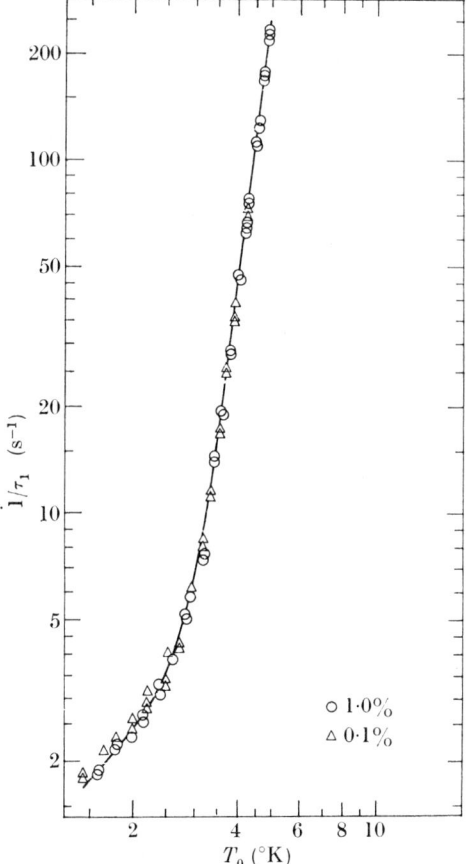

FIG. 1.24. Experimental measurements of spin-lattice relaxation time for Nd^{3+} in yttrium ethylsulphate. The results are fitted with a combination of direct and Raman terms

$$\frac{1}{\tau_1} = 1{\cdot}18 T_0 + 1{\cdot}59 \times 10^{-4}\, T_0^9\ (\mathrm{s}^{-1})$$

and show no significant dependence on the concentration of neodymium ions. The measurements were made by observation of the transient recovery in the magnetic resonance signal following saturation by a high power r.f. pulse. The steady magnetic field ($H = 3390$ G) is normal to the crystal axis, and the resonant frequency is 9·40 GHz (after Larson and Jeffries 1966a).

shows the first experimental confirmation of the exponential dependence characteristic of the Orbach process, determined by the classical method of measurement of the magnetic susceptibility at audio frequencies. Further experimental results are given in Chapter 10.

In this section we have discussed hitherto the rate at which the spin system relaxes to the phonons, with the implicit assumption that the

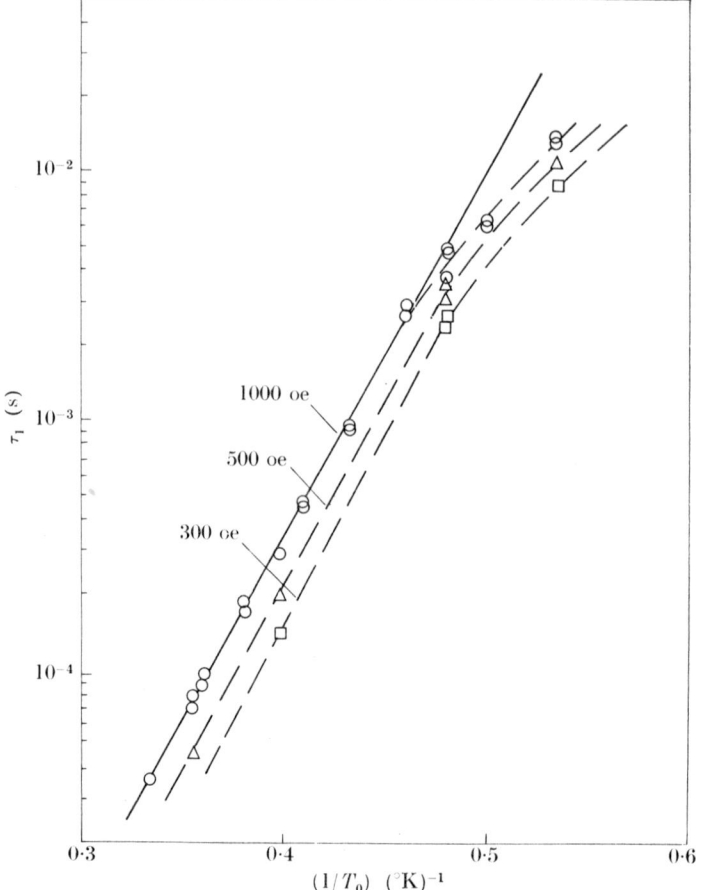

Fig. 1.25. Spin-lattice relaxation times for the ground doublet of cerium magnesium nitrate. The straight line represents the exponential temperature dependence

$$\tau_1 = 2 \times 10^{-10} \exp\left(\frac{34}{T_0}\right) \text{ (s)}, \quad \text{or} \quad \frac{1}{\tau_1} = 5 \times 10^{+9} \exp\left(-\frac{34}{T_0}\right) \text{ (s}^{-1})$$

typical of an Orbach process involving phonon excitation of the spin system to another crystal field level at 34°K. The measurements were made by determining the magnetic susceptibility in the audio-frequency range at frequencies of order $(2\pi\tau_1)^{-1}$ (after Finn, Orbach, and Wolf 1961).

phonons are always in thermal equilibrium with a bath at temperature T_0. However we have already seen that in the direct process the number of phonons on 'speaking terms' with the spins is small compared with the number of spins. Thus, unless these phonons can rapidly transfer energy to other phonons or to a surrounding bath of high heat capacity, the energy they receive from the spin system via the direct process will quickly heat the resonant phonons to a temperature close to that of the spin system. This is followed by a regime in which the combined (spin+phonon) system relaxes to the bath with a time constant τ_b which is much longer than that for the phonons alone, owing to the high heat capacity of the spin system. This phenomenon is known as the 'phonon bottle-neck' and was first discussed by Van Vleck (1941a,b); it is represented by a block diagram in Fig. 1.26.

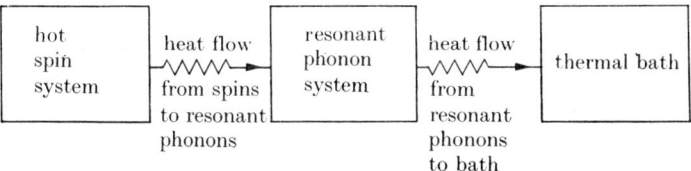

FIG. 1.26. The flow of heat from the spin system to the resonant phonons, and on to the thermal bath. If the heat capacity of the resonant phonon system is very small compared with that of the spin system, the initial process is a rapid heating of the phonons to a temperature close to that of the spin system. This is followed by a much slower process in which the combined systems (spins+phonons) relax to the bath with a characteristic time constant τ_b, which becomes the effective relaxation time constant for the spin system.

If the ratio (heat capacity of spin system/heat capacity of resonant phonon system) is denoted by b, a quantity that may be as high as 10^4 (see § 10.2), then the effective time constant for the initial (spin to phonon) transfer is of order (τ_{1d}/b), which is very much shorter than τ_{1d}, the ordinary spin-lattice relaxation time for the direct process. On the other hand, the effective relaxation time τ_b for the combined (spin+phonon) to bath transfer is of order $b\tau_{ph}$, where τ_{ph} is the normal relaxation time for transfer to the bath from the phonons alone. Because of the large value of b, it is the quantity τ_b that is observed to control the rate of change of the spin temperature, after the initial rapid heating of the phonon system in a time $\sim(\tau_{1d}/b)$. This results in a different temperature dependence for the apparent relaxation time, as demonstrated by Scott and Jeffries (1962). It may also become dependent on sample size, being shorter for a fine powder with a shorter value of τ_{ph} owing to the more intimate contact between the phonon system and the surrounding bath.

Under these conditions where b is very large, the resonant phonons will be heated almost to the same temperature as the spins following a saturating magnetic resonance pulse, and the temperature of the two systems then falls together towards the bath temperature. However it is possible instead to invert the populations in the spin system, i.e. to make $n_b > n_a$, giving effectively a 'negative temperature' for T_s (cf. eqn (1.128)). This is not possible for the phonon system, since the equally spaced quantum levels of a simple harmonic oscillator continue to infinity in energy. The transient decay of the negative temperature T_s of the spin system must therefore be different from that for a positive value of T_s, as demonstrated by Brya and Wagner (1965). Immediately after an inverted population is established (e.g. by the method of adiabatic rapid passage; see § 2.5), phonons will be emitted from the spin system mainly by spontaneous emission if $h\nu > kT_0$. These heat the resonant phonons, thus increasing their radiation

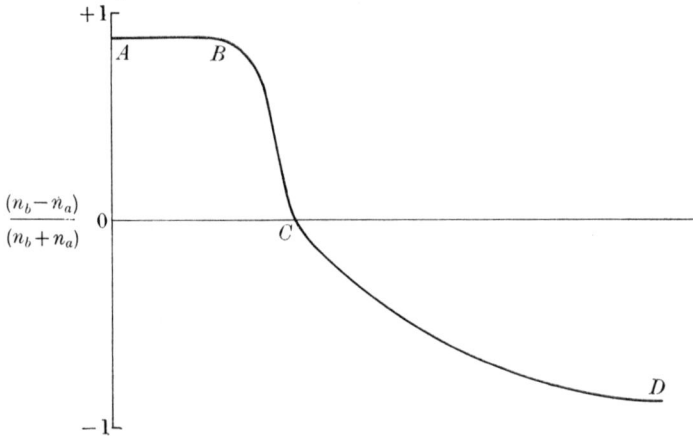

FIG. 1.27. The 'phonon avalanche' effect. After an inverted population ($n_b > n_a$) has been established in the spin system, the population difference decays only slowly in the region AB mainly through spontaneous emission of phonons if $h\nu \gg kT_0$. The resonant phonons become heated, increasing the phonon radiation density and the chance of stimulated emission of more phonons, giving the 'phonon avalanche' in the region BC. After the population has ceased to be inverted, the temperatures of spin and resonant phonon systems decay together in the region CD towards the thermal equilibrium value at a rate determined by the phonon-bath relaxation times τ_b.

density which in turn increases the rate of emission of further phonons by the process of induced emission. The results are cumulative, producing a 'phonon avalanche' that is particularly dramatic if initially $n_b \gg n_a$. In the avalanche the population of the upper spin level falls rapidly until it drops below that of the lower level, then reaching a positive value of T_s at which transient thermal equilibrium with the resonant phonons can be attained. The subsequent decay is governed by the value of τ_b, as illustrated in Fig. 1.27.

The hot phonons generated through a phonon avalanche have been directly detected by Shiren (1966). The phonon bottle-neck and avalanche effects are further discussed in § 10.6.

Although we have used the concept of spin temperature, defined for a simple two-level system by eqn (1.128), this is justified only if the spins are in thermal equilibrium amongst themselves. For spins with the same resonant frequency, such equilibrium is established through the mutual spin flips due to spin-spin interaction outlined in § 1.10; if equilibrium is disturbed, it is re-established in a time of order τ_2. The

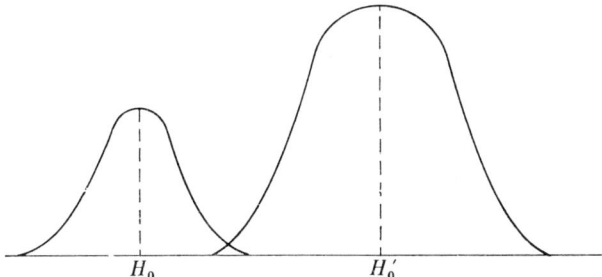

FIG. 1.28. Two overlapping lines, centered on fields H_0 and H_0'. In strong fields mutual spin-flips between two systems can occur through spin-spin interaction only in the region of overlap, where spin packets belonging to the two systems coincide in energy.

concept of a spin temperature is then valid if $\tau_2 \ll \tau_1$, the spin-lattice relaxation time, as is often the case. Similarly, if the spin temperature in one region only of a crystal is raised above the average value, energy will be transferred gradually through the spin-spin interaction to spins in other regions; this process is known as 'spin diffusion', and represents a flow of heat through the spin system which adds (on the macroscopic scale) to the thermal conductivity of the crystal. If we have two spin systems with different resonant frequencies, there is no obvious reason for their spin temperatures T_s, T_s' to be the same unless they are both in thermal equilibrium with a bath (e.g. at temperature T_0). The reason for this is that energy must be conserved in a spin-spin relaxation process, and this is not possible for two spins with different resonant frequencies. However this neglects the fact that the resonance lines have a finite width; if the tails of the two lines overlap, as in Fig. 1.28, mutual spin flips can take place (with conservation of energy) in the region of overlap. This process is known as 'cross relaxation', and since spin-spin relaxation times are often much shorter than spin-lattice relaxation times, this may be a more effective process in removing energy from a 'hot' spin system by transferring it to another 'cold' spin system than direct transfer to the lattice.

If both resonance lines are homogeneously broadened, with overlapping tails, spin-flips in the overlap region will affect all the spins. If the lines are inhomogeneously broadened, only those spin packets in the overlap region will be able to take part in cross-relaxation. However each spin-packet has a finite width associated with its real value of τ_2, and will be able to make spin-flips with neighbouring (and overlapping)

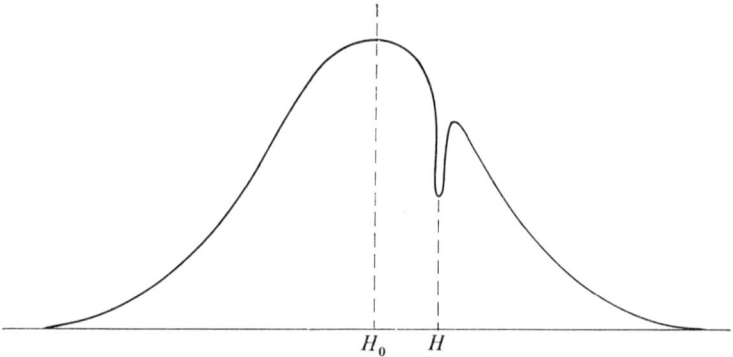

FIG. 1.29. 'Burning a hole in a line'. If the line is inhomogeneously broadened, and a strong saturating power is applied at a frequency such that $h\nu = g\beta H$, only the spin packets resonant at this frequency are saturated, while others are not. In consequence these spin packets give a smaller contribution to the line relative to the remaining packets, 'burning a hole' in the line as shown. This effect is not possible for a homogeneously broadened line, in which all spins absorb power from the applied saturating power.

spin-packets, thus transferring energy gradually throughout the inhomogeneously broadened lines. This is another aspect of cross relaxation, which is connected with the rather dramatically named process of 'burning a hole in a line'. If the line is inhomogeneously broadened, and microwave power is supplied at a single frequency sufficient to cause power saturation, only those spin packets at the resonant frequency will be saturated, so that a 'hole' appears in the resonance curve, as in Fig. 1.29. The saturation effect then spreads through the remainder of the line by cross-relaxation between adjacent spin packets, the degree to which it spreads being determined by the competition between such cross relaxation processes and other processes such as spin-lattice relaxation. On the other hand, in a homogeneously broadened line all spins are affected immediately by the saturating microwave power and the intensity falls throughout the line, no hole being produced.

ELECTRON PARAMAGNETIC RESONANCE

We conclude this section on spin-lattice relaxation by mentioning (and not discussing elsewhere) the solid-state maser. As pointed out initially, it follows from eqn (1.127) that a spin system with an inverted population ($n_b > n_a$) will emit instead of absorbing energy in a magnetic resonance experiment. The net induced emission transfers energy from the spin system to the electromagnetic field, and may thus be used to

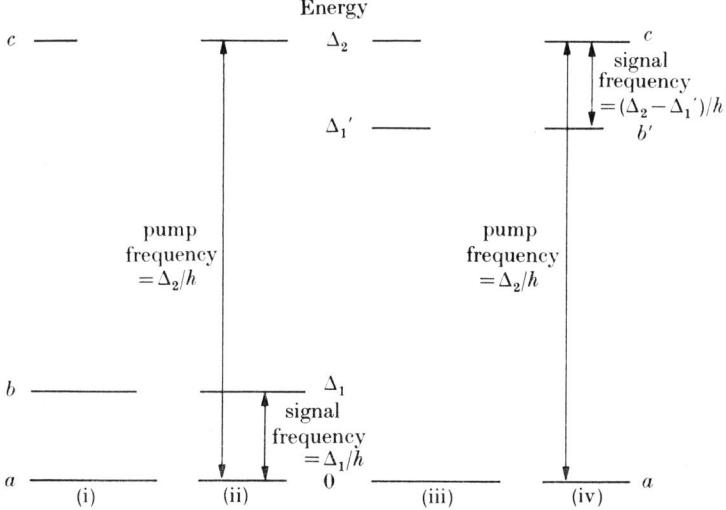

FIG. 1.30. The principle of the three level maser. In (i), (iii) the lengths of the three horizontal lines represent the populations of the levels in thermal equilibrium; in (ii) strong pumping equalizes the populations of levels (a) and (c), causing that of (a) to fall below that of (b). A signal causing transitions between (a) and (b) can then be amplified. The alternative situation, when (b') is nearer in energy to (c) than (a), is shown in (iii), (iv); in this case signal amplification is obtained through transitions between (c) and (b').
The figure is drawn to scale for $(\Delta_2/kT_0) = 1$, $(\Delta_1/kT_0) = 0.2$, $(\Delta_1'/kT_0) = 0.8$.

amplify a signal at the resonant frequency. The noise power inherent in the system is (in a band-width df)

$$P_n = 4k |T_s'| df \qquad (1.140)$$

and if the working substance is at helium temperatures, the equivalent noise temperature can be very low provided that the effective spin temperature is numerically small, even when negative. A negative temperature can be achieved in a two-level system by adiabatic rapid passage (§ 2.5), but this is necessarily a transient method. Steady-state working can be achieved with a spin system containing at least three levels (Bloembergen 1956), as illustrated in Fig. 1.30. The energy levels are at 0, Δ_1 (or Δ_1') and Δ_2, with populations proportional to 1, $\exp(-\Delta_1/kT_0)$ or $\exp(-\Delta_1'/kT_0)$ and $\exp(-\Delta_2/kT_0)$ respectively. If strong saturating microwave power is applied at frequency $\nu_p = \Delta_2/h$, the population difference between levels (a) and (c) will be diminished, and in the limit of infinite pump power the populations of the two levels will be equalized, if we neglect the possibility of relaxation transitions to and from level (b). This situation is shown in Fig. 1.30 (ii), where it is clear that the population of level (b) exceeds that of level (a), as a result of pumping out of (a) into (c). If a signal at frequency $\nu_s = \Delta_1/h$ is applied, it can be amplified through stimulated emission of photons corresponding to transitions between (b) and (a). In Fig. 1.30 (iv) the corresponding situation is shown for the case where the intermediate level is at Δ_1', nearer to (c) than to (a). In this case the pumping action between (a) and (c) produces an excess population in (c) over (b'), and amplification is possible at a signal frequency $\nu_s' = (\Delta_2 - \Delta_1')/h$.

1.12. Dynamic nuclear orientation

We now outline some interesting nuclear effects that can be produced using electron paramagnetic resonance, through the hyperfine interaction between the magnetic electrons and the nuclear spins.

At ordinary temperatures all nuclear sub-states I_z are substantially equally occupied, corresponding to random orientation of the nuclear spins. At sufficiently low temperatures, if kT can be made of the same order as the splitting of the nuclear levels, significant departures from equal occupation and random orientation can occur. These are measured by the following two quantities.

(a) Nuclear polarization is the first quantity, for which ideally all nuclear spins are in the substates $I_z = +I$ (or all in $I_z = -I$), so that all point in the same direction. The degree of polarization is defined by the magnitude of the quantity

$$P_1 = \langle I_z \rangle / I, \tag{1.141}$$

which reaches unity in the ideal case and is zero for random orientation. In a static magnetic field H the nuclear Zeeman interaction gives a level separation $g_I \beta H$; for protons this is about 400 MHz in a field of 100 kG, which is equal to kT only when $T \sim 0.02°K$. Thus significant nuclear polarization demands these somewhat extreme conditions (Simon 1939), and not surprisingly the method is commonly known as the 'brute force' method. It corresponds to reaching saturation in the nuclear paramagnetic susceptibility curve, similar to that common in electronic paramagnetism, but since for electrons $g \sim 10^3 g_I$, the latter can be readily attained in a field of 10 kG at a temperature of 1°K.

A less exacting method was proposed by Gorter (1948) and by Rose (1949), in which a relatively small field is applied to polarize the magnetic electrons in a paramagnetic substance, the nuclei then being polarized through the magnetic hyperfine interaction. In Fig. 1.10 we see that in the region indicated by the arrow marked GR, all the levels are separated by amounts $\sim A$, where A is the magnetic hyperfine constant. This region is reached in fields where $g\beta H \sim A$, and H must therefore be comparable with H_n in eqn (1.28), i.e. 10^2 to 10^3 G. Then at temperatures where $A \sim kT$ (i.e. $T \sim 10^{-2}$ to 10^{-1} °K), appreciable nuclear polarization will be set up through preferential population of the lowest hyperfine level. Such temperatures can be reached by partial adiabatic demagnetization from 1°K using values of $(H/T) \sim 10^4$ to 10^5 G per °K. This clever trick thus makes use of the fact that H_e, the electronic magnetic field at the nucleus (see eqn (1.28)), is of order 10^5 to 10^6 G, but is obviously limited to the nuclei of paramagnetic ions.

(b) Nuclear quadrupolarization is the second quantity, in which the nuclei are preferentially aligned (parallel or antiparallel) along an axis, or in the plane normal to this axis. Ideally all nuclei should be in the sub-states $|I_z| = I$, or in $|I_z| = 0$ or $\frac{1}{2}$ (according as I is integral or half-integral). The degree of quadrupolarization is defined by the magnitude of the quantity

$$P_2 = \frac{\langle 3I_z^2 - I(I+1) \rangle}{3I^2 - I(I+1)} = \frac{\langle 3I_z^2 - I(I+1) \rangle}{I(2I-1)}, \qquad (1.142)$$

which reaches $+1$ for $|I_z| = I$ and is negative if $|I_z| = 0$ or $\frac{1}{2}$. Clearly polarization implies quadrupolarization, but not vice versa. A direct method of quadrupolarization was suggested by Pound (1949), based on cooling a substance with a large nuclear electric quadrupole interaction

$$\mathscr{H} = P_\parallel \{I_z^2 - \tfrac{1}{3}I(I+1)\} \qquad (1.71)$$

down to temperatures such that $kT \approx P_\parallel$ (see Fig. 1.31).

This method has been used to produce appreciable nuclear quadrupolarization for uranium nuclei in $Rb(UO_2)(NO_3)_3$ (see § 6.5) which has a singlet electronic ground state. In normal paramagnetic substances with degenerate electronic ground states the nuclear electric quadrupole interaction is generally small compared with the magnetic hyperfine interaction. When the latter is anisotropic, as is frequently the case, it can be used to produce nuclear quadrupolarization by adiabatic demagnetization to zero magnetic field (Bleaney 1951a). From Fig. 1.10 it is clear that when there is no anisotropy there can be no nuclear

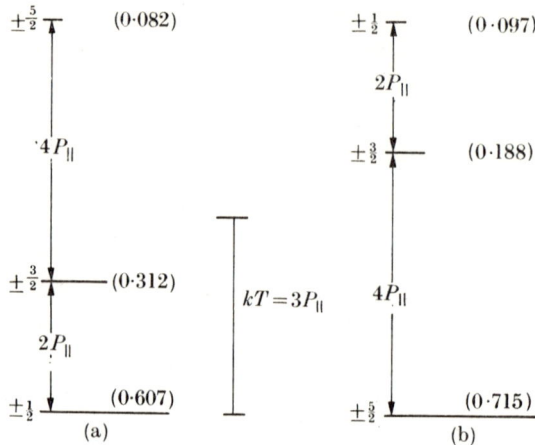

FIG. 1.31. Energy level diagram for a nuclear electric quadrupole splitting

$$P_{\parallel}\{I_z^2 - \tfrac{1}{3}I(I+1)\}$$

for a nucleus with $I = \tfrac{5}{2}$. The numbers in parentheses and lengths of horizontal line represent the populations (normalized to unity) of each level at a temperature such that $kT = 3P_{\parallel}$ (equal to half the overall splitting), indicated by the vertical line in the centre.
(a) $P_{\parallel}$ positive; (b) $P_{\parallel}$ negative.

polarization or quadrupolarization in zero magnetic field, since all substates m_F for each given value of F are degenerate and equally occupied; this result follows more generally from the fact that there is nothing to define a preferred axis in space. On the other hand, when the magnetic hyperfine interaction is anisotropic and, for example, of the form (with $A_{\parallel} \neq A_{\perp}$)

$$\mathcal{H} = A_{\parallel}\tilde{S}_z I_z + A_{\perp}(\tilde{S}_x I_x + \tilde{S}_y I_y) \tag{1.85}$$

there is a preferred axis. F is not a good quantum number (see § 3.9), though $m_F = M + m$ is when the magnetic quantum numbers are referred to the z-axis. The splitting pattern in zero field is altered; states of different values of m_F have different energies (cf. Fig. 3.12), so that those lowest in energy are preferentially populated, giving the possibility of a nuclear quadrupolarization. The limiting cases of anisotropy, one with $A_{\perp} = 0$ and the other with $A_{\parallel} = 0$, are illustrated in Fig. 1.32, and it is clear that they lead to opposite signs for the nuclear quadrupolarization. The first successful nuclear alignment experiments were carried out using this method (Daniels, Grace, and Robinson 1951), in which preferential emission of γ-rays was observed normal to the z-axis from ^{60}Co nuclei. Since parity is conserved in such a process, the

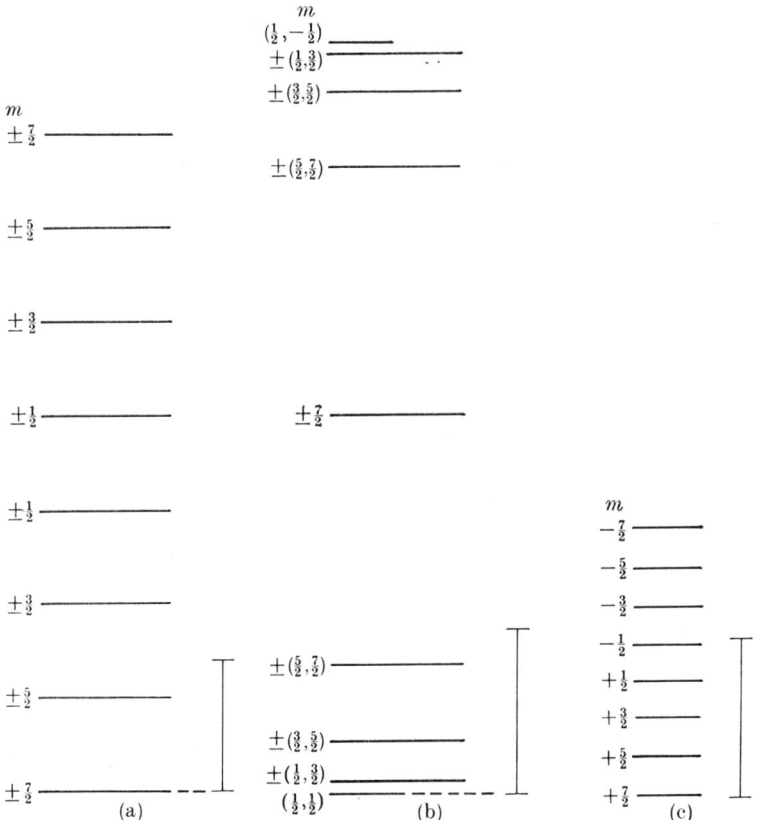

FIG. 1.32. Energy levels in zero magnetic field for $\tilde{S} = \frac{1}{2}$, $I = \frac{7}{2}$, for the extremes (a) $A_\perp = P_\parallel = 0$, (b) $A_\parallel = P_\parallel = 0$. Cobalt ammonium sulphate approximates to (a), and, for comparison, the nuclear levels of ^{59}Co in a diamagnetic cobalt compound in a field of 100 kG are shown to scale in (c). The vertical lines in (a) and (b) represent kT for the lowest temperatures which could be reached by demagnetization from 1°K, assuming that this is limited only by the hyperfine splittings. The vertical line in (c) represents kT for $T = 0.01°K$. Most states in (b) are admixtures of two values of $m = m_F \pm \frac{1}{2}$. In the actual experiment (see Bleaney, Daniels et al. (1954)) a diluted copper tutton salt containing a small quantity of Co^{2+} was used in order to attain the lowest possible temperature (after Bleaney 1951b).

angular distribution depends on the nuclear quadrupolarization and analogous higher even-order quantities. In β-emission, parity is not conserved, and the angular distribution depends on the nuclear polarization and higher-order quantities, both odd and even. Thus in the celebrated experiment of Wu et al. (1957), which showed that β-rays are preferentially emitted in one *direction* rather than in the reverse direction, an external magnetic field was applied as in the Gorter–Rose method in order to produce a nuclear polarization.

The experiments just described are all essentially static methods, in which any external field applied is a steady one. Magnetic resonance is not involved, except that discovery of the highly anisotropic nature of the hyperfine interaction suggested the first successful method, and detailed measurements indicated which salt to use. We consider below methods of 'dynamic nuclear polarization', in which magnetic resonance is directly involved, but first discuss the question of relaxation times. These are in fact an important parameter in the static experiments, since in the methods of 'brute force' and Pound only nuclear levels are involved, and thermal contact with the nuclear system is only established extremely slowly. In the other two methods using complete or partial demagnetization of a paramagnetic salt, the electronic and nuclear levels are so completely mixed by the hyperfine interaction that equilibrium is virtually immediate; the process involves only a change in the way in which the entropy (a constant in the adiabatic demagnetization process) is divided between the electronic and nuclear systems, at temperatures where the entropy of the lattice is in comparison negligible.

For simplicity we restrict the discussion to an example with axially symmetric magnetic hyperfine interaction as in eqn (1.85) with a strong magnetic field along the z-axis, so that the spin Hamiltonian is

$$\mathscr{H} = g_\| \beta H_z \tilde{S}_z + A_\| \tilde{S}_z I_z + \tfrac{1}{2} A_\perp (\tilde{S}_+ I_- + \tilde{S}_- I_+), \qquad (1.143)$$

where we have re-written the $A_\perp$ term using raising and lowering spin operators. For the case of $\tilde{S} = \tfrac{1}{2}$, $I = 1$ the energy levels are shown schematically in Fig. 1.33, together with the approximate values of the populations (unnormalized) when $h\nu/kT \approx g_\| \beta H_z/kT = \delta_e \ll 1$. We have assumed that $A_\|/kT$ is so small that we can neglect the population differences due to the hyperfine splittings, and retain only those due to the electronic splitting. The important relaxation processes are those associated with electronic transitions, and in order of importance (i.e. order of rapidity) they follow the order of intensity set out in (a), (b), (c) of § 1.6. The spin-lattice relaxation rate for the $\Delta M = \pm 1, \Delta m = 0$ transitions is just $1/\tau_1$; for the $\Delta M = \pm 1, \Delta m = \pm 1$ transitions the rate is much slower, approximately $(A_\perp/h\nu)^2(1/\tau_1)$, but for the spin Hamiltonian of (1.143) only the

$$\Delta M = \pm 1, \qquad \Delta m = \mp 1; \quad \text{or} \quad \Delta(M+m) = 0,$$

transitions (corresponding to 'flip-flop' transitions) are partially allowed, and have relaxation rates of this order. This is because the spin

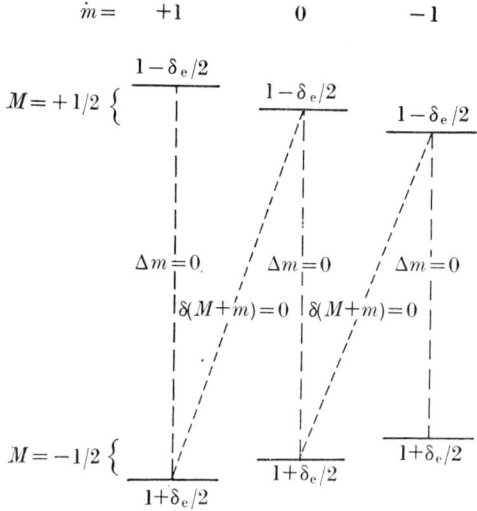

FIG. 1.33. The energy levels for the spin Hamiltonian (1.143) for $\tilde{S} = \frac{1}{2}$, $I = 1$ in strong magnetic field H_z. Main relaxation paths are shown by broken lines ($\Delta m = 0$), and subsidiary relaxation paths (less rapid by factors of order $(A_\perp/h\nu)^2$) by dotted lines ($\Delta(M+m) = 0$). Much slower relaxation paths are omitted. The relative populations of each state in thermal equilibrium are given in the approximation where $(A_\parallel/kT)$ is negligible and
$$\delta_e = (h\nu/kT) \ll 1$$

Hamiltonian contains terms of the form S_+I_-, S_-I_+ but not of the form S_+I_+ or S_-I_-. The remaining transitions are associated with much slower relaxation rates, which we neglect. The main 'relaxation paths' are indicated in Fig. 1.33, and maintain the population differences shown (in the above approximation), with no net nuclear polarization or quadrupolarization.

When an oscillatory magnetic field is applied at a magnetic resonance frequency, sufficiently strong to cause transitions at a rate much greater than $(1/\tau_1)$, saturation sets in, and the population of the two levels involved tends to become equalized. We simplify the algebra by assuming they become exactly equal. The relaxation processes then act to maintain population differences of order δ_e between other states with different values of M by means of the relaxation paths that are not in direct competition with the magnetic resonance transition. The resulting populations of the various states (under steady-state conditions) are shown in Fig. 1.34, where the magnetic resonance transition being saturated (that for which $m = +1$) is shown by a continuous line and the relaxation paths by broken lines. If we compare the populations of the various nuclear states it is clear that the population of the $m = +1$

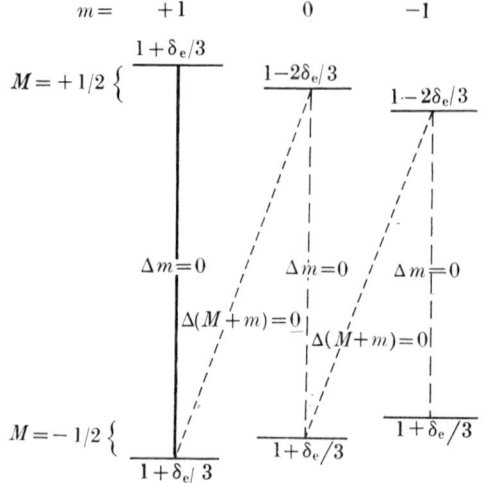

Fig. 1.34. Similar to Fig. 1.33, except that strong saturating magnetic resonance power is applied to the $\Delta m = 0$ line $(+\tfrac{1}{2}, +1) \leftrightarrow (-\tfrac{1}{2}, +1)$, which equalizes the populations of these two states. The populations of the other states are maintained by relaxation paths so that the upper states, apart from $(+\tfrac{1}{2}, +1)$, are always less in population than the lower states by an amount δ_e. The result is a nuclear polarization $P_1 = +\delta_e/6$, and a nuclear quadrupolarization $P_2 = +\delta_e/6$.

states has been enhanced in comparison with that of the $m = 0$ or -1 states. Thus a nuclear polarization has been set up, for which $P_1 = \langle I_z \rangle / I = +\delta_e/6$, while the nuclear quadrupolarization (see eqn (1.142)) is $P_2 = +\delta_e/6$, also.

The results of saturating any of the other $\delta m = 0$ transitions can readily be found (see Fig. 1.35). If the transition between the $m = -1$ states is saturated, the value of P_1 is the same as above, but $P_2 = -\delta_e/6$. If the transition between the $m = 0$ states is saturated, one obtains twice as much polarization, $P_1 = \delta_e/3$, but $P_2 = 0$. If all three of the $\delta m = 0$ transitions represented by vertical lines are saturated, the net polarization is $P_1 = 2\delta_e/3$, but the quadrupolarization $P_2 = 0$. These last results are the same as found by taking the sum of the values of P_1 or P_2 obtained by saturating individual $\delta m = 0$ lines, but this simple rule holds only for $\delta_e \ll 1$.

In this method of 'dynamic nuclear polarization', first analyzed by Abragam (1955), no nuclear polarization or quadrupolarization is set up except through the relaxation processes, which must therefore be allowed time to act. The controlling relaxation rate is that for the $\Delta(M+m) = 0$ transitions (represented by sloping broken lines), which is of order $(A_\perp/h\nu)^2(1/\tau_1)$, and in some cases may be quite long. This

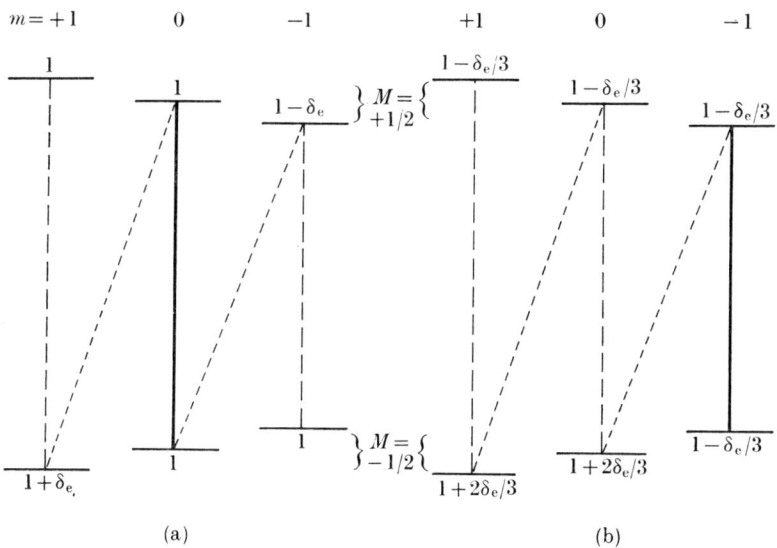

Fig. 1.35. Similar to Fig. 1.34, except that strong saturating electron resonance power is used to equalize the populations of two different hyperfine states. The populations of the other states are maintained by the relaxation paths shown. (a) Saturation of the $m = 0$ hyperfine line, giving nuclear polarization $P_1 = +\delta_e/3$, nuclear quadrupolarization $P_2 = 0$. (b) Saturation of the $m = -1$ hyperfine line, giving

$$P_1 = +\delta_e/6, \qquad P_2 = -\delta_e/6.$$

Whichever hyperfine line is saturated (see also Fig. 1.34), the nuclear polarization P_1 has the same sign because of the asymmetry in the direction of the sloping relaxation paths.

difficulty is avoided by an important modification due to Jeffries (1957), in which one of the $\Delta(M+m)$ transitions is saturated instead of one of the $\Delta m = 0$ transitions. This is represented schematically in Fig. 1.36, in which the $(+\frac{1}{2}, 0) \leftrightarrow (-\frac{1}{2}, +1)$ transition is saturated. The rapid relaxation paths are indicated by broken vertical lines, and there is one much slower relaxation path indicated by a sloping dotted line (other relaxation paths are again neglected). At a time τ after the saturating power is switched on, the populations are shown in Fig. 1.36(a) for $\tau_1 \ll \tau \ll (h\nu/A_\perp)^2 \tau_1$, and in Fig. 1.36(b) for $\tau \gg (h\nu/A_\perp)^2 \tau_1$. In the former case only the vertical relaxation paths have had time to operate, whereas in the latter the sloping relaxation path has also become fully effective. The resulting polarization parameters are

(a) $P_1 = -\delta_e/6, \qquad P_2 = -\delta_e/2,$

(b) $P_1 = -\delta_e/3, \qquad P_2 = -\delta_e/3.$

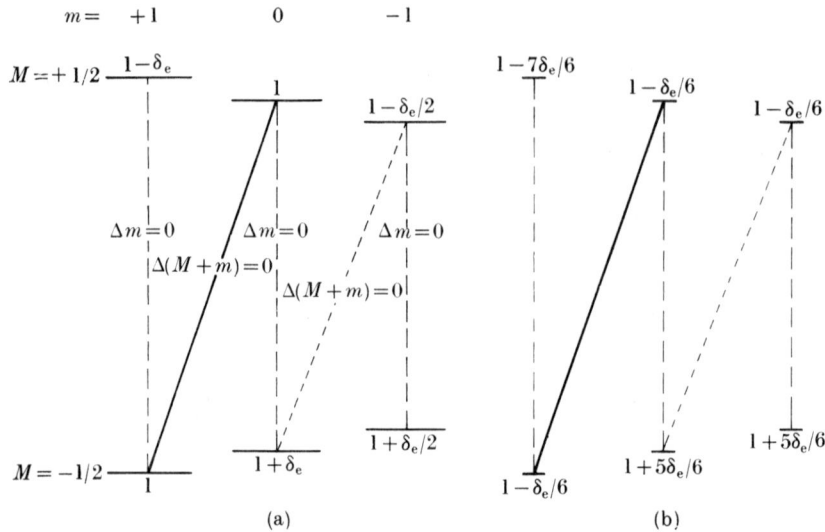

FIG. 1.36. Similar to the last two figures, except that a $\Delta(M+m) = 0$ line, $(+\tfrac{1}{2}, 0) \leftrightarrow (-\tfrac{1}{2}, +1)$ is saturated, equalizing the populations of these two states. The populations given in (a) assume that the sloping relaxation path (dotted line) is inoperative; those in (b) assume that this relaxation path is playing its full role. The nuclear parameters are (a) $P_1 = -\delta_e/6$, $P_2 = -\delta_e/2$; (b) $P_1 = -\delta_e/3$, $P_2 = -\delta_e/3$.

If the other transition, $(+\tfrac{1}{2}, -1) \leftrightarrow (-\tfrac{1}{2}, 0)$ is saturated, the values of P_1 are the same, while those of P_2 are reversed. A notable advantage of this method is that one can detect quickly when the right frequency is being applied to saturate one of the $\Delta(M+m) = 0$ transitions by observing the anisotropy in the γ-ray emission (or β-ray, but this is generally experimentally more complicated) so that the transitions detected are those of the radio-active isotope itself. With short-lived isotopes, quantities far below those necessary to give a paramagnetic resonance signal directly detected can be used, and the value of the hyperfine splitting for the radio-active isotope is found from the positions of the resonance lines thus detected. In principle, this is possible also by saturating the $\Delta m = 0$ transitions, but long relaxation times may make this difficult in practice. The nuclear spin can also be determined in either case, since the number of $\Delta(M+m) = 0$ transitions (for $S = \tfrac{1}{2}$) is just $2I$, while the number of $\Delta m = 0$ transitions is $2I+1$ (but note that no nuclear quadrupolarization is obtained by saturating the $m = 0$ transition); see, for example, Abraham, Kedzie, and Jeffries (1960).

In cases where relaxation times are unusually long, the technique of adiabatic rapid passage (see § 2.5) can be used to produce nuclear orientation or alignment (Feher 1956a), which endures until thermal equilibrium is re-established through relaxation effects. The method is illustrated in Fig. 1.37, where relaxation processes are assumed to be negligible and the relaxation paths have been omitted. The effect of adiabatic rapid passage is to invert the populations of a pair of states when the system is swept over the resonance transition between them. If first, an electronic rapid passage experiment is carried out for the

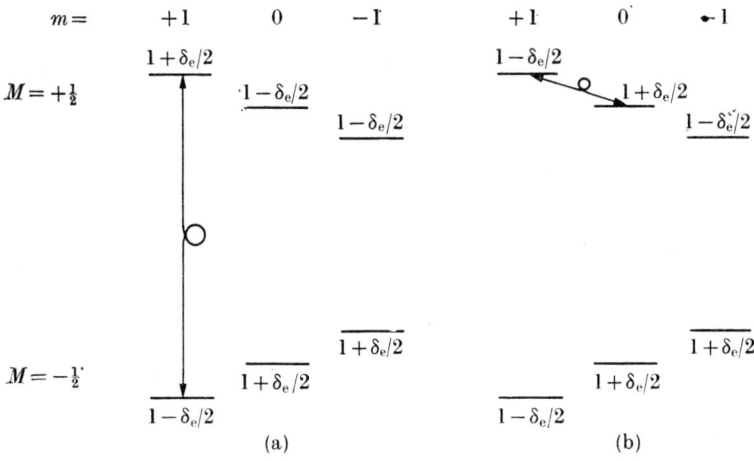

FIG. 1.37. Population inversion in pairs of states following adiabatic rapid passage over the transition between them. Relaxation mechanisms are omitted, since inversion of the population is independent of relaxation effects, though it lasts only for a time of the order of the relaxation time. (a) Populations after adiabatic rapid passage over the $m = +1$, $M = +\frac{1}{2} \longleftrightarrow -\frac{1}{2}$ transition, followed by (b) adiabatic rapid passage over the $M = +\frac{1}{2}$, $m = +1 \longleftrightarrow 0$ transition. (a) results in no nuclear alignment, but after (b) the nuclear polarization parameters are $P_1 = -\delta_e/6$, $P_2 = -\delta_e/2$.

$\Delta m = 0$ line $m = +1$, the populations of the two states $M = \pm\frac{1}{2}$, $m = 1$ are inverted, as shown in Fig. 1.37(a). It is obvious from the populations given in this diagram that no nuclear alignment is thus achieved. If now a nuclear rapid passage experiment is carried out for the transition $(+\frac{1}{2}, +1) \leftrightarrow (+\frac{1}{2}, 0)$, the populations become as shown in Fig. 1.37(b). Both a nuclear polarization, $P_1 = -\delta_e/6$, and a quadrupolarization $P_2 = -\delta_e/2$ are created. If a third adiabatic rapid passage experiment is carried out over the next nuclear transition $(+\frac{1}{2}, 0) \leftrightarrow (+\frac{1}{2}, -1)$, the nuclear parameters become $P_1 = -\delta_e/3$, $P_2 = 0$, so that the polarization has been doubled but the quadrupolarization reduced to zero. Feher and Gere (1956) demonstrated the success of such an experiment for phosphorus nuclei in silicon by observing the strength of the various resonance transitions after adiabatic rapid passage; Pipkin and Culvahouse (1957) observed the alignment of the radioactive isotope ^{76}As in silicon through the anisotropy in its γ-ray emission.

The methods of dynamic nuclear orientation outlined hitherto have assumed a well-resolved hyperfine structure, typically that due to the interaction between the magnetic electrons and the nucleus of the paramagnetic ion itself. We end this section by discussing briefly the dynamic orientation of nuclei of other (diamagnetic) ions through their interaction (mainly magnetic dipolar in nature) with a few paramagnetic ions present in low concentration, which can be subjected to electron paramagnetic resonance fields at power saturation levels. This subject is of importance, since it gives a means of producing highly polarized nuclear targets (usually of protons) for particle scattering experiments, without the extreme conditions required for a static method such as the 'brute force' method. The method was proposed and demonstrated in a solid by Abragam and Proctor (1958), and is often known as the 'effet solide' or 'solid effect', though it was simultaneously observed in a liquid (Erb, Motchane, and Uebersfeld 1958).

If the interaction between a neighbouring nuclear spin j and a paramagnetic ion i is just due to their two magnetic dipole moments, it is essentially similar to that illustrated in Fig. 1.17. We do not, however, need to assume that it is dipolar in character, but write the interaction in the form $(\tilde{\mathbf{S}}_i \cdot \mathscr{J}_{ij} \cdot \mathbf{I}_j)$, which is a shorthand notation similar to (1.62) or (1.126) involving all possible products of the components of $\tilde{\mathbf{S}}_i$ and $\mathbf{I}_j$. In particular, it contains products of the form $\tilde{S}_{iz}I_{j\pm}$ as well as $\tilde{S}_{i\pm}I_{j\pm}$, the former being more significant for the 'effet solide'.

On including the nuclear Zeeman term, the spin Hamiltonian is

$$\mathscr{H} = g\beta(\mathbf{H} \cdot \tilde{\mathbf{S}}_i) - g_I\beta(\mathbf{H} \cdot \mathbf{I}_j) + (\tilde{\mathbf{S}}_i \cdot \mathscr{J}_{ij} \cdot \mathbf{I}_j) \qquad (1.144)$$

where, for simplicity, we assume that g is isotropic. If we can neglect the third term in comparison with the second (and, *a fortiori*, in comparison with the first), the energy levels in a strong field H are as shown in Fig. 1.38 for the case $\tilde{S} = I = \frac{1}{2}$. The importance of the third term in (1.144) arises from the presence of products such as $\tilde{S}_z I_{\pm}$, which admix states $m = \pm\frac{1}{2}$ belonging to the same value of M, with amplitudes $\sim(\mathscr{J}_{ij}/g_I\beta H)$. This makes possible partially allowed transitions of the form $\Delta M = +1, \Delta m = +1$ or $\Delta M = -1, \Delta m = -1$ ('flip-flip' transitions) as well as $\Delta M = \pm 1, \Delta m = \mp 1$ ('flip-flop' transitions), with intensity $\sim(\mathscr{J}_{ij}/g_I\beta H)^2$ relative to the $\Delta m = 0$

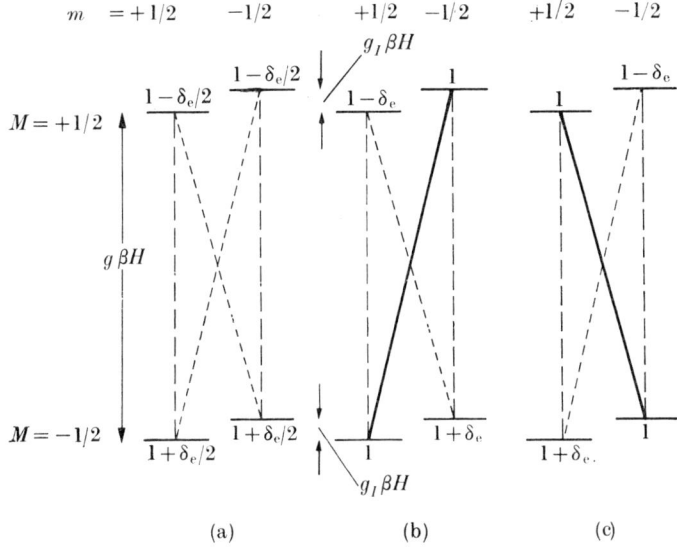

FIG. 1.38. The energy levels for the system $\tilde{S} = I = \frac{1}{2}$ corresponding to the spin Hamiltonian (1.144) with the third term neglected. The main relaxation paths are represented by the vertical broken lines, and less important paths by the sloping dotted lines; other relaxation paths are omitted. The level populations are shown for three cases: (a) thermal equilibrium; nuclear polarization $P_1 = 0$ as shown, but actually $P_1 = (g_I \beta H / 2kT)$ if the population differences due to the nuclear Zeeman term are included; (b) saturation of the $(+\frac{1}{2}, -\frac{1}{2}) \leftrightarrow (-\frac{1}{2}, +\frac{1}{2})$ satellite, giving $P_1 = -\delta_e/2$; (c) saturation of the $(+\frac{1}{2}, +\frac{1}{2}) \leftrightarrow (-\frac{1}{2}, -\frac{1}{2})$ satellite, giving $P_1 = +\delta_e/2$, where $\delta_e = (g\beta H/kT)$.

transitions. If $g_I \beta H$ is greater than the line width, these are observed as satellite lines at $h\nu = (g \pm g_I)\beta H$ on either side of the main line ($\Delta m = 0$) at $h\nu = g\beta H$, as shown in Fig. 1.39. The products $\tilde{S}_\pm I_\pm$, which also occur, only produce admixtures of amplitude $\sim (\mathscr{I}_{ij}/g\beta H)$, and are relatively less important. The important relaxation paths follow the allowed and partially allowed transitions, with similar relative importance.

These paths, together with the relative populations, are shown in Fig. 1.38(a), where we have neglected population differences of order $\delta_n = (g_I \beta H/kT)$ arising from the nuclear Zeeman interaction. If either of the allowed transitions $\Delta m = 0$ is saturated, no change in the nuclear polarization is induced, except through the residual relaxation paths that may produce changes of either sign, depending on which of the sloping relaxation paths is the more important (and this may differ for the various neighbouring nuclei). If the satellite transition

$$(+\tfrac{1}{2}, -\tfrac{1}{2}) \leftrightarrow (-\tfrac{1}{2}, +\tfrac{1}{2})$$

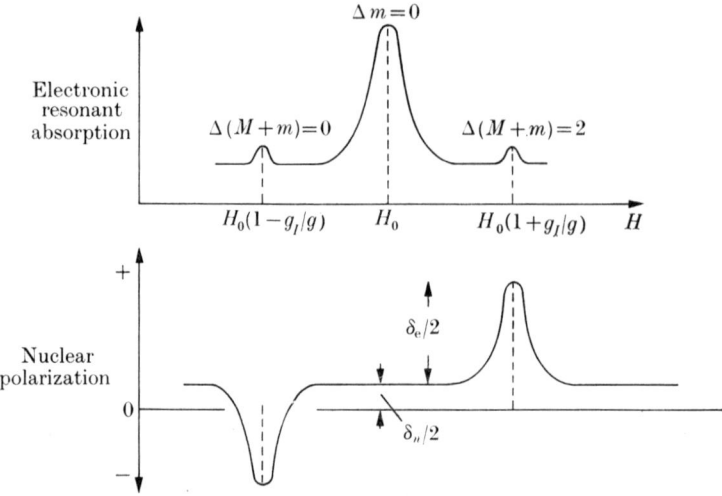

Fig. 1.39. Illustrating the electron paramagnetic resonance absorption and nuclear polarization for a system with $\tilde{S} = I = \frac{1}{2}$, and the spin Hamiltonian (1.144). Constant frequency is assumed, with scanning of the magnetic field, so that the satellite on the high field side is the $(+\frac{1}{2}, +\frac{1}{2}) \leftrightarrow (-\frac{1}{2}, -\frac{1}{2})$ transition, and that on the low field side is $(+\frac{1}{2}, -\frac{1}{2}) \leftrightarrow (-\frac{1}{2}, +\frac{1}{2})$.

is saturated, the populations become as shown in Fig. 1.38(b), with a net nuclear polarization $P_1 = -\delta_e/2$; if the other satellite

$$(+\tfrac{1}{2}, +\tfrac{1}{2}) \leftrightarrow (-\tfrac{1}{2}, -\tfrac{1}{2})$$

is saturated, the populations given in Fig. 1.38(c) show that the net nuclear polarization $P_1 = +\delta_e/2$ (see Fig. 1.39). Thus by saturating either satellite, nuclear polarizations of either sign can be produced which are enhanced over the value $P_1 = \delta_n/2$ appropriate to conditions of thermal equilibrium by factors of order $(\delta_e/\delta_n) = (g/g_I)$. If δ_e is not small compared with unity, the electronic polarization becomes $\tanh(\delta_e/2) = \tanh(h\nu/2kT)$, and this is the limiting value of the nuclear polarization which can be reached under ideal conditions. Obviously for $I = \frac{1}{2}$, the question of nuclear quadrupolarization does not arise.

The various methods of dynamic nuclear orientation discussed above essentially enhance nuclear polarizations from the thermal equilibrium value of order $\delta_n = (g_I \beta H/kT)$ to polarizations or quadrupolarizations of order $\delta_e = (g\beta H/kT)$, though these limits may not be reached because of competition from other relaxation processes which we have neglected in order to bring out the basic mechanisms on which the enhancement depends. The subject will not be discussed further, in view of the comprehensive reviews by Jeffries (1963) and Abragam and Borgini (1964), which also include the method of Overhauser (1953) for metals, the first method of enhancement proposed using electron paramagnetic resonance.

1.13. Endor

In the method of nuclear orientation of Feher (1956a) discussed in the last section, both electronic paramagnetic resonance and nuclear magnetic resonance (at a hyperfine frequency) are involved, using the technique of adiabatic rapid passage. In this section we outline an allied extension of electron paramagnetic resonance, also due to Feher (1956b), which has proved of outstanding importance. This is known generally by the convenient name of Endor, an acronym for Electron-Nuclear DOuble Resonance. In essence it makes possible the detection of nuclear magnetic resonance through its effect on the electronic magnetic resonance signal, thus making use of the high sensitivity associated with the latter. The nuclear magnetic resonance occurs at a hyperfine frequency, and is insufficient in intensity for direct detection. The Endor method thus offers a direct measurement of the hyperfine frequencies, with (as we shall see) much higher precision than can be obtained from a simple electronic magnetic resonance experiment.

We consider a system with $\tilde{S} = \frac{1}{2}$, $I = 1$, for which the level populations in thermal equilibrium (neglecting all differences except those between levels whose energy differs by an electronic quantum $h\nu = g\beta H$) are shown in Fig. 1.33. If low-level electron paramagnetic resonance power is applied to any of the $\Delta m = 0$ transitions (or $\Delta(M+m) = 0$ transitions), the level populations are not appreciably disturbed and the signal intensity is determined primarily by the difference in population between the lower and upper levels involved in the transition, such as that between the $(+\frac{1}{2}, +1)$ and $(-\frac{1}{2}, +1)$ levels. If at the same time we apply nuclear magnetic resonance power, for example, to the transition $(+\frac{1}{2}, +1) \leftrightarrow (+\frac{1}{2}, 0)$, the nuclear signal will have in comparison an extremely low intensity, partly because the quanta absorbed or emitted are so much smaller than in the electronic transition, and partly because the rate of absorption and emission is so much smaller. The n.m.r. signal cannot therefore be directly detected. Even if we supply sufficient power at this nuclear frequency to reach saturation, the most we can do is to equalize the populations of the $(+\frac{1}{2}, +1)$, $(+\frac{1}{2}, 0)$ levels, and since otherwise they differ only by $\delta_n = (h\nu_n/kT)$, a quantity neglected in Fig. 1.33, the biggest fractional effect we can produce on the intensity of the electronic signal will be of order (δ_n/δ_e).

The situation becomes significantly different if strong electron resonance power is applied, sufficient, for example, to saturate the $(+\frac{1}{2}, +1) \leftrightarrow (-\frac{1}{2}, +1)$ transition and equalize the populations of these two levels, as shown in Fig. 1.34. The populations of the remaining levels attain the values given in this figure through relaxation effects by the paths shown. In these circumstances, the populations of the two levels involved in the nuclear transition $(+\frac{1}{2}, +1) \leftrightarrow (+\frac{1}{2}, 0)$ differ by an amount of order δ_e instead of δ_n. The nuclear signal at this frequency will therefore be enhanced by a factor of order (δ_e/δ_n), though this is still generally insufficient to make it directly detectable in a dilute magnetic system. If, however, sufficient power can be applied at the nuclear frequency to saturate this transition, the population of the $(+\frac{1}{2}, +1)$ level will be reduced towards equality with the $(+\frac{1}{2}, 0)$ transition, thus affecting the intensity of the electronic signal corresponding to transitions between the $(+\frac{1}{2}, +1)$ and $(-\frac{1}{2}, +1)$ levels. We have thus achieved the desired effect, of being able to detect when the right frequency is applied to cause a nuclear transition, through its effect on the intensity of the much stronger electronic resonance signal.

A full analysis of the Endor process, including a calculation of the signal intensity, is very complex, depending on the relative importance of the transitions induced through the relaxation paths (including those omitted from Fig. 1.34) and through the applied saturating magnetic resonance power. We shall here carry the analysis only one stage further, making the simplifying assumptions that both the electronic and nuclear transitions are fully saturated, and retaining only the more important relaxation paths. This situation is illustrated in Fig. 1.40, which is similar to Fig. 1.34 except that a nuclear transition is saturated as well as an electronic transition. This means that all three levels involved in these two transitions are equalized in population; the remaining three levels attain steady-state populations relative to the first three via relaxation paths that connect them immediately or ultimately to the $|+\frac{1}{2}, 0\rangle$ state. Comparison of Figs. 1.34 and 1.40 shows that the populations of the two states $|+\frac{1}{2}, +1\rangle$ and $|-\frac{1}{2}, +1\rangle$, between which the electronic transitions are taking place, have altered by an amount $\sim \delta_e$. We cannot simply use this to find out how much the electronic resonance signal is altered, since in the limiting case of complete saturation that we have assumed for simplicity, this signal would apparently be zero because of the equal populations of the two

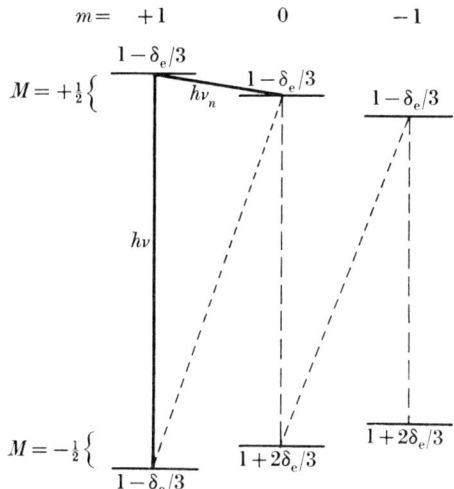

Fig. 1.40. Level populations in a system with $S = \frac{1}{2}$, $I = 1$ while saturating power is applied both to the electronic transition $(+\frac{1}{2}, +1) \leftrightarrow (-\frac{1}{2}, +1)$ and to the nuclear transition $(+\frac{1}{2}, +1) \leftrightarrow (+\frac{1}{2}, 0)$, thus equalizing the populations of three levels. This makes the populations of the $m = +1$ levels $(1-\delta_e/3)$ instead of the value $(1+\delta_e/3)$ which they have if only the $m = +1$ electronic resonance is saturated (see Fig. 1.34). The change in population shows that the intensity of the electronic resonance is changed by the application of power at the nuclear (hyperfine) frequency. The other nuclear transitions can be observed similarly if the other electronic transitions are saturated, as in Fig. 1.35.

states $|+\frac{1}{2}, +1\rangle$ and $|-\frac{1}{2}, +1\rangle$ in both Fig. 1.34 and Fig. 1.40. In practice it would neither be desirable nor readily possible to attain complete saturation, and a power level would be used which appreciably reduces the difference in population of these two states, while still allowing a reasonably strong electron resonance signal; this signal would still be significantly altered in magnitude when sufficient power is applied to the nuclear transition.

This problem is discussed further in § 4.6. We can, however, readily see from a comparison of Figs. 1.34 and 1.40 that the nuclear power must be strong enough to induce transitions between the $|+\frac{1}{2}, +1\rangle$ and $|+\frac{1}{2}, 0\rangle$ states at a rate comparable with, or faster than, thermalizing transitions through the relaxation path between the $|-\frac{1}{2}, +1\rangle$ and $|+\frac{1}{2}, 0\rangle$ levels. Hence the nuclear power must compete, not just with the nuclear relaxation time for $\Delta M = 0$ transitions, but with the electronic relaxation rate at least for the $\Delta(M+m) = 0$ transitions, which is of order $(A_\perp/h\nu)^2(1/\tau_1)$ for the spin Hamiltonian (1.143) (see § 1.12).

In a steady-state Endor experiment the actual signal intensity depends on the effects of all the relaxation paths competing with the saturating power, and the ultimate steady state is of course only reached in a time at least of order of the longest relevant relaxation time. An experiment carried out in a much shorter time must necessarily produce only transient effects, though if all relaxation times are long such effects may endure for correspondingly long times. The Endor process is much easier to analyze if relaxation effects can be neglected, as in the closely allied nuclear polarization experiment of Feher (1956a) described in § 1.12 and illustrated in Fig. 1.37. This made use of the technique of adiabatic rapid passage, but similar effects are obtained by applying a short saturating pulse. Figure 1.41 illustrates

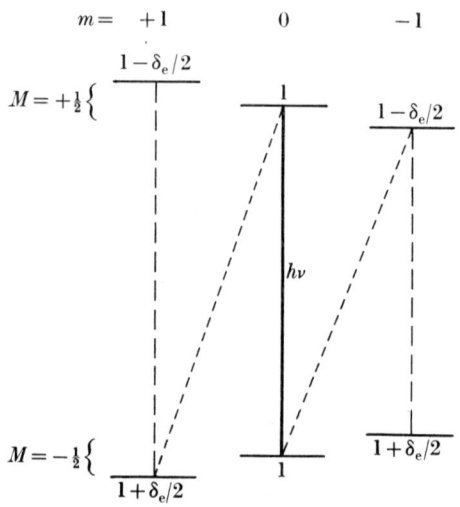

FIG. 1.41. Level populations in a system with $\tilde{S} = \frac{1}{2}$, $I = 1$ after saturating power is applied to the $(+\frac{1}{2}, 0) \leftrightarrow (-\frac{1}{2}, 0)$ transition but before relaxation processes have reacted to the disturbance. A population difference of order $|\delta_e/2|$ is created between any two adjacent nuclear levels, between which $\Delta M = 0$, $\Delta m = \pm 1$ transitions can be induced at a nuclear (hyperfine) frequency.

the level populations obtained by saturating the $(+\frac{1}{2}, 0) \leftrightarrow (-\frac{1}{2}, 0)$ transition for the system $\tilde{S} = \frac{1}{2}$, $I = 1$, before relaxation effects have had time to react to the disturbance from thermal equilibrium; the populations of the two $m = 0$ levels are equalized, while those of the other four levels are unaltered. There is now a population difference of order $|\delta_e/2|$ between all adjacent nuclear levels, so that by saturating any of the nuclear transitions $\Delta M = 0$, $\Delta m = \pm 1$ the populations of at least three levels including the two $m = 0$ levels will be changed. This alters the intensity of the electronic paramagnetic resonance absorption for the $(+\frac{1}{2}, 0) \leftrightarrow (-\frac{1}{2}, 0)$ transition, by means of which the nuclear transitions can be detected. Thus Endor methods are available both for materials with very long relaxation times and for those with comparatively short relaxation times within the limitation cited at the end of the previous paragraph.

The information content of the Endor signals is illustrated by taking the simple diagonal Hamiltonian

$$\mathscr{H} = g_\parallel \beta H_z \tilde{S}_z + A_\parallel \tilde{S}_z I_z + P_\parallel \{I_z^2 - \tfrac{1}{3} I(I+1)\} - g_I \beta H_z I_z \quad (1.145)$$

for the case $\tilde{S} = \frac{1}{2}, I = 1$. In a simple electron paramagnetic resonance experiment, the transition $(+\frac{1}{2}, m) \leftrightarrow (-\frac{1}{2}, m)$ requires a quantum

$$h\nu = |g_\parallel \beta H_z + A_\parallel m| \quad (1.146)$$

from which we can find $g_\parallel$ and $A_\parallel$, but not $P_\parallel$ or g_I. The accuracy is limited by the line width, and by the fact that $A_\parallel$ is determined just from the spacing between the hyperfine lines and hence from the *differences* in the values of H_z (at constant frequency) for the different values of m. In the Endor method we detect transitions such as $(M, m) \leftrightarrow (M, m-1)$, which requires a nuclear quantum

$$h\nu_n = |A_\parallel M + P_\parallel (2m-1) - g_I \beta H_z|. \quad (1.147)$$

The four such transitions in the $\tilde{S} = \frac{1}{2}, I = 1$ system are all of the type (with different combinations of signs)

$$h\nu_n = |\pm \tfrac{1}{2} A_\parallel \pm P_\parallel - g_I \beta H_z| \quad (1.148)$$

from which the three quantities $A_\parallel$, $P_\parallel$, g_I can be extracted with at least one check on the validity of the spin Hamiltonian and the precision of the measurements. Additional checks are obtained by working at different field strengths; this can be done at the same electronic frequency, simply by saturating either of the electronic $m = \pm 1$ hyperfine lines instead of the $m = 0$ line.

In addition to providing values of the hyperfine constants and of g_I (or, more correctly, of $g^{(I)}$, which includes the 'pseudo-nuclear Zeeman interaction'—see § 1.8), Endor also determines them with a precision immensely greater than can be obtained in a pure electron paramagnetic resonance experiment, provided that the lines are inhomogeneously broadened. If the line width in such an experiment is, say, 3 G, this corresponds to a frequency width in the electron paramagnetic resonance spectrum of 10 MHz; if the separation of the hyperfine lines is 300 G ($\sim$1000 MHz), and the centre of each line can be determined with a precision of $\frac{1}{10}$ of the line width, the resultant error in the hyperfine constant will be $\sim$1 MHz, or 0·1 per cent. In contrast, the nuclear Zeeman interaction at $H = 3000$ G (even for a rather large nuclear moment such as those of ^{1}H or ^{19}F), is of order 10 MHz, and for most nuclei it is considerably smaller. Thus the precision possible in determining g_I must be quite low, quite apart from the fact the nuclear Zeeman interaction does not shift the $\Delta m = 0$ lines, and can be observed only through $\Delta m \neq 0$ lines that are normally partially forbidden. However, if the effective line width for a nuclear transition is also only 3 G, corresponding to a frequency width of $\sim$10 kHz for such a transition, we can hope by using Endor to measure g_I with a precision of order 0·1 per cent, and the hyperfine constant to a precision of 0·001 per cent, or better. In the latter case the high precision results from the fact that the nuclear magnetic resonance transition is taking place, not just in the external field H, but also in the electronic magnetic field H_e (see § 1.2), which is commonly in the range 10^5–10^6 G.

This high precision is obtained only when the electron paramagnetic resonance line is inhomogeneously broadened, so that we are concerned at any one time only with a spin packet whose inherent line width is of order 10 to 100 kHz, corresponding to a value of τ_2 of order 10^{-5} to 10^{-6} s. Obviously τ_1 must also be at least as long as this in order that the spin packet be not broadened by spin-lattice relaxation, but long values of τ_1 are needed in any case in order to allow saturation of the electron paramagnetic resonance line (or of the spin packet), see § 2.8; and, more seriously, to make it possible for the nuclear magnetic resonance transition rate to compete with relaxation effects (see above). If the electron magnetic resonance line is homogeneously broadened, the hyperfine levels will have frequency widths similar to those observed in the electron resonance line, and an Endor experiment is generally impossible.

A fuller discussion of Endor is given in Chapter 4.

1.14. Experimental aspects

Experimental techniques of electron paramagnetic resonance will not be discussed in this book, except for an assessment of the limiting spectrometer sensitivity (see § 2.10). Most experiments are carried out at centimetre or millimetre wavelengths, partly for the historical reason that oscillators and detectors have been available; clearly many important experiments need and are carried out at still shorter wavelengths whenever feasible. In general, the intensity increases with frequency; from eqn (1.127) the power absorbed by the spin system in a transition between levels a and b is

$$\mathrm{d}W/\mathrm{d}t = w_\mathrm{e}(h\nu)(n_a - n_b) \qquad (1.127)$$

and if $(h\nu) \ll kT$ we have approximately $(n_a - n_b) = n_a(h\nu/kT)$, so that $\mathrm{d}W/\mathrm{d}t$ increases with the square of the frequency since w_e has no inherent frequency dependence. The factors arising from spectrometer sensitivity which influence the choice of frequency are also discussed in § 2.10. However the wavelength region 30 mm to 3 mm (10 to 100 GHz) is very convenient experimentally, for a number of reasons:

(a) Many initial splittings lie in this range;
(b) line widths in concentrated paramagnetic salts lie in the range 0.1 to 10 GHz
(c) cavity resonators and wave-guide components are of convenient size for crystals of 1 to 10 mm linear dimensions;
(d) magnetic fields up to 20 kG are readily achieved with iron-cored electromagnets.

Amongst the advantages of electron paramagnetic resonance are the following:

(i) high sensitivity—only a small concentration of paramagnetic species is required to give a good signal (see § 2.10); in fact the use of highly dilute samples is often essential in order to realize the following advantages;
(ii) detailed and accurate information is obtained about the nature of the paramagnetic species (hyperfine structure is often a salient clue in identification) and its immediate surroundings, which affect the anisotropy and fine and hyperfine splittings, and may provide a ligand hyperfine structure; an exception arises in structureless lines at the free spin value $g = 2 \cdot 0023$ due to defects or free radicals, though the intensity still provides a useful method of estimating concentrations;
(iii) high precision is possible in measurement of the spin Hamiltonian parameters, especially for the hyperfine parameters if Endor is possible;
(iv) spin-spin interactions between neighbouring paramagnetic ions can be estimated from line widths, and determined more precisely from the spectra of interacting pairs of ions in semi-dilute salts;
(v) spin-lattice relaxation times can be estimated from excess line widths, saturation effects, and transient recovery signals following saturation.

Some disadvantages are:
 (i) resonance may only be observable at low temperatures because of fast spin-lattice relaxation;
 (ii) resonance is usually limited to the ground state or levels within a few cm^{-1} of the ground state, unless special techniques such as optical pumping can be used;
 (iii) resonance is not always allowed within the ground state or low-lying excited states;
 (iv) (from the analytical point of view) good single crystals are generally required, except in cases of no anisotropy.

PART II

GENERAL SURVEY

2

THE RESONANCE PHENOMENON

2.1. Use of rotating coordinates

IT was shown in § 1.1 that when an atom with angular momentum **G** and magnetic moment $\boldsymbol{\mu} = \gamma \mathbf{G}$ is placed in a magnetic field **H**, the equation of motion is

$$d\mathbf{G}/dt = \boldsymbol{\mu} \wedge \mathbf{H} = \gamma \mathbf{G} \wedge \mathbf{H} \tag{2.1}$$

and the solution is one in which both **G** and $\boldsymbol{\mu}$ precess about **H** with angular velocity $\boldsymbol{\omega}_L = -\gamma \mathbf{H}$. The question arises, what will the motion appear to be to an observer in a rotating coordinate system, the axis of rotation being taken to lie along the field **H**? If this system is rotating with angular velocity $\boldsymbol{\omega}$ relative to a system stationary in the laboratory, then the rate of variation of a vector quantity such as **G** in the laboratory system (which we denote by $d\mathbf{G}/dt$) is related to its rate of variation in the rotating system (which we denote by $D\mathbf{G}/Dt$) through the equation

$$d\mathbf{G}/dt = D\mathbf{G}/Dt + \boldsymbol{\omega} \wedge \mathbf{G}. \tag{2.2}$$

This may be combined with the equation of motion (2.1) to obtain the equation of motion in the rotating system, which is then

$$D\mathbf{G}/Dt = d\mathbf{G}/dt - \boldsymbol{\omega} \wedge \mathbf{G} = \gamma \mathbf{G} \wedge \mathbf{H} + \mathbf{G} \wedge \boldsymbol{\omega}$$
$$= \gamma \mathbf{G} \wedge (\mathbf{H} + \boldsymbol{\omega}/\gamma). \tag{2.3}$$

This equation is similar to the original eqn (2.1) except that instead of **H** we have $(\mathbf{H} + \boldsymbol{\omega}/\gamma)$. Thus the motion relative to the rotating system will again be a precession, but with angular velocity

$$\boldsymbol{\omega}' = -\gamma \mathbf{H}' = -\gamma(\mathbf{H} + \boldsymbol{\omega}/\gamma)$$
$$= \boldsymbol{\omega}_L - \boldsymbol{\omega}. \tag{2.4a}$$

As would be expected, the apparent precession velocity is equal to the difference between $\boldsymbol{\omega}_L$, the angular velocity observed in the stationary system, and $\boldsymbol{\omega}$, the velocity of the rotating coordinate system relative to the fixed one. This is equivalent to saying that in the rotating system there is an effective magnetic field parallel to the z-axis

$$\mathbf{H}' = \mathbf{H}+\boldsymbol{\omega}/\gamma = \mathbf{H}-\mathbf{H}^*, \qquad (2.4\text{b})$$

where $\mathbf{H}^* = -\boldsymbol{\omega}/\gamma$. Clearly, if we choose $\boldsymbol{\omega}$ to be equal to $\boldsymbol{\omega}_L$, the precession vanishes and both $\mathbf{G}$ and $\boldsymbol{\mu}$ are at rest in the rotating system, corresponding to the effective field $\mathbf{H}'$ being zero.

2.2. Magnetic resonance

This result enables us to derive simply the motion of an atom under the combined action of a steady field $\mathbf{H}$, which we suppose to lie

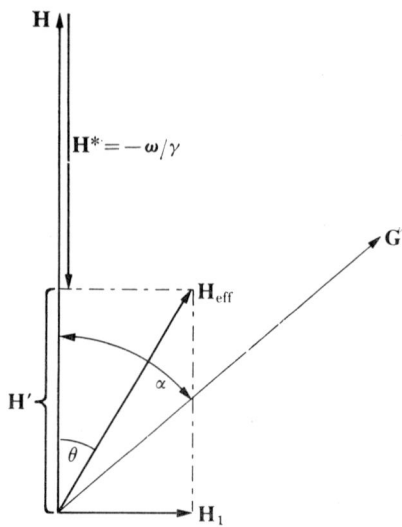

FIG. 2.1. Effective fields in the rotating coordinate system.

along the z-axis of a Cartesian coordinate system, and a field $\mathbf{H}_1$ rotating with angular velocity $\boldsymbol{\omega}$ about the z-axis. It is clear that if we change to a coordinate system that also rotates about the z-axis with angular velocity $\boldsymbol{\omega}$, then the field $\mathbf{H}_1$ will be represented in this system by a constant vector $\mathbf{H}_1$ normal to the z-axis, as in Fig. 2.1. At the same time the constant field $\mathbf{H}$ must be replaced by the effective field $\mathbf{H}' = \mathbf{H}+\boldsymbol{\omega}/\gamma$, which (see eqn (2.4b)) may conveniently be written as $\mathbf{H}' = \mathbf{H}-\mathbf{H}^*$, where $\mathbf{H}^* = -\boldsymbol{\omega}/\gamma$. The couple acting on the magnetic moment, referred to the rotating system, is now

$$D\mathbf{G}/Dt = \gamma \mathbf{G} \wedge (\mathbf{H}'+\mathbf{H}_1) = \gamma \mathbf{G} \wedge \mathbf{H}_\text{eff} \qquad (2.5)$$

where $\mathbf{H}_{\text{eff}}$ is the resultant of $\mathbf{H}'$ and $\mathbf{H}_1$, added vectorially as shown in Fig. 2.1. This equation of motion is again the same as the original eqn (2.1). Hence referred to the rotating system, $\mathbf{G}$ and $\boldsymbol{\mu}$ will undergo a precession about $\mathbf{H}_{\text{eff}}$ with angular velocity $-\gamma \mathbf{H}_{\text{eff}}$.

We consider first the particular case where $\omega = \omega_L$, so that $\mathbf{H}' = 0$ and $\mathbf{H}_{\text{eff}} = \mathbf{H}_1$, and assume that the rotating field $\mathbf{H}_1$ is switched on suddenly at time $t = 0$, when the magnetic moment $\boldsymbol{\mu}$ is aligned along $\mathbf{H}$. The motion of the magnetic moment is then a precession about $\mathbf{H}_1$ with angular velocity $-\gamma \mathbf{H}_1$, and every half-cycle of this motion it changes from being parallel to $\mathbf{H}$ to being anti-parallel and back again. In the rotating system this rotation takes place in the plane normal to $\mathbf{H}_1$. In general $H_1 \ll H$, so that the precession about $\mathbf{H}_1$ occurs at a much lower velocity than that at which $\mathbf{H}_1$ rotates in the laboratory system. Thus, in the latter system, the motion of $\mathbf{G}$ and $\boldsymbol{\mu}$ consists of a rapid rotation about $\mathbf{H}$ at an angle α to $\mathbf{H}$ which varies slowly from 0 to π and back again.

When the frequency of rotation of $\mathbf{H}_1$ is not equal to the natural precession frequency ω_L, the magnetic moment $\boldsymbol{\mu}$ precesses (in the rotating coordinate frame) about the field $\mathbf{H}_{\text{eff}}$. This field makes an angle θ with $\mathbf{H}$, where

$$\tan \theta = H_1/H' = H_1/(H-H^*) = H_1/(H+\omega/\gamma) \qquad (2.6)$$

and from simple geometrical considerations the value of the angle α at time t (assuming $\alpha = 0$ at $t = 0$) is found to be

$$\cos \alpha = \cos^2\theta + \sin^2\theta \cos(\gamma H_{\text{eff}} t)$$
$$= 1 - 2\sin^2\theta \sin^2(\tfrac{1}{2}\gamma H_{\text{eff}} t), \qquad (2.7)$$

where

$$\gamma H_{\text{eff}} = \{(\omega - \omega_L)^2 + \gamma^2 H_1^2\}^{\frac{1}{2}}. \qquad (2.8)$$

The maximum value attained by α is clearly 2θ, and if $H_1 \ll H$ (as is normally the case) this value is large only when we are correspondingly close to fulfilling the condition $\omega = \omega_L$. This is clearly a resonance effect, and the possibility of being able to change appreciably the orientation of the magnetic moment $\boldsymbol{\mu}$ with respect to the steady field $\mathbf{H}$ by the application of a relatively small rotating field $\mathbf{H}_1$ constitutes the phenomenon of 'magnetic resonance'. When $\omega = \omega_L$, the magnetic moment $\boldsymbol{\mu}$, assumed to be initially parallel to $\mathbf{H}$, can be completely reversed by application of the rotating field $\mathbf{H}_1$; in the ideal case this will occur no matter how small the value of $\mathbf{H}_1$, though the rate of reversal is of course proportional to $\mathbf{H}_1$.

Such a change in the orientation of the magnetic moment with respect to **H** (i.e. a change in the value of α) leads to the possibility of the detection of magnetic resonance in one of two ways. In the molecular beam method of Rabi for free particles the re-orientation is detected by the change in path of the molecule in an inhomogeneous field. For matter in the aggregated state, magnetic resonance is usually detected by an electromagnetic method. In a transient method energy must be supplied to or taken from the system of dipoles in order respectively to increase or decrease the average value of the angle α. This represents an interchange of stored energy with the source of electromagnetic radiation. In a steady-state method, power must be supplied in order to maintain the angle α at a value larger than the thermal equilibrium value. This power can be detected either through the extra dissipation it represents in the electromagnetic system, or through its appearance as heat resulting in a rise in temperature in the aggregated state of matter. A magnetic induction method has also been used (Bagguley 1955), in which the signal is induced by the oscillatory magnetic moment of the system arising from the forced precession about **H**, the oscillatory moment increasing as α increases towards $\pi/2$. In spin-echo methods, the observed signal is again induced by either a free or a forced precession of the magnetic moment.

2.3. Quantum-mechanical analysis

The discussion hitherto has been on a classical basis, and an outline of the quantum-mechanical approach will now be given.

When an atom with no nuclear hyperfine interaction is placed in a steady magnetic field **H** (which we take to be parallel to the z-axis), the interaction with its dipole moment is described by the Hamiltonian

$$\mathscr{H} = -\boldsymbol{\mu} \cdot \mathbf{H} = -\gamma\hbar(\mathbf{J} \cdot \mathbf{H}) = -\gamma\hbar H J_z. \qquad (2.9)$$

This equation holds also for the case where **J** reduces simply to **L** or **S**, and also for an atom with no electronic momentum but with nuclear angular momentum if we replace **J** by **I**, provided that the appropriate value of γ is inserted in each case. Hence in this section we may regard **J** as representing an angular momentum, either electronic or nuclear, for an atom with no hyperfine interaction.

The only matrix element of J_z is

$$\langle M| J_z |M \rangle = M, \qquad (2.10)$$

where M is the magnetic quantum number appropriate to J. Hence the energy in a stationary state is

$$W = -\gamma\hbar H \langle M| J_z |M \rangle = -\gamma\hbar H M \qquad (2.11)$$

as given by the classical theory.

If now an oscillatory magnetic field is applied in the plane normal to **H**, with components H_x, H_y, there will be an additional perturbation described by the Hamiltonian

$$\mathcal{H}' = -(H_x\mu_x + H_y\mu_y). \qquad (2.12)$$

If the oscillatory field consists of a rotating field with components $H_x = H_1 \cos \omega t$, $H_y = H_1 \sin \omega t$, this can be written as

$$\mathcal{H}' = -\tfrac{1}{2}H_1\{\mu_+ \exp(-i\omega t) + \mu_- \exp(i\omega t)\}, \qquad (2.13)$$

where

$$\mu_+ = \mu_x + i\mu_y, \qquad \mu_- = \mu_x - i\mu_y.$$

Since the perturbation is an oscillatory one we use time-dependent perturbation theory to evaluate its effect. If the overall wave-function is written as a summation over all states $|u_n\rangle$,

$$\psi = \sum_n a_n |u_n\rangle \exp(-iW_n t/\hbar),$$

where a_n is the amplitude and W_n the energy of the nth state, then the coefficients a_n obey the relation

$$\frac{da_k}{dt} = (i\hbar)^{-1} \sum_n \mathcal{H}_{kn} a_n \exp\{i(W_k - W_n)t/\hbar\}, \qquad (2.14)$$

where $\mathcal{H}_{kn}$ is the matrix element of the perturbation between the states k and n,

$$\mathcal{H}_{kn} = \langle u_k| \mathcal{H}' |u_n\rangle.$$

In our case the only matrix elements that arise are those between states M and $M \pm 1$, which are

$$\langle M+1| \mu_+ |M\rangle = \langle M+1| \gamma\hbar J_+ |M\rangle = \gamma\hbar\{J(J+1) - M(M+1)\}^{\frac{1}{2}}, \qquad (2.15)$$
$$\langle M-1| \mu_- |M\rangle = \langle M-1| \gamma\hbar J_- |M\rangle = \gamma\hbar\{J(J+1) - M(M-1)\}^{\frac{1}{2}},$$

which lead to the selection rule $M \leftrightarrow M \pm 1$, or $\Delta M = \pm 1$. If we assume that the system is initially in the state $J_z = M$, so that all the coefficients a_n are initially zero except a_M, which is unity, then since

$(W_{M\pm1}-W_M)/\hbar = \mp\gamma H$, for small values of the perturbation we have approximately

$$\frac{da_{M\pm1}}{dt} = \tfrac{1}{2}i\gamma H_1\langle M\pm1|\,J_\pm\,|M\rangle\exp[i\{(W_{M\pm1}-W_M)\mp\hbar\omega\}t/\hbar]$$
$$= \tfrac{1}{2}i\gamma H_1\langle M\pm1|\,J_\pm\,|M\rangle\exp\{\mp i(\gamma H+\omega)t\}. \qquad (2.16)$$

This is readily integrated, and yields for the probability $(aa^*)_{M\pm1}$ of finding the system in the states $M\pm1$ after a time t

$$(aa^*)_{M\pm1} = (\gamma H_1)^2\langle M\pm1|\,J_\pm\,|M\rangle^2\frac{\sin^2\tfrac{1}{2}(\gamma H+\omega)t}{(\gamma H+\omega)^2}$$
$$= (\gamma H_1)^2\langle M\pm1|\,J_\pm\,|M\rangle^2\frac{\sin^2\tfrac{1}{2}(\omega-\omega_L)t}{(\omega-\omega_L)^2}. \qquad (2.17)$$

The transition probability per unit time w is given by the time derivative $d(aa^*)/dt$ integrated over a distribution of Larmor frequencies:

$$w = (\gamma H_1)^2\langle M\pm1|\,J_\pm\,|M\rangle^2\int_{-\infty}^{+\infty}\frac{\sin(\omega-\omega_L)t}{2(\omega-\omega_L)}f(\omega_L)\,d\omega_L$$
$$= (\pi/2)(\gamma H_1)^2\langle M\pm1|\,J_\pm\,|M\rangle^2 f(\omega), \qquad (2.17a)$$

where it is assumed that $f(\omega_L)$ has a wide spectrum, but the only significant contribution to the integral comes from the region where $\omega_L \approx \omega$. The quantity $f(\omega)$, known as the 'shape function', is normalized by the condition

$$\int_0^\infty f(\omega)\,d\omega = 1, \qquad (2.17b)$$

and describes the distribution of energy levels between which the transition is taking place. For the case of $J = \tfrac{1}{2}$, Rabi (1937) has shown that an exact solution is

$$(aa^*) = \frac{(\gamma H_1)^2}{(\gamma H+\omega)^2+(\gamma H_1)^2}\sin^2[\tfrac{1}{2}\{(\gamma H+\omega)^2+(\gamma H_1)^2\}^{\tfrac{1}{2}}t] \qquad (2.18)$$

for the probability of finding the system in the state $M = -\tfrac{1}{2}$ at time t if it was in the state $M = +\tfrac{1}{2}$ at $t = 0$. Since for $J = \tfrac{1}{2}$ the matrix elements of $J_\pm$ are just unity, eqn (2.18) is similar to (2.17) except that an additional term $(\gamma H_1)^2$ is added to $(\gamma H+\omega)^2$.

It is clear from these equations that the probability of a transition is large only if $(\gamma H+\omega)$ is close to zero. This corresponds to the resonance condition deduced from the classical equations, but it is possible to make a much more detailed correlation between the classical and quantum mechanical approaches, as has been shown, for example, by Rabi, Ramsey, and Schwinger (1954). First, it will be shown that the classical equation of motion (2.1) is valid also in quantum mechanics.

The rate of variation of a quantity such as **G** is given by

$$-i\hbar(d\mathbf{G}/dt) = \mathscr{H}\mathbf{G} - \mathbf{G}\mathscr{H}, \tag{2.19}$$

where $\mathscr{H}$ is the Hamiltonian which for our system is

$$\mathscr{H} = -\boldsymbol{\mu}\cdot\mathbf{H} = -\gamma\mathbf{G}\cdot\mathbf{H}.$$

Hence

$$-i\hbar(d\mathbf{G}/dt) = -\gamma\{(\mathbf{G}\cdot\mathbf{H})\mathbf{G} - \mathbf{G}(\mathbf{G}\cdot\mathbf{H})\},$$

which does not vanish because the components of **G** do not commute with one another, though they do with **H**. The bracket on the right-hand side thus has the x-component

$$G_z H_z G_x - G_x G_z H_z + G_y H_y G_x - G_x G_y H_y$$
$$= (G_z G_x - G_x G_z)H_z - (G_x G_y - G_y G_x)H_y$$
$$= i\hbar(G_y H_z - G_z H_y)$$

by using the commutation relations for the components of **G**. The last expression is just $i\hbar(\mathbf{G}\wedge\mathbf{H})_x$, so that we obtain again the eqn (2.1),

$$d\mathbf{G}/dt = \gamma\mathbf{G}\wedge\mathbf{H}.$$

Thus we may expect that our classical analysis is valid also in quantum mechanics, and a simple comparison can again be made for the case of $J = \frac{1}{2}$.

The wave-function for this case will, in general, be a linear combination of the wave-functions for the states $|J_z\rangle = +\frac{1}{2}$ and $-\frac{1}{2}$, which can be written as just $|+\rangle$ and $|-\rangle$. A suitable combination with normalized coefficients is

$$\psi = \cos\tfrac{1}{2}\alpha\,|+\rangle + \sin\tfrac{1}{2}\alpha\,|-\rangle,$$

for which the z-component of the magnetic moment is

$$\mu_z = \langle\psi^*|\,\gamma\hbar J_z\,|\psi\rangle = \tfrac{1}{2}\gamma\hbar(\cos^2\tfrac{1}{2}\alpha - \sin^2\tfrac{1}{2}\alpha)$$
$$= \tfrac{1}{2}\gamma\hbar\cos\alpha = \mu\cos\alpha,$$

showing that our wave-function corresponds to the magnetic moment lying at an angle α to the z-axis. Thus the angle introduced to give normalized coefficients in the wave-function can be associated with a simple physical quantity.

If the system is in the $|+\rangle$ state at time $t = 0$, then α is initially zero, and the probability of a transition to the $|-\rangle$ state at time t is given by the square of the coefficient of the $|-\rangle$ state, i.e. by $\sin^2\tfrac{1}{2}\alpha$. From the classical eqn (2.7) one finds

$$\sin^2\tfrac{1}{2}\alpha = \tfrac{1}{2}(1-\cos\alpha)$$
$$= \sin^2\theta \sin^2(\tfrac{1}{2}\gamma H_{\text{eff}}t),$$

which on substitution from eqn (2.8) and use of the relation

$$\sin\theta = H_1/H_{\text{eff}} = (\gamma H_1)/\{(\gamma H+\omega)^2+(\gamma H_1)^2\}^{\tfrac{1}{2}}$$

is seen to be identical with the quantum-mechanical eqn (2.18).

2.4. Magnetic resonance in aggregated systems

It has now been shown that the classical equations of motion can be justified on quantum mechanics, and that they provide an easy physical approach to the phenomenon of magnetic resonance. This approach is valid for a simple system, such as an atom with an electronic but no nuclear magnetic moment, or vice versa, but not for a complex system such as an atom with hyperfine structure. This restriction follows from the use of the simple Hamiltonian of eqn (2.9) instead of the complex one (eqn (1.67)). Even in the absence of hyperfine structure, the simple Hamiltonian (2.9) may not be adequate when dealing with aggregated systems, for there may then be other interactions such as quadrupole splittings in the nuclear case, or its equivalent in the electronic case, the so-called fine structure terms (see §§ 3.3–3.6). The classical approach is not then valid, but there are many cases when it is, and it is particularly fruitful in providing a physical picture in many situations in which it is only approximately valid. We shall therefore investigate the solutions of the classical equation which describe the motion of the magnetization in bulk matter.

In an aggregated system of permanent magnetic dipoles it is convenient to think in terms of the magnetization **M**, or the magnetic moment of the whole system, rather than in terms of individual dipoles. Now the magnetic moment of the whole system is simply the vector resultant of the individual dipoles. Similarly, the angular momentum, **G**, of the whole system is the vector resultant of the individual momenta $\hbar\mathbf{J}$. Since each individual dipole, **μ**, is parallel and proportional to its own associated angular momentum vector ($\boldsymbol{\mu} = \gamma\hbar\mathbf{J}$), then for a system of particles which all have identical values of γ we have

$$\mathbf{M} = \gamma\mathbf{G} \tag{2.20}$$

and the system as a whole will obey the classical equation of motion

$$d\mathbf{G}/dt = \gamma \mathbf{G} \wedge \mathbf{H}$$

which, in order to deal in terms of **M** instead of **G**, can be multiplied by γ yielding

$$d\mathbf{M}/dt = \gamma \mathbf{M} \wedge \mathbf{H}. \tag{2.21}$$

With suitable modifications, this equation is widely used to treat the phenomenon of resonance in cooperative systems (ferro-, antiferro-, or ferri-magnetic). We are concerned with paramagnetic systems where the dipoles interact only weakly with one another; primarily this means systems for which we can write

$$\mathbf{M} = \chi_0 \mathbf{H} = N \frac{\gamma^2 \hbar^2 J(J+1)}{3kT} \mathbf{H} \tag{2.22}$$

in the high temperature limit where $(\mu H/kT) \ll 1$. Essentially our analysis will consist of finding solutions to eqn (2.21) which will yield the conditions under which magnetic resonance can be observed. As we should expect, this phenomenon is readily observable only when the resonance condition $\boldsymbol{\omega} = -\gamma \mathbf{H}$ is fulfilled. We can, however, vary either $\boldsymbol{\omega}$, the angular velocity of the applied rotating field $\mathbf{H}_1$, or the steady field **H**, to reach resonance. In practice, it is nearly always preferable experimentally to work at fixed frequency and vary the steady field **H**. This is also the simpler case to analyze since we can make use of rotating coordinate systems as has been done earlier in this chapter. There will be some restriction on the rate at which **H** is varied, however, as a rapidly changing field will have oscillatory components that introduce transitions between the atomic energy levels. When the field is changed slowly such transitions do not take place, and the change is said to be 'adiabatic'. The condition that this should hold can be found by means of the rotating coordinate system.

Instead of varying the magnitude of **H**, let us assume that **H** is constant in magnitude but changing in direction; for simplicity we assume that this change consists of a rotation about the x-axis, with angular velocity $\boldsymbol{\omega}'$. Then, if we change from the laboratory system to one rotating with this velocity, **H** will be a constant vector in the rotating system, but there will also be an apparent field $\mathbf{H}^* = -\boldsymbol{\omega}'/\gamma$ at right angles to **H**, as in Fig. 2.2. Then the effective field $\mathbf{H}_{\text{eff}}$ in the rotating system will be the vector sum of **H** and $\mathbf{H}^*$. If the magnetization was initially parallel to **H**, then it will start to precess about $\mathbf{H}_{\text{eff}}$ and hence will not remain closely parallel to **H** unless the angle between

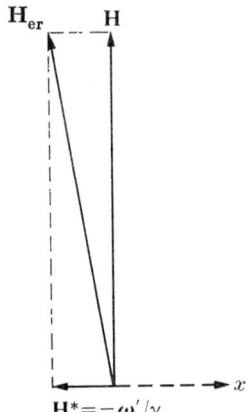

Fig. 2.2. Rotation of the field **H** about the x-axis with angular velocity $\boldsymbol{\omega}'$.

H and $\mathbf{H}_{\text{eff}}$ in the rotating system is very small. This implies that $H^* \ll H$, or

$$\omega' = |\gamma H^*| \ll |\gamma H|. \tag{2.23}$$

In other words, the Larmor precession frequency in the field H must be very much higher than the frequency of rotation of the field H itself. Then the precessional motion of the magnetization caused by the rotation of **H** will cause it to deviate from being parallel to **H** only by a very small angle, instead of the large angles possible when a rapidly rotating field at the Larmor resonance frequency is applied. Since characteristic electron resonance frequencies lie in the range 10^{10} to 10^{12} Hz, while modulation of the field **H** is seldom used at a frequency much above 10^5 Hz, the adiabatic condition is easily fulfilled under normal conditions.

2.5. Adiabatic rapid passage

We will now discuss a special solution of the equation of motion (2.21) and its implications, before proceeding to the more general case. As before, we assume that the system is subject to a steady magnetic field directed along the z-axis, together with a field H_1 precessing with angular velocity ω in the xy-plane, so that it has components $H'_x = H_1 \cos \omega t$, $H'_y = H_1 \sin \omega t$. A plausible solution might correspond to the magnetization precessing about the z-axis at an angle θ, so that it has components

$$\begin{aligned} M_x &= M \sin \theta \cos \omega t, \\ M_y &= M \sin \theta \sin \omega t, \\ M_z &= M \cos \theta. \end{aligned} \tag{2.24}$$

This corresponds to a steady-state solution, with forced precession at the same angular velocity as that of the applied field. To find whether this satisfies eqn (2.21) it is convenient to work in terms of the quantities

$$M_+ = M_x + iM_y = M \sin\theta \exp(i\omega t),$$
$$M_- = M_x - iM_y = M \sin\theta \exp(-i\omega t), \quad (2.25)$$
$$M_z = M \cos\theta.$$

In terms of these components the equations of motion become

$$\dot{M}_+ + i\gamma H M_+ = +i\gamma M_z H_1 \exp(i\omega t),$$
$$\dot{M}_- - i\gamma H M_- = -i\gamma M_z H_1 \exp(-i\omega t), \quad (2.26)$$
$$\dot{M}_z = \tfrac{1}{2}i\gamma\{M_+ \exp(-i\omega t) - M_- \exp(i\omega t)\}H_1.$$

The first two equations in (2.26) (which are complex conjugates) can be integrated at once if M_z is constant, as assumed in (2.25), and substitution from (2.25) into the equation for $\dot{M}_z$ shows that the solutions assumed for M_+, M_- do give $\dot{M}_z = 0$. Hence the equation of motion is satisfied by our solution, and integration gives

$$M_\pm = \frac{\gamma H_1 M_z}{(\gamma H + \omega)} \exp(\pm i\omega t). \quad (2.27)$$

These are of the form (2.25) provided that we take

$$\tan\theta = \gamma H_1/(\gamma H + \omega)$$
$$= H_1/(H - H^*). \quad (2.28)$$

The nature of this solution can readily be seen by working in a system of coordinates rotating about **H** with angular velocity **ω**. In this system $\mathbf{H}_1$ is a constant vector perpendicular to **H**, while parallel to **H** the effective field component is $\mathbf{H}' = \mathbf{H} + \boldsymbol{\omega}/\gamma = \mathbf{H} - \mathbf{H}^*$. The magnetization vector **M** is also constant in this system, and from (2.28) we see that it is parallel to $\mathbf{H}_{\text{eff}}$, the vector resultant of $(\mathbf{H} - \mathbf{H}^*)$ and $\mathbf{H}_1$.

Let us assume that the steady-state solution (2.24) is attained under the condition $(H - H^*) \gg H_1$, so that θ is initially close to zero, and **M** is almost parallel to **H**. Then, if H is diminished slowly, the angle θ increases and passes through $\pi/2$ at resonance when $H = H^*$, attaining the limiting value π when $H \ll H^*$ and again $\{H_1/(H - H^*)\} \to 0$. The magnetization **M** is now reversed in direction and is anti-parallel to **H**, while at intermediate times when $(H - H^*) \sim H_1$, there is a large precessing magnetization in the xy-plane in the laboratory system. If the specimen is inside a coil of area A which picks up a flux in the

y-direction, a voltage will be induced in the coil of magnitude

$$V = -(4\pi A/c)\dot{M}_y$$
$$= -\omega(4\pi A/c)M \sin\theta \cos\omega t. \qquad (2.29)$$

This is largest when $\theta = \pi/2$, corresponding to H being at the resonance value $H^* = -\omega/\gamma$. Detection of this voltage gives a method of detecting the magnetic resonance and is the essence of the method of nuclear induction.

The solution obtained above was based on the assumption that **H** was constant, but we have presumed that it still holds when **H** is varied 'slowly'. We note that **M** in our solution is always parallel to the effective field $\mathbf{H}_\mathrm{eff}$ in the rotating coordinate system, and that variation of **H** corresponds to a rotation of $\mathbf{H}_\mathrm{eff}$ as well as a change in its magnitude, its smallest value being just $\mathbf{H}_1$. From the 'adiabatic condition' discussed in § 2.4, we should expect that **M** will follow $\mathbf{H}_\mathrm{eff}$ at a small angle when $\mathbf{H}_\mathrm{eff}$ is rotated provided that the rate of rotation $(\mathrm{d}\theta/\mathrm{d}t)$ is small compared with $|\gamma H_1|$. Since $\cot\theta = H'/H_1$,

$$\mathrm{d}\theta/\mathrm{d}t = -\sin^2\theta\, \mathrm{d}(\cot\theta)/\mathrm{d}t$$
$$= -(H_1^2/H_\mathrm{eff}^2)\,\mathrm{d}(H'/H_1)/\mathrm{d}t$$
$$= -(H_1^2/H_\mathrm{eff}^2)\,\mathrm{d}(H/H_1)/\mathrm{d}t.$$

The maximum value of H_1/H_eff is unity, and hence the adiabatic condition is satisfied provided that

$$\frac{1}{H_1}|\mathrm{d}H/\mathrm{d}t| \ll \gamma H_1. \qquad (2.30)$$

In words, this means that any change in H in a time of the order $(\gamma H_1)^{-1}$ must be small compared to H_1. We can obtain some idea of the magnitude of this restriction by taking $H_1 = 1$ G. Then for protons, for which $\gamma = 2\cdot 6\times 10^4$ rad G^{-1}, $(\gamma H_1)^{-1} = 4\times 10^{-5}$ s, while for electron spins it is 6×10^{-8} s. Thus H must be varied in such a way that a change of 1 G must take place in a time long compared with either of these, a condition that is readily fulfilled. In terms of the solution of eqns (2.26), this corresponds to $(\mathrm{d}\theta/\mathrm{d}t)$ being so small that it can be neglected in the differentiation of eqns (2.25).

Since the equation of motion is linear in M, the magnetization does not become infinite during forced resonance, as is commonly the case in other systems without any damping. It means also that the size of M is indeterminate, though if the system of magnetic dipoles is allowed to come to thermal equilibrium at a constant field $\mathbf{H}_0$ by waiting a long time before commencing the sweep through resonance, then we should expect **M** to have its equilibrium value

$$\mathbf{M_0} = \chi_0 \mathbf{H_0}$$

with χ_0 given by eqn (2.22). Then **M** is initially parallel to **H**, and also parallel to $\mathbf{H'} = \mathbf{H} - \mathbf{H^*}$, as assumed above in taking $\theta = 0$ initially, provided that $\mathbf{H_0} > \mathbf{H^*}$. If, on the other hand, the system is allowed to reach thermal equilibrium in a field $\mathbf{H_0}$ that is less than $\mathbf{H^*}$, then **M** is initially anti-parallel to $\mathbf{H'} = \mathbf{H} - \mathbf{H^*}$, so that $\theta = \pi$ initially. Then on sweeping through resonance **M** is again reversed, but the change in θ is from π to 2π, and the sign of the voltage picked up in the coil (see eqn (2.29)) is the reverse of that in the former case. This is readily seen from the diagrams showing the relative positions of the vectors in the rotating coordinate system in Fig. 2.3.

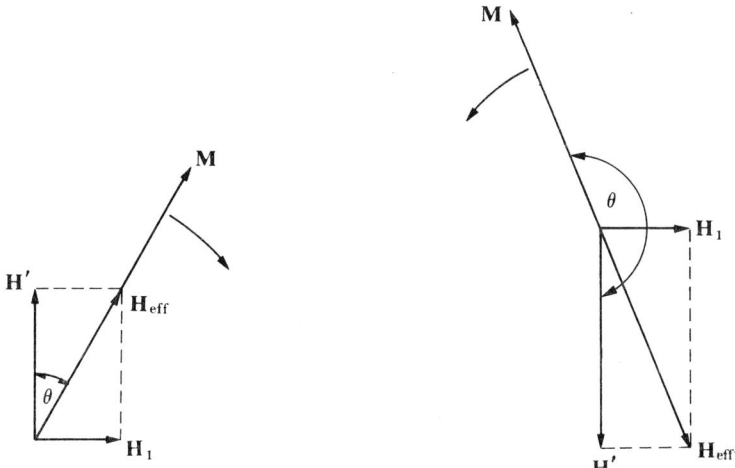

Fig. 2.3. Motion of the magnetic vector in the rotating coordinate system during adiabatic rapid passage.

In this discussion it has been necessary to determine **M** by reference to the initial conditions, assuming the system was initially in thermal equilibrium. Such equilibrium is attained by means of relaxation processes, which occupy a finite time. It is clear that if the sweep through resonance occupies a time long compared with the 'relaxation time', reversal of the magnetization will not be attained because the relaxation will tend to restore **M** to being parallel to **H**. Thus there is a lower limit to the rate of sweep through resonance as well as the upper limit given by the adiabatic condition (eqn (2.30)). When the rate lies between these limits the sweep through resonance is called 'adiabatic rapid passage' through resonance. When the relaxation time τ_1 is much longer than $(\gamma H_1)^{-1}$, [that is, $(\gamma H_1 \tau_1) \gg 1$], it is clearly possible to satisfy both conditions. We note also that this gives a lower limit for the size of H_1, unlike the analysis of § 2.2 in which no damping forces were present.

2.6. Relaxation effects

In view of the restriction that it has been necessary to introduce concerning relaxation effects, it is clearly essential to consider these in more detail before examining the case of slow passage through resonance. This amounts to abandoning the assumption that all changes in orientation or magnitude of the magnetization are due entirely to the external magnetic fields applied.

There are three major internal interactions to be considered. The first of these includes the ligand interaction, which may result simply in a modified form of the Zeeman interaction but may also include 'fine structure' effects (see Chapter 3); it also includes hyperfine interactions between the magnetic electrons and nuclear moments. If these interactions are comparable with or greater than the electronic Zeeman interaction energy, the macroscopic model is seldom applicable. If they are smaller than the electronic Zeeman energy, and only those parts of the interaction Hamiltonian are retained which are diagonal in the representation in which the electronic Zeeman energy has been diagonalized, the macroscopic analysis often gives a useful physical picture that forms a good approximation to the actual motion.

The second and third internal interactions are the effects of thermal agitation and of interaction between the dipoles themselves. These are usually much weaker than the Zeeman interaction with the external fields, but are of importance because of their cumulative effects over long periods of time. Both these effects represent random interactions with the various atomic dipoles, and to this extent our model becomes only 'semi-macroscopic' when they are introduced. However, there is a fundamental difference in that only thermal perturbations can change the energy of the total dipole system (the 'spin system' as it is generally called, though this must be taken to include orbital dipoles in the electronic case), whereas interactions within the spin system (i.e. between the individual dipoles) leave this energy unchanged.

The major part of the energy of the spin system, is, under the restrictions retained above, the Zeeman energy in the steady field **H**, which is just

$$W = -\mathbf{M}\cdot\mathbf{H} = -M_z H,$$

assuming **H** to be in the z-direction, as before. Major changes in the energy are therefore due to changes in M_z, for which thermal fluctuations must be responsible. If at any time the magnetization component M_z is not equal to its equilibrium value $M_0 = \chi_0 H$, it is not a bad approximation to assume that it will approach this equilibrium value exponentially according to the differential equation

$$\dot{M}_z = -(M_z - M_0)/\tau_1. \tag{2.31}$$

τ_1 is a characteristic time-constant, sometimes called the 'longitudinal' relaxation time, since it governs the changes in M_z, the component parallel to the steady magnetic field, and sometimes the 'spin-lattice' relaxation time, since it is connected with the exchange of energy between the spin system and the lattice in which the dipoles are embedded. We use τ_1 and (below) τ_2 instead of the more usual T_1, T_2 as a more logical nomenclature which avoids confusion with the temperature T.

We now consider the mutual interactions between the dipoles, which are conveniently, if somewhat loosely, described by means of 'local fields'. As above, we assume that the spin system is in an external magnetic field parallel to the z-axis which is much larger than the internal fields. Then the mutual interactions between the dipoles, or spin-spin interactions, cannot affect the longitudinal component M_z of the magnetization, but only the transverse components M_x, M_y.

Since the interactions between the dipoles do not affect the total energy of the system, in magnetic fields much larger than the local fields they can have no effect on the component M_z of the magnetization, but only on M_x, M_y. For simplicity of analysis it was assumed by Bloch that the effect of these mutual interactions could again be represented by a relaxation time, so that M_x, M_y obey the differential equations

$$\dot{M}_x = -M_x/\tau_2, \qquad \dot{M}_y = -M_y/\tau_2. \tag{2.32}$$

τ_2 is generally known as the 'transverse' relaxation time, and the sense in which the mutual interactions can be treated as giving rise to a relaxation effect will be discussed later.

These relaxation effects may be introduced into the equations of motion by combining eqns (2.26) with (2.31) and (2.32). Since M_+, M_-

will obey equations similar to (2.32), we have

$$\dot M_+ + i\gamma H M_+ + M_+/\tau_2 = i\gamma M_z H_1 \exp(i\omega t),$$
$$\dot M_- - i\gamma H M_- + M_-/\tau_2 = -i\gamma M_z H_1 \exp(-i\omega t),$$
$$\dot M_z + M_z/\tau_1 = \tfrac{1}{2} i\gamma\{M_+ \exp(-i\omega t) - M_- \exp(i\omega t)\} H_1 + M_0/\tau_1.$$
(2.33)

These equations were first given by Bloch (1946).

It is necessary to point out that while the rate eqn (2.31) defining τ_1 has a wide range of validity, the eqns (2.32) that define τ_2 are valid only for magnetic moments in fast relative motion with respect to each other. They may describe correctly the behaviour of the transverse magnetization of paramagnetic ions or free radicals in liquid solution, or that of conduction electrons in metals and semiconductors, but in general they will be grossly incorrect in insulating paramagnetic solids. A rate equation such as (2.32) implies that the time dependence of the decaying amplitude of the transverse magnetization is exponential. It can be shown under fairly general assumptions that this decay function is the Fourier transform of the resonance line shape observed with a vanishingly small r.f. field, since the Fourier transform of a decaying exponential $\exp(-t/\tau_2)$ is proportional to the Lorentzian function (where $\Delta\omega = \tau_2^{-1}$)

$$f(\omega) = \frac{1}{\pi}\frac{\tau_2}{1+(\omega-\omega_L)^2\tau_2^2} = \frac{1}{\pi}\frac{\Delta\omega}{(\Delta\omega)^2+(\omega-\omega_L)^2}. \qquad (2.34)$$

Thus the validity of eqns (2.32) would imply that resonance lines always have a Lorentzian shape, which is contradicted by experiment and not to be expected on theoretical grounds.

Whatever the line shape, one may of course somewhat loosely define a time τ_2' by the relation $\tau_2' = (\gamma\,\Delta H)^{-1}$ where ΔH is the line width, but it is preferable to introduce first a distinction between homogeneous and inhomogeneous broadening. We shall say that a line is inhomogeneously broadened if the line width is caused by a distribution across the sample of the Larmor frequencies of the various magnetic moments. The causes of such a distribution are manifold, ranging from a badly inhomogeneous external field to local variation of the gyromagnetic factor γ arising from local imperfections in the crystalline environment of the magnetic dipoles. Whatever the origin, such inhomogeneous broadenings have one feature in common: the loss of phase coherence caused by the fanning out in the equatorial plane of the individual precessing dipoles is not an irreversible phenomenon—there are ways, known as spin echoes (see § 2.7), whereby phase coherence can be restored.

The problem of phase coherence can be thought of in the following way. In thermal equilibrium the only steady component of the magnetization is M_z, since both M_x and M_y are zero on a time average, though they may fluctuate about this value. This does not imply that the individual dipoles are all necessarily parallel or anti-parallel to the z-axis, but that their components normal to this axis are random and average to zero. Suppose, however, that at some instance there exists a finite value of the magnetization in the equatorial plane. In the presence of a uniform field **H** this will precess about **H** with angular velocity $-\gamma \mathbf{H}$, and since the individual dipoles all precess with this velocity they remain in phase and the magnetization will retain its initial magnitude. This will only be true if the field acting on each dipole is exactly equal to **H**; if the applied field is inhomogeneous, the individual dipoles will precess at different velocities. The mutual interaction of the dipoles produces an effect similar to that of inhomogeneity in **H**, since the local field at individual dipoles then varies by an amount, ΔH, of the order μ/r^3. This gives rise to a spread in the precessional velocity of order $|\gamma| \Delta H$, which will cause the dipoles to get out of phase in a time of order

$$\tau_2' \sim (|\gamma| \Delta H)^{-1}. \qquad (2.35)$$

For nuclear moments of size $10^{-3}\beta$ at an average separation of 0·2 nm we have $\Delta H \sim 1$ G, and $\tau_2' \sim 10^{-4}$ s. For electronic moments in a moderately dilute paramagnetic salt such as a double sulphate the average distance between dipoles is about 0·6 nm, $\Delta H \sim 50$ G and

$$\tau_2' \sim 10^{-9} \text{ s}. \qquad (2.36)$$

By diluting the salt with an isomorphous diamagnetic salt the value of ΔH can be reduced to a few gauss, or even less than 1 G if the salt contains no nuclear dipole moments of appreciable size, and in this case the value of τ_2' may rise to about 10^{-7} s.

The effect of the mutual interactions cannot be represented entirely as giving rise to inhomogeneity in **H**, since the precessing components of the dipole field give rise to transitions ('spin flips') where one dipole gains energy and another loses it by exchange of a quantum of energy. This process is most effective when the dipoles precess at the same frequency. In such a mutual flip the total value of M_z is conserved, assuming that the external field is much larger than the local fields, so that the total energy of the system remains unchanged. However there is no such restriction on the total values of M_x or M_y, and the result of such spin flips is a gradual destruction of the phase coherence between the components of the individual dipoles in the xy-plane, so that any precessing magnetization in this plane will gradually decay to zero. In so far as this decay takes place exponentially in time, it can be represented by a transverse relaxation time in an equation such as (2.32). Thus the spin-flip process introduces a 'genuine' transverse relaxation time τ_2. In undiluted salts, the spin-flip process between identical spins increases the line width by a factor $\sim\frac{3}{2}$ (see § 9.7), so that τ_2 is not much greater than τ_2'. In diluted salts where the line width is mainly due to nuclear moments in the host lattice, the true value of τ_2 may be $\sim 10^{-5}$ s, corresponding to 'spin packet' widths of order 10^4 Hz.

In non-conducting solids τ_1 is generally much longer than τ_2 for nuclei, but this is not necessarily true for electronic dipoles that are in much more intimate contact with the lattice through the action of thermally modulated electric fields on the translational or orbital motion of the electrons. At low temperatures where thermal fluctuations die out τ_1 becomes large, but at room temperature it may be much shorter for electrons than the value of τ_2' given in eqn (2.36). This does not mean that τ_1 becomes shorter than τ_2, since thermal fluctuations that affect M_z also affect M_x, M_y and result in a τ_2 comparable with τ_1. When the whole of the line width is due to relaxation effects, the resonance line is 'homogeneously broadened'.

We return now to the consideration of the phenomenon of adiabatic rapid passage in the light of our remarks about relaxation effects. The essential feature of the process is that in the rotating frame the magnetization vector follows the effective field $\mathbf{H}_{\text{eff}}$, but since the magnetization is simply the vector sum of the individual dipoles it follows that they must all undergo a similar motion (this does not, of course, imply that they are each parallel to $\mathbf{H}_{\text{eff}}$). The presence of a local field $\Delta\mathbf{H}$ that varies from point to point in the sample means that the field seen by an individual dipole in the rotating frame is $\mathbf{H}_{\text{eff}} + \Delta\mathbf{H}$, and in view of the random nature of $\Delta\mathbf{H}$ the dipoles will not all undergo a similar motion unless $\Delta H \ll H_{\text{eff}}$ always, which means $\Delta H \ll H_1$. If ΔH were a purely longitudinal field, parallel to H, this restriction would not apply since the effect of ΔH would then be merely to change the value of $H' = H - H^*$, so that the individual dipoles pass through the equatorial plane at different times, but all would end anti-parallel (or parallel) to the external field. If, however, $\Delta \mathbf{H}$ has transverse components with a frequency dependent spectrum so that there is a genuine transverse relaxation time τ_2, then it is necessary to pass through resonance in a time short compared with τ_2 in order that the transverse magnetization shall not decay appreciably during passage, which must therefore be rapid. In fact spin-flip processes arise from local transverse magnetic fields near the resonance frequency, which appear as constant fields $\Delta \mathbf{H}'$ in the rotating system to be added vectorially to $\mathbf{H}_1$. We therefore do require $H_1 \gg \Delta H'$, or

$$\tau_2 = (|\gamma \, \Delta H'|)^{-1} \gg (|\gamma H_1|)^{-1},$$

for adiabatic rapid passage to be effective in reversing the magnetization, and it is thus possible only in heavily diluted salts, where $\tau_2 > 10^{-7}$ s, since at centimetre wavelengths oscillatory fields are not normally greater than about 1 G, for which $(|\gamma H_1|)^{-1}$ is about 10^{-7} s.

Similar restrictions apply to τ_1, but as remarked earlier, if τ_1 is not greater than τ_2 it can be made so by working at lower temperatures, so that the condition $|\gamma H_1 \tau_1| \gg 1$ is readily fulfilled. However, there remains the requirement that the sweep through resonance must be made in a time short compared with τ_1 and τ_2, which is difficult to fulfill unless both are rather long. In addition, any reversed magnetization set up will persist only for a time $\sim \tau_1$. The conditions are, of course, much more easily met in the case of nuclear dipoles.

2.7. Radio-frequency pulses and spin echoes

Reversal of the magnetization such as occurs in adiabatic rapid passage can also be effected by the application of a short pulse of the oscillatory field at the resonance frequency. The situation then is similar to that of the free atom case discussed in § 2.2. Suppose the system is initially in thermal equilibrium in a magnetic field $\mathbf{H}$, and a strong rotating field $\mathbf{H}_1$ at the exact resonance frequency is applied suddenly for a time δt. Then the magnetization $\mathbf{M}$ is initially parallel to $\mathbf{H}$, and in the rotating coordinate system of Fig. 2.1 is normal to $\mathbf{H}_1$. In this system there is no other field, and in time δt, $\mathbf{M}$ will precess about $\mathbf{H}_1$ through an angle $|\gamma| H_1 \, \delta t$. If δt is chosen so that $|\gamma| H_1 \, \delta t = \pi$, the magnetization will just reach the direction anti-parallel to $\mathbf{H}$ (a '180 degree pulse'), while if δt is half as long, $\mathbf{M}$ is turned to a position normal to $\mathbf{H}_1$ in the equatorial plane (a '90 degree pulse'). It is readily seen that relaxation effects will only be negligible provided that $\delta t \ll \tau_1, \tau_2$; thus the condition $\tau_1, \tau_2 \gg (|\gamma| H_1)^{-1}$ applies in this case also. If the transverse relaxation time is expressed as a spread in field $\Delta H = (|\gamma|\, \tau_2)^{-1}$, the second and more rigorous of these conditions is readily seen to be $H_1 \gg \Delta H$. In the rotating coordinate system this means that the effective z-component of the field $\mathbf{H} - \mathbf{H}^*$, which is not zero for all parts of the sample if we regard ΔH as being equivalent to an inhomogeneity in $\mathbf{H}$, is small everywhere compared with $\mathbf{H}_1$, so that the precession of $\mathbf{M}$ is about an effective field (the vector resultant of $\mathbf{H} - \mathbf{H}^*$ and $\mathbf{H}_1$) which does not deviate appreciably in direction and magnitude from $\mathbf{H}_1$. This condition is more stringent than that required for adiabatic rapid passage, where a static spread in field of ΔH ('inhomogeneous broadening') only results in dipoles passing through the equatorial plane at different instants of time.

Pulses of this kind are used in spin-echo techniques, originated by Hahn (1950). After the magnetization has reached thermal equilibrium in a field $\mathbf{H}$, so that $\mathbf{M}$, like $\mathbf{H}$, lies along the z-axis, a '90 degree pulse' is applied by means of an oscillatory field at the resonance frequency. If, in the rotating coordinate system, $\mathbf{H}_1$ is taken to lie along the x-axis, then after the pulse the magnetization vector $\mathbf{M}$ lies in the equatorial plane, parallel to the y-axis if γ is positive, and anti-parallel to the y-axis if γ is negative. There is now no field in the rotating frame, apart from a small spread in field due to the local field or other inhomogeneity which has a steady component ΔH in the rotating frame along the z-axis. The effect of this is to cause the dipoles to fan out in the equatorial plane with angular velocity $-\gamma\, \Delta H$, with resulting diminution in the net magnetization. If at a time τ after the first pulse is applied, a '180 degree pulse' of oscillatory field is turned on, the result is to turn all the individual dipoles in the equatorial plane through $180°$ about $\mathbf{H}_1$. Thus a dipole that had deviated in time τ by an angle $\delta' = -\gamma\, \Delta H \tau$ from the y-axis is turned to an angle

$$\pi - \delta' = \pi - (-\gamma\, \Delta H \tau) = \pi + \gamma\, \Delta H \tau.$$

Under the action of the local field it will subsequently precess in the equatorial plane with angular velocity $-\gamma\, \Delta H$, so that after a further time τ it will reach an angle $(\pi + \gamma\, \Delta H \tau) - \gamma\, \Delta H \tau = \pi$. Thus at time 2τ after the first pulse all dipoles are at the same angle, and a macroscopic magnetization is reconstructed in the equatorial plane which is observed as a pulse on the oscilloscope (see Fig. 2.4). The process can be repeated by applying further $180°$ pulses at times 3τ, 5τ, etc., so that pulsed signals appear at times 4τ, 6τ, etc. as well as 2τ. The amplitude of successive pulses decays because of relaxation effects, and the method

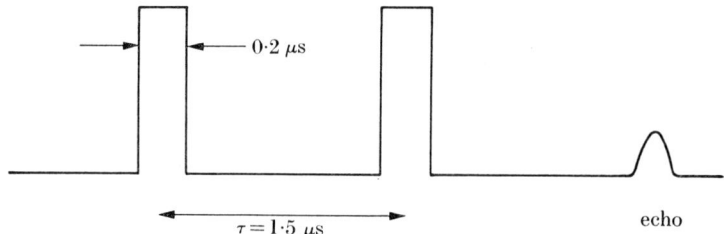

FIG. 2.4. Typical spin echo sequence for electron spin resonance (after Kiel and Mims 1967). Two pulses of r.f. power are applied at an interval of 1·5 μs. The r.f. field at the sample during each pulse is about 0·5 G; for $g = 2$, $\gamma H_1 \approx 10^7$ rad s^{-1} so that $\gamma H_1 t \approx 2$ for a pulse of duration 0·2 μs. The spin echo is detected by a superheterodyne receiver. Note that pulse lengths other than 90° followed by 180° (that described in the text) are often used in practice. In the work of Kiel and Mims the spin echo method is used as a sensitive detector in measurements of τ_1.

gives therefore a measurement of the 'genuine' relaxation time τ_2. It can also be applied to the measurement of τ_1.

2.8. Solution of the macroscopic equations for slow passage

In the case of very slow passage through resonance, which is the case commonly encountered in electronic magnetic resonance, another limiting solution of the Bloch equations can easily be found. Essentially we assume that passage is so slow that at all times steady-state conditions prevail, and again look for a solution of eqns (2.33) where $\dot{M}_z = 0$, so that M_z is constant. Then the equations are readily integrable, giving the complex conjugate solutions

$$\left. \begin{array}{l} M_+ = \dfrac{\gamma H_1 M_z \exp(+i\omega t)}{\omega + \gamma H - i/\tau_2}, \\[6pt] M_- = \dfrac{\gamma H_1 M_z \exp(-i\omega t)}{\omega + \gamma H + i/\tau_2}. \end{array} \right\} \qquad (2.37)$$

Substitution back into (2.33) gives the value of M_z:

$$\frac{M_z}{M_0} = \frac{1+(\omega-\omega_L)^2\tau_2^2}{1+(\omega-\omega_L)^2\tau_2^2+\gamma^2 H_1^2 \tau_1 \tau_2} \qquad (2.38)$$

whence

$$\frac{M_\pm}{M_0} = \frac{\{(\omega-\omega_L)\tau_2 \pm i\}\gamma H_1 \tau_2 \exp(\pm i\omega t)}{1+(\omega-\omega_L)^2\tau_2^2+\gamma^2 H_1^2 \tau_1 \tau_2}, \qquad (2.39)$$

where, as before, the symbol ω_L is used for the Larmor angular velocity $-\gamma H$. These equations show that because of the term $\{(\omega-\omega_L)\tau_2\}^2$ M_z remains practically equal to M_0 except at resonance, and even then the departure is small unless $\gamma^2 H_1^2 \tau_1 \tau_2 > 1$. The whole solution corresponds to the magnetization vector precessing about H with angular velocity ω, but tipped at an angle θ to H such that

$$\tan\theta = \frac{|M_\pm|}{M_z} = \frac{\gamma H_1 \tau_2}{\{1+(\omega-\omega_L)^2\tau_2^2\}^{\frac{1}{2}}}. \tag{2.40}$$

If, as in most experiments, H is varied slowly and ω is kept constant, it is convenient to express these quantities in terms of magnetic field. Using $H^* = -(\omega/\gamma)$ for the resonance field at frequency ω, and $\Delta H = -(\gamma\tau_2)^{-1}$, we have

$$\tan\theta = \frac{H_1}{\{(\Delta H)^2+(H-H^*)^2\}^{\frac{1}{2}}}, \tag{2.41}$$

while the equatorial component $M\sin\theta$ of the magnetization leads the rotating field H_1 by an angle ϵ such that

$$\tan\epsilon = \{\tau_2(\omega-\omega_L)\}^{-1} = \Delta H/(H^*-H). \tag{2.42}$$

We have already seen that in a paramagnetic substance with electronic dipoles ΔH is fairly large ($\sim 10^2$ G) unless the substance is heavily diluted. Thus, for ordinary values of H_1, $\tan\theta$ is very small even at resonance, for which $\tan\theta = (H_1/\Delta H)$, so that the angle θ at which **M** precesses round **H** is small. The value of ϵ varies from 0 for $H^* \ll H$, through $-\pi/2$ for $H^* = H$ to $-\pi$ for $H^* \gg H$. This behaviour is similar to that usually found for the phase angle in forced resonance, and we can express the oscillatory part of the magnetization in terms of a complex susceptibility χ such that

$$M_+ = \chi H_1 \exp(i\omega t). \tag{2.43}$$

Then, since $M_0 = \chi_0 H$,

$$\frac{\chi}{\chi_0} = \frac{\gamma H \tau_2\{(\omega-\omega_L)\tau_2+i\}}{1+(\omega-\omega_L)^2\tau_2^2+\gamma^2 H_1^2\tau_1\tau_2}. \tag{2.44}$$

The real and imaginary parts of $\chi = \chi'-i\chi''$ can be expressed in terms either of fields or of frequencies (writing $\Delta\omega = \tau_2^{-1}$), giving in the latter case (since $\gamma H = -\omega_L$)

$$\left.\begin{aligned}\frac{\chi'}{\chi_0} &= \frac{\omega_L(\omega_L-\omega)}{(\omega-\omega_L)^2+(\Delta\omega)^2+\gamma^2 H_1^2(\Delta\omega)\tau_1}, \\ \frac{\chi''}{\chi_0} &= \frac{\omega_L \Delta\omega}{(\omega-\omega_L)^2+(\Delta\omega)^2+\gamma^2 H_1^2(\Delta\omega)\tau_1}.\end{aligned}\right\} \tag{2.45}$$

Thus χ' is zero at resonance, while at other frequencies it is positive or negative (see Fig. 2.5) according to whether ω_L is greater or less than

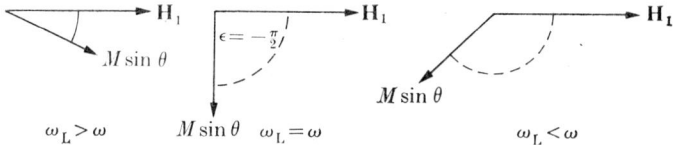

FIG. 2.5. Position of the equatorial component $M \sin \theta$ of the magnetization $\mathbf{M}$, relative to $\mathbf{H}_1$ in the rotating frame, on either side of resonance. When the natural resonance frequency ω_L is numerically larger than the applied frequency ω, the phase angle ϵ given by

$$\tan \epsilon = \Delta\omega/(\omega - \omega_L) \tag{2.42a}$$

lies between 0 and $-\pi/2$; at resonance it is just equal to $-\pi/2$, and $\chi' = 0$; as ω is increased above ω_L, the angle ϵ moves towards $-\pi$. In terms of magnetic field, the same pattern is followed, starting with $H > H^*$ and decreasing H through the resonance value H^*.

In eqn (2.42a), all angular velocities have the same sign (usually positive for electronic dipoles) provided that $\mathbf{H}_1$ rotates in the same sense as the natural sense of precession of the atomic dipoles. If it rotates in the opposite sense, then the quantities ω, ω_L add numerically and the value of ϵ is always close to zero. Note that $\mathbf{M}$ deviates in the direction corresponding to an incipient precession about $\mathbf{H}_1$.

the applied ω. The imaginary part of the susceptibility is a maximum at resonance, and provided that $\gamma^2 H_1^2 \tau_1 \ll \Delta\omega$, one has at resonance

$$\frac{\chi''}{\chi_0} = \frac{\omega_L}{\Delta\omega}, \tag{2.46}$$

showing that the imaginary part of the susceptibility is much greater than the static susceptibility provided that the line width is small compared with the resonance frequency, i.e. that the lines are fairly sharp.

The condition that was introduced in deducing eqn (2.46) involves the spin-lattice relaxation time τ_1, and is the same condition as is required to make the departure of M_z from M_0 negligible even at resonance. It is readily shown that at resonance

$$\frac{|M|}{M_0} = \frac{\{1 + \gamma^2 H_1^2 \tau_2^2\}^{\frac{1}{2}}}{1 + \gamma^2 H_1^2 \tau_1 \tau_2} \tag{2.47}$$

so that M can never be greater than M_0. It can, however, be significantly smaller than M_0 for large amplitudes of the rotating field H_1. This is because the spin system is absorbing energy from the rotating field at a rate

$$\frac{dW}{dt} = \left\{ H_x \frac{dM_x}{dt} + H_y \frac{dM_y}{dt} \right\}_{\text{av}} = \omega \chi'' H_1^2, \tag{2.48}$$

and this absorption of power raises the temperature of the spin system to the point where (dW/dt) is equal to the rate at which energy is transferred from the spin system to the lattice. Hence M falls below the equilibrium value M_0, which is appropriate to the spin system being at the same temperature as the lattice, and the temperature rise in the spin system with increasing H_1 will be greater, the longer τ_1 and τ_2. If $H_1 = 1$ G, and $\gamma = 10^7$ rad G^{-1}, $\tau_2 = 10^{-9}$ s as for an ordinary paramagnetic, then there will be no significant temperature rise in the spin system unless $\tau_1 \geqslant 10^{-5}$ s, a value that is ordinarily attained only at low temperatures.

This phenomenon is known as 'dynamic saturation'; in experiments where resonance is detected by means of the energy absorbed from the rotating field it limits the gain that can be achieved through increasing H_1. At resonance the size of the precessing component $\chi''H_1$ of the magnetization in the equatorial plane, which is $\pi/2$ out of phase with the rotating field, is given by

$$\frac{\chi'' H_1}{M_0} = -\frac{\gamma H_1 \tau_2}{1 + \gamma^2 H_1^2 \tau_1 \tau_2}, \qquad (2.49)$$

which passes through a maximum at $|\gamma H_1| = (\tau_1 \tau_2)^{-\frac{1}{2}}$. This maximum arises because the angle θ at which the vector $\mathbf{M}$ precesses about $\mathbf{H}$ increases as $\mathbf{H}_1$ increases (eqn (2.40)), but the magnitude of the vector shrinks as saturation sets in. The absorbed energy $\mathrm{d}W/\mathrm{d}t$ given by (2.48) can be written, using (2.49) and (2.38), as (at resonance)

$$\frac{\mathrm{d}W}{\mathrm{d}t} = \frac{M_0 H}{\tau_1}\left(\frac{\gamma^2 H_1^2 \tau_1 \tau_2}{1+\gamma^2 H_1^2 \tau_1 \tau_2}\right) = \frac{M_0 H}{\tau_1}\left(\frac{M_0 - M_z}{M_0}\right), \qquad (2.50)$$

where the quantity $s = (M_0 - M_z)/M_0$ is sometimes known as the 'saturation parameter'. As H_1 is increased without limit, the value of $\mathrm{d}W/\mathrm{d}t$ tends to the limiting value $(M_0 H/\tau_1)$, which is independent of H_1 and τ_2. This limit occurs because the temperature of the spin system rises and hence χ'' falls as H_1 increases, the temperature adjusting itself to a value such that the rate at which heat is absorbed is just equal to the rate at which it is transferred to the lattice; the latter becomes the controlling factor.

In the simple treatment of § 1.11 we saw that saturation sets in when $2w_e \tau_1 \approx 1$. We can compare this result with the treatment given in this section by using eqns (2.17a) and (2.34). For a system with $J = \frac{1}{2}$ the matrix element $\langle M \pm 1 | J_\pm | M \rangle = 1$, and exactly at resonance ($\omega = \omega_L$) the value of the shape function is $f(\omega) = \tau_2/\pi$. Hence

$$2w_e \tau_1 = 2(\pi/2)(\gamma H_1)^2 (\tau_2/\pi) \tau_1 = \gamma^2 H_1^2 \tau_1 \tau_2,$$

which is just the quantity that occurs in the saturation parameter at resonance (cf. eqn (2.50)).

It may also be helpful to give some numerical values. If we take $H_1 = 0\cdot1$ G as being typical of an electron resonance experiment in which $\gamma = 10^7$ rad G^{-1} with $J = \frac{1}{2}$, we find that saturation sets in when $\gamma^2 H_1^2 \tau_1 \tau_2 = 10^{12} \tau_1 \tau_2 \approx 1$. Thus in a concentrated salt for which $\tau_2 = 10^{-9}$ s ($\Delta H \approx 100$ G), saturation occurs only if $\tau_1 \geqslant 10^{-3}$ s. On the other hand, in an inhomogeneously broadened dilute paramagnetic for which $\tau_2 = 10^{-5}$ s in a spin packet, saturation requires only $\tau_1 \geqslant 10^{-7}$ s; since τ_2 cannot be longer than τ_1, saturation would set in at $\tau_2 = \tau_1 = 10^{-6}$ s. For a nuclear paramagnet $\gamma_n = 10^4$ rad G^{-1}, and if we take $H_1 = 1$ G in this case, and $\Delta H = (\gamma_n \tau_2)^{-1} = 10$ G in a solid lattice, saturation occurs if $\tau_1 \geqslant 10^{-3}$ s.

2.9. Intensity and line width

The classical analysis using the macroscopic equations gives an excellent approach to magnetic resonance, particularly because it gives a simple physical picture. It is rather less applicable to electronic paramagnetic than to nuclear paramagnetic substances because the case where the only major interaction is that with the external applied field **H** tends to be the exception rather than the rule. In a solid paramagnetic the energy levels in zero magnetic field are commonly split by crystal field effects as well as nuclear interactions, and even where these are absent, the behaviour of the energy levels in an applied field depends on the angle that this field makes with the crystal axes. In these circumstances a quantum-mechanical approach is essential, and the resonance spectrum can only be interpreted in terms of a 'spin Hamiltonian' where the effects of interaction between the paramagnetic ion and the crystal lattice are included. The relation between the spectrum and this spin Hamiltonian is considered in the next chapter, but it is appropriate to derive here a formula for the intensity of the lines by quantum-mechanical methods. This approach is similar to that given by Bloembergen, Purcell, and Pound (1948) for the nuclear case, with a slight extension.

The number of transitions per second between levels of energy W_i and W_j which are induced by an oscillatory field of amplitude H_1 and angular frequency ω is

$$w_{ij} = \frac{\pi H_1^2}{2\hbar^2} |\mu_{ij}|^2 f(\omega). \tag{2.51}$$

This equation is similar to (2.17a) except that a more general expression, $|\mu_{ij}|^2$, has been introduced for the square of the magnetic dipole matrix element between the states i and j. The 'shape function' $f(\omega)$ is a function of the applied frequency ω, the resonance frequency ω_0, and a parameter describing the line width; it is normalized as in eqn (2.17b).

The oscillating field induces transitions between states i and j with equal probability in either direction, and in each transition the energy change is $\hbar\omega$. There will, in general, be an absorption of energy since the state of lower energy normally has the greater population, and the total number of transitions in either direction is proportional to the population of the initial state. The net absorption of energy per second (assuming no significant departure from thermal equilibrium) is

$$\mathrm{d}W/\mathrm{d}t = w_{ij}(\hbar\omega)(N_i - N_j) = N_i w_{ij}(\hbar\omega)\{1 - \exp(-\hbar\omega/kT)\},$$

where N_i, N_j are the populations of lower and upper states respectively. From eqn (2.51) we have then

$$\frac{\mathrm{d}W}{\mathrm{d}t} = \frac{N_i \pi \omega H_1^2}{2\hbar} \{1 - \exp(-\hbar\omega/kT)\} |\mu_{ij}|^2 f(\omega). \qquad (2.52\mathrm{a})$$

For a simple ground doublet, the total number of ions is

$$N = N_i + N_j = N_i\{1 + \exp(-\hbar\omega/kT)\}$$

and

$$\frac{\mathrm{d}W}{\mathrm{d}t} = N \tanh(\hbar\omega/2kT) \frac{\pi\omega H_1^2}{2\hbar} |\mu_{ij}|^2 f(\omega). \qquad (2.52\mathrm{b})$$

In many experiments $(\hbar\omega/kT) \ll 1$ (it would be equal to unity for radiation of wave number 1 cm^{-1} at a temperature of 1·43°K), so that $\{1 - \exp(-\hbar\omega/kT)\} \approx \hbar\omega/kT$, and the power absorbed in a transition between states i, j is

$$\frac{\mathrm{d}W}{\mathrm{d}t} = N_i \frac{\pi\omega^2 H_1^2}{2kT} |\mu_{ij}|^2 f(\omega). \qquad (2.52\mathrm{c})$$

Circularly polarized radiation

Combining eqns (2.48) and (2.52c), we find for the imaginary part of the susceptibility, when circularly polarized radiation is used,

$$\chi''_{ij} = \frac{N_i \pi \omega |\mu_{ij}|^2 f(\omega)}{2kT}. \qquad (2.53)$$

This formula is a general one, and it may be compared with that derived from the macroscopic equations on a classical basis for the case of a rotating magnetic field. For this comparison we must assume that all the states have the same energy in zero magnetic field, so that for a system with angular momentum J the partition function

$$Z = \sum \exp(-W_j/kT)$$

may be approximated by $Z = 2J+1$, and $N_i = N/(2J+1)$, where N is the total number of spins. For a rotating field the matrix elements of μ_{ij} are $(\gamma\hbar)$ times those of $|J_\pm|$ between states $M\pm 1$ and M, as given in eqns (2.15), and hence we have for the transition $M \leftrightarrow M-1$

$$\chi''_{M,M-1} = \frac{N\pi\omega\gamma^2\hbar^2\{J(J+1)-M(M-1)\}f(\omega)}{2(2J+1)kT}. \tag{2.54}$$

The macroscopic equations are applicable to a substance where all the transitions occur at the same frequency, since the energy levels are separated by equal amounts $\gamma\hbar H$ as given by eqn (2.11). Hence (2.54) must be summed over all transitions; using the relation

$$\sum_{-(J-1)}^{J} \{J(J+1)-M(M-1)\} = \tfrac{2}{3}J(J+1)(2J+1)$$

and eqn (2.22), we have

$$\chi'' = \frac{N\pi\omega\gamma^2\hbar^2 J(J+1)f(\omega)}{3kT} = \pi\omega\chi_0 f(\omega). \tag{2.55}$$

With a suitable choice for $f(\omega)$, to which we shall return later, this is similar to eqns (2.45) derived from the macroscopic theory.

In practice it is seldom that a circularly polarized oscillatory field is used in electronic magnetic resonance apart from exceptional cases where the sign of the magnetogyric ratio γ is of importance in deciding the nature of the electronic ground state. We shall therefore consider briefly the modification to the theory required for a linearly polarized oscillatory field.

Linearly polarized oscillatory field

If the oscillatory magnetic field is of amplitude H_1 and is linearly polarized along a direction whose direction cosines are (l_1, m_1, n_1) with respect to the Cartesian axes x, y, z, the interaction with the magnetic dipole moment may be written as

$$\begin{aligned}\mathcal{H}_1 &= -(l_1\mu_x+m_1\mu_y+n_1\mu_z)H_1\cos\omega t \\ &= -\tfrac{1}{2}(l_1\mu_x+m_1\mu_y+n_1\mu_z)H_1\{\exp(i\omega t)+\exp(-i\omega t)\}. \end{aligned} \tag{2.56}$$

Then if μ_{ij} is the matrix element of the dipole moment

$$\mu_{ij} = \langle \mathbf{u}_j | (l_1\mu_x+m_1\mu_y+n_1\mu_z) | \mathbf{u}_i \rangle$$

between states i and j, the power absorbed (dW/dt) is given by eqn (2.52a). As before, we shall assume that the steady magnetic field is directed along the z-axis, so that the transition probabilities involve only the matrix elements of μ_+, μ_-. The dipole moment operator for the linear oscillatory field may be written as

$$\tfrac{1}{2}(l_1-im_1)\mu_+ + \tfrac{1}{2}(l_1+im_1)\mu_- + n_1\mu_z$$

and the square of the matrix element for the transition $M \leftrightarrow M-1$ is

$$|\mu_{ij}|^2 = \tfrac{1}{4}(l_1^2+m_1^2)\langle M-1| \,\mu_-\, |M\rangle^2. \tag{2.57}$$

It is clear that this will have its maximum value if $l_1^2+m_1^2 = 1$, that is, if the oscillatory field is in the xy-plane, normal to the direction of the steady field. If this condition is satisfied, and we choose the direction of the x-axis so that it coincides with that of H_1, the oscillatory Hamiltonian reduces to

$$\mathscr{H}_1 = -\tfrac{1}{4}\gamma\hbar H_1(J_+ + J_-)\{\exp(i\omega t)+\exp(-i\omega t)\}. \tag{2.58}$$

This differs from the equivalent eqn (2.13) for a rotating field in two respects: the operators J_+, J_- are each multiplied both by $\exp(i\omega t)$ and by $\exp(-i\omega t)$, so that transitions can be induced irrespective of the sign of γ, and the amplitude of the perturbation is only half as large as for a rotating field. These differences correspond just to the fact that a linearly polarized field $H_1 \cos \omega t$ can be decomposed into two circularly polarized fields each of amplitude $\tfrac{1}{2}H_1$ but rotating in opposite sense with the same numerical value of the angular velocity ω. Of these two, only the component that rotates in the same sense as that in which the magnetization is precessing will be effective in inducing transitions. Since the amplitude of this component is only half as great as before, we would expect the induced precessing magnetization to be only half as great, and the complex susceptibility likewise to be halved. This is confirmed by carrying through the computation of χ''.

In the case of a linearly polarized field we have, instead of eqn (2.48),

$$\frac{dW}{dt} = \tfrac{1}{2}\omega\chi''H_1^2, \tag{2.59}$$

so that from eqn (2.52c)

$$\chi''_{ij} = \frac{\pi N_i \omega \, |\mu_{ij}|^2 f(\omega)}{kT}, \tag{2.60}$$

where N_i is the number of atoms in the lower of the two states between which a transition is taking place. For the transition $M \leftrightarrow M-1$, on using eqns (2.57) or (2.58) and inserting the matrix element of μ_- or J_- from eqn (2.15), we have

$$\chi''_{M,M-1} = \frac{N_i \pi \omega \gamma^2 \hbar^2 \{J(J+1) - M(M-1)\} f(\omega)}{4kT}. \tag{2.61}$$

For an electronic transition, the value of γ is usually negative, so that the lower state is $M-1$, for which

$$N_i = N_{M-1} = N \exp(-W_{M-1}/kT)/Z. \tag{2.62}$$

For a simple ion, as considered earlier in this section, all the transitions are superposed, and at high temperatures where $\hbar\omega \ll kT$, the exponential has a value close to unity and $Z \to 2J+1$. Then, on summing over all transitions, we have

$$\chi'' = \tfrac{1}{2}\pi\omega\chi_0 f(\omega), \tag{2.63}$$

which is just half as great as the corresponding expression (2.55) for the case of a rotating field.

This expression for χ'' satisfies the Kramers–Kronig relations, one of which reduces to

$$\chi_0 = \frac{2}{\pi} \int_0^\infty \frac{\chi''(\omega)\, d\omega}{\omega}. \tag{2.64}$$

It is readily seen that this equation is satisfied through the normalization of $f(\omega)$ introduced by eqn (2.17b). In this connection it should be noted that the macroscopic eqns (2.45) do not satisfy the Kramers–Kronig relations except in the limit of small H_1 (see Portis 1953); this arises from the presence of the term in H_1 in the denominator of eqns (2.45). As already discussed in § 2.8, this term is connected with saturation effects in which the temperature of the spin system rises above that of the lattice because of the absorption of energy from the oscillatory field; this makes the magnetization non linear with applied oscillatory field, contrary to one of the assumptions on which the Kramers–Kronig relations are based. Saturation effects have, of course, been neglected in the derivation of formulae for χ'' in the present section (§ 2.9).

The question of an appropriate formula for the line shape in electron paramagnetic resonance will not be discussed in detail here. The problem of the line shape due to spin-spin interaction is considered in Chapter 9, but when the line shape is determined by spin lattice interaction (or 'collisions with phonons'), or another true relaxation process, an appropriate formula for $f(\omega)$ should be the collision broadening formula of Van Vleck and Weisskopf (1945),

$$f(\omega) = \frac{1}{\pi}\left\{\frac{\Delta\omega}{(\omega-\omega_\mathrm{L})^2+(\Delta\omega)^2} + \frac{\Delta\omega}{(\omega+\omega_\mathrm{L})^2+(\Delta\omega)^2}\right\}. \quad (2.65)$$

Here the second term (which is absent from eqn (2.34), but which is needed to make $\int_0^\infty f(\omega)\,\mathrm{d}\omega = 1$ exactly) can be regarded as arising from the presence of an oscillatory field component rotating in the opposite sense to that required for resonance; correspondingly, the formula allows for reversal of the applied magnetic field H (negative values of ω_L), when linearly polarized oscillatory fields are used. It also gives a suitable form for the line shape near $H = 0$ (compare Fig. 2.6), and in

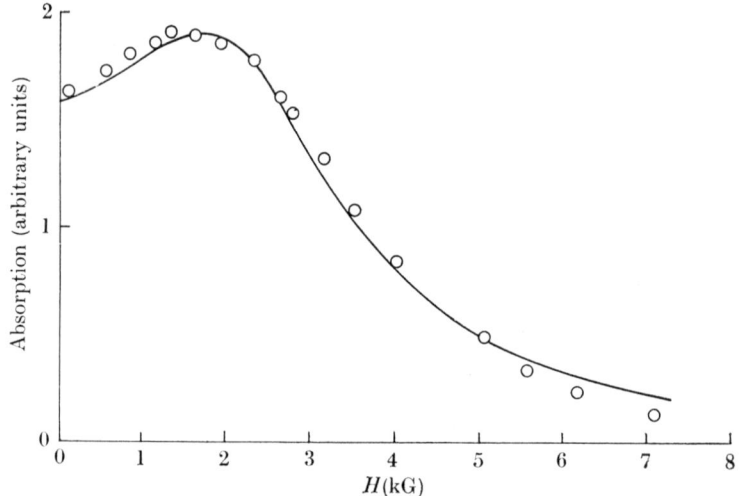

Fig. 2.6. Line shape in $\mathrm{Co(NH_4)_2(SO_4)_2, 6H_2O}$ at 90°K and $\bar{\nu} = 0\cdot3$ cm^{-1}.

the limit of $\omega_\mathrm{L} = 0$ (or $\omega_\mathrm{L} \ll \Delta\omega$, corresponding to $H \ll \Delta H$), the combination of eqns (2.63) and (2.65) gives

$$\chi'' = \chi_0 \frac{\omega\,\Delta\omega}{\omega^2+(\Delta\omega)^2}, \quad (2.66)$$

the formula used in low-frequency measurements on paramagnetic relaxation (see Gorter 1947).

In both eqns (2.65) and (2.66) the effects of saturation by a high-intensity oscillatory magnetic field have been omitted (cf. Karplus and Schwinger (1948) for the case of saturation in microwave gas spectroscopy).

2.10. Spectrometer sensitivity

We now consider in a simple way the practical aspects of the detection of magnetic resonance in an electronic paramagnetic salt, including the important question of spectrometer sensitivity. For simplicity we restrict the discussion to linearly polarized oscillatory fields, since these are almost invariably used in practice. If the paramagnetic sample is subjected to such a linearly polarized magnetic field $H_1 \cos \omega t$, the power absorbed by it is given by eqn (2.59),

$$dW/dt = \tfrac{1}{2}\omega \chi'' H_1^2, \tag{2.59}$$

where χ'' is the imaginary part of the complex susceptibility of the sample. To make H_1^2 as large as possible, the sample is placed in a tuned circuit or resonant cavity, in which the power dissipated through resistive and other non-magnetic losses may be written

$$\frac{1}{Q_0} \omega \frac{H_1^2}{8\pi} V, \tag{2.67}$$

where Q_0 is the 'quality factor' of the unloaded cavity. The quantity V is the 'effective volume' of the cavity, defined by the fact that $V(H_1^2/8\pi)$ is the total energy stored in the cavity, related to H_1, the amplitude of the oscillatory magnetic field at the sample. The important quantity for the detection of magnetic resonance is the ratio of the power absorbed by magnetic resonance in the sample to that otherwise dissipated in the cavity. From the above equations this ratio is

fractional power loss in magnetic resonance $= 4\pi\chi'' Q_0/V$. (2.68)

It is clear that this quantity can be maximized by making Q_0 as large as possible, and keeping the effective volume V as small as possible. The latter is achieved partly by making the actual cavity volume as small as possible, and partly by locating the sample at the position of maximum H_1. As far as the geometry of the cavity is concerned, higher values of (Q_0/V) are achieved by keeping the cavity small; for example, a cavity one-half wavelength long is better than a cavity $n/2$ wavelengths long, since the small gain in Q_0 in the larger cavity is much outweighed by the increase in V. A smaller value of V can be obtained by using a cavity filled with a low-loss dielectric of high permittivity, but this usually has practical difficulties.

The size of the effect on the cavity at resonance can be estimated as follows. At a frequency of 10 GHz (a wavelength of 3 cm), the values for a typical cavity resonator are $Q_0 = 5000$, $V = 3$ cm³. The power absorbed in the crystal due to magnetic resonance will then equal that dissipated in the cavity (i.e. the Q will be halved at resonance) by a sample for which χ'' is about 5×10^{-5}. This value can be related to the static susceptibility by using eqn (2.63) and taking $f(\omega) = (\pi \Delta\omega)^{-1}$ from eqn (2.65) for a reasonably narrow line at resonance. This gives

$$\chi'' = \chi_0\{\omega/(2\,\Delta\omega)\}. \qquad (2.69)$$

For a reasonably dilute paramagnetic salt such as a tutton salt (typical formula $M^{2+}K_2(SO_4)_2, 6H_2O$) the volume per gram-ion of paramagnetic ion M^{2+} is close to 200 cm³. If the magnetic ions have effective spin $S = \tfrac{1}{2}$ and $g = 2$, the line width ΔH is about 100 G and the resonant field H at 10 GHz is a little over 3000 G. Thus $(\omega/\Delta\omega) = (H/\Delta H)$ is roughly 30, and the gram-ionic susceptibility is

$$\chi_0 = 0\cdot 375/T \text{ (per gram-ion)}$$

so that at room temperature ($T = 300°K$) we have

$$\chi'' = 0\cdot 018 \text{ (per gram-ion)}.$$

Hence the value of $\chi'' = 5 \times 10^{-5}$ required above to halve the cavity Q needs about 3×10^{-3} gram-ion or 0·6 cm³ of salt at room temperature. Such a sample clearly provides an effect that is very readily detectable and increases inversely with the temperature of the sample. In fact serious detuning of the cavity will occur owing to the anomalous variation in χ' through resonance, and a distorted shape of resonance curve will result unless the cavity is continuously retuned.

Electromagnetic detection

A prime advantage of electron paramagnetic resonance is its ability to detect effects caused by rather small quantities of magnetic ions, and we now consider the ultimate sensitivity that can be achieved. This does not depend on the details of the spectrometer, and we therefore consider a simple lumped-parameter equivalent circuit. A power source of voltage V_1, with internal resistance R_1, is coupled through a mutual inductance M_1 to a tuned circuit of series resistance r, as in Fig. 2.7. The power level in the tuned circuit is detected by an output coupling of mutual inductance M_2 to a voltage detector of input resistance R_2, the voltage across which is V_2. When magnetic resonance occurs in a paramagnetic sample placed in the tuned circuit, the additional power loss is represented by an extra resistance δr. Since we are interested in

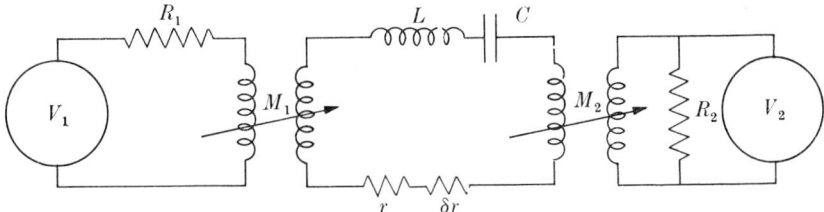

FIG. 2.7. Simple equivalent circuit for 'transmission' spectrometer. A voltage source V_1, with internal resistance R_1, is coupled through M_1 to a tuned circuit in which the magnetic resonance absorption contributes an extra resistance δr. The tuned circuit is coupled through M_2 to a detector of resistance R_2. When M_1, M_2 are adjusted to give the maximum sensitivity (largest change in V_2 for a given δr), the extra resistances coupled into the tuned circuit through input and output are together just equal to r, so that the loaded Q is just half the unloaded Q_0.

the ultimate sensitivity, we are concerned with values of $\delta r \ll r$, and we can write, from eqn (2.68),

$$\delta r/r = 4\pi\chi'' Q_0/V. \tag{2.70}$$

Our problem is to relate the change in V_2 to δr, assuming that the source voltage V_1 is constant. In fact we shall specify the source by its 'available power' $P_1 = V_1^2/4R_1$, which is the maximum power it can deliver into a load matched to its internal resistance R_1. If we write $r_1 = \omega^2 M_1^2/R_1$, $r_2 = \omega^2 M_2^2/R_2$, where r_1, r_2 are the resistances coupled into the tuned circuit, then at resonance in the tuned circuit it is easily shown that

$$\frac{V_2}{\sqrt{R_2}} = \frac{2\sqrt{(P_1)}\sqrt{(r_1 r_2)}}{(r_1+r_2+r)} \tag{2.71}$$

and hence that

$$\frac{|\delta V_2|}{\sqrt{R_2}} = \frac{2\sqrt{(P_1)}\sqrt{(r_1 r_2)}}{(r_1+r_2+r)^2} \delta r. \tag{2.72}$$

We wish to maximize δV_2 by adjusting the input and output couplings, which is equivalent to finding the optimum values of r_1, r_2. These occur at $r_1 = r_2 = r/2$, and the maximum value of δV_2 is then

$$\frac{|\delta V_2|}{\sqrt{R_2}} = \tfrac{1}{4}\sqrt{(P_1)}\frac{\delta r}{r}. \tag{2.73}$$

This maximum occurs when the loaded Q of the cavity is just half the unloaded Q_0, since it requires $(r_1+r_2+r) = 2r$. We can see how this maximum arises by re-writing (2.72) in terms of the power $P_2 = V_2^2/R_2$ reaching the detector. The value of $\sqrt{P_2}$ is given by (2.71), so that

(2.72) becomes

$$\frac{|\delta V_2|}{\sqrt{R_2}} = \sqrt{P_2}\frac{\delta r}{(r_1+r_2+r)} \quad \text{or} \quad \frac{|\delta V_2|}{V_2} = \frac{\delta r}{(r_1+r_2+r)}, \tag{2.74}$$

which shows that the fractional voltage change ($\delta V_2/V_2$) at the detector is just equal to the fractional change in the 'loaded Q', which is $\delta r/(r_1+r_2+r)$. As we increase the coupling, V_2 rises and so initially does δV_2, but ultimately the increase in V_2 becomes less important than the fall in the loaded Q because of the heavier coupling. By combining eqns (2.70) and (2.73) we find that under optimum coupling conditions

$$|\delta V_2|/\sqrt{R_2} = \sqrt{(P_1)(\pi\chi''Q_0/V)}. \tag{2.75}$$

If the detector has an equivalent noise temperature T_N and bandwidth df, the signal output will be equal to the noise output when

$$|\delta V_2|/\sqrt{R_2} = (kT_N\,\mathrm{d}f)^{\frac{1}{2}}, \tag{2.76}$$

and if this is regarded as the minimum signal detectable, we find that

$$(\chi'')_{\min} = \left(\frac{V}{\pi Q_0}\right)\left(\frac{kT_N\,\mathrm{d}f}{P_1}\right)^{\frac{1}{2}}. \tag{2.77}$$

To gain an idea of the order of magnitude implied by this equation we insert some numerical values. Taking (as before) $Q_0 = 5000$, $V = 3$ cm³, together with $P_1 = 40$ mW, $\Delta f = 1$ Hz, and $T_N = 3000°$K, we find

$$(\chi'')_{\min} = 2\times 10^{-13}. \tag{2.78}$$

When working at the limit of sensitivity we are likely to be dealing with highly diluted salts; for convenience we take $\Delta H = 1$ G (i.e. line width $= 2$ G right across at half intensity) and $H = 3000$ G (the resonance field for a salt with $g = 2$ at about 9 GHz frequency). From eqn (2.69) we then obtain

$$(\chi_0)_{\min} \approx 10^{-16}, \tag{2.79}$$

which for a system of spins $S = \frac{1}{2}$ requires some $3\times 10^{-16}T$ moles at temperature T, or about 10^9 spins at $4°$K.

In making this calculation we have assumed that the detector is not overloaded by the power P_2 incident on it, which under the optimum coupling conditions assumed above is $P_1/4$. If it is necessary to keep P_2 to some maximum value less than $P_1/4$, then obviously it is better to loosen the coupling to the cavity so as to increase the loaded Q rather than to insert attenuation by means of a lossy device before or after

THE RESONANCE PHENOMENON

the cavity since, from (2.74),

$$|\delta V_2|/V_2 = \delta(1/Q).$$

The means by which we can improve the sensitivity are rather obvious from eqn (2.77). Assuming T_N and P_1 already have the best attainable values, we can reduce the band width df. In considering the choice of frequency we combine (2.77) with (2.69) to obtain

$$(\chi_0)_{\min} = \left(\frac{2\Delta\nu}{\nu}\right)\left(\frac{V}{\pi Q_0}\right)\left(\frac{kT_N\,df}{P_1}\right)^{\frac{1}{2}}. \qquad (2.80)$$

If we can scale the cavity in proportion to the wavelength, V varies as ν^{-3} and Q_0 as $\nu^{-\frac{1}{2}}$, so that $(\chi_0)_{\min}$ varies as $\nu^{-\frac{7}{2}}$, giving a big improvement at higher frequencies, even allowing for rather poorer values of T_N and P_1. This improvement assumes that we do not simultaneously have to scale the size of the sample in order to accommodate it in the cavity. In more precise language, the improvement is real for a sample that is so small that it can be contained in the smallest wavelength cavity without affecting the scaling argument. If we can use a larger sample of the same material at longer wavelengths, so that we have a constant filling factor, it is more realistic to discuss the quantity $(\chi_0/V)_{\min}$, which refers to the minimum value of the volume susceptibility. This varies only as $\nu^{-\frac{1}{2}}$, so that allowing for poorer values of T_N and P_1 there is little gain in sensitivity at higher frequencies.

For a paramagnetic substance where resonance is observed in the ground levels we have χ'' proportional to T^{-1} so that we always gain by reducing the temperature provided that the spin-lattice relaxation time does not become so long that we must reduce P_1 in order to avoid dynamic saturation effects. When $h\nu$ becomes larger than kT, χ_0 approaches a constant value (equivalent to static saturation of the sample in the steady magnetic field), but this occurs only at temperatures where the spin-lattice relaxation time is also likely to be long because of the drop in the number of phonons of energy $h\nu$ (see Chapter 10).

Dynamic saturation effects set a more fundamental limit than the possible detector overload mentioned above, since the value of χ'' falls at high values of H_1^2. The optimum then occurs when $1 = \gamma^2 H_1^2 \tau_1 \tau_2$, (see eqn (2.49)), since from eqn (2.75) we need to maximize $\chi''\sqrt{P_1}$, which for a given coupling is equivalent to maximizing $\chi''H_1$. With the coupling conditions assumed in (2.73) the power dissipated in the cavity is $P_1/2$, and hence from (2.67) we have

$$P_1 = \frac{1}{Q_0} \omega \frac{H_1^2}{4\pi} V = \frac{\omega V}{4\pi Q_0} \frac{1}{\gamma^2 \tau_1 \tau_2}$$

$$= \frac{V}{4\pi Q_0} \frac{H \Delta H}{\tau_1}, \qquad (2.81)$$

since $|(\omega/\gamma)| = H$ and $|(1/\gamma\tau_2)| = \Delta H$.

Using the same numerical values as for eqn (2.79) we find

$$P_1 \text{ (erg s}^{-1}\text{)} = 0\cdot 15/\tau_1,$$

which means that P_1 must be limited to a smaller value than $40 \text{ mW} = 4 \times 10^5 \text{ erg s}^{-1}$ unless τ_1 is shorter than about 4×10^{-7} s. This is quite a severe restriction. For example, the value of $H_1 = 0\cdot 1$ G used in § 2.8 corresponds to $P_1 = 3$ mW, a level at which saturation sets in for $\Delta H = 1$ G unless τ_1 is shorter than about 5×10^{-6} s. If we have an inhomogeneously broadened line for which the true $\tau_2 = 10^{-5}$ s, corresponding to a spin packet width $\Delta H = 10^{-2}$ G, saturation of the spin packet would occur for a value of P_1 as small as 10^{-2} mW even if τ_1 were as short as τ_2.

The situation can be improved (but only marginally) by using a looser input coupling to the cavity to reduce H_1. In terms of the power P_c dissipated in the cavity, (2.74) becomes, since $P_c/P_2 = \sqrt{(r/r_2)}$,

$$\frac{|\delta V_2|}{\sqrt{R_2}} = \sqrt{P_c} \sqrt{\left(\frac{r_2}{r}\right)} \frac{1}{(r_1+r_2+r)} \delta r, \qquad (2.82)$$

which is optimized for a given value of P_c by making $r_2 = r+r_1$ (matching the detector to the loaded cavity) and then adjusting the input coupling to make r_1 as small as possible consistent with a given value of P_c. If then $r_1 \ll r$, the optimum output coupling condition becomes essentially $r_2 = r$, so that the detector is just matched to the unloaded cavity.

Bolometer detection

In the previous section we have assumed that the detector was linear in its response to the incident voltage (or amplitude of the microwave radiation). We consider now a bolometer system, in which the microwave power is absorbed in a suitable element and detected through the rise in temperature of this element, which is proportional to the magnitude of the incident power. The analysis is based on that of Schmidt and Solomon (1966).

If the bolometer has thermal capacity C and thermal time constant

τ, its response to an incident power P_2 is determined by the equation

$$C\{dT/dt+(T/\tau)\} = P_2. \qquad (2.83)$$

If P_2 is time-dependent (e.g. modulated) with frequency f, the bolometer will respond fully only if $f\tau \ll 1$. If the bolometer is in thermal equilibrium at temperature T_0, it undergoes temperature fluctuations δT, whose mean square value is

$$\langle \delta T^2 \rangle = (k/C)T_0^2, \qquad (2.84)$$

where k is Boltzmann's constant. In a narrow band df of frequencies near the modulation frequency f, the fluctuations have the mean square value

$$\langle \delta T^2 \rangle_f = (k/C)T_0^2 \frac{4\tau\,df}{1+f^2\tau^2} \qquad (2.85)$$

and when $f\tau \ll 1$ this can be written as

$$\langle \delta T^2 \rangle_f = (k/C)T_0^2(4\tau\,df). \qquad (2.86)$$

This temperature fluctuation is equivalent to a fluctuation P_n in the incident power of magnitude

$$P_n = (C/\tau)\langle \delta T^2 \rangle_f^{\frac{1}{2}}$$
$$= (4kCT_0^2\,df/\tau)^{\frac{1}{2}},$$

and this can be written in the form

$$P_n = (4kT_0P_0\,df)^{\frac{1}{2}}, \qquad (2.87)$$

where $P_0 = CT_0/\tau$ is a useful parameter with the dimensions of a power that characterizes the sensitivity of the bolometer. The value of P_n obviously sets a lower limit to the changes in incident power that can be detected by the bolometer. Very much lower values of P_n can be obtained by operating the bolometer below room temperature, partly because T_0 occurs in (2.87) but more particularly because P_0 is greatly reduced because of the rapid fall in the heat capacity C as the temperature is reduced. At 4°K it is not difficult to make $P_0 = 10^{-3}$ W, which with unit band width $df = 1$ Hz gives $P_n = 5 \times 10^{-13}$ W.

We take this value of the 'thermal noise power' P_n to be the smallest detectable amount of power P_2 incident on the bolometer. In an ideal system the spin-resonance sample acts as its own bolometer, the power absorbed through the magnetic resonance effect appearing as heat in the sample. The highest sensitivity is obtained by making the power absorption as large as possible by increasing H_1, and we must allow for saturation effects. From eqn (2.50) we have

$$P_2 = \frac{dW}{dt} = \frac{M_0 H}{\tau_1} \frac{\gamma^2 H_1^2 \tau_1 \tau_2}{1+\gamma^2 H_1^2 \tau_1 \tau_2}, \qquad (2.50)$$

which approaches a limiting value as we increase H_1. An experiment where this is advantageous is the spin resonance of conduction electrons in a semiconductor, where the spin system relaxes through the motion of the conduction electrons, and the magnetic absorption results in a heating of the conduction electrons which can be detected as a change in the resistance of the specimen. In such a system τ_1 is short, and $\tau_1 = \tau_2$. If we write $\gamma\tau_1 = \gamma\tau_2 = (\Delta H)^{-1}$, then since $M_0 = \chi_0 H$ we have

$$P_2 = \frac{\chi_0 H^2}{\tau_1}\left(\frac{H_1^2}{(\Delta H)^2 + H_1^2}\right). \qquad (2.88)$$

If we can make $H_1 \gg \Delta H$, the term in parentheses becomes unity, and on setting $P_2 = P_n$ we have for the limiting sensitivity

$$(\chi_0)_{\min} = \left(\frac{\tau_1}{H^2}\right)(4kT_0 P_0 \, df)^{\frac{1}{2}}. \qquad (2.89)$$

This assumes, as in (2.50), that we are using circularly polarized radiation. If, as is more likely, we use linearly polarized radiation, the value of (2.88) should be multiplied by a factor $\frac{1}{2}$, and (2.89) becomes

$$(\chi_0)_{\min} = \left(\frac{2\tau_1}{H^2}\right)(4kT_0 P_0 \, df)^{\frac{1}{2}}. \qquad (2.90)$$

In a typical example $\Delta H = 1$ G, corresponding to $\tau_1 = 5 \times 10^{-8}$ s. Then at a resonance frequency of 10 GHz, H is approximately 3000 G, and using the value $P_n = 5 \times 10^{-13}$ W given above for $T_0 = 4°K$ and $df = 1$ Hz, we have

$$(\chi_0)_{\min} = 10^{-20},$$

which for a sample with $g = 2$ at $4°K$ requires only about 10^5 spins. This very high sensitivity is obtained only for a system with a very short value of τ_1, such as conduction electrons. From eqns (2.89) or (2.90) we see that $(\chi_0)_{\min}$ varies as H^{-2} or as the inverse square of the resonance frequency, so that such a bolometer method of detection is advantageous at millimetre wavelengths but of little use for nuclear magnetic resonance.

3
THE SPIN HAMILTONIAN
AND THE SPECTRUM

3.1. The spin Hamiltonian

IN general a paramagnetic resonance spectrum is rather complex, with lines due to different electronic transitions, which may be further divided into sub-groups of lines by interaction with a nuclear moment. The magnetic fields at which these lines occur alter with the frequency of the applied radiation, and, if anisotropy is present they will also depend on the orientation of the external magnetic field with respect to the crystal axes. Measurement of the spectrum observed under varying conditions of frequency and orientation results in the accumulation of a mass of data, which is bulky and not very meaningful unless some simple interpretation can be found. The solution of this problem lies in the use of a 'spin Hamiltonian', whose form can often be guessed from considerations of crystal symmetry if it cannot be obtained from theory. Such a spin Hamiltonian will contain relatively few terms; its advantage is that a complete description of the experimental data can be presented by giving the size of the coefficients of these terms, together with the directions of the appropriate axes relative to the crystal axes where anisotropy is present. For substances where detailed crystallographic information exists about the paramagnetic ion and its surroundings in the crystal, it is often possible to construct a model and, by carrying through a full treatment of the problem, to derive the spin Hamiltonian, as described in Chapters 16–21. In general the information available is insufficient to give more than rough estimates of the size of the various terms. In other cases crystallographic data may be scanty, but it may be possible to guess a plausible spin Hamiltonian. In either case it is then necessary to relate the spectrum observed under various conditions to a spin Hamiltonian.

As a result of ligand field interaction with the neighbouring diamagnetic ions or atoms in a crystal, the ground state of a paramagnetic ion frequently consists of a group of electronic levels whose separation is a few wave numbers or less, while all other electronic levels lie considerably higher. The behaviour of this group can be represented by defining an 'effective spin' S, such that the total number of levels in the group is $2S+1$, the same as in an ordinary spin multiplet. It is further required that the matrix elements between the various states determined by the full Hamiltonian that describes the system shall be proportional to those of the effective spin. It is then possible to describe the behaviour of this group of levels by a spin Hamiltonian involving just the effective spin, and where hyperfine interactions are present, the nuclear spins. The use of the symbol S for effective spin may be somewhat misleading at first, but the spin Hamiltonian was initially used for certain cases where the effective spin was the same as the true electronic spin of the ion (for example, the introduction of terms such as S_z^2 to represent splittings due to crystal fields by Van Vleck (1940)). A general form of the spin Hamiltonian (see Chapter 19) for ions of the $3d$ group was derived later, and the use of the symbol S was thus gradually extended. In this chapter no confusion should arise, but in Chapter 1 and later chapters the symbol $\tilde{S}$ is used for the effective spin when it is necessary to distinguish it from the true spin.

The most general form of the spin Hamiltonian contains a large number of terms, representing the Zeeman interaction of the magnetic electrons with an external field, level splittings due to indirect effects of the crystal field (which we shall refer to as 'fine structure'), hyperfine structure due to the presence of nuclear magnetic dipole and electric quadrupole moments in the central ion or in ligand ions, and the Zeeman interaction of the nuclear moment with the external field (which may be modified by induced electronic moments—equivalent to a 'paramagnetic shift'). In this chapter we shall give a fairly detailed but not exhaustive treatment of the relationship between the various terms in the spin Hamiltonian and the observed spectrum, starting with the electronic Zeeman term and including progressively a number of other terms. To keep the complexity within reasonable bounds we shall generally assume some degree of symmetry, for example, that the principal axes of the 'g-tensor' and hyperfine 'tensor' coincide. The extent to which these quantities are justifiably called 'tensors' is discussed in Chapter 15; for simplicity we shall here continue to use the name tensor, omitting the inverted commas but retaining the mental reservations. For example, the electronic Zeeman interaction may be written as

$$\beta \mathbf{H} \cdot \mathbf{g} \cdot \mathbf{S} = \beta(g_{xx}H_xS_x + g_{yy}H_yS_y + g_{zz}H_zS_z + g_{xy}H_xS_y + g_{yx}H_yS_x + \\ + g_{yz}H_yS_z + g_{zy}H_zS_y + g_{zx}H_zS_x + g_{xz}H_xS_z),$$

but if we choose a system where the (x, y, z) axes are the principal axes this reduces to the form

$$\beta(g_x H_x S_x + g_y H_y S_y + g_z H_z S_z), \tag{3.1}$$

where for shortness g_x has been written for g_{xx}, etc.

3.2. The effect of anisotropy in the g-factor

In order to elucidate the behaviour of a resonance spectrum when anisotropy is present we shall consider first the case of an ion with an anisotropic g-factor, but without any initial splitting of the electronic levels and without any hyperfine structure. A suitable spin Hamiltonian then contains only the terms given in eqn (3.1). If the field **H** is in such a direction that it has direction cosines (l, m, n) with respect to the principal axes (x, y, z) of the g-tensor, the Hamiltonian is

$$\mathcal{H} = \beta H(l g_x S_x + m g_y S_y + n g_z S_z). \tag{3.2}$$

If g were isotropic we could change to a new set of axes (x_e, y_e, z_e) where z_e is parallel to the field **H**, and the Hamiltonian would reduce (using primes to denote components of **S** relative to the new axes) to

$$\mathcal{H} = g\beta H(l S_x + m S_y + n S_z) = g\beta H S_z' \tag{3.3}$$

where $g = g_x = g_y = g_z$. The energy levels are then clearly a set of $(2S+1)$ levels equally spaced in energy by $g\beta H$, and the allowed transitions are those in which the component S_z' changes by one unit, so that they require a quantum

$$h\nu = g\beta H. \tag{3.4}$$

When anisotropy is present, we cannot eliminate the terms off-diagonal in the energy matrix by changing to a set of axes where z_e is parallel to **H**. It is obvious, however, that our Hamiltonian can be written in the form

$$\mathcal{H} = g\beta H(l' S_x + m' S_y + n' S_z),$$

where $g l' = g_x l$, etc. If we now make $(l'^2 + m'^2 + n'^2) = 1$, the Hamiltonian is identical with that for an ion with a spectroscopic splitting factor g, in a field **H** with direction cosines (l', m', n'). Hence the energy levels will be just as given above, and the resonance condition will be given by eqn (3.4), where, from the normalization of the apparent direction cosines (l', m', n') we have

$$g^2 = l^2 g_x^2 + m^2 g_y^2 + n^2 g_z^2. \tag{3.5}$$

In essence, we have diagonalized the energy matrix by changing to a new set of axes where the z_e-axis has direction cosines (l', m', n') with respect to the principal axes (x, y, z) of the g-tensor.

Evaluation of the transition probability is rather more complicated than in the case of a simple ion with an isotropic g-factor. In the latter case it is possible to choose the z-axis as that of the steady magnetic field $\mathbf{H}$, and the analysis of § 2.9 can be applied, if $\gamma\hbar$ is replaced by $-g\beta$. Then the operator components in eqn (2.56) such as $-l_1\mu_x$ are replaced by $l_1 g\beta S_x$, etc., and for the transition $M \leftrightarrow M-1$ we have, instead of eqn (2.57),

$$|\mu_{M,M-1}|^2 = \tfrac{1}{4}(l_1^2+m_1^2)g^2\beta^2\langle M-1|\,S_-\,|M\rangle^2$$
$$= \tfrac{1}{4}g_1^2\beta^2\langle M-1|\,S_-\,|M\rangle^2 \qquad (3.6)$$

where

$$g_1^2 = (l_1^2+m_1^2)g^2. \qquad (3.7)$$

The value of g_1 is clearly a maximum if $\mathbf{H}_1$ is normal to $\mathbf{H}$, when $(l_1^2+m_1^2) = 1$ and $g_1 = g$.

To analyze the position when anisotropy is present, we assume that we have a linearly polarized oscillatory field $H_1 \cos \omega t$, which is in a direction with cosines (l_1, m_1, n_1) with respect to the principal axes of the g-tensor. The 'oscillatory Hamiltonian' is then

$$\mathscr{H}_1 = \beta(l_1 g_x S_x + m_1 g_y S_y + n_1 g_z S_z)H_1 \cos \omega t. \qquad (3.8)$$

This must be referred to the system of axes (x_e, y_e, z_e) in which the Hamiltonian (3.2) is diagonal. We can write the result formally as

$$\mathscr{H}_1 = \beta H_1(g_{1x}S'_x + g_{1y}S'_y + g_{1z}S'_z) \cos \omega t \qquad (3.8\text{a})$$
$$= \beta H_1\{\tfrac{1}{2}(g_{1x}-ig_{1y})S'_+ + \tfrac{1}{2}(g_{1x}+ig_{1y})S'_- + g_{1z}S'_z\} \cos \omega t. \qquad (3.8\text{b})$$

The square of the matrix element for the transition $M \leftrightarrow M-1$ is

$$|\mu_{M,M-1}|^2 = \tfrac{1}{4}g_1^2\beta^2\langle M-1|\,S'_-\,|M\rangle^2 \qquad (3.9)$$

where, after some reduction, it can be shown that one obtains a value of g_1 analogous to that given above (eqn (3.7)) given by

$$g_1^2 = (g_{1x})^2 + (g_{1y})^2 \qquad (3.10)$$
$$= (l_1^2 g_x^2 + m_1^2 g_y^2 + n_1^2 g_z^2) - g^{-2}(l_1 l g_x^2 + m_1 m g_y^2 + n_1 n g_z^2)^2, \qquad (3.10\text{a})$$

where g is given by eqn (3.5). An alternative form of this is

$$g_1^2 g^2 = g_x^2 g_y^2 (l_1 m - l m_1)^2 + g_y^2 g_z^2 (m_1 n - m n_1)^2 + g_z^2 g_x^2 (n_1 l - n l_1)^2. \qquad (3.10\text{b})$$

To find the optimum direction of $\mathbf{H}_1$ from these equations we note that, from eqn (3.10b), g_1 vanishes if $l_1 = l$, $m_1 = m$, $n_1 = n$; that is, if $\mathbf{H}_1$ is parallel to $\mathbf{H}$ (this follows also from the fact that (3.8) would then transform like (3.3) into a single term containing only S'_z). Thus we conclude that $\mathbf{H}_1$ should be normal to $\mathbf{H}$ to have the maximum effect, since any component parallel to $\mathbf{H}$ would be ineffective. (Note that this does not in general mean that the coefficient of $H_1 S'_z$ is then zero.) The value of g_1 will still depend on the orientation of $\mathbf{H}_1$ in the plane normal to $\mathbf{H}$, so that there will be a specific orientation of $\mathbf{H}_1$ which gives the greatest intensity.

This behaviour can be understood more clearly by considering a special case. Let us suppose that $\mathbf{H}$ is confined to the plane $y = 0$, and that it is at an angle θ with the z-axis. We retain the general condition that g_x, g_y, g_z are all unequal, but if $g_x = g_y = g_\perp$, so that the z-axis is a symmetry axis, the treatment will apply generally since we can then always choose our x-axis so that $\mathbf{H}$ lies in the zx-plane.

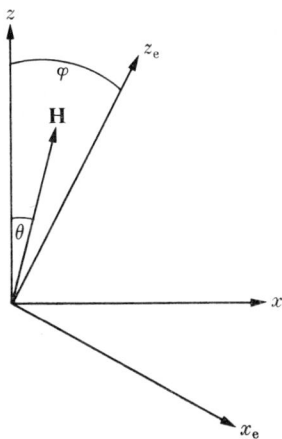

FIG. 3.1. Rotation of axes (z_e, x_e) from axes (z, x) through angle ϕ about y-axis, where $\tan \phi = (g_x/g_z) \tan \theta$. Note that unless g_x, g_z are both positive, z_e lies in a different quadrant from $\mathbf{H}$. The equations relating the two systems (where primes indicate components in the x_e, z_e system) are (see eqns (3.11), (3.13), (3.13a), (3.14)).

$$S_z = S'_z \cos\phi - S'_x \sin\phi = (g_z/g) S'_z \cos\theta - (g_x/g) S'_x \sin\theta,$$
$$S_x = S'_z \sin\phi + S'_x \cos\phi = (g_x/g) S'_z \sin\theta + (g_z/g) S'_x \cos\theta,$$
$$S_y = S'_y.$$

If we change to a set of axes (x_e, y_e, z_e) rotated about the y-axis by an angle ϕ (see Fig. 3.1), this requires the transformation

$$\begin{aligned} S_z &= S'_z \cos\phi - S'_x \sin\phi, \\ S_x &= S'_z \sin\phi + S'_x \cos\phi. \end{aligned} \quad (3.11)$$

Then
$$\mathscr{H} = \beta H(g_x S_x \sin\theta + g_z S_z \cos\theta)$$
$$= \beta H \begin{Bmatrix} S'_x(g_x \sin\theta \cos\phi - g_z \cos\theta \sin\phi) + \\ + S'_z(g_x \sin\theta \sin\phi + g_z \cos\theta \cos\phi) \end{Bmatrix} \quad (3.12)$$

and we can make the coefficient of S'_x vanish by choosing the angle ϕ such that
$$\tan\phi = (g_x/g_z)\tan\theta. \quad (3.13)$$

If we define g by the equation
$$g^2 = g_x^2 \sin^2\theta + g_z^2 \cos^2\theta, \quad (3.14)$$

which is clearly a special case of eqn (3.5), then we have
$$\sin\phi = (g_x/g)\sin\theta, \quad \cos\phi = (g_z/g)\cos\theta \quad (3.13a)$$

and the Hamiltonian reduces to the form $\mathscr{H} = g\beta H S'_z$.

Suppose now that $\mathbf{H}_1$ is applied in a direction making an angle η with the y-axis, and that its projection on the xz-plane makes an angle θ_1 with the z-axis. Then the components of $\mathbf{H}_1$ are $H_x = H_1 \sin\eta \sin\theta_1$, $H_y = H_1 \cos\eta$ and $H_z = H_1 \sin\eta \cos\theta_1$. On transforming to the (x_e, y_e, z_e) axes the oscillatory Hamiltonian becomes

$$\mathscr{H}_1 = \beta H_1 \cos\omega t \begin{Bmatrix} S'_x(g_x g_z/g)\sin\eta \sin(\theta_1 - \theta) + \\ + S'_y g_y \cos\eta + \\ + S'_z g^{-1}(g_x^2 \sin\theta \sin\theta_1 + g_z^2 \cos\theta \cos\theta_1)\sin\eta \end{Bmatrix}. \quad (3.15)$$

From this the value of g_1 is found to be given by
$$g_1^2 = (g_x g_z/g)^2 \sin^2\eta \sin^2(\theta_1 - \theta) + g_y^2 \cos^2\eta. \quad (3.16)$$

It is clear from this equation that the best value of θ_1 is $\theta \pm \pi/2$, i.e. the oscillatory field should be in the plane normal to $\mathbf{H}$. The best value of η is then either 0 or $\pi/2$, depending on whether $|g_y|$ is greater or less than $|(g_x g_z/g)|$. With axial symmetry ($g_x = g_y = g_\perp$, $g_z = g_\parallel$) this condition reduces to making $\eta = 0$ if $|g_\perp| > |g_\parallel|$, and $\eta = \pi/2$ if $|g_\perp| < |g_\parallel|$, since g always lies between $|g_\parallel|$ and $|g_\perp|$.

Note that if $g_x = g_y = 0$ then $g_1 = 0$, corresponding to the fact that there are no allowed transitions.

Determination of the sign of the g-values

Though g is essentially positive for a free electron, the signs of the principal g-values in the spin Hamiltonian using an effective spin are not known *a priori*, and the question arises, how far can they be determined experimentally? The spectrum is not changed by reversing the direction of **H**, and reversal of $\mathbf{H}_1$ is equivalent to a simple phase change if $\mathbf{H}_1$ is linearly polarized. If circularly polarized radiation is used, however, a comparison of the relative intensities of a given transition using right- and left-handed senses can be used to obtain information about signs. This will be discussed for the special case where **H** lies in the xz-plane at an angle θ to the z-axis just considered.

The oscillatory Hamiltonian for a field $\mathbf{H}_1$ (perpendicular to **H**) rotating in the right-handed sense about **H** is

$$\mathcal{H} = \beta H_1 \{(+g_x S_x \cos\theta - g_z S_z \sin\theta)\cos\omega t + g_y S_y \sin\omega t\}. \quad (3.17)$$

(By changing the sign of ω a field rotating in the opposite sense would be obtained.) On transforming to the (x_e, y_e, z_e) axes in which the static Zeeman energy is diagonalized, the oscillatory Hamiltonian becomes (using eqns (3.11), (3.13a) and $S'_\pm = S'_x \pm iS'_y$),

$$\mathcal{H} = \tfrac{1}{4}\beta H_1 \begin{bmatrix} S'_+\left\{\left(\dfrac{g_x g_z}{g} - g_y\right)e^{i\omega t} + \left(\dfrac{g_x g_z}{g} + g_y\right)e^{-i\omega t}\right\} + \\ +S'_-\left\{\left(\dfrac{g_x g_z}{g} + g_y\right)e^{i\omega t} + \left(\dfrac{g_x g_z}{g} - g_y\right)e^{-i\omega t}\right\} + \\ +S'_z\left\{\left(\dfrac{g_x^2 - g_z^2}{g}\right)\sin 2\theta(e^{i\omega t} + e^{-i\omega t})\right\} \end{bmatrix}. \quad (3.18)$$

We now assume that the sign of g is always positive, thereby resolving the sign ambiguity in eqns (3.5) and (3.14). (If either g_x, g_z or both have negative signs, this means that the z_e-axis lies in a different quadrant from **H** in Fig. 3.1.) It then follows that the level with $S'_z = M$ lies higher in energy than $M-1$, and on calculating the intensity of the transition between these two levels by the methods of § 2.3 it is found to be proportional to $\{(g_x g_z/g) + g_y\}^2$ for radiation that is right-hand circularly polarized about **H**, and to $\{(g_x g_z/g) - g_y\}^2$ for the opposite sense. Since g is taken to be always positive, experimental observation of which sense gives the higher intensity yields the sign of $g_x g_z$ relative to g_y, i.e. it determines whether the quantity $(g_x g_y g_z)$ is positive or negative. In § 15.6 it is shown that this quantity is invariant.

3.3. Multipole fine structure

When $S = 1$ or more, an additional splitting of the energy levels may appear due to indirect effects of the crystal field. This may be represented by adding to the spin Hamiltonian terms of higher powers in S_x, S_y, S_z. It is convenient to group such terms into combinations of spin operators, each such operator being the equivalent of a combination of spherical harmonics. This has the advantage that the appropriate spin Hamiltonian can generally be written down without detailed calculation, since it must reflect the symmetry of the crystal field. If the latter has trigonal symmetry, for example, it is only necessary to include the spin operators corresponding to spherical harmonics with threefold or higher symmetry about the trigonal axis. The number of such spin operators is further limited by the more general restriction that spin operators of odd degree are excluded because they are not invariant under time reversal. Also operators of degree higher than $2S$ can be omitted since they have zero matrix elements; thus for $S = 1$ or $\frac{3}{2}$, only terms of the second degree are required; for $S \geqslant 2$, terms of the fourth degree must be included, and for $S \geqslant 3$, terms of the sixth degree. In general, the spin Hamiltonian may be written (in the absence of hyperfine structure) as

$$\mathcal{H} = \beta(\mathbf{H} \cdot \mathbf{g} \cdot \mathbf{S}) + \sum_{k,q} B_k^q O_k^q, \qquad (3.19)$$

where O_k^q is a spin operator. The more important equivalent operators are listed in Table 16 in terms of angular momentum J; expressions identical with these, except for the replacement of J by S, are required in eqn (3.19).

The use of spin operators of this form has several advantages. The matrix elements of such operators are tabulated; these elements are diagonal for operators with $q = 0$, and off-diagonal when $q \neq 0$. Thus in a very strong magnetic field parallel to the z-axis of the spin operators only the matrix elements of the operators with $q = 0$ are required for a first order perturbation calculation. When the magnetic field is applied in a different direction, it is sometimes convenient to use the standard formulae for the rotation of axes of spherical harmonics to transform to a system where $\mathbf{H}$ is again along the 'polar' axis. This means that the corresponding combination of spherical harmonics can be transformed to the new axes, after which the new set of spin operators can be written down to correspond with the new combination of spherical harmonics (see, for example, Bleaney, Scovil, and Trenam (1954); Baker and Williams (1961)). (In this procedure it must be remembered that the spherical harmonics contain normalizing coefficients that unfortunately have not been included in a systematic way in the spin operators.) Then in strong magnetic fields the spectrum is determined by the new set of coefficients of the operators O_k^0, which are of course related to the original set of coefficients B_k^q in a known way. Thus by observation of the spectrum at a sufficient variety of angles, all the coefficients B_k^q can be determined.

In a case where a number of multipole spin operators are required, there will in general be more than one interpretation depending on the identification of the observed lines with the various allowed transitions. However, the fact that different transitions have different intensities is of considerable help in the identification. For example, in strong magnetic fields (Zeeman energy $\gg$ multipole energy), the strongly allowed transitions are those for which $\Delta M = \pm 1$, and the intensity of the $M \leftrightarrow M-1$ transition is proportional to $S(S+1) - M(M-1)$.

When measurements of the positions of the spectral lines as a function of angle have been completed, the *relative* signs of the coefficients B_k^q are determined as well as the magnitudes. There is, however, an ambiguity in the overall sign, which arises as follows. The reversal of the direction of flow of time ('time reversal') is an operation which does not change the energy spectrum of a Hermitian operator since, as shown in Chapter 15, time reversal is an antiunitary operator. However, under such an operation the spin components S_x, S_y, S_z change their sign and the Hamiltonian (3.19), in which all the indices k are even, becomes

$$\mathcal{H}' = -\beta(\mathbf{H} \cdot \mathbf{g} \cdot \mathbf{S}) + \sum_{k,q} B_k^q O_k^q. \tag{3.19a}$$

The Hamiltonian (3.19a) must have the same energy spectrum as (3.19) and, since in a resonance experiment only differences of energy are measured, (3.19a) must have the same *resonance* spectrum as

$$-\mathcal{H}' = \beta(\mathbf{H} \cdot \mathbf{g} \cdot \mathbf{S}) - \sum_{k,q} B_k^q O_k^q. \tag{3.19b}$$

A comparison of (3.19) and (3.19b), which differ only in the reversed signs of all the B_k^q but correspond to the same resonance spectrum, shows that the absolute signs of the B_k^q remain undetermined.

This ambiguity can be resolved by observation of the relative intensities of the different lines as the temperature is reduced towards the region where kT is comparable with the overall splitting of the energy levels. The population of the lower levels will then be increased at the expense of that of the higher levels, and transitions between the lowest levels will gain in intensity noticeably over transitions between the highest levels; from this the order of the levels can be found. With this information, the absolute signs of the spin operator terms can be found; an example of this is given in § 3.6.

3.4. Fine structure in cubic fields ($S = \frac{5}{2}, \frac{7}{2}$)

The highest degeneracy left in a crystal field even of cubic symmetry is usually only fourfold, corresponding to an effective spin of $S = \frac{3}{2}$. Hence magnetic resonance spectra corresponding to $S = 2$ or more occur normally only for ions with half-filled electron shells, where the free ion is in an S-state that is split in the solid state by rather small amounts through higher-order effects of the crystalline field. Examples are $3d^5$, $^6S_{\frac{5}{2}}$, (Mn^{2+}, Fe^{3+}) and $4f^7$, $5f^7$, $^8S_{\frac{7}{2}}$, (Eu^{2+}, Gd^{3+}, Cm^{3+}). In these cases g is isotropic and very close to the free-spin value; a special case of some interest is that of cubic symmetry, and this will be considered first.

Formulae for $S = \frac{5}{2}$

In terms of Cartesian coordinates, the potential term of the fourth degree with cubic symmetry that satisfies Laplace's equation is

$$x^4 + y^4 + z^4 - \tfrac{3}{5}r^4. \qquad (3.20)$$

This has the operator equivalent

$$\tfrac{1}{6}a\{S_x^4 + S_y^4 + S_z^4 - \tfrac{1}{5}S(S+1)(3S^2 + 3S - 1)\}. \qquad (3.21)$$

The coefficient a is that used in early work on paramagnetic salts with spin $S = \frac{5}{2}$. In terms of the operator equivalents listed in Table 16 the fourth degree cubic operator may be written as

$$B_4\{O_4^0 + 5O_4^4\} \qquad (3.22a)$$
$$= -\tfrac{2}{3}B_4\{O_4^0 + 20\sqrt{(2)}O_4^3\}, \qquad (3.22b)$$

where the first form is referred to the fourfold axes and the second to the threefold axes. The numerical relation between these and the earlier form is expressed by the relation

$$a = 120B_4.$$

The matrix elements of the spin operators are given in Table 17. For either (3.22a) or (3.22b) there is only one off-diagonal element, so that the energy levels are given by quadratic expressions. The matrix elements of the Zeeman interaction are diagonal if we take its direction to be the z-axis, and if the magnetic field is directed along either a fourfold or threefold axis we again have only one off-diagonal element

THE SPIN HAMILTONIAN AND THE SPECTRUM

Table 3.1

Energy levels and states for $S = \frac{5}{2}$, cubic field, magnetic field along fourfold axis

---

Energy matrix

$$\mathscr{H} = g\beta H S_z + (a/120)\{O_4^0 + 5O_4^4\} \quad (3.22c)$$

$$\begin{array}{c|cc}
\pm\frac{5}{2} & \pm\frac{5}{2}g\beta H + \frac{1}{2}a, & \frac{1}{2}\sqrt{(5)}a \\
\mp\frac{3}{2} & \frac{1}{2}\sqrt{(5)}a & \mp\frac{3}{2}g\beta H - \frac{3}{2}a \\
\pm\frac{1}{2} & \pm\frac{1}{2}g\beta H + a &
\end{array}$$

States	Energy levels
$\cos\alpha\,\lvert\pm\tfrac{5}{2}\rangle + \sin\alpha\,\lvert\mp\tfrac{3}{2}\rangle$	$\tfrac{1}{2}(\pm g\beta H - a) + \{(a\pm 2g\beta H)^2 + 5a^2/4\}^{\frac{1}{2}}$
$\sin\alpha\,\lvert\pm\tfrac{5}{2}\rangle - \cos\alpha\,\lvert\mp\tfrac{3}{2}\rangle$	$\tfrac{1}{2}(\pm g\beta H - a) - \{(a\pm 2g\beta H)^2 + 5a^2/4\}^{\frac{1}{2}}$
$\lvert\pm\tfrac{1}{2}\rangle$	$\pm\tfrac{1}{2}g\beta H + a$

Here the upper signs must be taken with $\tan 2\alpha = \sqrt{(5)}\,a/(2a+4g\beta H)$ and the lower signs with $\tan 2\alpha = \sqrt{(5)}\,a/(2a-4g\beta H)$.

of the cubic field, provided we use the operator expression whose 'polar' axis is the direction of **H**. The energy matrices, energy levels, and states are given in Tables 3.1 and 3.2.

In zero magnetic field the levels reduce to a quadruplet state at $+a$, and a doublet at $-2a$; these are just the fourfold Γ_8 and twofold Γ_7 levels into which a level with angular momentum $\tfrac{5}{2}$ splits in a cubic field (see § 14.4). We shall evaluate the zero-field states, from which we also obtain the first-order weak-field Zeeman effect for a magnetic field along the fourfold and threefold axes. A complete series expansion including the second-order Zeeman effect in weak fields $(g\beta H \ll a)$

Table 3.2

Energy levels and states for $S = \tfrac{5}{2}$, cubic field, magnetic field along threefold axis

---

Energy matrix

$$\mathscr{H} = g\beta H S_z - (a/180)\{O_4^0 + 20\sqrt{(2)}O_4^3\} \quad (3.22d)$$

$$\begin{array}{c|cc}
\pm\tfrac{5}{2} & \pm\tfrac{5}{2}g\beta H - \tfrac{1}{3}a & \mp\tfrac{1}{3}\sqrt{(20)}a \\
\mp\tfrac{1}{2} & \mp\tfrac{1}{3}\sqrt{(20)}a & \mp\tfrac{1}{2}g\beta H - \tfrac{2}{3}a \\
\pm\tfrac{3}{2} & \pm\tfrac{3}{2}g\beta H + a &
\end{array}$$

States	Energy levels
$\cos\alpha\,\lvert\pm\tfrac{5}{2}\rangle + \sin\alpha\,\lvert\mp\tfrac{1}{2}\rangle$	$\pm g\beta H - \tfrac{1}{2}a + \tfrac{1}{6}\{(a\pm 9g\beta H)^2 + 80a^2\}^{\frac{1}{2}}$
$\sin\alpha\,\lvert\pm\tfrac{5}{2}\rangle - \cos\alpha\,\lvert\mp\tfrac{1}{2}\rangle$	$\pm g\beta H - \tfrac{1}{2}a - \tfrac{1}{6}\{(a\pm 9g\beta H)^2 + 80a^2\}^{\frac{1}{2}}$
$\lvert\pm\tfrac{3}{2}\rangle$	$\pm\tfrac{3}{2}g\beta H + a$

Here the upper signs must be taken with $\tan 2\alpha = -\sqrt{(80)}\,a/(a+9g\beta H)$ and the lower signs with $\tan 2\alpha = \sqrt{(80)}\,a/(a-9g\beta H)$.

for an arbitrary direction of the magnetic field is given by Kronig and Bouwkamp (1939).

Weak-field Zeeman effect ($g\beta H \ll a$)

We consider first the Γ_7 doublet at energy $-2a$. The states referred to the fourfold and threefold axes, and the energy in a weak field along these axes are given in Table 3.3. We see at once that the Zeeman

TABLE 3.3

Energy levels and states for $S = \tfrac{5}{2}$, Γ_7 doublet in weak magnetic field ($g\beta H \ll a$)

States (fourfold axis)	Energy	States (threefold axis)
$\sqrt{(\tfrac{1}{6})}\lvert+\tfrac{5}{2}\rangle-\sqrt{(\tfrac{5}{6})}\lvert-\tfrac{3}{2}\rangle$	$-2a-\tfrac{5}{3}g\beta H$	$\sqrt{(\tfrac{4}{9})}\lvert-\tfrac{5}{2}\rangle+\sqrt{(\tfrac{5}{9})}\lvert+\tfrac{1}{2}\rangle$
$\sqrt{(\tfrac{1}{6})}\lvert-\tfrac{5}{2}\rangle-\sqrt{(\tfrac{5}{6})}\lvert+\tfrac{3}{2}\rangle$	$-2a+\tfrac{5}{3}g\beta H$	$\sqrt{(\tfrac{4}{9})}\lvert+\tfrac{5}{2}\rangle-\sqrt{(\tfrac{5}{9})}\lvert-\tfrac{1}{2}\rangle$

splitting is the same along the fourfold and threefold axes; in fact the Γ_7 doublet has an isotropic g-factor, as we would expect for a doublet in a cubic field. In representing the doublet by a spin-Hamiltonian with an effective spin $\tilde{S} = \tfrac{1}{2}$, we must identify the state $\tilde{S}_z = +\tfrac{1}{2}$ with that on the upper line, and $\tilde{S}_z = -\tfrac{1}{2}$ with that on the lower line, since we need the operator $\tilde{S}_-$ to take us from the former to the latter in the one representation and S_- in the other. This means that the effective g-factor for the weak-field Zeeman effect (and hence for the representation $\tilde{S} = \tfrac{1}{2}$) is $-\tfrac{5}{3}g$; this is a simple example of a case where the effective g-factor is negative.

We consider now the Γ_8 quadruplet at energy $+a$. The unusual behaviour of such a quadruplet will be discussed for the general case in Chapter 18, where it is shown that the fourfold manifold of states can be fitted with a spin Hamiltonian of the form

$$\mathcal{H} = g'\beta(H_x\tilde{S}_x+H_y\tilde{S}_y+H_z\tilde{S}_z)+g''\beta(H_x\tilde{S}_x^3+H_y\tilde{S}_y^3+H_z\tilde{S}_z^3) \quad (3.22e)$$

with $\tilde{S} = \tfrac{3}{2}$. This Zeeman interaction is a special case of the general vector **V** discussed in Chapter 18 where the matrix elements of V_z are

$$\langle+\tfrac{3}{2}\lvert V_z\rvert+\tfrac{3}{2}\rangle = -\langle-\tfrac{3}{2}\lvert V_z\rvert-\tfrac{3}{2}\rangle = P = \beta H_z(\tfrac{3}{2}g'+\tfrac{27}{8}g'')$$
$$\langle+\tfrac{1}{2}\lvert V_z\rvert+\tfrac{1}{2}\rangle = -\langle-\tfrac{1}{2}\lvert V_z\rvert-\tfrac{1}{2}\rangle = Q = \beta H_z(\tfrac{1}{2}g'+\tfrac{1}{8}g'').$$

Here the (x, y, z) axes are the fourfold cubic axes, so that the matrix elements of V_z give the Zeeman interaction for a field along one of the fourfold axes in the weak field case where $\beta H \ll a$. The energy levels of the Γ_8 quadruplet of $S = \tfrac{5}{2}$ for this case are shown in Fig. 3.2(a),

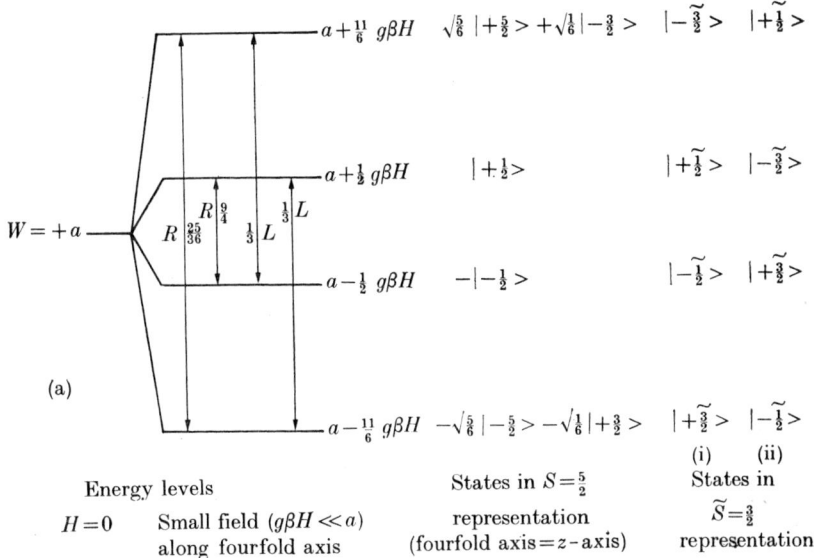

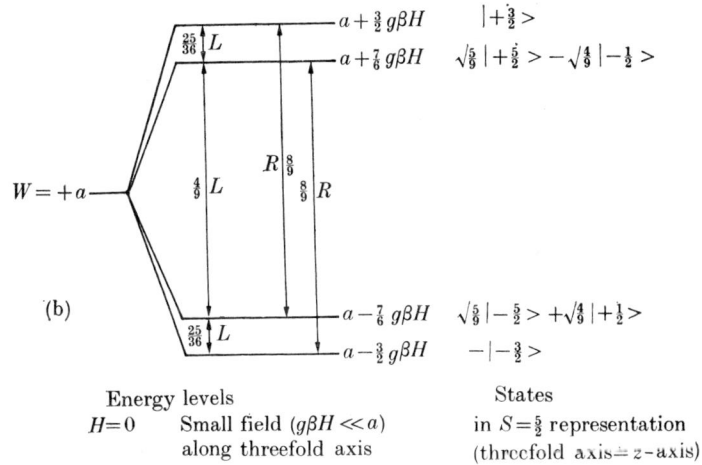

Fig. 3.2. (a) Energy levels in weak magnetic field along the fourfold axis for the Γ_8 quadruplet of $S = \frac{5}{2}$ in a strong cubic crystal field. The states have two alternative representations in $\widetilde{S} = \frac{3}{2}$; in the nomenclature of Chapter 18, eqn (18.16), the choice labelled (i) above requires $P = -(\frac{11}{6})g\beta H, Q = +\frac{1}{2}g\beta H$, while (ii) requires $P = -\frac{1}{2}g\beta H$, $Q = +(\frac{11}{6})g\beta H$. The symbol R by a transition indicates that it requires right-handed circular polarization, while L requires left-handed polarization, if circularly polarized radiation is used. The numbers indicate the relative intensities. (b) Energy levels in weak magnetic field along threefold axis for the Γ_8 quadruplet of $S = \frac{5}{2}$ in a strong cubic crystal field. The symbols R, L indicate whether right- or left-handed polarization is required, if circularly polarized radiation is used. The numbers indicate the relative intensities. The Zeeman splittings agree with those calculated from eqn (18.36), using either of the two sets of values for P, Q given above.

together with the states calculated from eqn (3.22c). These can be identified with the four states of $\tilde{S} = \frac{3}{2}$ in two ways, as shown on the diagram. These require either $P = -(\frac{11}{6})g\beta h$, $Q = +\frac{1}{2}g\beta H$ or $P = -\frac{1}{2}g\beta H$, $Q = +(\frac{11}{6})g\beta H$, where g is of course the parameter in eqn (3.22c) and not either of those in eqn (3.22e).

The Zeeman interaction is anisotropic; its value for an arbitrary direction of the magnetic field can be found from eqn (18.30) or from the expressions given by Kronig and Bouwkamp. For the particular case of a field along a threefold axis the levels are shown in Fig. 3.2(b), together with the states calculated from eqn (3.22d) where the threefold axis is taken as the z- or polar axis. It is readily verified that the Zeeman splittings agree with those calculated from eqn (18.36).

If circularly polarized radiation is used, some of the transitions require right-handed polarization and some left-handed polarization. In terms of the states formed from $S = \frac{5}{2}$, all transitions are of the type $\Delta M = \pm 1$, that is they are allowed (as we should expect) between levels whose states contain values of S_z differing by one unit. The polarization rule can be stated as follows; in a transition of the form $M \leftrightarrow M-1$, right-handed polarization is required if the level $M-1$ lies lower in energy than M, and left-handed polarization if it lies higher. The polarization rules and the relative intensities are also shown in Figs. 3.2(a) and 3.2(b).

Strong-field Zeeman effect ($g\beta H \gg a$)

To the second order in perturbation theory, the energy levels in a strong magnetic field (see also Kronig and Bouwkamp) with direction cosines (l, m, n) are given in Table 3.4, and the strong transitions in

Table 3.4

Energy levels for $S = \frac{5}{2}$, cubic field, with strong magnetic field $\mathbf{H}$ along direction (l, m, n) with respect to the fourfold axes

States	Energy levels
$\pm \frac{5}{2}$	$\pm \frac{5}{2}g\beta H + \frac{1}{2}pa \pm (212 - 24p - 113p^2)(a^2/240g\beta H)$
$\pm \frac{3}{2}$	$\pm \frac{3}{2}g\beta H - \frac{3}{2}pa \pm (12 + 8p - 15p^2)(a^2/16g\beta H)$
$\pm \frac{1}{2}$	$\pm \frac{1}{2}g\beta H + pa \pm (-2 - 3p + 5p^2)(a^2/3g\beta H)$

Table 3.5. In these tables $p = 1 - 5(l^2m^2 + m^2n^2 + n^2l^2)$; the quantum numbers labelling the states, and the relative intensities, are exact only in the strong field limit $(a/g\beta H) \to 0$.

TABLE 3.5

Resonance transitions ($\Delta M = \pm 1$) *and approximate relative intensities for* $S = \frac{5}{2}$, *cubic field, in strong magnetic field* ($g\beta H \gg a$)

Transition	Position	Relative intensity
$+\frac{5}{2} \leftrightarrow +\frac{3}{2}$	$h\nu = g\beta H + 2pa + (2-9p+7p^2)(a^2/15g\beta H)$	5
$+\frac{3}{2} \leftrightarrow +\frac{1}{2}$	$g\beta H - \frac{5}{2}pa + (68+72p-125p^2)(a^2/48g\beta H)$	8
$+\frac{1}{2} \leftrightarrow -\frac{1}{2}$	$g\beta H + (-2-3p+5p^2)(2a^2/3g\beta H)$	9
$-\frac{1}{2} \leftrightarrow -\frac{3}{2}$	$g\beta H + \frac{5}{2}pa + (68+72p-125p^2)(a^2/48g\beta H)$	8
$-\frac{3}{2} \leftrightarrow -\frac{5}{2}$	$g\beta H - 2pa + (2-9p+7p^2)(a^2/15g\beta H)$	5

In sufficiently strong fields, where terms of order $a^2/g\beta H$ can be neglected, the resonance spectrum consists just of the five lines for which $\Delta M = \pm 1$. Their relative intensities follow from the square of the matrix elements, which is just $S(S+1)-M(M-1)$ for the transition $M \leftrightarrow M-1$. The lines form a symmetrical pattern, as shown in Fig. 3.3(a); the spectrum retains the same form for all directions of **H**, but with varying displacements of the outer lines depending on the quantity p. This displacement varies with angle in the same way as the cubic potential, since p can be written in the form

$$2p/5 = l^4 + m^4 + n^4 - \tfrac{3}{5}.$$

The extreme values of p are $+1$ along a $\langle 100 \rangle$ axis and $-\tfrac{2}{3}$ along a $\langle 111 \rangle$ axis, with another stationary value of $-\tfrac{1}{4}$ along an $\langle 011 \rangle$ axis. In the important cube planes p varies as follows:

$\{111\}$ plane: $p = -\tfrac{1}{4}$ (invariant).

$\{001\}$ plane: $p = 1 - \tfrac{5}{4}\sin^2 2\theta$, where $\theta = 0$ along a $\langle 100 \rangle$ axis.

$\{01\bar{1}\}$ plane: $p = 1 - 5\sin^2\theta + \tfrac{15}{4}\sin^4\theta$, where $\theta = 0$ along a $\langle 100 \rangle$ axis.

The last plane is important, as it contains the $\langle 100 \rangle$, $\langle 111 \rangle$, and $\langle 011 \rangle$ axes.

Formulae for $S = \tfrac{7}{2}$

For ions where $S = 3$ or more, terms of the sixth degree must also be included. The full operator equivalent for the case of cubic symmetry then becomes

$$B_4(O_4^0 + 5O_4^4) + B_6(O_6^0 - 21O_6^4)$$
$$= -\tfrac{2}{3}B_4\{O_4^0 + 20\sqrt{(2)}O_4^3\} + (\tfrac{16}{9})B_6[O_6^0 - \{35/\sqrt{(8)}\}O_6^3 + (\tfrac{77}{8})O_6^6], \quad (3.23)$$

where the first form is that referred to the four fold axes and the second that referred to the three fold axes. The case most generally

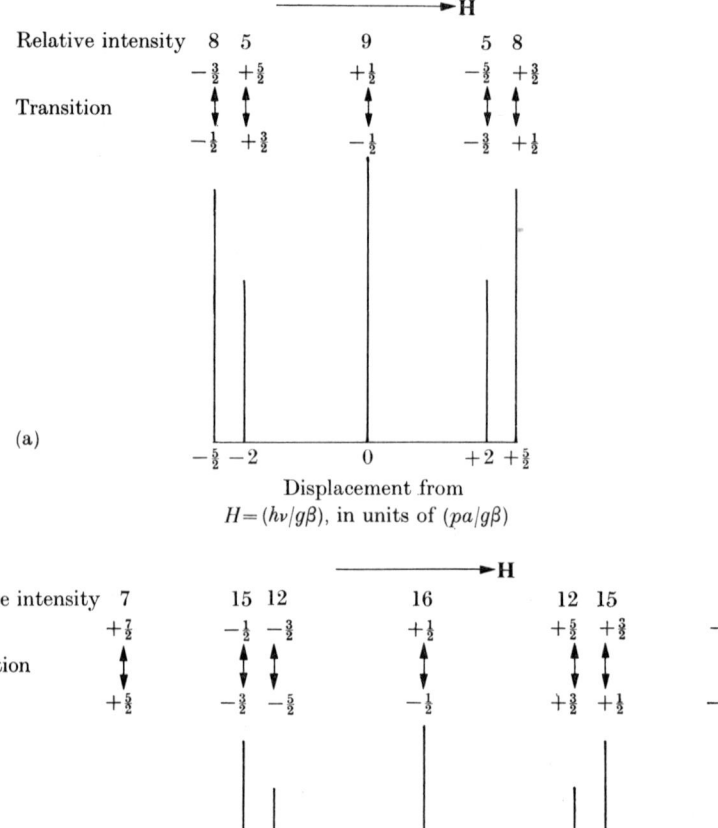

Fig. 3.3. (a) Spectrum for $S = \frac{5}{2}$, cubic field splitting, in strong magnetic field. (b) Spectrum for $S = \frac{7}{2}$ cubic crystal field splitting of fourth degree only, in strong magnetic field. In each case constant frequency and variable field are assumed.

encountered is that of $S = \frac{7}{2}$, for the ions with configuration f^7. For this case closed formulae for the levels can be given only when the magnetic field is directed along one of the $\langle 001 \rangle$ axes (see Table 3.6).

These results show that the values of α, β when $H = 0$ become independent of b_4, b_6 so that the states are uniquely determined. In a cubic field $S = \frac{7}{2}$ splits into two doublets Γ_6, Γ_7 and a quadruplet Γ_8;

Table 3.6

Energy levels and states for $S = \frac{7}{2}$, cubic field, with magnetic field **H** *along an $\langle 001 \rangle$ axis*

States	Levels
$\cos \alpha \lvert \pm \frac{7}{2} \rangle + \sin \alpha \lvert \mp \frac{1}{2} \rangle$	$\pm \frac{3}{2} g\beta H + 8b_4 - 2b_6 + \{(\pm 2g\beta H - b_4 + 3b_6)^2 + 35(b_4 - 3b_6)^2\}^{\frac{1}{2}}$
$\sin \alpha \lvert \pm \frac{7}{2} \rangle - \cos \alpha \lvert \mp \frac{1}{2} \rangle$	$\pm \frac{3}{2} g\beta H + 8b_4 - 2b_6 - \{(\pm 2g\beta H - b_4 + 3b_6)^2 + 35(b_4 - 3b_6)^2\}^{\frac{1}{2}}$
$\cos \beta \lvert \pm \frac{5}{2} \rangle + \sin \beta \lvert \mp \frac{3}{2} \rangle$	$\pm \frac{1}{2} g\beta H - 8b_4 + 2b_6 + \{(\pm 2g\beta H - 5b_4 - 7b_6)^2 + 3(5b_4 + 7b_6)^2\}^{\frac{1}{2}}$
$\sin \beta \lvert \pm \frac{5}{2} \rangle - \cos \beta \lvert \mp \frac{3}{2} \rangle$	$\pm \frac{1}{2} g\beta H - 8b_4 + 2b_6 - \{(\pm 2g\beta H - 5b_4 - 7b_6)^2 + 3(5b_4 + 7b_6)^2\}^{\frac{1}{2}}$

Here the abbreviations $b_4 = 60B_4$, $b_6 = 1260B_6$ have been used, while the values of α and β are given by

$$\tan 2\alpha = \sqrt{(35)}(b_4 - 3b_6)/(\pm 2g\beta H - b_4 + 3b_6),$$
$$\tan 2\beta = \sqrt{(3)}(5b_4 + 7b_6)/(\pm 2g\beta H - 5b_4 - 7b_6).$$

since these occur only once, the states are independent of the cubic field parameters. The energy levels in a weak field parallel to a fourfold axis, and the zero field states are given in Table 3.7.

More general expressions for the energy levels in a magnetic field, including series expansions for arbitrary directions of the magnetic field in both strong and weak magnetic fields, are given by Lacroix (1957). He uses a different nomenclature (as also does Low (1960)), whose connection with that used here can be seen from a comparison of the formulae for the levels in zero magnetic field given in Table 3.8.

In strong magnetic fields the only allowed transitions are those in which $\Delta M = \pm 1$, and the states are fairly accurately represented by single values of M, where M is the component of angular momentum about an axis parallel to the magnetic field. There are then seven

Table 3.7

Energy levels and states for $S = \frac{7}{2}$, cubic field, in a weak magnetic field **H** *along an $\langle 001 \rangle$ axis*

	States	Energy levels
Γ_6	$\sqrt{(\tfrac{5}{12})} \lvert \pm \tfrac{7}{2} \rangle + \sqrt{(\tfrac{7}{12})} \lvert \mp \tfrac{1}{2} \rangle$	$14 b_4 - 20 b_6 \pm \tfrac{7}{6} g\beta H$
Γ_7	$\sqrt{(\tfrac{3}{4})} \lvert \pm \tfrac{5}{2} \rangle - \tfrac{1}{2} \lvert \mp \tfrac{3}{2} \rangle$	$-18 b_4 - 12 b_6 \pm \tfrac{3}{2} g\beta H$
Γ_8	$\sqrt{(\tfrac{7}{12})} \lvert \pm \tfrac{7}{2} \rangle \quad \sqrt{(\tfrac{5}{12})} \lvert \mp \tfrac{1}{2} \rangle$	$2 b_4 + 16 b_6 \pm \tfrac{11}{6} g\beta H$
	$\tfrac{1}{2} \lvert \pm \tfrac{5}{2} \rangle + \sqrt{(\tfrac{3}{4})} \lvert \mp \tfrac{3}{2} \rangle$	$2 b_4 + 16 b_6 \mp \tfrac{1}{2} g\beta H$

In this Γ_8 quadruplet the values of P, Q are the same as for the Γ_8 quadruplet of $S = \tfrac{5}{2}$.

TABLE 3.8

Comparison of nomenclatures for $S = \frac{7}{2}$ in cubic field

	(This notation)	Lacroix	Low
Doublet$_6$	$32b_4 - 8b_6$	$(8+\epsilon)\delta$	$8c - 2d$
Quadruplet$_8$	$20b_4 + 28b_6$	5δ	$5c + 7d$
Doublet$_7$	0	0	0

allowed transitions, whose position and relative intensities are given in Table 3.9. Here the abbreviations $b_4 = 60B_4$, $b_6 = 1260B_6$ have again been used; p is as defined above for $S = \frac{5}{2}$, while q is the equivalent expression of the sixth degree

$$q = (\tfrac{2\,1}{2})\{11l^2m^2n^2 - (l^2m^2 + m^2n^2 + n^2l^2) + \tfrac{2}{21}\}$$

whose angular variation is similar to that of the sixth degree cubic potential

$$11x^2y^2z^2 - (x^2y^2 + y^2z^2 + z^2x^2)r^2 + 2r^6/21.$$

Values of q for specific directions are $+1$, $+\tfrac{16}{9}$ along $\langle 100 \rangle$ and $\langle 111 \rangle$ axes respectively. In a $\{111\}$ plane

$$q = \tfrac{77}{72} \cos 6\theta - \tfrac{5}{9}.$$

where θ is measured from a $\langle 11\bar{2} \rangle$ direction. Since p is invariant at $-\tfrac{1}{4}$ in this plane the sixth degree term can be measured directly by rotating H in a $\{111\}$ plane. These results follow also from eqn (3.23) since both O_4^3 and O_6^3 have zero expectation values in this plane, and if

TABLE 3.9

Resonance transitions ($\Delta M = \pm 1$) and relative intensities for $S = \frac{7}{2}$, cubic field, in a strong magnetic field $\mathbf{H}$ ($g\beta H \gg b_4, b_6$)

Transition	Position	Intensity
$+\tfrac{7}{2} \leftrightarrow +\tfrac{5}{2}$	$h\nu = g\beta H + 20pb_4 + 6qb_6$	7
$+\tfrac{5}{2} \leftrightarrow +\tfrac{3}{2}$	$g\beta H - 10pb_4 - 14qb_6$	12
$+\tfrac{3}{2} \leftrightarrow +\tfrac{1}{2}$	$g\beta H - 12pb_4 + 14qb_6$	15
$+\tfrac{1}{2} \leftrightarrow -\tfrac{1}{2}$	$g\beta H$	16
$-\tfrac{1}{2} \leftrightarrow -\tfrac{3}{2}$	$g\beta H + 12pb_4 - 14qb_6$	15
$-\tfrac{3}{2} \leftrightarrow -\tfrac{5}{2}$	$g\beta H + 10pb_4 + 14qb_6$	12
$-\tfrac{5}{2} \leftrightarrow -\tfrac{7}{2}$	$g\beta H - 20pb_4 - 6qb_6$	7

eqn (3.23) is rotated through 90° (cf. eqns (5.78), (5.79)) we have for the diagonal part just
$$-\tfrac{1}{4}B_4 O_4^0 + B_6 O_6^0(\tfrac{77}{72}\cos 6\theta - \tfrac{5}{9}). \qquad (3.23\mathrm{a})$$

In an {001} plane we have
$$q = 1 - \tfrac{21}{8}\sin^2 2\theta$$
and in an {01$\bar{1}$} plane (which contains the $\langle 100 \rangle$, $\langle 111 \rangle$, and $\langle 011 \rangle$ axes)
$$q = 1 - \tfrac{21}{8}(4\sin^2\theta - 14\sin^4\theta + 11\sin^6\theta),$$
the angle θ being measured in each case from the $\langle 100 \rangle$ axis.

Measurements of line positions in the spectrum do not yield the absolute signs of any of the splitting parameters, though the relative signs of b_4, b_6 can be found when both fourth and sixth degree terms are present. The absolute signs can, however, be determined by observing the change in relative intensities of the lines as the temperature is reduced, because of the depopulation of the states with positive values of M, just as in the case of a second degree splitting (see § 3.6). In practice the splitting due to b_6 is usually small compared with that due to b_4. The spectrum in a strong magnetic field ($g\beta H \gg b_4$) with $b_6 = 0$ is shown in Fig. 3.3(b).

The results quoted in this section are based on the assumption that the spectroscopic splitting factor g is isotropic; in a cubic field g would be expected to be rigorously isotropic. In fact with very few exceptions no anisotropy in g has been detected outside the experimental error for an ion in an S-state.

3.5. Electronic 'quadrupole' fine structure ($S = 1, \tfrac{3}{2}$)

When the symmetry is less than cubic, fine structure terms of the second degree will generally be present. This is an effect of more common occurrence than higher multipole fine structure, and hence we shall consider it separately, another reason for so doing being that it occurs in cases when g may be anisotropic.

In terms of the spin operators O_2^0, O_2^2 listed in Table 16, the second degree or 'quadrupole' terms are

$$B_2^0 O_2^0 + B_2^2 O_2^2 = B_2^0 \{3S_z^2 - S(S+1)\} + \tfrac{1}{2}B_2^2(S_+^2 + S_-^2). \qquad (3.24)$$

These may be expressed in an alternative form, which can be written as a single term $\mathbf{S} \cdot \mathbf{D} \cdot \mathbf{S}$ where $\mathbf{D}$ is a tensor quantity. Referred to the principal axes, this term is

$$D_x S_x^2 + D_y S_y^2 + D_z S_z^2, \qquad (3.25)$$

where it is convenient to take the sum of the three coefficients as zero, i.e. $D_x+D_y+D_z = 0$. If this sum is not zero, it can be made so by subtracting the quantity

$$\tfrac{1}{3}(D_x+D_y+D_z)(S_x^2+S_y^2+S_z^2) = \tfrac{1}{3}(D_x+D_y+D_z)S(S+1),$$

which is just a constant that moves all the levels up or down by the same amount, and so does not affect the resonance spectrum. The fact that one can set the sum of the three coefficients equal to zero means that there are really only two independent coefficients, as in eqn (3.24). The connection between the two forms is revealed by manipulation of eqn (3.25) as follows:

$$\begin{aligned} D_x S_x^2 &+ D_y S_y^2 + D_z S_z^2 \\ &= \tfrac{1}{2}(D_x+D_y)(S_x^2+S_y^2) + \tfrac{1}{2}(D_x-D_y)(S_x^2-S_y^2) + D_z S_z^2 \\ &= D\{S_z^2 - \tfrac{1}{3}S(S+1)\} + \tfrac{1}{2}E(S_+^2 + S_-^2), \end{aligned} \qquad (3.26)$$

where $D = \tfrac{3}{2}D_z = 3B_2^0$, $\tfrac{1}{2}(D_x - D_y) = E = B_2^2$.

The form (3.25) is often convenient when all the coefficients are unequal, since the energy levels can be computed for the case of a magnetic field along one axis, and the formulae for the other axes obtained by cyclic permutation of the subscripts. The permutation required is (moving from column to column):

g_x	g_y	g_z
$\tfrac{3}{2}D_x$	$\tfrac{3}{2}D_y$	$\tfrac{3}{2}D_z$
$= \tfrac{1}{2}(3E-D)$	$= -\tfrac{1}{2}(3E+D)$	$= D$
$\tfrac{1}{2}(D_y-D_z)$	$\tfrac{1}{2}(D_z-D_x)$	$\tfrac{1}{2}(D_x-D_y)$
$= -\tfrac{1}{2}(D+E)$	$= \tfrac{1}{2}(D-E)$	$= E$

When axial symmetry is present, $D_x = D_y$ and $E = 0$. The forms (3.24) or (3.26) are then to be preferred since they contain only one parameter. Even when $E \neq 0$ they have the advantage of containing only two parameters instead of three (which are not all independent).

Formulae for $S = 1$

The quadrupole terms have no effect on the energy of a doublet ($S = \tfrac{1}{2}$) but with larger values of S they produce a splitting of the levels when no magnetic field is present. For the cases of $S = 1$ and $S = \tfrac{3}{2}$, the energy levels can be expressed in a closed form when a magnetic field is applied, so long as it is parallel to one of the three principal axes of the tensor **D**. We shall examine the first of these ($S = 1$) fairly completely, as it illustrates in a simple fashion the effects that occur.

THE SPIN HAMILTONIAN AND THE SPECTRUM

The spin Hamiltonian, when an external magnetic field is applied along the z-axis, is

$$\mathcal{H} = g_z\beta H S_z + D\{S_z^2 - \tfrac{1}{3}S(S+1)\} + E(S_x^2 - S_y^2). \quad (3.27)$$

The energy matrix is

$$\begin{array}{c|ccc} & |+1\rangle & |0\rangle & |-1\rangle \\ \hline |+1\rangle & \tfrac{1}{3}D+G & 0 & E \\ |0\rangle & 0 & -\tfrac{2}{3}D & 0 \\ |-1\rangle & E & 0 & \tfrac{1}{3}D-G \end{array}$$

and the quantum states and energy levels can be written as

$$\left.\begin{array}{ll} |+\rangle = \cos\alpha\,|+1\rangle + \sin\alpha\,|-1\rangle, & W_+ = +\tfrac{1}{3}D + (G^2+E^2)^{\tfrac{1}{2}}, \\ |0\rangle = |0\rangle, & W_0 = -\tfrac{2}{3}D, \\ |-\rangle = \sin\alpha\,|+1\rangle - \cos\alpha\,|-1\rangle, & W_- = +\tfrac{1}{3}D - (G^2+E^2)^{\tfrac{1}{2}}, \end{array}\right\} \quad (3.28)$$

where $\tan 2\alpha = E/G$ and $G = g_z\beta H$. The allowed transitions between these levels have varying intensity according to the orientation of the oscillatory magnetic field, as shown in Table 3.10. The behaviour of

TABLE 3.10

Transitions and relative intensities for $S = 1$, rhombic field, $\mathbf{H}$ along z-axis

Transition	Quantum required	Intensity (value of $g_1^2\,|S_{ij}|^2$)		
		H_1 along x-axis	y-axis	z-axis
$|+\rangle \leftrightarrow |0\rangle$	$D + (G^2+E^2)^{\tfrac{1}{2}}$	$\tfrac{1}{2}g_x^2(1+\sin 2\alpha)$	$\tfrac{1}{2}g_y^2(1-\sin 2\alpha)$	0
$|0\rangle \leftrightarrow |-\rangle$	$-D + (G^2+E^2)^{\tfrac{1}{2}}$	$\tfrac{1}{2}g_x^2(1-\sin 2\alpha)$	$\tfrac{1}{2}g_y^2(1+\sin 2\alpha)$	0
$|+\rangle \leftrightarrow |-\rangle$	$2(G^2+E^2)^{\tfrac{1}{2}}$	0	0	$g_z^2\sin^2 2\alpha$

the energy levels for the special case of $E = \tfrac{1}{2}D$ are shown in Fig. 3.4, together with the transitions requiring a quantum equal to $\tfrac{5}{2}D$.

In zero magnetic field $\alpha = 45°$ and the three states are

$$2^{-\tfrac{1}{2}}\{|+1\rangle \pm |-1\rangle\}; \; |0\rangle,$$

each of which is an eigenstate with $\langle S_x\rangle = \langle S_y\rangle = \langle S_z\rangle = 0$. Thus the system has no permanent magnetic moment, and the line width due to spin-spin interaction is extremely small. This makes it possible to measure the zero-field transitions with very high accuracy, a result that has been applied in the measurement of the spectrum of the photoexcited triplet state of phenanthrene (see Brandon, Gerkin, and Hutchison (1962)). Another curiosity of the zero field spectrum is that

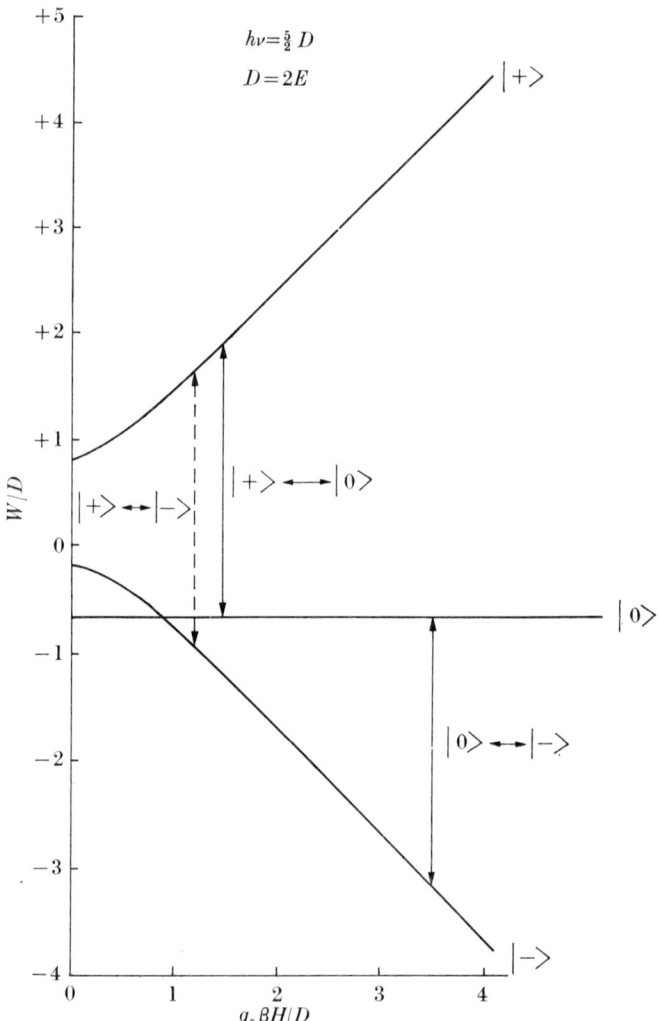

Fig. 3.4. Energy levels for $S = 1$ with rhombic splitting ($D = 2E$), in units of ($g_z\beta H/D$), with magnetic field along the z-axis. The vertical arrows indicate the allowed transitions for a quantum $h\nu = \frac{5}{2}D$.

each of the transitions is strongest for an oscillatory magnetic field linearly polarized along a different principal axis.

When a magnetic field of increasing size is applied along the z-axis, the value of α decreases towards zero; the three quantum states approach the simple description $|+1\rangle$, $|0\rangle$, and $|-1\rangle$, and the energy levels eventually diverge linearly with slopes $g_z\beta H$ times $+1$, 0, and -1 respectively. This shows that in strong fields the magnetic moment simply precesses about the axis determined by the magnetic field, whereas in zero field its motion can be regarded as consisting of three mutually perpendicular linear vibrations of different frequencies; in intermediate fields its motion is more complicated. In strong fields the transition $|+\rangle \leftrightarrow |-\rangle$ becomes smaller and smaller in intensity; it is often loosely referred to as a '$\Delta M = \pm 2$' transition, but this assumes that the states are labelled by their strong field description as $|+1\rangle$ and $|-1\rangle$. This nomenclature is only accurate in infinitely strong fields, and the two states $|+\rangle$, $|-\rangle$ at finite fields are each an admixture of the two states $|+1\rangle$, $|-1\rangle$. It would be more correct to call it a $\Delta M = 0$ transition, since it is allowed only because the same values of M occur in the two states. This agrees with the simple polarization rule, that $\Delta M = 0$ transitions require an oscillatory field parallel to the axis used to specify the components M of angular momentum while $\Delta M = \pm 1$ transitions in strong fields require an oscillatory field normal to this axis.

Although these formulae have been given only for a field along the z-axis, they can be applied for fields along the x- and y-axes by making the substitutions $D = \frac{3}{2}D_z$, $E = \frac{1}{2}(D_x - D_y)$ and using cyclic permutations. They can also be used for the case of an axial fine structure term with the magnetic field normal to the axis of symmetry by taking the field direction as the z-axis, and the fine structure term as

$$D'\{S_x^2 - \tfrac{1}{3}S(S+1)\} = \tfrac{1}{3}D'(2S_x^2 - S_y^2 - S_z^2)$$
$$= -\tfrac{1}{2}D'\{S_z^2 - \tfrac{1}{3}S(S+1)\} + \tfrac{1}{2}D'(S_x^2 - S_y^2),$$

which is equivalent to $D = -\tfrac{1}{2}D'$, $E = +\tfrac{1}{2}D'$.

Formulae for $S = \tfrac{3}{2}$

Using the spin Hamiltonian (3.27) but with $S = \tfrac{3}{2}$, the energy matrix is

$$\begin{array}{c|cc} & \pm\tfrac{3}{2} & \mp\tfrac{1}{2} \\ \hline \pm\tfrac{3}{2} & D \pm \tfrac{3}{2}G & \sqrt{(3)}E \\ \mp\tfrac{1}{2} & \sqrt{(3)}E & -D \mp \tfrac{1}{2}G \end{array}$$

from which the states and energy levels are

States	Energy
$\|\pm\tfrac{\overline{3}}{2}\rangle - \cos\alpha \, \|\pm\tfrac{3}{2}\rangle \| \sin\alpha \, \|\mp\tfrac{1}{2}\rangle$	$\pm\tfrac{1}{2}G + \{(D \pm G)^2 + 3E^2\}^{\tfrac{1}{2}}$
$\|\mp\tfrac{\overline{1}}{2}\rangle = -\sin\alpha \, \|\pm\tfrac{3}{2}\rangle + \cos\alpha \, \|\mp\tfrac{1}{2}\rangle$	$\pm\tfrac{1}{2}G - \{(D \pm G)^2 + 3E^2\}^{\tfrac{1}{2}}$

where $G = g_z \beta H$ and $\tan 2\alpha = \sqrt{(3)}E/(D \pm G)$. The nomenclature $|\bar{M}\rangle$ is a useful shorthand which is a correct description of the states only

TABLE 3.11

Transitions and relative intensities for $S = \frac{3}{2}$, rhombic field, $\mathbf{H}$ along z-axis

Transition	Orientation of H_1	Intensity (value of $g_1^2 \|S_{ij}\|^2$)
$\|+\tfrac{3}{2}\rangle \leftrightarrow \|-\tfrac{1}{2}\rangle$	z-axis	$g_z^2 \sin^2 2\alpha$
$\|-\tfrac{3}{2}\rangle \leftrightarrow \|+\tfrac{1}{2}\rangle$	z-axis	$g_z^2 \sin^2 2\alpha$
$\|+\tfrac{3}{2}\rangle \leftrightarrow \|-\tfrac{3}{2}\rangle$	x-axis y-axis	$g_x^2 \{\sin^2 \alpha \pm \sqrt{(3)} \cos \alpha \sin \alpha\}^2$
$\|+\tfrac{1}{2}\rangle \leftrightarrow \|-\tfrac{1}{2}\rangle$	x-axis y-axis	$g_x^2 \{\cos^2 \alpha \mp \sqrt{(3)} \cos \alpha \sin \alpha\}^2$
$\|+\tfrac{3}{2}\rangle \leftrightarrow \|+\tfrac{1}{2}\rangle$	x-axis y-axis	$\tfrac{1}{4}g_x^2 \{\sin 2\alpha \pm \sqrt{(3)} \cos 2\alpha\}^2$
$\|-\tfrac{3}{2}\rangle \leftrightarrow \|-\tfrac{1}{2}\rangle$	x-axis y-axis	$\tfrac{1}{4}g_x^2 \{\sin 2\alpha \pm \sqrt{(3)} \cos 2\alpha\}^2$

In the limit of strong magnetic fields $\alpha \to 0$ and only three transitions with relative intensity 3:4:3 are allowed, with the oscillatory field normal to the steady field.

in the limit of very strong magnetic field. The transition probabilities for various orientations of the oscillatory field are given in Table 3.11.

3.6. Electronic 'quadrupole' fine structure in a strong magnetic field

For values of $S > \tfrac{3}{2}$, or if the external field is in an arbitrary direction, calculation of the energy levels is cumbersome and no simple formulae can be given. In cases where the Zeeman energy is very large or very small compared with the fine structure energy, perturbation methods can be used. The former case is the one more commonly encountered in practice, and the following discussion is restricted to this case.

When the symmetry is rhombic, measurements at arbitrary field directions are useful in finding the positions of the principal axes, if not already known; precise measurements can then be made in the directions of the principal axes, where the theory is more amenable to simple calculation. For intermediate directions the formulae are therefore carried only as far as first order perturbation theory. On transforming to the new set of axes (x_e, y_e, z_e) to diagonalize the Zeeman energy, as in § 3.2, the quadrupole term

$$D_x S_x^2 + D_y S_y^2 + D_z S_z^2$$

transforms into a mixture of all possible products of the second degree in the spin components. Using the relation $D_x + D_y + D_z = 0$, the

diagonal term becomes
$$\tfrac{1}{2}D''\{S_z'^2 - \tfrac{1}{3}S(S+1)\}, \tag{3.29}$$
where
$$g^2 D'' = 3(l^2 g_x^2 D_x + m^2 g_y^2 D_y + n^2 g_z^2 D_z) \tag{3.30}$$
and g is given by eqn (3.5).

The transition $M \leftrightarrow M-1$ occurs at
$$h\nu = g\beta H + D''(M-\tfrac{1}{2}) + \ldots, \tag{3.31}$$
and, in this approximation where terms of order $D^2/h\nu$ are neglected, the spectrum consists of $2S$ lines equally spaced at intervals of D'' about the position corresponding to $D = 0$. The splitting interval D'' varies with angle, reaching extreme values along the principal axes, but becoming zero at certain intermediate directions since not all the coefficients D_x, D_y, D_z can have the same sign. This equal spacing is upset by the second-order terms, but these will be evaluated only for the case of axial symmetry.

Axial symmetry

When axial symmetry is present,
$$g_z = g_\parallel, \quad g_x = g_y = g_\perp, \quad D_x = D_y = -\tfrac{1}{2}D_z = -\tfrac{1}{3}D, \quad (E = 0).$$
The calculations will be carried to second order of perturbation theory, assuming that the Zeeman energy $\gg$ electronic quadrupole energy, and that the external field is at an angle θ with the z-axis in the xz-plane. Then the spin Hamiltonian becomes
$$\mathscr{H} = \beta H(g_\perp S_x \sin\theta + g_\parallel S_z \cos\theta) + D\{S_z^2 - \tfrac{1}{3}S(S+1)\}. \tag{3.32}$$
The Zeeman energy is first diagonalized as in § 3.2 by a rotation about the y-axis through an angle ϕ given by eqn (3.13). Then the spin Hamiltonian becomes
$$\mathscr{H} = g\beta H S_z' + \tfrac{1}{2}D\{S_z'^2 - \tfrac{1}{3}S(S+1)\}(3\cos^2\phi - 1) - \\ - D(S_x' S_z' + S_z' S_x')\cos\phi \sin\phi + \tfrac{1}{4}D(S_+'^2 + S_-'^2)\sin^2\phi. \tag{3.32a}$$
The last two terms are off-diagonal, and on evaluation by second-order perturbation theory the energy of the state $S_z' = M$ is found to be
$$W = g\beta H M + \tfrac{1}{2}D\{(3g_\parallel^2/g^2)\cos^2\theta - 1\}\{M^2 - \tfrac{1}{3}S(S+1)\} + \\ + (g_\parallel^2 g_\perp^2/g^4)(D^2 \cos^2\theta \sin^2\theta/2G)M\{8M^2 + 1 - 4S(S+1)\} + \\ + (g_\perp^4/g^4)(D^2 \sin^4\theta/8G)M\{2S(S+1) - 2M^2 - 1\}. \tag{3.33}$$
Here g is given by eqn (3.14), and $G = g\beta H$. The ratios $g_\perp/g$, $g_\parallel/g$

that occur arise from the relations between $\sin\phi$, $\cos\phi$ and $\sin\theta$, $\cos\theta$.

The diagonal term in eqn (3.33) is of course just a special case of (3.29) with
$$D'' = D\{(3g_\parallel^2/g^2)\cos^2\theta - 1\}, \qquad (3.34)$$
and to a first approximation the spectrum is given by eqn (3.31) with this special value of D''. The angular variation of D'' is just $(3\cos^2\phi - 1)$, corresponding to the fact that $3S_z'^2 - S(S+1)$ transforms like
$$3z_e^2 - r^2 = r^2(3\cos^2\phi - 1)$$
and ϕ is the angle made by z_e, the spin axis, with the symmetry axis. The variation with angle is shown in Fig. 3.5 for the case of isotropic

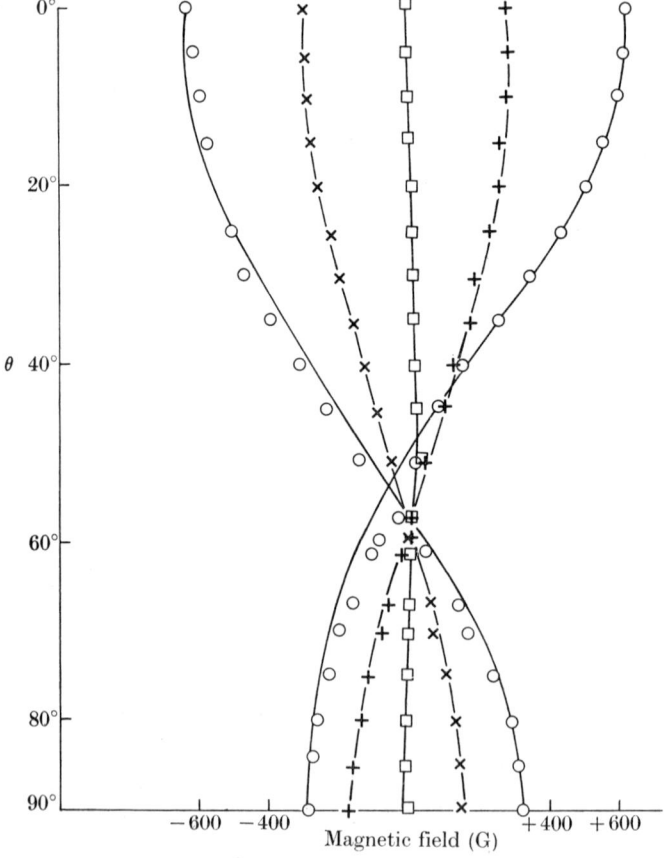

FIG. 3.5. Variation of the fine structure with angle (showing the experimental points and the calculated curves). Dilute manganese fluosilicate, $S = \frac{5}{2}$, at 3·3-cm wave-length, and 90°K (Bleaney and Ingram 1951a).

g and $S = \frac{5}{2}$. At $\theta = \pi/2$ the splitting is just half of that for $\theta = 0$, and the lines occur in the reverse order; the first-order splitting is zero at $3\cos^2\theta = 1$, or $\theta = 54° 44'$. If g is anisotropic, this zero occurs at $3\cos^2\phi = 1$.

The effect of off-diagonal terms in the energy matrix which were neglected in eqn (3.31) but included in eqn (3.33) is to give small shifts in the energy levels of order D^2/G. There are corresponding shifts in the lines, which for the transition $M \leftrightarrow M-1$ is given by adding to eqn (3.31) the quantity

$$(g_\parallel^2 g_\perp^2 /g^4)(D^2 \cos^2\theta \sin^2\theta/2G)\{24M(M-1)+9-4S(S+1)\}+$$
$$+(g_\perp^4/g^4)(D^2 \sin^4\theta/8G)\{2S(S+1)-6M(M-1)-3\}. \quad (3.35)$$

These shifts account for the departure from a simple $(3\cos^2\theta-1)$ variation shown in Fig. 3.5.

The off-diagonal matrix elements are of two kinds: elements between $|M\rangle$ and $|M-1\rangle$, whose magnitudes vary with $\sin 2\theta$ and which therefore vanish both parallel and perpendicular to the symmetry axis; elements between $|M\rangle$ and $|M-2\rangle$, which vary with $\sin^2\theta$ and hence vanish only parallel to the symmetry axis. Elements of these two kinds give rise to and determine the intensity of the so-called 'forbidden' transitions with $\Delta M = \pm 2$ or higher. When the external field is along the symmetry axis such transitions are forbidden entirely; when $\mathbf{H}$ is normal to this axis they are allowed only if the oscillatory field $\mathbf{H}_1$ is parallel to $\mathbf{H}$; but for intermediate directions they are allowed with $\mathbf{H}_1$ either parallel or perpendicular to $\mathbf{H}$. The transition $M+1 \leftrightarrow M-1$ occurs at

$$h\nu = 2G + 2D''M +$$
$$+(g_\parallel^2 g_\perp^2/g^4)(D^2 \cos^2\theta \sin^2\theta/G)\{24M^2+9-4S(S+1)\}+$$
$$+(g_\perp^4/g^4)(D^2 \sin^4\theta/4G)\{2S(S+1)-6M^2-3\}. \quad (3.36)$$

When the spectrum is observed at constant frequency these lines occur at roughly half the field required for the normal transitions but the average separation in field between successive lines is the same; the intensity is smaller by a factor of order $(D/G)^2$. Also, for intermediate directions of the magnetic field $\mathbf{H}$ the intensity of the normal transitions ($\Delta M = \pm 1$) does not vanish when $\mathbf{H}_1$ is parallel to $\mathbf{H}$ but falls only by a factor of order $(D/G)^2$.

The number of strong transitions of the form $\Delta M = \pm 1$ is just $2S$, and the intensity of the transition $M \leftrightarrow M-1$ is proportional to $\{S(S+1)-M(M-1)\}$ in the first approximation. Unlike the cases of higher multipole splitting discussed in § 3.4 the intensity is highest in the centre and smallest for the outside lines of the spectrum. This intensity variation makes it possible to identify the transition corresponding to a given line, except for an ambiguity in the sign of the values of M involved. This ambiguity can be resolved by observing the change in the relative intensity of the lines as the temperature is reduced as mentioned in § 3.3. The way in which this can be done is readily understood from the energy levels shown in Fig. 3.6 for the

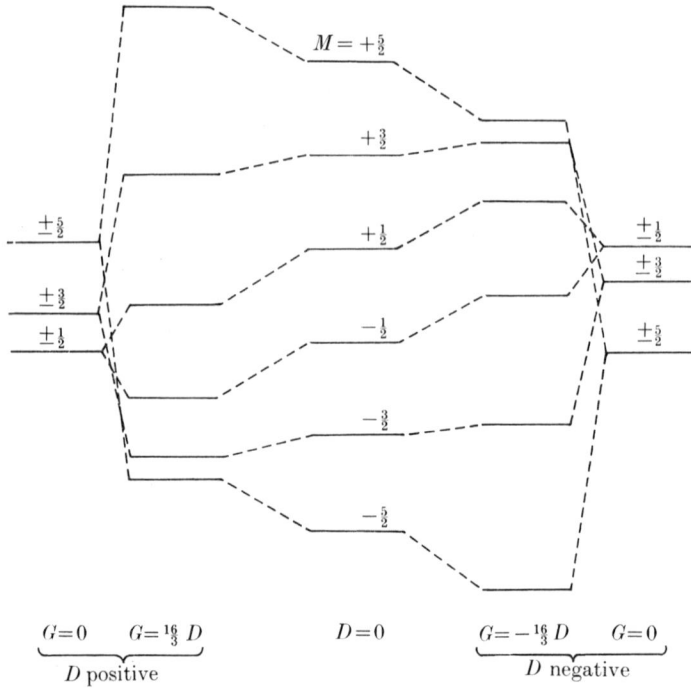

FIG. 3.6. Energy levels with electronic 'quadrupole' splitting, for $S = \frac{5}{2}$, with magnetic field H along symmetry axis ($G = g_z\beta H$), for the spin Hamiltonian,

$$\mathscr{H} = g_z\beta HS_z + D\{S_z^2 - \tfrac{1}{3}S(S+1)\}.$$

case of $S = \frac{5}{2}$ in a strong field. The diagram shows positive and negative values of D respectively and it will be seen that one set of levels is just inverted from the other. At low temperatures when kT is comparable with the overall splitting the population of the highest levels will be noticeably smaller than that of the lowest levels, and the intensity of the transitions between the higher levels will decrease relative to those between the lower levels. Comparison of the intensity of these transitions is easiest in strong fields working at constant frequency, when the different transitions can be traversed by varying the field. It is readily seen that if D is positive the transition between the highest two levels, which has the lower intensity at lower temperatures, comes at highest frequency in a given field or at lowest field for a given frequency; if D is negative the reverse is the case. In this way the order of the levels can be identified, and the sign of D established.

This is possible in measurements at zero field, for which the energy levels are shown to the right and left of the diagram, but then the splittings are much smaller and the difference in populations will be much lower than that for the transitions compared in strong fields. In addition it would be necessary to compare the relative strengths of transitions at different frequencies when working in zero field, which is experimentally much more difficult to do accurately.

The strength of this method is illustrated by inserting some numerical values. For an ion with $S = \frac{5}{2}$, $g = 2$ in a field of about 10 kG the allowed transitions occur near 1 cm wavelength. The intensity ratio between the extreme transitions $+\frac{5}{2} \leftrightarrow +\frac{3}{2}$ and $-\frac{3}{2} \leftrightarrow -\frac{5}{2}$ is very nearly $\exp(-4h\nu/kT)$. Even at 14°K, a temperature readily obtained with pumped hydrogen, this intensity ratio is about 1·5, which is easily discernible in the spectrum. This makes it possible to label the transitions and determine the sign of D; it should be noted that the method can be used for quite small values of D, the only limitation being that D must be larger than the line width so that the various transitions are resolved.

The method outlined above for determining the sign of D can be used for multipole splittings of any degree. Figure 3.7 shows an example where it is used for a splitting that is predominantly due to the cubic term of eqn (3.21). As remarked in § 3.3, measurements of the line positions in the spectrum at a sufficient number of directions will give the values of the coefficients of all the multipole splitting terms, together with their relative signs. Thus only one observation of the change of intensity with temperature is required to determine all the signs. This may be illustrated, using as an example, a rhombic splitting of the second degree. Measurements of the fine structure of the spectrum observed with a magnetic field along each of the three principal axes of the tensor **D** determine the three quantities $|D_x|$, $|D_y|$, $|D_z|$, but since $D_x + D_y + D_z = 0$ there is only one choice of the relative signs of the three coefficients which is possible. We may summarize this situation by saying that only the sign of the product $(D_x D_y D_z)$ is unknown,

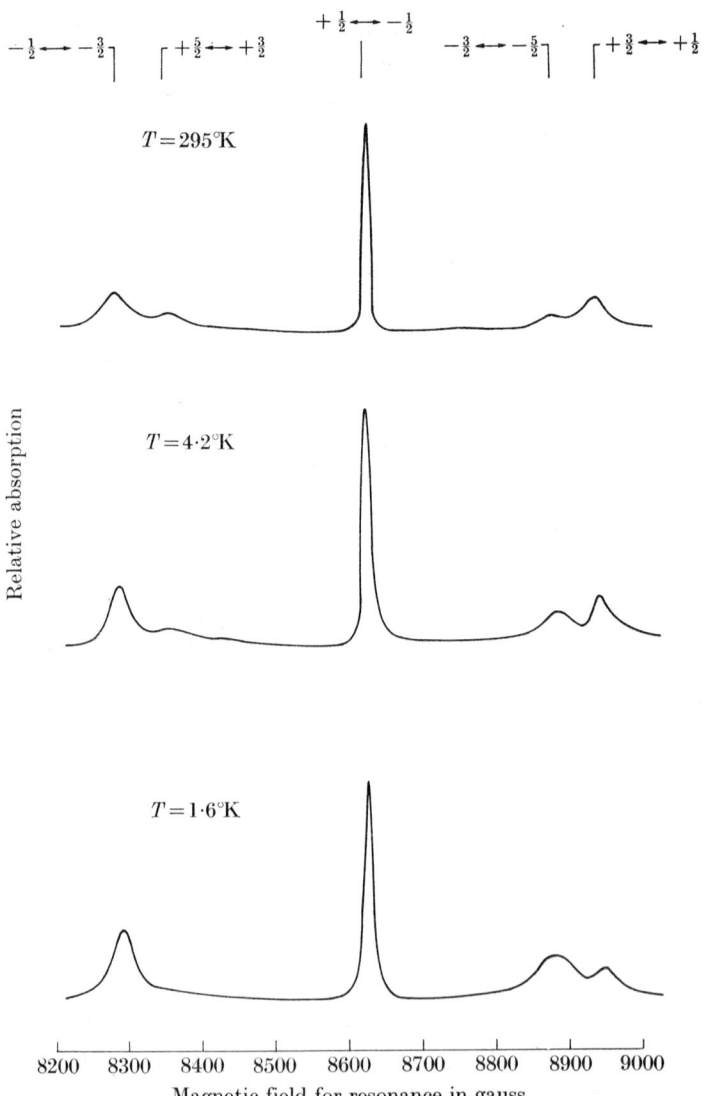

Fig. 3.7. Paramagnetic resonance at 24000 MHz of Fe^{3+} in rubidium aluminium sulphate with **H** parallel to $\langle 100 \rangle$ direction. The change of intensity at low temperatures shows that the cubic field parameter a is positive, since the transitions which gain in relative intensity must be those between the states of low Zeeman energy (negative values of M); cf. Fig. 3.3(a)). (Geschwind 1959).

THE SPIN HAMILTONIAN AND THE SPECTRUM

and a knowledge of the sign of any one of the three individual parameters is sufficient to determine the overall sign. In a strong field along the x-axis, for example, the energy levels will approximately be as shown in Fig. 3.6 except that the parameter D would be replaced by $3D_x/2$. Observation of the change of relative intensity with temperature would then give the sign of D_x. It may be remarked here that the change in population of the levels with temperature also gives a term in T^{-2} in the susceptibility whose sign is related to that of D (see eqn (A.34), Appendix A); hence in some cases the sign of D can be found from anisotropy measurements (see, for example, Bleaney and Ingram 1951a). This is not such a good method, since it can only be used when g is quite isotropic, and for non-cubic crystals.

These last limitations apply generally to susceptibility measurements as compared with paramagnetic resonance measurements. Also the information obtained from a resonance measurement in strong fields is very much greater than that in zero field. For example, in zero field for $S = \frac{3}{2}$, the energy levels reduce to two doublets separated by an energy $(D^2+3E^2)^{\frac{1}{2}}$, so that a measurement in zero field would yield a value only for this quantity. On the other hand, by measurements in a strong field one can find the values of D and E separately, and also determine the directions of the axes associated with these quantities, together with the principal g-values.

3.7. Hyperfine structure I (introductory remarks)

The effects of hyperfine structure in free atoms are well known, and can usually be described by a Hamiltonian of the form

$$\mathcal{H}_n = a_J \mathbf{J} \cdot \mathbf{I} + \frac{b}{2I(2I-1)J(2J-1)}\{3(\mathbf{I} \cdot \mathbf{J})^2 + \tfrac{3}{2}(\mathbf{I} \cdot \mathbf{J}) - I(I+1)J(J+1)\} - g_I \beta \mathbf{H} \cdot \mathbf{I}, \quad (3.37)$$

where the first term represents the interaction between the nuclear magnetic moment and the electronic magnetic field (whose steady component is parallel to the electronic angular momentum vector $\mathbf{J}$); the second term the interaction between the nuclear electric quadrupole moment and the electronic electric field gradient (also parallel to $\mathbf{J}$) at the nucleus, and the third the Zeeman interaction between the nuclear magnetic moment and the external magnetic field. In an external magnetic field we have also the electronic Zeeman interaction $\mathcal{H}_Z = g_J \beta \mathbf{H} \cdot \mathbf{J}$, but none of the fine structure terms discussed earlier in this chapter. In a strong field $(g_J \beta H \gg a_J)$, the allowed electronic transitions are those in which the nuclear magnetic quantum number m does not change, and the diagonal part of the quadrupole interaction term is

$$\frac{b}{4I(2I-1)J(2J-1)}\{3J_z^2-J(J+1)\}\{3I_z^2-I(I+1)\}, \qquad (3.37\text{a})$$

where the z-axis is parallel to the external field. The quantum required for a transition $(M,m) \leftrightarrow (M-1, m)$ is

$$h\nu = g_J\beta H + a_J m + \frac{3b}{4I(2I-1)J(2J-1)}(2M-1)\{3m^2-I(I+1)\} \qquad (3.38\text{a})$$

and by measuring a sufficient number of such transitions the parameters can be determined except g_I, which does not appear in eqn (3.38a).

In strong magnetic fields **L** and **S** may be partially decoupled, a result that is expressed quantum mechanically by the admixture of the first excited state $J \pm 1$ into the ground state J. Amongst other effects this modifies the apparent nuclear Zeeman interaction through cross terms of the form

$$\frac{\langle J|\mathcal{H}_z|J+1\rangle\langle J+1|\mathcal{H}_n|J\rangle + \langle J|\mathcal{H}_n|J+1\rangle\langle J+1|\mathcal{H}_z|J\rangle}{W_J-W_{J+1}}, \qquad (3.38\text{b})$$

where J is the ground state and $J+1$ the excited state. This may be written in the form

$$\frac{2a_{J,J+1}\beta H \langle J+1\|\Lambda\|J\rangle\{(J+1)^2-M^2\}m}{W_J-W_{J+1}}, \qquad (3.38\text{c})$$

where

$$a_{J,J+1} = 2g_I\beta^2 \langle r^{-3}\rangle \langle J+1\|N\|J\rangle$$

and $\langle J+1\|N\|J\rangle$, $\langle J+1\|\Lambda\|J\rangle$ are numbers such as those tabulated for the $4f$ group in Table 20. This gives an additional energy for the transition $(M, m) \leftrightarrow (M-1, m)$ of

$$-\frac{2a_{J,J+1}\beta H \langle J+1\|\Lambda\|J\rangle(2M-1)m}{W_J-W_{J+1}}, \qquad (3.38\text{d})$$

which gives useful additional information about the atomic structure (see, for example, Harvey (1965)). It will be seen that the additional energy given by eqn (3.38c) is of the form **H . I**; it is essentially the same effect as gives rise to the 'paramagnetic shift' in nuclear resonance. The atomic state is modified by the applied field by an amount which is linear in H, making the field seen by the nucleus different from the external field by a constant fraction varying as $(2M-1)$.

In a weak magnetic field ($g_J\beta H \ll a_J$) the off-diagonal terms in eqn (3.37) cannot be neglected and the behaviour of the energy levels is correspondingly somewhat more complicated. In zero field **I** and **J** may be vectorially combined to give a resultant **F**, whose motion in a weak magnetic field is a precession about **H**. This gives a weak field Zeeman interaction that can be expressed in terms of a magnetic quantum number $m_F = M+m$. We shall not consider the possible transitions in detail but note that transitions in which m_F changes by ± 1 require an oscillatory magnetic field perpendicular to the steady magnetic field, while those in which $\Delta m_F = 0$ require an oscillatory field parallel to the external field. In the strong magnetic field limit the latter have vanishing intensity, while the former fall into two classes: those in which $\Delta M = \pm 1$, which have an intensity determined by the electronic magnetic moment, and those in which $\Delta m = \pm 1$, whose intensity is determined by the nuclear magnetic moment and are therefore very weak in comparison.

The method of electron paramagnetic resonance has been applied to free atoms in a gaseous mixture, and has produced a number of significant results. These form rather a separate field, more related to the results of atomic beam and other methods of precision hyperfine structure measurement than to solid state paramagnetism, and we shall not discuss them further. Our main purpose in writing down the equations given above is for convenience in introducing the more complex subject of hyperfine interactions in the solid state, and later to draw some useful comparisons with atomic hyperfine structure.

In the solid state the electronic magnetic moment is often highly anisotropic, and we would expect its interaction with the nuclear moment also to be anisotropic. The electronic charge cloud is far from spherical in shape, and the electric field gradient at the nucleus must be expressed by a tensor whose principal axes are fixed in the crystal. Before considering the main results of these effects on the hyperfine structure, we mention the case of cubic symmetry. For a Γ_8 quadruplet the hyperfine magnetic dipole and electric quadrupole terms require drastic modification; this may also occur for Γ_4, Γ_5 triplet states, and even the Γ_3 doublet, as discussed in Chapter 18. Minor modifications may also be needed for ions with half-filled shells in S-states with $S = \frac{5}{2}, \frac{7}{2}$; the corrections are usually so small that they can be detected only in Endor experiments, and will be discussed in Chapter 5. The following discussion applies mainly to ions where the crystal field leaves a doublet ground state, or to states with a higher multiplicity where the complications just mentioned do not occur or are so small that they can be neglected.

In the presence of anisotropy the interaction between the magnetic moment of a nucleus of spin **I** with the electronic field of an ion whose effective spin is **S**, may be written in the form **S . A . I**. The nature of the quantity **A** is discussed in Chapter 15; for brevity we shall still refer to it as a 'tensor' with a mental reservation similar to that for the g 'tensor'. When referred to its principal axes the term takes the form

$$A_x S_x I_x + A_y S_y I_y + A_z S_z I_z, \tag{3.39a}$$

but when axial symmetry is present we may use the form

$$A_\perp (S_x I_x + S_y I_y) + A_\parallel S_z I_z \tag{3.39b}$$

(in the literature the nomenclature A, B has generally been used for $A_\parallel$, $A_\perp$).

When the nucleus has a spin $I \geqslant 1$, and there is also an electric field gradient at the nucleus, an interaction with the nuclear electric quadrupole moment may arise whose form for an ion with effective spin $S = \frac{1}{2}$ may be written as **I . P . I**. Here **P** is a tensor quantity. The electric field gradient is set up by the anisotropic distribution of electric charge on the paramagnetic ion and its immediate neighbours, and the tensor **P** generally has the same principal axes as the quantities **g**, **D**, **A**. The general form of the interaction is

$$\begin{aligned}\mathbf{I.P.I} = {} & P_{xx}I_x^2 + P_{yy}I_y^2 + P_{zz}I_z^2 + \\ & + P_{xy}I_xI_y + P_{yx}I_yI_x + \\ & + P_{yz}I_yI_z + P_{zy}I_zI_y + \\ & + P_{zx}I_zI_x + P_{xz}I_xI_z,\end{aligned} \tag{3.40a}$$

but when referred to its principal axes it has the simpler form

$$P_x I_x^2 + P_y I_y^2 + P_z I_z^2, \tag{3.40b}$$

which is similar to that of the tensor **D** (eqn (3.25)). Thus by analogous reasoning, we can set $P_x + P_x + P_z = 0$, so that there are only two independent parameters. Then we can transform (3.40b) into an expression analogous to (3.26):

$$P_\parallel [\{I_z^2 - \tfrac{1}{3}I(I+1)\} + \tfrac{1}{3}\eta(I_x^2 - I_y^2)], \tag{3.40c}$$

where $P_\parallel = 3P_z/2$, $\eta = (P_x - P_y)/P_z$. This form is particularly useful if axial symmetry is present, when $\eta = 0$.

Since the nomenclature adopted here is somewhat different from that in other branches of radio-frequency spectroscopy, we note that

$$P_\| = \frac{3eQq}{4I(2I-1)} = \frac{3eQ}{4I(2I-1)} \frac{\partial^2 V}{\partial z^2}, \tag{3.41}$$

where $q = (\partial^2 V/\partial z^2)$ is the electric field gradient at the nucleus.

Finally, we have a term $-\beta \mathbf{H} \cdot \mathbf{g}^{(I)} \cdot \mathbf{I}$, which represents the interaction between the nuclear magnetic moment and the external field $\mathbf{H}$. If this were a simple direct interaction, $\mathbf{g}^{(I)}$ would be a scalar quantity, equal to the ordinary nuclear g-factor. However, when an external magnetic field is applied the electronic wave-function is changed by an amount that is in first approximation proportional to H, and the electronic field at the nucleus is similarly modified, producing the equivalent of paramagnetic shielding (or anti-shielding). An atomic example of this is given in eqn (3.38c). In the solid state this effect can be quite large; when low-lying excited electronic states are present so that the electronic wave-function is appreciably modified in a magnetic field, and the (anti-)shielding correction may outweigh the direct interaction. For example, Lewis and Sabisky (1963) find that for the Ho^{2+} ion in CaF$_2$ the apparent value of $\mathbf{g}^{(I)}$ is forty times larger than the true value. Also, when the electronic ground state is anisotropic the (anti-)-shielding will be anisotropic, so that $\mathbf{g}^{(I)}$ becomes a 'tensor' usually with the same principal axes as the g-factor.

3.8. Hyperfine structure II—strong external field

In general the energy levels of a system with hyperfine structure must be evaluated either by numerical or by perturbation methods. In most cases the electron Zeeman energy is the largest term in the Hamiltonian, and perturbation methods can be used if this term is diagonalized first. These methods are adequate for many purposes, and also give qualitatively the behaviour of the spectrum in the general case.

The hyperfine terms in the spin Hamiltonian are

$$\mathcal{H}_n = \mathbf{S} \cdot \mathbf{A} \cdot \mathbf{I} + \mathbf{I} \cdot \mathbf{P} \cdot \mathbf{I} - \beta \mathbf{H} \cdot \mathbf{g}^{(I)} \cdot \mathbf{I}. \tag{3.42}$$

If the electronic Zeeman term $\beta \mathbf{H} \cdot \mathbf{g} \cdot \mathbf{S}$ is the largest interaction term, then it must be diagonalized first. This is accomplished by changing to new axes, as in § 3.2; if the magnetic field $\mathbf{H}$ has direction cosines (l, m, n) with respect to the principal axes of the g-tensor, then the z_e-axis of the new system must have direction cosines $(lg_x/g, mg_y/g, ng_z/g)$.

When the new system of axes is introduced for **S** in the magnetic hyperfine term **S . A . I**, where **A** is assumed to have the same principal axes as the g-tensor and to be a symmetric tensor (see below), terms involving products of all possible pairs of the components of **S** and **I** appear. Those containing S'_x, S'_y are off-diagonal and connect levels separated by the electronic Zeeman energy; they can be treated by second-order perturbation theory. The term in S'_z is

$$S'_z(lg_x A_x I_x + mg_y A_y I_y + ng_z A_z I_z)/g. \tag{3.43}$$

The coefficients of I_x, I_y, I_z, if suitably normalized, can be regarded as a set of direction cosines; this normalization requires that

$$g^2 A^2 = l^2 g_x^2 A_x^2 + m^2 g_y^2 A_y^2 + n^2 g_z^2 A_z^2. \tag{3.44}$$

Then by changing to a set of axes (x_n, y_n, z_n) for **I**, where z_n has the direction cosines $(lg_x A_x/gA)$, $(mg_y A_y/gA)$, $(ng_z A_z/gA)$, the first order hyperfine term given by eqn (3.43) reduces to

$$AS'_z I'_z, \tag{3.45}$$

where the primes indicate that S'_z refers to the 'electronic' coordinate system (x_e, y_e, z_e) and I'_z refers to the 'nuclear' coordinate system (x_n, y_n, z_n). From eqn (3.45) it is obvious that in the first order the energy of a state (M, m) is displaced by an amount AMm, where A is given by eqn (3.44). The strongly allowed transitions are of the type $(M, m) \leftrightarrow (M-1, m)$, and these are displaced in energy by an amount Am. Thus each electronic transition is divided into $2I+1$ hyperfine lines, which are equally spaced in this approximation (see Fig. 3.8). Since at ordinary temperatures all nuclear orientations are equally probable, the hyperfine lines in each electronic transition have equal intensity; this makes it readily possible to distinguish a hyperfine splitting from an electronic 'fine splitting', where the lines have unequal intensity (see § 3.3).

In physical terms these results arise as follows. In a sufficiently strong external magnetic field the magnetic electrons may be regarded as giving rise to a magnetic field H_e at the centre of the ion, whose direction and (to some extent) magnitude depend on the direction of the external field. H_e is usually much stronger than the external field (and here we have assumed that this is so); its direction defines the axis denoted by z_n above, so that it is natural to take this axis in labelling the $2I+1$ nuclear quantum states. In this approximation their energy is just $-g_I \beta H_e m$, where g_I is the true nuclear g-factor. In an electronic transition the value of H_e changes; it is proportional to M, and in the

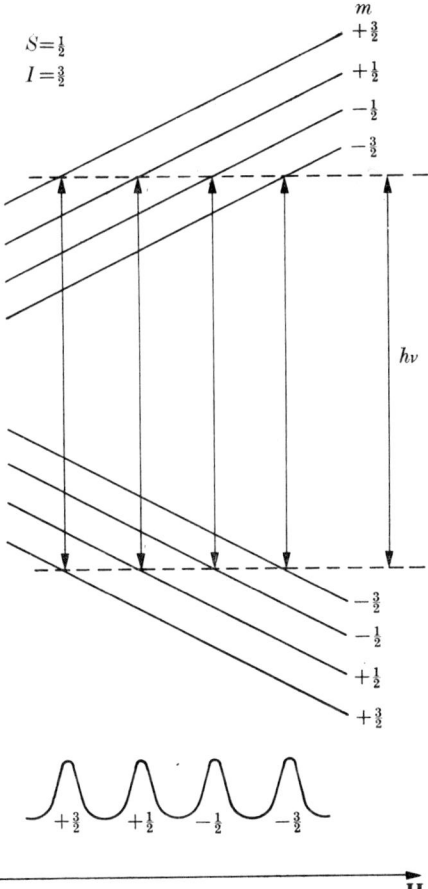

FIG. 3.8. Energy levels and allowed transitions for $S = \frac{1}{2}$, $I = \frac{3}{2}$ with magnetic hyperfine splitting, observed at constant frequency in strong magnetic field.

transition $M = +\frac{1}{2} \leftrightarrow -\frac{1}{2}$ it simply changes sign. The energy difference between each electronic state is therefore different for the $2I+1$ nuclear orientations, and this gives rise to $2I+1$ lines. Alternatively we may regard the electronic transition as taking place in a net field $H + H_n$, where H_n is the mean magnetic field produced by the nucleus and acting on the electronic magnetic moment; the value of H_n is proportional to the nuclear magnetic quantum number m.

In the treatment given above we have assumed that the quantity **A** has the same principal axes as **g**. This is not necessarily true in cases of low symmetry (see, for example, Bleaney and O'Brien (1956)) when, if x, y, z are the principal axes of the **g** 'tensor', the magnetic hyperfine structure term will have the general form

$$A_{xx}S_xI_x+A_{yy}S_yI_y+A_{zz}S_zI_z+A_{xy}S_xI_y+A_{yx}S_yI_x+$$
$$+A_{yz}S_yI_z+A_{zy}S_zI_y+A_{zx}S_zI_x+A_{xz}S_xI_z. \quad (3.45a)$$

Then on transforming to a new set of axes (x_e, y_e, z_e) where the z_e-axis has direction cosines $(g_x l/g, g_y m/g, g_z n/g)$, in order to diagonalize the electronic Zeeman interaction when a field **H** is applied with direction cosines (l, m, n), we obtain a term in S'_z

$$S'_z \begin{bmatrix} I_x\{A_{xx}(lg_x/g)+A_{yx}(mg_y/g)+A_{zx}(ng_z/g)\}+ \\ +I_y\{A_{xy}(lg_x/g)+A_{yy}(mg_y/g)+A_{zy}(ng_z/g)\}+ \\ +I_z\{A_{xz}(lg_x/g)+A_{yz}(mg_y/g)+A_{zz}(ng_z/g)\} \end{bmatrix} \quad (3.45b)$$

of which eqn (3.43) is clearly a special case. This can be diagonalized by changing to a set of axes (x_n, y_n, z_n) for **I**, where the direction cosines of z_n are just the coefficients of I_x, I_y, I_z in eqn (3.45b), whose normalization condition gives

$$g^2 A^2 = (lg_x A_{xx}+mg_y A_{yx}+ng_z A_{zx})^2+(lg_x A_{xy}+mg_y A_{yy}+ng_z A_{zy})^2+$$
$$+(lg_x A_{xz}+mg_y A_{yz}+ng_z A_{zz})^2. \quad (3.45c)$$

This is the equation of an ellipsoid in (l, m, n) space whose projection on a plane is an ellipse. If the quantity $g^2 A^2$ is plotted as a function of angle while the magnetic field is rotated in any given plane, a sinusoidal variation is obtained repeating every 180°. If the angle θ is measured from an axis corresponding to an extremum of $g^2 A^2$, then a plot of $g^2 A^2$ against $\cos^2\theta$ or $\sin^2\theta$ will give a straight line. If the plane contains one of the principal axes (assumed to coincide) of **g** and **A**, such a plot (cf. Pryce (1949)) can be used to extrapolate the hyperfine constants to axes along which resonance cannot be observed.

The method of diagonalization we have used above depends on the fact that we can use different sets of axes as a reference for **S** and for **I**, a technique that would not be possible if **A** were a genuine tensor quantity. This is a particular illustration of a result that is discussed more generally for **A** and for **g** in Chapter 15; in contrast, quantities like **D** and **P** are tensors whose components multiply two components of the same vector quantity, **S** and **I** respectively.

The results obtained above are not changed in the first approximation by the inclusion of the second and third terms in $\mathcal{H}_n$, eqn (3.42), provided that they are small compared with the first term, as is usually the case. (Other possibilities are considered in § 3.11.) When this condition is satisfied, we diagonalize the magnetic hyperfine structure by transforming to the axes (x_n, y_n, z_n) for $\mathbf{I}$; then the diagonal terms of $\mathcal{H}_n'$ are (assuming all 'tensor' quantities to have the same principal axes)

$$\mathcal{H}_n' = AS_z'I_z' + P\{I_z'^2 - \tfrac{1}{3}I(I+1)\} - G_I I_z' \; (+\text{second order terms}), \quad (3.46)$$

where

$$g^2 A^2 P = \tfrac{3}{2}\{l^2 g_x^2 A_x^2 P_x + m^2 g_y^2 A_y^2 P_y + n^2 g_z^2 A_z^2 P_z\} \quad (3.47)$$

and

$$gAG_I = \beta H(l^2 g_x A_x g_x^{(I)} + m^2 g_y A_y g_y^{(I)} + n^2 g_z A_z g_z^{(I)}), \quad (3.48)$$

while A is given by eqn (3.44). These diagonal terms in P and G_I change the energy levels but not, in a first approximation, the transitions, since only levels with the same value of m are involved in a strong transition, and these levels are displaced equally by such terms. Hence to the first order, the transition $|M, m\rangle \leftrightarrow |M-1, m\rangle$ is still displaced from the position at which the transition $M \leftrightarrow M-1$ would have occurred in the absence of the hyperfine structure by an amount Am. The quantity A, like g, must be taken to be positive.

The derivation of eqns (3.47), (3.48) is as follows. In the transformation to axes (x_n, y_n, z_n) the term $P_x I_x^2 + P_y I_y^2 + P_z I_z^2$ gives rise to a term

$$\{(lg_x A_x/gA)^2 P_x + (mg_y A_y/gA)^2 P_y + (mg_z A_z/gA)^2 P_z\} I_z'^2 = P_z' I_z'^2$$

together with terms in $I_x'^2$, $I_y'^2$ and cross terms. The diagonal terms are

$$P_z' I_z'^2 + \tfrac{1}{2}(P_x' + P_y')(I_x'^2 + I_y'^2) = \tfrac{1}{2}P_z'(2I_z'^2 - I_x'^2 - I_y'^2) = \tfrac{3}{2}P_z'\{I_z'^2 - \tfrac{1}{3}I(I+1)\}$$

since $P_x' + P_y' + P_z' = P_x + P_y + P_z = 0$. Similarly the term

$$\beta H(lg_x^{(I)} I_x + mg_y^{(I)} I_y + ng_z^{(I)} I_z)$$

transforms to

$$\beta H I_z'(l^2 g_x A_x g_x^{(I)} + m^2 g_y A_y g_y^{(I)} + n^2 g_z A_z g_z^{(I)})/(gA)$$

plus terms in I_x', I_y'. It will be seen that a general advantage of the method of approximate diagonalization used in this chapter is that the first-order terms can be extracted from the directions of z_e and z_n without determining the directions of the axes x_e, y_e or x_n, y_n.

Off-diagonal terms

The effect of the off-diagonal terms in the magnetic *hfs* is twofold: (1) they allow transitions in which the nuclear magnetic quantum

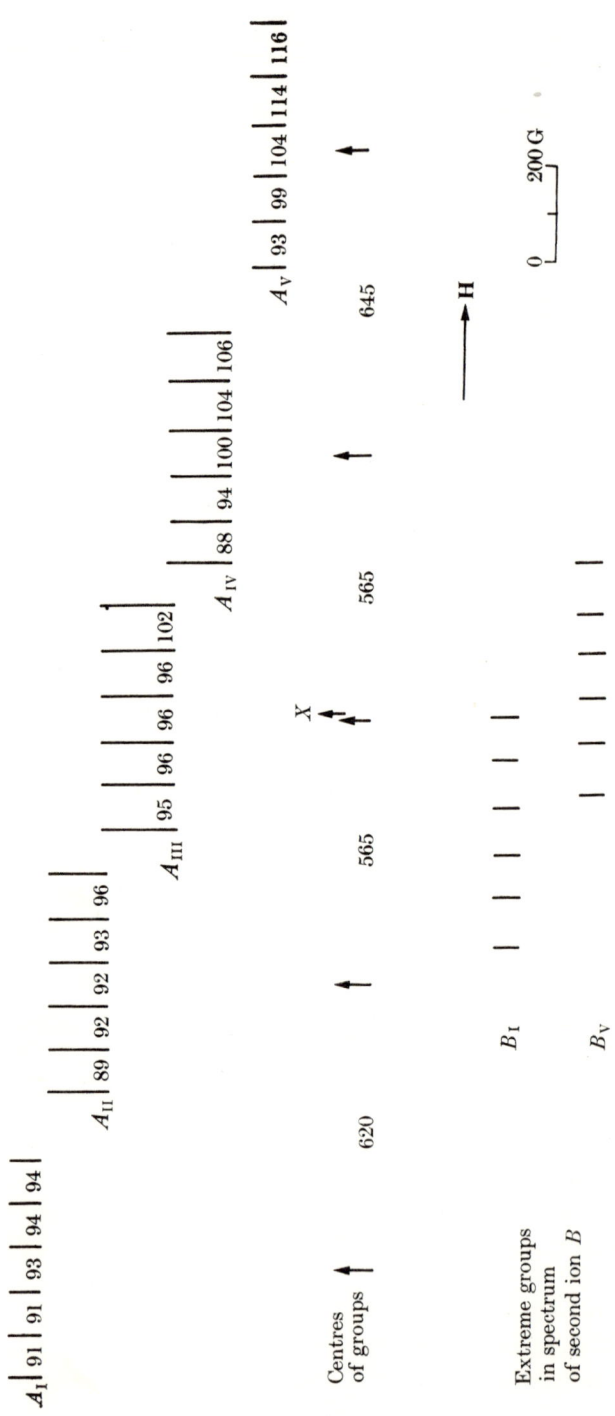

FIG. 3.9. Spectrum, at 20°K and 3·3 cm wavelength, of dilute manganese ammonium sulphate, with magnetic field parallel to the tetragonal axis of ion A. The hyperfine structure and fine-structure separations are given in gauss. Only two groups of lines for ion B are resolved; X is the narrow line of the organic radical used to determine g. Note that within each hyperfine group (I to V) splittings increase to the right (increasing magnetic field), and that different groups have different mean hyperfine splittings; the latter can be used for determining relative signs (see text) (Bleaney and Ingram 1951a).

number m changes by ± 1, smaller in intensity than the strong transitions by a factor $\sim(A/h\nu)^2$; (2) they introduce energy terms of order $A^2/(h\nu)$ which involve m^2, and hence make the hyperfine lines unequally spaced; (3) for $S > \frac{1}{2}$, they introduce terms in Mm which change the spacing of the hyperfine lines in different electronic transitions by amounts of the order $A^2/(h\nu)$. These effects are illustrated in Fig. 3.9. The effect of terms of this order can be eliminated by taking the average of all the separations between successive hyperfine lines to find A.

The effect of the off-diagonal terms arising from $\mathbf{I}\cdot\mathbf{P}\cdot\mathbf{I}$ is also to make transitions allowed in which m changes, smaller in intensity by a factor $\sim(P/A)^2$. The positions of the strong transitions are affected by amounts of order P^2/A. The off-diagonal terms in $\beta\mathbf{H}\cdot\mathbf{g}^{(I)}\cdot\mathbf{I}$ are usually too small to have an observable effect.

In general we shall evaluate these second-order effects only in the case of axial symmetry.

Axial symmetry—magnetic hyperfine interaction only

In order to look at the behaviour of hyperfine structure in some detail, it is more profitable to restrict the discussion to the case of axial symmetry, and to omit at first the nuclear electric quadrupole interaction as well as the electronic fine structure terms. With axial symmetry we can choose the direction of the x-axis so that the external magnetic field lies in the xz-plane. Then the Hamiltonian may be written as

$$\mathcal{H} = \beta H(g_\parallel S_z \cos\theta + g_\perp S_x \sin\theta) + \\ + A_\parallel S_z I_z + A_\perp (S_x I_x + S_y I_y) - \\ - \beta H(g_\parallel^{(I)} I_z \cos\theta + g_\perp^{(I)} I_x \sin\theta), \quad (3.49)$$

where the effective nuclear g-factor $\mathbf{g}^{(I)}$ is assumed also to have axial symmetry. The electronic Zeeman term, which we again assume to be much the largest, is first diagonalized, as in § 3.2, by changing to a new set of axes for $\mathbf{S}$ which are rotated from the original set by an angle ϕ about the y-axis. This leaves terms in the hyperfine structure involving $S_z' I_z$, as well as terms such as $S_x' I_x$ and $S_y' I_y$; the latter connect levels that differ in energy by $g\beta H$ and can be treated by second-order perturbation theory if $|A| \ll g\beta H$. However a term such as $S_z' I_x$ cannot, and must be eliminated by changing to a new set of axes for $\mathbf{I}$. If this system is rotated from the original system (as in Fig. 3.10) by an angle ψ about the y-axis, then the coefficient of $S_z' I_x'$ after this

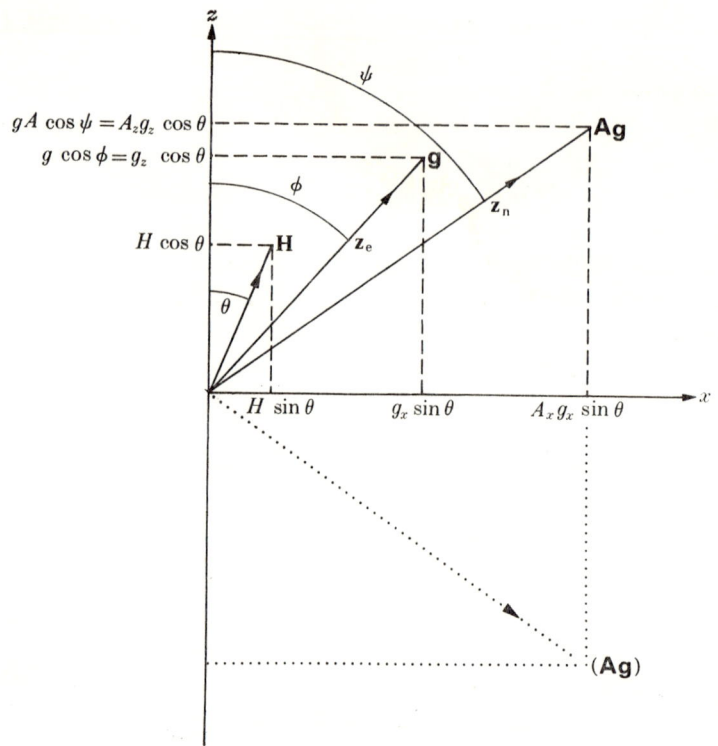

FIG. 3.10. Orientation of magnetic field **H**, electron spin (z_e) and nuclear spin (z_n) when anisotropy is present. The figure is drawn for $g_x > g_z$, $A_x > A_z$ (all positive). The dotted vector is appropriate to a negative value of A_z, the other parameters remaining positive. The equations relating the components of **I** in the two systems eqns (3.49), (3.51) and Fig. 3.1.) are (primes refer to the x_n, y_n, z_n system):

$$I_z = I'_z \cos \psi - I'_x \sin \psi,$$
$$I_x = I'_z \sin \psi + I'_x \cos \psi,$$
$$I_y = I'_y,$$

with

$$\cos \psi = (A_z/A) \cos \phi = (g_z A_z/gA) \cos \theta,$$
$$\sin \psi = (A_x/A) \sin \phi = (g_x A_x/gA) \sin \theta.$$

transformation is found to be

$$A_\perp \sin \phi \cos \psi - A_\parallel \cos \phi \sin \psi, \tag{3.50}$$

which vanishes if we choose ψ so that

$$\tan \psi = (A_\perp/A_\parallel)\tan \phi = (A_\perp g_\perp/A_\parallel g_\parallel)\tan \theta.$$

The term in $S'_z I'_z$ can now be written as $AS'_z I'_z$, where

$$g^2 A^2 = g_\parallel^2 A_\parallel^2 \cos^2\theta + g_\perp^2 A_\perp^2 \sin^2\theta. \tag{3.51}$$

This is clearly the special form of eqn (3.44) appropriate to axial symmetry.

The spin Hamiltonian now becomes

$$\mathcal{H} = g\beta H S'_z + A S'_z I'_z + \\
+ S'_x[(A_\parallel A_\perp/A)I'_x - \tfrac{1}{2}\{(A_\parallel^2 - A_\perp^2)/A\}(g_\parallel g_\perp/g^2)\sin 2\theta\, I'_z] + \\
+ A_\perp S'_y I'_y - \\
- \beta H \left[I'_z(g_\perp^{(I)}\sin\theta\sin\psi + g_\parallel^{(I)}\cos\theta\cos\psi) + \\
+ I'_x(g_\perp^{(I)}\sin\theta\cos\psi - g_\parallel^{(I)}\cos\theta\sin\psi) \right]. \quad (3.52)$$

If $g^{(I)}$ is isotropic, the last term reduces to

$$-g^{(I)}\beta H\{I'_z\cos(\theta-\psi) + I'_x\sin(\theta-\psi)\}, \quad (3.53)$$

corresponding to the fact that the angle between **H** and the z_n-axis is just $(\theta-\psi)$. The angle ψ can be eliminated using the relations

$$\sin\psi = (g_\perp A_\perp/gA)\sin\theta, \quad \cos\psi = (g_\parallel A_\parallel/gA)\cos\theta.$$

In most cases the nuclear Zeeman term is so small that we need retain only its diagonal part, and using first- and second-order perturbation theory the energy of the state (M, m) is found to be

$$W_{M,m} = g\beta HM + Am[M + (A_\parallel A_\perp^2/2A^2G)\{M^2 - S(S+1)\}] + \\
+ (A_\parallel^2 + A^2)(A_\perp^2/4A^2G)M\{I(I+1) - m^2\} - \\
- [\{(A_\parallel^2 - A_\perp^2)^2/8A^2G\}(g_\parallel g_\perp/g^2)^2\sin^2 2\theta]Mm^2 - G_I m, \quad (3.54)$$

where $G = g\beta H$ and G_I is given by the relation

$$gAG_I = \beta H(g_\parallel^{(I)}g_\parallel A_\parallel \cos^2\theta + g_\perp^{(I)}g_\perp A_\perp \sin^2\theta). \quad (3.55)$$

Hence the transition $(M, m) \leftrightarrow (M-1, m)$ requires a quantum

$$h\nu = g\beta H + Am\{1 + (A_\parallel A_\perp^2/2A^2G)(2M-1)\} + \\
+ (A_\parallel^2 + A^2)(A_\perp^2/4A^2G)\{I(I+1) - m^2\} + \\
+ [\{(A_\parallel^2 - A_\perp^2)^2/8A^2G\}(g_\parallel g_\perp/g^2)^2\sin^2 2\theta]m^2. \quad (3.56)$$

The significance of the angle ψ is similar to that of ϕ, and can be pictured as follows. The steady component of the electronic moment lies along z_e, and has components proportional to $\cos\phi = (g_\parallel/g)\cos\theta$ and $\sin\phi = (g_\perp/g)\sin\theta$ parallel and perpendicular to the symmetry axis respectively. Owing to the anisotropy in the hyperfine interaction, the corresponding components of the magnetic field H_e which the electrons produce at the nucleus are proportional to $A_\parallel(g_\parallel/g)\cos\theta$ and $A_\perp(g_\perp/g)\sin\theta$ respectively, and the nucleus therefore precesses about the direction z_n given by the resultant of these two components, in a field $\mathbf{H}_e$ proportional to $(g_\parallel^2 A_\parallel^2 \cos^2\theta + g_\perp^2 A_\perp^2 \sin^2\theta)^{\frac{1}{2}}/g$. It is convenient to call this direction the 'nuclear spin axis', though the nucleus does not precess in a simple manner about this axis.

The second-order terms vary as $(g\beta H)^{-1}$ or $(h\nu)^{-1}$ and so vanish in strong fields. They are of two kinds, and we consider first those which vary as m^2. These cause a departure from equal spacing in the hyperfine structure; under normal conditions the lines observed at lower field, when working at constant frequency, are closer together than those at higher field, as illustrated in Fig. 3.9. This may be reversed at certain angles if $|A_\parallel| > |A_\perp|$ because the term in $\sin^2 2\theta$ in eqn (3.56) has the opposite sign to the term in which θ does not occur explicitly (there is an implicit variation with angle in both terms through the variation in A). This effect is shown in Fig. 3.11.

The second type of second order term is linear in m, and has been

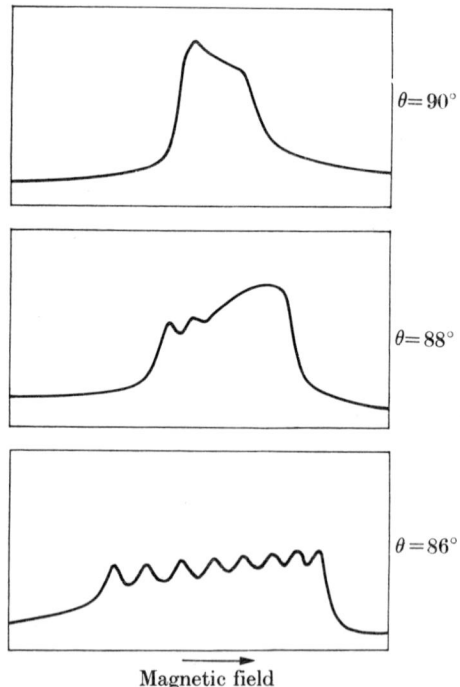

Fig. 3.11. Spectra of dilute cobalt ammonium sulphate near the perpendicular to the tetragonal axis at $T = 20°\text{K}$, $\tilde{\nu} = 0.8$ cm^{-1}. Normally the second-order terms make the hyperfine lines closer together at the high field end, but in cases of high anisotropy this may be reversed near $\theta = 90°$ by the term in $\sin^2 2\theta$ in eqn (3.56) (Bleaney and Ingram 1951b).

combined with the principal term Am. It is present only when $S > \tfrac{1}{2}$. Its effect is to produce a variation in the spacing of the hyperfine components in different electronic transitions. If these are not split by a fine structure interaction, then the hyperfine lines belonging to different electronic transitions are slightly split by an amount increasing with $|m|$; this is greatest at the two ends of the hyperfine structure pattern, and if not resolved, may appear as a broadening of the outside lines. If there is a fine structure, then the appropriate terms from §§ 3.3–3.6 must be included in the spin Hamiltonian and in a first approximation the consequent displacements of the lines can be added linearly to eqn (3.56). It is then possible to observe experimentally whether the transitions occurring at higher frequencies in a constant field (i.e. in lower fields at constant frequency) have greater or smaller mean separations between the hyperfine lines. From this some information can be obtained about the signs of A_x, etc., provided that the signs of the g-values and of the fine structure terms have been determined, as described earlier.

For simplicity we consider only a quadrupole-like electronic splitting, but the argument is readily extended; initially we also assume axial symmetry. Then, if **H** is along the axis of symmetry, it is readily seen from Fig. 3.6 that if D is positive, the transitions between levels with positive values of M occur at higher frequencies in constant field (lower fields at constant frequency), and vice versa if D is negative. Then from eqn (3.56) the hyperfine structures in the transitions with positive values of M have a larger mean spacing if $A_\parallel$ is positive; thus the sign of $A_\parallel$ relative to that of D is determined. If **H** is perpendicular to the symmetry axis, the electronic transitions occur in the reverse order (see Fig. 3.5) and so also does the change in the mean spacing. However it is clear from eqn (3.56) that this still yields only the sign of $A_\parallel$ relative to that of D, since $A_\perp$ occurs only in the square in the second-order terms.

If there is only rhombic symmetry, the coefficient of $(2M-1)$ in eqn (3.56) becomes $(A_x A_y A_z / 2A^2 G)$, where the value of A is A_x, A_y, A_z according as the external field is directed along the corresponding axis (the coefficient has not been evaluated for intermediate directions, though its form suggests that it is still correct if the appropriate value of A is used). Thus in this case observation of the variation of the mean hyperfine spacing between different electronic transitions determines the sign of $(A_x A_y A_z)$ relative to that of D_x, D_y, D_z if the external field is directed along the three axes in turn. The situation is rather similar to that for the g-values: with rhombic symmetry the use of a circularly polarized oscillatory field yields the sign of $(g_x g_y g_z)$, while from the fine structure one finds the sign of $(D_x D_y D_z)$ provided a relative intensity measurement can be made at low temperatures; from the hyperfine structure the sign of $(A_x A_y A_z)$ is determined if the signs of the principal values of the **D**-tensor are known (if not, only the sign of $(A_x A_y A_z)$ relative to that of $(D_x D_y D_z)$ can be found). With axial symmetry, circularly polarized radiation yields the sign of $g_\parallel$ uniquely, and the hyperfine structure gives the sign of $A_\parallel$ relative to that of D. No information about the signs of A_x, A_y, A_z is obtained if only one electronic transition is observed, as in the case of $S = \frac{1}{2}$.

At first sight it might appear that the argument given above might have to be reversed if one of the g-values were negative, since G appears in the denominator of the second-order terms in eqn (3.56). However the labelling of the transitions would also have to be reversed in sign, and the argument is still correct. This illustrates the advantage of always taking g and A to be positive quantities; the actual signs of the principal values of the **g**, **A** tensors then only affect the quadrants in which the angles ϕ, ψ lie, and appear automatically in equations such as (3.56) where functions of ϕ, ψ have been translated back into functions of θ.

3.9. Hyperfine structure III—nuclear electric quadrupole interaction

We discuss first a special case where explicit formulae can be obtained for the energy levels, and which shows some similarity to a free atom: an ion with $S = \frac{1}{2}$, and axial symmetry. If the external magnetic field **H** is applied along the axis of symmetry, the spin Hamiltonian is

$$\mathcal{H} = g_\parallel \beta H S_z + A_\parallel S_z I_z + A_\perp (S_x I_x + S_y I_y) + P_\parallel \{I_z^2 - \tfrac{1}{3}I(I+1)\} - g_\parallel^{(I)} \beta H I_z. \tag{3.57}$$

The effect of the term in $A_\perp$ is to admix the states $(M = +\tfrac{1}{2}, m)$ and $(M = -\tfrac{1}{2}, m+1)$, where m is the nuclear magnetic quantum number. These two states both have the same component of total angular momentum along the z-axis; this component is $m + \tfrac{1}{2}$ (in units of $\hbar$), which we may write as m_F, where $m_F = M + m$. It is similar to the magnetic quantum number m_F used in the hyperfine structure of free atoms (see Fig. 3.12), being the component of the total angular momentum along the z-axis, but with the difference that the symmetry axis of **A** and $\bar{\mathbf{P}}$ (the z-axis) is fixed in the crystal, whereas in the free

THE SPIN HAMILTONIAN AND THE SPECTRUM

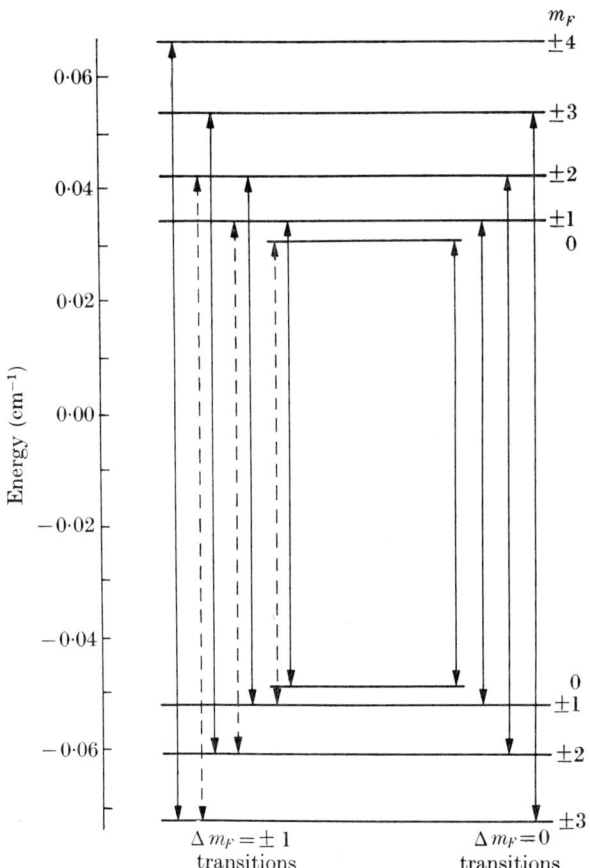

FIG. 3.12. Hyperfine levels of ^{143}Nd in lanthanum ethyl sulphate in zero magnetic field. The Nd^{3+} ion in this salt has axial symmetry, so that the levels $\pm m_F$ are degenerate, giving doublets except for the levels $m_F = 0$. The transitions corresponding to $\Delta m_F = \pm 1$ are shown on the left, the broken lines indicating rather weak transitions; the transitions $\Delta m_F = 0$ are shown on the right. If there were no anisotropy ($A_\parallel = A_\perp$, $P_\parallel = 0$), the upper and lower groups of levels would each be unsplit, corresponding to $F = 4$ and $F = 3$ respectively.

atom there is no restriction and the z-axis can be taken as that of the external magnetic field. In the present instance we restrict the discussion to the case where **H** is along the z-axis, in order to obtain a simple closed form for the energy levels. Then the energy of the two states with the same value of m_F is

$$W = -g_\parallel^{(I)}\beta H m_F - \tfrac{1}{4}A_\parallel + P_\parallel\{m_F^2 + \tfrac{1}{4} - \tfrac{1}{3}I(I+1)\} \pm$$
$$\pm \tfrac{1}{2}[\{\beta H(g_\parallel + g_\parallel^{(I)}) + (A_\parallel - 2P_\parallel)m_F\}^2 + A_\perp^2\{I(I+1) + \tfrac{1}{4} - m_F^2\}]^{\tfrac{1}{2}}, \quad (3.58)$$

where the positive sign must be taken for the state where the electronic spin moment is parallel to the field and the negative sign when it is anti-parallel.

The transitions that are allowed within this set of energy levels are of two kinds. If the oscillatory field $\mathbf{H}_1$ is normal to the z-axis, only transitions of the type $\Delta m_F = \pm 1$ are allowed, but we must distinguish between transitions of the form $|M, m\rangle \leftrightarrow |M \pm 1, m\rangle$ and of the form $|M, m\rangle \leftrightarrow |M, m \pm 1\rangle$. In strong external fields where $h\nu \gg A_\perp$ the admixture of states for a given value of m_F becomes very small, so that each state is rather accurately specified as just $|M, m\rangle$. Then the transitions in which $\Delta M = \pm 1$, $\Delta m = 0$ are strong ones, involving the electronic moments, and require a quantum of the order

$$|g_\| \beta H + A_\| m|.$$

The transitions in which $\Delta M = 0$, $\Delta m = \pm 1$ are weak transitions involving (in the strong field limit) only the nuclear moments, and require a quantum of the order $A_\| M + P_\|(2m \pm 1) \mp g_\|^{(I)} \beta H$.

In intermediate fields these latter transitions gain in intensity because some electronic magnetic moment is involved through the mixture of states.

If the oscillatory field is parallel to the z-axis, the selection rule is $\Delta m_F = 0$. Hence the only allowed transitions are those between the two energy levels with the same value of m_F, and from eqn (3.58),

$$h\nu = [\{\beta H(g_\| + g_\|^{(I)}) + (A_\| - 2P_\|)m_F\}^2 + A_\perp^2\{I(I+1) + \tfrac{1}{4} - m_F^2\}^2]^{\frac{1}{2}}. \tag{3.59}$$

The matrix element for these transitions has the squared value (neglecting the nuclear magnetic moment)

$$|\mu_{ij}|^2 = \tfrac{1}{4} g_\|^2 \beta^2 (A_\perp/h\nu)^2 \{I(I+1) + \tfrac{1}{4} - m_F^2\}. \tag{3.60}$$

It depends on the value of m_F, and as the frequency increases the transition probability becomes vanishingly small. From eqn (3.59) it follows that measurements on transitions of this kind yield the quantities $|(g_\| + g_\|^{(I)})|$, $|A_\| - 2P_\||$, and $|A_\perp|$. If these are combined with observation of transitions of the type $\Delta M = \pm 1$, $\Delta m = 0$, which yield different combinations of the parameters, we can find the quantities $|g_\|^{(I)}|$, $|A_\||$, $|P_\||$, the sign of $g_\|^{(I)}$ relative to $g_\|$, and that of $P_\|$ relative to $A_\|$. If the sign of $g_\|$ is known, then that of $g_\|^{(I)}$ is determined (and vice versa); similarly for $P_\|$ and $A_\|$. In practice it is not usually possible to find $g_\|^{(I)}$ because it is $\sim 10^{-3} g_\|$, and to obtain sufficiently

THE SPIN HAMILTONIAN AND THE SPECTRUM

accurate measurements observations must be made in fields such that $g^{(I)}\beta H$ is comparable with the line width; then the transitions involved in eqn (3.59) usually have insufficient intensity to be detectable.

Nuclear electric quadrupole interaction—perturbation theory

We now extend the treatment of § 3.8, hyperfine structure in strong external fields, to include the nuclear electric quadrupole interaction. The first-order effects of this interaction have already been evaluated (eqn (3.47)) and shown to produce no effect on the spectrum in this approximation. The second-order effects therefore become unusually important, since it is only through their effect on the spectrum that the size of the nuclear electric quadrupole interaction can be found. In general, the magnitude of this interaction is smaller than that of the magnetic hyperfine interaction, and we make this assumption in applying perturbation theory to treat the case of axial symmetry, without restriction on the value of S.

If an external magnetic field is applied at an angle θ to the symmetry axis (the z-axis) in the xz-plane, and the nuclear coordinate system (x_n, y_n, z_n) is introduced where the z_n-axis makes an angle ψ with the z-axis, the nuclear electric quadrupole interaction

$$P_{\parallel}\{I_z^2 - \tfrac{1}{3}I(I+1)\}$$

becomes

$$\tfrac{1}{2}P_{\parallel}\{I_z'^2 - \tfrac{1}{3}I(I+1)\}(3\cos^2\psi - 1) - $$
$$- P_{\parallel}(I_x'I_z' + I_z'I_x')\cos\psi\sin\psi +$$
$$+ \tfrac{1}{4}P_{\parallel}(I_+'^2 + I_-'^2)\sin^2\psi. \quad (3.61)$$

The last two terms are off-diagonal and give rise to matrix elements between the states $|m\rangle$ and $|m\pm 1\rangle$, $|m\pm 2\rangle$ belonging to the same value of the electronic magnetic quantum number M. This gives a complicated energy matrix in which perturbation methods can only be applied if $|P| \ll |AM|$, since AM is the energy difference between successive nuclear levels. If this inequality holds, the strongly allowed transitions are still those in which $\Delta m = 0$, $\Delta M = \pm 1$. The first (diagonal) term in eqn (3.61) adds to the energy of the state $|M, m\rangle$ in eqn (3.54) an amount

$$\tfrac{1}{2}P_{\parallel}\{m^2 - \tfrac{1}{3}I(I+1)\}\left\{3\frac{g_{\parallel}^2 A_{\parallel}^2}{g^2 A^2}\cos^2\theta - 1\right\}$$
$$= P\{m^2 - \tfrac{1}{3}I(I+1)\}, \quad (3.62)$$

which is just a special case of eqn (3.47). The off-diagonal terms, treated by second-order perturbation theory, add to the energy a further amount

$$(g_\| g_\perp A_\| A_\perp/g^2 A^2)^2 \{P_\|^2 \sin^2 2\theta/8AM\} m\{8m^2+1-4I(I+1)\}+$$
$$+(g_\perp A_\perp/gA)^4 \{P_\|^2 \sin^4\theta/8AM\} m\{2I(I+1)-2m^2-1\}. \quad (3.63)$$

The quantum of energy required for the transition $|M, m\rangle \leftrightarrow |M-1, m\rangle$ is increased by an amount

$$-(g_\| g_\perp A_\| A_\perp/g^2 A^2)^2 \{P_\|^2 \sin^2 2\theta/8AM(M-1)\} m\{8m^2+1-4I(I+1)\}-$$
$$-(g_\perp A_\perp/gA)^4 \{P_\|^2 \sin^4\theta/8AM(M-1)\} m\{2I(I+1)-2m^2-1\}, \quad (3.64)$$

where the term in P (eqn (3.62)) has vanished since it does not depend on M. The second-order terms, unlike those in eqn (3.56), do not vanish in strong fields, since the denominators contain A and not G. In addition to terms in m, which change the spacing of all the hyperfine lines equally, they contain terms in m^3 that make the spacing of the hyperfine lines greater at the outside than at the middle when $\theta = 90°$, with the reverse effect at intermediate angles. This effect vanishes at $\theta = 0°$, where there are no off-diagonal terms if there is axial symmetry. Thus quadrupole effects give a pattern that is symmetrical about the centre, unlike second-order effects of the magnetic hyperfine structure, which produce terms in m^2.

A more important and more easily detectable effect of the off-diagonal elements of the quadrupole interaction is that they give rise to transitions in which the nuclear magnetic quantum number changes by one or two units ($\Delta m = \pm 1, \pm 2$) that would otherwise be forbidden. In a first approximation, where energies of the order $(P/A)^2$ are neglected, these transitions occur as follows (in these equations the value of P, which occurs in the energy, is given by eqn (3.62)). Transitions $|M, k\pm\tfrac{1}{2}\rangle \leftrightarrow |M-1, k\mp\tfrac{1}{2}\rangle$:

$$\text{Hyperfine energy} = Ak \pm \{A(M-\tfrac{1}{2})+2Pk-G_I\}. \quad (3.65)$$

The intensities, relative to that of the $\Delta m = 0$ transitions, are:

$$(g_\| g_\perp A_\| A_\perp/g^2 A^2)^2 \{P_\| \cos\theta \sin\theta/AM(M-1)\}^2 \times k^2\{(I+\tfrac{1}{2})^2-k^2\}. \quad (3.66)$$

Transitions $|M, m\pm 1\rangle \leftrightarrow |M-1, m\mp 1\rangle$:

$$\text{Hyperfine energy} = Am \pm \{A(2M-1)+4Pm-2G_I\}. \quad (3.67)$$

The intensities, relative to that of the $\Delta m = 0$ transitions, are:

$$(g_\perp^2 A_\perp^2/g^2 A^2)^2 \{P_\| \sin^2\theta/8AM(M-1)\}^2 \times \{(I+1)^2-m^2\}\{I^2-m^2\}. \quad (3.68)$$

Here G_I is in general given by eqn (3.55), but if $g^{(I)}$ is isotropic the rather complicated angular dependence of G_I is simply $\cos(\theta-\psi)$, as can be seen from eqn (3.53). If the quantities $g_\| A_\|, g_\perp A_\perp$ are of opposite sign, G_I may pass through zero, since there is an angle at which $(\theta-\psi)$ is 90°, corresponding to a situation where the electronic magnetic field $\mathbf{H}_e$ at the nucleus is perpendicular to the external magnetic field $\mathbf{H}$.

The general behaviour of the energy levels and transitions is illustrated in Fig. 3.13.

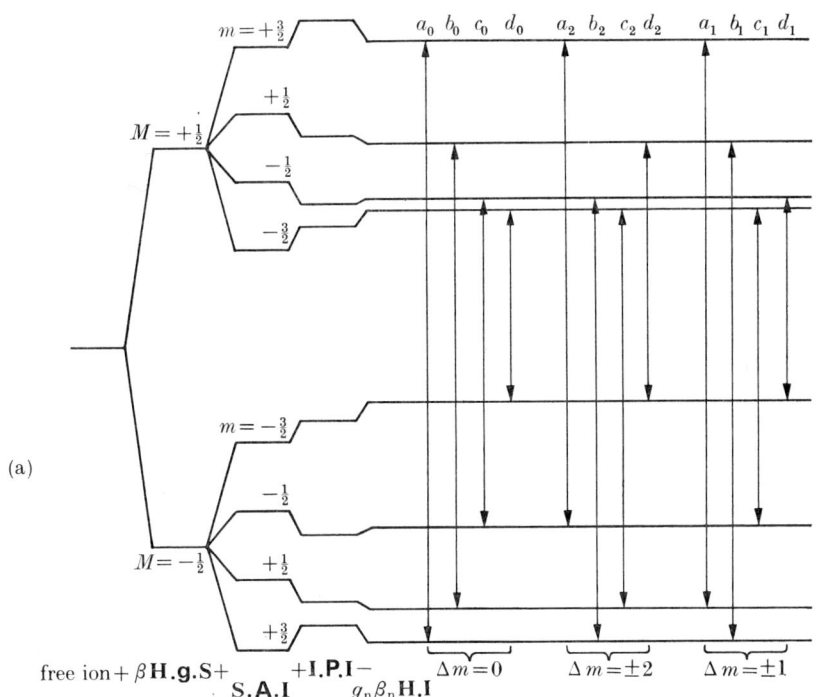

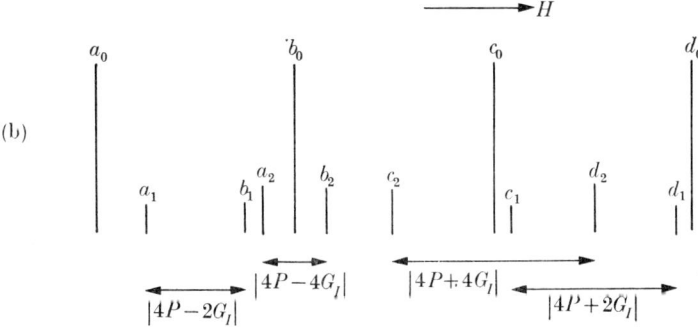

FIG. 3.13. (a) Schematic energy-level diagram for $S = \frac{1}{2}$, $I = \frac{3}{2}$. Of the transitions shown, $\Delta m = 0$ and $\Delta m = \pm 2$ are allowed if the external magnetic field is along a rhombic axis, while $\Delta m = \pm 1$ are allowed only in intermediate directions.
(b) Schematic representation of the spectrum corresponding to (a).
Second-order terms have been neglected; the intensities are not to scale.

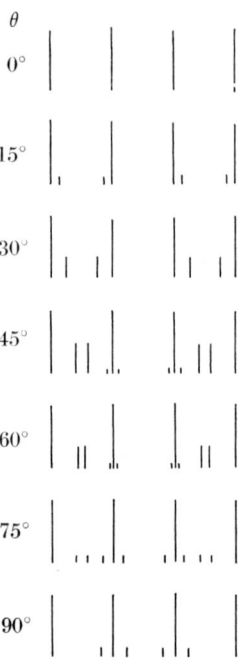

Fig. 3.14. Quadrupole effects in spectrum for $S = \frac{1}{2}$, $I = \frac{3}{2}$, with no anisotropy and $A_\parallel = A_\perp = 5P_\parallel$.

The behaviour of these lines can best be appreciated from a simple example. Figure 3.14 shows the spectrum at various angles for a hypothetical ion for which $S = \frac{1}{2}$, $I = \frac{3}{2}$, $A_\parallel = A_\perp = 5P_\parallel$, $g_\parallel = g_\perp$. At $\theta = 0$ only the four $\Delta m = 0$ lines are allowed. As θ increases the $\Delta m = \pm 1$ lines appear first, as doublets symmetrically situated about the midpoints between all the $m = 0$ lines except at the centre of the whole spectrum. From (3.65), since $(M - \frac{1}{2}) = 0$ in this case, the splitting of these doublets is $(4Pk - 2G_I)$ and so is greatest at the ends of the spectrum, but is asymmetrical because the terms $4Pk$ and $2G_I$ add at one end and subtract at the other. The intensities (eqn (3.66)) are greatest when $\theta = 45°$, and fall to zero again at $90°$. The behaviour of the $\Delta m = \pm 2$ lines is rather similar, except that the doublets have twice as great a splitting (see eqn (3.67)) and are centred on each of the $\Delta m = 0$ lines except the outside lines; the intensity (eqn (3.68)) increases to a maximum at $\theta = 90°$K. Figure 3.14 is drawn for a case of no anisotropy, but the behaviour is qualitatively similar when axial anisotropy is present. The nuclear quadrupole effects then essentially depend on the angle ψ, which may be very much closer than θ to 0 or π if $|(g_\| A_\|/g_\perp A_\perp)| \gg 1$, or closer to $\pm\pi/2$ if this quantity is less than unity. If $|(g_\| A_\|/g_\perp A_\perp)|$ is much greater than unity, the quadrupole lines appear only at values of θ close to $90°$, and the whole of the variation of P (which passes through zero at $\cos\psi = 3^{-\frac{1}{2}}$) occurs over a small range of θ close to $90°$. On the other hand, if $|(g_\| A_\|/g_\perp A_\perp)| \ll 1$, the quadrupole effects are spread over a wide range of the angle θ, and may appear with considerable intensity when θ is quite small.

In principle, observation of these quadrupole lines gives a good deal of extra information, since the doublet splitting yields both $|P|$, from which the nuclear quadrupole moment can be calculated if the electric field gradient is known, and $|G_I|$ from which the nuclear magnetic moment can be obtained directly. Determination of the relative signs of these quantities needs careful analysis.

Since both g and A are positive quantities, the lines for which m or k are positive occur always at the high frequency (low field) end of a group of hyperfine lines. When $S = \frac{1}{2}$, the extra lines due to quadrupole interaction fall into doublets from whose splitting the relative signs of P and G_I can be determined, since the splitting will be smallest at the high frequency (low field) end if P, G_I have the same signs, and vice versa (see Fig. 3.13(b)). Inspection of the expressions for P, G_I shows that observations at different angles then give the sign of $P_\|$ relative to the signs of $(A_\| g_\|/g_\|^{(I)})$, $(A_\perp g_\perp/g_\perp^{(I)})$ which occur in G_I (eqn (3.55)), together with the magnitudes of $|P_\||$, $|g_\|^{(I)}|$, $|g_\perp^{(I)}|$, $|A_\||$, $|A_\perp|$.

Further information about signs can be obtained from theory or from consideration of second-order effects:

(1) atomic theory of the electronic ground state may give the sign and magnitudes of the quantities $g_\|$, $g_\perp$, $(A_\|/g_\|^{(I)})$, $(A_\perp/g_\perp^{(I)})$, and (P/Q); then the signs of $(A_\| g_\|/g_\|^{(I)})$, $(A_\perp g_\perp/g_\perp^{(I)})$ are known and from the experiments one can deduce the sign of P and hence also of Q, the electric quadrupole moment of the nucleus;

(2) the second-order terms in eqn (3.64) are of fixed sign and if their effect is observable, the actual signs of P and G_I can be found;

(3) the second-order terms in eqn (3.56) are of fixed sign and by observing their effect in intermediate fields the sign of P, and hence of G_I (either directly or from the strong field experiments) can be found;

(4) if $S = 1$ or more, the signs of P and G_I are found from first-order effects through the terms in AM in eqns (3.65), (3.67), provided that the sign of M has been identified for the various electronic transitions; otherwise the first-order terms give only the signs of P, G_I relative to the fine structure terms.

In many cases sufficient resolution may not be available to determine all these quantities. The term G_I is usually rather small so that it cannot be determined accurately unless measurements are made at rather high fields.

It must be stressed that the perturbation methods used above are valid only if $|P/A|$ is quite small. From Fig. 3.14 it can be seen that even for $P/A = 0.2$ the $\Delta m = \pm 1$ transitions have an intensity about 0.4 that of the $\Delta m = 0$ transitions at some angles; this reflects the fact that the off-diagonal terms are by no means small compared with the diagonal terms in the energy matrix. The effects of G_I and terms of order A^2/G on the intensity can also be quite important, as is illustrated in Fig. 3.15.

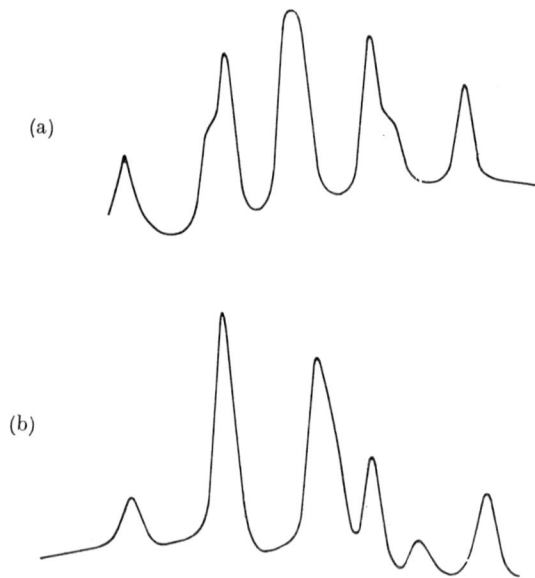

FIG. 3.15. 'Perpendicular' spectrum of $(Mg, Cu)_3La_2(NO_3)_{12}.24D_2O$ in a $\langle 110 \rangle$ direction relative to the local fourfold axes: (a) at 0.3 cm^{-1}, (b) at 0.8 cm^{-1}. The increased asymmetry at the higher frequency is due to the interaction of the larger external magnetic field with the nuclear magnetic moment (Bleaney, Bowers, and Trenam 1955).

3.10. 'Forbidden' hyperfine transitions

In the paramagnetic resonance spectra of ions which have a more than doubly degenerate ground state, split by the crystalline electric field by amounts of the order 10^{-2} to 10^{-1} cm^{-1}, and which also have a hyperfine structure, a number of extra hyperfine lines have often been observed which appear to correspond to transitions in which the nuclear magnetic quantum m changes by ± 1. These were first reported by Bleaney and Ingram (1951a) in manganese ammonium sulphate and manganese fluosilicate (both diluted); they appeared when the external magnetic field was at an angle between parallel and perpendicular to the crystal axis, and were inexplicably large in intensity. They arise from second-order effects due to cross terms in the spin Hamiltonian between the 'fine structure' splitting and the hyperfine structure splitting, as shown by Bleaney and Rubins (1961).

We consider an ion with spin $S(> \frac{1}{2})$ and a fine structure term of the second degree with axial symmetry; for simplicity we shall assume that the spectroscopic splitting factor and the hyperfine structure are both isotropic, as is closely the case for ions such as d^5 and f^7, or d^3 in an octahedral field, where such effects are usually observed. The spin Hamiltonian is then

$$\mathscr{H} = g\beta \mathbf{H} \cdot \mathbf{S} + D\{S_z^2 - \tfrac{1}{3}S(S+1)\} + A\mathbf{S} \cdot \mathbf{I} - g_I\beta \mathbf{H} \cdot \mathbf{I}. \quad (3.69)$$

If $\mathbf{H}$ is at an angle θ to the crystal axis, and the Zeeman energy is the largest term, the latter can be diagonalized by choosing the direction of $\mathbf{H}$ as the new z-axis. With axial symmetry this change can be treated as a rotation through an angle θ about the y-axis, and the Hamiltonian then becomes

$$\mathscr{H} = g\beta H S_z + \tfrac{1}{2}D\{S_z^2 - \tfrac{1}{3}S(S+1)\}\{3\cos^2\theta - 1\} +$$
$$+ D(S_zS_x + S_xS_z)\cos\theta\sin\theta + \tfrac{1}{4}D(S_+^2 + S_-^2)\sin^2\theta +$$
$$+ A\mathbf{S} \cdot \mathbf{I} - g_I\beta \mathbf{H} \cdot \mathbf{I}. \quad (3.70)$$

Zero-order wave-functions $|M, m\rangle$ (where M, m are the electronic and nuclear magnetic quantum numbers), which would be appropriate to an ion in a very large magnetic field, are admixed by the off-diagonal terms in this Hamiltonian. The effect on the energy of first-order admixtures due to terms in A and D has been evaluated earlier, but there are important second-order admixtures arising from cross products of the terms in D and A. These consist of products

$$(S_zS_+)(S_-I_+), \ (S_zS_-)(S_+I_-)$$

and other products formed by permutation of the components of S, whose energy coefficients are of order $(DA/\text{Zeeman energy})$. Such terms can be reduced to an equivalent operator

$$\tfrac{1}{8}DA\sin 2\theta(I_+ + I_-)\left[\frac{(2S_z-1)\{S(S+1)-S_z^2+S_z\}}{W_M - W_{M-1}} + \right.$$
$$\left. + \frac{(2S_z+1)\{S(S+1)-S_z^2-S_z\}}{W_M - W_{M+1}}\right] \quad (3.71)$$

and if $D \ll g\beta H$, we can take the energy denominators as
$$W_{M+1}-W_M = W_M-W_{M-1} = g\beta H,$$
when (3.71) reduces to
$$\frac{3DA \sin 2\theta}{4g\beta H}(I_++I_-)\{S_z^2-\tfrac{1}{3}S(S+1)\}. \tag{3.72}$$

The importance of the terms in (3.71) and (3.72) derives from the fact that they contain operators connecting hyperfine levels with the same M, which differ in energy only by AM. Hence a nuclear state $|m\rangle$ becomes admixed with states $|m\pm 1\rangle$ by amounts of order
$$3D \sin 2\theta/4g\beta H$$
and transitions of the type $\Delta M = \pm 1$, $\Delta m = \pm 1$ become allowed with intensities of the order of the square of this admixture coefficient, relative to the intensities of the ordinary transitions in which m does not change. Detailed evaluation of these gives for the relative intensity the expression
$$\left(\frac{3D\sin 2\theta}{4g\beta H}\right)^2\left\{1+\frac{S(S+1)}{3M(M-1)}\right\}^2\{I(I+1)-m^2+m\} \tag{3.73}$$
for the transitions
$$|M, m\rangle \leftrightarrow |M-1, m-1\rangle \text{ and } |M, m-1\rangle \leftrightarrow |M-1, m\rangle.$$

These lines are strongest in the electronic $|+\tfrac{1}{2}\rangle \leftrightarrow |-\tfrac{1}{2}\rangle$ transition, and their angular dependence is such that they vanish parallel or perpendicular to the symmetry axis, being strongest at $\theta = 45°$. Their intensity is surprisingly large, for at this angle they would rival the $\Delta m = 0$ transitions in intensity for a value of $D/g\beta H$ as small as $\tfrac{1}{12}$.

Under such conditions transitions with $\Delta m = \pm 2$ will also have appreciable intensity. It is readily seen that these arise from admixtures due to the repeated application of the cross-product terms already considered, together with cross-products of the form $(DS_+^2)(AS_-I_+)^2$ and $(DS_-^2)(AS_+I_-)^2$. By means of perturbation theory it can be shown that the relative intensity of the transitions
$$|M, m\pm 1\rangle \leftrightarrow |M-1, m\mp 1\rangle$$
is, assuming $D, A \ll g\beta H$,
$$\left[\frac{1}{2}\left(\frac{3D\sin 2\theta}{4g\beta H}\right)^2\left\{1+\frac{S(S+1)}{3M(M-1)}\right\}^2 \pm \frac{3DA\sin^2\theta}{16(g\beta H)^2}\left\{1+\frac{S(S+1)}{3M(M-1)}\right\}\right]^2 \times$$
$$\times\{I^2-m^2\}\{(I+1)^2-m^2\}. \tag{3.74}$$

Positions of lines

The energy of the state $|M, m\rangle$ due to the hyperfine terms is

$$AMm+m(A^2/2g\beta H)\{M^2-S(S+1)\}+M(A^2/2g\beta H)\{I(I+1)-m^2\}-$$
$$-g_I\beta Hm+(D\sin 2\theta/4g\beta H)^2(2Am/M)[\{M^2-S(S+1)\}^2-M^2]+$$
$$+(D\sin^2\theta/4g\beta H)^2(2AmM)\{2M^2+1-2S(S+1)\}, \quad (3.75)$$

where all second-order terms have been included, together with the important third-order terms arising from the cross products considered above. An idea of the appearance of the spectrum is obtained by omitting at first the energy terms which vanish at high frequencies; then for an ion with $S = \frac{3}{2}$, $I = \frac{3}{2}$ and D large enough to separate out the three electronic transitions, the spectrum will be roughly as in Fig. 3.16. For the central transition, the $\Delta m = \pm 1$ lines occur as

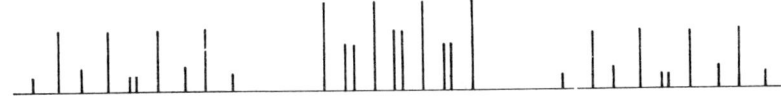

FIG. 3.16. Schematic plot of spectrum for $S = \frac{3}{2}$, $I = \frac{3}{2}$ showing the strong $\Delta m = 0$ lines and the extra $\Delta m = \pm 1$ lines.

doublets, with splitting $2g_I\beta H$, centred on the points midway between the main lines, and with greatest intensity at the centre. For the outer transitions, half of the $\Delta m = \pm 1$ lines are displaced by an amount A towards the higher frequencies, while the other half are equally displaced in the opposite sense; hence one line at each end of a hyperfine set falls outside the main hyperfine structure lines. For an ion with a higher value of S, the displacements are $\pm 2A$ for the $|\pm\frac{5}{2}\rangle \leftrightarrow |\pm\frac{3}{2}\rangle$ transitions, and $\pm 3A$ for the $|\pm\frac{7}{2}\rangle \leftrightarrow |\pm\frac{5}{2}\rangle$ transitions.

To this must be added the second-order displacements. As would be expected, the term in $A(D/g\beta H)^2$ is the most important if D is large. However, for the electronic $|+\frac{1}{2}\rangle \leftrightarrow |-\frac{1}{2}\rangle$ transition this term shifts the two components of the $\Delta m = \pm 1$ doublets equally, and so does not affect the splitting of each doublet, which is

$$|(A^2/g\beta H)\{S(S+1)-\tfrac{1}{4}\}+2g_I\beta H|. \quad (3.76)$$

Measurement of the splitting at different field strengths makes it possible to determine both quantities in (3.76). This is not usually a very precise method of determining g_I, but it may be of use in fixing its sign (strictly only its sign relative to g can be found). If there were a nuclear electric quadrupole interaction (with a corresponding term $P\{m^2-\tfrac{1}{3}I(I+1)\}$ as in eqn (3.62)), a term $2P(2m-1)$ would have to be included between the modulus signs in eqn (3.76). This would increase the splitting of the doublets at one end and decrease it at the other. Thus equal spacing in the doublets serves as rather a delicate test of such an interaction, though caution must be observed in using this test, since terms in m^2 which would give similar effects may arise from higher order effects neglected in treatment.

The positions of the $\Delta m = \pm 2$ lines can be easily seen. For the central electronic transition they form doublets spaced by twice the amount given by eqn (3.76), centred on all the main hyperfine lines except the outside ones. For the outer electronic transitions they will be displaced sideways, but by an amount twice that (in first order) for the $\Delta m = \pm 1$ transitions.

These sideways displacements give the sign of A relative to g_I, from observation whether the displacement is added to or subtracted from that due to $g_I \beta H$ in the outer transition with positive M. The sign of M is, however, only identified if that of D is known; otherwise only the sign of the product DAg_I is determined.

The general correctness of this theory has been verified by Bleaney and Rubins (1961) in the spectrum of a single crystal of $ZnSiF_6, 6H_2O$ containing about 1 per cent of V^{2+}. This ion, $3d^3$, has a singlet ground state with $S = \frac{3}{2}$, and the abundant stable isotope ^{51}V has $I = \frac{7}{2}$. The spin Hamiltonian (3.69) is applicable with the parameters:

$g = 1 \cdot 971 \pm 0 \cdot 002;$ $\qquad g_I = +0 \cdot 80 \times 10^{-3};$
$D = (+808 \pm 1) \times 10^{-4} \text{ cm}^{-1};$ $\qquad A = (-84 \cdot 2 \pm 0 \cdot 1) \times 10^{-4} \text{ cm}^{-1}.$

The value of D is so large that perturbation theory can be used only for angles where $\sin 2\theta$ is rather small; Fig. 3.17 shows the central electronic transition $|+\frac{1}{2}\rangle \leftrightarrow |-\frac{1}{2}\rangle$, with **H** at 78° to the crystal axis. At the shorter wavelength the $\Delta m = \pm 1$ lines are visible as unresolved doublets, whose intensity at the centre is nearly equal to that of the $\Delta m = 0$ lines. At the longer wavelength the intensity of the 'forbidden' transitions is considerably greater: both $\Delta m = \pm 1$ and ± 2 lines are visible, while the $\Delta m = 0$ lines have sunk to small intensity except at the two ends. Exact diagonalization of the energy matrix using a computer showed that a very complex spectrum would be expected; for example near $\theta = 45°$ the $\Delta m = \pm 3$ and ± 4 transitions should be the most intense.

Graphs showing the computed variation of intensity of the various 'forbidden' transitions with angle are given by Bleaney and Rubins; though applicable only to a special case, they illustrate the fact that the intensity can be remarkably large, so that the observed spectrum may bear little resemblance to that expected from simple theory. Obviously fine structure terms other than the simple DS_z^2 term can produce similar effects, and these have been analyzed for a cubic term (see, for example, Cavenett (1964) who has compared the theory with the experimental results for Mn^{2+} in cubic ZnSe).

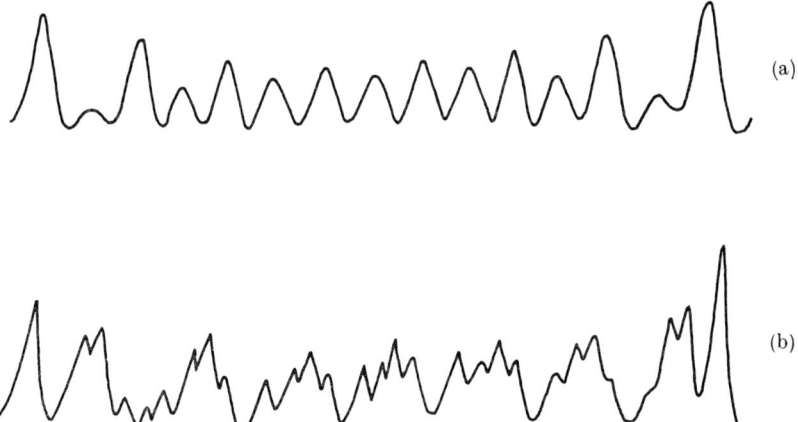

FIG. 3.17. Central $|+\tfrac{1}{2}\rangle \leftarrow \rightarrow |-\tfrac{1}{2}\rangle$ transition for dilute vanadium fluosilicate, with **H** at 78° from the crystal axis: (a) wavelength 1·3 cm ($H \sim 8200$ G), (b) wavelength 3·2 cm ($H \sim 3400$ G); the line at the extreme right is abnormally intense through an accidental coincidence with a line belonging to an outer electronic transition (Bleaney and Rubins 1961).

Although the appearance and intensity of the 'forbidden' transitions is their most striking feature, they are naturally accompanied by displacements of the $\Delta m = 0$ lines from their expected positions. These displacements must of course be taken into account in calculating values of the spin Hamiltonian parameters from experimental measurements on the spectrum; however, unless the orientation of the magnetic field with respect to the crystal axes is very accurately known it is generally best to make measurements in extreme directions where less accuracy in orientation is required. This rather elementary rule is quite general in its application, but we wish to emphasize that in Endor measurements, where hyperfine structure can be measured with great accuracy, the effects of the off-diagonal terms discussed in this section can become quite remarkably important. In many cases serious errors arising from their neglect can be avoided only by aligning the magnetic field along a crystal axis to a high degree of accuracy.

3.11. Ligand hyperfine structure

When the magnetic electrons are not entirely localized on the central 'magnetic' ion, and their wave-functions include orbitals belonging to the ligand ions, an additional splitting of the paramagnetic resonance lines is observed if the ligand nuclei possess nuclear spins. This ligand hyperfine structure (sometimes known as 'super' hyperfine structure) is described by additional terms in the spin Hamiltonian, which have axial symmetry about the 'bond' axis—the line through the centre of the ligand ion and the centre of the magnetic ion. If this line is the z-axis, the additional terms due to the one ligand ion L are

$$\mathscr{H}_L = A_\parallel^L I_z^L S_z + A_\perp^L (I_x^L S_x + I_y^L S_y) + \\ + P_\parallel^L \{(I_z^L)^2 - \tfrac{1}{3}I^L(I^L+1)\} - g_I^L \beta \mathbf{H} \cdot \mathbf{I}^L, \quad (3.77)$$

where $\mathbf{I}^L$ is the spin of the ligand nucleus, $\mathbf{S}$ the fictitious spin of the magnetic ion, and the parameters are analogous to those used to describe the hyperfine structure of the central ion.

In general the magnitude of these parameters is such that when an external field is applied large enough to bring the resonance spectrum into the microwave region, we can treat the ligand interaction by means of perturbation theory retaining only the first-order terms. Then the allowed electronic transitions are those in which the orientation of the ligand nucleus remains unchanged ($\Delta m^L = 0$), and if the external magnetic field makes an angle θ with the bond axis, the transition is displaced by an amount $A^L m^L$, where

$$(A^L)^2 = (A_\parallel^L)^2 \cos^2\theta + (A_\perp^L)^2 \sin^2\theta \quad (3.78)$$

provided that the central ion has an isotropic g-factor. If the g-factor is anisotropic, and its principal axes coincide with the x, y, z axes appropriate to eqn (3.77) and the z-axis is the bond axis, the value of A^L is given by an equation similar to (3.44),

$$g^2(A^L)^2 = (l^2 g_x^2 + m^2 g_y^2)(A_\perp^L)^2 + n^2 g_z^2 (A_\parallel^L)^2, \quad (3.79)$$

where (l, m, n) are the direction cosines of the magnetic field with respect to the x, y, z axes.

The effect of this additional interaction is to sub-divide each electronic transition, or each hyperfine line if the central ion has a nuclear spin, into $(2I^L+1)$ lines of equal intensity with separation A^L. If the number of ligands is n (all identical), then the sub-division is into $(2I^L+1)^n$ lines, which with an ion octahedrally coordinated to six ligand ions of nuclear spin $I^L = \tfrac{1}{2}$ would give $2^6 = 64$ lines.

THE SPIN HAMILTONIAN AND THE SPECTRUM   193

Fortunately the problem is not so serious as this calculation suggests, since many of the lines coincide. If the magnetic ion is at a centre of inversion symmetry, the ligand ions will always be identical in pairs, and the line displacement due to such a pair will be $A^L(m_1+m_2)$, where m_1, m_2 each take on their $(2I^L+1)$ possible values. The ligand hyperfine structure due to such a pair is shown in Fig. 3.18 for values of I^L

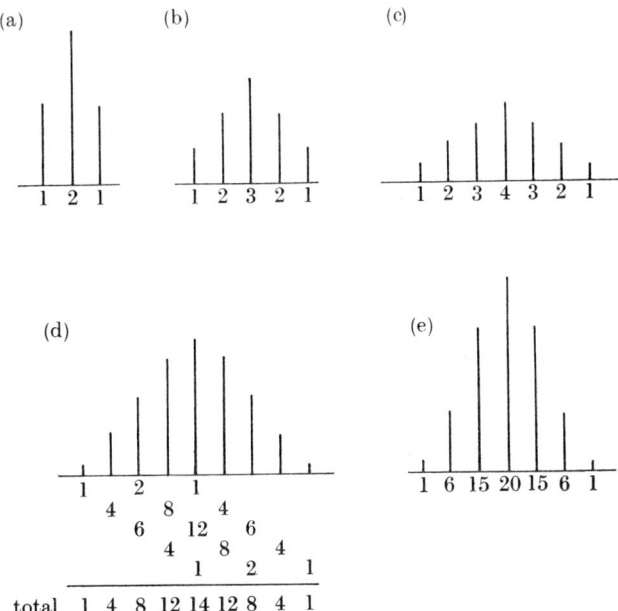

FIG. 3.18. (a)(b)(c) Hyperfine structure due to two identical ligand ions, with nuclear spins (a) $I = \frac{1}{2}$, (b) $I = 1$, (c) $I = \frac{3}{2}$. (d), (e) Hyperfine structure due to 6 ligand nuclei, all with $I = \frac{1}{2}$. In (d) the hyperfine constant is twice as large for two nuclei as for the other four. In (e) all hyperfine constants are identical.

equal to $\frac{1}{2}$, 1, and $\frac{3}{2}$. The structure consists of a set of equally spaced lines of separation A^L, but with an intensity which increases in arithmetical progression towards the centre owing to the number of ways in which (m_1+m_2) can sum to a given value. The total intensity in (a), (b), and (c) of Fig. 3.18 is drawn to be identical in each case; the higher the nuclear spin, the greater the number of lines between which the intensity must be divided, and the lower the intensity of individual lines.

Two examples of the ligand hyperfine structure due to six nuclei, where simple numerical relationships between the hyperfine parameters produce a simplification of the patterns, are also shown in Fig. 3.18.

In (d) it is assumed that the hyperfine splitting of two identical ions is twice that of the other four, so that lines from different ligands are superimposed. Each nucleus is assumed to have $I = \frac{1}{2}$, so that two identical ions produce a 1:2:1 intensity pattern, and four identical ions produce a 1:4:6:4:1 pattern. The resultant intensity pattern can be regarded either as a superposition of five 1:2:1 patterns, with relative intensity 1:4:6:4:1, or three 1:4:6:4:1 patterns with relative intensity 1:2:1. These can be identified visually by reading the intensity ratios below the figure horizontally and diagonally, respectively. The case where four ions are identical but different from the other two is typical of a regular octahedron of ligand ions, with the magnetic field in one of the planes of reflection symmetry normal to a twofold axis. The ratio of 2:1 in the hyperfine splittings may occur at some particular orientation of the magnetic field; if the coupling were purely dipolar, it would be numerically twice as large for the two ions whose bond axis is parallel to the magnetic field as for the four ions whose bond axes are normal to the field.

Naturally the simplest pattern is obtained when all ions give the same hyperfine splitting. This is shown in (e) for six ligand ions each with $I = \frac{1}{2}$. The intensity pattern is 1:6:15:20:15:6:1, the same as the binomial coefficients in the expansion of $(x+y)^6$. The rather large intensity ratios, typical of patterns due to a number of nuclei, often make it difficult to see the outermost lines in the spectrum (to bring the total intensity to be the same as in (a), (b), (c) the lines should be smaller by a factor 2 in (d) and (e)). Thus the outside lines in (e) have each an intensity which is only $\frac{1}{64}$ ($= (\frac{1}{2})^6$) of the total intensity associated with any given electronic transition (which may already be subdivided because of hyperfine structure due to a nuclear moment on the central ion). In the case of eightfold coordination to a regular cube of fluorines (for example, a paramagnetic ion replacing a Ca^{2+} ion in CaF_2) the intensity pattern is 1:8:28:56:70:56:28:8:1 if all ions give the same hyperfine splitting, so that each outside line would have only $\frac{1}{128} = (\frac{1}{2})^8$ of the total intensity.

A situation where all ligand hyperfine parameters are equal normally occurs only for certain directions of the external magnetic field, and also requires high symmetry in the complex. With a regular octahedron of ligand ions, the magnetic field must be directed along a threefold axis; with a regular cube or tetrahedron of ligand ions, the magnetic field must be along a fourfold axis. Of course it will also occur if the hyperfine parameters $|A_\parallel^L|$, $|A_\perp^L|$ in eqn (3.77) are equal, in which case the ligand hyperfine pattern will be independent of the orientation of the magnetic field. Since there are a number of contributions to the ligand hyperfine structure, the condition $A_\parallel^L = \pm A_\perp^L$ may be fortuitously obeyed; it will also be approximately correct if the contribution from s-electrons on the ligand ion is much larger than the contributions from any other source (including dipolar interaction).

The Hamiltonian for ligand hyperfine structure has been treated above by means of first-order perturbation theory. This implies that the hyperfine parameters are small compared with the electronic Zeeman energy, and this condition is normally fulfilled. The treatment given also implies that the magnetic hyperfine interaction be large compared with the other terms in eqn (3.77), since we have diagonalized the main portion of this term, as in § 3.8; then the allowed transitions are those in which $\Delta m^L = 0$, whose positions are not affected (in first order) by the nuclear electric quadrupole term or the nuclear Zeeman interaction with the external field. Obviously there will be second-order shifts of order $((A^L)^2/h\nu)$, but these are not usually important in an electron spin resonance experiment, though they may well be in an Endor experiment where the precision is much higher. For a nucleus whose spin I is greater than $\tfrac{1}{2}$, the electric quadrupole term $P_\parallel^L$ may be comparable with A^L, in which case transitions with $\Delta m^L \neq 0$ may have appreciable intensity; this gives a situation similar to that considered in § 3.9, but the hyperfine structure is more complex because of the number of nuclei involved. A further possibility is that the ligand hyperfine structure is rather small, so that A^L is comparable with, or smaller than, the nuclear Zeeman term; this needs a rather different treatment, which we give below.

For simplicity we consider only the case where the nuclear electric quadrupole interaction is zero. However our treatment is not necessarily confined to a hyperfine structure due to a ligand nucleus, so that we will omit the superscript L and the restriction to axial symmetry. The hyperfine Hamiltonian including the nuclear Zeeman energy is then

$$\mathscr{H}_n = A_x S_x I_x + A_y S_y I_y + A_z S_z I_z - \beta(g_x^{(I)} H_x I_x + g_y^{(I)} H_y I_y + g_z^{(I)} H_z I_z), \quad (3.80)$$

where the nuclear g-factor may be anisotropic owing to indirect effects as discussed in Chapter 18, but is assumed to have the same principal axes as the hyperfine 'tensor' A. Suppose the external magnetic field has direction cosines (l, m, n) with respect to these axes. Then the electronic Zeeman interaction is diagonalized as in § 3.2, and on retaining only the terms in S'_z in the magnetic hyperfine interaction the nuclear Hamiltonian becomes

$$\mathscr{H}_n = lI_x\left\{\frac{g_x A_x S'_z}{g} - g_x^{(I)}\beta H\right\} + mI_y\left\{\frac{g_y A_y S'_z}{g} - g_y^{(I)}\beta H\right\} +$$

$$+ nI_z\left\{\frac{g_z A_z S'_z}{g} - g_z^{(I)}\beta H\right\}. \quad (3.81)$$

The coefficients of I_x, I_y, I_z may be regarded as the components of an interaction which can be diagonalized by a suitable transformation of the nuclear axes to a set (x_n, y_n, z_n) where the z_n-axis has direction cosines proportional to the coefficients of I_x, I_y, I_z. Then the interaction may be written in the form KI'_z, where

$$K^2 = \sum_{x,y,z} l^2 \left\{\frac{g_x A_x S'_z}{g} - g_x^{(I)}\beta H\right\}^2 \quad (3.82)$$

and the hyperfine energy is simply Km, where m is the nuclear magnetic quantum number defined by the projection of the nuclear spin onto an axis parallel to the vector **K**. The vector **K** may be regarded as the sum of two vectors, one representing the interaction with the electronic magnetic field and the other the interaction with the external field (the latter vector is, of course, parallel to the external field if $g^{(I)}$ is isotropic), as shown in Fig. 3.19. The length of the former vector is proportional to S'_z, and is therefore different in different electronic states; thus the resultant vector **K** is different both in magnitude and direction for different electronic states. In a given state with $S'_z = M$, we denote it by $\mathbf{K}_M$. In a transition $(M, m) \leftrightarrow (M', m')$ the additional energy due to the hyperfine interaction will be $mK_M - m'K_{M'}$, assuming that the state M lies higher in energy than M'. Here there will in general be no restriction relating the allowed values of m' to m; in other words, changes in the nuclear magnetic quantum number are allowed. The reason for this is that m, m' represent the components of nuclear spin along the directions $\mathbf{K}_M$, $\mathbf{K}_{M'}$, which are differently oriented. If we take a given state m referred to the $\mathbf{K}_M$ axis, and transform it by a rotation that takes the z_n-axis from $\mathbf{K}_M$ to $\mathbf{K}_{M'}$, it will be represented by a linear combination of all possible states m' referred to the latter axis. Thus a transition in which the orientation of the nuclear magnetic moment remains fixed in space must be represented by a sum of transitions in which the nuclear magnetic quantum number m, as defined above, changes to all values m' without restriction, though in certain definite proportions that determine the

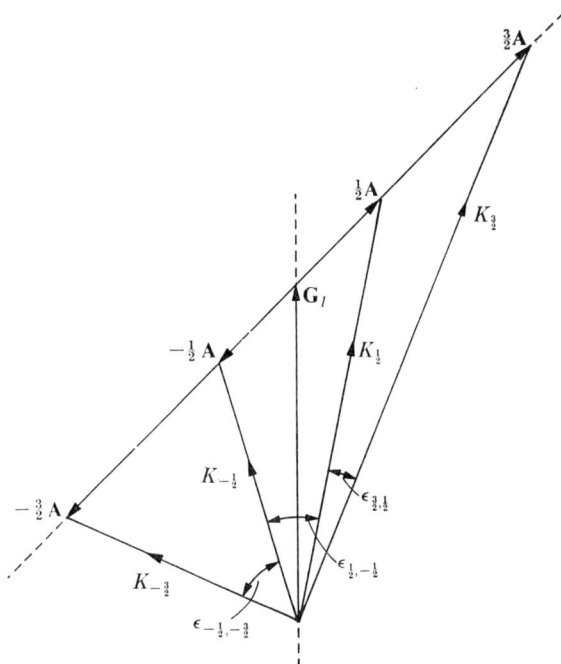

FIG. 3.19. Diagram illustrating the vector addition of the electronic interaction with the nucleus (AS'_z) and the nuclear Zeeman interaction $(\mathbf{G}_I)$ for the case of $S = \tfrac{3}{2}$. If the nuclear g-factor $g^{(I)}$ is isotropic the direction of $\mathbf{G}_I$ is the same as that of the external magnetic field. The net hyperfine interaction energy in a state (M, m) is mK_M.

relative amplitudes in the various transitions. These proportions are a function of the angle ϵ between $\mathbf{K}_M$ and $\mathbf{K}_{M'}$, and are shown for the case of $S = \tfrac{1}{2}$, $I = \tfrac{3}{2}$ in Fig. 3.20. From this figure it will be seen that when the angle $\epsilon = 0$, the only allowed transitions are those in which $m' = m$; this corresponds to the limit in which the electronic magnetic field at the nucleus is zero $(A = 0)$, and the only field is the external field, which remains fixed in direction during a transition. When the angle $\epsilon = 180°$, the only allowed transitions are those in which $m' = -m$; this corresponds (for $S = \tfrac{1}{2}$) to the case considered earlier, where the electronic magnetic field at the nucleus is much greater than the external field, so that to a good approximation the net field is just reversed in the transition $M = \tfrac{1}{2} \leftrightarrow -\tfrac{1}{2}$. (These allowed transitions are of course those which were labelled $\Delta m = 0$ in earlier sections, since the positive direction for m was then considered fixed, instead of reversing when M goes from $+\tfrac{1}{2}$ to $-\tfrac{1}{2}$, as in the present treatment.)

The hyperfine displacement $mK_M - m'K_{M'}$ vanishes for the transitions in which $m' = m$ if $K_M = K_{M'}$. The latter holds if $A = 0$, but it may

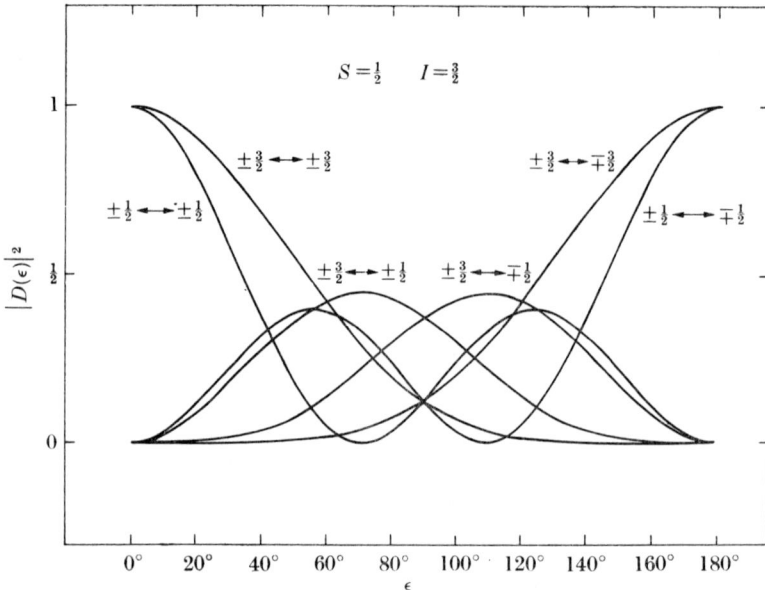

Fig. 3.20. Relative strengths of the various transitions $(+\tfrac{1}{2}, m) \leftrightarrow (-\tfrac{1}{2}, m')$ as a function of the angle ϵ between $\mathbf{K}_{\tfrac{1}{2}}$ and $\mathbf{K}_{-\tfrac{1}{2}}$ for $S = \tfrac{1}{2}$, $I = \tfrac{3}{2}$. $D_{mm'}(0, \epsilon, 0)$ is a matrix element of the three-dimensional rotation operator. (After Weil and Anderson 1961.)

also occur when $A \neq 0$ provided that the hyperfine energy is comparable with the nuclear Zeeman energy. For the transition $M = \tfrac{1}{2} \leftrightarrow -\tfrac{1}{2}$ this requires that the direction of $\mathbf{A}$ be normal to that of $\mathbf{G}_I$, as in Fig. 3.21; this will occur only at certain specific orientations of the external magnetic field, and also requires that the principal values $g_x A_x$, $g_y A_y$, $g_z A_z$ do not all have the same sign (or, a less probable alternative, that $g_x^{(I)}$, $g_y^{(I)}$, $g_z^{(I)}$ do not all have the same sign). Cases where the hyperfine structure passes through zero in this way have been observed by Weil and Anderson (1961) and Woodbury and Ludwig (1961).

When $A \gg G_I$ we have a hyperfine structure of the normal kind discussed in earlier sections where the extreme hyperfine lines are displaced by $\pm AI$ in any electronic transition. If transitions occur in which $\Delta m \neq 0$, these lie within these extremes when $S = \tfrac{1}{2}$. This is not necessarily true for the case discussed in this section, where A may be smaller than G_I. A clear example arises in the situation represented by Fig. 3.21, where the hyperfine transitions $(+\tfrac{1}{2}, m) \leftrightarrow (-\tfrac{1}{2}, m)$ are undisplaced, but a transition with $m' \neq m$ occurs at $K(m-m')$, where $K = |K_{\pm\tfrac{1}{2}}|$. Such transitions appear as 'satellites' either side of the main line; if $A \ll G_I$ they are displaced by an amount almost equal to

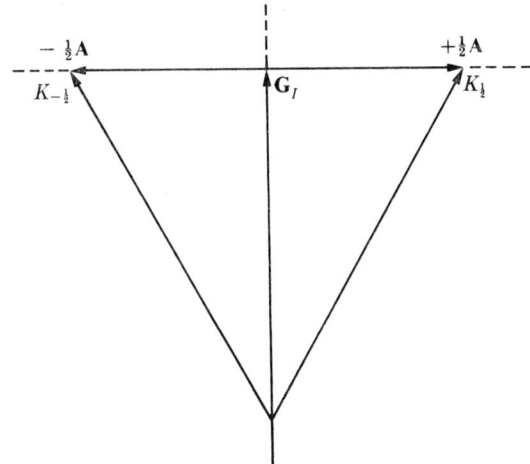

Fig. 3.21. When the vectors $\mathbf{A}$, $\mathbf{G}_I$ are perpendicular the values of $K_{\frac{1}{2}}$ and $K_{-\frac{1}{2}}$ are equal, so that the net hyperfine energy for the transition $(+\frac{1}{2}, m) \leftrightarrow (-\frac{1}{2}, m)$ vanishes. From eqn (3.82) this requires that

$$\sum l^2 \left\{ \frac{g_x A_x}{2g} - g_x^{(I)} \beta H \right\}^2 = \sum l^2 \left\{ \frac{g_x A_x}{2g} + g_x^{(I)} \beta H \right\}^2$$

or $(lg_x A_x/g)(lg_x^{(I)}) + (mg_y A_y/g)(mg_y^{(I)}) + (ng_z A_z/g)(ng_z^{(I)}) = 0$, or $\mathbf{A} \cdot \mathbf{G}_I = 0$; that is the vectors $\mathbf{A}$ (direction cosines $lg_x A_x/g$, etc.) and $\mathbf{G}_I$ (direction cosines $lg_x^{(I)}$, etc.) are mutually perpendicular.

the precession frequency of the nuclear moment in the external field (cf. Fig. 1.39), or a multiple thereof. Such transitions were first explained by Trammell, Zeldes, and Livingstone (1958) as due to 'spin flips' of the proton nuclear moments in the electron magnetic resonance spectrum of various irradiated acids. When the ligand nucleus has $I = \frac{1}{2}$ we would expect only one satellite line on each side of the main line, due to the nuclear magnetic quantum number changing from $+\frac{1}{2}$ to $-\frac{1}{2}$ or vice versa, but when a number of ligand nuclei have nuclear moments, further satellites of smaller intensity can be seen due to 'flips' of more than one nuclear moment. The theory of such transitions has been considered in some detail by Baker, Hayes, and O'Brien (1960), and compared with experiment for a number of lanthanide ions in CaF_2.

In conclusion, we point out that such satellite lines have a simple classical explanation. If the external magnetic field is directed along the z-axis, the Hamiltonian may contain terms such as $A_{zx} S_z I_x$. To take a concrete example, let the magnetic field be parallel to a fourfold axis in CaF_2, with a paramagnetic ion on a Ca^{2+} site surrounded by a cube of eight fluorine ions. The Ca—F bonds lie along the threefold axes, so that a dipole on a Ca^{2+} site parallel to the fourfold or z-axis produces a magnetic field at a F$^-$ nucleus which is normal to the z-axis. If the direction of this dipolar field at this site is taken as the x-axis, then the interaction of the fluorine nuclear moment with this field will give rise to a term of the form $A_{zx}S_zI_x$, since the x-component of the field is proportional to S_z. Conversely, the x-component of the nuclear moment of the fluorine ion will give rise to a magnetic field at the Ca^{2+} site in the z-direction. If the external field is much larger than the local field of the paramagnetic ion at the F$^-$ site, the fluorine nuclear moment will precess about the external field at a frequency $\nu_n = g_I\beta H/h$, and will give rise to a magnetic field ΔH parallel to H at the Ca^{2+} site which oscillates at this frequency. Thus the electron moment will precess about a field $H + \Delta H \cos 2\pi\nu_n t$, so that the precession may be regarded as frequency modulated, with a carrier frequency ν_e and sidebands separated from this frequency by $\pm \nu_n$ and multiples thereof, with decreasing intensity.

3.12. The spectrum of a powder

If a paramagnetic resonance spectrum is completely isotropic, it can be examined in a powder (under which term we include a polycrystalline specimen) without loss of resolution. For this purpose local cubic symmetry at the ion is a necessary but not a sufficient condition—for example, the spectrum of an ion with a cubic 'fine structure' term of the type discussed in § 3.4 is by no means isotropic. In a powder the spectrum of any ion that is anisotropic will naturally be spread out; to a considerable extent the detail will be lost, and the information extractable from the spectrum may be drastically reduced. Such information as can be retrieved comes usually from the limits of the spread-out spectrum, since these normally appertain to ions whose orientation is such that the magnetic field happens to lie along one of the principal axes of the tensor parts of the spin Hamiltonian.

Little (if any) analysis has been attempted of the powder spectrum of ions with less than axial symmetry, and we shall restrict discussion to this case. The number of ions for which the magnetic field makes an angle between θ and $\theta + d\theta$ with the unique axis is then proportional to $\sin\theta \, d\theta$, and since the spectrum is not affected by reversal of the magnetic field we may restrict the range of angles to be considered to lie between 0 and $\pi/2$, in which case the factor of proportionality is unity. Assuming the spectrum is observed at constant frequency and variable magnetic field, we need to relate the angle θ to the magnetic field at which the spectrum line for an ion at this angle would be observed. For this purpose we consider first an ion with an anisotropic g-factor of axial symmetry, for which (eqn (3.14))

$$g^2 = g_\parallel^2 \cos^2\theta + g_\perp^2 \sin^2\theta.$$

From this relation we have

$$\cos\theta = \left(\frac{g^2-g_\perp^2}{g_\parallel^2-g_\perp^2}\right)^{\frac{1}{2}} = \left(\frac{H^{-2}-H_\perp^{-2}}{H_\parallel^{-2}-H_\perp^{-2}}\right)^{\frac{1}{2}}, \tag{3.83}$$

where the symbols H, $H_\parallel$, and $H_\perp$ refer to the fields at which the lines appropriate to values g, $g_\parallel$, and $g_\perp$ would occur. Differentiation gives

$$\begin{aligned}\sin\theta\,d\theta &= \frac{H^{-3}\,dH}{\{(H^{-2}-H_\perp^{-2})(H_\parallel^{-2}-H_\perp^{-2})\}^{\frac{1}{2}}} \\ &= \frac{H_\perp^2 H_\parallel\,dH}{H^2\{(H^2-H_\perp^2)(H_\parallel^2-H_\perp^2)\}^{\frac{1}{2}}}.\end{aligned} \tag{3.84}$$

This distribution function allows only for the random orientation of the crystallites, and is equivalent to that given by Sands (1955). It must be modified to allow for the fact that the transition probability is also a function of angle (see Bleaney 1960); in other words, to allow for the anisotropy in g_1^2. Equation (3.16) shows that the intensity is a maximum when the microwave field is normal to the steady field, and we shall assume that it is so oriented. Then, in eqn (3.16), $\sin(\theta_1-\theta)=1$, and on averaging over η for a powder we have $\langle\sin^2\eta\rangle = \langle\cos^2\eta\rangle = \frac{1}{2}$, so that the average value of g_1^2 is

$$g_1^2 = \frac{g_\perp^2(g_\parallel^2+g^2)}{2g^2}. \tag{3.85}$$

This can be related as before to the field at which resonance occurs, and for the sake of obtaining an expression for the distribution of intensity which is non-dimensional we extract only the factor

$$(g_\parallel^2+g^2)/g^2 = (H^2+H_\parallel^2)/H_\parallel^2.$$

On multiplying this by eqn (3.84) we find for the intensity distribution,

$$\text{powder intensity} \propto \frac{H_\perp^2(H^2+H_\parallel^2)\,dH}{H_\parallel H^2\{(H^2-H_\perp^2)(H_\parallel^2-H_\perp^2)\}^{\frac{1}{2}}}. \tag{3.86}$$

The main feature of this function is the infinity at $H = H_\perp$ (see Fig. 3.22) which arises because no line width mechanism has been included. Attempts have been made to remedy this deficiency (see, for example, Searl, Smith, and Wyard 1961; Ibers and Swalen 1962) which give fair

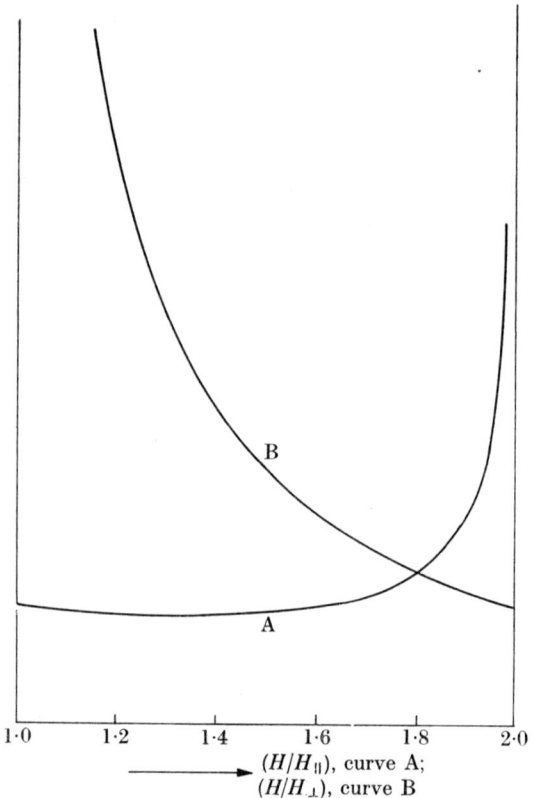

Fig. 3.22. Relative intensity distribution in the spectrum of a powder for an ion with $S = \frac{1}{2}$ and an axially symmetric g-factor, from eqn (3.86).
Curve A, $H_{\parallel} = H_{\perp}/2$; curve B, $H_{\parallel} = 2H_{\perp}$.

agreement with experiment for small anisotropy. With large anisotropy a proper knowledge of the line shape (which may be a function of angle) becomes much more important, particularly in concentrated substances.

In some cases effects due to hyperfine structure can be observed in a powder, as each hyperfine line has its own limits for maximum and minimum fields. An example is shown in Sands (1955) for the Cu^{2+} ion in glass, though it should be noted that his conclusion that $A_{\parallel}$ and $A_{\perp}$ have opposite signs cannot be justified from the observed spectrum, as the hyperfine lines do not invert in order (as assumed by him) in going from the parallel to the perpendicular direction (see § 3.8).

In his work on glass Sands found that a large number of samples showed lines corresponding to apparent anisotropic g-values of approximately 6 and 4·2, which later work (Castner, Newell, Holton, and Slichter 1960) suggests are due to Fe^{3+} ions with rather large crystal field splittings. If the term DS_z^2 is large compared with $h\nu$, we are in the weak Zeeman effect region. For an ion with $S = \frac{5}{2}$, the spectrum in zero field consists of three well-spaced doublets corresponding to $S_z = \pm\frac{5}{2}$, $\pm\frac{3}{2}$, and $\pm\frac{1}{2}$ respectively. In a weak magnetic field only resonance within the latter doublet is observable, and it behaves like an ion with an apparent $S = \frac{1}{2}$ and $g_\parallel = 2$, $g_\perp = 6$. The latter corresponds to the point of maximum intensity in a powder, and is presumed to give rise to the anisotropic line at $g = 6$. The line at $g = 4\cdot2$ has been ingeniously explained by Castner et al., who assume that it is due to a ferric ion with a fine structure term of the form $E(S_x^2 - S_y^2)$. In zero magnetic field this splits the $S = \frac{5}{2}$ levels into three equally spaced doublets, the wave-functions for the middle doublet being $(\frac{9}{14})^{\frac{1}{2}}|\pm\frac{5}{2}\rangle - (\frac{5}{14})^{\frac{1}{2}}|\mp\frac{3}{2}\rangle$. In a weak magnetic field this has an isotropic g-factor of $(15g_s/7) = \frac{30}{7} = 4\cdot28$, in good agreement with the value of 4·27 observed by Castner et al. This unexpected result of an isotropic weak field g-factor for an ion with a highly anisotropic Hamiltonian shows that great care is often required in the interpretation of a spin resonance spectrum.

For an ion whose effective spin is greater than $\frac{1}{2}$, a number of electronic transitions occur whose position is generally sharply dependent on orientation, owing to the 'fine structure' discussed in §§ 3.3–3.6, even when g is isotropic. A conspicuous exception in such cases is the $|+\frac{1}{2}\rangle \leftrightarrow |-\frac{1}{2}\rangle$ transition, which is not dependent on orientation as far as terms linear in the fine structure parameters, provided that g is isotropic (as is often true in such cases to a good approximation). Thus, if terms of order (fine structure parameter)$^2/g\beta H$ are relatively unimportant, the central transition $|+\frac{1}{2}\rangle \leftrightarrow |-\frac{1}{2}\rangle$ can be observed in a powder without much loss of resolution. If the source of such lines is unknown, the spectrum gives little help in identification unless a characteristic hyperfine structure is present; if g is isotropic the hyperfine structure is usually also isotropic and may be well resolved in the central transition. The spectrum of a sample of modelling clay is shown in Fig. 3.23; at a wavelength of 1·3 cm, six almost equally spaced lines are observed which may be rather confidently attributed to Mn^{2+}, with $g = 2\cdot0013 \pm 0\cdot0006$ and $|A| = (88\cdot7 \pm 0\cdot3) \times 10^{-4}$ cm^{-1}. At the longer wavelength of 3·2 cm, the spectrum shows a number of additional lines, which may be identified as the 'forbidden' hyperfine

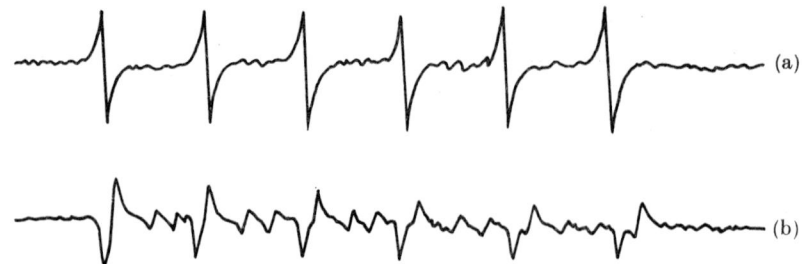

Fig. 3.23. Spectrum of modelling clay at (a) 1·3 cm, (b) 3·2-cm wavelength. The spectrum is attributed to the $|+\frac{1}{2}\rangle \leftrightarrow |-\frac{1}{2}\rangle$ transition of a trace of ^{55}Mn^{2+}. High field is towards the right (Bleaney and Rubins 1961).

lines discussed in the previous section. Their intensity may be used to estimate a value of D, which cannot be determined directly, by using eqn (3.73) averaged over all values of θ from 0 to $\pi/2$, with a weighting factor of $\sin\theta$. This results in $\sin^2 2\theta$ being replaced in eqn (3.73) by $\frac{8}{15}$.

An independent estimate of D in such a specimen or a powder can be obtained from the splitting of the hyperfine lines $\Delta m = 0$ shown in Fig. 3.23(b). This results from the combination of higher-order effects of two kinds: shifts of order $D^2/g\beta H$, which can be found from eqn (3.35) and are the same for all hyperfine lines; and shifts of order $D^2A/(g\beta H)^2$, which arise from the last terms in eqn (3.75), which are proportional to the distance of the hyperfine line from the centre of the spectrum. On combining all the terms that depend on the angle θ, we find that the position of the transition $|+\frac{1}{2}, m\rangle \leftrightarrow |-\frac{1}{2}, m\rangle$ is shifted by an amount

$$\Delta H = x(y\cos^2\theta - z\sin^2\theta)\sin^2\theta, \qquad (3.87)$$

with $x = (D^2/16g\beta H)\{4S(S+1)-3\}$ and $y = 8(1-\delta)$, where

$$\delta = (Am/4g\beta H)\{4S(S+1)+1\},$$

and $z = 1+(Am/g\beta H) = 1+\epsilon$. The number of ions whose spectrum lies in the range of field H to $H+\Delta H$ is proportional to

$$\Delta n = \sin\theta(d\theta/dH)\,\Delta H,$$

and intensity peaks occur in the line shape when $(\Delta n/\Delta H)$ is infinite. From eqn (3.87) the quantity $\sin\theta(d\theta/dH)$ is inversely proportional to $\cos\theta\{(y+2z)\sin^2\theta - y\cos^2\theta\}$ and thus becomes infinite at $\theta_1 = \pi/2$ and at θ_2 where $\tan^2\theta_2 = y/(y+2z)$. Peaks in the spectrum corresponding to the two values of θ will therefore occur at fields which differ by

$$\delta H = x\left\{z + \frac{y^2}{4(y+z)}\right\}. \qquad (3.88)$$

Since y is a function of m, the splitting of the hyperfine components varies within the central electronic transition. When δ is small, eqn (3.88) reduces to

$$\delta H = \left(\frac{25x}{9}\right)\left\{1 - \frac{32\delta - 13\epsilon}{45}\right\} \qquad (3.89)$$

and since δ, ϵ are proportional to m, the splitting varies linearly from one end of the hyperfine structure to the other. In fact the splitting is always greatest at the high field end (as in Fig. 3.23(b)), since this corresponds to negative values of m if A is positive (and vice versa), and δ, ϵ are proportional to (Am).

In making a similar calculation for the 'forbidden' hyperfine lines allowance must be made for the fact that the transition probability is proportional to $\sin^2 2\theta$ (see eqn (3.73)) so that there is no intensity peak at $\theta_1 = \pi/2$. Hence the forbidden lines are not split, though they will be broadened.

In the modelling clay specimen examined by Bleaney and Rubins, the average separation of the 'forbidden' $\Delta m = \pm 1$ lines at the shorter wavelength was 16 ± 1 G, in good agreement with the value calculated from (3.76) using the known nuclear moment of ^{55}Mn. The difference in splittings at either end was less than 2 G, suggesting that any nuclear electric quadrupole term P was less than 10^{-5} cm^{-1}. The value estimated for $|D|$ by the methods outlined above is about 0.01 cm^{-1}. This value was confirmed by the observation of two asymmetrical lines, assigned to the electronic transitions $|\pm\frac{3}{2}\rangle \leftrightarrow |\pm\frac{1}{2}\rangle$, which move as $\pm D(3\cos^2\theta - 1)$, with intensity peaks at $\theta = \pi/2$.

3.13. Effects of crystal imperfections

We have hitherto assumed that all parameters have sharp values, which might be true in a perfect crystal but is seldom the case in practice. In this section we consider the broadening of resonance lines due to crystal imperfections (broadening due to spin-spin interaction and spin-lattice relaxation is discussed in Chapters 9 and 10).

We distinguish between two effects:

(a) mosaic structure, in which the crystal is not truly single but consists of a multitude of small crystallites with slightly different orientations. The spin Hamiltonian parameters are the same in each crystallite, so that no broadening is observed in the resonance lines in zero magnetic field, but the lines become broadened in a magnetic field because (with a few obvious exceptions) the positions of the resonance transitions depend on the angle that **H** makes with the individual crystal axes. Such broadening is a minimum at directions (such as the principal axes) where the resonance lines reach extreme positions in the spectrum, and their first-order dependence on crystal mis-orientation by a small angle vanishes.

(b) impurities, dislocations, and other defects set up stresses within a single crystal, which may change the point symmetry at the paramagnetic ion, or preserve the point symmetry but change the crystal field parameters slightly. The changes in the parameters are not independent of each other, since for any given ion they are linked to a specific deviation in its ligand field. Simple examples illustrating this point abound in Chapter 5, where it is shown that under axial symmetry the values of $g_\parallel$, $g_\perp$ can often be related to a single parameter (see for example, eqn (5.12)); thus even a strain preserving axial symmetry may produce changes in $g_\parallel$, $g_\perp$ that are simply related to one another.

Whatever the cause, the effect of a variation in g differs from that of variations in other spin Hamiltonian parameters because $H = (h\nu/g\beta)$, so that at constant frequency ν we have

$$\langle \delta H^2 \rangle = (h\nu/g^2\beta)^2 \langle \delta g^2 \rangle, \tag{3.90}$$

and the root mean square width in field increases linearly with the resonance frequency. On the other hand, for any parameter that splits the spectrum lines as in fine or hyperfine structure effects, we have $h\nu = g\beta H + X$, or $H = (h\nu/g\beta) - (X/g\beta)$, and

$$\langle \delta H^2 \rangle = (1/g\beta)^2 \langle \delta X^2 \rangle, \tag{3.91}$$

showing that $\langle \delta H^2 \rangle$ from this cause will not be frequency dependent.

We now consider briefly the effect of mosaic structure. Variation of orientation of the crystallites means that instead of **H** being along a specific direction (l, m, n) with respect to the principal axes for all ions, we have a spread in the values of l, m, n. For small variations, since $(l^2 + m^2 + n^2) = 1$, we have

$$l\, \delta l + m\, \delta m + n\, \delta n = 0, \tag{3.92}$$

and if **H** is along a principal axis for the median point of the distribution (say along [100]) we have $m = n = 0$ and hence $\delta l = 0$. Then from a formula such as (3.5) we have

$$g\, \delta g = l\, \delta l g_x^2 + m\, \delta m g_y^2 + n\, \delta n g_z^2 \tag{3.93}$$

and clearly $\delta g = 0$ when **H** is along a principal axis. Since all spin Hamiltonian parameters such as D'', A, P (see eqns (3.30), (3.44), (3.47)) appropriate to a strong external field vary in a manner dependent on l^2, m^2, n^2, the spread in field $\langle \delta H^2 \rangle$ due to mosaic structure also vanishes in first order when **H** is along a principal axis.

For simplicity we now restrict discussion to the case of axial symmetry. Then, from eqn (3.14), writing $g_\parallel$, $g_\perp$ for g_z, g_x, we have

$$g(\partial g/\partial \theta) = (g_\perp^2 - g_\parallel^2)\sin \theta \cos \theta$$

and hence, using (3.90),

$$\langle \delta H^2 \rangle = (h\nu/g^3\beta)^2(g_\perp^2 - g_\parallel^2)^2 \sin^2\theta \cos^2\theta \langle \delta\theta^2 \rangle, \qquad (3.94)$$

which obviously vanishes always if $g_\parallel = g_\perp$ (no anisotropy), but otherwise only at the extrema $\theta = 0$ and $\pi/2$. Similarly we have for the second degree fine structure parameter in strong magnetic field $(g\beta H \gg D'')$,

$$D'' = D\{(3g_\parallel^2/g^2)\cos^2\theta - 1\}$$

from which

$$\partial D''/\partial \theta = -2D(3g_\parallel^2/g^2)\sin\theta \cos\theta, \qquad (3.95)$$

which again vanishes at $\theta = 0$ and $\pi/2$. For fine structure splittings of higher degree there may be intermediate positions where the splitting parameter is independent of small angular variations. For example, in a cubic field the splitting parameter $p = 1 - 5(l^2m^2 + m^2n^2 + n^2l^2)$ has stationary values (independent of small variations in three mutually perpendicular directions) along the fourfold, threefold, and twofold axes, as can be verified from the formulae in § 3.4. Hence we would expect spreading in the resonance line to vanish when **H** is along any of these axes, as far as small random variation in crystallite orientation is concerned. This has been verified by Shaltiel and Low (1961) for Gd^{3+} in ThO_2, where the splitting is predominantly due to the fourth-degree cubic field parameter (in fact the splitting due to the sixth-degree parameter has extrema along the same axes).

Whatever the cause of the variation in the fine structure parameters, the resulting line widths in strong external magnetic fields are largest for lines furthest from the centre of the spectrum, being just proportional to the distance from the centre (the same is true for hyperfine lines). Since the fine structure parameters involve only even powers of the magnetic quantum number M, a transition such as $|M\rangle \leftrightarrow |-M\rangle$ is not broadened in first order, though it may be in second or higher order. Thus for an ion with half-integral spin, the 'allowed' transition $|+\tfrac{1}{2}\rangle \leftrightarrow |-\tfrac{1}{2}\rangle$ is not broadened in first order, nor are 'forbidden' transitions such as $|+\tfrac{3}{2}\rangle \leftrightarrow |-\tfrac{3}{2}\rangle$; for integral spin this holds for a 'forbidden' transition such as $|+1\rangle \leftrightarrow |-1\rangle$. In either case $\Delta M = \pm 1$ transitions (except $|+\frac{1}{2}\rangle \leftrightarrow |-\frac{1}{2}\rangle$) are broadened in first order, and the variation in line width for different transitions within the spectrum gives a measure of the variation in the fine (or hyperfine) structure parameters.

Attempts have been made (see Feher 1964, McMahon 1964, Stoneham 1966) to relate the excess line width (which is not necessarily isotropic even in a cubic crystal such as MgO) to strain fields within the crystal. When present, excess line width due to variations in the fine structure parameters is generally larger than that due to variation in the g-value, though both are strongly dependent on the amount of orbital momentum in the ground state. Representative values of the excess width (which may vary of course from crystal to crystal) are given in Table 3.12, for the ions Mn^{2+}, Fe^{2+}, Ni^{2+} in MgO. In the latter two cases the

TABLE 3.12

Measured line widths for three ions in MgO, showing the variation between different lines for each ion, and the large variation from ion to ion. Since the excess line width is due to strains, variations between crystals are to be expected, and the values given here must only be regarded as representative order of magnitude results

Ion	g	Transition	Line width (G)	Reference
$Mn^{2+}, 3d^5, S=\frac{5}{2}$	2·0014	$\|+\frac{1}{2}\rangle \leftrightarrow \|-\frac{1}{2}\rangle$	0·7	Feher (1964)
		$\|\pm\frac{3}{2}\rangle \leftrightarrow \|\pm\frac{1}{2}\rangle$	0·9 to 2·4	
		$\|\pm\frac{5}{2}\rangle \leftrightarrow \|\pm\frac{3}{2}\rangle$	1·4 to 4·5	
$Ni^{2+}, 3d^8, S=1$	2·1728	$\|\pm 1\rangle \leftrightarrow \|0\rangle$	50	Orton et al. (1960a)
		$\|+1\rangle \leftrightarrow \|-1\rangle$	3	Smith et al. (1969)
$Fe^{2+}, 3d^6, \tilde{S}=1$	3·428	$\|\pm 1\rangle \leftrightarrow \|0\rangle$	10^3	McMahon (1964)
		$\|+1\rangle \leftrightarrow \|-1\rangle$	20	

$|+1\rangle \leftrightarrow |-1\rangle$ transition (which would be forbidden in the absence of strains) is much narrower than the $|\pm 1\rangle \leftrightarrow |0\rangle$ transitions; coincident with the latter a narrow line has also been observed which is attributed to a 'double quantum' transition. This is essentially a $|+1\rangle \leftrightarrow |-1\rangle$ transition (hence its small line width) which involves the absorption or emission of two quanta instead of one. Thus it occurs at the point where the $|\pm 1\rangle \leftrightarrow |0\rangle$ transitions would occur in the absence of strain, whereas the ordinary $|+1\rangle \leftrightarrow |-1\rangle$ 'forbidden' single quantum transition occurs at twice the frequency at constant field strength, or at half the field strength at constant frequency.

3.14. The weak-field Zeeman interaction for non-Kramers ions

Specific formulae for the weak-field Zeeman effect, where the Zeeman splitting is small compared with fine structure splittings, have not so far been considered except in the cases (see § 3.4) of ions with $S = \frac{5}{2}, \frac{7}{2}$ in a cubic field. In a ligand field of axial symmetry a manifold of $(2S+1)$ states will split into $(S+\frac{1}{2})$ doublets if S is half-integral, or S doublets and one singlet if S is integral. If the ligand field splitting is large compared with the Zeeman (and other, such as hyperfine) interactions, each doublet can be treated using an effective spin Hamiltonian with an effective spin $\tilde{S} = \frac{1}{2}$. For a set of Kramers doublets this was illustrated in § 1.8 by the example of a $4f^1$ ion in its ground state $J = \frac{5}{2}$. The position for non-Kramers ions is rather different, and we first illustrate the problem by means of another simple example.

Assume we have an ion with $S = 2$, subject to a strong axial splitting represented by a term $D\{S_z^2 - \frac{1}{3}S(S+1)\}$. Then, if a magnetic field **H** is applied at an angle θ to the symmetry axis (the z-axis) in the zx-plane, the spin Hamiltonian is

$$\mathcal{H} = D\{S_z^2 - \tfrac{1}{3}S(S+1)\} + g_\parallel \beta H S_z \cos\theta + g_\perp \beta H S_x \sin\theta. \quad (3.96)$$

Under the conditions $g\beta H \ll D$, the energy levels and states are in first order (see Fig. 3.24(a) and (b))

$$|\pm 2\rangle \quad W_{\pm 2} = +2D \pm 2g_\parallel \beta H \cos\theta,$$
$$|\pm 1\rangle \quad W_{\pm 1} = -D \pm g_\parallel \beta H \cos\theta,$$
$$|0\rangle \quad W_0 = -2D,$$

with no allowed transition in either doublet. In second-order perturbation theory the calculation is complicated because the states $|\pm 1\rangle$, which are degenerate when $\cos\theta = 0$, each has a matrix element of $g_\perp H S_x \sin\theta$ with the state $|0\rangle$. Similar difficulties arise if a small rhombic splitting term $E(S_x^2 - S_y^2) = \frac{1}{2}E(S_+^2 + S_-^2)$ is present; not only does this split the $|\pm 1\rangle$ states in first order, but it also splits the $|\pm 2\rangle$ states in second order through their matrix element with $|0\rangle$. A useful trick in such cases is to use as the basis states $|2^s\rangle, |2^a\rangle, |1^s\rangle, |1^a\rangle$ which are simple linear combinations of the form

$$|2^s\rangle = \frac{1}{\sqrt{2}}\{|+2\rangle + |-2\rangle\}, \quad |2^a\rangle = \frac{1}{\sqrt{2}}\{|+2\rangle - |-2\rangle\}$$

The energy matrix and energy levels for the spin Hamiltonian (3.96) together with the term $E(S_x^2 - S_y^2)$ are given in Table 3.13, correct to

210   THE SPIN HAMILTONIAN AND THE SPECTRUM

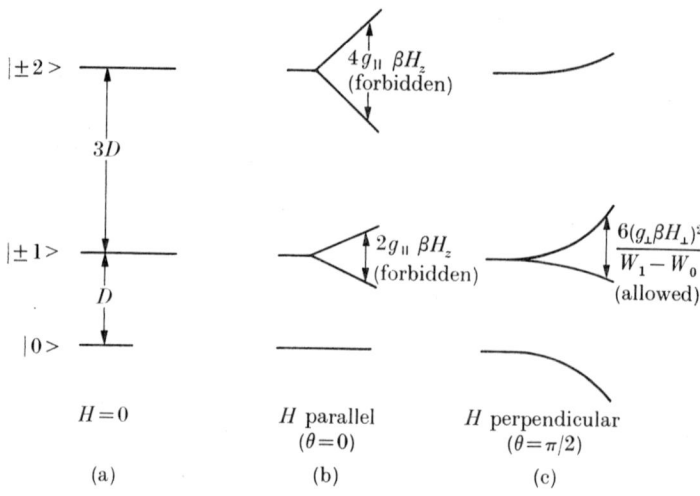

FIG. 3.24. Energy levels corresponding to $S = 2$ for the spin Hamiltonian (3.96) with (a) $H = 0$, (b) H parallel to z-axis (the symmetry axis), (c) H perpendicular to the z-axis.

second order in perturbation theory. This shows the following:

$E = 0$.

No allowed transition within the $|\pm 2\rangle$ doublet, but for the $|\pm 1\rangle$ doublet each state is an admixture of the form $|+1, 0, -1\rangle$ for an arbitrary value of θ, and a transition is weakly allowed except when **H** is along the z-axis ($\theta = 0$) through the second-order Zeeman effect, as shown in Fig. 3.24(c). (The splitting of the $|\pm 1\rangle$ doublet states due to the second-order Zeeman effect can also be seen in Fig. 1.7 for $S = 1$.)

$E \neq 0$.

The states are again admixed, but both for $|\pm 1\rangle$ and for $|\pm 2\rangle$. However, if either $\theta = 0$ or we can neglect the small splitting of order $(g_\perp \beta H \sin \theta)^2/(W_1 - W_0)$ of the $|\pm 1\rangle$ states, we have for both doublets energy levels of the form (see Fig. 3.25)

$$W_\pm = \pm \tfrac{1}{2}\{(\tilde{g}_\parallel \beta H \cos \theta)^2 + \Delta^2\}^{\tfrac{1}{2}}, \qquad (3.97)$$

where for the $|\pm 2\rangle$ doublet $\tilde{g}_\parallel = 4g_\parallel$, $\Delta_2 = 12E^2/(W_2 - W_0)$, and for the $|\pm 1\rangle$ doublet $\tilde{g}_\parallel = 2g_\parallel$, $\Delta_1 = 6E$, neglecting small corrections to $\tilde{g}_\parallel$ that arise from admixtures of the other states. In each case the states can be written in the form

$$\begin{aligned}|+'\rangle &= \cos \alpha \, |+\rangle + \sin \alpha \, |-\rangle, \\ |-'\rangle &= \sin \alpha \, |+\rangle - \cos \alpha \, |-\rangle, \end{aligned} \qquad (3.98)$$

TABLE 3.13

Energy matrix and energy levels for $S = 2$ with the spin Hamiltonian
$$\mathcal{H} = D\{S_z^2 - \tfrac{1}{3}S(S+1)\} + E(S_x^2 - S_y^2) + g_\parallel \beta H S_z \cos\theta + g_\perp \beta H S_x \sin\theta.$$
The perturbation denominators are written $(W_2 - W_1)$, etc., because this indicates the origin of the second-order terms and allows the displacement of the energy levels by a diagonal fourth degree term of the form $B_4^0 O_4^0$ to be easily included if present

Energy matrix

	$\|2^s\rangle$	$\|2^a\rangle$	$\|1^s\rangle$	$\|1^a\rangle$	$\|0\rangle$
$\|2^s\rangle$	$2D$	$2g_\parallel \beta H \cos\theta$	$g_\perp \beta H \sin\theta$	0	$\sqrt{(12)}E$
$\|2^a\rangle$	$2g_\parallel \beta H \cos\theta$	$2D$	0	$g_\perp \beta H \sin\theta$	0
$\|1^s\rangle$	$g_\perp \beta H \sin\theta$	0	$-D+3E$	$g_\parallel \beta H \cos\theta$	$\sqrt{(3)} g_\perp \beta H \sin\theta$
$\|1^a\rangle$	0	$g_\perp \beta H \sin\theta$	$g_\parallel \beta H \cos\theta$	$-D-3E$	0
$\|0\rangle$	$\sqrt{(12)}E$	0	$\sqrt{(3)} g_\perp \beta H \sin\theta$	0	$-2D$

Energy levels

$$W_{\pm 2} = 2D + \frac{(g_\perp \beta H \sin\theta)^2}{W_2 - W_1} + \tfrac{1}{2}\Delta_2 \pm \{(2g_\parallel \beta H \cos\theta)^2 + (\Delta_2/2)^2\}^{\frac{1}{2}},$$

$$W_{\pm 1} = -D + \frac{(g_\perp \beta H \sin\theta)^2}{W_1 - W_2} + \frac{3(g_\perp \beta H \sin\theta)^2}{2(W_1 - W_0)} \pm$$
$$\pm \left[(g_\parallel \beta H \cos\theta)^2 + \left\{3E + \frac{3(g_\perp \beta H \sin\theta)^2}{2(W_1 - W_0)}\right\}^2\right]^{\frac{1}{2}}$$

$$W_0 = -2D + \frac{12E^2}{W_0 - W_2} + \frac{3(g_\perp \beta H \sin\theta)^2}{W_0 - W_1}$$

where $\quad \Delta_2 = \dfrac{12E^2}{W_2 - W_0}.$

where $\tan 2\alpha = \Delta/(\tilde{g}_\parallel \beta H \cos\theta)$, and $|+\rangle$, $|-\rangle$ are the states in the limit $(\Delta/H) = 0$, while in the opposite limit of $(H/\Delta) = 0$ the states are $(1/\sqrt{2})\{|+\rangle \pm |-\rangle\}$. The states (3.98) have a matrix element linking them of $\langle +'| \tilde{g}_\parallel \beta S_z | -' \rangle = \tfrac{1}{2}\tilde{g}_\parallel \beta \sin 2\alpha$, so that transitions are allowed between the two levels for an oscillating magnetic field along the z-axis whose probability is given by (cf. eqn (2.51))

$$|\mu_{ij}|^2 = |\mu_z|^2 = \tfrac{1}{4}\tilde{g}_\parallel^2 \beta^2 \frac{\Delta^2}{(\tilde{g}_\parallel \beta H \cos\theta)^2 + \Delta^2}$$

$$= \tfrac{1}{4}\tilde{g}_\parallel^2 \beta^2 \frac{\Delta^2}{(\hbar\omega)^2} \tag{3.99}$$

and which require a quantum

$$\hbar\omega = h\nu = \{(\tilde{g}_\parallel \beta H \cos\theta)^2 + \Delta^2\}^{\frac{1}{2}}. \tag{3.100}$$

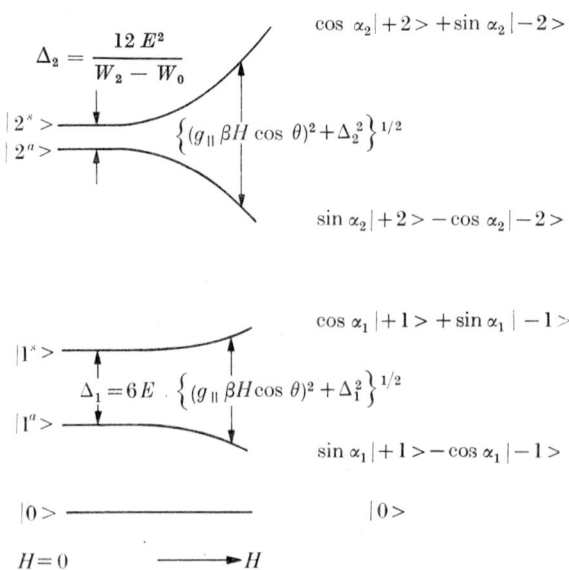

Fig. 3.25. Energy levels and states corresponding to Table 3.13, but neglecting terms of order $(g\beta H)^2/\Delta W$, where ΔW equals W_2-W_1 or W_1-W_0. The states in zero magnetic field are shown on the left, the modified states in a field **H**, with a component $H\cos\theta$ along the z-axis, on the right.

An example of such a transition occurs in the ground doublet ($S_z = \pm 2$) of Fe^{2+}, $3d^6$, $S = 2$ in ZnF_2 which has rhombic symmetry (Tinkham 1956a). Such transitions are allowed transitions, though from eqn (3.99) their probability diminishes with the inverse square of the frequency, and the line shape is normal. The value of Δ will obviously be larger for the $|\pm 1\rangle$ doublet than for the $|\pm 2\rangle$ doublet, depending on E in the former case and on $E^2/(W_2-W_0)$ in the second case. These conclusions are not substantially altered by the presence of fine-structure terms of the fourth degree, such as $B_4^0 O_4^0 + B_4^2 O_4^2 + B_4^4 O_4^4$. The first of these is diagonal, the second has matrix elements between states differing by 2 in S_z and thus behaves similarly to

$$B_2^2 O_2^2 = \tfrac{1}{2}B_2^2(S_+^2 + S_-^2) = E(S_x^2 - S_y^2),$$

while the last connects states differing by 4 in S_z and thus splits the $|\pm 2\rangle$ levels in first order.

When $E = 0$, and H is zero, we are in the anomalous position that if we take the doublet states as $|\pm 2\rangle$, $|\pm 1\rangle$ there appear to be no allowed transitions within either doublet, but if we take any linear combinations (which should be equally valid as basis states) such as

$|2^s\rangle$, $|2^a\rangle$, $|1^s\rangle$, $|1^a\rangle$ for which there are matrix elements

$$\langle 2^s| S_z |2^a\rangle = 2\langle 1^s| S_z |1^a\rangle$$

we expect to find allowed transitions if the oscillatory magnetic field is along the z-axis. In practice there will always be some perturbation, such as a component H_z due to the external or the local field, or a deviation from axial symmetry, which lifts the degeneracy and gives properly defined states.

In this situation strains that upset the axial symmetry (e.g. strains corresponding to $E \neq 0$) play a special role for non-Kramers doublets. In the absence of any strain, and neglecting second and higher-order Zeeman effects, we see that a field **H** with a component H_z along the z-axis splits such a doublet, giving states such as $|\pm 2\rangle$, $|\pm 1\rangle$ with no allowed transitions, corresponding to $\tilde{g}_\perp = 0$ in the effective spin Hamiltonian with $\tilde{S} = \tfrac{1}{2}$ for each doublet. A rhombic strain will give rise to a small splitting Δ when $H = 0$, and even for a finite H_z the states will be of the form (3.98) for which weak transitions are allowed. If the strains are random, we do not expect sharp energy levels and the resonance line will be broadened. A suitable spin Hamiltonian under these conditions (see § 18.5) is, with $\tilde{S} = \tfrac{1}{2}$ for each doublet,

$$\mathcal{H} = \tilde{g}_\| \beta H_z \tilde{S}_z + \Delta_x \tilde{S}_x + \Delta_y \tilde{S}_y \qquad (3.101)$$

for which the energy levels and transition frequency are given by eqns (3.97), (3.100) and the transition probability by (3.99), with

$$\Delta^2 = \Delta_x^2 + \Delta_y^2.$$

The resonance line lies wholly on the high-frequency side of $h\nu = \tilde{g}_\| \beta H_z$ at constant H, or on the low-field side of $H_0 = (h\nu/\tilde{g}_\| \beta)$ at constant frequency, the displacement being determined by the value of Δ^2. If the line shape is dominated by the spread in Δ^2 rather than by any other source of line width, we have, in place of eqn (2.51), for an oscillatory field $(H_1)_\|$ along the z-axis

$$w \, d(\Delta^2) = \frac{\pi}{2\hbar^2} \{|\mu_z|^2 (H_1)_\|^2\} P(\Delta^2) \, d(\Delta^2) \qquad (3.102)$$

where $P(\Delta^2)$ is the normalized probability distribution in Δ^2. If we assume a random Gaussian distribution for Δ_x, Δ_y we have

$$P(\Delta^2) \, d(\Delta^2) = \exp\left\{-\frac{\Delta^2}{\Delta_0^2}\right\} d\left(\frac{\Delta^2}{\Delta_0^2}\right), \qquad (3.103)$$

where Δ_0^2 is the most probable value of Δ^2. On inserting eqn (3.103) into (3.102) and using eqn (3.99) we see that the transition rate is zero for $\Delta = 0$ and for $\Delta = \infty$, with an asymmetrical line shape. If the resonance is observed at constant frequency and variable field, and $\Delta_0 \ll (\hbar\omega)$, we can write $H_z = H_0 - h$ where

$$\frac{\Delta^2}{(\hbar\omega)^2} = \frac{2\tilde{g}_\| \beta h}{(\hbar\omega)}, \qquad (3.104)$$

and instead of a frequency distribution we have

$$P(\Delta^2)\, \mathrm{d}(\Delta^2) = \exp(-h/h_0)\, \mathrm{d}(h/h_0), \qquad (3.105)$$

where $\tilde{g}_\| \beta h_0 = \tfrac{1}{2}\Delta_0^2/(\hbar\omega)$. Then the transition rate is, as a function of field strength,

$$w(h)\, \mathrm{d}h = \frac{\pi}{2\hbar^2}\{\tfrac{1}{2}\tilde{g}_\| \beta (H_1)_\|\}^2 \left(\frac{2\tilde{g}_\| \beta h}{\hbar\omega}\right)\exp\left(-\frac{h}{h_0}\right)\mathrm{d}\left(\frac{h}{h_0}\right), \qquad (3.106)$$

for which a typical line shape is shown in Fig. 3.26(a), with maximum intensity at $h = h_0$.

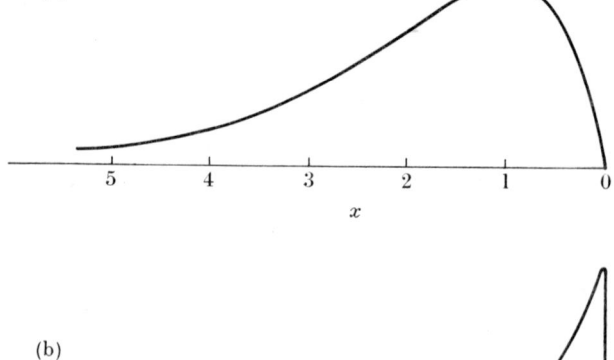

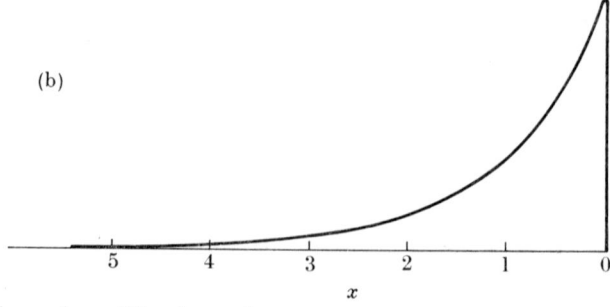

Fig. 3.26. Comparison of line shapes for resonance in a non-Kramers doublet (a) due to an oscillating magnetic field $(H_1)_\|$ along the symmetry axis, (b) due to an oscillating electric field $(E_1)_\perp$ normal to the symmetry axis. The quantity $x = (h/h_0)$, where h, h_0 are defined in the text; $h = H_0 - H_z = (\hbar\omega/\tilde{g}_\| \beta) - H_z$, so that increasing h (or x) means decreasing field strength H_z at constant frequency. The units for intensity are arbitrary and not the same in (a) and (b).

If the paramagnetic ions occupy sites whose point symmetry is such that there are one or more planes that are not planes of reflection symmetry, there may be an electric dipole moment normal to any such plane (see § 15.10). For a non-Kramers ion in a ligand field of axial symmetry, the important term in the (anisotropic) interaction of such an electric dipole moment with an applied electric field, which we may write in the form

$$\mathcal{H}_E = \beta(\tilde{\mathbf{S}} \cdot \mathbf{g}^{(E)} \cdot \mathbf{E}) \tag{3.107}$$

where $\mathbf{E}$ is the local applied field (which may be different from the external field), will be

$$g_\perp^{(E)} \beta(\tilde{S}_x E_x + \tilde{S}_y E_y). \tag{3.108}$$

Such a term is allowed if the site symmetry lacks reflection symmetry in planes containing the z-axis, and is thus possible, for example, for C_{3h} symmetry but not for C_{3v}. If an oscillatory electric field $(E_1)_\perp$ is applied normal to the z-axis, transitions can be induced within the doublet, and instead of (3.102) we have for the transition rate

$$w \, d(\Delta^2) = \frac{\pi}{2\hbar^2}\{|\mu_x|^2 + |\mu_y|^2\}(E_1)_\perp^2 P(\Delta^2) \, d(\Delta^2) \tag{3.109}$$

where

$$|\mu_x|^2 + |\mu_y|^2 = (\tfrac{1}{2}g_\perp^{(E)}\beta)^2\left\{1 - \left(\frac{\Delta}{\hbar\omega}\right)^2\right\} \tag{3.110}$$

This differs from (3.99) in the important respect that it is a maximum at $\Delta^2 = 0$, leading to a different line shape in which maximum intensity occurs at the undisturbed frequency $\hbar\omega = \tilde{g}_\parallel \beta H_0$, since a strain ($\Delta^2 \neq 0$) is no longer needed to produce an allowed transition. Under these circumstances we have instead of eqn (3.106) a transition rate as a function of magnetic field strength $h = H_0 - H_z$ at constant frequency given by

$$w(h) \, dh = \frac{\pi}{2\hbar^2}\{\tfrac{1}{2}g_\perp^{(E)}\beta(E_1)_\perp\}^2\left\{1 - \left(\frac{h}{H_0}\right)\right\}^2 \exp\left(-\frac{h}{h_0}\right) d\left(\frac{h}{h_0}\right) \tag{3.111}$$

The line shape as a function of h when transitions are due to a perpendicular oscillatory electric field only is shown in Fig. 3.26(b), the maximum intensity now falling at $h = 0$.

In a resonant cavity at microwave frequencies a sample can be placed in a position of maximum $(H_1)_\parallel$ or of maximum $(E_1)_\perp$, and (if sufficiently small) it can experience one field without the other. The results of an experiment (Williams (1967)) on yttrium ethylsulphate

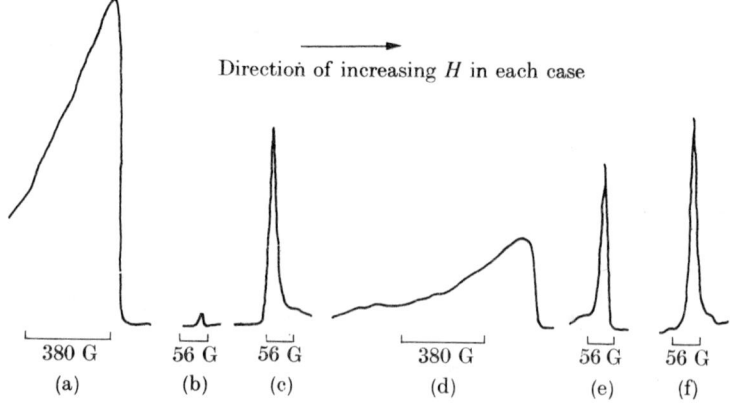

FIG. 3.27. Comparison of absorption intensities due to oscillatory 'electric' and 'magnetic' fields for the non-Kramers doublet ground state of Pr^{3+} in yttrium ethylsulphate. The purely 'magnetic' resonance of Nd^{3+} ions in the sample and in another crystal is used as a reference signal.

(a) Pr^{3+}
(b) Nd^{3+} } in sample at position of maximum $(E_1)_\perp$ and minimum $(H_1)_\parallel$;
(c) Nd^{3+} in second crystal used as reference signal for (a), (b);
(d) Pr^{3+}
(e) Nd^{3+} } in sample at position of minimum $(E_1)_\perp$ and maximum $(H_1)_\parallel$;
(f) Nd^{3+} in second crystal used as reference signal for (d), (e).
(After Williams (1967).)

(symmetry C_{3h}) containing small amounts of Pr^{3+} and Nd^{3+} ions, whose ground states are a non-Kramers and a Kramers doublet respectively, are shown in Fig. 3.27. This shows from the comparison of intensities that the transition in the non-Kramers doublet is primarily due to $(E_1)_\perp$ rather than to $(H_1)_\parallel$, with a corresponding line shape in which the maximum is close to $h = 0$.

4

ELECTRON-NUCLEAR DOUBLE RESONANCE (ENDOR)

4.1. Introduction

IN hydrated crystals such as many paramagnetic salts of the iron group, the magnetic ion is surrounded by a number of water molecules. The line width of the electron spin resonance spectrum, even in a highly diluted specimen at such low temperatures that the spin lattice relaxation time is too long to give an appreciable contribution, is of order 10 G, or about 30 MHz for $g = 2$. This line width is due mainly to the local magnetic fields of the proton nuclear moments in the water molecules. In a deuterated crystal this can be reduced by a factor of about $\frac{1}{3}$, because of the smaller nuclear magnetic moment of the deuteron. In oxides such as MgO, CaO, or ThO_2, where only a minority of the nuclei have magnetic moments, the line width may be about 1 G or smaller (dependent on crystal imperfections), corresponding to a few MHz in frequency units. This residual line width sets a limit to the accuracy with which measurements of the position of the line may be made, which in very round terms we may take to be of the order of 1 MHz.

In the determination of a hyperfine structure parameter from a spin resonance spectrum, measurements must be made of the positions of successive lines, and the parameter is deduced from the differences in these positions. When anisotropy is present, it is best to make measurements along each of the principal axes; not only does this give directly the 'diagonal' components of any 'tensor' interaction, but the positions of the spin resonance lines are then passing through extrema, and small inaccuracies in orientation of the external magnetic field are least important. The accuracy with which the magnetic hyperfine parameter A can be determined is then limited by the considerations of line width set out above. The other hyperfine parameters such as a nuclear electric quadrupole interaction or the nuclear Zeeman interaction with the external magnetic field are more difficult to determine because they are usually rather small, and their presence is revealed in the electron spin resonance spectrum only if the simple selection rule $\Delta m = 0$ is broken down. In many cases this means that measurements must be made with the external field at an angle to the principal axes of the hyperfine 'tensor', in order that otherwise 'forbidden' lines can be observed. Even in cases where this is possible, the accuracy of such measurements is rather restricted, and the determination of signs of the parameters is, to say the least, awkward.

It is true that if measurements are made in zero or rather small magnetic field, transitions of the type $\Delta m \neq 0$ have matrix elements comparable with those for which $\Delta m = 0$ (cf. § 3.9 and Fig. 3.12), and have been observed, for example, by Bleaney, Scovil, and Trenam (1954). Such transitions have line widths comparable with those quoted above, and they generally occur in frequency regions where the sensitivity of a magnetic resonance spectrometer is inherently much lower, with the additional drawback of having to sweep over a wide frequency range.

The accuracy of such hyperfine measurements, and the total information content, would be greatly improved if nuclear transitions could be observed **directly,** particularly in strong external magnetic fields where the nuclear Zeeman interaction is not unduly small. Unfortunately the matrix elements for such transitions involve (in the strong field limit of complete decoupling) only the nuclear magnetic moment, and the transitions are correspondingly weak, though there is some enhancement in practice of the oscillatory field at the nuclear frequency (see below and § 4.3). In this chapter we describe a technique by which these problems are solved. In §§ 4.2–4.8 we shall be primarily concerned with hyperfine interactions with the nuclear moments of the paramagnetic ion itself; these are usually associated with a resolved hyperfine structure in the electron resonance spectrum, and the approximation $A \gg g_I \beta H$ is generally valid. They can also be applied to unresolved hyperfine structure due to ligand nuclei (the cause of the line widths mentioned above); the relevant hyperfine parameters are then usually comparable with $g_I \beta H$. This rather different case is discussed in more detail in § 4.9.

The difficulties are overcome by the method of Electron-Nuclear DOuble Resonance, or Endor, invented by Feher (1956b). The method has been outlined in § 1.13, and here we illustrate the principles briefly by reference to a system with $S = \frac{1}{2}$, $I = \frac{1}{2}$. In the absence of anisotropy, the spin Hamiltonian is

$$\mathscr{H} = g\beta(\mathbf{H} \cdot \mathbf{S}) + A(\mathbf{S} \cdot \mathbf{I}) - g_I \beta(\mathbf{H} \cdot \mathbf{I}). \tag{4.1}$$

In a strong magnetic field ($g\beta H \gg A \gg g_I \beta H$) the levels are approximately as shown in Fig. 4.1. The electronic states $M = +\frac{1}{2}$ lie above

ELECTRON-NUCLEAR DOUBLE RESONANCE (ENDOR) 219

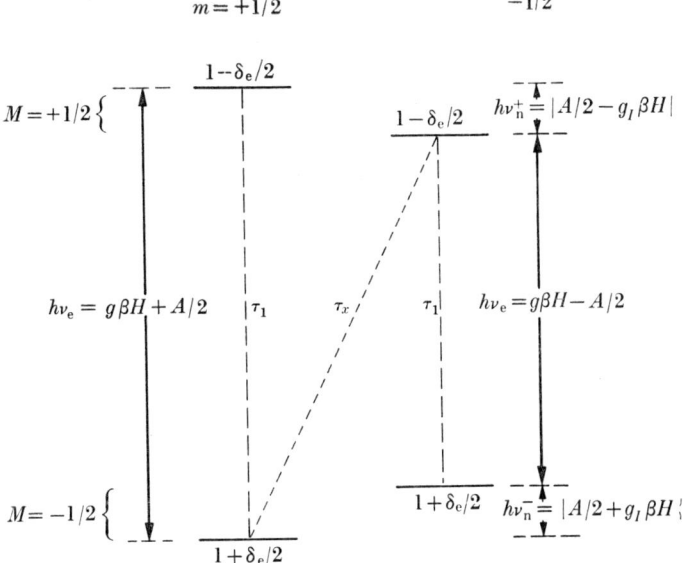

FIG. 4.1. The energy level scheme for $S = I = \frac{1}{2}$, showing the level populations in thermal equilibrium in the approximation where $\delta_e = (h\nu_e/kT) \ll 1$, and $\delta_n = (h\nu_n/kT)$ is neglected. The relaxation paths are indicated by broken lines, the vertical paths being the more rapid ($\tau_1 \ll \tau_x$); other, slower, relaxation paths are omitted.

the $M = -\frac{1}{2}$ states by about $g\beta H$, while the nuclear states with the same value of M are separated by $(\pm A/2 - g_I \beta H)$, if we ignore for the present terms of order $A^2/g\beta H$.

In thermal equilibrium the populations are approximately $1 \pm \delta_e/2$, where $\delta_e = (g\beta H/kT) = (h\nu_e/kT)$, provided that $\delta_e \ll 1$. Here $h\nu_e$ is the quantum required for electron magnetic resonance, and for the $\Delta m = 0$ transitions

$$h\nu_e = |g\beta H \pm A/2|, \tag{4.2}$$

but we ignore the correction to δ_e due to the term in A. This is equivalent to ignoring the small differences in population of order $\delta_n = (h\nu_n/kT)$, where $h\nu_n$ is the quantum required for a nuclear magnetic resonance transition, which is

$$\begin{aligned} h\nu_n^+ &= |+A/2 - g_I\beta H| \\ h\nu_n^- &= |-A/2 - g_I\beta H| \end{aligned} \tag{4.3}$$

for the $M = +\frac{1}{2}$, $m = +\frac{1}{2} \leftrightarrow -\frac{1}{2}$ and $M = -\frac{1}{2}$, $m = +\frac{1}{2} \leftrightarrow -\frac{1}{2}$ transitions respectively. The population differences are maintained by means of relaxation to the lattice through the paths marked by broken lines, the vertical paths being usually the more rapid in action.

If only a weak electron magnetic resonance driving field is applied to either of the $\Delta M = \pm 1$, $\Delta m = 0$ transitions, the populations will remain substantially unaltered; this is also true if we apply simultaneously a driving field at the nuclear magnetic resonance frequency, and since this does not appreciably affect the intensity of the electron magnetic resonance signal, the nuclear transition will remain undetected, being too weak to be observed directly. If, instead, a strong magnetic resonance driving field is applied to either of the $\Delta M = \pm 1$, $\Delta m = 0$ transitions, the difference between the populations of the two levels involved will be reduced and, in the limit of complete saturation, they will become equal. If the relaxation time for the sloping relaxation path is τ_x, the populations of the two $m = -\frac{1}{2}$ states will remain unaltered for a time short compared with τ_x if the $m = +\frac{1}{2}$ transition is saturated; and vice versa (see Fig. 4.2(a)). Thus we have a population

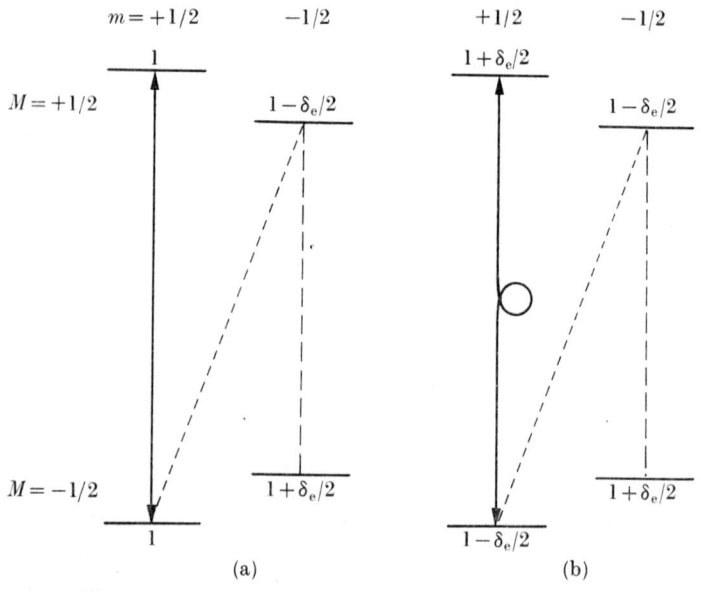

FIG. 4.2. The level populations in the system $S = I = \frac{1}{2}$, after (a) saturation of the electronic $(+\frac{1}{2}, +\frac{1}{2}) \leftrightarrow (-\frac{1}{2}, +\frac{1}{2})$ transition, (b) adiabatic rapid passage through the same transition, in each case before the populations of the $m = -\frac{1}{2}$ levels are affected through the sloping relaxation path. In (a) the populations of the two levels involved in the electronic magnetic resonance are equalized; in (b) the populations are inverted.

difference of order δ_e between adjacent nuclear levels $m = \pm\frac{1}{2}$ with the same value of M, where formerly the population difference was (in our approximation) zero. Under these conditions the effect of applying a nuclear magnetic resonance driving field to the $\Delta M = 0$,

$\Delta m = \pm 1$ transitions is enhanced, though not usually enough to make them directly detectable. However, if this driving field is strong enough to induce nuclear transitions at a rate faster than $1/\tau_x$, the population of the two $m = +\frac{1}{2}$ levels will be appreciably altered, with a resultant change in the intensity of the electron magnetic resonance absorption between these two levels. Obviously the effect is greatest when the nuclear driving field is at either of the resonant values given by eqn (4.3), so that we have a method of determining these two frequencies (and hence measuring A and $g_I \beta H$) through the effect on the electron paramagnetic resonance signal, with a sensitivity comparable with that in a simple electron paramagnetic resonance experiment.

When the relaxation times are exceptionally long, as in the original experiments of Feher on donors in silicon, it may be more convenient to use the technique of adiabatic rapid passage. This results in inversion, instead of equalization, of the populations of the two levels involved, as shown in Fig. 4.2(b). Clearly the situation as regards the difference in population between adjacent nuclear levels resembles that in Fig. 4.2(a) (actually the differences are twice as large), so that application of the appropriate nuclear resonance frequency will again affect the intensity of the electron resonance signal, as measured either by application of a weak driving field at the electron resonance frequency, or by a further traverse of the electronic line under adiabatic rapid passage conditions.

The major advance in an Endor type experiment is that, not only does it enable us to measure the hyperfine interaction constants and the nuclear Zeeman interaction directly, but it also gives them with much higher precision. In fact much greater (fractional) precision may be obtained in the measurement of A in an Endor experiment than of g in a simple electron paramagnetic resonance experiment. As indicated in § 1.13, this is possible only when we are dealing with inhomogeneously broadened lines. We shall not at this point repeat the reasons for the high precision obtainable when this requirement is satisfied (see § 1.13), believing that the experimental results reported later in this chapter will serve as an adequate demonstration. Essentially it stems from the fact that the nuclear resonance is taking place in the electronic field H_e (see § 1.2), which is of order 10^5 to 10^6 G, rather than just the applied field H, but the line width may still be only a few gauss.

This large electronic field also assists in enhancing the nuclear transition rate w_n. As indicated above, this rate must be at least comparable with $1/\tau_x$ in order to affect the level populations sufficiently to produce an observable change in intensity in the electron paramagnetic resonance signal. For a nucleus with $g_I \sim 10^{-3}$, and an r.f. field amplitude $H_1 = 1$ G, we have $w_n \sim 10^3$ s^{-1} if the Endor line width is 10 kHz, and 10^2 s^{-1} if it is 100 kHz; this is sufficient only if τ_x is greater than 10^{-3} or 10^{-2} s respectively, so that any enhancement in w_n may be very important. The enhancement produced by the electronic field H_e at the nucleus may be understood classically from Fig. 4.3.

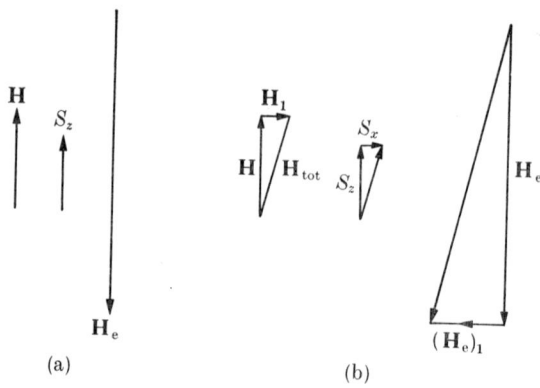

FIG. 4.3. (a) Direction of a steady magnetic field **H** along the z-axis, the component S_z of the electron spin, and the electronic field at the nucleus $H_e = -(AS_z/g_I\beta)$; (b) when a low-frequency oscillatory field H_1 is applied normal to **H**, both the electron spin and the electronic field H_e follow the resultant external field H_{tot}, setting up an additional oscillatory field $(H_e)_1$ at the nucleus, which is larger than H_1 by a factor $|(H_e/H)|$.

Suppose the steady external field **H** is applied in the z-direction; then, assuming no anisotropy, the electronic magnetic moment corresponding to a given value of S_z is also directed along the z-axis, and this sets up an electronic field H_e at the nucleus, corresponding to the term $AS_zI_z = -g_I\beta H_eI_z$, which is also along the z-axis, but oppositely directed to S_z if (A/g_I) is positive. The application of an oscillatory field H_1 normal to the z-axis gives a resultant field H_{tot} inclined at an angle $\theta \sim (H_1/H)$ (assumed small) to the z-axis. If H_1 is a rotating field, then H_{tot} rotates about the z-axis; if H_1 is a linearly polarized field along (say) the x-axis, then H_{tot} oscillates about the z-axis in the zx-plane. Provided that H_1 oscillates at a frequency low compared with the electronic Larmor frequency (see § 2.4), the direction of the electronic magnetization will follow H_{tot}, setting up an oscillatory field $(H_e)_1$ at the nucleus, as in Fig. 4.3(b). Although $(H_e)_1$ is here oppositely directed to H_1, it is of order $\theta H_e = (H_1/H)H_e = H_1(H_e/H)$, and so may be as much as 100 times or more larger than the externally applied field

H_1. As can be seen, the degree of enhancement is proportional to H_e and hence to $S_z = M$, and thus is largest for the largest values of $|M|$, vanishing for $M = 0$.

This enhancement process is simply a form of 'paramagnetic antishielding' (equivalent to an enormous 'chemical shift'), and is similar to that which occurs *within* a domain in a ferromagnetic substance. In a paramagnetic substance the situation may be complicated through the presence of anisotropy. A quantum-mechanical analysis is given in § 4.3.

In the following section (§ 4.2) we carry out a somewhat tedious but necessary analysis of the information that can be obtained from Endor measurements. Also, § 4.3 contains a rather detailed analysis of the enhancement of the nuclear transition probability, and the reader who wishes to obtain a general picture of the power of the Endor method may prefer to read first the outline of the elegant experiments of Feher, given in §§ 4.4 and 4.5.

4.2. The Endor spectrum

In general, the accuracy of Endor measurements is such that they must be fitted to a numerical solution of the spin Hamiltonian evaluated by means of a computer—approximate methods of solution using perturbation theory will not suffice. However, to discover what information can be obtained from the Endor spectrum it is convenient to make use of the formulae obtained in the previous chapter for the energy levels, using perturbation theory. An important question is—what information do the experiments give about the signs of the parameters in the spin Hamiltonian, as well as their numerical magnitudes. To simplify this problem we shall use the approximate formulae derived in the last chapter for the energy levels in a strong magnetic field with arbitrary orientation. As before, we assume that quantities such as $G = g\beta H$, A are necessarily positive, though the principal values of the **g** and **A** 'tensors' may be of either sign. As a further simplification, we consider first the case of an ion with effective spin $S = \frac{1}{2}$ and magnetic hyperfine structure only.

$S = \frac{1}{2}$, magnetic hyperfine structure only, axial symmetry

A typical Endor transition is that in which the nuclear magnetic quantum number changes by one unit while the electronic quantum number remains fixed. We shall therefore consider always the transition $|M, m\rangle \leftrightarrow |M, m-1\rangle$. From the formula for the energy levels given in eqn (3.54), this requires a quantum of energy

$$|h\nu_n| = AM - (A_\parallel A_\perp^2/4AG) - G_I -$$
$$- M(2m-1)\left\{(A_\parallel^2 + A^2)(A_\perp^2/4A^2G) + \frac{(A_\parallel^2 - A_\perp^2)^2}{8A^2G}\left(\frac{g_\parallel g_\perp}{g^2}\right)^2 \sin^2 2\theta\right\}. \quad (4.4)$$

This formula applies to the case where **H** is applied at an angle θ to the unique axis of an ion with axial symmetry, so that $G = g\beta H$ where g is defined by eqn (3.14), and A is defined by eqn (3.51). G_I is the diagonal part of the nuclear Zeeman term as defined by eqn (3.55), so that we are making the implicit assumption that $A \gg G_I$. If linearly polarized radiation is used to induce the nuclear Endor transition, we cannot tell experimentally which of the two levels involved is the higher and we therefore write $h\nu_n$ within modulus signs.

The terms contained in eqn (4.4) are of three types:

(i) AM, which has the values $+A/2$ or $-A/2$ according to the electronic energy level involved. Since the Endor transition is detected by the change in the intensity of the electronic transition $M \leftrightarrow M-1$, transitions in both the levels $M = +\frac{1}{2}$ and $-\frac{1}{2}$ will in general be observed while sitting on a given hyperfine line in the e.s.r. spectrum (see § 4.6);

(ii) the two terms $(A_\parallel A_\perp^2/4AG) + G_I$, which occur together and are independent of M and m. These two terms can be separated by measurements at different e.s.r. frequencies (different magnetic fields), since the term with G in the denominator varies as H^{-1} while G_I varies as H. To obtain good accuracy in measuring the two terms separately it is preferable to make measurements at widely different fields, but it may also be possible to calculate the first term with sufficient accuracy from the known values of the hyperfine parameters and g.

(iii) the last term, which involves both M and m in the form $M(2m-1)$. This term vanishes for $m = \frac{1}{2}$, and otherwise may be calculable with sufficient accuracy, as in (ii).

The different dependences on M and on m mean that by making sufficient measurements all three terms can be evaluated separately, while the two terms in (ii) can be separated using measurements at different electron resonance frequencies. This raises the general question, how many different measurements can be made? Clearly this number can be multiplied by making Endor measurements over a range of different electron resonance frequencies, and to define the position we restrict ourselves to just one such frequency.

In practice, the Endor transitions $|M, m\rangle \leftrightarrow |M, m\pm 1\rangle$ and $|M-1, m\rangle \leftrightarrow |M-1, m\pm 1\rangle$ can usually be detected while observing just the one electronic hyperfine line $|M, m\rangle \leftrightarrow |M-1, m\rangle$. This means that at a single value of H, four Endor frequencies can be measured except for the hyperfine lines $m = \pm I$, for each of which obviously only two are possible. Altogether, for an ion with effective spin S the number of Endor frequency measurements possible is $16IS$, and this usually means that the hyperfine parameters are overdetermined, giving a valuable check on internal consistency.

We consider now the question of the determination of signs. Since we can measure only $|h\nu_n|$, only $|(A_\parallel A_\perp^2/4AG)+G_I|$ can be found, but as pointed out above these two terms can be measured separately by observations at different field strengths. Their relative sign is also determined in this process, and since we take A, G to be positive quantities, we know the sign of G_I relative to $A_\parallel$. However, from eqn (3.55),

$$gAG_I = \beta H(g_\parallel^{(I)}g_\parallel A_\parallel \cos^2\theta + g_\perp^{(I)}g_\perp A_\perp \sin^2\theta),$$

so that if we measure in the direction $\theta = 0$ we find, relative to $A_\parallel$, the sign of $g_\parallel^{(I)}g_\parallel A_\parallel$, and the final result is a determination of the sign of $g_\parallel^{(I)}$ relative to that of $g_\parallel$. This is the same result as is obtained by observation of the 'forbidden' lines in the electron spin resonance spectrum itself, as outlined in § 3.8. Measurement at $\theta = 90°$ would give the sign of G_I (i.e. of $g_\perp^{(I)}g_\perp A_\perp$) relative to that of $A_\parallel$, but as the signs of $g_\perp$ and $A_\perp$ are indeterminate, no fresh information is obtained.

It is possible, of course, to use circularly polarized radiation for the Endor transitions. If the largest term in eqn (4.4) is AM, we would need right-handed polarization for the transitions $m \leftrightarrow m-1$ in the electronic level $M = +\tfrac{1}{2}$, and left-handed polarization for those in $M = -\tfrac{1}{2}$. For simplicity, we consider just the transitions $m = +\tfrac{1}{2}$ to $-\tfrac{1}{2}$. If those requiring right-handed polarization have the higher frequency, then $G_I+(A_\parallel A_\perp^2/4AG)$ is negative (and vice versa), so that we find the sign of $A_\parallel$ unambiguously, together with that of the product $(g_\parallel^{(I)}g_\parallel)$, but the latter still gives the sign of $g_\parallel^{(I)}$ only relative to that of $g_\parallel$.

$S = \tfrac{1}{2}$, axial symmetry, including electric quadrupole interaction

We now include a term $P_\parallel\{I_z^2-\tfrac{1}{3}I(I+1)\}$, which means that for an arbitrary orientation of the external magnetic field we must add to the energy in first order (i.e. if $P \ll A$) an amount

$$P\{m^2-\tfrac{1}{3}I(I+1)\}, \tag{3.62}$$

where P is given by the equation

$$P = \tfrac{1}{2}P_{\|}\left(3\frac{g_{\|}^2 A_{\|}^2}{g^2 A^2}\cos^2\theta - 1\right).$$

There are also second-order terms in m/M and m^3/M, whose coefficients involve only the squares of $g_{\|}$, $g_{\perp}$, $A_{\|}$, $A_{\perp}$ and are therefore fixed in sign; these terms are given in eqn (3.63). In an Endor transition for $S = \tfrac{1}{2}$ these give rise to terms that behave like M, Mm, and Mm^2, which can all be sorted out by their angular dependence and give no further information. They vanish if measurements are made with the magnetic field along the unique axis, and considerable complications are thus avoided.

Essentially therefore we have to add to $|h\nu_n|$ for the transition $m \leftrightarrow m-1$ an increment $P(2m-1)$. In its dependence on m alone it differs from any of the terms in eqn (4.4) and it can therefore be evaluated separately by making sufficient measurements of different hyperfine lines. Since A is positive, the e.s.r. lines for positive m come at the low-field end of the spectrum for constant frequency (and vice versa). If we take the mean of the Endor transitions for a given value of m (i.e., the mean of the measurements for $M = +\tfrac{1}{2}$ and $-\tfrac{1}{2}$), we obtain the quantity

$$|P(2m-1) - G_I - (A_{\|}A_{\perp}^2/4AG)|,$$

and since we know the sign of m, we can determine the sign of P relative to those of $A_{\|}$ and G_I. But the sign of G_I can be found relative to that of $A_{\|}$, as already shown, so that we can find the sign of P (and hence of $P_{\|}$) relative to that of $A_{\|}$.

The conclusions we have reached about signs for the case of $S = \tfrac{1}{2}$ agree with those stated by Halford (1962), and can be summarized as follows:

   (i) Endor gives the sign of $g_{\|}^{(I)}$ relative to that of $g_{\|}$;
   (ii) Endor gives the sign of $P_{\|}$ relative to that of $A_{\|}$.

In addition, if circularly polarized nuclear radiation is used,

   (iii) the sign of $A_{\|}$ (and hence of $P_{\|}$) can be found.

(The fact that the net oscillatory field at the nucleus may be oppositely directed to the external oscillatory field (see end of § 4.1) does not cause any complication, here, since both must rotate in the same sense.)

In cases where an ion with $S = \tfrac{1}{2}$ interacts with two nuclear spins, information concerning the relative signs of the two hyperfine parameters can be obtained by driving two Endor transitions simultaneously (see Cook and Whiffen 1964, 1965).

$S > \frac{1}{2}$, axial symmetry, magnetic dipole and electric quadrupole hyperfine structure

The more general case of $S > \frac{1}{2}$ can now be disposed of quite simply. From eqns (3.54) and (3.62) the Endor quantum for the transition $m \leftrightarrow m-1$ is

$$|h\nu_n| = AM + (A_\parallel A_\perp^2/2AG)\{M^2 - S(S+1)\} - G_I + P(2m-1) -$$
$$- M(2m-1)\left\{(A_\parallel^2 + A^2)(A_\perp^2/4A^2G) + \frac{(A_\parallel^2 - A_\perp^2)^2}{8A^2G}\left(\frac{g_\parallel g_\perp}{g^2}\right)^2 \sin^2 2\theta\right\}, \quad (4.5)$$

which differs only from our previous expressions in that the second term contains the quantity $\{M^2 - S(S+1)\}$, which is just $-\frac{1}{2}$ for $S = \frac{1}{2}$. This means that we can separate this term from G_I by making measurements using different electronic transitions, without needing to make measurements at two different e.s.r. frequencies. The Endor transitions for different values of $|M|$ come at very different frequencies, and there is no difficulty in identifying the value of $|M|$ concerned. As regards signs, the Endor spectrum yields (as before) the sign of $g_\parallel^{(I)}$ relative to that of $g_\parallel$, and the sign of $P_\parallel$ relative to that of $A_\parallel$. However the sign of the fine-structure terms can be found by measurements of the e.s.r. spectrum at low temperatures, and hence the sign of $A_\parallel$ from the second-order effects in the e.s.r. spectrum, as described in §§ 3.3, 3.6. In this case the sign of $P_\parallel$ can be established.

In eqn (4.5) the second-order terms in P^2/A have been omitted; in fact they vanish if measurements are made along the unique axis. It is almost essential to make Endor measurements along such an axis for reasons of precision. Not only are the off-diagonal terms tiresome to evaluate (though this can be done by a computer), but they cannot be evaluated correctly unless the exact orientation of the external magnetic field relative to the crystal axes is known (which cannot be found by a computer). When $S > \frac{1}{2}$, the second-order terms of the type considered in § 3.10, which are of order $(A/G) \times$ (fine structure parameter), become of tremendous importance and must be eliminated by careful alignment of the crystal so that the external magnetic field is exactly along a principal axis. Experimentally, the alignment of the crystal can be checked directly in the e.s.r. spectrum by observing the intensity of the 'forbidden' transitions due to these second-order terms, and adjusting the orientation to make their intensity as small as possible. With exact alignment they should vanish, and in practice the degree of mis-alignment can be estimated from their intensity, so that the resultant error in the Endor measurement can be gauged and, in favourable cases, reduced to an insignificant amount. For example, Baker and Hurrell (1963) estimate that misalignment by 0·1° could produce an unresolved splitting of 40 kHz in Endor of Eu^{2+} in CaF_2. A ligand Endor spectrum can also be used to obtain exact alignment (see Bessent and Hayes 1965, Davies and Hurrell 1968).

In this connection, measurement of the Endor transitions using the highest possible frequency for observation of the e.s.r. spectrum is clearly very desirable because

(i) second-order terms arising from matrix elements between different electronic levels become less important, since the energy in the denominators of a perturbation expansion is maximized; thus alignment of the crystal becomes less critical. In addition, line widths are reduced, as mentioned in § 4.10;

(ii) in many Endor experiments the prime motive is determination of $g^{(I)}$ through measurement of the energy G_I which is linearly proportional to H. The value of $g^{(I)}$ can therefore be measured with greater accuracy if a high e.s.r. frequency (larger magnetic field) is used. Obviously this does not help to separate the true g_I from the pseudo-nuclear Zeeman terms, and since these are largest when there are nearby electronic levels they can then be calculated accurately only if the electronic states and levels are known rather well.

The importance of the pseudo-nuclear Zeeman terms can be seen from a rough estimate. The change in frequency that they produce in a nuclear transition is of order $\nu_e(A/\Delta)$, where ν_e is the microwave frequency, A the hyperfine constant, and Δ the energy of the relevant excited state. Taking $\nu_e = 10^{10}$ Hz, we need (A/Δ) to be less than 10^{-6} if the shift is not to exceed the line width, which may be only 10^4 Hz. This inequality is fulfilled usually only for a half-filled shell such as d^5 or f^7, where A is unusually small and Δ exceptionally large. In an extreme case such as Ho^{2+} in CaF_2, which has a low-lying excited state at about 30 cm^{-1} (see Lewis and Sabisky (1963)), the apparent value of $g^{(I)}$ is some forty times larger than the true value of g_I.

4.3. Enhancement of the nuclear transition probability

We consider now the question of the nuclear transition probability, since this is important if the nuclear transition rate is to compete with spin-lattice relaxation processes that are electronic rather than nuclear in origin. In addition to the direct coupling of the oscillatory field with the nuclear magnetic moment, we have two other effects:

(a) the pseudo-nuclear Zeeman effect, discussed in § 1.8(e), § 3.7, and more fully in § 18.1; this arises from the admixture of excited states by the electronic Zeeman interaction, and in some cases can be much larger than the true nuclear Zeeman interaction (cf. §§ 4.2 and 4.8);

(b) modulation of the electronic hyperfine field by the applied oscillatory field, as outlined at the end of § 4.1.

In fact both these effects are essentially similar; we can incorporate (a) with the direct nuclear Zeeman interaction by using the same approach as in the static spin Hamiltonian, and follow this by a separate calculation of (b).

On this basis, the modified nuclear Zeeman effect is written in the form (cf. eqn (3.42))

$$\mathscr{H} = -\beta(\mathbf{H} \cdot \mathbf{g}^{(I)} \cdot \mathbf{I}) = -\beta(g_x^{(I)} H_x I_x + g_y^{(I)} H_y I_y + g_z^{(I)} H_z I_z), \quad (4.6)$$

where (x, y, z) are the principal axes of the nuclear 'g-tensor', which includes the direct interaction with the nuclear moment as well as the 'pseudo-nuclear' Zeeman interaction. For simplicity, we take not the most general case of an arbitrary orientation of the static field $\mathbf{H}$, but (as in § 3.2) assume $\mathbf{H}$ lies in the plane $y = 0$ at an angle θ to the z-axis (cf. Fig. 3.1). We assume that all tensor quantities have the same principal axes, and allow all principal values to be unequal; however, if the principal values for the x, y axes are equal, so that the z-axis is the unique symmetry axis, there is no loss of generality in our assumption about $\mathbf{H}$ since we can always choose the x-axis so that $\mathbf{H}$ lies in the xz-plane. Without appreciable danger of confusion, we call the nuclear oscillatory field $\mathbf{H}_1$, and assume it to be linearly polarized in a direction making an angle η with the y-axis, whose projection on the xz-plane makes an angle θ_1 with the z-axis (cf. § 3.2). Then the components of $\mathbf{H}_1$ are

$$\left. \begin{array}{l} H_x = H_1 \sin \eta \sin \theta_1, \\ H_y = H_1 \cos \eta, \\ H_z = H_1 \sin \eta \cos \theta_1. \end{array} \right\} \quad (4.7)$$

In a strong steady magnetic field $\mathbf{H}$, we diagonalize the static Hamiltonian as in Fig. 3.10 by rotating the nuclear axes about the y-axis through an angle ψ, a procedure that is appropriate if the magnetic hyperfine term dominates the nuclear Zeeman interaction; then

$$\left. \begin{array}{l} I_x = I'_x \cos \psi + I'_z \sin \psi, \\ I_y = I'_y, \\ I_z = -I'_x \sin \psi + I'_z \cos \psi, \end{array} \right\} \quad (4.8)$$

where the primes refer to the axes (x_n, y_n, z_n) and

$$\cos \psi = \left(\frac{g_z A_z}{gA}\right) \cos \theta, \quad \sin \psi = \left(\frac{g_x A_x}{gA}\right) \sin \theta. \qquad (4.9)$$

Our oscillatory nuclear Zeeman interaction then becomes, combining eqns (4.6)–(4.9),

$$-\beta H_1 \begin{bmatrix} I'_x\{(g_x^{(I)} g_z A_z/gA) \cos \theta \sin \theta_1 - (g_z^{(I)} g_x A_x/gA) \sin \theta \cos \theta_1\} \sin \eta + \\ +I'_y g_y^{(I)} \cos \eta + \\ +I'_z\{(g_x^{(I)} g_x A_x/gA) \sin \theta \sin \theta_1 + (g_z^{(I)} g_z A_z/gA) \cos \theta \cos \theta_1\} \sin \eta \end{bmatrix}. \qquad (4.10)$$

The rate at which nuclear transitions are induced is proportional to the sum of the squares of the coefficients of I'_x, I'_y so that by analogy with eqns (3.8a) and (3.16) we can define an oscillatory nuclear g-factor $g_1^{(I)}$ whose value is given by

$$(g_1^{(I)})^2 = \sin^2 \eta \left\{ \left(\frac{g_x^{(I)} g_z A_z}{gA}\right) \cos \theta \sin \theta_1 - \left(\frac{g_z^{(I)} g_x A_x}{gA}\right) \sin \theta \cos \theta_1 \right\}^2 + \\ + \cos^2 \eta (g_y^{(I)})^2. \qquad (4.11)$$

It is clearly best to make either $\sin \eta$ or $\cos \eta = 1$, according to which of the multiplying factors is the larger. In the former case the optimum value of θ_1 is given by

$$\tan \theta_1 = -\left(\frac{g_x^{(I)} g_z A_z}{g_z^{(I)} g_x A_x}\right) \cot \theta = -\left(\frac{g_x^{(I)}}{g_z^{(I)}}\right) \cot \psi, \qquad (4.12)$$

so that the optimum value of θ_1 is $\psi + \pi/2$ if there is no anisotropy in $g^{(I)}$, as we should expect, and $\theta + \pi/2$ if there is no anisotropy in $g^{(I)}$, g, or A.

We now consider the effect of $\mathbf{H}_1$ acting through the electronic magnetization and the hyperfine interaction. From eqns (3.8a) and (3.15) the electronic oscillatory spin Hamiltonian is

$$\mathcal{H}_1 = \beta H_1 (g_{1x} S'_x + g_{1y} S'_y + g_{1z} S'_z), \qquad (4.13)$$

where

$$\begin{aligned} g_{1x} &= (g_x g_z/g) \sin \eta, \quad g_{1y} = g_y \cos \eta, \\ g_{1z} &= \{(g_x^2 - g_z^2)/2g\} \sin 2\theta \sin \eta \end{aligned} \qquad (4.14)$$

if we assume that $\theta_1 = \theta + \pi/2$, so that $\mathbf{H}_1$ is normal to $\mathbf{H}$. The primes on the components of $\mathbf{S}$ indicate that these refer to the axes (x_e, y_e, z_e) in which the static electronic Zeeman interaction is diagonal (see § 3.2).

Since we are interested only in nuclear transitions that depend on the components I'_x, I'_y, we combine (4.13) with the relevant part of the hyperfine interaction, which from (3.52) is

$$(A_x A_z/A) S'_x I'_x + A_y S'_y I'_y. \tag{4.15}$$

In the combination we use the terms S'_+, S'_- from (4.13) and (4.15), and vice versa to obtain terms diagonal in S'_z, which are found to be altogether

$$\tfrac{1}{2}\beta H_1 \{I'_x g_{1x}(A_x A_z/A) + I'_y g_{1y} A_y\} \left\{ \frac{S(S+1)-M(M+1)}{W_M - W_{M+1}} + \right.$$
$$\left. + \frac{S(S+1)-M(M-1)}{W_M - W_{M-1}} \right\}, \tag{4.16}$$

and if we assume that $W_M - W_{M-1} = -(W_M - W_{M+1}) = g\beta H$, (4.16) reduces to

$$\mathcal{H}_1 = \frac{MH_1}{H} \left\{ I'_x \left(\frac{g_x g_z}{g^2}\right)\left(\frac{A_x A_z}{A}\right) \sin \eta + I'_y \left(\frac{g_y}{g}\right) A_y \cos \eta \right\} \tag{4.17}$$

on substituting for g_{1x}, g_{1y} and omitting the term in I'_z.

To be strictly logical we should add the coefficients of I'_x, I'_y in (4.17) to those in (4.10) (with $\theta_1 = \theta + \pi/2$), and take the sum of the squares of the resultant coefficients to give a net value of $(g_1^{(I)})^2$. However the two mechanisms are so disparate in size that this is hardly worth while. Clearly we should again make $\sin \eta$ or $\cos \eta = 1$ in (4.17), according to which coefficient is the larger, and with high anisotropy they can be quite different in magnitude. In the absence of anisotropy (4.17) and (4.10) with $\theta_1 = \theta + \pi/2$ add to

$$\mathcal{H}_1 = H_1\{-g^{(I)}\beta + (AM/H)\}(I_x \sin \eta + I_y \cos \eta), \tag{4.18}$$

where the coefficient $H_1(AM/H)$ is just equivalent to the classical value deduced at the end of § 4.1, which would correspond to an oscillatory Hamiltonian

$$\mathcal{H}_1 = -g_I \beta H_1 (H_e/H)(I_x \sin \eta + I_y \cos \eta), \tag{4.18a}$$

since $g_I \beta H_e = -AM$, by comparison with eqn (1.28).

If we refer to the formulae used to derive the static pseudo-nuclear Zeeman effect (e.g. eqn (1.97)), we can see that the mechanisms involved in deducing eqns (4.10) and (4.16) are basically similar. The oscillatory electronic Zeeman interaction induces an oscillatory component in the electronic magnetic moment (by admixture of other electronic states) proportional to H, which then interacts with the nuclear magnetic moment through the magnetic hyperfine interaction. In general, (4.16) gives the larger effect, because the admixed electronic state comes from a level only distant $g\beta H$ in energy, whereas (4.10) involves levels separated by a ligand field splitting (or other larger interactions). If the ligand field leaves low-lying excited states, and we apply a static field H such that $g\beta H$ becomes comparable with the energy of another ligand field state, the two expressions (4.10), (4.16) will become comparable in magnitude, since the energy denominators required in the perturbation theory will be of the same order. We have also inherently assumed that $H_e \gg H$, since essentially we diagonalized the hyperfine interaction (as far as possible) by choosing nuclear axes such that z_n is parallel to H_e. If H_e and H are comparable in magnitude (as often occurs in ligand hyperfine structure), we should use a more sophisticated approach analogous to that in § 3.11 following eqn (3.81). However the enhancement factor for the nuclear transition probability is then relatively small, and usually of insufficient importance to justify a more elaborate calculation. We content ourselves by remarking that the optimum direction of H_1 for an Endor nuclear transition, if $g^{(I)}$ is isotropic, is simply normal to the appropriate vector K_M in Fig. 3.19.

From eqn (2.51), the rate at which nuclear transitions are induced between levels m and $m \pm 1$ is

$$w_n = \frac{\pi}{2}\left(\frac{g_1^{(I)}\beta H_{1n}}{\hbar}\right)^2 \{I(I+1) - m(m\pm 1)\}f(\omega), \quad (4.19)$$

where we have written (H_{1n}) to remind ourselves that this must be at the nuclear (Endor) frequency, and $(g_1^{(I)})^2$ is given by (4.11) modified to include the analogous terms from (4.17). $f(\omega)$ is the appropriate shape function, which we take to be that of a spin packet; if the half width at half intensity of the latter is $(\Delta\omega/2\pi)$, then at the centre of the Endor line we have from (2.65) for a narrow line, $f(\omega) = 1/(\pi\Delta\omega) = \tau_2/\pi$.

Use of a circularly polarized nuclear field

When we have anisotropy and the steady field **H** is in an arbitrary direction, analysis of the Endor signal resulting from the use of a circularly polarized nuclear oscillatory field is exceedingly tedious, and not worth while in view of the fact (see § 4.2) that accurate measurements are made usually with **H** along a principal axis. We therefore assume that **H** is along the z-axis, and that $\mathbf{H}_{1n}$ is normal to it; however we do not assume that we have axial symmetry about the z-axis, since the analysis is then applicable to the x, y axes by using a cyclical interchange $x \to y \to z$. The oscillatory nuclear Hamiltonian is then, instead of eqn (4.6),

$$\mathcal{H}_{1n} = -\beta H_{1n}(g_x^{(I)} I_x \cos \omega t + g_y^{(I)} I_y \sin \omega t)$$

$$= -\tfrac{1}{4}\beta H_{1n}\begin{bmatrix} I_+\{(g_x^{(I)}-g_y^{(I)})e^{i\omega t}+(g_x^{(I)}+g_y^{(I)})e^{-i\omega t}\}+ \\ +I_-\{(g_x^{(I)}+g_y^{(I)})e^{i\omega t}+(g_x^{(I)}-g_y^{(I)})e^{-i\omega t}\} \end{bmatrix}, \quad (4.20a)$$

where we have used a transformation similar to that between eqns (3.17) and (3.18).

The operator I_- induces the transition $m \to m-1$, for which the required energy is (in first order)

$$W_m - W_{m-1} = A_z M - g_z^{(I)} \beta H + \tfrac{3}{2} P_z(2m-1),$$

which may be positive or negative. By an analysis similar to that following eqn (3.18) we find that the rates at which nuclear transitions are induced vary as follows.

Sign of	Transition rates are proportional to	
$A_z M - g_z^{(I)} \beta H + \tfrac{3}{2} P_z(2m-1)$	Right-hand sense	Left-hand sense
positive	$(g_x^{(I)}+g_y^{(I)})^2$	$(g_x^{(I)}-g_y^{(I)})^2$
negative	$(g_x^{(I)}-g_y^{(I)})^2$	$(g_x^{(I)}+g_y^{(I)})^2$

This table shows that the sign of $W_m - W_{m-1}$ (and hence of the largest term in this energy) can be found by observing the Endor intensities with right- and left-handed nuclear signals, provided the relative signs of $g_x^{(I)}$ and $g_y^{(I)}$ are known.

If the oscillatory field inducing nuclear transitions is due primarily to the effect of 'stirring' the hyperfine field rather than to direct interaction with the nuclear moment, the oscillatory Hamiltonian under the conditions assumed above is

$$\mathcal{H}_{1n} = (M/H)H_1\{I_x A_x(g_x/g_z) \cos \omega t + I_y A_y(g_y/g_z) \sin \omega t\}, \quad (4.20b)$$

from which a similar table is readily constructed, and it is apparent that we need to know the relative signs of $(g_x A_x)$ and $(g_y A_y)$.

If we have axial symmetry about the z-axis, then

$$(g_x^{(I)} \; g_y^{(I)}) - (g_x A_x - g_y A_y) = 0$$

and only one sense of the nuclear circularly polarized signal is effective. If $A_z M$ is the largest term in $W_m - W_{m-1}$, this gives the sign of $A_z = A_\parallel$ (those of $A_x = A_y = A_\perp$ are indeterminate).

4.4. Endor on donors in silicon

The method of Endor was invented by Feher (1956b), and it is fitting that we should start discussion of experimental work by outlining his comprehensive and elegant experiments on donors in silicon. As we do not discuss elsewhere shallow traps in semiconductors, we give here a very brief resumé of the essential features. The donors used are principally the group V atoms P, As, Sb, which have one electron more in their outer shells than silicon. When these atoms replace a silicon atom in the silicon lattice, four of the outer electrons are used in forming bonds with the four immediate silicon neighbours of the diamond-type lattice, as in the case of a silicon atom itself. The extra electron of the group V atom may remain attached to the donor, giving it one unpaired electron spin; or it may be excited in to the conduction band to act as a conduction electron, in which case the ionized impurity atom is left with one net positive charge and no unpaired electron spins. The 'ionization potentials' required for this process are given in Table 4.1;

TABLE 4.1

Data of Feher (1959) for three group V donors in silicon. V_i is the ionization potential to the conduction band; g the spin resonance g-factor relative to that $g_c = 1{\cdot}99875(10)$ for conduction electrons in a more highly doped sample, A is the magnetic hyperfine constant for the isotope whose mass and nuclear spin are also given

Donor atom	V_i meV	$g-g_c$	Isotope mass	$\|A\|$ (MHz)	Nuclear spin
Sb	39	$-1{\cdot}7(1)\times 10^{-4}$	121	186·802(5)	$\frac{5}{2}$
			123	101·516(4)	$\frac{7}{2}$
P	44	$-2{\cdot}5(1)\times 10^{-4}$	31	117·53(2)	$\frac{1}{2}$
As	49	$-3{\cdot}8(1)\times 10^{-4}$	76	198·35(2)	2

since for room temperature the equivalent voltage ($eV = kT$) is about 25 mV, the electrons will be practically all in the lower energy state, bound to their parent donor atoms, at helium temperatures. The substance is then a good electrical insulator so long as the donor concentrations are of order 10^{16} cm^{-3}, as used in Feher's experiments; conduction electrons occur in more highly doped samples, with donor concentrations $\sim 10^{18}$ cm^{-3}, and give a narrow homogeneously broadened electron spin resonance line.

The ground state of the extra electron attached to a donor is $^2S_{\frac{1}{2}}$, though its wave-function is very much more extended than in a free atom and has a finite density at many neighbouring silicon nuclei. The electron spin resonance spectrum occurs at a g-value very close to the free electron value, and shows hyperfine structure due to interaction both with the nuclear magnetic moment of its parent donor and with the nuclear moments of silicons on neighbouring lattice sites provided these are occupied by the odd isotope ^{29}Si, for which $I = \frac{1}{2}$, and whose natural abundance is about 5 per cent. The effective spin Hamiltonian for the electron bound to its parent donor is therefore ($S = \frac{1}{2}$)

$$\mathscr{H} = g\beta(\mathbf{H}\cdot\mathbf{S}) + A(\mathbf{S}\cdot\mathbf{I}) - g_I\beta(\mathbf{H}\cdot\mathbf{I}), \quad (4.21)$$

where g, A are isotropic and g_I is the true nuclear g-factor. To this we must add the interaction with neighbouring ^{29}Si nuclei for which, with $I = \frac{1}{2}$, we have

$$\mathscr{H}_L = \sum_L \{A^L(\mathbf{S}\cdot\mathbf{I}^L) - (g_I)^L\beta(\mathbf{H}\cdot\mathbf{I}^L)\} + \sum_L A_p^L(3S_{z'}I_{z'}^L - \mathbf{S}\cdot\mathbf{I}^L), \quad (4.22)$$

where the summation is over all lattice sites L occupied by ^{29}Si nuclei. The first term in (4.22) represents the interaction arising from the electron density of s-like character at the ^{29}Si nucleus due to the donor electron; the second is the ^{29}Si nuclear Zeeman interaction in the external field $\mathbf{H}$; and the third arises from magnetic dipole interaction between the ^{29}Si nuclear moment and the extended electron spin magnetization of the donor electron, the direction z' being that line joining the donor nucleus to the ^{29}Si nucleus at lattice site L. In a strong external magnetic field $\mathbf{H}$ the energy levels are approximately

$$W = g\beta HM + AMm - g_I\beta Hm + \\ + \sum_L [\{A^L M - (g_I)^L\beta H\}m^L + A_p^L M^L(3\cos^2\theta - 1)m^L], \quad (4.23)$$

where θ is the angle between H and the z'-axis, and the approximation we have made is that $A_p^L \ll A^L$.

In an ordinary electron resonance transition

$$(M, m, m^L) \leftrightarrow (M-1, m, m^L)$$

we have for $M = \frac{1}{2}$,

$$h\nu_e = g\beta H + Am + \sum_L \{A^L + A_p^L(3\cos^2\theta - 1)\}m^L, \quad (4.24)$$

in which the hyperfine splitting Am due to the donor nucleus is usually well resolved, while that from the remaining terms due to the ^{29}Si nuclei is not and gives an inhomogeneously broadened line. The nuclear

(Endor) transitions occur for $\Delta M = \Delta m^L = 0$, $\Delta m = \pm 1$, giving

$$h\nu_n^\pm = |A/2 \mp g_I \beta H| \qquad (4.25)$$

for the donor nucleus, and for the silicon nuclei at $\Delta M = \Delta m = 0$, $\Delta m^L = \pm 1$, giving

$$h\nu_L^\pm = |-(g_I)^L \beta H \pm A^L/2 \pm \tfrac{1}{2} A_p^L (3\cos^2\theta - 1)|, \qquad (4.26)$$

where the $\pm$ signs refer to the electronic states $M = \pm\tfrac{1}{2}$ respectively. The frequencies (4.25) are close to the values of A given in Table 4.1, while the frequencies (4.26) give two sets of lines around

$$(g_I)^L \beta H / h = 2 \cdot 6 \text{ MHz}$$

at $H = 3000$ G, and the A^L, A_p^L terms range from a few MHz for nearby ^{29}Si nuclei down to zero for the more distant nuclei.

Because of the very long spin-lattice relaxation times in silicon ($\tau_1 \sim 1$ h), it is possible to carry out either adiabatic rapid passage experiments or saturation experiments, both for the electronic and nuclear transitions (cf. §§ 1.12, 1.13). In the work of Feher (1959), a small part of the inhomogeneously broadened line was saturated by a high-power pulse, so that a 'hole was burnt' in the line, as shown in Fig. 4.4, since only the spin packets within that part of one hyperfine line at the resonance frequency were saturated. If the electron resonance signal at this same frequency is now monitored at a low level, the

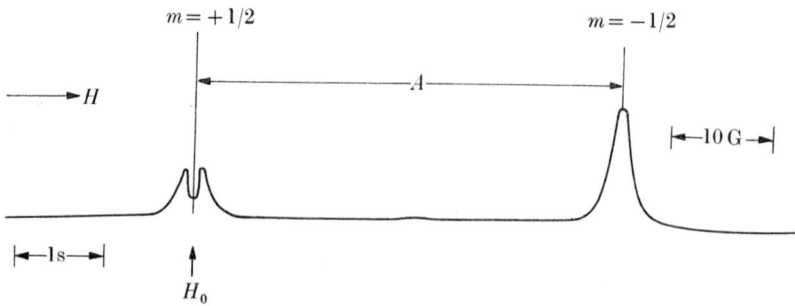

Fig. 4.4. 'Burning a hole' in the electronic transition $(+\tfrac{1}{2}, +\tfrac{1}{2}) \leftrightarrow (-\tfrac{1}{2}, +\tfrac{1}{2})$ for phosphorus donors in silicon. After a strong saturating pulse is applied at H_0, the electronic magnetic resonance signal is traversed at low power level, to show that only the spin packets at the resonance frequency for the saturating signal are affected. If the electron resonance signal at this frequency is monitored at low level, an increase in signal is observed when Endor transitions at either the donor or ^{29}Si frequencies are induced. A is the hyperfine splitting for the ^{31}P donor ($I = \tfrac{1}{2}$) nuclei; the inhomogeneous line width is due to interaction with the nuclear moments of ^{29}Si nuclei ($I = \tfrac{1}{2}$) on neighbouring lattice sites. The time scale for the traverse at low power level is indicated at bottom left, and the magnetic field scale at the right. (After Feher 1959.)

application of a strong oscillatory field at either ν_n or ν_L will cause a change in level. This will be of order $\delta_e/2$ if the populations of the two nuclear levels involved are inverted by an adiabatic rapid passage experiment, or of order $\delta_e/4$ if the populations are equalized through saturation. The two alternatives are illustrated in Fig. 4.5.

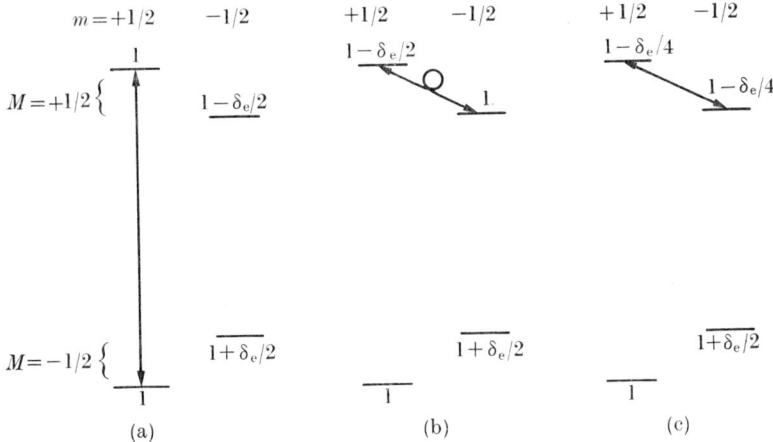

Fig. 4.5. Level populations in an Endor experiment before spin-lattice relaxation can act to restore thermal equilibrium.
(a) After the electronic transition $(+\tfrac{1}{2}, +\tfrac{1}{2}) \leftrightarrow (-\tfrac{1}{2}, +\tfrac{1}{2})$ has been saturated. (b) Inversion of the nuclear populations in the $M = +\tfrac{1}{2}$ states following adiabatic rapid passage through the Endor $(+\tfrac{1}{2}, +\tfrac{1}{2}) \leftrightarrow (+\tfrac{1}{2}, -\tfrac{1}{2})$ transition. (c) Equalization of the populations in the $M = +\tfrac{1}{2}$ states following saturation of the same Endor transition. The levels are drawn for the system $S = I = \tfrac{1}{2}$. Similar effects are obtained by saturating the other electronic transition followed by adiabatic rapid passage through or saturation of the other Endor transition.

In the process of burning a hole in the line we are saturating just those spin packets consisting of electron spins that experience a given local field from the neighbouring silicon nuclei. When Endor transitions are induced which change the orientation of these neighbouring nuclei, the local field will be altered both for the saturated packets and also for unsaturated spin packets consisting of other electron spins to which the same nuclei are also neighbours. Effectively this transfers neighbouring spin packets into and out of the hole, thus giving an increased electron resonance signal again. This is known as a 'packet-shifting' mechanism; from eqn (4.23) we see that an Endor transition $\Delta m^L = \pm 1$ just changes the local field at a donor relative to which the ^{29}Si nucleus occupies lattice site L by an amount

$$|M\{A^L + A_p^L(3\cos^2\theta - 1)\}/g\beta|, \qquad (4.27)$$

which is just the difference between the Endor frequency (4.26) ν_L and the ordinary nuclear resonance frequency $(g_I)^L \beta H/h$, divided by $(g\beta/h)$.

A typical Endor experiment of Feher (1959) is shown in Fig. 4.6, the electron resonance being observed at 9000 MHz and $T = 1\cdot 25°\text{K}$.

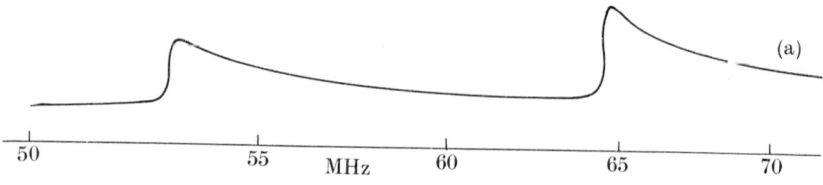

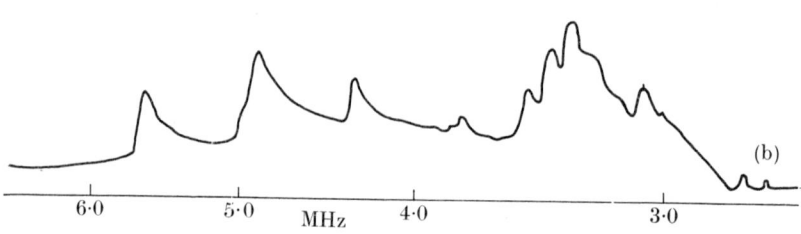

Fig. 4.6. Change in the electron resonance signal at 9000 MHz and $T = 1\cdot 25°\text{K}$ due to phosphorus (^{31}P) donors in silicon, upon application of power at the Endor frequencies. (a) Traverse through the ^{31}P Endor spectrum; the two frequencies correspond to those in eqn (4.25). (b) Traverse through the high frequency side of the ^{29}Si spectrum; the frequencies correspond to those in (4.26) with the lower $(-)$ sign before A^L. Another line is observed around $2\cdot 7$ MHz, due to more distant ^{29}Si for which A^L is small (after Feher 1959).

The upper diagram (Fig. 4.6a) shows the increase in electron spin resonance signal obtained by traversing the Endor transitions for the donor nucleus ^{31}P, which also has $I = \tfrac{1}{2}$. These transitions are given by eqn (4.25). Because the Endor line width is only 10 kHz, there is a sharp rise at the front, due to the sudden increase in population difference between the two previously saturated electron levels, followed by a slow recovery towards a steady-state condition. The second Endor transition is more intense for the following reason. From Fig. 4.5(b) it can be seen that adiabatic rapid passage through one Endor transition gives a population difference $\delta_e/2$ between the two electron levels for $m = +\tfrac{1}{2}$; if this is followed quickly by adiabatic rapid passage through the other Endor transition, the populations of the two $M = -\tfrac{1}{2}$ levels will be inverted, giving a net population difference between the $m = +\tfrac{1}{2}$ levels of δ_e. Thus the second Endor transition should be twice as intense as the first if they are traversed in a time too short for the populations to recover towards steady state values.

The hyperfine parameters for the donor nuclei are given in Table 4.1, and illustrate the accuracy that can be obtained in Endor experiments. The ratio of the two values of A for the isotopes 121, 123 of antimony is not quite equal to the ratio of the nuclear moments, and yields a hyperfine anomaly (cf. eqn 17.91, § 17.7) of $\Delta = -(0{\cdot}352 \pm 0{\cdot}005)$ per cent, where

$$\frac{A(121)}{A(123)} = \frac{g_I(121)}{g_I(123)}(1+\Delta). \qquad (4.28)$$

The ratio of the hyperfine constants was found to be $1{\cdot}84012(1)$ in the Endor experiments (Eisinger and Feher (1958)), while separate nuclear magnetic resonance measurements gave $1{\cdot}84661(1)$ for the ratio of the g_I values.

The g-values for the various donors were measured by comparison with that g_c arising from magnetic resonance of conduction electrons in a more heavily doped sample (3×10^{18} P/cm³), for which the value of $g_c = 1{\cdot}99875(10)$ was determined separately by comparison with a proton magnetic resonance signal. The value of $|(g-g_c)|$ increases with the value of V_i (see Table 4.1), where V_i is the 'ionization potential' required to lift the donor electron into the conduction band.

The Endor transitions due to ^{29}Si lattice sites are shown in Fig. 4.6(b). Here it is possible to identify the lattice sites involved through the anisotropy due to the A_p^L term in (4.26) by plotting the line frequency as a function of the orientation of the external field **H**, and using the relative intensity variation arising from the number of equivalent lattice sites giving an Endor transition at the same frequency. These experiments form an important method of checking the theory of shallow donors in semiconductors and measuring the parameters involved, details of which may be found in Feher (1959).

From the values of A^L, A_p^L determined for the various lattice points by means of Endor, it is possible to calculate both the shape and the width of the inhomogeneously broadened electron resonance line, and good agreement was obtained (Feher (1959)).

4.5. Endor on donors in silicon—relaxation effects

We now consider briefly relaxation effects in Endor, starting in this section with the experimental work of Feher and Gere (1959) on donors in silicon, in particular ^{31}P, which has $I = \frac{1}{2}$. Since the donor electron has $S = \frac{1}{2}$, we are dealing with the smallest possible number of energy levels, two electronic and two nuclear. The various relaxation processes that are operative in such a system are indicated in Fig. 4.9(a). The simple electron spin relaxation processes ($\Delta M = \pm 1$, $\Delta m = 0$) follow the broken vertical lines, with characteristic times τ_1 which are the same for either hyperfine line. One broken sloping line represents a process involving both an electronic 'flip' and a nuclear 'flop' such that $\Delta(M+m) = 0$, with a characteristic time τ_x. The other two broken sloping lines represent relaxation effects within the nuclear spin system for which $\Delta M = 0$, $\Delta m = \pm 1$, with characteristic time τ_n which is essentially the same for either nuclear transition. The remaining path for which $\Delta(M+m) = 2$ is not allowed when the hyperfine interaction is isotropic—see eqn (4.32).

In the experiments of Feher and Gere the level populations were disturbed from the values appropriate to thermal equilibrium, either by saturating one transition or more, or by adiabatic rapid passage through one line or more to reverse the populations. The latter proved more convenient when dealing with relaxation times longer than about a second. Special methods were devised of disturbing the equilibrium populations in different ways so as to disentangle the various relaxation processes (see below). The relaxation time was then determined by observing the exponential recovery of the populations towards the thermal equilibrium values, by monitoring the electron resonance signal at low level.

The first experiments were simple electron magnetic resonance experiments to determine τ_1, usually the fastest of all the relaxation times. It was found that the value of τ_1 was concentration dependent above about 10^{16} P/cm^3, and concentration independent below this level. In the latter case τ_1 was proportional to $(1/T)$, typical of a direct relaxation process (see § 1.11 and Chapter 10), below 2°K, the value of τ_1 lying in the range 10^{+3} to 10^{+4} s (about 1 h). A much faster rate of change of τ_1 with temperature set in above 2°K.

When the value of τ_x is very much longer than that of τ_1, it is necessary to adopt a method of measuring it that avoids the short-circuiting effect of the τ_1 paths. This method is to saturate both electronic transitions, so that all population differences of order δ_e are removed as shown in Fig. 4.7(a). If this situation is retained for a time of order τ_x, relaxation via the sloping relaxation path tends to re-establish the normal population difference between the $(+\frac{1}{2}, -\frac{1}{2})$ and $(-\frac{1}{2}, +\frac{1}{2})$ levels, the saturating signal causing the populations of the other two levels to follow, as in Fig. 4.7(b). If the saturating power is

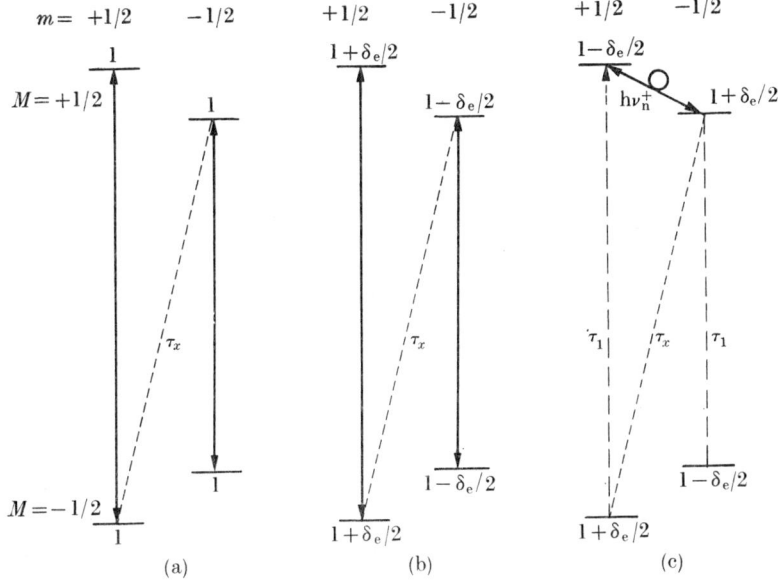

FIG. 4.7. Illustrating the method of Feher and Gere (1959) for measuring τ_x. (a) Both electronic transitions are saturated thus initially equalizing all the populations. (b) Saturation of both electronic transitions is maintained, but after a time $\gg \tau_x$ the sloping relaxation path establishes the normal population difference between the $(+\frac{1}{2}, -\frac{1}{2})$ and $(-\frac{1}{2}, +\frac{1}{2})$ levels. (c) Adiabatic rapid passage through the $h\nu_n^+$ transition inverts the populations of the $M = +\frac{1}{2}, m = \pm\frac{1}{2}$ levels. The electronic transitions (at low level) now have the normal intensity, except that the $m = -\frac{1}{2}$ line is reversed in sign. If (c) is carried out at various times after (a), the degree to which τ_x has re-established the population difference between the $(+\frac{1}{2}, -\frac{1}{2})$ and $(-\frac{1}{2}, +\frac{1}{2})$ levels can be found, thus giving the value of τ_x.

now removed, and the nuclear transition $h\nu_n^+$ induced by adiabatic rapid passage, the populations of the two $M = +\frac{1}{2}$ nuclear levels are inverted, as shown in Fig. 4.7(c). A quick sweep through the electronic resonance spectrum then shows one normal line ($m = +\frac{1}{2}$) and one inverted line ($m = -\frac{1}{2}$). By observing the strength of these last signals after waiting various intervals to allow the τ_x relaxation process to take effect, values of τ_x of order 30 h at $H = 3200$ G and 5 h at $H = 8000$ G were estimated, both at $T = 1.25°K$.

To estimate τ_n a different trick was employed, involving preferential population of the two nuclear $m = -\frac{1}{2}$ levels. If the system is allowed to reach thermal equilibrium in a high field (8000 G, for which $\delta_e' = h\nu_e'/kT$), and then adiabatic rapid passages are performed, first of the electronic line $m = +\frac{1}{2}$ and then on the nuclear line $M = +\frac{1}{2}$, the populations become as shown in Fig. 4.8(a) and 4.8(b). If the field

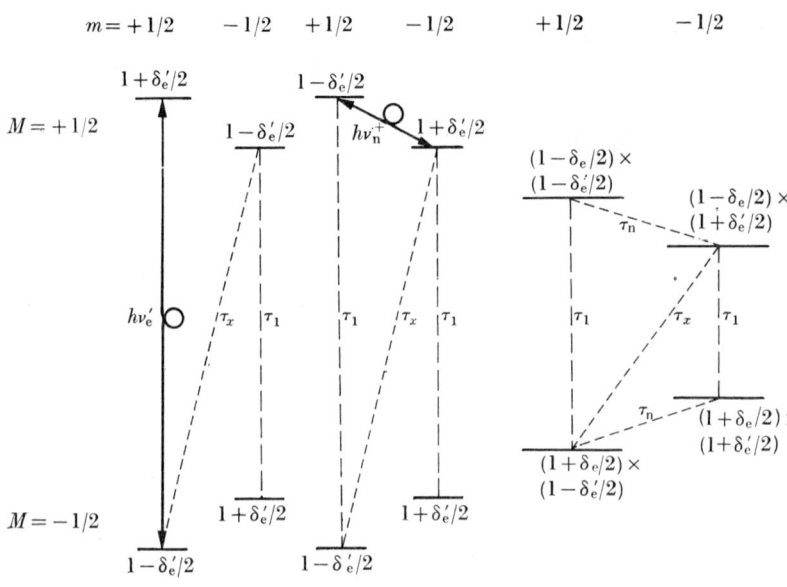

Fig. 4.8. Illustrating the method of Feher and Gere (1959) for measuring τ_n. (a) After thermal equilibrium is reached in a high field ($H = 8000$ G), adiabatic rapid passage of the electronic transition $m = +\frac{1}{2}$ inverts the population of the two $M = \pm\frac{1}{2}, m = +\frac{1}{2}$ levels. (b) Adiabatic rapid passage through the $h\nu_n^+$ transition then inverts the population of the $M = +\frac{1}{2}, m = \pm\frac{1}{2}$ levels. (c) In a lower field ($H = 3200$ G), relaxation through the rapid τ_1 paths modifies the populations of upper and lower electronic levels by factors $(1-\delta_e/2)$, $(1+\delta_e/2)$ respectively. This can be observed through the unequal strengths of the two electronic transitions which are in the ratio $(1-\delta'_e/2)/(1+\delta'_e/2)$. The relative strengths of these two transitions then return gradually to equality through the relaxation mechanisms τ_x, τ_n. The quantities $\delta'_e = (h\nu'_e/kT)$, $\delta_e = (h\nu_e/kT)$ where ν'_e, ν_e are the electron resonance frequencies in the higher and lower fields respectively.

is now dropped to $H = 3200$ G, for which $\delta_e = h\nu_e/kT$, relaxation through the more rapid τ_1 paths gives the populations shown in Fig. 4.8(c), and the intensity of the two electronic hyperfine lines is in the ratio $(1-\delta'_e/2)/(1+\delta'_e/2)$. This difference is then progressively eliminated as populations return to thermal equilibrium through the τ_x, τ_n processes. As τ_x had been measured separately, a lower limit of $\tau_n \approx 10$ h was established.

When free carriers are present, all these relaxation times are shorter through collisions with the conduction electrons, a process in which a mutual spin flip may occur between donor electron and conduction electron.

4.6. Relaxation effects in Endor—general

Endor experiments on donors in silicon have the unusual feature that the spin-lattice relaxation times are extremely long; in most paramagnetic systems relaxation times of a fraction of a second are the rule. Effects due to adiabatic rapid passage (even if this is feasible) will then vanish in a fraction of a second, and Endor experiments are carried out under steady-state conditions. Signals are somewhat smaller than immediately after a fast transient, but the analysis is rather complex and depends on competition between the various relaxation processes that are operative. An order of magnitude estimate can be obtained as follows for the size of the Endor signal. The relaxation paths for a system with $S = I = \frac{1}{2}$ and an isotropic hyperfine interaction are shown in Fig. 4.9(a). If saturating power is applied to the electronic $m = +\frac{1}{2}$ transition, as in Fig. 4.9(b), the net effect is to 'pump' spins from the lower to the upper of the two $m = +\frac{1}{2}$ levels; the spins can then return to the lower state by phonon emission, either following the direct vertical path τ_1, or by the indirect path via the $(+\frac{1}{2}, -\frac{1}{2})$ state which

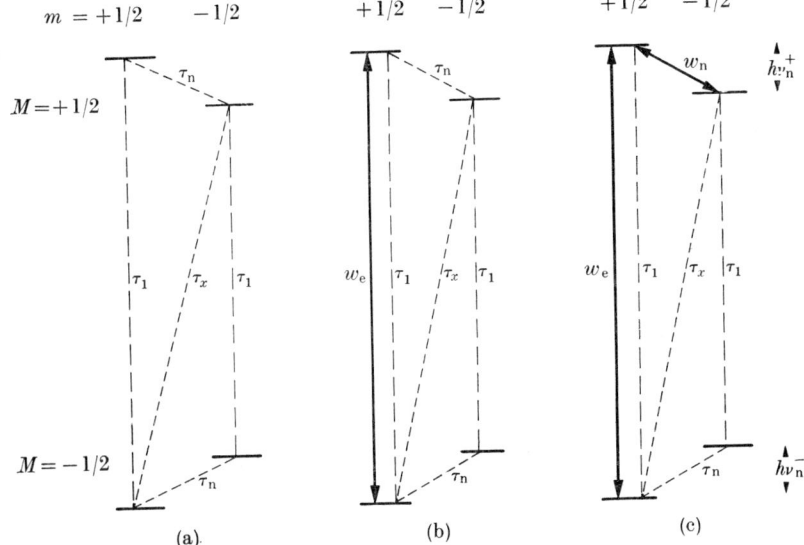

Fig. 4.9. Steady-state Endor signals. (a) Relaxation paths in a system $S = I = \frac{1}{2}$, with an isotropic hyperfine interaction. (b) As in (a), but with an electron resonance signal inducing transitions at a rate w_e on the $\Delta M = \pm 1$, $m = +\frac{1}{2}$ line. (c) As in (b), but with an Endor resonance signal also inducing transitions at a rate w_n on the $M = +\frac{1}{2}$, $\Delta m = \pm 1$ line. In general, $\tau_1 \ll \tau_x \ll \tau_n$, and to be effective in giving an Endor signal we need $w_n \tau_x \geqslant 1$, and not just $w_n \tau_n \geqslant 1$. The reason for this is that the relaxation path τ_n is essentially 'short-circuited' by the much faster indirect paths through τ_x and τ_1 (or w_e), for which the controlling rate is $1/\tau_x$.

involves both τ_n and τ_x and is therefore a very much slower process, controlled by τ_n if this is the longer relaxation time of the two. If power is now also applied to the transition $h\nu_n^+$, transitions to the $(+\frac{1}{2}, -\frac{1}{2})$ level are induced at a rate w_n, and if this is much faster than $1/\tau_x$, the factor controlling return to the $(-\frac{1}{2}, +\frac{1}{2})$ ground state will become τ_x. The net relaxation rate is thus changed from approximately $(1/\tau_1)+(1/\tau_n)$ to $(1/\tau_1)+(1/\tau_x)$, which is a change of order (τ_1/τ_x) in the effective value of τ_1 if $\tau_n \gg \tau_x$, i.e. $(\delta\tau_1/\tau_1) = -(\tau_1/\tau_x)$.

We now use this relation to make an estimate of the change in the electron resonance signal by using the macroscopic equations of Chapter 2 (with our usual reservations as to their validity when dealing with inhomogeneously broadened lines). From eqn (2.75) the electron resonance signal is proportional to $\chi''\sqrt{(P_1)}$, or $\chi'' H_1$, where P_1 is the applied power producing an oscillatory field strength H_1. From eqn (2.49) we have

$$\frac{\chi'' H_1}{M_0} = -\frac{\gamma H_1 \tau_2}{1+\gamma^2 H_1^2 \tau_1 \tau_2},$$

where for an inhomogeneously broadened line τ_2 is the true spin-spin relaxation time appropriate to the width of a spin-packet. The change in signal resulting from a change $\delta\tau_1$ in τ_1 is

$$\delta\left(\frac{\chi'' H_1}{M_0}\right) = \gamma H_1 \tau_2 \left\{ \frac{\gamma^2 H_1^2 \tau_1 \tau_2}{(1+\gamma^2 H_1^2 \tau_1 \tau_2)^2} \right\} \frac{\delta\tau_1}{\tau_1}, \qquad (4.29)$$

which has its maximum value when $\gamma^2 H_1^2 \tau_1 \tau_2 = 3$. Under this condition we find readily that

$$\frac{\delta(\chi'' H_1)}{(\chi'' H_1)} = -\frac{3}{4}\frac{\delta\tau_1}{\tau_1} = \frac{3}{4}\frac{\tau_1}{\tau_x}. \qquad (4.30)$$

The optimum condition $\gamma^2 H_1^2 \tau_1 \tau_2 = 3$ is somewhat higher than that which gives the maximum electron resonance signal (see eqn (2.49)), and H_1 should be adjusted to this optimum value.

So far we have considered cases where the only important cross-relaxation path is τ_x. This leads to a restriction on the number of Endor signals that we can expect to observe, the reason for which can be understood in two ways. (a) We consider the population differences of order δ_e, and neglect relaxation paths whose times are of order τ_n. Then under complete saturation of the electronic hyperfine transition for $m = +\frac{1}{2}$ we get the populations shown in Fig. 4.10(a), while if the $m = -\frac{1}{2}$ electronic transition is saturated, we get the populations shown in Fig. 4.10(b). A population difference of order δ_e exists in Fig.

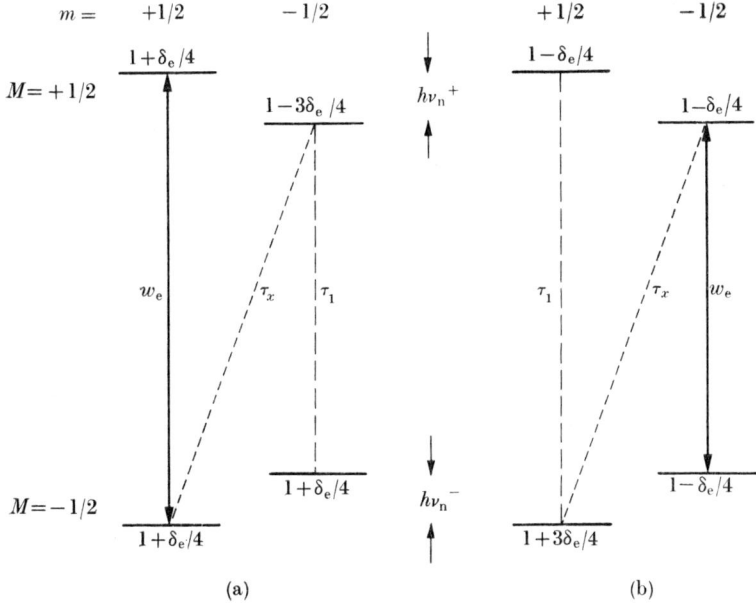

FIG. 4.10. Populations of the levels for the system $S = I = \frac{1}{2}$ in the steady state resulting from saturation of (a) the electronic hyperfine transition for $m = +\frac{1}{2}$, (b) the electronic hyperfine transition for $m = -\frac{1}{2}$. On this simple model, Endor transitions can be observed by applying nuclear power to the nuclear transitions between levels whose population differences are of order δ_e, i.e. to the upper nuclear transition $h\nu_n^+$ in (a) and the lower nuclear transition $h\nu_n^-$ in (b).

4.10(a) only between the nuclear levels for $M = +\frac{1}{2}$, and not between the nuclear levels for $M = -\frac{1}{2}$. Since application of nuclear power can only promote equalization of the populations of the two levels between which transitions are induced, it follows that an Endor signal will be observed only from the transition $h\nu_n^+$ in Fig. 4.10(a), and only from $h\nu_n^-$ in Fig. 4.10(b). (b) We consider the change in effective relaxation rate. In Fig. 4.9, saturation of the $h\nu_n^-$ transition instead of the $h\nu_n^+$ transition can only influence the return flow towards thermal equilibrium from the $M, m = +\frac{1}{2}, +\frac{1}{2}$ to the $-\frac{1}{2}, +\frac{1}{2}$ levels via the more indirect path

$$++ \xrightarrow{\tau_n} +- \xrightarrow{\tau_1} -- \xrightarrow{w_n} -+,$$

where the controlling rate will be the slowest relaxation rate $1/\tau_n$. Hence we can expect a fractional change in τ_1 only of order (τ_1/τ_n).

It is found in general that both Endor transitions can be observed (though often with unequal intensities) while saturating either hyperfine electronic line, showing that our model of the relaxation paths is

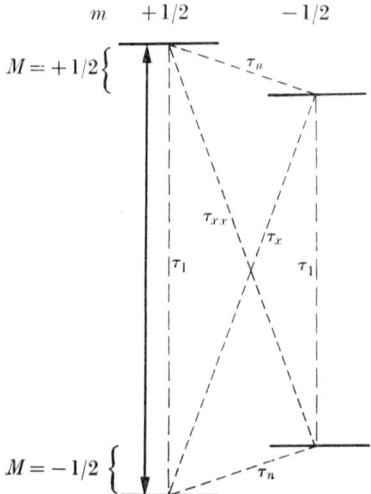

FIG. 4.11. Relaxation paths for a system $S = I = \tfrac{1}{2}$, with the spin Hamiltonian
$$\mathcal{H} = \beta(\mathbf{H}\cdot\mathbf{g}\cdot\mathbf{S}) + (\mathbf{S}\cdot\mathbf{A}\cdot\mathbf{I}).$$
The continuous vertical line represents the strong $(+\tfrac{1}{2}, +\tfrac{1}{2}) \leftrightarrow (-\tfrac{1}{2}, +\tfrac{1}{2})$ electron resonance transition. Relaxation via the path labelled τ_{xx} does not occur when the hyperfine interaction is isotropic. In general, $(1/\tau_1) \gg (1/\tau_x)$, $(1/\tau_{xx}) \gg (1/\tau_n)$.

over-simplified. We now consider a number of ways in which it can be modified.

When the magnetic hyperfine interaction is anisotropic, transitions between the $(+\tfrac{1}{2}, +\tfrac{1}{2})$ and $(-\tfrac{1}{2}, -\tfrac{1}{2})$ states are also partially allowed, and correspondingly there is an additional relaxation path labelled τ_{xx} in Fig. 4.11. When electron resonance power is applied to the $\Delta M = \pm 1$, $m = +\tfrac{1}{2}$ line, as shown in Fig. 4.11, spins are 'pumped' from the lower to the upper of these two states. In addition to the direct return flow through τ_1, there are additional return flows of the type

$$|{++}\rangle \xrightarrow{\tau_n} |{+-}\rangle \xrightarrow{\tau_x} |{-+}\rangle,$$
$$|{++}\rangle \xrightarrow{\tau_{xx}} |{--}\rangle \xrightarrow{\tau_n} |{-+}\rangle,$$

and

$$|{++}\rangle \xrightarrow{\tau_n} |{+-}\rangle \xrightarrow{\tau_1} |{--}\rangle \xrightarrow{\tau_n} |{-+}\rangle.$$

If the relaxation times are widely disparate, then the flow is controlled by the longest relaxation times, which are normally τ_n. If in the first two return paths, saturating power $(w_n \tau_x) > 1$ is applied at either of the nuclear frequencies, the effective value of τ_1 is altered by an amount of order (τ_1/τ_x) or (τ_1/τ_{xx}). It is true that if both nuclear frequencies are applied simultaneously at saturating level, we can also get a change through the third path, but the need for two frequencies to be at the correct value obviously creates difficulties when searching for an unknown Endor frequency.

We can obtain an estimate of the relative orders of magnitude of the various relaxation times by assuming that

(a) relaxation takes place by a 'direct' process (see § 1.11 and Chapter 10), involving the absorption and emission of phonons at the magnetic resonance frequencies;

(b) the only important mechanism is the interaction between the electron spin and a thermal 'noise spectrum' producing oscillatory magnetic fields of magnitude $\sim h_1$ such that $|h_1|^2$ is isotropic in space, and independent of frequency at least over a band of order (A/h), which includes all the hyperfine transitions in the vicinity of the electron resonance frequency ν.

On this basis the relative relaxation rates for different electronic transitions vary in the same way as the transition probabilities in the magnetic resonance spectrum. The sloping relaxation paths τ_x, τ_{xx} arise through transitions partly allowed by the terms in the static spin Hamiltonian

$$A'_x S'_x I'_x + A'_y S'_y I'_y$$
$$= \tfrac{1}{4}(A'_x + A'_y)(S'_+ I'_- + S'_- I'_+) + \tfrac{1}{4}(A'_x - A'_y)(S'_+ I'_+ + S'_- I'_-), \quad (4.31)$$

where the primes indicate that we have chosen coordinate systems for electron and nucleus such that the electronic Zeeman interaction and (as far as possible) the magnetic hyperfine interaction are diagonalized (cf. Chapter 3; for a typical example of such terms, see eqn (3.52)). These terms admix states $(M \pm 1, m \mp 1)$ and $(M \pm 1, m \pm 1)$ into (M, m), with amplitudes $\sim (A'_x + A'_y)/4h\nu_e$ and $(A'_x - A'_y)/4h\nu_e$ respectively. The transition probabilities are proportional to the squares of these amplitudes, giving

$$\frac{1}{\tau_x} \sim \frac{1}{\tau_1}\left(\frac{A'_x + A'_y}{4h\nu_e}\right)^2; \quad \frac{1}{\tau_{xx}} \sim \frac{1}{\tau_1}\left(\frac{A'_x - A'_y}{4h\nu_e}\right)^2. \quad (4.32)$$

Thus $(1/\tau_{xx})$ vanishes when the magnetic hyperfine interaction is isotropic, and in particular cases such that $A'_x = A'_y$. There is also a different dependence on electron resonance frequency for τ_1 and τ_x, τ_{xx}; for the direct process in a Kramers doublet $(1/\tau_1)$ varies as H^4 or $(h\nu_e)^4$ (see eqn (10.69)), and hence from (4.32) we would expect $1/\tau_x$, $1/\tau_{xx}$ to vary only as H^2.

In a more general spin Hamiltonian we may have terms of the form $S_z' I_\pm'$ which lend some electronic character to the $\Delta M = 0, \Delta m = \pm 1$ transitions. An effect of the same order of magnitude arises from the effect of a thermal low-frequency oscillatory field (normal to **H**) in driving the electronic magnetization, and hence producing an oscillatory component in the hyperfine field $\mathbf{H}_e$ at the nucleus (see § 4.3). These effects give matrix elements for the transitions $\Delta M = 0$ whose square is $(A/h\nu_e)^2 \times$ the square of the matrix elements for the electronic transitions $\Delta m = 0$. However in estimating relaxation times we must also remember that the phonon density at the nuclear frequency ν_n is smaller by a factor $(\nu_n/\nu_e)^2$ than that at the electron frequency ν. Hence we expect

$$\frac{1}{\tau_n} \sim \frac{1}{\tau_x}\left(\frac{\nu_n}{\nu_e}\right)^2, \tag{4.33}$$

which is of order

$$\frac{1}{\tau_n} \sim \frac{1}{\tau_1}\left(\frac{A}{h\nu_e}\right)^4, \tag{4.34}$$

since $(h\nu_n/h\nu_e)^2 \sim (A/h\nu_e)^2$ and, from eqn (4.32), $(1/\tau_x) \sim (1/\tau_1)(A/h\nu_e)^2$.

In addition to the relaxation processes considered above, which may be regarded as by-products of the simple electron spin-lattice relaxation mechanism, there are others that arise from thermal modulation of the hyperfine terms. Thermal modulation of the magnetic hyperfine interaction (Abragam 1955) has been considered by Jeffries (1960), who shows that it may contribute to all the relaxation paths shown in Fig. 4.11. Hence τ_{xx} may be finite even when the static Hamiltonian contains only an isotropic magnetic hyperfine interaction, since the thermal vibrations distort the lattice and give rise to all terms of the type given in eqn (1.62). Contributions to τ_1^{-1}, τ_x^{-1} and τ_{xx}^{-1} may all be of the same order, and unlike those in eqn (4.32), do not depend on the states being admixed by the static Hamiltonian. Thus at high static fields, modulation of the hyperfine interaction may make a more important contribution to τ_x^{-1}, τ_{xx}^{-1} than that indicated by eqn (4.32); the same is likely to be true for τ_n^{-1}. When $I > \frac{1}{2}$, we may have a further contribution to τ_n^{-1} through thermal modulation of the nuclear electric quadrupole coupling, which will produce terms of all the types in eqn (3.40a), even though the static quadrupole tensor vanishes (as for a Kramers doublet in cubic symmetry). Such terms will produce relaxation paths corresponding to the transitions $\Delta M = 0$, $\Delta m = \pm 1$ and $\Delta M = 0$, $\Delta m = \pm 2$ (cf. § 3.9). Yet another mechanism arises through the effect of hyperfine interaction in breaking time reversal symmetry in a Kramers doublet (cf. Fig. 10.9 and the reference given there).

We have so far assumed that the spin packet width is determined by a true spin-spin relaxation time τ_2. However the application of power at the electron resonance frequency will shorten the lifetime of the electron states, and we inquire now whether this will increase the effective width of the spin packet. As before, we use the macroscopic formulae to provide a convenient mode of analysis without implying that exact numerical interpretation is possible in practice. From eqn (2.51) the rate w_e at which electronic transitions are induced is (for spin $S = \tfrac{1}{2}$)

$$w_e = \frac{\pi}{2}\gamma^2 H_1^2 f(\omega), \qquad (4.35)$$

where H_1 is the amplitude of the oscillatory field at the electron resonance frequency. For a narrow spin packet whose width is determined by a true relaxation time τ_2 we have, from eqn (2.65), at exact resonance

$$f(\omega) = 1/(\pi\,\Delta\omega) = \tau_2/\pi,$$

so that we can write

$$w_e = \tfrac{1}{2}\gamma^2 H_1^2 \tau_2. \qquad (4.36)$$

We have seen from eqn (4.29) that the condition for optimum Endor signal is $\gamma^2 H_1^2 \tau_1 \tau_2 = 3$, in which case (4.36) becomes

$$2w_e \tau_1 = 3. \qquad (4.37)$$

(Note that this is just the condition that makes the denominator in eqn (1.134) equal to 4, so that the electron 'spin temperature' for the transition in question is raised to roughly four times the lattice temperature.) The electronic transition rate w_e changes $1/\tau_2$ by an amount $\sim w_e$. Hence if $\tau_1 \gg \tau_2$ we can satisfy (4.37) without increasing the Endor line width, but if τ_1 is so short that it is the determining factor in the spin packet width ($\tau_2 = \tau_1$), partial saturation of the electron resonance line will increase the Endor line width.

We are now in a position to make some rough estimates of the signal strength at the nuclear frequency needed to give an Endor signal. The rate at which nuclear transitions are induced is given by eqn (4.19),

$$w_n = \frac{\pi}{2}\left(\frac{g_1^{(I)}\beta H_{1n}}{\hbar}\right)^2 \{I(I+1) - m(m\pm 1)\} f(\omega). \qquad (4.19)$$

At the centre of a nuclear line the shape function

$$f(\omega) = 1/(\pi\Delta\omega) = 1/(2\pi^2\,\Delta\nu),$$

where $2\Delta\nu$ is the overall width of the nuclear line at half intensity. Hence at the centre, assuming $I = \frac{1}{2}$ for which

$$\{I(I+1)-m(m-1)\} = 1,$$

we have

$$w_n = \frac{\pi}{\Delta\nu}\left(\frac{g_1^{(I)}\beta H_{1n}}{h}\right)^2 \tag{4.38}$$

and to approach saturation of the nuclear transition we need

$$2w_n\tau_x \sim 1,$$

assuming τ_x to be the fastest relaxation time with which we have to compete.

In measurements on a ligand hyperfine structure there is little enhancement of the nuclear transition rate, and we may take $g_1^{(I)}$ to be roughly equal to g_I. If the nuclear signal strength is assumed to be $H_{1n} = 1$ G, and $\Delta\nu = 10$ kHz, we find for a proton or an ^{19}F ligand nucleus that

$$2w_n \sim 10^4 \text{ s}^{-1}.$$

Thus this value of H_{1n} is sufficient to produce appreciable saturation of the nuclear transition only if $\tau_x \geqslant 10^{-4}$ s, which is appreciably longer than the value of $\tau_2 = 1/(\Delta\nu) \approx 10^{-5}$ s, assumed for the spin packet width.

In considering Endor on the nucleus of the paramagnetic ion itself we expect considerable enhancement of the nuclear transition rate from 'stirring' the hyperfine field, as discussed in § 4.3. If for simplicity we assume **H** to be along an axis of symmetry (the z-axis) so that $A_x = A_y = A_\perp$, we have first from eqn (4.32)

$$\frac{1}{\tau_x} \sim \frac{1}{\tau_1}\left(\frac{A_\perp}{2h\nu_e}\right)^2. \tag{4.39}$$

Second, from eqn (4.18) we must now replace $g_1^{(I)}\beta H_{1n}$ by $(M/H)A_\perp H_{1n}$ so that for $I = \frac{1}{2}$ we have, instead of (4.38),

$$w_n = \frac{\pi}{\Delta\nu}\left(\frac{MA_\perp H_{1n}}{hH}\right)^2, \tag{4.38a}$$

and since $h\nu_e = g_\parallel \beta H$ we find that in this case

$$2w_n\tau_x = \frac{2\pi\tau_1}{\Delta\nu}\left(\frac{2Mg_\parallel\beta H_{1n}}{h}\right)^2, \tag{4.40}$$

independent of both $A_\perp$ and of ν_e, except that τ_1 may vary with ν_e. If

we take $|M| = \frac{1}{2}$, $g_\| = 2$ and $H_{1n} = 10^{-2}$ G, $\Delta = 10$ kHz, we obtain
$$2w_n\tau_x \sim 5\times 10^5 \tau_1,$$
so that even with this value of H_{1n} we can strongly saturate the nuclear transition provided that $\tau_1 \geqslant 10^{-5}$ s, a condition that must in any case be fulfilled if we are to keep the value of $\Delta\nu$ down to the region of 10 kHz.

In conclusion, we point out that saturation of the 'forbidden' transitions such as $\Delta(M+m) = 0$, which is a powerful tool in creating dynamic nuclear polarization (see § 1.12), is not an efficient method for Endor, since the nuclear transition rate w_n must then compete with the fastest relaxation rate $1/\tau_1$ in order to affect the intensity of the electronic transition that is being saturated.

4.7. The hyperfine structure of europium

As a second illustration of the power of the Endor technique we shall discuss the measurement of the hyperfine structure and nuclear moments of europium. The europium atom has the configuration $4f^7 6s^2$ and the dipositive ion Eu²⁺ has the configuration $4f^7$, so that both contain half-filled $4f$ shells and the ground spectroscopic state is $^8S_{\frac{7}{2}}$. The presence of the two extra electrons in the $6s$ shell of the atom makes little difference to the $4f$ electrons, but the magnetic hyperfine structure is distinctly larger in the ion. This difference is ascribed to core polarization, a conclusion that is supported by measurement of the hyperfine anomaly.

Europium has two stable isotopes, of mass 151 and 153, of roughly equal abundance. Each has nuclear spin $\frac{5}{2}$, but the nuclear magnetic dipole moments and nuclear electric quadrupole moments of the two isotopes each differ by a factor 2, though in opposite directions. The magnetic hyperfine anomaly is defined by the ratio

$$\frac{A(151)}{A(153)} = \frac{g_I(151)}{g_I(153)}(1+\Delta), \tag{4.41}$$

and Δ vanishes (see § 17.7) unless

(i) there is a finite density of magnetic electrons with a spatial variation within the nucleus (in practice this means that s-electrons must contribute to the magnetic hyperfine structure);

(ii) the nuclear magnetic dipole moment must be differently distributed within the nucleus for each isotope. (It was anticipated that this might be the case in view of the large differences in the nuclear moments quoted above.)

Obviously a hyperfine anomaly ($\Delta \neq 0$) can be observed only if the nuclear dipole moments and the magnetic hyperfine structure constants can be measured for both isotopes. This has been carried out for the ion by means of Endor in the solid state (Eu^{2+} in CaF_2), and for the atom by means of atomic beam triple resonance; nuclear magnetic resonance cannot be used to determine the nuclear moments because no configuration is available which does not have a resultant electronic angular momentum.

The atomic beam results for the atom are fitted to a Hamiltonian

$$\mathcal{H} = g\beta(\mathbf{H}\cdot\mathbf{S}) - g_I\beta(\mathbf{H}\cdot\mathbf{I}) + A(\mathbf{S}\cdot\mathbf{I}) + \frac{B}{2I(2I-1)2S(2S-1)} \times$$
$$\times \left[\{3S_z^2 - S(S+1)\}\{3I_z^2 - I(I+1)\} + \tfrac{3}{2}(S_zS_+ + S_+S_z)(I_zI_- + I_-I_z) + \right.$$
$$\left. + \tfrac{3}{2}(S_zS_- + S_-S_z)(I_zI_+ + I_+I_z) + \tfrac{3}{2}(S_+^2I_-^2 + S_-^2I_+^2) \right],$$
(4.42)

where the first two terms represent the Zeeman interactions, the third the magnetic hyperfine structure, and the last the electric quadrupole interaction. The same Hamiltonian is used for the ionic spectrum with some additional terms, which arise because of the cubic crystal field. These will be discussed in §§ 5.9, 5.10 and here we shall consider only the principal hyperfine terms.

The results of a number of measurements of the hyperfine constants are given in Table 4.2. We consider first the magnetic hyperfine interaction constant A. This may be regarded as a sum of contributions from

TABLE 4.2

Comparison of hyperfine measurements for the europium atom by atomic beam and for Eu^{2+} in CaF_2 by Endor. In each case the ground state is $4f^7$, ^8S, and the data are based on the work of Sandars and Woodgate (1960), Evans, Sandars, and Woodgate (1965), and Baker and Williams (1962). The values of g_I are uncorrected for diamagnetism, and are expressed in Bohr magnetons (eqn (4.42)); the nuclear magnetic moments are positive

	Eu atom	Eu^{2+} in CaF_2
g	1·9934(1)	1·9926(3)
^{151}A (MHz)	−20·0523(2)	−102·9069(13)
^{151}B (MHz)	−0·7012(35)	−0·7855(52)
$^{151}g_I$	+7·4921(13)×10^{-4}	+7·4969(20)×10^{-4}
$^{151}A/^{153}A$	+2·26498(8)	+2·25313(15)
$^{151}B/^{153}B$	+0·393(3)	+0·387(4)
$^{151}g_I/^{153}g_I$	+2·26505(42)	+2·2632(26)
Δ (%)	−0·003(20)	−0·53(2)

the 4f electrons, A_f, and from polarization of the core which we write as A_s because the dominant contribution must be from s-electrons through their finite density at the nucleus. Since only s-electrons are expected to contribute to the hyperfine anomaly, we can write, following Baker and Williams (1962),

$$A = A_f + A_s, \qquad A\Delta = A_s\Delta_s \qquad (4.43)$$

where Δ is the measured anomaly and Δ_s that due to s-electrons. We have two pairs of such equations, one pair for the atom and one pair for the ion. In each pair there are three unknowns, A_f, A_s, and Δ_s but only two measured quantities, A and Δ, so that some further assumptions must be made to obtain a solution. These are: (1) that Δ_s is independent of the principal quantum numbers of the s-electrons involved (see Pichanick, Sandars, and Woodgate (1960)), so that Δ_s can be taken to be the same for atom and ion; (2) that A_f (ion) can be plausibly related to A_f (atom).

We consider first the atom, for which the measured anomaly (see Table 4.2) is essentially zero, showing that we must take A_s(atom) = 0. We have then the problem of accounting for the observed value of A(atom), which on this basis must arise from the f-electrons. For a pure ^8S state A_f should be zero, since there is no orbital moment and the spherical distribution of spin density gives no field at the nucleus, provided there is no core polarization. However it is well known that the spin-orbit coupling tends to break down LS-coupling, giving a ground state that is of the form

$$(1+\alpha^2)^{-\frac{1}{2}}\{|^8S_{\frac{7}{2}}\rangle + \alpha\,|^6P_{\frac{7}{2}}\rangle + \ldots\}.$$

The difference between the measured g-value and the spin-only value can be explained on this basis, but it turns out that the contributions to the hyperfine constants A, B are both positive (the best estimates are $+7\cdot70$ MHz for ^{151}A and $+1\cdot87$ MHz for ^{151}B). It was suggested by Sandars and Woodgate (1960) that the discrepancy might be due to relativistic contributions. The theory of these effects has been worked out by Sandars and Beck (1965), and for europium Evans, Sandars, and Woodgate (1965) estimate that they could account for about half the remaining discrepancy in A_f and nearly all that in B. It may be that calculations with better wave-functions could remove the discrepancy entirely, but the important point is that the relativistic effects give contributions of the right sign, and that they involve the 4f electrons (not s-electrons) and therefore give rise to no hyperfine anomaly.

Similar effects will be present in the ion, but in the absence of more precise calculations we probably cannot do better than assume that the value of A_f is the same for the ion as the measured value of A in the atom. We then obtain $^{151}A_s$(ion) $= -83$ MHz, Δ_s(ion) $= -0.66$ per cent, which agrees well with the best value from the optical spectrum of $\Delta_s = -0.64(10)$ per cent deduced by Bordarier, Judd, and Klapisch (1965) from the measurements of Müller, Steudel, and Walther (1965).

The theory indicates that no anomaly is to be expected in the nuclear electric quadrupole interaction, so that the ratio of $B(151)/B(153)$ should be the same for atom and for ion; reference to Table 4.2 confirms that this prediction is fulfilled. However, as remarked above, the negative sign for B can only be explained by relativistic effects. For a pure ^8S state the electronic charge distribution is accurately spherical and no electric field gradient at the nucleus would be expected. The small break-down of LS-coupling which admixes $^6P_{\frac{7}{2}}$, etc., would give a positive value of B, estimated as $+1.87$ MHz for the isotope 151 (Evans, Sandars, and Woodgate 1965), but these authors find that relativistic effects contribute an amount -2.44 MHz, giving a total of -0.57 MHz in reasonable agreement with the experimental value of -0.70 MHz for the atom. Similar contributions would be expected in the ion, and it is reassuring to find that the observed value of -0.78 MHz is reasonably close to that for the atom.

We have described the results of Endor measurements on Eu^{2+} ions in CaF_2, and of atomic beam triple resonance measurements on the Eu atom, in some detail because they illustrate the following points.

(a) They confirm that most of the contribution to the isotropic magnetic hyperfine constant in the ion comes from core polarization, though relativistic effects are not negligible (and may be expected to be increasingly important in heavier ions);

(b) they show that Endor measurements of hyperfine constants can be made with an accuracy approaching that of magnetic resonance experiments on atoms using conventional atomic beam methods;

(c) they give measurements of the nuclear Zeeman interaction with an accuracy comparable with atomic beam triple resonance experiments.

The latter does not always mean that the nuclear magnetic moment can be determined with comparable accuracy. Both in atoms and in ions there may be an appreciable pseudo-nuclear Zeeman interaction, but this can generally be calculated more accurately for atoms because more exact information on excited states is available, in particular because crystal field interactions are absent. Obviously ions with half-filled shells are exceptional in this respect, and in the following section we discuss another type of ion where the interpretation of the Endor results involves a number of correction terms.

4.8. The Endor spectrum of Nd^{3+} in $LaCl_3$

We consider now the Endor spectrum of an ion where a number of important corrections must be included in the hyperfine structure owing to the presence of nearby excited electronic states. This ion is Nd^{3+}, $4f^3$, whose Endor spectrum has been analyzed very fully by Halford (1962). The ground state of this ion, which has $L = 6$, $S = \frac{3}{2}$, is $J = \frac{9}{2}$, which is split by the crystal field (see § 5.6) into five doublets. Resonance is observed in the lowest of these, and fitted to a spin Hamiltonian of the usual form for effective spin $S = \frac{1}{2}$ and axial symmetry

$$\mathcal{H} = g_{\parallel}\beta H_z S_z + g_{\perp}\beta(H_x S_x + H_y S_y) + \\ + A'_{\parallel} S_z I_z + A_{\perp}(S_x I_x + S_y I_y) + \\ + P'_{\parallel}\{I_z^2 - \tfrac{1}{3}I(I+1)\} - \\ - g_{\parallel}^{(n)}\beta_n H_z I_z - g_{\perp}^{(n)}\beta_n(H_x I_x + H_y I_y). \tag{4.44}$$

This is essentially the same as discussed in Chapter 3 except that primed coefficients have been used for $A'_{\parallel}$ and $P'_{\parallel}$ to denote the presence of correction terms, and the nuclear g-factor (here expressed in terms of nuclear magnetons) is allowed to be anisotropic because of contributions from the pseudo-nuclear Zeeman effect.

The electronic g-factor was determined from measurements on the single strong line in the electron spin resonance spectrum due to the even isotopes of Nd, the values obtained being

$$g_{\parallel} = +3 \cdot 9903(5), \qquad |g_{\perp}| = 1 \cdot 7635(12). \tag{4.45}$$

A large number of Endor measurements were made for the two odd isotopes ^{143}Nd, ^{145}Nd, each of which has $I = \frac{7}{2}$, and fitted by a least squares procedure using computer diagonalization of the Hamiltonian; the more accurate sets of results are shown in Table 4.3. The values of $A'_{\parallel}$ and $P'_{\parallel}$ determined with the magnetic field along the crystal axis are the more accurate, and were used in the subsequent analysis. As pointed out in § 4.2, the relative signs of $g_{\parallel}$ and $g_{\parallel}^{(n)}$, and of $A'_{\parallel}$ and $P'_{\parallel}$, can be determined from the Endor experiments. The crystal field analysis shows $g_{\parallel}$ to be positive, and that the sign of $A'_{\parallel}$ is the same as

Table 4.3

Results of Endor measurements on Nd^{3+} in $LaCl_3$ (from Halford (1962)). Column 2 gives the orientation of the external magnetic field $\mathbf{H}$ relative to the crystal axis

Isotope	H	$A'_\parallel$ (MHz)	$\|A_\perp\|$ (MHz)	$P'_\parallel$ (MHz)	$g^{(n)}_\parallel$	$\|g^{(n)}_\perp\|$
^{143}Nd	$\parallel$	$-1272 \cdot 1428(80)$	$499 \cdot 568(42)$	$+0 \cdot 0887(10)$	$-0 \cdot 543(3)$	
	$\perp$	$-1273 \cdot 036(240)$	$499 \cdot 532(40)$	$+0 \cdot 136(15)$		$0 \cdot 874(7)$
^{145}Nd	$\parallel$	$-790 \cdot 736(16)$	$310 \cdot 682(140)$	$+0 \cdot 0427(22)$	$-0 \cdot 347(18)$	
	$\perp$	$-790 \cdot 810(300)$	$310 \cdot 540(48)$	$+0 \cdot 035(12)$		$0 \cdot 538(9)$

that of g_n, thus establishing the signs given in the table, but only after a complete analysis has been used to relate $g^{(n)}_\parallel$ to g_n.

To make full use of the high experimental accuracy a rather extensive crystal field analysis is required to determine accurately the wave functions $|+\rangle$ and $|-\rangle$ spanning the ground doublet, together with an approximate determination of the energies of the excited levels. This was carried out by Halford including contributions from the $J = \frac{11}{2}$ and $\frac{13}{2}$ manifolds (those from $J = \frac{15}{2}$ were shown to be negligible) as well as the first-order contributions from $J = \frac{9}{2}$, and including corrections due to intermediate coupling, from a fit to the optical spectrum by Judd (1959a). The results were checked against the experimentally determined electronic g-values of the ground doublet, and a small correction made.

To proceed further, we write out the parameters of the experimental spin Hamiltonian in some detail, separating out the correction terms. This gives

$$\left.\begin{aligned} A'_\parallel &= 4g_n\beta\beta_n\langle r^{-3}\rangle\langle +|N_z|+\rangle + C + \Delta A_\parallel \\ A_\perp &= 4g_n\beta\beta_n\langle r^{-3}\rangle\langle +|N_x|-\rangle + C \end{aligned}\right\} \quad (4.46)$$

$$P'_\parallel = -\frac{3e^2Q}{4I(2I-1)}\langle r^{-3}\rangle\langle +|M|+\rangle + \Delta P_\parallel \quad (4.47)$$

$$\left.\begin{aligned} g^{(n)}_\parallel &= g_n + \Delta g^{(n)}_\parallel \\ g^{(n)}_\perp &= g_n + \Delta g^{(n)}_\perp \end{aligned}\right\}. \quad (4.48)$$

In eqns (4.46) the first term is the first-order magnetic hyperfine parameter; N is the operator defined in eqn (17.48), and an extra factor 2 appears because the parameters $A_\parallel$, $A_\perp$ refer to the effective spin Hamiltonian (4.44) for which $\langle +|S_z|+\rangle = \frac{1}{2}$. The term C was introduced by Halford to allow for a possible contact hyperfine interaction from unpaired s-electrons which he incorrectly assumed (see below) would give an isotropic contribution to the magnetic hyperfine structure. The extra term $\Delta A_\parallel$ is the second-order correction to the magnetic hyperfine structure given by eqn (18.9),

$$\Delta A_\parallel = -2u_{33} = -2d_3^{12},$$

which can be written according to eqn (18.8) for axial symmetry as

$$\Delta A_\parallel = 2(2g_n\beta\beta_n\langle r^{-3}\rangle)^2 \sum_n{}' \frac{\langle +|\,N_x\,|n\rangle\langle n|-iN_y\,|+\rangle}{W_0-W_n}. \quad (4.49)$$

It is readily verified that the corresponding correction $\Delta A_\perp$ which would involve terms

$$\frac{\langle +|\,N_x\,|n\rangle\langle n|\,N_z\,|-\rangle}{W_0-W_n} \quad (4.50)$$

vanishes and for that reason was omitted.

The quadrupole interaction (4.47) can be written as

$$P'_\parallel = P_\parallel + \Delta P_\parallel, \quad (4.51)$$

where $P_\parallel$ is the true quadrupole interaction, here assumed to arise only from the $4f$ electrons. In fact matrix elements between states of different J can be neglected, and the value of the matrix element is then

$$\langle +|\,M\,|+\rangle = \sum_J n_J^2 \langle J\|\,\alpha\,\|J\rangle\langle +|3J_z^2-J(J+1)\,|+\rangle, \quad (4.52)$$

where each term in the summation is weighted by n_J^2, the square of its amplitude in the ground state wavefunction.

The extra term $\Delta P_\parallel$ is the pseudo-quadrupolar interaction constant which, from eqn (18.8), can be written in the case of axial symmetry as

$$\Delta P_\parallel = \tfrac{3}{2}(2g_n\beta\beta_n\langle r^{-3}\rangle)^2 \sum_n{}' \frac{\langle +|\,N_z\,|n\rangle^2}{W_0-W_n}. \quad (4.53)$$

Both $\Delta A_\parallel$ and $\Delta P_\parallel$ are second-order effects of the magnetic hyperfine interaction, which admixes excited crystal field states into the ground states and thus changes (in second order) the hyperfine interactions.

Finally, $\Delta g_\parallel^{(n)}$ and $\Delta g_\perp^{(n)}$ are constants representing the pseudo-nuclear Zeeman effect whose values are, from eqn (18.2),

$$\left.\begin{aligned}
\Delta g_\parallel^{(n)} &= -4g_n\beta^2\langle r^{-3}\rangle \sum_n{}' \frac{\langle +|\,L_z+g_sS_z\,|n\rangle\langle n|\,N_z\,|+\rangle}{W_0-W_n}, \\
\Delta g_\perp^{(n)} &= -4g_n\beta^2\langle r^{-3}\rangle \sum_n{}' \frac{\langle +|\,L_x+g_sS_x\,|n\rangle\langle n|\,N_x\,|+\rangle}{W_0-W_n}.
\end{aligned}\right\} \quad (4.54)$$

Their interpretation is as follows: the electronic Zeeman interaction mixes excited crystal field states into the ground states and thus changes the magnetic hyperfine interaction by an amount proportional to the external magnetic field.

The value calculated for $\Delta A_\parallel$ is very small, being $+0.0351(10)$ MHz for isotope 143, and $+0.0136(4)$ MHz for isotope 145. The value of C can be found by comparing the ratio of $A'_\parallel - \Delta A_\parallel$ to $A_\perp$ found experimentally with that calculated for the ratio of $A_\parallel$ to $A_\perp$ from the crystal field states. C is found to be zero within the experimental error, but this does not imply that there is no core polarization effect. The matrix elements describing this effect are proportional to those of the vector $\mathbf{S}$; if the ground state wave-functions are accurately described by states belonging only to one manifold of given $\mathbf{J}$, the matrix elements both of $\mathbf{S}$ and of the Zeeman interaction $\mathbf{L} + g_s \mathbf{S}$ are each proportional to those of $\mathbf{J}$, and a core polarization contribution will not change the ratio $A_\parallel/A_\perp$. In so far as we cannot neglect admixtures from states of different $\mathbf{J}$, this proportionality will break down, and the ratio $A_\parallel/A_\perp$ will be affected, though the change cannot be represented by adding an isotropic contribution C. Since such admixtures in the present case are fairly small, and the expected contribution to the magnetic hyperfine structure from core polarization is probably only about 2 per cent (Bleaney (1964b)), the net effect is likely to be quite small.

We now consider the nuclear g-factors. The fact that $g_\parallel^{(n)}$ and $g_\perp^{(n)}$ are experimentally found to be quite different shows at once that the pseudo-nuclear Zeeman effect, which can be anisotropic, is playing an important role. In fact Halford's analysis gives $\Delta g_\parallel^{(n)} = -0.235(18)$, $\Delta g_\perp^{(n)} = -0.564(18)$ for isotope 143 which are of the same order as the value obtained for $g_n = -0.308(18)$. The latter gives a nuclear magnetic moment of $-1.079(60)$ n.m., in good agreement with the later result of Smith and Unsworth (1965) of $-1.063(5)$ n.m., obtained from atomic beam triple resonance measurements on the free atom.

Halford's results also yield the values:

isotope 143 $\Delta P_\parallel = +0.0302(8)$ MHz; $P_\parallel = +0.0585$ MHz
isotope 145 $\Delta P_\parallel = +0.0117(3)$ MHz; $P_\parallel = +0.0310$ MHz

(4.55)

where only the more accurate values of $P'_\parallel$ obtained from measurements with the magnetic field parallel to the crystal axis are used to find $P_\parallel$. The latter is the true quadrupole interaction, and its ratio for the two isotopes is 1·90, in very close agreement with the ratio of the quadrupole interaction constants for the free atom (Spalding (1963)). The pseudo-quadrupole effect $\Delta P_\|$ is of considerable importance, and since it is proportional to the square of the magnetic hyperfine constant and not to the quadrupole moment, the ratio of the constants $P'_\|$ for the two isotopes is quite different from that of the constants $P_\|$, which should also be the ratio of the quadrupole moments. The values of the quadrupole moments themselves, deduced by Halford are, however, quite different from those obtained from atomic beam measurements on the free atom, as follows:

	Halford (Nd^{3+} in $LaCl_3$)	Smith and Unsworth (Nd)
isotope 143	$Q = +0\cdot0206(30)$ barns	$Q = -0\cdot484(20)$ barns,
isotope 145	$Q = +0\cdot0105(20)$ barns	$Q = -0\cdot253(10)$ barns.

In each case a value of $\langle r^{-3} \rangle$ deduced from the magnetic hyperfine constant and the nuclear moment is used, but that for the atom is only a few per cent larger than that for the ion, as would be expected. We must therefore look elsewhere for an explanation of the discrepancy. Bleaney (1964b) has suggested that there is an appreciable contribution to $P_\|$ from the lattice, which, from the results of Edmonds (1963), might well be of the order of $+1$ MHz for isotope 143. From the atomic beam results we would expect the contribution from the $4f$ electrons for this isotope to be about $-1\cdot4$ MHz in Halford's experiment, so that his value of $P_\| = +0\cdot06$ MHz could be explained if the lattice contribution were just over $+1\cdot4$ MHz.

As discussed in § 5.5, the lattice contribution is enhanced by a large anti-shielding factor, while that of the $4f$ electrons is slightly diminished by shielding. In the compound used by Halford, the ground state wave-functions give an unusually small electric field gradient from the $4f$ electrons; in other cases the $4f$ electrons generally provide the dominant contribution to the electric field gradient. Nevertheless, the complications introduced by the crystal field make it unlikely that nuclear moment determinations by the Endor method in the solid state will approach in accuracy (except for ions with half-filled shells) those given by atomic beam methods for the free atoms of the $4f$ group.

4.9. Endor measurements of ligand hyperfine structure

Magnetic hyperfine structure due to interaction of the magnetic electrons with the nuclear magnetic moments of ligand ions is usually fairly small, particularly in the rare-earth group, and it can be measured only with limited accuracy in e.s.r. experiments. A further disadvantage of such experiments is that the allowed transitions are usually those in which the ligand nuclear magnetic quantum number does not change, so that no measure of any electric quadrupole interaction (such as may exist if the ligand ions are chlorine, with $I = \frac{3}{2}$) is obtained. If transitions violating this selection rule are observed, the hyperfine pattern is usually so complicated that accurate measurements are very difficult. This is just the situation in which the Endor technique can be used to good advantage, and as an example we shall discuss measurements of the fluorine hyperfine structure in the Eu^{2+} spectrum in CaF_2 (Baker and Hurrell (1963)).

The crystal structure of CaF_2 is outlined in § 5.1; the point symmetry of the Eu^{2+} ion (substituting for Ca^{2+}) is cubic O_h, while that of the nearest neighbour fluorine is trigonal C_{3v} about its bond axis (the line joining it to the Eu^{2+} ion), which is a $\langle 111 \rangle$ axis of the crystal. This symmetry restricts the form of the magnetic hyperfine interaction to

$$A_\parallel S_{z'} I_{z'} + A_\perp (S_{x'} I_{x'} + S_{y'} I_{y'}), \tag{4.56}$$

where the z'-axis lies along the bond direction. Using the substitutions

$$A_\parallel = A_s + 2A_p \quad \text{and} \quad A_\perp = A_s - A_p \tag{4.57}$$

this can be resolved into an isotropic term and one that has the same form as the point dipolar interaction. The complete spin Hamiltonian for each nearest-neighbour fluorine nucleus is then

$$\mathcal{H}_F = -g_I \beta \mathbf{H} \cdot \mathbf{I} + A_s \mathbf{S} \cdot \mathbf{I} + A_p (3 S_{z'} I_{z'} - \mathbf{S} \cdot \mathbf{I}), \tag{4.58}$$

where the z'-axis is along the appropriate bond axis and there are altogether eight such terms (with different bond axes, except that pairs of fluorine ions have the same bond axis because of the inversion symmetry about the Eu^{2+} site). The electronic Zeeman term is so large that to a good approximation off-diagonal terms in $\mathbf{S}$ can be neglected, and if $\mathbf{H}$ is applied along an axis (the z-axis) which makes an angle θ with the z'-axis, the relevant terms in the Hamiltonian are

$$\mathcal{H}_F = -g_I \beta H I_z + A_s S_z I_z + A_p (3 \cos^2\theta - 1) S_z I_z + 3 A_p \sin\theta \cos\theta S_z I_x, \tag{4.59}$$

which can be diagonalized (cf. § 3.11) by choosing a z_n-axis for the nuclear coordinates which makes an angle ϕ with the z-axis, such that

$$\tan\phi = \frac{3 A_p \sin\theta \cos\theta}{-g_I \beta H + \{A_s + A_p (3\cos^2\theta - 1)\} S_z}. \tag{4.60}$$

The energy then has the values $\pm\tfrac{1}{2}h\nu_\text{F}$, and the nuclear transition $\Delta I_z = \pm 1$ between these two states occurs at frequency ν_F where

$$h\nu_\text{F} = [\{-g_I\beta H + A_s S_z + (3\cos^2\theta - 1)A_p S_z\}^2 + 9\sin^2\theta\cos^2\theta A_p^2 S_z^2]^{\frac{1}{2}}. \tag{4.61}$$

Essentially this formula reflects the fact that the nuclear transition takes place in a net magnetic field which is the vectorial sum of the external magnetic field and the steady component of the local magnetic field due to the magnetic electrons; the terms that have been neglected correspond to the rapidly precessing components of the electronic magnetic field to which the fluorine magnetic moment cannot in a first approximation react, owing to its gyromagnetic properties. The steady magnetic field of the magnetic electrons has components both parallel and perpendicular to the external magnetic field which are each proportional to the value of S_z. As a result, the nuclear frequency given by eqn (4.61) has terms involving the zero, first, and second power of S_z, and by making Endor measurements on different electronic lines there is more than enough information to determine all the terms separately even while working only at one particular value of θ.

In practice all accurate measurements of the Endor transitions were made with **H** along a $\langle 100 \rangle$ direction where all eight nearest neighbour fluorines are equivalent with $\cos^2\theta = \tfrac{1}{3}$. Other measurements to confirm the sign of the interactions and the general form of eqn (4.58) were made in a $\langle 110 \rangle$ direction where four nearest neighbours have $\cos^2\theta = 0$ and four have $\cos^2\theta = \tfrac{2}{3}$. The advantages of such directions were: (i) the e.s.r. spectrum of eighty-four lines was almost resolved so that the line being saturated can be identified; (ii) the fluorine Endor spectrum is greatly simplified owing to the high degeneracy; (iii) this degeneracy enabled the crystal to be accurately aligned using the Endor spectrum.

Using a value of $H_1 \approx 0.1$ G for the nuclear driving field, the relative intensities of different nearest-neighbour transitions were found to vary by a factor of about 50, with a best signal/noise ratio of ~ 60 for a time constant of 1 s. The average line width was about 40 kHz with no resolved splitting, and this width could be explained by the accuracy of about $0.1°$ in orienting the crystal. An analysis of thirty-seven lines gave

$A_s = -2.23(1)$ MHz, $A_p = +4.01(1)$ MHz; (nearest neighbours).

The sign of A_p cannot be found from measurements in a $\langle 100 \rangle$ direction, since $(3\cos^2\theta - 1)$ is then zero, but is found relative to that of g_I from measurements in a ⟨110⟩ direction. Likewise the sign of A_s is obtained only relative to that of g_I, but the latter is known to be positive. These sign determinations are possible because $S > \frac{1}{2}$ and the signs of S_z are known, through the methods described in Chapter 3. The value of A_p is somewhat smaller than the value of $+5\cdot7$ MHz calculated for point dipoles, and suggests that some bonding occurs, since the alternative explanation of a reduction of 10 per cent in the bond length when Eu^{2+} is substituted for Ca^{2+} seems unlikely.

The analysis of twenty-eight next nearest neighbour lines gave $A_s = -7(5)$ kHz, $A_p = +785(5)$ kHz; (next nearest neighbours) which is consistent within the accuracy of the measurement and first-order theory with a purely point dipole interaction, for which

$$A_s = 0; \quad A_p = A_d = +0\cdot8 \text{ MHz}.$$

Endor measurements have also been made of the fluorine hyperfine interaction for two ions with the configuration $4f^{13}$ in CaF_2; these are Tm^{2+} (Bessent and Hayes 1965) and Yb^{3+} (Ranon and Hyde 1966). In each case the ion has cubic symmetry, and the ground state is a

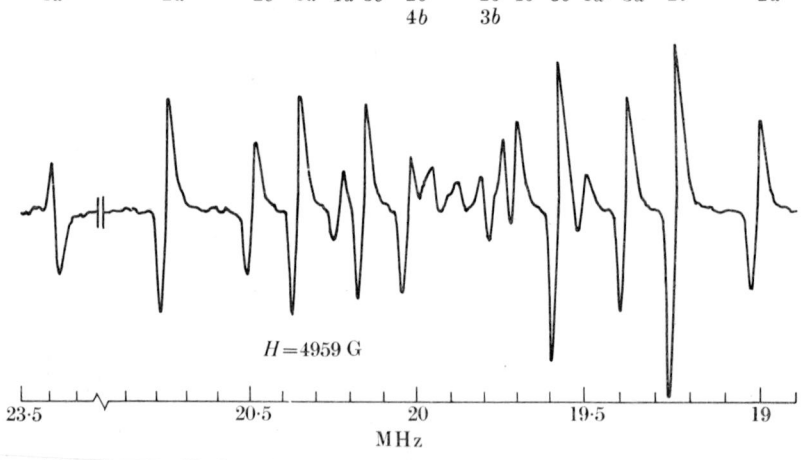

FIG. 4.12. Fluorine Endor spectrum for Tm^{2+} in CaF_2. The electron paramagnetic resonance is observed at about 24000 MHz and $H = 5000$ G, directed along the ⟨111⟩ axis, using magnetic field modulation at 115 kHz, giving an output after detection at this frequency. This output varies as a nuclear resonance oscillator (modulated at ~100 Hz) is swept through the Endor spectrum. After detection, the 115-kHz signal is amplified and passed to a synchronous detector at the second modulation frequency, whose output feeds (through a smoothing circuit with a 10 s time-constant) a chart recorder. The ordinate represents $\delta^2\chi''/\delta\nu\,\delta H$, the derivative with respect to frequency of the derivative $\delta\chi'/\delta H$ of the e.p.r. dispersion line. The lines labelled 1, 2, 3, 4 arise from fluorines in the first, second, third, and fourth shells. (After Bessent and Hayes 1965.)

Γ_7 doublet with an isotropic g-factor close to 3·45. A typical Endor spectrum is shown in Fig. 4.12 and the results are summarized in Table 4.4. For Tm^{2+}, as for Eu^{2+}, the interaction with more distant shells of fluorines is found to agree within experimental error with that

TABLE 4.4

Endor measurements of the fluorine hyperfine interaction of some rare-earth ions in CaF_2. The actual signs are experimentally determined only for Eu^{2+}, but for the other ions (A_s/A_p) is found to be positive. A_s, A_p are the parameters in the effective spin Hamiltonian (eqn (4.58)); A_d is the value that A_p would have (with $A_s = 0$) if the interaction were that between two point dipoles situated in an undistorted CaF_2 lattice

		Nearest fluorine neighbours			Second nearest fluorine neighbours			
		A_s (MHz)	A_p (MHz)	A_d (MHz)	A_s (MHz)	A_p (MHz)	A_d (MHz)	Reference
$4f^7$	Eu^{2+}	−2·23(1)	+4·01(1)	+5·7	−0·007(5)	+0·785	+0·81	Baker and Hurrell (1963)
$4f^{13}$	Tm^{2+}	2·584(10)	12·283(10)	+9·8	0·010(10)	1·386(5)	+1·40	Bessent and Hayes (1965)
$4f^{13}$	Yb^{3+}	1·67(5)	17·57(5)	+9·8				Ranon and Hyde (1966)

expected for a point dipole interaction; this is larger for Tm^{2+} than for Eu^{2+} in the ratio of the electronic g-values 3·45:2·00. The interaction with the nearest-neighbour fluorines is not point dipole, nor is it purely dipolar in character, since some s character is found. The sign of the interaction is not determined for the $4f^{13}$ ions where the electronic ground state is a doublet, but is assumed to be positive for both A_s and A_p (experimentally these are found to have the *same* sign), since otherwise the difference between A_p and the point dipole value A_d appears to be implausibly large.

To make a proper comparison of the non-dipolar part of the interaction for the different ions we must allow for the fact that the interaction for the $4f^{13}$ ions is expressed in terms of the effective spin of $\frac{1}{2}$ for the electronic state, which is not the same as the true spin. The

Γ_7 doublet states are (neglecting any admixture of $J = \tfrac{5}{2}$ into $J = \tfrac{7}{2}$)

$$|\pm\rangle = \pm\sqrt{(\tfrac{3}{4})}\,|J_z = \pm\tfrac{5}{2}\rangle \mp \tfrac{1}{2}|J_z = \mp\tfrac{3}{2}\rangle \tag{4.62}$$

$$= \pm\sqrt{(\tfrac{3}{4})}\{\sqrt{(\tfrac{6}{7})}\,|\pm 2, \pm\rangle + \sqrt{(\tfrac{1}{7})}\,|\pm 3, \mp\rangle\} \mp$$
$$\mp \tfrac{1}{2}\{\sqrt{(\tfrac{5}{7})}\,|\mp 1, \mp\rangle + \sqrt{(\tfrac{2}{7})}\,|\mp 2, \pm\rangle\}, \tag{4.63}$$

where the last representation is in terms of $|l_z, s_z\rangle$. From this it can be verified that the expectation values of the true spin component s_z are $\pm\tfrac{3}{14}$ in the two doublet states, and are thus a factor $(\tfrac{3}{7})$ smaller than the expectation values of the effective spin. Thus the non-point dipole parts, A_s and $(A_p - A_d)$, of the hyperfine interaction for the $4f^{13}$ ions must be increased by a factor $(\tfrac{7}{3})$ for comparison with Eu^{2+}. With this change one finds that the isotropic interaction is larger for both the $4f^{13}$ ions than for Eu^{2+}; so also is the adjusted value of $A_p - A_d$. The signs are also probably different for A_s, as well as for $A_p - A_d$; (if the signs of A_s are assumed to be all negative, the resultant value of $A_p - A_d$ becomes an order of magnitude larger for the $4f^{13}$ ions than for Eu^{2+}, though of the same sign). The rather larger value of A_p for Yb^{3+} compared with Tm^{2+} (both assumed positive) could be partly due to a larger point dipole interaction if the extra charge on the former ion pulls the nearest F$^-$ ions inwards by an appreciable amount.

The ligand hyperfine interaction discussed in this section and in § 4.10 is not essentially different from the interaction between the nuclear moments of ^{29}Si in the host lattice and the donor electron in silicon (§ 4.4), as can be seen from a comparison of eqns (4.22) and (4.58). If the magnetic electrons were wholly localized on the parent ion (or donor), interaction with a ligand nucleus would be purely 'point dipole', i.e. the same as that for two magnetic point dipoles. The additional ligand interaction comes from delocalization of the magnetic electrons. In silicon their wave-function spreads widely, and the resultant hyperfine interaction does not decrease monotonically with distance. In the $4f$ group bonding takes place only to the immediate ligand ions, and the degree of electron transfer is usually small. In the $3d$ group the transfer is usually larger, and may not be confined to the immediate ligand neighbours.

4.10. Endor line widths

To obtain narrow Endor lines the value of τ_2, the true spin-spin relaxation time, must be fairly long and in most cases this means that the main cause of broadening of the electron paramagnetic resonance line must be inhomogeneous. Then ideally one spin packet of width

$\Delta\omega \sim 1/\tau_2$ is saturated at the centre of the electron hyperfine transition $(M, m) \leftrightarrow (M-1, m)$, and only spins within this packet take part in the Endor transitions $(M, m) \leftrightarrow (M, m\pm 1)$, $(M-1, m) \leftrightarrow (M-1, m\pm 1)$. If $\tau_2 \sim 10^{-5}$ s, then the Endor line width parameter $\Delta\nu_n = 1/(2\pi\tau_2)$ should be of order 10 kHz. In practice the ideal situation is not always attained, and we now consider three reasons for this: inhomogeneous broadening, cross-relaxation within the electron spin system, and spin diffusion within the nuclear spin system.

Inhomogeneous broadening

When the effective spin $S \geqslant 1$ it is frequently observed that the electronic transitions are broadened owing to random variations in the crystal field from site to site. The electron resonance energy may be written

$$h\nu_e = g\beta H + g\beta\Delta H + (h\nu_e)_{fs} + (h\nu_e)_{hfs} + \qquad (4.64)$$

$$+ \delta g\beta H + \delta(h\nu_e)_{fs} + \delta(h\nu_e)_{hfs}, \qquad (4.65)$$

where ΔH is the random variation of local field due to magnetic dipole moments of neighbouring ions or nuclei. The terms $(h\nu_e)_{fs}$ and $(h\nu_e)_{hfs}$ represent the energy contributions from fine structure terms or 'initial splittings' (see §§ 3.3 to 3.6) and hyperfine terms respectively. The terms $\delta g\beta H$, $\delta(h\nu_e)_{fs}$ and $\delta(h\nu_e)_{hfs}$ represent respectively the effect of random variation in the crystal field giving rise to random variations in g, and in the fine structure and hyperfine structure contributions. The first of these may always occur, but the second is present only for $S \geqslant 1$. As discussed in § 3.13, for $S \geqslant \frac{3}{2}$ it is manifested in the electron resonance spectrum by a narrower line width for the $M = +\frac{1}{2} \leftrightarrow -\frac{1}{2}$ transition than for the outer transitions, since changes in the fine structure parameters affect the position of the central transition only in second order but the others in first order.

If random crystal field effects are absent, a fixed value of ν_e and a fixed value of H allow the resonance condition to be satisfied only for one value of ΔH. If, on the other hand, random crystal fields do contribute to the line width, then even for fixed ν_e and H, the term $g\beta\Delta H$ and those in (4.65) can vary at random so long as their sum is constant. However the terms in (4.65) do not necessarily vary independently of one another, as has been elegantly demonstrated by an Endor experiment of Locher and Geschwind (1963) on ^{61}Ni ($I = \frac{3}{2}$) in Al_2O_3, where the Ni^{2+} ion has an effective spin $S = 1$. The ion has axial symmetry about the z-axis, and for a magnetic field directed along this axis the diagonal parts of the spin Hamiltonian are

$$\mathcal{H} = g_\parallel \beta H S_z + D\{S_z^2 - \tfrac{1}{3}S(S+1)\} + \\ + A_\parallel S_z I_z + P_\parallel \{I_z^2 - \tfrac{1}{3}I(I+1)\} - g_\parallel^{(I)} \beta H I_z. \qquad (4.66)$$

The electron resonance condition for the transition $(M, m) \leftrightarrow (M-1, m)$ is

$$h\nu_e = |g_\parallel \beta H + D(2M-1) + A_\parallel m|, \qquad (4.67)$$

from which it is clear that a variation δD in D due to random crystal field effects must be compensated by a variation in H that will be of opposite sign for the two allowed transitions for which $(2M-1)$ has the values $+1$ and -1 respectively. Thus points within the electron

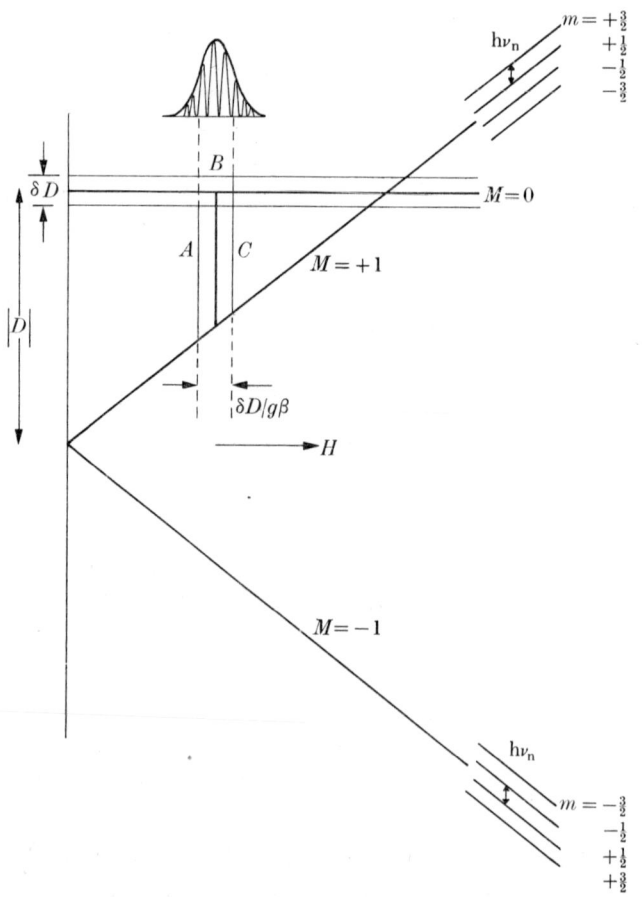

FIG. 4.13. Effect of inhomogeneity in the value of D for an ion such as Ni^{2+} ($S=1$) in Al$_2$O$_3$, with the spin Hamiltonian of eqn (4.66). The Endor transitions of energy $h\nu_n$ shown are for the isotope ^{61}Ni ($I = \tfrac{3}{2}$). (After Locher and Geschwind 1963.)

resonance line associated with different values of H correspond to different values of D, as illustrated in Fig. 4.13.

The Endor transition $(M, m) \leftrightarrow (M, m-1)$ requires a quantum

$$h\nu_n = |A_\| M + P_\|(2m-1) - g_\|^{(I)}\beta H| \tag{4.68}$$

and, in principle, the values of $A_\|$ and $P_\|$, and also of $g_\|^{(I)}$ if it contains an appreciable 'pseudo-nuclear Zeeman' contribution, may all vary in a manner linked to the crystal field, and hence to the value of D. In fact only the variation in $A_\|$ is significant in this case, and Locher and Geschwind showed that as H was varied through the resonance lines the value of $h\nu_n$ varied linearly not just as $-g^{(I)}\beta H$ but as

$$\delta(h\nu_n) = |(\delta A_\|/\delta H)M - g_\|^{(I)}\beta|\, \delta H, \tag{4.69}$$

as shown in Fig. 4.14. Here the actual rates of change $(\delta\nu_n/\delta H)$ are $+0\cdot 13\,\text{kHz/G}$ for the $M = +1$ Endor transitions, and $-0\cdot 65\,\text{kHz/G}$ when $M = -1$, showing that the first term in (4.69) is comparable with the second. The main change in $A_\|$ arises from the variation in the orbital contribution to the hyperfine field, which is linked to the residual unquenched orbital momentum ($g_\| = 2\cdot 195$) and hence to the splitting parameter D through the spin-orbit coupling.

In some cases such as ions with half-filled shells the hyperfine interactions are mainly due to effects such as core-polarization and relativistic effects (cf. § 4.7) which may be expected to be substantially independent of the crystal field. Thus variations in the latter which affect the fine structure parameters will have no corresponding effect on the hyperfine parameters. However there may still be some inhomogeneous broadening of the Endor line because of the inhomogeneous width ΔH of the electron resonance line, since the Endor frequency is not independent of H. For axial symmetry and an arbitrary direction of $\mathbf{H}$ the width of the Endor line can be found by differentiation of eqn (4.5) with respect to H, which is inherent both in G_I (see eqn (3.55)) and in $G = g\beta H$. For the special case where $\mathbf{H}$ is along the symmetry axis such a differentiation gives

$$\delta(h\nu_n) = \left| -g_\|^{(I)}\beta\, \Delta H + \left(\frac{A_\perp^2}{2h\nu_e}\right)\{S(S+1) - M^2 + M(2m-1)\}\frac{\Delta H}{H} \right|. \tag{4.70}$$

This expression has a complicated dependence on M and m, but is often smallest for the $M = +\tfrac{1}{2} \leftrightarrow -\tfrac{1}{2}$ transition simply because ΔH for the electron resonance line is also normally smallest for this transition.

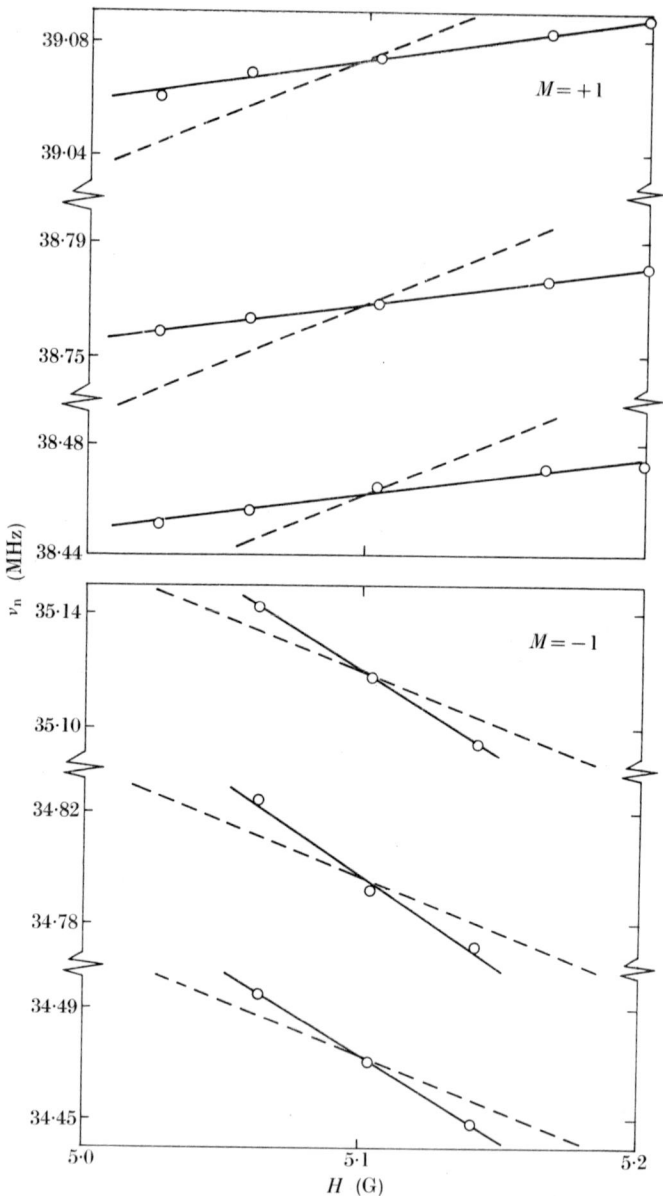

Fig. 4.14. The six Endor frequencies ν_n of ^{61}Ni observed in Al_2O_3 for $\mathbf{H} \parallel c$ axis at the $M = 0 \leftrightarrow +1$ e.p.r. line. H was varied within the broad e.p.r. line keeping the microwave frequency fixed. The points and the solid lines give the experimental results whereas the dashed lines indicate what would be expected if the hfs parameter, $A_\parallel$, were constant throughout the line. Note that the Endor transitions in the $M = -1$ states are also observed (though much more weakly than those in the $M = +1$ states) while saturating the $M = +1 \leftrightarrow 0$ transition (see § 4.11). (Locher and Geschwind 1963)

Since the effects discussed above depend on $g^{(I)}$ and on the magnetic hyperfine constant (either as A or A^2), and possibly also on the nuclear electric quadrupole parameter P, variation in these quantities will cause the inhomogeneous contribution to the Endor line width to vary from isotope to isotope of the same ion if the nuclear moments are different. There may also be a variation with m (cf. eqns (4.68) and (4.70)).

We conclude this discussion of inhomogeneous broadening of Endor lines by giving some numerical values. Locher and Geschwind (1963) deduced in their experiment on ^{61}Ni^{2+} in Al_2O_3 that on varying H by 200 G (going from A to C in Fig. 4.13) the variation in the hyperfine field H_e was some 130 G. Since the total orbital contribution to H_e is estimated as 170 kG, the change is fractionally very small though readily measurable; it corresponds to a frequency change of some 34 kHz whereas the half width at half intensity $2\delta\nu_n$ in the Endor lines was 23 kHz.

In contrast, the effect given by (4.70) can sometimes be quite small. In their experiments on Eu^{2+} Endor in CaF$_2$ (§ 4.7), Baker and Williams (1962) found line widths of 7 and 8 kHz respectively for the isotopes ^{153}Eu and ^{151}Eu, whose nuclear moments are in the ratio 1:2·27. In addition, the outer electron resonance lines gave the same Endor width as the $M = +\frac{1}{2} \leftrightarrow -\frac{1}{2}$ transition. These two facts indicate that inhomogeneous broadening of the Endor lines in this case was only very minor. We note also from (4.70) that the second term varies as ν_e^{-2} or H^{-2}, and its effect can be reduced by working at a higher electron resonance frequency. Thus Bessent and Hayes (1965) found for ^{169}Tm^{2+} in CaF$_2$ that the Endor width was reduced from 500 to 75 kHz on increasing the electron resonance frequency from 10 to 24 GHz. The second term in (4.70) is more important in this case because of the high value of $A = 1100$ MHz (isotropic), and the observed Endor widths correspond to $\Delta H \sim 30$ G, which is roughly the extent of the fluorine ligand hyperfine structure in the electron resonance line rather than the width of individual lines.

Cross-relaxation within the electron spin system

Cross-relaxation is due to the possibility of two nearby ions, some of whose resonance frequencies coincide, exchanging a quantum of energy and undergoing a 'flip-flop' transition. If one ion is subjected to an electron resonance driving field tending to equalize the populations of two specific levels, while the other belongs to another spin-packet which is not at resonance, the effect of cross-relaxation is partially to transfer saturation to the second ion. Conversely, if a nuclear driving field tends to diminish saturation (or merely change populations of levels) in this second ion, the effect will also be felt by the first ion, and will be manifested as an Endor signal for the first ion.

As a specific example we refer to Fig. 1.34, where the $\Delta M = \pm 1$, $m = +1$ hyperfine line is saturated in a system $S = \frac{1}{2}$, $I = 1$. Normally we would expect to observe at most the $m = +1 \leftrightarrow 0$ nuclear transitions in such an Endor experiment. However, if the ions can make a spin flip corresponding to the left-hand sloping transition

$$(-\tfrac{1}{2}, +1) \leftrightarrow (+\tfrac{1}{2}, 0)$$

with other ions not undergoing saturation, both the $m = +1$ level populations for the first set of ions and the $m = 0$ populations for the second set will be changed. As a result, nuclear power applied to the $m = 0$ to -1 transitions on the second set of ions may appear as an Endor transition while saturating the $m = +1$ line of the first set.

The chance of such a cross-relaxation process occurring is greatly increased if the concentration of paramagnetic ions is raised. In an Endor experiment on $^{59}\text{Co}^{2+}$ ($I = \tfrac{7}{2}$, $S = \tfrac{1}{2}$) in MgO, Fry and Llewellyn (1962) found that at a concentration of 0·1 per cent Co they observed just $16IS = 28$ strong Endor transitions by saturating in turn each of the eight hyperfine lines. However at 0·5 per cent concentration the total number of Endor lines was close to 100; as a specific example, all fourteen Endor transitions $M = \pm\tfrac{1}{2}$, $\Delta m = \pm 1$ were observed (with varying intensity) while saturating the $(+\tfrac{1}{2}, -\tfrac{7}{2}) \leftrightarrow (-\tfrac{1}{2}, -\tfrac{7}{2})$ electron resonance line. The higher concentration not only makes cross-relaxation more likely because on average the paramagnetic ions are closer, but also because the greater line width makes it easier to satisfy the conditions for mutual spin flips, which may involve any allowed transition and not just the type quoted above. Obviously the greater line width is not purely inhomogeneous in nature, and results in a decrease in the value of τ_2 and an increase in the spin-packet width, giving greater Endor line widths.

Cross-relaxation in the nuclear spin system

In addition to mutual spin flips involving electronic transitions we may have spin flips involving nuclear transitions. The rate at which these occur depends on the size of the spin-spin interaction between nuclei and is thus obviously very much smaller than in the electronic case. This is particularly true for the central nuclei of the paramagnetic ions themselves as they are comparatively far apart in a dilute crystal. On the other hand, when a ligand hyperfine structure is present, we are usually concerned with a number of nuclei of the same species separated by distances ≈ a few nm. For example, in CaF_2 the F^- ions lie on a simple cubic lattice of spacing 0·27 nm, and each Ca^{2+} is surrounded by a cube of eight F^- ions at a distance of 0·235 nm. When all F^- ions are equivalent, as in pure CaF_2, the spin-spin interaction between them gives an r.m.s. line width $2\delta\nu_n$ for the ^{19}F nuclear magnetic resonance varying from 12 to 29 kHz with the orientation of the external magnetic field (Van Vleck (1948), Pake and Purcell (1948)), which is partly due to mutual spin flip processes. We may therefore expect a line width of this order in the fluorine ligand Endor experiments discussed in § 4.9. When **H** is along a ⟨110⟩ direction, the value of $2\delta\nu_n$ is 20 kHz in the n.m.r. experiment, and Bessent and Hayes (1965) find a line width of 20 ± 12 kHz in the F^- endor in $Tm^{2+}:CaF_2$.

Although the n.m.r. line width should be equal to the Endor line width for more distant fluorine ions, whose immediate environment is similar to that in pure CaF_2, this is not necessarily the case for Endor of the F^- ions immediately adjacent to a paramagnetic ion. There are two reasons for this:

(a) such F^- ions experience a large local field from the paramagnetic ion, which even for a purely dipolar interaction is of order $\mu/r^3 = 10^3$ G if $S = \frac{1}{2}$, $g = 2$ and may be considerably larger. Hence the total field at these F^- nuclei differs considerably from that at more distant nuclei, thus eliminating the possibility of mutual spin flips except between the smaller number of nuclei in identical fields. This should give a smaller Endor line width for the immediate ligand nuclei of the paramagnetic ion:

(b) such F^- nuclear dipoles can, however, interact with one another indirectly via the paramagnetic ion, since terms of the form

$$\frac{\langle M, m^i|\, A_F S_+ I_-^i\, |M-1, m^i+1\rangle \langle M-1, m^j+1|\, A_F S_- I_+^j\, |M, m^j\rangle}{h\nu_e} \quad (4.71)$$

and similar terms give an $(\mathbf{I}^i \cdot \mathbf{I}^j)$ interaction of order $A_F^2/(h\nu_e)$. If $A_F \sim 10$ MHz, $\nu_e \sim 10$ GHz, this interaction constant is of order 10 kHz, and may be multiplied by an appreciable numerical factor for larger values of S (e.g., $S = \frac{7}{2}$ for Eu^{2+}). This indirect interaction will increase the Endor line width, accounting for the observed value ∼40 kHz for F^- in $Eu^{2+}:CaF_2$ (Baker and Hurrell (1963)). For a fuller discussion of the effects of such interactions, reference should be made to this paper.

4.11. 'Indirect' observation of Endor transitions

For an ion with $S = \frac{1}{2}$ the populations of the two electronic levels are both immediately affected by saturating an electronic transition. For $S \geqslant 1$, we have three or more electronic levels but the microwave power directly affects only two levels unless the resonance frequencies for several transitions coincide accidentally. Nevertheless relaxation proceeds between all the levels of a single ion at a rate $\sim 1/\tau_1$, so that the populations of all levels are indirectly affected by the saturating power after a time of order τ_1. This is illustrated in Fig. 4.15 for $S = 1$,

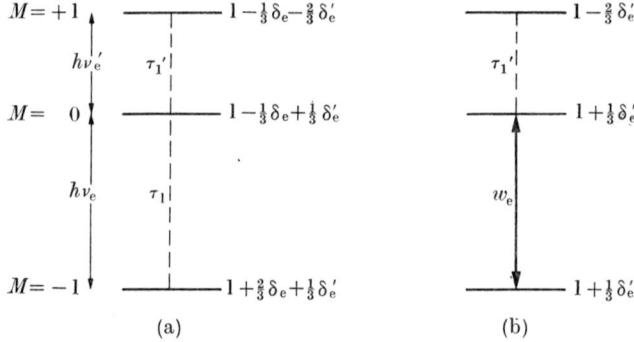

FIG. 4.15. Level populations within an electron spin system $S = 1$. (a) Levels in thermal equilibrium, with thermal populations according to their unequal spacing $\delta_e = (h\nu_e/kT)$, $\delta_e' = (h\nu_e'/kT)$. (b) Populations when the lower transition $h\nu_e$ is saturated, showing how the population of the top level is also changed following relaxation through the path labelled τ_1'.

where there are two allowed transitions at different frequencies ν_e and ν_e'. If the $M = 0 \leftrightarrow -1$ transition for one hyperfine line $I_z = m$ is saturated, the population of the $M = +1$, m state is altered through a τ_1 relaxation path, while those of the other $M = +1$ nuclear levels depend on other relaxation paths that are usually less effective. It follows conversely that nuclear power applied to nuclear transitions within the $M = +1$ level can change the populations of the $M = 0, -1$ levels and may thus give an Endor signal while the $M = 0 \leftrightarrow -1$ transition is observed. The lower three Endor transitions in Fig. 4.14 were detected 'indirectly' in this way; they belong to the $M = -1$ electronic state but are observed through the electronic

$M = +1 \leftrightarrow 0$ transition. This is a result of population adjustments within the levels of a single paramagnetic ion resulting from spin-lattice relaxation processes.

We have already mentioned in § 4.10 how cross-relaxation processes due to spin-spin interactions in the electron spin system may allow Endor transitions to be observed indirectly for ions with $S = \frac{1}{2}$, and such effects are equally possible for ions with $S \geqslant 1$. We now discuss similar effects arising from cross-relaxation within a nuclear spin system.

If we have a system of identical spins in which the level populations at one point in the specimen differ from those at other points, the spin flip process will act to make the population differences uniform throughout the specimen. This process is known as 'spin diffusion'; essentially it creates a uniform 'spin temperature' throughout the specimen, and without it the concept of a 'spin temperature' would be thermodynamically meaningless, or at least trivial as in the case of a single spin. It is particularly important in methods of dynamic nuclear polarization (cf. § 1.12) for ligand nuclei. If the ligand nuclei in the neighbourhood of a paramagnetic ion become polarized through interaction with this ion while a strong microwave field near its electron resonance frequency is being applied, the polarization is gradually transferred to more distant nuclei through spin diffusion. (There is of course a barrier to this for the closest ligand nuclei, because they experience different local fields due to their proximity to the paramagnetic ion and are thus not at the same resonance frequency as more distant ligand nuclei.) The process of polarizing more distant nuclei involves a flow of energy, whose ultimate source must be the electron resonance oscillatory field, through the paramagnetic ion. The electron resonance intensity of the latter can therefore be sensitive to population differences on quite distant nuclei (of which there are many), giving the possibility of an Endor experiment on such nuclei. Clearly this type of 'distant Endor' process is closely linked with that of dynamic nuclear polarization, as might be expected from similarities in the treatment of §§ 1.12 and 1.13. It is also linked with the fact that the principal agents in relaxing nuclear spins in a solid are very commonly paramagnetic ions, which are in very much more intimate contact with the lattice than the nuclear spin system.

A study of 'distant Endor' has been made for the system $Cr^{3+}:Al_2O_3$ (ruby) by Lambe, Laurance, McIrvine, and Terhune (1961). Essentially they found that

(a) the behaviour of the electron resonance signal is complicated; when nuclear power is applied to the ^{27}Al ($I = \frac{5}{2}$) system, the dispersion signal (χ') is decreased at points away from the centre of the electron resonance line, while the absorption signal (χ'') is increased at such points;

(b) such effects are observed when the nuclear frequency was that required for nuclear resonance on Al nuclei which are so far away that their nuclear resonance frequency is not shifted through interaction with the magnetic field of the Cr^{3+} ion;

(c) the recovery time of the electron resonance signal after removal of the nuclear power was comparable with the nuclear spin–lattice relaxation time ($\sim$10 s), whereas the Endor mechanisms discussed in §§ 4.5 and 4.6 would lead to a recovery time of order of the electronic spin–lattice relaxation time $\tau_1 \sim 0.1$ s.

The authors concluded that the distant Al nuclei were being polarized by the $S^{\pm}I^{\pm}$, $S^{\pm}I^{\mp}$ terms in their interaction with the paramagnetic ions (the mechanism discussed at the end of § 1.12), which are effective when the electron resonance frequency is on either side of its central frequency. Depolarization of the Al nuclei occurs when nuclear magnetic resonance power is applied to them, and conversely this affects the electron resonance signal in the wings, not in the centre, at the points where the electron resonance frequency has the required value to produce dynamic nuclear polarization (cf. Fig. 1.39).

4.12. Summary

We conclude the discussion of Endor by recapitulating some of the salient features of this type of experiment.

(a) *High sensitivity.* Nuclear magnetic resonance transitions are observed through their effect on the electron paramagnetic resonance signal, in many cases aided by the considerable enhancement of the nuclear transition probability brought about through 'stirring' of the hyperfine field by the nuclear oscillatory field.

(b) *Direct measurement of hyperfine frequencies with high accuracy.* For inhomogeneously broadened electron spin resonance lines, Endor line widths lie in the region 10–50 kHz, so that Endor frequencies even of order 10^3 MHz can be measured to a few kHz. This makes it possible to observe high-order effects in the hyperfine structure of the paramagnetic ion with its own nucleus, as well as ligand hyperfine structure associated with electron transfer and bonding with ligand ions. However in solids the nuclear Zeeman interaction can be drastically changed by the 'pseudo-nuclear Zeeman effect'; thus g_I is often not determined as accurately as is possible using triple resonance methods in atomic beam measurements on free atoms, where the corrections are smaller and can be calculated more accurately.

(c) *At least three mechanisms for the Endor process are possible—'packet shifting'* (§ 4.4), *change in rate of relaxation to lattice* (§ 4.6), *and 'distant Endor'* (§ 4.10).

THE LANTHANIDE (4f) GROUP

THE rare earth or 'lanthanide' series forms a distinctive group whose chemical properties are remarkably similar. The valency is usually 3, with occasional values of 2 or 4 which occur mostly for ions that thereby achieve an empty, half-full, or full shell. The significant feature of the lanthanide group is the filling of the $4f$ shell, apart from some exceptional cases where an electron enters the $5d$ rather than the $4f$ shell. The closed shells of electrons correspond to an xenon 'core' $1s^2 2s^2 2p^6 3s^2 3p^6 3d^{10} 4s^2 4p^6 4d^{10} 5s^2 5p^6$, which is the spectroscopic state of La^{3+} and Ce^{4+}. In the following ions electrons are added successively into the $4f$ shell, until this is full with fourteen electrons at Yb^{2+} and Lu^{3+}. Table 5.1 gives the basic information about the ground configuration of the tripositive ions.

A striking feature of the lanthanides is that most compounds exhibit sharp lines in their optical spectra, particularly at low temperatures, and much detailed information has been obtained for both ground and

TABLE 5.1

Ionic radii (in nanometres) and values of $\langle r^2 \rangle$, $\langle r^4 \rangle$, and $\langle r^6 \rangle$ (in atomic units). The ionic radii are from Evans (1964) and the other data from Freeman and Watson (1962)

Z	Ion	State	Ionic radius (nm)	$\langle r^2 \rangle$ (a.u.)	$\langle r^4 \rangle$ (a.u.)	$\langle r^6 \rangle$ (a.u.)
57	La^{3+}	$4f^0$	0·115			
58	Ce^{3+}	$4f^1$	0·102	1·200	3·455	21·226
59	Pr^{3+}	$4f^2$	0·100	1·086	2·822	15·726
60	Nd^{3+}	$4f^3$	0·099	1·001	2·401	12·396
61	Pm^{3+}	$4f^4$	0·098			
62	Sm^{3+}	$4f^5$	0·097	0·883	1·897	8·775
63	Eu^{3+}	$4f^6$	0·097			
64	Gd^{3+}	$4f^7$	0·097	0·785	1·515	6·281
65	Tb^{3+}	$4f^8$	0·100			
66	Dy^{3+}	$4f^9$	0·099	0.726	1·322	5·102
67	Ho^{3+}	$4f^{10}$	0·097			
68	Er^{3+}	$4f^{11}$	0·096	0·666	1·126	3·978
69	Tm^{3+}	$4f^{12}$	0·095			
70	Yb^{3+}	$4f^{13}$	0·094	0·613	0·860	3·104
71	Lu^{3+}	$4f^{14}$	0·093			
63	Eu^{2+}	$4f^7$		0·938	2·273	11·670

excited states by high resolution spectroscopy. An extensive discussion of the theory of the spectroscopic properties is given by Wybourne (1965), much of which is relevant to the paramagnetic resonance properties with which this chapter is concerned.

5.1. Lanthanide compounds

Lanthanide ions enter many chemical compounds but we shall confine discussion to a few types of compound where the local symmetry is at least axial and in which a wide range of lanthanide ions have been studied. The first of these is the series of lanthanide ethylsulphates, $Ln(C_2H_5SO_4)_3,9H_2O$, whose X-ray crystallography was investigated by Ketelaar (1937) and more recently by Fitzwater and Rundle (1959), who examined the compounds of Pr, Er, and Y. The latter workers show that all heavy atoms are found in the space group P^63/m, and that probably the hydrogen positions also conform to this space group. The point symmetry at the lanthanide ion is C_{3h}. This ion has nine water molecules as nearest neighbours; six form a triangular prism with three above and three below the mirror plane containing the other three water oxygens and the lanthanide ion (see Fig. 5.1). In the erbium compound the Er—O distances to the prism are 0·237 nm, and the remaining three distances are 0·252 nm. If all but the nearest oxygen positions are neglected, the symmetry about the lanthanide ion is almost D_{3h}. This structure has a vertical threefold axis of symmetry; if the structure were exactly D_{3h} there would also be both a vertical and a horizontal plane of reflexion symmetry. The compounds are isomorphous throughout the group from La^{3+} to Lu^{3+}, and also with Y^{3+}. Since the ionic size of Y^{3+} is 0·093 nm, the later ions of the group (see Table 5.1) probably fit better into yttrium ethylsulphate as a diluent than into lanthanum ethylsulphate.

A second series, in which the paramagnetic resonance results are remarkably similar to those in the ethylsulphates, is formed by the anhydrous trichlorides, type $LnCl_3$. $LaCl_3$ crystallizes in a structure whose space symmetry (see Hutchison and Wong 1958) is $C6\ 3/m$, the point symmetry at the La site being C_{3h}. The La^{3+} ion is surrounded by nine approximately equidistant nearest neighbours (chloride ions). Three of these are coplanar with the La^{3+} and at a distance of 0·297 nm; the other six lie, three in a parallel plane above and three in a parallel plane below, all six at a distance of 0·299 nm from the La^{3+} ion. All the lanthanide series enter $LaCl_3$ as a diluent with this structure, but most of the undiluted anhydrous trichlorides of the heavier ions have

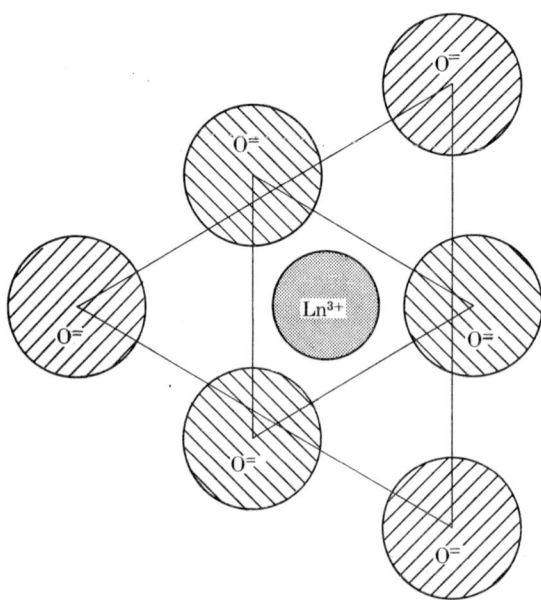

Fig. 5.1. Arrangement of the double triangular prism of water oxygen ions in the lanthanide ethylsulphates, in a plan view as seen along the crystallographic c-axis. In (R, θ, ϕ) coordinates, the Ln^{3+} ion is at the origin; there are three $O^=$ ions in the same plane at (0·252 nm, $\pi/2$, $\phi = (2n+1)\pi/3$), and six $O^=$ ions (above and below) at (0·237 nm, θ, $\phi = 2n\pi/3$) and (0·237 nm, $\pi-\theta$, $\phi = 2n\pi/3$) with θ approximately 40°. The values of R, θ vary slightly with the lanthanide ion; the values of R given are for Er^{3+}. The radii of the circles denoting ions are drawn on about half scale relative to the inter-ionic distances.

a different structure, similar to YCl_3. The anhydrous bromides and iodides form similar series, but less resonance work has been done on these.

A third series with somewhat different symmetry is formed by the double nitrates, formula type $Ln_2Mg_3(NO_3)_{12}$, $24H_2O$. The crystals are rhombohedral, space group $R\bar{3}$, and the hexagonal cell with $a = 1·1004(6)$ nm, $c = 2·4592(12)$ nm (values for the cerium salt) contains three formula units. The detailed X-ray crystallographic study has been carried out by Zalkin, Forrester, and Templeton (1963). It shows that the lanthanide ion is surrounded by twelve oxygen atoms at an average distance of 0·264 nm; these atoms, belonging to six nitrate ions, are at the corners of a somewhat irregular icosahedron. The Mg atoms are of two kinds, each surrounded by six water molecules whose oxygen atoms lie at the corners of an octahedron with an average Mg—O distance of 0·207 nm. Lanthanum is the usual diluent (yttrium does not fit into this lattice); bismuth has also been used but its crystals are not so stable chemically. Zinc or an ion of the $3d$ group can replace magnesium.

The X-ray results confirm a number of conjectures made about the structure from paramagnetic resonance studies. In particular, Judd (1957a) had deduced that the crystal field resembled closely that of an icosahedron. The presence of two types of sites for $3d$ ions, with roughly 2:1 abundance, was observed by Trenam (1953), and the spectra of these ions suggested they were octahedrally coordinated. In the copper salt Bleaney, Bowers, and Trenam (1955) found that the line width was reduced in the deuterated compound, suggesting that the copper ligands are water molecules, while the rather small line width in the spectrum of the lanthanide ion suggested that its immediate neighbours are predominantly nitrate groups (Bleaney, Hayes, and Llewellyn 1957). The site symmetry at the Ce ion found by Zalkin, Forrester, and Templeton (1963) is C_3, but the spectroscopic and resonance data have mostly been interpreted assuming the symmetry to be C_{3v} (see, however, Devine (1967)).

The arrangement of the nearest magnesium ions around a cerium ion is shown in Fig. 5.2; the nearest cerium neighbours are three at 0·856 nm and three at 0·859 nm, so that this compound is magnetically very dilute and very useful for magnetic cooling by adiabatic demagnetization. If the magnesium ions are replaced by paramagnetic ions of the $3d$ group, the compound is not particularly dilute, as can be seen from Fig. 5.2. The nearest neighbours to a Mg(1) site include six Ce at 0·698 nm and six Mg(2) at 0·715 nm, but the nearest neighbours to a Mg(2) site include another Mg(2) ion at only 0·499 nm.

In this series, and in the ethylsulphates and trichlorides, there is for magnetic purposes only one lanthanide ion per unit cell, at a site of threefold or sixfold symmetry. The ethylsulphates are also magnetically rather dilute, the nearest lanthanide neighbours being two at a distance of 0·71 nm along the c-axis. In the trichlorides the nearest lanthanide neighbours are again two along the c-axis, but at the much smaller distance of 0·43 nm.

Two other series of compounds have been examined in which threefold symmetry at the lanthanide site might be anticipated. These are LaF_3 and in contrast to this rather concentrated lanthanide compound, an unusually dilute series, lanthanum hexa-antipyrene iodide (HAPI), $La(C_{11}H_{12}ON_2)_6I_3$, (Baker and Rubins 1961, Baker and Williams 1961), doped with various lanthanide ions. The resonance results indicate a

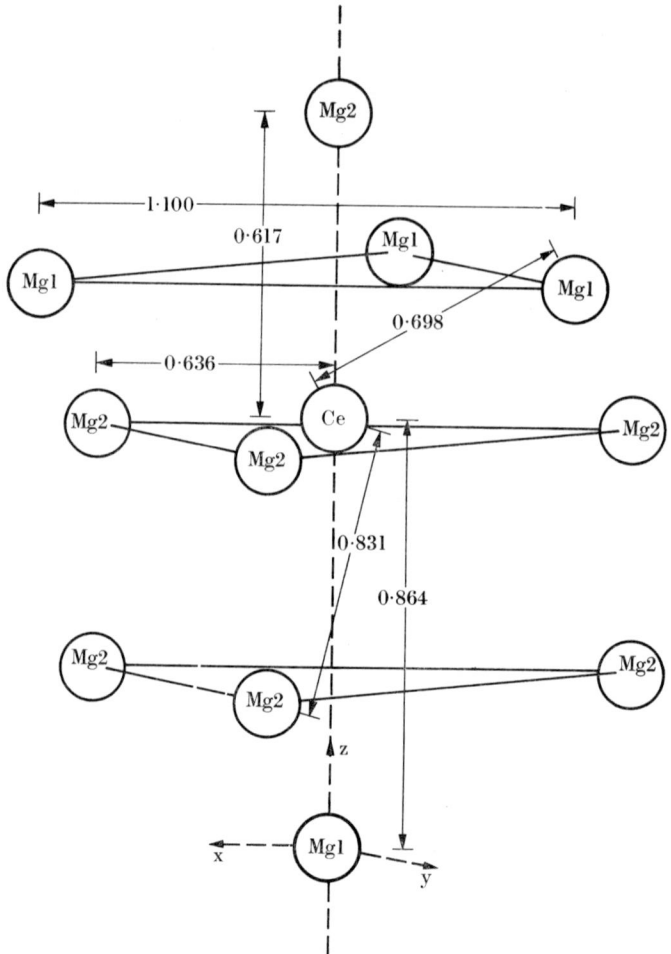

FIG. 5.2. Positions of the magnesium ions (distances in nanometres) relative to a cerium ion in cerium magnesium nitrate (Zalkin, Forrester, and Templeton 1963).

number of complexities in the crystal structure, including transitions at low temperatures, and we shall not discuss them further.

A rather different host lattice that has been used for both $4f$ and $3d$ ions is $CaWO_4$, and its isomorph, $SrWO_4$. These crystals (see Zalkin and Templeton 1964) are tetragonal, space group $I4_1/a$, with lattice constants $a = 0.5243$ nm, $c = 1.1376$ nm for the calcium salt. The point group at the Ca site is S_4; there are two calcium sites in the unit cell, but these are related by a reflection in the (001) plane, so that they are magnetically equivalent. Both tripositive and dipositive ions substitute for Ca^{2+}, and provided that charge compensation in the former case is remote, fourfold symmetry about the c-axis is maintained for the paramagnetic ion. The calcium ion has eight oxygen neighbours at an average distance of 0·246 nm, while the tungsten has tetrahedral coordination with a tetragonal distortion.

Paramagnetic resonance studies have also been made of lanthanide ions subject to crystal fields of octahedral or cubic symmetry. The former occurs in the compounds MgO and CaO, both of which form crystals of the NaCl type (space group O_h^5), with lattice constants 0·4203 and 0·4797 nm respectively. The cation lies at the centre of a regular octahedron of six O^{2-} ions, and lanthanide ions with three positive charges can be introduced as impurities in small concentration on cation sites. Their spectrum is observed to have cubic symmetry, and charge compensation must take place at remote sites. Divalent ions such as Eu^{2+} can also be introduced, and the undiluted compounds EuO, EuS, EuSe, EuTe have the same structure.

Trivalent lanthanide ions can be introduced into CdF_2, CaF_2, SrF_2, BaF_2, and $SrCl_2$, whose crystals also belong to the O_h^5 space group, with lattice constants as given in Fig. 5.3. The structure may be

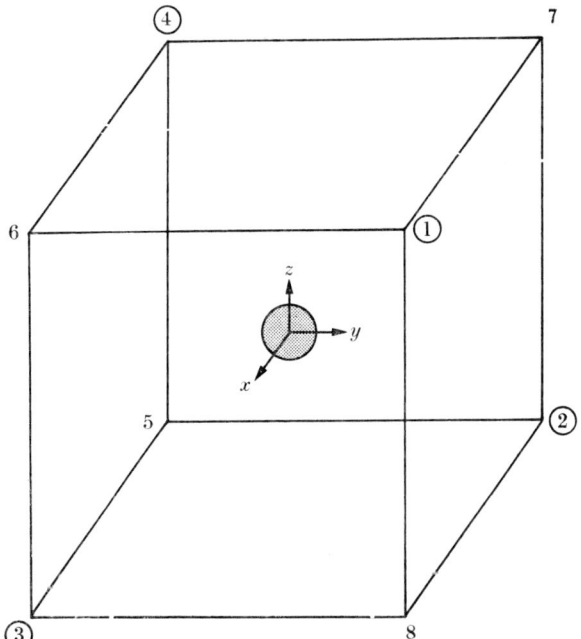

FIG. 5.3. Crystal structure of CaF_2 and isomorphous compounds. The Ca^{2+} ion is at the centre of a cube of eight F^- ions, and the (x, y, z) axes are the fourfold cubic axes. The lattice constants (twice the length of the cube edge in this diagram) are CdF_2, 0·540 nm; SrF_2, 0·586 nm; $SrCl_2$, 0·700 nm; CaF_2, 0·545 nm; BaF_2, 0·619 nm. The numbers 1 to 4, and 5 to 8, label the apices of the two tetrahedra which make up the cube.

regarded as a simple cubic array of fluorine ions with calcium ions at every other body centre. Each calcium is at the centre of a cube of eight fluorine ions (see Fig. 5.3), the Ca—F distance being 0·236 nm, and the lanthanide ions substitute for calcium. The tripositive ions are observed with cubic symmetry, or with tetragonal or trigonal symmetry. In the first case charge compensation must be remote; in the second an extra F$^-$ ion occupies the nearest vacant interstitial site, and in the third it is presumed that an O^{2-} ion replaces an F$^-$ ion on one apex of the immediate cube of fluorines (other possibilities may occur). A number of dipositive lanthanide ions have also been observed in CaF$_2$, occurring either naturally (such as Eu^{2+}, Sm^{2+}), or produced by irradiation.

Another crystal of the same basic structure is ThO$_2$ or CeO$_2$, where the cation is quadripositive and the anion dinegative. The lattice constants are 0·559 nm and 0·541 nm respectively, and crystals have been grown, both with four-valent lanthanide ions such as Tb^{4+} and with the more common tripositive ions, in which the paramagnetic resonance spectrum shows cubic symmetry.

5.2. The free ions

Outside the xenon core, the electrons in the lanthanide group of ions nearly all occupy either the $4f$ shell or the $6s$ shell. For the tripositive ions, the configuration is $4f^n$, where $n = Z-57$ and Z is the atomic number of lanthanum. For the dipositive ions the configuration is $4f^{n+1}$, and for the neutral atoms the ground state is $4f^{n+1}6s^2$, except for LaI which is $5d6s^2$ and GdI which is $4f^75d6s^2$. Since $6s^2$ is a closed shell the spectroscopic ground states of the atoms and dipositive ions are the same. For the few quadripositive ions measured the configuration is $4f^{n-1}$, and hence in nearly all the cases with which we shall be concerned the magnetic properties are decided by the number of electrons in the f-shell. The coupling between these electrons is not far from perfect LS-coupling, the ground state being determined by Hund's rules. There are, however, some appreciable corrections that arise from the fact that the spin-orbit coupling energy is not negligible in comparison with the Coulomb energy. For example, the splittings within the ground multiplet do not obey the Landé interval rule and corrections must be applied in deriving the value of the spin-orbit coupling parameter. These are summarized in a paper by Judd and Lindgren (1961) (see also Conway and Wybourne 1963), and their magnitude can be seen from Table 5.2, which gives the energy levels for the configuration $4f^6$. This table shows the energy levels calculated first, on pure LS-coupling, and second, after being corrected for second-order effects of the spin-orbit coupling, the spin-orbit parameter being chosen to give the best fit to the experimental values, for the

TABLE 5.2

Energy levels in the ground multiplet of $4f^6$, ^7F, ($L = 3$, $S = 3$, $J = 0$ to 6), in cm^{-1} relative to $J = 0$

J	SmI (free atom) Pure LS-coupling	Corrected	Experiment	Sm^{2+} in SrF_2	Sm^{2+} (calc)	Eu^{3+} in europium ethyl-sulphate	Eu^{3+} (calc)
6	3724	4021	4021	( )	3981	5000	5022
5	2660	3127	3125	(3100)	3099	3920	3921
4	1773	2275	2273	(2300)	2257	2880	2866
3	1064	1490	1490	1506	1481	1911	1888
2	532	812	812	805	808	1046	1063
1	177	291	293	263	291	376	374
0	0	0	0	0	0	0	0

The first three columns are for SmI; the first two are calculated by Conway and Wybourne (1963) using a value of $\zeta = 1064$ cm^{-1} chosen to fit the experimental values of Albertson (1937) in column 3. The effect of the second-order spin-orbit terms is shown by the difference between column 1 (pure LS-coupling) and column 2 (corrected for second-order terms). Column 4 gives values for Sm^{2+} (SmIII) in SrF_2, estimated from Wood and Kaiser (1962). Column 6 gives values for Eu^{3+} (EuIV) in europium ethylsulphate estimated from the analysis of Judd (1959b). Columns 5 and 7, calculated values for Sm^{2+} and Eu^{3+}, are from Ofelt (1963).

SmI spectrum. The corrections are quite considerable, especially for the intermediate levels. This table shows also estimated values for Sm^{2+} (in SrF_2) and for Eu^{3+} (in europium ethylsulphate). Both these ions are isoelectronic with neutral Sm, and a comparison of Sm and Sm^{2+} shows that the removal of the $6s^2$ electrons in the latter has only a small effect, considerably less than that of the extra nuclear charge in the case of Eu^{3+}.

The values of the spin-orbit coupling parameter ζ have been calculated using Hartree–Fock wave-functions by Blume, Freeman, and Watson (1964); the results are rather higher than the experimental values, as can be seen from the comparison in Table 5.3. This table gives also the ground term of the multiplet, and its value of g_J from a simple Landé formula, together with the energy of the first excited state of the multiplet. The values of g_J need a number of corrections. Besides the Schwinger electrodynamic correction for the spin g-factor

TABLE 5.3

Values of the spin orbit coupling parameter ζ for the tripositive ions of the 4f group, together with the Landé value of g_J for the ground state and the energy of the first excited state of the multiplet. The values of ζ are from Blume, Freeman, and Watson (1964). More accurate values of ζ and the energy of the first excited state, which vary slightly from salt to salt, are obtained from analyses of the optical spectra

Z		Ion	ζ (exp)	ζ (calc)	Term	g_J	Energy (cm^{-1})
58	$4f^1$	Ce^{3+}	640	740	$^2F_{\frac{5}{2}}$	$\frac{6}{7}$	0
					$^2F_{\frac{7}{2}}$		2200
59	$4f^2$	Pr^{3+}	750	878	$^3H_4$	$\frac{4}{5}$	0
					$^3H_5$		2100
60	$4f^3$	Nd^{3+}	900	1024	$^4I_{\frac{9}{2}}$	$\frac{8}{11}$	0
					$^4I_{\frac{11}{2}}$		1900
61	$4f^4$	Pm^{3+}			$^5I_4$	$\frac{3}{5}$	0
					$^5I_5$		1600
62	$4f^5$	Sm^{3+}	1180	1342	$^6H_{\frac{5}{2}}$	$\frac{2}{7}$	0
					$^6H_{\frac{7}{2}}$		1000
63	$4f^6$	Eu^{3+}	1360		$^7F_0$	0	0
					$^7F_1$	$\frac{3}{2}$	400
64	$4f^7$	Gd^{3+}		1717	$^8S_{\frac{7}{2}}$	2	0
					^6P		30000
65	$4f^8$	Tb^{3+}	1620	1915	$^7F_6$	$\frac{3}{2}$	0
					$^7F_5$	$\frac{3}{2}$	2000
66	$4f^9$	Dy^{3+}	1820	2182	$^6H_{\frac{15}{2}}$	$\frac{4}{3}$	0
					$^6H_{\frac{13}{2}}$		
67	$4f^{10}$	Ho^{3+}	2080	2360	$^5I_8$	$\frac{5}{4}$	0
					$^5I_7$		
68	$4f^{11}$	Er^{3+}	2470	2610	$^4I_{\frac{15}{2}}$	$\frac{6}{5}$	0
					$^4I_{\frac{13}{2}}$		6500
69	$4f^{12}$	Tm^{3+}	2750	2866	$^3H_6$	$\frac{7}{6}$	0
					$^3H_5$		†
70	$4f^{13}$	Yb^{3+}	2950	3161	$^2F_{\frac{7}{2}}$	$\frac{8}{7}$	0
					$^2F_{\frac{5}{2}}$	$\frac{6}{7}$	10000

† This level lies above $^3H_4$.

(which requires the use of 2·00232 for g_s instead of 2), and relativistic and diamagnetic corrections (usually of order 0·1 to 0·2 per cent), there are somewhat larger corrections due to admixtures of excited states through the spin-orbit coupling. For example, this makes the ground state of HoI (see Judd and Lindgren 1961)

$$0·9860\,|^4I_{\frac{15}{2}}\rangle - 0·1669\,|^2K_{\frac{15}{2}}\rangle + ..., \tag{5.1}$$

but fortunately the admixture only effects the Zeeman splitting in the square, i.e.

$$g_J(\text{corrected}) = (0\cdot9860)^2 g(^4I_{\frac{15}{2}}) + (0\cdot1669)^2 g(^2K_{\frac{15}{2}}) + ..., \qquad (5.2)$$

so that the effect is fairly small (usually less than 1 per cent). The overall effect of the various corrections can be assessed from Table 5.4, which summarizes some experimental results for the neutral atoms, mostly obtained by atomic beam measurements.

5.3. Crystal field theory—C_{3h} symmetry

Early measurements of the susceptibility of rare-earth salts in the vicinity of room temperature showed that it was generally rather close to that expected from an assembly of free ions. Curie's law was fairly well obeyed, and the magnitude of the susceptibility was close to that expected from values of g_J and J appropriate to the ground term of the multiplet. Since the energy of the first excited state of the multiplet is generally a few thousand cm^{-1}, which is large compared with room temperature ($300k \equiv 200$ cm^{-1}), contributions to the susceptibility from the excited states are negligible except for the ions $4f^5$ (Sm^{3+}) and $4f^6$ (Sm^{2+}, Eu^{3+}) where the first excited state is unusually low in energy (see, for example, Van Vleck (1932)).

The inference from these results is that the magnetic electrons are but little affected by the surroundings of the magnetic ion in a crystal, and since the $4f$ electrons form an 'inner shell' this is not surprising. (In fact $4f$ electrons have about the same radial extension as $3d$ electrons, but the diffuse outer electron shells $5s$, $5p$ keep the ligand ions away.) At temperatures well below room temperature departures from Curie's law appear, which are attributed to interactions with the ligand ions with energies of the order 10–100 cm^{-1}. Since bonding effects are expected to be very small or negligible for $4f$ electrons, the interaction with the ligands is treated by the approximation of a crystal potential V, set up by the electric charges on the ligands, in which the magnetic electrons move. Numerical calculation of this potential is difficult, but it must reflect the symmetry of the crystal at a rare-earth ion site. If, further, we confine ourselves to potential terms that have matrix elements within the manifold of $4f$ states, terms of odd parity can be omitted. Further it can be shown from the properties of spherical harmonics that terms of higher degree than the sixth need not be considered, since their matrix elements are zero for f-electrons.

To illustrate the theory it is convenient to take a concrete example, and for this purpose we choose a crystal potential appropriate to

TABLE 5.4

Spectroscopic data for the neutral atoms with $4f^n$ configurations from atomic beam measurements. A, B are the magnetic dipole and electric quadrupole hyperfine constants. The isotopes marked † are radioactive. For the nuclear moments, some data from other experimental methods are included (values marked ‡ include the diamagnetic correction)

	Atom	State	g_J	Isotope	I	A (MHz)	B (MHz)		μ_I (nuclear magnetons)	Q (barns)
$4f^3$	Pr	$^4I_{9/2}$	0·731044(33)	141	5/2	+926·20894(27)	−11·87375(543)		+4·25(5)	−0·0589(42)
				142†	2	67·5(5)	7·0(20)			
$4f^4$	Nd	$^5I_4$	0·6032(1)	143	7/2	−195·649(9)	+122·25(28)	{‡‡	−1·063(5)	−0·484(20)
									−1·085(60)	
				145	7/2	−121·627(27)	+64·60(37)	{‡‡	−0·654(4)	−0·255(10)
				147†	7/2	−95·531(19)	+60·65(70)		−0·675(40)	
$4f^5$	Pm	$^6H_{5/2}$		147†	7/2	447(9)	267(70)		+2·58(7)	+0·74(20)
		$^6H_{7/2}$	0·831(5)	151†	(5/2)	358(23)	777(94)			
$4f^6$	Sm	$^7F_0$		147	7/2	−33·4971(1)	−58·7543(5)		−0·8074(7)	−0·20(2)
		$^7F_1$	1·49840(5)	149	7/2	−27·6139(1)	+16·9195(2)		−0·665	+0·058(6)
				153†	3/2	−2·100(5)	+289·042(4)		−0·021(1)	+1·1(3)
$4f^7$	Eu	$^8S_{7/2}$	1·99340(7)	151	5/2	−20·0523(2)	−0·7012(35)	‡‡	+3·4630(6)	+1·16
				152†	3	9·345(6)	1·930(165)		1·912(4)	
				153	5/2	−8·8532(2)	−1·7852(35)	‡‡	+1·5292(8)	+2·9
	Gd			155	3/2				−0·2567(6)	
								‡‡	−0·2584(5)	+1·59(16)
				157	3/2				−0·3363(12)	+1·70(16)
$4f^9$	Tb	$^6H_{15/2}$	1·3225(30)	159	3/2	+667·9(16)	+1381(19)		+2·000(12)	+1·12(10)
									+1·994(4)	
$4f^{10}$	Dy	$^5I_8$	1·24166(7)	161	5/2	−115·8(10)	+1102(15)		−0·47(9)	+2·36(40)
				163	5/2	+162·9(6)	+1150(20)		+0·66(13)	+2·46(40)
$4f^{11}$	Ho	$^4I_{15/2}$	1·1951540(36)	165	7/2	+800·5844(4)	−1667·957(15)		+4·03(5)	+2·4
$4f^{12}$	Er	$^3H_6$	1·16377(17)	167	7/2	−120·4864(5)	−4552·959(23)	‡‡‡‡	−0·5647(24)	+2·827(12)
				169†	1/2	725·46(31)	—		0·513(25)	
$4f^{13}$	Tm	$^2F_{7/2}$	1·1411891(32)	169	1/2	−374·137641(10)		‡‡‡‡	−0·2310(15)	
				170†	1	200(3)	1010(15)		0·2448(36)	
				171†	1/2	372(4)			0·2277(36)	
	Yb			171	1/2			‡‡‡‡	+0·49188(2)	
				173	5/2				−0·67755(2)	+2·8(2)

a site of C_{3h} symmetry. In many ways this is one of the simplest, and historically it is the most important because of the experimental and theoretical work on the rare-earth ethylsulphates. Single crystals of these salts can be grown from aqueous solution for all the rare-earth ions, and were used for most of the early susceptibility measurements at low temperatures and for magnetic resonance work. The existence of experimental results for a homologous series was a vital touchstone in the development of a successful theoretical approach by Elliott and Stevens. The anhydrous chlorides, on which work was initiated later by Hutchison, have the same site symmetry, and the resonance results for this series are remarkably similar to those for the ethylsulphates. It is therefore convenient to consider them together.

For a site of C_{3h} (or D_{3h}) symmetry the crystal potential takes the form (omitting terms whose matrix elements are zero for f-electrons)

$$V = A_2^0(3z^2-r^2)+A_4^0(35z^4-30z^2r^2+3r^4)+ \\ +A_6^0(231z^6-315z^4r^2+105z^2r^4-5r^6)+ \\ +A_6^6(x^6-15x^4y^2+15x^2y^4-y^6). \quad (5.3)$$

This can be written in the abbreviated form

$$V = \sum V_k^q = A_2^0 P_2^0 + A_4^0 P_4^0 + A_6^0 P_6^0 + A_6^6 P_6^6 \quad (5.4)$$

by using the functions P_k^q listed in Table 15. (The terms P_4^3 and P_6^3 are proportional to odd powers of z and are missing because of reflection symmetry in the xy plane for C_{3h}; they will be needed when dealing with the double nitrates, which have C_{3v} symmetry). The first three terms are independent of x, y and are axially symmetric about the z-axis; the fourth term has sixfold symmetry about this axis. Its form depends on a particular choice for the x-axis relative to the crystal, and there are six such choices, related by rotations of 60° about the z-axis.

The effect of this potential can be found by integrating over the wave-function of each electron and then summing over all the magnetic electrons. As, however, neither the coefficients A_k^q nor the radial parts of the atomic wave-functions are known with any accuracy, only relative values due to the different angular dependences of the wave-functions are known precisely. The calculation of the latter is greatly simplified by use of the simple relations between the matrix elements of the potential operators and the appropriate 'spin' operators, as discussed in Chapter 16. Within a manifold of states for which J is constant the potential (5.3), for example, can be replaced by

$$A_2^0\langle r^2\rangle\langle J\|\ \alpha\ \|J\rangle O_2^0 + A_4^0\langle r^4\rangle\langle J\|\ \beta\ \|J\rangle O_4^0 +$$
$$+ A_6^0\langle r^6\rangle\langle J\|\ \gamma\ \|J\rangle O_6^0 + A_6^6\langle r^6\rangle\langle J\|\ \gamma\ \|J\rangle O_6^6. \quad (5.5)$$

Here $\langle r^k\rangle$ is the mean value of r^k averaged over the atomic wavefunctions; some calculated values are given in Table 5.1. $\langle J\|\ \alpha\ \|J\rangle$, etc., are numerical coefficients listed for each ion in Table 20, and the O_k^q are the spin operators listed in Table 16 whose matrix elements are given in Tables 17. Thus the whole calculation is reduced to a matter of looking up numerical values in the appropriate Tables, apart from the $A_k^q\langle r^k\rangle$ which are regarded as adjustable parameters.

The theory outlined above is based on the work of Stevens and Elliott (Stevens 1952a, Elliott and Stevens 1952, 1953a, 1953b), who attempted a systematic treatment of the rare-earth ethylsulphates. Within a manifold of given J the results can be illustrated rather simply. Spin operators of the type O_k^0 contain only J_z and hence have only diagonal matrix elements for the $2J+1$ states characterized by different values of the magnetic quantum number J_z or M. These diagonal matrix elements are the same for the states $+J_z$ and $-J_z$ since the operators contain only even powers of J_z. Since states with different values of $|J_z|$ will in general have different energy, a crystal potential containing only such terms will produce a series of doublets characterized by states of the type $|\pm M\rangle$, with one singlet state, $|0\rangle$, if J is integral.

Spin operators O_k^q for which m is not zero have only off-diagonal elements and therefore admix states of different M to give a resultant state of the type

$$\sum_M c_M |J, M\rangle \quad (5.6)$$

with, of course, $\sum c_M^2 = 1$ to satisfy the normalization condition. In any given admixture the values of M are those with successive differences equal to q. Thus inclusion of the last spin operator O_6^6 in (5.5) produces states of the form

$$c_{M+6}|J, M+6\rangle + c_M|J, M\rangle + c_{M-6}|J, M-6\rangle,$$

where the number of states admixed is not more than three because the maximum value of J in the ground state is 8.

So far we have considered just a first-order perturbation approach in which only matrix elements within a manifold of given J are included. In a second-order approach it is necessary to include matrix elements between states of different J and these can again be obtained by the method of operator equivalents. The most important elements are those between values of J which differ by unity, since these are adjacent in energy; they are given in eqns (16.33)–(16.36) for C_{3h} symmetry, from which it can be seen that they produce states of the form

$$\sum_M c_M |J, M\rangle + \sum_M c'_M |J', M\rangle, \qquad (5.7)$$

where the values of M in J' are the same as those in J except that if $J' > J$ there may be an additional value of M which is greater than J. For the sake of simplicity admixtures of states of different J will be mostly ignored in the following general discussion; their inclusion alters the numerical values, but (with a few exceptions) does not affect the basic properties.

In this discussion it is convenient to treat first the ions with Kramers degeneracy (half-integral values of J), for which the energy levels are degenerate in pairs, with states linked by time reversal. We shall, however, leave the ions with a half-filled shell until later, since their ground state is $^8S_{\frac{7}{2}}$, which is split by the crystal field only through higher-order perturbations. Secondly we shall discuss the position for non-Kramers ions, with integral values of J.

Ions with Kramers degeneracy

For ions with an odd number of electrons (half integral values of J), the crystal field lifts the degeneracy completely except for the necessary twofold degeneracy imposed by Kramers theorem, so that the levels consist of $(J+\frac{1}{2})$ doublets. Since for such ions the maximum value of J in the ground state is $\frac{15}{2}$, the wave-functions of the Kramers doublets are linear combinations of states $|\pm M\rangle$ of the form

$$\pm \tfrac{15}{2}, \pm \tfrac{3}{2}, \mp \tfrac{9}{2},$$
$$\pm \tfrac{13}{2}, \pm \tfrac{1}{2}, \mp \tfrac{11}{2}, \qquad (5.8)$$
$$\pm \tfrac{7}{2}, \mp \tfrac{5}{2}.$$

The actual energy levels and relative admixtures of the states depend of course on the size of the various parameters of the crystal field. In general, the doublets are separated in energy by 10–100 cm^{-1}, so that a magnetic resonance transition is observable in the microwave region only between the two components of a doublet. At temperatures where excited doublets are populated the lifetimes of the excited states are generally so short because of relaxation to the lattice that transitions between them are too broad to be observable, so that in practice resonance is almost always restricted to the ground doublet.

To find the resonance condition we must calculate the Zeeman effect for each doublet. In a first approximation, where we consider only matrix elements within the ground manifold of a given value of $\mathbf{J}$, the Zeeman operator $\mathbf{L}+2\mathbf{S}$ reduces to the simpler form $g_J\mathbf{J}$ or, in the notation of Elliott and Stevens and Chapter 16, $\langle J\| \Lambda \|J\rangle\mathbf{J}$. Thus calculation of the first-order Zeeman effect is reduced to finding the matrix elements of J_x, J_y, J_z within each doublet. Since we have axial symmetry, the matrix elements of J_x, J_y are equal, but in general different from those of J_z, so that the Zeeman effect within each doublet is described by a spin Hamiltonian with an effective spin $S = \frac{1}{2}$ and an anisotropic g-'tensor' with axial symmetry of the form

$$\mathcal{H} = g_\| \beta H_z S_z + g_\perp \beta (H_x S_x + H_y S_y) \tag{5.9}$$

where

$$\begin{aligned} g_\| &= 2\langle J\| \Lambda \|J\rangle\langle +| J_z |+\rangle, \\ g_\perp &= \langle J\| \Lambda \|J\rangle\langle +| J_+ |-\rangle. \end{aligned} \tag{5.10}$$

Here $|+\rangle$, $|-\rangle$ are the two components of the doublet given by one of the combinations (5.8). To preserve the correct sign of $g_\|$, which is important for some purposes, the two states $|+\frac{1}{2}\rangle$, $|-\frac{1}{2}\rangle$ of the effective spin should be chosen to satisfy the correspondence

$$g_\perp \langle +\tfrac{1}{2}| S_+ |-\tfrac{1}{2}\rangle \equiv \langle J\| \Lambda \|J\rangle\langle +| J_+ |-\rangle.$$

Physically this correspondence is required in order that the spin Hamiltonian shall give the correct transition intensities when circularly polarized radiation is used (see Chapter 3).

It is readily seen that there are no matrix elements of the operators J_+, J_- between the two states of the first doublet listed in (5.8), so that $g_\perp = 0$ and no transition is allowed within the doublet (cf. §§ 1.8, 3.2). For the other two doublets $g_\perp$ is finite and transitions are allowed; values of $g_\|$ and $g_\perp$ can then be measured in a resonance experiment with the external field parallel and perpendicular to the z-axis respectively. For the second doublet in (5.8) this means that all the admixture coefficients can be determined, since we have two pieces of experimental information and the normalizing relation for the coefficients means there are really only two unknowns though there are three coefficients. For the third doublet, which can be written in the form

$$\cos \theta\, |\pm\tfrac{7}{2}\rangle + \sin \theta\, |\mp\tfrac{5}{2}\rangle, \tag{5.11}$$

we have

$$\begin{aligned} g_\| &= \langle J\| \Lambda \|J\rangle(7\cos^2\theta - 5\sin^2\theta), \\ |g_\perp| &= \langle J\| \Lambda \|J\rangle\, 2\, |\cos\theta \sin\theta|\, \{(J+\tfrac{7}{2})(J-\tfrac{5}{2})\}^{\frac{1}{2}}. \end{aligned} \tag{5.12}$$

This gives a relation between $g_\parallel$ and $g_\perp$ which is exact only in so far as we can neglect admixtures of states of different J, and experimental measurement of the two g-values gives a simple test of the validity of this approximation.

In more accurate measurements it may be necessary to take account of the corrections to $\langle J \| \Lambda \| J \rangle$ discussed in § 5.2. The most important of these are the use of 2·0023 for g_s instead of 2, and the correction for the breakdown of LS-coupling, which for most purposes can be adequately allowed for by the use of a modified value of $\langle J \| \Lambda \| J \rangle$ or g_J such as that given by eqn (5.2) for a special case. However in general the important corrections to the Zeeman energy calculation arise through the admixture of states of different J (principally $J \pm 1$, the state adjacent in energy); if these admixtures are known the corrections to the Zeeman energy can be found using §§ 16.3, 16.4 and the values of $\langle J+1 \| \Lambda \| J \rangle$ listed in Table 20.

For smaller values of J not all the values of M shown in eqns (5.6) may appear. Thus for $J < \tfrac{11}{2}$ the second doublet is just $|\pm\tfrac{1}{2}\rangle$, for which the values of both $g_\parallel$ and $g_\perp$ are completely defined

$$g_\parallel = \langle J \| \Lambda \| J \rangle;\ g_\perp = (J+\tfrac{1}{2})\langle J \| \Lambda \| J \rangle \qquad (5.13)$$

if admixtures of different J are negligible. Similarly in this case the first doublet will span just the values $|\pm\tfrac{3}{2}\rangle$ and $|\mp\tfrac{9}{2}\rangle$ so that it can be written in a form similar to (5.11). Obviously for every doublet whose states involve two values of J_z there is another doublet with orthogonal states given by changing θ to $\theta+\pi/2$, which has a different energy due to the crystal field. Similarly, when three values of J_z are admixed there are three orthogonal doublets, all with different energy.

At field strengths where the Zeeman energy is not small compared with the crystal field splittings it may be necessary to include Zeeman effects of higher order than the first. The second-order Zeeman effect generally results in both levels of a doublet being displaced equally by an amount proportional to H^2; this does not change the frequency of a transition between the two states of a doublet as far as terms of order H^2 are concerned. The third-order Zeeman effect does affect the transition frequency, since it may be of opposite sign for the two components of a doublet and displace them in opposite directions. This can be seen in another way, as a by-product of the second-order Zeeman effect. This modifies the wave-functions of a doublet in such a way that instead of $|+\rangle$, $|-\rangle$ they take the form

$$(1+\alpha^2)^{-\frac{1}{2}}\{|+\rangle+\alpha\,|+'\rangle\} \qquad (5.14)$$
$$(1+\alpha^2)^{-\frac{1}{2}}\{|-\rangle+\alpha\,|-'\rangle\}$$

where α, the admixture coefficient for the other states, is proportional to H. If we now calculate the first-order Zeeman effect using the new states, its energy, which is proportional to H, will be modified by a fraction of order α^2, so that we have a correction to the energy separation of order H^3. This effect is most striking for a doublet for which $g_\perp = 0$, which is not split by a perpendicular magnetic field in first-order, but which may be split in third-order and give a weakly allowed transition at high magnetic fields. In the case of the first doublet listed in (5.8) this can be regarded as a result of admixing of states of the other two doublets by off-diagonal matrix elements of the Zeeman interaction.

Non-Kramers ions

For the rare-earth ions that have an even number of electrons in the $4f$ shell, the values of J are integral, with a maximum value of 8 in the ground state. Since the crystal potential for C_{3h} symmetry (eqn (5.4)) admixes states whose values of J_z differ by 6, it is readily found from the spin operators that its effect on a manifold of $2J+1$ states is to give a series of doublet and singlet energy levels. The doublets are of the form

$$\pm 8,\ \pm 2,\ \mp 4 \qquad (5.15)$$
$$\pm 7,\ \pm 1,\ \mp 5$$

while the singlets have admixtures of the form

$$|3^s\rangle = \frac{1}{\sqrt{2}}\{|+3\rangle+|-3\rangle\};\ |3^a\rangle = \frac{1}{\sqrt{2}}\{|+3\rangle-|-3\rangle\} \qquad (5.16)$$

or a linear combination of the form

$$+6,\ 0,\ -6. \qquad (5.17)$$

The singlets (5.16) arise in the following way. The states $M = \pm 3$ are degenerate except through the action of the V_6^6 term in the crystal field, which has a matrix element between them; this splits them by an amount directly proportional to the magnitude of the V_6^6 term, with states of the simple symmetrical and antisymmetrical form given in (5.16). It can be seen that the operator J_z then has a matrix element

$$\langle 3^s|\,J_z\,|3^a\rangle = 3\sqrt{2}$$

and it follows that an oscillatory magnetic field along the z-axis can produce transitions between them. Because of the magnitude of the V_6^6 term, this transition will usually be in the infrared.

Similarly, the states $M = \pm 6$ are degenerate except through the action of the V^6 term which has matrix elements

$$\langle +6|\, V_6^6\, |0\rangle = \langle 0|\, V_6^6\, |-6\rangle.$$

The energy matrix will thus be a 3×3 matrix, but we can simplify it by choosing as the basic states the linear combinations

$$|6^s\rangle = \frac{1}{\sqrt{2}}\{|+6\rangle+|-6\rangle\}\,;\quad |6^a\rangle = \frac{1}{\sqrt{2}}\{|+6\rangle-|-6\rangle\} \quad (5.18)$$

since the second state $|6^a\rangle$ then has no matrix element with $|0\rangle$ and is thus an eigenstate. The other eigenstates are found from the energy matrix

$$\begin{array}{c|cc} & |6^s\rangle & |0\rangle \\ \hline |6^s\rangle & 0 & \sqrt{(2)}c \\ |0\rangle & \sqrt{(2)}c & d \end{array}$$

where $d = \langle 0|\sum_k V_k^0 |0\rangle - \langle 6|\sum_k V_k^0 |6\rangle$ and $c = \langle 0|\, V_6^6\, |\pm 6\rangle$. The eigenstates are clearly of the form

$$p\,|6^s\rangle + q\,|0\rangle;\quad (p^2+q^2 = 1). \quad (5.19)$$

When $c \ll d$ the state that lies close to $|6^a\rangle$ in energy has approximately

$$q = -\sqrt{(2)}(c/d)$$

and is depressed below $|6^a\rangle$ by an amount $\Delta = 2c^2/d$.

The basic states for the ground doublet are thus

$$|\xi'\rangle = p\,|6^s\rangle + q\,|0\rangle$$
$$|\eta'\rangle = |6^a\rangle$$

where q is small and p is close to unity. In the presence of a magnetic field it is convenient to replace these states by the linear combinations

$$|\xi\rangle = \frac{1}{\sqrt{2}}(|\xi'\rangle+|\eta'\rangle) = \tfrac{1}{2}(1+p)\,|+6\rangle - \tfrac{1}{2}(1-p)\,|-6\rangle + (q/\sqrt{2})\,|0\rangle$$

$$|\eta\rangle = \frac{1}{\sqrt{2}}(|\xi'\rangle-|\eta'\rangle) = -\tfrac{1}{2}(1-p)\,|+6\rangle + \tfrac{1}{2}(1+p)\,|-6\rangle + (q/\sqrt{2})\,|0\rangle.$$

(5.19a)

The Zeeman interaction $Z = -\boldsymbol{\mu}\cdot\mathbf{H} = \Lambda\beta\mathbf{J}\cdot\mathbf{H}$ has no matrix elements between $|\xi\rangle$ and $|\eta\rangle$ and opposite expectation values in either state:

$$\langle\xi|\,Z\,|\xi\rangle = -\langle\eta|\,Z\,|\eta\rangle = 6\Lambda\beta p H_z \approx 6\Lambda\beta\left(1-\frac{q^2}{2}\right)H_z.$$

Using a fictitious spin $S = \tfrac{1}{2}$ the spin Hamiltonian can be written as

$$\mathscr{H} = g_{\parallel}\beta H_z S_z + \Delta_x S_x + \Delta_y S_y, \tag{5.20}$$

where approximately

$$g_{\parallel} = 12\Lambda\left(1-\frac{q^2}{2}\right) = 12\Lambda\left(1-\frac{c^2}{d^2}\right),$$
$$\Delta = (\Delta_x^2 + \Delta_y^2)^{\frac{1}{2}} = 2c^2/d. \tag{5.20a}$$

The individual values of Δ_x, Δ_y have no significance. Transitions occur at

$$\hbar\omega = \{(g_{\parallel}\beta H_z)^2 + \Delta^2\}^{\frac{1}{2}} \tag{5.21}$$

and are allowed when the oscillatory magnetic field is along the z-axis. The square of the magnetic dipole matrix element for the transition is

$$|\mu_z^2| = \tfrac{1}{4}(g_{\parallel}\beta\Delta/\hbar\omega)^2 \tag{5.22}$$

and grows progressively smaller as the frequency increases. This corresponds to a change from the states $|\xi'\rangle$, $|\eta'\rangle$ which are correct in zero magnetic field through intermediate states in a finite magnetic field to the states $|\xi\rangle$, $|\eta\rangle$ of eqn (5.19a), which are correct when the Zeeman interaction is very large compared with Δ.

We return now to the doublet levels whose states are of the form (5.15). Their behaviour is considered in detail in § 18.5 (see also § 3.14) and we give here only an outline. Operating with J_z gives diagonal matrix elements of opposite sign for the two states of a doublet so that they have a linear Zeeman splitting when a magnetic field is applied along the z-axis. There are no matrix elements between the two states, so there is no Zeeman splitting with a perpendicular magnetic field, nor will transitions be allowed between them through the action of an oscillatory perpendicular magnetic field. However the fact that we are dealing with non-Kramers ions means that the doublets can be split by a crystal field of lower symmetry, such as might arise through the Jahn–Teller effect or through crystal imperfections. Any such splitting that arises from a matrix element of the distorted crystal field between the two states of the doublet will admix the two states, so that a transition is now allowed between them with an oscillatory magnetic field along the z-axis, in the same way as for the levels considered earlier in this section. An appropriate 'spin Hamiltonian' for such a doublet ($S = \frac{1}{2}$) is again (5.20), but now Δ_x, Δ_y are allowed to have a distribution of values and Δ does not have just one value. The quantum of energy required for a transition is given by (5.21) and involves Δ, so that it will have a spread of values; the transition intensity also depends on Δ (eqn (5.22)) and vanishes for $\Delta = 0$, so that we expect a broad line, asymmetrical in shape. (This contrasts with the situation considered above, where the transition was allowed without any distortion of the crystal field, and therefore is narrow and normal in shape.)

When the point symmetry at the lanthanide site lacks a plane of inversion symmetry, a non-Kramers ion can have an electric dipole moment normal to this plane, as discussed in §§ 15.10, 18.5. In C_{3h} symmetry, there is only one plane of reflection symmetry, normal to the c-axis, so that an electric dipole moment can exist in a non-Kramers ion normal to, but not parallel to, the c-axis. If the effect of an electric field on this dipole moment is represented by an additional term (cf. eqns (3.108) or (18.67))

$$g_{\perp}^{(E)} \beta (E_x S_x + E_y S_y)$$

in the spin Hamiltonian, transitions will be allowed for an oscillatory electric field $(E_1)_\perp$ normal to the c-axis, whose intensity depends on $(g_{\perp}^{(E)} E_1)^2$. The transition rate is now given by eqns (3.109), (3.110).

If resonance is observed at constant frequency and variable magnetic field it will lie wholly at field strengths below the value $H_0 = (\hbar\omega/g_{\parallel}\beta)$ corresponding to $\Delta = 0$, and will be asymmetrical in shape. The exact line shape will depend on the fraction of ions with a given value of Δ^2 (see § 3.14), and will be different according to whether transitions are induced by $(H_1)_\parallel$ or by $(E_1)_\perp$. In particular, if the transition is due to $(H_1)_\parallel$ it will have zero intensity at $H_z = H_0$ while if it is due to $(E_1)_\perp$ it will have a finite intensity (in fact the maximum intensity) at $H_z = H_0$ (see Fig. 3.20).

Although we have not yet considered hyperfine structure in this chapter it is convenient to point out here that for a non-Kramers doublet the hyperfine Hamiltonian will have the simple form

$$A_{\parallel} S_z I_z + P_{\parallel} \{I_z^2 - \tfrac{1}{3} I(I+1)\} \tag{5.23}$$

and the only change required above is that for each hyperfine component $g_{\parallel} \beta H_z$ must be replaced by $(g_{\parallel} \beta H_z + A_{\parallel} m)$ in eqn (5.21), where $m = I_z$.

5.4. Magnetic hyperfine structure

A general expression for the hyperfine interaction is given in eqn (17.61). We shall consider first the magnetic interaction where the nuclear magnetic moment can be considered as interacting with a local magnetic field due to the surrounding electrons. This field, $\mathbf{H}_e$, arises partly from the orbital motion of the electrons and partly from their spin magnetization. For a free ion in LS-coupling this resultant electronic field precesses rapidly about the angular momentum vector $\mathbf{J}$, and to a good approximation only that component of the field that is parallel to $\mathbf{J}$ and hence a constant of the motion need be retained. This makes it possible to reduce the magnetic hyperfine Hamiltonian

$$\mathcal{H}_n = 2g_n\beta\beta_n\langle r^{-3}\rangle[\mathbf{L}\cdot\mathbf{I} + \xi\{L(L+1)-\kappa\}(\mathbf{S}\cdot\mathbf{I}) - \tfrac{3}{2}\xi\{(\mathbf{L}\cdot\mathbf{S})(\mathbf{L}\cdot\mathbf{I}) + (\mathbf{L}\cdot\mathbf{I})(\mathbf{L}\cdot\mathbf{S})\}]$$

$$= 2g_n\beta\beta_n\langle r^{-3}\rangle\,(\mathbf{N}\cdot\mathbf{I}) \quad (5.24)$$

to the simple form

$$\mathcal{H}_n = A_J\mathbf{J}\cdot\mathbf{I} \quad (5.25)$$

where

$$A_J = 2g_n\beta\beta_n\langle r^{-3}\rangle\langle J\|N\|J\rangle + A_s'. \quad (5.26)$$

In these formulae g_n is the nuclear magnetogyric ratio appropriate to the nuclear magnetic moment expressed in nuclear magnetons, and $\langle J\|N\|J\rangle$ is a numerical factor given by eqns (17.51), (17.52) and listed for each ion in Table 20. A_s' is discussed below.

In the ground states of the rare-earth ions (with the exception of the half filled shell $^8S_{\frac{7}{2}}$) the magnetic hyperfine structure is due mainly to the orbital rather than the spin contribution. To illustrate this we carry out the numerical calculation of $\langle J\|N\|J\rangle$ for the ground state of the Tb^{3+} ion, $^7F_6$ ($L=3$, $S=3$, $J=6$). From eqn (17.51) we have

$$\langle J\|N\|J\rangle = \frac{1}{J(J+1)}(\mathbf{L}\cdot\mathbf{J}) \text{ (orbital contribution)} +$$
$$+\frac{1}{J(J+1)}\xi\{L(L+1)(\mathbf{S}\cdot\mathbf{J}) - 3(\mathbf{L}\cdot\mathbf{J})(\mathbf{L}\cdot\mathbf{S})\} \text{ (spin contribution),}$$

$$(5.27)$$

where from eqn (17.52) for this ion, $(\mathbf{L}\cdot\mathbf{J}) = (\mathbf{S}\cdot\mathbf{J}) = \tfrac{1}{2}J(J+1)$ and $(\mathbf{L}\cdot\mathbf{S}) = \tfrac{3}{14}J(J+1)$ and from eqn (17.46) $\xi = -\tfrac{1}{135}$. Hence

$$\langle J\|N\|J\rangle = \tfrac{1}{2} + \tfrac{1}{18} = \tfrac{5}{9} \text{ for } Tb^{3+},\ ^7F_6,$$

where $\tfrac{1}{2}$ represents the orbital contribution and $\tfrac{1}{18}$ the spin contribution. The reason why the former dominates is that we have quite a large orbital momentum making its full contribution, while the spin magnetization is rather uniformly distributed at any given distance so that the net field it produces at the nucleus is small (for a distribution with spherical symmetry it would vanish).

For an ion with a half-filled shell ($^8S_{\frac{7}{2}}$ ground state) the spin magnetization has spherical symmetry, and the orbital momentum is zero; we should therefore expect the magnetic hyperfine constant A to be zero. In practice a finite value is obtained (see §§ 4.7, 5.9), and this is attributed mainly to the 'core-polarization' effect (see §§ 17.5, 17.6) which produces a non-zero spin density at the nucleus. This is represented by a term similar to that in eqn (17.60).

$$A_c(\mathbf{S}\cdot\mathbf{I}) = -2g_n\beta\beta_n\langle r^{-3}\rangle\kappa(\mathbf{S}\cdot\mathbf{I}). \tag{5.28}$$

In addition we have a contribution from relativistic effects (see below) which is also of the form $(\mathbf{S}\cdot\mathbf{I})$ so that it is simplest to lump the two effects together in a term $A_s(\mathbf{S}\cdot\mathbf{I})$. As above, we must project $\mathbf{S}$ onto $\mathbf{J}$ using eqns (17.52), (17.53) so that we obtain for the contribution A'_s in eqn (5.26)

$$A'_s = A_s\frac{(\mathbf{S}\cdot\mathbf{J})}{J(J+1)} = A_s(\langle J\|\Lambda\|J\rangle - 1). \tag{5.29}$$

We consider now an ion subjected to a crystal field. Within the approximation where we need include only matrix elements within a given manifold J, the magnetic hyperfine calculation is simple because, like the Zeeman interaction, it requires only the matrix elements of the operator $\mathbf{J}$. There is therefore a linear relation between the hyperfine and the Zeeman interaction. For any sub-space of the $2J+1$ manifold which we can represent by a fictitious spin $\mathbf{S}$, and for which the Zeeman interaction takes the form $\beta\mathbf{H}\cdot\mathbf{g}\cdot\mathbf{S}$, the hyperfine interaction will have the form $\mathbf{S}\cdot\mathbf{A}\cdot\mathbf{I}$, where $\mathbf{A}$ is a 'tensor' which obviously has the same principal axes as the g-'tensor'. The principal values of these tensors are connected by the relation

$$\frac{A_x}{g_x} = \frac{A_y}{g_y} = \frac{A_z}{g_z} = \frac{A_J}{g_J} \tag{5.30}$$

where $g_J = \langle J\|\Lambda\|J\rangle$. In measurements at constant frequency and varying field, this relation means that the hyperfine separation will be the same in field along each of the principal axes; except in the case of no anisotropy, this will not be true in other directions because of the different transformation properties of $\mathbf{g}$ and $\mathbf{A}$ as expressed in eqns (3.5) and (3.44). It follows also that we can find the value of A_J from a magnetic resonance experiment by using the experimental values of **g, A** along the principal axes without any knowledge of the crystal field.

These simple results no longer hold when the crystal field admixes considerable amounts of states of different J. The reason for this is that, although the matrix elements of the hyperfine operator can always be related to those of the Zeeman operator (see § 17.4), the constant of proportionality does not vary in the same way for the two operators, as can be seen from eqns (17.56), (17.57). Thus in practice the extent to which the relation (5.30) is satisfied serves as a measure of the validity of the approximation in which matrix elements with states of different J are neglected.

In the absence of direct measurements of the hyperfine interaction for a free tripositive ion the value of A_J can be deduced from measurements on salts if allowance is made for the admixture of states of different J through the action of the crystal field. A list of the values obtained in this way for the ground state J of the stable isotopes is given in Table 5.5. Before any direct measurements of the nuclear

TABLE 5.5

Values of the magnetic hyperfine constant A_J for the free tripositive ions, calculated from magnetic resonance measurements on salts. The contribution from core polarization is assumed to be given by the formula $A'_c = -(63\pm10)(\langle J\| \Lambda \|J\rangle - 1)g_n$ (MHz), while A'_s is similar but with the parameter $-(85\pm10)$. The value of A_J for Tm^{3+} is estimated from measurements of Tm and Tm^{2+} (Bleaney 1964b)

	Ion	Isotope	Abundance (%)	Nuclear spin (I)	A_J (MHz)	A'_c (MHz)	A'_s (MHz)
59	Pr^{3+}	141	100	$\frac{5}{2}$	$+1093(10)$	$+20(3)$	$+27(3)$
60	Nd^{3+}	143	12·3	$\frac{7}{2}$	$-220·3(2)$	$-5·3(8)$	$-7·2(8)$
		145	8·3	$\frac{7}{2}$	$-136·9(1)$	$-3·3(5)$	$-4·5(5)$
61	Pm^{3+}	147	radioactive	$\frac{7}{2}$	$(+)599(6)$	$(+)20(3)$	$(+)27(3)$
62	Sm^{3+}	147	15·0	$\frac{7}{2}$	$-240(3)$	$-10·7(17)$	$-15(2)$
		149	13·9	$\frac{7}{2}$	$-194(3)$	$-8·6(14)$	$-11·7(14)$
65	Tb^{3+}	159	100	$\frac{3}{2}$	$+530(5)$	$-39(6)$	$-54(6)$
66	Dy^{3+}	161	19·0	$\frac{5}{2}$	$-109·5(22)$	$+3·4(6)$	$+4·6(6)$
		163	24·9	$\frac{5}{2}$	$+152·4(30)$	$-5·0(8)$	$-6·8(8)$
67	Ho^{3+}	165	100	$\frac{7}{2}$	$+812·1(10)$	$-17·5(27)$	$-24(3)$
68	Er^{3+}	167	22·9	$\frac{7}{2}$	$-125·3(12)$	$+2·0(3)$	$+2·7(3)$
69	Tm^{3+}	169	100	$\frac{1}{2}$	$(-393·5)$	$+4·8(8)$	$+6·5(8)$
70	Yb^{3+}	171	14·4	$\frac{1}{2}$	$+887·2(15)$	$-9·0(15)$	$-12(2)$
		173	16·2	$\frac{5}{2}$	$-243·3(4)$	$+2·5(4)$	$+3·4(4)$

magnetic moments of the lanthanides were available, such values of A_J were used to calculate the nuclear moments by making estimates of $\langle r^{-3} \rangle$ and neglecting the core polarization effect. However, such estimates obtained in different ways showed as much as 25 per cent variation for the lighter rare earth ions. Following the introduction of Endor and atomic beam triple resonance methods, a number of the nuclear moments have been directly determined, thus affording a check on the estimates of $\langle r^{-3} \rangle$, if the contribution from core polarization is known. There is no experimental evidence about the magnitude of this contribution except from measurements on ions with the $4f^7$ configuration. If it is assumed that the magnetic field at the nucleus produced by core polarization is constant per unit of electron spin throughout the $4f$ group, as seems to be roughly true for the $3d$ group if bonding effects are small, we can estimate the contribution to the other ions from measurements on the $4f^7$ ions. On this basis one obtains (Bleaney 1964b)

$$A'_c = -(63 \pm 10)(\langle J \| \Lambda \| J \rangle - 1)g_n \text{ MHz} \qquad (5.31)$$

for the tripositive ions, where the factor $(\langle J \| \Lambda \| J \rangle - 1)$ allows for the projection of **S** onto **J** (see eqn (5.29)); this corresponds to a magnetic field of $-(83 \pm 13)S$ kG at the nucleus.

It has been suggested (see Bleaney 1967) that the relativistic contribution to A_s may also be roughly constant per unit of spin in the field it produces at the nucleus, and if this is included we obtain

$$A'_s = -85(\langle J \| \Lambda \| J \rangle - 1)g_n \text{ MHz} \qquad (5.32)$$

for the tripositive ions.

The values of A'_{core} and A'_s given by eqns (5.31), (5.32) are listed in Table 5.5. Obviously they may not be very accurate, but they represent a comparatively small correction in most cases, the largest being about 7 and 9 per cent respectively for Tb^{3+}. Subtraction of these from the value of A_J gives a hyperfine parameter from which $\langle r^{-3} \rangle$ can be calculated using eqn (5.26) if g_n is known. A list of such values (based on the quantity $A_J - A'_s$, with A'_s given by eqn (5.32)), is given in Table 5.6 for comparison with calculated values. Three sets of the latter are given: the Hartree–Fock values of Freeman and Watson (1962); the values of Judd (1963a), which are a modification of the former to allow for the effect of configuration interaction; and the values of Lindgren (1962) (see also Judd and Lindgren (1961)) obtained from the observed spin-orbit constants. The values of Judd give the best fit with the

TABLE 5.6

Values of $\langle r^{-3} \rangle$ in atomic units for the tripositive ions (after Judd 1963a). Experimental values marked † are calculated from $(A_J - A'_s)$ in Table 5.5 and the value of g_n (where known). The last column gives the experimental values after including the correction for the break-down of LS-coupling. Values in parentheses are interpolated

	Ion	\multicolumn{4}{c	}{Values of $\langle r^{-3} \rangle$ in atomic units}	Experimental (corrected)
		Lindgren	Hartree–Fock	Judd	Experimental	
58	Ce^{3+}	3·66	4·72	4·17		
59	Pr^{3+}	4·26	5·37	4·76	†5·0	
60	Nd^{3+}	4·86	6·03	5·35	5·64(6)	5·49(6)
61	Pm^{3+}	5·46	—	(5·95)		
62	Sm^{3+}	6·07	7·36	6·56	†6·5	
63	Eu^{3+}	6·70	—	(7·22)		
64	Gd^{3+}	7·35	8·84	7·92		
65	Tb^{3+}	8·03	—	(8·60)	†8·3	
66	Dy^{3+}	8·74	10·34	9·32		
67	Ho^{3+}	9·50	—	(10·18)	†9·7	
68	Er^{3+}	10·32	12·01	10·89	10·60(12)	11·07(12)
69	Tm^{3+}	11·20	—	(11·73)		
70	Yb^{3+}	12·18	13·83	12·63	12·50(8)	

experimental values; here it must be remembered that the values of A_J in Table 5.5 have not been corrected for deviations from LS-coupling, which may be as large as 4 per cent (see Judd 1963a) in some cases.

The magnetic hyperfine constants for the tripositive ions given in Table 5.5 may be compared with those measured for the free atoms (Table 5.4) to find how great is the difference in $\langle r^{-3} \rangle$ for the two cases. In this comparison it must be remembered that the spectroscopic state is different in the two cases, but this involves just the extraction of the factors $\langle J \| N \| J \rangle$. Allowance should also be made for a contribution from core polarization and relativistic effects in the free atom, but these appear to be rather small. If they are neglected, the value of $\langle r^{-3} \rangle_{\text{ion}} / \langle r^{-3} \rangle_{\text{atom}}$ can be easily calculated and appears to fall smoothly from about 1·17 for Pr to 1·05 for Yb; the ratio appears to be close to the ratio of the values of $\langle r^{-3} \rangle$ for ion and atom calculated by Judd and Lindgren (1961), though the values of $\langle r^{-3} \rangle$ themselves are higher.

Few experimental results are available for the hyperfine structure constants of the dipositive ions, but it seems likely that these are close to those of the free atoms which are in the same spectroscopic state. After allowing for core polarization Bleaney (1964a) has estimated that $\langle r^{-3}\rangle$ is $(1\cdot 4\pm 0\cdot 3)$ per cent larger for Ho²⁺ than for Ho, and $(2\cdot 2\pm 0\cdot 5)$ per cent larger for Tm²⁺ than for Tm.

5.5. Nuclear electric quadrupole interaction

This interaction is normally so small that we need consider only matrix elements within a given manifold J. Hence we can use the form of the interaction Hamiltonian appropriate to a free ion

$$\mathscr{H}_q = -\frac{e^2Q\langle r_q^{-3}\rangle}{I(2I-1)}\langle J\|\alpha\|J\rangle \times$$
$$\times \begin{bmatrix} \tfrac{1}{4}\{3J_z^2-J(J+1)\}\{3I_z^2-I(I+1)\}+ \\ +\tfrac{3}{8}\{(J_zJ_++J_+J_z)(I_zI_-+I_-I_z)+(J_zJ_-+J_-J_z)(I_zI_++I_+I_z)\}+ \\ +\tfrac{3}{8}(J_+^2I_-^2+J_-^2I_+^2) \end{bmatrix}. \quad (5.33)$$

Furthermore, the nuclear electric quadrupole interaction is normally so small that, except in rather precise Endor experiments, we need consider only matrix elements that are diagonal within the electronic states left as groups of sub-states by the crystal field. If the latter has threefold or higher symmetry about the z-axis, a given electronic state will consist of admixtures of $J_z = M$ differing by 3 or more. In eqn (5.33) only the operator $\{3J_z^2-J(J+1)\}$ will have non-zero matrix elements, so that the nuclear electric quadrupole operator will have axial symmetry and be of the form

$$P_\|\{I_z^2-\tfrac{1}{3}I(I+1)\}, \quad (5.34)$$

where

$$P_\| = -\frac{3e^2Q\langle r_q^{-3}\rangle}{4I(2I-1)}\langle J\|\alpha\|J\rangle\langle\ |3J_z^2-J(J+1)|\ \rangle \quad (5.35)$$

and $\langle\ |3J_z^2-J(J+1)|\ \rangle$ is the matrix element of this operator computed for the given electronic state. Clearly $P_\|$ has the same value for any pair of states differing only in the sign of J_z, and which form a degenerate doublet (Kramers or non-Kramers) in such a crystal field. If the crystal field has only rhombic symmetry, so that the z-axis is only a twofold axis, then the crystal field states will contain values of J_z differing by 2, so that the terms J_+^2, J_-^2 in eqn (5.33) will also have diagonal matrix elements, and the quadrupole interaction will be of the more general form (cf. eqn (3.40c))

$$P_\|[\{I_z^2-\tfrac{1}{3}I(I+1)\}+\tfrac{1}{3}\eta(I_x^2-I_y^2)] \quad (5.36)$$

where, since $(I_x^2-I_y^2) = \frac{1}{2}(I_+^2+I_-^2)$, the value of η is given by

$$\eta = \frac{3\langle\ |J_\pm^2|\ \rangle}{\langle\ |3J_z^2-J(J+1)|\ \rangle}. \tag{5.37}$$

The matrix elements of the operator $3J_z^2-J(J+1)$ are given in Table 17. Thus for the states $\cos\theta\,|\pm\frac{9}{2}\rangle + \sin\theta\,|\mp\frac{7}{2}\rangle$ of eqn (5.11) belonging to a manifold with $J = \frac{9}{2}$ we have

$$\langle\ |3J_z^2-J(J+1)|\ \rangle = 12\cos^2\theta - 6\sin^2\theta.$$

This result is sharply dependent on the value of θ (it changes sign at $\tan\theta = \sqrt{2}$), and is always small compared with the maximum value of $3J_z^2-J(J+1)$ which is obtained for a pure state $|\pm J_z\rangle = |J\rangle$ and equals 36 for $J = \frac{9}{2}$. It follows that the quadrupole term may often be rather small for the ground doublet of a rare-earth ion, and also that to evaluate the quantity $Q\langle r_q^{-3}\rangle$ from an experimental result we are dependent in first order on a knowledge of the crystal field.

To proceed further, and estimate Q, we need to know what value to use for $\langle r_q^{-3}\rangle$. The simplest assumption is that it has the same value as the $\langle r^{-3}\rangle$ that appears in the magnetic hyperfine structure, and hence can be determined from the latter if the value of the nuclear magnetic moment is known. However distortion of the charge cloud of the inner closed shells of electrons produces an appreciable change in the electric field gradient at the nucleus (Sternheimer shielding), which may be represented by writing $\langle r_q^{-3}\rangle = (1-R)\langle r^{-3}\rangle$. The contributions to R are both positive (shielding) and negative (anti-shielding), but it seems probable that the net effect is a small amount of shielding.

In addition to the electric field gradient at the nucleus due to the $4f$ electrons, there will be a contribution in the solid state from the lattice, associated with the term $A_2^0(3z^2-r^2)$ and other second-degree terms in the crystal field. Here again a shielding (anti-shielding) factor must be introduced, which is generally written as a multiplicative factor $(1-\gamma_\infty)$. In the case of axial symmetry the contribution to $P_\parallel$ from the lattice or crystal field is then

$$P_\parallel\ (\text{lattice}) = -\frac{3QA_2^0}{I(2I-1)}(1-\gamma_\infty), \tag{5.38}$$

which must be added to (5.35). The size of the lattice contribution has been measured directly by Edmonds (1963), using nuclear magnetic resonance for three lanthanum salts, where the La^{3+} ion has no $4f$ electrons. Using calculated values of A_2^0, and $Q = 0\cdot 27$ barns for ^{139}La, an estimate of around 20 was obtained for $(1-\gamma_\infty)$, the sign being probably positive. A number of theoretical calculations give even larger negative values; for example, Ghatikar, Raychaudhuri, and Ray (1965) find $(1-\gamma_\infty)$ to be $+81$ for Pr^{3+} and $+75$ for Tm^{3+}. On the other hand, their values of R are quite small ($+0.20$ and $+0.15$ for these two ions respectively). The large anti-shielding factors for the lattice contribution mean that this may predominate over the gradient from the $4f$ electrons in cases when the latter is exceptionally small (see § 4.8 for Nd^{3+} and below for Eu^{3+}). The relative importance of the two contributions has been measured directly in Mössbauer experiments on two salts of thulium by Barnes, Mössbauer, Kankeleit, and Poindexter (1964). A more general analysis is given by Blok and Shirley (1966).

In cases of cubic symmetry there is no lattice contribution but the appropriate form of the nuclear electric quadrupole interaction due to the $4f$ electrons is more complicated; it is discussed in §§ 4.7, 5.9, 5.10 and Chapter 18.

5.6. Experimental results for ethylsulphates and anhydrous chlorides

These two salts have C_{3h} point symmetry at the rare-earth site and it is convenient to consider them together; in fact the electron spin results in the two salts are remarkably similar. The ethylsulphates were the first rare-earth salts to be measured by resonance, and the theoretical interpretation by Elliott and Stevens was based mainly on the resonance results; with this limited range of experimental data they succeeded in making remarkably good estimates of the crystal field parameters. The application of high resolution optical and infrared spectroscopy has greatly changed the situation, giving directly and accurately many of the crystal field splittings in the ground state and in some of the excited states. From these the crystal field parameters have been derived for almost all the ethylsulphates and anhydrous chlorides, and these are quoted in Tables 5.7 and 5.8. In each case the V_6^6 term is much the largest; the general trends are that V_6^6 and V_6^0 fall as we proceed along the series to the heavy end, V_4^0 varies little (with perhaps a slight drop in the middle of the series), while V_2^0 increases noticeably in the ethylsulphates (less so in the chlorides). If the crystal field potential were the same throughout a series, the changes in V_k^q would be entirely due to changes in $\langle r^k \rangle$, which we should expect to diminish steadily towards the heavy end because the

TABLE 5.7

Crystal field parameters in cm^{-1} for the rare-earth ethylsulphates; the crystals are undiluted except where the symbol (La) occurs, meaning diluted in the lanthanum salt

	$A_2^0 \langle r^2 \rangle$ (cm^{-1})	$A_4^0 \langle r^4 \rangle$ (cm^{-1})	$A_6^0 \langle r^6 \rangle$ (cm^{-1})	$|A_6^6| \langle r^6 \rangle$ (cm^{-1})	Reference
Ce	+9	−42	−45	680	see Birgeneau (1967a)
Pr(La)	+15	−88	−49	548	Gruber (1963)
Pr	+23	−80	−44	695	Hüfner (1962)
Nd	+58	−68	−43	595	Gruber and Satten (1963)
Pm					
Sm	+60	−64	−40	575	Hill and Wheeler (1966)
	+78	−53			Hüfner (1962)
Sm(La)	+45	−25	−30	450	Larson and Jeffries (1966a)†
Eu	+80	−63	−39	510	Judd (1959b)
Gd					
Tb	+110	−75	−34	465	Hüfner (1962)
Dy	+124	−79‡	−31	492	Powell and Orbach (1961)
	+144	−85	−33	535	Hill and Wheeler (1966)
Ho	+125	−79	−30	391	Hüfner (1962)
Er	+126	−81	−31	387	Erath (1961)
	+119	−74	−30	376	Hill and Wheeler (1966)
Tm(La)	+130	−71	−29	433	Wong and Richman (1961)
Yb					

† Estimated values, which with $A_4^3 \langle r^4 \rangle \approx 360$ cm^{-1}, fit the g-values and splittings for the $J = \frac{5}{2}$ manifold.
‡ Corrected value, Hüfner (1962).

increasing nuclear charge pulls the electron orbits into smaller radii. As can be seen from Table 5.1, Hartree–Fock calculations suggest that as we go from Ce^{3+} to Yb^{3+} the values of $\langle r^2 \rangle$, $\langle r^4 \rangle$, and $\langle r^6 \rangle$ fall by factors of (roughly) 2, 4, and 7 respectively. The experimental results show that the crystal potential cannot be assumed to remain constant, presumably because the positions of the ligand ions change as the size of the central ion is reduced (small angular displacements may well be more important than small changes in distance). It is also possible that there are important shielding or anti-shielding effects due to distortion of the outer closed shells of electrons by the crystal field, which will vary throughout the series. The importance of shielding effects is a vexed question; Burns (1962) has argued that shielding effects are quite small, while Freeman and Watson (1964) take the opposite view. They also suggest that non-linear shielding may occur, i.e. that shielding effects are different for wave-functions with different angular distributions of the electronic charge (if this were appreciable,

Table 5.8

Crystal field parameters in cm^{-1} for the anhydrous rare earth chlorides ($LaCl_3$ structure, hexagonal). The symbol (La) indicates that the measurements were made in a sample diluted with $LaCl_3$. The energy levels in the ground manifold and the corresponding wave-functions are listed for most of the ions by Mikkelson and Stapleton (1965)

	$A_2^0 \langle r^2 \rangle$ (cm^{-1})	$A_4^0 \langle r^4 \rangle$ (cm^{-1})	$A_6^0 \langle r^6 \rangle$ (cm^{-1})	$\lvert A_6^6 \rvert \langle r^6 \rangle$ (cm^{-1})	Reference
Ce	+65	−41	−64	399	Hellwege, Orlich and Schaack (1965), Bagguley and Vella-Colleiro (1969)
Pr(La)	+50	−40	−39	397	Judd (1957b)
	+47	−41	−40	405	Margolis (1961) (see also Hutchings and Ray 1963)
Nd(La)	+104	−36	−45	426	Judd (1959a)
	+98	−39	−44	443	Eisenstein (1963b)
Pm					
Sm	+81	−23	−44	426	Axe and Dieke (1962)
Eu(La)	+89	−38	−51	495	de Shazer and Dieke (1963)
Gd(La)	+97	−42	−30	290	Piksis, Dieke, and Crosswhite (1967)
Tb(La)	+92	−40	−30	290	Thomas, Singh, and Dieke (1963)
Dy(La)	+90	−40	−23	253	Crosswhite and Dieke (1961)
Ho(La)	+122	−45	−28	280	Dieke and Pandey (1964)
	+114	−34	−28	277	Rajnak and Krupke (1967, 1968)
Er(La)	+94	−37	−27	265	Eisenstein (1963a)

it would not be possible to fit the crystal field levels observed experimentally with a single set of parameters V_k^q). Further calculations have been made by Sternheimer (1966) and Sternheimer, Blume, and Peierls (1968), which suggest that shielding may be much more important for V_2 than for V_4 or V_6. Finally, we note that some attempts have been made to compute the crystal potential and hence to find effective values of $\langle r^k \rangle$ by comparison with the spectroscopic crystal field parameters. A detailed calculation of this kind has been carried out for $PrCl_3$ and $PrBr_3$ by Hutchings and Ray (1963), with the surprising results that the effective values of $\langle r^2 \rangle$ are a good deal smaller (a factor of 10 to 30) than the Hartree–Fock values; on the other hand the effective value of $\langle r^4 \rangle$ is somewhat larger, and that of $\langle r^6 \rangle$ is about 10 times larger than the Hartree–Fock value. A similar result has been obtained for $\langle r^4 \rangle$, $\langle r^6 \rangle$ by Bleaney (1964a) for Tm^{2+} in CaF_2; Axe and Burns (1966) suggest this may be partly due to covalency and overlap effects (see also Ellis and Newman (1966), Watson and Freeman (1967b)).

The paramagnetic resonance data for the ethylsulphates are summarized in Table 5.9, and for the anhydrous chlorides in Table 5.10.

TABLE 5.9

Paramagnetic resonance data for the diluted ethylsulphates $Ln(C_2H_5SO_4)_3, 9H_2O$

Ion	Diluent	T (°K)	$g_\parallel$	$\|g_\perp\|$	Isotope	$A_\parallel$ (cm^{-1})	$\|A_\perp\|$ (cm^{-1})	$P_\parallel$ (cm^{-1})
Ce	La	4·2	(+)0·955(5)	2·185(10)		ground state		
			(−)3·72(1)	0·20(5)		excited doublet at 3(1) cm^{-1}		
Pr	Y	20	1·525(20)		141	0·0755(20)		
Nd	La	20	(+)3·535(1)	2·072(1)	143	(−)0·03803(1)	0·01989(5)	<10^{-4}
					145	(−)0·02364(1)	0·01237(5)	<10^{-4}
Pm	La		0·432(4)		147	0·01655(13)		
Sm	La	4·2	(+)0·596(2)	0·604(2)	147	(−)0·0060(1)	0·0251(1)	<4×10^{-4}
					149	(−)0·0049(1)	0·0205(1)	
Tb	Y	12	17·72(2)		159	0·209(2)		
Dy	Y	14·4	(10·8)	(0)		ground state—no resonance		
			(+)5·86(10)	8·4(5)		excited doublet at 15(4) cm^{-1}		
Ho	Y	13	15·41(1)		165	0·3354(4)		
Er	La	4	(+)1·47(3)	8·85(20)	167	excited singlet at 5·5(1) cm^{-1}		
Tm	Y	13	no resonance			(−)0·0052(1)	0·0314(1)	0·0030(3)
Yb			no resonance					

TABLE 5.10

Paramagnetic resonance data for the diluted trichlorides $LnCl_3$ *(Hutchison and Wong 1958)*

Ion	Diluent	T (°K)	$g_{\|}$	$\|g_\perp\|$	Isotope	$A_{\|}$ (cm^{-1})	$\|A_\perp\|$ (cm^{-1})	$P_{\|}$ (cm^{-1})
Ce	La	4·2	(−)4·0366(15)	0·17(8)				
Pr	La	4·2	1·035(5)		141	0·0502(3)		
Nd	La	4·2	(+)3·996(1)	1·763(1)	143	(−)0·0425(2)	0·0167(1)	<10^{-4}
					145	(−)0·0264(2)	0·0104(1)	<10^{-4}
Sm	La	4·2	(+)0·5841(3)	0·6127(6)	147	(−)0·00607(2)	0·0245(1)	
					149	(−)0·00499(2)	0·0202(1)	
Tb	La	4·2	17·78(1)		159	0·212(3)		
Dy	La	4·2	no resonance					
Ho	La	4·2	16·01(2)		165	0·351(7)		
Er	La	4·2	(+)1·989(1)	8·757	167	(−)0·00664(3)	0·0304(2)	(+)0·00086
Tm	La	4·2	no resonance					
Yb	La	4·2	no resonance					

These tables include both the non-Kramers and the Kramers ions, with the exception of the half-filled shell, Gd^{3+}, $4f^7$, $^8S_{\frac{7}{2}}$, which will be considered separately. In most cases resonance data are available only for the ground state, and we shall discuss briefly the interpretation for each ion on the lines adopted in the original theoretical papers, since this illustrates the way in which the problem would be tackled in the absence of detailed knowledge about the crystal field, apart from its symmetry.

Ce^{3+}, $4f^1$, ^2F ($S = \frac{1}{2}$, $L = 3$; ground state $J = \frac{5}{2}$, $\langle J\|\Lambda\|J\rangle = \frac{6}{7}$)

In the free ion the first excited state $^2F_{\frac{7}{2}}$ lies some 2200 cm^{-1} above the ground state $^2F_{\frac{5}{2}}$, which is very much greater than the splittings produced by the crystal field, so that in first approximation we take J as a good quantum number. Inside the manifold $J = \frac{5}{2}$ the sixth-degree terms in the crystal potential have no matrix elements, so that effectively we have just the simple operator

$$V = A_2^0 \langle r^2\rangle \langle J\|\alpha\|J\rangle O_2^0 + A_4^0 \langle r^4\rangle \langle J\|\beta\|J\rangle O_4^0, \qquad (5.39)$$

where the matrix elements of O_2^0, O_4^0 are given in Table 17 and the values of $\langle J\|\alpha\|J\rangle$, $\langle J\|\beta\|J\rangle$ in Table 20. The matrix elements are all diagonal in J_z, so that the crystal field would give three Kramers doublets whose wave-functions are just

$$|J_z = \pm\tfrac{1}{2}\rangle, \ |J_z = \pm\tfrac{3}{2}\rangle, \ |J_z = \pm\tfrac{5}{2}\rangle,$$

for which the principal g-values are, since $\langle J\|\Lambda\|J\rangle = \frac{6}{7}$ for the ground state $J = \frac{5}{2}$, (cf. eqns 1.80)

$$\left.\begin{array}{l} |\pm\tfrac{1}{2}\rangle \quad g_\| = \tfrac{6}{7} = 0\cdot86; \quad g_\perp = \tfrac{18}{7} = 2\cdot57;\\ |\pm\tfrac{3}{2}\rangle \quad g_\| = \tfrac{18}{7} = 2\cdot57; \quad g_\perp = 0;\\ |\pm\tfrac{5}{2}\rangle \quad g_\| = \tfrac{30}{7} = 4\cdot29; \quad g_\perp = 0. \end{array}\right\} \qquad (5.40)$$

Comparison with the experimental results in Tables 5.9 and 5.10 shows that there is no exact fit, but that the first-order theory is qualitatively correct if we assume that the ground doublet in the chloride and the low-lying excited doublet in the ethylsulphate are approximately $|\pm\tfrac{5}{2}\rangle$, and the ground doublet in the latter salt is approximately $|\pm\tfrac{1}{2}\rangle$. We therefore look to see whether quantitative agreement with experiment can be obtained if contributions from the state $J = \tfrac{7}{2}$ are taken into account.

At this point we must use the more general potential operator (5.4) or (5.5) rather than the truncated operator (5.39) which suffices within the $J = \frac{5}{2}$ manifold. For the case of $J_z = \pm\frac{1}{2}$, the terms V_k^0 have matrix elements between $|J, J_z\rangle = |\frac{5}{2}, \pm\frac{1}{2}\rangle$ and $|\frac{7}{2}, \pm\frac{1}{2}\rangle$ but V_6^6 has no matrix elements so that the most general form of the states is

$$|\xi\rangle = \cos\varphi\,|\tfrac{5}{2}, +\tfrac{1}{2}\rangle - \sin\varphi\,|\tfrac{7}{2}, +\tfrac{1}{2}\rangle,$$
$$|\bar{\xi}\rangle = \cos\varphi\,|\tfrac{5}{2}, -\tfrac{1}{2}\rangle + \sin\varphi\,|\tfrac{7}{2}, -\tfrac{1}{2}\rangle. \qquad(5.41)$$

However, since we are dealing with a one-electron system, this doublet is spanned in the $|l_z, s_z\rangle$ representation by the wave-functions

$$|\xi\rangle = \cos\theta\,|+1, -\rangle - \sin\theta\,|0, +\rangle,$$
$$|\bar{\xi}\rangle = -\cos\theta\,|-1, +\rangle + \sin\theta\,|0, -\rangle, \qquad(5.42)$$

which is simpler to handle. We can use real admixture coefficients, $\cos\theta$ and $\sin\theta$, because, in the expansion (equ (16.1)) of the crystal potential, all the B_k^q and therefore all the matrix elements of the crystalline field can be chosen real for symmetry C_{3h} as explained in § 16.1 (which incidentally justifies the use of the expansion (16.10) with the P_k^q, rather than the more general expansion (16.1)). The admixture coefficients $\cos\varphi$ and $\sin\varphi$ in (5.41) are also real because the Clebsch–Gordan coefficients relating the representations $|l_z, s_z\rangle$ and $|J, J_z\rangle$ are all real.

From (5.42) we find

$$g_\| = 2\langle\xi|\,l_z+2s_z\,|\xi\rangle = 2\sin^2\theta,$$
$$|g_\perp| = 2\langle\xi|\,l_x+2s_x\,|\bar{\xi}\rangle = 2\,|\sqrt{3}\sin 2\theta - \sin^2\theta|, \qquad(5.43)$$

from which we can check that the first-order approximation in which $\varphi = 0$ in (5.41) and $\sin\theta = (\tfrac{3}{7})^{\frac{1}{2}}$ in (5.42) gives the same g-values as in (5.40). If we eliminate θ between the two eqns (5.43) we find that $g_\|$ and $g_\perp$ should satisfy the relation

$$(g_\|+g_\perp)^2/12g_\|(2-g_\|) = 1. \qquad(5.44)$$

The experimental values give 0·82 for this ratio, showing that no value of θ in (5.43) will give exact agreement with experiment.

For a pure $J_z = \pm\tfrac{5}{2}$ doublet, $g_\perp$ is zero and no resonance is observable. The finite value of $g_\perp$ here arises solely because of the admixture of the $J = \tfrac{7}{2}$ states, which gives the wave-functions

$$|+\rangle = p_1\,|\tfrac{5}{2}, -\tfrac{5}{2}\rangle + q_1\,|\tfrac{7}{2}, -\tfrac{5}{2}\rangle + r_1\,|\tfrac{7}{2}, +\tfrac{7}{2}\rangle,$$
$$|-\rangle = -p_1\,|\tfrac{5}{2}, +\tfrac{5}{2}\rangle + q_1\,|\tfrac{7}{2}, +\tfrac{5}{2}\rangle + r_1\,|\tfrac{7}{2}, -\tfrac{7}{2}\rangle, \qquad(5.45)$$

or, in the $|l_z, s_z\rangle$ representation,

$$|+\rangle = p_2\,|-2, -\rangle + q_2\,|-3, +\rangle + r_2\,|+3, +\rangle,$$
$$|-\rangle = p_2\,|+2, +\rangle + q_2\,|+3, -\rangle + r_2\,|-3, -\rangle, \qquad(5.46)$$

with, of course, $p^2+q^2+r^2 = 1$ in each case. (Note that p_1, q_1, r_1 can be related directly to p_2, q_2, r_2 by using the expansions for $|J, J_z\rangle$ in terms of $|l_z, s_z\rangle$, which give $r_1 = r_2$ and

$$\begin{aligned} -p_1+\sqrt{(6)}q_1 &= \sqrt{(7)}p_2, \\ \sqrt{(6)}p_1+q_1 &= \sqrt{(7)}q_2; \end{aligned} \qquad (5.47)$$

these formulae can be used to check those for the g-values given below.) In both (5.45) and (5.46) we have chosen as the $|+\rangle$ state that which requires operation by l_-+s_- to generate the $|-\rangle$ state, so as to preserve the correct handedness in the effective spin Hamiltonian if circularly polarized radiation is used; the correct signs in the $|-\rangle$ state are related to those in the $|+\rangle$ state by the time reversal formula (15.29).

The formulae for $g_\parallel$ and $g_\perp$ are found, using $\langle \tfrac{7}{2} \| \Lambda \| \tfrac{7}{2} \rangle = \tfrac{8}{7}$, eqns (16.39), (16.40), and Table 20 applied to (5.45), to be

$$\begin{aligned} g_\parallel &= \tfrac{2}{7}(-15p_1^2+2\sqrt{(6)}p_1q_1-20q_1^2+28r_1^2), \\ g_\perp &= \left| \frac{2}{\sqrt{7}} r_1(\sqrt{(6)}p_1+8q_1) \right|, \end{aligned} \qquad (5.48)$$

while from (5.46) we have

$$\begin{aligned} g_\parallel &= 2(-3p_2^2-2q_2^2+4r_2^2), \\ g_\perp &= |2r_2(\sqrt{(6)}p_2+2q_2)|. \end{aligned} \qquad (5.49)$$

It is clear that when r is small compared with the other coefficients, $g_\parallel$ will be negative, because the $|+\rangle$ state will have a negative Zeeman energy; we have therefore inserted a negative sign in parentheses before the corresponding entries in Tables 5.9 and 5.10, although these signs have not been verified experimentally.

For (La, Ce)Cl$_3$ all the $4f$ levels have been observed by optical spectroscopy, and are shown in Fig. 5.4. The calculated best fit to these levels with a single set of crystal field parameters (those given in Table 5.8) is also shown, a value of $\zeta = 626.5 \pm 1.6$ cm^{-1} being used for the spin-orbit coupling parameter. This value is smaller by a factor 0.9745 than that for the free Ce^{3+} ion, and Hellwege, Orlich, and Schaak (1965) find that a slightly better fit to the resonance value of $g_\parallel$ for the ground doublet is obtained by using an orbital reduction factor k assumed equal to 0.9745. The calculated value of $|g_\parallel|$ is thereby reduced from 4.222 to 4.078, the measured value being 4.0366 (but see below).

We return now to the question of the g-values of the ground doublets in cerium ethylsulphate. Elliott and Stevens (1962) pointed out that,

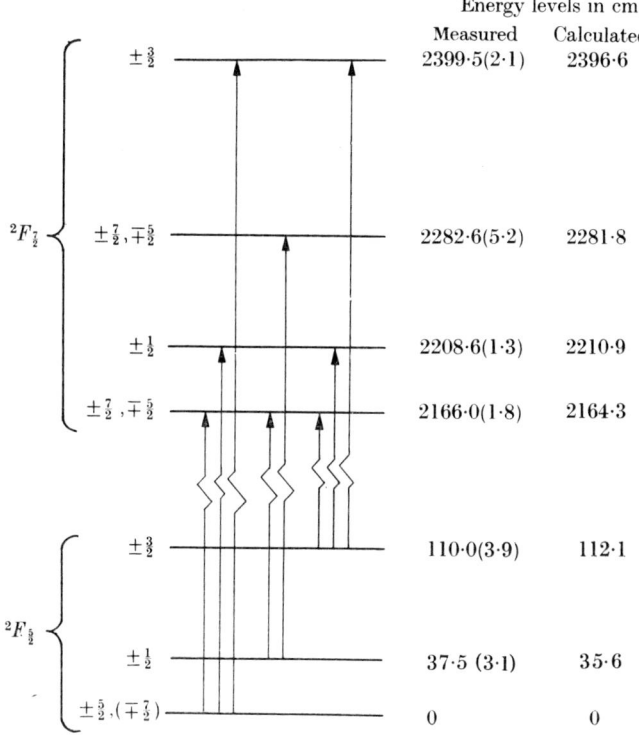

FIG. 5.4. Levels and allowed optical transitions for (Ce, La)Cl$_3$. The states are characterized by their values of $\pm J_z$, the lowest having an admixture of ($\mp \frac{7}{2}$) from the $J = \frac{7}{2}$ manifold. The measured energy values are compared with those calculated from the crystal field parameters given in Table 5.8 together with a spin-orbit coupling parameter $\zeta = 626 \cdot 5 \pm 1 \cdot 6$ cm^{-1}. (After Hellwege, Orlich, and Schaack 1965.)

because of the proximity of the two low-lying doublets, the theory is very sensitive to any small departure from reflection symmetry in the xy-plane. This would introduce terms $V_4^{\pm 3}$, $V_6^{\pm 3}$ into the crystal potential which couple into the ground doublet $|J_z = \pm \frac{1}{2}\rangle$ admixtures (which may have complex amplitudes) of $|J_z = \mp \frac{5}{2}\rangle$. It is then possible to fit simultaneously the g-values for both the low-lying doublets, and the temperature variation of the susceptibility. The latter is for the undiluted salt, where the two low-lying doublets are inverted in energy, the $|J_z = \pm \frac{1}{2}\rangle$ doublet being in this case the lower of the two. This change requires, however, only a small alteration in the crystal field coefficients, since the two doublets lie so close together only through the accidental cancellation of large terms.

If this hypothesis were correct, it would lead to g-values that are very sensitive to the splitting between the two doublets, as pointed out by Birgeneau (1967a), since this splitting appears as an energy denominator in the perturbation theory for the admixtures. In fact, experiments with different host lattices (La, Ce, Sm, and Y ethylsulphates) in which the energy splitting $W_{\pm\frac{1}{2}} - W_{\pm\frac{5}{2}}$ varies from $-3 \cdot 9$ to $+17 \cdot 4$ cm^{-1} show only very small changes in the g-values for the $|\pm\frac{5}{2}\rangle$ doublet. The value of $g_\perp$ is constant within the experimental error; $g_\parallel$ increases only from 3·70 to 3·81, whereas the theoretical values for $g_\parallel$ are all close to 4·14.

Since the correction to $g_\parallel$ for the $|\pm\frac{5}{2}\rangle$ doublet is proportional to the square of the admixture of the $|\pm\frac{1}{2}\rangle$ states, it should vary as $(W_{\pm\frac{1}{2}} - W_{\pm\frac{5}{2}})^{-2}$ on a static crystal field theory. If the admixture is due to a dynamic crystal field in which lattice vibrations introduce crystal field terms of the necessary lower symmetry, the energy denominator in the admixture contains a phonon energy as well as the static crystal field splitting. The admixtures and the corrections to the g-values will thus be much less dependent on $(W_{\pm\frac{1}{2}} - W_{\pm\frac{5}{2}})$ if the main phonon energies involved are large compared with this splitting. In this way Birgeneau (1967a) has succeeded in explaining quantitatively the low values of $g_\parallel$ for the $|\pm\frac{5}{2}\rangle$ doublet in the ethylsulphate. Bagguley and Vella-Colleiro (1969) have applied a simple correction of the same kind to the value of $g_\parallel$ for the analogous doublet in (Ce, La)Cl$_3$, as well as to the $|\pm\frac{1}{2}\rangle$ doublet of the $J = \frac{7}{2}$ level (see Fig. 5.4).

As pointed out by Birgeneau, corrections of this type (which may be positive or negative), originally proposed by Inoue (1963), give a clear-cut explanation of the rather large corrections to the g-values, whereas a covalent orbital reduction would here require overlap and covalent contributions to the static crystal field far larger than the actual crystal field.

Pr^{3+}, $4f^2$, $^3H_4$ ($S = 1$, $L = 5$; ground state $J = 4$, $\langle J \| \Lambda \| J \rangle = \frac{4}{5}$)

With two electrons in the $4f$ shell we have a non-Kramers ion whose ground state is $J = 4$. In C_{3h} symmetry we see from eqns (5.15)–(5.17) that the crystal field splits the ninefold $J = 4$ manifold into three singlet states and three doublets, the latter being of the form $|4, \pm 1\rangle$,

$$|\xi\rangle = \sin\theta\,|4, +2\rangle + \cos\theta\,|4, -4\rangle,$$
$$|\bar\xi\rangle = \sin\theta\,|4, -2\rangle + \cos\theta\,|4, +4\rangle, \qquad (5.50)$$

and

$$|\xi'\rangle = \cos\theta\,|4, +2\rangle - \sin\theta\,|4, -4\rangle,$$
$$|\bar\xi'\rangle = \cos\theta\,|4, -2\rangle - \sin\theta\,|4, +4\rangle, \qquad (5.51)$$

the two states in each case being conjugated by time reversal. The two sets of states are mutually orthogonal, and (5.51) can be derived from (5.50) by replacing θ by $\theta+\pi/2$. In each case the Zeeman Hamiltonian has opposite expectation values in the two conjugate states, separating them by an amount $g_\parallel \beta H_z$ which for (5.51) has the value

$$g_\parallel \beta H_z = 2\beta \langle J\| \Lambda \|J\rangle H_z(2\cos^2\theta - 4\sin^2\theta) \tag{5.52}$$

with $\langle J\| \Lambda \|J\rangle = \tfrac{4}{5}$ for this ion. In the ethylsulphate the value of $g_\parallel = 1\cdot52_5$ observed for the ground state can be fitted using $\theta = 24°$ in (5.51), with an appreciable correction from admixture of the next excited state $J = 5$; the rather smaller value of $g_\parallel = 1\cdot03_5$ in the anhydrous chloride requires a somewhat larger value of θ.

From the resonance results alone we have no means of telling whether the state $|\xi'\rangle$ or $|\bar\xi'\rangle$ has the positive Zeeman energy, and clearly the experimental results could be fitted by a different value of θ if $g_\parallel$ were taken as negative in (5.52); it is easily seen that this alternative does not correspond to using (5.50) instead of (5.51). In any given case the choice between the two alternatives must be made on other evidence; usually thermal, magnetic, or spectroscopic experiments will give some information that makes one choice plausible, if not decisive.

Since we are dealing with a non-Kramers doublet for which $g_\perp = 0$, resonance is observed only through distortions of the crystal field which admix the two conjugated states, giving an asymmetrical line whose intensity is greatest when the oscillatory magnetic field is also along the crystal axis. The theory of this is given in detail in § 18.5 and summarized in §§ 3.14, 5.3. The value of Δ_0 or h_0 (eqns (3.103), (3.105)) deduced from the experimental results on the assumption of magnetic dipole transitions varies considerably. For Pr^{3+} in yttrium ethylsulphate Baker and Bleaney (1958) found $\Delta_0 = 0\cdot11 \pm 0\cdot04$ cm^{-1}, with a larger value (0·19 cm^{-1}) in the lanthanum ethylsulphate; in the undiluted praseodymium ethylsulphate it is much larger still. For Pr^{3+} in $LaCl_3$ Hutchison and Wong (1958) find the much smaller value $\Delta_0 \approx 0\cdot02$ cm^{-1}, or $h_0 \approx 37$ G. A detailed comparison of the experimental and theoretical line shapes has been made by Williams (1967), and shows (see Fig. 3.27), that a much better fit is obtained for (Pr, Y) ethylsulphate using (3.111) assuming $\Delta_0 = 0\cdot15 \pm 0\cdot04$ cm^{-1} and that the transitions are primarily due to the perpendicular component of the oscillatory *electric* field.

Nd^{3+}, $4f^3$, $^4I_{\frac{9}{2}}$ ($S = \tfrac{3}{2}$, $L = 6$; *ground state* $J = \tfrac{9}{2}$, $\langle J\| \Lambda \|J\rangle = \tfrac{8}{11}$)

Inside the ground manifold $J = \tfrac{9}{2}$, for C_{3h} symmetry, the possible wave-functions for the Kramers doublets have the forms

(a) $|\pm\tfrac{1}{2}\rangle$, for which $g_\perp = 5g_\parallel = 5\langle J\|\Lambda\|J\rangle$;
(b) two doublets with admixtures $|\pm\tfrac{9}{2}, \mp\tfrac{3}{2}\rangle$, for which $g_\perp = 0$;
(c) two doublets of the form of eqn (5.11), whose g-values are given by (5.12).

Only the latter can fit the observed values for the ground state, and we can test the accuracy of the fit by calculating the value of the quantity

$$9g_\perp^2/4(g_\parallel + 5\Lambda)(7\Lambda - g_\parallel),$$

which is obtained by eliminating θ from eqns (5.12) and which should equal unity if the first-order theory is correct. Inserting

$$\Lambda = \langle J\|\Lambda\|J\rangle = \tfrac{8}{11}$$

and the experimental g-values we obtain 0·86 for the ethylsulphate and 0·84 for the trichloride, showing that the first-order theory is not a bad approximation. The best value for the ethylsulphate is $\theta = 24°$, which gives (cf. Table 5.9)

$$g_\parallel = 3\cdot 65,\ g_\perp = 2\cdot 16.$$

Neodymium has two odd isotopes of mass 143 and 145, both with nuclear spin $I = \tfrac{7}{2}$ (incidentally these were the first nuclear spins in the 4f group to be determined by electron paramagnetic resonance). From eqn (5.30) we should expect, on first-order theory, that

$$(g_\parallel/g_\perp)/(A_\parallel/A_\perp) = 1;$$

the experimental values for both ethylsulphate and trichloride are both 0·89, showing again that the first-order theory is approximately correct.

In a second approximation the crystal field terms will introduce into the ground doublet admixtures from the next excited level $J = \tfrac{11}{2}$ with the same values of J_z. By varying the amplitudes of these admixtures it is possible to improve the agreement between the calculated and experimental values of $g_\parallel$, $g_\perp$, and $A_\parallel/A_\perp$. The most extensive calculations of this kind have been carried out by Halford (1962) in order to interpret Endor measurements on the Nd^{3+} isotopes in $LaCl_3$. The complete crystal field and spin-orbit coupling energy matrix for the $^4I_{\tfrac{9}{2}}$, $^4I_{\tfrac{11}{2}}$, and $^4I_{\tfrac{13}{2}}$ manifolds was computed and diagonalized numerically, using the crystal field parameters obtained by Judd (1959a) with corrections for intermediate coupling. These parameters were obtained by fitting the levels observed in the optical fluorescence spectrum, and they lead to g-values for the ground doublet of

$$g_{\parallel} = +4\cdot 1016, \ |g_{\perp}| = 1\cdot 7925,$$

which are in good agreement with Halford's measurements of

$$g_{\parallel} = 3\cdot 9903(5), \ |g_{\perp}| = 1\cdot 7635(12),$$

which are more accurate than those of Hutchison and Wong (1958) given in Table 5.10.

Details of the Endor spectrum, including discussion of a number of important corrections, are given in § 4.8.

Pm^{3+}, $4f^4$, $^5I_4$ ($S = 2$, $L = 6$; ground state $J = 4$, $\langle J \| \Lambda \| J \rangle = \frac{3}{5}$)

This ion has no radioactively-stable isotopes, but resonance has been observed for the isotope ^{147}Pm, which has a half-life of 2·64 years and is a beta-emitter. The magnetic ground state of Pm^{3+} incorporated in lanthanum ethylsulphate is a non-Kramers doublet which is fitted by a spin Hamiltonian of the type given in eqn (5.20) with the addition of a hyperfine term $A_{\parallel}S_zI_z$. Stapleton, Jeffries, and Shirley (1961) observed eight hyperfine lines, corresponding to $I = \frac{7}{2}$, with an asymmetric shape, but the line width appeared to increase rapidly with time after crystal growth, presumably owing to radiation damage of the crystal. Their results for $g_{\parallel}$ and $A_{\parallel}$ are listed in Table 5.9, and are fitted by wave-functions of the form (5.51) with $\theta \sim 38°$ (and admixtures of order 0·03 from the $J = 5$ functions) which correspond to a negative value of $g_{\parallel}$ in eqn (5.52), though this of course has no physical significance.

Sm^{3+}, $4f^5$, $^6H_{\frac{5}{2}}$ ($S = \frac{5}{2}$, $L = 5$; ground state $J = \frac{5}{2}$, $\langle J \| \Lambda \| J \rangle = \frac{2}{7}$)

In the ground state the small value of $J = \frac{5}{2}$ means that, as in the case of cerium, the first-order theory is very simple, and only the doublet $|\pm\frac{1}{2}\rangle$ has $g_{\perp}$ different from zero. However it would have $g_{\parallel} = \frac{2}{7}, g_{\perp} = \frac{6}{7}$, which are very different from the observed values (see Tables 5.9, 5.10) that lie close to 0·6 with remarkably little anisotropy. Samarium has two odd isotopes, each with $I = \frac{7}{2}$, whose magnetic hyperfine constants have been measured, and these give for the ratio $(g_{\parallel}/g_{\perp})/(A_{\parallel}/A_{\perp})$ the values 0·24 for the ethylsulphate and 0·26 for the chloride. This shows that the first-order theory is completely inadequate for Sm^{3+}, and in view of the fact that the first excited state $J = \frac{7}{2}$ lies at about 1000 cm^{-1} this is not too surprising.

In the next order of perturbation theory the doublet becomes

$$\cos\theta \, |\tfrac{5}{2}, \pm\tfrac{1}{2}\rangle \pm \sin\theta \, |\tfrac{7}{2}, \pm\tfrac{1}{2}\rangle \tag{5.53}$$

for which

$$g_{\parallel} = \tfrac{2}{7}(\cos^2\theta + \tfrac{26}{9}\sin^2\theta + 3\sqrt{(10)}\sin 2\theta),$$
$$g_{\perp} = |\tfrac{2}{7}(3\cos^2\theta - \tfrac{104}{9}\sin^2\theta - \tfrac{3}{2}\sqrt{(10)}\sin 2\theta)|, \tag{5.54}$$

and

$$A_{\parallel} = 2g_n\beta\beta_n\langle r^{-3}\rangle\left(\frac{488}{315}\cos^2\theta + \frac{5056}{4725}\sin^2\theta - \frac{437\sqrt{(10)}}{525}\sin 2\theta\right),$$

$$A_{\perp} = \left|2g_n\beta\beta_n\langle r^{-3}\rangle\left(\frac{488}{105}\cos^2\theta - \frac{20224}{4725}\sin^2\theta + \frac{437\sqrt{(10)}}{1050}\sin 2\theta\right)\right|, \tag{5.55}$$

where the factors used are given in Table 20 except $\langle\tfrac{7}{2}\|\Lambda\|\tfrac{7}{2}\rangle = \tfrac{52}{63}$ and $\langle\tfrac{7}{2}\|N\|\tfrac{7}{2}\rangle = \tfrac{5056}{4725}$. With the relatively small admixture of the $J = \tfrac{7}{2}$ state given by $\theta = 0.07$ rad the correct value of $(g_{\parallel}/g_{\perp})/(A_{\parallel}/A_{\perp})$ can be obtained for the ethylsulphate, and the theoretical g-values are then $g_{\parallel} = 0.66$, $g_{\perp} = 0.65$, which is as good as can be expected from a second-order theory that makes such drastic changes. It can readily be seen that these are due to the terms in $\sin 2\theta$ that arise from the off-diagonal elements between $J = \tfrac{5}{2}$ and $J = \tfrac{7}{2}$; their coefficients are relatively large, particularly for the g-values, and, classically, this corresponds to a situation where there are large components of **L** precessing about **J** as can be seen in a vector diagram.

Eu^{3+}, $4f^6$, $^7F_0$ ($S = 3$, $L = 3$; ground state $J = 0$)

Obviously, no electron paramagnetic resonance can be expected for an ion with $J = 0$. Elliott (1957) has considered the possibility of a nuclear magnetic resonance experiment in the ground state, and has shown that this is considerably affected by off-diagonal effects from the low-lying $J = 1$ level at about 300 cm^{-1}. The magnetic hyperfine structure produces a pseudo-nuclear quadrupole splitting, and heavy magnetic 'shielding' (i.e. a pseudo-nuclear Zeeman effect of opposite sign to the true nuclear Zeeman interaction) in the $J = 0$ state. Judd, Lovejoy, and Shirley (1962) have observed the quadrupole splitting in the ethylsulphate using nuclear alignment techniques for ^{152}Eu and ^{154}Eu, and find it to be considerably larger and of opposite sign to that predicted by Elliott. Edmonds (1963) has shown that this can be ascribed to the electric field gradient at the nucleus due to the crystal field V_2^0, assuming it to be enhanced by the same large negative anti-shielding effects as observed for ^{139}La.

Gd^{3+}, $4f^7$, $^8S_{\frac{7}{2}}$ ($S = \frac{7}{2}$, $L = 0$); see § 5.9

Tb^{3+}, $4f^8$, $^7F_6$ ($S = 3$, $L = 3$; ground state $J = 6$, $\langle J \| \Lambda \| J \rangle = \frac{3}{2}$)

This is a non-Kramers ion where resonance is observed in the ground state when the oscillatory magnetic field is parallel to the crystal axis. The lines are however narrow and of normal shape; measurements at different frequencies show that the behaviour is that of a doublet with a small initial splitting of 0·2 to 0·3 cm^{-1}, and a value of $g_\parallel$ close to 18. These results identify the ground doublet as consisting principally of the states $J_z = |\pm 6\rangle$, split by a small amount and slightly admixed with the $J_z = |0\rangle$ state through the term V_6^6 in the crystal potential. It is fitted with a spin Hamiltonian of the type given by eqn (5.20) with a single value of $\Delta = (\Delta_x^2 + \Delta_y^2)^{\frac{1}{2}}$, with the addition of hyperfine terms, so that the transitions occur at

$$\hbar\omega = \{(g_\parallel \beta H_z + A_\parallel m)^2 + \Delta^2\}^{\frac{1}{2}}. \qquad (5.56)$$

Baker and Bleaney (1958) attempted to deduce values of the crystal field parameters from the experimental results for Δ and $(18 - g_\parallel)$, but even with corrections for intermediate coupling (which affects $g_\parallel$) and for an admixture of the state $J = 5$, $J_z = 0$ through the V_6^6 term (which affects Δ) they arrived at values considerably different from the optical values in Table 5.7 found later by Hüfner (1962) for Tb^{3+} in the ethylsulphate. In fact the difference between $g_\parallel$ and 18 is predominantly due to intermediate coupling, and the correction to Δ is of the same order as the experimental values, so that such a discrepancy in the crystal field parameters is hardly surprising. Hüfner's crystal field parameters give $c = 5\cdot 7$ cm^{-1}, $d = 111$ cm^{-1}, and (from eqns (5.20a) with the corrections mentioned above) $\Delta = 0\cdot 2$ cm^{-1}, $g_\parallel = 17\cdot 8$, in reasonable agreement with the resonance results $\Delta = 0\cdot 387$ cm^{-1}, $g_\parallel = 17\cdot 72$.

Dy^{3+}, $4f^9$, $^6H_{\frac{15}{2}}$ ($S = \frac{5}{2}$, $L = 5$; ground state $J = \frac{15}{2}$, $\langle J \| \Lambda \| J \rangle = \frac{4}{3}$)

No resonance has been observed in the ground state of either ethylsulphate or trichloride, indicating that $g_\perp = 0$. This is confirmed by susceptibility measurements (Cooke, Edmonds, McKim, and Wolf 1959) in the ethylsulphate, which give $g_\parallel = 10\cdot 8$, $g_\perp = 0$. Magnetic resonance has been observed in the first excited doublet by Baker and Bleaney (1958), who estimated it to lie at 15 ± 3 cm^{-1} from relative intensity measurements, with $g_\parallel = 5\cdot 86$, $g_\perp = 8\cdot 4$. Optical measurements of Gramberg (1960) give $16\cdot 0$ cm^{-1} for this splitting, with another doublet at 21 cm^{-1}. These results made it possible for Powell and Orbach (1961) to determine the four constants of the crystal potential (see Table 5.7), and hence the energies and wave-functions of the $J+\frac{1}{2}$ = eight doublets into which the crystal field splits the multiplet $J = \frac{15}{2}$. The ground doublet is a $\pm\frac{15}{2}$, $\pm\frac{3}{2}$, $\mp\frac{9}{2}$ combination, which has $g_\perp = 0$, and the first excited doublet is a $\pm\frac{7}{2}$, $\mp\frac{5}{2}$ combination of the type (5.11).

A weak resonance, corresponding to $g_\perp \approx 10^{-2}$ in the ground doublet of the ethylsulphate, has been reported by Gill (1963), while Brower and Stapleton (1967) have similarly observed resonance in the ground doublet of (Dy, La)Cl$_3$. These results are attributed to small distortions of the crystal field. In the latter salt resonance was also observed from a slightly split quartet of states at about 9·9 cm^{-1} above the ground doublet, essentially consisting of two doublets that accidentally lie very close together (only 0·135 cm^{-1} apart in zero magnetic field).

Ho^{3+}, $4f^{10}$, $^5I_8$ ($S = 2$, $L = 6$; *ground state* $J = 8$, $\langle J \| \Lambda \| J \rangle = \frac{5}{4}$)

The ground state in both the ethylsulphate and the chloride is a non-Kramers doublet in which the transitions can be described by eqn (5.56) with $g_\parallel$ about 16 and $A_\parallel$ about 0·35 cm^{-1} for the single stable isotope ^{165}Ho, $I = \frac{7}{2}$. The lines are asymmetric in shape and fairly narrow, so that the hyperfine components, which are over 450 G apart, are well resolved. Other sets of lines (see Fig. 5.5) can be observed at high fields, as well as some intermediate lines, in the ethylsulphate. It turns out that in this case the ground doublet, which is of the type

$$\left| \begin{matrix} \xi \\ \eta \end{matrix} \right\rangle = \pm\{\alpha \left| \pm 7 \right\rangle + \beta \left| \pm 1 \right\rangle + \gamma \left| \mp 5 \right\rangle\} \qquad (5.57)$$

for either salt, lies a little below the singlet

$$\zeta = \cos\delta \left| 6^s \right\rangle + \sin\delta \left| 0 \right\rangle \qquad (5.58)$$

and it is possible to observe transitions between ξ and η with the oscillatory magnetic field parallel to the crystal axis, and also transitions between ζ and ξ or η with the oscillatory magnetic field perpendicular to the crystal axis. The resonance data can be described by means of a fictitious spin $S = 1$, where ξ and η are eigenstates $S_z = \pm 1$, and ζ is the eigenstate $S_z = 0$. The spin Hamiltonian is then

$$\mathscr{H} = D\{S_z^2 - \tfrac{1}{3}S(S+1)\} + g_\parallel \beta H_z S_z + g_\perp \beta(H_x S_x + H_y S_y) + \\ + A_\parallel S_z I_z + A_\perp(S_x I_x + S_y I_y) + \\ + \{(P_1 - P_0)S_z^2 + P_0\}\{I_z^2 - \tfrac{1}{3}I(I+1)\}, \qquad (5.59)$$

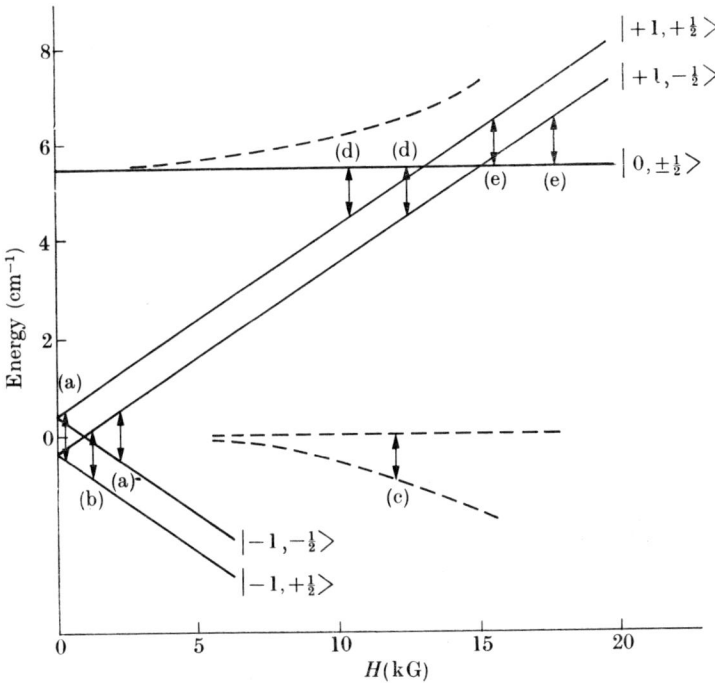

FIG. 5.5. Schematic representation of the three lowest energy levels of diluted holmium ethylsulphate. Continuous lines: behaviour of the energy levels when H is parallel to the crystal axis (for clarity only two hyperfine components are shown); Broken lines: behaviour when H is normal to the crystal axis (hyperfine components omitted). The various transitions observed are (in terms of $S = 1$): (a) $|+1, m\rangle \leftrightarrow |-1, m\rangle$ (H parallel), (b) $|+1, m\rangle \leftrightarrow |-1, m+1\rangle$ (H parallel), (c) $|+1, m\rangle \leftrightarrow |-1, m\rangle$ (H perpendicular), (d), (e) $|+1, m\rangle \leftrightarrow |0, m\rangle$ (H parallel). (After Baker and Bleaney 1958.)

where $D = -5.5$ cm^{-1} is the energy by which the doublet states lie above into singlet (in this case the negative sign means that the doublets lie below the singlet). The other constants are (Baker and Bleaney 1958)

$$g_\parallel/\Lambda = 7\alpha^2 + \beta^2 - 5\gamma^2,$$
$$g_\perp/\Lambda = \{\sqrt{(\tfrac{15}{2})}\alpha + \sqrt{(\tfrac{21}{2})}\gamma\}\cos\delta + 6\beta\sin\delta, \qquad (5.60)$$
$$\frac{A_\parallel}{g_\parallel} = \frac{A_\perp}{g_\perp} = 2g_n\beta\beta_n\langle r^{-3}\rangle\frac{\langle J\| N \|J\rangle}{\langle J\| \Lambda \|J\rangle},$$

where $\langle J\| N \|J\rangle = \tfrac{23}{60}$, $\langle J\| \Lambda \|J\rangle = \Lambda = \tfrac{5}{4}$. P_1 and P_0 are the strengths of the quadrupole coupling in the doublet and singlet respectively.

It should be noticed that in using the $S = 1$ formalism the experimental values of $g_\parallel$ and $A_\parallel$ are just half what they would be in the spin $S = \tfrac{1}{2}$ formalism, apart from a correction to $A_\|$ of order $A_\perp^2/D$. In Tables 5.9 and 5.10 the $S = \tfrac{1}{2}$ formalism is used for the doublet; for the ethylsulphate the constants in the spin Hamiltonian (5.59) are

$$g_\| = 7\cdot 71, \; |g_\perp| \sim 4,$$
$$A_\| = 0\cdot 167 \text{ cm}^{-1}, \; |A_\perp| \sim 0\cdot 09 \text{ cm}^{-1}.$$

Er^{3+}, $4f^{11}$, $^4I_{\frac{15}{2}}$ ($S = \tfrac{3}{2}$, $L = 6$; ground state $J = \tfrac{15}{2}$, $\langle J \| \Lambda \| J \rangle = \tfrac{6}{5}$)

The first excited state $J = \tfrac{13}{2}$ is some 8000 cm^{-1} away and first-order theory should be quite accurate. This is confirmed by observation of the hyperfine structure due to the odd isotope ^{167}Er, which yields a ratio $(g_\|/g_\perp)/(A_\|/A_\perp) = 1\cdot 00$ in the ethylsulphate and $1\cdot 04$ in the chloride. The ground doublet is of the form (5.11); its g-values could be fitted also by the combination $\pm\tfrac{13}{2}$, $\pm\tfrac{1}{2}$, $\mp\tfrac{11}{2}$ but for this to be the ground doublet would require rather unreasonable values for the crystal field parameters. The theoretical g-values given by eqn (5.12) with $J = \tfrac{15}{2}$ yield the relation

$$\left(\frac{g_\| - \Lambda}{6\Lambda}\right)^2 + \frac{1}{55}\left(\frac{g_\perp}{\Lambda}\right)^2 = 1 \tag{5.61}$$

and this relation is well obeyed both by the ethylsulphate and by the chloride.

Tm^{3+}, $4f^{12}$, $^3H_6$ ($S = 1$, $L = 5$; ground state $J = 6$, $\langle J \| \Lambda \| J \rangle = \tfrac{7}{6}$)

No resonance has been observed for this ion in either the ethylsulphate or the chloride, and the crystal field parameters predict that the ground state should be a singlet.

Yb^{3+}, $4f^{13}$, $^2F_{\frac{7}{2}}$ ($S = \tfrac{1}{2}$, $L = 3$; ground state $J = \tfrac{7}{2}$, $\langle J \| \Lambda \| J \rangle = \tfrac{8}{7}$)

Here again no resonance has been observed. The crystal field theory predicts a ground doublet $|J_z = \pm\tfrac{3}{2}\rangle$ (in keeping with susceptibility measurements) for which obviously $g_\perp = 0$.

5.7. Experimental results for the double nitrates, $Ln_2Mg_3(NO_3)_{12}, 24H_2O$

Magnetic resonance results for the rare-earth double nitrates were interpreted by Judd (1955), with the aid of some data from optical spectroscopy, on the basis of a crystal field of C_{3v} symmetry, a result that has been confirmed by X-ray crystallography (see § 5.1). The crystal potential then has the form (5.3) with the additional terms

$$A_4^3 z(x^3 - 3xy^2) + A_6^3 (11z^3 - 3zr^2)(x^3 - 3xy^2), \tag{5.62}$$

which have operator equivalents

$$A_4^3 \langle r^4 \rangle \langle J \| \beta \| J \rangle O_4^3 + A_6^3 \langle r^6 \rangle \langle J \| \gamma \| J \rangle O_6^3, \quad (5.63)$$

where the O_k^q are given in Table 16 and their matrix elements in Table 17. The extra terms correspond to the site having threefold instead of sixfold symmetry about the crystal axis. It appears that they are rather large, as can be seen from Table 5.11, which gives the parameters

TABLE 5.11

Crystal field parameters for the rare-earth double nitrates,
$Ln_2Mg_3(NO_3)_{12}, 24H_2O$

	$A_2^0 \langle r^2 \rangle$ (cm^{-1})	$A_4^0 \langle r^4 \rangle$ (cm^{-1})	$A_4^3 \langle r^4 \rangle$ (cm^{-1})	$A_6^0 \langle r^6 \rangle$ (cm^{-1})	$A_6^3 \langle r^6 \rangle$ (cm^{-1})	$|A_6^6| \langle r^6 \rangle$ (cm^{-1})	Reference
Ce	−70	−30	∓900	−60	±2800	850	Judd (1955)
	−155	+6	∓1780	−43	±4960	2390	Leask, Orbach, Powell, and Wolf (1963)
Pr	−70	−20	∓420	−50	±2300	700	Judd (1955)
Nd	−50	−30	∓400	−40	±1950	700	Judd (1955)
	−73	−18	∓790	−70	±1760	1330	Tinsley (1963)
Pm							
Sm	−30	−30	∓400	−40	±1800	700	Judd (1955)
	−14	−15	∓1440				Friederich, Hellwege, and Lämmerman (1960)

assumed by Judd (1955). Here only the relative signs and not the absolute signs of A_4^3 and A_6^3 are given, since they change sign on rotation by 180° about the crystal axis, unlike the remaining terms. The actual signs thus depend on the orientation chosen for the x-axis. It was pointed out by Judd (1957a) that these values lie fairly close to those appropriate to a crystal field of icosahedral symmetry, for which

$$|A_6^3/A_6^0| = 14\sqrt{5}, \quad |A_6^6/A_6^0| = 14, \quad (5.64)$$

all the other A_k^q being zero. This point has also been confirmed by the X-ray analysis, which shows that the lanthanide ion is surrounded by twelve oxygens at the corners of a somewhat irregular icosahedron. An important result is that for pure icosahedral symmetry, levels with J less than 3 are not split, while remarkable degeneracies occur for higher values of J. For example, $J = 3$ splits into a triplet and a quartet, whereas even in a pure cubic field it splits into three levels (two triplets and a singlet). The splittings for all manifolds with integral or half-integral values of J up to 8 are given by Judd (1957a) and

TABLE 5.12

Paramagnetic resonance data for the double nitrates $Ln_2Mg_3(NO_3)_{12}, 24H_2O$. Taken from the tables of Bowers and Owen (1955) and Orton (1959), except for (a) Stapleton, see Ruby, Benoit, and Jeffries (1962); (b) Judd and Wong (1958)

Ion	Diluent	T (°K)	$g_\parallel$	$\|g_\perp\|$	Isotope	$A_\parallel$ (cm^{-1})	$\|A_\perp\|$ (cm^{-1})	$P_\parallel$ (cm^{-1})
Ce	La	4·2	(±)0·25(5)	1·84(2)				
			(a) (±)0·032(68)	1·8264(13)				
Pr	La	4·2	1·55(2)					
Nd	La	4·2	(−)0·45(5)	2·72(2)	141	0·077(2)		
					143	(+)0·0052(5)	0·0312(1)	
					145	(+)0·0032(3)	0·0194(1)	
Pm								
Sm	Sm	4·2	(−)0·76(1)	0·40(5)	147	(+)0·0346(5)	<0·010	
					149	(+)0·0287(5)	<0·010	
Dy	La		(+)4·281(6)	8·923(160)	161	(−)0·01161(7)	0·02463(15)	−0·00142(10)
					163	(+)0·01622(7)	0·03415(15)	−0·00168(15)
Er	La		(b) (+)4·21(1)	7·990(10)	167	(−)0·0142(1)	0·0274(1)	+0·0013

Wybourne (1965), with their group-theoretical representations. Further details can be found in McLellan (1961).

The main effect of the extra terms (5.63) is to admix states whose values of J_z differ by 3, so that the states are more complicated than in the case of C_{3h} symmetry where only states whose values of J_z differ by 6 are admixed. The experimental magnetic resonance results are listed in Table 5.12, and we give here only a brief discussion.

Ce^{3+}, $4f^1$, $^2F_{\frac{5}{2}}$

First-order theory gives wave-functions for the ground doublet of the type
$$\cos\theta\,|\tfrac{5}{2},\,\pm\tfrac{1}{2}\rangle \pm \sin\theta\,|\tfrac{5}{2},\,\mp\tfrac{5}{2}\rangle \tag{5.65}$$
for which
$$\begin{aligned}g_\parallel &= \tfrac{6}{7}(\cos^2\theta - 5\sin^2\theta),\\ g_\perp &= \tfrac{6}{7}(3\cos^2\theta).\end{aligned} \tag{5.66}$$

These are connected by the relation $2g_\perp - g_\parallel = \tfrac{3\cdot 0}{7} = 4\cdot 29$, but the experimental values (taking $g_\parallel$ as zero) give 3·68 showing that the first-order theory is inadequate.

In second order no less than three admixtures from the $J = \tfrac{7}{2}$ state have to be included, with $J_z = \pm\tfrac{7}{2}$, $\pm\tfrac{1}{2}$, and $\mp\tfrac{5}{2}$; formulae for the g-values in this case are given by Judd (1955). The resonance results can then be fitted within experimental error, using the crystal field parameters given in Table 5.11.

Cerium magnesium nitrate is very useful in adiabatic demagnetization and as a magnetic thermometer; all its magnetic ions are equivalent, the g-tensor is highly anisotropic, and interactions between the ions are very small. It is also one of the first substances in which an Orbach relaxation process was observed, from which an excited doublet is found to lie at 34°K (Finn, Orbach, and Wolf (1961)). This splitting has been verified by magnetic susceptibility and calorimetric measurements, and has been observed directly by far infrared spectroscopy by Thornley (1963) who finds it to be 36·25±0·4°K in cerium magnesium nitrate and 30·2±0·4°K in the corresponding zinc salt. Leask, Powell, Orbach, and Wolf (1963) have attempted to compute the crystal field parameters from susceptibility measurements, obtaining the values given in Table 5.11. They worked in the L_z, S_z representation, and quote the eigenstates in this form; they show also that if a small orbital reduction ($k = 0\cdot 95$) is assumed, crystal field parameters some 10 to 20 per cent different are required to fit the experimental results.

Pr^{3+}, $4f^2$, $^3H_4$

Optical spectroscopic results of Hellwege and Hellwege (1953) and Brochard and Hellwege (1953) were used by Judd (1955) in the analysis for this ion. The results predict a ground doublet of the non-Kramers type

$$\alpha|\pm 4\rangle \pm \beta|\pm 1\rangle + \gamma|\mp 2\rangle \tag{5.67}$$

for which $g_\parallel = \frac{8}{5}(4\alpha^2 + \beta^2 - 2\gamma^2)$ and $g_\perp = 0$. Values of $\alpha = 0.458$, $\beta = -0.756$, $\gamma = 0.467$ give $g_\parallel = 1.56$ in good agreement with the resonance result. Note that the reversed signs (which follow from time reversal, eqn (15.29)) of the coefficient of $|\pm 1\rangle$ in the two substates lead to $g_\perp = 0$, as expected for a non-Kramers doublet.

Nd^{3+}, $4f^3$, $^4I_{\frac{9}{2}}$

First-order theory gives for the ground doublet

$$\pm \cos\theta \cos\varphi \, |\tfrac{9}{2}, \pm\tfrac{7}{2}\rangle + \sin\theta \cos\varphi \, |\tfrac{9}{2}, \mp\tfrac{5}{2}\rangle + \sin\varphi \, |\tfrac{9}{2}, \pm\tfrac{1}{2}\rangle,$$

which with the parameters given by Judd (see Table 5.11) leads to

$$g_\parallel = -0.52, \; |g_\perp| = 2.91$$

but with rather too low a value of $A_\perp$. Second-order theory gives four extra admixed states from $J = \tfrac{11}{2}$, with $J_z = \pm\tfrac{7}{2}, \pm\tfrac{1}{2}, \mp\tfrac{5}{2}$, and $\mp\tfrac{11}{2}$, for which the detailed formulae are given by Judd (1955), who obtains the calculated values

$$g_\parallel = -0.32, \; |g_\perp| = 2.72 \; \text{and} \; |A_\perp| = 0.0184 \; \text{cm}^{-1}$$

in good agreement with experiment. The rather different crystal field parameters of Tinsley (1963) are based on an analysis of the optical spectrum observed by Dieke and Heroux (1956).

Sm^{3+}, $4f^5$, $^6H_{\frac{5}{2}}$

Since the ground state has $J = \tfrac{5}{2}$ and the first excited state is $J = \tfrac{7}{2}$, the theory is similar to that for Ce^{3+} except that the second-order terms are more important (see Judd (1955)). Excited doublets of the $J = \tfrac{5}{2}$ manifold have been observed at $46.5 \; \text{cm}^{-1}$ and $68.9 \; \text{cm}^{-1}$ by means of optical spectroscopy (Friederich, Hellwege, and Lämmermann 1960), and the crystal field parameters attributed to them in Table 5.11 represent a fit to these splittings using first-order theory only (i.e. neglecting admixtures of the $J = \tfrac{7}{2}$ state); see Scott and Jeffries (1962).

Dy^{3+}, $4f^9$, $^6H_{15/2}$ and Er^{3+}, $4f^{11}$, $^4I_{15/2}$

It is convenient to take these two ions together because each has ground state $J = \tfrac{15}{2}$, and the theoretical interpretation uses the icosahedral crystal field as a starting point. In such a field the $J = \tfrac{15}{2}$ manifold breaks up into two sextet states and one quartet (Γ_8); for both ions the latter is expected to lie lowest. When a perturbation of lower (C_{3v}) symmetry is included, the quartet breaks up into two doublets, one of which would be

$$\mp|\pm\rangle = \sqrt{(\tfrac{91}{1215})}|\pm\tfrac{13}{2}\rangle \mp \sqrt{(\tfrac{121}{243})}|\pm\tfrac{7}{2}\rangle - \sqrt{(\tfrac{77}{405})}|\pm\tfrac{1}{2}\rangle \pm$$
$$\pm \sqrt{(\tfrac{55}{243})}|\mp\tfrac{5}{2}\rangle + \sqrt{(\tfrac{13}{1215})}|\mp\tfrac{11}{2}\rangle, \quad (5.68)$$

where the coefficients are those appropriate to a pure icosahedral field, and are uniquely defined in a group-theoretical treatment since the quartet (Γ_8) occurs only once in the reduction of $J = \tfrac{15}{2}$. For this doublet

$$g_\parallel = g_\perp/2 = (\tfrac{17}{5})\langle J \| \Lambda \| J \rangle,$$

which gives the values

$$g_\parallel = 4{\cdot}53, \ |g_\perp| = 9{\cdot}07 \text{ for } Dy^{3+} \quad \text{(Judd 1957a)}$$

and

$$g_\parallel = 4{\cdot}08, \ |g_\perp| = 8{\cdot}16 \text{ for } Er^{3+} \quad \text{(Judd and Wong 1958)}$$

in remarkably good approximation to the observed values (Table 5.12).

Relaxation measurements (Larson and Jeffries 1966a) indicate the presence of another doublet at 15 cm^{-1} in the case of Dy^{3+}.

5.8. Lanthanide ions in cubic symmetry

As mentioned in Chapter 18 and in § 5.1, the resonance spectra of a number of lanthanide ions, both di- and tri-positive, have been observed on sites of cubic symmetry in crystals such as MgO, CaO (octahedral symmetry) and CaF_2, SrF_2, BaF_2, ThO_2 (cubic, eightfold coordination). In cubic symmetry the way in which a manifold of given J is split is readily found from group theory, and the results are summarized in Table 5.13. Manifolds with J less than 2 are not split; half-integral values of J give rise to doublets (Γ_6, Γ_7) and quartets (Γ_8); integral values of J give singlets (Γ_1, Γ_2), doublets (Γ_3) and triplets (Γ_4, Γ_5). When a given representation occurs only once, its wave-functions are uniquely determined from symmetry alone and a number are listed in Tables 4 and 9. When a given representation occurs more than once, the wave-functions depend on the value of x (see eqns (5.72) or (18.14)). The doublets Γ_6, Γ_7 can be described by spin

TABLE 5.13

Group theoretical representation of the decomposition of manifolds of various J in a cubic field, showing how often the various representations occur

J	Γ_6 (doublet)	Γ_7 (doublet)	Γ_8 (quartet)
$\frac{1}{2}$	1		
$\frac{3}{2}$			1
$\frac{5}{2}$		1	1
$\frac{7}{2}$	1	1	1
$\frac{9}{2}$	1		2
$\frac{11}{2}$	1	1	2
$\frac{13}{2}$	1	2	2
$\frac{15}{2}$	1	1	3

J	$\Gamma_1(A_1)$ (singlet)	$\Gamma_2(A_2)$ (singlet)	$\Gamma_3(E)$ (doublet)	$\Gamma_4(T_1)$ (triplet)	$\Gamma_5(T_2)$ (triplet)
1				1	
2			1		1
3		1		1	1
4	1		1	1	1
5			1	2	1
6	1	1		1	2
7		1	1	2	2
8	1		2	2	2

Hamiltonians with $S = \frac{1}{2}$ and have isotropic g-values; where they occur in the ground manifolds $J = \frac{5}{2}, \frac{7}{2}, \frac{9}{2}$, or $\frac{15}{2}$ of the lanthanide ions, the wave-functions are uniquely determined and so are the g-factors, a list of which (in first-order theory) is given in Table 22. Apart from small effects induced by the external magnetic field (see § 5.10), the magnetic hyperfine interaction is isotropic, and the nuclear electric quadrupole interaction is zero. The Γ_8 quartets can be described by a spin Hamiltonian with $S = \frac{3}{2}$, but they have rather peculiar properties which are discussed in Chapter 18; in particular, the Zeeman interaction and magnetic hyperfine interaction are not isotropic; there may be a nuclear electric quadrupole coupling whose form is somewhat similar to that of a free ion with $J = \frac{3}{2}$ (but needing two constants instead of one), and a complicated 'pseudo-nuclear' Zeeman coupling.

In the non-Kramers cases (integral J), the singlet states are of course of no interest in electron paramagnetic resonance, though in nuclear magnetic resonance they may show an isotropic 'pseudo-Zeeman' interaction as well as a true nuclear Zeeman interaction, with no nuclear electric quadrupole splitting. The Γ_3 doublets have no first-order Zeeman interaction (though a second-order splitting has been observed in an electron magnetic resonance experiment for Dy^{2+} in CaF_2; see below), and similarly no magnetic hyperfine coupling, though they may have a nuclear electric quadrupole coupling (see Chapter 18). The Γ_4, Γ_5 triplets can be represented by a spin Hamiltonian with $S = 1$, with isotropic Zeeman interaction and magnetic hyperfine interaction, but the nuclear electric quadrupole coupling can be represented by a formula similar to that for Γ_8 quartets (with $S = 1$ instead of $S = \frac{3}{2}$) in which again two constants are needed.

A crystal field interaction of cubic symmetry is conveniently expressed in the operator equivalent form

$$\mathscr{H} = B_4 O_4 + B_6 O_6, \tag{5.69}$$

where

$$O_4 = O_4^0 + 5O_4^4; \quad O_6 = O_6^0 - 21 O_6^4; \tag{5.70}$$

and

$$B_4 = A_4 \langle r^4 \rangle \langle J \| \beta \| J \rangle; \quad B_6 = A_6 \langle r^6 \rangle \langle J \| \gamma \| J \rangle. \tag{5.71}$$

Since we have only two parameters in (5.69) the energy levels and eigen functions can be computed numerically, on a reduced energy scale, for various values of the ratio of the two parameters. This has been carried out by Lea, Leask, and Wolf (1962); for numerical convenience they write (see eqns (18.14), (18.15))

$$B_4 F_4 = Wx; \quad B_6 F_6 = W(1 - |x|) \tag{5.72}$$

so that the Hamiltonian becomes

$$\mathscr{H} = W\{x(O_4/F_4) + (1-|x|)(O_6/F_6)\}, \tag{5.73}$$

where F_4, F_6 are numbers listed by them, and which are mostly (though not invariably) the same as those in the columns headed F in Tables 17. With these definitions $B_4/B_6 = 0$ for $x = 0$ and $\pm \infty$ for $x = \pm 1$, so that the whole range is contained between $x = -1$ and $+1$.

The values of A_4 and A_6 based on a point charge model and including only the nearest neighbours are given for a regular octahedron and cube by eqns (16.15) and (16.16). The lattice sums converge rapidly and for the CaF_2 lattice the inclusion of more distant neighbours gives values of A_4 and A_6 that differ by -8.7 and $+8.8$ per cent respectively from those in (16.16). If Ca^{2+} is replaced by a dipositive ion, calculation of the crystal potential should be rather accurate if no distortion takes place of the crystal lattice. On this supposition Bleaney (1964a) has shown from the optical spectrum of Tm^{2+} in CaF_2 that the effective values of $\langle r^4 \rangle$ and $\langle r^6 \rangle$ are some 4 and 10 times larger respectively than the values calculated by Freeman and Watson (Table 5.1, and interpolation). If Ca^{2+} is replaced by a tripositive ion without local charge compensation, so that cubic symmetry is retained, the induced dipoles on nearby lattice sites give roughly a 20 per cent increase in the crystal potential, and on applying this to the crystal field parameters estimated by Bierig and Weber (1963) for Dy^{3+} in CaF_2 one again finds that the effective values of $\langle r^4 \rangle$, $\langle r^6 \rangle$ are some 3 and 8 times larger respectively than those of Freeman and Watson. The results of Kiss (1965) show that the crystal field parameters of Dy^{2+} in CaF_2 are very similar to those for Dy^{3+}. From a comparison of Dy^{2+} in CaF_2, SrF_2, and BaF_2 Kiss finds that B_4 varies approximately as R^{-2} rather than R^{-5}, where R is the lattice spacing, showing that the crystal field model is inadequate unless considerable local distortion of the lattice is assumed.

Also, the effect of uniaxial stress on optical transitions in $Tm^{2+}:CaF_2$ and $Tm^{2+}:SrF_2$ is found by Axe and Burns (1966) to be larger than that predicted by a point charge model, and their calculations (see also Watson and Freeman 1967b) suggest a sizable covalent contribution to the ligand field splitting.

The possibility that CaF_2 with suitable lanthanide doping would provide a good laser material has led to intensive investigation of its spectra. The following examples illustrate some points associated with the electron magnetic resonance spectra.

$4f^3$, $^4I_{9/2}$. Pr^{2+} (cubic)

The paramagnetic resonance spectrum of this ion in CaF_2 with cubic symmetry has been observed by Merritt, Guggenheim, and Garrett (1966). They find the ground state to be a Γ_8 quartet, with a large hyperfine structure. The Zeeman splittings are fitted with a value of -0.55 for the parameter x of Lea, Leask, and Wolf (1962), which is close to the value of $|x|$ needed to fit the spectra of a number of lanthanide ions in CaF_2.

$4f^9$, $^6H_{15/2}$. Dy^{3+} (cubic)

In CaF_2 Bierig and Weber (1963) find the ground state to be a Γ_8 quartet, with a Γ_7 doublet as the first excited state at 8.5 ± 1 cm^{-1} higher in energy. The doublet has an isotropic $g = 7.52(5)$, which agrees well with the predicted value of $(\frac{17}{3})\langle J \| \Lambda \| J \rangle = 7.55$. When the magnetic field is parallel to a $\langle 100 \rangle$ axis, three resonance lines are observed within the Γ_8 quartet at $2.63(5)$, $5.48(15)$, and $14(1) \times (\beta H)$, which can be fitted with calculated values of 2·57, 5·36, and 13·3 using $x = +0·6$ (eqn (5.72)). This is also the value of x at which Γ_7 and Γ_8 states lie lowest and very close together in the calculations of Lea, Leask, and Wolf (1962). Bierig and Weber find that the results can be fitted with the approximate crystal field parameters of

$$A_4\langle r^4\rangle \sim -200 \text{ cm}^{-1}, A_6\langle r^6\rangle \sim +30 \text{ cm}^{-1}.$$

They have verified that the Zeeman effect in the Γ_8 quartet has the correct angular dependence for $x = 0·6$ when **H** is rotated in a {100} plane, and that it is independent of orientation when **H** is in a {111} plane, as expected from eqn (18.28), since the quantity

$$\{1-5(n_1^2 n_2^2+n_2^2 n_3^2+n_3^2 n_1^2)\}$$

is invariant in this plane.

Low and Rubins (1963b) have observed an isotropic line at $g = 6·60(5)$ for Dy^{3+} in CaO, which they identify as the Γ_6 doublet that lies lowest in an octahedral field when x lies between -1 and $-0·48$. Its theoretical g-value is 6·67, and they suggest that there may be an orbital reduction factor $k = 0·98$.

$4f^{10}$, $^5I_8$. Dy^{2+} (*cubic*)

The spectrum of Dy^{2+} in CaF_2 has been extensively investigated. The ground state is a Γ_3 doublet, with a Γ_4 triplet at 4·863 cm^{-1} (determined by a direct measurement; Mergerian, Stombler, and Harrop 1965), and a Γ_5 triplet at 28·6 cm^{-1} (Kiss, Anderson, and Orbach 1965). The remaining levels, two triplets, a doublet and a singlet lie above 312 cm^{-1}. Such a grouping of levels occurs near $x = -0·46$ for $J = 8$ (see Lea, Leask, and Wolf 1962). The calculated g-value for the Γ_4 triplet is 6·84, but because of the cluster of low-lying states the levels are strongly perturbed when a magnetic field is applied (cf. Fig. 5.6). Their behaviour has been studied by Sabisky (1964), who finds $g = 6·76(2)$ after allowing for a quadratic Zeeman effect. The Γ_3 doublet is also split by the quadratic Zeeman effect, and transitions become allowed between the two components of the doublet when **H** is not along an $\langle 001\rangle$ axis. This occurs then because the magnetic field admixes some of the Γ_4 states characterized by J_z values (± 5, ± 1, ∓ 3, ∓ 7) into the Γ_3 states which are characterized by (± 6, ± 2) and (± 8, ± 4, 0), so that transitions of the type $\Delta J_z = 0$, ± 1 become allowed. Obviously transitions between the Γ_3 doublet and Γ_4 triplet are allowed in zero field for the same reason, and from

330　　　　　　THE LANTHANIDE (4f) GROUP

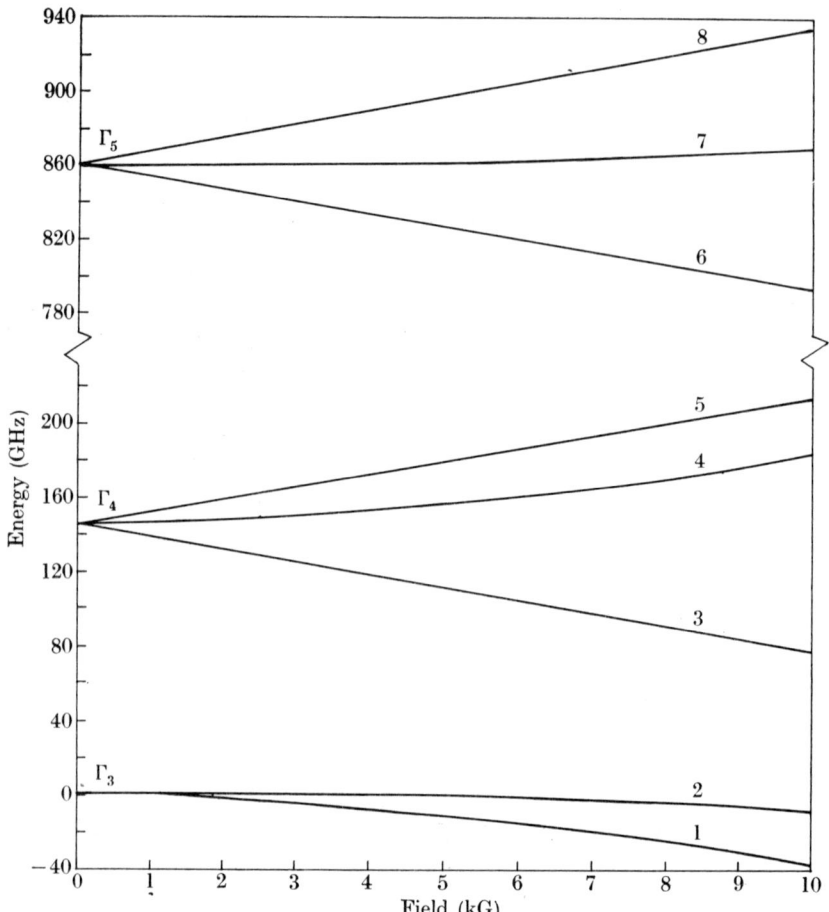

FIG. 5.6. Theoretical calculations of the Zeeman splitting of the lowest eight levels of Dy^{2+} ($4f^{10}$, $^5I_8$) in CaF_2 with **H** parallel to a $\langle 100 \rangle$ axis. (After Mergerian, Harrop, Stombler, and Krikorian 1967.)

a study of such transitions Mergerian, Harrop, Stombler, and Krikorian (1967) have determined the $\Gamma_4 - \Gamma_3$ splitting to be 145·89 GHz.

$4f^{11}$, $^4I_{\frac{15}{2}}$. Er^{3+}, Ho^{2+} (*cubic*)

The resonance results for these ions are summarized in Table 5.14.

The theoretical g-value for the Γ_7 doublet is $(+)6·8$ (or 6·77 allowing for intermediate coupling) while that for Γ_6 is $(-)6$; the rather low value for Ho^{2+} suggests that a small amount of orbital reduction is present (see Bleaney 1964a), which may be partly due to spin-phonon interactions (Inoue 1963). The fact that the ground states are different

TABLE 5.14

Resonance data for ground states of Ho^{2+}, Er^{3+} $(4f^{11}, {}^4I_{15/2})$ in a cubic field. The hyperfine constants are for ^{165}Ho and ^{167}Er $(I = \frac{7}{2}$ in both cases).

		g	A(MHz)	Reference
Eightfold coordination				
Ho^{2+} in CaF_2 doublet	Γ_6	(−)5·912(3)	(−)3921(1)	Lewis and Sabisky (1963)
Er^{3+} in CaF_2 doublet	Γ_7	(+)6·785(2)	(−)677(10)	Ranon and Low (1963)
CdF_2 dcublet	Γ_7	(+)6·758(10)	(−)693(10)	Zverev, Kornienko, Prokhorov, and Smirnov (1962)
ThO_2 dcublet	Γ_7	(+)6·752(5)	(−)697·8(9)	Abraham, Weeks, Clark, and Finch (1965)
CeO_2 dcublet	Γ_7	(+)6·747(6)	(−)696·3(9)	Abraham, Weeks, Clark, and Finch (1966)
Fourfold coordination				
Er^{3+} in ZnSe(I) dcublet†	Γ_7	(−)5·950(5)		Kingsley and Aven (1967)
quartet	Γ_8	at 6·5±1 cm^{-1} above doublet ground state		
(II) quartet	Γ_8	ground state		
Sixfold coordination				
Er^{3+} in MgO quartet	Γ_8			Descamps and Merle D'Aubigné (1964)
				Belorizky, Ayant, Descamps, and Merle D'Aubigné (1966)

† Note that the conventional nomenclatures for doublets Γ_6, Γ_7 in a tetrahedral field are interchanged from those for cubic (eightfold) field (see Kingsley and Aven (1967)).

for Er^{3+} and Ho^{2+} in CaF_2 indicates that we are near the point $x = -0.45$, where these two states cross, the value of $|x|$ being slightly greater in the latter case. The calculations of Lea, Leask, and Wolf (1962) show that there should also be a low-lying Γ_8 quartet. Lewis and Sabisky (1963) have observed 'forbidden' transitions in the hyperfine structure of Ho^{2+} in the Γ_6 doublet which show a very large 'pseudo-nuclear' Zeeman effect, from whose magnitude they deduce that the Γ_8 quartet lies at 32 ± 4 cm^{-1}, in good agreement with optical results of Weakliem and Kiss (1967) who find a Γ_8 quartet at 30.1 cm^{-1} and a Γ_7 doublet at 33.5 cm^{-1}.

The results for Er^{3+} in MgO are discussed in detail in Chapter 18 in connection with the general treatment of the Γ_8 quartet. Ayant and Belorizky (1964) find that the results are fitted best with a value of $\langle J\| \Lambda \|J\rangle$ which is 1.8 per cent lower than the simple free ion value of $\frac{6}{5} = 1.2$. Of this, 0.4 per cent is due to the admixture of states with different values of J, and if the remainder is ascribed to a reduction of the orbital k-value one obtains $k = 0.979$. Allowing for these effects, the results correspond to $x = 0.714$; the results of Low and Rubins for the same ion in CaO can be interpreted similarly with $x = 0.73$ (Descamps and Merle D'Aubigné 1964). A detailed study of the hyperfine structure of Er^{3+} in MgO by Belorizky, Ayant, Descamps, and Merle D'Aubigné (1966) confirms the expected form of the interactions, and gives a quadrupole coupling in good agreement with that calculated from atomic beam measurements on the free atom, the lattice contribution being of course zero in cubic symmetry.

The spectrum of Er^{3+} in cubic symmetry has also been observed for (Er, Zn)Se by Kingsley and Aven (1967). Two sets of spectra are found, believed to be associated with

(I) Er^{3+} substituting for Zn^{2+} (surrounded by a tetrahedron of Se^{2-} ions);

(II) Er^{3+} at an interstitial site (normally surrounded by a tetrahedron of 4 Zn^{2+} sites and an octahedron of 6 Se^{2-} ions).

In spectrum I, the ground state is a Γ_7 doublet, with a Γ_8 quartet 6.5 ± 1 cm^{-1} higher in energy. In spectrum II the ground state is a Γ_8 quartet.

$4f^{13}$, $^2F_{\frac{7}{2}}$. Yb^{3+}, Tm^{2+} (*cubic*)

A number of experimental resonance results are summarized in Table 5.15.

TABLE 5.15

Experimental results for Yb^{3+} *and* Tm^{2+} $(4f^{13})$ *in cubic symmetry*

			g	$^{171}A\ (I = \tfrac{1}{2})$ (MHz)	$^{173}A\ (I = \tfrac{5}{2})$ (MHz)	Reference
Yb^{3+}	CaO	Γ_6	$(-)2\cdot 585(3)$	$(-)2094(18)$		Low and Rubins (1963b)
	CaF_2	Γ_7	$(+)3\cdot 498(2)$	$(+)2638\cdot 70(5)$	$(-)727\cdot 094(60)$	Baker, Blake, and Copland (1969)
	CdF_2	Γ_7	$(+)3\cdot 4359(8)$	$(+)2649(5)$	$(-)726(1)$	Konyukhov, Pashinin, and Prokhorov (1962)
	ThO_2	Γ_7	$(+)3\cdot 423(1)$	$(+)2631(3)$	$(-)725(1)$	Abraham, Weeks, Clark, and Finch (1965)
	CeO_2	Γ_7	$(+)3\cdot 424(1)$	$(+)2631(3)$	$(-)726(1)$	Abraham, Weeks, Clark, and Finch (1966)
			g	$^{169}A\ (I = \tfrac{1}{2})$ (MHz)		
Tm^{2+}	CaF_2	Γ_7	$(+)3\cdot 443(2)$	$(-)1101\cdot 376(4)$	$J = \tfrac{7}{2}$	Bessent and Hayes (1965)
		Γ_7'	$(-)1\cdot 453(2)$	$(+)1160(5)$	$J = \tfrac{5}{2}$	Sabisky and Anderson (1966)

In eightfold coordination the Γ_7 doublet should be the ground state except for values of x less than about $+0.2$; in sixfold coordination (x negative) the Γ_6 doublet always lies lowest. The spectrum of Tm^{2+} in CaF_2 has been studied in considerable detail. From the optical spectrum of Kiss (1962), values of the crystal field were deduced by Bleaney (1964a), who interpreted the g-factor in the ground doublet as giving evidence for an orbital factor $k = 0.991$. Later, paramagnetic resonance measurements on the $J = \frac{5}{2}$, Γ_7 doublet were made by Sabisky and Anderson (1966) using an optical pumping technique to populate this doublet indirectly. In this case the deviation of the g-factor requires $k = 0.986$, so that $(1-k)$ is here significantly larger than in the ground doublet. One explanation of the deviation in the g-factors is that it is due to electron–phonon interactions, for which Inoue (1963) has estimated $\Delta g = -0.01$, a result that is rather smaller than the observed deviation of -0.025. However, measurements of the fluorine hyperfine structure (Bessent and Hayes 1965) show that the interaction, which is point dipole for more distant shells of fluorines, is larger than point dipole for the first shell, suggesting that there may also be a contribution from covalent bonding. For the isoelectronic ion $Yb^{3+}:CaF_2$ (cubic), a value of $k = 0.984$ is found for the Γ_7 ground doublet (Baker, Blake, and Copland 1969); this is a larger value of $(1-k)$ than for the ground doublet of $Tm^{2+}:CaF_2$, and it may be significant that in the fluorine ligand hyperfine structure (see Table 4.4) the value of A_p is appreciably larger for Yb^{3+} than for Tm^{2+}. On the other hand, Axe and Burns (1966) estimate that for Tm^{2+} covalency gives a value for $(1-k)$ of order 10^{-3} rather than 10^{-2}.

For the Γ_6 doublet of Yb^{3+} in CaO (where the analysis is simpler because it is not admixed with any $J = \frac{5}{2}$ state), Low and Rubins (1963b) find that the rather low value of g, which on simple theory ($k = 1$) should be $-\frac{7}{3}\langle J\|\Lambda\|J\rangle = -2.667$, corresponds to a value of $k = 0.959(1)$.

A comparison of magnetic hyperfine measurements on two doublets, such as given in Table 5.15 for Tm^{2+} in CaF_2, makes it possible to extract the $A_s(\mathbf{S} . \mathbf{I})$ contribution to the hyperfine structure. If no corrections are applied for electron–phonon or covalent effects in the hyperfine structure, a value is found (Sabisky and Anderson 1966) of $(+)87(13)$ MHz, which is equivalent to a coefficient of $-190(30)$ MHz in eqn (5.31) or (5.32), between 2 and 3 times the numerical values assumed in these equations. However this result depends on, and is rather sensitive to, the assumption that $\langle r_l^{-3}\rangle = \langle r_{sC}^{-3}\rangle$ (see § 17.7), and that these quantities are the same for doublets belonging to the $J = \frac{7}{2}$ and $J = \frac{5}{2}$ manifolds.

The hyperfine structure of Yb^{3+} in CaF_2 is discussed in § 5.10.

5.9. Ions with a half-filled $4f$-shell, $4f^7$, $^8S_{\frac{7}{2}}$. Eu^{2+}, Gd^{3+}, Tb^{4+}

If the ground state of the $4f^7$ ions were a pure $^8S_{\frac{7}{2}}$ state, the only non-zero interaction in a magnetic resonance experiment would be the Zeeman interaction

$$\mathscr{H} = g\beta \mathbf{H}.\mathbf{S}, \qquad (5.74)$$

where g would be isotropic and very close to the free spin value of 2·00232 (corrections for diamagnetic and relativistic effects, estimated by Judd and Lindgren (1961), reduce this to 2·0005 for the Eu atom, whose configuration is $4f^7 6s^2, {}^8S_{\frac{7}{2}}$). However, even for a free atom or ion, there is a significant amount of intermediate coupling, which on including a first-order correction makes the ground state

$$(1-\alpha^2)^{\frac{1}{2}}\,{}^8S_{\frac{7}{2}}+\alpha\,{}^6P_{\frac{7}{2}}(+\text{etc.}) \qquad (5.75)$$

whose g-value is

$$g = (1-\alpha^2)g({}^8S_{\frac{7}{2}})+\alpha^2 g({}^6P_{\frac{7}{2}}). \qquad (5.76)$$

For the Eu atom the best estimates of α give a result in good agreement with the experimental value of 1·9935(3). Less information is available for the ions in the solid state on which to base an estimate of α, but the measured g-values of Eu^{2+} and Gd^{3+} given in Table 5.16 (which

TABLE 5.16

Values of g and the cubic splitting parameters b_4, b_6 for a number of $4f^7$ ions in cubic fields. A number of the original sign determinations have been corrected. (See Shuskus 1962, and Low and Rubins 1963a)

	g	$b_4 = 60B_4$ (MHz)	$b_6 = 1260B_6$ (MHz)	Reference
Eu^{2+} in CaO	1·9914(10)	−77·1(15)	+4·5(15)	Shuskus (1962)
Eu^{2+} in CaF_2	1·9926(3)	−176·12(2)	+0·78(2)	Baker and Williams (1962)
Eu^{2+} in $SrCl_2$	1·995(1)	39		Low and Rosenburger (1959)
Gd^{3+} in CaO	1·9925(10)	−36·6(3)	+3·6(3)	Shuskus (1962)
Gd^{3+} in CaF_2	1·991(2)	(−)138(4)	(+)3(2)	Low (1958)
Gd^{3+} in ThO_2	1·991(1)	−108·75(60)	−1·28(60)	Low and Shaltiel (1958) Hurrell (1965)
Gd^{3+} in $SrCl_2$	1·9906(10)	(±)29·7(1)	(∓)0·2(1)	Low and Rosenburger (1959)
Tb^{4+} in ThO_2	2·0146(4)	−2527·53(10)	−24·84(4)	Baker, Chadwick, Garton, and Hurrell (1965)

are typical of many experimental results) seem to be explicable on this basis, while the value of 2·0146(4) for Tb^{4+} in ThO_2 is not.

In the solid state where an ion is subjected to a ligand field the Zeeman interaction may be expected to be more complicated, with terms linear in **H** but involving higher powers of the components of **S**. The form of these terms (which are experimentally very small) will be considered in § 5.10, together with comparable terms in the magnetic hyperfine structure. The latter have been determined with much greater certainty, because of the higher accuracy possible in Endor measurements (see Chapter 4).

The main result of the ligand field is to produce a splitting of the electronic states in zero magnetic field, with a consequent 'fine structure' in the resonance spectrum. This is possible only through perturbations of high order, so that the splitting is usually rather small, whereas for the lanthanide ions which are not in S-states the comparable splitting is of order 10^2 cm^{-1}. Fortunately the experimentalist can include such effects in his spin Hamiltonian by the simple device of introducing spin operators which reflect the symmetry of the site on which the ion is placed in the solid. Thus for cases of cubic symmetry we add to eqn (5.74) the general cubic terms

$$B_4(O_4^0+5O_4^4)+B_6(O_6^0-21O_6^4), \tag{5.77}$$

whose effects on the spectrum are discussed in § 3.4. B_4, B_6 are regarded as parameters to be determined experimentally, and a representative list of results is given in Table 5.16. For numerical convenience the modified parameters $b_4 = 60B_4$, $b_6 = 1260B_6$ are used, where the numerical multipliers are those listed in the column F for $J = \frac{7}{2}$ in Tables 17. For comparison with the nomenclature of Low (1960) we give the equivalence $c = 4b_4$, $d = 4b_6$ (see also Table 3.8).

The experimental results do not give any very firm guide to the theoretical interpretation. The values of b_4 appear to be negative both in octahedral and cubic (eightfold coordinated) sites, suggesting that perturbation mechanisms linear in the crystal field can be ruled out, but the magnitude of b_4 does not seem to be correlated in any obvious way with the strength of the crystal field estimated on a point charge model. The unusually large value of b_4 for Tb^{4+} in ThO_2 (all transitions observed come at intermediate fields—see Fig. 5.7), together with other abnormalities (see Baker, Chadwick, Garton, and Hurrell 1965) suggests that bonding effects are important in this case, but b_4 is not particularly large for Eu^{2+} in CaO (Shuskus 1962) though the magnetic

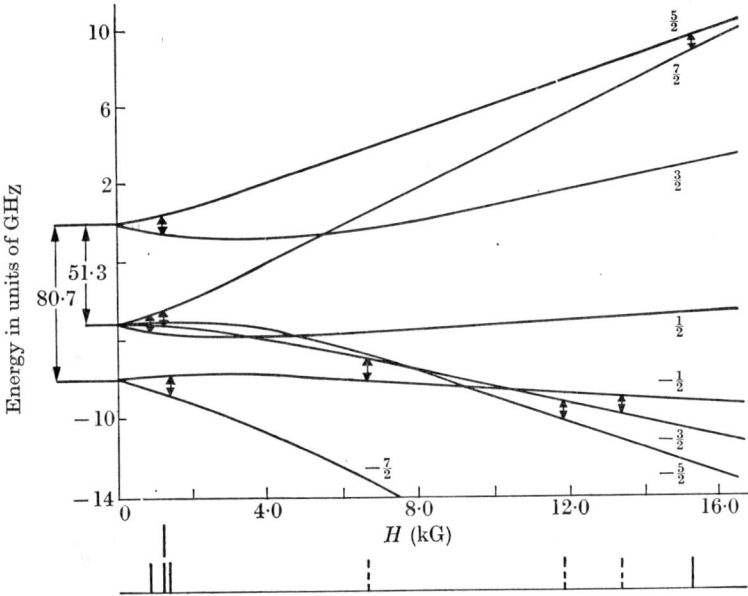

FIG. 5.7. Energy levels of Tb^{4+} in ThO_2 plotted as a function of a magnetic field applied in a $\langle 100 \rangle$ direction, showing the transitions observed at a frequency of ~ 9 GHz. The position of the lines is shown below, the broken lines being those on which Endor was observed. (After Baker and Hurrell 1965.)

hyperfine structure is distinctly smaller than for Eu^{2+} in CaF_2, a result that in $3d^5$ ions usually indicates a greater degree of covalent bonding. Furthermore, the value of b_4 for Eu^{2+} changes sign in the series of host lattices CaO, SrO, BaO (see Overmeyer and Gambino 1964, Miyata and Argyle 1967).

The value of b_6 always seems to be small compared with b_4, but because of the numerical multipliers perhaps the most direct comparison can be made by considering the crystal field splittings in zero magnetic field. The ^8S state then divides into two doublets (Γ_6, Γ_7) and a quartet (Γ_8), the ratio of the separations being

$$\frac{W(\Gamma_6)-W(\Gamma_8)}{W(\Gamma_8)-W(\Gamma_7)} = \frac{12b_4-36b_6}{20b_4+28b_6},$$

so that if $b_6 = 0$ the ratio is just 0·6. The deviation from this value is only a few per cent for the eightfold coordinated ions in Table 5.16, which are all in CaF_2 type lattices. On the other hand, for Tm^{2+} in CaF_2, whose ground state is $J = \frac{7}{2}$, the comparable ratio is 0·06. This seems to be in line with the suggestion (Lacroix 1957) that for $4f^7$, ^8S

states the crystal field produces the b_6 term only through higher-order perturbation mechanisms than are required for b_4.

In compounds where the symmetry is less than cubic a larger number of spin operators will be required, but these must still reflect the symmetry of the site. Thus for C_{3h} symmetry (LaCl$_3$, the ethyl-sulphates) we need to add to (5.74) the terms

$$B_2^0 O_2^0 + B_4^0 O_4^0 + B_6^0 O_6^0 + B_6^6 O_6^6. \qquad (5.78)$$

In experiments where these terms are small compared with the Zeeman interaction (as is often the case), the first three terms, which are diagonal, can be determined at once from the positions of the transitions when the external magnetic field is applied along the axis. The last term then enters only in second order, and can be found more accurately by measurements with the magnetic field in the plane normal to the symmetry axis. If the direction of the magnetic field is then taken as the z-axis of the spin operators, the Hamiltonian can be transformed, using the formulae in Table 5 of Jones, Baker, and Pope (1959), to (H at angle $\theta = \pi/2$, ϕ to z-axis)

$$\mathscr{H} = g\beta H S_z - \tfrac{1}{2} B_2^0 O_2^0 + \tfrac{3}{8} B_4^0 O_4^0 + \tfrac{1}{16} O_6^0 (B_6^6 \cos 6\phi - 5 B_6^0), \qquad (5.79)$$

where only the diagonal terms have been retained. These are similar to those in (5.78) except for numerical changes, but the presence of $\cos 6\phi$ in the last term means that the line spacings will vary as $\cos 6\phi$ as the magnetic field is rotated in the plane normal to the crystal axis. The value of B_6^6 can be obtained directly from this variation.

When the symmetry is only C_3, or C_{3v}, we must add to (5.78) the terms

$$B_4^3 O_4^3 + B_6^3 O_6^3, \qquad (5.80)$$

which affect the transition frequencies only in second order when the magnetic field is along the threefold axis. When the magnetic field is normal to this axis, the expectation values of O_4^3 and O_6^3 are zero, so that as H is rotated in this plane the variation in the spectrum reflects only the O_6^6 term. Thus to first order the behaviour of the spectrum under these conditions is the same as for C_{3h} symmetry, and measurements must be made at intermediate directions to determine the terms in (5.80), as is discussed for Gd^{3+} in Al$_2$O$_3$ by Geschwind and Remeika (1961), and in the hexa-antipyrene iodide by Baker and Williams (1961).

The results for Gd^{3+} measured in four compounds of three or sixfold symmetry are given in Table 5.17. The main feature of the results is

TABLE 5.17

'Fine structure' splitting parameters for Gd^{3+} in four compounds. They are somewhat temperature sensitive; in the first three cases the temperature for which values are given is 90°K, in the last it is 4.2°K. For Gd^{3+} in Al_2O_3 the additional parameters $\bar{b}_4^3 = 3B_4^3 = 18\cdot3(10)$, $\bar{b}_6^3 = 36B_6^3 \approx 1\cdot0$ (in units of 10^{-4} cm^{-1}) have been determined

		g	$b_2^0 = 3B_2^0$	$b_4^0 = 60B_4^0$	$b_6^0 = 1260B_6^0$	$\lvert b_6^6 = 1260B_6^6 \rvert$	Reference
			in units of 10^{-4} cm^{-1} (2·998 MHz)				
$LaCl_3$	C_{3h}	1·991(1)	+16·0(2)	+2·13(5)	+0·25(5)	1·4(3)	Hutchison, Judd, and Pope (1957)
Lanthanum ethylsulphate	C_{3h}	1·990(2)	+204·7(20)	−3·96(30)	+0·63(60)	3·5(5)	Bleaney, Scovil, and Trenam (1954)
Bismuth magnesium nitrate	C_{3v}	1·992(2)	+124(1)	+0·9(1)	+0·6(1)	12(1)	Trenam (1953)
Al_2O_3	C_3	1·9912(5)	+1032·9(20)	+26·0(10)	+1·0(5)	5·0(5)	Geschwind and Remeika (1961)

TABLE 5.18

Principal hyperfine constants of some ions with $4f^7$, $^8S_{\frac{7}{2}}$ ground states, as measured by Endor. The column $-AS/g_I\beta$ gives the magnetic field at the nucleus in the fully polarized state $S_z = S$. The following columns give the estimated contributions from f-electrons (A_f/g_n) including relativistic effects, and from core polarization (A_c/g_n). A, B are the magnetic dipole and electric quadrupole parameters (see eqn (4.42)); g_I, g_n are the nuclear g-factors referred to Bohr magnetons and nuclear magnetons respectively

		I	A (MHz)	μ_I (nuclear magnetons)	$-AS/g_I\beta$ (kG)	A_f/g_n (MHz)	A_c/g_n (MHz)	B (MHz)	Q (barns)	B/Q (MHz/barn)
CaF_2	Eu^{2+} 151	5/2	$-102\cdot907(1)$	$+3\cdot4413(9)$				$-0\cdot7855(52)$	$+1\cdot16(8)$	
	Eu^{2+} 153	5/2	$-45\cdot673(2)$	$+1\cdot5196(13)$				$-2\cdot0294(68)$	$+2\cdot92(20)$	$-0\cdot68(5)$
	Eu^{2+}				340	-15	-60			
CeO_2	Gd^{3+} 155	3/2	$(+)12\cdot048(3)$	$(-)0\cdot2567(6)$						
	Gd^{3+} 157	3/2	$+15\cdot809(5)$	$-0\cdot3357(6)$						
ThO_2	Gd^{3+} 157	3/2	$+15\cdot768(3)$	$-0\cdot3371(6)$				$-0\cdot687(18)$	$+1\cdot70(16)$	
	Gd^{3+}				320	-18	-52			$-0\cdot40(4)$
ThO_2	Tb^{4+} 159	3/2	$-73\cdot891(23)$	$+1\cdot994(4)$	250	-22	-33	$+6\cdot194(38)$	$1\cdot23(10)$	$+5\cdot0(4)$

References: Eu^{2+}, Baker and Williams (1962).
Gd^{3+}, Hurrell (1965), Baker, Copland, and Wanklyn (1969).
Tb^{4+}, Baker, Chadwick, Garton, and Hurrell (1965).

that, except for Gd^{3+} in $LaCl_3$, the 'fine structure' is dominated by the 'quadrupole' term B_2^0. This suggests that such a term is produced by lower order perturbation mechanisms than the other terms, since the B_2^0 term is far from dominant in the splittings of other lanthanide ions that are not in S-states. Wybourne (1966) has shown that when relativistic effects are included, a term of the right order of magnitude (but of the wrong sign) can be produced by a second-order process involving the first power of the spin-orbit coupling and of the crystal field.

We consider now the principal hyperfine interactions, listed in Table 5.18 for some $4f^7$ ions in cases of cubic symmetry. We can isolate the electronic effects by dividing through by g_I (or g_n) and by Q. The column headed $-AS/g_I\beta$ gives the electronic magnetic field at the nucleus in the state $S_z = S$, and is of interest in a comparison with the corresponding field in a substance (such as the metals) which enter a magnetically co-operative state. This field shows a small but significant decrease as the valency of the ion increases. The following two columns give the contributions (A_f/g_n) from f-electrons including relativistic effects, and (A_c/g_n), from core polarization, as estimated by Baker, Chadwick, Garton, and Hurrell (1965), and show that the latter decreases markedly for Tb^{4+}, a result again consistent with increased covalent bonding. The column (B/Q) shows a more dramatic difference in that this quantity is much larger and of opposite sign for Tb^{4+} than for the two other ions.

In these and a number of other cases small higher-order terms in the spin Hamiltonian have been determined experimentally. Some of these are outlined in the following section.

5.10. Higher-order terms in the spin Hamiltonian

In the spin Hamiltonian the allowed terms have the dimensions $H^r S^s I^t$, where $r+s+t$ must be even to satisfy time reversal invariance, and where $s \leqslant 2S$, $t \leqslant 2I$ because operators containing higher powers have zero matrix elements. The chief terms in the spin Hamiltonian are those for which $r+s+t = 2$, except in the case $r = t = 0$, when powers of S up to 4 are encountered for d-electrons and 6 for f-electrons. These are the 'fine structure' terms that occur for d^5 (see § 7.12) and for f^7 (see §§ 5.9, 6.2, 6.3), and whose effect on the spectrum has been discussed for the case of cubic symmetry in § 3.4. Fine structure terms with $r = t = 0$, $s = 2$ occur for other configurations, and are similarly discussed in §§ 3.5, 3.6.

The case $r = s = 0$ arises only for the nuclear electric quadrupole interaction, which under non-cubic symmetry may be expressed in a form with $t = 2$ for a Kramers doublet and sometimes also for non-Kramers doublets (an exception to the latter is discussed in § 18.6). Occasionally such a form suffices for manifolds of higher degeneracy, but then we may need terms for which $s = t = 2$. We have already met such terms in § 4.7, where a nuclear electric quadrupole interaction of the same form as for a free atom was assumed for Eu^{2+}, $4f^7$ in CaF_2. This is not strictly correct, since even in cubic symmetry two parameters rather than one are needed in this interaction (see §§ 18.4 and 18.6), but the error is not likely to be serious for Eu^{2+} where the size of the interaction is close to that for the free Eu atom.

Fine structure terms are often comparable with or larger in energy than the electronic Zeeman interaction. In contrast, the nuclear electric quadrupole interaction, although one of the two principal hyperfine interactions, is usually rather small. The remaining terms with which we shall be concerned in this section are also rather small, with one exception. For an ion with appreciable 'unquenched' orbital momentum, such as Co^{2+} in a tetrahedral field or a rare-earth ion in a cubic field (cf. § 5.8), groups of states with degeneracies up to four may occur, as in the Γ_8 quartet. The spin Hamiltonian for this is discussed in some detail in Chapter 18, using an effective spin $S = \frac{3}{2}$. Terms of the form HS^3 may then be as important as the simple Zeeman term HS, but the reason for this is that in using $S = \frac{3}{2}$ to describe the quartet we are labelling the states by quite fictitious quantum numbers that have no simple relation to the actual angular momenta. It would be incorrect to ascribe terms of dimensions HS^3 to an electronic magnetic octupole moment, since this would interact only with a non-uniform magnetic field which has an octupole-like distribution over the region of the electronic magnetic moment, whereas H is a uniform field. Terms in IS^3 arise in a similar way from the magnetic field of the nucleus, but the fact that this field is non-uniform over the electronic distribution does not play a significant role.

In considering other terms, we confine ourselves to cases of cubic symmetry. Nevertheless, we can perhaps further emphasize our point about the effective spin by referring to § 18.6, where it is shown that terms of degree SI^2 are allowed for a Γ_3 doublet in a cubic field. This is possible because for a non-Kramers doublet (see § 18.5) the components of the effective spin S may not be time-odd operators, whereas the condition that $r+s+t$ must be even rests on the assumption that

THE LANTHANIDE (4f) GROUP

the components of S *are* time-odd operators. General discussions of the allowed terms are given by Koster and Statz (1959) and Tuhina Ray (1963), and below we illustrate the situation by outlining two sets of experimental results.

For an ion with configuration $4f^7$, whose ground state $S = \tfrac{7}{2}$ has an eightfold multiplicity, we may anticipate a number of higher-order terms. In the electronic Zeeman interaction, additional terms that have been explored are

$$\mathcal{H}'_{\text{Zeeman}} = g_1\beta H_z\{1/\sqrt{(10)}\}\{5S_z^3 - 3S_zS(S+1) + S_z\} + g_2\beta H_z\{1/3\sqrt{(56)}\} \times$$
$$\times\{63S_z^5 - 70S_z^3S(S+1) + 105S_z^3 + 15S_zS^2(S+1)^2 - 50S_zS(S+1) + 12S_z\} +$$
$$+ g_3\beta H_z\{5/8\sqrt{(10)}\}\{S_z(S_+^4 + S_-^4) + (S_+^4 + S_-^4)S_z\}, \quad (5.81)$$

where the last term applies when H_z is along a cube axis of fourfold symmetry. (The parameter g_1 in (5.81) is of course quite different from that in eqn (3.9).) Similarly, in addition to the principal hyperfine terms in eqn (4.42), we have

$$\mathcal{H}'_{\text{hfs}} = A'(S_x^3I_x + S_y^3I_y + S_z^3I_z) +$$
$$+ A''(S_x^5I_x + S_y^5I_y + S_z^5I_z) + B'\beta H_zS_zI_z^2. \quad (5.82)$$

Here we have retained the nomenclature of Baker and Williams (1962); more symmetrical forms of the terms in S^3 (which have the advantage that they vanish for $S < \tfrac{3}{2}$) are given in eqn (18.21)).

Some experimental results are listed in Table 5.19. As we would expect for ions where the orbital angular momentum is very small, the terms of higher degree become rapidly smaller. They arise from the

Table 5.19

Electron paramagnetic resonance and Endor data for some $4f^7$ ions. A, g are the usual parameters; the others are defined by eqns (5.77), (5.81), and (5.82). For references, see Table 5.18

	Eu²⁺ in CaF₂	Gd³⁺ in ThO₂	Tb⁴⁺ in ThO₂
$60B_4$ (MHz)	−176·12(2)	−168·75(60)	−2527·53(10)
$1260B_6$ (MHz)	+0·78(2)	−1·28(60)	−24·84(4)
g	1·9926(3)	1·991(1)	2·0146(4)
g_1	0(1)×10⁻⁵	—	0(2)×10⁻⁵
g_2	0(1)×10⁻⁶	—	2·1(4)×10⁻⁶
g_3	—	—	1·1(5)×10⁻⁶
	Isotope 151	Isotope 157	Isotope 159
A (MHz)	−102·907(1)	+15·7679(33)	−73·891(23)
A' (kHz)	+13·77(46)	−3·05(30)	+103·3(23)
A'' (kHz)	+0·084(31)	—	—
B'	—	—	+1·02(14)×10⁻⁶

small admixture of excited states through the spin-orbit coupling and the ligand field. This means that we might expect the ratio A'/A to be of the same order as $g_1/\Delta g$ rather than g_1/g, where $\Delta g = g_s - g$ is a measure of the orbital admixture. The measurements of g_1 are not sufficiently accurate to check this expectation, and the fact that (in contrast) fairly accurate values of A'/A are listed reflects the fact that Endor measurements are often higher in precision than electron paramagnetic resonance measurements (cf. § 1.13 or § 4.1).

An interesting contrast is provided by Endor measurements on Yb^{3+} (cubic) in CaF$_2$, where the ground state is an electronic Kramers doublet Γ_7 (see Table 5.15). The Endor spectrum is fitted to the spin Hamiltonian (with $S = \tfrac{1}{2}$)

$$\begin{aligned}\mathscr{H} =\ & g\beta \mathbf{H}\cdot\mathbf{S} + A\mathbf{I}\cdot\mathbf{S} - g^{(I)}\beta \mathbf{H}\cdot\mathbf{I} + \\
& + \beta h\{S_x I_x^2 H_x + S_y I_y^2 H_y + S_z I_z^2 H_z - \tfrac{1}{3}I(I+1)(\mathbf{S}\cdot\mathbf{H})\} + \\
& + \beta q\{(S_x H_y + S_y H_x)(I_x I_y + I_y I_x) + (S_y H_z + S_z H_y)(I_y I_z + I_z I_y) + \\
& \qquad + (S_z H_x + S_x H_z)(I_z I_x + I_x I_z)\} + \\
& + C[I_x^3 S_x + I_y^3 S_y + I_z^3 S_z - \tfrac{1}{5}\{3I(I+1) - 1\}(\mathbf{S}\cdot\mathbf{I})] + \\
& + D\{35 I_z^4 - 30 I_z^2 I(I+1) + 25 I_z^2 - 6I(I+1) + 3I^2(I+1)^2 + \\
& \qquad + \tfrac{5}{2}(I_+^4 + I_-^4)\}, \end{aligned} \qquad (5.83)$$

where the nomenclature (apart from the pseudo-nuclear Zeeman term) is that of Baker, Blake, and Copland (1969). For the isotope ^{171}Yb, $I = \tfrac{1}{2}$, and terms in higher powers of I than the first are allowed only

TABLE 5.20

Nuclear and hyperfine parameters for Yb^{3+} *(cubic) in* CaF$_2$, *for the* Γ_7 *Kramers doublet ground state, with* $g = +3\cdot4380(5)$. *The Endor spectrum is fitted to the spin Hamiltonian of eqn (5.83), and the data are from Baker, Blake, and Copland (1969), except for the values of* g_I *which are those of Olschewski and Otten (1967)*

	^{171}Yb ($I = \tfrac{1}{2}$) $\langle 100 \rangle$ axis	^{173}Yb ($I = \tfrac{5}{2}$) mean of $\langle 100 \rangle$ and $\langle 110 \rangle$ axes
A (MHz)	$+2638\cdot70(5)$	$-727\cdot094(60)$
$g^{(I)}$	$+8\cdot702(50)\times 10^{-4}$	$-2\cdot396(12)\times 10^{-4}$
g_I	$+5\cdot35816\times 10^{-4}$	$-1\cdot4761\times 10^{-4}$
βh (HzG^{-1})		$+49\cdot6(7)$
βq (HzG^{-1})		$+65\cdot2(10)$
C (kHz)		$+12(5)$
D (kHz)		$+133(33)$
$g^{(I)}/g_I$	$+1\cdot6240$	$+1\cdot6231$

for the other stable isotope ^{173}Yb, for which $I = \frac{5}{2}$. The measured parameters are given in Table 5.20, those for ^{173}Yb being a weighted mean of measurements in $\langle 100 \rangle$ and $\langle 110 \rangle$ directions. The external magnetic field admixes the excited Γ_8 quartet states into the ground Γ_7 doublet, and this accounts for the difference between the effective value $g^{(I)}$ and the true nuclear g-factor g_I measured by optical pumping experiments (Olschewski and Otten 1967) in the free atom. From the difference, the energy of the Γ_8 level is found to be 604 ± 10 cm^{-1}.

This same admixture is responsible for the terms in βh and βq; physically they correspond to interactions between the nuclear electric quadrupole moment and an electric field gradient arising from a distortion of the electronic charge cloud whose magnitude is proportional to the applied magnetic field. The same admixture contributes to the terms in C and D, which are essentially second-order effects of the nuclear electric quadrupole interaction, together with (in C) cross terms between the nuclear magnetic dipole and electric quadrupole interactions. In addition, such second-order effects admix the excited Γ_6 states, giving further contributions to C, D; finally, there is a small contribution to A from such terms which similarly involves admixtures from both the Γ_8 and Γ_6 levels. When allowance is made for this, the following corrected values of A were obtained:

$$^{171}A = +2638 \cdot 84(5) \text{ MHz}, \qquad ^{173}A = -726 \cdot 91(5) \text{ MHz}.$$

The size of the correction can be seen by comparing these values with those in Table 5.20.

Although the various nuclear parameters in eqn (5.83) were determined from the Endor spectrum as six independent constants, the results were also successfully analyzed by Baker, Blake, and Copland in terms of just the two principal free ion hyperfine parameters (magnetic dipole and electric quadrupole), together with two energy differences, those of the Γ_8, Γ_6 levels relative to the ground Γ_7 doublet.

6
THE ACTINIDE (5*f*) GROUP

6.1. Ions and compounds of the actinide group

ALTHOUGH Bohr had suggested as early as 1923 that another paramagnetic group might occur at the upper end of the periodic table, with electrons in the $5f$ shell, it was not until much later that the point at which such a series started became clear. In a review of the evidence Seaborg (1949) postulated that the $5f$ series starts with actinium as the analogue of lanthanum, thorium that of cerium, etc., the closed shells of electrons corresponding to a radon core. Later Dawson (1952), on the basis of magnetic data, suggested that at the start of the series the magnetic electrons occupied $6d$ states rather than $5f$, but it is now accepted that this interpretation is incorrect (see, for example, Bleaney (1955)). The number of unpaired electron spins is shown in Table 6.1 for the ions where the evidence is most complete. These electrons occupy the $5f$ states, and at 7 electrons (Am^{2+}, Cm^{3+}) we reach the half-filled shell; presumably the shell would be complete with 14 electrons at the tripositive ion of element 103. Measurements on the heavier ions are, of course, very difficult because of the high level of radioactivity, and magnetic resonance measurements have been confined to the lighter members of the group. Since most of the isotopes are α-particle emitters, radiation damage is heavy, causing magnetic resonance spectra to deteriorate rapidly in a single crystal, or even to be unobservable immediately after crystal growth. Ionic radii and some other data for the first half of the actinide group are given in Table 6.2.

In the solid state the tripositive ions have been studied mainly in the compounds $LaCl_3$ and $La(C_2H_5SO_4)_3, 9H_2O$, which have been extensively used in work on the lanthanide group. The site symmetry is C_{3h} (see the discussion in the previous chapter) and analysis of the results is similar to that for the $4f$ group. Uranium ions have been incorporated into CaF_2, and its analogues, where charge compensation results in magnetic resonance spectra that show tetragonal or trigonal symmetry as well as cubic symmetry.

The actinide ions in higher valency states have been studied in compounds that are, in the main, different from those of the $4f$ group. The two sets of compounds of greatest importance have regular octahedral symmetry at the actinide ion site. In the gas phase the molecules UF_6, NpF_6, and PuF_6 have a structure in which the metal

TABLE 6.1

Number of unpaired electron spins for the principal ions of the actinide group. In addition the complex ions $(UO_2)^{2+}$, $(NpO_2)^{2+}$, $(PuO_2)^{2+}$ have 0, 1, and 2 unpaired electron spins respectively. For these ions, as for those in the table, the unpaired electrons occupy the 5f states

	0	1	2	3	4	5	6	7
Thorium	Th^{4+}							
Protoactinium	Pa^{5+}	Pa^{4+}						
Uranium	U^{6+}		U^{4+}	U^{3+}	U^{2+}			
Neptunium		Np^{6+}		Np^{4+}	Np^{3+}			
Plutonium			Pu^{6+}		Pu^{4+}	Pu^{3+}		
Americium							Am^{3+}	Am^{2+}
Curium								Cm^{3+}

atom is at the centre of a regular octahedral arrangement of six fluorine atoms, the metal to fluorine distances being 0·1996, 0·1981, and 0·1971 nm respectively. A magnetic resonance experiment shows that cubic symmetry is preserved in the solid state for NpF_6 (Hutchison and Weinstock 1960). Some quadripositive ions have been studied in the diluted state in the compounds Cs_2ZrCl_6 and $[(CH_3)_4N]_2PtCl_6$, in each

TABLE 6.2

Some data for ions in the first half of the actinide series

	Tripositive ions			Quadripositive ions		
Z	Ion	Ionic radius (nm)	Spin-orbit coupling (cm^{-1})	Ion	Ionic radius (nm)	Spin-orbit coupling (cm^{-1})
90	Th^{3+}, $5f^1$	0·108		Th^{4+}, $5f^0$	0·095	
91	Pa^{3+}, $5f^2$	0·106		Pa^{4+}, $5f^1$	0·091	1490 ASJ
92	U^{3+}, $5f^3$	0·104	1666 CW 1659 GCCN	U^{4+}, $5f^2$	0·089	1800 JSSW
93	Np^{3+}, $5f^4$	0·102	2070 CW 1969 GCCN	Np^{4+}, $5f^3$	0·088	
94	Pu^{3+}, $5f^5$	0·101	2292 CW 2260 GCCN	Pu^{4+}, $5f^4$	0·086	
95	Am^{3+}, $5f^6$	0·100	2548 CW 2605 C	Am^{4+}, $5f^5$	0·085	
96	Cm^{3+}, $5f^7$		2948 CW 2842 GCCN			

The ionic radii are from Evans (1964); the other references are
ASJ   Axe, Stapleton, and Jeffries (1961).
CW   Carnall and Wybourne (1964), Wybourne (1964).
GCCN Gruber, Cochran, Conway, and Wong (1966).
JSSW Johnston, Satten, Schreiber, and Wong (1966).
C   Conway (1964).

of which the metal ion is at the centre of a regular octahedron of chlorine ions.

The actinide series differs also from the lanthanide series in the formation of complex ions such as the uranyl ion $(UO_2)^{2+}$ and its analogues $(NpO_2)^{2+}$ and $(PuO_2)^{2+}$. These ions are linear, of the form O—U—O, with a rather short distance for the U—O bond which provides evidence for a covalent type of bonding. Uranyl sodium acetate, $(UO_2)Na(C_2H_3O_2)_3$ is cubic, with four (UO_2) groups per unit cell whose axes lie along the threefold axes of the cube; paramagnetic resonance has been observed in the plutonyl salt (Hutchison and Lewis 1954). A salt that has been used more extensively is uranyl rubidium nitrate, $(UO_2)Rb(NO_3)_3$. This salt is hexagonal and all the (UO_2) groups lie parallel to the c-axis, surrounded by three nitrate groups in the equatorial plane, making the c-axis a threefold axis for each (UO_2) group. Paramagnetic resonance has been observed in the neptunyl and plutonyl ions present in dilute form in uranyl rubidium nitrate (Bleaney, Llewellyn, Pryce, and Hall 1954a, b) and the properties of these ions have been the subject of a detailed theoretical study by Eisenstein and Pryce (1955, 1956).

6.2. Tripositive actinide ions

Optical studies have been made of a number of the tripositive ions of the actinide group, mostly incorporated in crystals of $LaCl_3$. An analysis of these results together with data obtained on ions in solution shows that the behaviour is similar to that of the $4f$ group except that the spin-orbit coupling is considerably larger, as can be seen from a comparison of Tables 6.2 and 5.3. Two sets of values are given for the tripositive ions, which illustrate the spread in the experimental values.

The electrostatic parameters such as F_2 are rather smaller than in the $4f$ group, as can be seen from Tables 6.18, 6.19 in Wybourne (1964), so that departures from LS-coupling are much more pronounced. For example, Conway (1964) finds that in the ground multiplet $5f^6$, ^7F of Am^{3+} the level $J = 1$ lies 2720 cm^{-1} above $J = 0$, and $J = 6$ lies 12350 cm^{-1} above $J = 0$. Thus these two splittings are in the ratio 2:9·1 instead of the ratio 2:42 that is appropriate to pure LS-coupling, a departure that is far greater than those given in Table 5.2 for the corresponding $4f^6$ ions. It follows that intermediate coupling is an important factor in interpreting any resonance results; however no analysis of the crystal field will be attempted here and the resonance results given below summarize mainly the spin Hamiltonian parameters.

$5f^3$, U^{3+} in $LaCl_3$

The resonance results for this substance have been fitted to the usual spin Hamiltonian for effective spin $S = \tfrac{1}{2}$ with axial symmetry; they are summarized in Table 6.3.

The g-values are remarkably close to those ($g_\parallel = 4.00$, $g_\perp = 1.76$) measured for the corresponding ion $4f^3$, Nd^{3+} in $LaCl_3$, a result which in itself strongly suggests that the resonance is due to a doublet split off by the crystal field from a ground state characterized fairly accurately

TABLE 6.3

Spin Hamiltonian parameters for U^{3+} in $LaCl_3$

$\|g_\parallel\| = 4.153(5)$, $\|g_\perp\| = 1.520(2)$		Hutchison, Llewellyn, Wong, and Dorain (1956)
	Isotope 233 ($I = \tfrac{5}{2}$)	Isotope 235 ($I = \tfrac{7}{2}$)
$\|A_\parallel\|$	$378.6(12) \times 10^{-4}$ cm^{-1}	$176(1) \times 10^{-4}$ cm^{-1}
$\|A_\perp\|$	$123.6(10) \times 10^{-4}$ cm^{-1}	$58.5(5) \times 10^{-4}$ cm^{-1}
$\|P_\parallel\|$	$9.9(10) \times 10^{-4}$ cm^{-1}	$5.5(5) \times 10^{-4}$ cm^{-1}
	Bleaney, Hutchison, Llewellyn, and Pope (1956)	Dorain, Hutchison, and Wong (1957)

as $5f^3$, $^4I_{\tfrac{9}{2}}$. The ratio $(A_\parallel/A_\perp)/(g_\parallel/g_\perp)$ is 1·12, which is also very close to that for Nd^{3+} in $LaCl_3$, and suggests that crystal field strength and spin-orbit coupling are roughly in the same ratio as for Nd^{3+}. The nuclear spins of $\tfrac{5}{2}$ for isotope 233 and $\tfrac{7}{2}$ for isotope 235 are clearly established from the hyperfine structure, the ratio of the nuclear moments being $\mu_I^{235}/\mu_I^{233} = 0.66$. The quadrupole interaction is measured by observation of the 'forbidden' hyperfine lines that appear when the external magnetic field is not along the symmetry axis (cf. § 3.9). The values of $P_\parallel$ are very much larger than in the case of Nd^{3+}, and the correction for the pseudo-quadrupolar interaction $\Delta P_\parallel$ which was quite important for Nd^{3+} (§ 4.8) should be quite small for U^{3+}. Thus the ratio of the quadrupole moments, $Q^{235}/Q^{233} = 1.17(20)$ should be quite accurate. As in the case of Nd^{3+}, the lattice may make a substantial contribution to the electric field gradient at the nucleus, since the anti-shielding factor γ_∞ is likely to be larger for U^{3+} as well as $\langle r^{-3} \rangle$, which would increase the gradient due to the f-electrons.

$5f^7$, Cm^{3+} in $LaCl_3$, and $La(C_2H_5SO_4)_3$, $9H_2O$

This ion is expected to be in an $^8S_{\tfrac{7}{2}}$ state, like Gd^{3+}, where crystal field effects produce splittings only through higher-order effects. Abraham, Judd, and Wickman (1963) have shown that some spectra attributed earlier to Cm^{3+} must be due to Gd^{3+}; they find that only a single anisotropic resonance line is visible, with the following g-values:

$$\text{Cm}^{3+} \text{ in } \text{La}(\text{C}_2\text{H}_5\text{SO}_4)_3, 9\text{H}_2\text{O} \quad |g_\parallel| = 1\cdot 925(2) \quad |g_\perp| = 7\cdot 73(2),$$
$$\text{Cm}^{3+} \text{ in } \text{LaCl}_3 \quad |g_\parallel| = 1\cdot 925(2) \quad |g_\perp| = 7\cdot 67(2).$$

Since $g_\perp$ is very close to $4g_\parallel$, this must correspond to the transition $|+\tfrac{1}{2}\rangle \leftrightarrow |-\tfrac{1}{2}\rangle$ of $S = \tfrac{7}{2}$ split by an axial term $D\{S_z^2 - \tfrac{1}{3}S(S+1)\}$ (or $B_2^0\{3S_z^2 - S(S+1)\}$) which is large compared with the frequency at which resonance is observed ($\sim 0\cdot 3$ cm^{-1}). The value of $g_\parallel$ should be equal to the g-value of the free ion, and its rather low value is due to a fair degree of intermediate coupling, being in reasonable agreement with that calculated from the wave-functions

$$0\cdot 8884\,|^8S\rangle + 0\cdot 4197\,|^6P\rangle - 0\cdot 0909\,|^6D\rangle + \ldots$$

suggested by Carnall and Wybourne (1964) (see also Gruber, Cochran, Conway, and Nicol (1966)). The larger degree of intermediate coupling means that we should also expect the crystal field splittings to be much larger for Cm^{3+} than for Gd^{3+}, in agreement with experiment. A proportionate increase in line-width for the other transitions due to crystal imperfections would also be expected, and the damage due to the high radioactivity would readily make such lines too broad to be observed.

6.3. Actinide ions in CaF_2

The electron spin resonance and optical spectra of several actinide ions have been studied in CaF_2 and isomorphic crystals, partly in a search for suitable laser crystals.

$5f^4$, U^{2+}; $5f^3$, U^{3+} and $5f^2$, U^{4+} in CaF_2

Magnetic resonance was observed by Bleaney, Llewellyn, and Jones (1956) due to uranium in CaF_2 and SrF_2; the spectra showed axial symmetry about a fourfold axis with the following g-values:

	Uranium		Neodymium	
	$|g_\parallel|$	$|g_\perp|$	$|g_\parallel|$	$|g_\perp|$
CaF_2	3·501	1·866	4·412	1·301
SrF_2	3·433	1·971	4·289	1·505

(Errors: $\pm 0\cdot 008$ in $g_\parallel$, $\pm 0\cdot 002$ in $g_\perp$.)

The similarity of the results suggests that the ions are U^{3+}, Nd^{3+}. A fluorine hyperfine structure was observed in the U^{3+} spectrum, but not in the Nd^{3+} spectrum, suggesting that the $5f$ electrons take part in an appreciable amount of covalent bonding.

Title et al. (1962) have reported resonance lines whose angular variation can be fitted to that of a Γ_8 quartet in cubic symmetry, which would be expected as the ground state for $5f^3$ in the cubic field of CaF_2; however the observed g-values have not been interpreted, and do not seem to correspond to those expected from simple crystal field theory using the calculations of Lea, Leask, and Wolf (1962), though this is perhaps not surprising.

A number of asymmetrical lines typical of non-Kramers doublets with $g_\perp < 0.1$ and axial symmetry about a threefold axis have also been observed in CaF_2 doped with uranium and, for comparison, with praseodymium. The values of $g_\parallel$ are

(a) $g_\parallel = 5.65(1)$ McLaughlan (1966) Pr^{3+}, $4f^2$, $^3H_4$.
(b) $g_\parallel = 5.83(1)$ McLaughlan (1966) Pr^{3+}, $4f^2$, $^3H_4$.
(c) $g_\parallel = 5.66(2)$ McLaughlan (1966) U^{4+}, $5f^2$, $^3H_4$.
(d) $g_\parallel = 4.02(1)$ McLaughlan (1966) U^{4+}, $5f^2$, $^3H_4$.
(e) $g_\parallel = 3.238(5)$ Yariv (1962); Title et al. (1962).

The first three lines are so similar in the value of $g_\parallel$ that they are assigned to f^2 ions, where the $^3H_4$ manifold is split by the trigonal field to give as ground state the non-Kramers doublet (cf. eqn (5.67))

$$\alpha |\pm 4\rangle \pm \beta |\pm 1\rangle + \gamma |\mp 2\rangle.$$

For this doublet the maximum possible value of $g_\parallel$ is $8\langle J\| \Lambda \|J\rangle$ (corresponding to $\alpha = 1$, $\beta = \gamma = 0$); this maximum is 6.4 for f^2, $^3H_4$ but only 4.8 for f^4, $^5H_4$, so that the lines cannot be due to $5f^4$, U^{2+} ions. The last value (e), also ascribed to $5f^2$, U^{4+} ions by the authors quoted, is assigned to $5f^4$, U^{2+} by Hargreaves (1967).

In the absence of local charge compensation, an f^2, $^3H_4$ ion in the purely cubic field of CaF_2 may have as ground state (see Lea, Leask and Wolf (1962)) a Γ_5 triplet with an isotropic $\tilde{g} = 2$ and an effective spin $\tilde{S} = 1$. Such an isotropic line has been observed by McLaughlan (1967), and the line (d) above is attributed to a $|+\tilde{1}\rangle \leftrightarrow |-\tilde{1}\rangle$ transition for such a triplet split by a trigonal distortion, the $|\tilde{0}\rangle$ level lying higher by some 7 cm^{-1} (McDonald 1969; Wetsel and Donoho 1965).

$5f^7$, Am^{2+} and Cm^{3+} in CaF_2

The magnetic resonance spectra of these ions at sites of cubic symmetry in CaF_2 have been observed by Edelstein and Easley (1968). At low field strengths a single resonance line with an isotropic g-value of $(-)4.492$ is observed for $^{244}Cm^{3+}$, but at higher field strengths (frequencies of 35 GHz) some anisotropy is found. Similar results are

TABLE 6.4

Magnetic resonance results for the Γ_6 doublet of the $^8S_{\frac{7}{2}}$ state split by a cubic field in CaF_2 (Edelstein and Easley 1968). Similar but slightly smaller values of $(-)4\cdot484(2)$ and $(-)4\cdot475(2)$ have been obtained for $^{244}Cm^{3+}$ in ThO_2 and CeO_2 respectively (Abraham, Finch, and Clark 1968)

Ion	$g(\Gamma_6)$	I	$|A|$ (cm^{-1})
$^{244}Cm^{3+}$	$(-)4\cdot492(2)$	0	0
$^{241}Am^{2+}$	$(-)4\cdot490(2)$	$\frac{5}{2}$	$0\cdot01837(2)$
$^{243}Am^{2+}$	$(-)4\cdot490(2)$	$\frac{5}{2}$	$0\cdot01821(2)$

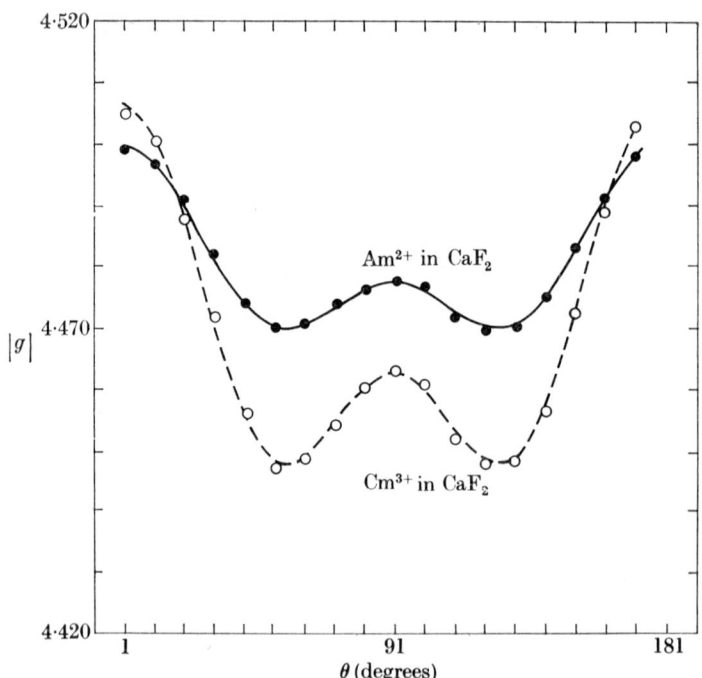

FIG. 6.1. The apparent g-factor of the Γ_6 doublet of two $5f^7$ ions on cubic sites in CaF_2, as measured at 35 GHz. The magnetic field is rotated in an $\langle 011 \rangle$ plane, and $\theta = 0$ corresponds to a $\langle 100 \rangle$ axis. The anisotropy is due to admixture of the excited Γ_8 states by the Zeeman interaction (Edelstein and Easley 1968).

—●—●— Am^{2+}, $W(\Gamma_8) - W(\Gamma_6) = 18\cdot6$ cm^{-1};
—○—○— Cm^{3+}, $W(\Gamma_8) - W(\Gamma_6) = 13\cdot4$ cm^{-1}.

THE ACTINIDE (5f) GROUP

```
                                    Γ₇(doublet), if
                                    b₆ = −0·1 cm⁻¹
                            (58·4) ─────

   Γ₇(doublet), if          (49·6) ─────
   b₆ = −0·1 cm⁻¹                  Γ₇(doublet), if
(45·1) ─────                       b₆ = 0


(35·7) ─────
       Γ₇(doublet), if
       b₆ = 0

                                      Γ₈ (quartet)
                            18·6 (5) ─────────

        Γ₈ (quartet)
13·4(5) ─────────


         Γ₆(doublet)                  Γ₆(doublet)
      0 ─────────                  0 ─────────
Energy          Cm³⁺          Energy          Am²⁺
(cm⁻¹)                        (cm⁻¹)
```

FIG. 6.2. Splitting of the ground $J = \frac{7}{2}$ manifold of two $5f^7$ ions on sites of cubic symmetry in CaF_2. The energy of the Γ_8 quartet level is deduced from the anisotropy shown in Fig. 6.1. Two calculated energies are given for the Γ_7 doublet, using two sets of values for the splitting parameters $b_4 = 60B_4$, $b_6 = 1260B_6$, either of which gives the correct energy for the Γ_8 level. These sets are (Edelstein and Easley 1968):

	b_4 (cm⁻¹)	b_6 (cm⁻¹)	$W(\Gamma_7)$ (cm⁻¹)
Cm³⁺	−1·115	0	35·7
	−1·433	−0·1	45·1
Am²⁺	−1·549	0	49·6
	−1·849	−0·1	58·4

obtained for Am^{2+}, with a hyperfine structure due to the odd isotopes 241, 243, each of which has $I = \frac{5}{2}$. The resonance is attributed to the Γ_6 doublet of an $^8S_{\frac{7}{2}}$ state with a large cubic field splitting; the effective g-value (see Table 22) is $(-\frac{7}{3})$ times the true g-value for $S = \frac{7}{2}$ at low fields, while at higher fields the anisotropy arises from admixture of the excited Γ_8 quartet by the Zeeman interaction. The energy of the Γ_8 level in zero field can be deduced from this anisotropy.

The results are listed in Table 6.4; the anisotropy is shown in Fig. 6.1 and the energy levels in Fig. 6.2. For the $S = \frac{7}{2}$ configuration the value of g_J is deduced to be 1·9261(10) for Cm^{3+} and 1·9258(10) for Am^{2+}; the latter is slightly smaller than the value $g_J = 1·93788(7)$ measured by Armstrong and Marrus (1966) for the free atom Am, which has the configuration $5f^7 7s^2$, $^8S_{\frac{7}{2}}$, but both are in good agreement with those for Cm^{3+} in C_{3h} symmetry (see § 6.2). The cubic field splittings (see Fig. 6.2) are enormously greater than those listed in Table 5.19 for the corresponding $4f^7$ ions.

6.4. Actinide ions in octahedral symmetry

We consider first the quadripositive ions. Optical measurements on Pa^{4+}, $5f^1$ in Cs_2ZrCl_6 (Axe, reported by Raubenheimer, Boesman, and Stapleton (1965)) yields the value $\zeta = 1490$ cm^{-1} for the spin-orbit parameter. This is in line with values for the tripositive ions (Table 6.2) and for U^{4+}, $5f^2$ (see below). On the other hand, the fourth degree term in the crystal potential is much larger than that for tripositive actinide or lanthanide ions. For Pa^{4+} the above authors give $A_4 \langle r^4 \rangle = +888$ cm^{-1}, $A_6 \langle r^6 \rangle = +42$ cm^{-1}, and similar values are found for U^{4+} (see below).

Magnetic resonance has been observed for Pa^{4+} in Cs_2ZrCl_6 by Axe, Stapleton, and Jeffries (1961) who also measured the hyperfine structure of ^{231}Pa ($I = \frac{3}{2}$) by Endor. In an octahedral field the ground manifold of $5f^1$, $^2F_{\frac{5}{2}}$ is split into a doublet (Γ_7) and a quartet (Γ_8), the former lying lower (from optical measurements) by about 1900 cm^{-1}. The magnetic resonance spectrum of the doublet was fitted to the usual spin Hamiltonian

$$\mathscr{H} = g\beta \mathbf{H} \cdot \mathbf{S} + A\mathbf{I} \cdot \mathbf{S} - g^{(I)}\beta \mathbf{H} \cdot \mathbf{I} \tag{6.1}$$

with the isotropic values $|g| = 1·1423(14)$, $|A| = 1578·6(14)$ MHz $|g^{(I)}| = 7·74 \times 10^{-4}$. After applying an 8 per cent correction for the pseudo-nuclear Zeeman effect, a value of 1·96 nuclear magnetons is obtained for the nuclear moment. The experiments show that g has the opposite sign to $g^{(I)}$, and hence also to g_I; since the nuclear moment is expected to be positive this gives a negative sign for g (and also for A). For a pure $J = \frac{5}{2}$ manifold split by a small cubic field the g-value for the Γ_7 doublet should be $(-\frac{5}{3}) \times \langle J \| \Lambda \| J \rangle = -\frac{10}{7}$ for f^1 (cf. § 18.2 and Table 22); the observed value is rather smaller because the crystal field admixes an appreciable amount of the $J = \frac{7}{2}$, Γ_7 states into the ground doublet. In general, one can write the ground doublet as

$$\cos \alpha \, |J = \tfrac{5}{2}, \Gamma_7\rangle + \sin \alpha \, |J = \tfrac{7}{2}, \Gamma_7\rangle \tag{6.2}$$

for which
$$g = -\frac{10}{7}\cos^2\alpha + \frac{16}{7\sqrt{3}}\cos\alpha\sin\alpha + \frac{24}{7}\sin^2\alpha, \quad (6.3)$$
showing that g can be positive or negative according to the value of α. The hyperfine constant is
$$A = 2g_n\beta\beta_n\langle r^{-3}\rangle\left\{-\frac{16}{7}\cos^2\alpha - \frac{8}{7\sqrt{3}}\cos\alpha\sin\alpha + \frac{16}{7}\sin^2\alpha - \right.$$
$$\left. -\kappa\left(\frac{5}{21}\cos^2\alpha + \frac{16}{7\sqrt{3}}\cos\alpha\sin\alpha + \frac{3}{7}\sin^2\alpha\right)\right\} \quad (6.3a)$$
where the term in κ allows for the possibility of a contribution from core polarization. An alternative approach is outlined below, in the theory for NpF_6.

Measurements of the susceptibility of $[(CH_3)_4N]_2UCl_6$, where the U^{4+} ion is in an octahedral environment, show that it has a small temperature-independent susceptibility up to 300°K (Hutchison and Candela 1957). This is consistent with a ground term $5f^2$, $^3H_4$, which in an octahedral field of the expected sign gives a Γ_1 singlet as the lowest level; the small constant susceptibility shows that the crystal field splittings are large compared with room temperature (200 cm^{-1}). Analysis of the optical spectrum (see, for example, Satten, Schreiber, and Wong (1965)) gives values of $A_4\langle r^4\rangle \approx 900$ cm^{-1}, $A_6\langle r^6\rangle \approx 56$ cm^{-1}, $\zeta = 1796$ cm^{-1}; the first excited state (the Γ_4 triplet of $^3H_4$) lies at about 900 cm^{-1}.

The magnetic resonance spectrum of NpF_6 at dilutions of 2 and 4 per cent in UF_6 has been observed at 4°K by Hutchison and Weinstock (1960). The results are fitted to the spin Hamiltonian (6.1); g is found to be isotropic with the value $g = -0.604(3)$, the sign being determined using circularly polarized radiation (cf. § 3.2). The hyperfine structure, also isotropic, is due to the isotope ^{237}Np ($I = \frac{5}{2}$), and its value is $|A| = 0.0665(5)$ cm^{-1}. These results are consistent with a structure in which the neptunium ion is surrounded by a regular octahedron of fluorine ions, with a single electron in the $5f$ shell. This is similar to that of Pa^{4+} in Cs_2ZrCl_6 (see above, but the much smaller g-value shows that a weak crystal field approach is an even poorer approximation in this case).

The theory of NpF_6 has been considered in some detail by Eisenstein and Pryce (1960). In a cubic field the sevenfold orbital degeneracy of an f-state is lifted, giving a singlet (a_2 or Γ_2), and two triplets (t_1 or Γ_4, and t_2 or Γ_5); the wave-functions are listed in Table 6.5. In an octahedral

TABLE 6.5

Splitting of the orbital states of an f-electron in a cubic field (Eisenstein and Pryce 1960). The wave-functions to the right are in terms of $|l_z\rangle$, and are real linear combinations of those given in Fig. 7.5. (A common factor of $\sqrt{7}$ has been omitted from the functions given in Cartesian coordinates as compared with Eisenstein and Pryce). If the orbital splitting is related to the spin operator form $B_4(O_4^0+5O_4^4)+B_6(O_6^0-21O_6^4)$, then $V=10b_4+84b_6$; $V'=18b_4+28b_6$, and for a single f-electron $B_4=b_4/60=2A_4\langle r^4\rangle/(11\times 45)$, $B_6=b_6/180=-4A_6\langle r^6\rangle/(11\times 13\times 27)$

Symmetry	Energy	Wave-functions
t_1 (Γ_4)	V'	$\|\delta_2\rangle \equiv \tfrac{1}{2}x(5x^2-3r^2) \equiv \dfrac{\sqrt{(3)}}{4}\{\|+1\rangle-\|-1\rangle\} - \dfrac{\sqrt{(5)}}{4}\{\|+3\rangle-\|-3\rangle\}$
		$\|\delta_3\rangle \equiv \tfrac{1}{2}y(5y^2-3r^2) \equiv -i\dfrac{\sqrt{(3)}}{4}\{\|+1\rangle+\|-1\rangle\} - i\dfrac{\sqrt{(5)}}{4}\{\|+3\rangle+\|-3\rangle\}$
		$\|\delta_1\rangle \equiv \tfrac{1}{2}z(5z^2-3r^2) \equiv \|0\rangle$
t_2 (Γ_5)	V	$\|\epsilon_2\rangle \equiv \tfrac{1}{2}\sqrt{(15)}x(y^2-z^2) \equiv \dfrac{\sqrt{(5)}}{4}\{\|+1\rangle-\|-1\rangle\} + \dfrac{\sqrt{(3)}}{4}\{\|+3\rangle-\|-3\rangle\}$
		$\|\epsilon_3\rangle \equiv \tfrac{1}{2}\sqrt{(15)}y(z^2-x^2) \equiv i\dfrac{\sqrt{(5)}}{4}\{\|+1\rangle+\|-1\rangle\} - i\dfrac{\sqrt{(3)}}{4}\{\|+3\rangle+\|-3\rangle\}$
		$\|\epsilon_1\rangle \equiv \tfrac{1}{2}\sqrt{(15)}z(x^2-y^2) \equiv \dfrac{1}{\sqrt{(2)}}\{\|+2\rangle+\|-2\rangle\}$
a_2 (Γ_2)	0	$\|\beta\rangle \equiv \sqrt{(15)}xyz \equiv \dfrac{1}{\sqrt{(2)}}\{\|+2\rangle-\|-2\rangle\}$

field the a_2 state is lowest, since it has zero density along each fourfold axis and so avoids the negatively charged ligand ions as much as possible; the t_2 states lie next, and the t_1 states highest. In crystal field theory this is equivalent to assuming that the fourth degree terms in the octahedral potential outweigh the sixth degree terms. In terms of bonding, the a_2 orbital is non-bonding, the t_1 orbitals can form σ-bonds, and the t_2 orbitals π-bonds. If we assume that the magnetic electron occupies an anti-bonding orbital, and that σ-bonding is stronger than

TABLE 6.6

Basis functions of f^1 including spin in a cubic field (Eisenstein and Pryce 1960)

			m_J
Γ_7	$\|\bar{A}\rangle = i\|\bar{\beta}\rangle$ $\|\bar{B}\rangle = (\frac{1}{3})^{\frac{1}{2}}[-\|\bar{\epsilon}_1\rangle + \|\epsilon_2\rangle - i\|\epsilon_3\rangle]$		$+\frac{3}{2}$
	$\|A\rangle = i\|\beta\rangle$ $\|B\rangle = (\frac{1}{3})^{\frac{1}{2}}[\|\epsilon_1\rangle + \|\bar{\epsilon}_2\rangle + i\|\bar{\epsilon}_3\rangle]$		$-\frac{3}{2}$
Γ_8	$\|\bar{C}\rangle = (\frac{1}{6})^{\frac{1}{2}}[-2\|\bar{\epsilon}_1\rangle - \|\epsilon_2\rangle + i\|\epsilon_3\rangle]$ $\|\bar{D}\rangle = -(\frac{1}{2})^{\frac{1}{2}}[\|\delta_2\rangle + i\|\delta_3\rangle]$		$+\frac{3}{2}$
	$\|C'\|\rangle = (\frac{1}{2})^{\frac{1}{2}}[\|\bar{\epsilon}_2\rangle - i\|\bar{\epsilon}_3\rangle]$ $\|D'\rangle = (\frac{1}{6})^{\frac{1}{2}}[2\|\delta_1\rangle - \|\bar{\delta}_2\rangle - i\|\bar{\delta}_3\rangle]$		$+\frac{1}{2}$
	$\|\bar{C}'\rangle = -(\frac{1}{2})^{\frac{1}{2}}[\|\epsilon_2\rangle + i\|\epsilon_3\rangle]$ $\|\bar{D}'\rangle = (\frac{1}{6})^{\frac{1}{2}}[2\|\bar{\delta}_1\rangle + \|\delta_2\rangle - i\|\delta_3\rangle]$		$-\frac{1}{2}$
	$\|C\rangle = (\frac{1}{6})^{\frac{1}{2}}[-2\|\epsilon_1\rangle + \|\bar{\epsilon}_2\rangle + i\|\bar{\epsilon}_3\rangle]$ $\|D\rangle = (\frac{1}{2})^{\frac{1}{2}}[\|\bar{\delta}_2\rangle - i\|\bar{\delta}_3\rangle]$		$-\frac{3}{2}$
Γ_6	$\|E\rangle = (\frac{1}{3})^{\frac{1}{2}}[\|\delta_1\rangle + \|\bar{\delta}_2\rangle + i\|\bar{\delta}_3\rangle]$		$+\frac{1}{2}$
	$\|\bar{E}\rangle = (\frac{1}{3})^{\frac{1}{2}}[-\|\bar{\delta}_1\rangle + \|\delta_2\rangle - i\|\delta_3\rangle]$		$-\frac{1}{2}$

The orbital functions are given in Table 6.5; the notation is that β is an orbital function combined with 'up' spin ($m_s = +\frac{1}{2}$) and $\bar{\beta}$ is an orbital function combined with 'down' spin ($m_s = -\frac{1}{2}$). The values in the column headed m_J are the values of $l_z + s_z$.

π-bonding, this gives the same order of levels as assumed above, the anti-bonding σ-orbitals t_1 lying highest, and the non-bonding orbital a_2 lowest.

When the electron spin is introduced the product wave-functions can be classified as Γ_6 (doublet), Γ_7 (doublet), and Γ_8 (quartet), where Γ_6 occurs only once, but Γ_7 and Γ_8 occur twice, as given in Table 6.6. From the energy matrices given in Table 6.7 it is clear that the ground state will be a Γ_7 doublet, which can be written as

$$|+\rangle = \cos\theta\,|A\rangle - \sin\theta\,|B\rangle \qquad (6.4)$$
$$|-\rangle = \cos\theta\,|\bar{A}\rangle - \sin\theta\,|\bar{B}\rangle$$

Table 6.7

Energy matrices for the Γ_7 (doublets), Γ_8 (quartets) and Γ_6 (doublet) states of NpF_6. The quantities V, V' are the energies of the t_2 and t_1 levels respectively (see Table 6.5); $\sqrt{k'}$, $\sqrt{k}$ are factors by which the amplitude of the 5f wave-functions are reduced in the t_1 and t_2 orbitals respectively. t_1 orbitals are σ-bonding, t_2 orbitals π-bonding and the remaining a_2 orbital is non-bonding (Eisenstein and Pryce 1960)

		A	B			C	D
Γ_7	A	0	$(3k)^{\frac{1}{2}}\zeta$	Γ_8	C	$V+\frac{1}{4}k\zeta$	$\frac{3}{4}(5kk')^{\frac{1}{2}}\zeta$
	B	$(3k)^{\frac{1}{2}}\zeta$	$V-\frac{1}{2}k\zeta$		D	$\frac{3}{4}(5kk')^{\frac{1}{2}}\zeta$	$V'-\frac{3}{4}k'\zeta$

		E
Γ_6	E	$\lvert V'+\frac{3}{2}k'\zeta \rvert$

whose g-value is

$$g = 2\cos^2\theta - 4(k/3)^{\frac{1}{2}}\sin 2\theta - \tfrac{2}{3}(1-k)\sin^2\theta \tag{6.5}$$

if an orbital reduction factor $k^{\frac{1}{2}}$ is assumed to allow for any π-bonding in the t_2 orbitals (note that the nomenclature used by Eisenstein and Pryce differs from that normally used for bonding effects with d-orbitals). Equation (6.5) with $k = 1$ is equivalent to eqn (6.3), as can be verified by noting that in the limit of a weak crystal field the two Γ_7 states belonging to $J = \tfrac{5}{2}$ and $J = \tfrac{7}{2}$ have the form

$$\begin{aligned} J = \tfrac{7}{2}, \Gamma_7; \quad &\sqrt{(\tfrac{4}{7})}\lvert A\rangle + \sqrt{(\tfrac{3}{7})}\lvert B\rangle \\ J = \tfrac{5}{2}, \Gamma_7; \quad &\sqrt{(\tfrac{3}{7})}\lvert A\rangle - \sqrt{(\tfrac{4}{7})}\lvert B\rangle \end{aligned} \tag{6.6}$$

and

$$\begin{aligned} \cos\theta &= \sqrt{(\tfrac{3}{7})}\cos\alpha + \sqrt{(\tfrac{4}{7})}\sin\alpha \\ \sin\theta &= \sqrt{(\tfrac{4}{7})}\cos\alpha - \sqrt{(\tfrac{3}{7})}\sin\alpha. \end{aligned} \tag{6.6a}$$

In their analysis for NpF_6 Eisenstein and Pryce attempted to fit the g-value, the optical spectrum, and the temperature-independent term in the magnetic susceptibility. With $k = 1$ (and also no orbital reduction factor for the t_1 orbitals), they adopted the values

$$\zeta = 2405 \text{ cm}^{-1}, \quad V = 5442 \text{ cm}^{-1}, \quad V' = 22220 \text{ cm}^{-1},$$

which give a temperature-independent susceptibility term of 158×10^{-6} per mol, in good agreement with the value of 165×10^{-6} obtained by Hutchison, Tsang, and Weinstock (1962) for NpF_6 in UF_6, extrapolated to infinite dilution. These parameters are considerably larger than those for the single f-electron of Pa^{4+} in Cs_2ZrCl_6 of

$$\zeta = 1490 \text{ cm}^{-1}, \quad V = 1496 \text{ cm}^{-1}, \quad V' = 3656 \text{ cm}^{-1}$$

(see Raubenheimer, Boesman, and Stapleton 1965). The corresponding crystal field parameters are

Pa^{4+} in Cs_2ZrCl_6 $A_4\langle r^4\rangle = 888$ cm^{-1}, $A_6\langle r^6\rangle = 41.9$ cm^{-1}
Np^{6+} in UF_6 $A_4\langle r^4\rangle = 5700$ cm^{-1}, $A_6\langle r^6\rangle = 320$ cm^{-1},

which showed a marked rise with increasing valency. Both parameters are positive, as would be expected for an octahedral configuration.

The magnetic hyperfine constant is

$$A = -2g_n\beta\beta_n\langle r^{-3}\rangle\left\{\frac{4}{\sqrt{3}}\sin 2\theta + \frac{1}{3}\kappa(1+2\cos 2\theta)\right\} \quad (6.7)$$

and it can be verified by using eqns (6.6a) that this is identical with eqn (6.3a). The value of 2θ required to fit the experimental g-value of NpF_6 is 63°, for which

$$A = -2g_n\beta\beta_n\langle r^{-3}\rangle(2\cdot 06 + 0\cdot 63\kappa). \quad (6.7a)$$

6.5. Neptunyl and plutonyl ions

As mentioned in § 6.1, ions of the type $(UO_2)^{2+}$ are linear. Uranyl salts have only a temperature-independent susceptibility, suggesting that there are no unpaired electrons. The free uranium atom has a closed radon core, and six valence electrons, two of which have been lost in $(UO_2)^{2+}$. Eisenstein and Pryce (1955) postulate that the remaining electrons are used to form covalent bonds with the oxygen atoms. In a simple model which uses only σ-bonding, the six valence electrons of each oxygen are assumed to be in non-bonding orbitals formed from the $2s$, $2p$ states on the oxygens, which are thus filled. For the central uranium ion, bonding orbitals of σ-type can be formed from $5f$, $6d$, $7s$ (and perhaps $7p$) using the substates with $m_l = 0$; taking the O—U—O axis as the z-axis, this means the f, δ_1 orbital in Table 6.5, whose wavefunction $z(5z^2-3r^2)$ is strongly directed along the z-axis, and similarly the $(3z^2-r^2)$ orbital belonging to the d-states. The two orbitals forming a bond with the inwardly directed sp_z orbital on each oxygen can accommodate just four electrons, which are assumed to be those originally associated with the uranium. The situation is therefore different from any considered hitherto for compounds of ions containing

f-electrons, in that we have strong bonding involving f-orbitals. In many respects it resembles a linear molecule (cf. Elliott 1953).

We now consider such 'actinyl' groups in the hexagonal compound $(UO_2)Rb(NO_3)_3$, where the axes of all the groups are parallel to the unique crystal axis.

$(UO_2)^{2+}$ in $(UO_2)Rb(NO_3)_3$

A strong nuclear electric quadrupole interaction with axial symmetry would be expected in this compound; this has been verified by measurements of anisotropy in the α-particle emission from ^{233}U, ^{235}U at temperatures below 1°K by Dabbs, Roberts, and Parker (1958).

$(NpO_2)^{2+}$ in $(UO_2)Rb(NO_3)_3$

The paramagnetic resonance spectrum of this compound has been measured by Bleaney, Llewellyn, Pryce, and Hall (1954b). As would be expected from the foregoing discussion, it shows strong axial symmetry, and the results have been fitted to a spin Hamiltonian with effective spin $S = \frac{1}{2}$ of the usual form with axial symmetry. The measured parameters are (see Pryce 1959)

$$g_\| = -3 \cdot 405(8), \qquad |g_\perp| = 0 \cdot 205(6);$$
$$A_\| = -0 \cdot 16547(5) \text{ cm}^{-1}, \qquad |A_\perp| = 0 \cdot 01782(3) \text{ cm}^{-1};$$
$$P_\| = +0 \cdot 03015(5) \text{ cm}^{-1}.$$

The hyperfine parameters are for the isotope ^{237}Np, whose nuclear spin is $I = \frac{5}{2}$. The actual signs are not determined experimentally, except for the result that $A_\|$, $P_\|$ are of opposite sign. Pryce (1959) has pointed out that theoretically $g_\|$ is negative (see below), and that $A_\|$ should be negative provided that the nuclear magnetic moment is (as expected) positive. A negative value of $A_\|$ and a positive value of $P_\|$ have been confirmed by Hanauer, Dabbs, Roberts, and Parker (1961) from the temperature dependence of the anisotropy in the emission of α-particles in a nuclear alignment experiment.

The concentration of negative charge along the axis of the O—U—O complex means that any additional electron such as the extra electron in the neptunyl ion will be strongly repelled, and in the ground state it would occupy an orbital whose charge distribution is concentrated in the equatorial plane. The $5f$ orbitals will therefore have very different energy, according to their charge distribution. The states for which $l_z = \pm 3$ will be lowest, followed by the $l_z = \pm 2$, ± 1 and the antibonding $l_z = 0$ states in that order. The excited states are estimated by Eisenstein and Pryce to lie at 20000–30000 cm^{-1}, as also are the $6d$

states. These splittings are large compared with the spin–orbit coupling, of order 2000 cm^{-1}, so that in a first approximation we retain only the diagonal term $\zeta l_z s_z$. Since ζ is positive for a single electron, the four states with $l_z = \pm 3$, $s_z = \pm \frac{1}{2}$ are split into two doublets, the lower doublet being that for which the axial components of orbit and spin are anti-parallel. These two states can be denoted by $j_z = l_z + s_z = |\pm \frac{5}{2}\rangle$, the other two states $|j_z = \pm \frac{7}{2}\rangle$ lying higher in energy by 3ζ. Use of the Zeeman operator $l_z + 2s_z$ shows that in this approximation the ground doublet would have $g_\| = 4$, $g_\perp = 0$. Such a doublet would not show a magnetic resonance line, though the similarity of the g-values to the experimental results suggests that the basic theory is correct.

We now allow for the effect of the crystal field due to other constituents, mainly the three nitrate ions symmetrically disposed in the equatorial plane. These produce a crystal field of threefold symmetry about the z-axis, of whose components V_6^6 is of vital importance because it admixes states with $l_z = \pm 3$. The ground doublet is then no longer exactly characterized by $j_z = \pm \frac{5}{2}$, but contains an admixture of the $j_z = \pm \frac{7}{2}$ states, making a resonance transition allowed. In addition off-diagonal elements of the spin–orbit coupling will admix some of the $l_z = \pm 2$, $j_z = \pm \frac{5}{2}$ states, but since the spin–orbit coupling is rather weak relative to the splitting due to the axial charge concentration, their amplitude is rather small. These two perturbations can be taken into account by writing the wave-functions of the ground doublet in terms of $|l_z, s_z\rangle$ as

$$|+\rangle = (1+p^2+q^2)^{-\frac{1}{2}}(|-3, +\tfrac{1}{2}\rangle + p\,|+3, +\tfrac{1}{2}\rangle - q\,|-2, -\tfrac{1}{2}\rangle)$$
$$|-\rangle = (1+p^2+q^2)^{-\frac{1}{2}}(|+3, -\tfrac{1}{2}\rangle + p\,|-3, -\tfrac{1}{2}\rangle - q\,|+2, +\tfrac{1}{2}\rangle) \quad (6.8)$$

where q is about 0·1. Use of the Zeeman operator $l_z + 2s_z$ then gives

$$g_\| = -(4 - 8p^2 + 6q^2)(1+p^2+q^2)^{-1}$$
$$g_\perp = |p\{4 - 2\sqrt{(6)}q\}(1+p^2+q^2)^{-1}|. \quad (6.9)$$

If q is fixed (and its effect is anyway rather small), the values of $g_\|$ and $g_\perp$ are related, depending only on the parameter p, and it is not possible to fit the experimental values of $g_\|$ and $g_\perp$ simultaneously. If $g_\perp$ is fitted, the value of $g_\|$ given by these equations is very close to -4; since most of the contribution to $g_\|$ comes from the orbit, this suggests that an orbital reduction factor k should be introduced to allow for some covalent bonding. The g-values then become

$$g_\| = -\{k(6 - 6p^2 + 4q^2) - 2(2 + 2p^2 - 2q^2)\}(1+p^2+q^2)^{-1}$$
$$g_\perp = |p\{4 - 2\sqrt{(6)}kq\}(1+p^2+q^2)^{-1}|. \quad (6.10)$$

The experimental g-values can then be fitted using $k = 0·9$, $p = 0·056$.

In eqns (6.8) we have chosen as the $|+\rangle$ state that which gives a 1:1 correspondence between the matrix elements of the effective spin and of the true angular momentum $\mathbf{l}+\mathbf{s}$. This makes the sign of $g_\parallel$ negative because it is dominated by the contribution from the orbit, which is opposed to the smaller spin contribution.

The theoretical expressions for the hyperfine constants given by Eisenstein and Pryce (1955) are

$$A_\parallel = -2g_n\beta\beta_n\langle r^{-3}\rangle$$
$$\{\tfrac{20}{3}+\tfrac{2}{3}\sqrt{(6)}q-\tfrac{16}{3}p^2+4q^2+\kappa(1+p^2-q^2)\}(1+p^2+q^2)^{-1}$$
$$A_\perp = |2g_n\beta\beta_n\langle r^{-3}\rangle p\{\tfrac{2}{3}-\tfrac{5}{3}\sqrt{(6)}q-2\kappa\}(1+p^2+q^2)^{-1}| \tag{6.11}$$

(here the sign of $A_\parallel$ has been reversed from the original, like that of $g_\parallel$; see Pryce (1959)). With the value $q = 0\cdot1$ assumed above, and the value $p = 0\cdot056$ required to fit the g-values, these equations can be solved for κ and $g_n\langle r^{-3}\rangle$ using the measured hyperfine constants. To fit the anisotropy, a value of about $-3\cdot1$ is required for κ, which then gives

$$g_n\langle r^{-3}\rangle = +13\cdot7.$$

(An alternative solution, $\kappa = +120$, which leads to $g_n\langle r^{-3}\rangle = 0\cdot4$ seems quite inadmissible.) Even if we take a value of $g_n = +2$, which corresponds to a magnetic moment close to the Schmidt limit of 4·8 n.m. for $I = \tfrac{5}{2}$, this result seems to give rather a large value for $\langle r^{-3}\rangle$, especially if some bonding occurs, which would tend to reduce $\langle r^{-3}\rangle$.

This result may be compared with that for NpF_6, where a similar value of $g_n\langle r^{-3}\rangle$ might be expected. If the measured value of A is there taken to be negative, eqn (6.7a) yields the result

$$g_n\langle r^{-3}\rangle = 21/(2\cdot06+0\cdot63\kappa),$$

which would be equal to 13·7 if we take $\kappa = -0\cdot75$. Thus to reconcile this result with that for the neptunyl ion we again need a negative value of κ, though not so large.

In the 3d and 4f groups the observed values of κ are positive but we shall not hazard a prediction of its sign for the 5f group. The theory given above for the magnetic hyperfine constants of the neptunyl ion should be accepted with reserve, since Eisenstein and Pryce (1956) have pointed out that the crystal field may contain a V_3^3 term, which could admix s-states directly to the 5f, $l_z = \pm 3$ states. This would affect the hyperfine structure in a way that cannot be represented by the addition of a κ-term, and the formulae (6.11) would no longer be correct.

The theory of Eisenstein and Pryce (1955) gives a nuclear electric quadrupole interaction

$$P_\| = -\frac{e^2 Q}{I(2I-1)}\{(\tfrac{4}{5}\alpha-\tfrac{1}{2})\langle r^{-3}\rangle_{5f}+\tfrac{6}{7}\alpha'\langle r^{-3}\rangle_{6d}\}, \qquad (6.12)$$

where α and α' are the probabilities that the bonding electrons are in the neptunyl $5f$ and $6d$ states rather than on the oxygen atoms. Since $\langle r^{-3}\rangle_{6d}$ is expected to be considerably larger than $\langle r^{-3}\rangle_{5f}$, this leads to a negative nuclear quadrupole moment for a positive value of $P_\|$, the field gradient at the nucleus being due mainly to the σ-bonding electrons. If there were an appreciable amount of π-bonding, this would give a contribution to $P_\|$ of opposite sign. Shielding corrections, and lattice contributions have been omitted from (6.12).

$(PuO_2)^{2+}$ in $(UO_2)Rb(NO_3)_3$

Paramagnetic resonance has been observed for this ion by Bleaney, Llewellyn, Pryce, and Hall (1954a). The line shape is asymmetrical, as expected for resonance due to a non-Kramers ion; the spin Hamiltonian parameters for a non-Kramers doublet with an effective spin of $\tfrac{1}{2}$ are found to be

$$|g_\|| = 5\cdot 32(2)$$

Isotope 239($I = \tfrac{1}{2}$) $|A_\|| = 0\cdot 0862(5)$ cm^{-1}
Isotope 241($I = \tfrac{5}{2}$) $|A_\|| = 0\cdot 0609(4)$ cm^{-1}.

The theory is discussed by Eisenstein and Pryce (1956). Exchange interaction between the two electrons outside the $(UO_2)^{2+}$ type core is assumed to be strong, so that they will have parallel spin and hence must occupy different orbital states. There are then two possibilities: (a) they occupy the states $l_z = +3$ and $l_z = -3$, giving a resultant $L_z = 0$ and $S_z = \pm 1$; (b) they occupy the states $l_z = \pm 3$, $l_z = \pm 2$ with parallel spin, giving four possible states $L_z = \pm 5$, $S_z = \pm 1$, which are then split by spin-orbit coupling into two doublets, that where the spin and orbital components are anti-parallel having the lower energy. Because of the Coulomb repulsion between the two electrons, the ground state is (b), where the electrons occupy orbitals with different values of l_z. The values of $L_z + 2S_z$ for the two states of the ground doublet are then $\pm(5-2)$, so that we have $g_\| = 6$, $g_\perp = 0$. In the next approximation, the crystal field and spin-orbit coupling will admix some other states into the ground state wave-functions, but the amplitude of the admixtures is expected to be small and does not seriously change the value of $g_\|$, while $g_\perp$ remains zero. To explain the observed value of $g_\|$ it seems necessary to introduce a reduction factor k for the orbital moment, which then gives

$$|g_\||= 2(5k-2), \qquad g_\perp = 0. \tag{6.13}$$

The measured value of $g_\| = 5\cdot 32$ can then be fitted with a value of k of about 0·93, which is not inconsistent with that needed to fit the corresponding neptunyl compound. Magnetic resonance has been observed by Hutchison and Lewis (1954) in another plutonyl compound, plutonyl sodium acetate; they find $g_\| = 5\cdot 92$, which would require a value of $k = 0\cdot 99$.

The magnetic hyperfine interaction arises from the 5f electrons, together with a contribution of s-like character due to core polarization and the fact that the V_3^3 term in the crystal field can directly admix s-states into the $l_z = \pm 3$ states. The two mechanisms cannot be differentiated in the present case, so that both can be combined in a single term, giving

$$A_\| = 2g_n\beta\beta_n\langle r^{-3}\rangle(\tfrac{32}{3}+2\kappa) \tag{6.14}$$

with, of course, $A_\perp = 0$. Using the value of $-0\cdot 200(4)$ n.m. (Faust, Marrus, and Nierenberg 1965) for the nuclear moment of ^{239}Pu, a value of about 6 a.u. is found for $\langle r^{-3}\rangle$ if $\kappa = 0$, and a rather smaller value if κ is positive. A strong nuclear electric quadrupole interaction may be expected for isotopes with $I > \tfrac{1}{2}$, but cannot be observed in an electron spin resonance experiment. An f-electron with $l_z = \pm 2$ has

$$3l_z^2 - l(l+1) = 0,$$

so the extra electron in the plutonyl ion gives no contribution to the nuclear electric quadrupole interaction, and eqn (6.12) should apply also to the plutonyl ion.

7
IONS OF THE 3d GROUP
IN INTERMEDIATE LIGAND FIELDS

7.1. Introduction

As soon as the investigation of paramagnetic ions by means of electron paramagnetic resonance had begun (Zavoisky 1945) it became clear that single crystals were needed. In the iron group a number of these can be grown fairly readily from aqueous solution, such as copper sulphate, the alums, and the tutton salts. Salts of the last two classes were already in use for magnetic cooling by means of adiabatic demagnetization, where a fairly high degree of magnetic dilution is an advantage in obtaining the lowest possible temperatures. Such dilution is also advantageous in reducing line-width and thus obtaining higher resolution in magnetic resonance experiments, so that these salts were both the natural ones to choose for resonance experiments and of interest in other fields. The need for still higher resolution in order to examine the finer detail of the spectrum led to the use of diluted salts, where most of the paramagnetic ions are replaced by diamagnetic ions; the alums and tutton salts, and other hydrated salts, lend themselves readily to this modification. An immediate result was the discovery of hyperfine structure in a paramagnetic copper salt by Penrose (1949); other paramagnetic ions with nuclear magnetic moments were quickly investigated, with results that led to the first successful nuclear alignment experiments. A subsidiary aim of magnetic resonance experiments became to find suitable salts for magnetic cooling and nuclear alignment, one of the most important series in this respect being the double nitrates, which act as host both to divalent iron group ions and to trivalent rare-earth ions.

Ligand hyperfine structure was first observed by Griffiths and Owen (1954) in complex chlorides of $4d$, $5d$ group ions, and later by Tinkham (1956a, b) for the fluorine ligands of several $3d$ group ions in ZnF_2. Almost simultaneously with Tinkham's work, nuclear magnetic resonance from the ^{19}F ion in MnF_2 was observed by Shulman and Jaccarino (1956). Fluorine has the advantage over other halides of a fairly large nuclear magnetic moment and small nuclear spin ($I = \frac{1}{2}$) which, together with the absence of nuclear electric quadrupole interaction, combine to give well-resolved ligand hyperfine structure in electron resonance experiments and signals of good intensity in nuclear resonance experiments. A considerable amount of work has therefore been carried out on iron group fluorides; in particular, salts of the type $KMgF_3$, which have cubic symmetry and in which the divalent cation is surrounded by a regular octahedron of fluorines, each fluorine being bonded to two collinear divalent cations, have been attractive for magnetic resonance work because of their simple structure. The presence of cubic symmetry reduces the unknown parameters in the theory to a number where most of them can be determined experimentally, and greatly enhances the possibility of an unambiguous interpretation of the results. In general single crystals of such fluorides must be grown from the melt or from a high melting point flux, and have become available only since about 1955. In view of the complexity of the spectrum in

TABLE 7.1

Formula	Typical host crystal	Local symmetry
(a) Crystals with approximately octahedral symmetry around the paramagnetic ion		
$M''O$	MgO, CaO	Regular octahedron
$KM''F_3$	$KMgF_3$	Regular octahedron
$M'F$	NaF (NaCl, etc.)	Regular octahedron
$KM'''(SO_4)_2, 12H_2O$	alums	Distorted octahedron with threefold symmetry axis
$La_2M''_3(NO_3)_{12}, 24H_2O$	'double nitrates'	Distorted octahedron with threefold symmetry axis
$M''SiF_6, 6H_2O$	$ZnSiF_6, 6H_2O$	Distorted octahedron with threefold symmetry axis
$M''(BrO_3)_2, 6H_2O$	$Zn(BrO_3)_2, 6H_2O$	Distorted octahedron with threefold symmetry axis
$M''Cl_2$	$MgCl_2$	Distorted octahedron with threefold symmetry axis
M'''_2O_3	Al_2O_3	Distorted octahedron with threefold symmetry axis
$K_2M''(SO_4)_2, 6H_2O$	tutton salts	Distorted octahedron with lower symmetry
$M''F_2$	ZnF_2	Distorted octahedron with lower symmetry
$K_4M''(CN)_6, 3H_2O$	$K_4Fe(CN)_6, 3H_2O$	Distorted octahedron with lower symmetry
$K_3M'''(CN)_6$	$K_3Co(CN)_6$	Distorted octahedron with lower symmetry
(b) Crystals with approximately tetrahedral symmetry around the paramagnetic ion (see also Table 7.24)		
$M''S, M''Se$, etc.	ZnS(cubic), CdSe, etc.	Regular tetrahedron
$M''S, M''Se$, etc.	ZnS(hexagonal), etc.	Distorted tetrahedron with threefold symmetry axis

such compounds, it was perhaps fortunate that most early work was carried out on hydrates, where no resolved ligand hyperfine structure is visible; the main features of the paramagnetic resonance spectrum of the central ion were thus discovered and interpreted before the complexity of ligand hyperfine structure was introduced.

A list of the principal types of compounds in which paramagnetic resonance of iron group ions has been studied is given in Table 7.1. In describing the experimental results we shall confine ourselves mainly but not exclusively to cases of cubic symmetry; we shall not attempt to mention all the experimental results (a catalogue is seldom readable) but shall single out those that illustrate the main aspects of the theory. At the same time we shall endeavour to give a reasonably complete picture of the behaviour of each ion. The ions that have been investigated by paramagnetic resonance are listed in Table 7.2, which shows the configuration and the ground state of the free ion. In this table and throughout this chapter (except in § 7.17) we assume that the paramagnetic ion is co-ordinated to six negatively charged ligand ions, and that the sign of the octahedral (cubic) crystal potential is the same as that calculated for a cage of negative ions on a point charge model.

In octahedral coordination the positively charged paramagnetic ion is surrounded by a cage of six negatively charged ions, as shown in Fig. 7.1. If these six ligand ions form a regular octahedron, the symmetry is exactly cubic, but in many cases small distortions occur. In general, two sets of reference axes are used:

(i) the fourfold axes, parallel to the lines joining the central ion to each pair of ligand ions, are taken as Cartesian axes (x, y, z). The six ligand ions of charge q at points $(0, 0, \pm R)$, etc., give a potential at a point (x, y, z) at a distance $r \ll R$ near the centre

$$V = \frac{6q}{R} + \frac{35}{4} \frac{q}{R^5} \left(x^4 + y^4 + z^4 - \frac{3}{5} r^4 \right),$$

where the first term has no angular variation and just changes the energy of all d-electrons on the central ion (assuming them to have identical radial wave-functions) by the same amount, while the second term gives rise to a spin operator term of the form $B_4\{O_4^0 + 5O_4^4\}$ in the ligand field interaction. If one pair of ligand ions (along the axis that we take to be the z-axis) is at a slightly different distance from the other four, we have a tetragonal distortion that produces extra potential terms of the form $(3z^2 - r^2)$, $(35z^4 - 30z^2r^2 + 3r^4)$. In terms of spin

TABLE 7.2

Configuration	d^1	d^2	d^3	d^4	d^5	d^6	d^7	d^8	d^9
Examples of ions	V^{4+} Ti^{3+}	Cr^{4+} V^{3+}	Mn^{4+} Cr^{3+} V^{2+}	Mn^{3+} Cr^{2+}	Fe^{3+} Mn^{2+} Cr^+	Co^{3+} Fe^{2+} Mn^+	Ni^{3+} Co^{2+} Fe^+	Cu^{3+} Ni^{2+} Co^+	Cu^{2+} Ni^+
Ground term of free ion	^2D	^3F	^4F	^5D	^6S	^5D	^4F	^3F	^2D
Total spin S	$\tfrac{1}{2}$	1	$\tfrac{3}{2}$	2	$\tfrac{5}{2}$	2	$\tfrac{3}{2}$	1	$\tfrac{1}{2}$
Ground term of ion in weakly bound octahedral complex { Orbital degeneracy (octahedral field)	3	3	1	2	1	3	3	1	2
Orbital degeneracy (tetrahedral field)	2	1	3	3	1	2	1	3	3
Ground term of ion in strongly bound octahedral complex { Total spin S	$\tfrac{1}{2}$	1	$\tfrac{3}{2}$	1	$\tfrac{1}{2}$	0	$\tfrac{1}{2}$	1	$\tfrac{1}{2}$
Orbital degeneracy	3	3	1	3	3	1	2	1	2

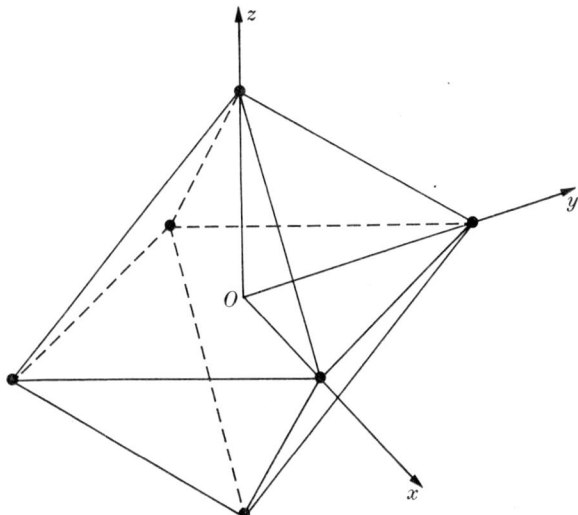

Fig. 7.1. A paramagnetic ion at the centre O of a regular octahedron of negative charges, showing one set of fourfold axes. A tetragonal distortion preserves the fourfold symmetry about one of these axes, assumed to be the z-axis. Typical inter-ionic distances (paramagnetic ion to ligand ion) are 0·207 nm when ligand is O^{2-} ion of H_2O in hydrated crystal; 0·210 nm for O^{2-} in MgO; and 0·201 nm when ligand is F^- ion in KM″F_3.

operators O_k^q these are conveniently introduced in the form given in eqn (7.2a) (see § 7.2). If each pair of ligand ions is at a different distance from the other pairs, we have a rhombic distortion, introducing further potential terms of the type (x^2-y^2), $(7z^2-r^2)(x^2-y^2)$, corresponding to spin operators O_2^2, O_4^2. Angular distortions of the octahedron may give rise to terms of still lower symmetry.

(ii) a threefold axis, a line drawn from the centre through the midpoint of one of the faces of the octahedron, whose direction cosines are [111], [11$\bar{1}$], etc. with respect to the fourfold axes, is taken as the z-axis. This is particularly convenient when a trigonal distortion of the octahedron is present, as illustrated in Fig. 7.2, giving a ligand field interaction in spin operator form of the type in eqn (7.2b).

Terms of higher degree in the ligand potential are of course present, but within a manifold of d-electrons their matrix elements are all zero. Within this manifold terms of odd degree also have zero matrix elements; such terms exist for tetrahedral coordination (see § 7.17), and other ligand arrangements lacking inversion symmetry.

In principle a potential function, representing the electrostatic potential of the cage of ligand ions at a point near the centre of the cage, can be written down in terms of the charges on the ligand ions

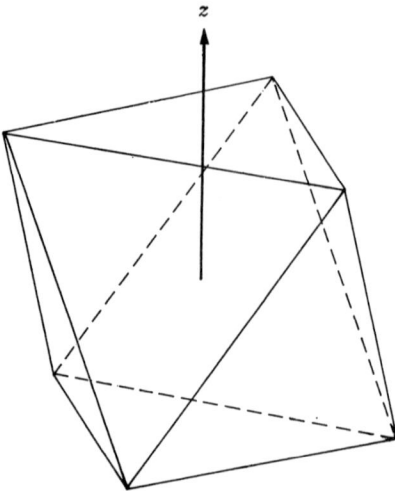

Fig. 7.2. A paramagnetic ion at the centre of a regular octahedron of negative charges, showing one of the threefold axes. A trigonal distortion preserves the threefold symmetry about the z-axis.

and their positions. This is not a very fruitful approach in the $3d$ group for a number of reasons:

(i) the charge distribution on the ligands and their exact positions are often not known sufficiently precisely;

(ii) the charge distribution (i.e. the radial wave-functions) of the magnetic electrons on the central ion is not known sufficiently precisely;

(iii) the electrostatic energy change for the magnetic electrons is by no means a small perturbation compared with other energy terms;

(iv) the wave-functions of the magnetic electrons on the central ion and of electrons on the ligand ions overlap appreciably, so that we ought to consider the complex consisting of the central ion in its cage of ligand ions as a whole.

In this situation numerical calculations are difficult and often unrealistic. Nevertheless the simple electrostatic approach has been surprisingly rewarding, particularly in the earlier days of the theory of paramagnetic ions in solids, using empirical values of the energy parameters. This is due to the fact that we are principally concerned with the way in which the energy levels of the d-electrons split under the action of the ligand field, and to a large part this is determined by symmetry considerations. In the following sections we shall derive these splittings, using a ligand field interaction in spin-operator form

$$\mathscr{H}_{\text{ligand}} = \sum_{k=2,4} \sum_{q=-k}^{k} B_k^q O_k^q$$

applied to the magnetic electrons as a perturbation. This is similar to that used for $4f$, $5f$ electrons (except that we shall mostly be concerned with complexes of nearly cubic symmetry), but differs in the way it is applied as a perturbation because of its relative magnitude. In the $4f$ group, its energy is small compared with the spin-orbit coupling energy, so that it is applied as a perturbation to the manifold of $(2J+1)$ states for a given term J of the free ion. In the $3d$ group the ligand field energy is larger than the spin-orbit coupling, and comparable with electrostatic interactions within the ion which give rise to LS-coupling. We therefore adopt first (§ 7.2) an 'intermediate ligand field' approach in which (L, S) wave-functions are used, and find how the manifold of $(2L+1)$ states is split by the ligand field. In the following section (§ 7.3) we consider the 'strong ligand field' approach in which single electron functions (l, s) are used to discuss the way in which the $(2s+1)(2l+1) = 2(2l+1)$ levels are filled in the presence of the ligand field splitting. In § 7.4 we outline the rather different approach using orbitals which are linear combinations of $3d$ and ligand ion wave-functions. The forms of the Zeeman and hyperfine interactions under these conditions are considered in §§ 7.5–7.7.

In these sections a static ligand field is assumed, but whenever this leaves a degenerate orbital ground state the celebrated theorem of Jahn and Teller (1937) shows that this is unstable against small displacements of the ligands. The complications introduced thereby (which may be static or dynamic in nature) are discussed in Chapter 21, and are mentioned only *in passim* in this chapter. Our primary concern is to indicate the principal features of the magnetic resonance spectra of ions of the $3d$ group, with a brief outline of the theoretical background needed for their interpretation. An 'intermediate ligand field' approach is assumed; compounds for which a 'strong ligand field' is needed are treated in Chapter 8 with ions of the $4d$, $5d$ groups.

In §§ 7.8–7.16 each $3d^n$ ion in sixfold coordination is considered in turn, while $3d^n$ ions in fourfold and eightfold coordination are briefly discussed in § 7.17. Some representative data are given in each case; usually this is limited to axial symmetry, for which the resonance data can be fitted to the spin Hamiltonian

$$\mathcal{H} = g_{\|}\beta H_z S_z + g_{\perp}\beta(H_x S_x + H_y S_y) + D\{S_z^2 - \tfrac{1}{3}S(S+1)\} + \\ + A_{\|}S_z I_z + A_{\perp}(S_x I_x + S_y I_y) + P_{\|}\{I_z^2 - \tfrac{1}{3}I(I+1)\} - \\ - g_{\|}^{(I)}\beta H_z I_z - g_{\perp}^{(I)}\beta(H_x I_x + H_y I_y), \quad (7.1)$$

though extra terms may be needed on occasion (e.g. for ions with $S = 2$ or $\tfrac{5}{2}$).

7.2. The intermediate crystal field approach

Apart from the half-filled shell, d^5, ^6S, the ground orbital states of free ions of the d-shell are either D or F. In a majority of cases the ligand field has predominantly cubic symmetry, with small distortions, in the sense that the splitting of the orbital states due to the cubic field is larger than that due to terms of lower symmetry. We shall diagonalize the matrix of the cubic field splitting first, and then evaluate the effect of terms of tetragonal or trigonal symmetry on the assumption that these can be treated by first-order perturbation theory. The calculation can be carried out rather neatly using the 'spin-operator technique', where (in spite of the nomenclature) the operators involve the orbital angular momentum.

D-states ($L = 2$)

In a cubic field with a tetragonal distortion the equivalent operator Hamiltonian is

$$B_4\{O_4^0 + 5O_4^4\} + B_2^0 O_2^0 + B_4^0 O_4^0, \quad (7.2a)$$

where the magnitude of the cubic field is denoted by B_4 and the other two terms represent the tetragonal distortion of second and fourth degree in the potential respectively. (Since O_4^0 occurs twice the separation into two fourth-degree terms is not unique, but it is convenient to use the form given in (7.2a) since the small terms are then diagonal.) The matrix elements of (7.2a) can be written down from Table 17, and the solution is straightforward. The energy levels and wave-functions are given in Fig. 7.3.

If the symmetry is predominantly cubic but with a trigonal distortion the equivalent operator Hamiltonian is

$$\mathcal{H} = -\tfrac{2}{3}B_4\{O_4^0 + 20\sqrt{(2)}O_4^3\} + B_2^0 O_2^0 + B_4^0 O_4^0, \quad (7.2b)$$

where the introduction of the fraction $(-\tfrac{2}{3})$ makes the parameter B_4

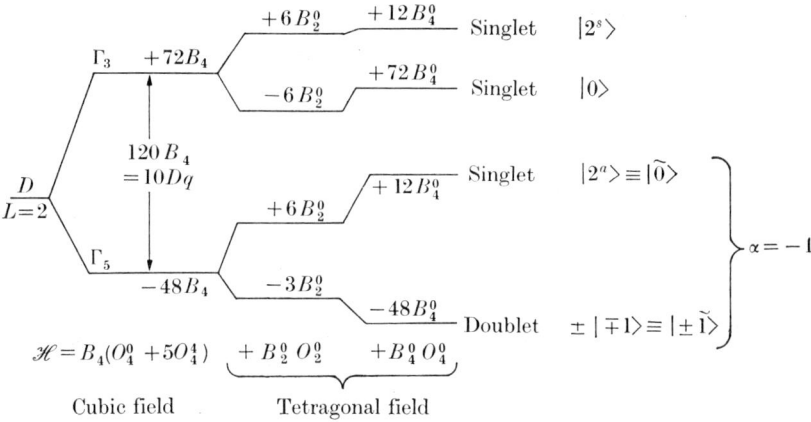

Fig. 7.3. Splitting of a D state in a cubic field and tetragonal fields of the second and fourth degree. On the right the states are given in terms of $|L_z\rangle$, and also for the triplet state in terms of $|\tilde{l}_z\rangle$, where $\tilde{l}$ is the fictitious orbital momentum. There are no off-diagonal elements of the tetragonal field between the orbital states which result from diagonalization of the cubic field.

of the cubic field operator identical with that in eqn (7.2a). The energy levels and wave-functions are given in Fig. 7.4.

The wave-functions given for the states in Fig. 7.3 are the same as those listed in Table 4, and on the extreme right of the figure the triplet Γ_5 states are labelled in terms of the fictitious angular momentum $\tilde{l} = 1$ by which it is often convenient to describe them. The matrix elements of orbital momentum within these three states are the same as those for $l = 1$, but with a change of sign ($\alpha = -1$), as shown in Chapter 14. On the other hand, the wave-functions listed in Fig. 7.4 are not the same as those given in Table 5, and differ by more than a simple change of overall sign. In fact the wave-functions in Table 5 are those appropriate to a cubic field operator $\{O_4^0 - 20\sqrt{(2)}O_4^3\}$; the difference is equivalent to a rotation of the coordinate system by 180° around the threefold axis, and either representation is equally valid. However it would be inconsistent (and lead to wrong answers) to use the wave-functions in Table 5 in calculations based on the Hamiltonian (7.2b).

An advantage of the equivalent operator language is that it gives a single representation for all states with $L = 2$. However it must be remembered that for a given crystalline potential the magnitude of the coefficients will depend on the configuration; thus for $L = 2$ they will differ in sign (and magnitude) for the configurations d^1, d^4, d^6, and d^9, all of which have a ground term with $L = 2$. We list below the sign of

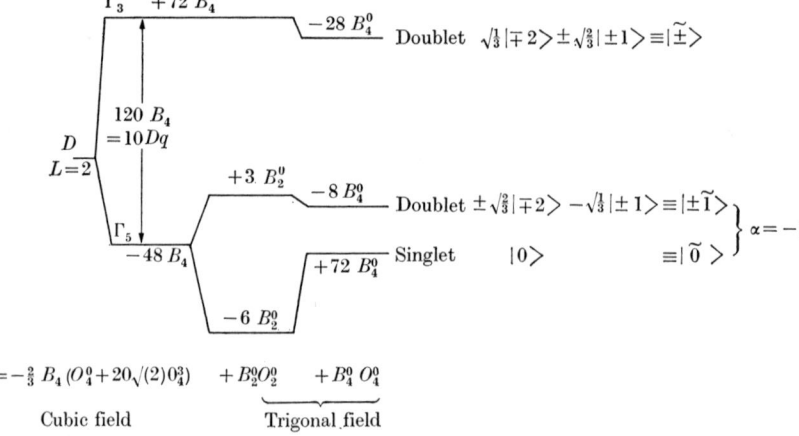

Fig. 7.4. Splitting of a D state in a cubic field and trigonal fields of the second and fourth degree. On the right the states are given in terms of $|L_z\rangle$, and also for the triplet state in terms of $|\tilde{l}_z\rangle$. The states shown are those resulting from diagonalization of the cubic field, and there is an off-diagonal element of the trigonal field

$$\langle \Gamma_3, \tilde{\pm}|\, B_2^0 O_2^0 + B_4^0 O_4^0 \,|\Gamma_5, \pm\tilde{1}\rangle = \pm\sqrt{(2)}(3B_2^0 + 20B_4^0).$$

B_4 for each of these configurations, in octahedral (sixfold) and in tetrahedral (fourfold) or cubic (eightfold) coordination with negatively charged ligand ions.

	B_4 positive	B_4 negative
Octahedral	d^1, d^6	d^4, d^9
Tetrahedral, cubic	d^4, d^9	d^1, d^6

F-states ($L = 3$)

In a cubic field with a tetragonal distortion the Hamiltonian (7.2a) leads to the wave-functions and energy levels given in Fig. 7.5. Similarly, in a cubic field with trigonal distortion the Hamiltonian (7.2b) gives the results shown in Fig. 7.6.

As before, the wave-functions in the tetragonal case are identical with those given in Table 4, but in the trigonal case there are differences corresponding to the wave-functions in Table 5 being associated with a cubic field operator $\{O_4^0 - 20\sqrt{(2)}O_4^3\}$. The signs of B_4 for various configurations whose ground state is F are listed below.

	B_4 positive	B_4 negative
Octahedral	d^3, d^8	d^2, d^7.
Tetrahedral (cubic)	d^2, d^7	d^3, d^8.

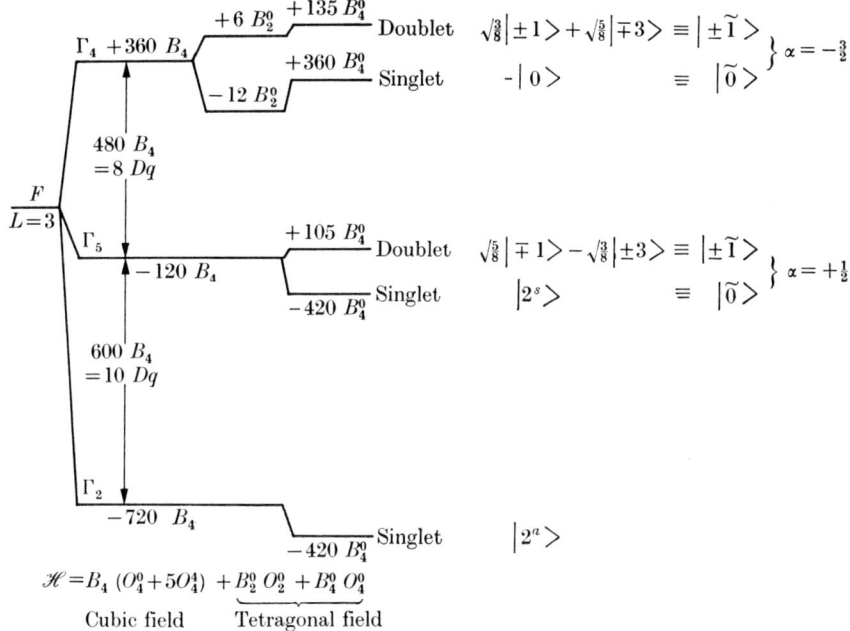

Fig. 7.5. Splitting of an F term in a cubic field and tetragonal fields of the second and fourth degree. On the right the states are given in terms of $|L_z\rangle$, and also for the triplet states in terms of $|\tilde{l}_z\rangle$. The states shown result from diagonalization of the cubic field, and there is an off-diagonal tetragonal field matrix element

$$\langle \Gamma_4, \pm\tilde{1}|\, B_2^0 O_2^0 + B_4^0 O_4^0\, |\Gamma_5, \mp\tilde{1}\rangle = -\sqrt{(15)}(3B_2^0 + 15B_4^0).$$

The states given in Figs. 7.3–7.6 are those appropriate to a cubic field, and are therefore only correct in some cases for the axial fields in so far as these are small compared with the cubic field. The axial field splittings have one or two peculiarities; for a D-state ion the doublet Γ_3 is not split by a trigonal field, a result that has important effects on the resonance spectrum of an ion for which this is the ground state, such as Cu^{2+} in an octahedral field. For an F-state ion the Γ_5 triplet is not split by a tetragonal field of the second degree, but this has only an indirect effect on the ground state. Axial field splittings do not displace the centre of gravity of the cubic field triplets or doublet from which they are formed so far as the second-degree terms $B_2^0 O_2^0$ are concerned; this does not hold for the fourth-degree terms $B_4^0 O_4^0$, though the latter still leave the centre of gravity of the whole manifold undisplaced.

To compare the splittings for different ions in a given crystal field we need the formulae deduced in Chapter 16, viz.

$$B_4 = \langle L\|\,\beta\,\|L\rangle A_4\langle r^4\rangle, \quad B_2^0 = \langle L\|\,\alpha\,\|L\rangle A_2^0\langle r^2\rangle, \quad B_4^0 = \langle L\|\,\beta\,\|L\rangle A_4^0\langle r^4\rangle,$$

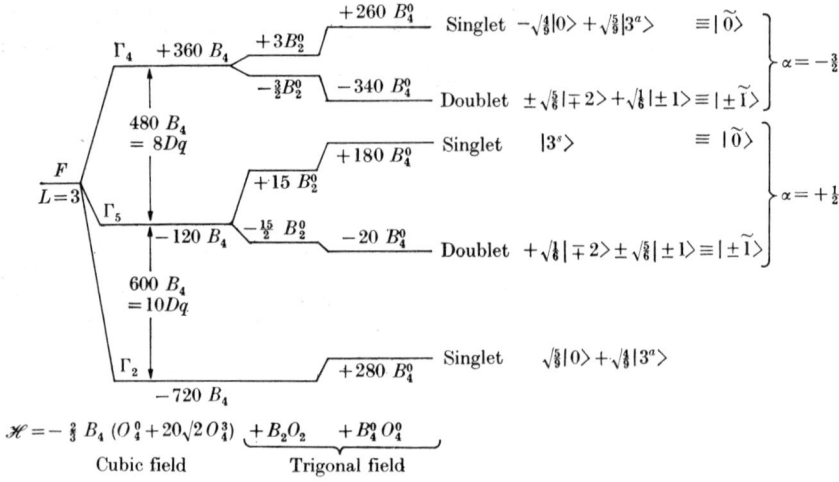

Fig. 7.6. Splitting of an F term in a cubic field and trigonal fields of the second and fourth degree. On the right the states are given in terms of $|L_z\rangle$, and also for the triplet states in terms of $|\tilde{l}_z\rangle$. The states shown result from diagonalization of the cubic field, and there are off-diagonal trigonal field matrix elements

$$\langle \Gamma_4, \tilde{0} | B_2^0 O_2^0 + B_4^0 O_4^0 | \Gamma_2 \rangle = \sqrt{(5)}(6B_2^0 - 40B_4^0);$$

$$\langle \Gamma_4, \pm\tilde{1} | B_2^0 O_2^0 + B_4^0 O_4^0 | \Gamma_5, \pm\tilde{1} \rangle = \pm\sqrt{(5)}(\tfrac{3}{2}B_2^0 - 80B_4^0).$$

where A_k^q denotes the strength of the potential, $\langle r^2 \rangle$ and $\langle r^4 \rangle$ are the mean second and fourth powers of the radius of the electronic charge distribution, and $\langle L \| \alpha \| L \rangle$, $\langle L \| \beta \| L \rangle$ are numerical coefficients given in Table 19. (Their value for an F-state is $\tfrac{1}{5}$ that for a D-state, but the signs depend on the number of electrons.) We shall not write them all out, but consider only the cubic field. Following the pioneer work of Van Vleck, the cubic field splitting for a single d-electron is frequently denoted in the literature by the symbol $10Dq$. From Figs. 7.3, 7.4 we see that this equals $120\,B_4$, or $(\tfrac{80}{21})A_4\langle r^4 \rangle$ for d^1; the splitting is the same for d^4, d^6, d^9 since these can be considered as an electron or hole plus a half-filled or filled shell, but the energy diagram for a hole is inverted from that for an electron. Since the value of $\langle L \| \beta \| L \rangle$ for an F-state is $\tfrac{1}{5}$ that of a D-state, the cubic field splittings for an F-state are $600\,B_4 = 10Dq$ between Γ_2 and Γ_5, and $480\,B_4 = 8Dq$ between Γ_5 and Γ_4, making the overall splitting equal to $18Dq$.

Once again, the inversion of the energy diagrams for d^3, d^8 compared with those for d^2, d^7 follows from the fact that the former can be regarded as a half- or filled shell with two holes, the latter as an empty or half-filled shell plus two electrons. Values of the cubic splitting parameter Dq can be deduced from the optical absorption spectrum; a typical set of values for hydrated ions in solution is given in Table 7.3 (slightly different values have been given by other authors). The values for hydrated ions in crystals do not appear to be significantly different, as can be readily seen from the fact that the colour is not materially changed. This table also gives the values of the one-electron spin-orbit coupling parameter ζ, as deduced from the spectrum of the free ion. In many cases, however, we shall use the spin-orbit coupling in the form $\lambda(\mathbf{L}\cdot\mathbf{S})$ rather than $\zeta(\mathbf{l}\cdot\mathbf{s})$; the two parameters are related by the equation (valid for a ground state obeying Hund's rules)

$$\lambda = \pm\zeta/2S, \tag{7.3}$$

where the $+$ sign is required for a shell that is less than half-filled, and the $-$ sign for a shell that is more than half-filled.

The spin operator technique we have used so far in this chapter represents interactions only within the ground term (D or F in the iron group). Except for the configurations d^1 and d^9, there are, however, excited terms belonging to the same configuration, and in general there will exist matrix elements of the crystal field between the ground term and some of the excited terms. In many compounds the cubic component of the crystal field is much larger than the other components and it is a good approximation to neglect all matrix elements of the crystal field to excited terms except those due to the cubic field. With this restriction there is only one such matrix element, that between the orbital triplet $|F, \Gamma_4\rangle$ and the triplet P, which also belongs to Γ_4. The effect is to depress in energy the Γ_4 triplet with respect to the other terms Γ_2, Γ_5 belonging to the F manifold, and to admix the states so that they take the form (with $\epsilon^2 + \tau^2 = 1$)

$$\epsilon\,|F, \Gamma_4\rangle + \tau\,|P, \Gamma_4\rangle.$$

This orbital triplet can still be represented by a fictitious angular momentum $\tilde{l} = 1$, but the fictitious orbital g-factor associated with it becomes

$$\tilde{g}_l = -\tfrac{3}{2}\epsilon^2 + \tau^2 = -\tfrac{1}{2}(3 - 5\tau^2). \tag{7.4}$$

The spin-orbit coupling will also be modified (see § 19.5).

7.3. The strong crystal field approach

The intermediate crystal field approach is a useful introduction to paramagnetic resonance of the iron group, and explains to a good approximation many of the experimental results. However the cubic

TABLE 7.3

Values of cubic field splitting parameter (Dq) for hydrated ions in solution, and of the one-electron spin-orbit parameter ζ. These values are based on Tables VII and VIII of McClure (1959). The values of B,C for the free dipositive ions are from Table VI of McClure (1959)

	$3d^1$	$3d^2$	$3d^3$	$3d^4$	$3d^5$	$3d^6$
Ion	Ti^{3+}	V^{3+}	Cr^{3+}	Mn^{3+}	Fe^{3+}	Co^{3+}
Dq (cm^{-1})	2030	1800	1760	2100	1400	1910
ζ (cm^{-1})	154	209	276	360		

	$3d^1$	$3d^2$	$3d^3$	$3d^4$	$3d^5$	$3d^6$	$3d^7$	$3d^8$	$3d^9$
Ion	Sc^{2+}	Ti^{2+}	V^{2+}	Cr^{2+}	Mn^{2+}	Fe^{2+}	Co^{2+}	Ni^{2+}	Cu^{2+}
Dq (cm^{-1})		120	1180	1400	750	1000	1000	860	1260
ζ (cm^{-1})	79		168	236	335	404	528	644	829
B (cm^{-1})		694	755	810	860	917	971	1030	
C (cm^{-1})		2910	3257	3565	3850	4040	4497	4850	
C/B		4·19	4·31	4·40	4·48	4·41	4·63	4·71	

crystal field is by no means a small perturbation, as can be seen from the fact that in many salts the overall cubic splitting of the F term is larger than the separation between the ground F term and the excited P term. An alternative procedure is the so-called 'strong crystal field' approach, in which the crystal field is diagonalized before the Coulomb interaction. Then we no longer consider terms such as ^4F, ^4P, etc., which represent the groups of energy levels characterized by given values of L, S into which the configuration is split by the Coulomb interaction, but return to the single electron states. We shall restrict the discussion to the case of a field of octahedral symmetry.

In such a field the five orbital states of a d-electron split into a lower triplet and an upper doublet, with separation $10Dq$. This is, of course, the same as shown in Figs. 7.3 and 7.4, but the nomenclature used for the symmetry classification is different: the triplet states are known as $d\epsilon$ states, the doublet states as $d\gamma$. However one reason for returning to the single electron states is to introduce the modifications due to bonding with the ligand ions; the $d\epsilon$ states take part in π-bonding, the $d\gamma$ states in σ-bonding. The nomenclature $d\epsilon$, $d\gamma$ refers to pure d-electron states; the augmented states formed by admixing orbitals belonging to the ligand ions have the more general symmetry classification t_2 (triplet), and e (doublet), and we shall anticipate the discussion of bonding by using the appropriate nomenclature immediately.

For configurations containing more than one d-electron, we have a choice of states for the various electrons. Obviously the crystal field energy is minimized by placing as many electrons as possible in the lower t_2 states, but we must take into account the powerful coupling between the electron spins which is expressed in the first of Hund's rules, that the ground state is the one of maximum spin. We shall begin by assuming that the spin coupling energy is larger than the crystal field energy; then only three electrons can be placed in the t_2 orbitals with parallel spin because of the exclusion principle, but two further electrons can be accommodated in the e-orbitals with parallel spin. This maximum of five electrons with parallel spin corresponds to a half-filled shell with $S = \frac{5}{2}$; since there is only one such arrangement, the overall state is an orbital singlet $L = 0$, or $^6S_{\frac{5}{2}}$. The ground state of any given configuration can quickly be constructed by following these rules, whereby the crystal field energy is minimized subject to maximizing the spin and the restrictions imposed by the exclusion principle. The results are shown in Fig. 7.7; the orbital multiplicity of the ground state is given by the number of ways in which the electrons can be

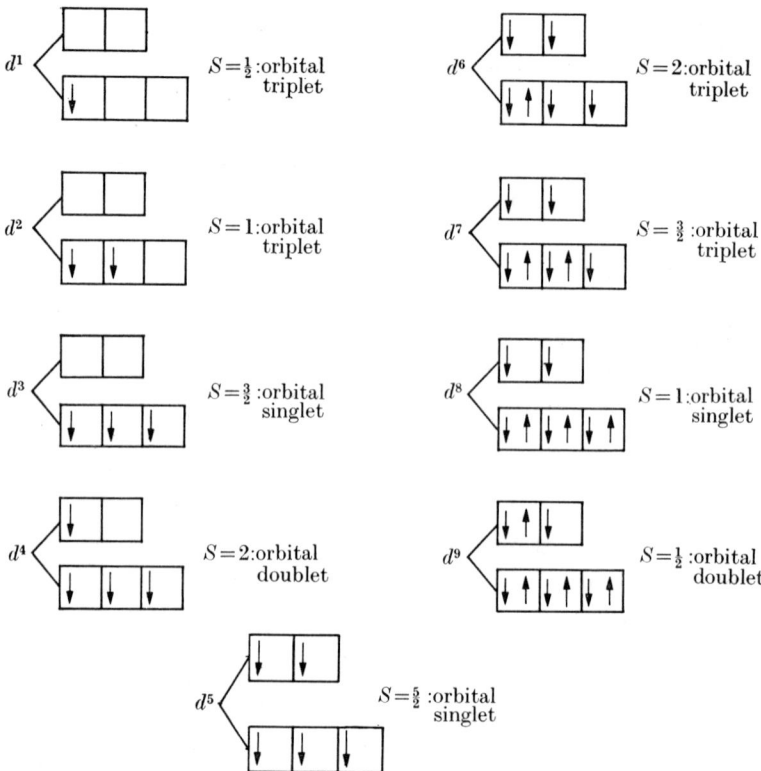

FIG. 7.7. Construction of the ground states of the d-configurations in an octahedral field using the strong crystal field approach, but assuming the spin coupling to be stronger than the crystal field energy.

distributed between orbitals of the same energy. These agree with the ground states in an octahedral field obtained by the intermediate crystal field method, and listed in Table 7.2; a similar procedure can be carried out for a tetrahedral field, placing the doublet e states below the triplet t_2 states instead of above.

In some compounds such as the complex cyanides of the $3d$ group the magnetic behaviour suggests that the crystal field energy is much larger, and sufficient to outweigh the spin coupling. We can construct the appropriate ground states for this situation by placing electrons as far as possible in the t_2 orbitals. For configurations containing up to three electrons the ground states are the same as those in Fig. 7.7, but the remainder are different (with the exception of d^8 and d^9), as shown in Fig. 7.8. The maximum spin multiplicity ($S = \frac{3}{2}$) is obtained for d^3, corresponding to a half-filled t_2 shell, and then decreases as far

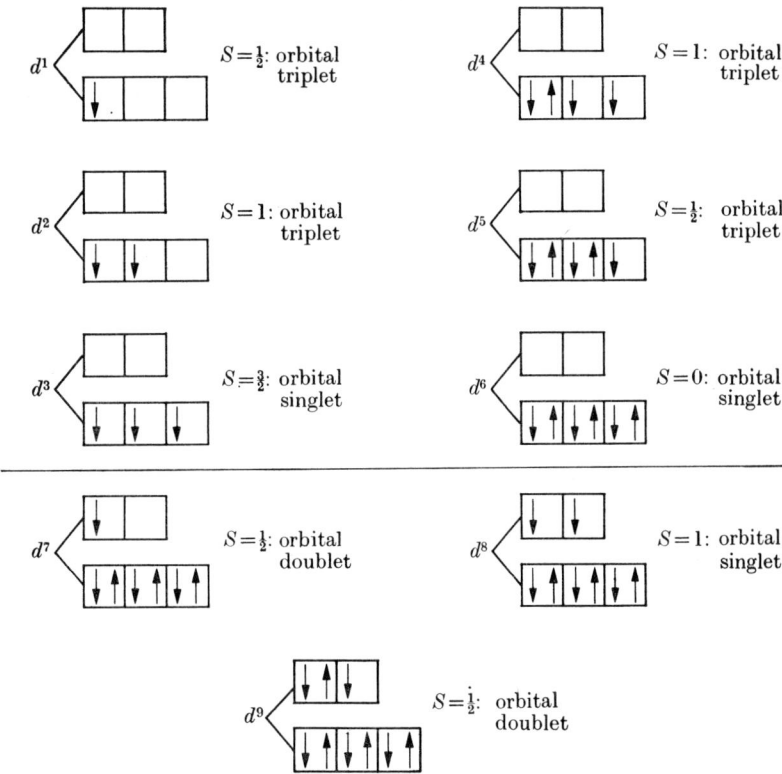

FIG. 7.8. Construction of the ground states of the d-configurations in an octahedral field using the strong crystal field approach, and assuming the crystal field energy to be larger than the spin coupling energy.

as d^6, where the t_2 shell is filled. Further electrons must go into the e-shell, which is rather high in energy so that compounds of this kind tend to be chemically unstable and occur rather rarely. We shall discuss the experimental results for compounds of this type, which include most paramagnetic salts of the $4d$ and $5d$ groups as well as the complex cyanides of the $3d$ group, separately in Chapter 8.

We consider now the excited states. For d^1 this is a trivial problem, the only excited state being that where the electron is placed in the e-shell, its energy being then $10Dq$ higher than in the t_2 shell. If at first we include only the crystal field energy, then for two electrons (d^2) there will be states t_2e and e^2 lying at $10Dq$, $20Dq$ relative to the ground state t_2^2 and formed by promoting first one and then two electrons to the e-shell. In all of these states we can keep the spins parallel, giving spin triplet states with $S = 1$, though there will of course be spin singlet states ($S = 0$) generally comparable in energy. If we confine ourselves to the spin triplet states, the orbital multiplicity of t_2^2 is three and that of e^2 is one, while the t_2e states have sixfold multiplicity since one electron can be in any three of the t_2 states while the other can be in either of the e-states. This gives in all ten states, the same total number of spin triplet states as in the free ion where they are distributed between the ^3F and ^3P states, with sevenfold and threefold orbital degeneracy respectively. Before introducing the Coulomb energy, we classify the strong crystal field states according to their symmetry properties, which we quote without proof. They are

$$^3T_1(t_2^2); \qquad ^3T_2(t_2e); \qquad ^3T_1(t_2e); \qquad ^3E(e^2),$$

of which the first three are threefold degenerate and the last is a singlet (apart, of course, from the spin multiplicity of three). The advantage of this procedure is that the Coulomb interaction $\sum e^2/r_{ij}$ has matrix elements only between states with the same total spin S and with the same symmetry classification. In the present case the only off-diagonal matrix element is that between the two $^3T_1$ states; when all the

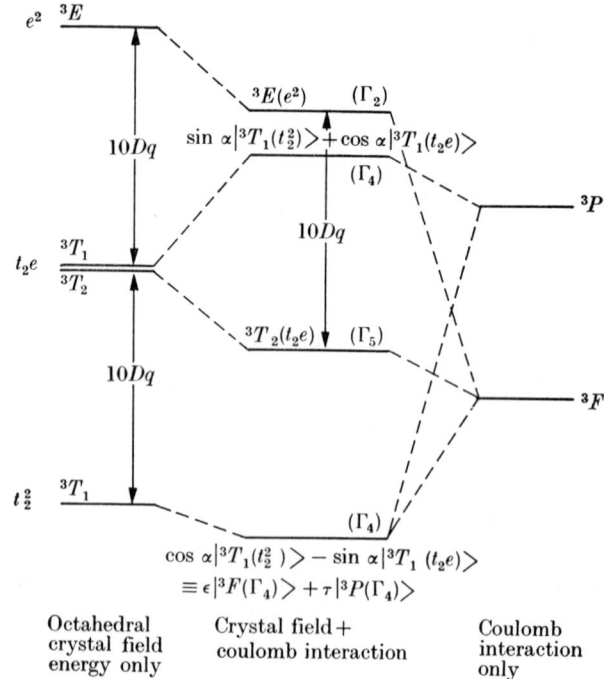

Fig. 7.9. The spin triplet states of d^2 in an octahedral crystal field, showing their derivation from the weak and strong crystal field approaches.

Coulomb energy terms are included, the energy matrix, in which B is
the radial integral parameter introduced by Racah, is

$$\begin{array}{c|cc} {}^3T_1(t_2^2) & 3B & 6B \\ {}^3T_1(t_2e) & 6B & 12B+10Dq \end{array},$$

which on diagonalization gives for the energy

$$W = \tfrac{1}{2}(15B+10Dq) \pm \{\tfrac{1}{4}(9B+10Dq)^2 + 36B^2\}^{\tfrac{1}{2}},$$

while the states ${}^3T_2(t_2e)$ and ${}^3E(e^2)$ remain at 10 and $20Dq$ respectively. In all we obtain three orbital triplets and one singlet (each with three-fold spin degeneracy), as shown in Fig. 7.9. In the limit of zero crystal field two triplets and the singlet collapse into the term ${}^3F$, and the remaining triplet becomes ${}^3P$, lying higher in energy than ${}^3F$ by $15B$.

To give an impression of the sizes of the various terms we mention that in the free V^{3+} ion (d^2) the separation ${}^3P-{}^3F$ is some 13000 cm^{-1}, while in aqueous solution $10Dq$ is about 18000 cm^{-1}. For the free ion the Racah parameter B is about 860 cm^{-1}, so that (Dq/B) is a little greater than 2. Reference to the energy level diagram of Tanabe and Sugano (1954), reproduced as Fig. 7.10, shows that a number of the singlet states have energies comparable with the excited triplet states at this value of (Dq/B). However such singlet states have little effect on the paramagnetic resonance spectrum as they are not appreciably admixed into the ground state by crystal fields of lower symmetry (assumed small in magnitude) or by the spin-orbit coupling.

The lowest triplet may be described in the two ways (see Fig. 7.9)

$$\cos\alpha\,|{}^3T_1(t_2^2)\rangle - \sin\alpha\,|{}^3T_1(t_2e)\rangle \equiv \epsilon\,|{}^3F(\Gamma_4)\rangle + \tau\,|{}^3P(\Gamma_4)\rangle,$$

where $\tan 2\alpha = 12B/(9B+10Dq)$, and $\epsilon^2+\tau^2 = 1$, and from the equivalence we find that $\sqrt{(5)}\tau = \cos\alpha - 2\sin\alpha$. The triplet may be described as before by a fictitious angular momentum $\tilde{l} = 1$, with orbital g-factor (cf. eqn (7.4))

$$\tilde{g}_l = -\tfrac{1}{2}(3-5\tau^2)$$
$$= -\cos^2\alpha - 2\cos\alpha\sin\alpha + \tfrac{1}{2}\sin^2\alpha. \quad (7.4a)$$

In the limit where $Dq \to 0$, the value of $\cos\alpha$ tends to $2/\sqrt{5}$ and $\sin\alpha$ to $1/\sqrt{5}$, and $\tilde{g}_l$ to $-\tfrac{3}{2}$. Likewise the separation between the Γ_4 and Γ_5 triplets belonging to ${}^3F$ tends to $8Dq$, as shown in Figs. 7.5 and 7.6. These results are valid also for the Γ_4 triplet that forms the ground state of d^7 in an octahedral field (compare § 20.4).

A similar procedure can be carried out for d^3 in an octahedral field.

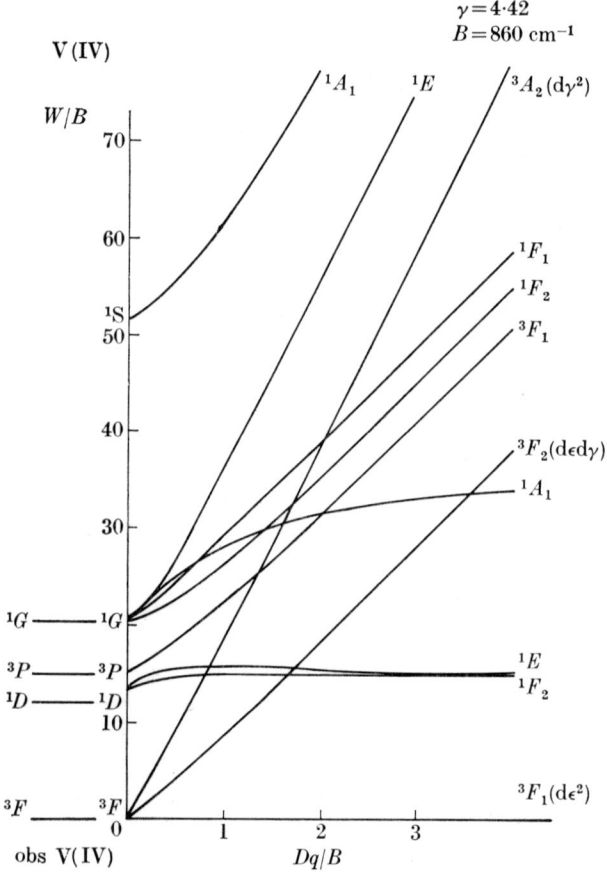

Fig. 7.10. Splitting of states of the d^2 configuration by an octahedral field (Tanabe and Sugano 1954). The observed levels of V^{+++} are shown to the left for comparison with the calculated positions of ^1D, ^3P, and ^1G as shown in the diagram. The value of $\gamma = C/B$ used in constructing the diagram is 4·42, the best value for V^{+++}. The importance of interference from the next higher configuration of the same parity as $3d^2$ viz. $3d4s$, may be judged from the fact that its lowest level ^3D is at $111B$ on the energy scale of the diagram. At the right of the diagram one notes that there are three sets of lines each characterized by a certain slope. These correspond to the three strong-field states at 0, 1, 2 × (10Dq). The notation used in these figures is that of Tanabe and Sugano; its relation to the notation used here is as follows:

$d\epsilon \to t$ $^3F_1, {}^3F_2,$ etc. $\to {}^3T_1, {}^3T_2,$ etc.
$d\gamma \to e$ V(IV), etc. $\to$ V^{+++}, etc.

In a strong crystal field without the Coulomb interaction the spin quartet states are $^4A_2(t_2^3)$, an orbital singlet, which we take to be at zero in energy; $^4T_1(t_2^2e)$ and $^4T_2(t_2^2e)$, two orbital triplets at $10Dq$; $^4T_1(t_2e^2)$, an orbital triplet at $20Dq$. Still higher in energy is e^3, at $30Dq$, but one of the spins must then be reversed because the e-states can only

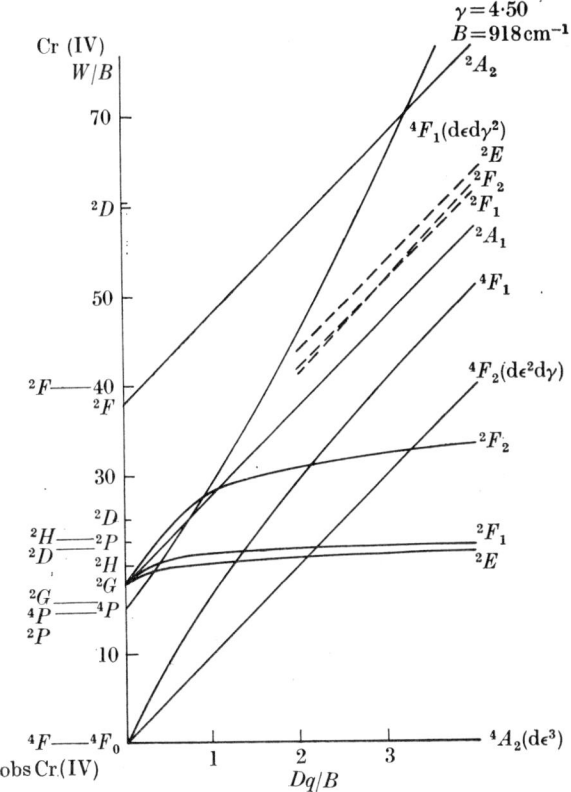

FIG. 7.11. Splitting of states of the d^3 configuration by an octahedral field (Tanabe and Sugano 1954). The ^4F state of the next excited configuration $3d^2(^3F)4s$ lies at $W/B = 113$.

accommodate two electrons with parallel spin, so that this must be a spin doublet state. Introduction of the Coulomb energy raises both the $^4T_1$ states and admixes them, giving one orbital triplet whose states are denoted by $\sin \alpha \, |^4T_1(t_2^2 e)\rangle + \cos \alpha \, |^4T_1(t_2 e^2)\rangle$; in the small crystal field limit $\cos \alpha$ tends to $2/\sqrt{5}$, and the energy to $18Dq$, as in Figs. 7.5 and 7.6. The orthogonal state is the orbital triplet corresponding to ^4P in the weak field limit.

When we come to d^4 there are only two possibilities of spin quintet states. These are $t_2^3 e$, which is an orbital doublet, and $t_2^2 e^2$, which is an orbital triplet lying higher in energy by $10Dq$. Other states such as $t_2 e^3$, t_2^4 must be states of lower spin. The latter makes it impossible for there to be matrix elements of the Coulomb interaction between such states and the spin quintet states, so that the latter are unadmixed and their relative separation is unchanged. For d^5 there is only one spin

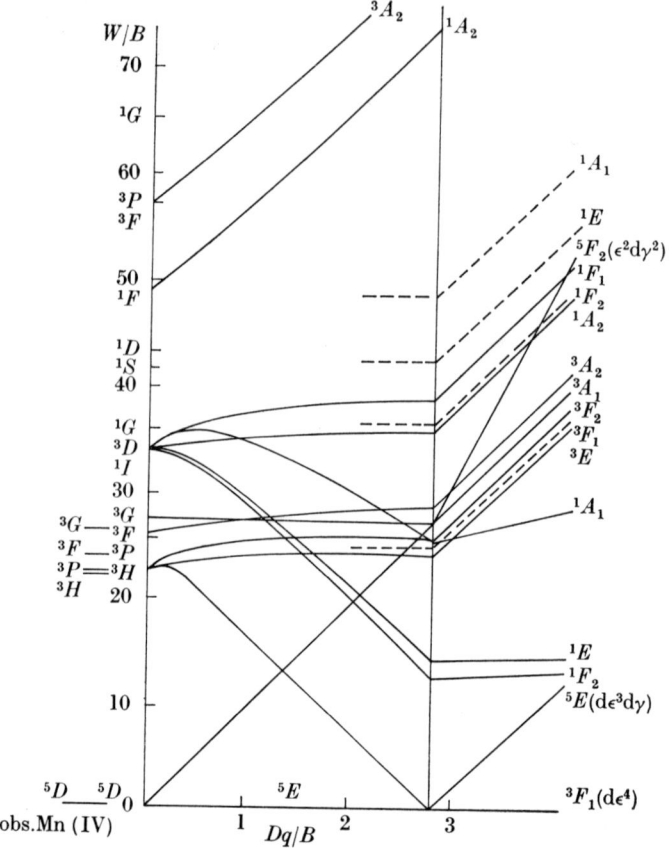

FIG. 7.12. Splitting of the states of the d^4 configuration in an octahedral field (after Tanabe and Sugano 1954). The observed levels of the free Mn IV (Mn^{3+}) ion are shown on the left. For this ion $C/B = 4.61$, $B = 965$ cm^{-1} and the next ^5F state from the excited configuration $3d^3(^4F)4s$ lies at $W/B \approx 116$.

sextet state, and the result is straightforward. Configurations with more electrons can be treated in the same way by considering holes in a filled shell instead of electrons, remembering of course that the states of higher energy are formed by displacing the holes downwards.

A complete solution of the problem including both Coulomb and crystal field energy involves three parameters, the two Racah parameters B and C and the crystal field energy Dq. Such solutions have been given by Tanabe and Sugano (1954); by assuming that the ratio C/B is the same as in the free ions, where it varies between about 4·2 to 4·9, they were able to draw reduced energy level diagrams for each configuration. A series of these are reproduced in Figs. 7.10–7.16; the

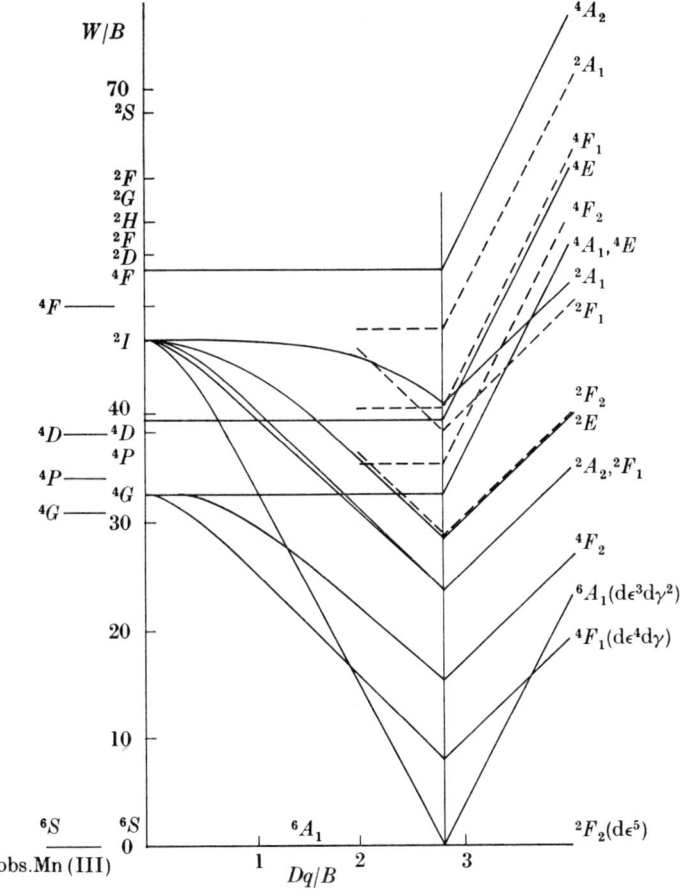

Fig. 7.13. Splitting of the states of the d^5 configuration by an octahedral field (Tanabe and Sugano 1954). For the free III (Mn^{2+}) ion, $C/B = 4.48$ and $B = 860$ cm^{-1}. The observed levels for this ion are shown on the left. The next lowest sextet state belongs to the configuration $3d^4(^5D)4s$, and lies at $W/B \approx 73$.

energy, relative to the Racah parameter B, is plotted against the reduced crystal field energy (Dq/B). At $(Dq/B) = 0$, the states group themselves into the free ion terms, but in the opposite limit of very large (Dq/B) the levels diverge linearly as sets of lines whose slope $(W/10Dq)$ tends to 0, 1, 2, etc., corresponding to the limit of strong crystal field. For d^2, d^3, and d^8 the ground term remains the same for all (Dq/B), but for the others there is a cross-over at a certain value of (Dq/B) lying between 2 and 3, above which the ground state becomes one of lower spin. This corresponds to the point at which it becomes energetically more advantageous to put electrons into the t_2 states,

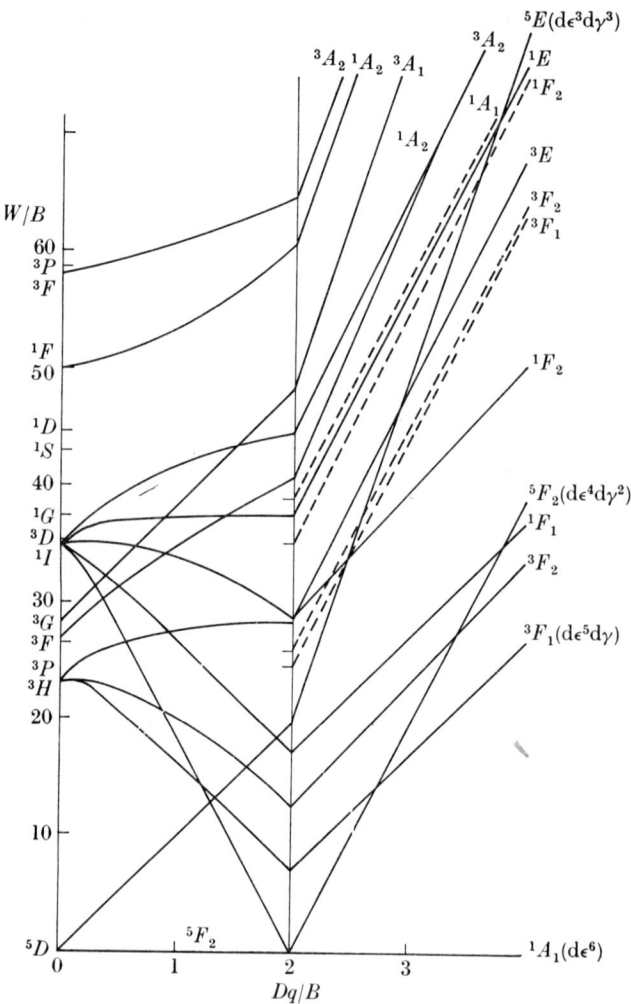

Fig. 7.14. Splitting of the states of the d^6 configuration in an octahedral field (Tanabe and Sugano 1954). For the free Co IV (Co^{3+}) ion $C/B = 4\cdot 81$, $B = 1065$ cm^{-1}.

which reduces the crystal field energy, than to conserve spin pairing energy by forming states of maximum spin. The reader may verify that in Figs. 7.12–7.15 the ground state for sufficiently high values of (Dq/B) is that corresponding to Fig. 7.8 instead of Fig. 7.7. Thus for d^4 the ground state becomes a spin triplet, for d^5 a spin doublet, and for d^6 a singlet, with a doublet again for d^7; although the spin quartet states mostly lie below the spin doublet states in d^5 for the free ion, there is no value of (Dq/B) at which a quartet state becomes the ground state in

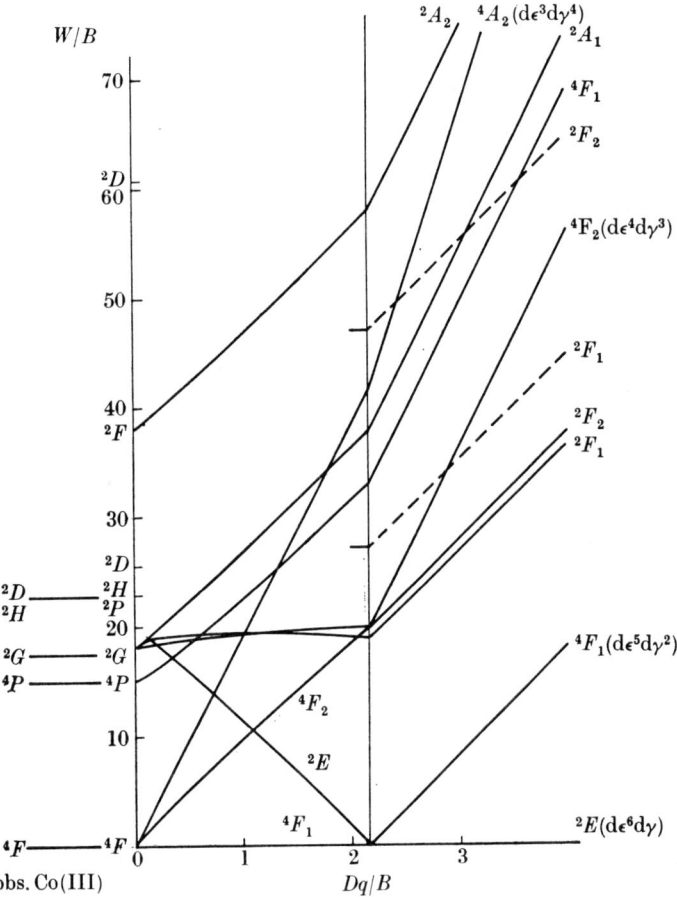

FIG. 7.15. Splitting of the states of the d^7 configuration by an octahedral field (Tanabe and Sugano 1954). The levels for the free Co III (Co^{2+}) ion are shown on the left; for this ion $C/B = 4.63$, $B = 971$ cm^{-1}. The excited level $3d^6(^5D)4s$, ^6D lies at $(W/B) \approx 48$.

an octahedral field, though a slightly split quartet ground state has been observed in an Fe^{3+} complex of C_{2v} symmetry (Wickman et al, 1967).

The optical spectra of a number of tripositive ions in corundum (Al_2O_3) have been studied by McClure (1962) over the temperature range 4·2–1200°K, and have been interpreted in terms of a strong crystal field or single electron approach. The values of the cubic field and trigonal field parameters are given in Table 7.4. In comparison with the nomenclature of our Fig. 7.4 (which is appropriate to a single d-electron in trigonal symmetry), the cubic field splitting $10Dq$ is taken as the separation between the upper orbital doublet Γ_3 or $E(e)$ and the

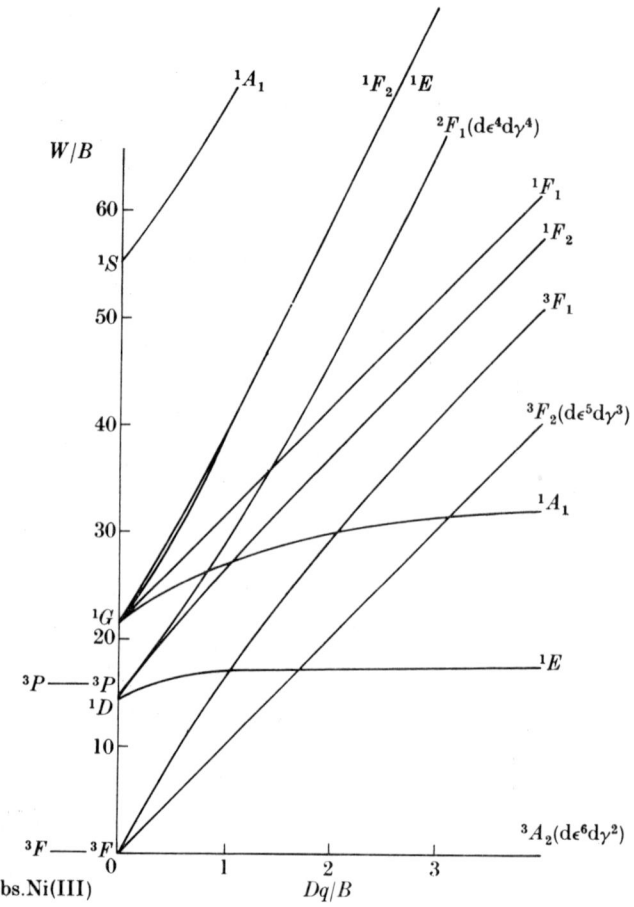

Fig. 7.16. Splitting of the states of the d^8 configuration by an octahedral field (Tanabe and Sugano 1954). The levels for the free Ni III (Ni^{2+}) ion are shown on the left; for this ion $(C/B) = 4\cdot 71$, $B = 1030$ cm^{-1}. The excited configuration $3d^7(^4F)4s$, ^5F lies at $(W/B) \approx 51$.

centre of gravity of the lower split triplet Γ_5 or $A_1(t)$, $E(t)$. These are displaced by the $B_4^0 O_4^0$ trigonal field term, so that in terms of Fig. 7.4 McClure's parameters are

$$\left.\begin{aligned}
10Dq &= 120B_4 - (\tfrac{140}{3})B_4^0, \\
v &= -9B_2^0 + 80B_4^0, \\
v' &= \pm\sqrt{(2)}(3B_2^0 + 20B_4^0),
\end{aligned}\right\} \quad (7.5)$$

where v, v' are the diagonal and off-diagonal elements of the trigonal field.

TABLE 7.4

Values of the cubic field parameter (Dq) and the trigonal field parameter v for tripositive ions in Al_2O_3. The values are taken from McClure (1962), except for (a) Pryce and Runciman (1958), (b) Blumberg, Eisinger, and Geschwind (1963). In terms of the nomenclature of Fig. 7.4, appropriate to a single d-electron in trigonal symmetry, the parameters are (see eqn (7.5)) $10Dq = 120B_4 - (140/3)B_4^0$; $v = -9B_2^0 + 80B_4^0$. The values of B, C for the free ion are from Table VI of McClure (1959). The values of ζ for the free ion are from Blume and Watson (1963)

Ion	$3d^1$	$3d^2$	$3d^3$	$3d^4$	$3d^5$	$3d^6$	$3d^7$	$3d^8$
	Ti^{3+}	V^{3+}	Cr^{3+}	Mn^{3+}	Fe^{3+}	Co^{3+}	Ni^{3+}	Cu^{3+}
Dq (cm^{-1})	1905	1750$^{(a)}$	1815	1947	1650	1830	1000	2100$^{(b)}$
v (cm^{-1})	>0	960	1425	1950				
ζ (cm^{-1})	159	219	292	380	486	720		
B (cm^{-1})		861	918	965	1015	1065	1115	
C (cm^{-1})		3814	4133	4450	4800	5120	5450	
C/B		4·43	4·50	4·61	4·73	4·81	4·89	

In this section we have attempted only an outline of the strong-field representation, and in the next we attempt the same for covalent bonding. A more general discussion is given in Chapter 20; for a still more complete account (including the Racah parameters B, C that are a measure of the mutual electrostatic repulsion energy of the electrons) a book such as Griffith (1961) may be consulted. For the magnetic properties that are our main interest we have drawn widely on the review article of Owen and Thornley (1966).

7.4. The effects of bonding

In considering the effects of bonding we make use of real wave-functions. For the d-wave-functions on the central ion these are related to the states specified by their angular momentum component along the z-axis as follows:

$$
\begin{aligned}
(16\pi/5)^{\frac{1}{2}} |0\rangle &= 3z^2 - r^2 \\
(16\pi/5)^{\frac{1}{2}} |2^s\rangle = (8\pi/5)^{\frac{1}{2}}\{|+2\rangle + |-2\rangle\} &= \sqrt{(3)}(x^2 - y^2)
\end{aligned}\Bigg\} e_g,
$$
$$
\left.\begin{aligned}
i^{-1}(16\pi/5)^{\frac{1}{2}} |2^a\rangle = i^{-1}(8\pi/5)^{\frac{1}{2}}\{|+2\rangle - |-2\rangle\} &= 2\sqrt{(3)}xy \\
i(16\pi/5)^{\frac{1}{2}} |1^s\rangle = i(8\pi/5)^{\frac{1}{2}}\{|+1\rangle + |-1\rangle\} &= 2\sqrt{(3)}yz \\
-(16\pi/5)^{\frac{1}{2}} |1^a\rangle = -(8\pi/5)^{\frac{1}{2}}\{|+1\rangle - |-1\rangle\} &= 2\sqrt{(3)}zx
\end{aligned}\right\} t_{2g}.
$$
(7.6)

In discussing the orbitals on the ligand ions in the complex XY_6 we follow the conventions adopted in Chapter 20. The six ligand ions are placed at equal distances along each of the x, y, z axes, the numbering being 1, 4 on the x-axis; 2, 5 on the y-axis; 3, 6 on the z-axis, with the smaller numeral on the positive side in each case. The ligand orbitals, consisting of s- and p-states, are divided into two classes: (1) σ-bonding orbitals, whose maximum charge density is along the bond-axis, and whose wave-functions are symmetrical about that axis; such orbitals are normally s-p hybrids, and suitable combinations belong to the symmetry class e_g; (2) π-bonding orbitals, whose maximum charge density is directed normal to the bond axis; these are pure p-states, and are denoted by X, Y, or Z, where an X-orbital is symmetrical about a line parallel to the x-axis, etc.; in suitable combinations their symmetry class is t_{2g}.

The correct combinations of orbitals on the central ion and on the ligand ions are derived in Chapter 20 and are listed in Table 24. Only central ion and ligand ion orbitals belonging to the same symmetry classes can be combined, with the result that e_g orbitals take part only

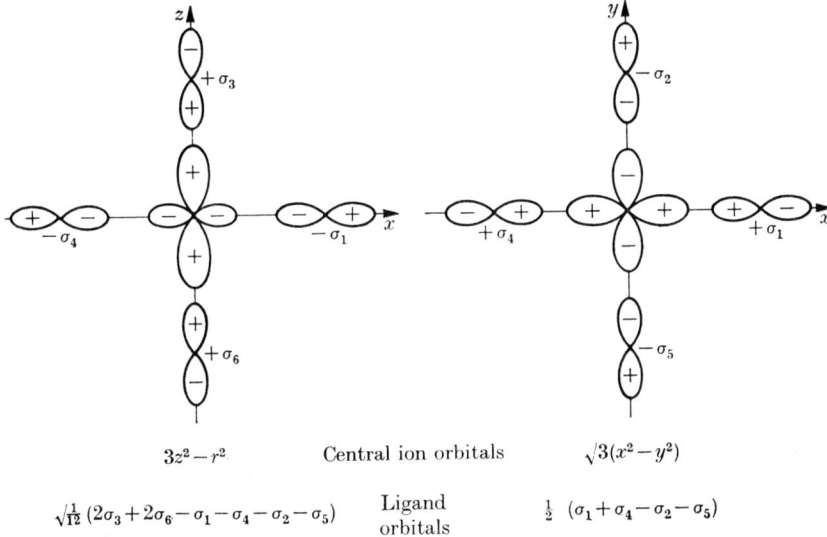

Fig. 7.17. Schematic representation of the central ion and ligand ion orbitals of symmetry e_g which are involved in σ-bonding. The ligand orbitals are drawn with equal lobes but in general they will be hybrid s-p orbitals with unequal lobes, for which we write $\sigma = \sigma_s + \sigma_p$.

in σ-bonding, and t_{2g} orbitals only in π-bonding. The way in which combinations of the correct symmetry are formed can be understood from Figs. 7.17 and 7.18, which show the two e_g and one t_{2g} combination respectively. The phases of the ligand orbitals are chosen to match the symmetry of the central ion orbital; the conventions of Chapter 20 are that the σ-orbitals on the ligands have their positive lobes facing inwards, and the π-orbitals have their positive lobes in the positive direction of the axis to which they are parallel. The correct signs in the combinations can then be inferred from the figures; the double amplitude assigned to σ_3 and σ_6 corresponds to the amplitude of $3z^2 - r^2$ being twice as great along the z-axis as along the x- and y-axes.

The augmented states listed in Table 24, consisting of linear combinations of central ion and ligand orbitals, are five in number, two of symmetry e_g and three of symmetry t_{2g}. The latter are lower in energy than the former by an amount $10Dq$, or Δ, where the change of nomenclature symbolizes that the energy splitting may be of different origin. On a pure crystal field model the energy is associated with the electrostatic repulsion between electrons on the central ion and the negatively charged ligand ions; the $e_g(d\gamma)$ orbitals have higher energy because their wave-functions have maxima along the axes and approach

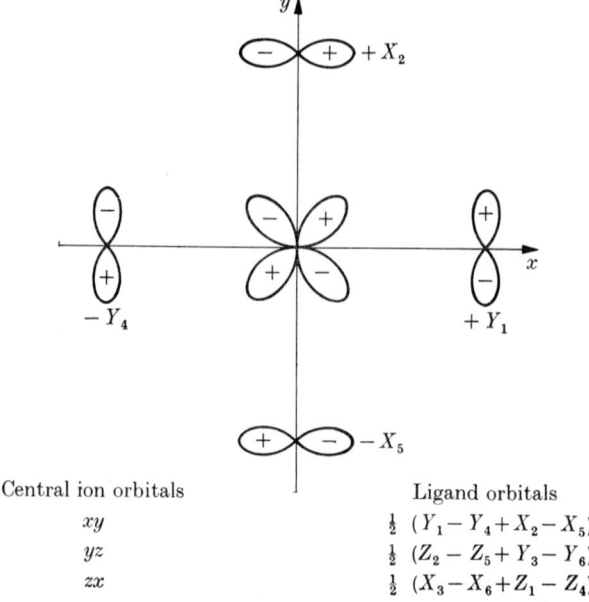

Fig. 7.18. Schematic representation of the central ion and ligand ion orbitals of symmetry t_{2g} which are involved in π-bonding. Only one set is shown; the other two are similar and can be derived by permuting the axes.

the ligand ions more closely than do the $t_{2g}(d\epsilon)$ states. However attempts to calculate the cubic field splitting from first principles on this basis have failed badly, the results not always being even of the correct sign. It is clear that overlap and bonding make a considerable contribution to the energy splitting, but although much interest attaches to this theoretical problem, the value of Δ must generally be regarded as a parameter to be determined from experimental results (cf. Owen and Thornley 1966).

In building up the states for a configuration d^n, electrons can be placed in the augmented states of symmetry e_g and t_{2g} in the same way as outlined in the last section for the strong crystal field states to which the augmented states reduce in the limit where they become pure central ion states. The resulting many-electron states are labelled by their spin multiplicity, the symmetry (A_2, T_1, T_2, etc.), and the single electron states (t_2^3, $t_2^2 e$, etc.) of which they are composed. States of the same spin and symmetry may be admixed by the Coulomb interaction, though we may expect the latter to be modified in magnitude for the augmented orbitals. Tanabe and Sugano (1954) assumed that the Racah parameter C was the same as in the free ion; and that the

Racah parameter B was similarly unaltered for the dipositive ions, but considerably reduced ($\sim$20 per cent) for the tripositive ions. Attempts have been made to derive values by fitting the optical absorption spectrum, though approximations such as $B = 4C$ are often made to reduce the labour of analysis (see, for example, Sugano and Peter 1961). Experimental evidence for a reduction of B, C in octahedral complexes is summarized in Table 7.5.

TABLE 7.5

Values of the Racah parameters B, C for the free ions and of the reduced values B', C' for the same ions in octahedral symmetry. Note that the values of B, C come from a different source from those in Tables 7.3, 7.4. This table is based on Table 5 of Owen and Thornley (1966), where references to the original papers are given

Ion	V^{2+}	Cr^{3+}	Mn^{2+}	Co^{2+}	Ni^{2+}
d^n	d^3	d^3	d^5	d^7	d^8
B (cm^{-1})	766	830	960	1115	1084
C (cm^{-1})	2855	3430	3325	4366	4831
Host crystal	MgO	MgO	MgO	MgO	KMgF$_3$
B' (cm^{-1})	550	688	786	845	955
C' (cm^{-1})	2475	3095	3210	3803	4234
B'/B	0·72	0·83	0·82	0·76	0·88
C'/C	0·87	0·90	0·97	0·87	0·88

The use of augmented orbitals introduces changes in the matrix elements of some other operators. The matrix elements of angular momentum for the central ion are multiplied by a factor that is written $k_{\pi\pi}$ for matrix elements between t_2 electrons and $k_{\pi\sigma}$ between t_2 and e-electrons; there are no matrix elements between e-electrons so that the factor $k_{\sigma\sigma}$ that we might have expected does not arise. For convenience we repeat the expressions for $k_{\pi\pi}$ and $k_{\pi\sigma}$ given in eqns (20.31) and (20.40).

$$k_{\pi\pi} = 1 - \tfrac{1}{2}\lambda_t^2 N_t^{-1},$$
$$k_{\pi\sigma} = 1 - \tfrac{1}{2}\left(\lambda_t^2 + \lambda_{\sigma p}^2 + \lambda_{\sigma s}^2 + \lambda_t\lambda_{\sigma p} + \lambda_t\lambda_{\sigma s}a\langle p_y|\frac{\partial}{\partial y}|s\rangle\right). \quad (7.7)$$

The matrix elements of the spin-orbit coupling are similarly multiplied by factors (see eqns 20.43–20.44), but there is some uncertainty as to the causes of the changes (see the discussion by Marshall and Stuart (1961)), and it is often simplest to regard the spin-orbit parameter as reduced in the solid compared with the free ion but by an amount to be found by comparison with experimental results. Different values may be expected for the effective spin-orbit parameter when only t_2 electrons are involved from that where matrix elements between t_2 and e-electrons arise, but there are seldom sufficient data to determine both.

The most direct manifestation of the fact that the magnetic electrons must be described by augmented wave-functions containing ligand orbitals comes from the observation of hyperfine structure due to the nuclear moments of the ligand ions. The magnetic hyperfine structure of each ligand ion is symmetrical about the bond axis, and may be represented by the spin Hamiltonian (20.47)

$$\mathscr{H}_L = A_\parallel^L S_z I_z^L + A_\perp^L (S_x I_x^L + S_y I_y^L) \tag{7.8}$$

for ligands $L = 3, 6$ which lie along the z-axis, with corresponding expressions for the other ligands with suitable permutations of the axes. Here S is the effective spin used in the spin Hamiltonian of the central ion. There are a number of contributions to the magnetic hyperfine structure, the most obvious being the simple dipolar interaction between the magnetic moment of the ligand nucleus and the electronic magnetic moment of the central ion. In a first approximation the latter may be regarded as localized at the centre, so that the interaction is that between two point dipoles, for which the contribution to A^L is $2A_d$, while that to $A_\perp^L$ is $-A_d$, where

$$A_d = g\beta\hbar\gamma_n R^{-3} = g\beta g_n \beta_n R^{-3} \tag{7.9}$$

where R is the distance between the centres of the central ion and the ligand ion. The non-spherical distribution of the magnetic dipole density on the central ion means that some corrections need to be applied to this formula (see Chapter 20) but these amount usually to no more than a few per cent and are normally neglected. A typical value of A_d is that for an impurity ion in $KMgF_3$, which is 3.1×10^{-4}cm^{-1} for $g = 2$. This is the one contribution to the ligand hyperfine structure which can be accurately calculated, and is present whether the magnetic electrons are in augmented orbitals or in purely central ion orbitals.

That part of the ligand hyperfine structure which arises because the magnetic electrons are in augmented orbitals containing ligand orbitals is considerably more complicated; its calculation is outlined in Chapter 20, and we shall give here only a summary. It is usual to analyze the ligand hyperfine structure into components as follows:

$$\begin{aligned} A_\parallel^L &= (A_s)_\parallel + 2A_d + (A_p)_\parallel + (\delta A_s)_\parallel + (\delta A_p)_\parallel \\ A_\perp^L &= (A_s)_\perp - A_d + (A_p)_\perp + (\delta A_s)_\perp + (\delta A_p)_\perp. \end{aligned} \tag{7.10}$$

Here the main components (apart from A_d) are those which would arise from electrons in s and p functions on the ligand ions, assuming the ground state to be one formed of augmented orbitals without any admixtures of excited states through the spin-orbit coupling; the correction terms δA arise through such admixtures, and are normally small. For ions such as d^3, d^5, d^8 where the ground term is an orbital singlet, $(A_s)_\| = (A_s)_\perp$ and $(A_p)_\| = -2(A_p)_\perp = 2A_p$. If the correction terms are neglected (or estimated), values of A_s and A_p can be found from the experimental measurements of $A_\|^L$ and $A_\perp^L$, using the calculated value of A_d. The results can then be related to the 'spin densities' defined by eqn (20.51):

$$f_{\sigma s} = \tfrac{1}{3}\lambda_{\sigma s}^2 N_\sigma^{-1}; \qquad f_{\sigma p} = f\tfrac{1}{3}\lambda_{\sigma p}^2 N_\sigma^{-1}; \qquad f_t = \tfrac{1}{4}\lambda_t^2 N_t^{-1}, \qquad (7.11)$$

using the hyperfine parameters

$$a_s = \frac{16\pi}{3}\beta\hbar\gamma_\mathrm{n}\,|\psi_s(0)|^2 = \frac{16\pi}{3}\beta\beta_\mathrm{n} g_\mathrm{n}\,|\psi_s(0)|^2, \qquad (7.12)$$
$$a_p = \tfrac{4}{5}\beta\hbar\gamma_\mathrm{n}\langle r^{-3}\rangle_p = \tfrac{4}{5}\beta\beta_\mathrm{n} g_\mathrm{n}\langle r^{-3}\rangle_p.$$

Since we shall be dealing mainly with fluorine ions, we give here the hyperfine constants calculated by Shulman and Sugano (1963) for the $2s$ and $2p$ states of $^{19}\mathrm{F}^-(I = \tfrac{1}{2})$:

$$a_{2s} = 1\cdot503\text{ cm}^{-1} = 34\cdot96\text{ GHz},$$
$$a_{2p} = 0\cdot0439\text{ cm}^{-1} = 1\cdot316\text{ GHz}.$$

Then for the ions d^3, d^5, d^8 with orbital singlet ground states we have

$$A_s = (f_s/2S)a_s; \qquad A_\sigma = (f_\sigma/2S)a_p; \qquad A_\pi = (A_t/2S)a_p, \qquad (7.13)$$

where S is the true spin of the central ion ground state, and the values of A_p are respectively:

$$d^3,\, A_p = -A_\pi; \qquad d^5,\, A_p = A_\sigma - A_\pi; \qquad d^8,\, A_p = A_\sigma. \qquad (7.14)$$

Here the negative sign that appears with A_π reflects the fact that the magnetic field produced at the nucleus by an electron is oppositely directed for an electron in a p_π orbital to that from an electron in a p_σ orbital. For d^3 only A_π appears because the ground state electrons are all in the π-bonding t_{2g} orbitals; conversely for d^8 only A_σ appears because the two holes are in the σ-bonding e_g states; while for d^5 both A_σ and A_π are present because every orbital contains one electron.

For ions in which the ground state is not an orbital singlet the calculations are considerably more complicated, and an example (for the ion d^7) is given in Chapter 20. An important difference is that $(A_p)_\parallel$ is not just equal to $-2(A_p)_\perp$, but each contains f_σ and f_t in different proportions so that in principle it is possible to determine them separately if the relation between $(A_s)_\parallel$ and $(A_s)_\perp$ is known.

7.5. The electronic spin Hamiltonian

In the previous sections we have considered the interaction of the magnetic ions with the ligand ions. This is primarily an interaction involving the orbital momenta of the magnetic electrons, and the electron spin entered the discussion only through the need to satisfy the over-riding requirements of the exclusion principle. We now introduce the interactions involving the electron spin and carry the theory a stage further, primarily in order to obtain an electronic 'spin Hamiltonian' that describes the behaviour of the ground manifold of electron states within which paramagnetic resonance is observed. In doing so we revert to the 'intermediate ligand field' approach, assuming LS- coupling to be valid; results for compounds where a 'strong ligand field' approach is required are considered in Chapter 8. General formulae can be deduced readily only for ions where the cubic ligand field results in the ground orbital state being a singlet, and the effective spin $\tilde{S}$ in the spin Hamiltonian is then generally the same as the true spin S, making a distinction between $\tilde{S}$ and S unnecessary. When the orbital ground state is not a singlet, the situation is more complex, and is outlined in later sections where the experimental results are considered ion by ion.

The interactions that we now introduce are as follows:

(a) the spin-orbit interaction

$$\mathcal{H}_{so} = \sum_i \zeta_i(\mathbf{l}_i \cdot \mathbf{s}_i) = \lambda(\mathbf{L} \cdot \mathbf{S}) \qquad (7.15)$$

where $\lambda = \pm(\zeta/2S)$ from eqn (7.3), the $+$ sign being required for a shell that is less than half-filled, and the $-$ sign for a shell that is more than half-filled;

(b) the spin-spin interaction (see § 16.5 or eqns 1.104, 1.105)

$$\mathcal{H}_{ss} = -\rho\{(\mathbf{L} \cdot \mathbf{S})^2 + \tfrac{1}{2}(\mathbf{L} \cdot \mathbf{S}) - \tfrac{1}{3}L(L+1)S(S+1)\} \qquad (7.16)$$

in the form appropriate to LS- coupling;

(c) the electronic Zeeman interaction

$$\mathcal{H}_z = \beta \sum_i \mathbf{H} \cdot (\mathbf{l}_i + g_s \mathbf{s}_i) = \beta \mathbf{H} \cdot (\mathbf{L} + g_s \mathbf{S}). \qquad (7.17)$$

These interactions are all small compared with the cubic ligand field

TABLE 7.6

Free ion values of the radial averages $\langle r^{-3}\rangle$, $\langle r^2\rangle$, $\langle r^4\rangle$, and of the spin-orbit parameter λ and the spin-spin parameter ρ, for ions of the 3d group. The radial averages $\langle r^n\rangle$ are calculated, from tables given by Burns (1962), Freeman and Watson (1965), Tucker (1966). The calculated values of λ are based on Blume and Watson (1963) and Dunn (1961). The experimental values of λ are obtained by setting the overall multiplet splitting in the free ion equal to $\lambda S(2L+1)$, without further correction. The calculated values of ρ are from Watson and Blume (1965)

		Ion	$\langle r^{-3}\rangle$ (a.u.)	$\langle r^2\rangle$ (a.u.)	$\langle r^4\rangle$ (a.u.)	λ (calc) (cm^{-1})	λ (exp) (cm^{-1})	ρ (cm^{-1})
$3d^1$	^2D	Sc^{2+}				86	79	
		Ti^{3+}	2·552	1·893	7·071	159	154	
		V^{4+}	3·684	1·377	3·593	255	248	
$3d^2$	^3F	Sc$^+$					35	
		Ti^{2+}	2·133	2·447	13·17	61	60	0·16
		V^{3+}	3·217	1·643	5·447	106	104	0·26
		Cr^{4+}	4·484	1·227	2·906	163	164	
$3d^3$	^4F	Ti$^+$	1·706	3·508	31·62		29	
		V^{2+}	2·748	2·070	9·605	57	55	0·11
		Cr^{3+}	3·959	1·447	4·297	91	91	0·17
		Mn^{4+}	5·361	1·104	2·389	135	134	
$3d^4$	^5D	V$^+$	2·289	2·819	20·71		34	
		Cr^{2+}	3·451	1·781	7·211	59	58	0·12
		Mn^{3+}	4·790	1·286	3·446	87	88	0·18
		Fe^{4+}	6·332	1·000	1·986	125	129	0·25
$3d^5$	^6S	Cr$^+$	2·968	2·319	14·14			
		Mn^{2+}	4·250	1·548	5·513			
		Fe^{3+}	5·724	1·150	2·789			
		Co^{4+}	7·421	0·9080	1·659			
$3d^6$	^5D	Mn$^+$	3·683	2·026	10·87	−64	−64	
		Fe^{2+}	5·081	1·393	4·496	−114	−103	0·18
		Co^{3+}	6·699	1·049	2·342	(−145)		
		Ni^{4+}	8·552	0·8371	1·423	(−197)		
$3d^7$	^4F	Fe$^+$		1·774	8·385	−115	−119	
		Co^{2+}	6·035	1·251	3·655	−189	−178	0·24
		Ni^{3+}	7·790	0·9582	1·971	(−272)		
		Cu^{4+}	0·814	0·7710	1·221	(320)		
$3d^8$	^3F	Co$^+$	5·388	1·576	6·637	−228	−228	
		Ni^{2+}	7·094	1·130	3·003	−343	−324	0·53
		Cu^{3+}	9·018	0·8763	1·662	(−438)		
$3d^9$	^2D	Ni$^+$		1·401	5·264	−605		
		Cu^{2+}	8·252	1·028	2·498	−830	−830	

splitting, which is of order 10^4 cm^{-1}. Values of ζ are given for the free ions in Tables 7.3 and 7.4, and lie in the range 10^2 to 10^3 cm^{-1}. Values of λ and other parameters for the 3d group are listed in Table 7.6. The spin-spin interaction is roughly of order 1 cm^{-1} in magnitude; some values of the constant ρ are listed in Table 7.6 for a number of ions (note that higher-order effects of spin–orbit coupling may produce contributions comparable to ρ; see Watson and Blume (1965)). The electronic Zeeman interaction is of order 1 cm^{-1} in a field of 10kG.

As indicated in § 7.4, we cannot expect the effective values of these interaction parameters to be the same for a magnetic ion in a complex as for the free ion. Experimental evidence for a reduction in the effective value λ' of the spin-orbit parameter is given in Table 7.7,

TABLE 7.7

Values of the spin-orbit coupling parameter (a) λ for free ion (from Landé interval rule) (b) λ' for ion in octahedral complex, found from paramagnetic resonance g value using the relation $g = g_s - 8\lambda'/\Delta$ and experimental value of cubic field splitting parameter Δ. The data are from Table 4 of Owen and Thornley (1966), where references to the sources are given

Ion	Ligand	λ (free ion) (cm^{-1})	λ' (complex) (cm^{-1})	λ'/λ
Cr^{3+}, $3d^3$	H$_2$O in Cr alum	91	57	0·63
Cr^{3+}, $3d^3$	O^{2-} in MgO	91	46	0·51
V^{2+}, $3d^3$	O^{2-} in MgO	55·5	36	0·65
Ni^{2+}, $3d^8$	H$_2$O in tutton salt	−324	−270	0·83
Ni^{2+}, $3d^8$	F$^-$ in KMgF$_3$	−324	−258	0·79
Ni^{2+}, $3d^8$	O^{2-} in MgO	−324	−242	0·75

taken from Owen and Thornley (1966) where an extensive discussion is given. The effective value of λ' depends on whether we are considering matrix elements between σ-orbitals, π-orbitals or between σ- and π-orbitals. In addition λ' may be anisotropic in a complex whose symmetry is less than cubic. Bonding will also affect the spin-spin interaction by changing the distribution of electronic magnetization, but this interaction is significant in so few cases that we shall not pursue this point. Much more important is the modification of the effective orbital moment through bonding to the ligands, which in octahedral (or cubic) symmetry we can introduce by replacing (7.17) by

$$\mathcal{H}_Z = \beta \mathbf{H} \cdot (k\mathbf{L} + g_s \mathbf{S}) \tag{7.18}$$

where $k \leqslant 1$. The value of k again depends on whether we are dealing with σ- or π-orbitals, and it may be anisotropic in complexes of less than cubic symmetry.

A further complication in interpreting experimental resonance

Application to singlet orbital ground state

We outline now the results of including these interactions in considering the case of a singlet orbital ground state, a full discussion being given in Chapter 19.

For such an orbital singlet state all diagonal components of the orbital angular momentum are zero, a result that we write in the form (see § 11.7)

$$\langle 0| L_p |0\rangle = 0 \qquad (7.19)$$

where $|0\rangle$ represents the singlet orbital state, and $p = x, y, z$. However for interactions such as the spin-spin interaction, which can be written in the form (see eqn (19.6))

$$\mathscr{H}_{SS} = -\rho \sum_{p,q} \{\tfrac{1}{2}(L_p L_q + L_q L_p) - \tfrac{1}{3}L(L+1)\delta_{pq}\} S_p S_q, \qquad (7.20)$$

where $p, q = x, y, z$, we meet the quadratic combinations of the orbital components whose values can be written (see eqn 19.15)

$$\tfrac{1}{2}\langle 0| (L_p L_q + L_q L_p) |0\rangle - \tfrac{1}{3}L(L+1)\delta_{pq} = l_{pq}. \qquad (7.21)$$

This completes the diagonal terms involving the orbital momentum, so that in first order we have just $g_s\beta(\mathbf{H}.\mathbf{S}) = g_s\beta H_p S_p$ for the Zeeman interaction and $-\rho l_{pq} S_p S_q$ for the spin-spin interaction. This means that in this approximation we have 'spin-only' g-values, but when $S \geqslant 1$ the levels may be split in zero magnetic field through spin-spin interaction. The latter arises only when the distribution of spin magnetization (determined by the geometrical 'shape' of the square of the orbital wave-function) is non-cubic, so that the self-energy of the magnetization depends on the orientation of the spin with respect to its anisotropic distribution in space.

We now consider second-order effects arising from matrix elements of the orbital angular momentum between the ground orbital and excited orbital states. Their effect on the manifold of spin states in the ground orbital level can be written

$$-\lambda\{\lambda\Lambda_{pq} + \rho(\Lambda'_{pq} + \Lambda'_{qp})\}S_p S_q, \qquad (7.22)$$

$$-2\lambda\beta\Lambda_{pq}H_p S_q, \qquad (7.23)$$

where the quantities

$$\Lambda_{pq} = \Lambda_{qp} = \sum_{n\neq 0} \frac{\langle 0| L_p |n\rangle\langle n| L_q |0\rangle}{W_n - W_0} \qquad (7.24)$$

and

$$\Lambda'_{pq} = -\tfrac{1}{2}i \sum_{t,r} \epsilon_{ptr} \sum_{n\neq 0} \frac{\langle 0| L_r |n\rangle\langle n| L_q L_t + L_t L_q |0\rangle}{W_n - W_0} \quad (7.25)$$

are tensor quantities formed from the matrix elements of L connecting the ground orbital state $|0\rangle$ of energy W_0 with excited orbital states $|n\rangle$ of energy W_n. From (7.24) we see that Λ_{pq} is a symmetric tensor; it can be shown (see § 19.2) that Λ'_{pq} is also symmetric provided that the symmetry of the ligand complex is no lower than rhombic. The first term in (7.22) in λ^2 comes from using the off-diagonal terms of the spin-orbit coupling twice, and the second from the cross term between the spin-orbit coupling and the spin-spin interaction. Essentially they arise from a second-order change in shape of the magnetization in the ground state, giving rise to changes in the self energy of the magnetization dependent on the orientation of the magnetization. Equation (7.23) comes from cross terms between the spin-orbit coupling and the Zeeman interaction, and represents the fact that the orbital momentum has a second-order contribution which is not exactly zero.

Collecting both the first- and second-order contributions we have altogether for the effective spin Hamiltonian

$$\mathcal{H} = \{-\lambda^2 \Lambda_{pq} - \rho l_{pq} - \rho\lambda(\Lambda'_{pq} + \Lambda'_{qp})\} S_p S_q \quad (7.26)$$

$$+ \beta\{\delta_{pq} g_s - \lambda \Lambda_{pq}\} H_p S_q, \quad (7.27)$$

where the cross term in $\rho\lambda$ in (7.26) can generally be neglected. (It should be noted—see § 7.10—that third-order contributions to the effective spin-spin interaction cannot always be neglected.) In the notation of Chapter 3 we can write the effective spin Hamiltonian as

$$\mathcal{H} = (\mathbf{S} \cdot \mathbf{D} \cdot \mathbf{S}) + \beta(\mathbf{H} \cdot \mathbf{g} \cdot \mathbf{S}), \quad (7.28)$$

where the first term gives rise to 'fine structure' splittings and the second to g-factors that differ from the free-spin value and may be anisotropic. In calculations based on (7.26), (7.27) we must remember that the quantities λ, Λ_{pq}, Λ'_{pq} may have to be modified in the presence of bonding effects.

Doublet orbital ground state

From Figs. 7.3, 7.4 we see that for an ion in a D-state the orbital levels in a cubic field split into a triplet and a doublet, the latter being the ground state in a number of cases. The doublet is peculiar in that there are no matrix elements of the orbital angular momentum between the two components of the doublet, and the diagonal elements are also zero. In a ligand field of tetragonal, rhombic, or lower symmetry the doublet splits into two singlets whose separation is usually large compared with kT at ordinary temperatures. An effective spin Hamiltonian can then be derived for the lower orbital singlet as in the previous section, the energy denominators in (7.24), (7.25) being determined primarily by the cubic field splitting.

In a cubic or trigonal field the two components of the orbital doublet do not split. However a distortion of lower symmetry will split them, giving one state of lower energy, and according to the theorem of Jahn and Teller (1937) such a distortion may be expected to occur spontaneously. This gives a singlet ground orbital state, as above, but when there are a number of distortions that give the same energy change, the system may resonate between them, giving interesting dynamic Jahn–Teller effects. These are discussed in Chapter 21.

Triplet orbital ground state

If the cubic ligand field has the right sign, an orbital triplet may be the ground state, either Γ_5 for an ion in a D-state or Γ_4 for an ion in an F-state. If the ligand field has an axial distortion (tetragonal or trigonal) the triplet splits into a doublet and a singlet, as in Figs. 7.5, 7.6; with lower symmetry the degeneracy is lifted completely. Obviously Jahn–Teller effects may drastically change the picture, as discussed in §§ 21.9 to 21.12. They may produce a static distortion; but in some cases the spin-orbit coupling produces splittings large enough to stabilize the system against such distortions, though a partial quenching of the orbital momentum may still be present (with an apparent reduction in the spin-orbit coupling parameter, and also in any trigonal distortion).

In cubic symmetry, if we represent the orbital triplet by a fictitious angular momentum $\tilde{l} = 1$, then under the action of spin-orbit coupling the $(2\tilde{l}+1)(2S+1) = 3(2S+1)$ manifold of states splits into a series of levels characterized by $\tilde{J} = S+1, S, |S-1|$ as shown in Figs. 7.19, 7.20. The energy separations are just those that would be produced by an effective spin-orbit parameter which we write as $\tilde{\lambda}$, i.e. by

$$\mathscr{H}'_{SO} = \tilde{\lambda}(\tilde{\mathbf{l}} \cdot \mathbf{S}), \tag{7.29}$$

so that they obey the interval rule

$$W(\tilde{J}) = \tfrac{1}{2}\tilde{\lambda}\{\tilde{J}(\tilde{J}+1) - \tilde{l}(\tilde{l}+1) - S(S+1)\}. \tag{7.30}$$

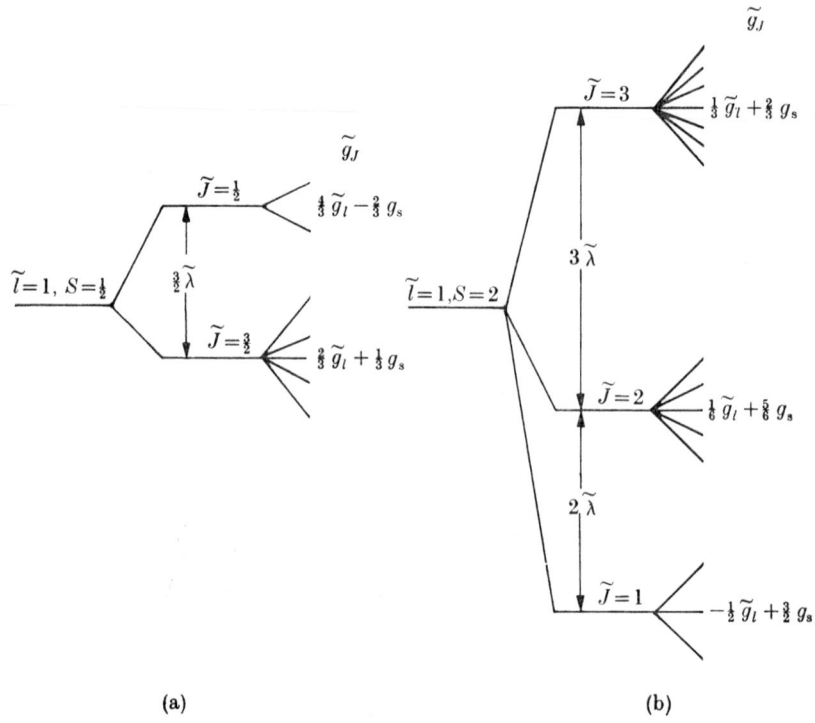

Fig. 7.19. Triplet orbital state for a D-ion in a cubic field, showing splitting of levels for (a) $\tilde{l}=1$, $S=\tfrac{1}{2}$, (b) $\tilde{l}=1$, $S=2$ under spin-orbit coupling (eqn (7.29)) and first-order Zeeman effect (eqn (7.31)). In (a), $\tilde{\lambda}$ is assumed negative (appropriate to d^1 in octahedral field); in (b), $\tilde{\lambda}$ is assumed positive (appropriate to d^6 in octahedral field). With the levels inverted, the diagrams are appropriate to (a) d^9 in tetrahedral symmetry, (b) d^4 in tetrahedral symmetry.

The levels have first-order Zeeman splittings of the form

$$\mathcal{H}'_Z = \tilde{g}_J \beta (\mathbf{H}.\tilde{\mathbf{J}}) \qquad (7.31)$$

in which, writing $\tilde{g}_l$ for the effective orbital g-factor,

$$\tilde{g}_J = \tfrac{1}{2}(\tilde{g}_l + g_s) + \frac{\tilde{l}(\tilde{l}+1) - S(S+1)}{2\tilde{J}(\tilde{J}+1)}(\tilde{g}_l - g_s) \qquad (7.32)$$

where, of course, $\tilde{l} = 1$ in both (7.30) and (7.32).

The fictitious angular momentum $\tilde{\mathbf{J}}$ has been introduced here in a manner that demonstrates its resemblance to the real total angular momentum $\mathbf{J} = \mathbf{L} + \mathbf{S}$ used in ordinary atomic theory. It is an example of the effective spin $\tilde{S}$ used in the effective spin Hamiltonian, though obviously it is not the same as the true spin S and the g_J-factor given

IN INTERMEDIATE LIGAND FIELDS

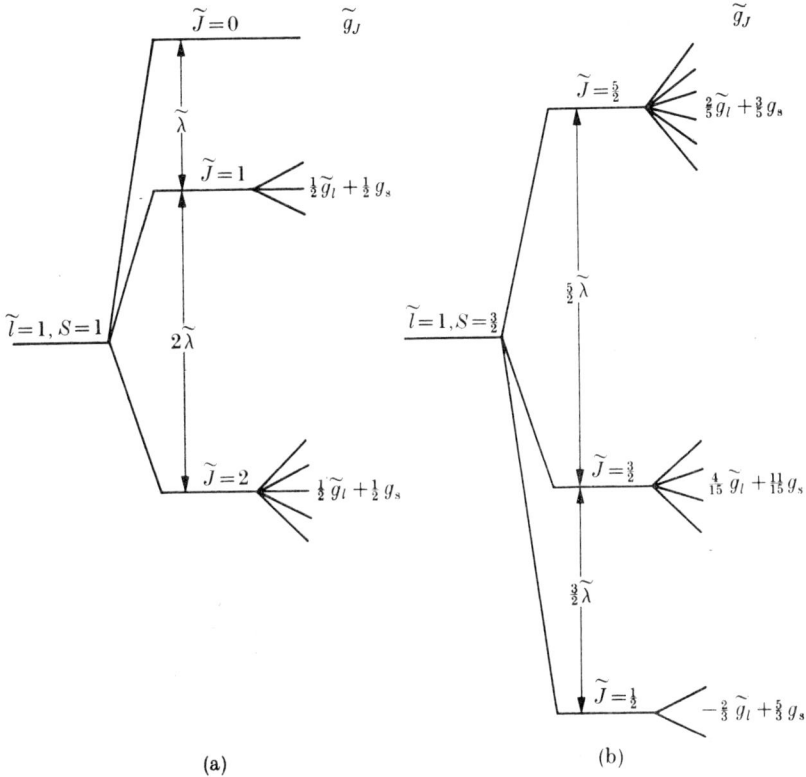

FIG. 7.20. Triplet orbital state for F-ion in cubic field, showing splitting of levels for (a) $\tilde{l} = 1, S = 1$; (b) $\tilde{l} = 1, S = \frac{3}{2}$ under spin-orbit coupling (eqn (7.29)) and first-order Zeeman effect (eqn (7.31)). In (a), $\tilde{\lambda}$ is assumed negative (appropriate to d^2 in octahedral field); in (b), $\tilde{\lambda}$ is assumed positive (appropriate to d^7 in octahedral field). With the levels inverted, the diagrams are appropriate to (a) d^8 in tetrahedral symmetry, (b) d^3 in tetrahedral symmetry.

by (7.32) is nothing like the free spin value except in accidental cases.

The simple picture given above must be modified to allow for a number of effects.

(i) As for orbital singlet states, second-order effects involving coupling through the spin-orbit interaction to excited orbital states change the effective value of $\tilde{g}_J$ (though it remains isotropic); together with the spin-spin interaction, they also give rise to small splittings of states with $\tilde{J} \geqslant 2$ in a manner consistent with group theoretical requirements for cubic symmetry.

(ii) In the presence of small departures from cubic symmetry, the $\tilde{g}_J$-factors become anisotropic, and there will be splittings of the manifolds for which $\tilde{J} \geqslant 1$, but not for $\tilde{J} = \frac{1}{2}$ because of the requirements of Kramers degeneracy. If the ligand field interaction of less than cubic symmetry is small compared with the spin-orbit splitting, its effect can be expressed by introducing spin operators $\sum_{k,q} O_k^q$ operating within each manifold of given $\tilde{J}$. For $\tilde{J} = 1$ or $\frac{3}{2}$, only operators with $k = 2$ are needed, but for larger values of $\tilde{J}$, operators with $k = 4$ will also be required.

(iii) If the departures from cubic symmetry give ligand field interactions comparable in energy with $\tilde{\lambda}$, the resultant energy matrices including both them and the spin-orbit coupling must be diagonalized, following which the first-order Zeeman effect can be calculated. Again modifications will result from matrix elements of the spin-orbit coupling with excited states, and from spin-spin interaction.

(iv) If the non-cubic ligand field interaction is much greater than the spin-orbit coupling, the splitting of the orbital levels is computed first as in Figs. 7.3–7.6, after which the other interactions (spin-orbit coupling, spin-spin and Zeeman interactions) are introduced.

(v) A static model may not be applicable because of the Jahn–Teller effect.

It is clear that no useful general formulae can be given at this stage, and the results for individual ions will be presented in §§ 7.8–7.16, where the effects of bonding are also discussed. An exception may arise for an ion with an odd number of electrons, where a ground state of twofold degeneracy may result (a 'Kramers doublet'), which can be represented by a spin Hamiltonian with $\tilde{S} = \frac{1}{2}$

$$\mathscr{H} = \beta(\mathbf{H} \cdot \tilde{\mathbf{g}} \cdot \tilde{\mathbf{S}}), \tag{7.33}$$

where along a principal axis

$$g_p = 2\langle \ |(\tilde{g}_l)_p \tilde{l}_p + g_s S_p|\ \rangle \tag{7.33a}$$

evaluated within the two states of the doublet. Here the notation $(\tilde{g}_l)_p$ indicates that the effective orbital g-factor $\tilde{g}_l$ may itself be anisotropic.

7.6. Magnetic hyperfine interaction

The magnetic hyperfine interaction is often thought of as an interaction between the nuclear magnetic moment $\boldsymbol{\mu}_I$ and a magnetic field $\mathbf{H}_e$ at the nucleus generated by the magnetic electrons; it can equally well be regarded as an interaction between the distributed electronic magnetic moment $\boldsymbol{\mu}_e$ and the average magnetic field $\mathbf{H}_n$ generated by the nuclear dipole moment. We have thus

$$\mathcal{H}_n^{mag} = -(\boldsymbol{\mu}_I \cdot \mathbf{H}_e) = -(\boldsymbol{\mu}_e \cdot \mathbf{H}_n), \tag{7.34a}$$

where the definitions of $\mathbf{H}_e$, $\mathbf{H}_n$ are rather more general than in eqns (1.28), (1.30) in that precessing components are included as well as steady components. For a free ion we can write $\boldsymbol{\mu}_I = g_n \beta_n \mathbf{I}$, and $\mathbf{H}_e = -(A/g_n \beta_n)\mathbf{J}$, so that (7.34a) becomes just

$$\mathcal{H}_n^{mag} = A(\mathbf{J} \cdot \mathbf{I}). \tag{7.34b}$$

For positive values of (A/g_n) we see that $\mathbf{H}_e$ is oppositely directed to $\mathbf{J}$; because of the negative sign of the electronic charge, this means that $\mathbf{H}_e$ is in the same direction as the electronic magnetization.

In the weak ligand field approximation which is appropriate to $4f$, $5f$ ions the magnetic hyperfine interaction was introduced in the form (7.34b), giving rather simple relations (see § 5.4) between the free ion value of A and the principal values of the tensor $\mathbf{A}$ in the interaction ($\tilde{\mathbf{S}} \cdot \mathbf{A} \cdot \mathbf{I}$) of the effective spin Hamiltonian. When the ligand field is intermediate or strong a rather different approach is necessary, in which we consider separately the various contributions to the magnetic hyperfine interaction. These arise partly from the orbital motion of the electrons, and partly from their spin magnetization.

Neglecting any hyperfine anomaly, we regard the nucleus as having a point dipole moment, while the electronic charge and spin magnetization are distributed in space. For an element of electronic charge dq at distance $\mathbf{r}$ from the nucleus, moving with velocity $\mathbf{v}$, we can write down the magnetic field at the nucleus, giving on integration

$$\mathbf{H}_e^l = \int \frac{(\mathbf{r} \wedge \mathbf{v})}{cr^3} dq = \int \frac{\mathbf{G}}{r^3} \frac{dq}{mc}, \tag{7.35}$$

where $\mathbf{G}$ is the angular orbital momentum and m the electronic mass. If $\mathbf{G}_i = \mathbf{1}_i \hbar$ for the ith electron of charge $-e$, we can write

$$\int r_i^{-3} dq = -e \langle r_i^{-3} \rangle_i \tag{7.36}$$

and, on summing over all electrons, we find

$$\mathbf{H}_e^l = (e\hbar/mc) \sum_i (-\mathbf{1}_i)\langle r_i^{-3} \rangle_i$$
$$= 2\beta \sum_i (-\mathbf{1}_i)\langle r_i^{-3} \rangle_i, \tag{7.37}$$

where β is the Bohr magneton.

In LS- coupling $\sum_i \mathbf{1}_i = \mathbf{L}$, and if $\langle r_i^{-3} \rangle_i$ is the same for all electrons in the open shell, we have

$$\mathbf{H}_e^l = 2\beta \langle r_i^{-3}\rangle (-\mathbf{L}) \tag{7.38}$$

where the minus sign shows that the orbital contribution to $\mathbf{H}_e$ is always in the same direction as the orbital magnetic moment. If we write $\boldsymbol{\mu}_I = (\mu_I/I)\mathbf{I}$, which avoids the choice between expressing the nuclear moment in nuclear or Bohr magnetons, we have for the orbital contribution to the magnetic hyperfine Hamiltonian

$$\mathscr{H}_n^l = \sum_i 2\beta(\mu_I/I)\langle r_i^{-3}\rangle_i (\mathbf{l}_i \cdot \mathbf{I}) \tag{7.39}$$

$$= 2\beta(\mu_I/I)\langle r_i^{-3}\rangle (\mathbf{L} \cdot \mathbf{I}), \tag{7.40}$$

where the first form is needed when we have to consider single electrons (as in strong ligand fields) and the second is suitable for intermediate ligand fields if LS-coupling is applicable.

The contributions to $\mathbf{H}_e$ associated with the electron spin can be divided into two; one of these arises from the spin magnetization 'outside' the nucleus, and the other from the magnetization 'inside' the nucleus. The former gives a simple dipolar contribution

$$\mathbf{H}_e^{sC} = -\int \frac{\{\mathbf{M}_s - 3(\mathbf{M}_s \cdot \mathbf{r}_0)\mathbf{r}_0\}}{r^3} d\tau \tag{7.41}$$

where $\mathbf{M}_s$ is the spin magnetization density at a point $\mathbf{r}$, and $\mathbf{r}_0$ is a unit vector parallel to $\mathbf{r}$. We can write

$$\mathbf{M}_s = -g_s\beta \sum_i \mathbf{s}_i \rho_{si} \tag{7.42}$$

where $\mathbf{s}_i$ is the spin of the ith electron and ρ_{si} its spin density (i.e. the square of the amplitude of its wave-function at the point $\mathbf{r}$). Then

$$\mathbf{H}_e^{sC} = +g_s\beta \sum_i \{\mathbf{s}_i - 3(\mathbf{s}_i \cdot \mathbf{r}_0)\mathbf{r}_0\}\langle r_{sC}^{-3}\rangle_i, \tag{7.43}$$

where

$$\langle r_{sC}^{-3}\rangle_i = \int r_i^{-3}\rho_{si}\, d\tau. \tag{7.44}$$

Here $g_s = 2\cdot 0023$ is the anomalous g-factor of the electron spin, and the negative sign in (7.42) arises from the negative charge on the electron. Both here and in the orbital contribution there are small corrections due to diamagnetic effects and the relativistic increase of mass. It is generally convenient to absorb these, and also the difference between g_s and 2, into effective values of $\langle r_i^{-3}\rangle$ and of $\langle r_{sC}^{-3}\rangle$, so that the 'spin-dipolar' part of the hyperfine interaction becomes

$$\mathscr{H}_n^{sC} = -2\beta(\mu_I/I)\sum_i \{\mathbf{s}_i \cdot \mathbf{I} - 3(\mathbf{s}_i \cdot \mathbf{r}_0)(\mathbf{r}_0 \cdot \mathbf{I})\}\langle r_{sC}^{-3}\rangle_i. \tag{7.45}$$

This term is more complicated than the orbital interaction because it involves components of the unit vector $\mathbf{r}_0$. These can be replaced by components of the orbital angular momentum, just as in the equivalent operator method for representing the effect of the ligand field interaction. In doing so we must allow for the fact that components of $\mathbf{r}_0$ commute while those of $\mathbf{l}$ or $\mathbf{L}$ do not. Thus for a single electron whose orbital quantum number is l the operator

$$\{\mathbf{s} \cdot \mathbf{I} - 3(\mathbf{s} \cdot \mathbf{r}_0)(\mathbf{r}_0 \cdot \mathbf{I})\} \tag{7.46}$$

can be replaced by

$$\sqrt{(10)}(sC^2)^1 =$$
$$= -\frac{2}{(2l-1)(2l+3)}\{l(l+1)(\mathbf{I} \cdot \mathbf{s}) - \tfrac{3}{2}(\mathbf{l} \cdot \mathbf{I})(\mathbf{l} \cdot \mathbf{s}) - \tfrac{3}{2}(\mathbf{l} \cdot \mathbf{s})(\mathbf{l} \cdot \mathbf{I})\}, \tag{7.47}$$

and in LS-coupling the sum of the terms (7.46) for the several electrons in the partly-filled shell can be replaced by

$$\sqrt{(10)}(sC^2)^1_L = -\xi\{L(L+1)(\mathbf{I} \cdot \mathbf{S}) - \tfrac{3}{2}(\mathbf{L} \cdot \mathbf{I})(\mathbf{L} \cdot \mathbf{S}) - \tfrac{3}{2}(\mathbf{L} \cdot \mathbf{S})(\mathbf{L} \cdot \mathbf{I})\} \tag{7.48}$$

where

$$\xi = \frac{2l+1-4S}{S(2l-1)(2l+3)(2L-1)}. \tag{7.49}$$

The Hamiltonian for the 'spin-dipolar' part of the magnetic hyperfine interaction thus becomes either

$$\mathscr{H}_\mathrm{n}^{sC} = -2\beta(\mu_I/I) \sum_i \langle r_{sC}^{-3} \rangle_i \sqrt{(10)}(\mathbf{s}C^2)^1_i \tag{7.50}$$

in terms of single electrons, or in LS- coupling (assuming the values of $\langle r_{sC}^{-3} \rangle_i$ to be the same for all electrons in the open shell)

$$\mathscr{H}_\mathrm{n}^{sC} = -2\beta(\mu_I/I)\langle r_{sC}^{-3} \rangle \sqrt{(10)}(\mathbf{s}C^2)^1_L. \tag{7.51}$$

Here the notation $(sC^2)^1$ gives a useful shorthand for the operators in (7.47), (7.48).

An important case where this interaction vanishes is that where the electron spin density has spherical symmetry; classically it is easily shown that the magnetic field at the centre of a thin uniformly magnetized spherical shell is zero, since it involves the integral

$$\int_0^\pi (1 - 3\cos^2\theta) \sin\theta \, d\theta = 0, \tag{7.52}$$

where θ is the angle between $\mathbf{M}_s$ and $\mathbf{r}_0$ in eqn (7.41). It follows that the interaction (7.45) vanishes for a closed shell of electrons, and also, in LS-coupling, for a half-filled shell; in particular, it vanishes for a single s-electron, but in this case we must consider also the electron magnetization 'inside' the nucleus. Conceptually it is easiest at this point to regard the nuclear moment as due to an elementary current loop immersed in a magnetized medium arising from the spin magnetization density $\mathbf{M}_0$ at the nucleus. The latter gives rise to a magnetic field

$$\mathbf{B}_e = +\frac{8\pi}{3}\mathbf{M}_0 \qquad (7.53)$$

where for a single s-electron

$$\mathbf{M}_0 = -g_s\beta\mathbf{s}_i\rho_i^{(0)}, \qquad (7.54)$$

where $\rho_i^{(0)}$ is the spin density of the ith s-electron at the nucleus and is equal to the square of its wave-function at the origin. On summing over all s-electrons we obtain the Hamiltonian for this 'contact' interaction

$$\mathcal{H}_n^c = \frac{8\pi}{3}g_s\beta(\mu_I/I)\sum_i \rho_i^{(0)}(\mathbf{s}_i \cdot \mathbf{I}). \qquad (7.55)$$

In a many-electron atom the resultant hyperfine structure is the sum of the contributions from the individual electrons, but since we are summing vector quantities, the net effect depends on the balance between the positive and the negative contributions. For example, in a closed electron shell we have

$$\sum_i \mathbf{l}_i = \sum_i \mathbf{s}_i = 0 \qquad (7.56)$$

but it does not necessarily follow that the sums

$$\sum_i \mathbf{l}_i \langle r_l^{-3}\rangle_i = 0; \qquad \sum_i \rho_i^{(0)}\mathbf{s}_i = 0 \qquad (7.57)$$

if different electrons in the same shell have different values of $\langle r_l^{-3}\rangle_i$ or of $\rho_i^{(0)}$. If all electron shells are closed, the summations in eqns (7.39), (7.50), (7.57) are zero, so that there is no hyperfine structure. If the atom has an open shell in addition to a number of closed shells, then the electrostatic repulsion between the electrons ('configuration interaction') gives rise to energy differences for electrons in the same shell which result in changes in their radial wave-functions and hence in the individual values of $\langle r^{-3}\rangle_i$. For example, the electrostatic repulsion gives rise (through the exclusion principle) to an exchange interaction between electrons in the open shell and electrons in the closed shells. If the net spin **S** of the electrons in the open shell is 'up', then electrons in the closed shells whose individual spins are also 'up' will have a slightly different energy from those whose spins are 'down', and hence slightly different values of $\langle r^{-3} \rangle$.

The differences between $\langle r_l^{-3} \rangle$ and $\langle r_{sC}^{-3} \rangle$ which result from these effects cannot easily be calculated. Experimentally, differences of order 10 per cent between $\langle r_l^{-3} \rangle$ and $\langle r_{sC}^{-3} \rangle$ have been found by Harvey (1965) for the ground terms of the oxygen atom ($2p^4$, ^3P) and the fluorine atom ($2p^5$, ^2P), but these may be exceptionally large. For the samarium atom ($4f^6$, ^7F) the difference is only 2 per cent (Woodgate 1966). In the solid state there is no experimental evidence, and the possibility of a difference between $\langle r_l^{-3} \rangle$ and $\langle r_{sC}^{-3} \rangle$ is usually ignored (for a discussion, see Watson and Freeman (1967a)).

A much more dramatic effect is the appearance of a 'contact' term in cases where it would not be expected, such as open shells of p, d, f electrons which have no finite density at the nucleus. This is mainly due to the effect of the exchange interaction on the s-shells, giving different radial distributions of magnetization (and charge) for electrons with spin 'up' and with spin 'down', which results in a non-zero value for the second sum in (7.57). The effect is especially important for s-electrons because a small degree of unbalance can give a rather large contribution to the hyperfine structure, and because the contribution is different in form from the orbital and spin-dipolar terms. Since it arises from the exchange-induced magnetization of the closed shells or 'core electrons', the effect (see §§ 17.5, 17.6) is known as 'core polarization'. The most striking effect occurs for half-filled shells, such as Mn^{2+}, $3d^5$, ^6S, where zero hyperfine interaction would be expected because $L = 0$, the spin-dipolar interaction vanishes because of spherical symmetry, and no unpaired s-electrons are present. Nevertheless a hyperfine interaction of the form

$$\mathcal{H}_n = A_s(\mathbf{S} \cdot \mathbf{I}) \tag{7.58}$$

is observed whose size is not greatly different from that of other $3d^n$ ions.

As mentioned in § 4.7 and discussed in § 17.7, relativistic effects can also give a contribution of the form (7.58), and can be included in A_s. Since the quantity $\rho^{(0)}$ in eqn (7.55) has the dimensions r^{-3}, we can eliminate the factors $(8\pi/3)$ and $(g_s/2)$ in that equation by writing it in the form

$$\mathcal{H}_n^s = -2\beta(\mu_I/I)\langle r_s^{-3} \rangle(\mathbf{S} \cdot \mathbf{I}), \tag{7.59}$$

where we have used s rather than c as superscript and subscript to denote that all contributions of the form $(\mathbf{S}.\mathbf{I})$ are included. The quantity $\langle r_s^{-3} \rangle$ has no simple relation to the other parameters $\langle r_l^{-3} \rangle$, $\langle r_{sC}^{-3} \rangle$, and may be of either sign, though for $3d^n$ ions it is found experimentally to be always positive. With the notation

$$\mathscr{P} = 2\beta(\mu_I/I)\langle r^{-3} \rangle \tag{7.60}$$

and using the approximation that $\langle r_l^{-3} \rangle = \langle r_{sC}^{-3} \rangle$ we can write the full magnetic hyperfine interaction as

$$\mathscr{H}_n^{\mathrm{mag}} = \mathscr{P}\{(\mathbf{L}.\mathbf{I}) - \kappa(\mathbf{S}.\mathbf{I}) - \sqrt{(10)}(sC^2)_L^1\}, \tag{7.61}$$

where $\kappa = \langle r_s^{-3} \rangle / \langle r^{-3} \rangle$, and the third tensor operator is given by eqns (7.48) and (7.49).

Orbital singlet ground state

In applying these formulae to the case of a singlet orbital ground state resulting from the action of the ligand field we need just the quantities l_{pq}, Λ_{pq}, Λ'_{pq} already defined in eqns (7.21)–(7.25).

For the orbital contribution we have, from (7.19), no diagonal term, but from the cross term with the spin-orbit coupling we obtain

$$\mathscr{H}_n^l = -2\lambda\mathscr{P}\Lambda_{pq}S_pI_q = \mathscr{P}\Delta g_{pq}S_pI_q \tag{7.62}$$

where Δg_{pq} is the orbital contribution to g_{pq}. If the external field is along one of the principal axes of the ligand field interaction, such as the z-axis, (7.62) is equivalent to a component of the electronic magnetic field at the nucleus

$$(H_e^l)_z = -2\beta\langle r^{-3} \rangle \Delta g_z = -125\langle r^{-3} \rangle_{\mathrm{a.u.}}\Delta g_z \tag{7.63}$$

in kilogauss if $\langle r^{-3} \rangle_{\mathrm{a.u.}}$ is expressed in atomic units.

For the spin-dipolar contribution we obtain

$$\mathscr{H}_n^{sC} = 3\xi\mathscr{P}\{-l_{pq} + \lambda\Lambda'_{pq}\}S_pI_q, \tag{7.64}$$

where the first term l_{pq} needs a comment. If the cubic field itself gives a singlet orbital ground state, as for Cr^{3+}, $3d^3$ in an octahedral field, then all elements $l_{pq} = 0$, and the spin-dipolar field at the nucleus vanishes because the orbital wave-function and hence also the distribution of spin magnetization has cubic symmetry. The situation is different for an ion such as Cu^{2+}, $3d^9$ where an octahedral field leaves a ground orbital doublet that may be split by departures from exact octahedral symmetry. The two components of the orbital doublet, with wave-functions $(3z^2 - r^2)$, $(x^2 - y^2)$ are by no means cubic (see Fig. 7.17) in their distribution of charge and spin magnetization, which are concentrated along the z-axis in the former case and in the xy-plane in the latter. In first order the components of orbital momentum are zero, giving no orbital contribution to the hyperfine field, though there will be a second-order effect given by eqn (7.62). However, for either orbital there will be a first-order, anisotropic, spin-dipolar contribution, because of the anisotropic distribution of spin magnetization. For the orbital $(3z^2-r^2)$, $l_{zz} = -2$, $l_{xx} = l_{yy} = +1$, while the values for the orbital (x^2-y^2) are numerically the same but reversed in sign, thus giving the appropriate cancellation for cubic symmetry, when the two orbitals are degenerate and equally occupied, but not when they are split by a departure from cubic symmetry.

The term in $(\mathbf{S}.\mathbf{I})$, due almost entirely to core polarization, gives simply

$$\mathscr{H}_\mathrm{n}^s = -\mathscr{P}\kappa\delta_{pq}S_pI_q. \tag{7.65}$$

Often another parameter χ is used, such that

$$\kappa\langle r^{-3}\rangle = -\tfrac{2}{3}\chi \tag{7.66}$$

so that

$$\mathscr{H}_\mathrm{n}^s = \tfrac{4}{3}\beta(\mu_I/I)\chi\delta_{pq}S_pI_q. \tag{7.67}$$

The dimensions of χ are r^{-3}, and it is related to the electronic field at the nucleus due to core polarization by

$$H_\mathrm{e}^s = -\tfrac{4}{3}\beta\chi S = -83\cdot4(\chi S)\text{ kG} \tag{7.68}$$

when χ is in atomic units. Experimentally, it appeared for some time that χ was substantially constant (apart from covalent bonding effects) throughout the iron group at about -3 a.u., giving a core polarization field of 250 kG per unit of electron spin. Later measurements have shown that χ does increase with atomic number (by about 25 per cent between $3d^2$ and $3d^8$), in accordance with Hartree–Fock calculations (see Watson and Freeman 1967a). There are, nevertheless, some surprising consistencies for ions of a given configuration $3d^n$ (an isoelectronic sequence) in the same compound, and for individual ions in a given coordination but with varying ligand distance (see Geschwind (1967); §7.10 and Table 7.21.) In general, χ depends on the nature of the ligand, falling with increasing covalency (cf. Table 7.17), a result first noted experimentally by Van Wieringen (1955) for Mn^{2+} compounds, $3d^5$, where the hyperfine interaction is virtually entirely due to core polarization.

Orbital triplet ground state

As for the purely electronic interactions, it is difficult to give any general formulae (for a fairly detailed discussion, see Chapter 19). If the ground state is a Kramers doublet, represented by a spin Hamiltonian with effective spin $\tilde{S} = \frac{1}{2}$, the magnetic hyperfine interaction will be of the form

$$\mathscr{H}_\mathrm{n}^\mathrm{mag} = (\tilde{\mathbf{S}} \cdot \mathbf{A} \cdot \mathbf{I}) \tag{7.69}$$

where along a principal axis

$$A_p = 2\mathscr{P}\langle \ | -(\tilde{g}_l)_p \tilde{l}_p + \{\xi L(L+1) - \kappa\}S_p - \\ -\tfrac{3}{2}\xi\{L_p(\mathbf{L} \cdot \mathbf{S}) + (\mathbf{L} \cdot \mathbf{S})L_p\} | \ \rangle, \tag{7.70}$$

the matrix elements being evaluated within the two states of the doublet, as in eqn (7.33a).

In concluding this outline of the effects of magnetic hyperfine interaction we note that a term of the form (7.69) should express the interaction, not only for a Kramers doublet, but also (to a good approximation) for a singlet ground orbital state with a more than twofold spin degeneracy. For this the terms discussed earlier can be summarized in the form

$$A_{pq} = \mathscr{P}\{-2\lambda\Lambda_{pq} + 3\xi(-l_{pq} + \lambda\Lambda'_{pq}) - \kappa\delta_{pq}\}, \tag{7.71}$$

but for manifolds of fourfold or greater degeneracy ($S \geqslant \tfrac{3}{2}$) further terms may be required, as discussed in § 4.7 and Chapter 18. Finally, we mention that the values of $\langle r^{-3} \rangle$ are not necessarily the same as those for a free ion, being generally reduced by covalency in a similar manner to the reduction in the spin-orbit coupling.

7.7. Nuclear electric quadrupole and nuclear Zeeman interaction

For the intermediate ligand field approach where we work in terms of S,L the nuclear electric quadrupole interaction is conveniently written in the form

$$\mathscr{H}_\mathrm{n}^\mathrm{eq} = -\frac{e^2 Q}{I(2I-1)} \langle r_q^{-3} \rangle \langle L \| \alpha \| L \rangle \times \\ \times \tfrac{1}{2}\{3(\mathbf{L} \cdot \mathbf{I})^2 + \tfrac{3}{2}(\mathbf{L} \cdot \mathbf{I}) - L(L+1)I(I+1)\}. \tag{7.72}$$

This operator is similar in form to that for the electronic spin-spin interaction, eqn (7.16), except that $\mathbf{S}$ is replaced by $\mathbf{I}$. In practice the nuclear electric quadrupole interaction is so small in magnitude ($\sim 10^{-3}$ cm^{-1} or less) that we need only retain the diagonal terms, which give therefore (cf. eqns (7.20), (7.21)) for an orbital singlet state

$$\mathscr{H}_n^{eq} = -\frac{3e^2Q}{2I(2I-1)}\langle r_q^{-3}\rangle\langle L\|\alpha\|L\rangle l_{pq}I_pI_q, \qquad (7.73)$$

which is obviously of the form (**I . P . I**) used in Chapter 3.

When the cubic field leaves an orbital singlet ground state, as for Cr^{3+}, $3d^3$ in an octahedral field, we have $l_{pq} = 0$, reflecting the fact that in this approximation the electronic charge distribution has cubic symmetry and thus has no electric field gradient at the nucleus. On the other hand, in Cu^{2+}, $3d^9$, each of the two Γ_3 orbital states gives a first-order contribution to the nuclear electric quadrupole interaction because of their non-cubic electronic charge distribution, in the same way as the non-cubic spin magnetization contributes a first-order spin-dipolar magnetic hyperfine interaction.

As discussed in § 5.5 and § 17.7, the value of $\langle r_q^{-3}\rangle$ cannot be identified with $\langle r^{-3}\rangle$ because of electrostatic shielding effects. Similarly there may be a direct contribution to the nuclear electric quadrupole interaction from the ligand field if this is not exactly cubic, in which shielding effects must also be allowed for. It is likely that shielding effects will not be so important in the $3d$ as in the $4f$ group, since they increase considerably with atomic number (cf. Watson and Freeman 1967a).

For a ground electronic manifold of single or double degeneracy an electric quadrupole interaction of the form (7.73) will suffice, but for manifolds of three or more states with appreciable amounts of orbital momentum an operator similar to that for a free ion may be required. In such an operator, **L** (or **J**) is replaced by the effective spin $\tilde{\mathbf{S}}$, but two parameters rather than one may be needed (see the discussion in Chapter 18). An example of such an interaction is given for Fe^{2+}, $3d^6$, in an octahedral field, in § 7.13.

The effective value of $\langle r_q^{-3}\rangle$ will also be modified in the presence of covalent bonding. It has sometimes been assumed that there will be a reduction both in $\langle r_q^{-3}\rangle$ and in $\langle r^{-3}\rangle$ for the magnetic hyperfine interaction of the same magnitude as that in the spin-orbit coupling, but in a proper treatment the various hyperfine interactions must be evaluated using the appropriate bonding orbitals.

The nuclear Zeeman interaction

$$\mathscr{H}_n^L = -(\boldsymbol{\mu}_I\cdot\mathbf{H}) = -(\mu_I/I)(\mathbf{I}\cdot\mathbf{H}) \qquad (7.74)$$

is usually of order 10^{-4} cm^{-1}. In addition to this direct interaction there may be additional terms of the form H_pI_q arising from cross terms between the electronic Zeeman interaction and the hyperfine interaction. Essentially they arise from the modification of the electronic wave-functions of the ground state by the action of the electronic Zeeman interaction in admixing excited states, thus changing the magnetic hyperfine interaction by an amount proportional to **H**. For an orbital singlet ground state such terms are produced only by the off-diagonal elements of the orbital momentum, giving a 'pseudo-nuclear' Zeeman interaction

$$\mathcal{H}_n^{Z'} = \mathcal{P}\beta\Lambda_{pq}H_pI_q. \tag{7.75}$$

This must be added to the direct term (7.74), giving a resultant equivalent to that used in Chapter 3

$$\mathcal{H}_n^Z = \beta(\mathbf{H}\cdot\mathbf{g}^{(I)}\cdot\mathbf{I}) \tag{7.76}$$

where $\mathbf{g}^{(I)}$ is now a tensor.

For electronic manifolds of higher degeneracy, such as Γ_8 quartets, other additional terms of different form may be needed (see Chapter 18), but these are likely to be small. The latter is true also of

TABLE 7.8

Atoms of the 3d group, and nuclear properties of the stable isotopes that have nuclear spin not equal to zero. The nuclear moments are quoted to four or five significant figures (where justified from nuclear magnetic resonance) and include the diamagnetic correction

Z		Mass number	Abundance (%)	Nuclear spin I	Nuclear magnetic moment (n.m.)		Nuclear electric quadrupole moment Q(barns)	
21	Sc	45	100	$\tfrac{7}{2}$	+4·7564	+4·7492	−0·22	−0·22
22	Ti	47	7·28	$\tfrac{5}{2}$	−0·7884	−0·78710		
		49	5·51	$\tfrac{7}{2}$	−1·1040	−1·1022		
23	V	50	0·24	6	+3·3470	+3·3413		
		51	99·76	$\tfrac{7}{2}$	+5·148	+5·139	−0·052	−0·04
24	Cr	53	9·55	$\tfrac{3}{2}$	−0·4744	−0·47354	−0·03	
25	Mn	55	100	$\tfrac{5}{2}$	+3·443	+3·444	+0·4	+0·35
26	Fe	57	2·19	$\tfrac{1}{2}$	+0·0903	+0·09024		
27	Co	59	100	$\tfrac{7}{2}$	+4·649	+4·6163	+0·36	+0·40
28	Ni	61	1·19	$\tfrac{3}{2}$	±0·75	−0·74868	(+)0·16	
29	Cu	63	69·09	$\tfrac{3}{2}$	+2·226	+2·2206	−0·18	−0·16
		65	30·91	$\tfrac{3}{2}$	+2·385	+2·3789	−0·19	−0·15
					(a)	(b)	(a)	(b)

(a) Based on *Nuclear moments* (Fuller and Cohen 1965), *Table of rounded-off values*, with some later corrections and additions.
(b) Based on *A table of nuclear spins, moments and magnetic resonance frequencies* (Lee and Anderson 1967).

'pseudo-nuclear electric quadrupole' terms produced by second-order effects of the magnetic hyperfine interaction.

A list of nuclear data for ions of the 3d group is given in Table 7.8.

7.8. $3d^1$. Ti^{3+}, V^{4+} in an octahedral field. ^2D, $L = 2$, $S = \frac{1}{2}$

For these chemically rather unstable ions there is little experimental information. One of the few compounds measured is caesium titanium alum, Cs Ti $(SO_4)_2$, $12H_2O$, which has the usual alum structure, space group Pa3, with four molecules in unit cell. The electron paramagnetic resonance spectrum observed at helium temperatures in the undiluted salt is consistent with this structure, the four ions each having $|g_\parallel| = 1 \cdot 25$, $|g_\perp| = 1 \cdot 14$ (accurate to $\pm 0 \cdot 02$), the axis of symmetry being one of the four threefold axes of the cube in each case.

X-ray crystallography shows that the Ti^{3+} ions are coordinated to six water molecules in the form of a nearly regular octahedron, with a small distortion symmetrical about the threefold axis. The crystal field splittings would therefore be expected to be as shown in Fig. 7.4. We must, however, include the spin-orbit coupling, and the fact that the g-factors are so far from the spin-only value suggests that this may well be comparable with the trigonal field splitting of the ground triplet. On the other hand, the spin-orbit parameter ζ has the value $+154$ cm^{-1} for the free Ti^{3+} ion, while optical data suggest that the cubic field splitting (120 B_4) is some 20000 cm^{-1}, so that any effect due to admixture of the Γ_3 states by the spin-orbit coupling should be rather small. The trigonal field also has matrix elements (see Fig. 7.4) between the ground Γ_5 triplet and the excited Γ_3 doublet, but if the trigonal field splitting is comparable with the spin-orbit coupling these can be similarly neglected, except in the somewhat unlikely case of large second and fourth degree components that accidentally cancel as far as the splitting of the Γ_5 triplet is concerned (they then add in the off-diagonal components). A tetragonal field has no such off-diagonal matrix elements.

We shall see later that infrared measurements on Ti^{3+} and V^{4+} ions in Al_2O_3 have clarified the situation, and that a dynamic Jahn–Teller effect is responsible for many features of the observed spectrum. In this section we present only the static theory; this may be applicable in some cases, but primarily it illustrates the chronological approach to the problem, and the difficulties encountered in explaining just the single early magnetic resonance measurement of the g-values for caesium titanium alum.

TABLE 7.9

States and energy levels for the Γ_5 triplet of d^1 in an octahedral field, with splitting due to trigonal field and spin-orbit coupling. The orbital triplet has an effective momentum $\tilde{\mathbf{l}} = 1$, with effective orbital g-factor $\tilde{g}_l = -1$ in the column labelled 'isotropic'. In the column labelled 'anisotropic' both the spin-orbit coupling (and $\tilde{g}_l$) are allowed to be anisotropic. The states can be described similarly in each case, but with different formulae for $\tan 2\delta$. In terms of this parameter, the energy levels can be written in the form

$$W_{A \atop B} = (\lambda_\perp/\sqrt{2})(\cot 2\delta \pm \operatorname{cosec} 2\delta); \qquad W_C = \lambda_\perp \sqrt{2} \cot 2\delta - \lambda_\parallel$$

states	'isotropic' levels	'anisotropic' levels
$A^+ = \cos\delta \lvert +\tilde{1}^-\rangle - \sin\delta \lvert \tilde{0}^+\rangle$	$W_A = \tfrac{1}{2}\Delta + \tfrac{1}{4}\lambda + \tfrac{1}{2}(\Delta^2 + \Delta\lambda + \tfrac{9}{4}\lambda^2)^{\tfrac{1}{2}}$	$\tfrac{1}{2}\Delta + \tfrac{1}{4}\lambda_\parallel + \tfrac{1}{2}(\Delta^2 + \Delta\lambda_\parallel + \tfrac{1}{4}\lambda_\parallel^2 + 2\lambda_\perp^2)^{\tfrac{1}{2}}$
$A^- = \cos\delta \lvert -\tilde{1}^+\rangle - \sin\delta \lvert \tilde{0}^-\rangle$		
$B^+ = \sin\delta \lvert +\tilde{1}^-\rangle + \cos\delta \lvert \tilde{0}^+\rangle$	$W_B = \tfrac{1}{2}\Delta + \tfrac{1}{4}\lambda - \tfrac{1}{2}(\Delta^2 + \Delta\lambda + \tfrac{9}{4}\lambda^2)^{\tfrac{1}{2}}$	$\tfrac{1}{2}\Delta + \tfrac{1}{4}\lambda_\parallel - \tfrac{1}{2}(\Delta^2 + \Delta\lambda_\parallel + \tfrac{1}{4}\lambda_\parallel^2 + 2\lambda_\perp^2)^{\tfrac{1}{2}}$
$B^- = \sin\delta \lvert -\tilde{1}^+\rangle + \cos\delta \lvert \tilde{0}^-\rangle$		
$C^+ = \lvert +\tilde{1}^+\rangle$	$W_C = \Delta - \tfrac{1}{2}\lambda$	$\Delta - \tfrac{1}{2}\lambda_\parallel$
$C^- = \lvert -\tilde{1}^-\rangle$		
Value of δ	$\tan 2\delta = \sqrt{(2)}\lambda/(\Delta + \tfrac{1}{2}\lambda)$	$\tan 2\delta = \sqrt{(2)}\lambda_\perp/(\Delta + \tfrac{1}{2}\lambda_\parallel)$

IN INTERMEDIATE LIGAND FIELDS

In the following treatment we neglect any admixture of the Γ_3 doublet, and operate just within the manifold of the Γ_5 triplet with its twofold spin degeneracy. We label the three orbital states by their fictitious angular momentum components $|\pm\tilde{1}\rangle$, $|\tilde{0}\rangle$, and introduce just the spin-orbit coupling, plus an axial field which lifts the $|\pm\tilde{1}\rangle$ levels above $|\tilde{0}\rangle$ by an amount Δ (where Δ is not to be confused with the $\Delta = 10\,Dq$ mentioned in §7.4) which may be of either sign. The axial field may be either tetragonal or trigonal (in the latter case $\Delta = -v$ in the nomenclature of Table 7.4). For the matrix elements of the fictitious orbital momentum we must include the factor $\alpha = -1$ (see Table 5), and the energy matrix is then

$$\begin{array}{c|ccc|c} |-\tilde{1}^-\rangle & \Delta-\lambda/2 & 0 & 0 & |+\tilde{1}^+\rangle \\ |+\tilde{1}^-\rangle & 0 & \Delta+\lambda/2 & -\lambda/2 & |-\tilde{1}^+\rangle \\ |\ \tilde{0}^+\rangle & 0 & -\lambda/2 & 0 & |\ \tilde{0}^-\rangle \end{array} \qquad (7.77)$$

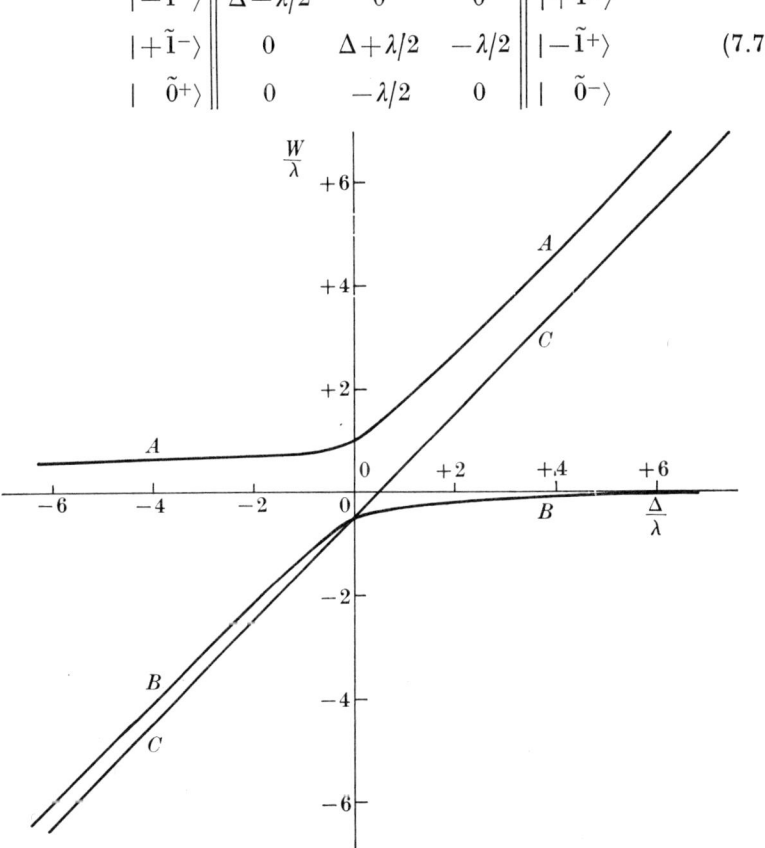

FIG. 7.21. Plot of the reduced energy (W/λ) against the reduced trigonal or tetragonal field splitting (Δ/λ) for the three Kramers doublets of the Γ_5 orbital triplet of d^1 in a nearly octahedral field.

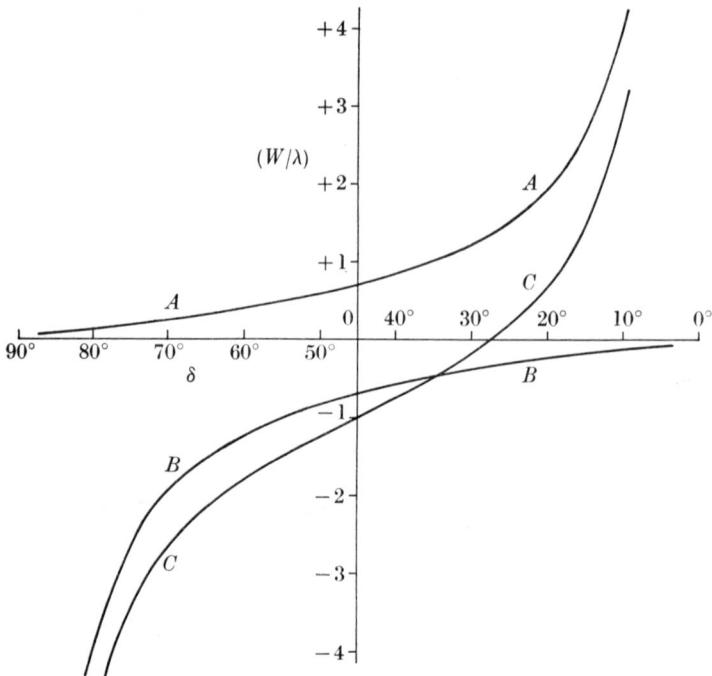

FIG. 7.22. Plot of the reduced energy parameter (W/λ) as a function of δ for the 'isotropic' equations given in Table 7.9. Positive values of δ have been plotted towards the left to make the figure correspond to Fig. 7.21. Cubic symmetry occurs at $\cos 2\delta = \frac{1}{3}$, or $\delta = 35.5°$.

where there are two sets of states (Kramers conjugates) shown to the left and to the right with the same energy matrix. Table 7.9 gives the general solution, consisting of three energy levels, each a Kramers doublet. The behaviour of these three energy levels is shown in Fig. 7.21 as a plot of the reduced energy (W/λ) against the reduced axial field splitting (Δ/λ), and in Fig. 7.22 as a function of the angle δ defined in Table 7.9. If λ is positive, as it is for d^1, level A is always uppermost; levels B, C intersect at the point $(\Delta/\lambda) = 0$, or $\cos 2\delta = \frac{1}{3}$, which corresponds to exact cubic symmetry. The level C is higher than B for positive values of (Δ/λ), and just lower than B for negative values.

In magnetic fields such that the Zeeman energy is small compared with the splittings between the three levels, each doublet is characterized by an anisotropic g-factor, whose values are given in Table 7.10. The result that $g_\| = g_\perp = 0$ for level C is rather surprising, but results from a cancellation that follows immediately from the states that compose the doublet C. Remembering that the fictitious angular

TABLE 7.10

Values of $g_\parallel$, $g_\perp$ for the three doublets of Γ_5 orbital triplet with $S = \frac{1}{2}$ of d^1 in an octahedral field, with splitting by trigonal field and spin-orbit coupling. The two sets of values correspond to those given in Table 7.9: 'isotropic' means both λ and the effective orbital g-factor $\tilde{g}_l = -1$ are assumed to be isotropic; 'anisotropic' means that parameters $\lambda_\parallel$, $\lambda_\perp$ are introduced, together with parallel and perpendicular components of $\tilde{g}_l = -k_\parallel$, $-k_\perp$; in the latter case we have not made the approximation $g_s = 2$, so that the spin and orbital contributions are immediately apparent. Note that the g-values for the doublet A are obtained from those for the doublet B by changing the signs of $\cos 2\delta$, $\sin 2\delta$; this is equivalent to increasing δ by $\pi/2$, and corresponds to the requirement that the two sets of states given for doublets A, B in Table 7.9 must be orthogonal

Doublet energy	W_A	W_B	W_C
	$-1 - 3\cos 2\delta$	$-1 + 3\cos 2\delta$	0
'isotropic' $\begin{cases} g_\parallel \\ g_\perp \end{cases}$	$\|1 - \cos 2\delta + \sqrt{(2)}\sin 2\delta\|$	$\|1 + \cos 2\delta - \sqrt{(2)}\sin 2\delta\|$	0
'anisotropic' $\begin{cases} g_\parallel \\ g_\perp \end{cases}$	$-g_s \cos 2\delta - k_\parallel(1 + \cos 2\delta)$	$g_s \cos 2\delta - k_\parallel(1 - \cos 2\delta)$	$g_s - 2k_\parallel$
	$\|\tfrac{1}{2}g_s(1 - \cos 2\delta) + \sqrt{(2)}k_\perp \sin 2\delta\|$	$\|\tfrac{1}{2}g_s(1 + \cos 2\delta) - \sqrt{(2)}k_\perp \sin 2\delta\|$	0

momentum $l = 1$ for the Γ_5 triplet is associated with the factor $\alpha = -1$, we see that

$$\langle +\tilde{1}^+| L_z + 2S_z |+\tilde{1}^+\rangle = \langle +\tilde{1}^+| -\tilde{l}_z + 2S_z |+\tilde{1}^+\rangle = -1 + 2(\tfrac{1}{2}) = 0, \tag{7.78}$$

so that $g_\parallel = 0$, while $g_\perp$ is also zero because the matrix elements of $L_\pm$, $S_\pm$ are each separately zero between the two states $|+1^+\rangle$, $|-1^-\rangle$.

For doublets A, B the values of $g_\parallel$, $g_\perp$ may be expressed in terms of a single angular parameter 2δ, and a plot of $g_\parallel$ against $g_\perp$ has the form of an ellipse, as shown in Fig. 7.23. The points on this ellipse include all possible g-values for either doublet, since the ellipse is completed by allowing 2δ to take all values from 0 to 2π, and the g-values for doublet A may be regarded as corresponding to a value of 2δ that differs by π from those for doublet B. At this point we see that some care must be exercised in using this ellipse as regards the signs of the

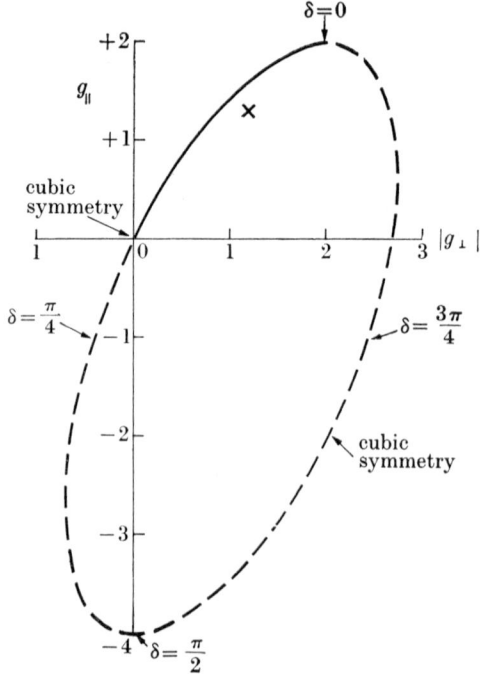

Fig. 7.23. Plot of $g_\parallel$ against $g_\perp$ for the doublets A, B, of Table 7.10 'isotropic'. The left-hand portion ($\delta = 0$ to $\pi/2$) corresponds to doublet B, the right-hand portion ($\delta = \pi/2$ to π) to doublet A. The thick portion indicates the range over which doublet B is the ground doublet, lying below doublet C. ×—experimental values of $(g_\parallel, g_\perp)$ for caesium titanium alum. For doublet A it would be better to plot $g_\perp$ as a negative quantity, so that the point for cubic symmetry would lie at $(-2, -2)$; see § 8.8.

g-values. The sign of $g_\parallel$ is a physically real quantity, which can be determined by experiment, while that of $g_\perp$ is not, so that there will be an ambiguity in plotting a point corresponding to an experimental determination of $(g_\parallel, g_\perp)$ in Fig. 7.23. (If the sign of $g_\parallel$ is not determined experimentally, there will be a double ambiguity.)

Before applying our results to a special case, we make a few remarks about some special features of the ellipse shown in Fig. 7.23. In order to identify points on it unambiguously, we assume that $g_\perp$ is for this purpose taken as positive towards the right (this corresponds to using the expression for $g_\perp$ given in the Table 7.10 above for the doublet B rather than that for the doublet A). The ellipse passes through the point $(g_\parallel, g_\perp) = (+2, +2)$, corresponding to the solution $\delta = 0$ appropriate to a positive infinite value of Δ. The doublet B is then lowest, as can be seen at once from the energy matrix (7.77) or from Fig. 7.22, and its two states are simply $|\tilde{0}^+\rangle$, $|\tilde{0}^-\rangle$, whose magnetic moment is entirely due to the spin, giving an isotropic g-value of 2. For this value of Δ the doublet A is the highest of the three doublets, and its g-values correspond to the point $(-4, 0)$ on the ellipse.

As the value of Δ diminishes from plus infinity the value of 2δ increases from zero, and the plot of $(g_\parallel, g_\perp)$ for the doublet B (which is still the ground doublet) moves anti-clockwise round the top left-hand part of the ellipse back towards the point $(0, 0)$ which corresponds to $\Delta = 0$ and cubic symmetry. From Fig. 7.21 we see that energy levels B and C cross at this point, and for negative values of Δ the ground doublet is C; since resonance is very rarely observed in an excited state, we shall not consider the behaviour of the g-values of the doublet B in this region. We can however check our results for the case of cubic symmetry, for then the Γ_5 triplet with its twofold spin degeneracy is split by the spin-orbit coupling into a doublet and a quadruplet, which we can denote by fictitious angular momenta $(\Gamma_5, \tfrac{1}{2})$ and $(\Gamma_5, \tfrac{3}{2})$ respectively. The g-factors of doublet and quadruplet are of course isotropic when the symmetry is cubic, and can be calculated using the modified Landé formula given in eqn (7.32). On inserting the values $\tilde{l} = 1$, $S = \tfrac{1}{2}$; $g_l = \alpha = -1$, $g_s = 2$, we get

$$g(\Gamma_5, \tilde{J}) = \frac{1}{2} - \frac{15}{8\tilde{J}(\tilde{J}+1)} \qquad (7.79\text{a})$$

and hence

$$g(\Gamma_5, \tfrac{1}{2}) = -2, \qquad g(\Gamma_5, \tfrac{3}{2}) = 0. \qquad (7.79\text{b})$$

The value of δ appropriate to cubic symmetry is that for which $\cos 2\delta = \frac{1}{3}$, $\sin 2\delta = \sqrt{(8)}/3$, for which the formulae in Table 7.10 lead to isotropic values of -2 and 0 in agreement with (7.79b).

From this result we see that the quadruplet state has $g = 0$; in fact it is formed by the crossing at $\Delta = 0$ of the two doublets B and C, of which the latter always has $g_\| = g_\perp = 0$, while the former has $g_\| = g_\perp = 0$ at the moment of crossing corresponding to the point $(0, 0)$ in Fig. 7.23. For the doublet A the fact that g, though isotropic, is equal to -2, not to $+2$, reminds us that the orbital moment must be contributing, a point to which we shall return later. (Unfortunately this result corresponds to the point $(-2, +2)$ in Fig. 7.23, because we have made a choice of sign in labelling the value of $g_\perp$ more appropriate to the doublet B than to the doublet A; for this reason we shall later redraw the ellipse (see Fig. 8.2) when considering the doublet A, which is the ground state if the spin-orbit coupling parameter λ is negative.)

We now return to the measured values $g_\| = 1\cdot25$, $g_\perp = 1\cdot14$ for caesium titanium alum. Though the sign of $g_\|$ has not been experimentally determined, there is no ambiguity because the doublet C has $g_\| = g_\perp = 0$, and the measured values must therefore belong to the doublet B. They are plotted in Fig. 7.23, as a point which lies clearly well inside the ellipse of allowed values; we must therefore conclude that the approximations on which the ellipse was constructed are inadequate. One such approximation is neglect of admixtures from the Γ_3 states, but these are unlikely to affect the g-values by more than a few per cent, and an exact treatment (Bleaney, Bogle, et al. 1954) shows that no acceptable fit can be obtained in this way.

A plausible alternative is that the simple crystal field theory must be abandoned, and allowance must be made for the effects of bonding to the ligand ions. As discussed in § 7.4, and shown in Chapter 20, this leads to a reduction of the orbital moment; in the case of d^1, if we confine ourselves to the Γ_5 states which can only take part in π-bonding, this means that the matrix elements of orbital momentum should be multiplied by a factor $k_{\pi\pi}$, which for brevity we shall here write as k. This factor is isotropic in the case of cubic symmetry, but not otherwise; however to avoid introducing too many parameters we shall assume that k is isotropic even in the presence of a small trigonal distortion. Then for the doublet B

$$\left.\begin{aligned} g_\| &= (2+k)\cos 2\delta - k, \\ g_\perp &= |1 + \cos 2\delta - \sqrt{(2)}k \sin 2\delta|. \end{aligned}\right\} \quad (7.80)$$

For a given value of k (< 1) this gives an ellipse that passes (at $\delta = 0$) through the point ($+2$, $+2$), which corresponds to spin-only magnetism, but otherwise lies inside the ellipse shown in Fig. 7.23. However it proves necessary to take rather a low value of k, about 0·7, in order to obtain g-values near those measured experimentally. The fit is not exact, and in any case corrections due to admixtures of the Γ_3 states have been neglected; however it is not worth while carrying the analysis further in the absence of experimental information for caesium titanium alum about the positions of the levels A, B.

An equally difficult problem for simple ligand field theory is posed by the experimental results for Ti^{3+} in the trigonal field of Al$_2$O$_3$, summarized in Table 7.11. The resonance results for the ground doublet

TABLE 7.11

Measured energy levels and g-values from infrared spectroscopy for the three Kramers doublets of the Γ_5 orbital triplet of two $3d^1$ ions in Al$_2$O$_3$. Data from Nelson, Wong, and Schawlow (1967); Wong, Berggren, and Schawlow (1968); Joyce and Richards (1969). Magnetic resonance has been observed in the ground doublet of Ti^{3+} in Al$_2$O$_3$, with $|g_\parallel| = 1\cdot07$, $|g_\perp| < 0\cdot1$, by Kask, Kornienko, Mandel'stam, and Prokhorov (1964)

Ti^{3+} in Al$_2$O$_3$			V^{4+} in Al$_2$O$_3$		
Energy (cm^{-1})	$g_\parallel$	$|g_\perp|$	Energy (cm^{-1})	$g_\parallel$	$|g_\perp|$
107·5	1·9 (calc)		52·6	1·8 (calc)	
37·8	−2·00(6)	<0·1	28·0	−1·43(4)	<0·2
0	+1·11(3)	<0·1	0	+1·43(4)	<0·2

cannot sensibly be fitted to the ($g_\parallel$, $g_\perp$) curve of Fig. 7.23 for doublets A, B. For the doublet C we have $g_\perp = 0$, giving no allowed transition and we can only account for the low value of $g_\parallel = 1\cdot07$ by using the formula $g_\parallel = g_s - 2k_\parallel$ of Table 7.10 with the very low value $k_\parallel = 0\cdot47$. Since the actual value of $g_\perp$ is not known, it is possible that resonance is allowed through a third-order Zeeman effect arising through admixtures of the low-lying excited state at 38 cm^{-1}.

The positions of the excited doublets are also difficult to fit to Fig. 7.21, where it is obvious that there is no negative value of (Δ/λ), for which alone C is the lowest doublet, where ($W_A - W_B$) is roughly twice ($W_B - W_C$). In addition, the minimum overall splitting of the three doublets is 1·46λ; thus the measured overall splitting of 108 cm^{-1}

leads to $\lambda \sim 74$ cm^{-1}, less than half the free ion value. This result is in line with the low value of $k_\parallel$, but both are remarkably small.

We can extend the theory by allowing both λ and k to be 'anisotropic', obtaining the results thus labelled in Tables 7.9, 7.10. The spin-orbit coupling and Zeeman interactions then have the form

$$-\lambda_\parallel \tilde{l}_z S_z - \lambda_\perp(\tilde{l}_x S_x + \tilde{l}_y S_y) + \\ + \beta(-k_\parallel \tilde{l}_z + g_s S_z)H_z - k_\perp \beta(\tilde{l}_x H_x + \tilde{l}_y H_y) + g_s \beta(S_x H_x + S_y H_y),$$

in which the negative signs arise as before in using a fictitious orbital momentum $\tilde{l} = 1$ for the Γ_5 triplet. In the energy matrix (7.77) the only change is that $\lambda_\parallel$ appears in the diagonal elements and $\lambda_\perp$ in the off-diagonal elements, so that the states given in Table 7.9 are still valid, but with the formulae for the energy levels and g-factors labelled 'anisotropic' in Tables 7.9, 7.10. The observed energy levels and value of $g_\parallel$ for the ground doublet are insufficient to determine all the parameters, but require $\lambda_\perp \leqslant 50$ cm^{-1}, $\lambda_\parallel > 73$ cm^{-1}; $k_\parallel = 0\cdot47$. The value of the trigonal field splitting $\Delta = -v$ must be quite small, in contrast to the relatively large values found for other ions in Al$_2$O$_3$, listed in Table 7.4.

For the isoelectronic ion V^{4+} in Al$_2$O$_3$ the splittings (Table 7.11) are even smaller than for Ti^{3+}, though the free ion value of the spin-orbit parameter is larger. A static ligand field model is clearly inadequate, and Macfarlane, Wong, and Sturge (1968) (see also Joyce and Richards (1969)) have shown that the main features of the results can be accounted for in terms of a dynamic Jahn–Teller effect. This produces a dynamic quenching not only of the spin-orbit coupling and the orbital momentum in the Γ_5 triplet, but also of the splitting due to the trigonal component of the ligand field. The theory allows values for the latter of $v = 700$ to 1000 cm^{-1}, which are in line with the values given for other ions in Al$_2$O$_3$ in Table 7.4. In addition, a considerable reduction in the value of k ($\approx 0\cdot8$) due to covalency is required. For further discussion see Chapter 21 and Ham (1969), or the papers quoted above.

7.9. $3d^2$. V^{3+}, Cr^{4+} in an octahedral field. ^3F, $L = 3$, $S = 1$

Again this ion is chemically rather unstable, and there is little experimental information. One of the few compounds examined is vanadium ammonium alum, V(NH$_4$)(SO$_4$)$_2$,12H$_2$O, which is expected to have the usual alum structure, the local symmetry being octahedral with a small trigonal distortion; we shall assume that this is the case. (This guarded statement reflects the fact that X-ray crystallographic analysis shows this to be correct at room temperature, but a number of alums have been found to undergo crystallographic changes associated with ferro-electric effects at low temperatures.) In fact the main results we shall discuss are not from magnetic resonance, but the susceptibility measurements of van den Handel and Siegert (1937), which can be interpreted in terms of a ground state whose Hamiltonian is

$$\mathscr{H} = D\{S_z^2 - \tfrac{1}{3}S(S+1)\} + g_\parallel \beta H_z S_z + g_\perp \beta (H_x S_x + H_y S_y) \quad (7.81)$$

with $S = 1$, $D = +4\cdot 8$ cm^{-1}, $g_\parallel = 1\cdot 96$, $g_\perp = 1\cdot 82$. No other levels are appreciably populated below room temperature. Thus the position is that we appear to have a ground state that is an orbital singlet, with threefold spin degeneracy that is lifted by a rather large initial splitting D, and a somewhat anisotropic g-factor.

The ground state of a d^2 ion is ^3F, whose orbital splitting in a cubic field with a trigonal distortion is shown in Fig. 7.6. The cubic field parameter B_4 is negative for a d^2 ion in an octahedral field, so that the Γ_4 triplet lies lowest. This is split into a singlet and a doublet by the trigonal distortion, and in view of the results quoted above we expect the singlet to lie well below the doublet. This orbital singlet has a threefold spin degeneracy, which is split through second-order effects of the spin-orbit coupling, with a small contribution from spin–spin interaction (eqn (7.16)).

We shall give an approximate analysis, using only the manifold of Γ_4 states, which is based on the work of Chakravarty (1959), which follows the theory of Abragam and Pryce (1951a). The manifold is treated as an orbital triplet with a fictitious angular momentum $\tilde{l} = 1$, with effective orbital g-values α, α' for the components of orbital momentum parallel and perpendicular to the trigonal axis respectively. We assume that the effect of the trigonal field splitting is to raise the states $|+\tilde{1}\rangle$, $|-\tilde{1}\rangle$ by an amount Δ above the state $|\tilde{0}\rangle$, so that the effective Hamiltonian operating within the Γ_4 manifold with its threefold spin degeneracy ($S = 1$) is

$$\mathscr{H}(\Gamma_4) = \Delta(\tilde{l}_z^2) + \alpha \lambda \tilde{l}_z S_z + \alpha' \lambda'(\tilde{l}_x S_x + \tilde{l}_y S_y) + \\ + \beta(\alpha \tilde{l}_z + 2S_z)H_z + \beta(\alpha' \tilde{l}_x + 2S_x)H_x + \beta(\alpha' \tilde{l}_y + 2S_y)H_y, \quad (7.82)$$

where the reader should note that the signs of Δ, α, and α' are the reverse of those used by Chakravarty. Taking Δ and the spin-orbit parameters λ, λ' to be positive, one obtains a set of three low-lying states corresponding to the ground 'spin' triplet of eqn (7.81). In the absence of a magnetic field these three states are described by the following functions:

States in terms of $|\tilde{l}_z, S_z\rangle$ .... Energy

$(1+\delta^2)^{-\frac{1}{2}}\{|\tilde{0}, +1\rangle + \delta\,|+\tilde{1}, 0\rangle\}$ .... $\frac{1}{2}(\Delta - S_1)$

$(1+2\epsilon^2)^{-\frac{1}{2}}\{|\tilde{0}, 0\rangle + \epsilon\,|+\tilde{1}, -1\rangle + \epsilon\,|-\tilde{1}, +1\rangle\}$ .... $\frac{1}{2}(\Delta - \alpha\lambda - S_0)$

$(1+\delta^2)^{-\frac{1}{2}}\{|\tilde{0}, -1\rangle + \delta\,|-\tilde{1}, 0\rangle\}$ .... $\frac{1}{2}(\Delta - S_1)$

where $\delta = (\Delta - S_1)/2\alpha'\lambda'$, $\epsilon = (\Delta - \alpha\lambda - S_0)/4\alpha'\lambda'$
and
$$S_1 = (\Delta^2 + 4\alpha'^2\lambda'^2)^{\frac{1}{2}}, \qquad S_0 = \{(\Delta - \alpha\lambda)^2 + 8\alpha'^2\lambda'^2\}^{\frac{1}{2}}.$$

The first and third states (corresponding to $S_z = \pm 1$ in the spin Hamiltonian for the ground triplet) are raised in energy above the second state ($S_z = 0$) by an amount

$$D = \tfrac{1}{2}(S_0 - S_1 + \alpha\lambda).$$

In the limit where the spin-orbit coupling is small compared with Δ, both δ and ϵ tend to the limit $\delta = \epsilon = -\alpha'\lambda'/\Delta$, and D becomes $\alpha'^2\lambda'^2/\Delta = -\alpha'\lambda'\delta$. On evaluating the Zeeman energy as a perturbation we find

$$g_\| = (2 + \alpha\delta^2)(1+\delta^2)^{-1} \approx 2 - (2-\alpha)\delta^2$$
$$g_\perp = \{2 + 2\delta\epsilon + \alpha'(\delta+\epsilon)\}\{(1+\delta^2)(1+2\epsilon^2)\}^{-\frac{1}{2}}$$
$$\approx 2 + 2\alpha'\delta - \delta^2$$

where in the approximate formulae we have assumed $\epsilon = \delta$.

We see at once that the theory leads in the right direction in that it predicts a first-order deviation from the spin-only value of 2 for $g_\perp$, but only a second-order deviation for $g_\|$, and also gives a positive value for D. Making the assumption that $\lambda = \lambda'$, Chakravarty fitted the susceptibility of vanadium ammonium alum from room temperature down to liquid helium temperatures, using the values

$$\Delta = 1390 \text{ cm}^{-1}, \qquad \lambda = +64 \text{ cm}^{-1}$$
$$\alpha = -1\cdot 10, \qquad \alpha' = -1\cdot 35.$$

This value of λ is markedly smaller than the value $+104$ cm^{-1} for the free ion, a result also obtained by Siegert (1937), and suggests there is considerable covalent bonding. However no reduction in the orbital moment was made by Chakravarty, except in so far as this may be disguised in the low values of α, α', which were ascribed to admixtures of the ^3P state.

Magnetic resonance has been observed for the d^2 ions V^{3+} and Cr^{4+} in α-Al_2O_3, and the results are summarized in Table 7.12. The most accurate values are those obtained for V^{3+} by Joyce and Richards (1969) using far-infrared Fourier-transform spectroscopy to observe the $|\pm1\rangle \leftrightarrow |0\rangle$ transitions in a magnetic field. The following discussion (cf. § 3.14) illustrates the approach needed in the work of Zverev and Prokhorov (1960, 1961) at lower frequencies where the resonance observed is the $S_z = |+1\rangle$ to $|-1\rangle$ transition. In zero magnetic field, with a field of trigonal symmetry, these two levels would be degenerate,

TABLE 7.12

Magnetic resonance and infrared spectroscopy results for some $3d^2$ ions in α-Al_2O_3

Ion	D (cm^{-1})	$g_\parallel$	$g_\perp$	$A_\parallel$ (10^{-4} cm^{-1})	Reference
$^{51}V^{3+}$	$+7.0(3)$	$1.915(2)$	$1.63(5)$	$95.9(5)$	Zverev and Prokhorov (1960, 1961)
	$8.29(2)$				Sauzade, Pontnau, Lesas, and Silhouette (1966)
	$8.25(2)$	$1.92(3)$	$1.74(2)$		Joyce and Richards (1969)
Cr^{4+}	$+7$	1.90			Hoskins and Soffer (1964)

and if a field is applied parallel to the trigonal axis, their separation (including hyperfine structure) would be $2g_\parallel\beta H + 2A_\parallel m$. This transition has been observed with the radio-frequency field parallel to the trigonal axis; in a field of pure trigonal symmetry the transition would be forbidden, and the observation is attributed to a term of the form $E(S_x^2 - S_y^2)$ which is present due to strains and crystal defects. The value of D is found from the temperature dependence of the intensity below 4°K, which falls sharply as the temperature is reduced and the populations of the excited $|\pm1\rangle$ states decline. If the magnetic field is inclined at an angle θ to the trigonal axis, the transitions move to higher fields since in a first approximation the splitting of the ±1 states varies as the component $H\cos\theta$ along the trigonal axis. However Zverev and Prokhorov have observed a broad transition centred on 19800 G at 9300 MHz for V^{3+} when the field is normal to the trigonal axis. Under such conditions the two levels which are degenerate when $E = 0$ are split by an amount $(g_\perp\beta H)^2/D$, and the transition is partially allowed with the oscillatory field along the trigonal axis. This experiment gave the first value of $g_\perp$ listed in Table 7.12; the hyperfine structure was not resolved. Zverev and Prokhorov interpreted the results in terms of the theory outlined above with $\lambda = +38$ cm^{-1}, $\alpha = -1\cdot 40$, $\alpha' = -1\cdot 17$, and a trigonal splitting of 280 cm^{-1} of the Γ_4 orbital triplet. These values differ considerably from those of Pryce and Runciman (1958), who obtained a trigonal splitting of 1200 cm^{-1} and $\lambda \approx 70$ cm^{-1} from the optical spectrum and fitting a ground state splitting of 8 cm^{-1}.

In the formulae given above we have neglected the contribution of the spin-spin interaction, which when $\Delta \gg \lambda$ adds an amount $3\rho/2$ to D. However ρ is estimated to be about $0\cdot 26$ cm^{-1} (see Table 7.6), so that the contribution is not important.

7.10. $3d^3$. V^{2+}, Cr^{3+}, Mn^{4+} in an octahedral field. ^4F, $L = 3$, $S = \frac{3}{2}$

Extensive magnetic resonance and optical investigations have been carried out on the Cr^{3+} ion in a predominantly octahedral field, and a fair amount of data is available for V^{2+} and Mn^{4+}. The resonance spectrum observed in the ground state is generally that expected for a spin quadruplet, with g-value close to that of a free electron spin, and a small initial splitting. In terms of a spin Hamiltonian

$$\mathscr{H} = \beta(\mathbf{H}\cdot\mathbf{g}\cdot\mathbf{S})+(\mathbf{S}\cdot\mathbf{D}\cdot\mathbf{S})+(\mathbf{S}\cdot\mathbf{A}\cdot\mathbf{I})+(\mathbf{I}\cdot\mathbf{P}\cdot\mathbf{I}) \qquad (7.83)$$

g is found to be about 1·98, with little or no anisotropy, and the initial splitting parameter D is of order 0·1 cm^{-1}. The magnetic hyperfine structure is practically isotropic, and the electric quadrupole interaction is very small.

These results point to the ground state of the ion being an orbital singlet, with all excited states lying higher in energy by amounts large compared with the spin-orbit coupling. This is consistent with crystal field theory, which for octahedral coordination predicts a singlet orbital ground state for d^3, as can be seen from Figs. 7.5, 7.6 for positive values of B_4. Thus, neglecting spin-orbit coupling, we would expect the ground state to be an unsplit spin quadruplet whose g-value is equal to the free spin value. There are however matrix elements of the spin-orbit coupling between the ground orbital singlet Γ_2 and the excited triplet Γ_5 (but not between Γ_2 and the second excited triplet Γ_4). If we label the triplet Γ_5 states in terms of a fictitious angular momentum $|\tilde{l}\rangle$, and use second-order perturbation theory, the results for the ground spin quadruplet can be given in a form applicable to a distortion of the octahedron with axial symmetry about either a tetragonal or a trigonal axis.

For this purpose we assume that the Γ_5, $\tilde{l}_z = \pm 1$ states lie above the ground orbital singlet by an amount Δ_1, and the Γ_5, $\tilde{l}_z = 0$ state lies above the singlet by an amount Δ_0. The difference between Δ_1 and Δ_0 is equal to the splitting of the Γ_5 triplet by a tetragonal or trigonal distortion. The off-diagonal matrix elements of the spin-orbit coupling then give the states for the ground 'spin' quartet listed in Table 7.13. This quartet can be fitted to the spin Hamiltonian (7.81) with $S = \tfrac{3}{2}$ and

$$g_\| = g_s - 8\lambda/\Delta_0, \quad g_\perp = g_s - 8\lambda/\Delta_1. \tag{7.84a}$$

Correct to second order in perturbation theory, the energy levels are

$$\begin{aligned} W_{\pm\frac{3}{2}} &= -9\lambda^2/\Delta_0 - 6\lambda^2/\Delta_1, \\ W_{\pm\frac{1}{2}} &= -\lambda^2/\Delta_0 - 14\lambda^2/\Delta_1, \end{aligned} \tag{7.84b}$$

which gives

$$W_{\pm\frac{3}{2}} - W_{\pm\frac{1}{2}} = 2D = -8\lambda^2/\Delta_0 + 8\lambda^2/\Delta_1 \tag{7.84c}$$

but there is a third-order correction to D which is really of the same size (see below) in the case of a trigonal distortion.

TABLE 7.13

States for ground spin triplet $S = \tfrac{3}{2}$ of orbital singlet Γ_2 of d^3 ion in octahedral ligand field, with tetragonal or trigonal distortion

$$|+\tfrac{3}{2}\rangle = |\Gamma_2, +\tfrac{3}{2}\rangle - \frac{3\lambda}{\Delta_0}|\Gamma_5(\tilde{0}), +\tfrac{3}{2}\rangle + \frac{\sqrt{(6)}\lambda}{\Delta_1}|\Gamma_5(+\tilde{1}), +\tfrac{1}{2}\rangle$$

$$|+\tfrac{\tilde{1}}{2}\rangle = |\Gamma_2, +\tfrac{1}{2}\rangle - \frac{\lambda}{\Delta_0}|\Gamma_5(\tilde{0}), +\tfrac{1}{2}\rangle + \frac{2\sqrt{(2)}\lambda}{\Delta_1}|\Gamma_5(+\tilde{1}), -\tfrac{1}{2}\rangle - \frac{\sqrt{(6)}\lambda}{\Delta_1}|\Gamma_5(-\tilde{1}), +\tfrac{3}{2}\rangle$$

$$|-\tfrac{\tilde{1}}{2}\rangle = |\Gamma_2, -\tfrac{1}{2}\rangle + \frac{\lambda}{\Delta_0}|\Gamma_5(\tilde{0}), -\tfrac{1}{2}\rangle - \frac{2\sqrt{(2)}\lambda}{\Delta_1}|\Gamma_5(-\tilde{1}), +\tfrac{1}{2}\rangle + \frac{\sqrt{(6)}\lambda}{\Delta_1}|\Gamma_5(+\tilde{1}), -\tfrac{3}{2}\rangle$$

$$|-\tfrac{3}{2}\rangle = |\Gamma_2, -\tfrac{3}{2}\rangle + \frac{3\lambda}{\Delta_0}|\Gamma_5(\tilde{0}), -\tfrac{3}{2}\rangle - \frac{\sqrt{(6)}\lambda}{\Delta_1}|\Gamma_5(-\tilde{1}), -\tfrac{1}{2}\rangle$$

If the ion is in a field of cubic symmetry, $\Delta_0 = \Delta_1$, so that g is isotropic and we have an unsplit spin quartet with a g-value slightly less than the free spin value. This is observed for Cr^{3+} and Mn^{4+} ions in MgO with no local charge compensation, as well as for V^{2+} (see Table 7.14). Similar g-values are observed in hydrated salts, and there is evidence that the departure of the g-value from the free spin value is less than predicted by the simple formula. Optical absorption bands give a value of about 17500 cm^{-1} for Δ for the complex $[Cr(H_2O)_6]^{3+}$, so that from the free ion value of λ we should expect $g - 2 = -8\lambda/\Delta$ to be about -0.04 rather than the value of -0.02 generally observed; the discrepancy can

Table 7.14

Experimental results for $3d^3$ ions in near octahedral symmetry. The spin Hamiltonian is that of eqn (7.1). The hyperfine interaction is almost entirely due to core polarization, which produces a field ≈ -192 kG at the nucleus, with a small orbital contribution of -4 to -6 kG. The results for the ions in Al_2O_3 are from Laurance and Lambe (1963)

		Cubic symmetry		
Host	Ion	g	A (cm^{-1})	Reference
MgO	^{51}V^{2+}	1·9803(5)	0·00743(2)	Low (1956)
MgO	^{53}Cr^{3+}	1·9800(5)	0·00162(3)	Wertz and Auzins (1957)
MgO	^{55}Mn^{4+}	1·9941(2)	0·007082(10)	Davies, Smith and Wertz (1969)

		Trigonal symmetry					
Host	Ion	$g_{\parallel}$	$g_{\perp}$	D (MHz)	$A_{\parallel}$ (MHz)	$A_{\perp}$ (MHz)	$P_{\parallel}$ (MHz)
Al$_2$O$_3$	^{51}V^{2+}	1·991	$= g_{\parallel}$	$-4803·6$	$-220·62$	$-222·9$	$-0·02$
Al$_2$O$_3$	^{53}Cr^{3+}	1·984(1)	1·9867(10)	$-5723·5$	$+48·5$	$= A_{\parallel}$	$-0·105$
Al$_2$O$_3$	^{55}Mn^{4+}	1·993	$= g_{\parallel}$	-5868	$-208·83$	$-211·4$	$+0·069$

be attributed to bonding effects which produce a reduction in the effective value of the spin-orbit coupling and also in the orbital moment.

The substance most extensively studied (because of its importance as a laser material) is Al_2O_3, where there is trigonal symmetry at the aluminium site. The electron paramagnetic resonance results are given in Table 7.14, where for the Cr^{3+} ion the sign of D is given by measurements on the optical spectrum (Sugano and Tanabe 1958). At first sight the theoretical expression for D can be written down from the second-order shifts of the Γ_2 levels as in Table 7.13, but since Δ_0 only differs from Δ_1 because of the trigonal field splitting of the Γ_5 levels, this expression is actually of the third order, $\sim \lambda^2(\Delta_{\text{trig}}/\Delta^2)$. We must therefore be careful to include all third-order terms, and Sugano and Tanabe (1958) have pointed out that a term comparable in magnitude arises from an admixture through the trigonal field of the $|\Gamma_4, \tilde{0}\rangle$ state into Γ_2 (cf. Fig. 7.6). The amplitude of this admixture, δ, is $\sim(\Delta_{\text{trig}}/\Delta)$, and since the spin-orbit coupling has matrix elements between Γ_4 and Γ_5, there is a further contribution to D of order $\lambda^2\delta/\Delta \approx \lambda^2(\Delta_{\text{trig}}/\Delta^2)$.

By using simple crystal field theory explicit expressions for these third-order terms could be written down in terms of B_2^0, B_4^0, and Δ (or B_4). In practice this would be pointless for a number of reasons. Not only is the crystal field so strong that we must allow for the presence of

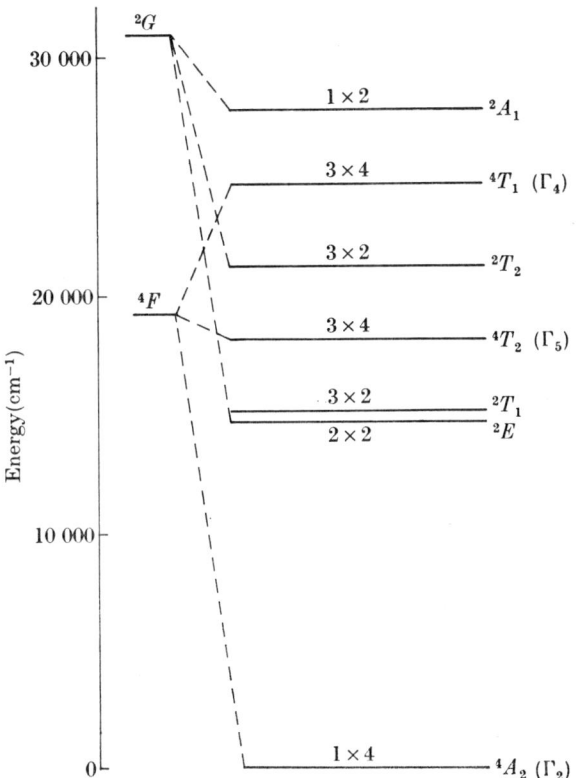

Fig. 7.24. The lowest energy levels of Cr^{3+} in a cubic field; the splittings are those appropriate to Cr^{3+} in Al_2O_3 with neglect of the trigonal distortion. The multiplicities shown are those of orbit × spin.

excited doublet levels, some of which actually fall below the excited triplet states belonging to the ^4F ground state (see Fig. 7.24), but it has been shown (Sugano and Peter 1961) that configurational interaction cannot be neglected; bonding effects must also be included. Any reasonably complete theory therefore contains a number of adjustable parameters, and some approximations must be made to reduce these to manageable proportions. Though an exact fit to the optical and magnetic resonance data has not been obtained, it is clear (see Sugano and Peter) that a considerable degree of covalency must be ascribed to both the t_2 and e-electrons, as well as a considerable reduction in the spin-orbit coupling. This gives a reasonable approximation to the value of $g_\parallel$, but to obtain a value of D as large as that observed experimentally it may be necessary to allow the spin-orbit coupling to be anisotropic.

This corresponds to using different values of λ in the two terms with denominators Δ_0, Δ_1 for the g-values and second-order energy displacements of the ground quartet in eqns (7.84). However, later calculations (Rimmer and Johnston 1966) suggest that such refinements may not be necessary.

The Endor results for $^{51}V^{2+}$, $^{53}Cr^{3+}$, $^{55}Mn^{4+}$ in Al_2O_3 given in Table 7.14 make it possible to extract a value for the core-polarization field for this sequence (Geschwind 1967). After allowing for a small orbital contribution of -4 to -6 kG, calculated from $g_\parallel$ using eqn (7.63), the field at the nucleus H_e^s (eqn 7.68) is found to lie in the range $-192 \cdot 3 \pm 1$ kG per unit of electron spin for all three ions. The negative sign means that $\mathbf{H}_e^s$ is oppositely directed to the spin magnetization, and parallel to the electron spin $\mathbf{S}$. For $^{53}Cr^{3+}$ in MgO (cubic) some higher-order terms, including a nuclear electric quadrupole interaction of the form given in eqn (18.50), have been measured in the Endor spectrum by Woonton and Dyer (1967).

7.11. $3d^4$. Cr^{2+} in an octahedral field. ^5D, $L = 2$, $S = 2$

One of the few salts of this ion in which resonance has been observed is $CrSO_4, 5H_2O$, which has the same crystal structure as copper sulphate. The resonance results of Ono, Koide, Sekiyama, and Abe (1954) show that the ground state is a 'spin' quintuplet described by the Hamiltonian

$$\mathscr{H} = D\{S_z^2 - \tfrac{1}{3}S(S+1)\} + E(S_x^2 - S_y^2) +$$
$$+ \beta(g_x H_x S_x + g_y H_y S_y + g_z H_z S_z), \quad (7.85)$$

with $S = 2$, $|D| = 2 \cdot 24$ cm^{-1}, $|E| = 0 \cdot 10$ cm^{-1}, and g very close to 2.

From Fig. 7.3, using the negative sign of B_4 appropriate to a d^4 ion in an octahedral field, we see that the Γ_3 states are lower and that these are split by a tetragonal field. If an orthorhombic distortion is also present, the two Γ_3 states are not only split but also admixed, so that the ground state will be an orbital singlet of the form

$$\cos \delta \, |2^s\rangle + \sin \delta \, |0\rangle$$

with a fivefold spin degeneracy. The latter will be slightly lifted by matrix elements of the spin-orbit coupling, which connect with the Γ_5 states but not with the other Γ_3 orbital state. These also cause the g-factor to fall below the free spin value, and on including the spin-spin interaction, second-order perturbation theory gives the following

relations:
$$D = -3(\rho+\lambda^2/\Delta)\cos 2\delta,$$
$$E = -\sqrt{(3)}(\rho+\lambda^2/\Delta)\sin 2\delta,$$
$$g_x = 2-2(\lambda/\Delta)(\cos\delta-\sqrt{(3)}\sin\delta)^2, \quad (7.86)$$
$$g_y = 2-2(\lambda/\Delta)(\cos\delta+\sqrt{(3)}\sin\delta)^2,$$
$$g_z = 2-8(\lambda/\Delta)\cos^2\delta.$$

Here the orbital corrections to the g-factors are given only to the first order in (λ/Δ), where λ is the spin-orbit splitting parameter and Δ the amount by which the Γ_5 states (whose splitting is neglected) lie above the ground state; ρ is the constant of the spin-spin interaction (see eqns (16.43), (16.44)).

These formulae show that $E/D = 3^{-\frac{1}{2}}\tan 2\delta$, and the experimental result that $|E|$ is small compared with $|D|$ means that δ must be close to 0 or $\pi/2$. The correct choice depends on the sign of D, but unfortunately this was not determined experimentally. However in the isomorphous crystal $CuSO_4$, $5H_2O$ the results point unambiguously to a value of δ close to zero, and it may be assumed that this is the correct choice also for $CrSO_4$, $5H_2O$. Using $\lambda = 58$ cm^{-1}, the results can be fitted with $\Delta = 10000$ cm^{-1}, which is a reasonable order of magnitude for $10Dq$ (see Table 7.3). We then expect $g_z = 1\cdot95$, $g_x \approx g_y = 1\cdot99$.

A d^4 ion in a purely octahedral field presents a situation of some interest. The orbital states $|2^s\rangle$, $|0\rangle$ are then degenerate; when we include the spin $S = 2$ we have a manifold of 10 states that are split by second-order effects of the spin-orbit coupling (of order λ^2/Δ) and the spin-spin interaction. In zero magnetic field the states and energy levels are given in Table 7.15, where we have used the abbreviations

$$|2^s\rangle = (1/\sqrt{2})\{|+2\rangle+|-2\rangle\}, \quad |2^a\rangle = (1/\sqrt{2})\{|+2\rangle-|-2\rangle\}$$

for the spin states as well as for the orbital states. Here the zero of energy has been taken to lie at the centre of gravity of the states. The Zeeman effect in such a manifold is obviously rather complicated. In weak fields, such that $\beta H \ll (\rho+\lambda^2/\Delta)$, only the triplets have a first-order Zeeman effect, and each behaves like a triplet with zero initial splitting and an isotropic $g = 1$, if we neglect corrections of order λ/Δ. Since the separation between successive energy levels in zero field would be 2–3 cm^{-1}, the weak field condition is the one most likely to be encountered in a magnetic resonance experiment. These calculations may often prove to be no more than a pedantic curiosity (but see §7.17, $3d^6$ in a tetrahedral field). Certainly this approach is quite unrealistic

Table 7.15

Energy levels and states for the orbital doublet Γ_3 with spin $S = 2$, which is the ground state of d^4 in an exactly octahedral field. The ground orbital states are $|0\rangle$, $|2^s\rangle = (1/\sqrt{2})\{|+2\rangle+|-2\rangle\}$; second-order admixtures of states through spin-orbit coupling to the excited Γ_5 orbital triplet are omitted, but their contribution to the energy splitting is included. The nomenclature $|2^s\rangle$, $|2^a\rangle$ is also used for the spin states, with $|2^a\rangle = (1/\sqrt{2})\{|+2\rangle-|-2\rangle\}$.

Energy	Multiplicity	States
$W = +12\left(\rho+\dfrac{\lambda^2}{\Delta}\right)$	Singlet	$\dfrac{1}{\sqrt{(2)}}\{|0, 2^s\rangle-|2^s, 0\rangle\}$
$W = +6\left(\rho+\dfrac{\lambda^2}{\Delta}\right)$	Triplet	$\sqrt{(\tfrac{3}{4})}\,|2^s, +1\rangle-\tfrac{1}{2}|0, -1\rangle$ $|0, 2^a\rangle$ $-\sqrt{(\tfrac{3}{4})}\,|2^s, -1\rangle+\tfrac{1}{2}|0, +1\rangle$
$W = 0$	Doublet	$\dfrac{1}{\sqrt{(2)}}\{|0, 2^s\rangle+|2^s, 0\rangle\}$ $\dfrac{1}{\sqrt{(2)}}\{|2^s, 2^s\rangle-|0, 0\rangle\}$
$W = -6\left(\rho+\dfrac{\lambda^2}{\Delta}\right)$	Triplet	$-\tfrac{1}{2}|2^s, -1\rangle-\sqrt{(\tfrac{3}{4})}\,|0, +1\rangle$ $|2^s, 2^a\rangle$ $\tfrac{1}{2}|2^s, +1\rangle+\sqrt{(\tfrac{3}{4})}\,|0, -1\rangle$
$W = -12\left(\rho+\dfrac{\lambda^2}{\Delta}\right)$	Singlet	$\dfrac{1}{\sqrt{(2)}}\{|2^s, 2^s\rangle+|0, 0\rangle\}$

for Cr^{2+} in MgO, where the ion is so strongly coupled to the lattice that the energy levels are completely altered by the Jahn–Teller effect and by lattice strains. This problem is discussed by Fletcher and Stevens (1969) in connection with acoustic paramagnetic resonance measurements of Marshall and Rampton (1968).

7.12. $3d^5$. Cr^+, Mn^{2+}, Fe^{3+} in an octahedral field. $^6S_{\frac{5}{2}}$, $L = 0$, $S = \frac{5}{2}$.

If the ground state of these ions were a pure $^6S_{\frac{5}{2}}$, the magnetic resonance spectrum would be remarkably simple. The sextet would have no splitting other than that due to the Zeeman interaction, and a single line would be observed at the free spin value $g = 2\cdot 0023$. In practice the spectrum is rather complex, with both 'fine structure' and hyperfine structure. The fine structure can be represented by a series of spin operators O_k^q, where k is even and cannot have a value greater than 4. In the case of predominantly cubic symmetry with a small axial distortion about the tetragonal or trigonal axis, the appropriate forms of the spin Hamiltonian are

Tetragonal: $\mathcal{H} = \beta(\mathbf{H}\cdot\mathbf{g}\cdot\mathbf{S}) + B_4(O_4^0 + 5O_4^4) + B_2^0 O_2^0 + B_4^0 O_4^0$ (7.87)

Trigonal: $\mathcal{H} = \beta(\mathbf{H}\cdot\mathbf{g}\cdot\mathbf{S}) - \tfrac{2}{3}B_4(O_4^0 + 20\sqrt{(2)}O_4^3) + B_2^0 O_2^0 + B_4^0 O_4^0.$ (7.88)

In most published work the results have been expressed in terms of the equivalent spin Hamiltonian

$$\mathcal{H} = \beta(\mathbf{H}\cdot\mathbf{g}\cdot\mathbf{S}) + \tfrac{1}{6}a\{S_\xi^4 + S_\eta^4 + S_\zeta^4 - \tfrac{1}{5}S(S+1)(3S^2+3S-1)\} +$$
$$+ D\{S_z^2 - \tfrac{1}{3}S(S+1)\} +$$
$$+ \tfrac{1}{180}F\{35S_z^4 - 30S(S+1)S_z^2 + 25S_z^2 - 6S(S+1) + 3S^2(S+1)^2\},$$ (7.89)

where the coordinate system $\xi\eta\zeta$ refers to three mutually perpendicular axes which are the fourfold axes of the crystal field, and the z-axis lies along the trigonal axis of the crystal, which is also the [111] axis of the $\xi\eta\zeta$ system, or along a tetragonal axis, in which case the $\xi\eta\zeta$ system can be taken to coincide with the xyz system of axes. The relations between the parameters in the two types of Hamiltonian are:

$$a = 120B_4, \qquad D = 3B_2^0, \qquad F = 180B_4^0.$$

The g-value would be expected also to have axial symmetry, but in practice any anisotropy is generally within the experimental error.

The energy levels and strongly allowed transitions have been evaluated (Bleaney and Trenam 1954), assuming the Zeeman interaction to be dominant, by perturbation theory, including second-order terms due to a and D but only first-order terms from F, which is generally small. The strongly allowed transitions occur at fields:

$M = \pm\tfrac{5}{2} \leftrightarrow \pm\tfrac{3}{2}: \quad g\beta H = g\beta H_0 \mp \{2D(3\cos^2\theta - 1) + 2pa + \tfrac{1}{6}Fq\} -$
$$- 32\delta_1 + 4\delta_2 + \epsilon_1;$$

$\pm\tfrac{3}{2} \leftrightarrow \pm\tfrac{1}{2}: \quad g\beta H = g\beta H_0 \mp \{D(3\cos^2\theta - 1) - \tfrac{5}{2}pa - \tfrac{5}{24}Fq\} +$
$$+ 4\delta_1 - 5\delta_2 + \epsilon_2;$$

$+\tfrac{1}{2} \leftrightarrow -\tfrac{1}{2}: \quad g\beta H = g\beta H_0 + 16\delta_1 - 8\delta_2 + \epsilon_3.$ (7.90)

In these formulae H is the field at which a line is observed at the frequency ν of the applied radiation, and $H_0 = h\nu/g\beta$ is the field at which it would have occurred if all the fine-structure terms had been zero. The parameter p which appears in the terms due to the cubic field is given (see Kronig and Bouwkamp 1939) by the equation $p = (1-5\phi)$, where $\phi = (l^2m^2+m^2n^2+n^2l^2)$, (l, m, n) being the direction cosines of **H** referred to the axes of the cubic crystal field. The second-order terms due to the cubic field are

$$\epsilon_1 = \frac{a^2}{g\beta H_0}\{\tfrac{5}{3}\phi(1-7\phi)\},$$

$$\epsilon_2 = -\frac{a^2}{g\beta H_0}\{\tfrac{5}{48}(3+178\phi-625\phi^2)\}, \qquad (7.91)$$

$$\epsilon_3 = \frac{a^2}{g\beta H_0}\{\tfrac{10}{3}\phi(7-25\phi)\}.$$

The angle that **H** makes with the trigonal axis is denoted by θ, and the following abbreviations have been used:

$$q = (35\cos^4\theta - 30\cos^2\theta + 3), \qquad \delta_1 = (D^2/g\beta H_0)\cos^2\theta\sin^2\theta,$$

$$\delta_2 = (D^2/4g\beta H_0)\sin^4\theta.$$

In zero magnetic field the levels fall into three Kramers doublets, whose energy and states are given in Table 7.16.

For the case of strong external magnetic field it is readily seen from

TABLE 7.16

Energy levels and states in zero magnetic field for $S = \tfrac{5}{2}$, with fine structure Hamiltonian given by eqn (7.87) for a tetragonal splitting and (7.88) for a trigonal splitting, in each case combined with a cubic splitting. The parameters used are

$$a = 120 B_4, \quad D = 3B_2^0, \quad F = 180 B_4^0$$

Tetragonal distortion

Energy	State
$-\tfrac{1}{2}(a+\tfrac{2}{3}F)+\tfrac{2}{3}D+\{(a+\tfrac{2}{3}F+2D)^2+\tfrac{5}{4}a^2\}^{\frac{1}{2}}$	$\cos\alpha\,\lvert\pm\tfrac{5}{2}\rangle + \sin\alpha\,\lvert\mp\tfrac{3}{2}\rangle$
$-\tfrac{1}{2}(a+\tfrac{2}{3}F)+\tfrac{2}{3}D-\{(a+\tfrac{2}{3}F+2D)^2+\tfrac{5}{4}a^2\}^{\frac{1}{2}}$	$\sin\alpha\,\lvert\pm\tfrac{5}{2}\rangle - \cos\alpha\,\lvert\mp\tfrac{3}{2}\rangle$
$a+\tfrac{2}{3}F-\tfrac{8}{3}D$	$\lvert\pm\tfrac{1}{2}\rangle$

where $\tan 2\alpha = \dfrac{\sqrt{(5)}\,a}{2(a+\tfrac{2}{3}F)+4D}$

Trigonal distortion

Energy	State
$-\tfrac{1}{2}(a-F)+\tfrac{1}{3}D+\tfrac{1}{6}\{(a-F+18D)^2+80a^2\}^{\frac{1}{2}}$	$\cos\alpha\,\lvert\pm\tfrac{5}{2}\rangle \mp \sin\alpha\,\lvert\mp\tfrac{1}{2}\rangle$
$-\tfrac{1}{2}(a-F)+\tfrac{1}{3}D-\tfrac{1}{6}\{(a-F+18D)^2+80a^2\}^{\frac{1}{2}}$	$\sin\alpha\,\lvert\pm\tfrac{5}{2}\rangle \pm \cos\alpha\,\lvert\mp\tfrac{1}{2}\rangle$
$(a-F)-\tfrac{2}{3}D$	$\lvert\pm\tfrac{3}{2}\rangle$

where $\tan 2\alpha = \dfrac{\sqrt{(80)}\,a}{a-F+18D}$

the formulae given above that the different parameters representing the fine structure terms are associated with different angular variations, so that they can be determined uniquely (except for the overall sign) by making measurements in a sufficient number of directions. In the case of trigonal symmetry the first-order terms in a, F occur always in combination as an integral or half-integral multiple of $(pa+\frac{1}{12}Fq)$; along the trigonal axis $(pa+\frac{1}{12}Fq)$ takes the value $-\frac{2}{3}(a-F)$, while for any direction perpendicular to the trigonal axis its value is $-\frac{1}{4}(a-F)$. Thus measurements with the magnetic field parallel and normal to the trigonal axis are not sufficient to determine a and F separately, and only the values of D and $(a-F)$ can be found. The determination of a and F separately requires measurements in some intermediate direction, a fourfold axis of the cube often being the most convenient. A further result is that in a crystal such as $CaCO_3$, which has D_{3d}^6 symmetry with two ions in the unit cell which are related by a reflection in the plane normal to the trigonal axis, the spectra of the two Mn^{2+} ions which substitute for Ca^{2+} are identical if the magnetic field lies in this plane or parallel to the trigonal axis. Since the value of a is rather small, the fact that there are two ions with different cubic axes shows only as a small doubling of the resonance lines at intermediate directions of the magnetic field.

A wide range of experimental magnetic resonance data exists for $3d^5$ ions, of which a representative selection is given in Table 7.17. This table contains the values of g, the cubic splitting parameter a, and the hyperfine parameter A (where measured) for the isotopes ^{53}Cr, ^{55}Mn, and ^{57}Fe in the ions Cr^+, Mn^{2+}, and Fe^{3+} respectively. In a number of cases the sign of a has been determined from the change in the relative intensities at low temperatures between different fine structure lines, and the sign of A is then found from the variation in hyperfine spacing due to second-order effects (see Chapter 3). In other cases only the relative signs of a and A have been determined, and the signs given in parentheses are based on the plausible assumption that A is negative, since no instances to the contrary are known. When the symmetry is not exactly tetrahedral, octahedral or cubic (4-, 6-, or 8-fold coordination), terms in D and F are needed in the spin Hamiltonian, but the values of these parameters (which vary rather widely) have not been included. In such circumstances a small anisotropy in g and A would be expected; the latter has been observed in the more accurate measurements, but seldom exceeds 1 or 2 per cent and has been omitted in the table.

TABLE 7.17

Representative values of g, the cubic splitting parameter a and the hyperfine parameter A for $3d^5$, ^6S ions. The values of a, A are in units of 10^{-4} cm^{-1}, which can be converted into MHz by multiplying by 2.998

Ion	Host	g	a (10^{-4} cm^{-1})	A (10^{-4} cm^{-1})	Symmetry
^{53}Cr$^+$	NaF	2·001(1)	2·2		octahedral
	KMgF$_3$	2·0005(5)	4·5		octahedral
	ZnS	1·9995(5)	3·9		tetrahedral
^{55}Mn^{2+}	KMgF$_3$	2·0015(5)	(+)6·5	(−)91	octahedral
	MgO	2·0014(5)	+18·6	−81·5(2)	octahedral
	CaO	2·0011(5)	(+)5·9	(−)81·640(1)	octahedral
	SrO		1	−80·9(2)	octahedral
	CdF$_2$	2·0026(6)	4	(−)93	cube
	ZnO	2·001(1)	(+)2 to 5	(−)75	tetrahedral (distorted)
	ZnS	2·0025	(+)7·8	−64	tetrahedral
	ZnSe	2·0069(2)	19·3	−62	tetrahedral
	ZnTe	2·0105	30·0	−56	tetrahedral
	CdTe	2·0075(10)	(+)28·0	(−)57	tetrahedral
^{57}Fe^{3+}	KMgF$_3$	2·0031(2)	51		octahedral
	KCdF$_3$	2·0027(2)	53		octahedral
	MgO	2·0037(7)	+205	10·059(1)	octahedral
	CaO	2·0059(6)	+65	10·05(5)	octahedral
	AlRb(SO$_4$)$_2$, 12H$_2$O	2·003(2)	+†134		octahedral (distorted)
	AlK(SeO$_4$)$_2$, 12H$_2$O	2·003(1)	+†127		octahedral (distorted)
	Yttrium iron garnet				octahedral (distorted)
	Yttrium gallium garnet }	2·003(1)	+185	10·083(10)	octahedral (distorted)
		2·0047(5)	+62		tetrahedral (distorted)
	ZnS	2·019(1)	+127	7·7	tetrahedral

† Sign corrected.

Measurements of the nuclear moments of ^{57}Fe and ^{55}Mn, more accurate than previous determinations because of the negligible size of the pseudo-nuclear Zeeman effect, have been made by Endor on Fe^{3+} (cubic) in MgO (Locher and Geschwind 1965), on Mn^{2+} in CaO, ZnS (Mims, Devlin, Geschwind, and Jaccarino 1967), and on Mn^{2+} in MgO (Dyer and Woonton 1967). Some small higher-order terms have also been detected.

Table 7.17 shows that the g-values are very close to the free spin value of 2·0023 (for this reason, anisotropy, which can only appear in the deviation from the free spin value, is very difficult to observe). To a good approximation the g-value would be the same as for the free ion; the spin-orbit coupling produces a slight breakdown of LS- coupling, introducing a small admixture of the excited state $^4P_{\frac{3}{2}}$ into the ground state $^6S_{\frac{5}{2}}$, but the expected deviation $\Delta g = g - g_s$ is only about $-0\cdot0003$ for Fe^{3+} and $-0\cdot0004$ for Mn^{2+} (Watanabe 1957). The observed deviations are of either sign, the larger ones being positive rather than negative. The latter are observed in compounds where a fair amount of covalent bonding (or a decrease in 'ionicity') would be expected. Fidone and Stevens (1959) have suggested an explanation of this which has been elaborated by Watanabe (1964). In the presence of covalent bonding excited sextets of symmetry $^6T_{1g}$ are present, whose single electron states will tend to be full or empty according to whether electrons are transferred to or from the central ion by the bonding action. A second-order shift (linear instead of quadratic in spin-orbit coupling) is then possible in the g-value, whose sign depends on the direction of electron transfer.

Table 7.17 shows also that in the cases where the sign of the cubic splitting parameter a has been determined it has been found to be positive, both for fourfold and for sixfold coordination. The cubic field splitting arises through perturbation of the free ion state by interactions involving both the cubic crystalline potential and the spin-orbit coupling; since the former is of opposite sign for fourfold and sixfold coordination the results suggest that a varies as an even power of the crystal field. There is, however, no theoretical reason why odd powers of the crystal field should be excluded, as pointed out by Powell, Gabriel, and Johnston (1960). The configuration d^5 may be regarded as made up either of five electrons or five holes; energy splittings due to either crystal fields or spin-orbit coupling are of opposite sign for the configurations d^n and d^{10-n}, so that for d^5 we can only conclude that the sum of the powers of crystal field and spin-orbit coupling must be even.

Assuming that the nucleus and the 'argon' core of electrons in closed shells can be regarded as a source of a rigid spherically symmetrical scalar potential V, the motion of the five d-electrons is described by the Hamiltonian

$$\mathscr{H} = T+V+ \sum_{i<j} e^2/r_{ij}+\sum_{i} V_i+\zeta \sum_{i} (\mathbf{l}_i \cdot \mathbf{s}_i)+W_{ss}, \qquad (7.92)$$

where V_i denotes the crystal field seen by the ith electron, and W_{ss} is the spin-spin interaction, which makes a small contribution. Watanabe (1957) treated the case where V_i has cubic symmetry by means of a perturbation expansion based on the spin-sextet and spin-quartet states of the free $3d^5$ configuration, using V_i as a perturbation relative to the $\sum e^2/r_{ij}$ term. It was shown, however, by Powell, Gabriel, and Johnston (1960) that the spin-doublet contribution to the ground state manifold, via the spin-orbit interaction, is of major importance; using a value of $\zeta = 400$ cm^{-1}, they found that for a range of values of Dq between $+1200$ and -1200 cm^{-1}, the cubic splitting parameter a varied very widely but was positive except for a narrow range of Dq between 0 and about $+200$ cm^{-1}. In a later and more detailed paper (Gabriel, Johnston, and Powell 1961) it was found possible to fit both the observed value of a for Mn^{2+} in MgO and the optical spectrum of MnO, using rather smaller (and probably more realistic) values of $\zeta < 300$ cm^{-1}. (A value for ζ of 320 ± 10 cm^{-1} has been obtained from the optical spectrum of Mn^{2+} in RbMnF$_3$ (Mehra and Venkateswarlu 1967).) In the range $+1000$ cm$^{-1} \leqslant Dq \leqslant +1500$ cm^{-1} the parameter a was found to be a sensitive function of ζ and Dq, varying approximately as ζ^4 and $(Dq)^n$, where n varies between 3·5 and 6.

A distinctive feature of the results for Mn^{2+} is the sharp rise in the value of the cubic parameter a with the atomic number of the anion in the II–VI compounds with fourfold coordination, as can be seen in the progression ZnO, ZnS, ZnSe, ZnTe. This is accompanied by a significant rise in the g-value above the free spin value, and a fall in the value of the hyperfine constant A. Each of these two effects would be expected to be associated with a progressive decrease of 'ionicity' of the Mn^{2+} ion, and the results suggest that we may also associate the rise in a with this decrease. A noticeable increase in the value of a occurs generally in the progression Cr$^+$, Mn^{2+}, Fe^{3+}; this may be partly due to the fact that the increase in spin-orbit coupling with higher nuclear charge may outweigh any reduction in Dq due to shrinkage of the electronic orbits, but a change in ionicity (e.g. in ZnS) may also play a role.

7.13. $3d^6$. Fe^{2+} in an octahedral field. ^5D, $L = 2$, $S = 2$

In an octahedral field the Γ_5 triplet is the orbital ground state, corresponding to the fact that d^6 can be regarded as a single electron outside a half-filled shell. We consider first the case of a purely octahedral field, with no distortion. Under the influence of the spin-orbit coupling, the orbital triplet with its fivefold spin degeneracy ($S = 2$), breaks up into a triplet, quintet, and septet, as shown in Fig. 7.19(b). The triplet is the lowest level, and its wave-functions in terms of $|L, S\rangle$ functions are

$$|\tilde{J}_z\rangle = |+\tilde{1}\rangle = \sqrt{(\tfrac{3}{5})} |1, 2\rangle + \sqrt{(\tfrac{3}{10})} |2^a, 1\rangle - \sqrt{(\tfrac{1}{10})} |-1, 0\rangle,$$
$$|\tilde{J}_z\rangle = |\tilde{0}\rangle = \sqrt{(\tfrac{3}{10})} |1, 1\rangle + \sqrt{(\tfrac{2}{5})} |2^a, 0\rangle - \sqrt{(\tfrac{3}{10})} |-1, -1\rangle,$$
$$|\tilde{J}_z\rangle = |-\tilde{1}\rangle = \sqrt{(\tfrac{1}{10})} |1, 0\rangle + \sqrt{(\tfrac{3}{10})} |2^a, -1\rangle - \sqrt{(\tfrac{3}{5})} |-1, -2\rangle.$$
(7.93)

In computing the Zeeman interaction within this triplet, we anticipate that it will be necessary to allow for some covalent bonding by including a reduction factor for the orbital moment. Since the Γ_5 orbital wave-functions take part only in π-bonding, the appropriate factor is $k_{\pi\pi}$, making the Zeeman operator equal to $k_{\pi\pi}\mathbf{L} + g_s\mathbf{S}$. Evaluation of the matrix elements within the triplet functions given in eqn (7.93) shows that the g-factor is isotropic and equal to $\tfrac{1}{2}(3g_s + k_{\pi\pi})$. (The phases of the wave-functions in eqn (7.93) have been chosen to make the off-diagonal components of the Zeeman operator positive, like the diagonal components.)

These results can be obtained easily by using the fact that the Γ_5 triplet may be described by an effective angular momentum $l = 1$, with $\alpha = -1$ (see § 7.5). Then the 'term' with $\tilde{l} = 1, S = 2$ breaks up through the spin-orbit coupling into states with effective angular momenta $\tilde{J} = 1, 2$, and 3. Since the spin-orbit parameter λ is negative for d^6 we would have expected an inverted multiplet, but because $\alpha = -1$ we in fact get a normal multiplet, with the triplet state lowest. The Zeeman splitting within each component of the multiplet can be described by means of an effective Landé factor, writing $\tilde{g}_l = -k_{\pi\pi}$ in order to include both the factor $\alpha = -1$ and the reduction due to covalency. The results are

$$\tilde{J} = 1, \quad g = \tfrac{1}{2}(3g_s + k_{\pi\pi}),$$
$$\tilde{J} = 2, \quad g = \tfrac{1}{6}(5g_s - k_{\pi\pi}), \quad (7.94)$$
$$\tilde{J} = 3, \quad g = \tfrac{1}{3}(2g_s - k_{\pi\pi}).$$

In passing we note that the states whose real orbital angular momentum component is specified by $|+1\rangle$, $|2^a\rangle$, $|-1\rangle$ must be identified in terms of the fictitious angular momentum l as $|-\tilde{1}\rangle$, $|\tilde{0}\rangle$, $|+\tilde{1}\rangle$ respectively. If we apply this change of nomenclature to eqns (7.93), we see that each state $|\tilde{J}_z\rangle$ is a mixture of states $|\tilde{l}_z, S_z\rangle$ such that for each $\tilde{J}_z = \tilde{l}_z + S_z$, as is the case in the coupling of real angular momenta in multiplet states.

We now consider the effect of off-diagonal elements of the spin-orbit coupling, which admix some of the excited Γ_3 states into the Γ_5 states (more exactly, since we shall allow for bonding effects, we should speak of admixing some of the e-states into the t_2 states). This will produce downward shifts in the $\Gamma_5(t_2)$ energy levels; in the case of $\tilde{J} = 1$ the shift is the same for all three states so the triplet remains unsplit, but in the cases of $\tilde{J} = 2$ and $\tilde{J} = 3$ there will be a splitting (to which the spin-spin interaction may also be expected to contribute). As these are excited states in which it is unlikely that resonance will be observed, we shall not evaluate such splittings in detail, but note that (assuming cubic symmetry) we expect the $\tilde{J} = 2$ state to split into a doublet and a triplet, and the $\tilde{J} = 3$ state into two triplets and a singlet. (We may also anticipate that such splittings can be represented by cubic spin operators of the fourth degree.)

In addition to the energy shifts and splittings, there will be corrections to the $\tilde{g}$ values $\sim(\lambda/\Delta)$, where Δ is the energy of the excited orbital doublet Γ_3. These arise from cross terms of the form

$$-\frac{\langle t_2 | e \rangle \langle e | t_2 \rangle}{\Delta}, \qquad (7.95)$$

where one matrix element comes from the spin-orbit coupling and the other from the orbital angular momentum. From Chapter 20 we find that the matrix elements of the former involve the parameter $(\lambda_{\pi\sigma} k_{\pi\sigma})$ instead of λ, and for the latter we have $k_{\pi\sigma}$ instead of unity, in order to allow for bonding effects. Hence for the ground triplet, on evaluating the matrix elements, we find the result

$$\tilde{J} = 1, \qquad \tilde{g} = \tfrac{3}{2}g_s + \tfrac{1}{2}k_{\pi\pi} - \frac{18}{5}\frac{\lambda_{\pi\sigma}}{\Delta}k_{\pi\sigma}^2. \qquad (7.96)$$

The paramagnetic resonance spectrum and also the optical spectrum of the Fe^{2+} ion in MgO have been studied by Low and Weger (1960) and others. The $\tilde{g}$-value of the ground triplet is found to be 3·428, while $\Delta = 10000$ cm^{-1}. The value of λ for the free Fe^{2+} ion is -100

cm^{-1}, and if we make the approximation that $(\lambda_{\pi\sigma}/\lambda) \sim k_{\pi\sigma} \sim k_{\pi\pi}$ in evaluating the small correction term of order $\lambda/\Delta = 0\cdot 01$, we find we must assume that $k_{\pi\pi} = 0\cdot 80 \pm 0\cdot 02$. The g-value of Fe^{2+} in NaF with cubic symmetry (Hall, Hayes, Stevenson, and Wilkens 1963) has been found to be 3·420, and although the size of the cubic field splitting in this case is not known it is clear that the value of $k_{\pi\pi}$ must be about the same as in MgO.

The only stable isotope of iron with non-zero nuclear spin is ^{57}Fe, and a number of studies of hyperfine structure in the paramagnetic state have been made using the Mössbauer effect. For Fe^{2+} in an exactly octahedral field, the triplet electronic ground state $\tilde{J} = 1$ will have a magnetic hyperfine interaction of the simple form

$$\mathscr{H}_n^{\text{mag}} = A(\tilde{\mathbf{J}} \cdot \mathbf{I}) \tag{7.97}$$

where

$$A = \mathscr{P}(\tfrac{1}{2} - \tfrac{3}{2}\kappa + \tfrac{3}{70}), \tag{7.98}$$

in which the three contributions come from the orbit, the core polarization, and the spin-dipolar interaction respectively.

Since the ground nuclear state of ^{57}Fe is $I = \tfrac{1}{2}$, it will have no nuclear electric quadrupole interaction, but for the excited nuclear state $I = \tfrac{3}{2}$ we have a term of the form given in eqn (18.50) with $\tilde{J} = 1$

$$\mathscr{H}_n^{\text{eq}} = -\frac{e^2 Q \langle r_q^{-3} \rangle}{I(2I-1)} \left[\tfrac{1}{6} m \sum_{x,y,z} \{3\tilde{J}_x^2 - \tilde{J}(\tilde{J}+1)\}\{3I_x^2 - I(I+1)\} + \right. \\ \left. + \tfrac{3}{4} n \sum_{x \neq y, \text{etc.}} (\tilde{J}_x \tilde{J}_y + \tilde{J}_y \tilde{J}_x)(I_x I_y + I_y I_x) \right], \tag{7.99}$$

where

$$m = \langle L \| \alpha \| L \rangle \langle +\tilde{1} | \, 3L_z^2 - L(L+1) \, | +\tilde{1} \rangle,$$
$$\sqrt{(2)} n = \langle L \| \alpha \| L \rangle \langle +\tilde{1} | \, L_z L_+ + L_+ L_z \, | \tilde{0} \rangle. \tag{7.100}$$

On evaluating the matrix elements we find $m = -n = +1/35$. This differs from the usual result for a free ion in that the latter has $n = m$; this emphasizes the point that, since the sign of n depends on the phases given to the wave-functions in (7.93), it is essential to choose these so that the off-diagonal elements of $\tilde{\mathbf{J}}$ are consistent in sign with those of $\mathbf{L}, \mathbf{S}$.

In relating the hyperfine Hamiltonian to the observed Mössbauer spectrum, allowance must be made for the presence of small static strains which broaden the resonance lines (see Ham 1967). The quadrupole splitting disappears above 14°K, and Ham suggested this arose from spin-lattice relaxation through an Orbach process (see

§ 10.4) involving an excited level at about 100 cm^{-1}. A level at 105 cm^{-1} (Wong 1968) has been observed for Fe^{2+} in MgO, which is a good deal lower than the value $|2\lambda| \approx 200$ cm^{-1} expected from static crystal field theory. This is ascribed to quenching of the spin-orbit coupling by a dynamic Jahn–Teller effect (Ham, Schwarz, and O'Brien 1969); the orbital momentum is not quenched so effectively in this case as it is for Fe^{2+} in CaO (see § 21.12), although covalent bonding is likely to be smaller in the latter compound which has a larger interionic spacing.

7.14. $3d^7$. Fe$^+$, Co^{2+}, Ni^{3+} in an octahedral field. ^4F, $L = 3$, $S = \tfrac{3}{2}$

For a d^7 ion in an octahedral field the Γ_4 triplet is lowest; this is split by the spin-orbit coupling and, if the octahedron is distorted, by crystal field terms of tetrahedral, trigonal, or lower symmetry. In most hydrated salts the latter are comparable in magnitude with the spin-orbit coupling, a problem that has been treated in some detail by Abragam and Pryce (1951c). We shall restrict our discussion almost entirely to the case of undistorted octahedral symmetry.

As shown in §14.2, the orbital triplet Γ_4 formed from ^4F can be assigned a fictitious angular momentum $\tilde{l} = 1$, with which must be associated a value of $\alpha = -\tfrac{3}{2}$. There is however a low lying ^4P term (some 14000 cm^{-1} higher in the free Co^{2+} ion), also arising from the d^7 configuration, which is also a Γ_4 state, and which is admixed to the ^4F, Γ_4 state by an octahedral crystal field. The ground triplet will therefore be of the form

$$\epsilon\,|^4F,\,\Gamma_4\rangle + \tau\,|^4P,\,\Gamma_4\rangle$$

for which the value of α is

$$\alpha = -\tfrac{3}{2}\epsilon^2 + \tau^2 = -\tfrac{1}{2}(3 - 5\tau^2).$$

In the presence of weak covalent bonding there will be a further reduction in the orbital moment, whose effect is allowed for by the introduction of a factor k. Thus the effective orbital g-factor is $\tilde{g}_l = \alpha k$.

On including the spin we have a twelvefold multiplicity in the ground state, corresponding to $\tilde{l} = 1$, $S = \tfrac{3}{2}$. On introducing the spin-orbit coupling this splits into a doublet, a quartet, and a sextet as shown in Fig. 7.20(b), to which may be assigned fictitious angular momenta $\tilde{J} = \tfrac{1}{2}, \tfrac{3}{2}$, and $\tfrac{5}{2}$ respectively. The energy splittings are shown in Fig. 20(b); the doublet $\tilde{J} = \tfrac{1}{2}$ is lowest because within the orbital triplet the effective spin-orbit parameter $\tilde{\lambda}$, which is $\alpha\lambda$ apart from any reduction due to covalent bonding, is positive because both α and λ are negative. The energy matrices for the spin-orbit interaction, and the transformation matrices which diagonalize them, are given in Table 7.18, where λ' is the ^4F spin-orbit parameter modified by admixture of the ^4P state and any covalent bonding, but not including the factor $\alpha = -\frac{3}{2}$. The 3×3 matrices belong to the states $\tilde{J}_z = \pm\frac{1}{2}$, the 2×2 matrices to the states $\tilde{J}_z = \pm\frac{3}{2}$, and the 1×1 matrix to $\tilde{J}_z = \pm\frac{5}{2}$. From the matrices we see that the states belonging to the ground doublet $\tilde{J} = \frac{1}{2}$ are

$$\sqrt{(\tfrac{1}{2})} \,|\mp\tilde{1}, \pm\tfrac{3}{2}\rangle - \sqrt{(\tfrac{1}{3})}\,|\tilde{0}, \pm\tfrac{1}{2}\rangle + \sqrt{(\tfrac{1}{6})}\,|\pm\tilde{1}, \mp\tfrac{1}{2}\rangle. \qquad (7.101)$$

Evaluation of the Zeeman effect within this doublet using the operator $\tilde{g}_l\mathbf{l} + g_s\mathbf{S}$ shows that it has the isotropic $\tilde{g}$-factor

$$\tilde{g} = \tfrac{5}{3}g_s - \tfrac{2}{3}\tilde{g}_l, \qquad (7.102a)$$

which for $g_s = 2$, $\tilde{g}_l = -\tfrac{3}{2}$ becomes $4 \cdot 33$; these results agree with the simple calculation given in § 19.3.

Second-order effects arise from the matrix elements of the spin-orbit coupling between Γ_4 and Γ_5. They give rise to: (1) a depression in energy of the Γ_4 levels; (2) a splitting of the $\tilde{J} = \tfrac{5}{2}$ sextet into a doublet and a quartet; (3) a small change in the $\tilde{g}$-values, of which only this last is important as far as resonance in the ground doublet is concerned. If in evaluating the correction we neglect the admixture of ^4P into the ^4F, Γ_4 triplet and assume that the Γ_5 triplet lies an amount Δ higher in energy than the ground doublet, we obtain

$$\tilde{g} = \tfrac{5}{3}g_s - \tfrac{2}{3}\tilde{g}_l - \frac{15\lambda''k'}{2\Delta}, \qquad (7.102b)$$

where λ'' and k' refer to the reduced values of the spin-orbit parameter and orbital k-factor appropriate to matrix elements between Γ_4 and Γ_5.

Endor measurements on $^{59}\mathrm{Co}^{2+}$ ($I = \tfrac{7}{2}$) in MgO give $\tilde{g} = 4 \cdot 280(2)$, $A = (+)290 \cdot 55(6)$ MHz, and $g^{(I)}/g_I = 1 \cdot 39(1)$ (Fry and Llewellyn 1962). In an appendix to this paper Pryce gives the formulae (where the sign of α has been reversed to agree with that adopted above)

$$A = 2g_\mathrm{n}\beta\beta_\mathrm{n}\langle r^{-3}\rangle\{-\tfrac{2}{3}\alpha - \tfrac{5}{3}\kappa + \tfrac{1}{63}(1 - 15\tau^2)\}, \qquad (7.103a)$$

$$\frac{g^{(I)}}{g_I} = \frac{1 + 20(2-\alpha)\beta^2\langle r^{-3}\rangle}{9\delta}\{-\alpha - \kappa + \tfrac{13}{420}(1 - 15\tau^2)\}, \qquad (7.103b)$$

where δ is the energy of the $\tilde{J} = \tfrac{3}{2}$ level, assumed to be ≈ 305 cm^{-1} above the ground $\tilde{J} = \tfrac{1}{2}$ doublet. The results are interpreted using the

TABLE 7.18

Matrices of spin-orbit coupling within ^4F, Γ_4, $\tilde{l}=1$, $S=\frac{3}{2}$

Energy matrix			Transformation matrix			Diagonal states	Energy
$\|\mp\tilde{1},\pm\tfrac{3}{2}\rangle$	$\|\tilde{0},\pm\tfrac{1}{2}\rangle$	$\|\pm\tilde{1},\mp\tfrac{1}{2}\rangle$					
$\tfrac{9}{4}\lambda'$	$-3\sqrt{(\tfrac{3}{8})}\lambda'$	0	$\sqrt{\tfrac{1}{2}}$	$-\sqrt{\tfrac{1}{3}}$	$\sqrt{\tfrac{1}{6}}$	$\|\tilde{J}=\tfrac{1}{2},\tilde{J}_z=\pm\tfrac{1}{2}\rangle$	$\tfrac{15}{4}\lambda'$
$-3\sqrt{(\tfrac{3}{8})}\lambda'$	0	$-3\sqrt{(\tfrac{5}{8})}\lambda'$	$\sqrt{\tfrac{2}{5}}$	$\sqrt{\tfrac{1}{15}}$	$-\sqrt{\tfrac{8}{15}}$	$\|\tilde{J}=\tfrac{3}{2},\tilde{J}_z=\pm\tfrac{1}{2}\rangle$	$\tfrac{3}{2}\lambda'$
0	$-3\sqrt{(\tfrac{5}{8})}\lambda'$	$\tfrac{3}{4}\lambda'$	$\sqrt{\tfrac{1}{10}}$	$\sqrt{\tfrac{3}{5}}$	$\sqrt{\tfrac{3}{10}}$	$\|\tilde{J}=\tfrac{5}{2},\tilde{J}_z=\pm\tfrac{1}{2}\rangle$	$-\tfrac{9}{4}\lambda'$

$\|\pm\tilde{1},\pm\tfrac{1}{2}\rangle$	$\|\tilde{0},\pm\tfrac{3}{2}\rangle$					
$-\tfrac{3}{4}\lambda'$	$-3\sqrt{(\tfrac{3}{8})}\lambda'$	$\sqrt{\tfrac{2}{5}}$	$-\sqrt{\tfrac{3}{5}}$		$\|\tilde{J}=\tfrac{3}{2},\tilde{J}_z=\pm\tfrac{3}{2}\rangle$	$\tfrac{3}{2}\lambda'$
$-3\sqrt{(\tfrac{3}{8})}\lambda'$	0	$\sqrt{\tfrac{3}{5}}$	$\sqrt{\tfrac{2}{5}}$		$\|\tilde{J}=\tfrac{5}{2},\tilde{J}_z=\pm\tfrac{3}{2}\rangle$	$-\tfrac{9}{4}\lambda'$

$\|\pm\tilde{1},\pm\tfrac{3}{2}\rangle$						
$-\tfrac{9}{4}\lambda'$					$\|\tilde{J}=\tfrac{5}{2},\tilde{J}_z=\pm\tfrac{5}{2}\rangle$	$-\tfrac{9}{4}\lambda'$

values $\alpha = -1{\cdot}406$, $\tilde{g}_l = k\alpha = -1{\cdot}22$, $\Delta = 8400$ cm^{-1}, $k = k' = 0{\cdot}82$ and $-15\lambda''k'/2\Delta = 0{\cdot}160$. The effective value of $\langle r^{-3}\rangle$ is 5·12 a.u., which is close to k times the Freeman and Watson value (Table 7.6), and $\kappa = 0{\cdot}320$.

The absence of any appreciable dynamic Jahn–Teller quenching of the orbital momentum for this ion is discussed in § 21.12.

In conclusion, we digress a little to show why cobaltous salts commonly show very high anisotropy. If the octahedron has a small tetragonal or trigonal distortion, the effect is to add terms on the diagonal of the energy matrices given in Table 7.18; identical matrices are obtained for either type of distortion provided of course that the basis functions are defined by taking the polar axis as a fourfold or threefold axis of the octahedron respectively. The effect of the additional terms is to split the quartet and sextet, so that J is no longer a good quantum number, giving a number of Kramers doublets for which $\tilde{J}_z = \pm\tfrac{1}{2}$, $\pm\tfrac{3}{2}$, $\pm\tfrac{5}{2}$. Provided that the additional terms are small compared with the spin-orbit coupling the ground doublet may be written in the form

$$|\tilde{J} = \tfrac{1}{2}, \tilde{J}_z = \pm\tfrac{1}{2}\rangle + a\,|\tfrac{3}{2}, \pm\tfrac{1}{2}\rangle + b\,|\tfrac{5}{2}, \pm\tfrac{1}{2}\rangle \qquad (7.104)$$

where a, and still more so b, are small compared with unity. The Zeeman operator has no matrix elements between $\tilde{J} = \tfrac{1}{2}$ and $\tilde{J} = \tfrac{5}{2}$, and to the approximation where terms of order a^2, b^2 are neglected we have

$$\left.\begin{array}{l} g_\| = \tfrac{5}{3}g_s - \tfrac{2}{3}\tilde{g}_l + (4\sqrt{(5)}a/3)(2g_s - \tilde{g}_l), \\ g_\perp = \tfrac{5}{3}g_s - \tfrac{2}{3}\tilde{g}_l - (2\sqrt{(5)}a/3)(2g_s - \tilde{g}_l), \end{array}\right\} \qquad (7.105)$$

showing that $g_\| + 2g_\perp \approx 5g_s - 2\tilde{g}_l \approx 13$. This relation is roughly obeyed by the results for a number of hydrated cobalt salts. The degree of anisotropy is illustrated by taking $a = 0{\cdot}2$; then for $g_s = 2$, $\tilde{g}_l = -\tfrac{3}{2}$ we have $g_\| = 6{\cdot}42$, $g_\perp = 3{\cdot}27$.

7.15. $3d^8$. Co$^+$, Ni^{2+}, Cu^{3+} in an octahedral field. ^3F, $L = 3$, $S = 1$

In an octahedral field the orbital F state splits in the same way as for d^3 (Cr^{3+}, etc.), with the singlet Γ_2 lowest. The paramagnetic resonance spectrum is that expected for a spin triplet, but because the spin-orbit coupling has the reverse sign from that for d^3, and is rather larger, the g-value is appreciably greater than the free spin value, being commonly in the range 2·15 to 2·35. In compounds such as the tutton salts, where the symmetry is not exactly octahedral, the spin triplet is split through the action of the non-cubic part of the crystal field and the spin-orbit coupling; the effective spin Hamiltonian contains a term of the form (**S . D . S**), where D may be of order of a few cm^{-1}.

The simple crystal field theory follows very closely that for d^3 in an octahedral field, except that $S = 1$ instead of $\frac{3}{2}$, and the results for the ground spin triplet can again be given in a form applicable to a distortion of the octahedron with axial symmetry about either a tetragonal or a trigonal axis. The spin-orbit coupling only couples the ground orbital singlet Γ_2 to the lower orbital triplet Γ_5, and using the same nomenclature as in Table 7.13, the states and energies for the ground spin triplet listed in Table 7.19 are obtained.

TABLE 7.19

States and energy levels for ground spin triplet state $(\Gamma_2, S = 1)$ for d^8 in a nearly octahedral crystal field, with tetragonal or trigonal distortion. The Γ_5, $\tilde{l}_z = \pm\tilde{1}$ orbital states are assumed to lie at an energy Δ_1 above Γ_2, and the Γ_5, $\tilde{l}_z = 0$ orbital state at an energy Δ_0; they are admixed through the spin-orbit coupling $\lambda(\mathbf{L . S})$

$$|\overline{+1}\rangle = |\Gamma_2, +1\rangle - \frac{2\lambda}{\Delta_0}|\Gamma_5(\tilde{0}), +1\rangle + \frac{2\lambda}{\Delta_1}|\Gamma_5(+\tilde{1}), 0\rangle$$

$$|\overline{0}\rangle = |\Gamma_2, 0\rangle - \frac{2\lambda}{\Delta_1}|\Gamma_5(-\tilde{1}), +1\rangle + \frac{2\lambda}{\Delta_1}|\Gamma_5(+\tilde{1}), -1\rangle$$

$$|\overline{-1}\rangle = |\Gamma_2, -1\rangle + \frac{2\lambda}{\Delta_0}|\Gamma_5(\tilde{0}), -1\rangle - \frac{2\lambda}{\Delta_1}|\Gamma_5(-\tilde{1}), 0\rangle$$

$$W_{\pm 1} = -\frac{4\lambda^2}{\Delta_0} - \frac{4\lambda^2}{\Delta_1}, \qquad W_0 = -\frac{8\lambda^2}{\Delta_1},$$

giving

$$W_{\pm 1} - W_0 = D = -4\lambda^2/\Delta_0 + 4\lambda^2/\Delta_1,$$

but there is a third-order correction to D of comparable size in the case of a trigonal distortion (see $3d^3$, §7.10).

In this table it has been assumed that (λ/Δ) is so small that quantities of order $(\lambda/\Delta)^2$ can be neglected. In this approximation the Zeeman effect in the Γ_2 spin triplet is characterized by formulae which are the same as for d^3;

$$g_\| = g_s - \frac{8\lambda}{\Delta_0}, \qquad g_\perp = g_s - \frac{8\lambda}{\Delta_1}. \tag{7.106}$$

If the symmetry is accurately cubic, then $g_\| = g_\perp$. In practice anisotropy in g has rarely been detected even in cases of lower symmetry, presumably because sufficiently accurate measurements of g are difficult to make when the spin triplet has an initial splitting of appreciable magnitude. For the tutton salts (Griffiths and Owen 1952a) the average value of g is close to 2·25, while the value of Δ for the $[\text{Ni}(\text{H}_2\text{O})_6]^{2+}$ complex in solution is found from optical absorption to be about 8400 cm^{-1}; this leads to an apparent value of λ of about -270 cm^{-1} (see Owen 1955), which is noticeably smaller than the free ion value. The discrepancy is attributed to bonding effects.

As in the case of d^3, the splitting of the triplet states is really of third order $\lambda^2(\Delta_1-\Delta_0)/\Delta^2$, and in the trigonal case there may be other contributions of the same order (see § 7.10), which should be included in any comparison of the observed value of D with the difference between $W_{\pm 1}$ and W_0 as listed in Table 7.19. In practice no such comparisons have been made except in Al_2O_3 (see below), and even here the splitting of the excited states has not been measured. The observed values of D for trigonal compounds of Ni^{2+} vary widely. In $\text{Zn}_3\text{La}_2(\text{NO}_3)_{12}$, $24\text{H}_2\text{O}$ there are two types of sites in which Ni^{2+} ions can replace Zn^{2+}; in one of these D is about $+0\cdot05$ cm^{-1}, in the other about $-2\cdot2$ cm^{-1} (Culvahouse 1962). D also varies with temperature, a conspicuous example being NiSiF_6, $6\text{H}_2\text{O}$, where it falls from $-0\cdot5$ cm^{-1} at room temperature to $-0\cdot12$ cm^{-1} at 20°K and below (see Penrose and Stevens 1950). The value of D has also been measured as a function of pressure (Walsh and Bloembergen 1957), and of uniaxial stress (Walsh 1958).

The most detailed studies of d^8 ions in a trigonal field have been made using Al_2O_3 as a host lattice; the results are given in Table 7.20. The crystal field splitting for Cu^{3+} is found from the optical absorption, and is over twice as large as that estimated for Ni^{2+} in the same lattice. This partly accounts for the g-values of Cu^{3+} being much closer to the free spin values than those of Ni^{2+}, but from eqns (7.106) for the g value (ignoring the small anisotropy) Blumberg, Eisinger, and Geschwind (1963) find that the effective values of the spin-orbit coupling parameter λ need to be taken as -214 cm^{-1} and -240 cm^{-1} for Cu^{3+} and Ni^{2+} respectively, which amount only to about 50 and 75 per cent respectively of the free ion values. As in the case of Cr^{3+} in Al_2O_3, the anisotropy of the g-values and the value of D are difficult to explain on the simple theory given above; in particular they appear to require a sign for the trigonal field splitting $(\Delta_0-\Delta_1)$ of the Γ_5 triplet different from that observed for other ions in Al_2O_3.

TABLE 7.20

Data for $3d^8$, Cu^{3+} and Ni^{2+} in Al_2O_3, fitted to the spin Hamiltonian (7.1) with $S = 1$

		Cu^{3+}		Ni^{2+}
$\Delta = 10Dq$		21000 cm^{-1}		10000 cm^{-1}(a)
$g_\parallel$		2·0788(5)		2·1948(10)
$g_\perp$		2·0772(5)		2·1853(10)
$g_\parallel - g_\perp$		0·0016		0·0095
D		$-0·1884$ cm^{-1}		$-1·312$ cm^{-1}
$A_\parallel$	(^{63}Cu)	$-192·947(1)$ MHz	(^{61}Ni)	$(+)36·767(2)$ MHz
	(^{65}Cu)	$-206·679(1)$ MHz		
$A_\perp$	(^{63}Cu)	$-180·10(5)$ MHz	(^{61}Ni)	$(+)32·55(4)$ MHz
	(^{65}Cu)	$-192·92(5)$ MHz		
$P_\parallel$	(^{63}Cu)	$-0·042(1)$ MHz	(^{61}Ni)	$-0·170(2)$ MHz
	(^{65}Cu)	$-0·036(1)$ MHz		

Notes. (a) Estimated by comparison with MgO:Ni^{2+}.
References
Cu^{3+}. Blumberg, Eisinger, and Geschwind (1963).
Ni^{2+}. Marshall, Kikuchi, and Reinberg (1962); Locher and Geschwind (1963).

Endor measurements have been made to determine accurately the hyperfine constants of the stable isotopes ^{63}Cu, ^{65}Cu, and ^{61}Ni (in the last case considerably enhanced in abundance above the natural value of 1·25 per cent). Each of these isotopes has spin $I = \frac{3}{2}$, and the anisotropic values of the magnetic hyperfine constants ($A_\parallel$ and $A_\perp$) and the quadrupole interaction parameter $P_\parallel$ ($= \frac{1}{4}e^2qQ$ for $I = \frac{3}{2}$) are given in Table 7.20. Comparison of the ratio of the values of $A_\parallel$ for the two isotopes of copper with the known nuclear moment ratio gives a hyperfine anomaly of about 0·015 per cent, very close to that found in atomic beam experiments for the $4s$ state of unionized copper. This suggests that the main contribution to the magnetic hyperfine structure comes from s-electrons through the core-polarization effect, the d-electron contribution being rather small. The latter would be expected because the electron density distribution in the Γ_2 state has almost cubic symmetry, so that the spin density would give zero magnetic field at the centre, and the orbital momentum in a pure Γ_2 state is quenched. The residual orbital momentum due to the admixture of the Γ_5 states produces a field at the nucleus which cannot be neglected, however, being positive and in the case of Ni^{2+} about twice as large as the observed field, which is negative. The various contributions to the magnetic hyperfine structure are summarized in Table 7.21 for Cu^{3+},

TABLE 7.21

Hyperfine fields for $3d^8$ ions, from Locher and Geschwind (1963) and ref. (d)

						Hyperfine field at nucleus (kG)			
								Calculated	
$3d^8$-ion	Compound	$H \parallel$ or $\perp$ to c-axis of Al_2O_3	$\Delta g = g - g_s$	$\langle r^{-3} \rangle^{(c)}_{a.u.}$	Measured total	Orbit	Spin dipolar	Core polarization
$Co^{2+(a)}$	MgO		0.1705	5.35	(−)160 (1)	+114	≈0	−274
$Co^{2+(d)}$	CaO		0.2720	5.35	(−)93 (1)	+182	≈0	−276
$Ni^{2+(a)}$	MgO		0.2122	7.09	(−)63 (3)	+188	≈0	−251
$Ni^{2+(d)}$	CaO		0.327	—	—	—	—	—
Ni^{2+}	Al_2O_3	$\parallel$	0.1925	7.09	−96.8	+171	−12	−256
Ni^{2+}	Al_2O_3	$\perp$	0.1830	7.09	−85.7	+162	+6	−254
$Cu^{3+(b)}$	Al_2O_3	$\parallel$	0.0765	9.02	−171.0	+86	−8	−249
$Cu^{3+(b)}$	Al_2O_3	$\perp$	0.0749	9.02	−159.6	+84	+4	−248

The (−) sign indicates that the sign is assumed, not measured.
(a) Orton, Auzins, and Wertz (1960b); Orton, Auzins, Griffiths, and Wertz (1961).
(b) Blumberg, Eisinger, and Geschwind (1963).
(c) Free ion values, from Freeman and Watson (1965); see Table 7.6.
(d) Low and Suss (1965).

Ni^{2+}, and Co^+, and it will be seen that the estimated core polarization contribution is remarkably constant.

The Endor line widths observed in these experiments were of order 20 and 60 kHz, but the e.s.r. spectrum line width was 180 and 20 G (in each case for Ni^{2+} and Cu^{3+} respectively). The width in the latter case is ascribed to inhomogeneity in the value of D, and Locher and Geschwind (1963) were able to show (see § 4.10) by measuring the Endor frequency as a function of position in the microwave line that the variation in D also entailed a variation in the orbital contribution to the g-value and hence in the orbital contribution to the hyperfine field. Similar large line widths have been observed for d^8 ions in cubic fields (e.g. in MgO and CaO), where D should be zero, and attributed to local departures from cubic symmetry. As pointed out by Orton, Auzins, Griffiths, and Wertz (1961) this interpretation, which assumes a distribution of zero field splittings centred upon zero, is supported by two pieces of evidence.

(1) A resonance line is observed at approximately half the field of the main line, whose shape is asymmetrical, with a sharp cut-off on the high-field side and a much more gradual tail on the low-field side. This line corresponds to single quantum transitions between the $|+\tilde{1}\rangle$ and $|-\tilde{1}\rangle$ levels, whose spacing is just $2g\beta H$ in energy and independent of D in first order, unlike the main line whose frequency depends directly on D (cf. eqns (3.28)). The transition probability for the half-field line is zero in an exactly cubic field, and any departure from cubic symmetry moves the half-field line to lower fields, thus accounting for the sharp cut-off on the high-field side. These features of the half-field line are similar to those observed for non-Kramers doublets (see § 3.14 or § 18.5).

(2) A much more striking feature is the occurrence of a single sharp line (divided into $2I+1$ sharp lines if a hyperfine structure is present) at the centre of the broad main line. This is attributed to a 'two-quantum' transition, which can only be observed in a three-level system if the three levels are nearly equally separated. The latter occurs in a cubic field, so that the two-quantum transition is confined to those ions for which the departure from cubic symmetry is virtually zero, and the line width is correspondingly narrower. The rather precise values of g and the hyperfine constant A for ions in MgO and CaO in Table 7.21 have each been obtained from the e.s.r. spectrum using the precision available through these two-quantum transitions. Orton, Auzins, and Wertz (1960b) have shown that such two-quantum transitions can be

observed for Ni^{2+} in MgO using two separate frequencies, which implies that the level spacings differ through the existence of a zero-field splitting. They have also shown (Orton, Auzins, Griffiths, and Wertz 1961) that the intensity of absorption falls off with increasing difference between the two applied frequencies at a rate consistent with the observed shape of the broad (single-quantum transition) main line.

7.16. $3d^9$. Ni$^+$, Cu^{2+} in an octahedral field. ^2D, $L = 2$, $S = \frac{1}{2}$

In an octahedral field the orbital D state splits into a Γ_3 doublet and a Γ_5 triplet, the former being the lower. The pattern for d^9 is thus inverted from that for d^1, corresponding to the fact that we are now dealing with a hole in a closed shell rather than an electron. The Γ_3 doublet is not split by a trigonal distortion, resulting in some unusual properties which will be discussed after we have considered the case of a tetragonal distortion, which does split the Γ_3 orbital doublet. If the second-degree tetragonal field has the same sign as the fourth-degree field (i.e. $B_2^0/B_4^0 > 0$) the sign of the second-degree field determines which of the two orbital states, $|2^s\rangle$ or $|0\rangle$, will be the lower, as can be seen from Fig. 7.3. If the two orbital states are well separated, the lower singlet with its twofold spin degeneracy will behave in a first approximation like a spin doublet. However the spin-orbit coupling mixes some of the Γ_5 states into the ground state, making the g-values depart from the free spin value and introducing some anisotropy. In Table 7.22 the formulae given are correct to order (λ/Δ), where λ is the spin-orbit parameter and Δ is mainly determined by the cubic field splitting; in most cases (λ/Δ) is of order -0.1, or less, -0.05 being a typical value. The formulae in Table 7.22 are given for both of the possible ground states, $|2^s\rangle$ and $|0\rangle$; the parameters are those for the spin Hamiltonian (with $S = \frac{1}{2}$, and $I = \frac{3}{2}$ for each of the stable isotopes 63, 65)

$$\mathcal{H} = g_\parallel \beta H_z S_z + g_\perp \beta (H_x S_x + H_y S_y) + \\ + A_\parallel S_z I_z + A_\perp (S_x I_x + S_y I_y) + P_\parallel \{I_z^2 - \tfrac{1}{3} I(I+1)\}. \quad (7.107)$$

Formulae to the next order $(\lambda/\Delta)^2$, for the more general case of a rhombic distortion which admixes the $|2^s\rangle$ and $|0\rangle$ orbital states, are given in Bleaney, Bowers, and Pryce (1955), from which the formulae in this table can be obtained as special cases. With a typical value of $(\lambda/\Delta) \sim -0.05$, corrections $\sim(\lambda/\Delta)^2$ are small but not negligible.

Apart from a few salts with trigonal symmetry, the majority of the

TABLE 7.22

Formulae for ground state of d^9 in octahedral field with tetragonal distortion

(a) *Ground state* $|2^s\rangle$, *with twofold spin degeneracy.* The states $|\widetilde{+}\rangle, |\widetilde{-}\rangle$ are

$$|\widetilde{+}\rangle = |2^s, +\rangle - \frac{\lambda}{\Delta_0}|2^a, +\rangle - \frac{\lambda}{\sqrt{(2)}\Delta_1}|-1, -\rangle$$

$$|\widetilde{-}\rangle = |2^s, -\rangle + \frac{\lambda}{\Delta_0}|2^a, -\rangle + \frac{\lambda}{\sqrt{(2)}\Delta_1}|+1, +\rangle$$

$$g_\| = 2 - \frac{8\lambda}{\Delta_0}$$

$$g_\perp = 2 - \frac{2\lambda}{\Delta_1}$$

$$A_\| = 2g_n\beta\beta_n\langle r^{-3}\rangle\left\{-\kappa - \frac{4}{7} - \frac{6\lambda}{7\Delta_1} - \frac{8\lambda}{\Delta_0}\right\}$$

$$A_\perp = 2g_n\beta\beta_n\langle r^{-3}\rangle\left\{-\kappa + \frac{2}{7} - \frac{11\lambda}{7\Delta_1}\right\}$$

$$P_\| = -\frac{3e^2Q}{7I(2I-1)}\langle r_q^{-3}\rangle$$

The orbital states $|2^s\rangle, |2^a\rangle, |+1\rangle, |-1\rangle, |0\rangle$ are those shown in Fig. 7.25. Δ_0 is the energy by which $|2^a\rangle$ lies above $|2^s\rangle$, and Δ_1 the energy by which the $|\pm 1\rangle$ states lie above $|2^s\rangle$.

(b) *Ground state* $|0\rangle$, *with twofold spin degeneracy.* The states $|\widetilde{+}\rangle, |\widetilde{-}\rangle$ are

$$|\widetilde{+}\rangle = |0, +\rangle - \sqrt{(\tfrac{3}{2})}\frac{\lambda}{\Delta_2}|+1, -\rangle$$

$$|\widetilde{-}\rangle = |0, -\rangle - \sqrt{(\tfrac{3}{2})}\frac{\lambda}{\Delta_2}|-1, +\rangle$$

$$g_\| = 2$$

$$g_\perp = 2 - \frac{6\lambda}{\Delta_2}$$

$$A_\| = 2g_n\beta\beta_n\langle r^{-3}\rangle\left\{-\kappa + \frac{4}{7} + \frac{6\lambda}{7\Delta_2}\right\}$$

$$A_\perp = 2g_n\beta\beta_n\langle r^{-3}\rangle\left\{-\kappa - \frac{2}{7} - \frac{45\lambda}{7\Delta_2}\right\}$$

$$P_\| = +\frac{3e^2Q}{7I(2I-1)}\langle r_q^{-3}\rangle$$

Δ_2 is the energy by which the $|\pm 1\rangle$ states lie above $|0\rangle$.

hydrated salts of copper contain a $[\text{Cu}(\text{H}_2\text{O})_6]^{2+}$ complex with a distortion of tetragonal or somewhat lower symmetry in which typical g-values are $g_\| \sim 2\cdot 4, g_\perp \sim 2\cdot 1$. This shows that in a tetragonal approximation, the orbital ground state must be $|2^s\rangle$. In aqueous solution the optical absorption spectrum of the cupric ion shows one broad peak at

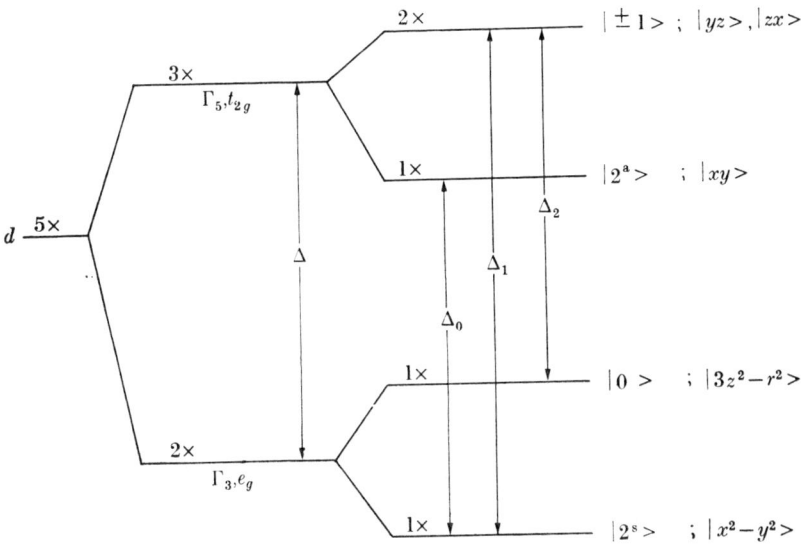

FIG. 7.25. Orbital energy levels and states for a d^9 ion in a crystal field of octahedral symmetry with a tetragonal distortion.

about 12300 cm^{-1} (Dreisch and Trommer 1937), which may be interpreted as arising from transitions from the ground orbital state to the Γ_5 triplet. If this value is taken as the mean value of Δ_0, Δ_1 then the observed g-values can be fitted if λ is assigned a value of about -710 cm^{-1} (see Bleaney, Bowers, and Pryce 1955), which is considerably smaller than the free ion value of about -830 cm^{-1}. The difference is ascribed to bonding effects.

The hyperfine structure of copper is of interest as being the first to be discovered in the solid state by the method of electron paramagnetic resonance. In a diluted tutton salt Penrose (1949) found that the four-line hyperfine structure was markedly anisotropic, being a maximum when the external field was directed along the axis in which the g-value was a maximum (i.e. the parallel direction for a $|2^s\rangle$ ground orbital state). It was pointed out by Abragam and Pryce (1949) that this anisotropy was in the reverse direction from that predicted by a first-order calculation of the contributions from electron spin and orbit. The latter is almost 'quenched' and contributes amounts proportional to $(g_\parallel-2)$ and $(g_\perp-2)$ in the parallel and perpendicular directions respectively. With the same factor of proportionality, the spin-dipolar contributions are $-\frac{4}{7}$ and $+\frac{2}{7}$ respectively, so that the spin and orbital contributions oppose one another in the parallel direction, and add in the perpendicular direction. This discrepancy between theory and experiment was overcome (Abragam 1950, Abragam and Pryce 1951b) where the importance of the unpaired spin density at the nucleus was pointed out; this effect is represented by the additional parameter κ which occurs in the formulae for $A_{\parallel}$ and $A_{\perp}$ given in Table 7.22. The small terms of order (λ/Δ) in these formulae are partly due to the residual orbital contribution, and partly to the modification of the spin contribution through the admixture of other states by the spin-orbit coupling, which alters the shape of the electron distribution in the ground state.

In the expressions for the magnetic hyperfine structure the small terms involving (λ/Δ) can be evaluated from measurements of the g-values. Measurements of the values of $A_{\parallel}$ and $A_{\perp}$ can then be regarded as determining two unknowns, $\langle r^{-3} \rangle$ and the core polarization parameter κ, since g_n is known accurately from nuclear magnetic resonance in diamagnetic cuprous salts. From the spectrum of the Cu^{2+} ion in $La_2Mg_3(NO_3)_{12}, 24D_2O$ (which is discussed further below) Bleaney, Bowers, and Pryce (1955) obtained the values 0.32_5 for κ (with slightly smaller values for other salts), and 6.3 a.u. for $\langle r^{-3} \rangle$. These give a value of -250 kG for the core polarization field per unit of electron spin, which is very close to that found for other neighbouring ions of the $3d$ group (see Table 7.21). The value of $\langle r^{-3} \rangle$ is rather smaller than that of 7.5 a.u. estimated for the free cupric ion; the diminution (which is of the same order as that in λ) is again attributed to bonding effects.

The electron magnetic resonance spectrum of the cupric ion is also noteworthy as the first example in which an effect due to the nuclear electric quadrupole interaction was observed (Ingram 1949). This has been analyzed in some detail (see, in particular, Bleaney, Bowers, and Ingram (1955) and Bleaney, Bowers, and Trenam (1955)), and the value of $P_{\parallel}$ in both tutton salts and lanthanum magnesium nitrate is found to be about 0.0011 cm^{-1} or 33 MHz (this is a mean value for the two isotopes 63, 65, which differ only by 8 per cent). Values of the nuclear electric quadrupole moments of -0.15_9 and -0.14_7 barns have been estimated (see Bleaney, Bowers, and Pryce 1955) for the isotopes 63 and 65 respectively, using the same value of 6.3 a.u. for $\langle r_q^{-3} \rangle$ as that deduced from the magnetic hyperfine structure. The spin Hamiltonian parameters for Cu^{2+} in $La_2Mg_3(NO_3)_{12}, 24H_2O$ (or D_2O), including some later and more accurate values, are listed in Table 7.23.

The d^9 ion in an octahedral field with a small trigonal distortion (or no distortion) is of particular interest because the two ground orbital

TABLE 7.23

Measured parameters for Cu^{2+} *in* $La_2Mg_3(NO_3)_{12}$, $24D_2O$. *The hyperfine constants are weighted mean values for the two stable isotopes* ^{63}Cu, ^{65}Cu. *The data at* $90°K$ *are from Bleaney, Bowers, and Trenam (1955); those at* $20°K$ *are from Breen, Krupka, and Williams (1969)*

Temperature	g-values	Hyperfine constants (in units of 10^{-4} cm^{-1})
90°K	$g_\| = 2 \cdot 219(3)$	$\|A_\|\| = 29 \cdot 0(5)$
	$g_\perp = 2 \cdot 218(3)$	$\|A_\perp\| = 27 \cdot 5(5)$
20°K	$g_\| = 2 \cdot 465(1)$	$A_\| = -111 \cdot 7(5)$
	$g_\perp = 2 \cdot 099(1)$	$\|A_\perp\| = 16 \cdot 0(5)$
	$\frac{1}{3}(g_\| + 2g_\perp)$	$\|\frac{1}{3}(A_\| + 2A_\perp)\|$
	$= 2 \cdot 221(1)$	$= 26 \cdot 6(5)$
		$P_\| = +10 \cdot 5(5)$

levels of the Γ_3 doublet are not split (see Fig. 7.4) by the ligand field. In addition the spin-orbit coupling has no matrix elements within the doublet, even when the twofold spin degeneracy is included, and a magnetic field has no matrix elements between the two orbital states. Thus the ground state of a d^9 ion in a trigonal field has a fourfold degeneracy of an unusual kind. As we have seen above, a distortion of tetragonal or rhombic (or lower) symmetry will split the orbital doublet, and the system is peculiarly susceptible to strain. We should not therefore be surprised to observe a spectrum appropriate to an effective spin of $\frac{1}{2}$ with anisotropic g-values and hyperfine structure corresponding to some symmetry lower than trigonal. If there is little or no strain, we may expect a static or dynamic Jahn–Teller effect; the theory of this is rather complex (see §§ 21.3 to 21.8). We shall give below only a rather over-simplified presentation in an effort to interpret the observed spectra.

Experimentally the spectrum of a salt containing a d^9 ion in a site of trigonal symmetry generally has the following characteristics (the parameters are appropriate to the Cu^{2+} ion):

(a) at high temperatures the spectrum is isotropic, with a g-value of about 2·2 and a rather small magnetic hyperfine structure ($|A| \approx 0.002$ cm^{-1}), and no observable quadrupole splitting;

(b) as the temperature is lowered a transition region sets in where the nature of the spectrum changes with temperature reversibly. The temperature region in which this occurs varies from salt to salt, being between 20°K and 60°K for the diluted salts

$(Cu,Mg)SiF_6,6H_2O$ and $(Cu,Mg)_3Bi_2(NO_3)_{12},24H_2O$, and between 20°K and 12°K in $(Cu, Zn)(BrO_3)_2,6H_2O$ (in all of which trigonal symmetry at the Cu^{2+} site would be expected), and below 4°K in $(Cu, Mg)O$ (where the symmetry should be cubic);

(c) at still lower temperatures (except in MgO) three spectra are observed with anisotropic g-values, magnetic hyperfine structure, and a nuclear electric quadrupole interaction typical of an ion subject to a tetragonal distortion from octahedral symmetry $(g_\| \sim 2\cdot 4,\ g_\perp \sim 2\cdot 1;\ A_\| \sim 0\cdot 01\ \mathrm{cm}^{-1}$ with a smaller value of $0\cdot 001$ to $0\cdot 003\ \mathrm{cm}^{-1}$ for $A_\perp;\ P_\| \sim 0\cdot 001\ \mathrm{cm}^{-1})$, of such a sign that the ground state is the $|2^s\rangle$ orbital singlet (see Fig. 7.25). The three spectra have axial symmetry about each of the three mutually perpendicular fourfold axes of the XY_6 complex.

Attention was first drawn by Van Vleck (1939) to the peculiar features which might be expected for a d^9 ion in a crystal field of octahedral symmetry with a trigonal distortion as a result of the Jahn–Teller effect. The Γ_3 orbital states $|2^s\rangle$ and $|0\rangle$ may be written in the form $|x^2-y^2\rangle$ and $|3z^2-r^2\rangle$, and it can be shown that they are sensitive to distortions of the octahedron with similar symmetry, i.e. to distortions of the type associated with the two normal modes of vibration Q_2 and Q_3 defined by Fig. 7.26. If the calculation of the Jahn–Teller effect is carried only to the second order (i.e. to order Q^2), it is found that the kinetic energy depends on $Q_2^2+Q_3^2$, and not on the ratio of Q_2 to Q_3, so that there are an infinite number of possible displacements which will yield the same energy. If we write (see eqn (21.15))

$$Q_3 = \rho \cos \phi, \qquad Q_2 = \rho \sin \phi \qquad (7.108)$$

then the kinetic energy is independent of ϕ, and is proportional to ρ^2.

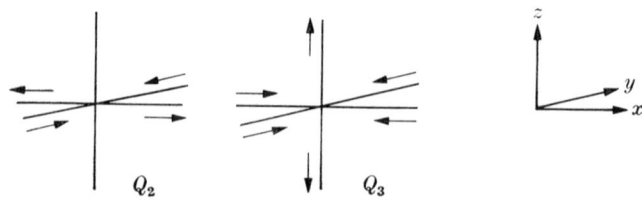

Fig. 7.26. The two normal vibrations of an octahedron that take part in the Jahn–Teller distortion for a d^9 ion. If Cartesian axes are taken through the ligands at the vertices of the octahedron, and if X is the outwards displacement of a ligand on the positive or negative x-axis and so on, then Q_2 and Q_3 are defined by

$$X = \tfrac{1}{2}\!\left(Q_2-\tfrac{1}{\sqrt{3}}Q_3\right),\quad Y = -\tfrac{1}{2}\!\left(Q_2+\tfrac{1}{\sqrt{3}}Q_3\right),\quad Z = Q_3/\sqrt{3}$$

if all the other normal coordinates are zero. (After O'Brien, 1964.)

Since the 'potential energy' associated with the coupling between the magnetic electrons and the ligand nuclei depends (in a first approximation) linearly on ρ, there is a minimum energy known as the Jahn–Teller energy W_{JT} (see § 21.4) whose locus is a circle in the Q_2, Q_3 plane. The corresponding eigenstates are

$$\cos \tfrac{1}{2}\phi \, |x^2-y^2\rangle + \sin \tfrac{1}{2}\varphi \, |3z^2-r^2\rangle, \qquad (7.109a)$$

which is the ground state for Cu^{2+}, together with the orthogonal excited state

$$\sin \tfrac{1}{2}\phi \, |x^2-y^2\rangle - \cos \tfrac{1}{2}\phi \, |3z^2-r^2\rangle, \qquad (7.109b)$$

which lies higher in energy by $4W_{JT}$.

A given value of ϕ corresponds in general to a rhombic distortion whose principal g-values are, to order λ/Δ and neglecting the difference between the various values of Δ in Fig. 7.25,

$$\begin{aligned} g_1 &= 2-(2\lambda/\Delta)(\cos \tfrac{1}{2}\phi - \surd(3) \sin \tfrac{1}{2}\phi)^2, \\ g_2 &= 2-(2\lambda/\Delta)(\cos \tfrac{1}{2}\phi + \surd(3) \sin \tfrac{1}{2}\phi)^2, \qquad (7.110) \\ g_3 &= 2-(8\lambda/\Delta) \cos^2 \tfrac{1}{2}\phi. \end{aligned}$$

The directions of g_1, g_2, g_3 are the three mutually perpendicular fourfold axes of the XY_6 complex. Tetragonal distortions are included as special cases; thus values of $\phi = 0$, $2\pi/3$, $4\pi/3$ correspond to tetragonal distortions about each of the three tetragonal axes in which the complex is elongated along the tetragonal axis, and values of $\phi = \pi$, $5\pi/3$, $\pi/3$ to similar distortions in which the complex is compressed along each of the three tetragonal axes. Since all values of ϕ give the same energy, and are equally probable, the complex moves continuously through the whole range of distortions corresponding to all values of ϕ. Since the vibrational frequency (of order 10^{13} Hz) is large compared with the frequency at which spin resonance is observed (of order 10^{10} Hz), only time average values are observed. Setting

$$\langle \cos^2 \tfrac{1}{2}\phi \rangle = \langle \sin^2 \tfrac{1}{2}\phi \rangle = \tfrac{1}{2}, \quad \langle \cos \tfrac{1}{2}\phi \sin \tfrac{1}{2}\phi \rangle = 0$$

in eqns (7.110), one obtains the result

$$g_1 = g_2 = g_3 = 2-(4\lambda/\Delta) = \tfrac{1}{3}(g_\parallel + 2g_\perp), \qquad (7.111)$$

which corresponds to the experimental observation in the high temperature spectrum (Abragam and Pryce 1950). This result can also be written in the form

$$g = \tfrac{1}{2}g_{x^2-y^2} + \tfrac{1}{2}g_{3z^2-r^2}, \qquad (7.112)$$

showing that the g-value in the high temperature spectrum is just the average of those appropriate to tetragonal distortions of opposite sign, as can be verified from Table 7.22 on putting $\Delta_0 = \Delta_1 = \Delta_2 = \Delta$. Similarly one finds

$$A = \tfrac{1}{2}A_{x^2-y^2} + \tfrac{1}{2}A_{3z^2-r^2} \tag{7.113}$$

and

$$P_{\parallel} = \tfrac{1}{2}P_{x^2-y^2} + \tfrac{1}{2}P_{3z^2-r^2}, \tag{7.114}$$

which give an isotropic value of

$$A = 2g_n\beta\beta_n\langle r^{-3}\rangle\left\{-\kappa - \frac{4\lambda}{\Delta}\right\} = \tfrac{1}{3}(A_{\parallel} + 2A_{\perp}) \tag{7.115}$$

to order (λ/Δ), and $P_{\parallel} = 0$. These two results correspond to what would be expected in cubic symmetry, where the electric field gradient at the nucleus vanishes, as also does the magnetic field at the nucleus due to the electron spin apart from the core-polarization contribution (the term in λ/Δ is here due entirely to the residual orbital contribution). It will be seen also that these values of g, A satisfy the relations $g = \tfrac{1}{3}(g_{\parallel} + 2g_{\perp})$, $A = \tfrac{1}{3}(A_{\parallel} + 2A_{\perp})$ for either of the ground states in Table 7.22; a comparison with experiment is given in Table 7.23.

Though this theory apparently gives a satisfactory explanation of the high temperature spectrum, it does not account for the changes observed at lower temperatures. Also, the result that the energy is independent of the value of ϕ holds only to the second degree in Q, and corresponds to complete axial symmetry whereas we have only three-fold symmetry. Öpik and Pryce (1957) showed that on expanding the energy in a Taylor's series in the distortions, a term varying as $\cos 3\phi$ is obtained, whose coefficient is plausibly expected to be negative for a complex such as $[Cu(H_2O)_6]^{2+}$. The states of lowest energy are then those for which $\phi = 0$, $2\pi/3$, $4\pi/3$ corresponding to the stable state being one in which the octahedron suffers an elongation of tetragonal symmetry, about one of the fourfold axes. This is clearly the position observed in the low temperature spectrum, and corresponds to a 'static' Jahn–Teller effect.

Even at 0°K the picture we have drawn is an over-simplification for three reasons: tunnelling between the three potential wells; zero point energy of vibration within a well; and the effects of strain. Of these, the first two are features of the 'dynamic' Jahn–Teller effect; the third results from imperfections within the crystal, and has the drawback that it obscures the effects in the magnetic resonance spectrum which might be expected from the tunnelling mechanism.

In such a triple potential well, vibrational states in each well are mutually orthogonal and have the same energy only if the potential barrier is infinitely high. With a finite potential barrier the quantum-mechanical tunnelling effect lifts the degeneracy, giving (on including the twofold spin degeneracy) a ground quartet (Γ_8) and an excited doublet with a separation in zero magnetic field which is small in the ground state but large in excited vibrational states. When a magnetic field is applied, the magnetic resonance spectrum will be rather complex, except in the limiting cases where the splitting due to tunnelling is either very small or very large compared with the difference in Zeeman energy $(g_\parallel - g_\perp)\beta H$, where $g_\parallel$ and $g_\perp$ are the principal g-values appropriate to a static tetragonal distortion. In the former case the spectrum should be similar to that of an assembly of complexes with static tetragonal distortions where the axes are distributed randomly between the three fourfold axes of the octahedron. However, some extra lines should be observable when the magnetic field is not along the threefold or any of the fourfold axes; careful measurements by Breen, Krupka, and Williams (1969) reveal no such effects for Cu^{2+} in lanthanum magnesium nitrate, and it is probable that the tetragonal distortions are stabilized by strains. This will be the dominant effect if the strain energy is large compared with the energy splitting due to tunnelling.

The effect of zero-point motion within a potential well means that the angle ϕ is not localized at the points 0, $2\pi/3$, $4\pi/3$ (or π, $5\pi/3$, $7\pi/3$) but oscillates about these mean values. If the electronic state is assumed to have the form given in eqn (7.109a),

$$\cos \tfrac{1}{2}\phi \, |x^2-y^2\rangle + \sin \tfrac{1}{2}\phi \, |3z^2-r^2\rangle,$$

where ϕ is a function of time because the distortion that couples the states is an oscillating one, then the mean value of ϕ may be (for example) zero, when the equilibrium position is at $\phi = 0$, but it will have a mean square value that is not zero. As in the explanation given earlier for the high temperature spectrum, the observed g-values and other constants will be the time averages, since the vibrational frequency (in this case of the zero point motion) is large compared with the electron spin resonance frequency. Since $\langle \cos \tfrac{1}{2}\phi \sin \tfrac{1}{2}\phi \rangle = 0$, the observed constants will be

$$g = \langle \cos^2 \tfrac{1}{2}\phi \rangle g_{x^2-y^2} + \langle \sin^2 \tfrac{1}{2}\phi \rangle g_{3z^2-r^2}, \tag{7.116}$$

$$A = \langle \cos^2 \tfrac{1}{2}\phi \rangle A_{x^2-y^2} + \langle \sin^2 \tfrac{1}{2}\phi \rangle A_{3z^2-r^2}, \tag{7.117}$$

with a similar expression for P.

The presentation of the Jahn–Teller effect given above is based mainly on the work of O'Brien (1964). This paper gives also an explanation of some effects observed in the spectrum of the Ni^+ ion in LiF by Hayes and Wilkens (1964). The impurity ion enters the lattice as Ni^{2+}; after irradiation Ni^+ ions are formed and occupy cubic lattice sites, substituting for Li^+ in LiF (and for Na^+ in NaF). Thus the Ni^+ ion with its d^9 configuration is situated at the centre of a regular octahedron of six F^- ions, on a site of cubic symmetry. In fact a number of spectra are observed, depending on the conditions of irradiation and the temperature; the majority of these show tetragonal symmetry, but with g-values corresponding to distortions of either sign. Typical results in LiF are

(a) $g_\parallel$, 2·53 to 2·62; $g_\perp$, 2·10 to 2·11;

(b) $g_\parallel$, 2·06 to 2·08; $g_\perp \approx 2·35$.

The orbital contributions to the g-values are larger than for most cupric salts because the cubic field splitting is smaller ($\Delta = 10Dq$ is estimated as 6200 cm^{-1} for Ni^+ in LiF and 5500 cm^{-1} in NaF). The spectra characterized by the g-values in (a) clearly belong to the orbital state $|2^s\rangle = (|x^2-y^2\rangle)$ while those in (b) are attributed to the orbital state $|0\rangle = (|3z^2-r^2\rangle)$. However the observed value of $g_\parallel$ in the latter case is greater than 2, while theory gives just 2 to order (λ/Δ) and somewhat less than 2 to order $(\lambda/\Delta)^2$, as can be seen from Table 7.22(b) on introducing the factors $(1+3\lambda^2/2\Delta_2^2)^{-\frac{1}{2}}$ required to normalize the states. The experimental result can be explained on the basis of zero-point motion in the Jahn–Teller effect, since the observed g-values will then be given by eqn (7.116). By taking

$$\langle \cos^2 \tfrac{1}{2}\phi \rangle = 0·10, \qquad \langle \sin^2 \tfrac{1}{2}\phi \rangle = 0·90$$

this gives $g_\parallel = 2·05$, $g_\perp = 2·36$ instead of the values $g_\parallel = 1·98$, $g_\perp = 2·42$ which would be obtained with no vibrational admixture. Rather similar but less detailed results are obtained for Ni^+ in NaF; in this case an isotropic spectrum was also observed above 130°K, associated with one of the anisotropic spectra.

Since nickel consists almost entirely of isotopes with no nuclear spin, no hyperfine structure due to the nickel nucleus is observed in these experiments, and this makes it easier to observe and measure the hyperfine structure due to the fluorine ligands. As shown in Chapter 20 (see also Fig. 7.17), the $|3z^2-r^2\rangle$ orbital bonds to all six ligands in octahedral symmetry (though not in equal amounts), while the $|x^2-y^2\rangle$ orbital can bond only to the four ligands along the x-, y-axes, apart from a weak interaction with the two ligands on the z-axis which arises from the admixture of π-bonding (t_{2g}) states through the spin-orbit coupling. Thus with a static tetragonal distortion which makes $|x^2-y^2\rangle$ the ground state only a very small hyperfine structure due to the ligands on the z-axis should be observed, apart from simple dipolar interaction. However the effect of the zero point motion in the Jahn–Teller effect is to admix the $|x^2-y^2\rangle$ and $|3z^2-r^2\rangle$ states, so that some bonding with all six ligands becomes possible. This is observed in the experiments of Hayes and Wilkens (1964).

We now return briefly to the high temperature spectrum. As the temperature rises, excited vibrational states become occupied which should have different magnetic resonance spectra because of their much larger splittings due to tunnelling. The rise in temperature, however, also results in a rapid increase in the rate at which transitions between the various vibronic levels are induced by the phonon spectrum. These may be accompanied by spin-flips in the ground state, and Breen Krupka, and Williams (1969) have found that the spin-lattice relaxation time at helium temperatures is some four orders of magnitude faster for Cu^{2+} in lanthanum magnesium nitrate than in tutton salts where the ligand field has tetragonal or lower symmetry. This result has been related by Williams, Krupka, and Breen (1969) to the rate $(1/\tau)$ at which relaxation occurs between equivalent distortions without re-orientation of the electron spin. The latter is comparable in its effect on the magnetic resonance spectrum with motion of the nuclei in a nuclear magnetic resonance. If the relaxation rate is fast compared with the frequency difference between resonance lines from the different states, these lines disappear and are replaced by a single line at the average frequency. It is probable (see §§ 21.7, 21.8) that such motional narrowing is responsible for the single isotropic line observed at high temperatures. It means that the resonance lines from the different levels will be replaced by a single line with a g-value equal to the mean of all the g-values, as in the original theory of Abragam and Pryce outlined earlier in this section.

In the absence of other forms of broadening, the width of this averaged line should be approximately equal to $\tau\{(g_\parallel - g_\perp)\beta H/h\}^2$, where τ is the relaxation time for phonon-induced transitions between equivalent distortions without re-orientation of the electron spin. If a hyperfine structure is present, the width of each hyperfine line under such conditions should be, in frequency,

$$\tau\{(g_\parallel - g_\perp)\beta H + (A_\parallel - A_\perp)m\}^2/h^2, \qquad (7.118)$$

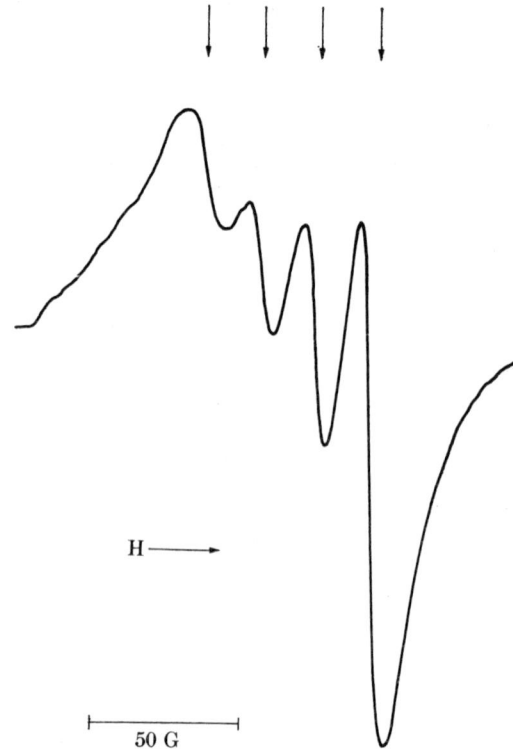

FIG. 7.27. Derivative of the Cu^{2+} spectrum in MgO at 77°K, showing decrease in width, and consequent increase in amplitude, of the hyperfine components towards higher magnetic field. (Orton, Auzins, Griffiths and Wertz (1961); see also Coffman (1968a).)

showing that the line width should depend on the nuclear magnetic quantum number m. An effect of this kind has been observed by Orton, Auzins, Griffiths, and Wertz (1961) in the spectrum of Cu^{2+} in MgO, as shown in Fig. 7.27; as pointed out by these authors, similar observations have been reported for Cu^{2+} ions in solution (McGarvey 1956), for which an explanation was proposed by McConnell (1956) on the basis of a rapid tumbling motion that in many ways resembles the Jahn–Teller effect. A feature of the spectra of both Cu^{2+} and the isoelectronic ion Ni^+ in MgO is that the isotropic spectrum persists down to 4°K, but at 1·2°K is replaced by an anisotropic spectrum; thus in these cases the transition occurs at an unusually low temperature. Details of the spectra of Cu^{2+} in MgO and CaO at 1·2°K, which are quite different from those discussed above, are given in papers by Coffman (1968) and Coffman, Lyle, and Mattison (1968); the results are briefly discussed in § 21.8.

7.17. 3d ions in tetrahedral symmetry

Under the title of 'tetrahedral' symmetry (fourfold coordination) we include 'cubic' symmetry (eightfold coordination). In the latter case we have a cage of eight ligand ions at the corners of a regular cube, with the paramagnetic ion at the body centre of the cube (cf. Fig. 5.3); in tetrahedral symmetry ligand ions at alternate corners are missing, so that the ligands consist of four ions in a regular tetrahedron. The host lattices of these types for the transition ions most studied are

(i) the cubic zinc blende structure, space group T_d^2, in which each dipositive ion is surrounded by four dinegative ions in a regular tetrahedron. The lattice constant increases with atomic number of either cation or anion, as shown in Table 7.24;

(ii) the wurtzite structure, with space group C_{4v}^6, which is similar to the preceding class except that the tetrahedron is distorted but retains trigonal symmetry (C_{3v} point group) about a line joining the paramagnetic ion to one of the four ligand ions; some typical crystals are also listed in Table 7.24;

TABLE 7.24

Space group and lattice constants of various compounds used as host lattices for paramagnetic ions in fourfold and eight-fold coordination

Type	Space group	Host lattice	Lattice constant c (nm)	a (nm)
Cubic	T_d^2	ZnS(β)	0·543	
(fourfold		ZnSe	0·565	
coordination)		ZnTe	0·607	
		CdO	0·469	
		CdS(β)	0·582	
		CdTe	0·641	
Cubic	O_h^5	CaF$_2$	0·545	
(eightfold		SrF$_2$	0·586	
coordination)		BaF$_2$	0 618	
Hexagonal	C_{4v}^6	BeO	0·439	0·270
(fourfold		ZnO	0·519	0·324
coordination)		ZnS(α)	0·628	0·384
		CdS(α)	0·672	0·414
		CdSe	0·702	0·430
Tetragonal	$I4/mcm$	CoCs$_3$Cl$_5$	1·455	0·922
(fourfold coordination)		CoCs$_3$Br$_5$	1·516	0·962
Tetragonal	$I4_1/a$	CaWO$_4$	1·138	0·524
(eightfold coordination)		SrWO$_4$	1·190	0·540

(iii) the fluorite structure CaF_2, etc., O_h^5, in which the Ca^{2+} ion is eightfold coordinated to the cube of F^- ions (see also § 5.1). Distorted tetrahedral/cubic sites occur in spinels, garnets and in crystals isomorphous with $CaWO_4$ (see also § 5.1); in the last case the distortion is tetragonal, as is also true for crystals of the type $CoCs_3Cl_5$.

On a point charge model (see eqns (16.15)–(16.17)) the size of the cubic field potential of the fourth degree is $-\tfrac{8}{9}$ for eightfold coordination and $-\tfrac{4}{9}$ for fourfold coordination of that for octahedral coordination to six ligand ions of the same charge and at the same distance from the paramagnetic ion. On this basis we would expect a cubic splitting $10Dq$ of opposite sign to that discussed previously in this chapter, and noticeably smaller in numerical value for tetrahedral coordination; this is confirmed by a number of optical measurements (see Table 7.25).

TABLE 7.25

Values of the cubic splitting parameter Dq (in cm^{-1}) for some $3d^n$ ions in tetrahedral coordination. The values are those quoted by Slack, Ham, and Chrenko (1966) and have been deduced from optical measurements, except that the last figure for Cu^{2+} in ZnO is from Dietz, Kamimura, Sturge, and Yariv (1963). Typical anion-cation distances are $0{\cdot}189\,nm$ in ZnO, $0{\cdot}252\,nm$ in CdS, $0{\cdot}236\,nm$ in CaF_2; these distances are $\sqrt{(3)}/4$ times the lattice constant of the cubic crystals listed in Table 7.24

Ion	Host lattice	ZnO	ZnS	CdTe	$MgAl_2O_4$
$3d^6$	Fe^{2+}		−340	−248	−447
$3d^7$	Co^{2+}	−390	−340		−400
$3d^8$	Ni^{2+}	−405			
		−420			
$3d^9$	Cu^{2+}	−480			
		−500			
		−570			

The important result of the change in sign of $10Dq$ compared with octahedral coordination is the inversion of the splitting of the orbital levels, which means that the pattern for d^n in tetrahedral coordination is the same as that for d^{10-n} in octahedral coordination. However we cannot take over the results obtained previously for octahedral compounds completely, because the sign of the spin-orbit parameter λ is also opposite for d^{10-n} to that for d^n. This is most important in triplet orbital ground states, where the splitting into J states is inverted (see

Figs. 7.19, 7.20); in orbital singlet states the orbital contribution to the g-value will be of opposite sign. The smaller overall cubic splitting makes the use of LS- coupling less of an approximation in crystal field theory, but it results in larger second-order splittings ($\sim\lambda^2/10Dq$) in ground states, and in larger departures ($\sim\lambda/10Dq$) in the g-values from the values given by first-order theory.

A detailed discussion of each ion on the lines of that for octahedral symmetry in §§ 7.8–7.16 will not be attempted here. Instead we give just a brief résumé, with a reference to a typical result (usually for undistorted tetrahedral/cubic symmetry) where available.

$3d^1$, ^2D; *lowest orbital level, doublet* Γ_3; $S = \frac{1}{2}$

Sc^{2+} in CaF_2, SrF_2 (Höchli and Estle 1967) shows a dynamic Jahn–Teller effect (cf. Cu^{2+} in octahedral symmetry) that is discussed in § 21.8.

$3d^2$, ^3F; *lowest orbital level, singlet* Γ_2; $S = 1$

A narrow double quantum line is observed (cf. Ni^{2+} in octahedral symmetry, § 7.15) within the spin triplet, split only by strains in cubic $ZnS(\beta)$. Typical results are:

Ti^{2+}; $g = 1 \cdot 9280(5)$ $|^{47}A| = 12 \cdot 8(1) \times 10^{-4}$ cm^{-1} Schneider and
$|^{49}A| = 12 \cdot 7(1) \times 10^{-4}$ cm^{-1} Räuber (1966)

V^{3+}; $g = 1 \cdot 9433$ $|^{51}A| = 63 \cdot 0 \quad \times 10^{-4}$ cm^{-1} Holton, Schneider, and Estle (1964)

$3d^3$, ^4F; *lowest orbital level, triplet* Γ_4; $S = \frac{3}{2}$

The triplet $l = 1$, $S = \frac{3}{2}$ manifold is split into $\tilde{J} = \frac{1}{2}$, $\tilde{J} = \frac{3}{2}$, $\tilde{J} = \frac{5}{2}$ as for Co^{2+} in an octahedral field, but the sextet $\tilde{J} = \frac{5}{2}$ is lowest. In an exactly cubic field it will split into a doublet and triplet, separation $\sim \lambda^2/(10Dq)$. However, for $^{51}V^{2+}$ in cubic ZnS, spectra with symmetry about the four $\langle 111 \rangle$ axes are attributed to a Jahn–Teller distortion (Schneider, Dischler, and Räuber 1967). The spin Hamiltonian constants are $(S = \frac{3}{2}, I = \frac{7}{2})$: $g_\parallel = 1 \cdot 9613(10)$, $g_\perp = 1 \cdot 9133(10)$, $|D| = 2 \cdot 02$ cm^{-1}, $|A_\parallel| = 120 \cdot 9(10)$ MHz, $|A_\perp| = 147 \cdot 0(10)$ MHz.

$3d^4$, ^5D; *lowest orbital level, triplet* Γ_5; $S = 2$

The $l = 1$, $S = 2$ ground manifold is split by the spin-orbit coupling into states with $\tilde{J} = 1$, $\tilde{J} = 2$, $\tilde{J} = 3$ as for Fe^{2+} in octahedral symmetry, but with the $\tilde{J} = 3$ state lowest. In a purely cubic field this will split by amounts $\sim \lambda^2/(10Dq)$ into two triplets and a doublet.

In hexagonal CdS(α) a spectrum corresponding to a slightly split non-Kramers doublet of Cr^{2+} is observed by Morigaki (1964a) and Estle, Walters, and de Wit (1963), and is attributed to a rhombic distortion (static Jahn–Teller effect).

$3d^5$, ^6S; *lowest orbital level, singlet* $(L = 0)$; $S = \frac{5}{2}$

Representative values for g and the cubic field splitting of the ^6S state for Cr^+, Mn^{2+}, Fe^{3+} in tetrahedral symmetry are listed in Table 7.17.

$3d^6$, ^5D; *lowest orbital level, doublet* Γ_3; $S = 2$

In an exactly tetrahedral field the manifold of $2(2S+1) = 10$ states is split into five equally spaced levels just like $3d^4$ in an exactly octahedral field (see Table 7.15). The splitting between successive levels is found to be $15.2(4)$ cm^{-1} for Fe^{2+} in ZnS(β), $10(2)$ cm^{-1} for Fe^{2+} in CdTe, and $13(2)$ cm^{-1} for Fe^{2+} in MgAl$_2$O$_4$ (Slack, Ham, and Chrenko 1966; Slack, Roberts, and Ham 1967).

$3d^7$, ^4F; *ground orbital state, singlet* Γ_2; $S = \frac{3}{2}$

The ground spin quartet is similar to that for Cr^{3+} in octahedral symmetry, but with a g-value larger than g_s. Typical g-values for Co^{2+} range from $2\cdot248$ in cubic ZnS(β) to $2\cdot309$ in CdTe, with small extra terms typical of a Γ_8 quartet (Chapter 18); these were the first such terms to have been observed (Ham, Ludwig, Watkins, and Woodbury 1960).

In trigonal symmetry the spin quartet splits into two Kramers doublets, split by $1\cdot34(2)$ cm^{-1} for Co^{2+} in CdS(α) (Morigaki 1964b); in tetragonal symmetry rather larger splittings of $2D = -8\cdot60(8)$ cm^{-1} and $-10\cdot7$ cm^{-1} for Co^{2+} in CoCs$_3$Cl$_5$ and CoCs$_3$Br$_5$ respectively are quoted by Wielinga, Blote, Röest, and Huiskamp (1967).

$3d^8$, ^3F; *ground orbital state, triplet* Γ_4; $S = 1$

The ground manifold of $l = 1$, $S = 1$ splits through spin-orbit coupling into states $\tilde{J} = 0$, $\tilde{J} = 1$, and $\tilde{J} = 2$, the singlet level $\tilde{J} = 0$ being lowest.

$3d^9$, ^2D; *ground orbital state, triplet* Γ_5; $S = \frac{1}{2}$

The ground manifold $l = 1$, $S = \frac{1}{2}$ is the same as that for d^1 in octahedral coordination (see § 7.8), but with the reversed sign of the spin-orbit coupling. This makes the doublet labelled A in Figs. 7.21, 7.22 the ground state, whatever the sign of a trigonal or tetragonal distortion.

Resonance has been observed by Dietz, Kamimura, Sturge, and Yariv (1963) for Cu^{2+} in hexagonal ZnO, with the following parameters for the doublet ground state, Γ_6 (more accurate values are given by Hausmann and Schreiber (1969)):

$$g_\parallel = -0.74; \quad |g_\perp| = 1.531,$$
$$|A_\parallel| = 195 \times 10^{-4} \text{ cm}^{-1}; \quad |A_\perp| = 231 \times 10^{-4} \text{ cm}^{-1}.$$

The hyperfine constants are for the unresolved structure due to the two isotopes 63, 65 each with $I = \frac{3}{2}$; the hyperfine interaction is attributed mainly to the orbital contribution. If the g-values are plotted at the point $(-0.74, +1.531)$ on Fig. 7.23 they are found to lie well within the ellipse, suggesting considerable orbital reduction. This is consistent with the g-values for a doublet belonging to the excited states at $10Dq$, which they find to be $g_\parallel = 1.63$, $g_\perp = 0$ instead of the simple crystal field values of $g_\parallel = 2$, $g_\perp = 0$. The theory is discussed by the authors in some detail; they find a reasonable interpretation using molecular orbitals in which the hole in the $t_2(\Gamma_5)$ orbital triplet spends only some 60 per cent of its time on the Cu^{2+} ion, and the t_2 wave-functions (but not the e-functions) are expanded radially relative to the d-electron wave-function in the free ion.

Rather different values are obtained for Cu^{2+} in hexagonal BeO by de Wit and Reinberg (1967). They are:

$$|g_\parallel| = 1.709(2); \quad |g_\perp| = 2.379(1),$$
$$|^{63}A_\parallel| = 50(1) \times 10^{-4} \text{ cm}^{-1}; \quad |^{63}A_\perp| = 108(1) \times 10^{-4} \text{ cm}^{-1},$$
$$|^{63}P_\parallel| = 11(1) \times 10^{-4} \text{ cm}^{-1}.$$

The results are interpreted with a positive sign for $g_\parallel$, giving a point near the top of the $(g_\parallel, g_\perp)$ ellipse in Fig. 7.23. A number of other spin resonance spectra for $3d^9$ ions (including Ni^+) in tetrahedral coordination have been measured (see references in de Wit and Reinberg 1967).

8
IONS OF THE d-GROUPS IN STRONG LIGAND FIELDS

8.1. The ions and their compounds

THE palladium ($4d$) and platinum ($5d$) groups are transition groups of the same general type as the $3d$ group, but on which very much less experimentation has been carried out, mainly because their compounds are less easy to prepare. The ions take up a number of valencies, corresponding to varying numbers of electrons in the partly-filled d-shells that lie 'outside' a core of closed shells. For the $4d$ group the core is like that of the krypton atom with configuration

$$1s^2 2s^2 2p^6 3s^2 3p^6 3d^{10} 4s^2 4d^6 \text{ (36 electrons)}.$$

Thus, for the $4d$ group the position resembles that of the $3d$ group, where the open $3d$ shell lies outside an argon core, $1s^2 2s^2 2p^6 3s^2 3p^6$ (18 electrons), but for the $5d$ group the position is more complicated. The core is xenon-like, with configuration

$$1s^2 2s^2 2p^6 3s^2 3p^6 3d^{10} 4s^2 4p^6 4d^{10} 5s^2 5p^6 \text{ (54 electrons)}$$

only for the ion La^{2+} for which, like the neutral atom La, it is energetically favourable for the odd electron to enter the $5d$ shell rather than the $4f$ shell. The main group of $5d$ ions have a core corresponding to Lu^{3+}, Hf^{4+}, Ta^{5+}, or W^{6+}, with configuration

$$1s^2 2s^2 2p^6 3s^2 3p^6 3d^{10} 4s^2 4p^6 4d^{10} 4f^{14} 5s^2 5p^6 \text{ (68 electrons)}$$

in which the lanthanide $4f$ shell is also filled.

The stable isotopes of the main $4d$, $5d$ groups (together with the long lived isotope 99 of technetium, which has no stable isotopes) are listed in Tables 8.1 and 8.2, together with their nuclear spins and moments. The ionic species principally studied in magnetic resonance are given in Table 8.3, together with a number of $3d$ ions which form 'strong ligand field' compounds in octahedral coordination. The ionic size for $4d$ ions, and more so for $5d$ ions, tends to be somewhat larger than that of comparable $3d$ ions in the same charge state, but it will be seen from Table 8.3 that the stable ions tend to carry a rather larger positive charge, so that their size is comparable with that of $3d$ ions with the same number of electrons in the d-shell.

TABLE 8.1

Atoms of the 4d group, and nuclear properties of the stable isotopes that have nuclear spin not equal to zero. Technetium has no stable isotopes, and the one radioactive isotope quoted is that for which magnetic resonance has been observed. The nuclear moments are quoted to four significant figures (where justified from nuclear magnetic resonance), and include the diamagnetic correction

Z		Mass number	Abundance (%)	Nuclear spin I	Nuclear magnetic moment (n.m.)	Nuclear electric quadrupole moment Q (barns)
39	Y	89	100	$\frac{1}{2}$	-0.1373	
40	Zr	91	11.23	$\frac{5}{2}$	-1.303	
41	Nb	93	100	$\frac{9}{2}$	$+6.167$	-0.2
42	Mo	95	15.72	$\frac{5}{2}$	-0.9135	0.12
		97	9.46	$\frac{5}{2}$	-0.9327	1.1
43	Tc	99	(radioactive)	$\frac{9}{2}$	$+5.680$	$+0.3$
44	Ru	99	12.7	$\frac{5}{2}$	-0.63	
		101	17.1	$\frac{5}{2}$	-0.69	
45	Rh	103	100	$\frac{1}{2}$	-0.0883	
46	Pd	105	22.2	$\frac{5}{2}$	-0.6015	
47	Ag	107	51.35	$\frac{1}{2}$	-0.1136	
		109	48.65	$\frac{1}{2}$	-0.1305	

TABLE 8.2

Atoms of the 5d group, and nuclear properties of the stable isotopes that have nuclear spin not equal to zero. The nuclear moments are quoted to four significant figures (where justified from nuclear magnetic resonance), and include the diamagnetic correction, except for the two iridium isotopes

Z		Mass number	Abundance (%)	Nuclear spin, I	Nuclear magnetic moment (n.m.)	Nuclear electric quadrupole moment Q (barns)
74	W	183	14.4	$\frac{1}{2}$	$+0.117$	
75	Re	185	37.1	$\frac{5}{2}$	$+3.172$	$+2.6$
		187	62.9	$\frac{5}{2}$	$+3.204$	$+2.6$
76	Os	187	1.64	$\frac{1}{2}$	$+0.067$	
		189	16.1	$\frac{3}{2}$	$+0.6566$	0.8
77	Ir	191	37.3	$\frac{3}{2}$	$+0.1440$	$+1.5$
		193	62.7	$\frac{3}{2}$	$+0.1568$	$+1.5$
78	Pt	195	33.8	$\frac{1}{2}$	$+0.6060$	
79	Au	197	100	$\frac{3}{2}$	$+0.1449$	$+0.58$

TABLE 8.3

Ions of the 4d and 5d groups, together with ions of the 3d group known to occur in 'strong crystal field' octahedral coordination

	d^1	d^2	d^3	d^4	d^5	d^6	d^7	d^8	d^9
Ions of the 3d group			Mn^{4+} Cr^{3+} V^{2+}	Mn^{3+}	Fe^{3+} Mn^{2+}	Co^{3+} Fe^{2+}	Ni^{3+}	Cu^{3+}	
Ions of the 4d group	Mo^{5+} Nb^{4+} Zr^{3+} Y^{2+}	Mo^{4+} Nb^{3+} Zr^{2+}	Tc^{4+} Mo^{3+} Nb^{2+}	Ru^{4+} Tc^{3+} Mo^{2+}	Rh^{4+} Ru^{3+} Tc^{2+} Mo^{+}	Rh^{3+} Ru^{2+}	Pd^{3+} Rh^{2+}	Ag^{3+} Pd^{2+} Rh^{+}	Ag^{2+} Rh^{0}
Ions of the 5d group	La^{2+}		Re^{4+}	Os^{4+}	Ir^{4+}	Pt^{4+} Ir^{3+}	Pt^{3+}		

TABLE 8.4

Free ion values of $\langle r^{-3} \rangle$, $\langle r^2 \rangle$, $\langle r^4 \rangle$ for ions of the 4d group, from Freeman and Watson (1965); values of ζ from Blume, Freeman and Watson (1964), with additional values of ζ_{exp} from estimates of Dunn (1961)

		$\langle r^{-3} \rangle$ (a.u.)	$\langle r^2 \rangle$ (a.u.)	$\langle r^4 \rangle$ (a.u.)	ζ (exp) (cm^{-1})	ζ (calc) (cm^{-1})
$4d^1$	Y^{2+}	2·034	5·588	59·00	300	312
	Zr^{3+}	3·160	3·857	25·33	500	507
	Nb^{4+}				750	
	Mo^{5+}				1030	
$4d^2$	Zr^{2+}	2·706	4·526	37·86	425	432
	Nb^{3+}	3·913	3·308	18·60	670	644
	Mo^{4+}				950	
$4d^3$	Nb^{2+}	3·414	3·829	26·98	555	560
	Mo^{3+}	4·707	2·905	14·39	800	812
	Tc^{4+}				(1150)	
$4d^4$	Mo^{2+}	4·175	3·319	20·22	695	717
	Tc^{3+}				(990)	
	Ru^{4+}				(1350)	
$4d^5$	Mo^{+}	3·662	3·954	32·98	(630)	
	Tc^{2+}	5·015	2·903	15·41	(850)	
	Ru^{3+}	6·496	2·313	9·17	(1180)	1197
	Rh^{4+}				(1570)	
$4d^6$	Ru^{2+}	5·858	2·628	12·87	1000	1077
	Rh^{3+}	7·447	2·117	7·79	1400	1416
$4d^7$	Rh^{2+}	6·804	2·374	10·60	1220	1291
	Pd^{3+}	8·487	1·939	6·59	1640	1664
$4d^8$	Pd^{2+}	7·814	2·158	8·83	1600	1529
	Ag^{3+}	9·611	1·782	5·61	1930	1940
$4d^9$	Ag^{2+}	8·905	1·972	7·41	1840	1794

The wave-functions for $4d$, $5d$ electrons have the same angular dependence as for $3d$ electrons, but their radial wave-functions differ in that the $4d$ functions have one node, and the $5d$ two nodes, while the $3d$ have none. Near the nucleus the radial wave-functions have an amplitude which increases rapidly with the atomic number Z, so that larger values of the hyperfine interactions, particularly in the $5f$ group, are to be expected than for $3d$ ions. This can be seen from Table 8.4, by comparing the values of $\langle r^{-3} \rangle$ obtained by Freeman and Watson (1965) by Hartree–Fock calculations for the $4d$ group, with the corresponding values for $3d$ ions in Table 7.6. Values of the spin-orbit coupling parameter ζ, which is similarly dependent on the amplitude of the wave-functions near the nucleus, are also noticeably larger for $4d$ ions, the values given in Table 8.4 being due to Blume, Freeman, and Watson (1964) and Dunn (1961). No detailed calculations have been made for $5d$ ions, but values of ζ in the range 2000–3000 cm^{-1} have been estimated, and correspondingly large values of $\langle r^{-3} \rangle$ for the hyperfine interaction must be expected.

Amongst the host lattices used for electron paramagnetic resonance experiments on these ions are MgO(cubic), Al_2O_3(trigonal), and $CaWO_4$(tetragonal), which have also been used extensively for other transition ions. For strong ligand field octahedral complexes the configuration $(d\epsilon)^6$ behaves as a 'diamagnetic' ion (though it has a small temperature independent susceptibility) with a closed shell (see § 8.2). Examples of this kind which have been used in host lattices are as follows:

$K_4Fe(CN)_6, 3H_2O$ for V^{2+}, Mn^{2+},
$K_3Co(CN)_6$ for Cr^{3+}, Fe^{3+},
$Co(NH_3)_6Cl_3$ for Ru^{3+},
$(NH_4)_2PtCl_6$, K_2PtCl_6 for Ir^{4+}, Tc^{4+}.

The point symmetry at the magnetic ion in the platinum compounds is exactly octahedral, the Pt^{4+} ion being surrounded by a regular octahedron of chlorine ions. In the other host lattices the point symmetry is rhombic or lower, and this is true also of a number of other substances that have been used as host lattices.

We shall discuss the theory and results for the octahedral complexes first, followed by one case of eightfold coordination ($5d^1$, La^{2+} in CaF_2) in § 8.13. Historically the explanation of the magnetic properties of the iron group cyanides in terms of a strong ligand field was first given by Van Vleck (1935). The contributions from the orbital magnetic moments, including the effects of spin-orbit splitting, were calculated by Kotani (1949) for octahedral symmetry. The theory has been extended by Kamimura (1956), Bleaney and O'Brien (1956), Kotani (1960), Tanabe (1960), and Sugano (1960); and, with particular reference to $4d$, $5d$ compounds, by Kamimura, Koide, Sekiyama, and Sugano (1960).

8.2. The strong ligand field octahedral complex

The strong crystal field approach using single electron states has been outlined in § 7.3, and the effects of bonding in § 7.4. The most important parameter is the octahedral ligand field splitting, which is denoted by $10Dq$ in the crystal field approach or, more generally, by Δ. A number of calculations indicate that the effect of covalent bonding is greatly to increase Δ over the value that is obtained in a purely 'ionic' approach; certainly, the latter gives a value that is only about one-third of that observed experimentally for Ni^{2+} in $KNiF_3$. The theoretical situation is outlined by Owen and Thornley (1966), who also quote experimental results showing how Δ increases for different ligand ions for the transition ions Cr^{3+}, $3d^3$ and Ni^{2+}, $3d^8$ (see Table 8.5). In general, the value of Δ appears to be larger by

TABLE 8.5

Variation of the octahedral crystal field splitting Δ (in units of 10^3 cm^{-1}) for various ligands and two different ions of the 3d group (quoted by Owen and Thornley (1966) from Jorgensen (1962))

Ion	Ligand	Br$^-$	Cl$^-$	F$^-$	H$_2$O	NH$_3$	CN$^-$
	Cr^{3+}, $3d^3$		13·8	15·2	17·4	21·6	26·7
	Ni^{2+}, $3d^8$	7·0	7·2	7·3	8·5	10·8	

a factor of about 2 for the tripositive ions compared with the dipositive ions (cf. also Cu^{3+} and Ni^{2+}, $3d^8$ in Al_2O_3, Table 7.20). In the $3d$ group the cyanides have unusually large values of Δ, values over 30000 cm^{-1} being reported for $Fe^{3+}(CN^-)_6$ by Naiman (1961) and for $Fe^{2+}(CN^-)_6$ by Jorgensen (1962). Since the values of the Racah parameter B are probably less than 1000 cm^{-1} for these ions in the solid state (see Table 7.5), it may be expected that such values of Δ ($= 10Dq$) are sufficient to make the configurations $(d\epsilon)^n$ the ground states (see Figs. 7.13, 7.14).

Calculations of Δ for the $4d$, $5d$ groups will obviously be much more

IONS OF THE d-GROUPS IN STRONG LIGAND FIELDS

difficult than for the $3d$ group. However, the values of $\langle r^4 \rangle$ for the $4d$ group in Table 8.4 are very much larger than for analogous ions of the $3d$ group (Table 7.6), so that a larger ionic contribution to Δ would be expected. It is not surprising that, experimentally, all ions studied in both the $4d$, $5d$ groups are found to belong to the 'strong ligand field' configurations.

Whatever the origin of Δ, its magnitude relative to the Racah parameter B determines the order in which the d-orbitals are filled. As can be seen from a comparison of Fig. 7.7 (applicable to small Δ) and Fig. 7.8 (applicable to large Δ), the configurations are the same for d^1, d^2, d^3, d^8, and d^9, and we shall expect only rather small differences between the results of paramagnetic resonance experiments for these ions in weak and strong ligand fields. The main differences occur for the ions d^4 to d^7; in a strong ligand field it is energetically preferable for the electrons first to occupy the $d\epsilon$ or t_2 shell, so that filling of this shell continues subject to the restraints imposed by the Pauli principle. This gives a closed $(d\epsilon)^6$ or $(t_2)^6$ sub-shell at d^6, and such ions have singlet ground states, both for orbit and spin, giving only a small temperature independent susceptibility. This makes it possible to use them as diluents in a number of strong ligand field complexes containing other paramagnetic ions in electron paramagnetic resonance experiments. The ion $(d\epsilon)^5$ resembles $(d\epsilon)^1$, except that it has one 'hole' instead of one electron in the $(d\epsilon)$ shell; similarly, $(d\epsilon)^4$ resembles $(d\epsilon)^2$, with two holes instead of two electrons. The remaining ion $(d\epsilon)^3$ has an orbital singlet state, with $S = \frac{3}{2}$ which is the maximum spin-multiplicity allowed in the $(d\epsilon)$ shell; its magnetic properties are virtually 'spin-only', as for the half-filled shell d^5 in a weak ligand field complex.

The four ions $(d\epsilon)^1$, $(d\epsilon)^2$, $(d\epsilon)^4$, and $(d\epsilon)^5$ all have triplet orbital ground states that can be regarded as manifolds with $\tilde{l} = 1$ and $\tilde{g}_l = -1$, though the latter may be reduced by a factor k to allow for bonding effects. The effective spin-orbit coupling is reversed in sign for the 'hole' configurations $(d\epsilon)^4$, $(d\epsilon)^5$ from that for the 'electron' configurations $(d\epsilon)^1$, $(d\epsilon)^2$, and in an exactly octahedral field the states $\tilde{J}$ into which the $(\tilde{l}, S)$ manifold splits are shown in Fig. 8.1. Again, the value of the spin-orbit coupling is likely to be lower than the free ion value, because of bonding effects. Within the $(d\epsilon)$ manifold, only π-bonding orbitals can be formed, so that the value of k above should be just $k_{\pi\pi}$, as given by eqn (7.7).

For d^7 in a strong octahedral field we have just one electron in a

478 IONS OF THE d-GROUPS IN STRONG LIGAND FIELDS

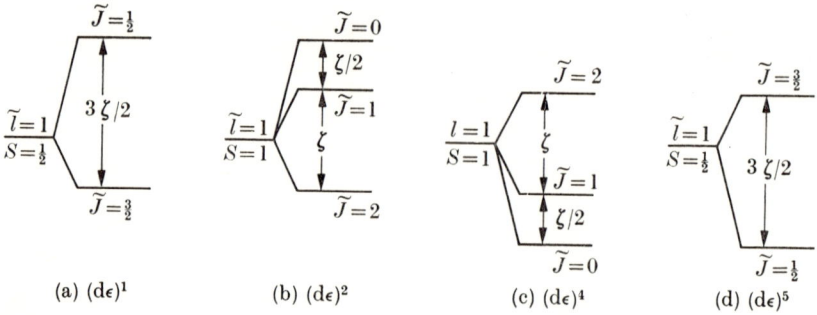

(a) $(d\epsilon)^1$ (b) $(d\epsilon)^2$ (c) $(d\epsilon)^4$ (d) $(d\epsilon)^5$

FIG. 8.1. Splittings of the $(\tilde{l} = 1, S)$ manifolds under spin-orbit coupling $\zeta(\mathbf{l}\cdot\mathbf{s})$ for the strong octahedral configurations $(d\epsilon)^1$, $(d\epsilon)^2$, $(d\epsilon)^4$, and $(d\epsilon)^5$, all of which have a triplet orbital ground state $\tilde{l} = 1$. The effective spin-orbit coupling is $\tilde{\lambda}(\tilde{\mathbf{l}}\cdot\mathbf{S})$, with $\tilde{\lambda} = -\zeta$ for $(d\epsilon)^1$ and $+\zeta$ for $(d\epsilon)^5$, and $\tilde{\lambda} = \pm\zeta/2S$ is $-\zeta/2$ for $(d\epsilon)^2$ and $+\zeta/2$ for $(d\epsilon)^4$. The splittings shown for $(d\epsilon)^2$, $(d\epsilon)^4$ assume that ζ is small compared with the electrostatic repulsion between the electrons, measured by the Racah parameter $(3B+C)$; exact formulae are given by Kamimura, Koide, Sekiyama, and Sugano (1960).

$d\gamma$ orbital 'outside' the closed sub-shell $(d\epsilon)^6$, and the behaviour of this ion (with $S = \frac{1}{2}$) is more akin to that of d^9, with one hole in the $(d\gamma)$ shell, than to that of d^7 in a weak octahedral field.

8.3. Hyperfine interaction

The hyperfine interaction must also be calculated using one-electron wave-functions, as in eqns (7.37), (7.45), or (7.47), and (7.55) for the magnetic interaction. Values for $\langle r^{-3} \rangle$ computed for the free $4d$ ions are given in Table 8.4, but smaller values are probably required for the ion in a magnetic complex owing to bonding effects. Values of χ (see eqn (7.66)) have been computed (see Watson and Freeman 1967a) for a number of $4d$ ions; they are found to lie in the range -8 to -9 a.u., which is about three times as large as for the $3d$ ions, but again smaller values are likely to hold for the complexes. For the nuclear electric quadrupole interaction an appropriate form is found by combining eqns (17.10) and (17.12), with

$$\langle l \| \alpha \| l \rangle = -\frac{2}{(2l-1)(2l+3)} = -\frac{2}{21}$$

for a d-electron (see Table 18).

8.4. d^1 in a strong octahedral field; $(d\epsilon)^1$, (t_2), $S = \frac{1}{2}$

The theory for this ion is essentially similar to that for d^1 in an intermediate ligand field, discussed in § 7.8.

Measurements have been reported on $4d^1$, Mo^{5+} in $K_3(InCl_6),3H_2O$

by Owen and Ward (1956); in TiO_2 by Kyi (1962) and Chang (1964); and in K_2SnCl_6 by Low (1967). Ignoring small departures from axial symmetry, we can summarize the results by saying that $g_\parallel$ lies between 1·9 and 2, with $g_\perp$ somewhat lower. This indicates that distortions of the octahedron give an axial field that splits the orbital triplet by an amount large compared with the spin-orbit coupling, making the ground state a Kramers doublet that is mainly $|\tilde{0}^\pm\rangle$ in the nomenclature of § 7.8 (see Tables 7.9, 7.10). We can then expand the formulae (7.80) for the g-values in the limit of small δ, giving

$$\begin{aligned} g_\parallel &= 2 - 2\delta^2(2+k), \\ g_\perp &= 2 - \sqrt{(8)}k\delta - 2\delta^2, \end{aligned} \quad (8.1)$$

which shows that we would expect a larger departure of $g_\perp$ than of $g_\parallel$ from the free spin value, since the former depends on δ and the latter only on δ^2. The same result is also clearly to be expected from Fig. 7.23, since we are near the top of the ellipse where $g_\parallel$ has its maximum value.

8.5. d^2 in a strong octahedral field; $(d\epsilon)^2$, $(t_2)^2$, $S = 1$

The theory for the ground state of this ion is similar to that for d^2 in intermediate octahedral ligand fields. In exact octahedral symmetry the splitting of the $(\tilde{l}, S)$ manifold by the spin-orbit coupling is shown in Fig. 8.1(b). However, for the $5d$ group the spin-orbit coupling may be so large that it competes with the electrostatic repulsion between the two electrons; in putting the two electrons into different $(d\epsilon)$ orbitals with parallel spin we have assumed the latter is overwhelming, giving a modified form of Hund's rule. Explicit formulae for the energy levels including both electrostatic repulsion and spin-orbit coupling can be written down in terms of the two parameters ζ and $(3B+C)$, where B, C are the Racah parameters (see Kamimura, Koide, Sekiyama, and Sugano 1960).

8.6. d^3 in a strong octahedral field; $(d\epsilon)^3$, $(t_2)^3$, $S = \frac{3}{2}$

The ground state is an orbital singlet with fourfold spin degeneracy, the same as for d^3 in intermediate octahedral fields (cf. Fig. 7.11). Typical g-values are 1·992 for V^{2+}, Cr^{3+} ($3d^3$) in the complex cyanides (Baker, Bleaney, and Bowers 1956); 1·9896(5) for Tc^{4+} ($4d^3$) in K_2PtCl_6 (Low and Llewellyn 1957). These are closer to the free spin value than the corresponding g-value for $3d^3$ ions in hydrated salts because the orbital contribution is more effectively quenched by the stronger cubic field. The magnetic hyperfine constants are almost isotropic, and arise mainly from the core polarization effect. For V^{2+}, Cr^{3+} the value of A is about 80 per cent of that observed in hydrated salts, the reduction being attributed to the effect of covalent bonding. In Tc^{4+} the large hyperfine constant $|A| = 0 \cdot 0135$ cm^{-1} for the isotope ^{99}Tc, $I = \frac{9}{2}$, leads to a value of about -5 a.u. for the core polarization parameter χ, rather smaller than the values of 8 to 9 a.u. calculated by Watson and Freeman (1967a).

The resonance spectrum of Re^{4+}, $5d^3$ has been observed in K_2PtCl_6 by Rahn and Dorain (1964). Only the $M = +\frac{1}{2} \leftrightarrow -\frac{1}{2}$ transition was visible, but assuming exact cubic symmetry they deduced the values $g = 1 \cdot 815(1)$, $|A| = 389(1) \times 10^{-4}$ cm^{-1} (unresolved value for isotopes 185, 187, each with $I = \frac{5}{2}$), together with appreciable higher-order terms of the form given in eqn (18.21) for a Γ_8 quartet. However the observed anisotropy may be due to small distortions from octahedral symmetry (see Dorain and Wheeler (1966)). The low value of g illustrates the considerable role played by intermediate coupling in the $5d$ group.

8.7. d^4 in a strong octahedral field; $(d\epsilon)^4$, $(t_2)^4$, $S = 1$

This ion resembles $(d\epsilon)^2$, except that with two holes instead of two electrons in the $d\epsilon$ shell the effective sign of the spin-orbit coupling is reversed. In the rather low symmetry of the Mn^{3+} ion in the $K_3Mn(CN)_6$ complex the energy levels are split into a number of singlets, but the susceptibility (Cooke and Duffus 1955) follows roughly the variation with temperature predicted for octahedral symmetry. The latter gives a ground orbital triplet state $l = 1$ with $\tilde{g}_l = -1$, which combines with the spin $S = 1$ through the spin-orbit coupling to give states $\tilde{J} = 0$, 1, and 2, the singlet state $\tilde{J} = 0$ being lowest as shown in Fig. 8.1. A rather better fit to the temperature variation of the susceptibility is obtained using a value of ζ somewhat smaller (60 to 80 per cent) than the free ion value. At very low temperatures only a temperature independent susceptibility due to the $\tilde{J} = 0$ state is observed, and the salt has been used as a diluent for $K_3Co(CN)_6$ by Baker, Bleaney, and Bowers (1956).

For an ion such as Os^{4+}, $5d^4$, the spin-orbit coupling is so large that only the singlet ground state is occupied at room temperature. The effect of its temperature independent susceptibility in shifting the nuclear magnetic resonance of ^{19}F in K_2OsF_6 through the formation of bonds has been studied theoretically, and compared with experiment,

IONS OF THE d-GROUPS IN STRONG LIGAND FIELDS

by Greenslade and Stevens (1967). The temperature variation of the susceptibility of K_2RuCl_6 and K_2OsCl_6 has been fitted theoretically by Kamimura, Koide, Sekiyama, and Sugano (1960).

8.8. d^5 in a strong octahedral field; $(d\epsilon)^5$, $(t_{2g})^5$, $S = \frac{1}{2}$

This ion can be regarded has having just one hole in the $(d\epsilon)$ shell, so that the spin $S = \frac{1}{2}$; this corresponds in Fig. 7.13 to a cubic field so strong $(Dq/B > 2\cdot9)$ that a doublet state attains a lower energy than the sextet state $^6S_{\frac{5}{2}}$, which is the ground state in weak and intermediate cubic fields. For one hole in the $(d\epsilon)$ shell we have again a triplet orbital ground state $l = 1$, with an effective $\tilde{g}_l = -1$; however the effective spin-orbit coupling constant will be reversed in sign to that for $(d\epsilon)^1$, so that within the triplet we have just $\zeta(\tilde{l} \cdot S)$. It has been shown by Bleaney and O'Brien (1956) that if all matrix elements to $(d\gamma)$ states are neglected, an energy matrix can be obtained somewhat similar to (7.77), irrespective of the symmetry of any distortion of the octahedron, by a suitable choice of axes. In general a set of three Kramers doublets is then obtained under the combined effect of the distortion and the spin-orbit coupling. In the nomenclature of Bleaney and O'Brien the three orbital states are $|b_+\rangle$, $|a\rangle$, and $|b_-\rangle$; the ground doublet can be written (with superscripts for the spin orientation) as

$$|+\rangle = \cos\theta[\sin\alpha\,|a^+\rangle + \cos\alpha\,|b_+^-\rangle] + \sin\theta\,|b_-^-\rangle \\ |-\rangle = -\cos\theta[\sin\alpha\,|a^-\rangle - \cos\alpha\,|b_-^+\rangle] + \sin\theta\,|b_+^+\rangle \quad (8.2)$$

and the principal g-values, using an orbital reduction factor k that is of course $k_{\pi\pi}$ in the nomenclature of § 7.4, are given by

$$g_z = \cos^2\theta\{g_s\sin^2\alpha - (g_s+2k)\cos^2\alpha\} + \sin^2\theta(2k-g_s), \\ \tfrac{1}{2}(g_x+g_y) = -\cos^2\theta\{g_s\sin^2\alpha + 2\sqrt{(2)}k\cos\alpha\sin\alpha\}, \\ \tfrac{1}{2}(g_x-g_y) = \sin 2\theta\{g_s\cos\alpha + \sqrt{(2)}k\sin\alpha\}. \quad (8.3)$$

Here the identification of the states $|+\rangle$, $|-\rangle$ has been chosen to make g_x, g_y have the same sign; for a given value of α this fixes the sign of g_z, since the sign of $(g_xg_yg_z)$ is invariant (see § 15.6).

Corresponding formulae for the magnetic hyperfine structure are more complicated because they are not independent of the symmetry of the distortion away from octahedral symmetry. Formulae are given by Bleaney and O'Brien (1956) for two cases, one in which the z-axis is along a fourfold axis of the octahedron and the other in which it is along a threefold axis; even so the magnetic hyperfine term $(\tilde{\mathbf{S}} \cdot \mathbf{A} \cdot \mathbf{I})$

with $\tilde{S} = \frac{1}{2}$ for the ground doublet is one in which **A** is not necessarily a 'symmetric tensor' (see the discussion in §§ 15.6 to 15.8).

$3d^5$ ions, Mn^{2+}, Fe^{3+}, in the complex cyanides

Paramagnetic resonance measurements have been made by Baker, Bleaney, and Bowers (1956) on the $3d^5$ ions, Mn^{2+} in $K_4Fe(CN)_6$, $3H_2O$ and Fe^{3+} in $K_3Co(CN)_6$. The results are summarized in Table 8.6. These results are characteristically different from those for $3d^5$ in the weak field state $^6S_{\frac{5}{2}}$ (see Table 7.17) in that there is no sign of any fine structure (i.e. $\tilde{S} = \frac{1}{2}$), and the g-values are markedly anisotropic, diverging widely from the free spin value. If we take $g_s = 2$, $k = 1$, we obtain from (8.3) the following simple relationship between the principal g-values,

$$\frac{(g_x - g_y)}{2\sqrt{(2)}\tan\theta} = \frac{(g_x + g_y)}{-2\sin\alpha} = \frac{g_z}{\sin\alpha - \sqrt{(2)}\cos\alpha}, \quad (8.4)$$

showing that both the unknown constants θ and α can then be found just from the ratios of the g-values. However, if this procedure is followed for $K_3Fe(CN)_6$ (diluted), one obtains values for the constants which give g-values in the correct proportions but whose actual values are about 8 per cent too high. This is a simple demonstration of the fact that the strong crystal field model cannot be used without making allowance for a reduction in the orbital moment because of bonding. If all the g-values are assumed to be negative, they can be fitted using $\cot\alpha = 1\cdot01$, $\tan\theta = -0\cdot028$, $k = 0\cdot87_5$, and a reasonable fit can then be obtained for the temperature variation of the susceptibility of a single crystal of $K_3Fe(CN)_6$ using a value of about 280 cm^{-1} for $|\zeta|$, which is a good deal smaller than the free ion value of about 486 cm^{-1} (see Table 7.4).

An alternative set of parameters is obtained if we assume that g_z is positive, and g_x, g_y both negative, but these require a very low value of $k = 0\cdot56$ and give a less good fit to the susceptibility.

$4d^5$ ions, Ru^{3+}

Not many results are available for $4d^5$, a typical set being shown in Table 8.6. These are for Ru^{3+} in $Co(NH_3)_6Cl_3$, which is monoclinic and shows three pairs of magnetically inequivalent complexes (Griffiths, Owen, and Ward 1953; Griffiths, O'Brien, Owen, and Ward 1955). The results resemble those for the $3d^5$ ions in the same table. The hyperfine constants are an average for the two isotopes 99,101, which each

TABLE 8.6

Representative values of g and A (signs undetermined) for a number of ions in $(d\epsilon)^5$ or $(t_2)^5$ configurations. The hyperfine constants given for Ru and Ir are from the unresolved structures due to the pairs of odd isotopes ^{99}Ru, ^{101}Ru and ^{191}Ir, ^{193}Ir respectively. In each case the spins of the pair are the same, and the nuclear magnetic moments differ by less than 10 per cent. The results for Ir^{4+} are from Thornley (1968), where further values are given.

	Ion and host lattice	$\|g_x\|$	$\|g_y\|$	$\|g_z\|$	$\|A_x\|$	$\|A_y\|$ (in units of 10^{-4} cm^{-1})	$\|A_z\|$
$3d^5$	Mn^{2+} $K_4Fe(CN)_6 \cdot 3H_2O$	2·624(8)	2·182(8)	0·63(10)	84·5(5)	46·5(5)	104(20)
	Fe^{3+} $K_3Co(CN)_6$	2·35(2)	2·10(2)	0·915(10)			
$4d^5$	Ru^{3+} $Co(NH_3)_6Cl_3$	2·06(1)	2·02(1)	1·72(1)	48(2)	48(2)	49(2)
		1·80(1)	1·90(1)	2·06(1)	48(2)	48(2)	50(2)
		1·15(1)	1·84(1)	2·66(1)	45(2)	41(2)	54(2)
$5d^5$	Ir^{4+} $Na_2PtCl_6 \cdot 6H_2O$	2·168(5)	2·078(5)	1·050(3)	25·5(10)	25·5(10)	24·9(5)
	Ir^{4+} (cubic)	(−)1·786(4), isotropic			26·3(6), isotropic		
	$(NH_4)_2PtCl_6$						

have $I = \frac{5}{2}$ and whose nuclear magnetic moments differ only by about 9 per cent (see Table 8.1). The rather small variation in the hyperfine constants suggests the main contribution comes from core polarization, though even the latter does not give an isotropic contribution because of the varying way in which spin and orbit are admixed in different directions. Although the hyperfine constants appear to be of the same order as those for ^{55}Mn in the same table, the nuclear moments of the two ruthenium isotopes are only about $\frac{1}{5}$ of that of ^{55}Mn (for which also $I = \frac{5}{2}$), so that the electronic parameters in the hyperfine structure for $4d^5$ are considerably larger than those for $3d^5$, as would be expected.

The spectrum of Ru^{3+} in Al_2O_3 has been observed by Geschwind and Remeika (1962), who find

$$g_{\parallel} < 0{\cdot}06, \quad |g_{\perp}| = 2{\cdot}430, \quad |^{101}A_{\perp}| = 0{\cdot}0043 \text{ cm}^{-1}.$$

The g-values can be fitted with $k = 0{\cdot}837$ in eqns (8.3).

$5d^5$ *ions*, Ir^{4+}

Octahedral complexes of the type $[IrCl_6]^{2-}$, $[IrBr_6]^{2-}$ containing the ion $5d^5$, Ir^{4+} have been studied extensively by Griffiths and Owen, following the early discovery of complex hyperfine structure due to the ligand ions (Owen and Stevens 1953). Some representative results are included in Table 8.6, those for the complex in $Na_2[PtCl_6],6H_2O$ which is not exactly octahedral being again rather similar to the results discussed above for the $3d^5$, $4d^5$ ions. The g-values for a number of salts (taking $g_{\perp}$ as an average of the values of g_x, g_y) can be fitted onto an ellipse similar to that in Fig. 7.23, but with a reduced value of $k = 0{\cdot}83$, as shown in Fig. 8.2.

The case of cubic symmetry is of particular interest. If in eqns (8.3) we put $\theta = 0$, $\sin \alpha = (\frac{1}{3})^{\frac{1}{2}}$, $\cos \alpha = (\frac{2}{3})^{\frac{1}{2}}$ we obtain

$$g_x = g_y = g_z = g = -\tfrac{1}{3}g_s - \tfrac{4}{3}k. \quad (8.5)$$

The same result follows from Table 7.10 on putting $\cos 2\delta = \frac{1}{3}$ and taking $g_{\perp}$ to be negative for the doublet A, which is always the lowest doublet for $(d\epsilon)^5$ because of the change of sign in the spin-orbit coupling; see Fig. 7.22. Here we have again identified the spin contribution by writing g_s and not just 2. If $k = 1$ we obtain $g = -2$, where the negative sign again emphasizes the fact that there is a large orbital contribution, and that this is not a spin-only value. If $k < 1$, we can fit the measured value of $|g| = 1{\cdot}786$ for Ir^{4+} in the exact octahedral symmetry of $(NH_4)_2PtCl_6$ by again taking $k = 0{\cdot}84$, and this point is also included in Fig. 8.2.

IONS OF THE d-GROUPS IN STRONG LIGAND FIELDS 485

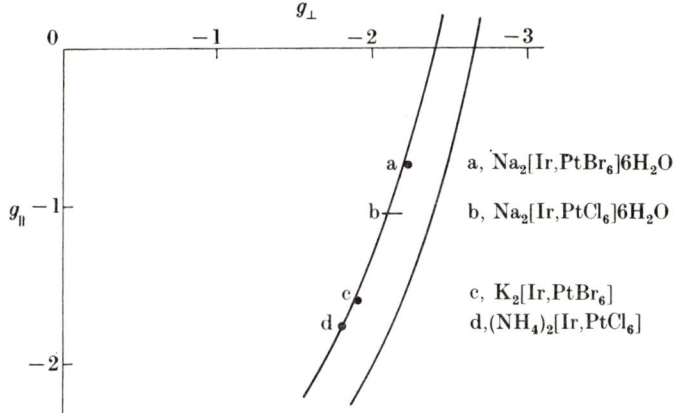

FIG. 8.2. Plot of $g_\parallel$ against $g_\perp$ for a number of iridium $(5d^5)$, $(d\epsilon)^5$ compounds. The curves correspond to eqns (8.3) with $g_\perp = \frac{1}{2}(g_x+g_y)$, (i.e. $\theta = 0$). The outer curve is drawn for $k = 1$, and is the same as that in Fig. 7.23 for a range of values between $\delta = \pi/2$ and π. The inner curve is drawn using a value of $k = 0.83$. (After Griffiths and Owen 1954).

In exact octahedral symmetry the magnetic hyperfine constant A is of course also isotropic, and from the formulae of Bleaney and O'Brien (1956) it is given by

$$A = 2g_n\beta\beta_n\langle r^{-3}\rangle(-\tfrac{8}{7}+\tfrac{1}{3}\kappa). \tag{8.6}$$

In exact octahedral symmetry the spin-dipolar contribution vanishes, and the first term in (8.6) is the orbital contribution and the second that from core-polarization plus other effects that may contribute to a term of the form $A_s(\mathbf{S} \cdot \mathbf{I})$. In terms of the parameter $\chi = -\tfrac{3}{2}\kappa\langle r^{-3}\rangle$ of eqn (7.66), and using a mean value of $g_n = 0.10$ for the two odd iridium isotopes 191, 193, for both of which $I = \tfrac{3}{2}$ and whose moments differ by less than 10 per cent (see Table 8.2), so that the hyperfine parameter in Table 8.6 is the mean for the unresolved structure, we find

$$|\chi+\tfrac{3.6}{7}\langle r^{-3}\rangle| = 37 \text{ a.u.} \tag{8.7}$$

Since both χ and $\langle r^{-3}\rangle$ are likely to be fairly large for ions in the $5d$ group, it is difficult to find a reasonable hypothesis about the relative sizes of the two contributions with which to extract further information.

Analysis of the complex hyperfine structure (cf. Fig. 1.16) due to both chlorine and bromine ligands has been attempted by Griffiths and Owen (1954). The t_2 orbitals on the central ion can bond only to p_π electrons on the ligand ions, so that eqn (7.10) reduces to

$$A_\parallel^\mathrm{L} = 2A_\mathrm{d}+(A_p)_\parallel, \qquad A_\perp^\mathrm{L} = -A_\mathrm{d}+(A_p)_\perp. \tag{8.8}$$

In cubic symmetry the ligand hyperfine parameter A_p is isotropic, its value being

$$A_p = -2g_n^L \beta \beta_n \langle r_L^{-3} \rangle \tfrac{16}{15} f_t \qquad (8.9)$$

where, from eqns (7.7), (7.11), $f_t = \tfrac{1}{4}\lambda_t^2 N_t^{-1} = \tfrac{1}{2}(1-k)$ is a measure of the strength of the π-bonding (in octahedral symmetry it is the probability of finding the magnetic electron on one particular ligand ion if the region where overlap of a ligand p-orbital onto a central ion d-orbital is assumed to be small). The isotropic ligand hyperfine structure $|A^L| = 26(1)$ MHz measured by Griffiths and Owen (1954) for Ir^{4+} in $(NH_4)_2[PtCl_6]$ suggests that A_d can be neglected, and gives a value of $f_t = 0\cdot 066$ in reasonable agreement with the value of $0\cdot 080$ obtained from the measured g-value. For bromine ligands the value of A^L is about 5 times larger, consistent with the increased value of $\langle r^{-3} \rangle$ for $4p$ orbits on a bromine ion and the larger nuclear magnetic moment.

The theory of the $(IrX_6)^{2-}$ complexes has been extended by Thornley (1968), who finds that the inclusion of admixtures from excited states appreciably affects the analysis given above of the experimental results, leading to a larger covalent character than previously assumed.

8.9. d^6 in a strong octahedral field; $(d\epsilon)^6$, $(t_2)^6$, $S = 0$

At this point the $(d\epsilon)$ shell is filled, giving as ground state an orbital and spin singlet when the octahedral ligand field strength is large enough to outweigh the effects of the electronic electrostatic repulsion (see Fig. 7.14). Under these conditions only a small temperature-independent susceptibility is observed for the ground state, and $(d\epsilon)^6$ ions have been used as diluents for other paramagnetic ions in strong octahedral field complexes.

8.10. d^7 in a strong octahedral field; $(d\epsilon)^6(d\gamma)$, $(t_2)^6 e$, $S = \tfrac{1}{2}$

With one electron more than can be accommodated in the $(d\epsilon)$ shell, the extra electron must enter the $(d\gamma)$ or e-shell, giving a net spin $S = \tfrac{1}{2}$. Since matrix elements of the spin-orbit coupling vanish within the e-manifold, the configurations $(d\epsilon)^6(d\gamma)$ or $(t_2)^6 e$ with one electron in the e-shell and $(d\epsilon)^6(d\gamma)^3$ or $(t_2)^6 e^3$ with one hole in the e-shell have similar ground states and are experimentally difficult to distinguish. The g-values are close to the free spin value, lying above it only through spin-orbit coupling to excited $(d\epsilon)^5(d\gamma)^2$ states. As can be seen from the experimental results in Table 8.7, a Jahn–Teller effect is to be expected

TABLE 8.7

Experimental results for d^7, $(d\epsilon)^6(d\gamma)$ or $(t_2)^6 e$, in strong octahedral coordination. In two cases all three g-values are unequal, and an average is given for $g_\perp$. Values of $\bar{g}$ are for the isotropic spectrum observed at higher temperatures in cubic or trigonal symmetry. The resonance values given here for Ni^{3+} in MgO, CaO were attributed in the references quoted to Ni^+, $3d^9$, but are ascribed to Ni^{3+} by Höchli, Müller, and Wysling (1965)

Ion	Host lattice	$\lvert g_\parallel \rvert$	$\lvert g_\perp \rvert$	$\lvert \bar{g} \rvert$	Reference
$3d^7$, Ni^{3+}	MgO			2·1693(5)	Orton, Auzins, Griffiths, and Wertz (1961).
	CaO			2·282(2)	Low and Suss (1963).
	Al_2O_3	2·045(5) (powder)		2·146(2)	Lacroix, Höchli, and Müller (1964).
$4d^7$, Pd^{3+}	Al_2O_3			2·163(1)	Lacroix, Höchli, and Müller (1964).
$4d^7$, Rh^{2+}	$ZnWO_4$	2·375(2)	2·12 (average)		Whiffin and Orton (1965).
$5d^7$, Pt^{3+}	Al_2O_3	2·011(6)	2·328(4)	2·220(1)	Geschwind and Remeika (1962).
	Yttrium Aluminium Garnet	1·962	2·39 (average)		Hodges, Serway, and Marshall (1966).
	$BaTiO_3$	1·950(5)	2·459(3)		Simanek, Sroubek, Zdansky, Kaczer, and Novak (1966).

in cubic or trigonal symmetry, giving a change from an isotropic to an anisotropic spectrum as the temperature is lowered.

8.11. d^8 in a strong octahedral field; $(d\epsilon)^6(d\gamma)^2$, $(t_2)^6 e^2$, $S = 1$

As can be seen from Fig. 7.16, the ground state of this ion is the same in a weak or a strong octahedral field; it is an orbital singlet with spin $S = 1$, for which formulae are given in Table 7.19. The g-value is almost isotropic even in a trigonal environment such as Al_2O_3, but a strong octahedral splitting Δ reduces the departure of g from the free spin value, and the splitting of the triplet state. This is exemplified in the comparison of results for Cu^{3+} and Ni^{2+} in Al_2O_3, given in Table 7.20.

8.12. d^9 in a strong octahedral field; $(d\epsilon)^6(d\gamma)^3$, $(t_2)^6 e^3$, $S = \tfrac{1}{2}$

Again, the properties of this ion in the ground state are similar irrespective of the strength of the octahedral field, so that the discussion of § 7.16 is valid.

TABLE 8.8

Resonance parameters for two $4d^9$ ions. The values of $A_\parallel$, $A_\perp$ are mean values for the unresolved hyperfine structure of the two isotopes ^{107}Ag, ^{109}Ag (both $I = \frac{1}{2}$; see Table 8.1); no hyperfine parameters for Rh° (^{103}Rh, $I = \frac{1}{2}$) are given

Ion		Host Lattice	$\|g_\parallel\|$	$\|g_\perp\|$	$\|A_\parallel\|$	$\|A_\perp\|$	Reference
					(in units of 10^{-4} cm^{-1})		
$4d^9$	Ag^{2+}	KCl (77°K)	2·193(1)	2·035(1)	40·9(5)	30·3(5)	Delbecq, Hayes, O'Brien, and Yuster (1963); Sierro (1967).
		(300°K)	2·09(1) (isotropic)				
	Rh°	AgCl (110°K)	2·015(2)	2·417(2)			Wilkens, de Graag, and Helle (1965).
		(160°–350°K)	2·436(5)	2·207(5)			
		(200°–430°K)	2·287(5) (isotropic)				

TABLE 8.9

Resonance parameters for some $4d^1$, $5d^1$ ions in eightfold and fourfold coordination. The nuclear spins are $\frac{5}{2}$ for both Mo isotopes, and $\frac{7}{2}$ for ^{139}La

Ion	Host lattice	Temperature (°K)	$\|g_\|\|$	$\|g_\perp\|$	$\|A_\|\|$	$\|A_\perp\|$	Isotope
					(in units of 10^{-4} cm^{-1})		
$4d^1$, Y^{2+}	CaF$_2$	4·2	2·00	1·958			
		20	1·971 (isotropic)				
		20	1·948	1·987			
$4d^1$, Mo^{5+}	CaWO$_4$	4·2	1·987	1·887	8·39	41·18	(isotope 95)
					8·64	42·52	(isotope 97)
$5d^1$, La^{2+}	CaF$_2$	4	2·00(1)	1·904(2)	37(9)	62·6(10)	(isotope 139)
		20	1·937(2) (isotropic)		52(1) (isotropic)		

Refs: Y^{2+}, O'Connor and Chen (1964).
Mo^{5+}, Azarbayejani and Merlo (1965).
La^{2+}, Hayes and Twidell (1963).

Resonance parameters for Ag^{2+} and Rh^0 (both $4d^9$) ions are given in Table 8.8. These are in sites of cubic symmetry, with static Jahn–Teller distortions at low temperatures; in each case an isotropic g-value resulting from the dynamic Jahn–Teller effect is observed at higher temperatures. Very similar results are obtained for Ag^{2+} in LiCl and NaCl (Sierro 1967).

8.13. d^1 in cubic (eightfold) coordination

The spectrum of the ion La^{2+}, $5d^1$, replacing Ca^{2+} in CaF_2 has been observed by Hayes and Twidell (1963). Because the sign of the cubic field is reversed in eightfold compared with sixfold coordination, the single d-electron is in one of the $(d\gamma)$ or e-states that form the ground state, and the behaviour of La^{2+} is similar to that of a d^9 ion in an octahedral field, except that its spin-orbit coupling parameter is positive, so that the g-values are smaller than the free spin value. At 4°K an anisotropic spectrum characteristic of a static Jahn–Teller effect with symmetry about a fourfold axis of the cube is observed; at 20°K this is replaced by an isotropic spectrum characteristic of a dynamic Jahn–Teller effect, whose resonance parameters are equal to the mean values of the previous spectrum. The resonance results are listed in Table 8.9, together with values for $4d^1$, Y^{2+} in CaF_2. For the latter, one set of results is very similar to those for La^{2+}, where the value of $g_{\parallel} = 2 \cdot 00$ in the anisotropic spectrum indicates that the orbital $|3z^2 - r^2\rangle$ has the lower energy. However, for Y^{2+} a spectrum with $g_{\parallel} < g_{\perp}$ could also be observed which must be ascribed to the $|x^2 - y^2\rangle$ orbital (cf. Table 7.22(a) and (b)); it is suggested that this is stabilized by a tetragonal distortion caused by an interstitial compensating ion.

The tetragonal spectra of $4d^1$, Mo^{5+} in $CaWO_4$ observed by Azarbayejani and Merlo (1965) presumably originate from a Mo^{5+} ion replacing a W^{6+} ion in a WO_4 distorted tetrahedron; the g-values again suggest one d-electron in a $|3z^2 - r^2\rangle$ orbital.

9
SPIN-SPIN INTERACTION

9.1. Introduction

IN previous chapters we have been concerned primarily with the interactions of paramagnetic ions with external magnetic fields, treating them as single ions and ignoring any mutual interaction between the ions. We now consider the nature of such mutual interactions, and their effect on the paramagnetic resonance spectrum.

Since we are concerned with magnetic dipoles, the simplest and most obvious interaction between them is magnetic dipole-dipole interaction. If we know the location of each ion in the crystal lattice, and the magnetic properties of each ion, the magnetic dipole-dipole interaction between any pair can be calculated quite accurately. Unfortunately this is seldom true for interactions of other types, which may be of comparable or greater importance.

Since our magnetic moments are associated with electrically charged particles, there exists also an electrostatic interaction. It is well known that the quantum-mechanical treatment of this interaction leads to terms of two types, one of which is similar in form to the Coulomb interaction of classical physics, and which we shall call for brevity the 'Coulomb' interaction. The best known example of this is the 'electric quadrupole-quadrupole' interaction. Such interactions are of interest to us only in so far as they give an interaction energy that depends on the orientation of the magnetic dipoles. This requires that the magnetic moment be due in part to the orbital motion of the atomic electrons, since only the sub-states corresponding to different orientations of the orbital magnetic moment have different distributions of electric charge. The spin sub-states are involved only in so far as they are strongly coupled to the orbital sub-states by the spin-orbit coupling. Calculation of such 'Coulomb' interactions is handicapped by uncertainty in the magnitude of shielding (or anti-shielding) effects due to intervening ions, and by the lack of good radial electronic wave-functions. The same considerations apply to another type of multipole interaction in which one ion is coupled to another through the medium of the 'phonon radiation field' instead of the 'photon radiation field'.

Similar problems arise for a better known type of interaction, the 'exchange interaction', which involves the spin more directly. In most cases of interest to us this interaction arises through partial electron transfer to intervening ('ligand') ions, and in magnitude the exchange interaction varies widely. If its energy is large or comparable with the microwave quantum involved in a magnetic resonance transition, the spectrum may be altered quite drastically. If the exchange energy is small compared with the microwave quantum, it affects primarily the shape and width of the resonance line. The latter is true generally for the magnetic and Coulomb interactions mentioned above, which seldom exceed 10^{-1} cm^{-1} in energy.

In §§ 9.2–9.4 we consider the form of these interactions in more detail. In the later sections we shall restrict the discussion to interactions of the form

$$\mathscr{H} = \mathbf{S}_i \cdot \mathscr{J} \cdot \mathbf{S}_j = \sum_{p,q} \mathscr{J}_{p,q} S_{ip} S_{jq}, \tag{9.1}$$

where $p, q = x, y, z$ and S_i, S_j are the effective spins for ions i, j. Obviously $\mathscr{J}_{pq}$ may be different for every pair of ions, and suffixes (i, j) are needed but were omitted here for simplicity. The magnitude of $\mathscr{J}$ embraces all contributions to spin-spin interaction which take the form (9.1), whatever their source.

9.2. Magnetic dipole-dipole interaction

In an ordinary paramagnetic crystal the magnetic ion is surrounded by other magnetic ions at minimum distances $\sim$0·3 to 0·8 nm. There is therefore an appreciable interaction between them, since the field acting on one ion due to its neighbour is of order βr^{-3}, which may be 10^2 G or so. This interaction is known as dipole-dipole coupling, since it is in a first approximation the same as the coupling between two localized magnets. From classical theory the energy of two point magnetic dipoles $\mathbf{m}_i$ and $\mathbf{m}_j$, a distance r apart, is

$$W = r^{-3}\{\mathbf{m}_i \cdot \mathbf{m}_j - 3r^{-2}(\mathbf{m}_i \cdot \mathbf{r})(\mathbf{m}_j \cdot \mathbf{r})\}. \tag{9.2a}$$

For a pair of isotropic electronic dipoles we can write

$$\mathbf{m}_i = -g_i \beta \mathbf{S}_i, \qquad \mathbf{m}_j = -g_j \beta \mathbf{S}_j,$$

so that (9.2a) becomes, in operator form,

$$\mathscr{H} = g_i g_j \beta^2 r^{-3}\{\mathbf{S}_i \cdot \mathbf{S}_j - 3r^{-2}(\mathbf{S}_i \cdot \mathbf{r})(\mathbf{S}_j \cdot \mathbf{r})\}. \tag{9.2b}$$

If the direction cosines of $\mathbf{r}$ are (l, m, n), this can be expanded:

$$\mathscr{H} = g_i g_j \beta^2 r^{-3}\{S_{ix}S_{jx}(1-3l^2) + S_{iy}S_{jy}(1-3m^2) + S_{iz}S_{jz}(1-3n^2) -$$
$$- (S_{ix}S_{jy} + S_{iy}S_{jx})3lm - (S_{iy}S_{jz} + S_{iz}S_{jy})3mn -$$
$$- (S_{iz}S_{jx} + S_{ix}S_{jz})3nl\}, \tag{9.3}$$

SPIN-SPIN INTERACTION

which is a symmetric tensor type interaction. Here it has been assumed that the dipoles are isotropic, but electronic dipoles in a solid are often anisotropic. We then require an expression for the energy of interaction of one anisotropic magnet in the field of the other (anisotropic) magnet. For a field component H_x in the x-direction the energy of such a magnet is

$$-m_x H_x = \beta H_x(g_{xx}S_x + g_{xy}S_y + g_{xz}S_z)$$

and it can be shown that (9.2a) is correct provided that for a component of a dipole moment such as m_x we write

$$m_x = -\beta(g_{xx}S_x + g_{xy}S_y + g_{xz}S_z). \tag{9.4}$$

The resultant formula can be written in the form

$$\mathcal{H} = \beta^2 r^{-3} S_{ip} S_{jq} \{g_{ips} g_{jqs} - (3st/r^2) g_{ips} g_{jqt}\}, \tag{9.5}$$

where each of the suffixes p, q, s, t takes the values x, y, z, and the usual summation rules are to be observed whenever a suffix occurs twice. This is not in general a symmetric tensor, though it may be symmetric in special cases. Two such examples are:

(a) the two ions are identical and the principal axes of their g-tensors are parallel. The tensor is symmetric provided that the coefficient of $S_{ip}S_{jq}$ is the same as that of $S_{iq}S_{jp}$; since S_i, S_j commute, interchange of (p, q) is equivalent to interchange of (i, j), and provided that all components $g_i = g_j$, such an interchange has no effect on the coefficients of each term. This result can also be obtained directly by taking the principal axes as the (x, y, z) axes, in which case the components in (9.2a) are all of the simple form $m_x = -\beta g_x S_x$, etc., giving

$$\mathcal{H} = \beta^2 r_{ij}^{-3} \begin{Bmatrix} (1-3l^2)g_{ix}g_{jx}S_{ix}S_{jx} + (1-3m^2)g_{iy}g_{jy}S_{iy}S_{jy} + \\ +(1-3n^2)g_{iz}g_{jz}S_{iz}S_{jz} - \\ -3lm(g_{ix}g_{jy}S_{ix}S_{jy} + g_{iy}g_{jx}S_{iy}S_{jx}) - \\ -3mn(g_{iy}g_{jz}S_{iy}S_{jz} + g_{iz}g_{jy}S_{iz}S_{jy}) - \\ -3nl(g_{iz}g_{jx}S_{iz}S_{jx} + g_{ix}g_{jz}S_{ix}S_{jz}) \end{Bmatrix} \tag{9.5a}$$

which is clearly symmetric provided that $g_{ix} = g_{jx}$, etc.

(b) the two ions have parallel principal axes for their g-tensors, and the line joining them lies along one of these common principal axes. Taking the latter as the z-axis, we have $l = m = 0$, $n = 1$, and (9.5a) becomes

$$\mathcal{H} = \beta^2 r_{ij}^{-3} \{g_{ix}g_{jx}S_{ix}S_{jx} + g_{iy}g_{jy}S_{iy}S_{jy} - 2g_{iz}g_{jz}S_{iz}S_{jz}\} \tag{9.5b}$$

where the g-values are not necessarily the same for the two ions.

It is readily verified that (9.3) is a traceless tensor, but in general (9.5), (9.5a), and (9.5b) are not.

So far we have considered only the interaction between point dipoles, but in a magnetic ion the dipole moment is distributed, and in a magnetic complex it may be distributed over the ligands. If the distribution is spherically symmetric, as in the spin magnetism of an ion in the S-state of a half-filled shell, simple magnetostatic theory shows that its magnetic potential at exterior points is the same as that of a point dipole at the centre. If it does not have spherical symmetry, the magnetic potential can be expanded in a power series where the leading term corresponds to the potential of the equivalent point dipole. If the magnetic moment distribution has a centre of symmetry, terms in the expansion of the *potential* with odd powers in r will vanish, and the next most important term after the dipolar term in the magnetic *energy* of interaction between two ions will vary as r^{-5}. At small inter-ionic distances where this term might be significant the magnetic interaction energy is generally unobservable because of exchange interaction.

Hitherto we have assumed that the magnetic moment of an ion is given by $\mathbf{m} = -\beta(\mathbf{g} \cdot \mathbf{S})$ of which a typical component is given in eqn (9.4). This assumption is justified if only the first-order Zeeman interaction of an ion with a magnetic field is important. More generally the Zeeman interaction can be written as a power series of the form (for $S = \frac{1}{2}$)

$$W_{\pm} = \pm \tfrac{1}{2}g\beta H - \alpha_1 H^2 \pm \alpha_2 H^3 + \text{etc.}$$

In a resonance experiment we have for the $|+\rangle \leftrightarrow |-\rangle$ transition

$$h\nu = g\beta H + 2\alpha_2 H^3 + O(H^5)$$

and the third and higher odd-order Zeeman terms appear as small corrections to the g-factor. However the effective magnetic moment at field H is

$$-(\mathrm{d}W/\mathrm{d}H) = \mp \tfrac{1}{2}g\beta + 2\alpha_1 H \mp 3\alpha_2 H^2, \text{ etc.}$$

and in an experiment at high magnetic field the second term may well be significant, giving changes of opposite sign in the apparent magnetic moment for the $|+\rangle$ and $|-\rangle$ states. An extreme case is the Γ_3 doublet in a cubic field, such as the ground state of Dy^{2+} in CaF_2; this is non-magnetic ($g = 0$) at $H = 0$, but the presence of low-lying excited states gives the rather complicated Zeeman splitting shown in Fig. 5.6. Clearly the magnetic dipolar interaction of such an ion with another ion is more complex than (9.3), and in general it may be necessary to allow for a field dependence in the ionic magnetic moment, with different effective moments in reversed orientations in the presence of an external magnetic field. Of course similar effects arise from the mutual magnetic interaction of two ions, but the magnetic field of one ion acting on a neighbouring ion is $\sim 10^2$ G, while the external field may be 10^4 to 10^5 G.

9.3. Exchange interaction

The theory of exchange interaction in insulators has been extensively reviewed by Anderson (1963), and the complexities of this subject will not be explored here. A common assumption is that the main interaction is of the form

$$\mathcal{H}_{ex} = \mathcal{J}_{ij}(\mathbf{S}_i \cdot \mathbf{S}_j), \qquad (9.6)$$

where $\mathbf{S}_i$, $\mathbf{S}_j$ are the true spins and not the effective spins. When $S > \frac{1}{2}$, this equation may only be the leading term of a series expansion in which higher terms (Stevens 1953b) such as $(\mathbf{S}_i \cdot \mathbf{S}_j)^2$ occur; the effect of such a term is considered briefly at the end of § 9.10.

In most paramagnetic crystals exchange interaction is not of the simple type arising from direct overlap of the electronic wave-functions discussed by Heisenberg and Dirac, but is due to super-exchange. The theory of Anderson (1959) has been extended by Moriya (1960) to include spin-orbit coupling, and the exchange energy then no longer takes the simple scalar form of eqn (9.6) ('isotropic exchange') but becomes 'anisotropic'. Moriya shows that in addition to the scalar term there may be other terms of the form

$$\mathbf{D} \cdot [\mathbf{S}_i \wedge \mathbf{S}_j] \qquad (9.7)$$

and

$$\mathbf{S}_i \cdot \mathcal{J}_{ij} \cdot \mathbf{S}_j, \qquad (9.8)$$

where the second term has the nature of a symmetrical tensor while the first is an antisymmetric tensor, with components of the form

$$D_x(S_{iy}S_{jz} - S_{iz}S_{jy}).$$

It is obvious that the form (9.1) can include both (9.6) and (9.7) if we allow $\mathcal{J}_{ij}$ to have the form of a general tensor that is not necessarily symmetric nor traceless.

The term (9.7) was proposed on symmetry grounds by Dzialoshinski (1958) and vanishes for pairs of ions embedded in a lattice under certain symmetry conditions. Moriya showed that it is of order $(\Delta g/g)$ times the isotropic super-exchange energy, while the magnitude of the interaction (9.8) is of order $(\Delta g/g)^2$ times the isotropic exchange, on the assumption that the ground state is an orbital singlet and the spin-orbit coupling is small compared with the crystal field splitting of the orbital states. This latter condition means that the departure Δg of the g-value from the spin-only value is rather small. However in many salts the orbital moment is far from quenched, and in these cases the anisotropic exchange interaction, expressed in terms of the effective spins, can easily become as important as the isotropic exchange, or even predominate. This can easily be seen by taking two examples.

Example 1: a rare-earth salt

We assume an isotropic interaction of the form (9.6) between the true spins. If admixture of states of different **J** can be neglected, so that only matrix elements within the ground manifold with a single value of **J** need be considered, the spin **S** can be projected onto **J**. From the equivalences $\mathbf{L+S} = \mathbf{J}$, $\mathbf{L+2S} = g_J\mathbf{J}$, this projection is $(g_J-1)\mathbf{J}$, and for a pair of identical ions the interaction may be written as

$$\mathcal{H}_{\text{ex}} = (g_J-1)^2 \mathscr{J}_{ij}(\mathbf{J}_i \cdot \mathbf{J}_j). \tag{9.9}$$

Suppose that the crystal-field splitting leaves ground states of effective spin **S**, with an anisotropic splitting factor **g**, whose principal values are (g_x, g_y, g_z). Then since we have the equivalence between the matrix elements

$$g_x \tilde{S}_x \equiv g_J J_x, \text{ etc.,}$$

the exchange interaction between a pair of identical ions in the ground doublet becomes, in terms of the effective spins,

$$\mathcal{H}_{\text{ex}} = \mathscr{J}_x \tilde{S}_{ix}\tilde{S}_{jx} + \mathscr{J}_y \tilde{S}_{iy}\tilde{S}_{jy} + \mathscr{J}_z \tilde{S}_{iz}\tilde{S}_{jz} \tag{9.10}$$

where

$$\mathscr{J}_x = (g_x/g_J)^2 (g_J-1)^2 \mathscr{J}_{ij}, \text{ etc.} \tag{9.11}$$

This interaction is of the form of eqn (9.1). It has the same principal axes as the g-tensor, but can be highly anisotropic; for example, in cases with axial symmetry, such as the ethylsulphates, we may have $g_x = g_y = 0$, while g_z^2 may be of order 100.

Example 2; a cobaltous salt

Similar relations can be deduced for salts of the iron group when the orbital momentum is not completely 'quenched'. We take as an example a cobaltous ion, Co^{2+}, $3d^7$, in an octahedral crystal field, where the orbital degeneracy of the $L = 3$ manifold is partially lifted, leaving a Γ_4 orbital triplet as the lowest level. This triplet behaves like a manifold with effective orbital momentum $\tilde{l} = 1$ (see §§ 7.14, 19.3), and an effective orbital g-factor $\tilde{g}_l$, which for our purposes we can assume to have the uncorrected value $-\frac{3}{2}$. We have also a fourfold spin degeneracy, with $S = \frac{3}{2}$.

Under the action of spin-orbit coupling, the orbital triplet ($\tilde{l} = 1$) with $S = \frac{3}{2}$ splits into a doublet, a quartet, and a sextet with effective angular momenta $\tilde{J} = \frac{1}{2}, \frac{3}{2}$, and $\frac{5}{2}$ respectively. The doublet is lowest, and its Zeeman splitting corresponds to a g-factor given by an equivalent Landé formula

$$g = \tfrac{1}{2}(\tilde{g}_l + g_S) + \frac{(\tilde{g}_l - g_S)\{\tilde{l}(\tilde{l}+1) - S(S+1)\}}{2\tilde{J}(\tilde{J}+1)}. \qquad (9.12)$$

To find the effective exchange interaction for a pair of ions both in this doublet state, we project the true spin S onto $\tilde{J}$, using the relations

$$\tilde{g}_l \tilde{\mathbf{l}} + g_S \mathbf{S} \equiv g\tilde{\mathbf{J}}; \qquad \tilde{\mathbf{l}} + \mathbf{S} \equiv \tilde{\mathbf{J}}, \qquad (9.13)$$

from which

$$\tilde{\mathbf{S}} \equiv \{(g - \tilde{g}_l)/(g_S - \tilde{g}_l)\}\mathbf{J}. \qquad (9.14)$$

Using the values $\tilde{g}_l = -\frac{3}{2}$, $g_S = 2$ we find that $g = \frac{13}{3}$ and $\mathbf{S} \equiv (\frac{5}{3})\tilde{\mathbf{J}}$. The latter factor enters twice in evaluating the exchange interaction for a pair of ions, and on reverting to our usual nomenclature of effective spin $\tilde{\mathbf{S}}$ instead of $\tilde{\mathbf{J}}$ we find that we have the equivalence

$$\mathscr{J}(\mathbf{S}_i \cdot \mathbf{S}_j) \equiv (\tfrac{5}{3})^2 \mathscr{J}(\tilde{\mathbf{S}}_i \cdot \tilde{\mathbf{S}}_j). \qquad (9.15)$$

Thus the effective exchange interaction for the ground doublet is isotropic, as we should expect from our initial assumption of cubic (octahedral) symmetry and an isotropic exchange interaction between the true spins. The latter is not, however, necessarily correct (see below).

A multiplying factor that is so simply related to the g-factor is no longer to be expected in the presence of lower symmetry. If there is a small distortion of axial symmetry, we may write the ground state as a Kramers doublet of the general form

$$a\,|\mp \tilde{1}, \pm \tfrac{3}{2}\rangle + b\,|\tilde{0}, \pm\tfrac{1}{2}\rangle + c\,|\pm \tilde{1}, \mp\tfrac{1}{2}\rangle, \qquad (9.16)$$

where each state $|\tilde{l}_z, S_z\rangle$ belongs to the Γ_4 triplet with effective orbital angular momentum $l = 1$ and $S = \frac{3}{2}$. The Zeeman interaction within the doublet can be evaluated using either the usual operator $\mathbf{L}+g_S\mathbf{S}$ or its equivalent form for the Γ_4 triplet, $\tilde{g}_l\tilde{\mathbf{l}}+g_S\mathbf{S}$, with $\tilde{g}_l = -\frac{3}{2}$ and $g_S = 2$ as above. In terms of an effective spin $\tilde{S} = \frac{1}{2}$, the doublet then has the principal values for its anisotropic g-factor

$$g_\| = 9a^2+2b^2-5c^2,$$
$$g_\perp = 4\sqrt{(3)}ac+4b^2-3\sqrt{(2)}bc. \qquad (9.17)$$

However to convert an exchange interaction $\mathscr{J}(\mathbf{S}_i \cdot \mathbf{S}_j)$ into the equivalent interaction in terms of the effective spins $\tilde{\mathbf{S}}$, the matrix elements of $\mathbf{S}$ are needed. It is easily found from (9.16) that we have then

$$\mathscr{J}(\mathbf{S}_i \cdot \mathbf{S}_j) \equiv \mathscr{J}\{(3a^2+b^2-c^2)^2\tilde{S}_{iz}\tilde{S}_{jz}+$$
$$+(2\sqrt{(3)}ac+2b^2)^2(\tilde{S}_{ix}\tilde{S}_{jx}+\tilde{S}_{iy}\tilde{S}_{jy})\}, \qquad (9.18)$$

from which it is clear that the coefficients are not simply related to the g-values, unlike the formulae (9.11) for the rare-earth salt. The case of cubic symmetry is, of course, an exception; the coefficients then are $a = 1/\sqrt{2}$, $b = -1/\sqrt{3}$, $c = 1/\sqrt{6}$, for which it can be verified that (9.18) reduces to (9.15).

These two examples illustrate how an exchange interaction, isotropic in terms of the true electron spins, can become highly anisotropic when expressed in terms of the effective spins provided that the orbital momentum is not completely quenched. Under the latter circumstances it is not necessarily true that the exchange interaction is isotropic even when expressed in terms of the true spins, as can readily be seen by considering the case of direct exchange due to overlap of the orbital wave-functions of two ions. Anisotropy arises when the spin and orbit are closely coupled, so that a given component of the spin may be associated with a given orbital wave-function that has a lobe extending in the direction of a neighbouring ion, while a different spin component may be associated with an orbital wave-function with minimal extension in this particular direction. It is clear that the overlap, and hence the exchange integral itself, can then change markedly according to the orientation of the true spin, and the exchange interaction will be anisotropic in the true spins, as in eqn (9.8).

Similar remarks hold for indirect exchange. It has been shown theoretically (Judd 1959c) and experimentally (Griffiths, Owen, Park, and Partridge 1959) that, even for magnetic complexes that individually have octahedral symmetry, the exchange interaction is not

SPIN-SPIN INTERACTION

necessarily isotropic in terms either of the true or the effective spins when considerable orbital momentum is involved. The only symmetry restriction is that the exchange interaction must have a form compatible with the overall symmetry, which is necessarily less than cubic for an assembly of two complexes.

9.4. Multipole interactions

In addition to magnetic dipole and exchange interactions between a pair of ions, there may also be a 'Coulomb' interaction between them. Suppose we have two ions, i, j separated by a distance R, as in Fig. 9.1. Then each electron on ion i will be subjected to an electric field set up by the electronic and nuclear charges on ion j. The electrostatic energy of a pair of electrons, one at a point (x_i, y_i, z_i) relative to the centre of ion i, the other at a point (x_j, y_j, z_j) relative to the centre of ion j, is

$$\frac{e^2}{\epsilon\{|(x_i-x_j)^2+(y_i-y_j)^2+(R+z_i-z_j)^2|\}^{\frac{1}{2}}} \quad (9.19)$$

if for simplicity we take the line joining the centres of the two ions to lie along the z-axis of the coordinate system. In this equation ϵ is a 'dielectric constant' that allows for shielding effects due to the induced charges on intervening ions; it may be <1, corresponding to an anti-shielding effect. If the 'medium' is anisotropic the 'dielectric constant' may also be anisotropic but this is a complication that we shall neglect.

If $R \gg x_i, x_j$, etc., we can expand (9.19) in inverse powers of R. We are, of course, interested only in those terms that vary with the magnetic sub-state of each ion. If the charge distribution on each ion does not have inversion symmetry, there may also be an electric dipole moment and an electric-dipole interaction. As an example, we take a pair of ions with non-Kramers doublets as ground states, such as Pr^{3+} in $LaCl_3$. For nearest neighbours along the symmetry axis, the net interaction may be of the form

$$\mathscr{J}_\perp(S_{ix}S_{jx}+S_{iy}S_{jy})+\mathscr{J}_\parallel S_{iz}S_{jz},$$

where the first term is due to electric dipole-dipole interaction and the last to magnetic dipole and/or exchange interaction.

The next term in the expansion of (9.19) is the quadrupole-quadrupole

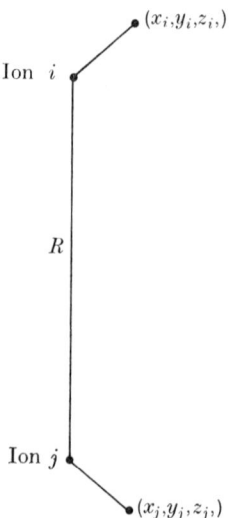

Fig. 9.1. Coordinate systems used in evaluating the electrostatic interaction between a pair of ions i, j whose centres are a distance R apart. The coordinate systems (x_i, y_i, z_i), (x_j, y_j, z_j) have parallel axes but their origins are at the centre of each ion respectively; the z-axes are taken to be parallel to the line joining the two centres.

interaction

$$\frac{3e^2}{8\epsilon R^5}\begin{bmatrix} 4(3z_i^2-r_i^2)(3z_j^2-r_j^2) - \\ -16z_i z_j\{(x_i+iy_i)(x_j-iy_j)+(x_i-iy_i)(x_j+iy_j)\} + \\ +(x_i+iy_i)^2(x_j-iy_j)^2+(x_i-iy_i)^2(x_j+iy_j)^2 \end{bmatrix}. \quad (9.20)$$

Here the interaction energy has been written in a form such that the equivalent quantum-mechanical expression can be found by replacing the spatial coordinates by angular momentum operators. For a pair of rare-earth ions, where for present purposes it is sufficient to work in terms of the ground manifold J for each ion, the Hamiltonian equivalent is

$$\mathcal{H} = A\{4O_{2i}^0 O_{2j}^0 - 16(O_{2i}^{+1}O_{2j}^{-1}+O_{2i}^{-1}O_{2j}^{+1})+(O_{2i}^{+2}O_{2j}^{-2}+O_{2i}^{-2}O_{2j}^{+2})\}, \quad (9.21)$$

where

$$O_2^0 = 3J_z^2 - J(J+1), \qquad O_2^{\pm 1} = \tfrac{1}{2}(J_z J_\pm + J_\pm J_z), \qquad O_2^{\pm 2} = J_\pm^2. \quad (9.22)$$

The value of the coefficient A is

$$A = \frac{3e^2 \langle r_i^2 \rangle \langle r_j^2 \rangle \langle J_i \| \alpha \| J_i \rangle \langle J_j \| \alpha \| J_j \rangle}{8\epsilon R^5}, \quad (9.23)$$

where $\langle r^2 \rangle$ is the mean square radius of the $4f$ ions on each ion, and $\langle J \| \alpha \| J \rangle$ is the numerical coefficient of Elliott and Stevens listed in Table 20.

The terms given above and higher-order terms, arising either from exchange or from other multipole interactions, are conveniently summarized in spin operator form by Birgeneau, Hutchings, Baker, and Riley (1969).

The terms in (9.20), (9.21) do not change sign under the parity operation, or under time reversal, and are thus essentially different from the interactions considered in §§ 9.2 and 9.3. They give rise to no first-order effect for ions whose states are Kramers doublets, or non-Kramers doublets in which each state has the same electric charge distribution. On the other hand, a first-order effect may occur for the non-Kramers doublet Γ_3, where the two states have electric quadrupole moments of opposite sign. First-order effects may arise also between ions in states of higher degeneracy, such as triplet states (Γ_4, Γ_5) or quadruplet (Γ_8) states in ions with cubic symmetry. Possible examples arise in the lanthanide nitrides and analogous compounds with the NaCl structure; UO_2, NpO_2, and similar compounds of the $5f$ group with the CaF_2 structure; and ions of the $3d$ group with unquenched orbital angular momentum such as Fe^{2+} in MgO. In accidental cases of near degeneracy, such as cerium ethylsulphate with two low-lying doublet states, the quadrupole-quadrupole interaction energy may be of the same order as the separation between the two doublets, giving first-order effects (Bleaney 1961).

Second-order effects may arise in ions whose ground states are doublets, but which have fairly low-lying excited states, provided that (9.21) has matrix elements between the ground and excited states. The resultant interaction is similar in form to anisotropic exchange interaction, making it difficult to distinguish unambiguously between the two (Baker 1964b). At first sight it would seem possible to calculate accurately the size of the coefficient A in eqn (9.21), but this is hampered by lack of reliable values of $\langle r^2 \rangle$ for the $4f$ ions, and of shielding (or anti-shielding) effects due to the intervening ions.

All the spin-spin interactions considered hitherto have arisen through electro-magnetic interactions between the ions. Sugihara (1959) has suggested that another type of spin-spin interaction may arise by coupling through the 'phonon field', corresponding to the virtual emission and absorption of phonons. When the coupling between the magnetic ions and the lattice waves or phonon field is quadrupolar in form, the 'virtual phonon interaction' may resemble a quadrupole-quadrupole interaction, and McMahon and Silsbee (1964) have shown that for a pair of Fe^{2+} ions in MgO the interaction Hamiltonian has the same form as eqn (9.21). A further theoretical treatment of the virtual phonon interaction is given by Orbach and Tachiki (1967).

Although we have primarily considered above the quadrupole-quadrupole interaction, in the case of ions (such as the $4f$, $5f$ ions) with large amounts of unquenched orbital momentum there may well be interactions involving other multipole moments (including those of odd degree). These may result from the 'Coulomb' interaction (see, for instance, Wolf and Birgeneau (1968)) or from the 'exchange' interaction (Levy 1964, 1968; Elliott and Thorpe 1968). The experimental problem of distinguishing between interactions that have the same form in the effective spin Hamiltonian but arise from different sources (including virtual phonon effects) is illustrated in the work of Wickersheim and White (1962), Baker and Mau (1966), Allen (1968), and Baker, Birgeneau, Hutchings and Riley (1968).

9.5. Interaction between a pair of similar ions

The general problem of magnetic resonance in a system of many interacting ions is extremely complicated and before discussing it in §§ 9.7–9.11 we consider the much simpler problem of a pair of interacting ions. Not only does this yield some useful insight into the more general problem, but the spectra of such pairs are themselves a direct method of investigating the nature of the spin-spin interaction, and almost the only method by which quantitative measurements of this interaction can be obtained. In this section we consider an isolated pair of identical ions, each with effective spin $S_i = S_j = \frac{1}{2}$ and with g-tensors

TABLE 9.1

Normalized states and energy levels for a pair of identical interacting ions with the spin Hamiltonian (9.24). The normalization coefficients are given by $\tan 2\alpha = (\mathscr{J}_x - \mathscr{J}_y)/4G_z$

Symmetric states	Energy levels
$\cos \alpha \,\|++\rangle + \sin \alpha \,\|--\rangle$	$+\tfrac{1}{4}\mathscr{J}_z + \{G_z^2 + \tfrac{1}{16}(\mathscr{J}_x - \mathscr{J}_y)^2\}^{\frac{1}{2}}$
$(1/\sqrt{2})\,\|+-\rangle + (1/\sqrt{2})\,\|-+\rangle$	$-\tfrac{1}{4}\mathscr{J}_z + \tfrac{1}{4}(\mathscr{J}_x + \mathscr{J}_y)$
$\sin \alpha \,\|++\rangle - \cos \alpha \,\|--\rangle$	$+\tfrac{1}{4}\mathscr{J}_z - \{G_z^2 + \tfrac{1}{16}(\mathscr{J}_x - \mathscr{J}_y)^2\}^{\frac{1}{2}}$
Antisymmetric state	Energy level
$(1/\sqrt{2})\,\|+-\rangle - (1/\sqrt{2})\,\|-+\rangle$	$-\tfrac{1}{4}(\mathscr{J}_x + \mathscr{J}_y + \mathscr{J}_z)$

whose principal values and principal axes are the same. We further simplify the problem by assuming that the spin-spin interaction has the simple form given in eqn (9.10), whose principal axes (x, y, z) are the same as the principal axes of the g-tensors. If the system is subjected to an external field **H** directed along one of these principal axes (which we take to be the z-axis), the appropriate Hamiltonian for the pair of ions is

$$\mathcal{H} = g_z\beta H_z(S_{iz}+S_{jz}) + \mathcal{J}_x S_{ix}S_{jx} + \mathcal{J}_y S_{iy}S_{jy} + \mathcal{J}_z S_{iz}S_{jz}. \quad (9.24)$$

This Hamiltonian gives the energy matrix

$$\begin{array}{c|cccc}
|++\rangle & +G_z+\tfrac{1}{4}\mathcal{J}_z, & 0 & , & 0 & , & \tfrac{1}{4}(\mathcal{J}_x-\mathcal{J}_y) \\
|+-\rangle & 0 & , & -\tfrac{1}{4}\mathcal{J}_z & , & \tfrac{1}{4}(\mathcal{J}_x+\mathcal{J}_y), & 0 \\
|-+\rangle & 0 & , & \tfrac{1}{4}(\mathcal{J}_x+\mathcal{J}_y), & -\tfrac{1}{4}\mathcal{J}_z & , & 0 \\
|--\rangle & \tfrac{1}{4}(\mathcal{J}_x-\mathcal{J}_y), & 0 & , & 0 & , & -G_z+\tfrac{1}{4}\mathcal{J}_z
\end{array} \quad (9.25)$$

in which $G_z = g_z\beta H_z$, and a state such as $|+-\rangle$ is one in which ion i is in the $S_{iz} = +\tfrac{1}{2}$ state and ion j in the $S_{jz} = -\tfrac{1}{2}$ state, etc. The diagonal states and energy levels are given in Table 9.1. This table shows that the four states of the two ions can be classified into three symmetric states and one antisymmetric state; it is convenient to refer to these as the 'triplet' and 'singlet' respectively, though the triplet levels are not degenerate in zero magnetic field unless the exchange interaction is isotropic. Magnetic resonance transitions are allowed only between states belonging to the triplet, and no singlet-triplet transitions are allowed, because the perturbation Hamiltonian is symmetric with respect to interchange of the two ions; that is,

$$\mathcal{H} = \beta\{\mathbf{H}_1 \cdot \mathbf{g} \cdot (\mathbf{S}_i + \mathbf{S}_j)\} \quad (9.26)$$

where $\mathbf{H}_1$ is the amplitude of an oscillatory magnetic field used to produce resonance transitions.

The division into symmetric and anti-symmetric states suggests that the system can be regarded as having two spectroscopic states, one a triplet with total effective spin $T = 1$, the other a singlet with total spin $T = 0$. This can be shown formally to be correct as follows.

Let T_x, T_y, T_z be the components along each of the three axes of the total effective spin of the system, so that $(S_{ix}+S_{jx}) = T_x$, etc. Then each of the terms in the exchange interaction can be transformed in the following way:

$$\begin{aligned}
\mathcal{J}_x S_{ix}S_{jx} &= \tfrac{1}{2}\mathcal{J}_x\{(S_{ix}+S_{jx})^2 - S_{ix}^2 - S_{jx}^2\} \\
&= \tfrac{1}{2}\mathcal{J}_x T_x^2 - \tfrac{1}{4}\mathcal{J}_x,
\end{aligned} \quad (9.27)$$

where we have used the fact that the values of S_{ix}^2, S_{jx}^2 are each $\tfrac{1}{4}$; similar expressions hold for the y, z components. Hence the Hamiltonian becomes

$$\mathcal{H} = g_z\beta H_z T_z + \tfrac{1}{2}(\mathcal{J}_x T_x^2 + \mathcal{J}_y T_y^2 + \mathcal{J}_z T_z^2) - \tfrac{1}{4}(\mathcal{J}_x + \mathcal{J}_y + \mathcal{J}_z). \quad (9.28)$$

If we now write

$$\mathcal{J}_x = \mathcal{J}_x' + \mathcal{J}, \text{ etc., where } \mathcal{J} = \tfrac{1}{3}(\mathcal{J}_x + \mathcal{J}_y + \mathcal{J}_z), \quad (9.29)$$

we can regard $\mathcal{J}$ as the 'isotropic' part of the interaction, and $\mathcal{J}'$ the 'anisotropic' part, for which the trace is zero:

$$\mathcal{J}_x' + \mathcal{J}_y' + \mathcal{J}_z' = 0. \quad (9.30)$$

Then, since $T_x^2 + T_y^2 + T_z^2 = T(T+1)$, the Hamiltonian becomes

$$\mathcal{H} = g_z\beta H_z T_z + \tfrac{1}{2}(\mathcal{J}_x' T_x^2 + \mathcal{J}_y' T_y^2 + \mathcal{J}_z' T_z^2) + \tfrac{1}{2}\mathcal{J}\{T(T+1) - \tfrac{3}{2}\}. \quad (9.31)$$

The allowed values of T are $T = 0$, a singlet level at $-3\mathcal{J}/4$, and a triplet $T = 1$ whose behaviour is identical with that of a spin triplet, the mean energy of the three levels being $+\mathcal{J}/4$.

If the exchange is isotropic, $\mathcal{J}_x' = \mathcal{J}_y' = \mathcal{J}_z' = 0$ and the second term in (9.31) vanishes. All three triplet levels have the same energy $(+\mathcal{J}/4)$ in zero magnetic field, and are equally spaced by $g_z\beta H_z$ when a field is applied, as shown in Fig. 9.2. The allowed transitions within the triplet are found from the oscillatory Hamiltonian

$$\mathcal{H} = \beta(\mathbf{H}_1 \cdot \mathbf{g} \cdot \mathbf{T}), \quad (9.32)$$

which is obtained from (9.26) by writing $\mathbf{T} = \mathbf{S}_i + \mathbf{S}_j$. The allowed transitions are of the usual type $\Delta T_z = \pm 1$, and occur at

$$h\nu = g_z\beta H_z. \quad (9.33)$$

This is independent of the value of $\mathcal{J}$, showing that isotropic exchange has no effect on the spectrum except at temperatures where $kT \sim \mathcal{J}$, when the intensity will no longer vary inversely as the absolute temperature because of the triplet-singlet splitting. If $\mathcal{J}$ is positive, and $\mathcal{J} \gg g_z\beta H_z$, the triplet states lie higher in energy and their populations will tend to zero at sufficiently low temperatures where $kT \ll \mathcal{J}$, so that the spectrum will then decline in intensity and ultimately vanish as the temperature is lowered. If $\mathcal{J}$ is negative, the singlet state is the higher in energy, and will become depopulated at sufficiently low temperatures; if $g_z\beta H_z \ll kT \ll \mathcal{J}$, the intensity of the spectrum line at $h\nu = g_z\beta H_z$ will be greater by a factor $\tfrac{4}{3}$ when the singlet state is completely depopulated than it would have been in the absence of the

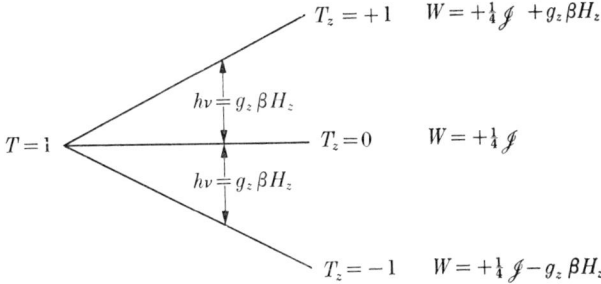

FIG. 9.2. Energy levels and allowed transitions for a pair of ions each with $S = \frac{1}{2}$ and coupled by isotropic exchange interaction:

$$\mathscr{H} = g_z\beta H_z(S_{iz}+S_{jz}) + \mathscr{J}(\mathbf{S}_i \cdot \mathbf{S}_j).$$

exchange interaction. Thus intensity measurements in the region where $kT \sim \mathscr{J}$ can be used to estimate the value of $\mathscr{J}$.

When the exchange interaction contains anisotropic components, the second term in (9.31) does not vanish and gives an initial splitting of the triplet levels in zero field similar to the 'fine structure' splitting discussed in § 3.5. This splitting arises only from the anisotropic part of the exchange, and is the same as that given by eqn (3.25) if we write

$$D_x = \mathscr{J}'_x/2, \text{ etc.} \tag{9.34}$$

If $g_z\beta H_z \gg \mathscr{J}'$, the allowed transitions are just those within the triplet for which $\Delta T_z = \pm 1$, and occur at (in first approximation)

$$h\nu = g_z\beta H_z \pm \tfrac{3}{4}\mathscr{J}'_z. \tag{9.35}$$

Hence the magnitude of $\mathscr{J}'_z$ can be found from the separation of the two resonance lines under these conditions, and similarly $\mathscr{J}'_x$, $\mathscr{J}'_y$ can be found from measurements with the external magnetic field along the axes x, y. If $g_z\beta H_z$ is comparable with $\mathscr{J}$, the spectrum is more complicated in its behaviour as discussed in § 3.5, but the values of $\mathscr{J}'_x$, $\mathscr{J}'_y$, $\mathscr{J}'_z$ can be found by fitting it to the Hamiltonian (9.31). The value of the isotropic part of the interaction $\mathscr{J}$ can be found only by intensity measurements as a function of temperature in the region where $kT \sim \mathscr{J}$, as mentioned above. Thus the isotropic and anisotropic parts of the spin-spin interaction produce significantly different effects in the spectrum of a pair of identical ions.

The spectrum of copper acetate

A spectrum of this type was discovered in copper acetate monohydrate, $Cu(CH_3COO)_2, H_2O$ by Bleaney and Bowers (1952a, b) and Abe and Shimada (1953, 1957). The cupric ion, $3d^9, {}^2D$ is expected to have a doublet ground state, but the spectrum observed corresponded to that for an effective spin of 1. It was suggested by Bleaney and Bowers that the copper ions must occur in isolated pairs in this salt, and this was later verified by an X-ray determination of the structure by van Niekerk and Schoening (1953), which showed that the copper-copper distance in a pair is only 0·264 nm.

The constants of the triplet spectrum in copper acetate and of the doublet spectrum in zinc-doped copper acetate are given in Table 9.2. The 'fine structure' constants D, E follow the customary nomenclature for a spin 1 spectrum (eqn (3.26)), and are related to the anisotropic parts of the exchange interaction discussed above:

$$D = 3\mathscr{J}'_z/4; \qquad E = \tfrac{1}{4}(\mathscr{J}'_x - \mathscr{J}'_y). \tag{9.36}$$

The intensity of the triplet spectrum passes through a maximum, and declines rapidly at temperatures below 90°K. From this intensity

TABLE 9.2

Parameters of the triplet spectrum of copper acetate and of the doublet spectrum in zinc-doped copper acetate

	$Cu_2Ac_4, 2H_2O$		$ZnCuAc_4, 2H_2O$
Effective spin	$T = 1$	$T = 1$	$S = \tfrac{1}{2}$
Temperature	90°K	300°K	77°K
g_x	2·08(3)	2·053(5)	2·052(7)
g_y	2·08(3)	2·093(5)	2·082(7)
g_z	2·42(3)	2·344(10)	2·344(5)
$\lvert D \rvert$ (cm^{-1})	0·34(3)	0·345(5)	—
E (cm^{-1})	0·010(5)	0·005(3)	—
A_x (cm^{-1})	<0·001		<0·0018
A_y (cm^{-1})	<0·001		<0·0023
A_z (cm^{-1})	0·008		0·0147(6)
Ref.	BB, 1952	AS, 1957	KAG, 1965

References: BB, Bleaney and Bowers (1952b).
AS, Abe and Shimada (1957).
KAG, Kokoszka, Allen, and Gordon (1965).

variation an estimate of (260 ± 50) cm^{-1} was obtained for the singlet-triplet splitting, which is equal to $\mathscr{J}$. A more accurate value of (310 ± 15) cm^{-1} is obtained by fitting the susceptibility measurements

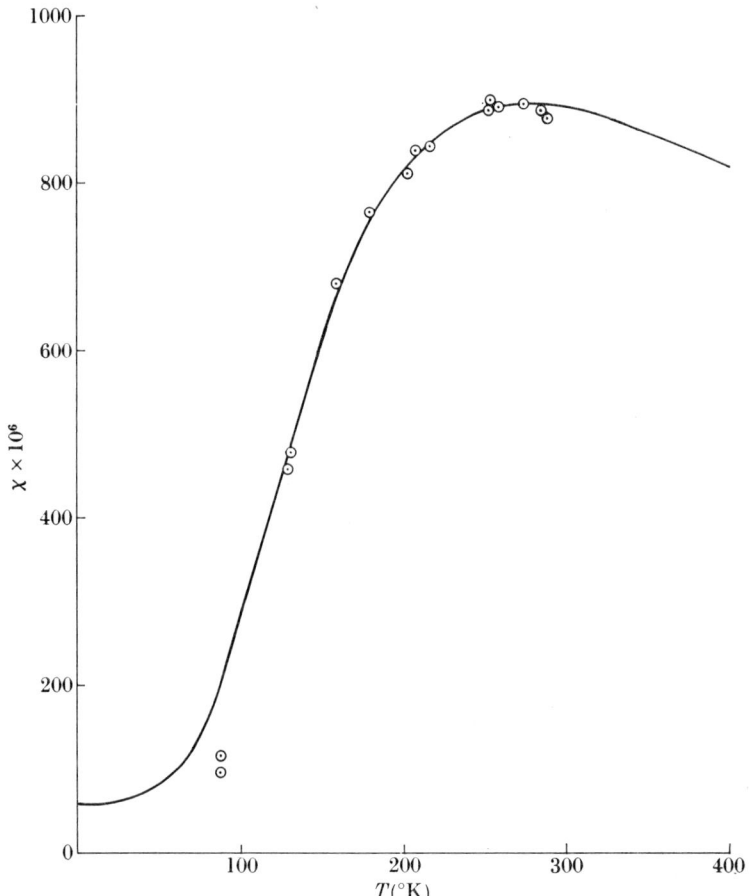

Fig. 9.3. The susceptibility of copper acetate monohydrate (per g mol of copper). The continuous line is the function

$$\chi = \frac{4C}{T\{3+\exp(\mathscr{J}/kT)\}}+\alpha,$$

where C is the Curie constant and α the temperature-independent susceptibility. The experimental points are those of Foex, Karantissis, and Perakis (1953).

(see Fig. 9.3). Thus the anisotropic part of the exchange interaction (measured by D, E) is very small compared with the isotropic part.

The g-values of Abe and Shimada (which are almost certainly more accurate than those of Bleaney and Bowers) for the triplet state agree very closely with those of isolated copper ions in copper-zinc pairs obtained from zinc-doped copper acetate by Kokoszka, Allen, and Gordon (1965). This supports a treatment (Bleaney and Bowers 1952b) in which the copper ions are assumed to be subjected to a normal type of crystal field, together with an exchange interaction $\mathscr{J}(\mathbf{S}_i \cdot \mathbf{S}_j)$ which is isotropic in the true spins. If the same tetragonal crystal field is assumed for each ion, then an anisotropic exchange interaction between the effective spins results when the spin-orbit interaction is introduced into the theory. Its size can be linked to the g-values:

$$D_{\text{ex}} = 3\mathscr{J}'_z/4 = \tfrac{1}{8}\mathscr{J}\{\tfrac{1}{4}(g_z-2)^2-(g_\perp-2)^2\}. \tag{9.37}$$

If we take $\mathscr{J} = 310$ cm^{-1}, $g_z = 2\cdot 344$ and $g_\perp = 2\cdot 073$ (the mean of g_x, g_y) we find $D_{\text{ex}} = +0\cdot 95$ cm^{-1} and, on the assumption of tetragonal symmetry, $E = 0$.

An additional contribution to D comes from magnetic dipole-dipole interaction between the two copper ions, which on a point dipole model gives

$$D_{\text{dip}} = -(2g_z^2+g_\perp^2)\beta^2/2r^3. \tag{9.38}$$

With the interionic distance $r = 0\cdot 264$ nm, and the measured values of g_z, $g_\perp$ this gives $D_{\text{dip}} = -0\cdot 19$ cm^{-1}, so that the net calculated value of D is $D_{\text{ex}}+D_{\text{dip}} = +0\cdot 76$ cm^{-1}. This is over twice the measured value. The discrepancy is most likely due to use of the value $\mathscr{J} = 310$ cm^{-1} in eqn (9.37). This value, obtained from the singlet-triplet splitting, represents the exchange interaction between two cupric ions in the ground orbital states produced by the splitting due to the tetragonal crystal field. However it is easily shown (see Bleaney and Bowers 1952b) that the splitting of the triplet depends on the exchange interaction between one cupric ion in the ground state and the other cupric ion in an excited orbital; there are in fact two such excited orbitals involved, which have matrix elements to the ground orbital through the spin-orbit coupling. If we allow for different values of the exchange interaction when an excited orbital is involved, we should write

$$D_{\text{ex}} = \tfrac{1}{8}\{\tfrac{1}{4}\mathscr{J}_1(g_z-2)^2 - \mathscr{J}_2(g_\perp-2)^2\}. \tag{9.39}$$

There is no evidence as to the values of $\mathscr{J}_1$, $\mathscr{J}_2$ and a meaningful comparison with experiment (where the sign of D has not been determined) is impossible.

A partially resolved hyperfine structure is observed (see Fig. 9.4) when the external magnetic field is along the z-axis. The electronic transition is split into components displaced from the centre by an

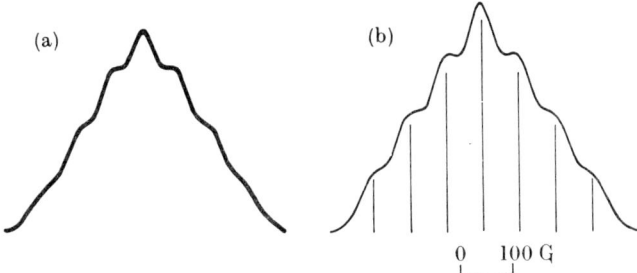

FIG. 9.4. Hyperfine structure due to coupled pairs of Cu^{2+} ions in cupric acetate, $Cu_2Ac_4, 2H_2O$. (a) Oscillogram showing hyperfine structure with magnetic field parallel to z-axis. (b) Calculated hyperfine structure, assuming Gaussian line shape and half width at half intensity of 35 G. (Bleaney and Bowers 1952b.)

amount $\tfrac{1}{2}A(m_i+m_j)$, where A is the hyperfine parameter for an isolated copper ion, and m_i, m_j the nuclear magnetic quantum numbers for the two copper ions of the pair. Since $I_i, I_j = \tfrac{3}{2}$, the sum m_i+m_j can take all integral values from $+3$ to -3, but the intermediate values can be obtained in more than one way, giving a set of 7 equally spaced lines with relative intensity $1:2:3:4:3:2:1$. The experimental line shape is compared with its reconstructed shape in Fig. 9.4. The values of A_x, A_y, A_z given in the first column of Table 9.2 for the triplet spectrum refer to the separation of successive lines, and are close to half the values in column three for the isolated ion. The factor $\tfrac{1}{2}$ occurs because we are dealing with two identical ions for which the hyperfine interaction is much smaller than the strong exchange interaction, so that we must project the spin vector of each ion onto the total spin $\mathbf{T}$. The factor is only $\tfrac{1}{2}$ if each ion has the same value of S, as can be immediately understood from a vector coupling model in which $\mathbf{S}_i$, $\mathbf{S}_j$ precess about the total spin $\mathbf{T}$.

9.6. Interaction between a pair of dissimilar ions

The discussion in the last section of the energy levels and spectrum of a pair of interacting ions with effective spin $S = \tfrac{1}{2}$ was confined to the case of a pair of similar ions. We now consider the case of a pair of dissimilar ions, again each with $S = \tfrac{1}{2}$. For the sake of simplicity we assume that the principal axes of the two g-tensors are identical but the principal g-factors themselves are not. Initially we allow the spin-spin interaction to be anisotropic, with the same form as in eqn (9.24). If an external magnetic field H_z is applied along the z-axis (a principal axis for both the g-tensor and the interaction tensor), the

Hamiltonian is

$$\mathscr{H} = \beta H_z(g_{iz}S_{iz}+g_{jz}S_{jz})+\mathscr{J}_x S_{ix}S_{jx}+\mathscr{J}_y S_{iy}S_{jy}+\mathscr{J}_z S_{iz}S_{jz}. \quad (9.40)$$

The energy matrix is similar to (9.25); writing

$$g_z = \tfrac{1}{2}(g_{iz}+g_{jz}), \qquad \delta g_z = \tfrac{1}{2}(g_{iz}-g_{jz}),$$

we obtain

$\|++\rangle$	$+g_z\beta H_z + \tfrac{1}{4}\mathscr{J}_z$,	0 ,	0 ,	$\tfrac{1}{4}(\mathscr{J}_x-\mathscr{J}_y)$
$\|+-\rangle$	0 ,	$+\delta g_z\beta H_z-\tfrac{1}{4}\mathscr{J}_z$,	$\tfrac{1}{4}(\mathscr{J}_x+\mathscr{J}_y)$,	0
$\|-+\rangle$	0 ,	$\tfrac{1}{4}(\mathscr{J}_x+\mathscr{J}_y)$,	$-\delta g_z\beta H_z-\tfrac{1}{4}\mathscr{J}_z$,	0
$\|--\rangle$	$\tfrac{1}{4}(\mathscr{J}_x-\mathscr{J}_y)$,	0 ,	0 ,	$-g_z\beta H_z+\tfrac{1}{4}\mathscr{J}_z$

(9.41)

in which the labelling of the states $|+-\rangle$, etc. is the same as in (9.25). The important difference between (9.41) and (9.25) is that the diagonal matrix elements for the states $|+-\rangle$ and $|-+\rangle$ are no longer identical, and this is most significant when the difference in Zeeman energy $\delta g_z\beta H_z$ is large compared with the interaction between the two ions, which appears in the off-diagonal matrix elements between these two states. For simplicity we consider first the case where this inequality is so well satisfied that we can use first-order perturbation theory and neglect the off-diagonal terms. In this approximation the states and energy levels are as shown in Fig. 9.5; the allowed transitions are just those in which the z-component of the spin of one ion is reversed, and occur at

$$h\nu = g_{iz}\beta H_z \pm \tfrac{1}{2}\mathscr{J}_z, \qquad h\nu = g_{jz}\beta H_z \pm \tfrac{1}{2}\mathscr{J}_z. \quad (9.42)$$

This spectrum consists of two pairs of lines, one pair centred on the point at which the transition for ion i would occur in the absence of any interaction, the other similarly centred for ion j. The separation of the lines in each pair is just $\mathscr{J}_z$, and the important difference between this and the separation in eqn (9.35) is that whereas the separation for similar ions depends only on the anisotropic part $\mathscr{J}'$ of the exchange interaction, the separation for dissimilar ions (eqn (9.42)) gives the component $\mathscr{J}_z$ of the entire interaction. Obviously similar measurements with the magnetic field along the x, y axes will yield the quantities $\mathscr{J}_x$, $\mathscr{J}_y$.

We now relax the restriction that the spin-spin interaction be small compared with the difference in Zeeman energy and consider a general case. However we reduce the number of parameters involved by assuming that the interaction is a simple isotropic interaction, so that $\mathscr{J}_x = \mathscr{J}_y = \mathscr{J}_z = \mathscr{J}$. This enables us to express the energy

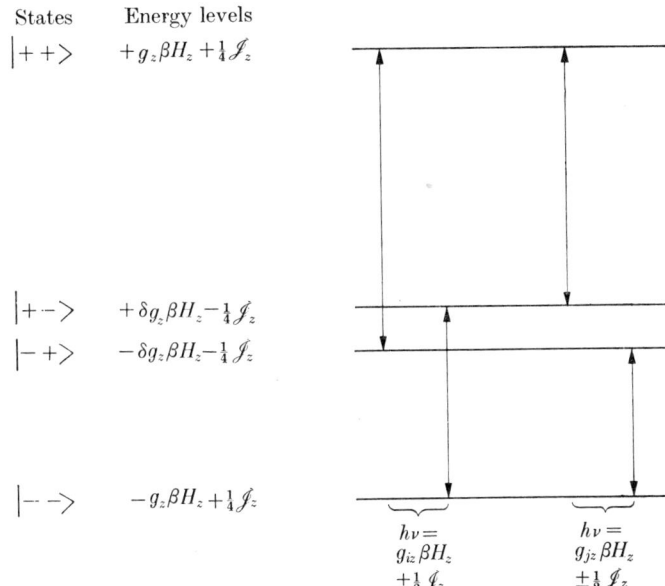

Fig. 9.5. Approximate states and energy levels for a pair of dissimilar ions for which the difference in Zeeman energy $(g_{iz}-g_{jz})\beta H_z = 2\delta g_z \beta H_z$ is large compared with the spin-spin interaction energy $\frac{1}{4}(\mathscr{J}_x+\mathscr{J}_y)$. The allowed transitions (in first approximation) correspond to the z-component of one spin being reversed under the combined effect of the external field plus the steady component $\mathscr{J}_z$ of the interaction due to the other spin.

levels in terms of a parameter $x = \frac{1}{4}\mathscr{J}/(\delta g_z \beta H_z)$, and the spectrum can be studied as x is allowed to increase from zero (no interaction) to very large values (strong interaction). The states and energy levels are given in Table 9.3.

When $x = 0$ (no interaction), the allowed transitions are $a \leftrightarrow c$, $b \leftrightarrow d$, which both occur at $h\nu = g_{iz}\beta H_z$; and $a \leftrightarrow b$, $c \leftrightarrow d$, which both occur at $h\nu = g_{jz}\beta H_z$; this is, of course, just the case of two

TABLE 9.3

States and energy levels for a pair of ions of spin $S = \frac{1}{2}$, which are dissimilar (different g-values), subject to isotropic exchange interaction. Here $\tan 2\alpha = 2x = \frac{1}{2}\mathscr{J}/(\delta g_z \beta H_z)$

States	Energy levels
(a) $\|++\rangle$	$+g_z\beta H_z + (\delta g_z \beta H_z)x$
(b) $\cos\alpha\|+-\rangle + \sin\alpha\|-+\rangle$	$(\delta g_z \beta H_z)\{-x+(1+4x^2)^{\frac{1}{2}}\}$
(c) $\sin\alpha\|+-\rangle - \cos\alpha\|-+\rangle$	$(\delta g_z \beta H_z)\{-x-(1+4x^2)^{\frac{1}{2}}\}$
(d) $\|--\rangle$	$-g_z\beta H_z + (\delta g_z \beta H_z)x$

isolated ions with different values of g_z. When x is small, these transitions split into two doublets, the components of each doublet being separated by an amount $\mathscr{J}$ in first approximation; this is the case treated earlier in this section, where it was shown more generally that if the interaction is anisotropic, the doublet separation is $\mathscr{J}_z$ ($=\mathscr{J}$ in this case) when the external field is applied along the z-axis.

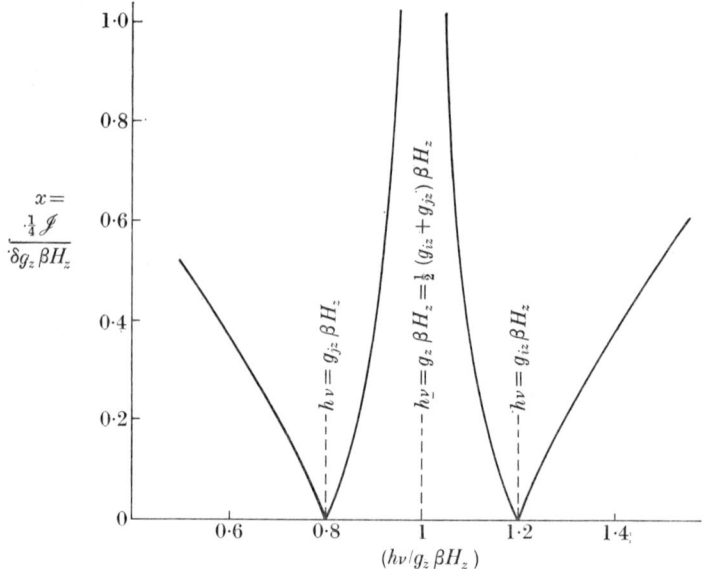

FIG. 9.6. Spectrum of a pair of dissimilar ions with isotropic exchange interaction corresponding to the Hamiltonian ($S_i = S_j = \frac{1}{2}$),

$$\mathscr{H} = \beta H_z(g_{iz}S_{iz}+g_{jz}S_{jz})+\mathscr{J}(\mathbf{S}_i \cdot \mathbf{S}_j).$$

The states and energy levels are given in Table 9.3 in terms of the parameter $x = \frac{1}{4}\mathscr{J}/(\delta g_z \beta H_z)$. At $x = 0$ (no interaction), the transitions occur at $h\nu = g_{iz}\beta H_z$, $g_{jz}\beta H_z$ appropriate to the isolated ions. As x increases the outer transitions diverge (and become steadily weaker); the inner transitions converge on the mean frequency $h\nu = g_z\beta H_z = \frac{1}{2}(g_{iz}+g_{jz})\beta H_z$ (and become stronger). The diagram is drawn for

$$\delta g_z/g_z = 0\cdot 2.$$

As x increases, the two transitions $a \leftrightarrow c$, $c \leftrightarrow d$, which are diverging in frequency (see Fig. 9.6), gradually lose intensity, while the transitions $a \leftrightarrow b$, $b \leftrightarrow d$, which are converging towards the point

$$h\nu = g_z\beta H_z = \tfrac{1}{2}(g_{iz}+g_{jz})\beta H_z,$$

become rather stronger. When x is extremely large, only these two latter have appreciable intensity, and they lie close to a frequency which is midway between the frequencies appropriate to the isolated ions. When x approaches infinity, the value of α approaches $\pi/4$ and the energy levels can be divided into a triplet a, b, d whose states are symmetric, and a singlet c whose state is antisymmetric. This is similar to the position considered in § 9.4, where transitions within the triplet only are allowed, but here transitions between the triplet and singlet are also allowed when x is small but become increasingly forbidden as $x \to \infty$.

The essential difference between the energy matrices (9.25) and (9.41) is that in the former the diagonal elements for the states $|+-\rangle$ and $|-+\rangle$ are the same, whereas in the latter they are different. In the former case the off-diagonal elements produce a splitting of the two states in first order, while in the latter the off-diagonal elements produce a first-order effect only when they are of the same order as the difference between the diagonal elements. Physically the situation can be interpreted as follows. When the ions are identical, they precess at the same rate in the external magnetic field. When spin-spin interaction is present, each ion experiences also a field set up by the other ion which has both steady and oscillatory components. (In the simple case of magnetic dipole interaction this is just the local magnetic field of the other ion.) For identical ions this oscillatory component is just at the right frequency to cause a spin 'flip', thus giving an additional interaction because of the resonance effect whereby ion 'i' moves from the 'up' to the down position while ion 'j' moves from 'down' to 'up'. This corresponds to the exchange of a quantum of energy and does not change the total Zeeman energy of the pair in the external field.

When the ions are not identical they precess at different velocities in the external magnetic field and there is no resonance effect if the interaction between them is small compared with the difference in velocities of precession. If this inequality is reversed, and the interaction is large, there is a high rate of mutual spin 'flipping' whose frequency is of order $\frac{1}{4}(\mathscr{J}_x+\mathscr{J}_y)/h$, which is rapid compared with the difference of precession frequencies in the external field. This field therefore sees only an average of the two ions, and the spectral line is observed at $\frac{1}{2}(g_{iz}+g_{jz})\beta H_z = g_z \beta H_z$ instead of at or near the individual frequencies $g_{iz}\beta H_z$, $g_{jz}\beta H_z$.

Figure 9.6 shows how the line positions vary as the relative sizes of the spin-spin interaction and $\delta g_z \beta H_z$ vary from one extreme to the other. In practice the spin-spin interaction for a given system is fixed in size but we can achieve a variation in $x = \frac{1}{4}\mathscr{J}/(\delta g_z \beta H_z)$ by measuring at different magnetic fields, i.e. at different frequencies. In copper sulphate, $CuSO_4, 5H_2O$, an effect of this kind for the two types of Cu^{2+}

ion was reported by Bagguley and Griffiths (1948) and Pryce (1948); this was the first direct observation of exchange effects in paramagnetic resonance. At lower frequencies ($\sim$10 GHz) only a single resonance line is observed, but at higher frequencies (40 to 60 GHz) two lines are resolved (Bagguley and Griffiths 1950, Ono and Ohtsuka 1958). The g-tensors of the two Cu^{2+} ions are similar but the principal axes are differently oriented, so that the phenomenon can also be studied as a function of δg by varying the orientation of the magnetic field with respect to the crystal axes.

Although we have treated above only the case of a pair of ions that are dissimilar in the principal values of their g-factors, the distinction between 'similar' and 'dissimilar' ions is more general. A pair of ions may be identical except that their nuclei are isotopes of the same element with different hyperfine structures (see, for example, Baker (1964a)). If the ions have effective spins greater than $S = \frac{1}{2}$ they may have 'fine structures' with differently oriented principal axes, as in a chromic alum. The essential distinction is: in the energy matrix of a pair of ions do two states have the same energy in the absence of spin-spin interaction, or do their energies differ by an amount large compared with the spin-spin interaction? In the latter case we need include only the diagonal components of their mutual interaction; in the former the off-diagonal components will be equally important. For the simple examples treated in this and the previous section, the interaction splits each spectrum line into a doublet; for dissimilar ions with a weak interaction, the doublet splitting (for a magnetic field along the z-axis) is $\mathscr{J}_z$, whereas for similar ions the corresponding splitting is

$$\mathscr{J}_z - \tfrac{1}{2}(\mathscr{J}_x + \mathscr{J}_y) = \mathscr{J}'_z - \tfrac{1}{2}(\mathscr{J}'_x + \mathscr{J}'_y) = \tfrac{3}{2}\mathscr{J}'_z, \qquad (9.43)$$

where we have used the fact that $\mathscr{J}'_x + \mathscr{J}'_y + \mathscr{J}'_z = 0$ for the components of the anisotropic interaction. We shall meet these two quantities again in the following sections on line broadening.

In conclusion it should be noted that when the interaction tensor contains off-diagonal terms such as $S_{ix}S_{jz}$, etc. (and these are generally present in magnetic dipole interaction), transitions between the 'triplet' and 'singlet' states are no longer forbidden either for pairs of similar or dissimilar ions.

9.7. Line broadening by spin–spin interaction

In the previous sections we have considered only isolated pairs of magnetic ions; but in a paramagnetic salt each ion is surrounded by a number of other ions at varying distances. The interaction between each pair is a function of the distance between them and the angle which the line joining their centres makes with the crystal axes; with indirect exchange it may also depend on the arrangement of diamagnetic ions in between. The effect of this interaction is to displace the resonance line from that of a single ion by an amount dependent on the strength of the interaction, and the intensity at a given displacement depends on the number of pairs with a given size of interaction. If we consider in turn pairs of ions in a crystal with increasing separation we may expect their interaction to diminish in size but the number of such pairs increases. Hence lines that are displaced by a small amount will have greater intensity, while lines displaced by a larger amount will have a smaller intensity. With only a small number of ions, the absorption curve may show some structure, but with a crystal of ordinary size the number of interacting ions is so large that a broad absorption curve with no structure is to be expected. Exceptions to this occur if the interaction of one ion with two of three of its neighbours is much greater than its interaction with other neighbours. One such example, cupric acetate, has already been described in § 9.5, where given pairs of ions are strongly coupled by exchange interaction; a less dramatic example is neodymium ethylsulphate (see Fig. 9.9) where interaction (mainly magnetic dipole in origin) with two nearest neighbours predominates sufficiently to give a structure to the absorption curve, while more distant neighbours give only an unresolved broadening.

The complete Hamiltonian for a set of interacting ions consists of (a) a set of terms $\mathcal{H}_i$, the separate spin Hamiltonian for each ion, summed over all ions; and (b) a set of terms $\mathcal{H}_{ij}$, the interaction Hamiltonian for each pair of ions, summed over all pairs. Formally we may write this as

$$\mathcal{H} = \sum_i \mathcal{H}_i + \sum_{j>i} \mathcal{H}_{ij}, \qquad (9.44)$$

where the summation nomenclature $j > i$ reminds us that each pair must be counted only once. Exact diagonalization of this Hamiltonian, which would yield all the eigenvalues and hence the details of the width and the shape of the absorption curve, is clearly impossible. An alternative approach, originally due to Waller (1932), is the method of 'moments'. The nth moment of an absorption line is defined by the relation

$$\langle \Delta \nu^n \rangle = \frac{\int (\nu - \nu_0)^n f(\nu) \, d\nu}{\int f(\nu) \, d\nu}$$
$$= \int (\nu - \nu_0)^n f(\nu) \, d\nu, \qquad (9.45)$$

where $f(\nu)$ is the value of the 'shape factor' at frequency ν, (see § 2.3), which is normalized so that $\int f(\nu)\,\mathrm{d}\nu = 1$ when the integral is taken over all frequencies, and ν_0 is the frequency at which the line would occur in the absence of any interaction. If the absorption curve (more strictly, the shape factor) is symmetrical about this frequency ν_0, then clearly all the odd moments (including the first, $n = 1$) will vanish, so that we need consider only the even moments. In principle these moments give information about the line shape in an indirect way, which is progressively more exact the higher the number of moments computed. Although these moments can be computed without having to diagonalize exactly the whole matrix corresponding to eqn (9.44), the labour is excessive for the higher moments. In practice only the second moment $\langle \Delta\nu^2 \rangle$ is easily found, and the fourth moment $\langle \Delta\nu^4 \rangle$ has been calculated only for a few special cases. This does not represent a serious drawback for comparison with experiment, because calculation of the higher moments from measured absorption curves is very inaccurate, the major contribution coming from the extreme wings of the curve where the absorption is small and difficult to measure exactly. In any case the theoretical computations can only be carried through if the way in which the interactions vary is known for all pairs of ions; this condition is fulfilled for a crystal of known structure if only magnetic dipole interactions are involved, but seldom holds if exchange or other interactions are present.

An extensive discussion of dipolar broadening (including the effect of isotropic exchange) can be found in Van Vleck (1948). In view of the limitations mentioned above, we give only a sketch of the method using the nomenclature adopted in this chapter. We begin by restricting the discussion to a set of identical spins, with no fine structure or hyperfine structure, subjected to an external magnetic field $\mathbf{H}$ whose direction we take to be the z-axis of a system of Cartesian coordinates. Then the general Hamiltonian (9.44) will have the form

$$\mathscr{H} = g\beta H \sum_i S_{iz} + \sum_{j>i} (\mathbf{S}_i \cdot \mathscr{J}_{ij} \cdot \mathbf{S}_j), \qquad (9.46)$$

where the interaction term embraces all interactions that are linear in the spin coordinates and expressible in tensor form. That is, each interaction term has the form (dropping the subscripts i, j on $\mathscr{J}$ for convenience)

$$\begin{aligned}(\mathbf{S}_i \cdot \mathscr{J} \cdot \mathbf{S}_j) = \mathscr{J}(\mathbf{S}_i \cdot \mathbf{S}_j) &+ \mathscr{J}'_{xx} S_{ix} S_{jx} + \mathscr{J}'_{yy} S_{iy} S_{jy} + \mathscr{J}'_{zz} S_{iz} S_{jz} + \\ &+ \mathscr{J}'_{xy} S_{ix} S_{jy} + \mathscr{J}'_{yx} S_{iy} S_{jx} + \\ &+ \mathscr{J}'_{yz} S_{iy} S_{jz} + \mathscr{J}'_{zy} S_{iz} S_{jy} + \\ &+ \mathscr{J}'_{zx} S_{iz} S_{jx} + \mathscr{J}'_{xz} S_{ix} S_{jz}.\end{aligned} \qquad (9.47)$$

Here the form of the first term is that of an isotropic exchange interaction while the remainder is that of an 'anisotropic' exchange interaction, in which

$$\mathscr{J}'_{xx}+\mathscr{J}'_{yy}+\mathscr{J}'_{zz} = 0. \tag{9.48}$$

This distinction is similar to that made in § 9.5, though the assumption was made there that the interaction could be reduced to the simple form of eqn (9.24). Here the z-axis is chosen to be that of the external magnetic field, which is not necessarily a principal axis of the interaction tensor, nor is the latter necessarily a symmetrical tensor. Equation (9.48) is valid whatever system of axes is used; it makes the trace of the anisotropic tensor zero, while the coefficient $\mathscr{J}$ of the term $\mathscr{J}(\mathbf{S}_i \cdot \mathbf{S}_j)$ is independent of the choice of axes.

Of the terms in (9.47) we now pick out those for which the total z-component of the spin $(S_{iz}+S_{jz})$ is constant. For any given pair these are

$$\mathscr{J}(\mathbf{S}_i \cdot \mathbf{S}_j)+\mathscr{J}'_{zz}S_{iz}S_{jz}+\tfrac{1}{2}(\mathscr{J}'_{xx}+\mathscr{J}'_{yy})(S_{ix}S_{jx}+S_{iy}S_{jy})$$
$$= (\mathscr{J}-\tfrac{1}{2}\mathscr{J}'_{zz})(\mathbf{S}_i \cdot \mathbf{S}_j)+(3\mathscr{J}'_{zz}/2)S_{iz}S_{jz}, \tag{9.49}$$

where we have used eqn (9.48). The reason why we select these terms can readily be seen by considering the energy matrix of a pair of similar ions. The Hamiltonian, including the Zeeman interaction and the 'truncated' Hamiltonian (9.49), is

$$\mathscr{H} = g\beta H(S_{iz}+S_{jz})+(\mathscr{J}-\tfrac{1}{2}\mathscr{J}'_{zz})(\mathbf{S}_i \cdot \mathbf{S}_j)+(3\mathscr{J}'_{zz}/2)S_{iz}S_{jz} \tag{9.50}$$

and, if the basis states (cf. § 9.5) are taken as

$$|++\rangle, \quad \frac{1}{\sqrt{2}}(|+-\rangle \pm |-+\rangle), \quad |--\rangle,$$

the energy matrix is diagonal. The allowed transitions are all of the form $\Delta M = \pm 1$, where $M = (S_{iz}+S_{jz})$ is the total z-component of the spin, and occur near $h\nu = g\beta H$ provided that $\mathscr{J}'_{zz} \ll g\beta H$. The terms omitted are all off-diagonal terms, whose effect is to admix the basis states and allow other transitions of the type $\Delta M = 0, \pm 2$, which occur near $h\nu = 0, 2g\beta H$. Similar considerations hold when we consider a large number of interacting similar ions, for which the truncated Hamiltonian is

$$\mathscr{H} = g\beta H \sum_i S_{iz}+\sum_{j>i} \{(\mathscr{J}-\tfrac{1}{2}\mathscr{J}'_{zz})(\mathbf{S}_i \cdot \mathbf{S}_j)+(3\mathscr{J}'_{zz}/2)S_{iz}S_{jz}\}. \tag{9.51}$$

The z-component $M = \sum_i S_{iz}$ is a constant of the motion, and the allowed transitions are all of the type $\Delta M = \pm 1$ and occur near $h\nu = g\beta H$. The interaction terms omitted from the truncated Hamiltonian (9.51) introduce matrix elements of the type $\Delta M = \pm 1, \pm 2$, and make transitions of the form $\Delta M = 0, \pm 2, \pm 3$ weakly allowed (higher values of ΔM also occur with rapidly decreasing intensity). The subsidiary line near $h\nu = 0$ corresponds to the non-resonant spin-spin relaxation phenomenon observed at radio-frequencies of order $\mathscr{J}'/h$; the line at $h\nu = 2g\beta H$ has been observed experimentally (see Fig. 9.7) by Bleaney and Ingram (1951) in a cobaltous salt where

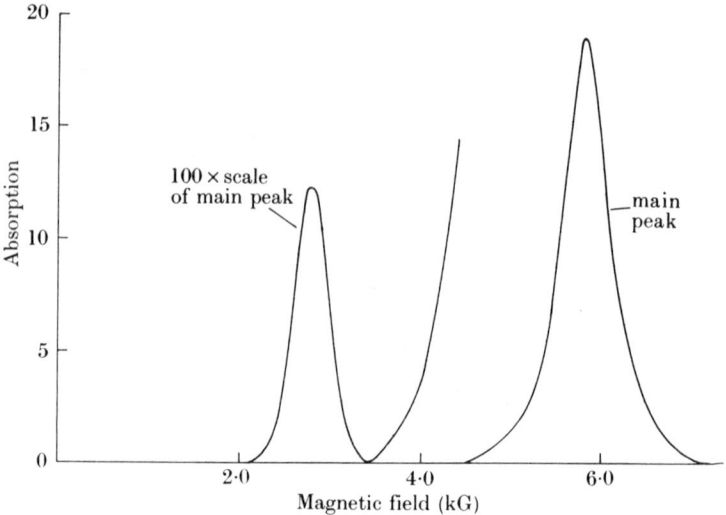

FIG. 9.7. Spectrum of concentrated cobalt ammonium sulphate along the K_2 axis at 20°K and $\bar{\nu} = 0.84$ cm^{-1}. The line at half-field is shown on a scale 100 × that of the main line (its height is 0.6 per cent of that of the main line). The line shape indicates that there is a considerable amount of exchange narrowing (Bleaney and Ingram 1951).

the effective spin of the Co^{2+} ion is $\tilde{S} = \frac{1}{2}$, and such a line cannot be a single ion transition.

We now consider the problem of calculating the second moment of an absorption line centred on frequency ν_0, and for which the shape factor at frequency ν has the value $f(\nu)$. Since we can write

$$\nu = \nu_0 + (\nu - \nu_0),$$

we have

$$\nu^2 = \nu_0^2 + 2\nu_0(\nu - \nu_0) + (\nu - \nu_0)^2$$

and

$$\int \nu^2 f(\nu)\, d\nu = \int \nu_0^2 f(\nu)\, d\nu + \int 2\nu_0(\nu - \nu_0)f(\nu)\, d\nu + \int (\nu - \nu_0)^2 f(\nu)\, d\nu.$$

SPIN-SPIN INTERACTION 519

If the line shape is symmetrical about ν_0, the second term on the right-hand side vanishes and, since ν_0 is a constant and $\int f(\nu)\,\mathrm{d}\nu = 1$, we have

$$\langle \nu^2 \rangle = \nu_0^2 + \langle \Delta \nu^2 \rangle. \tag{9.52}$$

The conditions under which the assumption of a line shape symmetrical about ν_0 is valid will be discussed later (§ 9.11).

Equation (9.52) enables us to find $\langle \Delta \nu^2 \rangle$ using a quantum-mechanical formula for $\langle \nu^2 \rangle$. Suppose transitions are induced by an oscillatory magnetic field polarized along the x-axis. Then, if there is no anisotropy in g, the transition probabilities are found from the matrix elements of the operator $S_x = \sum_i S_{ix}$. Let $(S_x)_{nn'}$ be the matrix element between two eigenstates of (9.51) denoted by n and n', for which the corresponding frequency is $\nu_{nn'} = (\mathscr{H}_n - \mathscr{H}_{n'})/h$. Then the mean square absorption frequency (in which each frequency is weighted with the square of the appropriate amplitude) is

$$\langle \nu^2 \rangle = \frac{\sum\limits_{n,n'} \{\nu_{nn'}^2 \, |(S_x)_{nn'}|^2\}}{\sum\limits_{n,n'} |(S_x)_{nn'}|^2}. \tag{9.53}$$

Both numerator and denominator can be expressed as diagonal sums, giving

$$h^2 \langle \nu^2 \rangle = -\frac{\mathrm{Tr}\,[\mathscr{H} S_x - S_x \mathscr{H}]^2}{\mathrm{Tr}\,[S_x]^2}, \tag{9.54}$$

which has the advantage that the spur or trace (Tr) is invariant and can be computed without having to diagonalize the Hamiltonian (9.51), and finding the individual eigenvalues. Evaluation of the traces (cf. Van Vleck 1948) gives

$$h^2 \langle \Delta \nu^2 \rangle = \tfrac{1}{3} S(S+1) \sum_{j>i} (3 \mathscr{J}'_{zz}/2)_{ij}^2. \tag{9.55}$$

The striking feature of this result is that it contains only the anisotropic part of the interaction, and is independent of the isotropic term $(\mathbf{S}_i \cdot \mathbf{S}_j)$. This is in line with the result found in § 9.5 for the spectrum of a pair of similar ions, which consists of a single line at $h\nu = g\beta H$ if only an isotropic interaction is present, but otherwise contains several lines split by an amount dependent on the size of the anisotropic part of the interaction. For a system of many interacting ions we may picture the broadened line as consisting of a multiplicity of such split pair spectra, though this is, of course, an over-simplification since we have to picture each ion as being paired off with every other ion.

In many paramagnetic crystals there may be more than one set of paramagnetic ions, such as two sets of electronic spins with different g-values, or one set of electronic spins and one set of nuclear spins. We must therefore carry out a calculation similar to that above, but in which we treat the interactions between similar spins all belonging to the same set S separately from the interactions between a spin S of one set and a spin S' of the other set. We assume that the two sets of spins have different g-values g and g', and give two distinct resonance lines at $h\nu = g\beta H$ and $g'\beta H$ respectively whose separation is large compared with their widths; this implies that the spin-spin interaction between spins of the different species is small compared with the difference $(g-g')\beta H$ in their Zeeman energies (cf. § 9.6). Suppose we wish to find the second moment of one of the lines (that of the first set of spins). For interactions with other spins of the same set we proceed exactly as above, but for interactions between dissimilar spins we retain in our truncated interaction Hamiltonian for each pair only the term

$$\mathscr{J}_{zz} S_{iz} S'_{jz}. \qquad (9.56)$$

Comparison with (9.49) shows that we have excluded terms of the type $S_{i+}S'_{j-}$, $S_{i-}S'_{j+}$, which would correspond to mutual spin flips between the two sets of ions; these terms are associated with a subsidiary absorption line at $h\nu = (2g-g')\beta H$ which, under the condition given above, is distinct from the main line at $g\beta H$. The formula for the second moment then becomes

$$h^2 \langle \Delta \nu^2 \rangle = \tfrac{1}{3} S(S+1) \sum_{j>i} (3\mathscr{J}'_{zz}/2)^2_{ij} + \tfrac{1}{3} S'(S'+1) \sum_{j',i} (\mathscr{J}^2_{zz})_{ij'}, \qquad (9.57)$$

where the first summation is over pairs of ion i with all similar ions of the first set only, while the second is over all pairs of ion i with dissimilar ions belonging to the second set only. An important difference between the two summations is that in the second case the quantity $\mathscr{J}_{zz}$ is involved and not just the term $\mathscr{J}'_{zz}$ belonging to the anisotropic part of the interaction. This means that the isotropic interaction contributes to the second moment for dissimilar ions, but not for similar ions, corresponding to the differences already found for single pairs of interacting ions in §§ 9.5 and 9.6. A further correspondence appears in the factor $(\tfrac{3}{2})$ in the term $(3\mathscr{J}'_{zz}/2)$ for similar ions arising from the resonance effect that enhances the interaction between them. This factor is absent from the term $\mathscr{J}_{zz}$ for dissimilar ions.

9.8. Line shape due to dipolar spin-spin interaction

As noted in the last section, computations of line width and moments can be carried out only if the interaction constants are known for all pairs of spins in a crystal. In general, this restricts the computations to cases where the interaction is entirely due to magnetic dipole interaction between the spins, a situation that is rather rare if the magnetic moments are electronic rather than nuclear in origin. For this reason the following discussion will in the main be qualitative rather than quantitative.

If the interaction is dipolar, and the g-factors of the two spins involved are each isotropic, the form of the interaction is given by eqn (9.3). This is a particular form of the more general tensor-type interaction assumed in eqn (9.46), and since (9.3) has zero trace the 'isotropic' part of the interaction as defined by eqns (9.47), (9.48) vanishes; that is, $\mathscr{J} = 0$, and the distinction between quantities $\mathscr{J}_{xx}$ and $\mathscr{J}'_{xx}$, etc., is unnecessary. (This does not hold if the g-factors are anisotropic, since the sum $\mathscr{J}_{xx} + \mathscr{J}_{yy} + \mathscr{J}_{zz} = 0$ for eqn (9.3) but not for (9.5a) or (9.5b).) If there are two sets of spins, S and S', with isotropic g-factors g, g', the quantities needed for the calculation of the second moment given by (9.57) are

$$(\mathscr{J}'_{zz})_{ij} = g^2\beta^2 r_{ij}^{-3}(1-3n_{ij}^2) \text{ for similar pairs,} \tag{9.58a}$$

and

$$(\mathscr{J}_{zz})_{ij'} = gg'\beta^2 r_{ij'}^{-3}(1-3n_{ij'}^2) \text{ for dissimilar pairs.} \tag{9.58b}$$

Here r_{ij} is the distance between the centres of spins i, j and $n_{ij} = \cos\theta_{ij}$ where θ_{ij} is the angle that the line joining the centres makes with the external magnetic field H (i.e. with the z-axis). The value of $\mathscr{J}_{zz}$ is just the interaction between the z-component of spin S_i and the z-component of the local field at its centre, that is produced by the z-component of spin S'_j. If mutual spin flips among the spins S' can be neglected, this component is a constant of the motion, and contributes just a static component to the local field, so that broadening of the line due to interaction with dissimilar spins is essentially similar to that which would be produced by inhomogeneities in the external magnetic field. If no other form of broadening is present the line is said to be 'inhomogeneously broadened'. In the extreme case the wave train emitted or absorbed by a spin is infinite in length, but is at a slightly different frequency from that of other spins subjected to different local fields (cf. § 2.6).

When the interacting spins are identical there is the additional resonance interaction due to the precessing components of the spins, which in eqn (9.57) makes the relevant quantity $(3\mathscr{J}'_{zz}/2)$ rather than $\mathscr{J}_{zz}$. Since in the present case there is no distinction between $\mathscr{J}'_{zz}$ and $\mathscr{J}_{zz}$, the second moment for similar spins is just a factor $(\frac{9}{4})$ greater than that obtained simply by including only the static components of the interaction. (Obviously the factor $(\frac{9}{4})$ is correct only if the g-factor is isotropic, since otherwise $\mathscr{J}'_{zz} \neq \mathscr{J}_{zz}$.) This resonance interaction between identical spins shortens the lifetime of the individual spin states through mutual spin flips. In the over-simplified model discussed in Chapter 2 the average lifetime of a spin state is described by a parameter τ_2, the spin-spin or transverse relaxation time. The line in such a case is said to be 'homogeneously broadened'; the spins emit or absorb wave trains of finite length whose mean duration in time is τ_2, and whose probability distribution is of the form $\exp(-t/\tau_2)$. The line shape should be given by the Lorentz factor

$$f(\nu) = \frac{2\tau_2}{1+4\pi^2(\nu-\nu_0)^2\tau_2^2}$$
$$= \frac{1}{\pi}\frac{\Delta\nu}{(\Delta\nu)^2+(\nu-\nu_0)^2}, \qquad (9.59)$$

where $\Delta\nu = 1/(2\pi\tau_2)$. For such a line shape, the intensity falls to one-half the maximum value when the frequency deviates by $\pm\Delta\nu$ from the central frequency ν_0. If the second moment of such a line is computed by eqn (9.45), it is found to be infinite; since we know from eqn (9.55) that the second moment is finite, it follows that the line shape cannot be Lorentzian, and the absorption in the wings must fall more rapidly than predicted by eqn (9.59). Again, since the second moment given by (9.59) is infinite, there is no way of relating the line breadth parameter $\Delta\nu$ in (9.59) to the second moment computed from eqns (9.57) and (9.58).

In practice it is often useful to have a simple analytical form for the line shape, even if only approximately correct. An alternative assumption for the shape factor is that of a Gaussian error function

$$f(\nu) = \frac{1}{(2\pi\langle\Delta\nu^2\rangle)^{\frac{1}{2}}}\exp\left\{-\frac{(\nu-\nu_0)^2}{2\langle\Delta\nu^2\rangle}\right\}, \qquad (9.60)$$

which is normalized, and contains no arbitrary parameter, since it gives a second moment just equal to $\langle\Delta\nu^2\rangle$. The higher moments can also be readily calculated, and (9.60) gives for the fourth moment

$\langle \Delta \nu^4 \rangle = 3(\langle \Delta \nu^2 \rangle)^2$, or
$$(\langle \Delta \nu^4 \rangle)^{\frac{1}{4}} = 1 \cdot 32 (\langle \Delta \nu^2 \rangle)^{\frac{1}{2}}. \qquad (9.61)$$

An explicit calculation of $\langle \Delta \nu^4 \rangle$ for a simple cubic lattice of spins by Van Vleck (1948, 1957) by the rigorous method of moments gives a check on the validity of the Gaussian assumption. For $S = \frac{1}{2}$, and an external field directed along an $\langle 001 \rangle$ axis, he finds the factor to be 1·25 rather than the 1·32 given in (9.61). For a field directed along a $\langle 111 \rangle$ axis, or if $S > \frac{1}{2}$, the factor is even nearer to 1·32.

An experimental determination of line shape is shown in Fig. 9.8; the experimental points agree well with the Gaussian shape whose second moment has been computed by the diagonal sum method (with

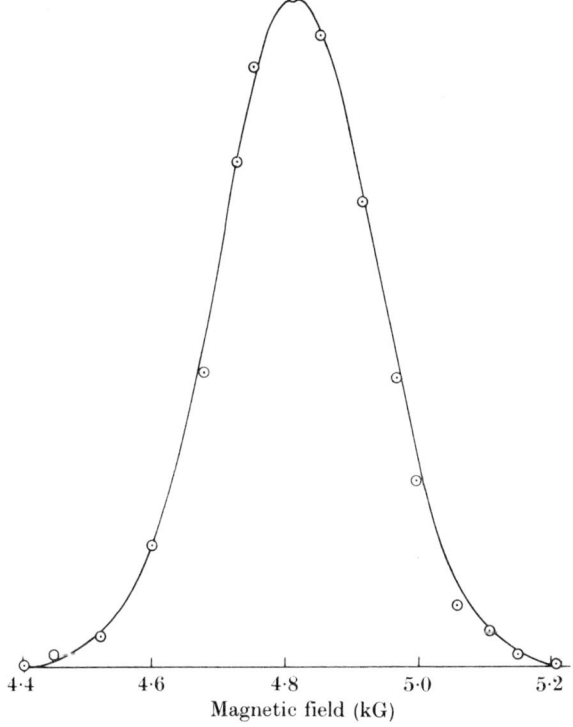

Fig. 9.8. The $-\frac{3}{2} \leftrightarrow -\frac{1}{2}$ transition in caesium chrome alum, $CsCr(SO_4)_2$, $12H_2O$, for which the single ion spin Hamiltonian is

$$\mathscr{H} = g\beta(\mathbf{H} \cdot \mathbf{S}) + D\{S_z^2 - \tfrac{1}{3}S(S+1)\}$$

with $S = \frac{3}{2}$. There are four ions in unit cell, and in the figure **H** is along the z-axis (a $\langle 111 \rangle$ axis) for one ion only. The continuous line is computed using a Gaussian shape and a root mean square width of 118 G, the calculated value for dipolar interaction assuming that ions whose trigonal axes are differently oriented are to be counted as dissimilar ions. This gives a good fit to the experimental points (Bleaney 1950).

one simplifying assumption—see legend to Fig. 9.8). In this substance (as in all the alums) the magnetic ions lie on a face-centred cubic lattice, and each ion is surrounded by twelve equidistant nearest neighbours. Although this substance is cubic, the line width is not isotropic because of the angular factors in eqns (9.58a), (9.58b). The variation in line width is most dramatic in cases where the dipolar contribution from all the nearest neighbours vanishes at certain orientations of the external field. If g is isotropic, this requires that $(1-3\cos^2\theta_{ij}) = 0$, or $\theta_{ij} = 54° 44'$, where θ_{ij} is the angle between the external field and the lines joining nearest neighbours. This condition is fulfilled in two cases of cubic symmetry:

(a) If the magnetic ions lie on a simple cubic lattice, so that the six nearest neighbours lie in pairs along [001], [010], and [100], and **H** is directed along a $\langle 111 \rangle$ axis;

(b) in a lattice of the CaF_2 type, with nearest neighbour magnetic ions along the $\langle 111 \rangle$ axes, when **H** is along an $\langle 001 \rangle$ axis.

These conditions, with isotropic g-factor and purely dipolar interaction, are more readily achieved with nuclear than with electronic dipoles. A special case of (b) is CaF_2 with an isolated electronic magnetic ion on a Ca site, when the first-order interaction with the eight nearest neighbour F^- nuclear moments vanishes for **H** along an $\langle 001 \rangle$ axis provided the interaction is entirely dipolar in form.

The assumption of a Gaussian line shape is most likely to be a good approximation when the number of equidistant nearest neighbours is fairly large. In other cases, when the number of nearest neighbours is rather small, the Gaussian assumption may be seriously wrong. A notable example is neodymium ethylsulphate, in which the Nd^{3+} ion has two equidistant nearest neighbours at 0·7 nm along the c-axis, which give a partly resolved structure (see Fig. 9.9) when the external field is along the c-axis. In a first approximation the line consists of a triplet with a 1:2:1 intensity distribution, corresponding to the chances of neighbouring spins being both 'up'; one 'up' and one 'down' (which can arise in two ways); and both 'down' respectively. The six next-nearest neighbours are about 0·88 nm away in directions such that when the external field is along the c-axis they contribute only a rather small dipolar field. Since all ions have symmetry about the c-axis, it is readily found from eqn (9.5a) that when **H** is along this axis

$$\mathscr{J}_{zz} = g_{iz}g_{jz}(1-3n_{ij}^2)\beta^2 r_{ij}^{-3} \tag{9.62}$$

and

$$\mathscr{J}'_{zz} = \tfrac{1}{3}(2g_{iz}g_{jz}+g_{i\perp}g_{j\perp})(1-3n_{ij}^2)\beta^2 r_{ij}^{-3}. \tag{9.63}$$

SPIN-SPIN INTERACTION 525

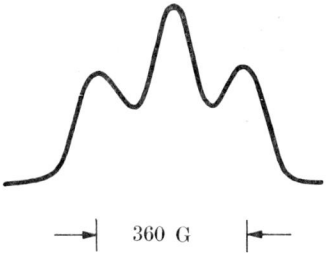

FIG. 9.9. Oscillogram of the spectrum of neodymium ethylsulphate. Above, triplet structure observed with magnetic field parallel to the crystal (hexagonal) axis; below, spectrum with magnetic field normal to crystal axis (Bleaney, Elliott, and Scovil 1951).

Thus the first-order dipolar contribution from neighbouring ions vanishes in this case if their directions are such that $(1-3n_{ij}^2) = 0$, irrespective of whether they are identical or not, and irrespective of the anisotropy in their g-tensors provided they have axial symmetry ($g_x = g_y = g_\perp$). In the ethylsulphates the directions of the next nearest neighbours are near to fulfilling the condition $(1-3n_{ij}^2) = 0$.

The angular dependence just derived applies to the case of **H** along the unique axis and neighbouring ions not along this axis. The converse case, of a pair of ions along the c-axis (such as the nearest neighbours in the ethylsulphates) and **H** at an angle to this axis, does not give the same angular dependence except when the g-values are isotropic. For such a pair of ions, if **H** is at an angle θ to the z-axis, which is the symmetry axis and also the line joining the ions, we have

$$\mathscr{H}_i = \beta H(S_{iz}g_{i\|} \cos\theta + S_{ix}g_{i\perp} \sin\theta) \qquad (9.64)$$

$$\mathscr{H}_j = \beta H(S_{jz}g_{j\|} \cos\theta + S_{jx}g_{j\perp} \sin\theta) \qquad (9.65)$$

where, because of the axial symmetry, we can assume that **H** lies in the xz-plane. The interaction Hamiltonian would be just (9.5b) if the interaction were purely dipolar, but we may take the rather more general form

$$\mathscr{H}_{ss} = \mathscr{J}_\perp(S_{ix}S_{jx} + S_{iy}S_{jy}) + \mathscr{J}_\| S_{iz}S_{jz}, \qquad (9.66)$$

which includes other possible contributions axially symmetric in form. If the Zeeman interactions are much larger than the spin-spin interaction, they must be diagonalized first; this can be achieved by a rotation of axes about the y-axis (as in § 3.2) through angles of ϕ_i and ϕ_j for spins S_i and S_j respectively, where $\tan \phi_i = (g_{i\perp}/g_{i\parallel})\tan\theta$ and $\tan \phi_j = (g_{j\perp}/g_{j\parallel})\tan\theta$. Then the interaction Hamiltonian becomes

$$\mathscr{H}''_{ss} = (\mathscr{J}_\perp \cos\phi_i \cos\phi_j + \mathscr{J}_\parallel \sin\phi_i \sin\phi_j) S'_{ix} S'_{jx} + \mathscr{J}_\perp S_{iy} S_{jy} +$$
$$+ (\mathscr{J}_\perp \sin\phi_i \sin\phi_j + \mathscr{J}_\parallel \cos\phi_i \cos\phi_j) S'_{iz} S'_{jz} + \ldots, \quad (9.67)$$

where we have retained only terms of the form

$$\mathscr{H}''_{ss} = \mathscr{J}_x S'_{ix} S'_{jx} + \mathscr{J}_y S'_{iy} S'_{jy} + \mathscr{J}_z S'_{iz} S'_{jz}. \quad (9.67a)$$

If the ions are dissimilar we need for the second moment calculation just the coefficient

$$\mathscr{J}_z = \mathscr{J}_\perp \sin\phi_i \sin\phi_j + \mathscr{J}_\parallel \cos\phi_i \cos\phi_j$$
$$= (g_{i\perp} g_{j\perp}/g_i g_j) \mathscr{J}_\perp \sin^2\theta + (g_{i\parallel} g_{j\parallel}/g_i g_j) \mathscr{J}_\parallel \cos^2\theta, \quad (9.68)$$

in which we have, as usual,

$$g_i^2 = g_{i\perp}^2 \sin^2\theta + g_{i\parallel}^2 \cos^2\theta,$$
$$g_j^2 = g_{j\perp}^2 \sin^2\theta + g_{j\parallel}^2 \cos^2\theta. \quad (9.69)$$

If the ions are similar, their g-factors are the same and $\phi_i = \phi_j = \phi$. Then in (9.67a) we have

$$\left.\begin{array}{l}\mathscr{J}_x = \mathscr{J} + \mathscr{J}'_x = \mathscr{J}_\perp \cos^2\phi + \mathscr{J}_\parallel \sin^2\phi, \\ \mathscr{J}_y = \mathscr{J} + \mathscr{J}'_y = \mathscr{J}_\perp, \\ \mathscr{J}_z = \mathscr{J} + \mathscr{J}'_z = \mathscr{J}_\perp \sin^2\phi + \mathscr{J}_\parallel \cos^2\phi,\end{array}\right\} \quad (9.70)$$

in which we have made the usual separation into an interaction with trace $\mathscr{J}'_x + \mathscr{J}'_y + \mathscr{J}'_z = 0$, and an 'isotropic' part whose coefficient is

$$\mathscr{J} = \tfrac{1}{3}(\mathscr{J}_x + \mathscr{J}_y + \mathscr{J}_z) = \tfrac{1}{3}(\mathscr{J}_\parallel + 2\mathscr{J}_\perp). \quad (9.71)$$

In computing the second moment (cf. eqn (9.57)) we need the quantity

$$\tfrac{3}{2}\mathscr{J}'_z = \tfrac{1}{2}(\mathscr{J}_\parallel - \mathscr{J}_\perp)(3\cos^2\phi - 1)$$
$$= \tfrac{1}{2}(\mathscr{J}_\parallel - \mathscr{J}_\perp)\left(3\frac{g_\parallel^2}{g^2}\cos^2\theta - 1\right). \quad (9.72)$$

If the interaction is purely dipolar in origin, we have from (9.5b),

$$\left.\begin{array}{l}\mathscr{J}_\parallel + 2\mathscr{J}_\perp = -2(g_\parallel^2 - g_\perp^2)\beta^2 r_{ij}^{-3}, \\ \mathscr{J}_\parallel - \mathscr{J}_\perp = -(2g_\parallel^2 + g_\perp^2)\beta^2 r_{ij}^{-3},\end{array}\right\} \quad (9.73)$$

from which we see that the tensor interaction (9.67a) is traceless ($\mathscr{J} = 0$) only if g is isotropic.

SPIN-SPIN INTERACTION 527

A perhaps surprising feature of (9.72) is that whatever values the coefficients $\mathscr{J}_{\|}$, $\mathscr{J}_{\perp}$ may have, the angular variation of $\mathscr{J}'_z$ given by the last bracket in (9.72) is from a maximum of $+2$ when **H** is parallel to the symmetry axis ($\theta = 0$) to -1 when **H** is perpendicular ($\theta = \pi/2$). The anisotropy in g affects only the angular variation at intermediate values of θ; in particular, $\mathscr{J}'_z$ passes through zero at $\tan^2\theta = (2g_{\|}^2/g_{\perp}^2)$. In the case of neodymium ethylsulphate referred to earlier in this section, the triplet separation is proportional to $\mathscr{J}'_z$ and is only well resolved (cf. Fig. 9.9) when $\mathscr{J}'_z$ has its maximum value, i.e. when **H** is along the symmetry axis.

9.9. Effect of exchange interaction on line shape

In the previous section we considered the width and shape of a line when the spin-spin interaction is wholly dipolar in origin. This is seldom the case for a system of electronic dipoles, and in many paramagnetic substances exchange interaction may be more important than the dipolar interaction.

Initially we consider again the simple case of a set of identical ions with isotropic g-factors and no 'fine' or hyperfine splittings, coupled together by dipolar interaction and isotropic exchange interaction. As we have seen already from § 9.7, eqn (9.55), the second moment is unaltered by an isotropic exchange term. However Van Vleck (1948) has shown that the formula for the fourth moment contains the isotropic exchange energy, so that the fourth moment is larger than it would be in the absence of such exchange interaction. This information is obviously insufficient to determine the line shape, but it is readily seen that the behaviour of the two moments implies that the line must be narrowed in the centre and extended in the wings in order to increase the fourth moment while keeping the second moment constant. This is the phenomenon of 'exchange narrowing', first predicted by Gorter and Van Vleck (1947). Physically it can be regarded as an effect akin to motional narrowing of a magnetic resonance line in a liquid. In order that the local field experienced by a given ion shall shift its precession frequency, and hence broaden the resonance for an assembly of ions, the local field must persist for a time long compared with the duration of the 'wave train' absorbed by the ion. If the local field fluctuates rapidly during this time, its effect tends to be averaged out. When the exchange energy is large, the orientation of neighbouring spins is being changed at a rate of order $|(\mathscr{J}/h)|$ through mutual spin flips, so that the local dipolar field fluctuates at a similar rate and tends to be averaged out. For this to be effective, we require

$$|(\mathscr{J}/h)| \gg (\langle \Delta \nu^2 \rangle)^{\frac{1}{2}},$$

where $\langle \Delta \nu^2 \rangle$ is the second moment due to dipolar interaction.

The problem of the shape of an exchange-narrowed line has been considered in more detail by Anderson and Weiss (1953), using a mathematical model in which the dipolar interaction is assumed to produce a Gaussian distribution of internal fields and, in addition, this field is time-modulated in a random way. If the above inequality is satisfied, this theory gives a line shape where the wings of the line fall away exponentially as for a Gaussian shape, but near the centre the line is Lorentzian in shape, with a half-width

$$(\Delta \nu)_{\frac{1}{2}} \approx (\langle \Delta \nu^2 \rangle)/(\mathscr{J}/h). \tag{9.74}$$

Here $\langle \Delta \nu^2 \rangle$ is the second moment due to dipolar interaction, and $(\mathscr{J}/h)$ is the exchange energy in frequency units. However, when this frequency exceeds the resonance frequency, the truncated Hamiltonian (9.55) should not be used in computing $\langle \Delta \nu^2 \rangle$, but all the $\mathscr{J}'$ terms in (9.47). For a set of identical spins with isotropic g-factors, this gives a value for $\langle \Delta \nu^2 \rangle$ which is larger by a factor $\frac{10}{3}$ than that obtained from (9.55) and (9.58a) (see, for instance, Abragam (1961)). The increase in line width at lower resonance frequencies, and its reduction at higher frequencies, has been measured for $K_2CuCl_4, 2H_2O$ (see, for example, Henderson and Rogers (1966)), and for $Cu(NH_3)_4SO_4, H_2O$ by Rogers, Carboni, and Richards (1967).

Equation (9.74) is only valid and exchange narrowing will only be appreciable provided that $|\mathscr{J}/h| \gg (\langle \Delta \nu^2 \rangle)^{\frac{1}{2}}$. Here $\mathscr{J}$ must be the value of the isotropic exchange interaction expressed in terms of the *effective spins*. We have seen (§ 9.3) that when there is considerable magnetic anisotropy, the exchange interaction expressed in terms of effective spins may be quite anisotropic, even if we start from an exchange interaction that is isotropic when expressed in terms of the *real spins*. In this case the appropriate value for $\langle \Delta \nu^2 \rangle$ is given by eqn (9.55), which includes all contributions to the anisotropic part of the spin-spin interaction; clearly, when $\mathscr{J}'$ is of the same order as $\mathscr{J}$, there is no appreciable narrowing, and if $\mathscr{J}'$ is mainly due to exchange its effect is to broaden rather than to narrow the line. Thus the narrowing of resonance lines by exchange interaction is most likely to be effective when the dipole moments are almost wholly due to electron spin, as in a half-filled shell such as $3d^5$, Mn^{++}, Fe^{3+} in a weak crystal field.

The phenomenon of exchange narrowing depends on the property that isotropic exchange interaction does not contribute to the second moment when we have a system of identical ions. This is no longer true when the ions are not identical, and isotropic exchange then makes a contribution to the second moment. Apart from the obvious case of ions of different species, ions of the same species are essentially not identical under conditions such as the following:

(a) the axes of their g-tensors are differently oriented, e.g. $CuSO_4$, $5H_2O$; see § 9.6;

(b) they possess a hyperfine structure, e.g. neodymium ethyl-sulphate; see Baker (1964a);

(c) they possess a 'fine structure', e.g. nickel salts; see Ishiguro, Kambe, and Usui (1951), Stevens (1952b).

Basically the essential criterion is that ions are counted as dissimilar if they give resolved resonance lines; then in calculating the second moment we take as the centre point the position of the unbroadened line. When the exchange interaction becomes comparable with the separations between the lines the situation becomes complicated, and if the exchange interaction is still larger the lines become merged. Exchange narrowing then sets in, with half-width given by eqn (9.74), provided that we include in $\langle \Delta \nu^2 \rangle$ the contributions from the line splittings.

The general question of methods of calculating moments when fine structure is present is discussed by Pryce and Stevens (1950), Kambe and Usui (1952), and McMillan and Opechowski (1960, 1961). The last authors also treat particularly the changes in moments which occur at low temperatures where departures from random orientation in the spin system cannot be neglected, a subject that is briefly discussed in § 9.11.

9.10. Magnetic dilution, and the spectra of pairs

In most salts that are paramagnetic down to fairly low temperatures the distances between the nearest paramagnetic ions lie in the range 0·5–0·8 nm, and the line widths due to magnetic dipolar interaction between the ions are of order 10^2–10^3 G. This is sufficient to obscure many details of the resonance line, such as hyperfine structure. In more concentrated salts the lines may be narrowed by exchange interaction, but at the same time small splittings due to fine or hyperfine effects tend to be averaged out. Since spin-spin interaction falls off rapidly with increasing inter-ionic distance, considerable reductions in line width can be achieved by the use of 'magnetically dilute' salts, in which the majority of the paramagnetic ions are replaced by suitable diamagnetic ions in an isomorphic crystal. For hydrated salts where the waters of hydration are the immediate neighbours of the paramagnetic ion, a residual width of order 10 G is attained at a relative concentration of paramagnetic ions in the range 0·1 to 1 per cent. This residual width is due to the nuclear magnetic moments of the protons, which are typically at a distance of a little over 0·2 nm; an $M(6H_2O)$ complex characteristic of the iron group contains twelve such protons, while a lanthanide ethylsulphate group contains eighteen. This residual width can be reduced by a factor of about one-third by deuteration, because of the smaller nuclear moment of the deuteron.

Considerable narrower lines, with widths less than 1 G in favourable circumstances, can be achieved by the use of oxides such as MgO, ThO_2 where the only isotopes with nuclear moments (^{25}Mg, ^{17}O) are in low abundance. Clearly such narrow lines can only be attained if the concentration of paramagnetic ions is made sufficiently small. Paradoxically, quite 'narrow' lines can also be observed in substances such as fluorides (ZnF_2, CaF_2) where the ^{19}F ions with their large nuclear moments are the immediate neighbours of the paramagnetic ion; this occurs when electron transfer between the paramagnetic ion and the nearest F^- ions is sufficient to give a resolved fluorine hyperfine structure. The width of the individual lines then arises only from interaction with more distant magnetic moments.

If the replacement of paramagnetic by diamagnetic ions is a purely random process, the chance of a paramagnetic ion occupying a given site is just equal to the fractional concentration c. In the calculation of the second moment $\langle \Delta \nu^2 \rangle$, the sums involved in, for example, eqn (9.57), are each proportional to c since, for any given paramagnetic ion, the chance of any given neighbouring site being occupied by another paramagnetic ion is just proportional to c. At first sight this would suggest that the line width would fall on dilution only as $c^{\frac{1}{2}}$, but in practice the line width falls faster than this. The reason for this is that dilution does not reduce the magnitude of the interaction between any given pair of ions, but only the chance that such a pair will occur. Thus a pair of near-by ions with (presumably) a large spin-spin interaction will give rise to lines well out in the wings of the main line, and as the line width is reduced these appear as satellites to the main line. In the sums involved in computing the second (and higher) moments the furthest out satellites make a more than average contribution;

such contributions keep the second moment proportional to the concentration, but the fallacy in using the root mean second moment ($\langle \Delta \nu^2 \rangle)^{\frac{1}{2}}$ as a measure of line width is that this incorrectly assumes that the line shape remains unaltered. At sufficiently low concentrations the dominant contribution to all moments $\langle \Delta \nu^{2n} \rangle$ comes from sums over pairs of ions, and is thus proportional to c, showing that the line shape must change (Kittel and Abrahams 1953). The arguments using the method of moments set out by these authors agree with the analysis of Anderson (1951) that at concentrations below about 0·1 the centre of the line is Lorentzian in shape, with a half-width roughly proportional to the concentration.

At relative concentrations of order $c = 10^{-2}$–10^{-1} the satellite lines are sufficiently intense and well resolved that their spectrum can be observed and fitted to a spin Hamiltonian with fair accuracy. The centre portion of the spectrum is obscured by the main line due to 'isolated' ions, which at low concentrations are relatively more abundant than 'pairs' of ions, and a rather careful and painstaking analysis is needed to identify with what type of pair a given set of satellites is associated. When the effective spin of the ions is $S = \frac{1}{2}$, the energy levels split into singlet and triplet as discussed in §§ 9.5–9.6. When the ions are identical, the 'isotropic' part of the interaction determines the singlet-triplet splitting whose size can be found from intensity measurements as a function of temperature, while the 'anisotropic' part is found from the splitting of the triplet as measured by the separation of the satellites as a function of orientation of the external magnetic field. An example of an extensive analysis of this type is the work on $K_2(Ir, Pt)Cl_6$ and related compounds by Griffiths, Owen, Park, and Partridge (1959). Here the Ir^{4+}, $5d^5$, ions are subject to a strong ligand field (see Chapter 8) of exact octahedral symmetry, with effective spin $S = \frac{1}{2}$ and $g = 1\cdot79$ (isotropic); the Pt^{4+}, $5d^6$, ions with a closed $d\epsilon$ shell act as diluent. The spin-spin interaction is assumed to have the form

$$\mathcal{H}_{SS} = \mathcal{J}(\mathbf{S}_i \cdot \mathbf{S}_j) + \mathcal{J}'_x S_{ix} S_{jx} + \mathcal{J}'_y S_{iy} S_{jy} + \mathcal{J}'_z S_{iz} S_{jz}, \qquad (9.75)$$

where the anisotropic terms can also be written in the form

$$D_e \{ 2 S_{iz} S_{jz} - S_{ix} S_{jx} - S_{iy} S_{jy} \} + E_e (S_{ix} S_{jx} - S_{iy} S_{jy}) \qquad (9.76)$$

$$= D_e \{ 3 S_{iz} S_{jz} - (\mathbf{S}_i \cdot \mathbf{S}_j) \} + E_e (S_{ix} S_{jx} - S_{iy} S_{jy}) \qquad (9.77)$$

with (since $\mathcal{J}'_x + \mathcal{J}'_y + \mathcal{J}'_z = 0$)

$$D_e = \tfrac{1}{2} \mathcal{J}'_z, \qquad E_e = \tfrac{1}{2}(\mathcal{J}'_x - \mathcal{J}'_y). \qquad (9.78)$$

The crystal structure is face-centred cubic, so that each Ir^{4+} ion has twelve nearest neighbours (nn) and six next nearest neighbours (nnn). The interaction parameters are given in Table 9.4.

TABLE 9.4

Interaction parameters between Ir^{4+} ions in two salts. After Table 1 (see also Table 8) of Harris and Owen (1965)

		$K_2(Ir, Pt)Cl_6$	$(NH_4)_2(Ir, Pt)Cl_6$
Nearest neighbours	$\mathscr{J}$ (cm^{-1})	$+8 \cdot 0 \pm 0 \cdot 8$	$+5 \cdot 2 \pm 0 \cdot 8$
	$\mathscr{J}$ (°K)	$+11 \cdot 5 \pm 1$	$+7 \cdot 5 \pm 1$
	D_e (cm^{-1})	$+0 \cdot 45 \pm 0 \cdot 01$	$+0 \cdot 42 \pm 0 \cdot 01$
	E_e (cm^{-1})	$-0 \cdot 18 \pm 0 \cdot 01$	$-0 \cdot 22 \pm 0 \cdot 01$
Next-nearest neighbours	$\mathscr{J}$ (cm^{-1})	$+0 \cdot 38 \pm 0 \cdot 03$	$+0 \cdot 27 \pm 0 \cdot 03$

The interactions are antiferromagnetic ($\mathscr{J}$ positive) in character, the triplet level lying above the singlet. Comparisons have been made with the magnetic behaviour (Cooke, Lazenby, et al. 1959) of the undiluted salts, which become antiferromagnetic below 3·05 and 2·15°K respectively. These transition temperatures are smaller than the Curie–Weiss constants by a factor of about $\frac{1}{10}$; this is related to the small size of the isotropic interaction between next nearest neighbours, which has been measured indirectly from the spectra of triads of three coupled ions (Harris and Owen 1965). The magnitude of the exchange interaction and its dependence on the inter-ionic distance has been discussed using a simple model by Griffiths et al. (1959), and in more detail by Judd (1959c). The problem of magnetic order in the co-operative state is considered in the papers quoted, and is discussed using spin-wave theory by Lines (1963).

The observation of pair spectra is considerably more difficult for ions whose effective spins are greater than $\frac{1}{2}$ because the spectrum of each single ion may be complicated through the presence of fine and hyperfine structure. A brief survey of some experiments is given by Owen (1961). Suppose the spectrum of a single ion i is given by the Hamiltonian

$$\mathscr{H}_i = g\beta(\mathbf{H} \cdot \mathbf{S}_i) + D_c\{S_{iz}^2 - \tfrac{1}{3}S_i(S_i+1)\} + \\ + E_c(S_{ix}^2 - S_{iy}^2) + A(\mathbf{S}_i \cdot \mathbf{I}_i) \quad (9.79)$$

with an identical Hamiltonian for ion j. The fine structure terms in D_c, E_c are labelled with subscript c, and have their origin in higher-order effects of the crystal field. Two similar ions are then coupled by an interaction of the type given by eqns (9.75), (9.77). We assume that the axes (x, y, z) for D_e, E_e are the same as those for D_c, E_c; usually this requires that the z-axis is not only the line joining the ions but also a principal axis of the ligand interaction (in other cases this assumption is equivalent to neglect of terms of the form $S_{ix}S_{iz}$, etc. in (9.79) or terms of the form $S_{ix}S_{jz}$, etc. in (9.75)).

If g, A are isotropic, and the term in $\mathscr{J}$ is much larger than any of the others, the spins $\mathbf{S}_i$, $\mathbf{S}_j$ couple to form states of total spin $\mathbf{S}$ whose values are (S_i+S_j), (S_i+S_j-1), ..., 0, in terms of which the spin Hamiltonian may be written

$$\mathscr{H} = g\beta(\mathbf{H}\cdot\mathbf{S}) + \tfrac{1}{2}\mathscr{J}\{S(S+1)-S_i(S_i+1)-S_j(S_j+1)\}+ \quad (9.80)$$
$$+D_S\{S_z^2-\tfrac{1}{3}S(S+1)\}+E_S(S_x^2-S_y^2)+ \quad (9.81)$$
$$+\tfrac{1}{2}A\{\mathbf{S}\cdot(\mathbf{I}_i+\mathbf{I}_j)\}. \quad (9.82)$$

Here (9.81) is the reduced value of the 'fine' structure for the state of total spin S. The values of D_S, E_S are given by

$$D_S = (3\alpha_S D_e + \beta_S D_c); \qquad E_S = (\alpha_S E_e + \beta_S E_c) \quad (9.83)$$

where

$$\left.\begin{aligned}\alpha_S &= \tfrac{1}{2}\left\{\frac{S(S+1)+4S_i(S_i+1)}{(2S-1)(2S+3)}\right\}, \\ \beta_S &= \frac{3S(S+1)-3-4S_i(S_i+1)}{(2S-1)(2S+3)}.\end{aligned}\right\} \quad (9.84)$$

The hyperfine constant in eqn (9.82) is halved because each ion is identical (see end of § 9.5). The overall width of the hyperfine structure is unaltered, but the components have maximum intensity in the middle and minimum at the ends of the hyperfine structure as in Fig. 9.4.

The major term (9.80) represents a set of states of total spin S whose energies follow the Landé interval rule. If $\mathscr{J}$ is positive (antiferromagnetic), the state $S = 0$ is lowest, with $S = 1$ higher in energy by $\mathscr{J}$, and $S = 2$ higher than the latter by energy $2\mathscr{J}$, etc. The Zeeman interaction splits each state of total spin S into $2S+1$ substates, the separation of consecutive substates being just $g\beta H$, independent of S. However the spectrum of a state of total spin S may be identified (a) by its fine structure (resulting from (9.81)) of $2S$ transitions allowed by the selection rule $\Delta S_z = \pm 1$, which is valid so long as $g\beta H \gg D_S$, E_S; (b) by the variation of intensity with temperature. If $\mathscr{J}$ is positive, a state of total spin S is higher in energy than the ground state $S = 0$

by an amount $W_S = \frac{1}{2}\mathscr{J}S(S+1)$, and the population of each of its substates is proportional to the quantity

$$I_S = \frac{\exp(-W_S/kT)}{\sum_S (2S+1)\exp(-W_S/kT)} \quad (9.85)$$

which (apart from temperature-independent factors) gives the temperature variation of the intensity of a transition within a given manifold S relative to that for a normal paramagnetic ion with no excited states. The form of the function I_S is shown in Fig. 9.10 for the case

$$S_i = S_j = \tfrac{5}{2}$$

and a simple isotropic exchange interaction for which $W_S = \frac{1}{2}\mathscr{J}S(S+1)$.

As mentioned in § 9.3, extra terms may occur in the exchange interaction for values of S_i, S_j greater than $\frac{1}{2}$. If we include a simple biquadratic term, making the exchange Hamiltonian

$$\mathscr{H}_{\text{ex}} = \mathscr{J}(\mathbf{S}_i \cdot \mathbf{S}_j) + \mathscr{J}^{(2)}(\mathbf{S}_i \cdot \mathbf{S}_j)^2, \quad (9.86)$$

the separation between successive manifolds of total spin S, $S-1$ becomes

$$W_S - W_{S-1} = \mathscr{J}S + \mathscr{J}^{(2)}S\{S^2 - S_i(S_i+1) - S_j(S_j+1)\}, \quad (9.87)$$

showing that the Landé interval rule is no longer obeyed. If the energy

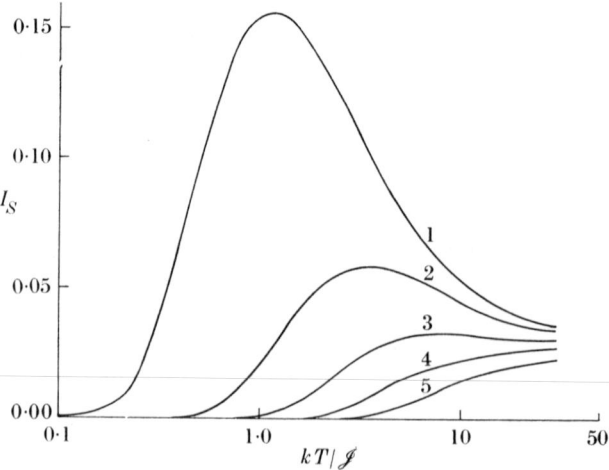

FIG. 9.10. The relative intensity, as measured by the function I_S (eqn (9.85)), of transitions for a pair of spins $S_i = S_j = \frac{5}{2}$, coupled by an exchange interaction $\mathscr{J}(\mathbf{S}_i \cdot \mathbf{S}_j)$. The numbered curves give the intensity for the different manifolds of total spin $S = 1, 2, 3, 4$, and 5, whose energy is $W_S = \frac{1}{2}\mathscr{J}S(S+1)$ above the ground state $S = 0$, plotted against the reduced temperature $(kT/\mathscr{J})$. (After Owen 1961.)

of more than one excited state can be found experimentally, the values of both $\mathscr{J}$ and $\mathscr{J}^{(2)}$ are determined. Work on Mn^{2+} pairs in MgO suggests that $\mathscr{J}^{(2)}$ is about a few per cent of $\mathscr{J}$, though the biquadratic term may be partly due to exchange-induced distortion of the lattice (Harris and Owen 1963; Rodbell, Jacobs, Owen, and Harris 1963).

9.11. Temperature-dependent effects

Apart from the excited states in pair spectra, the possibility of temperature-dependent effects in spin-spin interaction has not so far been considered. The Hamiltonian for spin-spin interaction does not in itself contain any explicit temperature-dependence, and the width and shape of lines resulting from spin-spin interaction might therefore be expected to be constant. This is true over a wide temperature range in most substances, but in this section we discuss two effects that modify this result, one at high temperatures, the other at low temperatures.

As pointed out in § 9.9, the full effects of spin-spin interaction in line broadening appear only if the local field due to neighbouring ions persists for a time long compared with a characteristic time τ_2 whose reciprocal is immediately related to the line width due to spin-spin interaction only when the resonance line is 'homogeneously broadened'. If the line width is predominantly due to 'inhomogeneous broadening', it is better to regard τ_2 as representing the average duration of the wave train emitted or absorbed by the spin system. If the local field is due to another species of ions whose spin-lattice relaxation time is τ_1, the local field will persist only for a time of order τ_1, and over longer periods the local field will be 'averaged out'. This is a form of motional narrowing which can be observed in a compound that contains two paramagnetic species, one with an abnormally short value of τ_1, the other with a long value of τ_2 (and hence also of τ_1), such as neodymium ethylsulphate containing a small fraction of gadolinium ions (Bleaney, Elliott, and Scovil 1951). At a temperature of 90°K, no resonance due to the Nd^{3+} ions was observed because of broadening due to the very small value of τ_1 ($< 10^{-11}$ s), while the Gd^{3+} resonance was observed with a line width substantially the same as in a diamagnetic host lattice such as lanthanum ethylsulphate. At 20°K, where the value of τ_1 for the Nd^{3+} ions is at least 10^{-8} s, the full effect of their magnetic dipolar field is observed in the Gd^{3+} spectrum. The contrast in resolution is shown in Fig. 9.11. The effective spin-lattice relaxation time for the Gd^{3+} ions was found to be much shorter than in a diamagnetic host owing to coupling with the rapidly fluctuating

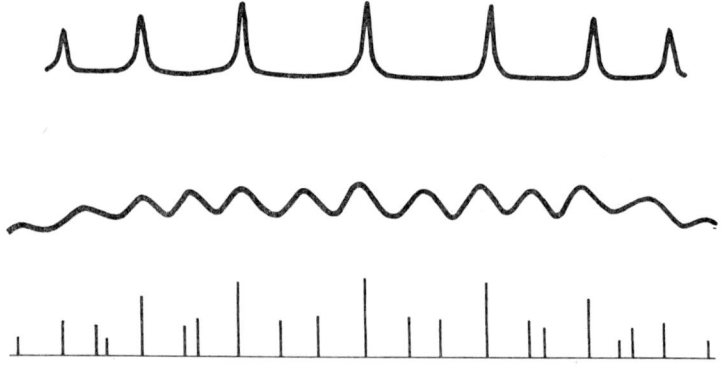

FIG. 9.11. The spectrum of Gd^{3+}, $4f^7$, $^8S_{\frac{7}{2}}$ present in low concentration in a single crystal of neodymium ethylsulphate. Above, spectrum at 90°K, showing absence of broadening, owing to rapid relaxation of the Nd^{3+} ions. Centre, spectrum at 20°K, broadened by dipolar field of Nd^{3+} ions. Below, triplet structure on each Gd^{3+} line, calculated assuming dipolar interaction with nearest pair of Nd^{3+} ions (cf. Fig. 9.9). The external magnetic field is parallel to the c-axis of the crystal, and the two nearest Nd^{3+} ions lie along this axis. (After Bleaney, Elliott, and Scovil 1951.)

fields produced by the Nd^{3+} ions, which provides a path by which energy can be rapidly transferred to the lattice. As in the case of nuclear magnetic resonance in a liquid, the narrowing effect for the Gd^{3+} ions is most pronounced the shorter the spin-lattice relaxation time of the neodymium ions, while the reduction of τ_1 for the Gd^{3+} ions should be most pronounced when τ_1^{-1} for the Nd^{3+} ions is of the same order as the resonant frequency of the Gd^{3+} ions (i.e. when the Fourier components at the Gd^{3+} resonance frequency of the fluctuating local field due to the Nd^{3+} ions have their maximum amplitude).

True motional narrowing requires much shorter correlation times for electron than for nuclear magnetic resonance, and (apart from conduction electron spin resonance) is correspondingly less important. In liquids it is associated with some unusual effects on the hyperfine structure (see, for example, McConnell (1956); McGarvey (1956)); in ionic solids it occurs only at temperatures where rapid ionic diffusion becomes possible.

The second (low temperature) effect that we now discuss does not depend on the intervention of another phenomenon such as spin-lattice relaxation, but is a consequence simply of the changing Boltzmann population of the different magnetic levels. In considering broadening due to spin-spin interaction we have so far assumed that all orientations of neighbouring dipoles are equally probable. This is of course an approximation that is not seriously in error at high temperatures such that $kT \gg g\beta H$, but as the temperature falls and we approach the region where kT is of the same order as $g\beta H$ the population of the spin-orientation states of higher energy becomes substantially less than that of the low-energy spin orientations. If (as we are assuming) the energy difference is mainly due to the Zeeman energy in an external field, the population shift is from the orientations where the magnetic dipoles are antiparallel to the field (if the gyromagnetic ratio is negative, as for free electrons, this means the spin orientations parallel to the field) into the dipole parallel (spin antiparallel) positions. For convenience we refer to this as a shift from the 'up' positions to the 'down' positions. When $kT \ll g\beta H$, the spins are almost entirely in the 'down' position and the static magnetization of the sample approaches its saturation value.

We now consider the local field at a given ion. If the neighbouring spins are predominantly in the 'up' position, the local field that they set up at the given ion will be (say) positive, while if they are predominantly 'down' the local field will be negative. The line broadening arises from the random statistical variation in the chances of any given neighbour being 'up' or 'down', but if on average these chances are equal the centre of gravity of the line will not be shifted from the resonance position for a single ion. If the chances are unequal, one wing of the broadened line will have higher intensity and the other lower intensity (see Fig. 9.12), so that the centre of gravity will be shifted. The effect becomes progressively greater in proportion to the amount by which the mean expectation value of 'down' spin predominates over 'up' spin. The size of the shift is determined by the mean value of the local field, which is proportional to the mean magnetization of the spin system(s) responsible for the local field.

As the temperature is reduced so that kT becomes much smaller than $g\beta H$, the spins fall more and more into the 'down' state and the local field approaches its maximum value corresponding to the saturation magnetization of the paramagnetic system. At the same time the shift in the centre of gravity of the resonance line approaches its maximum value. This is accompanied by a decrease in the randomness of spin orientation (decrease in entropy) so that the resonance line becomes less broad and, ideally, when all spins are completely aligned, the width should approach zero. However this assumes that the local field (and hence the line shift) is the same in all parts of the sample, and, because of demagnetizing fields, this is true only for certain sample shapes such as a sphere or ellipsoid.

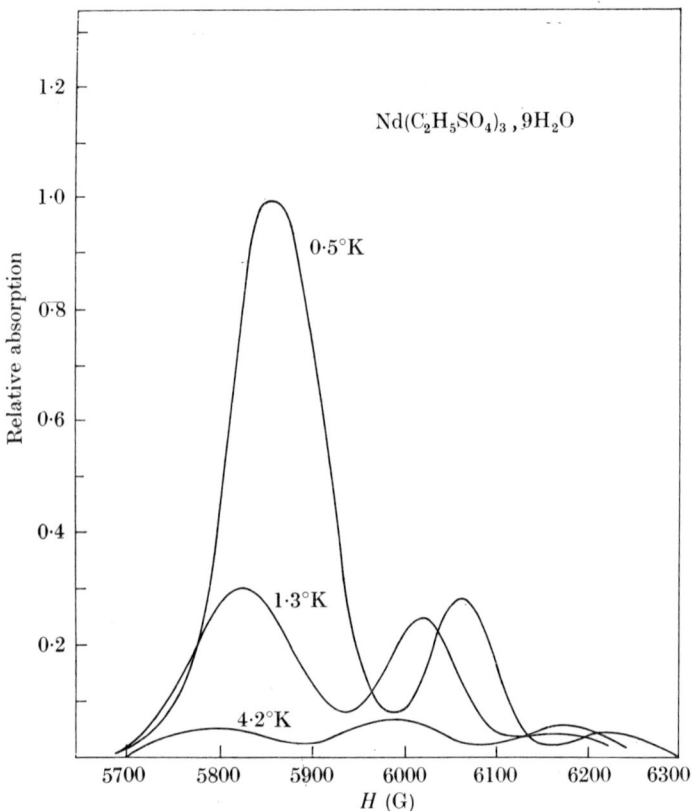

Fig. 9.12. Resonance absorption at 30 GHz in neodymium ethylsulphate at very low temperatures (magnetic field parallel to hexagonal axis). As the temperature is lowered the absorption strength increases and shifts progressively to one side. The spin-spin interaction is mainly due to the dipolar field of nearest neighbours along the hexagonal axis; when these dipoles are aligned in the external field their local field adds to the external field, so that at very low temperatures the intensity is concentrated in the line on the low field side when observed at constant frequency. (After Svare and Seidel 1964.)

For a system of spins $S = \tfrac{1}{2}$ the nature of the temperature dependence proves to be rather simple. If the Zeeman Hamiltonian for a magnetic field parallel to the z-axis is

$$\mathscr{H} = g\beta H S_z$$

the mean expectation value of S_z is

$$\langle S_z \rangle = -\tfrac{1}{2}\tanh(g\beta H/2kT). \tag{9.88}$$

The shift in the centre of gravity of the resonance line with temperature is measured by the first moment

$$\langle \Delta \nu \rangle = \langle \nu \rangle - \nu_0, \tag{9.89}$$

where ν_0 is the centre frequency of the unshifted line. At any given temperature we should expect the shift $\langle\Delta\nu\rangle$ to be proportional to the value of $\langle S_z\rangle$. From the method of moments the actual relations are found to be

$$h\langle\Delta\nu\rangle = -\tfrac{1}{2}\tanh\left(\frac{g\beta H}{2kT}\right)\sum_j(3\mathscr{J}'_{zz}/2)_{ij} \qquad (9.90)$$

if all the spins are identical, and

$$h\langle\Delta\nu\rangle = -\tfrac{1}{2}\tanh\left(\frac{g'\beta H}{2kT}\right)\sum_{j'}(\mathscr{J}_{zz})_{ij'} \qquad (9.91)$$

for interaction with a system of non-identical spins $S' = \tfrac{1}{2}$ denoted by j', whose g-factor for a field along the z-axis is g'.

In calculating the second moment we are interested in its value relative to the centre of gravity $(\nu_0+\langle\Delta\nu\rangle)$ of the shifted line, i.e. in the quantity

$$\langle\Delta\nu^2\rangle_c = \langle\{\nu-(\nu_0+\langle\Delta\nu\rangle)\}^2\rangle$$
$$= \langle\Delta\nu^2\rangle-(\langle\Delta\nu\rangle)^2 \qquad (9.92)$$

rather than the second moment $\langle\Delta\nu^2\rangle$ measured relative to the frequency ν_0 of the unshifted line. In the high temperature limit assumed in §§ 9.7–9.10 the value of $\langle\Delta\nu\rangle$ goes to zero, and the distinction between $\langle\Delta\nu^2\rangle_c$ and $\langle\Delta\nu^2\rangle$ is unnecessary. This is no longer true at low temperatures. It turns out that for a set of identical spins $S = \tfrac{1}{2}$ the formula for the second moment measured from the centre of the shifted line is

$$h^2\langle\Delta\nu^2\rangle_c = \tfrac{1}{4}\left\{1-\tanh^2\left(\frac{g\beta H}{2kT}\right)\right\}\sum_j(3\mathscr{J}'_{zz}/2)^2_{ij}, \qquad (9.93)$$

while for a spin i surrounded by a set of non-identical spins j', also with $S' = \tfrac{1}{2}$, we have

$$h^2\langle\Delta\nu^2\rangle_c = \tfrac{1}{4}\left\{1-\tanh^2\left(\frac{g'\beta H}{2kT}\right)\right\}\sum_{j'}(\mathscr{J}_{zz})^2_{ij'}. \qquad (9.94)$$

The temperature-dependent factors in curly brackets in (9.93), (9.94) reduce to unity in the high temperature limit (in agreement with eqn (9.57), specialized to $S = S' = \tfrac{1}{2}$) and vanish as $T \to 0$.

Experimental determinations of line shift and shape at very low temperatures have been made by Svare and Seidel (1964), who also discuss the effects of demagnetizing fields and non-uniform precession within the sample (magnetostatic modes). The classification into 'identical' and 'non-identical' spins is not straightforward when the resonance frequencies are shifted by appreciably different amounts through interaction with neighbouring spins, as in Fig. 9.12. When $S > \frac{1}{2}$, there may be the additional complication of crystal field splittings. The theory is discussed in some detail by McMillan and Opechowski (1960, 1961), and compared with experiment by Svare and Seidel (1964).

10
SPIN-PHONON INTERACTION

IN this chapter we shall be primarily concerned with the mechanisms by which energy is exchanged between the paramagnetic ions (or spin system) and thermal reservoirs such as electromagnetic radiation, the lattice vibrations (or phonon radiation) within the paramagnetic crystal, and the surroundings of the paramagnetic crystal, such as a bath of liquid helium. These mechanisms are fairly well established; they give a good qualitative, and sometimes a reasonably quantitative explanation of the way in which the spin-lattice relaxation time varies, particularly with temperature. Some related effects, such as the phonon bottle-neck and phonon avalanche that result from the comparatively small heat capacity at low temperatures of the phonons resonant with the spins, are outlined in the last section.

The spin-phonon interaction also affects the parameters in the static spin Hamiltonian, giving rise to shifts in the g-value, the fine structure splitting, and the hyperfine interaction. The phonons may also induce a 'spin-spin' interaction between ions, as mentioned in § 9.4. Such effects arise at 0°K from the zero-point vibrations of the lattice, and though they increase at finite temperatures, they are often rather small and we shall not discuss them further. In addition we shall be mainly concerned with fairly dilute paramagnetic substances, thus avoiding the complexities that arise in more concentrated compounds from the interplay of spin-spin interaction and spin-lattice relaxation. The excellent review by Stevens (1967) covers similar ground; its different approach strongly recommends it as companion reading.

10.1. The attainment of thermal equilibrium

If we have a system whose temperature is not the same as that of its surroundings, it cannot be in thermal equilibrium with these surroundings, and provided there is thermal contact between the two, their temperatures will approach a common value at a rate determined by the heat capacities of the system and the surroundings together with the rate at which energy can be transferred from one to the other. If the surroundings have a heat capacity that is immensely greater than the heat capacity of the system, the temperature of the latter will approach the temperature of the surroundings at a rate determined only by the heat capacity of the system and the rate of heat transfer.

We start by assuming that the surroundings are essentially infinite in their heat capacity, and that they are in thermal equilibrium with a bath of electromagnetic radiation whose density ρ_{em} is that appropriate to a temperature T_0, which is thereby defined as the temperature of the surroundings.

For an atomic system, energy transfer occurs through transitions between the atomic energy levels, such transitions being accompanied by the emission and absorption of quanta of electromagnetic radiation. For simplicity we consider first a system of just two energy levels a, b in which a is the lower level and b the upper level, for which the difference in energy is

$$W_b - W_a = \hbar\omega. \tag{10.1}$$

A simple example would be a spin system $S = \tfrac{1}{2}$ which is subjected to an external field that separates the two levels by an energy $\hbar\omega$. Let the populations of the lower and upper levels respectively be n_a and n_b at any instant. Then owing to interaction with the radiation bath, transitions will be taking place between the two levels in either direction, the number leaving the lower state being proportional to n_a and the number leaving the upper state being proportional to n_b. Thus the populations of the two levels obey the differential equation

$$-\frac{dn_a}{dt} = \frac{dn_b}{dt} = w_\uparrow n_a - w_\downarrow n_b, \tag{10.2}$$

where $w_\uparrow$, $w_\downarrow$ are the rates at which ions make transitions from the lower to the upper state, and vice versa. If the spin system is in thermal equilibrium with a radiation bath at temperature T_0, the time differentials are zero and the populations of the two states will have their equilibrium values N_a, N_b. Thus

$$0 = w_\uparrow N_a - w_\downarrow N_b \tag{10.3}$$

and hence

$$N_a/N_b = w_\downarrow/w_\uparrow. \tag{10.4}$$

The transition rates are given by the Einstein coefficients of absorption and emission

$$w_\uparrow = B\rho_{\text{em}}, \tag{10.5}$$

$$w_\downarrow = A + B\rho_{\text{em}} = B\rho_{\text{em}} \exp(\hbar\omega/kT_0), \tag{10.6}$$

where ρ_{em} is the radiation density, B the coefficient of stimulated emission or absorption, and A the coefficient of spontaneous emission. Hence we have

$$N_a/N_b = (A + B\rho_{\text{em}})/(B\rho_{\text{em}}) = \exp(\hbar\omega/kT_0), \tag{10.7}$$

showing that the transition rates must differ just by the factor required to maintain the ratio of the populations in accordance with Boltzmann's law.

We return now to the case where the spin system is not in equilibrium with the radiation bath. By a simple algebraic manipulation we have from (10.2)

$$\begin{aligned}\frac{\mathrm{d}(n_a-n_b)}{\mathrm{d}t} &= -2(w_\uparrow n_a - w_\downarrow n_b) \\ &= (n_a+n_b)(w_\downarrow - w_\uparrow) - (w_\downarrow + w_\uparrow)(n_a - n_b) \\ &= (w_\downarrow + w_\uparrow)\{(N_a - N_b) - (n_a - n_b)\},\end{aligned} \quad (10.8)$$

where we have used the fact that the total number of spins N is constant; i.e.,

$$n_a + n_b = N_a + N_b = N \quad (10.9)$$

and the relation

$$(w_\downarrow - w_\uparrow)/(w_\downarrow + w_\uparrow) = (N_a - N_b)/(N_a + N_b)$$

from (10.4). The solution of (10.8) is

$$(n_a - n_b) = (N_a - N_b) + \{(n_a - n_b)_0 - (N_a - N_b)\}\exp(-t/\tau_1), \quad (10.10)$$

where $(n_a - n_b)_0$ is the population difference at $t=0$, and the time constant τ_1 is given by the relation

$$\frac{1}{\tau_1} = (w_\downarrow + w_\uparrow). \quad (10.11)$$

On substituting from (10.5), (10.6) this becomes

$$\frac{1}{\tau_1} = A + 2B\rho_{\text{em}} = B\rho_{\text{em}}\{\exp(\hbar\omega/kT_0)+1\}. \quad (10.12)$$

If now we insert the electromagnetic (photon) radiation density

$$\rho_{\text{em}}\,\mathrm{d}\omega = \frac{\hbar\omega^3}{\pi^2 c^3}\frac{\mathrm{d}\omega}{\exp(\hbar\omega/kT_0)-1} \quad (10.13)$$

we have, using the well known relation between the coefficients A, B

$$\frac{1}{\tau_1} = \frac{\hbar\omega^3}{\pi^2 c^3} B \coth\left(\frac{\hbar\omega}{2kT_0}\right) = A \coth\left(\frac{\hbar\omega}{2kT_0}\right). \quad (10.14)$$

Equation (10.11) gives a quite general relation for the relaxation time for a two-level system. If $\hbar\omega \ll kT_0$, so that the difference between $w_\downarrow$ and $w_\uparrow$ can be neglected, we have simply $1/\tau_1 = 2w$. Further, since the coefficient B is not temperature-dependent, we find from (10.14) that in this region τ_1^{-1} varies as T_0, corresponding to the classical limit where stimulated emission and absorption predominate, and the rate at which transitions occur at a given frequency is just proportional to the radiation energy density at this frequency, which varies as kT_0. In the opposite extreme where $\hbar\omega \gg kT_0$, τ_1 becomes independent of temperature, being determined entirely by A, the rate of spontaneous emission from the upper state, since in this limit $\coth(\hbar\omega/2kT_0)$ tends to 1, and $A \gg B\rho_{\rm em}$ in eqn (10.12).

In ordinary laboratory optical spectroscopy we have generally $\hbar\omega \gg kT_0$, though this may be far from true in high temperature plasmas. In the radio-frequency region, we have generally $\hbar\omega \ll kT_0$. In centimetre/millimetre wave-spectroscopy at liquid-helium temperatures, we are just in the transition region, since $(\hbar\omega/kT_0) = 1$ at $T_0 = 1\cdot438°$K for a frequency $(\omega/2\pi) = 30$ GHz corresponding to a wavelength of 1 cm. Hence in electron paramagnetic resonance we shall frequently need the more general expression for the dependence on temperature of the relaxation time.

We now identify our two-level system with an assembly of magnetic ions for each of which $S = \tfrac{1}{2}$, subjected to a field H that splits the levels by an amount $\hbar\omega = g\beta H$. Then the magnetization is given by

$$M = \tfrac{1}{2}g\beta(n_a - n_b), \qquad M_0 = \tfrac{1}{2}g\beta(N_a - N_b), \qquad (10.15)$$

where M_0 is the thermal equilibrium value corresponding to temperature T_0. Substituting in (10.8) and using (10.11) we obtain

$$\frac{dM}{dt} = \frac{1}{\tau_1}(M_0 - M), \qquad (10.16)$$

which is identical with eqn (2.31) if H is along the z-axis. On this basis we recognize τ_1 as the 'spin-lattice relaxation time' introduced into the classical equations of motion to describe the way in which the magnetization returns from a non-equilibrium value to the thermal equilibrium value. Our treatment indicates that it is a special case of a general process involving the flow of energy between an assembly of atoms and ions with discrete energy levels and a thermal reservoir (the radiation field). In magnetic terms the flow of energy arises as follows. If the magnetization M in a constant field H changes by an amount dM, the magnetic energy of the system changes by $-H\,dM$; this magnetic energy is associated with the spin system, which we regard as possessing internal energy that may change through a transfer to or from another thermal reservoir.

SPIN-PHONON INTERACTION 545

The concept of thermal equilibrium involves that of temperature. For our system of ions with two levels separated in energy by $\hbar\omega$, we can define a temperature for the spin system known for brevity as the 'spin temperature' T_s by means of the relation

$$n_a/n_b = \exp(\hbar\omega/kT_s), \qquad (10.17)$$

which is analogous to eqn (10.7). Since the magnetization of such an assembly of N magnetic ions of spin $S = \tfrac{1}{2}$ is

$$M = \tfrac{1}{2}Ng\beta \tanh(\hbar\omega/2kT_s)$$

we can write eqn (10.16) in the form

$$\frac{d}{dt}\{\tanh(\hbar\omega/2kT_s)\} = \frac{1}{\tau_1}\{\tanh(\hbar\omega/2kT_0) - \tanh(\hbar\omega/2kT_s)\}. \qquad (10.18a)$$

Using eqn (10.14) this becomes

$$\frac{d}{dt}\{\tanh(\hbar\omega/2kT_s)\} = A\left\{1 - \frac{\tanh(\hbar\omega/2kT_s)}{\tanh(\hbar\omega/2kT_0)}\right\}. \qquad (10.18b)$$

In the high-temperature region where $(\hbar\omega/kT) \ll 1$, Curie's law holds, so that M, M_0 vary as T_s^{-1}, T_0^{-1} respectively. Using the nomenclature $\beta_s = (1/kT_s)$, $\beta_0 = (1/kT_0)$, we can write (10.18a) in the approximate form

$$\frac{d\beta_s}{dt} = \frac{1}{\tau_1}(\beta_0 - \beta_s). \qquad (10.19)$$

In the classical thermodynamical approach to spin-lattice relaxation, a specific heat C_H is associated with the spin system at constant field, and it is assumed that the rate of energy transfer between spin system and lattice is proportional to the difference in their temperatures. Then $-C_H(dT_s/dt) = \alpha(T_s - T_0)$, where α is a 'coefficient of thermal contact', or

$$dT_s/dt = (\alpha/C_H)(T_0 - T_s).$$

We can relate this to the magnetization using Curie's law and then, using eqn (10.16), we obtain

$$\frac{dT_s}{dt} = \left(\frac{dM}{dt}\right) \bigg/ \left(\frac{dM}{dT_s}\right) - \frac{1}{\tau_1}\left(\frac{T_s}{T_0}\right)(T_0 - T_s).$$

Comparison of the two equations for (dT_s/dt) shows that we can write

$$\tau_1 = C_H/\alpha \quad \text{(cf. Gorter 1947, Cooke 1950)}$$

only if $|T_0-T_s| \ll T_0$; this is a more stringent condition than the approximation that leads to eqn (10.19), and reflects the fact that both C_H and α are temperature-dependent.

For a system of more than two levels, a spin temperature can be defined provided that the ratio of the populations of *any* two levels is given by (10.17), and the condition for this is that the spin system must be in internal equilibrium. Such internal equilibrium is reached through the spin-spin interaction in a time τ_2, and if $\tau_2 \ll \tau_1$ (as is frequently the case for a magnetically concentrated system) the concept of a spin temperature is valid in dealing with the relatively slow process of spin-lattice relaxation. In a magnetically dilute system, on the other hand, there will be $(2S+1)$ rate equations describing the way in which the populations of the $(2S+1)$ levels change with time, whose solutions are not necessarily describable in terms of a single spin-lattice relaxation time.

We now return to the simple two-level system to obtain an estimate of the relaxation time τ_1 associated with transitions induced by the electromagnetic radiation field. In eqn (10.14) we need the value of B, to obtain which we make use of the standard formula for the induced transition probability rate

$$w_{ij} = \frac{2\pi}{\hbar^2} |\langle i| \mathscr{H}'_0 |j\rangle|^2 f(\omega), \qquad (10.20)$$

where $\langle i| \mathscr{H}'_0 |j\rangle$ is the matrix element of the perturbing Hamiltonian $\mathscr{H}' = \mathscr{H}'_0 \exp(\pm i\omega t)$, and $f(\omega)$ is the line shape function introduced in § 2.9. For a linearly polarized electromagnetic field of amplitude H_1, we have from eqns (2.56), (2.57), for the transition $M \leftrightarrow M-1$,

$$|\mathscr{H}'_0|^2 = (\tfrac{1}{4}H_1)^2(l_1^2+m_1^2)\langle M-1| \mu_- |M\rangle^2$$

and since the mean values of l_1^2, m_1^2 for random polarization are each $(\tfrac{1}{3})$, and $(H_1^2/8\pi) = \rho_{\text{em}}\,\mathrm{d}\omega$, while for a transition between the two states of a doublet $\langle M-1| \mu_- |M\rangle = \gamma\hbar$, we find

$$w = \left(\frac{\pi}{12}\right)\gamma^2 H_1^2 f(\omega) = (2\pi^2/3)\gamma^2 \rho_{\text{em}} f(\omega)\,\mathrm{d}\omega. \qquad (10.21)$$

If we assume that the line width is so small that we can neglect the variation in the energy density ρ_{em} over the width of the line, then by integrating over the line shape and using the relation $\int f(\omega)\,\mathrm{d}\omega = 1$ we have

$$B = (2\pi^2/3)\gamma^2 \qquad (10.22)$$

and hence

$$\frac{1}{\tau_1} = \left(\frac{2\gamma^2}{3}\right)\left(\frac{\hbar\omega^3}{c^3}\right)\coth\left(\frac{\hbar\omega}{2kT_0}\right) \tag{10.23}$$

for interaction with the electromagnetic radiation field. For single electron spins ($J = \frac{1}{2}$, $g = 2$) the value of B is $(2\pi^2/3)(e/mc)^2$, and at a frequency of 3×10^{10} Hz and a temperature of $1°K$ the value of τ_1 given by eqn (10.23) is about 10^{10} s. In a solid paramagnetic substance, however, equilibrium is established not through interaction with the thermal electromagnetic radiation field but through interaction with the lattice vibrations. These may be regarded as giving a 'phonon radiation field' in the solid whose energy density, apart from the complication arising from the presence of longitudinal and transverse phonons, is higher at a given frequency than that given by eqn (10.13) by a factor of order $(c/v)^3 \sim 10^{15}$, where v is the velocity of sound. Thus unless the coefficient B for stimulated emission and absorption of phonons by the magnetic ions is less than 10^{-15} of that for photons, the former process will dominate the latter.

The very long value of τ_1 obtained above is due to the small radiation density at the resonant frequency. The latter means also that the electromagnetic radiation field has a very small heat capacity compared with the spin system, so that we ought to take into account the rate at which heat flows from the surroundings of high heat capacity into the electromagnetic radiation bath. There is no point in pursuing this here, but in the next section we shall see that even for the phonon radiation bath, with its enormously higher energy density, its heat capacity may still be small compared with that of the spin system. The consequences of this are discussed in § 10.6.

10.2. The phonon radiation bath

The energy density in the phonon radiation bath can be written down quite simply by analogy with that in the photon radiation bath, remembering that as well as two transverse polarizations we have in addition a longitudinal wave motion. If the transverse and longitudinal wave velocities are v_t, v_l we have

$$\rho_{\text{ph}}\,d\omega = \frac{\hbar\omega^3}{2\pi^2}\left(\frac{2}{v_t^3} + \frac{1}{v_l^3}\right)\frac{d\omega}{\exp(\hbar\omega/kT_{\text{ph}})-1}. \tag{10.24}$$

Since we shall only be making order of magnitude calculations we shall ignore the difference between v_t and v_l (as well as any dispersion in the wave velocity), obtaining

$$\rho_{\text{ph}}\,d\omega = \frac{3\hbar\omega^3}{2\pi^2 v^3}\frac{d\omega}{\exp(\hbar\omega/kT_{\text{ph}})-1} \qquad (10.25)$$

for the phonon energy density in the frequency range ω to $\omega+d\omega$. It is readily seen that this is the product of the quantity

$$\Sigma = \frac{3\omega^2}{2\pi^2 v^3}\,d\omega, \qquad (10.26)$$

which is the number of phonon modes per unit volume in the frequency range just quoted, and $\bar{p}$, the phonon 'occupation number' given by

$$\bar{p} = \frac{1}{\exp(\hbar\omega/kT_{\text{ph}})-1}, \qquad (10.27)$$

multiplied by the energy quantum $(\hbar\omega)$, since the quantity $\bar{p}\,(\hbar\omega)$ is just the average energy per mode. In these equations we have introduced a 'phonon temperature' T_{ph} (see § 10.6), which for a given phonon frequency is essentially defined by eqn (10.27).

The processes by which transitions can be induced in the spin system require the presence of oscillatory electromagnetic fields at the resonance frequency of the spins. The phonons can give rise to such oscillatory electromagnetic fields through the motion with respect to the paramagnetic ion of other ions carrying either electric charges (or electric dipole moments) or magnetic dipole moments. This relative motion is caused by the lattice vibrations, and is determined by the lattice strain ϵ, which we must therefore relate to the phonon energy density. For a vibrational mode with wave-vector $q = \omega/v$, the matrix elements for phonon absorption and emission can be related to a strain ϵ_q, where for a crystal of unit volume and density ρ,

$$\epsilon_q = q\left(\frac{\hbar}{2\rho\omega}\right)^{\frac{1}{2}} p^{\frac{1}{2}}$$

for absorption of a lattice quantum that reduces the vibrational quantum number from p to $(p-1)$, and

$$\epsilon'_q = q\left(\frac{\hbar}{2\rho\omega}\right)^{\frac{1}{2}}(p+1)^{\frac{1}{2}}$$

for emission of a quantum raising the vibrational quantum number from p to $p+1$. On averaging over all values of p, and summing over all vibrational modes in unit volume in the appropriate frequency range, we have, since $q = \omega/v$,

$$\epsilon^2 = \left(\frac{\omega}{v}\right)^2 \left(\frac{\hbar}{2\rho\omega}\right) \bar{p} \Sigma \qquad (10.28)$$

for phonon absorption, while the corresponding value for phonon emission is greater by a factor (see eqn (10.27))

$$(\bar{p}+1)/\bar{p} = \exp(\hbar\omega/kT_{\text{ph}}), \qquad (10.29)$$

which is just the ratio of emission to absorption rates given in eqn (10.7). Hence from (10.28)

$$2\rho v^2 \epsilon^2 = (\hbar\omega)\bar{p} \sum = \rho_{\text{ph}} \, d\omega \qquad (10.30)$$

where ρ is the crystal density.

The phonon radiation bath differs from the electromagnetic radiation bath in that it is limited by the crystal boundaries, and for an isolated crystal *in vacuo* phonon radiation energy cannot readily flow into or out of the crystal, unlike electromagnetic radiation. This makes it easy to compare the heat capacities of the two systems, the assembly of spins and the phonon radiation. Retaining for simplicity a simple two-level system, such as a spin $S = \tfrac{1}{2}$, whose two levels have energy difference $(\hbar\omega)$, we find that the specific heat per spin is, at constant field H (since $\hbar\omega = g\beta H$), and at temperature T,

$$C_H/k = \left(\frac{\hbar\omega}{2kT}\right)^2 \text{sech}^2\left(\frac{\hbar\omega}{2kT}\right), \qquad (10.31a)$$

while the specific heat per vibrational mode at the same frequency is

$$C_{\text{ph}}/k = \left(\frac{\hbar\omega}{2kT}\right)^2 \text{cosech}^2\left(\frac{\hbar\omega}{2kT}\right). \qquad (10.31b)$$

The two functions are plotted in Fig. 10.1 as a function of the reduced parameter $(1/x)$, where $x = (\hbar\omega/kT)$. For high values of $(1/x)$, or high temperatures, C_{ph}/k approaches unity while C_H/k falls as T^{-2}. For low values of $(1/x)$, or low temperatures, the two specific heats become identical since we can neglect the population of all but the lowest two vibrational levels, and both vanish exponentially as $T \to 0$. Hence, except when $(\hbar\omega/kT) \ll 1$, there is no great difference between the specific heat per spin and per vibrational mode.

The situation is very different when we include the relative numbers of spins and of vibrational modes. For an undiluted tutton salt, whose molar volume is about 200 cm³, the number of paramagnetic ions per

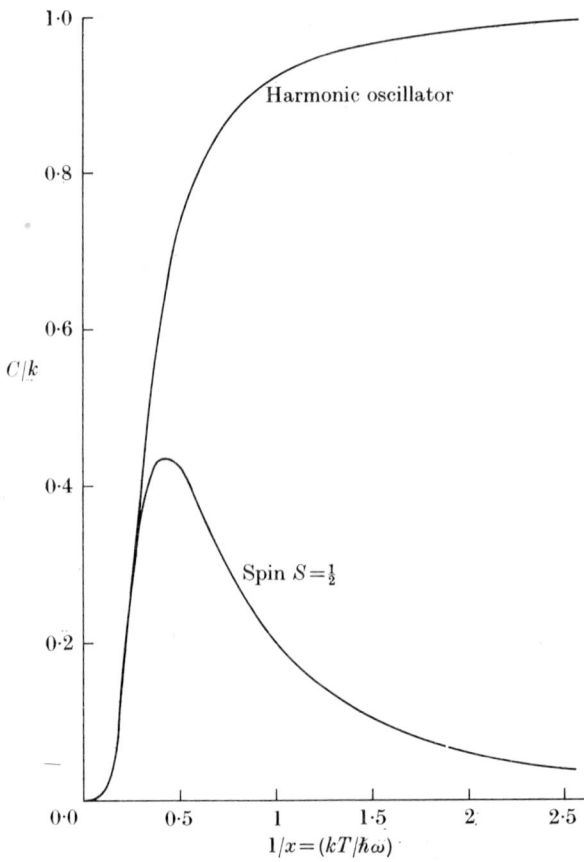

Fig. 10.1. Specific heats of a spin $S = \frac{1}{2}$ with splitting $\hbar\omega$ and of a harmonic oscillator of the same frequency.

cm³ is about 3×10^{21} per cm³. For the number of vibrational modes given by eqn (10.26) which are in resonance with the spins, we need to know the value of $d\omega$, which is roughly equal to the width of the resonance line. If we take a line width $\Delta\nu = 0.1$ cm⁻¹, corresponding to a spread in field ≈ 1 kG in the resonance line, we have $d\omega = 2\pi(3 \times 10^9)$, and for a resonance frequency of 1 cm⁻¹ we find $\sum = 4 \times 10^{15}$ per cm³. Even if we take a well-diluted salt such as a tutton salt in which only 1 in 10^4 of the dipositive ions are paramagnetic, we still have 3×10^{17} spins per cm³, while with a line width of say 10 G (due to the nuclear moments of the protons), we have some 4×10^{13} vibrational modes per cm³. Hence at temperatures in the region of a few °K the heat capacity of the spin system will far outweigh that of the vibrational modes in resonance with the spins, a point we shall return to in § 10.6. In the immediately following sections we shall ignore this point, and calculate the spin-lattice relaxation time on the assumption of a constant phonon temperature T_0. Clearly this assumption is only realistic if the resonant phonons are in close thermal contact with a third system of high thermal capacity at the temperature T_0 which we shall refer to as the 'bath temperature'.

In the estimate of $\sum$ given above we have used the convenient value of 3×10^3 ms^{-1} for the phonon velocity v. Few measurements of v exist for crystals of the type normally used in paramagnetic resonance experiments: for LaCl$_3$ an estimate of 2×10^3 ms^{-1} has been obtained by Mikkelson and Stapleton (1963), while Shiren (1962) has measured a velocity of $9 \cdot 25 \times 10^3$ ms^{-1} for longitudinal waves of frequency 10 GHz along the $\langle 100 \rangle$ axis of MgO. (The velocity for transverse waves is only $6 \cdot 5 \times 10^3$ ms^{-1}.) For simplicity we shall continue to use the value 3×10^3 ms^{-1} for later calculations in this chapter. For acoustic waves of frequency 30 GHz (which corresponds to an electromagnetic wavelength of 1 cm), the phonon wavelength is $\lambda \approx 10^{-5}$ cm $= 10^2$ nm, which is large compared with inter-ionic distances in a crystal. This justifies the approximation that ϵ_q is proportional to $(2\pi/\lambda) = q = (\omega/v)$ which appears as the first factor in equations for the strain such as (10.28). At liquid-helium temperatures phonons of wavelengths near the inter-ionic distance (i.e. phonons near the upper frequency limit corresponding to the Debye temperature) are no longer excited, and this simplifies a number of integrals over all phonon modes and frequencies that we shall encounter later. These are commonly of the form

$$I_n = \int_0^\infty \frac{x^n e^x}{(e^x - 1)^2} \, \mathrm{d}x \approx I'_n = \int_0^\infty x^n e^{-x} \, \mathrm{d}x = n! \, . \quad (10.32)$$

For example the exact value of $I_4 = 4\pi^4/15 = 26 \cdot 0$ while $I'_4 = 4! = 24$. The fractional error is smaller for larger values of n.

10.3. Spin-lattice relaxation by phonons—Waller processes

The relaxation process involves the emission or absorption of a quantum by the spin system, and we consider now how this quantum can be absorbed or emitted by the lattice vibrations (the phonon field). A transition between two levels of the spin system can only be induced by an oscillatory electromagnetic field of the right frequency, and it follows that some mechanism is required whereby the mechanical vibrations of the lattice can produce such an oscillatory electromagnetic field. In the earliest theory, that of Waller (1932), the mechanism involved was modulation of the spin-spin interaction by the lattice waves; that is, the local magnetic field, which exists at one ion because of the magnetic dipole on a neighbouring ion, fluctuates because the distance between the two ions fluctuates under the action of the lattice vibrations.

It might be thought that modulation of the exchange interaction, which is strongly dependent on distance, would provide a mechanism for spin-lattice relaxation. This is not so for isotropic exchange, since a change in its size merely alters the singlet-triplet splitting in Fig. 9.2 and does not introduce any new matrix elements (in quantum-mechanical language, the isotropic exchange operator and the Zeeman operators commute). Anisotropic exchange, whose operator is formally similar to that for magnetic dipole interaction, does provide such a mechanism (for a discussion, see Harris and Yngvesson (1968)).

The magnetic field of one dipole μ at a neighbouring ion, distance r, is of order μ/r^3. As a result of a lattice strain fluctuating at angular frequency ω, we have $r = r_0(1+\epsilon \cos \omega t)$ and

$$r^{-3} = r_0^{-3}(1-3\epsilon \cos \omega t + \ldots). \tag{10.33}$$

An oscillating magnetic field of frequency ω is thus set up whose mean square value is

$$H_1^2 = (3\epsilon)^2 H_i^2 \tag{10.34}$$

where H_i^2 is the mean square value of the static local field, given by

$$H_i^2 = 2\mu^2 \sum r_0^{-6} = \alpha n^2 \mu^2 \tag{10.35}$$

where α is a constant somewhat greater than unity and n is the number of magnetic ions of moment μ per unit volume. Using eqn (10.30) we have

$$H_1^2 = \frac{9H_i^2}{2\rho v^2} \rho_{\text{ph}} \, d\omega, \tag{10.36}$$

whereas the equivalent value for electromagnetic radiation would be $H_1^2 = 8\pi \rho_{\text{em}} \, d\omega$. The Waller mechanism is therefore faster by a factor

$$\left(\frac{9H_i^2}{16\pi \rho v^2}\right)\left(\frac{3c^3}{2v^3}\right)$$

so that instead of eqn (10.23) we have

$$\frac{1}{\tau_1} = \frac{9H_i^2\gamma^2\hbar\omega^3}{16\pi\rho v^5} \coth\left(\frac{\hbar\omega}{2kT_0}\right) \quad (10.37)$$

$$= \left(\frac{9\alpha}{16\pi}\right)\left(\frac{\gamma^2 n^2\mu^2\hbar}{\rho v^5}\right)\omega^3 \coth\left(\frac{\hbar\omega}{2kT_0}\right). \quad (10.38)$$

To obtain an order of magnitude estimate we take the numerical factor $(9\alpha/16\pi)$ as unity, $\rho = 2$ g-cm^{-3} and $n = 3 \times 10^{21}$ cm^{-3} (which is typical of an undiluted double sulphate of the iron group). Then with $\mu = 1$ Bohr magneton, the value of τ_1 due to this mechanism at a frequency of 30 GHz and a temperature of 1°K would be about 10^3 s. This value, though much less than that found in § 10.1, is still too high; it would take about 20 min for the heat of magnetization to be transferred to the lattice in a field of 10 kG at 1°K. In highly dilute salts the value of τ_1 would be much longer because of the smaller values of H_i^2, which decrease linearly with the concentration for random dilution; in practice no concentration dependence of τ_1 of this order is found.

The processes considered so far are all 'direct' processes, in which a single quantum is exchanged between the spin system and the radiation field. They result in long relaxation times because the energy density in the radiation field at the resonance frequency is very small. In the electromagnetic case the scattering of a non-resonant photon, combined with a transition in the spin system, is a higher-order process that can be neglected. It was pointed out by Waller that the analogous process for phonons, in which we consider one phonon to be absorbed and another emitted, can be very important. This mechanism can be discussed in very simple terms by considering the modulation of the internal field. Suppose that two strains are present of different frequencies, which we write as $\epsilon_1 \cos \omega_1 t$ and $\epsilon_2 \cos \omega_2 t$. Then in the expansion of r^{-3} we have

$$r^{-3} = r_0^{-3}(1 + \epsilon_1 \cos \omega_1 t + \epsilon_2 \cos \omega_2 t)^{-3}$$
$$= r_0^{-3}[1 - 3\epsilon_1 \cos \omega_1 t - 3\epsilon_2 \cos \omega_2 t$$
$$+ 3\epsilon_1\epsilon_2\{\cos(\omega_1 - \omega_2)t + \cos(\omega_1 + \omega_2)t\}\ldots], \quad (10.39)$$

from which it is apparent that in second order there will be fluctuations at the frequencies $(\omega_1 - \omega_2)$ and $(\omega_1 + \omega_2)$. This means that a fluctuating magnetic field at the frequency ω which is required to cause transitions within the spin system will be set up by the combined effect of the strains at frequencies ω_1 and ω_2 provided that $\omega = \omega_1 \pm \omega_2$. In practice (see below) we shall be concerned with the difference frequency, and we can regard the process as consisting of the absorption of a phonon of energy $\hbar\omega_1$ by the spin system combined with the emission of a phonon of energy $\hbar\omega_2$, the difference in energy $\hbar\omega$ being supplied by the spin system, so that in the final state energy is conserved (see Fig. 10.2). This process is analogous to the Raman effect in optical

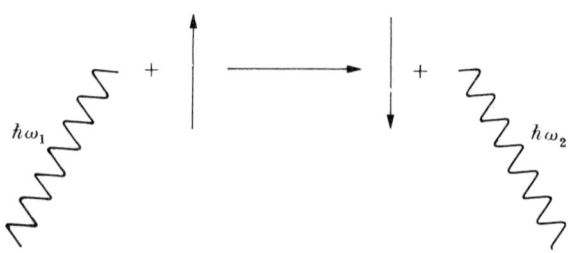

Fig. 10.2. Schematic representation of a 'Raman process', in which a 'spin-flip' occurs through the scattering of a phonon with a change in the phonon frequency. It may be regarded as the virtual absorption of a phonon of energy $\hbar\omega_1$ and the emission of a phonon of energy $\hbar\omega_2$, energy being conserved provided that $\hbar\omega_2 - \hbar\omega_1 = \hbar\omega$, where $\hbar\omega$ is the change in the energy of the spin. Obviously, if the latter is positive, $\hbar\omega_2 < \hbar\omega_1$, and vice versa.

spectroscopy, and is often referred to as a 'Raman process'. Its importance as a mechanism for spin-lattice relaxation can be seen from a plot of the energy density in the phonon system, Fig. 10.3. In the direct process only phonons of frequency ω can take part, and these are very few in number. In the Raman process any two phonons can take part provided that their frequency difference is equal to the resonance frequency of the spin system, so that the much more abundant phonons near the peak of the energy distribution are also available. At high temperatures this more than compensates for the fact that a second-order process involves a much weaker coupling mechanism than a first-order process.

It is obvious from Fig. 10.3 that the number of phonons for which $\omega_1 + \omega_2 = \omega$ is negligible compared with those for which $\omega_1 - \omega_2 = \omega$, and this is the reason why we disregard contributions from the former.

From the expansion of r^{-3} we find that the mean square magnetic field fluctuating at the frequency $\omega = \omega_1 - \omega_2$ is

$$H_1^2 = (3\epsilon_1\epsilon_2)^2 H_i^2 = 9\alpha n^2 \mu^2 \epsilon_1^2 \epsilon_2^2, \tag{10.40}$$

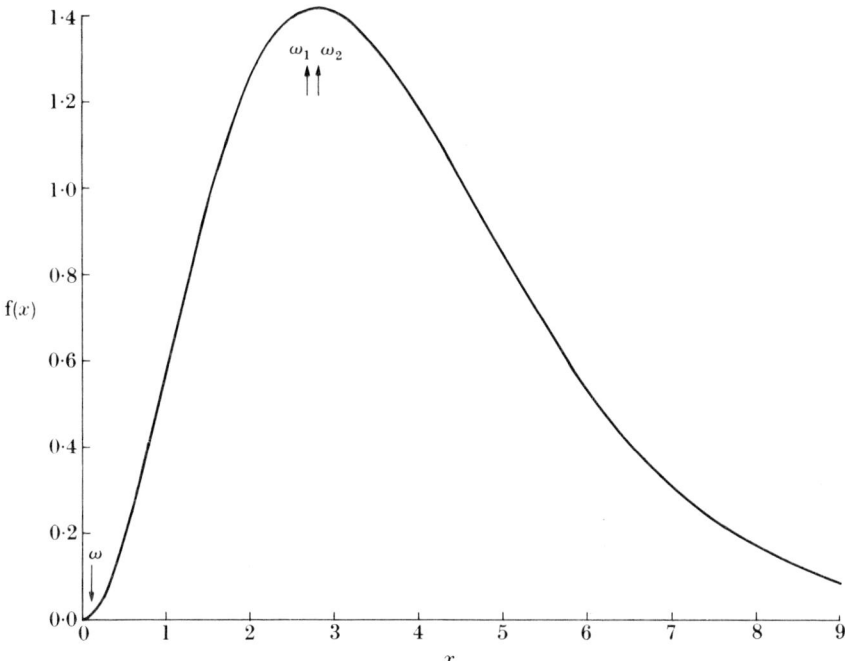

FIG. 10.3. Plot of $f(x) = x^3/(e^x-1)$ showing the energy density of phonons as a function of $x = \hbar\omega/kT_0$. The arrows indicate the energy density of phonons available for direct processes (ω), at $(\hbar\omega/kT_0) = 0\cdot 1$, compared with the densities at (ω_1, ω_2) available for Raman processes, which may be anywhere in the spectrum.

and from eqn (10.21) the rate at which transitions occur in the spin system is

$$w = \frac{\pi}{12}\gamma^2 H_1^2 f(\omega).$$

In relating ϵ_1^2, ϵ_2^2 to the energy density in the phonon field we must remember that ϵ_2 refers to a quantum being emitted by the spin system into the phonon field, and ϵ_1 to a quantum absorbed by the spin system from the phonon field so that (cf. eqn (10.30))

$$\left.\begin{aligned}\rho v^2 \epsilon_1^2 &= \tfrac{1}{2}\rho_1\, d\omega_1 \\ \rho v^2 \epsilon_2^2 &= \tfrac{1}{2}\rho_2\, d\omega_2 \exp(\hbar\omega_2/kT_0)\end{aligned}\right\} \quad (10.41)$$

so that

$$w = \frac{3\pi a n^2 \mu^2 \gamma^2}{16\rho^2 v^4}\iint f(\omega)\rho_1\rho_2 \exp(\hbar\omega_2/kT_0)\, d\omega_1\, d\omega_2. \quad (10.42)$$

In evaluating the double integral we must remember that the two integrations are not independent because $|\omega_1-\omega_2| = \omega$; and since in general $\omega \ll \omega_1, \omega_2$, we can replace the double integral above by

$$\int \rho_{ph}^2 \exp(\hbar\omega_l/kT_0)\, d\omega_l \int f(\omega)\, d\omega$$

$$= \int \frac{9\hbar^2 \omega_l^6 \exp(\hbar\omega_l/kT_0)\, d\omega_l}{4\pi^4 v^6 \{\exp(\hbar\omega_l/kT_0) - 1\}^2}$$

$$= \frac{9\hbar^2}{4\pi^4 v^6}\left(\frac{kT_0}{\hbar}\right)^7 \int_0^{x_m} \frac{x^6 e^x\, dx}{(e^x - 1)^2} = \frac{9\hbar^2}{4\pi^4 v^6}\left(\frac{kT_0}{\hbar}\right)^7 I_6, \quad (10.43)$$

where ω_l denotes a phonon frequency and we have used the fact that $\int f(\omega)\, d\omega = 1$. The integral over the lattice frequencies is carried from 0 to the Debye frequency but at low temperatures where such frequencies are not excited there is no significant error in replacing the upper limit by infinity. Then the integral I_6 is approximately (see eqn (10.32))

$$\int_0^\infty x^6 e^{-x}\, dx = 6!$$

and we have

$$\frac{1}{\tau_1} = 2w = 6!\left(\frac{27\alpha}{32\pi^3}\right)\frac{\gamma^2 \hbar^2 n^2 \mu^2}{\rho^2 v^{10}}\left(\frac{kT_0}{\hbar}\right)^7. \quad (10.44)$$

Using the same parameters as before this gives

$$1/\tau_1 \sim 10^{-13} T_0^7,$$

which is insignificant at 1°K but is more important than the Waller direct process above 50°K.

The integral I_6 can also be evaluated in the high temperature limit where kT_0 is large compared with the maximum phonon energy ($\hbar\omega_m$), though this is a situation usually of less practical interest. Then since $x \ll 1$,

$$I_6 = \int_0^{x_m} \frac{x^6 e^x\, dx}{(e^x - 1)^2} \sim \int_0^{x_m} x^4\, dx = \tfrac{1}{5}x_m^5 = \tfrac{1}{5}\left(\frac{\hbar\omega_m}{kT_0}\right)^5$$

and hence

$$\frac{1}{\tau_1} = \left(\frac{27\alpha}{160\pi^3}\right)\frac{\gamma^2 n^2 \mu^2}{\rho^2 v^{10}}\omega_m^5 k^2 T_0^2, \quad (10.45)$$

where ω_m is the maximum phonon frequency. Taking the Debye temperature to be about 100°K, which is typical of many paramagnetic salts, ω_m is about 10^{13}, and using the same parameters as before we find

$$\frac{1}{\tau_1} \sim 10^{-6} T_0^2,$$

so that τ_1 would still be about 10 s at 300°K. This is in violent contrast with experimental results, which show that τ_1 varies from 10^{-6} s to less than 10^{-12} s at room temperature.

10.4. Spin-lattice relaxation by modulation of the ligand field

The relaxation processes considered in the previous section, involving modulation of the internal dipolar field, are clearly insufficiently rapid to explain the experimental results, at any rate in magnetically dilute salts. A more potent mechanism was suggested by Heitler and Teller (1936); the theory was elaborated by Kronig (1939) and Van Vleck (1940), who in a long and detailed paper obtained relaxation times of the right order of magnitude. This mechanism consists of modulation of the crystal electric field or ligand field through motion of the electrically charged ions under the action of the lattice vibrations. Obviously this produces a fluctuating *electric* field that is essentially a dynamic orbit-lattice interaction, with no direct interaction with the electron spin. However, just as a static ligand field splits the orbital states of a magnetic ion and indirectly influences the spin levels through the spin-orbit coupling, so a fluctuating ligand field can cause transitions between these levels. In considering such processes we must of course take account of restrictions on the way in which electric fields can interact with magnetic ions, the most important of these being that the matrix elements of an electric perturbation vanish between a pair of Kramers conjugate states.

A numerical calculation of the effects of lattice vibrations in distorting the crystalline electric field is faced with the same computational difficulties (or more) that arise in calculating the static crystal field (or ligand field). This problem has been side-stepped by Orbach (1961a) by expanding the crystalline electric potential V in powers of the strain:

$$V = V^{(0)} + \epsilon V^{(1)} + \epsilon^2 V^{(2)} + \ldots, \quad (10.46)$$

where the first term on the right is just the static term. The second and third terms represent the additional electric potential generated by the strain in first- and second-order respectively. To find their effect on the magnetic ion, we must include also the spin-orbit coupling, and use the weak, intermediate, or strong crystal field representation appropriate to the static interaction. On this basis we may regard (10.46) as the 'magnetic ion-lattice interaction'.

If ϵ is a fluctuating strain caused by the lattice vibrations, the second and later terms give the dynamic part of the magnetic ion-lattice interaction. A reasonable assumption (which is considered in more detail by Scott and Jeffries (1962)) is that each of the terms $V^{(n)}$ is of the same order as $V^{(0)}$, though they will of course contain terms of lower symmetry than $V^{(0)}$ because the strain reduces the local symmetry. This assumption makes it possible to obtain some order of magnitude estimates of the spin-lattice relaxation time, and once more we consider first the so-called 'direct process', in which one phonon is emitted or absorbed in a given relaxation transition.

The direct process

From eqns (10.20), (10.46) we obtain

$$w_{ij} = (2\pi/\hbar^2)\epsilon^2 |\langle i| V^{(1)} |j\rangle|^2 f(\omega) \qquad (10.47)$$

and on using (10.30) to relate ϵ^2 to the energy density of phonons, and integrating over the line shape, we find

$$B = \frac{\pi}{\hbar^2 \rho v^2} |V^{(1)}|^2 \qquad (10.48)$$

for the Einstein coefficient of induced emission or absorption of phonons. Hence, using (10.25) for the phonon energy density, we obtain

$$\frac{1}{\tau_1} = \frac{3}{2\pi\hbar\rho v^5} |V^{(1)}|^2 \omega^3 \coth\left(\frac{\hbar\omega}{2kT_0}\right). \qquad (10.49)$$

In the case of a non-Kramers ion the ground states generated by the static crystal potential may have different orbital states between which the matrix elements of $V^{(1)}$ are finite and independent of frequency, in which case (10.49) has the same frequency and temperature dependence as for photons (10.23). If the levels are degenerate in zero magnetic field we have $\hbar\omega = g\beta H$, so that

$$1/\tau_1 \text{ varies as } H^3 \coth(g\beta H/2kT_0) \qquad (10.50)$$

$$\text{or as } H^2 T_0 \text{ when } (g\beta H/kT_0) \ll 1. \qquad (10.51)$$

In the case of a Kramers doublet, because of time reversal, the matrix elements of an electric potential such as $V^{(1)}$ vanish in zero magnetic

field. However the presence of an external magnetic field produces admixtures of excited states of order $(g\beta H/\Delta)$ or $(\hbar\omega/\Delta)$, where Δ is a crystal field splitting. To this order the two states of the doublet are no longer exactly time-reversed states, and we must replace $V^{(1)}$ by roughly
$$(\hbar\omega/\Delta)V^{(1)}.$$
Hence
$$\frac{1}{\tau_1} = \frac{3\hbar}{2\pi\rho v^5} \frac{1}{\Delta^2} |V^{(1)}|^2 \omega^5 \coth\left(\frac{\hbar\omega}{2kT_0}\right). \quad (10.52)$$

For the field and temperature dependence this gives

$1/\tau_1$ varies as $H^5 \coth(g\beta H/2kT_0)$ (10.53)

or as $H^4 T_0$ when $(g\beta H/kT_0) \ll 1$. (10.54)

In the rare-earth group $|V^{(1)}|^2$ for non-Kramers ions is of order 10 to 100 (cm^{-1})^2; thus taking $3V = \Delta = 30$ cm^{-1} makes τ_1 about 10^{-3} s if the levels are split by 1 cm^{-1} at 1°K, while for Kramers doublets $(\hbar\omega/\Delta)^2$ produces an extra factor $\sim 10^{-3}$ in τ_1^{-1} and $\tau_1 \sim 1$ s. In the iron group the matrix elements are rather greater and the relaxation times correspondingly shorter.

The need to break the time-reversal symmetry of a Kramers doublet in order to produce relaxation by the process discussed above was pointed out by Van Vleck (1940). The factor that it introduces is closely related to the temperature-independent (or Van Vleck) term in the magnetic susceptibility, and like it can be quite anisotropic. It follows that for a given crystal orientation we get an ω^5 or H^5 dependence, but if measurements are made at constant frequency, and H is varied as the crystal is rotated, we cannot expect a variation in $(1/\tau_1)$ that follows a simple $H^2\omega^3$ dependence.

A magnetic hyperfine field will also be effective in breaking the time-reversal symmetry of a Kramers doublet, producing admixtures of excited electronic states that are of two kinds. First, the diagonal part of the magnetic hyperfine interaction can be combined with the Zeeman term to give $\hbar\omega = g\beta(H+H_n)$, where H_n is proportional to the nuclear magnetic quantum number m (cf. eqn (1.30) or § 1.6); thus if measurements are made at constant frequency eqn (10.52) still holds, though at low fields ($H < H_n$) the effect of H_n will be more important than that of H. Second, the off-diagonal terms in the magnetic hyperfine interaction involving $I_\pm$ produce admixtures of excited states of order (A'/Δ') which are independent of ω, so that we get a frequency dependence of τ_1^{-1} that is similar to (10.49) rather than

(10.52), but that is numerically smaller than (10.49) by a factor $\sim (A'/\Delta')^2$. Here A' is the matrix element of the magnetic hyperfine interaction between the ground doublet and an excited state at Δ' (which may not be the same as Δ). This provides a relaxation path through the forbidden transitions $\Delta m = \pm 1$, whose rate in strong fields is slower by a factor $\sim (A'/\hbar\omega)^2$, if we neglect the difference between Δ and Δ', than that for the allowed transitions $\Delta m = 0$ given by eqn (10.52). The importance of these extra relaxation paths has been demonstrated, for example, by Baker and Ford (1964), and more extensively by Larson and Jeffries (1966b).

The two-phonon Orbach process

In 1961 an important two-phonon process suggested by Orbach was verified experimentally by Finn, Orbach, and Wolf (1961) (see also Manenkov and Prokhorov (1962)). Suppose the magnetic ion has a set of energy levels such as that shown in Fig. 10.4(i), where there are two

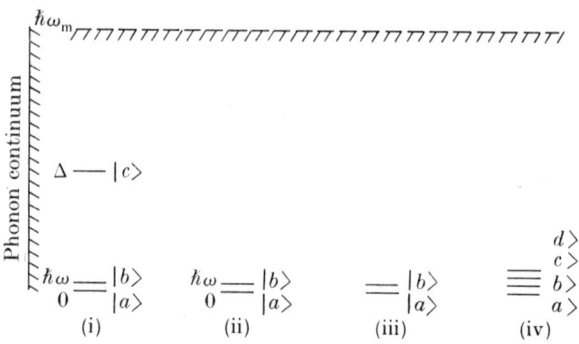

FIG. 10.4. Level schemes for spin-lattice relaxation theory. (i) Orbach process, (ii) Raman process (non-Kramers), (iii) Raman process (Kramers ion), (iv) Orbach–Blume process (ground multiplet).

low-lying states (such as a Kramers doublet) and an excited state $|c\rangle$ whose energy is less than the maximum phonon energy. It is then possible for an ion in, say, state $|b\rangle$ to absorb a phonon of the appropriate frequency by a direct process, and be excited to the state $|c\rangle$. In this state it emits a second phonon by spontaneous or induced emission and falls down to state $|a\rangle$. This gives an indirect transfer of ions from state $|b\rangle$ to state $|a\rangle$, and constitutes a relaxation process that may be faster than the direct transfer from state $|b\rangle$ to $|a\rangle$ because of the much higher density of phonons of energy Δ. If the direct process between states $|a\rangle$ and $|b\rangle$ is negligible, the rate equations become

$$dn_a/dt = -w_{ac}n_a + w_{ca}n_c,$$
$$dn_b/dt = -w_{bc}n_b + w_{cb}n_c,$$

where the transition rates now refer to transfers from the lower states to the excited state $|c\rangle$. In general, we are concerned with effects that primarily disturb the population of the states $|a\rangle$, $|b\rangle$, such as saturation of the transition $|a\rangle \leftrightarrow |b\rangle$, but which do not in first order change n_c. Hence if we neglect the difference between w_{ac} and w_{bc}, writing them simply as $w_\uparrow$, we have

$$d(n_a - n_b)/dt = -w_\uparrow(n_a - n_b)$$

from which

$$\frac{1}{\tau_1} = w_\uparrow = \frac{3}{2\pi\hbar^4 \rho v^5}|V^{(1)}|^2 \Delta^3 \frac{1}{\exp(\Delta/kT_0)-1} \quad (10.55)$$

where $V^{(1)}$ is a product of matrix elements between the states $|a\rangle$, $|b\rangle$ and the state $|c\rangle$. This gives a relaxation process that is extremely temperature-dependent since, in the limit where $\Delta \gg kT_0$,

$$1/\tau_1 \text{ varies as } \Delta^3 \exp(-\Delta/kT_0) \quad (10.56)$$

independent of an applied magnetic field except in so far as this may alter the value of Δ.

Using the same parameters as before for the rare-earth group, one finds

$$1/\tau_1 \sim 10^4 \Delta^3 \exp(-\Delta/T_0)$$

if Δ is measured in °K; the experimental results lie roughly in the range $(10^3 \text{ to } 10^5) \Delta^3 \exp(-\Delta/T_0)$.

In any such two-stage process we may expect to find more than one relaxation rate, but if the rates are materially different the controlling factor is the slower rate. In the discussion above, if $\Delta \gg kT_0$, the slower rate is that for induced phonon absorption raising the ion to state $|c\rangle$, the return to the ground state being primarily due to spontaneous phonon emission that is faster under this condition. Thus the controlling rate is $w_\uparrow$, and this is the rate that appears in (10.55).

If w_{ca}, w_{cb} are not equal we may not get a simple one-exponential relaxation rate. A more general discussion (including the case when the states $|a\rangle$, $|b\rangle$ lie higher in energy than $|c\rangle$) is given by Harris and Yngvesson (1968).

The two-phonon Raman process

We consider now processes involving two phonons which are Raman-type processes similar to those introduced by Waller, but in which the interaction mechanism is provided by the ligand field rather than the magnetic spin-spin interaction. As before, the magnetic ion makes a transition from state $|b\rangle$ to state $|a\rangle$, or vice versa, accompanied by the virtual absorption of a phonon of angular frequency ω_1 and the emission of a phonon of frequency ω_2. The condition for conservation of energy in the final state is again $\omega_2 - \omega_1 = \pm\omega$. Physically this requires that lattice distortions oscillating at frequencies ω_1, ω_2 can combine to generate a crystal potential fluctuating at frequency $\omega = |\omega_2 - \omega_1|$. Like the magnetic spin-spin interaction mechanism discussed in § 10.3, the crystal potential depends on inverse powers of the ligand distance, so that non-linear terms in ϵ^2 (and higher powers) are to be expected, as indicated in eqn (10.46).

We distinguish between two types of Raman process, which we call 'first order' and 'second order' according to the nature of the perturbation involved:

(i) a first-order Raman process arises when the crystal potential $V^{(2)}$ in (10.46) has matrix elements between the two states $|a\rangle$, $|b\rangle$ of the magnetic ion. The corresponding transition probability is then

$$w = (2\pi/\hbar^2)\,|\epsilon_1\epsilon_2 V^{(2)}|^2 f(\omega); \qquad (10.57a)$$

(ii) a second-order Raman process in which one phonon causes a virtual transition from one of the ground states to an excited state $|c\rangle$, followed by another virtual transition induced by the second phonon in which the magnetic ion returns from $|c\rangle$ to the other ground state. In this case the transition probability is

$$w = (2\pi/\hbar^2)\left|\frac{\epsilon_1 V_1^{(1)} \epsilon_2 V_2^{(1)}}{\Delta}\right|^2 f(\omega), \qquad (10.57b)$$

where $V_1^{(1)}$, $V_2^{(1)}$ are matrix elements of $V^{(1)}$ between the states $|a\rangle$, $|b\rangle$ and the excited state $|c\rangle$. The reason why only 'virtual' transitions are involved (unlike the Orbach process) is that the state $|c\rangle$ has an energy Δ that lies outside the continuum of allowed phonon frequencies, as indicated in Fig. 10.4(ii).

As before, each of these mechanisms gives transition rates that depend on the product $\epsilon_1^2 \epsilon_2^2$, where the squares of the strains can be related to the energy densities at the phonon frequencies ω_1, ω_2 by eqns (10.28)–(10.30). The calculation proceeds as in § 10.3 and, on integrating over all phonon frequencies ω_l, we obtain

$$\frac{1}{\tau_1} = 2w = \frac{9}{4\pi^3\rho^2 v^{10}}\left\{(V^{(2)})^2 + \frac{(V^{(1)})^4}{\Delta^2}\right\} \times$$

$$\int_0^{\omega_m} \frac{\omega_l^6 \exp(\hbar\omega_l/kT_0)}{\{\exp(\hbar\omega_l/kT_0)-1\}^2}\, d\omega_l$$

$$= \frac{9}{4\pi^3\rho^2 v^{10}}\left\{(V^{(2)})^2 + \frac{(V^{(1)})^4}{\Delta^2}\right\}\left(\frac{kT_0}{\hbar}\right)^7 I_6. \quad (10.58)$$

At low temperatures the integral can, as before, be carried to infinity and approximated by $I_6' = 6!$, giving

$$\frac{1}{\tau_1} = \frac{9(6!)}{4\pi^3\rho^2 v^{10}}\left\{(V^{(2)})^2 + \frac{(V^{(1)})^4}{\Delta^2}\right\}\left(\frac{kT_0}{\hbar}\right)^7. \quad (10.59)$$

In an approximation where $V^{(1)} \sim V^{(2)} \sim \Delta$, the two contributions from the first- and second-order processes are clearly comparable, and it is difficult to separate them experimentally. A rough estimate suggests that the Raman processes will be more rapid than the direct process (eqn (10.49)) by a factor of order $(10^{-7}$ to $10^{-6})T_0^6$, so that it may become the more important process above about 10°K. Equations (10.58), (10.59) are valid, however, only for non-Kramers ions, and we must consider the case of ions with Kramers degeneracy separately.

We have seen that in the direct process the matrix elements of $V^{(1)}$ vanish between two time-conjugate states of a Kramers ion, and the same is true for $V^{(2)}$. However the application of a magnetic field destroys the time-conjugate nature of the doublet to order $(g\beta H/\Delta')$ or $(\hbar\omega/\Delta')$; the first-order Raman process is therefore allowed at a reduced rate which, by a straightforward modification of (10.59), is

$$\frac{1}{\tau_1} = \frac{9(6!)}{4\pi^3\rho^2 v^{10}}\left(\frac{\hbar\omega}{\Delta'}\right)^2 (V^{(2)})^2 \left(\frac{kT_0}{\hbar}\right)^7. \quad (10.60)$$

Here we have written Δ' rather than Δ because the Zeeman interaction has a matrix element with an excited state (of energy Δ') that is not necessarily the same state as that to which the crystal field $V^{(1)}$ has a matrix element from the ground state.

In the second-order Raman process for a Kramers ion a similar difficulty arises, but in a less direct way. The energy levels of such an ion generated by the static crystal field are always degenerate in pairs, so that the excited state $|c\rangle$ is accompanied by its time-conjugate state $|d\rangle$, as in Fig. 10.4(iii). The matrix elements that occur in eqn

(10.57b) must be summed over both states $|c\rangle$ and $|d\rangle$, and are of the form

$$\frac{\langle a|\ V_2^{(1)}\ |c\rangle\langle c|\ V_1^{(1)}\ |b\rangle}{\hbar\omega_1-\Delta}+\frac{\langle a|\ V_1^{(1)}\ |d\rangle\langle d|\ V_2^{(1)}\ |b\rangle}{-\hbar\omega_2-\Delta}. \qquad (10.61)$$

The products of the matrix elements in the two numerators are equal but of opposite sign, a result often referred to as the 'Van Vleck (1940)' cancellation, and which can be shown to follow generally from time-reversal symmetry (see § 15.4). It follows that the sum of the two terms in (10.61) vanishes except for the difference in the denominators, which arises from the presence of the phonon energy $\hbar\omega_l$ with opposite signs, corresponding respectively to absorption and emission. The net result is a residuum of order

$$\frac{2\hbar\omega_l}{\Delta}\frac{V^2}{\Delta}$$

where ω_l is a lattice frequency, if $\Delta \gg (\hbar\omega_l)$.

Using this matrix element for the transition probability in the second-order Raman process we find

$$\frac{1}{\tau_1} = \frac{9\hbar^2}{\pi^3\rho^2 v^{10}}\left(\frac{V^4}{\Delta^4}\right)\int_0^{\omega_m}\frac{\omega_l^8\exp(\hbar\omega_l/kT_0)}{\{\exp(\hbar\omega_l/kT_0)-1\}^2}\,d\omega_l$$

$$= \frac{9\hbar^2}{\pi^3\rho^2 v^{10}}\left(\frac{V^4}{\Delta^4}\right)\left(\frac{kT_0}{\hbar}\right)^9 I_8, \qquad (10.62)$$

where

$$I_8 = \int_0^{x_m}\frac{x^8 e^x}{(e^x-1)^2}\,dx.$$

At low temperatures this integral can be approximated by $I_8' = 8!$, giving

$$\frac{1}{\tau_1} = \frac{(9!)\hbar^2}{\pi^3\rho^2 v^{10}}\left(\frac{V^4}{\Delta^4}\right)\left(\frac{kT_0}{\hbar}\right)^9. \qquad (10.63)$$

With the same parameters as before, a rough estimate gives

$$(1/\tau_1) \sim 10^{-5}T_0^9,$$

so that this process would predominate over the direct process for a Kramers doublet already above about 5°K. At still higher temperatures the approximations inherent in our treatment of the phonon spectrum mean that we cannot necessarily expect a simple T_0^9 law to hold (see Kiel and Mims (1967)).

The process just considered allows relaxation to take place within a Kramers doublet at a rate independent of the applied magnetic field, in contrast with the process described by (10.60) which depends on the application of a magnetic field to destroy the time-conjugate nature of the doublet states. If we set $V^{(1)} = V^{(2)} = \Delta = \Delta'$, a very rough estimate shows that (10.60) is slower than (10.63) by a factor $\sim 10^{-2}(\hbar\omega/kT_0)^2$, making it usually unimportant except at extremely low temperatures where, unlike other Raman processes, it gives an H^2 dependence of $1/\tau_1$ for a doublet whose splitting is $\hbar\omega = g\beta H$. At such temperatures the direct process will probably predominate.

Multiplet ground state

Another important case, considered by Blume and Orbach (1962), is that of a multiplet ground state; that is, the case (see Fig. 10.4(iv)) where the states $|c\rangle$, $|d\rangle$ have virtually the same energy as the states $|a\rangle$, $|b\rangle$. Then, in eqn (10.61), Δ is very small and can be neglected in comparison with the phonon energy $\hbar\omega_1$, $\hbar\omega_2$. Since the two numerators are of opposite sign, the two terms now reinforce each other instead of cancelling, giving a matrix element of order $(V^2/\hbar\omega_l)$. Instead of eqn (10.58) we have

$$\frac{1}{\tau_1} = \frac{9}{4\pi^3\rho^2v^{10}}\left(\frac{V^4}{\hbar^2}\right)\int_0^{\omega_m}\frac{\omega_l^4\exp(\hbar\omega_l/kT_0)}{\{\exp(\hbar\omega_l/kT_0)-1\}^2}\,d\omega_l$$

$$= \frac{9}{4\pi^3\rho^2v^{10}}\left(\frac{V^4}{\hbar^2}\right)\left(\frac{kT_0}{\hbar}\right)^5 I_4. \tag{10.64}$$

At low temperatures the integral can be taken to infinity, when $I_4 \approx I_4' = 4!$, and

$$\frac{1}{\tau_1} = \frac{9(3!)}{\pi^3\rho^2v^{10}}\left(\frac{V^4}{\hbar^2}\right)\left(\frac{kT_0}{\hbar}\right)^5. \tag{10.65}$$

For non-Kramers doublets Walker (1968) has described a mechanism that also gives a T_0^5 dependence and can operate between states of a multiplet that are not time-conjugate.

10.5. Summary and comparison with experiment

We now present a summary of the formulae for the spin-lattice relaxation time derived in the previous section, together with a comparison with some experimental results. We confine ourselves to situations in which the results do not appear to be affected by any phonon bottle-neck, reserving discussion of the latter to § 10.6. We consider first non-Kramers ions, then Kramers ions, followed by ions with $S > \frac{1}{2}$, all in the low-temperature region; finally the high-temperature region will be briefly discussed.

The principal experimental techniques for the measurement of τ_1 may be classified into two main types: steady-state methods and transient methods. The former includes (a) the classical low-frequency method (Gorter 1947, Cooke 1950) and (b) a resonance method in which the degree of saturation is observed as a function of oscillatory power level, but this requires knowledge of the transition probability, the line width and the oscillatory signal strength (cf. §§ 2.8, 2.9). Transient methods include (c) observation of the recovery of the paramagnetic resonance signal at low incident power level following the application of a saturating pulse, a method that has the advantages that each resolved resonance line can be studied separately, and that the presence of more than one relaxation time can be detected; (d) spin-echo methods, which have the further advantage over (c) of better sensitivity because observation of the transient recovery is not restricted to low power level; and (e) observation of the recovery of the steady-state magnetization following a saturating pulse, either by direct measurement or observation of some quantity (e.g. Faraday rotation) that is proportional to the magnetization. Methods (a) and (e) measure an average over all magnetic sub-states, while (b) is determined by the combined action of all relaxation paths. In general, (c) or (d) is the most accurate method, and can readily be used over a wide range of relaxation times. (d) is particularly sensitive and (like (b)) can be applied to very dilute samples.

Non-Kramers ions

For a resonance frequency $(\omega/2\pi)$ our formulae for the relaxation time can be summarized as

$$\left. \begin{aligned} \frac{1}{\tau_1} &= R_\mathrm{d}(\hbar\omega)^3 \coth(\hbar\omega/2kT_0) \text{ (direct process, (10.49))} + \\ &\quad + R_\mathrm{or}\,\Delta^3\{\exp(\Delta/kT_0)-1\}^{-1} \text{ (Orbach process, (10.55))} + \\ &\quad + R_\mathrm{r} T_0^7 \text{ (Raman process, (10.59))}, \end{aligned} \right\} \quad (10.66)$$

where R_d, R_or, R_r are parameters in an obvious notation, and Δ is the energy of an excited electronic state lying within the phonon energy band.

An experimental determination of the direct and Raman terms is shown in Fig. 10.5 for Tb^{3+} in yttrium ethylsulphate. This non-Kramers ion (see § 5.6) has a doublet ground state, slightly split by a crystal-field effect; the measurements shown are in zero external field, when there are two allowed electronic transitions for the hyperfine states $|m| = \frac{3}{2}$ and $\frac{1}{2}$ respectively. In these measurements $\hbar\omega \ll kT_0$, so that the first term in (10.66) can be approximated by $R_\text{d}(\hbar\omega)^2(2kT_0)$, giving the term linear in T_0. When the initial splitting of a non-Kramers

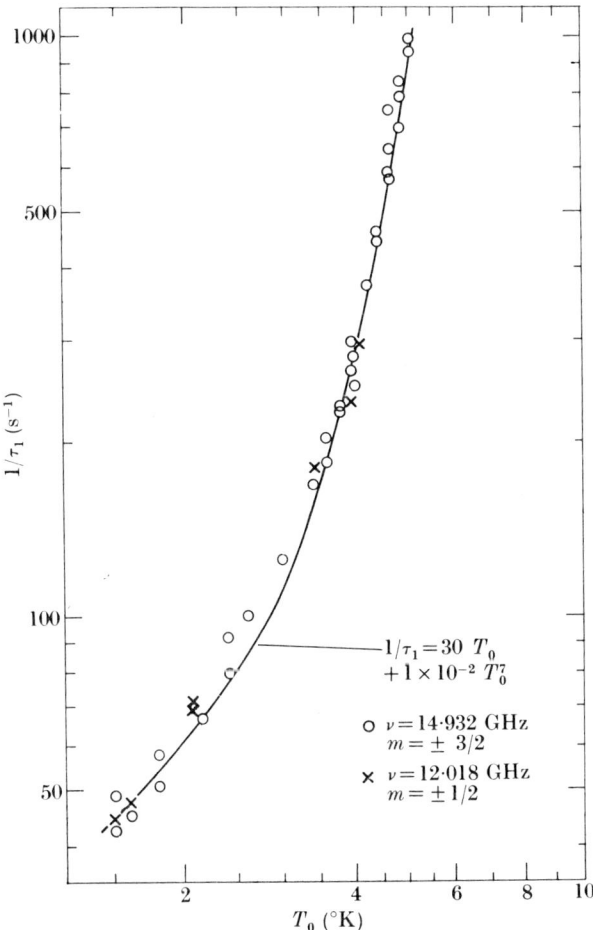

FIG. 10.5. Spin lattice relaxation rate $1/\tau_1$ as a function of temperature T_0 for 1 per cent Tb^{3+} in yttrium ethylsulphate at zero magnetic field. (After Larson and Jeffries 1966a.)

doublet is small compared with the Zeeman splitting the direct process gives $(1/\tau_1)$ varying as $H^2 T_0$.

As a very rough order of magnitude, in the rare-earth group when $\hbar\omega = g\beta H \equiv 1 \text{ cm}^{-1}$, we may expect

$$1/\tau_1 = 10^3 T_0 + (10^3 \text{ to } 10^5)\, \Delta^3 \exp(-\Delta/T_0) + (10^{-2} \text{ to } 10^2) T_0^7, \quad (10.67)$$

where τ_1 is in seconds and Δ in °K, assuming that $(T_0/\Delta) \ll 1$.

Kramers doublets

The formulae we have obtained can be summarized as follows.

$$\begin{aligned}
\frac{1}{\tau_1} = {} & R_d(\hbar\omega)^5 \coth(\hbar\omega/2kT_0) \text{ (direct process, (10.52))} + \\
& + R_{or}\, \Delta^3 \{\exp(\Delta/kT_0) - 1\}^{-1} \text{ (Orbach process, (10.55))} + \\
& + R_r T_0^9 + R_r'(\hbar\omega/k)^2 T_0^7 \text{ (Raman processes, (10.63)} \\
& \text{and (10.60)).}
\end{aligned} \quad (10.68)$$

In the region where $g\beta H = \hbar\omega \ll kT_0 \ll \Delta$, this approximates to

$$\frac{1}{\tau_1} = R_d' H^4 T_0 + R_{or}\, \Delta^3 \exp(-\Delta/kT_0) + R_r T_0^9 + R_r'(g\beta H/k)^2 T_0^7. \quad (10.69)$$

In the direct process it must be remembered that one factor $(\hbar\omega)^2$ or H^2 arises from the need to break the time-conjugate nature of the states, and that R_d, R_d' are dependent on the orientation of the magnetic field. Similar considerations would apply to R_r' (see Kiel and Mims (1967)). As a very rough order of magnitude in the rare-earth group, we have, when $g\beta H = \hbar\omega \equiv 1 \text{ cm}^{-1}$,

$$\begin{aligned}
\frac{1}{\tau_1} = {} & T_0 + (10^3 \text{ to } 10^5)\, \Delta^3 \exp(-\Delta/T_0) + (10^{-5} \text{ to } 10^{-1})\, T_0^9 + \\
& + (10^{-5} \text{ to } 10^{-1})\, T_0^7,
\end{aligned} \quad (10.70)$$

where τ_1 is in seconds and Δ is in °K.

Experimental results confirming the terms in T_0 and T_0^9 have already been presented in Fig. 1.24 for Nd^{3+} in yttrium ethylsulphate. Measurements on the same ion in lanthanum magnesium nitrate are shown in Fig. 10.6; in this case the predominant term is an Orbach term together with a direct process $\coth(\hbar\omega/2kT_0)$ measured into the region where $(\hbar\omega/2kT_0)$ is roughly 0.6. No differences are observed for samples containing 1 and 5 per cent Nd^{3+}, confirming that the relaxation times are independent of concentration in this range.

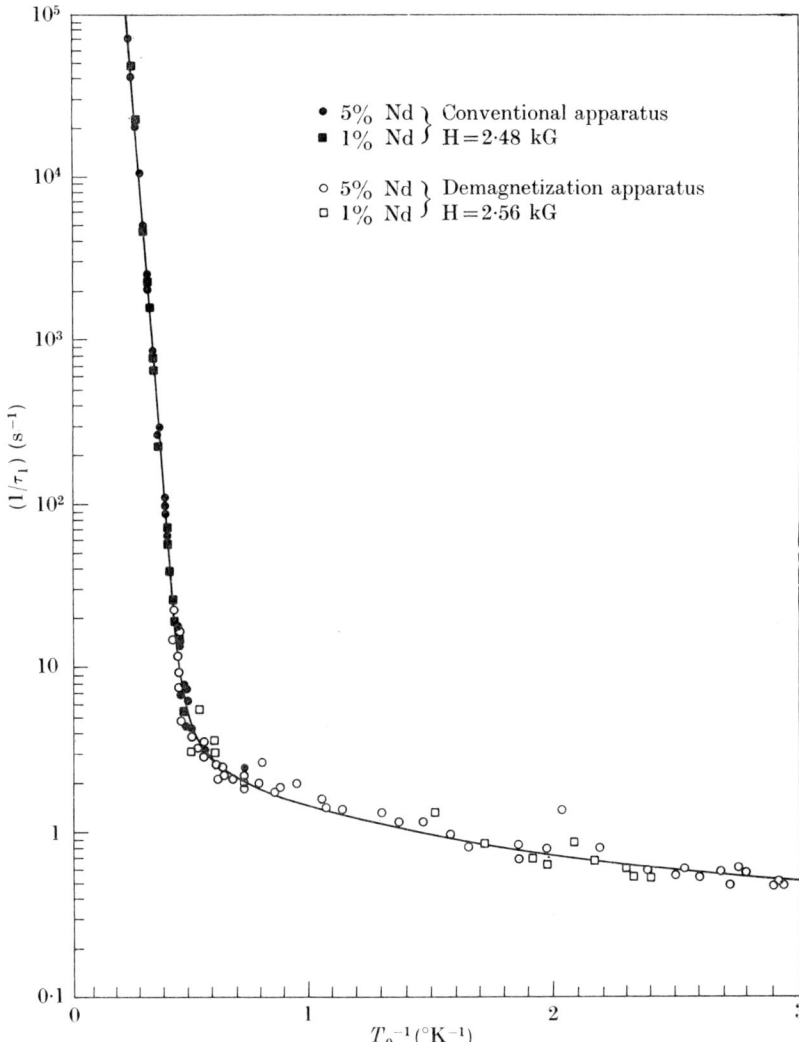

FIG. 10.6. Experimental results for the spin-lattice relaxation rate $1/\tau_1$ for Nd^{3+} in $La_2Mg_3(NO_3)_{12}, 24H_2O$ with H perpendicular to the c-axis in the temperature range from $0.3°K$ to $4.3°K$. The data show no concentration dependence, and are fitted accurately by the formula

$$1/\tau_1 = 6.3 \times 10^9 \exp(-47.6/T_0) + 0.3 \coth(\hbar\omega/2kT_0) \text{ (s}^{-1}),$$

showing that at higher temperatures relaxation is dominated by an Orbach process and at lower temperatures by the direct process. At the lowest temperatures the relaxation rate is chiefly due to the spontaneous emission of phonons. (After Ruby, Benoit, and Jeffries 1962.)

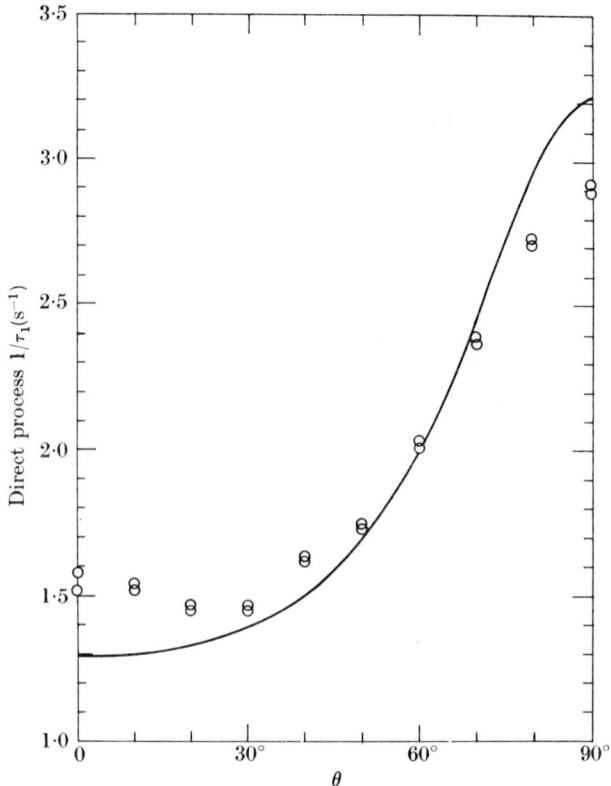

Fig. 10.7. Variation of $(1/\tau_1)$ at 9·41 GHz with the angle θ made by the external magnetic field H with the symmetry axis for 0·1 % Nd^{3+} in yttrium ethylsulphate at a temperature $T_0 = 1\cdot52°K$ where the direct process predominates. The full line represents the variation predicted by an approximate theoretical calculation. (Larson and Jeffries 1966b.)

The Nd^{3+} ion in these two substances also illustrates a number of other features of the relaxation process. Figure 10.7 shows how $(1/\tau_1)$ varies with the angle θ which the external field H makes with the crystallographic axis of symmetry at a temperature of 1·52°K where the direct process is dominant, while at 4·21°K where the Raman process is much the more important, the contribution to $(1/\tau_1)$ from the Raman process is independent of angle (Fig. 10.8). Measurements on a sample of 0·5 per cent Nd^{3+} in lanthanum magnesium nitrate, enriched to 91·3 per cent in the isotope ^{143}Nd ($I = \tfrac{7}{2}$), shown in Fig. 10.9, demonstrate the importance of extra paths involving the magnetic hyperfine interaction. These give an additional direct process

$$\frac{1}{\tau_1} = R_d^{hfs}(\hbar\omega)^3 \coth(\hbar\omega/2kT_0) \qquad (10.71)$$

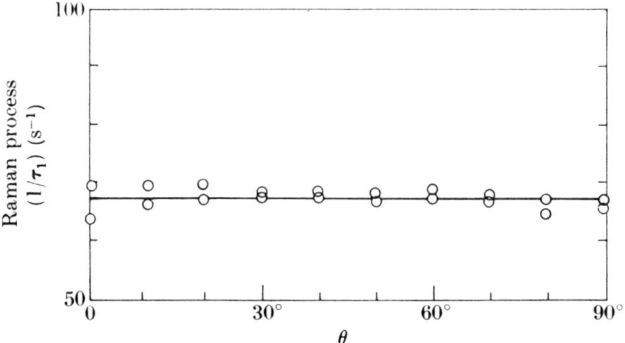

FIG. 10.8. Measurements on the same sample as in Fig. 10.7 but at a higher temperature $T = 4 \cdot 21°K$. The contribution to $(1/\tau_1)$ from the Raman process is found to be isotropic. This agrees with eqn (10.63); the Raman rate is found to vary as T_0^9 (see Fig. 1.24). (Larson and Jeffries 1966b.)

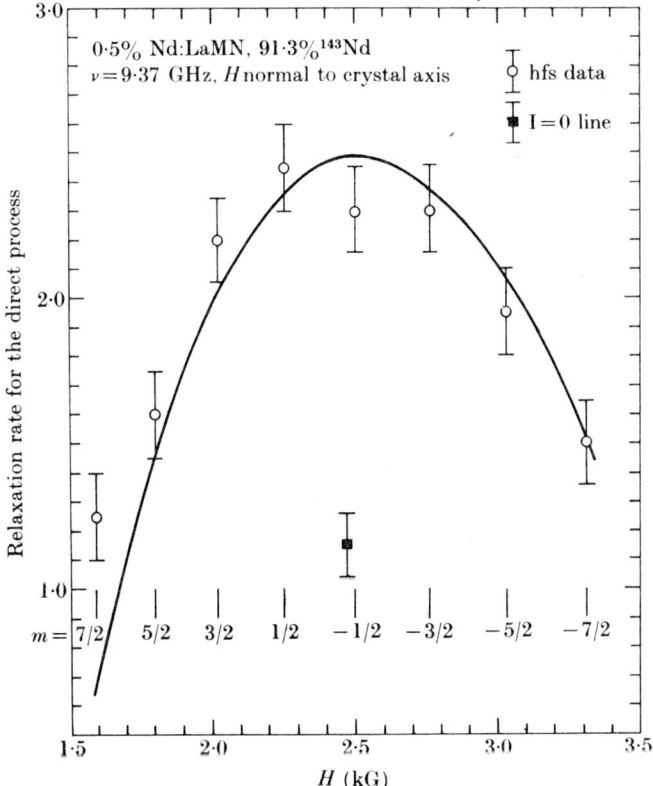

FIG. 10.9. The effect of extra paths involving the hyperfine interaction in 0·5% Nd^{3+} in lanthanum magnesium nitrate. The full line represents the empirical relation $(\tau_1 T_0)^{-1} = 0 \cdot 12\{I(I+1)-m^2\} + 1 \times 10^{-7} H^2$ for the direct process, where the first term arises from the effect of the hyperfine interaction and the second from the external field in breaking the time-conjugate nature of the doublet states. (Larson and Jeffries 1966b.)

where in this case the coefficient

$$R_d^{hfs} \text{ is proportional to } \{I(I+1)-m^2\} \tag{10.72}$$

where m is the nuclear magnetic quantum number.

Multiplets $(S > \tfrac{1}{2})$

In a system with more than two electronic levels the situation is complicated because of the possibility of a number of transitions within a given ion, not all at the same frequency. The coth formula is not appropriate, but when $(\hbar\omega/2kT_0) \ll 1$ the direct process should give $(1/\tau_1)$ varying as T_0. We have then

$$\left. \begin{array}{l} \dfrac{1}{\tau_1} = R_d T_0 \quad \text{(direct process)} \\ \quad + R_r T_0^5 \quad \text{(Raman process, eqn (10.65))} \end{array} \right\} \tag{10.73}$$

and there may also be an Orbach process if other low-lying levels are present. Equation (10.65) gives a value of order 10^{-3} s^{-1} °K^{-5} for R_r, and Chao-Yuan Huang (1965) finds for the $+\tfrac{1}{2} \leftrightarrow -\tfrac{1}{2}$ transitions of Eu$^{2+}(4f^7, {}^8S_{\tfrac{7}{2}})$ in CaF$_2$

$$1/\tau_1 = 12T_0 + 5\cdot 3 \times 10^{-4} T_0^5 \text{ (s)}^{-1}. \tag{10.74}$$

The fit to the experimental results is shown in Fig. 10.10. A T_0^5 term with a coefficient of the same order of magnitude has been found for the iso-electronic ion Gd^{3+} in CaF$_2$ by Bierig, Weber, and Warshaw (1964). A considerably faster T_0^5 process has been observed for Mn^{2+} in BaF$_2$ and SrF$_2$ by Horak and Nolle (1967).

Raman processes in the high-temperature limit

In the high temperature limit where $(kT_0) \gg \hbar\omega_m$, in which $\hbar\omega_m = k\theta_D$ and θ_D is the Debye temperature, all the Raman processes approximate in their temperature dependence to

$$\begin{aligned} \frac{1}{\tau_1} &= R_r(kT_0/\hbar)^{n+1} I_n = R_r(kT_0/\hbar)^{n+1} \int_0^{x_m} x^{n-2} \, dx \\ &= R_r(kT_0/\hbar)^{n+1} \{x_m^{n-1}/(n-1)\} \\ &= R_r' T_0^2 \end{aligned} \tag{10.75}$$

since $x_m = (\hbar\omega_m/kT_0) = (\theta_D/T_0) \ll 1$. Such a dependence on temperature just reflects the fact that two phonons are involved, each with average energy (kT_0) in the classical limit. In some exceptional cases where the vibrations of a defect are involved rather than a second phonon, a

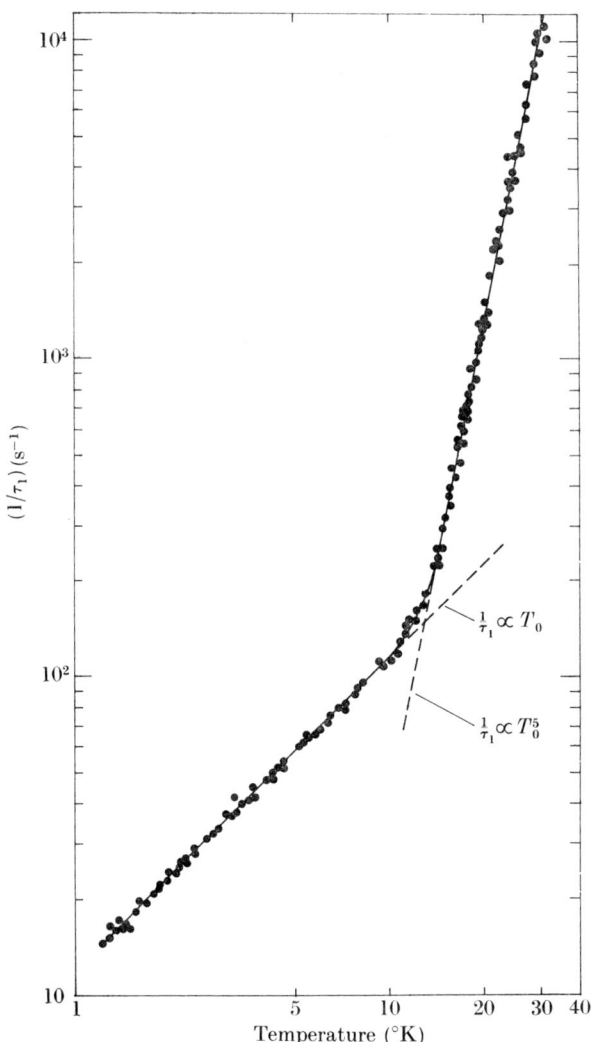

FIG. 10.10. Experimental results for $(1/\tau_1)$ as a function of temperature for Eu^{2+} in CaF_2 at 0·0047% concentration. The measurements are on the $M = +\tfrac{1}{2} \leftrightarrow -\tfrac{1}{2}$ transition of the $^8S_{\tfrac{7}{2}}$ ground state, which has a small initial splitting, at a field of 3·27 kG along a $\langle 100 \rangle$ axis and a frequency of about 10 GHz. The data for this concentration, and also for concentrations of 0·0074% and 0·18%, are well fitted by eqn (10.74).

(After Chao-Yuan Huang 1965.)

different temperature-dependence is observed (see, for example, Murphy (1966), Feldman, Castle, and Wagner (1966)).

For a few ions (such as S-state ions) where the orbital momentum is rather fully 'quenched', values of τ_1 of order 10^{-6} s at room temperature have been measured by paramagnetic relaxation experiments (Gorter

1947). For other ions τ_1 is very much shorter, and spin-lattice relaxation frequently becomes the dominant process in broadening an electron resonance line at temperatures above 100°K (or even less). Values of τ_1 can then be estimated from the line width, the shape being Lorentzian (eqn (2.65)) with $\Delta\omega = 1/\tau_1$. Šroubek and Zdansky (1966) report that a T_0^2 variation of τ_1^{-1} for Cu^{2+} in $CdWO_4$ has been verified in this way between 95 and 300°K.

10.6. The phonon 'bottle-neck' and phonon 'avalanche'

So far we have assumed that the spins relax to the phonons, and that the latter correspond to a radiation field whose temperature T_0 remains constant and equal to that of the 'bath' (for example, the bath of liquid helium in which the paramagnetic sample is immersed). This implies either that the heat capacity of the phonons is very much larger than that of the spins, so that the energy transferred from the spins to the phonons is insufficient to raise the temperature of the phonons appreciably, or that the phonons are in such intimate thermal contact with the bath that heat received from the spins is transferred instantaneously (or at any rate in a time very much less than τ_1) to the bath. We shall consider this last point first, remembering that in any steady-state experiment the temperature of the spin system will ultimately depend on the rate at which heat is transferred from it to the bath.

In a reasonably perfect crystal at liquid helium temperatures it is known from thermal conductivity data that the mean free path of the phonons is mainly determined by the crystallite size, showing that scattering of the phonons occurs principally through collisions with the boundaries. The latter gives a mean lifetime τ_{ph} for the phonons of order $L/2v$, where L is the linear dimension of the crystallites. For $L = 6$ mm, this gives a value τ_{ph} of order 10^{-6} seconds from boundary wall scattering. However there is usually a large acoustic impedance mis-match at the boundary, so that we cannot assume that every wall collision is efficient in transferring energy to the bath and restoring thermal equilibrium in the phonon system, and the time of 10^{-6} s must be regarded very much as a lower limit.

In a 'direct' process, energy is transferred from the spins only to those modes which (within the resonance line width) have the same frequency as the resonance frequency of the spins. We have seen in § 10.2 that the number of such lattice modes is very much smaller than the number of spins. For an undiluted tutton salt with some

3×10^{21} spins per cm³, the number of resonant lattice modes (even for a splitting ~ 1 cm⁻¹) is smaller by a factor of roughly 10^6. If we regard the spin-phonon relaxation (with characteristic time τ_1) as due to 'collisions' between spins and phonons, then for every collision experienced by a spin there must be some 10^6 collisions experienced by a phonon, so that the mean life time for a phonon between spin collisions is only $10^{-6}\tau_1$. This will lie in the range 10^{-6}–10^{-9} s at helium temperatures, corresponding to mean free paths for the phonons which may be much smaller than the crystallite size. This phenomenon is similar to the optical 'imprisonment of resonance radiation', and the phonons can only 'diffuse' to the walls, greatly increasing the effective phonon-bath relaxation time. If the rate at which phonons collide with spins is greater than $\Delta\omega = 1/\tau_2$, the line width assumed for the spin resonance, the phonon line width would appear to be greater than $1/\tau_2$, and it is not obvious which value we should take in estimating Σ, the number of phonons on 'speaking terms' with the spins. In fact under these circumstances the spins and phonons constitute strongly coupled systems, and we cannot correctly speak of either 'spins' or 'phonons' nor discuss their properties separately. The properties of the coupled spin-phonon system have been considered by Jacobsen and Stevens (1963), and the implications for paramagnetic relaxation discussed by Giordmaine and Nash (1965).

We now consider the transfer of energy from the spin system to the bath via the phonons as a simple classical problem in heat transfer. We represent the situation schematically in Fig. 10.11, where C_H is the heat capacity of the spin system given for $S = \tfrac{1}{2}$ by eqn (10.31a) multiplied by the number of spins N, and C_{ph} is the heat capacity of the resonant phonons given by (10.31b) multiplied by Σ, the number of phonons on speaking terms with the spins. The rates of heat transfer are characterized by time constants τ_1 for transfer from spins to

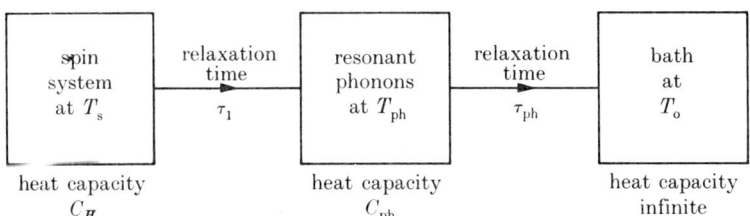

Fig. 10.11. Schematic representation of the flow of energy from the spin system through the resonant phonons to the bath.

phonons, and τ_{ph} from phonons to bath. We assume that we can speak of a spin temperature T_s and a phonon temperature T_{ph}. Then writing $T'_s = (T_s - T_0)$, $T'_{\text{ph}} = (T_{\text{ph}} - T_0)$, where T_0 is the bath temperature, we have the differential equations

$$\frac{dT'_s}{dt} = -\frac{1}{\tau_1}(T'_s - T'_{\text{ph}}), \tag{10.76}$$

$$C_{\text{ph}}\frac{dT'_{\text{ph}}}{dt} = +\frac{C_H}{\tau_1}(T'_s - T'_{\text{ph}}) - \frac{C_{\text{ph}}}{\tau_{\text{ph}}}T'_{\text{ph}}. \tag{10.77}$$

These two equations give differential equations for T'_s, T'_{ph} which are each of the form

$$\frac{d^2 T'}{dt^2} + \frac{dT'}{dt}\left\{\frac{1}{\tau_{\text{ph}}} + \frac{1}{\tau_1}\left(1 + \frac{C_H}{C_{\text{ph}}}\right)\right\} + T'\left\{\frac{1}{\tau_1 \tau_{\text{ph}}}\right\} = 0. \tag{10.78}$$

The solutions of this equation are of the form

$$T' = a_1 \exp\left(-\frac{t}{\tau'_1}\right) + a_2 \exp\left(-\frac{t}{\tau_b}\right), \tag{10.79}$$

and in the approximation where $C_H \gg C_{\text{ph}}$ we have

$$\tau'_1 = \frac{C_{\text{ph}}}{C_{\text{ph}} + C_H}\tau_1, \tag{10.80}$$

$$\tau_b = \tau_1 + \tau_{\text{ph}}\left(\frac{C_{\text{ph}} + C_H}{C_{\text{ph}}}\right). \tag{10.81}$$

The first of these time constants is very short, $\tau'_1 \approx \tau_1(C_{\text{ph}}/C_H)$, and corresponds to an initial rapid transfer of energy from the spins to the phonons, heating the latter quickly to the same temperature as the former because of their relatively small heat capacity. The second time constant τ_b corresponds to the much slower rate at which the combined systems (spins+phonons) relax to the bath, a time constant that is greater than τ_{ph} in the ratio $(C_H + C_{\text{ph}})/C_{\text{ph}}$, which is the ratio of the combined heat capacities to that of the phonons alone. For a spin system with $S = \frac{1}{2}$, we have, using eqns (10.31a, b),

$$\frac{C_H}{C_{\text{ph}}} = \frac{N}{\Sigma}\tanh^2\left(\frac{\hbar\omega}{2kT_0}\right), \tag{10.82}$$

where we have written T_0 on the assumption that the spin and phonon temperatures do not differ greatly from the bath temperature. This assumption is clearly essential for the validity of (10.76), (10.77).

A similar result is obtained from a microscopic approach on the same

SPIN-PHONON INTERACTION

lines as that used in § 10.1 for the spin system alone. The temperature $T_{\rm ph}$ of the phonon system is related to the phonon occupation number $\bar{p}$ by (10.27); for the bath temperature T_0

$$\bar{p}_0 = \frac{1}{\exp(\hbar\omega/kT_0)-1} \tag{10.83}$$

and for a two-level spin system as in § 10.1 we have in thermal equilibrium at temperature T_0

$$\frac{N_a}{N_b} = \frac{\bar{p}_0+1}{\bar{p}_0} = \exp\left(\frac{\hbar\omega}{kT_0}\right). \tag{10.84}$$

The microscopic differential equations are

$$-\frac{{\rm d}n_a}{{\rm d}t} = +\frac{{\rm d}n_b}{{\rm d}t} = B' \sum \{n_a(\bar{p})-n_b(\bar{p}+1)\}, \tag{10.85}$$

$$\frac{{\rm d}\bar{p}}{{\rm d}t} = \frac{1}{\tau_{\rm ph}}(\bar{p}_0-\bar{p})-B'\{n_a(\bar{p})-n_b(\bar{p}+1)\}, \tag{10.86}$$

where the ratio $(\bar{p}+1)/\bar{p}$ represents the ratio of the probability of emission of a phonon to that of absorption by the spin system. The constant B' is related to $(1/\tau_1)$ by

$$B' \sum (2\bar{p}_0+1) = \frac{1}{\tau_1}, \tag{10.87}$$

as can be verified by comparison with eqns (10.2) and (10.11). In terms of the parameters

$$x = \frac{n_a-n_b}{N_a-N_b}, \qquad y = \frac{\bar{p}-\bar{p}_0}{\bar{p}_0+\tfrac{1}{2}} \tag{10.88}$$

the differential equations can be written

$$\left.\begin{array}{l}\dfrac{{\rm d}x}{{\rm d}t} = \dfrac{1}{\tau_1}(1-x-xy),\\[2mm]\dfrac{{\rm d}y}{{\rm d}t} = -\dfrac{1}{\tau_{\rm ph}}y+b\dfrac{{\rm d}x}{{\rm d}t}\end{array}\right\} \tag{10.89}$$

where, with $N = N_a+N_b = n_a+n_b$,

$$b = \frac{N}{\sum (2\bar{p}_0+1)^2} = \frac{N}{\sum}\tanh^2\left(\frac{\hbar\omega}{2kT_0}\right) = \frac{C_H}{C_{\rm ph}} \tag{10.90}$$

from eqn (10.82). The solutions of these equations are discussed by Faughnan and Strandberg (1961), who plot curves for various ranges of b and τ_{ph}. When recovery is nearly complete ($x \to 1$, $y \to 0$) a simple exponential is obtained with time constant

$$\tau_{\text{b}} = \tau_1 + (1+b)\tau_{\text{ph}}$$

in agreement with eqn (10.81).

When $b\tau_{\text{ph}}$ is the largest term in this equation for τ_{b} we expect quite a different temperature and frequency dependence from the ordinary direct process. We have then to a good approximation

$$\frac{1}{\tau_{\text{b}}} = \frac{1}{\tau_{\text{ph}}}\left(\frac{3\omega^2 \, d\omega}{2\pi^2 v^3 N}\right)\coth^2\left(\frac{\hbar\omega}{2kT_0}\right). \tag{10.91}$$

An experimental result of this type is shown in Fig. 10.12. The results fit very well to a $\coth^2(\hbar\omega/2kT_0)$ dependence, corresponding to a temperature-independent value of τ_{ph} which would be expected if a relation such as $\tau_{\text{ph}} = L/2v$ is valid. Figure 10.12 shows also a concentration dependence, but R_{b} is not simply proportional to N^{-1}, presumably because $d\omega$ (and perhaps also τ_{ph}) in eqn (10.91) are concentration (or sample) dependent.

At temperatures where $(\hbar\omega/2kT_0) \ll 1$, $\coth^2(\hbar\omega/2kT_0)$ is approximately $(2kT_0/\hbar\omega)^2$, and $1/\tau_{\text{b}}$ is found to vary as T_0^2. This corresponds to a region where C_H varies as T_0^{-2}, while C_{ph} is independent of temperature, so that $(1/b)$ varies as T_0^2. Thus instead of a spin relaxation rate $(1/\tau_1)$ varying as T_0 in the direct process, we have a variation of $(1/\tau_{\text{b}})$ as T_0^2, indicating that the temperature dependence is determined just by the heat capacity of the spin system.

The difficulty in transferring energy from the spins to the bath which arises from the presence of relatively few phonons on speaking terms with the spins is known as the 'phonon bottle-neck' and was first recognized by Van Vleck (1941a,b). If instead of just heating the spin system by a saturating pulse, we go further and invert the populations of the spin levels by means of adiabatic rapid passage or a 180° high-power pulse (cf. §§ 2.5–2.7), we have an interesting situation because the phonons, unlike the spins, cannot attain a 'negative' temperature because their energy levels are not finite in number and continue to infinity. This is reflected in the different way in which the parameter y is defined in eqns (10.88); its value is zero in thermal equilibrium, and can rise to large positive values, bounded by $+\infty$; (negative values of y correspond to the phonons being *colder* than the bath). On the other hand, for a two-level spin system, $x = +1$ at thermal equilibrium, falls to zero when the populations of the two levels are equal (corresponding to $T_{\text{s}} = \infty$), and becomes negative when the level populations

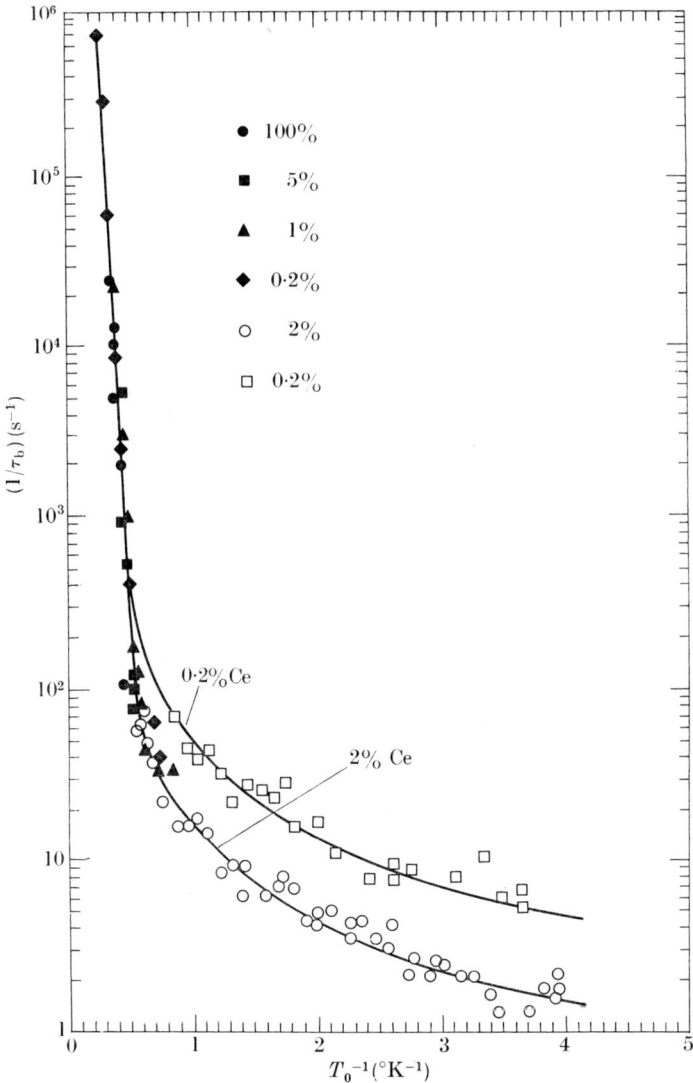

FIG. 10.12. Experimental data for the spin-bath relaxation rate $1/\tau_b$ for Ce^{3+} in $La_2Mg_3(NO_3)_{12}, 24H_2O$. The data are fitted accurately by the formula

$$1/\tau_b = 2{\cdot}7 \times 10^9 \exp(-34/T_0) + R_b \coth^2(\hbar\omega/2kT_0) \text{ (s}^{-1}),$$

where the value of R_b is concentration dependent, being 2·4 for 0·2 percent Ce and 0·8 for 2 per cent Ce. Thus the relaxation rate to the bath is dominated by an Orbach process at the higher temperatures and by the phonon bottleneck (see eqn (10.91)) at lower temperatures. (After Ruby, Benoit, and Jeffries 1962.)

are inverted. If we write $u = -x$, then exact population inversion corresponds to u changing from -1 to $+1$, which will be followed first by a fall in u back to 0 (equal populations), and then by a return through negative values to -1. In this second region the spin temperature is high but positive, so that the position is similar to that analyzed above, and we discuss now just the region where u is positive.

In terms of the parameter $u = -x$, eqns (10.89) become

$$\left.\begin{aligned}\frac{du}{dt} &= -\frac{1}{\tau_1}(1+u+uy), \\ \frac{dy}{dt} &= -\frac{1}{\tau_{ph}}y - b\frac{du}{dt}.\end{aligned}\right\} \quad (10.92)$$

These rate equations are non-linear and have been solved by Brya and Wagner (1967) for particular conditions using numerical techniques, representative curves being reproduced in Fig. 10.13. Initially u changes rather slowly, because we start with $y = 0$, and from the first equation in (10.92) the initial rate of change of u is just characterized by the normal spin-phonon relaxation time τ_1. For large values of b, which from eqn (10.90) and the earlier discussion may easily be in the region of 10^4, the second equation in (10.92) shows that y will rise very much faster than u falls, because of the relatively small heat capacity of the phonons. In turn, the rapid increase in the phonon energy density stimulates the emission of further phonons by the spin system, the deluge rapidly becoming catastrophic and resulting in a 'phonon avalanche'. This region is represented by the rapid fall in u and large rise in y shown in Fig. 10.13, the effective relaxation time in this region becoming τ_f where approximately

$$\frac{1}{\tau_f} = (bu_i)\frac{1}{\tau_1}, \quad (10.93)$$

u_i being the initial value of u. Equation (10.92) shows also that at $u = 0$, $(du/dt) = -(1+u)/\tau_1$, so that at this point again u changes at a rate given by τ_1, since there is no net stimulated emission (or absorption) of phonons from the spin system when the populations of the two spin levels are just equal. Thus the value of u changes rather slowly both initially and as u approaches 0, with a very rapid change in the 'avalanche' region. Similarly y rises slowly from zero initially, then very fast to a high value in the avalanche region, followed by a rather swift drop at a rate primarily determined by $1/\tau_{ph}$ to comparatively small values. It can be seen from Fig. 10.13 that the maximum

SPIN-PHONON INTERACTION

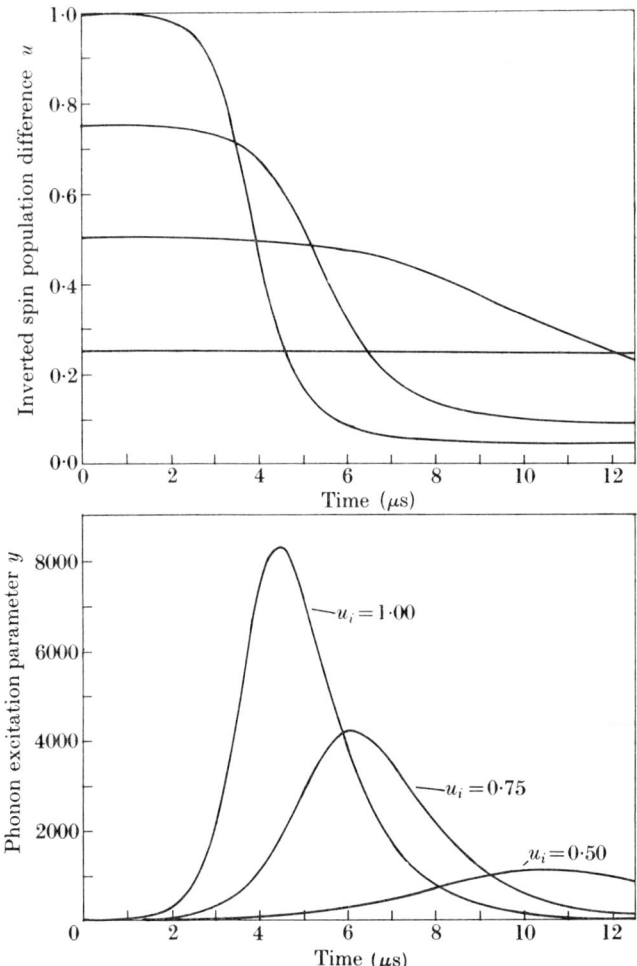

Fig. 10.13. Numerically computed solutions to eqns (10.92) for various initial values of the inverted spin population parameter u, assuming the numerical values $b = 2\cdot 5 \times 10^4$, $\tau_{ph} = 10^{-6}$ s, $\tau_1 = 10^{-9}$ s, where b is the ratio of the specific heat of the spin system to that of the resonant phonons at temperature T_0 (the bath temperature). u_i is the value of u at $t = 0$; note that the scale for y is larger than that for u by a factor of 10^4. (After Brya and Wagner 1967.)

value of y attained depends rather sharply on the value of u_i, the initial degree of inversion, and it is clear that much higher values of phonon temperature are generated in the avalanche than can be produced by saturation of the spin system, for which u_i is negative, and at most zero. The experiments of Brya and Wagner using dilute cerium magnesium nitrate give general confirmation of the theory. Similar results have been obtained for Ni^{2+} and Fe^{2+} in MgO by Shiren (1967), who has succeeded in detecting the phonons generated in an avalanche at one end of an MgO rod doped with Fe^{2+} by their effect on the spin resonance of Fe^{2+} ions (with uninverted level populations) at the other end. In this case the characteristic time constant for the avalanche was 4×10^{-8} s, in contrast with the normal spin-phonon relaxation time τ_1 of 4×10^{-4} s. The phonons generated were predominantly transverse, because the transition probability varies as v^{-5} and in MgO (§ 10.2)

$$(v_t/v_l) \sim 0{\cdot}7.$$

In this section so far we have obviously been concerned just with the 'direct process', and the phenomena of the phonon bottle-neck and avalanche depend on the very high ratio of the heat capacity of the spin system to that of the resonant phonon modes. It is possible that analogous effects might occur by means of an Orbach process, which also involves a narrow band of phonons, in this case of energy Δ, though the number of such modes will be higher by a factor $(\Delta/\hbar\omega)^2$. In the Raman process all lattice modes are involved so that the specific heat ratio will be much lower than for the direct process; however it should be noticed that a spin system $S = \tfrac{1}{2}$ in an undiluted tutton salt, with a splitting equal to 1 cm^{-1}, will have a magnetic specific

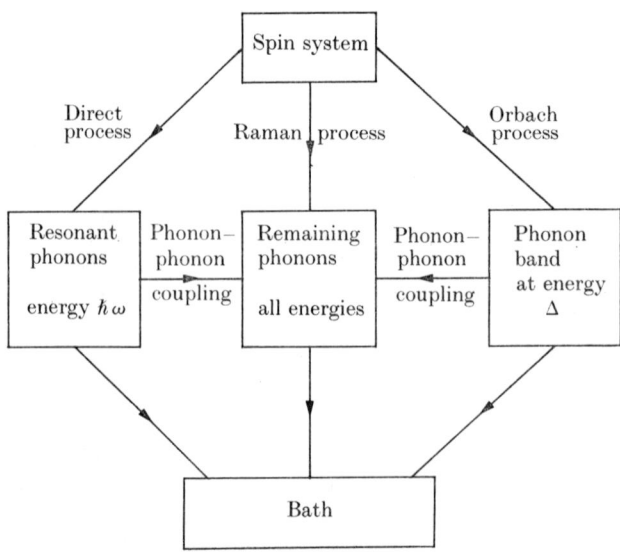

FIG. 10.14. Schematic representation of the various paths for spin-phonon relaxation, phonon-phonon coupling and phonon-bath relaxation. The arrows indicate the direction of energy flow assuming that the spin system is 'hot' and the bath 'cold'.

heat that is larger than the Debye specific heat (i.e. the specific heat of all lattice modes) even up to 4°K, if we assume the Debye θ_D to be about 100°K.

In considering the direct process we have neglected coupling between lattice modes, though this seems a reasonable approximation in a good crystal at helium temperatures if the mean free path of the phonons (in the absence of spins) is limited by crystallite size rather than by scattering through collisions with other phonons. A block diagram that includes all the above relaxation paths, and phonon-phonon coupling, is given in Fig. 10.14. An analysis of a less general system, involving just three 'thermal reservoirs', has been given by Petersen (1965, 1967); this uses rate equations for the quantity $\beta_i = (kT_i)^{-1}$, (cf. eqn (10.19)) which are valid provided that each reservoir i is in internal thermal equilibrium so that it can be characterized by a temperature T_i. This is basically similar to the analysis of Fig. 10.11 given at the beginning of this section; however it is valid over a wide temperature range because population differences (e.g., $n_a - n_b$) are proportional to $(1/T)$ and not to T, and it thus avoids the approximation of small temperature changes which is frequently inherent in the use of classical thermal analysis (cf. eqns (10.76), (10.77)).

PART III

THEORETICAL SURVEY

11

THE ELECTRONIC ZEEMAN INTERACTION

11.1. The interaction between electrons and a magnetic field

ELECTRONIC magnetic resonance occurs when an oscillating magnetic field induces transitions between two separated, discrete atomic levels. The most common case is one in which the levels are split through the application of a steady external magnetic field, but it sometimes happens that the two levels may be separated even in the absence of such a field. Nevertheless, it may still be convenient to apply to the system a magnetic field that will control within certain limits the spacing between the two levels.

An important problem, then, is to determine the position and the nature of the levels of an electronic system in the presence of an applied magnetic field $\mathbf{H}_0$. If $\mathscr{H}_0$ is the Hamiltonian of the system in the absence of the field, it is known from electromagnetic theory that the effect of a uniform field $\mathbf{H}_0$ is described by replacing in $\mathscr{H}_0$ the momentum $\mathbf{p}_i$ of each electron i of charge $-e$ by the quantity

$$\mathbf{p}_i + (e/c)\mathbf{A}(\mathbf{r}_i),$$

where $\mathbf{A}(\mathbf{r}_i) = \frac{1}{2}(\mathbf{H}_0 \wedge \mathbf{r}_i)$ is the value at the electron position $\mathbf{r}_i$ of the vector potential from which the field $\mathbf{H}_0$ is derived.

If the kinetic energy of the electrons has the correct relativistic form given by Dirac, the coupling of $\mathbf{H}_0$ with both the orbital magnetic moment and the spin magnetic moment of the electrons is obtained directly. On the other hand, if the non-relativistic form of the kinetic energy

$$T = \frac{1}{2m} \sum_i \mathbf{p}_i^2$$

is being used, the spin magnetism does not appear and the orbital magnetism is obtained as follows: the substitution

$$\mathbf{p} \to \mathbf{p} + (e/c)\mathbf{A}$$

yields

$$T = \frac{1}{2m} \sum_i \left(\mathbf{p}_i + \frac{e}{c}\mathbf{A}(\mathbf{r}_i)\right)^2. \tag{11.1}$$

THE ELECTRONIC ZEEMAN INTERACTION

The extra terms containing the magnetic field can be written:

$$\mathcal{H}_m = \frac{e}{2mc}\mathbf{H}_0 \cdot \sum_i (\mathbf{r}_i \wedge \mathbf{p}_i) + \frac{e^2}{8mc^2}\sum_i (\mathbf{r}_i \wedge \mathbf{H}_0)^2. \quad (11.2)$$

If we now introduce the quantity

$$\sum_i (\mathbf{r}_i \wedge \mathbf{p}_i) = \hbar\mathbf{L}, \quad (11.3)$$

where $\mathbf{L}$ is the orbital momentum of the system with respect to the origin of coordinates (the origin of the vectors $\mathbf{r}_i$), and the Bohr magneton $\beta = e\hbar/2mc$, eqn (11.2) becomes

$$\mathcal{H}_m = \beta\mathbf{H}_0 \cdot \mathbf{L} + \frac{e^2}{8mc^2}\sum_i (\mathbf{r}_i \wedge \mathbf{H}_0)^2. \quad (11.4)$$

The first term in (11.4) represents the orbital magnetism. It shows that an electronic system with an orbital momentum $\hbar\mathbf{L}$ possesses a magnetic moment $\boldsymbol{\mu}_L = -\beta\mathbf{L}$. The second term is the so-called diamagnetic term.

In the expression for the vector potential we did not specify the origin of the vectors $\mathbf{r}_i$. It is clear that the separation of (11.4) into a paramagnetic and a diamagnetic part depends on the choice of this origin. On the other hand, the principle of gauge invariance in electromagnetism tells us that this choice should not affect the result of the calculation of any physical quantity that can be observed experimentally. In this chapter, where we assume that the electrons responsible for the magnetic properties of the system are localized on a single atom, we shall choose as origin for the vector potential $\mathbf{A}$ and the orbital momentum $\mathbf{L}$, the nucleus of this atom. It is then easily shown that for the values of the field $\mathbf{H}_0$ that are currently used in laboratories the diamagnetic term is much smaller than the paramagnetic one.

The diamagnetic term can be written

$$\tfrac{1}{4}\beta H_0 \frac{e^2}{\hbar c}\sum_i \left\{\frac{r_i^2}{e}\sin^2\theta_i\right\}H_0, \quad (11.5)$$

where θ_i is the angle between the vector $\mathbf{r}_i$ and the applied field. In order of magnitude, if we assume that we can take $\mathbf{L}$ to be of order unity, this term is smaller than the paramagnetic term by a factor of $\sim \tfrac{1}{4}\tfrac{1}{137}(\langle r^2\rangle/e)H_0$ where $\langle r^2\rangle$ is the mean square radius of the atom. For $H_0 \sim 10^4$ e.m.u., and atomic dimensions of the order of 0·1 nm, the ratio of the diamagnetic to the paramagnetic term is of order 10^{-6}. This is comparable with the ratio between the diamagnetic and the paramagnetic susceptibility measured at very low temperatures when the electron magnetic moments are almost completely aligned in the applied field and $\langle L \rangle$ can indeed be of order unity.

Another reason why we shall often be able to disregard the diamagnetic term is that it has the same expectation value for both levels $|a\rangle$ and $|b\rangle$ between which the resonance is observed and thus does not affect the resonance frequency.

In the foregoing non-relativistic treatment the spin magnetism must be introduced phenomenologically by assigning to the electron an intrinsic spin magnetic moment $\boldsymbol{\mu}_s = -g_s\beta\mathbf{s}$, which has the same position as the electron charge. It is well known (and will be proved (in § 17.3)) that reduction of the Dirac equation to the non-relativistic form due to Pauli leads to a value of 2 for g_s. However, both accurate experiments and a more refined theory that takes into account the interaction of the electron with the fluctuations of the radiation field lead to a slightly different value of 2·0023 for g_s. We are thus led to describe the magnetic coupling of an atom (or an ion) with an applied field by means of a Hamiltonian

$$Z = \beta \mathbf{H}_0 \cdot (\mathbf{L} + g_s \mathbf{S}), \qquad (11.6)$$

where $\mathbf{S}$ is the total spin of the atom. In some experiments of very high accuracy a few small corrections, most of them of a relativistic nature, have to be applied to (11.6), but we shall disregard them with one or two possible exceptions. Frequently we shall even neglect the difference $g_s - 2$.

11.2. The Zeeman effect in a free atom (or ion)

Whatever the interactions between the electrons of a free atom or ion, in the absence of external fields the magnitude J of its total angular momentum $\mathbf{J}$ is a good quantum number and each energy level corresponding to a given value of J will have a degeneracy of $2J+1$.

The behaviour of the energy levels of the atom, when a steady magnetic field is applied to it, will depend on whether the Zeeman coupling (11.6) is large or small compared to the spacings of the atomic energy levels between which it possesses non-vanishing matrix elements. For magnetic fields currently used in laboratories, Z will in general be much smaller than these spacings. If we neglect the effects of the Zeeman coupling between different J levels, we may calculate the splitting of any given J level by first-order perturbation theory, using only the matrix elements of Z inside the manifold of degeneracy $(2J+1)$.

A fundamental theorem of group theory, the Wigner–Eckart theorem, of which we shall have more to say in § 13.5, tells us that inside the manifold J all vector operators have the same matrix elements within a constant factor. Within the approximation of the first-order perturbation theory we shall then write

$$\mathbf{L} + g_s \mathbf{S} = g_J \mathbf{J} \tag{11.7}$$

$$Z = g_J \beta \mathbf{H}_0 \cdot \mathbf{J}, \tag{11.8}$$

where g_J is a constant, called the Landé factor. The manifold J is split by a magnetic field into $2J+1$ equidistant levels each corresponding to a different value of J_z, where z is the direction of the applied field. As magnetic resonance transitions are allowed only between successive levels for which J_z differs by one unit there is a single resonance frequency,

$$\nu = g_J \beta H_0/h. \tag{11.9}$$

Whereas the proof of (11.7) is very general, the actual calculation of g_J requires more detailed hypotheses.

11.3. LS-coupling and the Landé formula

In a non-relativistic theory the interactions inside an atom can be divided into two parts: one, a spin-dependent part and two, a spin-independent part.

Quite generally (with the possible exception of the heaviest atoms) spin-dependent forces are much weaker than purely orbital forces and can be neglected in a first approximation. The description of atomic states in the framework of this approximation is known as the Russell–Saunders or LS-method of coupling. In this description it is well known that the magnitude L of the total orbital momentum is a good quantum number (this is a result of rotational invariance in the free atom and will be discussed in Chapter 13 from the viewpoint of group theory).

A second good quantum number is the total spin S. It is worth pointing out that, in contradistinction to L, the fact that S is a good quantum number has nothing to do with rotational invariance and stems simply from the fact that spin-dependent forces have been neglected. As Dirac (1930) has shown in his chapter on identical particles, the quantum number S is related to the type of symmetry of the electronic wave-function with respect to a permutation of the orbital coordinates of the various atomic electrons. Indeed one could (and did in the early days of quantum mechanics) classify atomic energy levels, without mentioning the spin at all, by using the language of the theory of the permutation group. It is just the fact that the electronic wave-function is antisymmetric with respect to the permutation of *both* the orbital and the spin coordinates of two electrons that enables one to replace, at least partially, this complicated description by the introduction of the total spin S as a good quantum number.

In atomic spectroscopy an energy level specified by the values of L and S is called a term. In practice, L and S alone will often not be sufficient to specify a term and extra parameters have to be introduced. A given term (L, S) has a degeneracy $(2L+1)\times(2S+1)$. We consider now the effect on such a term of spin-dependent forces that will partially lift this degeneracy. Since the magnitude J of the total angular momentum $\mathbf{J} = \mathbf{L}+\mathbf{S}$ is a good quantum number, the term (L, S) will be split into a number of submultiplets (J, L, S), where each multiplet has still a degeneracy of order $(2J+1)$ and where J ranges from $L+S$ to $|L-S|$. The representation of each multiplet by (J, L, S) implies that, according to our assumptions, the spin-dependent forces that split the (L, S) term are small compared to the distance between two neighbouring terms (L, S) and (L', S'), and in a multiplet (J, L, S) the admixture of wave-functions originating in another term (L', S') is small.

We have made no assumptions so far about the nature of the spin-dependent forces and thus can tell nothing of the spacings between the various multiplets (J, L, S). If these spacings are large compared to the Zeeman coupling Z we can compute the Landé factor g_J of formula (11.7) as follows. On multiplying each side of (11.7) by $\mathbf{J}$ we have

$$\mathbf{L}.\mathbf{J}+g_s\mathbf{S}.\mathbf{J} = g_J J(J+1) \tag{11.10}$$

and from squaring the relations $\mathbf{S} = \mathbf{J}-\mathbf{L}$ and $\mathbf{L} = \mathbf{J}-\mathbf{S}$ we find

$$\left.\begin{array}{l}2\mathbf{L}.\mathbf{J} = L(L+1)+J(J+1)-S(S+1)\\ 2\mathbf{S}.\mathbf{J} = S(S+1)+J(J+1)-L(L+1)\end{array}\right\} \tag{11.11}$$

whence

$$g_J = \frac{1}{2J(J+1)}[(g_s+1)J(J+1)-(g_s-1)\{L(L+1)-S(S+1)\}]. \tag{11.12a}$$

In particular, if we make the approximation $g_s = 2$, eqn (11.12a) becomes

$$g_J = 1+\frac{J(J+1)+S(S+1)-L(L+1)}{2J(J+1)}. \tag{11.12b}$$

The formulae (11.12a) or (11.12b) become particularly simple if J has the maximum possible value $J = L+S$, for then the substate

$J_z = J$ is also the state $L_z = L$, $S_z = S$ and the z component of the vector equation (11.7) can be written as the scalar relation

$$L + g_s S = g_J J,$$

whence

$$g_J = \frac{L + g_s S}{J} = \frac{L + g_s S}{L + S}, \qquad (11.13)$$

which could of course be obtained directly from (11.12a) by writing $J = (L+S)$.

It is worth emphasizing once more the extreme generality of the assumptions made so far. Equation (11.7) has simply implied that the Zeeman coupling Z is small compared to the spacings between the various levels of the free atom. The more detailed relations (11.12a) and (11.12b) are based on the further assumption that spin-dependent forces are small compared to orbital forces. The great simplicity of the Zeeman splitting in the free atom is related to the fact that the applied field $\mathbf{H}_0$ is uniform and that its effects are not sensitive to the detailed structure of the wave-functions. To a lesser extent this holds for a paramagnetic ion embedded in a crystal potential and the main features of the Zeeman splitting in such an ion could be derived without introducing any further details about the electronic structure of the free atom. However this is unadvisable for the following reasons.

(a) While the foregoing simple assumptions of rotational symmetry of the atom and of relative weakness of the spin-dependent forces have demonstrated the existence of (L, S) levels, they have provided no indication of their number or of their relative positions. Far more detailed predictions for these levels can be obtained.

(b) Furthermore, in addition to the effects of a uniform magnetic field we shall have to consider those of the very inhomogeneous fields produced by the nuclear magnetic dipole and electric quadrupole moments that give rise to the hyperfine structure of the resonance lines; these are very sensitive to the details of the atomic wave-function.

11.4. Self-consistent field configurations

The spin-independent part of the Hamiltonian of an atom can be written

$$\mathcal{H} = \frac{1}{2m} \sum_i p_i^2 - \sum_i \frac{Ze^2}{r_i} + \sum_{i<k} \frac{e^2}{r_{ik}} \qquad (11.14)$$

(in which we have neglected small orbital magnetic interactions between electrons).

For any atom beyond hydrogen, the problem of finding the eigenstates and eigenvalues of (11.14) cannot be solved exactly. The best, and in fact the only, approximation used to solve this problem is that of the self-consistent field suggested by Hartree and improved by Fock.

In Hartree's method it is assumed that each electron moves independently of the others in a central potential $V(r)$, which represents the combined effects of the attraction by the nuclear charge and a suitably averaged repulsion by the other electrons. In such a field, electrons have individual energy states and individual wave-functions, specified by the usual four quantum numbers n, l, m_l, m_s, where, because of rotational symmetry, the energy of each level is independent of the last two numbers. Thus at this stage of the approximation each atomic energy level is specified by a set of values for the quantum numbers n and l of each electron, where the restrictions introduced by the Pauli principle have been taken into account. Such a set is called a configuration.

Hartree's method has the following drawback. Consider two individual wave-functions φ_i and φ_j. The average potential $V(r)$ 'seen' by an electron with wave-function φ_i is that produced by all the electrons except itself. It will therefore be different from the potential 'seen' by the electron φ_j. The wave-functions φ_i and φ_j, being eigenfunctions of two slightly different one-electron Hamiltonians, will not be exactly orthogonal, which is a nuisance. Moreover, the fact that the electrons are indistinguishable is not taken properly into account in this method.

The answer to these difficulties is to be found in Fock's variational method. In this method one seeks the 'best' wave-function

$$\Psi(r_1, r_2, \ldots, r_n),$$

which is a product of individual wave-functions $\varphi_1(r_1)\varphi_2(r_2) \ldots \varphi_n(r_n)$ or, more exactly, the antisymmetrized product

$$\Psi = \frac{1}{\sqrt{n!}} \begin{vmatrix} \varphi_1(r_1) & \ldots & \varphi_1(r_n) \\ \varphi_2(r_1) & \ldots & \varphi_2(r_n) \\ \vdots & & \vdots \\ \varphi_n(r_1) & \ldots & \varphi_n(r_n) \end{vmatrix}, \quad (11.15)$$

which is called a Slater determinant. The individual or trial functions φ_i obey the orthogonality conditions $(\varphi_i \cdot \varphi_k) = \delta_{ik}$ and are determined by the requirement that $\langle \mathcal{H} \rangle = \int \Psi^* \mathcal{H} \Psi \, d\tau_1 \ldots d\tau_n$ be an extremum.

In practice, for free atoms, the trial functions are taken of the form

$$\varphi_i = P_{nl}(r) Y_l^{m_l}(\theta, \varphi) \chi_{m_s}(s),$$

where only the radial parts are unknown. The condition $\delta \langle \mathcal{H} \rangle = 0$ leads to a set of integro-differential equations for the $P_{nl}(r)$, which are then solved numerically. The energy level that corresponds to a configuration is still highly degenerate since the quantum numbers m_l and m_s remain unspecified and can take all the values compatible with the exclusion principle.

Consider then two Slater determinants $\Psi_{A'}$ and $\Psi_{A''}$ belonging to a configuration A of energy W_A, and two other determinants $\Psi_{B'}$ and $\Psi_{B''}$ belonging to a different configuration B of energy W_B. If these Slater determinants were exact eigenfunctions of $\mathcal{H}$ the following would hold:

$$(\Psi_{A'}| \mathcal{H} |\Psi_{A'}) = (\Psi_{A''}| \mathcal{H} |\Psi_{A''}) = W_A,$$
$$(\Psi_{B'}| \mathcal{H} |\Psi_{B'}) = (\Psi_{B''}| \mathcal{H} |\Psi_{B''}) = W_B, \quad (11.16)$$
$$(\Psi_{A'}| \mathcal{H} |\Psi_{B'}) = (\Psi_{A''}| \mathcal{H} |\Psi_{B''}) = 0.$$

In fact these equalities will be only approximate: the diagonal matrix elements $(\Psi_{A'}| \mathcal{H} |\Psi_{A'})$ and $(\Psi_{A''}| \mathcal{H} |\Psi_{A''})$ will be slightly different and will partially lift the degeneracy of configuration A. Off-diagonal matrix elements $(\Psi_{A'}| \mathcal{H} |\Psi_{B'})$ will admix into each configuration wave-functions originating in another configuration, an effect known to the spectroscopists as configuration interaction. The usefulness of the self-consistent field approximation resides in the fact that off-diagonal matrix elements of $\mathcal{H}$, linking two different configurations A and B, are small compared to the energy difference $W_A - W_B$ between these configurations. Configuration interaction can then be neglected and the 'correct' eigenfunctions $\Phi_{A'}$ of $\mathcal{H}$ are given by first-order perturbation theory as linear combinations of Slater determinants from a single configuration. The choice of the 'correct' linear combinations Φ_A is greatly simplified through L and S being good quantum numbers. As is well known in spectroscopy, and as will appear in more detail later on in connection with a group-theoretical discussion, the determination of the 'correct' linear combination will involve solving a secular equation only if more than one term (L, S) can be constructed from a given configuration. For instance, two distinct ^2D terms can be constructed from the configuration d^3; a quadratic secular equation must be solved in order to determine the structure of the wave-functions and the energies of the two ^2D terms.

Thanks to a rule originally due to Hund, the ground state of an atom (which is with few exceptions, the one of widest interest in paramagnetic resonance) has a structure much simpler than that of the excited states. Hund's rule states that the ground state has the maximum possible total spin S and, among states of maximum S, the greatest orbital momentum L. For instance, in a configuration d^4 the ground state will have $S = 2$, $L = 2$; in a configuration f^3 the ground state is $S = \frac{3}{2}$, $L = 6$, etc. For such a state, the magnetic substate $L_z = L$, $S_z = S$ is a single Slater determinant where all the electrons have the maximum values m_s and m_l compatible with the exclusion principle.

The low energy of a state with maximum S corresponds to the fact that, being completely symmetrical with respect to spin variables, the wave-function is completely antisymmetrical with respect to space variables. This keeps the electrons apart and reduces the positive contribution to the energy arising from their electrostatic repulsion. Unfortunately no such simple interpretation can be given for the rule for maximum L.

11.5. Spin-orbit coupling

Among spin-dependent interactions the most important by far is the spin-orbit coupling; for each electron this can be written as $\zeta(r)\mathbf{l} \cdot \mathbf{s}$. A simple physical (and somewhat incorrect) model of this interaction obtains as follows: the electron moves with a velocity $\mathbf{v} = \mathbf{p}/m$ in an electrostatic field $\mathbf{E}$, which is the gradient of an electrostatic central potential $-V/e$. The electron 'sees' a magnetic field

$$\mathbf{H} = -(\mathbf{v}/c) \wedge \mathbf{E} \qquad (11.17)$$

where $\mathbf{E}$ is related to the potential energy V by

$$\mathbf{E} = -\nabla\left(-\frac{V}{e}\right) = \frac{\mathbf{r}}{r}\frac{d}{dr}\left(\frac{V}{e}\right).$$

The magnetic coupling $2\beta\mathbf{s} \cdot \mathbf{H}$ between the field (11.17) and the spin magnetic moment $-2\beta\mathbf{s}$ of the electron can be written

$$-2\beta\mathbf{s} \cdot \left(\frac{\mathbf{p}}{mc} \wedge \frac{\mathbf{r}}{r}\right)\frac{d(V/e)}{dr} = \frac{\hbar^2}{m^2c^2}\frac{1}{r}\frac{dV}{dr}(\mathbf{l} \cdot \mathbf{s}) \qquad (11.18)$$

A correct calculation from Dirac's equation yields a result smaller by a factor 2 leading to

$$\zeta(r) = \frac{\hbar^2}{2m^2c^2}\frac{1}{r}\frac{dV}{dr}, \qquad (11.19)$$

where $\zeta(r)$ and $V(r)$ have both the dimensions of an energy and $(\hbar/mc)^2$, the square of the Compton wavelength, the dimensions of an area.

The assumption of LS-coupling is that spin-orbit coupling is small compared to the energy difference between two neighbouring terms, and this makes it possible to neglect off-diagonal matrix elements of $\sum_i \zeta(r_i)\mathbf{l}_i \cdot \mathbf{s}_i$ between two such terms. Inside each term the Wigner–Eckart theorem permits the replacement of $\sum_i \zeta_i \mathbf{l}_i \cdot \mathbf{s}_i$ by a single scalar product $\lambda \mathbf{L} \cdot \mathbf{S}$. From this approximation it follows immediately that the energies of the various (J, L, S) multiplets originating from a given term (L, S) vary as $(\lambda/2)\{J(J+1)\}$, a result known as the Landé interval rule.

If, as happens in heavy atoms, the off-diagonal matrix elements of spin-orbit coupling between two different terms are not negligibly small, L and S are no longer good quantum numbers and both the Landé formula for the g-factor and the interval rule become more or less incorrect. This situation is called intermediate coupling. The extreme case of j-j coupling, where the spin-orbit coupling is so strong that it is a better approximation to couple first $\mathbf{l}$ and $\mathbf{s}$ to a resultant $\mathbf{j}$ for each electron and then to couple the various $\mathbf{j}$ to a resultant $\mathbf{J}$, is never realized in practice.

Besides the spin-orbit coupling, by far the most important spin-dependent interaction, there are others much smaller, namely the spin-spin coupling between two electrons and the coupling of the spin of one electron with the orbit of another. These are discussed in some detail, together with the relation between λ and ζ_i by Blume and Watson (1962, 1963) and Watson and Blume (1965).

11.6. Matrix elements between Slater determinants

We summarize here without proof (Condon and Shortley 1935) a few simple rules relating to the calculation of matrix elements. We assume that the wave-function is a single determinant of the form (11.15), which we shall write for brevity:

$$\Psi = (\varphi_1, \varphi_2, \ldots, \varphi_n). \tag{11.20}$$

The extension to a linear combination of such determinants is straightforward.

In transition elements we shall specify a Slater determinant by the values of m_l and m_s for electrons outside closed shells. For instance, in the configuration d^3, $\Psi = (2^+, 1^-, 0^+)$ is a state where the quantum numbers of the three d-electrons are $m_l = 2, m_s = \frac{1}{2}; m_l = 1, m_s = -\frac{1}{2};$

$m_l = 0$, $m_s = +\tfrac{1}{2}$. From well known properties of the determinants it follows that the interchange of two one-electron states leads to a change of sign of Ψ:

$$(\varphi_1, \varphi_2, \varphi_3) = -(\varphi_2, \varphi_1, \varphi_3). \tag{11.21}$$

The diagonal matrix element $(\Psi|\ A\ |\Psi)$

The operators A that we shall deal with are either one-electron operators

$$A = \sum_i a(i), \tag{11.22}$$

such as the electrostatic coupling of an electron with the nucleus or with an external crystalline or magnetic field, etc., or two-electron operators (electrostatic repulsion, spin-spin interaction, etc. ...) of the form

$$A = \sum_{i<j} a(i,j). \tag{11.23}$$

For a one-electron operator the following rule holds:

$$(\Psi|\ A\ |\Psi) = \sum_p (\varphi_p(1)|\ a(1)\ |\varphi_p(1)). \tag{11.24}$$

For a two-electron operator the rule is

$$(\Psi|\ A\ |\Psi) = \sum_{p<q} (\varphi_p(1)\varphi_q(2)|\ a(1,2)\ |\varphi_p(1)\varphi_q(2)) -$$
$$- \sum_{p<q} (\varphi_p(1)\varphi_q(2)|\ a(1,2)\ |\varphi_q(1)\varphi_p(2)). \tag{11.25}$$

The second term of (11.25) is the so-called exchange term. In particular, if the operator $a(1, 2)$ is spin-independent the exchange term vanishes unless φ_p and φ_q correspond to the same spin orientation.

Off-diagonal matrix elements

Consider two different Slater determinants

$$\Psi = (\varphi_1, \ldots, \varphi_p), \quad \Psi'' = (\varphi_1', \ldots, \varphi_p').$$

The matrix elements $(\Psi|\ A\ |\Psi'')$ are then given by the following rules.

(a) *One-electron operators*

If Ψ and Ψ'' differ by more than one one-electron state, $(\Psi|\ A\ |\Psi'') = 0$. If they differ by a single individual state, which is φ in Ψ and φ' in Ψ'',

$$(\Psi|\ A\ |\Psi'') = \pm(\varphi|\ a\ |\varphi'). \tag{11.26}$$

The sign in (11.26) is that of the permutation that has to be performed among the states of Ψ'' in order to bring φ' into the place corresponding

THE ELECTRONIC ZEEMAN INTERACTION 595

to that which φ has in Ψ', all other identical individual states appearing in the same order in both Ψ' and Ψ''.

For instance,

$$(2^+, 1^-, 0^+|\, A\, |2^+, 0^+, -2^-) = -(2^+, 1^-, 0^+|\, A\, |2^+, -2^-, 0^+)$$
$$= -(1^-|\, a\, |-2^-).$$

(b) *Two-electron operators*

If Ψ' and Ψ'' differ by more than two individual states, $(\Psi'|\, A\, |\Psi'') = 0$.

If Ψ' and Ψ'' differ by two individual states, φ and χ for Ψ' and φ', χ' for Ψ''

$$(\Psi'|\, A\, |\Psi'') = \pm\{(\varphi(1)\chi(2)|\, a(1, 2)\, |\varphi'(1)\chi'(2)) -$$
$$-(\varphi(1)\chi(2)|\, a(1, 2)\, |\chi'(1)\varphi'(2))\}. \quad (11.27)$$

The choice of sign is determined by the permutation on the states of Ψ'' that will bring φ' in front of φ and χ' in front of χ.

If Ψ' and Ψ'' differ by a single state, namely φ and φ',

$$(\Psi'|\, A\, |\Psi'') = \pm \sum_{\varphi_i \neq \varphi, \varphi'} \{(\varphi(1)\varphi_i(2)|\, a(1, 2)\, |\varphi'(1)\varphi_i(2)) -$$
$$-(\varphi(1)\varphi_i(2)|\, a(1, 2)|\varphi_i(1)\varphi'(2))\}. \quad (11.28)$$

The choice of sign is determined as before.

The simplicity of the formulae (11.24)–(11.28) rests entirely on the assumption that the one-electron wave-functions φ_i are orthogonal to each other. This is always the case if one-electron Hartree–Fock wave-functions of a single atom are used. On the other hand, if we allow the magnetic electrons to be spread over several atoms, as will be the case when we shall consider the effects of covalent bonding, one-electron orbitals belonging to different atoms are no longer mutually orthogonal and the calculation of matrix elements of the energy becomes far more complicated.

11.7. Introduction of the crystal field

In the years from 1929 onwards it was suggested by a number of authors that the paramagnetic properties of an ion should be greatly affected by its electrostatic interactions with its surroundings. These interactions, at least in a first approximation, could be described by introducing an electrostatic potential in the region occupied by the electrons of the paramagnetic ion, produced by the neighbouring atoms or ions. Van Vleck (1932) and Bethe (1929) put this description on a quantitative basis, developing what is known as the crystal field theory, of paramount importance for our understanding of magnetism in general and more specifically of the results of paramagnetic resonance experiments. The first success of crystal field theory was the explanation of the 'quenching' of the orbital momentum in the iron group. It had been found that experimental values of the susceptibilities of a good many salts of the iron group could be made to agree with the theoretical formulae if in the latter the orbital contribution to the magnetic moment were equated to zero; that is, if Ψ is the wavefunction of the ion, one had to assume that

$$\langle \Psi | L_x | \Psi \rangle \equiv \langle L_x \rangle = 0 = \langle L_y \rangle = \langle L_z \rangle. \qquad (11.29)$$

Van Vleck has shown that the absence of orbital degeneracy is a sufficient condition for this quenching of the orbital momentum, his proof being as follows. Let Ψ be an eigenstate of the system, assumed non-degenerate. If spin-dependent forces can be neglected, the Hamiltonian $\mathscr{H}$, the sum of the kinetic and potential energy of the electrons, is a real operator. We can then assume that Ψ is real; for, if it were complex, of the form $\Psi_1 + i\Psi_2$, $\mathscr{H}$ being real, Ψ_1 and Ψ_2 would be separately eigenfunctions of $\mathscr{H}$ with the same energy and the level would be degenerate, in contradiction with the initial assumption. However the angular momentum operator equivalent to (11.3) is

$$\mathbf{L} = \frac{1}{i}(\mathbf{r} \wedge \mathbf{\nabla}), \qquad (11.30)$$

which is purely imaginary, so that the expectation value of any of its components taken over a real wave-function is imaginary. On the other hand, since $\mathbf{L}$ is Hermitian, this expectation value must be real. It must therefore vanish.

We shall present in § 15.4 a more general proof of the foregoing statement. Here we note that the quenching of the orbital momentum is a prime instance of the influence of the surroundings on the magnetic properties of an ion and more specifically of the close connection between magnetism and degeneracy. The rotational symmetry enjoyed by a free ion corresponds to a degeneracy that, in surroundings of the lower symmetry encountered in a crystal, may be lifted partially or even totally, leading to a change of the magnetic properties of the ion.

In order to estimate the effects of the crystal potential, the most important question is its magnitude relative to the energy spacings encountered in the free ion. These spacings can be classified according to their order of magnitude into three different categories:

(a) the energy separation C between two neighbouring configurations, of the order of the excitation energy of an individual electron;

(b) the separation T between two terms originating in the same configuration; as stated above the usefulness of the central field approximation resides in the condition $T \ll C$;

(c) spin-orbit coupling energies λ, which are smaller than T in LS-coupling, comparable to it in intermediate coupling and larger in j-j coupling.

In the following we shall always assume $\lambda \leq T$ and often $\lambda \ll T$.

We now compare the magnitude V of the crystal potential with the energy spacings of the free ion. Three extreme situations can be used as starting points for a quantitative description of the properties of an ion embedded in a crystal field, all of which are actually encountered in transition elements.

(1) *Weak field case*

The crystal field energy is smaller than the distance between two spin-orbit multiplets (J, L, S) and (J', L, S). It lifts partially or wholly the degeneracy $2J+1$ of each multiplet, which at least in a first approximation can be studied by neglecting the admixture from neighbouring multiplets so that J is approximately a good quantum number.

The weak field case is a good approximation for the rare-earth group where the spin-orbit coupling is larger and the crystal fields as a rule are smaller than in the iron group.

It is sometimes stated that the crystal fields are small because the $4f$ electrons of the rare-earth ions are shielded by the $5s$ and $5p$ electrons of external closed shells, but there is some doubt as to the reality of this shielding effect. Rather the result is due to the ligand ions being pushed further away by the extended $5s$, $5p$ orbitals.

(2) *Intermediate case*

Here the multiplet structure disappears completely and J ceases to be a good quantum number, even approximately.

Since perturbation theory requires that the various terms of the Hamiltonian should be introduced in order of decreasing magnitude, one must consider first the effect of the crystal potential on an (L, S) term and only afterwards examine how spin-orbit coupling affects the wave-functions and the energy levels modified by the crystal potential. In first approximation the admixtures from neighbouring terms (L', S')

may be neglected in the study of the splitting of a given term (L, S) so that L, and to a better approximation S, remain good quantum numbers; in more refined calculations these admixtures have to be considered.

This is the situation for many salts of the iron group. It often happens that the crystal potential V can be split into terms of decreasing symmetry V_1, V_2, etc. of which only the first is much larger than λ, which is a complication. One may have to calculate first the effect of V_1, then apply to the resulting structure simultaneously V_2 and the spin-orbit coupling as a perturbation, etc.

(3) *Strong field case*

The crystal field energy is stronger than the electronic repulsion energy responsible for the LS- coupling.

The term structure is washed out and L itself ceases to be even an approximately good quantum number. The correct treatment is then to go back to the configuration to see how each one-electron orbit is modified by the crystal potential and how the degeneracy of individual electron levels is lifted. The various electrons are then placed in the new orbits and a new configuration of minimum energy is built up. We shall call it a 'crystal field configuration'. One can then switch on as a perturbation the electron-electron interaction; this will partially lift the degeneracy of the crystal field configuration, introducing new levels that we may call by analogy 'crystal field terms' with new good quantum numbers that replace the total orbital momentum L. The correct interpretation of these new quantum numbers will appear in the group-theoretical discussion to follow. This situation is encountered mainly in the $4d$ and $5d$ group (palladium and platinum groups) and also for certain complexes in the iron group. Actually we shall see that the so-called strong field case is generally associated with covalent bonding where the description of the magnetic behaviour of an ion in terms of wave-functions strictly localized on this ion becomes inadequate. Nevertheless, the main features of the magnetic properties of the ion can still be understood in the framework of the crystal field theory. This is due to the fact that many of these properties are already determined by the symmetry of the surroundings without any detailed assumptions about their interactions with the paramagnetic ion.

In order to obtain a detailed quantitative description of the magnetic properties of a paramagnetic ion, in principle the following information should be available:

(i) The wave-functions and the energy levels of the free ion must be known. Fortunately, very often the knowledge of the ground term (L, S) is sufficient and, as mentioned earlier, this term being of maximum S and L has a simpler structure than that of excited states. If the crystal field is sufficiently weak, as in the case of the rare earths, it may be enough to know the ground spin-orbit multiplet (J, L, S).

(ii) One must be able to calculate the crystal field acting on the ion. Even if the positions of all neighbouring nuclei are known accurately from X-ray analysis, this is a very difficult problem fraught with considerable uncertainty. We shall discuss it in some detail in Chapter 16.

(iii) The matrix elements of the crystal field between the various states of the free ion must be computed. This involves in particular the knowledge of the radial parts of the one-electron Hartree–Fock wave-function of the free ion. These are matrix elements of one particle operators, and the only ones that appear for weak and intermediate crystal fields. For strong fields one has also to compute matrix elements of the electron–electron repulsion, a two-particle operator.

(iv) With the matrix elements thus calculated, secular equations must be set up and solved. It is particularly important to extract any multiple roots that correspond to degenerate energy levels. Wave-functions for the paramagnetic ion, which are linear combinations of the wave-functions for the free ion, are also obtained.

(v) With these functions, expectation values and, if necessary, off-diagonal matrix elements are computed for the physical quantities of interest in paramagnetic resonance: components of the magnetic moment, electron spin-spin interaction, magnetic dipole and electric quadrupole hyperfine interaction, etc. The hyperfine structure is very sensitive to the details of the wave-functions and configuration interaction must often be taken into account in its evaluation.

This is a rather formidable programme and it is neither possible nor fortunately always necessary to determine quantitatively all the elements listed above.

12
GROUP THEORY—AN OUTLINE

IT is difficult to exaggerate the role played in paramagnetic resonance by symmetry considerations, from which alone a tremendous amount of theoretical information can be obtained. Fortunately there exists a mathematical tool made to measure, as it were, for the task of extracting this information and translating it into quantitative results; this tool is the theory of group representations. Considered not so long ago as a somewhat sophisticated branch of mathematics, this theory has become accessible to physicists in a number of books written for them rather than for mathematicians. This poses the problem of on what level group theory should be treated in a book on paramagnetic resonance, in which perhaps the only certainty is that it cannot be ignored altogether.

Group-theoretical considerations enter the subject in both rather simple and rather complicated ways, as well as lying behind a number of statements contained in elementary books on atomic theory. For example, when it is said that the orbital angular momentum is a good quantum number, or that **L** and **S** are coupled together to give a resultant **J**, use is being made (consciously or unconsciously) of the theory of the representations of the rotation group, which we shall consider in the next chapter. On a simpler level, group theory is hardly needed to predict that in a field of cubic symmetry the three wavefunctions xy, yz, zx will correspond to states of the same energy; or, that if the field has a distortion along the z-axis, the function xy will have an energy different from the other two. On the other hand, without any knowledge of group theory it would hardly be possible to predict that a level with $J = \tfrac{15}{2}$ would split in a cubic field into two doublets and three quartets.

Given that group theory cannot be excluded, one may either assume that the reader already has a working knowledge of the subject, or attempt to give sufficient information to enable those unfamiliar with group theory to follow its applications in the field of paramagnetic resonance. Not without some hesitation, we have decided on the latter course. Apart from our desire to make this book reasonably self-contained, we felt (perhaps mistakenly) that, being ourselves more interested in paramagnetic resonance than in group theory *per se*, our account of the latter might be useful to readers sharing the same interest. In this chapter we shall therefore attempt to outline the main results of group theory relevant to our subject, and in subsequent chapters to apply them first to the free atom or ion and later more specifically to the conditions of lower symmetry experienced by paramagnetic ions in solids.

12.1. Invariance and degeneracy

The definition of a group is well known and will not be reproduced here. Consider a system defined by a set of dynamical variables $x_1, \ldots, x_n$, and assume that its Hamiltonian $\mathscr{H}$ is unchanged by all the transformations $\mathscr{R}$ of a group $\mathscr{G}$ which take a point of the set of variables $\mathbf{x} = (x_1, \ldots, x_n)$ into a point $\mathbf{x}' = \mathscr{R}\mathbf{x} = (x'_1, \ldots, x'_n)$. Let $\Psi(\mathbf{x})$ be a function of $(x_1, \ldots, x_n)$ and define $\Psi''(\mathbf{x})$ as the function

$$\Psi''(\mathbf{x}) = R\Psi = \Psi(\mathscr{R}^{-1}\mathbf{x}), \qquad (12.1)$$

that is, the function that takes a value equal to $\Psi(\mathbf{x})$ at the point $\mathbf{x}'$ into which $\mathscr{R}$ transforms $\mathbf{x}$. We have written symbolically $\Psi'' = R\Psi$, the meaning of this symbol being given by (12.1). The transformation R is a linear operator acting on the function Ψ and it is easy to see that the operators R form a group G that is isomorphous with $\mathscr{G}$. The relationship between G and $\mathscr{G}$ is a little more complicated when the electron spin is introduced, as will appear later on. Since $\mathscr{H}$ is invariant under G, $R(\mathscr{H}\Psi) = \mathscr{H}(R\Psi)$ for all Ψ and hence $\mathscr{H}$ commutes with all operators of G. This invariance entails the following fundamental property: unless all the operators R of the group commute with each other (abelian group) the system necessarily has degenerate levels: the proof of this is as follows.

Let R and S be two non-commuting operators of G, and let Ψ be an eigenfunction of $\mathscr{H}$ with eigenvalue W. Then

$$\mathscr{H}(R\Psi) = R\mathscr{H}\Psi = RW\Psi = W(R\Psi),$$
$$\mathscr{H}(S\Psi) = S\mathscr{H}\Psi = SW\Psi = W(S\Psi),$$

from which it is clear that $R\Psi$ and $S\Psi$ are eigenstates of $\mathscr{H}$ with the same eigenvalue W. If the system has no degenerate levels the three functions Ψ, $R\Psi$ and $S\Psi$ will represent the same state and differ by constant phase factors only. It follows that $[R, S]\Psi = 0$ for all eigenfunctions of the system. Since these functions form a complete set, $[R, S] = 0$, in contradiction with the initial assumption.

12.2. Linear representations, equivalence, and irreducibility

Let $\Psi_1, \Psi_2, \ldots, \Psi_p$ be a set of orthogonal, normalized eigenstates spanning the manifold of a p-fold degenerate energy level.

As we have seen, any state $\Phi_i = R\Psi_i$ is also an eigenstate of $\mathscr{H}$ with the same energy W, and it must therefore be a linear combination of the Ψ_k which we write as

$$R\Psi_i = \sum_k \Psi_k D_{ki}(R). \tag{12.2}$$

The linear operators $\mathscr{D}(R)$ represented by the matrices $D_{ki}(R)$ form what is known as a linear representation of the group G. We shall review briefly the properties of the linear representations that are of importance for the understanding of properties of paramagnetic ions in crystals. The groups of interest to us are certain finite groups, that is groups with a finite number of elements and also one infinite group, the group of spatial rotations.

A linear, p-dimensional representation of a group G establishes a correspondence between every element R of the group and a matrix $D(R)$, with a non-vanishing determinant, that transforms any vector $\mathbf{X}$ of a p-dimensional vector space $\mathscr{E}$ into a vector $\mathbf{Y} = D(R)\mathbf{X}$ of the same space through the formula

$$Y_i = \sum_j X_j D_{ji}(R), \tag{12.3}$$

with the isomorphism condition

$$D(R_1 R_2) = D(R_1) D(R_2). \tag{12.4}$$

If in the space $\mathscr{E}$ we make a change of coordinate axes $\mathbf{X} = S\mathbf{X}'$ and $\mathbf{Y} = S\mathbf{Y}'$, the relationship $\mathbf{Y} = D(R)\mathbf{X}$ becomes

$$\mathbf{Y}' = D'\mathbf{X}' = S^{-1}DS\mathbf{X}'. \tag{12.5}$$

The two representations $\mathscr{D}'$ and $\mathscr{D}$ related by the similarity transformation (12.5) are said to be equivalent. For all finite groups and for the rotation group it can be shown (and we assume it without proof) that every linear representation has an equivalent representation that is unitary, i.e. formed of unitary matrices. From now on we shall consider unitary representations only.

A unitary representation $\mathscr{D}$ of dimension p is said to be reducible if through a similarity transformation all its matrices D can be transformed to the quasi-diagonal form

$$\begin{pmatrix} D_1 & 0 \\ 0 & D_2 \end{pmatrix}, \tag{12.6}$$

where the submatrices D_1 and D_2 have dimensions p_1 and p_2, with $p_1+p_2 = p$. Geometrically, the reducibility of the representation $\mathscr{D}$ corresponds to the fact that in the representation space $\mathscr{E}$ there are two invariant orthogonal subspaces $\mathscr{E}_1$ and $\mathscr{E}_2$ with dimensions p_1 and p_2, such that every vector of $\mathscr{E}_1$ is transformed by $\mathscr{D}$ into another vector of $\mathscr{E}_1$, and similarly for $\mathscr{E}_2$. A representation that cannot be reduced to the form (12.6) by a similarity transformation is said to be irreducible. If the matrices D_1 and/or D_2 are reducible the process can be carried further until all the matrices D are of the form:

$$\begin{pmatrix} D_1 & & & \\ & D_2 & & \\ & & \cdot & \\ & & & \cdot \\ & & & & D_q \end{pmatrix}$$

where all the submatrices are irreducible.

12.3. Orthogonality relations, characters, and classes

We shall make use of the following theorems, most of which are stated without proof.

(a) If a matrix M commutes with all the matrices of an irreducible representation it is a multiple of the unit matrix.

(b) For two irreducible non-equivalent representations $\mathscr{D}_1$ and $\mathscr{D}_2$ of a finite group G the following orthogonality relation holds whatever the values of the indices i, j, k, l,

$$\sum_S D_{1,ik}(S) D^*_{2,jl}(S) = 0, \tag{12.7}$$

where the summation is over all the elements S of the group.

Each group possesses as one irreducible representation the unit representation where to each operation of the group corresponds the multiplication by the number 1. Hence if in (12.7) we take $\mathscr{D}_2$ to be the unit representation we get

$$\sum_S D_{1,ik}(S) = 0. \tag{12.8}$$

(c) For every irreducible representation $\mathscr{D}$ of order p the following holds:

$$\sum_S D_{ik}(S) D^*_{jl}(S) = \frac{m}{p} \delta_{ij} \delta_{kl}, \tag{12.9}$$

where m is the number of elements of the group, also called the order of the group. For the rotation group, an infinite group, the orthogonality relations (12.7) and (12.9) are still valid provided the summation over the group elements is replaced by a suitable integration. It can be shown that if we define a rotation R by the three Euler angles α, β, γ the relations (12.7) and (12.9) adapted for the rotation group can be rewritten as

$$\int_0^{2\pi} d\alpha \int_0^{2\pi} d\gamma \int_0^{\pi} D_{ik}^*(\alpha, \beta, \gamma) D_{jl}(\alpha, \beta, \gamma) \sin \beta \, d\beta = \delta_{ij}\delta_{kl}\frac{8\pi^2}{p}. \quad (12.10)$$

(d) The traces of the matrices of a linear representation are called its characters and are denoted by χ. Since the trace of a matrix is invariant under a similarity transformation such as (12.5), two equivalent representations have the same set of characters. From (12.7) and (12.9) we obtain immediately the relations

$$\sum_S \chi_1(S)\chi_2^*(S) = 0,$$
$$\sum_S \chi_1(S)\chi_1^*(S) = m. \quad (12.11)$$

The concept of character is intimately connected with that of class: two elements A and B of a group are said to belong to the same class if the group contains a third element C such that

$$B = CAC^{-1}. \quad (12.12)$$

For instance, two rotations A and B through the same angle but around two different axes X and Y, belong to the same class provided that the element C in (12.12) is the rotation that converts X into Y. From the definition (12.12) it follows that the unit element of the group is in a class by itself. It also follows from that definition that all the matrices of a linear representation that belong to the same class have the same character.

If the group contains r different classes $C^{(k)}$ with g_k elements per class the relations (12.11) can be rewritten as

$$\sum_{k=1}^r \chi_1^{(k)} \chi_2^{(k)*} g_k = 0,$$
$$\sum_{k=1}^r \chi_1^{(k)} \chi_1^{(k)*} g_k = m. \quad (12.13)$$

(e) A problem in the theory of linear representations of a finite group of paramount importance, for quantum mechanics in general and for paramagnetic resonance in particular, is the discovery of all the irreducible representations of a group G. Let $\mathscr{D}_1, \mathscr{D}_2, \ldots, \mathscr{D}_l$ be all the non-equivalent irreducible representations of a finite group G containing r classes. The number l is finite, as will appear presently. In a space $\mathscr{E}$ with r dimensions consider l vectors $\mathbf{X}_1, \mathbf{X}_2, \ldots, \mathbf{X}_l$, the r components of the vector $\mathbf{X}_i$ being the r numbers, with $k = 1, 2, \ldots, r$:

$$\chi_i^{(k)} \sqrt{\left(\frac{g_k}{m}\right)}.$$

It follows from (12.13) that these vectors are orthogonal and therefore linearly independent. Their number l cannot exceed the number r of dimensions of the space $\mathscr{E}$, and we must have $l \leqslant r$. It can also be shown (and we assume it without proof) that $l \geqslant r$ and thus $l = r$, which means that the number of non-equivalent irreducible representations of a finite group is equal to the number of its classes.

Let $l_1, l_2, \ldots, l_r$ be the dimensions of the r irreducible representations

$$\mathscr{D}_1, \mathscr{D}_2, \ldots, \mathscr{D}_r.$$

Consider one such representation $\mathscr{D}_p$ and the m matrix elements $D_{p,ij}(S)$ that correspond to the m elements S of the group. We can regard these numbers as the m components of a vector $\mathbf{X}_{p,ij}$ in a space $\mathscr{E}$ with m dimensions. The formula (12.7) shows that these vectors are all orthogonal and therefore their number $l_1^2 + l_2^2 + \ldots + l^2$ cannot be larger than the number of dimensions m of the space $\mathscr{E}$, which is the order of the finite group G.

Here again it can be shown that the following equality holds:

$$l_1^2 + l_2^2 + \ldots + l_r^2 = m, \tag{12.14}$$

which once more we assume to hold without proof.

12.4. Reduction of a representation and calculation of the characters

Let us assume that we know the characters of all the irreducible representations of a group G and let $\mathscr{D}$ be a reducible representation. We want to know the numbers of times $a_1, a_2, \ldots, a_r$ that each irreducible representation $\mathscr{D}_i$ is contained in $\mathscr{D}$. Let $\chi(S)$ be the character of the transformation S in the representation $\mathscr{D}$ and $\chi_1(S), \ldots, \chi_r(S)$ be the characters of S in the irreducible representations $\mathscr{D}_i$. From the very definition of reducibility and of the characters it is clear that

$$\chi(S) = \sum_{p=1}^{r} a_p \chi_p(S). \tag{12.15}$$

Multiplying each side of (12.15) by $\chi_q^*(S)$, where q is one of the indices $1, 2, \ldots, r$, and summing over all the group elements S, we obtain from (12.11)

$$a_q = \frac{1}{m} \sum_{S=1}^{m} \chi(S) \chi_q^*(S), \tag{12.16}$$

where a_q is necessarily an integer.

The prediction of the breakdown of a given representation into its irreducible components is therefore straightforward if the characters of all the irreducible representations are known. A straightforward answer to the question of how to find the characters of irreducible representations is to look them up in a table. Nevertheless we will indicate briefly the principle of their calculation.

If we consider the elements of the group as linear operators as explained in § 12.1 it follows from the definition (12.12) that the operator $\mathscr{C}^{(g)} = R_1 + R_2 + \ldots + R_g$, the sum of all elements of a given class (g) of the group, commutes with all the elements of the group and conversely that every operator that commutes with all the elements of the group is a sum of operators $\mathscr{C}^{(g)}$. It is clear, therefore, that the product $\mathscr{C}^{(p)} \mathscr{C}^{(q)}$ is also a sum of operators $\mathscr{C}$ which can be written as

$$\mathscr{C}^{(p)} \mathscr{C}^{(q)} = \sum_s a_{pq}^s \mathscr{C}^{(s)}. \tag{12.17}$$

The coefficients a_{pq}^s are obtained in a straightforward manner if the multiplication table of the group is known. Consider now an irreducible representation $\mathscr{D}_i$ of order l_i. The matrices representing the operators $\mathscr{C}^{(p)}$ commute with all the matrices of this representation and must therefore be multiples of the unit matrix. The operator $\mathscr{C}^{(p)}$ is represented by the matrix

$$C_i^{(p)} = \mathscr{C}_i^{(p)} I, \tag{12.18}$$

where I is the unit matrix and $\mathscr{C}_i^{(p)}$ a number.

Taking the trace of each side of (12.18) we get

$$\mathscr{C}_i^{(p)} = \chi_i^{(p)} \frac{g_p}{l_i} = \chi_i^{(p)} \frac{g_p}{\chi_i^{(1)}}, \tag{12.19}$$

where $\chi_i^{(1)} = l_i$ is the character of the class corresponding to the unit operator. Inserting this result into (12.17) we find

$$\frac{\chi^{(p)}}{\chi^{(1)}} \frac{\chi^{(q)}}{\chi^{(1)}} g_p g_q = \sum_{s=1}^{r} a_{pq}^s \frac{\chi^{(s)}}{\chi^{(1)}} g_s, \tag{12.20}$$

where we have dropped the index i of the particular representation $\mathscr{D}_i$. With the notation $y^{(s)} = \chi^{(s)}/\chi^{(1)}$ where $y^{(1)} = 1$, (12.20) becomes

$$y^{(p)}y^{(q)}g_p g_q = \sum_{s=1}^{r} a_{pq}^s y^{(s)} g_s, \qquad (12.21)$$

which can be solved for the values of y as follows. Consider a linear function

$$B = \sum_{s=1}^{r} \alpha_s y^{(s)}, \qquad (12.22)$$

where we choose arbitrarily the coefficients α_s. According to (12.21) the successive powers $B^2, \ldots, B^r$ will also be linear functions of the $y^{(s)}$ of the form

$$\begin{cases} B^2 = \sum_{s=1}^{r} \beta_s y^{(s)} \\ \text{\hspace{2cm}} \\ B^r = \sum_{s=1}^{r} \rho_s y^{(s)} \end{cases}, \qquad (12.23)$$

where the $\beta_s, \ldots, \rho_s$ are deduced from the α_s by means of (12.21). From the $r+1$ linear equations (12.22), (12.23) and $y^{(1)} = 1$ we can eliminate the $y^{(s)}$, obtaining a secular equation of order r for B:

$$B^r + C_1 B^{r-1} + \ldots + C_r = 0, \qquad (12.24)$$

where the coefficients $C_1, \ldots, C_r$ are known.

For each of the r roots of (12.24) we get from the sets of linear equations (12.22), (12.23) a set of values for the $y^{(s)}$ and thus also a set of characters if we know the $\chi_i^{(1)}$; that is, the dimensions l_i of the irreducible representation. The latter are often determined uniquely from the relation (12.14): $l_1^2 + \ldots + l_r^2 = m$.

We are now in a position to enumerate the irreducible representations of a group, to calculate their characters, and to break down every reducible representation into its irreducible components. These methods enable us to make predictions about the splitting of an energy level by a perturbation of known symmetry.

12.5. Splitting of a degenerate level by a perturbation of lower symmetry

Consider a Hamiltonian $\mathscr{H}_0$ which is invariant under all the transformations of a group G_0, and a level W_0 of that Hamiltonian with p-fold degeneracy. For instance, for a given level J of a free ion, the group G_0 will be the rotation group and the degeneracy is $p = 2J+1$. Under the operations of the group G_0, the p eigenfunctions Ψ_i that span the manifold $\mathscr{E}_0$ of the wave-functions with eigenvalue W_0 are transformed into each other as explained in § 12.2, and provide a linear representation $\mathscr{D}_0$ of order p of the group G_0, a representation that we shall initially assume to be irreducible. We introduce a perturbing Hamiltonian V of lower symmetry than $\mathscr{H}_0$. What is meant by lower symmetry is that only a fraction of the transformations belonging to G_0, namely a subgroup G of G_0, will leave V invariant. The linear transformation of the functions Ψ_i under the more restricted set of operations belonging to G, provides for this latter group a representation $\mathscr{D}$ that may not be irreducible. For, reducing the representation $\mathscr{D}_0$ of G_0 consisted of bringing by a similarity transformation all the matrices of $\mathscr{D}_0$ to the quasi-diagonal form (12.6), and our assumption of the irreducibility of $\mathscr{D}_0$ meant that this could not be done. On the other hand, since the representation $\mathscr{D}$ of G contains only a *fraction* of all the matrices of G_0, we have to 'quasi-diagonalize' fewer matrices by a similarity transformation and this may not be impossible.

We can also say that no sub-manifold $\mathscr{E}'$ of the manifold $\mathscr{E}_0$ was left invariant by all the transformations of the group G_0 but that it is not inconceivable that a fraction G of these transformations might leave such a sub-manifold invariant.

Let us assume then that $\mathscr{D}$, a linear representation of G, is reducible. We can reduce it by means of the formulae (12.15) and (12.16), which tell us how many times a given irreducible representation, say $\mathscr{D}_i$, of G is contained in $\mathscr{D}$. Assume for the sake of argument that $\mathscr{D}$ contains two irreducible representations $\mathscr{D}_1$ and $\mathscr{D}_2$ of dimensions p_1 and p_2 with $p_1 + p_2 = p$. We can show that the level W will be split into two sublevels W_1 and W_2 with respective degeneracies of order p_1 and p_2. Let φ_i ($i = 1, 2, \ldots, p_1$) be p_1 functions of the manifold $\mathscr{E}$ that transform by $\mathscr{D}_1$, and χ_j ($j = 1, 2, \ldots, p_2$) those that transform by $\mathscr{D}_2$, and calculate the following matrix elements:

$$A_{ij} = (\varphi_i|\ V\ |\chi_j),$$
$$B_{ii'} = (\varphi_i|\ V\ |\varphi_{i'}),$$
$$C_{jj'} = (\chi_j|\ V\ |\chi_{j'}).$$

(Here of course χ_j has nothing to do with the quantity χ used to denote the character of a representation in § 12.3.)

The matrix element A_{ij} can be written

$$A_{ij} = \int \varphi_i^*(\mathbf{x}) V(\mathbf{x}) \chi_j(\mathbf{x})\ \mathrm{d}\tau. \qquad (12.25)$$

GROUP THEORY—AN OUTLINE 609

Let R be an operation of group G that leaves V invariant, i.e.

$$A_{ij} = \int \varphi_i^*(\mathscr{R}^{-1}\mathbf{x}) V(\mathscr{R}^{-1}\mathbf{x}) \chi_j(\mathscr{R}^{-1}\mathbf{x}) \, d\tau \qquad (12.26)$$

(this involves just a change of variable in the integration). Making use of the definition (12.1) we find that

$$A_{ij} = \int (R\varphi_i)^* V(\mathbf{x})(R\chi_j) \, d\tau = \frac{1}{m} \sum_R \int (R\varphi_i)^* V(\mathbf{x})(R\chi_j) \, d\tau, \qquad (12.27)$$

where m is the number of elements in the group.

From the formulae (12.2) and the orthogonality relations (12.7) we have

$$A_{ij} = \frac{1}{m} \sum_{k,l} \left(\int \varphi_k^* V \chi_l \, d\tau \right) \sum_R \mathscr{D}_{1,ki}^*(R) \mathscr{D}_{2,lj}(R) = 0. \qquad (12.28)$$

In particular, if in (12.25) we replace V by unity we find

$$\int \varphi_i^* \chi_j \, d\tau = 0. \qquad (12.29)$$

Thus two such functions belonging to two non-equivalent irreducible representations are orthogonal.

In the same way, with the help of (12.9), we find

$$B_{ii'} = (\varphi_i | V | \varphi_{i'}) = \frac{\delta_{ii'}}{p_1} \sum_{k=1}^{p_1} (\varphi_k | V | \varphi_k),$$

$$C_{jj'} = (\chi_j | V | \chi_{j'}) = \frac{\delta_{jj'}}{p_2} \sum_{l=1}^{p_2} (\chi_l | V | \chi_l). \qquad (12.30)$$

Equations (12.28) and (12.30) express precisely the fact that the functions φ_i and χ_j are the 'correct' zero-order wave-functions for the perturbation V, and that the level W is split into two levels W_1 and W_2 of respective multiplicity p_1 and p_2.

This may be generalized immediately to the case where $\mathscr{D}$ is reduced into q irreducible representations $\mathscr{D}_1, \ldots, \mathscr{D}_q$ of dimensions $p_1, \ldots, p_q$, all different. The level W is split into q sublevels $W_1, \ldots, W_q$ of multiplicity $p_1, \ldots, p_q$, and the correct zero-order wave-functions of the level W_q form a basis for the representation $\mathscr{D}_q$ and are determined from symmetry considerations alone. However, if a given irreducible representation occurs more than once in $\mathscr{D}$, the situation is rather more complicated.

Assume for the sake of argument that an irreducible representation $\mathscr{D}_1$ occurs twice in $\mathscr{D}$, spanned respectively by a set of functions

$\varphi_1, \ldots, \varphi_n$ and $\chi_1, \ldots, \chi_n$. We assume further that the φ_i and χ_j transform *exactly* by the same matrices D_1 and not only within a similarity transformation. We can establish as before that $B_{ii'}$ and $C_{jj'}$ vanish for $i \neq i'$ and $j \neq j'$ but we can no longer state that $(\varphi_i|\ V\ |\chi_i) = 0$ and indeed this cross-element will not, in general, be zero. φ_i and χ_i are not 'correct' zero-order wave-functions any more and the correct linear combinations as well as the energy splittings caused by the perturbation V are obtained from the secular equation

$$\begin{vmatrix} (\varphi_1|\ V\ |\varphi_1) - W & (\varphi_1|\ V\ |\chi_1) \\ (\chi_1|\ V\ |\varphi_1) & (\chi_1|\ V\ |\chi_1) - W \end{vmatrix}. \quad (12.31)$$

It follows from (12.30) that

$$\begin{aligned}(\varphi_1|\ V\ |\varphi_1) &= (\varphi_i|\ V\ |\varphi_i) \\ (\chi_1|\ V\ |\chi_1) &= (\chi_i|\ V\ |\chi_i).\end{aligned} \quad (12.32)$$

The same argument shows that

$$(\varphi_1|\ V\ |\chi_1) = \frac{1}{p} \sum_{l=1}^{p} (\varphi_l|\ V\ |\chi_l) = (\varphi_i|\ V\ |\chi_i). \quad (12.33)$$

We would therefore have obtained the same secular equation by choosing two other functions φ_i and χ_i instead of φ_1 and χ_1. The 'correct' zero-order wave-functions will be first a set of linear combinations

$$\alpha'\varphi_1 + \beta'\chi_1,\ \alpha'\varphi_2 + \beta'\chi_2,\ \ldots,\ \alpha'\varphi_p + \beta'\chi_p$$

spanning a first manifold of order p, with an energy W' that will be one of the roots of (12.31), and a second set orthogonal to it,

$$\alpha''\varphi_1 + \beta''\chi_1,\ \ldots,\ \alpha''\varphi_p + \beta''\chi_p,$$

with an energy W'' that will be the second root of (12.31). The 'correct' zero-order wave-functions can no longer be obtained from symmetry considerations alone but involve the perturbing potential V explicitly.

More generally, if a representation $\mathscr{D}_q$ of dimension p_q occurs a_q times in $\mathscr{D}$, the original level W will be split into a_q sublevels $W_1, \ldots, W_a$ each of degeneracy p_q (and also into other sublevels). The energies $W_1, \ldots, W_a$ and the 'correct' zero-order wave-functions are obtained by solving a secular equation of order a_q.

In the foregoing discussion use has been made of first-order perturbation theory and one might ask whether a higher-order calculation might not lead to a further removal of degeneracy. The answer is no. Let us consider a total Hamiltonian $\mathscr{H}(\varepsilon) = \mathscr{H}_0 + \varepsilon V$, where W_0 is a

p-fold degenerate level of $\mathscr{H}(0) = \mathscr{H}_0$ spanned by p eigenfunctions Ψ_i, which provide an irreducible representation $\mathscr{D}$ of the group G that leaves $\mathscr{H}_0$ invariant. We assume that G also leaves V invariant. We saw that first-order perturbation theory predicts then no splitting for the level W, and we shall show that this statement is correct to all orders. We assume that W_0 is split into two levels W_1 and W_2 spanned respectively by p_1 eigenfunctions φ_j and p_2 eigenfunctions χ_k. The p_1 functions φ_j provide a representation $\mathscr{D}'$ of G, irreducible or not, but which can neither contain $\mathscr{D}$ nor coincide with it since $p_1 < p$, nor be contained in it since $\mathscr{D}$ is irreducible. From the orthogonality relations it follows that the Ψ_i are orthogonal to all the φ_j and also to all the χ_k. The manifold of the functions (φ_j, χ_k) being orthogonal to the manifold Ψ_i cannot be made to coincide with it in a continuous manner when $\varepsilon \to 0$, and the two levels W_1 and W_2, if they do exist, do not originate from a splitting of W_0.

This takes us back to the assumption made at the beginning of this section (§ 12.5) that the representation $\mathscr{D}_0$ of the group G_0, which left $\mathscr{H}_0$ invariant and which can be represented by the eigenfunctions of a level W_0, was irreducible. Had it been reducible into, say, two representations $\mathscr{D}_0'$ and $\mathscr{D}_0''$, a perturbation of the same *symmetry* as $\mathscr{H}_0$ could have split W_0 into two levels W_0' and W_0''. A degeneracy of this type is called accidental and will be disregarded from now on.

12.6. The direct product of two representations

We consider now two dynamic systems S_1 and S_2, and to start with we assume them to have no mutual interaction, so that their Hamiltonians $\mathscr{H}_1$ and $\mathscr{H}_2$ commute with each other and are each invariant under operations of the same group G. Examples of this type are two electrons moving in the central potential of the same type of atom, or two nuclear spins in an applied magnetic field.

Let W_1 be a level of $\mathscr{H}_1$ spanned by p_1 functions ψ_i and W_2 a level of $\mathscr{H}_2$ spanned by p_2 functions χ_j. The φ and the χ provide two representations $\mathscr{D}_1$ and $\mathscr{D}_2$, each of which, barring accidental degeneracy, we assume to be irreducible. It is clear that the products $\varphi_i \chi_j$ also provide a representation of G, which we shall denote as $\mathscr{D}_1 \times \mathscr{D}_2$ and shall call the direct product of the two representations $\mathscr{D}_1$ and $\mathscr{D}_2$. A matrix element of this representation is defined by the relations

$$R(\varphi_i \chi_j) = \sum_{kl} \varphi_k \chi_l D_{1,ki}(R) D_{2,lj}(R),$$
$$(D_1 \times D_2)_{kl,ij} = D_{1,ki} D_{2,lj}.$$
(12.34)

The degeneracy of the level $W = W_1+W_2$ of the combined system S_1+S_2 is at most of order $p_1 p_2$. (It may be less because of the limitations due to the Pauli principle.) If we introduce an interaction $\mathscr{H}_{12}$, between the two systems, invariant by the same group G, the degeneracy of the level W may be partially lifted, for $\mathscr{D}_1 \times \mathscr{D}_2$ will not in general be an irreducible representation of the group G. We shall be able to bring it to a reduced form by means of the relations (12.15) and (12.16), since the character of a matrix

$$(D_1 \times D_2)_{ij,kl} = D_{1,ik} D_{2,jl}$$

is clearly

$$\chi(D_1 \times D_2) = \chi(D_1)\chi(D_2). \tag{12.35}$$

The problem of reduction of direct products of representations is closely connected with the calculation of matrix elements of various physical operators. A special case is the establishment of selection rules that correspond to the vanishing of certain matrix elements.

Consider first a function $\Psi_\alpha(\mathbf{r})$, which belongs to an irreducible representation $\mathscr{D}$ of a group G that is not the unit representation. It can be seen that $\int \Psi_\alpha(\mathbf{r})\, d\tau = 0$. Indeed, following the argument that led to the relation (12.28), we have, because of the relation (12.8),

$$\int \Psi_\alpha(\mathbf{r})\, d\tau = \frac{1}{m} \sum_R \int R\Psi_\alpha(\mathbf{r})\, d\tau = \frac{1}{m} \sum_R \sum_\beta D_{\beta\alpha}(R)\left(\int \Psi_\beta(\mathbf{r})\, d\tau\right) = 0. \tag{12.36}$$

Consider now a matrix element of the form

$$(\Psi_\alpha|\, V_\beta\, |\Phi_\gamma) = \int \Psi_\alpha^*(\mathbf{r}) V_\beta(\mathbf{r}) \Phi_\gamma(\mathbf{r})\, d\tau, \tag{12.37}$$

where Ψ_α, V_β, and Φ_γ transform according to irreducible representatins $\mathscr{D}$, $\mathscr{D}'$, $\mathscr{D}''$ of a certain group. The product $\Psi_\alpha^* V_\beta \Phi_\gamma$ transforms according to the representation $\mathscr{D}^* \times \mathscr{D}' \times \mathscr{D}''$ which is, in general, a reducible representation of G, and the matrix element (12.37) will be different from zero only if the representation $\mathscr{D}^* \times \mathscr{D}' \times \mathscr{D}''$ contains the unit representation at least once.

Practically all the group representations that we shall encounter have real characters and are therefore equivalent to their complex conjugates; if this is the case for $\mathscr{D}$, the direct product $\mathscr{D}^* \times \mathscr{D}' \times \mathscr{D}''$ can be replaced by $\mathscr{D} \times \mathscr{D}' \times \mathscr{D}''$. We can show that, if all three representations $\mathscr{D}$, $\mathscr{D}'$, $\mathscr{D}''$ have real characters, the condition that $\mathscr{D} \times \mathscr{D}' \times \mathscr{D}''$ contains the unit representation can be replaced by the requirement that $\mathscr{D} \times \mathscr{D}'$ contains $\mathscr{D}''$. The proof is as follows. The number of times, n_1, that the unit representation is contained in the direct product

$\mathscr{D}_a \times \mathscr{D}_b$ of two irreducible representations $\mathscr{D}_a$ and $\mathscr{D}_b$ is unity if $\mathscr{D}_a$ and $\mathscr{D}_b$ are equivalent and zero if they are not, for n_1 is given by the following relation derived from (12.16) and (12.35):

$$n_1 = \frac{1}{m}\sum_R \chi^{(a\times b)}(R)\chi^1(R) = \frac{1}{m}\sum_R \chi^{(a)}(R)\chi^{(b)}(R) = \delta_{a,b}. \quad (12.38)$$

Thus $(\mathscr{D} \times \mathscr{D}') \times \mathscr{D}''$ contains the unit representation if and only if $(\mathscr{D} \times \mathscr{D}')$ contains $\mathscr{D}''$.

Let us assume now that Ψ_α and Φ_γ of (12.37) belong to the same representation $\mathscr{D} = \mathscr{D}''$. It is easily shown that the trace

$$T_\beta = \sum_\alpha (\Psi_\alpha|\, V_\beta\, |\Psi_\alpha)$$

vanishes unless V_β belongs to the unit representation. We can write

$$T_\beta = \frac{1}{m}\sum_{R,\alpha}(R\Psi_\alpha|\, RV_\beta\, |R\Psi_\alpha)$$

$$= \frac{1}{m}\sum_{R,\alpha,\alpha',\alpha'',\beta'} D^*_{\alpha'\alpha}(R)D_{\alpha''\alpha}(R)D'_{\beta'\beta}(R)(\Psi_{\alpha'}|\, V_{\beta'}\, |\Psi_{\alpha''}). \quad (12.39)$$

Since the matrices D are unitary, $\sum_\alpha D^*_{\alpha'\alpha}(R)D_{\alpha''\alpha}(R) = \delta_{\alpha'\alpha''}$ and (12.39) can be rewritten:

$$T_\beta = \frac{1}{m}\sum_{R,\beta'} D'_{\beta'\beta}(R)T_{\beta'}, \quad (12.40\text{a})$$

which vanishes because of (12.8) unless V_β belongs to the unit representation. Let $\delta W_1^\beta, \ldots, \delta W_p^\beta$ be the energy changes produced in the degenerate level spanned by a set of functions Ψ_α belonging to the representation $\mathscr{D}$, by the perturbation V_β. Because of the invariance of the trace we have

$$\sum_i \delta W_i^\beta = T_\beta = 0, \quad (12.40\text{b})$$

which means that the mean position of the energy levels is unchanged by the perturbation that lifts the degeneracy.

If in the direct product (12.34) the two representations $\mathscr{D}_1$ and $\mathscr{D}_2$ of order p are the same $\mathscr{D}_1 = \mathscr{D}_2 = \mathscr{D}$, the set of functions $\varphi_k\chi_l$ that span the reducible representation $(\mathscr{D})^2 = \mathscr{D} \times \mathscr{D}$ can be broken up into a symmetrical subset $\frac{1}{2}\{\varphi_k\chi_l + \varphi_l\chi_k\}$ that contains $\frac{1}{2}p(p+1)$ terms, and into an antisymmetrical subset $\frac{1}{2}\{\varphi_k\chi_l - \varphi_l\chi_k\}$ that contains $\frac{1}{2}p(p-1)$ terms. It is clear that in the reduction of $(\mathscr{D})^2 = \mathscr{D} \times \mathscr{D}$ the two subsets do not mix and that in the decomposition

$$(\mathscr{D})^2 = \mathscr{D}_a + \mathscr{D}_b + \ldots + \mathscr{D}_l + \ldots + \mathscr{D}_r, \quad (12.41)$$

where the $\mathscr{D}_l$ are irreducible representations, each of them belongs either to the symmetrical or to the antisymmetrical part of $(\mathscr{D})^2$.

The following relations for the characters of the symmetrical and antisymmetrical parts of $(\mathscr{D})^2$ are easily established:

$$\chi^{\mathscr{D}^2}(R) = \sum_{i,j} D_{ii}(R)D_{jj}(R) = \{\chi^{\mathscr{D}}(R)\}^2, \qquad (12.42)$$

$$\begin{cases} \chi_S^{\mathscr{D}^2}(R) = \tfrac{1}{2} \sum_{i,j} \{D_{ii}(R)D_{jj}(R) + D_{ij}(R)D_{ji}(R)\} = \tfrac{1}{2}\{\chi^{\mathscr{D}}(R)\}^2 + \tfrac{1}{2}\{\chi^{\mathscr{D}}(R^2)\}, \\ \chi_A^{\mathscr{D}^2}(R) = \tfrac{1}{2} \sum_{i,j} \{D_{ii}(R)D_{jj}(R) - D_{ij}(R)D_{ji}(R)\} = \tfrac{1}{2}\{\chi^{\mathscr{D}}(R)\}^2 - \tfrac{1}{2}\{\chi^{\mathscr{D}}(R^2)\}. \end{cases}$$
$$(12.43)$$

In (12.42), (12.43), R is any transformation of the group G and the suffixes S and A stand for the symmetrical and antisymmetrical parts of $\mathscr{D}^2$.

13
GROUP THEORY—THE ROTATION GROUP

Few subjects in modern physics have been more extensively studied and have a wider field of applications than the theory of representations of the group of spatial rotations, and the reader is likely to be already familiar with the subject. Two truly great books, *Group theory* (Wigner 1959) and *Irreducible tensorial sets* (Fano and Racah 1958), contain all the information on this subject that we might require. Excellent books such as *Quantum mechanics* (Messiah 1961) and more specifically *Group theory* (Tinkham 1964) and *Angular momentum in quantum mechanics* (Edmonds 1957), to name but three, also give a clear and comprehensive review of the subject. The importance of the group of spatial rotations for the study of paramagnetic resonance stems from the fact that although a paramagnetic ion embedded in bulk matter does not enjoy full rotational symmetry, the starting point for its study is the free ion. The wave-functions of the free ion do transform according to representations of the rotation group and the calculation of matrix elements between two such functions is greatly simplified by this fact. The sole aim of the developments to be given below is to remind the reader of those properties that are of direct interest for the calculation of wave-functions of low-lying energy levels of paramagnetic ions in the surroundings in which they are usually found in nature. Far more sophisticated developments based chiefly on the work of Racah and his school are utilized in atomic spectroscopy (see, for example, Judd (1963b)) but are not of immediate concern to us.

13.1. Angular momentum

Consider an electronic system with a Hamiltonian $\mathscr{H} = \mathscr{H}(\mathbf{p}_i, \mathbf{r}_i, \mathbf{s}_i)$ which may contain forces dependent on the spins $\mathbf{s}_i$ of the individual electrons. It is well known that the total angular momentum of the system, $\mathbf{J}$, is a Hermitian vector operator defined by

$$\begin{cases} \hbar \mathbf{J} = \hbar \mathbf{L} + \hbar \mathbf{S} = \sum_i (\hbar \mathbf{l}_i + \hbar \mathbf{s}_i), \\ = \sum_i \mathbf{r}_i \wedge \mathbf{p}_i + \hbar \sum_i \mathbf{s}_i. \end{cases} \quad (13.1)$$

If the system is in a state described by a wave-function $\Psi(\mathbf{r}_i, \sigma_i)$ where the σ_i are the spin variables, the definition (13.1) enables one to calculate the wave-function resulting from the operation on Ψ of any of the components J_x, J_y, J_z of $\mathbf{J}$. It is also well known that the components of $\mathbf{J}$ obey the commutation rules $[J_x, J_y] = \mathrm{i} J_z$, etc.

From these commutation rules it can be deduced that the operator $\mathbf{J}^2 = J_x^2 + J_y^2 + J_z^2$ commutes with the three components of $\mathbf{J}$ and that it has eigenvalues of the form $j(j+1)$ where j is an integer or half integer. For brevity we shall call (somewhat incorrectly) j an eigenvalue of the operator $\mathbf{J}$. For each value of j there are $(2j+1)$ states such that $\mathbf{J}^2 = j(j+1)$.

Since J_z commutes with $\mathbf{J}^2$, these states can be specified by the eigenvalues of $J_z = m$ and it can be shown that m can take the values $m = j, j-1, \ldots, -j$. If we take as basis functions $\Psi_{j,m} \equiv |j, m\rangle$, which are eigenfunctions of $\mathbf{J}^2$ and J_z, it is possible to choose their phases in such a way that

$$\langle j, m|\, J_\pm\, |j, m\mp 1\rangle = \{j(j+1) - m(m\mp 1)\}^{\frac{1}{2}}, \qquad (13.2\mathrm{a})$$

where $J_\pm = J_x \pm \mathrm{i} J_y$, and also

$$\langle j, m|\, J_z\, |j, m\rangle = m. \qquad (13.2\mathrm{b})$$

The commutation relations are a consequence of the fact that under infinitesimal rotation by an angle ε around an axis, say Oz, the wavefunction Ψ is changed into a function

$$\Phi = (1 - \mathrm{i}\varepsilon J_z)\Psi. \qquad (13.3)$$

Indeed, although the definition (13.1) for $\mathbf{J}$ is the more usual one, it would be more logical to derive it as a consequence of (13.3). It follows from (13.3) that a finite rotation by an angle φ around an axis of unit vector $\mathbf{n}$, transforms Ψ into a function

$$\Phi = \mathrm{e}^{-\mathrm{i}\varphi(\mathbf{n}.\mathbf{J})}\Psi. \qquad (13.4)$$

Since $\mathbf{J}^2$ commutes with the three components of $\mathbf{J}$ it also commutes with all rotational operators $R = \exp\{-\mathrm{i}\varphi(\mathbf{n}.\mathbf{J})\}$. It follows that if a function $\Psi_{j,m}$ is an eigenstate of $\mathbf{J}^2$ so is $R\Psi_{j,m}$ and, since there are $(2j+1)$ linearly independent functions $\Psi_{j,m}$,

$$R\Psi_{j,m} = \sum_{m'=-j}^{j} D^j_{m'm}(R)\Psi_{j,m'}. \qquad (13.5)$$

The $2j+1$ functions $\Psi_{j,m}$, where m takes the values $j, j-1, \ldots, -j$, provide a representation of the rotation group of order $2j+1$ which is denoted as $\mathscr{D}^j$. It can be shown that these representations are irreducible and that the rotation group has no other representations. It is often convenient to specify a rotation by the well-known Euler angles, already mentioned in § 12.3.

It can be shown by a straightforward geometrical reasoning (see for instance Edmonds (1957)) that to a rotation specified by these three angles there corresponds an operator

$$R(\alpha, \beta, \gamma) = e^{-i\alpha J_z} e^{-i\beta J_y} e^{-i\gamma J_z}. \tag{13.6}$$

13.2. The irreducible representations

Consider more particularly the case $j = \tfrac{1}{2}$ when (13.6) can be written

$$R(\alpha, \beta, \gamma) = e^{-i(\alpha/2)\sigma_z} e^{-i(\beta/2)\sigma_y} e^{-i(\gamma/2)\sigma_z}, \tag{13.7}$$

where σ_x, σ_y, σ_z are the well-known Pauli matrices. Inside the manifold $\Psi_{j,m}$ where $j = \tfrac{1}{2}$, $m = \pm\tfrac{1}{2}$ it is easy to show that the matrix representing (13.7) is

$$D^{\tfrac{1}{2}}(\alpha, \beta, \gamma) = \begin{pmatrix} e^{-i\{(\alpha+\gamma)/2\}} \cos\dfrac{\beta}{2} & -e^{-i\{(\alpha-\gamma)/2\}} \sin\dfrac{\beta}{2} \\ e^{-i\{(\alpha-\gamma)/2\}} \sin\dfrac{\beta}{2} & e^{-i\{(\alpha+\gamma)/2\}} \cos\dfrac{\beta}{2} \end{pmatrix}. \tag{13.8}$$

This follows immediately from (13.7) through the relation

$$e^{-i(\beta/2)\sigma_y} = \cos\dfrac{\beta}{2} - i\sigma_y \sin\dfrac{\beta}{2}. \tag{13.9}$$

An inspection of the matrix (13.8) shows that it is unitary (as indeed it must be since it represents a unitary operator) and that its determinant is equal to unity. Conversely it can be shown that any unitary matrix u with determinant equal to unity can be cast in the form (13.8). It is also often written in the form

$$\begin{pmatrix} a & b \\ -b^* & a^* \end{pmatrix} \text{ with } aa^* + bb^* = 1, \tag{13.10}$$

which exhibits its properties; (13.10) can be rewritten as

$$\begin{pmatrix} \alpha' + i\beta' & \gamma' + i\delta' \\ -\gamma' + i\delta' & \alpha' - i\beta' \end{pmatrix} \tag{13.11}$$

with $\alpha'^2 + \beta'^2 + \gamma'^2 + \delta'^2 = 1$, or in the operator form

$$\alpha' + i\beta'\sigma_z + i\delta'\sigma_x + i\gamma'\sigma_y. \tag{13.12}$$

Unitary matrices u with a determinant equal to unity clearly form a group U that is called the unitary unimodular group. We can thus say that U and $\mathscr{D}^{\frac{1}{2}}$ are the same group. The correspondence between this group and the group of spatial rotations is established as follows: every two-dimensional Hermitian matrix h with zero trace can be written

$$h = \mathbf{r} \cdot \boldsymbol{\sigma} = x\sigma_x + y\sigma_y + z\sigma_z$$
$$= \begin{pmatrix} z & x-iy \\ x+iy & -z \end{pmatrix}, \tag{13.13}$$

where the coefficients x, y, z are real and its determinant is $-(x^2+y^2+z^2)$. Consider the matrix

$$h' = uhu^+ \tag{13.14}$$

where u is unitary and unimodular. h' is clearly Hermitian and also has trace zero. It can therefore be written as

$$h' = x'\sigma_x + y'\sigma_y + z'\sigma_z = \mathbf{r}' \cdot \boldsymbol{\sigma},$$

where x', y', z' are real; as is apparent from (13.14) they are linear functions of x, y, z with coefficients that are necessarily real and depend on the elements of the matrix u in a manner we need not specify. The determinant of h', $-(x'^2+y'^2+z'^2)$ is equal to that $-(x^2+y^2+z^2)$ of h. It is therefore clear that the relation resulting from (13.14), between x', y', z' and x, y, z or $\mathbf{r}' = R_u \mathbf{r}$, is a real orthogonal matrix. It will be a spatial rotation if its determinant is $+1$, which is easy to check by a direct calculation.

From the relations

$$h = \mathbf{r} \cdot \boldsymbol{\sigma}, \qquad h' = uhu^+ = \mathbf{r}' \cdot \boldsymbol{\sigma}, \qquad r' = R_u \mathbf{r},$$

we have

$$h'' = vh'v^+ = (vuhu^+v^+) = \mathbf{r}'' \cdot \boldsymbol{\sigma},$$

or

$$\mathbf{r}'' = R_v \mathbf{r}' = R_v R_u \mathbf{r},$$

and it follows that if to the matrix u there corresponds the rotation R_u and to the matrix v the rotation R_v, to the product uv there corresponds the rotation $R_u R_v$. On the other hand, the correspondence between the matrix u and the rotation R_u is not one-to-one for it appears from the relation (13.14), which actually specifies R_u, that a change of u into $-u$ does not change h' and therefore leaves R_u unchanged. Hence to each rotation R_u there correspond two matrices u and $-u$. In particular, to the element unity of the rotation group there correspond the two matrices $\begin{pmatrix} 1 & \\ & 1 \end{pmatrix}$ and $\begin{pmatrix} -1 & \\ & -1 \end{pmatrix}$. It follows that $\mathscr{D}^{\frac{1}{2}}$ is not strictly speaking a representation of the group of spatial rotations: it is what is called a two-valued representation. This appears very clearly in formula (13.8) since an increase of, say, α by an angle 2π, which brings the system (or the coordinate axes) back into the original position, changes $D^{\frac{1}{2}}$ into $-D^{\frac{1}{2}}$. The fact that a rotation by an angle 2π leads to a change in the sign of the wave-function is not a paradox. The value Ψ of the wave-function is not a physically observable quantity as contrasted for instance with $|\Psi|^2$, which is observable and indeed remains unchanged under a rotation through 2π. The same two-valuedness applies to all representations D^j when j is half an odd integer (we shall say for brevity half-integer). This is seen most easily by considering rotations around the z-axis when the rotation operator can be written $e^{-i\varphi J_z}$. This operator acting upon a wave-function $|j, m\rangle$ multiplies it by $e^{-i\varphi m}$, which is -1 for $\varphi = 2\pi$ if j is half integer.

It is thus more correct to say that the representations D^j are representations of the group U, which is the group of the matrices $u = D^{\frac{1}{2}}$. To each matrix u there corresponds a single spatial rotation R_u and a single matrix $D^j(R_u)$ for each j. On the other hand, corresponding to two opposite matrices u and $-u$ we have the same rotation R_u, the single matrix $D^j(R_u)$ for j integer, and two opposite matrices $\pm D^j(R_u)$ if j is half-integer.

It is useful to know the matrix elements $D^j_{mm'}(R)$ which, according to (13.5), determine the transformation of the wave-functions $\Psi_{j,m} = (j, m)$, which for a free ion are eigenstates of the energy and thus form a convenient basis for the study of the bound ion. For $j = \frac{1}{2}$ they are given by (13.8). For higher values of j they have no simple analytical form and the usual advice of looking them up in tables still holds. At a pinch they can be calculated using the following remarks. The relation

$$\begin{aligned} D^j_{m'm}(\alpha, \beta, \gamma) &= \langle jm'| \, e^{-i\alpha J_z} e^{-i\beta J_y} e^{-i\gamma J_z} \, |jm\rangle \\ &= e^{-i(\alpha m' + \gamma m)} \langle jm'| \, e^{-i\beta J_y} \, |jm\rangle \\ &= e^{-i(\alpha m' + \gamma m)} D^j_{mm'}(0, \beta, 0), \end{aligned} \quad (13.15)$$

shows that only the simpler quantity $d^j_{mm'}(\beta) = D^j_{mm'}(0, \beta, 0)$ need be computed. Its evaluation is greatly facilitated by the following property that we quote without proof. Let ξ and η be the two components of a two-dimensional vector that transform through the $D^{\frac{1}{2}}$ matrices. It can be shown that the $2j+1$ quantities

$$\frac{\xi^{j+m}\eta^{j-m}}{\sqrt{\{(j+m)!\,(j-m)!\}}}$$

with $m = j, j-1, \ldots, -j$ transform according to the matrices $D^j_{mm'}$. This provides a method of calculating the latter, knowing $D^{\frac{1}{2}}$.

The characters of the representation D^j are easiest to calculate by considering the rotations through an angle φ around the z-axis. Rotations through the same angle φ around another axis belong to the same class and have the same characters. The operator R takes the form $e^{-i\varphi J_z}$ and

$$D^j_{m'm}(R) = \delta_{mm'}e^{-im\varphi},$$

whence

$$\chi^j(\varphi) = \sum_{m=-j}^{j} e^{-im\varphi} = \frac{\sin(j+\tfrac{1}{2})\varphi}{\sin(\tfrac{1}{2}\varphi)}. \tag{13.16}$$

13.3. The coupling of angular momenta

Clebsch–Gordan and Wigner coefficients

The coupling of angular momenta is one of the best known laws of physics. Given two angular momenta $\mathbf{j}_1$ and $\mathbf{j}_2$ with eigenvalues j_1, j_2 the vector $\mathbf{j} = \mathbf{j}_1 + \mathbf{j}_2$ has eigenvalues $j = j_1+j_2, j_1+j_2-1, \ldots, |j_1-j_2|$. In particular, the value $j = 0$ appears only if $j_1 = j_2$.

From the viewpoint of group representation it has the following meaning. Consider two sets of wave-functions $\Psi_{j_1 m_1} = |j_1 m_1\rangle$ and $\Psi_{j_2 m_2} = |j_2 m_2\rangle$, which are respectively bases for the representations $\mathscr{D}^{j_1}$ and $\mathscr{D}^{j_2}$. The set of all the products $\Psi_{j_1 m_1}\Psi_{j_2 m_2}$ is a basis for the representation $\mathscr{D}^{j_1} \times \mathscr{D}^{j_2}$. This representation is reducible and contains the following irreducible representations:

$$\mathscr{D}^{j_1} \times \mathscr{D}^{j_2} = \mathscr{D}^{j_1+j_2} + \mathscr{D}^{j_1+j_2-1} + \ldots + \mathscr{D}^{|j_1-j_2|}. \tag{13.17}$$

An important point is that every representation $\mathscr{D}^j$ with

$$|j_1-j_2| < j < j_1+j_2$$

is contained only once in the decomposition (13.17).

The equation (13.17) has a symbolical character. Its matrix transcription is

$$S(D^{j_1} \times D^{j_2})S^{-1} = D^{j_1+j_2} + \ldots + D^{|j_1-j_2|}, \tag{13.18}$$

where S is a unitary matrix that relates the basic states $|j_1 m_1\rangle|j_2 m_2\rangle$ of $D^{j_1} \times D^{j_2}$ to the basic states $|j_1 j_2 j m\rangle$ with respect to which the matrices of the representation $\mathscr{D}^{j_1} \times \mathscr{D}^{j_2}$ take the reduced quasi-diagonal form

$$\begin{pmatrix} D^{j_1+j_2} & & \\ & \cdot & \\ & & \cdot \\ & & & D^{|j_1-j_2|} \end{pmatrix} \qquad (13.19)$$

The elements of the matrix S are represented by the self-evident symbols $\langle j_1 m_1, j_2 m_2 | j_1 j_2, jm \rangle$, which are known in the literature as Clebsch–Gordan coefficients. Their properties have been studied extensively and tables of their numerical values have been prepared; we shall therefore be very brief on this subject. According to the vector coupling law they vanish unless $|j_1-j_2| \leq j \leq j_1+j_2$ and also if $m \neq m_1+m_2$. The latter follows from the fact that a rotation by an angle φ around the z-axis multiplies $|j_1 m_1\rangle|j_2 m_2\rangle$ by $e^{-i(m_1+m_2)\varphi}$ and $|j, m\rangle$ by $e^{-im\varphi}$.

The phases of the various basis vectors can be chosen so that the Clebsch–Gordan coefficients are real, a point that is by no means trivial. From the unitary character of the S matrix, obvious orthogonality relations for the Clebsch–Gordan coefficients can be deduced such as, for instance,

$$\sum_{m_1 m_2} \langle j_1 j_2 m_1 m_2 | j_1 j_2 jm \rangle \langle j_1 j_2 m_1 m_2 | j_1 j_2 j'm \rangle = \delta_{jj'}. \qquad (13.20)$$

Clebsch–Gordan coefficients possess somewhat complicated symmetry properties that are best exhibited by introducing related coefficients called Wigner coefficients or $3j$-symbols denoted by

$$\begin{pmatrix} j_1 & j_2 & j_3 \\ m_1 & m_2 & m_3 \end{pmatrix}$$

with $m_1+m_2+m_3 = 0$ and defined by the relation

$$\begin{pmatrix} j_1 & j_2 & j_3 \\ m_1 & m_2 & m_3 \end{pmatrix} = \frac{(-1)^{j_1-j_2-m_3}}{\sqrt{(2j_3+1)}} \langle j_1 m_1 j_2 m_2 | j_1 j_2; j_3, -m_3 \rangle. \qquad (13.21)$$

Wigner coefficients have very simple symmetry properties that are easy to memorize. An interchange of two columns multiplies a Wigner coefficient by $(-1)^{j_1+j_2+j_3}$ and so does the change of m_1, m_2, m_3 into $-m_1$, $-m_2$, $-m_3$. The more complicated symmetry properties of

Clebsch–Gordan coefficients can be obtained from these, using (13.21). It is worth noting that

$$\langle j_1 m_1 j_2 m_2 | j_1 j_2 j m \rangle \neq \langle j_2 m_2 j_1 m_1 | j_2 j_1 j m \rangle. \tag{13.22}$$

The following relations can be established between $|j_1 m_1\rangle |j_2 m_2\rangle$ and $|j_1 j_2 j m\rangle$.

$$|j_1 j_2 j m\rangle = (-1)^{j_2 - j_1 - m} \sum_{m_1 m_2} \sqrt{(2j+1)} \begin{pmatrix} j_1 & j_2 & j \\ m_1 & m_2 & -m \end{pmatrix} |j_1 m_1\rangle |j_2 m_2\rangle, \tag{13.23}$$

$$|j_1 m_1\rangle |j_2 m_2\rangle = \sum_{jm} (-1)^{j_2 - j_1 - m} \sqrt{(2j+1)} \begin{pmatrix} j_1 & j_2 & j \\ m_1 & m_2 & -m \end{pmatrix} |j_1 j_2 j m\rangle. \tag{13.24}$$

13.4. Multiple vector coupling and Racah symbols

Given three vectors $\mathbf{j}_1$, $\mathbf{j}_2$, $\mathbf{j}_3$ there is more than one way of coupling them to a value $\mathbf{j}$. Consider first the case when $\mathbf{j} = 0$. According to the laws of vector coupling this can only be achieved by coupling $\mathbf{j}_1$ and $\mathbf{j}_2$ to a value $\mathbf{j}_{12}$ equal to $\mathbf{j}_3$, or $\mathbf{j}_1$ and $\mathbf{j}_3$ to $\mathbf{j}_2$, or $\mathbf{j}_2$ and $\mathbf{j}_3$ to $\mathbf{j}_1$. The important point is that all these coupling schemes, within a phase factor ± 1, lead to the same non-degenerate state, invariant under spatial rotation.

The symbol $|j_1 j_2 j_3, 00\rangle$ represents the state of zero angular momentum where $\mathbf{j}_1$ and $\mathbf{j}_2$ are coupled to a value $\mathbf{j}_{12} = \mathbf{j}_3$ and $\mathbf{j}_{12}$ is then coupled to $\mathbf{j}_3$ to give $\mathbf{j} = 0$. It can be shown that the following relation holds:

$$|j_1 j_2 j_3, 00\rangle = \sum_{m_1 m_2} \begin{pmatrix} j_1 & j_2 & j_3 \\ m_1 & m_2 & m_3 \end{pmatrix} |j_1 m_1\rangle |j_2 m_2\rangle |j_3 m_3\rangle. \tag{13.25}$$

The symmetry of $|j_1 j_2 j_3, 00\rangle$ with respect to an interchange of two indices follows immediately from that of the Wigner coefficients: an interchange of two indices multiplies the result by $(-1)^{j_1 + j_2 + j_3}$.

We consider now the general case when the sum $\mathbf{J}$ of the three vectors $\mathbf{j}_1$, $\mathbf{j}_2$, $\mathbf{j}_3$ is different from zero. We wish to calculate the scalar product of the two states

$$\begin{cases} |a\rangle = |(j_1 j_2) j_{12}, j_3; JM\rangle, \\ \text{and} \\ |b\rangle = |j_1, (j_2 j_3) j_{23}; JM'\rangle, \end{cases} \tag{13.26}$$

where $|a\rangle$ is obtained by coupling $\mathbf{j}_1$ and $\mathbf{j}_2$ to a value $\mathbf{j}_{12}$ and the resulting vector $\mathbf{j}_{12}$ to $\mathbf{j}_3$ to give $\mathbf{J}$ and $|b\rangle$ by first coupling $\mathbf{j}_2$ and $\mathbf{j}_3$ to $\mathbf{j}_{23}$. Since $|a\rangle$ and $|b\rangle$ are two basis states that both belong to the same irreducible

GROUP THEORY—THE ROTATION GROUP

representation D^J, their scalar product vanishes if $M \neq M'$ and is independent of M if $M = M'$. This is a special case of eqn (12.33) where V is taken equal to unity. We can then drop the index M and write the product $\langle a \mid b \rangle$ as

$$\langle a \mid b \rangle = \langle (j_1 j_2) j_{12}, j_3, J \mid j_1, (j_2 j_3) j_{23}, J \rangle. \tag{13.27}$$

Racah introduces a function of six variables

$$\begin{pmatrix} j_1 & j_2 & j_{12} \\ j_3 & J & j_{23} \end{pmatrix}$$

called the six-j symbol which is essentially just this scalar product apart from a numerical factor. By definition of the six-j symbol, we have

$$\langle (j_1 j_2) j_{12}, j_3, J \mid j_1, (j_2 j_3) j_{23}, J \rangle$$
$$\equiv (-1)^{j_1+j_2+j_3+J} \sqrt{\{(2j_{12}+1)(2j_{23}+1)\}} \begin{pmatrix} j_1 & j_2 & j_{12} \\ j_3 & J & j_{23} \end{pmatrix}. \tag{13.28}$$

We shall not attempt to give an analytical expression for the six-j symbol, which is rather complicated. Numerical values have been tabulated by Rotenberg, Bivins, Metropolis, and Wooten (1959). The six-j symbol

$$\begin{pmatrix} a & b & c \\ d & e & f \end{pmatrix}$$

has some remarkable symmetry properties that are easily memorized by means of the so-called tetrahedron diagram. Construct a tetrahedron and call the three edges of any one plane a, b, c and the three opposite edges respectively d, e, f. The symbol

$$\begin{pmatrix} a & b & c \\ d & e & f \end{pmatrix}$$

has the symmetry of this figure. For instance, selecting another plane, say aef, we obtain the symbol

$$\begin{pmatrix} a & e & f \\ d & b & c \end{pmatrix},$$

which is equal to the former.

From the definition (13.28) of the six-j symbol it appears that they can be expressed in a straightforward manner as a sum of products of four Clebsch–Gordan coefficients. We shall not write down these expressions explicitly.

13.5. Irreducible tensor operators, the Wigner-Eckart theorem, and equivalent operators

Consider an operator T that acts on a wave-function Ψ, the result being another function Φ as symbolized by the equation

$$\Phi = T\Psi. \tag{13.29}$$

Let R be a transformation belonging to a group G that transforms Ψ into $\overline{\Psi}$ and Φ into $\overline{\Phi}$ according to $\overline{\Psi} = R\Psi$, $\overline{\Phi} = R\Phi$. From (13.29) we obtain for the relationship between $\overline{\Phi}$ and $\overline{\Psi}$,

$$\overline{\Phi} = RTR^{-1}\overline{\Psi} = \overline{T}\overline{\Psi}. \tag{13.30}$$

The operator $\overline{T} = RTR^{-1}$ is said to be the transform of T under the transformation R.

A very important class of operators are the irreducible tensors. By definition, a tensor operator T_k of order k is a set of $2k+1$ operators T_k^q, where $q = k, k-1, \ldots, -k$, which under the group of spatial rotations transform according to the representation $\mathscr{D}^k$; i.e.,

$$RT_k^q R^{-1} = \sum_{q'} D_{q'q}^k(R) T_k^{q'}. \tag{13.31}$$

The simplest tensor operator is a scalar, invariant under spatial rotations, with a single component. Next come vectors with three components, etc.

If we take for R the infinitesimal rotations $R = 1 - i\varepsilon J_z$ and $R = 1 - i\varepsilon J_\pm$, we obtain from (13.31),

$$(1-i\varepsilon J_z)T_k^q(1+i\varepsilon J_z) = \sum_{q'} (kq'|\,1-i\varepsilon J_z\,|kq)T_k^{q'}, \tag{13.32}$$

$$(1-i\varepsilon J_\pm)T_k^q(1+i\varepsilon J_\pm) = \sum_{q'} (kq'|\,1-i\varepsilon J_\pm\,|kq)T_k^{q'}. \tag{13.33}$$

Expansion of these relations, neglecting terms of order ε^2, gives, using equations (13.2),

$$\begin{cases} [J_z, T_k^q] = qT_k^q, \\ [J_\pm, T_k^q] = \{k(k+1)-q(q\pm 1)\}^{\frac{1}{2}} T_k^{q\pm 1}. \end{cases} \tag{13.34}$$

The importance of irreducible tensors stems from the following theorem known in the literature as the Wigner–Eckart theorem. Given a tensor operator T_k^q, consider the set of all the matrix elements $(\alpha jm|\,T_k^q\,|\alpha'j'm')$ where α and α' stand for any extra parameters that may be necessary to specify the states $|jm\rangle$ and $|j'm'\rangle$. For fixed values of α, j, k, α', j', k' there are altogether $(2j+1)(2k+1)(2j'+1)$ such matrix elements. The Wigner–Eckart theorem states that all these matrix elements are uniquely determined within a single multiplying factor. The proof is as follows. Let us for brevity represent the matrix element $(\alpha jm| T_k^q |\alpha'j'm')$ as f_i, the index i standing for the indices m, q, m'. Under a rotation of the coordinate axes, f_i transforms according to the triple direct product $D^{j^*} \times D^k \times D^{j'}$. We can reduce this product by a similarity transformation to a sum of irreducible representations D^J by introducing a set of symbols $g_M^{\beta J}$ which transform according to D^J and are linear combinations of the f_i,

$$\begin{cases} f_i = \sum_{J,\beta} (i \mid \beta JM) g_M^{\beta J}, \\ g_M^{\beta J} = \sum_i (\beta JM \mid i) f_i. \end{cases} \quad (13.35)$$

The index β takes into account the fact that in the reduction of the product $D^{j^*} \times D^k \times D^{j'}$ to $\sum D^J$, a given value J may appear more than once. The value $J = 0$, however, can appear only once since there is only one way of coupling three angular momenta to zero and the first formula (13.35) can be rewritten:

$$f_i = \sum_{\beta, J \neq 0} (i \mid \beta JM) g_M^{\beta J} + (i \mid 00) g_0^0. \quad (13.36)$$

The coefficients $(i \mid \beta JM)$ depend solely on the properties of the rotation group and are independent of the tensor T_k^q and of the parameters α, α' needed to specify the states $|\alpha jm\rangle$ and $|\alpha'j'm'\rangle$. In particular, the coefficient $(i \mid 00)$ is, according to eqn (13.25) for the invariant triple product, proportional to the Wigner coefficient

$$\begin{pmatrix} j & k & j' \\ -m & q & m' \end{pmatrix}$$

times a phase factor $(-1)^{j-m}$. Here the change of m into $-m$ in the Wigner coefficient, and the phase factor, are due to the fact that in the matrix element $(jm| T_k^q |j'm')$ it is the complex conjugate function Ψ_j^{m*} that appears.

Using the symmetry properties of the Wigner coefficients and the definition (13.21), we find

$$(-1)^{j-m} \begin{pmatrix} j & k & j' \\ -m & q & m' \end{pmatrix} = (-1)^{j-m}(-1)^{j+k+j'} \begin{pmatrix} j' & k & j \\ m' & q & -m \end{pmatrix}$$
$$= (-1)^{j-m+j+k+j'+j'-k+m} \frac{\langle j'm'kq \mid j'qjm \rangle}{\sqrt{(2j+1)}} = \frac{\langle j'm'kq \mid j'qjm \rangle}{\sqrt{(2j+1)}}. \quad (13.37)$$

On the other hand, a matrix element such as f_i does *not* depend on the orientation of the coordinate axes since it is an integral over space (and possibly spin) variables and a rotation of the axes is simply a change of integration variables. This is only possible if in (13.35) and (13.36) all the $g_M^{\beta J}$ with $J \neq 0$ vanish, whence $f_i = (i \mid 00)g_0^0$ which establishes the theorem, since a single constant g_0^0 distinguishes the set of matrix elements of T_k^q from those of another tensor operator of the same order k.

The gist of the proof is that in the reduction of the triple product $D^{j*} \times D^k \times D^{j'}$ the unit representation D^0 appears only once. We shall see later that for groups of lower symmetry this is not always true and that matrix elements of the form $(\Psi_\alpha' \mid V_\beta \mid \Psi_\gamma)$, where Ψ_α', V_β, Ψ_γ belong to three representations of, say, the cubic group, may require more than one constant.

The Wigner–Eckart theorem is usually expressed in the following form, consistent with (13.37),

$$(\alpha j m \mid T_k^q \mid \alpha' j' m') = \frac{1}{\sqrt{(2j+1)}} (\alpha j \| T_k \| \alpha' j') \langle j'm', kq \mid j'k; jm \rangle, \quad (13.38)$$

where $(\alpha j \| T_k \| \alpha' j')$ is independent of the magnetic quantum numbers m', q, m. The factor $1/\sqrt{(2j+1)}$ could equally well be included in its definition; we conform to the usual practice in leaving it out. Once a single matrix element $(\alpha j m_0 \mid T_k^{q_0} \mid \alpha' j' m_0')$ has been calculated for one set of magnetic quantum numbers m_0, q_0, m_0' (this is the hard part), all the other matrix elements can be computed from (13.38) if a numerical table of the relevant Clebsch–Gordan coefficients is available.

A slightly different way of applying the Wigner–Eckart theorem, sometimes more convenient than the direct use of formula (13.38), especially for the calculation of diagonal matrix elements $j = j'$, $\alpha = \alpha'$, is the so-called method of operator equivalents. Consider a component S_k^q of a tensor operator whose matrix elements are for some reason particularly easy to compute such as, for instance,

$$S_2^0 = 3J_z^2 - J(J+1)$$

for which

$$(\alpha j m \mid S_2^0 \mid \alpha j m') = \delta_{mm'}\{3m^2 - J(J+1)\}. \quad (13.39)$$

If we are interested in, say, the operator $T_2^0 = \sum_p (3z_p^2 - r_p^2)$ where the summation is over all the electrons of the ion, the Wigner–Eckart theorem enables us to write that inside the manifold $\mid \alpha, j \rangle$.

$$T_2^0 = \beta S_2^0,$$

Here the constant β may be obtained by calculating the expectation value of T_2^0 over a selected substate $|\alpha, j, m_0\rangle$ and writing

$$\beta\{3m_0^2 - j(j+1)\} = (\alpha, j, m_0|\, T_2^0\, |\alpha, j, m_0)$$

$$= \int \Psi^*(\mathbf{r}) \left|\sum_p (3z_p^2 - r_p^2)\right| \Psi(\mathbf{r})\, d\tau. \quad (13.40)$$

The method of operator equivalents is also convenient when a non-standard set of components such as, for instance, Cartesian tensor components are used for the tensor operators. As an example, consider a vector with three Cartesian components V_x, V_y, V_z. The three linear combinations of these components which transform under rotation in a standard way by means of the matrices $D^1_{mm'}$ are

$$V_1^0 = V_z, \qquad V_1^1 = -\frac{V_x + iV_y}{\sqrt{2}}, \qquad V_1^{-1} = \frac{V_x - iV_y}{\sqrt{2}}. \quad (13.41)$$

Inside the manifold J we can write

$$V_1^0 = \beta J_z, \qquad V_1^1 = -\frac{\beta}{\sqrt{2}} J_+, \qquad V_1^{-1} = \frac{\beta}{\sqrt{2}} J_-, \quad (13.42)$$

which we can express in the condensed form $\mathbf{V} = \beta \mathbf{J}$ and use any set of components we please to express this relation in a coordinate form. (Anticipating the Wigner–Eckart theorem we already made use of this result in the calculation of the Landé factor in § 11.2.)

Similarly from the components of the vector $\mathbf{J}$ we can build up an irreducible tensor S_2 with components S_2^q given by

$$\begin{cases} S_2^0 = 3J_z^2 - J(J+1), \\ S_2^{\pm 1} = \dfrac{\sqrt{6}}{2}(J_z J_\pm + J_\pm J_z), \\ S_2^{\pm 2} = \dfrac{\sqrt{6}}{2} J_\pm^2, \end{cases} \quad (13.43)$$

and write that inside the manifold J all second-order irreducible tensors T_2^q have matrix elements proportional to those of S_2^q.

Here too it may be more convenient to introduce S_2 as a symmetrical traceless Cartesian tensor with components

$$S_{ik} = \tfrac{3}{2}(J_i J_k + J_k J_i) - J(J+1)\delta_{ik}, \quad (13.44)$$

and state that inside the manifold J any other symmetrical traceless tensor T_{ik} will be of the form $T_{ik} = \beta S_{ik}$. For instance the tensor T, for which

$$T_{ik} = \sum_p \{3x_i^p x_k^p - (r^p)^2 \delta_{ik}\}, \quad (13.45)$$

where the x_i^p are the coordinates of the electrons of an atom, is called the electric quadrupole tensor of that atom and within a manifold J is proportional to the tensor (13.44).

The most cogent reason for preferring the equivalent operator formulation of the Wigner–Eckart theorem to the direct use of formula (13.38) is that most of the pioneer work in the field of magnetic resonance (including extensive and very useful numerical tables) is couched in that language. Despite some unfortunate inconsistencies of notation, such as the use of unnormalized spherical harmonics, we shall often use equivalent operators, pointing out occasionally the relationship with the more consistent notation of (13.38). A detailed study of the applications of this method to paramagnetic resonance is deferred until Chapter 16.

It may be useful to the reader to point out that Buckmaster (1962) has defined 'Racah operators' denoted by $\tilde{O}_k^q$ which correspond directly to the reduced spherical harmonics

$$C_k^q(\theta, \phi) = \{4\pi/(2k+1)\}^{\frac{1}{2}} Y_k^q(\theta, \phi)$$

and thus possess the consistency lacking in the operators defined by Stevens (1952a). Smith, and Thornley (1966) give a complete list of such operators up to $k = q = 6$, and their matrix elements are listed in Fortran notation by Birgeneau (1967b).

14

THE CUBIC GROUP
AND SOME OTHER GROUPS

14.1. The cubic group

We shall study this group in much greater detail than any other finite group, not only because cubic symmetry occurs quite often but even more so because it is the only type of symmetry for which the theory of group representations is really indispensable in paramagnetic resonance. This theory can often be dispensed with for environments of lower symmetry. Many of the results derived can be extended in a straightforward manner to other groups.

A level of a free ion with a given value J of its angular momentum and therefore a degeneracy of order $(2J+1)$ is spanned by a representation D^J of the rotation group with characters given by (13.16):

$$\chi^J(\varphi) = \sin\{(J+\tfrac{1}{2})\varphi\}/\sin(\tfrac{1}{2}\varphi).$$

In order to discover how this level is split in an environment of cubic symmetry we must obtain the irreducible representations of the cubic group O. We consider first the case when J is an integer. As we saw earlier, there is then a one-to-one correspondence between a rotation R and a matrix $D^J(R)$.

The cubic group O is the group of rotations that leaves invariant a cube or a regular octahedron. It contains the following classes:

E, the identity operation (1 element),

C_2, rotations through an angle π around the three axes perpendicular to the faces of the cube (3 elements),

C_4, rotations through the angles $\pm\pi/2$ around the same axes (6 elements),

C_2', rotations through π around the 6 axes passing through the centre points of opposite edges; these axes are parallel to the face-diagonals (6 elements),

C_3, rotations through angles $\pm(2\pi/3)$ around the 4 body diagonals (8 elements).

We have altogether 24 elements and 5 classes. In contrast to the situation in the full rotation group, two rotations through the same angle but around two different axes X and Y do not belong to the same class if the rotation C that brings X to Y is not an element of the group. This is why the two rotations through an angle π, C_2, and C', belong to two different classes.

Since a regular octahedron is a figure obtained by joining the centres of the six faces of a cube it is clear that it is left invariant by the same group. Note also that each operation of O amounts to a permutation of the four body-diagonals. Hence the cubic group is isomorphic to the group of permutations S_4.

Since there are 5 classes and 24 elements, the cubic group has 5 irreducible representations with dimensions $l_1, \ldots, l_5$ which obey the relation

$$l_1^2 + l_2^2 + l_3^2 + l_4^2 + l_5^2 = 24. \tag{14.1}$$

It is easy to check that the only solution of (14.1) is the set of 5 integers 1, 1, 2, 3, 3. The corresponding representations are denoted in the literature by the symbols Γ_1, Γ_2, Γ_3, Γ_4, Γ_5 or A_1, A_2, E, T_1, T_2. The characters can be obtained by the method described in paragraph 12.4 and their values are listed in Table 1.

We shall see shortly that it is possible to choose the basis functions of the irreducible representations of O to be real. Then the unitary matrices of these representations become real orthogonal matrices (that the characters are all real in Table 1 is only a necessary condition). The calculation of matrix elements of the form $(\Psi_\alpha| \, V \, |\Psi_{\alpha'})$, where Ψ_α and $\Psi_{\alpha'}$ belong to representations Γ and Γ' of O, is facilitated by the knowledge of the reducibility of the direct products $\Gamma^* \times \Gamma'$ or, since all the Γ can be made real, of $\Gamma \times \Gamma'$. The results are given in Table 2. For diagonal direct products $\Gamma \times \Gamma$ we have indicated by a superscript S or A whether an irreducible representation Γ_i belongs to the symmetric or antisymmetric part of $\Gamma \times \Gamma$. This can be established by using the formulae (12.43) for the characters.

The formula (13.16) yields the following values for the characters of $D^J(R)$ where R is a rotation belonging to the cubic group,

$$\begin{aligned} \varphi &= 0 & \chi(E) &= 2J+1 \\ \varphi &= \pi & \chi(C_2) &= \chi(C_2') = (-1)^J \\ \varphi &= \pm\frac{\pi}{2} & \chi(C_4) &= (-1)^{I(J/2)} \\ \varphi &= \pm\frac{2\pi}{3} & \chi(C_3) &= \frac{\sin\left(\frac{2J+1}{3}\pi\right)}{\sin(\pi/3)} \end{aligned} \tag{14.2}$$

where the symbol $I(J/2)$ means the integer part of $J/2$. With the help of

(14.2) and of the formulae (12.16), we obtain Table 3 for the reduction of D^J by the cubic group for integer values of J up to 10.

All free ions with an even number of electrons naturally have an integral value of J. However, even for an odd number of electrons, we saw earlier that in fields of intermediate strength such as occur in the iron group, the relevant quantum number of the free ion was not the total angular momentum J but the total orbital momentum L, which is always an integer. The wave-functions that span a representation D^J, where J is an integer, transform as the usual spherical harmonics. We observe in Table 3 that, up to and including $J = 4$, no representation appears more than once in D^J. The 'correct' zero-order wave-functions that span the various representations Γ_i are linear combinations of eigenfunctions of the free ion and can be obtained without explicit knowledge of the cubic Hamiltonian V for $J \leqslant 4$.

They are determined most conveniently by considering the functions

$$(p, q, r) = r^{2l+1} \frac{\partial^l (1/r)}{\partial x^p \, \partial y^q \, \partial z^r}, \tag{14.3}$$

with $p+q+r = l$. These functions are homogeneous polynomials of degree l which satisfy the Laplace equation. Equation (14.3) defines $(l+1)(l+2)/2$ such polynomials but the Laplace equation introduces $l(l-1)/2$ relations between them, leaving $(2l+1)$ independent polynomials.

$l = 1$

The three functions corresponding to (14.3) are (100), (010), (001), which are proportional to x, y, z. They clearly span a three-dimensional representation of the cubic group. A rotation of $\pi/2$ around Oz changes them into $-(010)$, (100), (001), and the trace of the transformation matrix is $\chi(C_4) = 1$, which from Table 1 shows that we are dealing with the representation Γ_4 and not with Γ_5. This can be seen directly from Table 3 since the representation D^1 reduces to Γ_4 under cubic symmetry.

$l = 2$

The functions (011), (101), (110) proportional to yz, zx, xy provide a three-dimensional representation of O that can only be Γ_5 since D^2 is split by O into Γ_3 and Γ_5.

The two-dimensional representation Γ_3 is spanned by the two orthogonal combinations (002) and $(1/\sqrt{(3)})\{(200)-(002)\}$ proportional to $3z^2-r^2$ and $\sqrt{(3)}(x^2-y^2)$. (The $\sqrt{3}$ factor ensures the same normalization for both functions.)

It is quite easy to see how the three wave-functions $\eta_x \propto yz$, $\eta_y \propto zx$, $\eta_z \propto xy$ transform under the cyclic permutations of (x, y, z), R and R^2; that is, under rotations through angles $2\pi/3$ and $4\pi/3$ respectively round the threefold axis (111). It is less easy to see how $\theta \propto (3z^2-r^2)$ and $\varepsilon \propto \sqrt{(3)}(x^2-y^2)$ transform under these conditions; we therefore quote the results

$$R\theta = -\frac{1}{2}\theta + \frac{\sqrt{3}}{2}\varepsilon; \qquad R^2\theta = -\frac{1}{2}\theta - \frac{\sqrt{3}}{2}\varepsilon;$$
$$R\varepsilon = -\frac{\sqrt{3}}{2}\theta - \frac{1}{2}\varepsilon; \qquad R^2\varepsilon = \frac{\sqrt{3}}{2}\theta - \frac{1}{2}\varepsilon. \qquad (14.4)$$

The first of these can easily be verified, for example, by noting that $(3z^2-r^2) = (2z^2-x^2-y^2)$ transforms into

$$(2x^2-y^2-z^2) = \tfrac{3}{2}(x^2-y^2) - \tfrac{1}{2}(2z^2-x^2-y^2).$$

$l = 3$

(111) proportional to (xyz) provides the unidimensional representation Γ_2. Γ_4 is spanned by the three functions (300), (030), (003) proportional to $x(3y^2+3z^2-2x^2)$, etc.

Γ_5 is spanned by the three orthogonal functions (102)–(120), (210)–(012), (021)–(201), proportional respectively to $x(y^2-z^2)$, etc.

$l = 4$

Here we quote just the results:

$\Gamma_1 \{(x^4+y^4+z^4-\tfrac{3}{5}r^4)$,

$$\Gamma_3 \begin{cases} [z^4 - \tfrac{1}{2}(x^4+y^4) - \tfrac{6}{7}r^2\sqrt{(5)}\{z^2-\tfrac{1}{2}(x^2+y^2)\}] \\ \dfrac{\sqrt{3}}{2}\left\{x^4-y^4-\dfrac{6}{7}\dfrac{\sqrt{(15)}}{2}(x^2-y^2)r^2\right\}, \end{cases}$$

$\Gamma_5 \left\{xy\left(z^2-\dfrac{r^2}{7}\right)\right.$ and two cyclic permutations,

$\Gamma_4 \{xy(x^2-y^2)$ and two cyclic permutations.

14.2. The fictitious angular momentum

Let **V** be a vector and consider the set of matrix elements $(\xi_i|\,V_k\,|\xi_j)$ where ξ_x, ξ_y, ξ_z are the three wave-functions spanning a representation Γ_4, which transform under O like x, y, z.

The vector components V_x, V_y, V_z that transform under rotations

THE CUBIC GROUP AND SOME OTHER GROUPS 633

like x, y, z, also transform according to Γ_4 under O. Since the reduction of $\Gamma_4 \times \Gamma_4$ contains Γ_4 only once (see Table 2), it follows that all the matrix elements $(\xi_i| V_k |\xi_j)$ are uniquely determined within a proportionality factor. The same is true of the matrix elements $(\eta_i| V_k |\eta_j)$ where η_x, η_y, η_z span Γ_5 and transform like yz, zx, xy, since $\Gamma_5 \times \Gamma_5$ also contains Γ_4 only once. If we introduce the new functions $|\tilde{m}\rangle$ by the formulae

$$|\pm \tilde{1}\rangle = \mp \frac{\xi_x \pm i\xi_y}{\sqrt{2}}, \qquad |\tilde{0}\rangle = \xi_z,$$

or (14.5)

$$|\pm \tilde{1}\rangle = \mp \frac{\eta_x \pm i\eta_y}{\sqrt{2}}, \qquad |\tilde{0}\rangle = \eta_z,$$

the matrix elements of $\langle \tilde{m}| A_k |\tilde{m}'\rangle$ will be proportional to those of an angular momentum $\tilde{l}$ with $\tilde{l} = 1$, sometimes called the fictitious angular momentum, that is,

$$\langle m| A_k |\tilde{m}'\rangle = \alpha \langle \tilde{l}, m| \tilde{l}_k |\tilde{l}, m'\rangle. \qquad (14.6)$$

Table 2 shows that the reduction of the direct product $\Gamma_3 \times \Gamma_3$ does not contain Γ_4 and therefore a vector has no matrix elements within the doublet Γ_3, a fact sometimes expressed by saying that this doublet is non-magnetic. As above, we shall denote by θ and ε the two functions that span Γ_3 and transform under O like $3z^2 - r^2$ and $\sqrt{(3)}(x^2 - y^2)$.

For many purposes it is convenient to express the various states $|\tilde{m}\rangle$ that originate in the decomposition of a representation D^J as a linear combination of eigenstates $|J, m\rangle$, where $m = J_z$, the axis Oz being one of the C_2 axes of the cube. Table 4 gives these expressions up to $J = 4$. We have written $|m\rangle$ instead of $|J, m\rangle$ for brevity, since the value of J is indicated unambiguously in the table. Besides each representation Γ_4 and Γ_5 we have written the value α of the proportionality coefficient between $\mathbf{J}$ and the fictitious moment $\tilde{\mathbf{l}}$: that is, $\mathbf{J} = \alpha \tilde{\mathbf{l}}$.

14.3. The multiplets Γ_4 and Γ_5 in trigonal axes

If the cubic environment of an ion is distorted along the direction of one of the C_2 axes of the cube, say Oz, it is fairly obvious that the triplet Γ_4 will be split into a doublet spanned by ξ_x, ξ_y (or $|\tilde{1}\rangle$ and $|-\tilde{1}\rangle$) and a singlet $\xi_z = |\tilde{0}\rangle$. The same will be true for Γ_5, with the functions η instead of ξ. The functions ξ_i for Γ_4 and η_i for Γ_5 (or the functions $|\tilde{m}\rangle$ for both) are thus the 'correct' zero-order wave-functions for this distortion of cubic symmetry. On the other hand, for a trigonal distortion, that is a distortion along a body diagonal, it is preferable to use the following functions which will be the 'correct' zero-order functions:

$$|\tilde{1}\rangle_T = a(\xi_x + e^{2\pi i/3}\xi_y + e^{4\pi i/3}\xi_z),$$
$$|\tilde{0}\rangle_T = a(\xi_x + \xi_y + \xi_z),$$
$$|-\tilde{1}\rangle_T = a(\xi_x + e^{-2\pi i/3}\xi_y + e^{-4\pi i/3}\xi_z), \tag{14.7}$$

where a is a normalization constant (for Γ_5, we use η_x, η_y, η_z instead of ξ_x, ξ_y, ξ_z). The functions (14.7) can be obtained from Table 4, but it is preferable then to quantize J_z along a body diagonal. The results for this are given in Table 5 for the representation Γ_5 contained in $J = 2$, and for both Γ_4 and Γ_5 originating in $J = 3$.

14.4. The double cubic group

In order to find how a half-integer J-level of the free ion splits in an environment of cubic symmetry we must introduce the so-called double cubic group O^+. We recall that D^J with J half-integer is not strictly speaking a representation of the spatial rotation group G but rather a representation of the unimodular group $U = D^{\frac{1}{2}}$, and that to each rotation belonging to G there correspond *two* matrices $\pm u$ of U. The double cubic group is then defined unambiguously as follows: consider the 24 rotations in G that belong to O, subgroup of G; to these correspond twice as many, namely 48, matrices of U, which form a subgroup o^+ of U. The abstract group that has the same multiplication table as o^+ is, by definition, the double cubic group O^+. The group O^+ is obtained by adding to O the element R which is represented in o^+ by the matrix $\begin{pmatrix} -1 & \\ & -1 \end{pmatrix}$. R commutes with all the other elements of O^+ and its square is equal to the unit element: $R^2 = E$.

If A is an element of O, every element of O^+ is either A or $RA = AR$. This definition of the double group O^+ can clearly be extended to any finite rotation group.

The irreducible representations Γ_1 to Γ_5 of O are also irreducible representations of O^+. They are representations where R is represented by the unit matrix, but there are others. To elucidate them we must find the classes of O^+. One might think naïvely that as O^+ has twice as many elements as O it has twice as many classes, those already listed for O plus

$$R = RE, RC_2, RC_2', RC_3, RC_4.$$

Actually this is not so: C_2 and RC_2 belong to the same class and similarly C_2' and RC_2'. This can be shown as follows (Opechowski 1940).

Since the two elements C and RC of a double group have characters of opposite sign it is a necessary condition that these must vanish if C and RC belong to the same class. From (13.16) the rotation that corresponds to these two elements is therefore through an angle π. Let us choose its axis as the z-axis. According to (13.8) with $\alpha = \pi$, $\beta = \gamma = 0$ we can represent C by the matrix $c = \begin{pmatrix} -i & 0 \\ 0 & i \end{pmatrix}$ and RC by $-c$. If there is in the group a twofold axis perpendicular to the z-axis let us choose it as the y-axis, the corresponding rotation $\alpha = \gamma = 0$, $\beta = \pi$ according to (13.8) can be represented by

$$b = \begin{pmatrix} 0 & -1 \\ +1 & 0 \end{pmatrix}$$

(or by $-b$) and one verifies immediately that $bcb^{-1} = -c$, which means that c and $-c$ belong to the same class.

We have thus shown that in a double group two elements, C and RC, corresponding to a rotation π around a certain axis Z, will belong to the same class provided that the group contains another twofold axis perpendicular to Z. This is clearly the case for the elements of O^+ belonging to C_2 or C_2'. O^+ has then 8 classes and 8 irreducible representations. Since we already know five of them, there must be three more with dimensions l_6, l_7, l_8 given by the relation

$$l_6^2 + l_7^2 + l_8^2 = 48 - (l_1^2 + \ldots + l_5^2) = 24, \tag{14.8}$$

which admits of a single solution,

$$l_6 = 2, \qquad l_7 = 2, \qquad l_8 = 4.$$

These representations, known in the literature as Γ_6, Γ_7, Γ_8 or E', E'', U (and possibly many other names), and sometimes called specific representations, differ from the first 5 through the fact that the element R is represented by the negative of the unit matrix. It is therefore clear that D^J for half-integral values of J contains specific representations only. Their characters are given in Table 6. The breakdown of D^J into specific representations is computed in the usual manner and the results are given for $J = \frac{1}{2}$ up to $J = \frac{15}{2}$ in Table 7.

From the character tables we can calculate the decomposition of the direct products. Table 8 contains results for the whole of the double group O^+ and thus incorporates Table 2.

We have not written out explicitly the matrices of the different representations Γ_1 to Γ_8. For the non-specific representations Γ_1 to Γ_5

the fact that xyz is a basis function for Γ_2, x, y, z for Γ_4, xy, yz, zx for Γ_5, $3z^2-r^2$ and $\sqrt{(3)}(x^2-y^2)$ for Γ_3, makes it easy to write them down if desired, using simple geometrical considerations. For the specific representations, we see from Table 7 that $D^{\frac{1}{2}}$ reduces just to Γ_6. The matrices of Γ_6 are thus simply the 48 matrices of $D^{\frac{1}{2}}$ that correspond to the 24 rotations of the cubic group. We shall represent the basis functions of Γ_6 as $|\pm\tilde{\tfrac{1}{2}}\rangle$, the tilde being there to remind us that these are *not* actually, except for $J = \tfrac{1}{2}$, the states $|J, \pm\tfrac{1}{2}\rangle$. Similarly since $D^{\frac{3}{2}}$ reduces to Γ_8 we can take as the basis functions of Γ_8, the functions $|\pm\tilde{\tfrac{3}{2}}\rangle$, $|\pm\tilde{\tfrac{1}{2}}\rangle$ that will transform under the rotations S of the cubic group in the same way as the functions $|\tfrac{3}{2}, \pm\tfrac{3}{2}\rangle, |\tfrac{3}{2}, \pm\tfrac{1}{2}\rangle$ would transform under $D^{\frac{3}{2}}(S)$.

The nature of Γ_7 is less obvious. Since $\Gamma_7 = \Gamma_6 \times \Gamma_2$ let us introduce the functions

$$|\alpha\rangle = |+\tilde{\tfrac{1}{2}}\rangle\rho,$$
$$|\beta\rangle = |-\tilde{\tfrac{1}{2}}\rangle\rho,$$
(14.9)

where ρ transforms according to Γ_2, that is as the function xyz, and let the functions $|\pm\tilde{\tfrac{1}{2}}\rangle$ transform according to Γ_6. α and β will then transform according to Γ_7. The matrices of Γ_7 are $\pm$ those of Γ_6. The matrices of Γ_6, Γ_7, Γ_8 cannot all be made real (see § 15.9, following eqn (15.53)). However, as appears from Table 6, their characters are all real and the Γ_i ($i = 6, 7, 8$) are equivalent to their complex conjugates. This is why in Table 8 we are able to consider direct products $\Gamma_i \times \Gamma_j$ rather than $\Gamma_i^* \times \Gamma_j$.

We see in Table 7 that, in the decomposition of D^J, Γ_6 and Γ_7 appear only once up to $J = \tfrac{15}{2}$, except for $J = \tfrac{13}{2}$ where Γ_7 appears twice, and that Γ_8 appears only once up to $J = \tfrac{7}{2}$. We can therefore write *a priori* the corresponding 'good' zero-order wave-functions as linear combinations of $|J, M\rangle$. This is done in Table 9. ($J = \tfrac{11}{2}$ and $\tfrac{13}{2}$ are omitted for they do not occur in practice as the ground states of paramagnetic ions.)

The reader perusing Tables 4 and 9 may well ask from what hat these rabbits have been extracted. A brief indication of the method used to obtain these tables will now be given. Consider a ket $|J, M\rangle$ belonging to a representation D^J of the rotation group. If S is an operator belonging to O (or O^+) we know how to calculate the ket $S |J, M\rangle$; it will be given by

$$S |J, M\rangle = \sum_{M'} |J, M'\rangle D^J_{M'M}(S),$$
(14.10)

where the $D^J_{M'M}(S)$ are certain matrix elements of the representation D^J of the rotation group, which are known in principle (see § 13.2). We also know in principle the matrices $A^k_{\mu\nu}$ of the various irreducible representations Γ_1 to Γ_8 of the cubic group, as explained above. In the cubic group k stands for the index 1 to 8 of the irreducible representation concerned and μ and ν refer to the various basis functions inside each representation Γ_k. Each representation Γ_k is spanned by l_k functions Ψ^k_μ which, under an operation S of the group, will transform accordingly to

$$S\Psi^k_\mu = \sum_\nu \Psi^k_\nu A^k_{\nu\mu}(S). \tag{14.11}$$

The various Ψ^k_ν will be the linear combinations of the $|J, M\rangle$, given in Tables 4 and 9 for the cubic group, which we wish to obtain. Conversely each ket $|J, M\rangle$ will be a linear combination of the Ψ^k_μ

$$|J, M\rangle = \sum_{k',\mu'} \Psi^{k'}_{\mu'} C^{k'}_{\mu'}(J, M). \tag{14.12}$$

Let us define now the operator

$$P^k_\mu = \frac{l_k}{g} \sum_S A^{k*}_{\mu\mu}(S) \cdot S, \tag{14.13}$$

where g is the number of elements of the group (48 for O^+) and the sum $\sum_S$ is over all the operations of the group. According to (14.10) we know how to express $P^k_\mu |J, M\rangle$ as a sum of $|J, M'\rangle$. On the other hand, if we take for $|J, M\rangle$ the expression (14.12), we get

$$P^k_\mu |J, M\rangle = \frac{l_k}{g} \sum_{S,k',\mu'} A^{k*}_{\mu\mu}(S)(S\Psi^{k'}_{\mu'})C^{k'}_{\mu'}(J, M) \tag{14.14}$$

or, using (14.11),

$$P^k_\mu |J, M\rangle = \frac{l_k}{g} \sum_{S,k',\mu',\nu'} A^{k*}_{\mu\mu}(S) \Psi^{k'}_{\nu'} A^{k'}_{\nu'\mu'}(S) C^{k'}_{\mu'}(J, M) \tag{14.15}$$

which, taking into account the orthogonality relations (12.9), reduces to

$$P^k_\mu |J, M\rangle = \Psi^k_\mu C^k_\mu(J, M). \tag{14.16}$$

We can thus generate all the Ψ^k_μ of Tables 4 and 9 (apart from normalization factors). The operation (14.14) will yield zero if $C^k_\mu(J, M) = 0$, that is if the expansion (14.12) of the ket $|J, M\rangle$ does not contain the function Ψ^k_μ. One then starts from another ket $|J, M'\rangle$ until all the Ψ^k_μ are obtained.

In practice our previous advice is the best: as somebody has prepared the tables, why not use them?

14.5. Groups of lower symmetry

The *tetragonal group*, also known in the literature as the group D_4, is the group that leaves invariant a cube or an octahedron that has been distorted along a C_2 axis called the tetragonal axis.

It contains the following elements and classes:

E, the identity operation,

C_2, rotation by an angle π around the tetragonal axis Oz (1 element),

C_4, rotation by $\pm\pi/2$ around the same axis (2 elements),

C_2', rotation by π around two axes Ox, Oy perpendicular to Oz (2 elements),

C_2'', rotation by π around two axes OX, OY at angles $\pi/4$ to Ox and Oy (2 elements).

Altogether we have 8 elements and 5 classes; whence (see § 12.3) there will be just 5 irreducible representations which we call Γ_1^t, Γ_2^t, Γ_3^t, Γ_4^t, Γ_5^t (the index t is a weak attempt to avoid confusion with the representations of the cubic group), and which are also known in the literature by other names. From eqn (12.14) it is readily apparent that they are all unidimensional except the last, which has two dimensions. The characters are given in Table 10.

We shall not write down the decomposition of D^J into the irreducible representations Γ_k^t of D_4 but be content to indicate the correspondence between the representations Γ_k of the cubic group and those Γ_k^t of the tetragonal group that is a subgroup of O. Using the character tables, it is possible to find, with the help of (12.16), that

$$\begin{cases} \Gamma_1 = \Gamma_1^t, & \Gamma_2 = \Gamma_3^t, & \Gamma_3 = \Gamma_3^t + \Gamma_1^t, \\ \Gamma_4 = \Gamma_2^t + \Gamma_5^t, & \Gamma_5 = \Gamma_4^t + \Gamma_5^t. \end{cases} \quad (14.17)$$

It is apparent that there is no logic whatsoever in this choice of indices (due to Bethe) but we shall not add to the confusion by proposing a new notation of our own differing from the many that already exist.

As far as the 'good' zero-order wave-functions are concerned we can see the following (taking the z-axis to be the tetragonal axis):

For $J = 1$, the cubic triplet Γ_4 spanned by x, y, z is split into the singlet Γ_2^t spanned by z and the doublet Γ_5^t spanned by x and y.

For $J = 2$, the cubic triplet Γ_5 spanned by xy, yz, zx is split into the singlet Γ_4^t spanned by xy and the doublet Γ_5^t spanned by yz and zx.

The cubic doublet Γ_3 is split into the singlet Γ_1^t spanned by $3z^2 - r^2$ and the singlet Γ_3^t spanned by $x^2 - y^2$.

For $J = 3$, the cubic singlet Γ_2 becomes the singlet Γ_3 spanned by xyz, the cubic triplet Γ_5 becomes the doublet Γ_5^t spanned by $x(y^2 - z^2)$

and $y(z^2-x^2)$, and the singlet Γ_4^t spanned by $z(x^2-y^2)$; the cubic triplet Γ_4 becomes the singlet Γ_2^t spanned by $z(3x^2+3y^2-2z^2)$, and a doublet Γ_5^t spanned by $x(3y^2+3z^2-2x^2)$ and $y(3z^2+3x^2-2y^2)$.

It will be noticed from (14.17) that already in the decomposition of D^3, Γ_5 appears twice and that the correct zero-order functions spanning Γ_5^t cannot be obtained without explicit knowledge of the tetragonal potential. The off-diagonal matrix elements of a tetragonal field between the cubic manifolds Γ_4 and Γ_5 are given explicitly in Fig. 7.5.

The properties of the *double tetragonal group* are obtained in the same manner as for the cubic group. It has 16 elements, twice as many as the simple group, but two more classes only, R and RC_4, since the classes RC_2, RC_2', RC_2'' are not distinct from C_2, C_2', C_2'' in accordance with the theorem in § 14.4. There are two specific representations Γ_6^t and Γ_7^t with dimensions, l_6 and l_7, such that

$$l_6^2 + l_7^2 = 16 - 8 = 8$$

whence $l_6 = l_7 = 2$ showing that both Γ_6 and Γ_7^t are bidimensional. We omit their character table and shall be content to notice that the decomposition of the specific representations of O^+ into those of the double tetragonal group is

$$\Gamma_6 = \Gamma_6^t, \qquad \Gamma_7 = \Gamma_7^t, \qquad \Gamma_8 = \Gamma_6^t + \Gamma_7^t. \tag{14.18}$$

The *rhombic group* or D_2 contains, besides the unit operation, just three binary orthogonal axes x, y, z and thus four classes and four unidimensional representations. For an integral value of J there is no degeneracy left in a rhombic field.

The double group has 8 elements and 5 classes, R being the only extra one. It has one specific representation of dimension 2. A multiplet J of the free ion with half-integral J will thus split into $J+\frac{1}{2}$ doublets.

The *trigonal group* or D_3 leaves invariant a cube distorted along a body-diagonal. Its elements and classes are as follows:

the identity E,

the rotations C_3 of angle $\pm 2\pi/3$ around the threefold axis Oz (2 elements),

the rotations C_2' of angle π around three axes perpendicular to Oz (3 elements).

Altogether we have 3 classes and 6 elements, which shows that there must be two one-dimensional representations that we call Γ_1^T and Γ_2^T and one bidimensional representation Γ_3^T. The characters are given in Table 11. The decomposition of D^J by the trigonal group is given in Table 12.

The connection between cubic and trigonal groups is as follows:
$$\Gamma_1 \to \Gamma_1^T, \quad \Gamma_2 \to \Gamma_2^T, \quad \Gamma_3 \to \Gamma_3^T,$$
$$\Gamma_4 \to \Gamma_2^T + \Gamma_3^T, \quad \Gamma_5 \to \Gamma_1^T + \Gamma_3^T.$$

It is worth noticing that the degeneracy of the cubic doublet Γ_3 is *not* lifted by a trigonal distortion though it is by a tetragonal distortion.

The double trigonal group has 12 elements and 6 classes. (C_2' and RC_2' are distinct since there is no binary axis perpendicular to the binary axes of C_2'.) We find readily that there are two unidimensional specific representations Γ_4^T, Γ_5^T, and one bidimensional Γ_6^T. Their characters are given in Table 13. The decomposition of D^J by the trigonal group for half integer J is given in Table 14.

As the representations Γ_4^T and Γ_5^T are unidimensional one might think that the decomposition of a half-integer J could contain singlets. We shall soon see (§ 15.4), that there is a very general theorem due to Kramers which proves that this can never occur. Therefore the two representations Γ_4^T and Γ_5^T (which are complex conjugates) must always correspond to the same energy of the system.

The decomposition of specific representations of O^+ into those of the trigonal group is the following:

$$\Gamma_6 = \Gamma_6^T, \quad \Gamma_7 = \Gamma_6^T, \quad \Gamma_8 = \Gamma_4^T + \Gamma_5^T + \Gamma_6^T. \tag{14.19}$$

14.6. Improper rotations

We have only considered so far symmetry groups containing pure rotations. Another type of symmetry element that occurs in nature is the improper rotation. An improper rotation is a symmetry element resulting from the combination of a rotation with an inversion with respect to a centre situated on the axis of rotation. For instance, reflection in a plane is the product of a rotation by an angle π with an inversion. In ordinary space an inversion corresponds to the reversal of the signs of the three space coordinates and an improper rotation is therefore represented by a real orthogonal matrix with determinant -1. It follows that the product of two improper rotations is a proper rotation and that the product of a proper and improper rotation is an improper rotation. A group G_i that contains at least one improper rotation must therefore contain as many proper as improper rotations and can be represented as the combination of a subgroup G_p of proper rotations with a set of improper rotations of the form $g_i G_p$ where g_i is an improper rotation. We write symbolically

$$G_i = G_p + g_i G_p. \tag{14.20}$$

THE CUBIC GROUP AND SOME OTHER GROUPS 641

We shall for brevity call a group such as (14.20) a group of improper rotations, although it contains naturally proper rotations as well. A finite group of improper rotations may or may not contain the inversion operation itself. If it does, we can use for the improper rotation g_i in (14.20) the inversion I itself and write

$$G_i = G_p + IG_p. \qquad (14.21)$$

Since I commutes with all the elements of the group G_i, and its square I^2 is just the identity operation, the matrix representing I in an irreducible representation of G_i can be either the unit matrix, for so-called even representations, or its negative for odd representations.

To each representation $\mathscr{D}_p$ of G_p correspond two representations $\mathscr{D}_i^\pm$ of G_i such that, if g_p is a proper rotation of G_p,

$$\begin{cases} \mathscr{D}_i^\pm(g_p) = \mathscr{D}_p(g_p) \\ \mathscr{D}_i^\pm(Ig_p) = \pm\mathscr{D}_p(g_p). \end{cases} \qquad (14.22)$$

In particular, if $\mathscr{D}_p$ is irreducible so is $\mathscr{D}_i^\pm$.

If $\mathscr{D}_i$ is a *reducible* representation of G_i of given parity there corresponds to it by (14.22) a single reducible representation $\mathscr{D}_p$ of G_p. If we know how to reduce $\mathscr{D}_p$ into its irreducible parts

$$\mathscr{D}_p = \sum a_k \mathscr{D}_p^{(k)}, \qquad (14.23)$$

the reduction of $\mathscr{D}_i$ will be

$$\mathscr{D}_i = \sum a_k \mathscr{D}_i^{(k)} \qquad (14.24)$$

with the same coefficients a_k.

As an example of a group of improper rotations that contains I, we may take the cubic group of improper rotations O_h defined in accordance with (14.21),

$$O_h = O + IO. \qquad (14.25)$$

On the other hand, the tetrahedral group T_d, which is the group of improper rotations that transform a regular tetrahedron into itself, does not contain the inversion I. Since each operation of T_d amounts to a permutation of the vertices of the tetrahedron, T_d is isomorphic with the permutation group S_4 and thus also with O. It has the same set of characters and the same representations as O. Geometrically, however, it is a different group. By adding to T_d the inversion operation we obtain again the full cubic group

$$O_h = T_d + IT_d, \qquad (14.26)$$

showing that T_d and O are two isomorphic subgroups of O_h. It should be noticed that in contradistinction to (14.25), (14.26) *is not* a relationship of the type (14.20) since T_d contains improper rotations as well as proper ones.

We examine now how the introduction of improper rotations modifies the splitting of a degenerate level of the free ion by the crystal potential, as studied in the previous sections. If the group of improper rotations G_i which describes the environment of the bound ion does contain the inversion element, the crystal field splitting is the same as calculated for the pure rotation group G_p associated with G_i by (14.21). This follows from the fact that the level J of the free ion has a definite parity and that the wave-functions that span this level provide a reducible representation $\mathscr{D}_i(G_i)$ of given parity which, as explained earlier, is reduced in the same manner as the representation $\mathscr{D}_p(G_p)$ in the absence of inversion. Thus the splitting pattern of a level of the free ion will be the same in a field of symmetry O_h as in symmetry O.

If the group G_i does not contain the inversion I we can still predict the splitting pattern as follows. Let us call $V(\mathbf{r})$ the crystalline potential invariant by G_i which represents the effects of the environment of the bound ion. We can write

$$V(\mathbf{r}) = \tfrac{1}{2}\{V(\mathbf{r})+V(-\mathbf{r})\}+\tfrac{1}{2}\{V(\mathbf{r})-V(-\mathbf{r})\} = V_{\text{even}}+V_{\text{odd}}. \quad (14.27)$$

Since the states of the free ion have a definite parity the matrix elements of V_{odd} vanish inside their manifold and we can in first approximation replace V by V_{even}. But V_{even} is invariant through I and thus through the group

$$G'_i = G_i+IG_i,$$

which can always be rewritten according to (14.20) as

$$G'_i = G'_p+IG'_p,$$

where G'_p is a certain pure rotation group. The splitting of D^J by a potential of symmetry G_i will then be the same as by one with symmetry G'_i and therefore, as we saw earlier, the same as by a potential of symmetry G'_p. As an example, from (14.26) and (14.25) we deduce that a level D^J will be decomposed in the same manner in a field of tetrahedral symmetry T_d as in a cubic field O and we need not repeat the detailed study of the last section.

The foregoing is valid provided we assume that the effects due to the environment are small compared with the distance between two configurations of the free ion with opposite parity.

The absence of a centre of inversion in the environment also modifies drastically the response to an applied electric field as we shall see in § 15.10.

15

TIME REVERSAL AND KRAMERS DEGENERACY

15.1. Operations involving the time

IN the previous sections we have considered the effects of certain purely spatial operations, namely proper and improper rotations, on the eigenstates of a Hamiltonian invariant with respect to these transformations. We consider now two new operations that involve the time. The first is the displacement in time D_t which relates the wavefunction Ψ at time 0 to the wave-function $\Psi(t)$ at time t. For a time-independent Hamiltonian $\mathscr{H}$, $D_t = \exp(-i\mathscr{H}t/\hbar)$ which, applied to an eigenstate of $\mathscr{H}$ of energy W, yields

$$D_t\Psi = \exp\left(-\frac{iW}{\hbar}t\right)\Psi. \tag{15.1}$$

The second operation, usually called time reversal but which would be better called reversal of direction of motion, is an operation where all velocities (including those associated with the spinning of the electrons) are reversed. If this operation, which we shall denote by the letter θ, leaves the Hamiltonian $\mathscr{H}$ invariant and if Ψ is an eigenstate of $\mathscr{H}$ with an energy W, $\theta\Psi$ is obviously also an eigenstate with the same energy. The invariance of the Hamiltonian of a system by time reversal has far-reaching consequences for its magnetic properties. The formalism involved is rather different from the usual calculations of quantum mechanics and as it is sometimes misunderstood it warrants a rather detailed discussion.

The following relation is a straightforward consequence of the definitions of θ and D_t:

$$D_t(\theta\varphi) = \theta D_{-t}(\varphi). \tag{15.2}$$

This equation states in effect that if two states of a system have opposite velocities at time zero they have opposite velocities at opposite times.

Since time reversal, as we shall continue to call it, is a symmetry operator it follows from general principles of quantum mechanics that it cannot change the transition probability between two states Ψ and Φ so that we have

$$|(\Psi \mid \Phi)| = |(\theta\Psi \mid \theta\Phi)|. \tag{15.3}$$

It can be shown (Wigner, *Group theory*, Chap. 26) that if a physical relationship expressed by an operator θ is such that (15.3) holds, it is always possible to re-define the phases of all wave-functions in such a way that θ is either a linear unitary operator or an antilinear anti-unitary one.

For convenience we re-state here that under a unitary transformation we have for a linear unitary operator A

$$A(C\Psi) = C(A\Psi), \qquad (15.4)$$

whereas for an antilinear operator we have

$$A(C\Psi) = C^*(A\Psi). \qquad (15.5)$$

An antilinear operator that satisfies (15.3) is further said to be antiunitary. The first alternative (15.4) implies

$$\begin{cases} \theta(C_1\Psi_1 + C_2\Psi_2) = C_1\theta\Psi_1 + C_2\theta\Psi_2, \\ (\theta\Psi \mid \theta\Phi) = (\Psi \mid \Phi). \end{cases} \qquad (15.6)$$

The second (15.5) implies that

$$\begin{cases} \theta(C_1\Psi_1 + C_2\Psi_2) = C_1^*\theta\Psi_1 + C_2^*\theta\Psi, \\ (\theta\Psi \mid \theta\Phi) = (\Psi \mid \Phi)^* = (\Phi \mid \Psi). \end{cases} \qquad (15.7)$$

To establish the category to which θ belongs, consider the relation (15.2) where we expand φ into a sum of eigenstates $\varphi = \sum a_k \Psi_k$ of a Hamiltonian $\mathcal{H}$. If θ is linear and (15.6) holds, the left-hand side of (15.2) can be written $\sum_k a_k e^{-iW_k t/\hbar}(\theta\varphi_k)$, while the right-hand side becomes $\sum_k a_k e^{+iW_k t/\hbar}(\theta\varphi_k)$, which is contradictory. On the other hand, if θ is antilinear, both sides of (15.2) yield $\sum a_k^* e^{-iW_k t/\hbar}(\theta\varphi_k)$. The time-reversal operator must thus be antilinear and antiunitary. Some caution must be exercised in the manipulation of antilinear operators. In particular, the customary notation $(\Phi \mid A \mid \Psi)$ for the matrix element of an operator A between two states Φ and Ψ is ambiguous if A is antilinear. We shall write instead $(A\Phi, \Psi)$ or $(\Phi, A\Psi)$, as the case may be.

15.2. Complex conjugation

One of the simplest antilinear operators is that which converts a wave-function Ψ into its complex conjugate Ψ^*. We denote this operator by K_0 and we use here the word wave-function in a rather general sense. For instance, dealing with angular momenta, if a state

TIME REVERSAL AND KRAMERS DEGENERACY 645

$|\xi\rangle$ can be expanded into a sum of state vectors

$$|\xi\rangle = \sum_{J,M} C_{J,M} |J, M\rangle, \tag{15.8}$$

what we call the wave-function of $|\xi\rangle$ is the set of numbers $C_{J,M}$; the state vector $K_0 |\xi\rangle$ will then have as wave-function the set of numbers $C_{J,M}^*$,

$$K_0 |\xi\rangle = \sum_{J,M} C_{J,M}^* |J,M\rangle. \tag{15.9}$$

Clearly K_0 is an antiunitary operator since

$$(K_0\Psi, K_0\Phi) = (\Psi, \Phi)^* = (\Phi, \Psi);$$

and it obeys the obvious relation

$$K_0^2 = 1. \tag{15.10}$$

An operator $K_0^{-1}AK_0$, the complex conjugate of an operator A, is by definition an operator whose matrix elements are the complex conjugates of those of A. It is clear that the definition of K_0 depends on the basis states chosen to define the representation. For instance, in the representation **r**, the complex conjugate of the operator

$$L_z = \frac{1}{\mathrm{i}}\left(x\frac{\partial}{\partial y} - y\frac{\partial}{\partial x}\right)$$

is $-L_z$, but it is $+L_z$ in the representation $|L, M_L\rangle$. In the representation **r** the operator $p_x = (\hbar/\mathrm{i}) \partial/\partial x$ changes its sign under conjugation, but not in the representation **p** where it is diagonal. We cannot therefore identify θ with K_0 in all representations since θ should be independent of the representation chosen, while K_0 clearly is not. There is, however, a relationship between θ and K_0 that we establish now, by introducing the operator $U = \theta K_0$ or, since $K_0^2 = 1$, $\theta = UK_0$.

We can show immediately that $U = \theta K_0$ is unitary; since θ and K_0 are both antilinear, $U = \theta K_0$ is linear, and since both θ and K_0 leave invariant the absolute value of the scalar product $|(\Phi, \Psi)|$ so does U. Hence it is unitary. We can also show that $\theta^2 = \pm 1$. The operator θ applied twice leaves the system in its original state and one must have

$$\theta^2 = UK_0 \cdot UK_0 = UK_0^{-1}UK_0 = UU^* = c \cdot 1, \tag{15.11}$$

where c is a number of modulus unity and 1 the unit operator. From

(15.11) we find:

$$U = cU^{*-1} \text{ or, since } U \text{ and } U^* \text{ are unitary,}$$

$$U = c(U^*)^\dagger = c\tilde{U}, \text{ the transposed of which is}$$

$$\tilde{U} = cU \text{ whence } U = c^2 U \text{ and}$$

$$c^2 = 1, \quad c = \pm 1.$$

It follows that the square of the time-reversal operator is

$$\theta^2 = \pm 1. \tag{15.12}$$

15.3. Determination of the time-reversal operator

We begin by disregarding the spin. The dynamical variables are then the coordinates x and momenta p_x, which under time reversal should transform as

$$\theta^{-1} x \theta = x; \quad \theta^{-1} p_x \theta = -p_x,$$

or

$$x\theta = \theta x, \quad p_x \theta = -\theta p_x. \tag{15.13}$$

Hence, on writing $\theta = UK_0$, we have

$$xUK_0 = UK_0 x, \quad p_x UK_0 = -UK_0 p_x,$$
$$xU = UK_0 x K_0^{-1}, \quad p_x U = -UK_0 p_x K_0^{-1}. \tag{15.14}$$

Let us choose the **r** representation where x is real; i.e., $x = K_0 x K_0^{-1}$ and p_x is purely imaginary so that $p_x = -K_0 p_x K_0^{-1}$. Then (15.14) yields

$$xU = Ux, \quad p_x U = Up_x. \tag{15.15}$$

Thus the operator U commutes with all the dynamical variables x and p_x and must therefore be a constant that we can choose equal to unity. In the absence of spin and in the **r** representation (but not in the **p** representation !) the time-reversal operator θ is represented by the complex conjugation operator K_0.

We now include the spin and consider first a single electron. The three components s_q of the spin operator change their sign under time-reversal and must satisfy the relations

$$s_q = -\theta s_q \theta^{-1}, \quad s_q \theta = -\theta s_q, \tag{15.16}$$

giving

$$s_q UK_0 = -UK_0 s_q,$$

or

$$s_q U = -UK_0 s_q K_0^{-1}, \tag{15.17}$$

so that

$$s_q U = -U s_q^*. \tag{15.18}$$

In the usual representation where s_x and s_z are real matrices and s_y is a purely imaginary one, U must anticommute with s_x and s_z and commute with s_y. We may take for U the matrix $\sigma_y = 2s_y$ which is unitary and answers our requirements, or better still $i\sigma_y$ which has the advantage of being a real matrix so that it commutes with K_0. The latter will be our choice. To recapitulate, in a representation where $\mathbf{r}$ is diagonal (the $\mathbf{r}$ representation) and s_x and s_z real, the time-reversal operator for a single electron can be represented by

$$\theta = i\sigma_y K_0. \tag{15.19}$$

Its square is given by

$$\theta^2 = (i\sigma_y)K_0(i\sigma_y)K_0 = (i\sigma_y)^2 K_0^2 = -1. \tag{15.20}$$

For n electrons $1, 2, \ldots, n$ the time-reversal operator θ will be the product

$$\theta = \prod_1^n \theta_p. \tag{15.21}$$

Its square will be ± 1 depending on whether the number n of electrons is even or odd, a particularly important result.

15.4. Kramers degeneracy

Consider a system with an odd number of electrons for which $\theta^2 = -1$ and assume that its Hamiltonian commutes with θ. This will be the case in the absence of a magnetic field for then the kinetic and potential energy as well as spin-spin and spin-orbit interactions are invariant through time reversal. Let Ψ be an eigenstate of the system with energy W; then the function $\Phi = \theta\Psi$ is also an eigenstate with the same energy. Consider the scalar product $(\Psi, \Phi) = (\Psi, \theta\Psi)$. Since θ is antilinear,

$$(\Psi, \Phi) = (\Psi, \theta\Psi) = (\theta\Psi, \theta^2\Psi)^* = (\theta^2\Psi, \theta\Psi) = -(\Psi, \theta\Psi) = 0,$$

showing that Φ is orthogonal to Ψ and is naturally distinct from it. It follows that the eigenvalue W is at least twofold degenerate. This is Kramers theorem. We shall call the state $\Phi = \theta\Psi$ the Kramers-conjugate of Ψ and represent it by the symbol $\overline{\Psi}$.

A few useful relations are connected with the properties of θ. We shall call an operator O time-odd if $\theta O \theta^{-1} = -O^\dagger$ and similarly an operator time-even if $\theta O \theta^{-1} = +O^\dagger$. If O is Hermitian, $O^\dagger = O$ but it is sometimes convenient to use non-Hermitian operators such as $L_\pm$ or $J_\pm$.

It should be noticed in this connection that the product of two non-commuting Hermitian operators A and B of well-defined time-parity (say for instance both even) does not have a well-defined time-parity but is the sum of a time-even and a time-odd operator:

$$AB = \tfrac{1}{2}(AB+BA)+\tfrac{1}{2}(AB-BA) = C+D$$

where C is Hermitian and D anti-Hermitian, and

$$\theta AB\theta^{-1} = \theta C\theta^{-1}+\theta D\theta^{-1} = C+D = C^\dagger - D^\dagger.$$

Hence C is time-even but D is time-odd!

Spin-spin and spin-orbit interactions, which are *Hermitian* products of time-odd operators, are time-even.

If $\theta^2 = -1$, the following theorems hold.

(a) A time-even operator has no matrix elements between two Kramers conjugate states, since

$$(\Psi|\, O\, |\bar{\Psi}) = (\Psi, O\theta\Psi) = (\theta O\theta\Psi, \theta\Psi) = -(\theta O\theta^{-1}\Psi, \theta\Psi)$$
$$= -(O^\dagger\Psi, \theta\Psi) = -(\Psi|\, O\, |\bar{\Psi}) = 0. \qquad (15.22a)$$

(b) A time-even operator has the same expectation value in two Kramers conjugate states since

$$(\Psi|\, O\, |\Psi) = (\Psi, O\Psi) = (\theta O\Psi, \theta\Psi) = (\theta O\theta^{-1}\theta\Psi, \theta\Psi)$$
$$= (O^\dagger\theta\Psi, \theta\Psi) = (\theta\Psi, O\theta\Psi) = (\bar{\Psi}|\, O\, |\bar{\Psi}). \qquad (15.22b)$$

(c) A time-odd operator has opposite expectation values in two Kramers conjugate states. The proof is similar to that in (b).

If $\theta^2 = +1$, time reversal plays a much less important role because a state Ψ can now coincide with its time-reversed state $\bar{\Psi} = \theta\Psi$. The analogues of the previous results (a), (b), (c) are the following:

(a') A time-odd operator has no matrix elements between two Kramers conjugate states. The proof is the same as for (a).

(b'), (c') The theorems (b), (c) do not depend on the value of θ^2 and are therefore also valid for $\theta^2 = +1$.

The result (a') leads to a generalization of Van Vleck's theorem on the quenching of the orbital momentum proved in § 11.7. Consider a time-even Hamiltonian $\mathscr{H}$ with a *non-degenerate* eigenstate Ψ, which necessarily implies $\theta^2 = +1$. The state $\bar{\Psi} = \theta\Psi$ is an eigenstate of $\mathscr{H}$ with the same energy and since Ψ is non degenerate $\bar{\Psi}$ must coincide with Ψ within a phase-factor. The matrix elements of time-odd operators between Ψ and $\bar{\Psi}$ which vanish according to (a') are just the expectation values in the state Ψ. It follows that all components of the magnetic moment, being time-odd operators have vanishing expectation values in non-degenerate states of time-even Hamiltonians, which is a generalization of Van Vleck's theorem on the quenching of the orbital momentum.

Invariance through time reversal is also connected with the so-called Van Vleck cancellation that is responsible for the T^9 temperature variation of the Raman relaxation rate for Kramers ions (see Chapter 10). What is meant by the Van Vleck cancellation is the *near* cancellation of two second-order probability amplitudes, caused by the *exact* cancellation

$$\langle \bar{a}| \ V' \ |c\rangle\langle c| \ V \ |a\rangle + \langle \bar{a}| \ V \ |\bar{c}\rangle\langle \bar{c}| \ V' \ |a\rangle = 0, \quad (15.23\text{a})$$

where V and V' are two time-even non-Hermitian operators (actually lattice-orbit coupling operators for emission or absorption of a phonon) and $|\bar{a}\rangle$, $|a\rangle$ and $|\bar{c}\rangle$, $|c\rangle$ are two pairs of Kramers conjugate states (see Chapter 10). The proof of (15.23a) is as follows (most of the proofs given in the literature, except that of Van Vleck, are incomplete or incorrect). The first term of (15.23a) can be written

$$(\theta a, \ V'c)(c, \ Va) = (\theta V'c, \ \theta^2 a)(\theta Va, \ \theta c) = (\theta V'\theta^{-1}\theta c, \ \theta^2 a)(\theta V\theta^{-1}\theta a, \ \theta c)$$
$$= -(V'^{+}\bar{c}, \ a)(V^{+}\bar{a}, \ \bar{c}) = -(\bar{c}| \ V' \ |a)(\bar{a}| \ V \ |\bar{c}), \quad (15.23\text{b})$$

which proves (15.23a).

Since Kramers degeneracy is directly related to invariance with respect to time reversal one may ask whether this is or is not a new degeneracy superimposed on that connected with the spatial symmetry of the environment. There is no single answer to that question since examples of either situation can be found. Thus, for instance, in an environment of cubic symmetry, for an odd number of electrons all representations are at least two-dimensional and no extra degeneracy is brought about by the invariance of the Hamiltonian with respect to time reversal. On the other hand, the double trigonal group (§ 14.5) has two unidimensional representations, Γ_4^T and Γ_5^T, which must correspond to the same energy if Kramers' theorem is to be obeyed. The mathematical criteria for the occurrence of either situation are discussed in Wigner's book but we shall have no occasion to make use of them.

15.5. Time-reversal operator in the $|J, M\rangle$ representation

Since, as has already been stated, eigenstates $|J, M\rangle$ of the angular momentum provide a convenient basis for expanding eigenstates of the bound ion, it is useful to be able to write directly the effect of time reversal on a state vector

$$|\xi\rangle = \sum_{J,M} C_{J,M} |J, M\rangle. \qquad (15.24)$$

Since **J** is a time-odd operator, we must have

$$\theta \mathbf{J} \theta^{-1} = -\mathbf{J}$$
$$\theta \mathbf{J} = -\mathbf{J}\theta$$
$$UK_0\mathbf{J} = -\mathbf{J}UK_0$$
$$UK_0\mathbf{J}K_0^{-1} = -\mathbf{J}U$$

or

$$U\mathbf{J}^* = -\mathbf{J}U. \qquad (15.25)$$

With the usual choice of phases, J_z and J_x are real operators, in the sense that all their matrix elements are real, whereas J_y is purely imaginary. According to (15.25) we need for U a unitary operator that commutes with J_y and anticommutes with J_x and J_z. The operator $e^{i\pi J_y}$ satisfies these requirements. This is obvious for J_y. For J_x, $e^{i\pi J_y}J_x = -J_x e^{i\pi J_y}$ implies $e^{-i\pi J_y}J_x e^{i\pi J_y} = -J_x$ which is satisfied since, as we saw in § 13.1, $e^{-i\pi J_y}$ expresses a rotation by an angle π around Oy which must change J_x into $-J_x$. The same is true for J_z. We can thus take for θ the form

$$\theta = e^{i\pi J_y}K_0 = K_0 e^{i\pi J_y}. \qquad (15.26)$$

For $J = \tfrac{1}{2}$

$$e^{i\pi J_y} = e^{(i\pi/2)\sigma_y} = \cos\frac{\pi}{2} + i\sigma_y \sin\frac{\pi}{2} = i\sigma_y, \qquad (15.27)$$

which is consistent with (15.19).

The only non-vanishing matrix elements of $e^{i\pi J_y}$ can be shown to be

$$\langle -M| e^{i\pi J_y} |M\rangle = (-1)^{J-M}, \qquad (15.28)$$

which gives for $\bar{\xi} = \theta \xi$ the expression

$$|\bar{\xi}\rangle = \theta |\xi\rangle = \sum_{J,M} C^*_{J,M}(-1)^{J-M} |J, -M\rangle. \qquad (15.29)$$

The reader may verify that this sign rule is consistent with the expansion of the various kets in Tables 4 and 9.

15.6. The 'Spin Hamiltonian' for a Kramers doublet

Consider a paramagnetic ion with an odd number of electrons in an environment of symmetry sufficiently low for the ground level to have only Kramers degeneracy. This will be the case for any symmetry lower than cubic, and also for the cubic levels Γ_6 and Γ_7.

This degenerate Kramers doublet is spanned by two kets $|\xi\rangle$ and $|\bar{\xi}\rangle = \theta |\xi\rangle$. There is an arbitrariness in the choice of these basis states. We can always choose two other states by a substitution of the form

$$\begin{cases} |\xi'\rangle = a |\xi\rangle + b |\bar{\xi}\rangle \\ |\bar{\xi}'\rangle = \theta |\xi'\rangle = -b^* |\xi\rangle + a^* |\bar{\xi}\rangle, \end{cases} \quad (15.30)$$

with $aa^* + bb^* = 1$. The substitution (15.30) is not only unitary but also unimodular and, as we saw in Chapter 13 in considering the rotation group, we can associate with it a certain rotation R by means of eqn (13.14) and the formulae following it in § 13.2.

The degeneracy of the Kramers doublet can only be lifted by a magnetic field, which may be either an applied external field or the field produced by the magnetic moment $\mu_I = -\gamma_n \hbar \mathbf{I}$ of the nucleus if it is not zero. The corresponding energies can be written $-\mu_e \cdot \mathbf{H}$ and $-\mu_I \cdot \mathbf{H}_e$ where vector operators μ_e and $\mathbf{H}_e$ represent respectively the magnetic moment of the ion and the magnetic field produced by its electrons at the position of the nucleus. If we assume that either interaction is much smaller than the distance between the doublet and excited levels, we need only their matrix elements within the ground manifold. It is well known that any Hermitian two-by-two matrix can be expressed as a linear combination with real coefficients of the three Pauli matrices and of the unit matrix. Since all components of μ and $\mathbf{H}_e$ are time-odd operators, their expectation values in the states $|\xi\rangle$ and $|\bar{\xi}\rangle$ have opposite values (theorem (c) of § 15.4) and we can write for the components μ_q and H_{eq}

$$\begin{cases} \mu_q = -\dfrac{\beta}{2} \sum_\alpha g_{q\alpha} \sigma_\alpha \\ H_{eq} = \sum_\alpha a_{q\alpha} \sigma_\alpha, \end{cases} \quad (15.31)$$

with no contribution from the unit matrix. The index q refers to the components x, y, z of the vectors μ and $\mathbf{H}_e$ and α to the Pauli matrices σ_1, σ_2, σ_3. The set of matrices $\sigma_1/2$, $\sigma_2/2$, $\sigma_3/2$ is often referred to as the components of the fictitious spin $\mathbf{s}$ and the sets of real numbers $g_{q\alpha}$ and $a_{q\alpha}$ are called the magnetogyric tensor and the hyperfine structure tensor. The deeply ingrained habit of calling the $g_{q\alpha}$ and the $a_{q\alpha}$ tensors, proceeds from a confusion between the fictitious spin $\sigma/2$ of the Kramers doublet and a real electronic spin and is a serious misnomer in general as will now appear.

There are two operations that can modify the set of numbers $g_{q\alpha}$ (and $a_{q\alpha}$). The first is a spatial rotation S of coordinate axes whereby the components μ_q are replaced by another set

$$\mu'_q = \sum_p S_{qp}\mu_p \tag{15.32}$$

and consequently

$$g'_{q\alpha} = \sum_p S_{qp}g_{p\alpha}. \tag{15.33}$$

The second is the substitution (15.30) of the basis states $|\xi\rangle$ and $|\bar{\xi}\rangle$ which, because of the *mathematical* connection between a matrix u of the unitary group U and a real orthogonal matrix R of the rotation group, results in the substitution

$$\sigma'_\gamma = \sum_\alpha R_{\gamma\alpha}\sigma_\alpha \quad \text{or} \quad \sigma_\alpha = \sum (R^{-1})_{\alpha\gamma}\sigma'_\gamma = \sum R_{\gamma\alpha}\sigma'_\gamma, \tag{15.34}$$

which leads to the substitution

$$g''_{p\gamma} = \sum R_{\gamma\alpha}g_{p\alpha}. \tag{15.35}$$

If we decided, quite arbitrarily, always to associate with a rotation S of the coordinate axes, expressed by (15.32), a substitution (15.30) of the basis states, such that the corresponding substitution R for the σ_α, expressed by (15.34), would coincide with the spatial rotation S, then and only then would the $g_{q\alpha}$ (and $a_{q\alpha}$) transform as the components of a tensor. A true tensor G_{pq} does, however, appear when we look for the eigenvalues of the Zeeman Hamiltonian $-\boldsymbol{\mu}\cdot\mathbf{H}$. Let λ_x, λ_y, λ_z be the cosines of the direction of $\mathbf{H}$. Then from (15.31)

$$-\boldsymbol{\mu}\cdot\mathbf{H} = \sum_{q,\alpha}\frac{\beta}{2}H_q g_{q\alpha}\sigma_\alpha = \sum_{q,\alpha}\frac{\beta H}{2}\lambda_q g_{q\alpha}\sigma_\alpha = \frac{\beta H}{2}\sum_\alpha \sigma_\alpha f_\alpha \tag{15.36}$$

where $f_\alpha = \sum_q \lambda_q g_{q\alpha}$.

The eigenvalues of (15.36) result from the well known properties of the Pauli matrices,

$$W_\pm = \pm\frac{\beta H}{2}\left(\sum_\alpha f_\alpha^2\right)^{\frac{1}{2}} = \pm\frac{\beta H}{2}\left(\sum_{p,q}\lambda_p\lambda_q\sum_\alpha g_{p\alpha}g_{q\alpha}\right)^{\frac{1}{2}}$$

$$= \pm\frac{\beta H}{2}\left(\sum_{p,q}\lambda_p\lambda_q G_{pq}\right)^{\frac{1}{2}}. \tag{15.37}$$

The set of numbers

$$G_{pq} = \sum_\alpha g_{p\alpha}g_{q\alpha} \tag{15.38}$$

is indeed a symmetric tensor that has the correct transformation properties under spatial rotations and can always be made diagonal by a proper choice of coordinate axes. Its eigenvalues are all positive and their square roots give the three principal values of the Larmor frequency of the electron spin in the applied field **H**. A tensor $A_{pq} = \sum_\alpha a_{p\alpha} a_{q\alpha}$ can be similarly defined.

From what has been said about the 'tensor' $g_{q\alpha}$ it follows that the question of whether it is symmetric is of little importance. Since we have six parameters at our disposal, three for the spatial rotation S in (15.33), and three for the fictitious rotation R in (15.35), we have enough freedom to make the 'tensor' $g_{q\alpha}$ diagonal. Once the 'tensor' g has been reduced to a diagonal form g_1, g_2, g_3 it is always possible to reverse the sign of two of the components, say g_1 and g_3, by a rotation of π around the y-axis (or by a change of σ_1 and σ_3 into $-\sigma_1$ and $-\sigma_3$ by an interchange of the basic states). On the other hand, the product $g_1 g_2 g_3$ which is the determinant of the $g_{q\alpha}$, is an invariant.

When $g_{p\alpha}$ is diagonal, so naturally is G_{pq} according to (15.38), since

$$G_{xx} = g_{x1}^2, \qquad G_{yy} = g_{y2}^2, \qquad G_{zz} = g_{z3}^2. \tag{15.38'}$$

In the same manner, with two other rotations R' and S' we can make the hyperfine 'tensor' $a_{p\alpha}$ diagonal. The relevant question is then whether both 'tensors' $g_{q\alpha}$ and $a_{q\alpha}$ be made diagonal simultaneously with the same choice of rotations R and S, that is with the same choice of coordinate axes and basis states.

We have so far said nothing about the spatial environment of our ion. If we make some restrictive assumptions on its symmetry we shall find situations where the 'tensors' $g_{q\alpha}$ and $a_{q\alpha}$ can be diagonalized simultaneously. For example, if the strength of the interactions due to the environment is small compared to the distance between two J multiplets of the free ion, we may assume that the wave-functions of our Kramers doublet can be constructed from the wave-functions of a single J level of the free ion (this may occur in the rare-earth group but not in general in the iron group). Then, according to the Wigner–Eckart theorem, all matrix elements within the doublet of the components of any vector, and in particular of μ and $\mathbf{H}_e$, will be proportional to the corresponding matrix elements of the vector $\mathbf{J}$. The 'tensors' $g_{q\alpha}$ and $a_{q\alpha}$ will be proportional to each other and can be diagonalized simultaneously.

15.7. The rhombic group

Suppose that the environment of the ion has two twofold symmetry axes, the z- and y-axes (it then has necessarily a third, the x-axis). We now show that in this case $g_{q\alpha}$ and $a_{q\alpha}$ can simultaneously be made diagonal.

We can expand one of the kets spanning the Kramers doublet into eigenstates $|J, M\rangle$ of the free ion (where $i \equiv \alpha, J, M$)

$$|\xi\rangle = \sum_{\alpha, J, M} C_i |\alpha, J, M\rangle, \qquad (15.39)$$

where $J_z = M$ is quantized along the binary z-axis. The values of M in (15.39) are half integer and differ by at least two units because of binary symmetry. This prevents the values M and $-M$ from being simultaneously present in the expansion (15.39).

The matrix elements $\langle \xi | \mu_x | \xi \rangle = g_{x3}$ and $\langle \xi | \mu_y | \xi \rangle = g_{y3}$ vanish because μ_x and μ_y obey the selection rule $|\Delta M| = 1$. To find the other matrix elements let R be the rotation $R = \mathrm{e}^{\mathrm{i}\pi J_y}$. Then from eqns (15.26), (15.29) the Kramers conjugate $|\bar{\xi}\rangle$ is given by

$$|\bar{\xi}\rangle = \theta |\xi\rangle = RK_0 |\xi\rangle = \sum_{\alpha, J, M} C_i^* (-1)^{J-M} |\alpha, J, -M\rangle. \qquad (15.40)$$

It is clear that $\langle \xi | \mu_z | \bar{\xi} \rangle = g_{z1} - \mathrm{i}g_{z2}$ vanishes since μ_z obeys the selection rule $|\Delta M| = 0$. Since the y-axis is a binary axis, $R |\xi\rangle$ is also an eigenstate with the same energy as $|\xi\rangle$ and clearly orthogonal to it. It must therefore coincide with $|\bar{\xi}\rangle = RK_0 |\xi\rangle$ within a phase factor and all the C_i in (15.39) can be chosen to be real. We then have the relations

$$R |\xi\rangle = \theta |\xi\rangle = |\bar{\xi}\rangle, \qquad \langle \bar{\xi}| = \langle \xi | R^\dagger$$

and hence

$$\langle \xi | \mu_x | \bar{\xi} \rangle = -\langle \xi | R^{-1} \mu_x R | \bar{\xi} \rangle = -\langle \xi | R^\dagger \mu_x R^2 | \xi \rangle = \langle \bar{\xi} | \mu_x | \xi \rangle$$

since $R^2 = -1$. Hence $\langle \xi | \mu_x | \bar{\xi} \rangle$ must be real and since it is equal to $g_{x1} - \mathrm{i}g_{x2}$, it follows that g_{x2} vanishes. Similarly it is found that

$$g_{y1} - \mathrm{i}g_{y2} = \langle \xi | \mu_y | \bar{\xi} \rangle = -\langle \bar{\xi} | \mu_y | \xi \rangle.$$

$\langle \xi | \mu_y | \bar{\xi} \rangle$ is imaginary and g_{y1} vanishes. The 'tensor' $g_{q\alpha}$ is thus diagonal with our choice of eigenstates and coordinate axes. Since we made no assumptions about $\boldsymbol{\mu}$ apart from its being a time-odd vector, this is clearly also true for the 'tensor' $a_{q\alpha}$ associated with $\mathbf{H}_\mathrm{e}$.

15.8. Threefold symmetry

The trigonal symmetry C_3 provides an example of a situation where it may be impossible to diagonalize $g_{q\alpha}$ and $a_{q\alpha}$ simultaneously.

Consider a system that possesses no symmetry beyond a threefold z-axis and assume that the expansion (15.39) of $|\xi\rangle$ contains the value $M = +\frac{1}{2}$. The other values of M in the expansion will be $+\frac{1}{2}+3p$ with p integer. Similarly the expansion of $|\bar{\xi}\rangle$ will contain the values $M = -\frac{1}{2}+3p'$. This has the following consequences for the 'tensor' g.

$$\begin{cases} \langle\xi| \mu_x |\xi\rangle = g_{x3} = 0, \\ \langle\xi| \mu_y |\xi\rangle = g_{y3} = 0, \\ \langle\xi| \mu_z |\xi\rangle = g_{z3} = g_\|, \end{cases} \tag{15.41}$$

and

$$\begin{cases} \langle\xi| \mu_z |\bar{\xi}\rangle = g_{z1}-ig_{z2} = 0, \\ \langle\xi| \mu_- |\bar{\xi}\rangle = \langle\xi| (\mu_x-i\mu_y) |\bar{\xi}\rangle, \\ \quad = g_{x1}-ig_{x2}-i(g_{y1}-ig_{y2}) = 0. \end{cases} \tag{15.42}$$

From these last equations (15.42) we have

$$\begin{aligned} g_{z1} &= g_{z2} = 0, \\ g_{x1} &= g_{y2} = g', \\ g_{x2} &= -g_{y1} = g''. \end{aligned} \tag{15.43}$$

Thus the 'tensor' g has the form

$$\begin{pmatrix} g' & g'' & 0 \\ -g'' & g' & 0 \\ 0 & 0 & g_\| \end{pmatrix}, \tag{15.44}$$

which can be brought into diagonal form by a rotation S of the coordinate axes around Oz through an angle Φ such that $\tan \Phi = -g''/g'$, giving

$$\begin{pmatrix} g_\perp & 0 & 0 \\ 0 & g_\perp & 0 \\ 0 & 0 & g_\| \end{pmatrix}, \tag{15.45}$$

where $g_\perp^2 = g'^2+g''^2$.

Similarly the hyperfine structure 'tensor' will initially have the form

$$\begin{pmatrix} a' & a'' & 0 \\ -a'' & a' & 0 \\ 0 & 0 & a_\| \end{pmatrix}, \tag{15.46}$$

which can be reduced by a rotation of the coordinates about the axis

Oz through an angle Ψ such that $\tan \Psi = -a''/a'$ to a diagonal form

$$\begin{pmatrix} a_\perp & 0 & 0 \\ 0 & a_\perp & 0 \\ 0 & 0 & a_\parallel \end{pmatrix}. \qquad (15.47)$$

However there is no *a priori* reason why $\tan \Psi$ should be equal to $\tan \Phi$, and this implies that we have to use two different sets of axes: Z, x, y for the components of the external magnetic field $\mathbf{H}$ and Z, X, Y for the components of the nuclear magnetic moment $\mathbf{\mu}_I = \gamma_n \hbar \mathbf{I}$. The spin Hamiltonian can then be written as

$$\tfrac{1}{2}\beta g_\parallel H_Z \sigma_3 + \tfrac{1}{2}\beta g_\perp (H_x \sigma_1 + H_y \sigma_2) - \gamma_n \hbar a_\parallel I_Z \sigma_3 - \gamma_n \hbar a_\perp (I_X \sigma_1 + I_Y \sigma_2). \qquad (15.48)$$

The fact that the components of $\mathbf{I}$ are not measured along the same axes as those of $\mathbf{H}$ would not be observable if there did not exist another term in the Hamiltonian, the direct coupling of the nuclear moment with the applied field $\mathbf{H}$, which is

$$-\gamma_n \hbar (\mathbf{I} \cdot \mathbf{H}) = -\gamma_n \hbar (I_x H_x + I_y H_y + I_z H_z). \qquad (15.49)$$

Although this term is very small, it can be measured with reasonable precision in Endor experiments (cf. Chapter 4). In general the hyperfine term will be much larger, so that we should use the (X, Y, Z) axes for the components of $\mathbf{I}$. Then if $\mathbf{H}$ is, for example, along the x-axis, the diagonal term in (15.49) becomes $-\gamma_n \hbar I_X H_X \cos(\Psi - \Phi)$. Thus in principle the difference in orientation that may occur in C_3 symmetry for the axes OX, OY and Ox, Oy could be detected, although it does not appear to have been observed in practice.

15.9. Selection rules related to time reversal

Let us consider a system S with a Hamiltonian $\mathscr{H}_0$ invariant through the transformations of a spatial group G, for instance, a paramagnetic ion whose environment has a symmetry describable by that group. As we have seen in § 12.6, a perturbation expressed by an operator V_β belonging to a representation Γ' of G cannot lift the p-fold degeneracy of an energy level of S, spanned by the wave-functions Ψ_α that belong to a p-dimensional representation Γ of G, unless the direct product $\Gamma^* \times \Gamma \times \Gamma'$ contains the unit representation of G. If this condition is not satisfied all the matrix elements $(\Psi_\alpha| V_\beta |\Psi_\gamma)$ vanish. It was shown in § 12.6 that if the representations Γ and Γ' are equivalent to their complex conjugates, respectively Γ^* and Γ'^*, that is if all their characters are real (an assumption that we retain in the following discussion), this condition is equivalent to the requirement that $(\Gamma \times \Gamma)$ contains Γ'. Furthermore, we shall show that if the Hamiltonian $\mathscr{H}_0$ is invariant through the time-reversal operation θ, the condition for the lifting of the degeneracy is more restrictive than the one stated above.

Let us define two symbols ε_V and ε_θ equal to ± 1: ε_V is equal to $+1$ if the operator V_β is time-even, and -1 if it is time-odd; ε_θ is equal to the square θ^2 of the time-reversal operator: it is $+1$ if the number of electrons is even and -1 if the number of electrons is odd (eqn (15.21)). The result that we shall establish can be stated as follows. In order for the operator V_β, belonging to Γ', to have non-vanishing matrix elements inside the manifold Γ the following are necessary conditions: if $\varepsilon_V \varepsilon_\theta > 0$, Γ' must be contained in the symmetric direct product $[\Gamma \times \Gamma]_S$; if $\varepsilon_V \varepsilon_\theta < 0$, Γ' must be contained in the antisymmetric direct product $[\Gamma \times \Gamma]_A$. The characters of these products are given in eqn (12.43). These conditions are more restrictive than the requirement obtained in § 12.6 prior to the introduction of time reversal, namely that Γ' must be contained in the direct product $\Gamma \times \Gamma$.

The proof is as follows. Under an operation R belonging to G the functions Ψ_α transform according to the law

$$R\Psi_\alpha = \sum_\gamma \Psi_\gamma D_\Gamma(R)_{\gamma\alpha}. \tag{15.50}$$

Let us introduce the set of functions

$$\overline{\Psi}_\alpha = \theta \Psi_\alpha, \tag{15.51}$$

which are generated from the Ψ_α through time reversal. Under an operation R the functions $\overline{\Psi}_\alpha$ transform as

$$R\overline{\Psi}_\alpha = R\theta\Psi_\alpha = \theta R\Psi_\alpha = \theta \sum_\gamma D_\Gamma(R)_{\gamma\alpha}\Psi_\gamma$$
$$= \sum_\gamma D_\Gamma^*(R)_{\gamma\alpha}\theta\Psi_\gamma = \sum_\gamma D_\Gamma^*(R)_{\gamma\alpha}\overline{\Psi}_\gamma, \tag{15.52}$$

where we have used the facts that the time-reversal operator θ commutes with the operations R of the group G and that it is antilinear. Equation (15.52) shows that the functions $\overline{\Psi}_\alpha$ afford for the group G a representation D_Γ^* which, according to our assumption of real characters, is equivalent to D_Γ; hence there must exist a unitary matrix U

relating the two representations D_Γ and D_Γ^* through the equations

$$D_\Gamma^*(R) = UD_\Gamma(R)U^{-1}. \qquad (15.53)$$

The fact that the two representations D_Γ^* and D_Γ are equivalent does not necessarily mean that they can be made to coincide, that is in the present case made real, by a similarity transformation as described by eqn (12.5). In fact it can be shown that the matrices D_Γ can always be made real by a similarity transformation if the square θ^2 of the time-reversal operation is $+1$; on the other hand, D_Γ and D_Γ^*, although equivalent, are always different if $\theta^2 = -1$ (see, for instance, Fano and Racah 1959, Griffith 1961) but we shall not make use of these results in the following.

Instead of the set of matrix elements $(\Psi_\alpha| V_\beta |\Psi_\gamma) = Y_{\alpha\gamma,\beta}$ it is convenient to introduce the mixed matrix elements $(\overline{\Psi}_\alpha| V_\beta |\Psi_\gamma) = Z_{\alpha\gamma,\beta}$. Since the $\overline{\Psi}_\alpha$ form a set of normalized orthogonal functions, spanning the same manifold as the Ψ_α, it is clear that the vanishing of all the $Y_{\alpha\gamma,\beta}$ implies that of the $Z_{\alpha\gamma,\beta}$ and vice versa. Let us then transform the expression $Z_{\alpha\gamma,\beta}$ as follows:

$$\begin{aligned}Z_{\alpha\gamma,\beta} &= (\overline{\Psi}_\alpha| V_\beta |\Psi_\gamma) = (\theta\Psi_\alpha, V_\beta\Psi_\gamma) = (\theta V_\beta\Psi_\gamma, \theta^2\Psi_\alpha) \\ &= (\theta V_\beta\theta^{-1}\theta\Psi_\gamma, \theta^2\Psi_\alpha) = \epsilon_\theta\epsilon_V(V_\beta^\dagger\overline{\Psi}_\gamma, \Psi_\alpha) \\ &= \epsilon_\theta\epsilon_V(\overline{\Psi}_\gamma, V_\beta\Psi_\alpha) = \epsilon_\theta\epsilon_V(\overline{\Psi}_\gamma| V_\beta |\Psi_\alpha) = \epsilon_\theta\epsilon_V Z_{\gamma\alpha,\beta}. \quad (15.54)\end{aligned}$$

We can write now:

$$Z_{\alpha\gamma,\beta} = \tfrac{1}{2}(Z_{\alpha\gamma,\beta} + \epsilon_V\epsilon_\theta Z_{\gamma\alpha,\beta}). \qquad (15.55)$$

It is clear from (15.55) that, depending on the sign of $\epsilon_V\epsilon_\theta = \pm 1$, $Z_{\alpha\gamma,\beta}$ belongs to the representation $[\Gamma \times \Gamma]_S \times \Gamma'$ or $[\Gamma \times \Gamma]_A \times \Gamma'$.

In order for this set and therefore also for the set $Y_{\alpha\gamma,\beta} = (\Psi_\alpha| V_\beta |\Psi_\gamma)$ not to vanish identically the unit representation must be contained either in the product $[\Gamma \times \Gamma]_S \times \Gamma'$ or $[\Gamma \times \Gamma]_A \times \Gamma'$ (depending on the sign of $\epsilon_\theta\epsilon_V$), rather than simply in the product $(\Gamma \times \Gamma) \times \Gamma'$ as proved in § 12.6. This in turn implies that Γ' must be contained either in $[\Gamma \times \Gamma]_S$ or $[\Gamma \times \Gamma]_A$ depending on the sign of $\epsilon_\theta\epsilon_V$.

Let us illustrate these results with a few examples. In § 13.5 we used the Wigner–Eckart theorem to prove that inside a manifold J every vector $\mathbf{V}$ is proportional to the angular momentum $\mathbf{J}$:

$$\mathbf{V} = \alpha\mathbf{J}, \qquad (15.56)$$

where the constant α must be real if $\mathbf{V}$ is Hermitian. As the vector $\mathbf{J}$ is time-odd, $\mathbf{V}$ must also be time-odd and $\epsilon_V = -1$. Since $\mathbf{V}$ transforms according to the representation D^1 of the rotation group, the condition for a matrix element $(JM|\,V_\beta\,|JM')$ to be different from zero is that $D^J \times D^J$ must contain D^1, which of course it does, if $J \neq 0$, as shown by the relation (13.17). However, our theorem tells us something more: if J is half integer, $\varepsilon_\theta = -1$, $\varepsilon_V \varepsilon_\theta = +1$ and D^1 belongs to $[D^J \times D^J]_S$; if J is an integer, $\varepsilon_\theta = +1$, $\varepsilon_V \varepsilon_\theta = -1$ and D^1 belongs to $[D^J \times D^J]_A$.

A less trivial example is that of the fictitious angular momentum introduced in § 14.2. There again the Wigner–Eckart theorem predicts that, inside a cubic triplet Γ_4 or Γ_5, the matrix elements of the components of a vector $\mathbf{V}$ are determined within a proportionality constant, because (see Table 2) the direct product $\Gamma_4 \times \Gamma_4$ or $\Gamma_5 \times \Gamma_5$ contains the vector representation Γ_4 only once. Moreover, as shown in this Table, it is the antisymmetric product $[\Gamma_4 \times \Gamma_4]_A$ or $[\Gamma_5 \times \Gamma_5]_A$ that contains Γ_4 and the product $\varepsilon_V \varepsilon_\theta$ must be equal to -1. Since the representations Γ_4 and Γ_5 arise only for an even number of electrons, $\varepsilon_\theta = +1$ and ε_V must be -1. Therefore only time-odd vectors have non-vanishing matrix elements inside the triplet Γ_4 or Γ_5. We anticipated this result when we introduced the fictitious angular momentum (necessarily time-odd) to which every vector must be proportional within the triplets Γ_4 or Γ_5.

15.10. The effect of an applied electric field on a paramagnetic ion

The potential energy of an ion, due to the presence of a homogeneous electric field $\mathbf{E}$ can be written:

$$V_E = -(-e)\mathbf{E} \cdot \sum_p \mathbf{r}_p = -\mathbf{E} \cdot \mathbf{p}_e, \qquad (15.57)$$

where $\sum_p \mathbf{r}_p$ is a summation over the positions of the electrons of the ion and $\mathbf{p}_e = -e \sum_p \mathbf{r}_p$ is here the operator for the electric dipole moment of the ion.

The energy changes induced by the electric perturbation (15.57) can be separated into first-order effects that are linear in $\mathbf{E}$ and higher-order effects, predominantly quadratic in the field strength, known as polarization effects. The quadratic polarization effects give rise to a change in energy expressed by the usual second-order perturbation formula:

$$\Delta_2 W_E = \sum_n{}' \frac{\langle 0|\,\mathbf{E}\cdot\mathbf{p}_e\,|n\rangle\langle n|\,\mathbf{E}\cdot\mathbf{p}_e\,|0\rangle}{W_0 - W_n}. \qquad (15.58)$$

The electric fields E that can be conveniently obtained in laboratories are at most of the order of a few thousand volts per millimeter and the corresponding matrix elements $\langle 0|\, \mathbf{p}_e \cdot \mathbf{E}\, |n\rangle$ expressed in wavenumbers turn out to be of the order of one cm^{-1} or so, that is, very much less than the energy separations $W_n - W_0$. The first-order effect of (15.57)

$$\Delta_1 W_E = -\mathbf{E} \cdot \langle \mathbf{p}_e \rangle, \qquad (15.59)$$

where the expectation value $\langle \mathbf{p}_e \rangle$ is to be taken over the ground state, or more generally inside the ground manifold if the latter is degenerate, will thus be very much greater than $\Delta_2 W_E$ given by (15.58), unless it vanishes because of selection rules. It is these selection rules that we wish to investigate using a few specific examples.

If the environment and therefore also the Hamiltonian of the paramagnetic ion is invariant through inversion, the ground manifold of the ion has a definite parity positive or negative and in either case $\langle \mathbf{p}_e \rangle$ vanishes because $\mathbf{p}_e$ is a polar (or space-odd) vector which is changed into $-\mathbf{p}_e$ through inversion. On the other hand, if the Hamiltonian of the ion is *not* invariant through inversion, $\langle \mathbf{p}_e \rangle$ *may* be different from zero.

If the starting point of our description of the ion is the free ion placed in a crystal potential $V(\mathbf{r})$, the lack of a centre of symmetry in the environment implies the existence in $V(\mathbf{r})$ of a part V_{odd} that changes its sign through inversion. The energy change linear in $\mathbf{E}$ appears then as a second-order term, cross product between $-(\mathbf{p} \cdot \mathbf{E})$ and V_{odd}. Writing c.c. for the complex conjugate, we have

$$\Delta_1 W_E = -\sum_n{}' \frac{\langle 0|\, \mathbf{p} \cdot \mathbf{E}\, |n\rangle \langle n|\, V_{\text{odd}}\, |0\rangle + \text{c. c.}}{W_0 - W_n}, \qquad (15.60)$$

where the states $|0\rangle$, $|n\rangle$ and the energies W_0, W_n are those of a free ion. Internal crystal electric fields are usually very much stronger than applied electric fields; the matrix element $\langle n|\, V_{\text{odd}}\, |0\rangle$ is therefore very much larger than $\langle n|\, \mathbf{p} \cdot \mathbf{E}\, |0\rangle$, and correspondingly $\Delta_1 W_E$ as given by (15.60), is very much larger than the polarization term $\Delta_2 W_E$ given by (15.58).

A different approach is first to diagonalize the Hamiltonian of the ion embedded in the crystal and then to calculate the expectation value

$$\Delta_1 W_E = -\langle 0|\, \mathbf{p} \cdot \mathbf{E}\, |0\rangle. \qquad (15.61)$$

The state $|0\rangle$ in (15.61) is the ground state of the *bound* ion which contains both odd and even admixtures. This approach is more convenient when covalent rather ionic bonding exists between the ion and its surroundings (see Chapter 20).

We shall now show that the absence of a centre of inversion is only a necessary condition for a change of energy linear with **E** and that selection rules connected with time reversal may also forbid such an effect. The first and very general example is that of a system, whose Hamiltonian is invariant through rotation without being necessarily invariant through inversion (as for instance that of a free molecule). Its energy eigenstates are degenerate manifolds spanned by eigenstates of $\mathscr{J}$. Inside each manifold $\mathscr{J}$, according to eqn (15.56), only time-odd vectors have non-vanishing matrix elements and the expectation value of the dipole moment $\mathbf{p}_e$, which is a time-even vector, vanishes (barring accidental degeneracy).

A second example is an environment of cubic symmetry *without* inversion symmetry that is invariant through the group O (but not O_h!). We have already shown in § 15.9 that only time-odd vectors have non-vanishing matrix elements inside the triplets Γ_4 or Γ_5, and this again precludes a finite expectation value for $\mathbf{p}_e$. It is easy to show that the same is true if the ion has an odd number of electrons, for which the specific representations are Γ_6, Γ_7, Γ_8. In that case, for the time-even vector $\mathbf{p}$ the product $\varepsilon_\Gamma \varepsilon_\theta$ defined in § 15.9 is equal to -1, which means that the vector representation Γ_4 should be contained in one of the antisymmetric products $[\Gamma_6 \times \Gamma_6]_A$, $[\Gamma_7 \times \Gamma_7]_A$, $[\Gamma_8 \times \Gamma_8]_A$ for the matrix elements of $\mathbf{p}_e$ to be non-zero.

We see, however, in Table 8, that none of these products contain Γ_4, and we can state the result that in a cubic environment an electric field cannot produce a first-order change in energy. To dispel a possible misunderstanding, let us emphasize again that it is invariance through time reversal rather than through space inversion which is responsible for this selection rule. This misunderstanding arises sometimes because a cubic environment is usually represented by drawing a cube or an octahedron, a figure that does possess a centre of symmetry, and this may create the mistaken impression that cubic symmetry does imply inversion symmetry. As a counter example one can think, for instance, of a cube with a left hand at each apex; such a figure is invariant through the cubic group of rotations but not through inversion.

At the risk of increasing the confusion a little further we shall show that if the environment of the ion has tetrahedral symmetry and therefore is invariant through the group of improper rotations T_d, which is isomorphic to O, and does not possess a centre of inversion, an applied electric field *does* produce a linear change in the energy of the ion. The temptation to associate the absence of a linear effect in O and its presence in T_d which is isomorphic to O, with the mistaken notion that O has a centre of inversion, must be resisted.

The gist of the proof for T_d is as follows. The components of a polar vector, which transform like the functions x, y, z under proper or improper rotations and which therefore in the group O span the representation Γ_4, in the group T_d span the representation Γ_5. This can be checked as follows. We have already stated respectively in §§ 14.1 and 14.6 that O and T_d are isomorphic to the group of permutations S_4; now the rotations of O permute among themselves the four body-diagonals of the cube, while the improper rotations of T_d permute the four apexes of the tetrahedron. Let us then consider for instance the permutation (12): for O it means a permutation of the body diagonals 1 and 2, that is a rotation C_2' around one of the six twofold axes passing through the centre points of opposite edges; for T_d it means a reflection in a plane passing through the edge 34 and the centre point of the edge 12. It is easy to check that the matrix for the transformation of (x, y, z) through the transformation (12) has a trace equal to -1 for O and $+1$ for T_d. From the characters in Table 1 (which is the same for isomorphic operations of O and T_d) we conclude that the components x, y, z of a polar vector must span the representation Γ_5 of T_d rather than Γ_4.

We see in Table 8 that Γ_5 is contained once in $[\Gamma_4 \times \Gamma_4]_S$, $[\Gamma_5 \times \Gamma_5]_S$, $[\Gamma_8 \times \Gamma_8]_A$. Therefore in all these multiplets a linear splitting is induced by an applied electric field. In the same way as we replaced a vector inside a multiplet Γ_4 or Γ_5 by a fictitious angular momentum $\tilde{l}$, within the multiplets Γ_4, Γ_5 and Γ_8 of T_d we can replace the three components of a polar vector, which transform through T_d according to the representation Γ_5, by the three operators

$$\tilde{\mathscr{J}}_y\tilde{\mathscr{J}}_z+\tilde{\mathscr{J}}_z\tilde{\mathscr{J}}_y, \quad \tilde{\mathscr{J}}_z\tilde{\mathscr{J}}_x+\tilde{\mathscr{J}}_x\tilde{\mathscr{J}}_z, \quad \tilde{\mathscr{J}}_x\tilde{\mathscr{J}}_y+\tilde{\mathscr{J}}_y\tilde{\mathscr{J}}_x. \quad (15.62)$$

In (15.62) $\tilde{\mathscr{J}}$ is an axial (or space-even) time-odd vector whose components are transformed by T_d according to the representation Γ_4 (in contrast to a polar vector whose components are transformed by T_d as Γ_5). Inside Γ_4 and Γ_5, $\tilde{\mathscr{J}}$ coincides with the fictitious angular momentum l introduced in § 14.2. Inside Γ_8, as will be explained in more detail in § 18.3, the matrix elements of $\tilde{\mathscr{J}}$ are those of a fictitious angular momentum $\tilde{\mathscr{J}} = \frac{3}{2}$. The splitting of each of the multiplets Γ_4, Γ_5, Γ_8

TIME REVERSAL AND KRAMERS DEGENERACY 663

of T_d by an electric field **E** can thus be expressed by a spin Hamiltonian:

$$\mathscr{H}_E = a\{E_x(\tilde{\mathscr{I}}_y\tilde{\mathscr{I}}_z+\tilde{\mathscr{I}}_z\tilde{\mathscr{I}}_y)+E_y(\tilde{\mathscr{I}}_z\tilde{\mathscr{I}}_x+\tilde{\mathscr{I}}_x\tilde{\mathscr{I}}_z)+E_z(\tilde{\mathscr{I}}_x\tilde{\mathscr{I}}_y+\tilde{\mathscr{I}}_y\tilde{\mathscr{I}}_x)\}. \tag{15.63}$$

The combined effect of an electric and a magnetic field

Finally we investigate changes in the energy of the ion which are bilinear in the components of an applied electric field **E** and an applied magnetic field **H**, that is changes in the 'g- tensor'.

For brevity we limit ourselves to the triplets Γ_4 and Γ_5 of the groups T_d and O. (For Γ_8 the situation becomes rather complicated.) The general expression for an operator responsible for such a change should be of the form

$$V_{EH} = \sum_{p,q} H^p E^q \mu_e^{\prime p} p_e^{\prime q}. \tag{15.64}$$

Here H^p, etc., are components of **H**, etc., and the vectors $\boldsymbol{\mu}_e'$ and $\mathbf{p}_e'$, which depend on the electronic variables of the ion, have respectively the time and space properties of the magnetic moment $\boldsymbol{\mu}_e$ and of the electric dipole moment $\mathbf{p}_e$; namely, $\boldsymbol{\mu}_e$ is time-odd and space-even, whereas $\mathbf{p}_e$ is time-even and space-odd.

We begin with the group O, within which μ_e' and p_e' each transform as Γ_4. V_{EH} is a time-odd operator that transforms through the group O (see Table 2) according to

$$\Gamma_4 \times \Gamma_4 = \Gamma_1+\Gamma_3+\Gamma_4+\Gamma_5. \tag{15.65}$$

Since the operator V_{EH} is time-odd, it will have non-vanishing matrix elements inside the triplets Γ_4 or Γ_5 of the group O only if at least one of the representations on the right-hand side of (15.65) is contained in the antisymmetric products $[\Gamma_4 \times \Gamma_4]_A$ or $[\Gamma_5 \times \Gamma_5]_A$, ($\varepsilon_\theta \varepsilon_V = -1$). We see in Table 2 that $[\Gamma_4 \times \Gamma_4]_A = [\Gamma_5 \times \Gamma_5]_A = \Gamma_4$. Therefore the only part of the operator V_{EH} given by (15.64) which has non-vanishing matrix elements both in Γ_4 and Γ_5 is that which contains the three combinations of the bilinear products $\mu_e^{\prime p} p_e^{\prime q}$ which transform according to Γ_4. It can be shown that these three combinations are

$$\mu_e^{\prime y}p_e^{\prime z}-\mu_e^{\prime z}p_e^{\prime y}, \quad \mu_e^{\prime z}p_e^{\prime x}-\mu_e^{\prime x}p_e^{\prime z}, \quad \mu_e^{\prime x}p_e^{\prime y}-\mu_e^{\prime y}p_e^{\prime x}. \tag{15.66}$$

Within the manifolds Γ_4 or Γ_5 of O they are proportional respectively to $\tilde{\mathscr{I}}_x, \tilde{\mathscr{I}}_y, \tilde{\mathscr{I}}_z$ and V_{EH} can be represented within Γ_4 or Γ_5 of O as

$$V_{EH} = a\{(H_yE_z-H_zE_y)\tilde{\mathscr{I}}_x+(H_zE_x-H_xE_z)\tilde{\mathscr{I}}_y+(H_xE_y-H_yE_x)\tilde{\mathscr{I}}_z\} \tag{15.67}$$

or in a vector notation as
$$V_{EH} = a(\mathbf{H} \wedge \mathbf{E}) \cdot \tilde{\mathscr{J}}.$$

We see from eqn (15.67) that for the group O, an electric field gives rise to a skew-symmetric contribution to the g-tensor.

We consider now the tetrahedral group T_d: here the space-even vector $\boldsymbol{\mu}'_e$ transforms like Γ_4, that is in the same way as through O, but the polar vector $\mathbf{p}'_e$ transforms like Γ_5 and the operator V_{EH} given by (15.64) transforms like

$$\Gamma_4 \times \Gamma_5 = \Gamma_2 + \Gamma_3 + \Gamma_4 + \Gamma_5. \tag{15.68}$$

For T_d, as for O, only the representation Γ_4 on the right-hand side of (15.68) is contained in $[\Gamma_4 \times \Gamma_4]_A$ and $[\Gamma_5 \times \Gamma_5]_A$. However, the three combinations of $\mu'^p_e p'^q_e$, which transform like Γ_4 through T_d and which inside the manifold Γ_4 or Γ_5 of T_d can be taken as proportional to $\tilde{\mathscr{J}}_x, \tilde{\mathscr{J}}_y, \tilde{\mathscr{J}}_z$, are different from (15.66). In this case they are

$$\mu'^y_e p'^z_e + \mu'^z_e p'^y_e, \quad \mu'^z_e p'^x_e + \mu'^x_e p'^z_e, \quad \mu'^x_e p'^y_e + \mu'^y_e p'^x_e. \tag{15.69}$$

Within the manifolds Γ_4 or Γ_5 of T_d the operator V_{EH} can therefore be represented as

$$V_{EH} = a\{(H_y E_z + H_z E_y)\tilde{\mathscr{J}}_x + (H_z E_x + H_x E_z)\tilde{\mathscr{J}}_y + (H_x E_y + H_y E_x)\tilde{\mathscr{J}}_z\} \tag{15.70}$$

and we see that the contribution to the g-tensor is symmetrical.

The present discussion is entirely based on symmetry considerations and no attempt is made to investigate the actual physical mechanism responsible for the g-shifts induced by an electric field. (It leans heavily on a set of unpublished lectures given by F. S. Ham in 1967 at the Nato School in Ghent.) Let us say only that an expression such as (15.64) for V_{EH} results in second order from the perturbing electric and magnetic Hamiltonians $-(\boldsymbol{\mu}_e \cdot \mathbf{H})$ and $-(\mathbf{p}_e \cdot \mathbf{E})$ and clearly cannot arise unless the system has states of mixed parity and therefore lacks an inversion centre.

Both of the effects studied above, namely a linear splitting of the energy levels and a linear g-shift, induced by an applied electric field, have been observed on transition ions of the iron group embedded in tetrahedral coordination in silicon (Ludwig and Woodbury 1961, Ham 1961, Ludwig and Ham 1962).

16
ELEMENTARY THEORY OF THE CRYSTAL FIELD

In this chapter we gather together some mathematical results useful for the calculation of the matrix elements of the crystal field and the determination of the wave-functions of the bound ion, in the framework of the elementary crystal-field theory. It should be made clear at the outset that this theory is a model that may or may not be an adequate description of the physical reality. It is probably quite good for rare-earth salts and rather bad for some complexes of the iron group, but we shall not discuss this point now.

16.1. The crystal field (or crystal potential)

Once the assumptions describing the model are stated, the rest is algebra. These assumptions can be formulated as follows. Let $\mathcal{H}_0$ be the Hamiltonian of the free ion. The Hamiltonian $\mathcal{H}$ of the bound ion is then assumed to be of the form $\mathcal{H} = \mathcal{H}_0 + V$ where V is an electrostatic potential energy that satisfies the Laplace equation $\Delta V = 0$ and has the symmetry of the surroundings of the ion. The potential V can then be expanded as a sum of spherical harmonics,

$$V = \sum_{k=0}^{\infty} \sum_{q=-k}^{k} B_k^q r^k Y_k^q(\theta, \varphi) = \sum V_k^q, \tag{16.1}$$

with the sign convention, $Y_k^{q*} = (-1)^q Y_k^{-q}$. Certain limitations are imposed on the B_k^q by the symmetry of the surroundings. For instance, if there is a centre of inversion there will be no harmonics of odd k in the expansion (16.1); also the reality of V requires $B_k^q = (-1)^q B_k^{-q*}$.

The potential energy V is assumed to arise from the electronic and nuclear charges of the atoms and ions surrounding the paramagnetic ion under study. To calculate this potential the simplest method is the so-called point charge approximation where each ion is replaced by a point charge equal to the total charge of the ion and located at its centre. For a monatomic ion the position of the centre is naturally that of its nucleus but it is not so clear what its position should be for, say, a negative ion such as $(NO_3)^-$.

As soon as one tries to improve on this approximation, difficulties arise. Even if one does take into account the spatial extension of the charge distribution producing V, there are still polarization effects, large and difficult to estimate. This comes from the fact that the closed shells of the paramagnetic ion are distorted by the potential V and under this distortion create a modified potential 'seen' by the magnetic electrons of the unfilled shells, which does not have the spherical symmetry of the self-consistent Hartree potential of the free ion anymore. Be it as it may, we start from (16.1), whatever its origin, and proceed to carry out some algebra with it, postponing discussion of the physics until later.

We are not interested in the absolute displacements of the energy levels but only in their splittings by the crystal potential. We can then without loss of generality omit the term $k = 0$ in the expansion (16.1).

The most important approximation we make is to take the electronic configuration as defined in § 11.4, as a good quantum number. This has the following implications. The wave-functions Ψ between which we calculate matrix elements of the crystal potential will be Slater determinants or linear combinations of Slater determinants. Each determinant will be of the form $(\chi_1, \ldots, \chi_N, \varphi_1, \ldots, \varphi_p)$ in which the first N one-electron functions $\chi_1, \ldots, \chi_N$ represent closed shells and unless stated otherwise will be the same for all the Slater determinants appearing in the expansion of the states Ψ of the paramagnetic ion. The remaining one-electron wave-functions $\varphi_1, \ldots, \varphi_p$ are those of the magnetic electrons of the unfilled shells, d- or f-electrons, depending on the transition group considered; they are therefore of the form $|l, m_l, m_s\rangle$ where $l = 2$ for d-electrons and 3 for f-electrons. The crystal potential V is a sum of one-electron operators $V = \sum V_p$ where V_p is obtained by replacing in (16.1) r, θ, φ by the coordinates r_p, θ_p, φ_p of the pth electron. According to the rules given in § 11.6, any matrix element $(\Psi| V |\Psi')$ will be a sum of one-electron matrix elements of the form $(\psi_a| V |\psi_b)$, where ψ_a and ψ_b are functions χ from the filled shells and φ from the unfilled shells. The contribution of the closed shells to a matrix element $(\Psi| V |\Psi')$ will be $\sum_{i=1}^{N} (\chi_i| V |\chi_i)$, which vanishes if the term $k = 0$ is omitted in the expansion (16.1). We can then omit closed shells altogether and write our Slater determinant as $(\varphi_1, \ldots, \varphi_p)$, building it solely from wave-functions of magnetic electrons (it should not be forgotten that we are only able to do this because we deal with one-electron operators such as V; it would not be quite correct for two-electron operators). A one-electron matrix element then takes the form

$$\sum_{k,q} \langle l, m_l, m_s| V_k^q |l, m_l', m_s'\rangle = \sum_k \langle r^k \rangle \sum_q B_k^q \langle l, m_l| Y_k^q |l, m_l'\rangle \delta(m_s, m_s'),$$

(16.2)

where

$$\langle r^k \rangle = \int_0^\infty |f_l(r)|^2 \, r^k r^2 \, \mathrm{d}r \tag{16.3}$$

and $f_l(r)$ is the radial wave-function of the Hartree–Fock approximation. V_k^q is a component of an irreducible tensor and the corresponding matrix element vanishes unless

$$k \leqslant 2l \text{ and } m_l = q + m_l'. \tag{16.4}$$

The selection rule $k \leqslant 2l$ decreases considerably the number of parameters that are needed for the description of the crystal potential. Furthermore, even in the absence of a centre of symmetry, terms V_k^q with odd k can be omitted because the corresponding matrix elements vanish. (These terms will, however, admix excited configurations.)

The number of terms in (16.1) is further reduced by the symmetry of the surroundings. Only even values of q are admissible if the surroundings have either a twofold axis of symmetry parallel to the axis of quantization or a plane of symmetry perpendicular to it. If there is a threefold axis, only values of q that are multiples of 3 appear.

Pairs of terms in (16.1) with values of q of opposite sign can be rewritten in the form

$$B_k^{|q|} Y_k^{|q|} + B_k^{|q|*} Y_k^{|q|*} = \frac{(B_k^{|q|} + B_k^{|q|*})}{2}(Y_k^{|q|} + Y_k^{|q|*}) - \frac{(B_k^{|q|} - B_k^{|q|*})}{2i} \frac{(Y_k^{|q|} - Y_k^{|q|*})}{i}$$

$$= f(\theta) \cos\{|q|\, \varphi\} + g(\theta) \sin\{|q|\, \varphi\}. \tag{16.5}$$

If there is a plane of symmetry passing through the axis of quantization we take it as the plane xOz; the term $\sin\{|q|\,\varphi\}$ must vanish and $B_k^{|q|}$ is real.

If there is a plane of symmetry perpendicular to a threefold axis of quantization (symmetry C_{3h}), $B_k^{|q|}$ can also be made real by a suitable choice of the xOz plane. For symmetry C_{3h} there is no term with $q \neq 0$ for $l = 2$ and a single one $B_6^6 Y_6^6 + B_6^{-6} Y_6^{-6}$ for $l = 3$ where it is always possible to make $B_6^6 - B_6^{6*}$ vanish by a rotation around the z-axis.

By a parallel argument similar relations are obtained for symmetry C_{4h} or D_4, where only $|q| = 0$ and $|q| = 4$ are possible.

We consider finally a potential of cubic symmetry. Being invariant under all transformations of the cubic group, this potential provides the representation Γ_1 of this group. We see in Table 3 that this representation appears once for $k = 4$ and once for $k = 6$. Hence there must be a single combination of spherical harmonics of fourth-order invariant by the cubic group and a single one of sixth order.

For $k = 4$ it is just the function corresponding to $J = 4$, Γ_1 in Table 4, that is,

$$V_4 = b_4 r^4 \left\{ \frac{\sqrt{(14)}}{\sqrt{(24)}} Y_4^0 + \frac{\sqrt{(5)}}{\sqrt{(24)}} (Y_4^4 + Y_4^{-4}) \right\}. \qquad (16.6)$$

For $k = 6$ it can be computed in the same way and is given by

$$V_6 = b_6 r^6 \left\{ \frac{2}{\sqrt{(32)}} Y_6^0 - \frac{\sqrt{(14)}}{\sqrt{(32)}} (Y_6^4 + Y_6^{-4}) \right\}. \qquad (16.7)$$

The coefficients b_4 and b_6 are defined in such a way that V_4/b_4 and V_6/b_6 should be normalized to unity over the unit sphere.

The reader will observe that there is a positive sign within the bracket in (16.6) but a negative sign in (16.7); this is not an accident. It is important, in calculations where both occur, to preserve these signs, since only then will (16.7) correspond to the use of the same set of axes as (16.6). A rotation through an angle ϕ about the polar axis multiplies a spherical harmonic Y_k^q by $\exp(-iq\phi)$; thus a rotation through $\pi/4$ just changes all the signs of the $Y_k^{\pm 4}$ terms giving

$$V_4 = b_4 r^4 \left\{ \frac{\sqrt{(14)}}{\sqrt{(24)}} Y_4^0 - \frac{\sqrt{(5)}}{\sqrt{(24)}} (Y_4^4 + Y_4^{-4}) \right\} \qquad (16.6a)$$

$$V_6 = b_6 r^6 \left\{ \frac{2}{\sqrt{(32)}} Y_6^0 + \frac{\sqrt{(14)}}{\sqrt{(32)}} (Y_6^4 + Y_6^{-4}) \right\}. \qquad (16.7a)$$

These are equally valid with (16.6) and (16.7), but one must not take a combination of one of these and one of (16.6) or (16.7).

These equations express the cubic potential in spherical harmonics such that the polar axis is one of the fourfold axes of the cube. It is sometimes convenient to have the corresponding expressions in the form of a set of spherical harmonics whose polar axis is one of the threefold axes (a body diagonal of the cube). With the same values of b_4 and b_6, the corresponding expressions for V_4 and V_6 are then

$$V_4 = -b_4 r^4 \{\sqrt{(\tfrac{7}{27})} Y_4^0 + \sqrt{(\tfrac{10}{27})}(Y_4^3 - Y_4^{-3})\} \qquad (16.8)$$

$$V_6 = +b_6 r^6 \frac{\sqrt{(2)}}{18} \{8 Y_6^0 + \sqrt{(\tfrac{70}{3})}(Y_6^3 - Y_6^{-3}) + \sqrt{(\tfrac{77}{3})}(Y_6^6 + Y_6^{-6})\}. \qquad (16.9)$$

Note that rotation of the coordinate system through an angle ϕ about the threefold axis multiplies each spherical harmonic Y_k^q by $\exp(-iq\phi)$, so that rotation through π changes the signs of the coefficients of the $Y_k^{\pm 3}$ but not those of the $Y_k^{\pm 6}$.

Although an expansion such as (16.1) for the crystal potential is the more natural one, it has been customary in the literature to expand it into homogeneous polynomials of degree k, each one a certain combination of spherical harmonics, without paying much attention to the normalization of these polynomials. When the symmetry is such that the coefficients B_k^q of the expansion (16.1) are real, this new expansion is of the form

$$V = \sum_{k,q>0} A_k^q P_k^q(x, y, z), \qquad (16.10)$$

where the P_k^q are unnormalized homogeneous polynomials proportional to $r^k(Y_k^q + Y_k^{q*})$ but which, in spite of the symbol, should not be confused with Legendre polynomials. A list of those that occur most frequently in the literature are given in Table 15 together with the relationship between the A_k^q of (16.10) and the B_k^q of (16.1).

For cubic symmetry one introduces the polynomials

$$P_4 = P_4^0 + 5P_4^4 = 20(x^4 + y^4 + z^4 - \tfrac{3}{5}r^4), \qquad (16.11)$$

$$P_6 = P_6^0 - 21P_6^4 = -14 \times 16 \times$$
$$\times \{x^6 + y^6 + z^6 + \tfrac{15}{4}(x^4y^2 + y^4x^2 + x^4z^2 + z^4x^2 + y^4z^2 + z^4y^2) - \tfrac{15}{14}r^6\}$$
$$(16.12)$$

and the cubic potential can be written

$$V \text{ (cubic)} = A_4 P_4 + A_6 P_6, \qquad (16.13)$$

where A_4 and A_6 are related to b_4 and b_6 of (16.6) and (16.7) by

$$\begin{cases} A_4 = \dfrac{1}{\sqrt{(2\pi)}} \dfrac{3}{16} \sqrt{\left(\dfrac{7}{6}\right)} b_4, \\ A_6 = \dfrac{1}{\sqrt{(2\pi)}} \dfrac{\sqrt{(13)}}{64} b_6. \end{cases} \qquad (16.14)$$

In the point charge approximation the coefficients A_4 and A_6 can be computed by an expansion of $\sum_i |\mathbf{r} - \mathbf{R}_i|^{-1}$ where $\mathbf{R}_i$ are the positions of the charges that produce the potential. The electrostatic energy of an electron of charge $-e$ in the potential produced by 6 charges $-Ze$ situated at the six vertices of a regular octahedron is given by (16.11)–(16.14) with

$$\begin{cases} A_4 \text{ (octahedron)} = \dfrac{7}{16} \dfrac{Ze^2}{R^5}, \\ A_6 \text{ (octahedron)} = \dfrac{3}{64} \dfrac{Ze^2}{R^7}. \end{cases} \qquad (16.15)$$

Here R is the distance of each charge $-Ze$ from the centre of the octahedron.

For 8 charges at the vertices of a cube one finds

$$\begin{cases} A_4 \text{ (cube)} = -\dfrac{7}{18}\dfrac{Ze^2}{R^5}, \\ A_6 \text{ (cube)} = \dfrac{1}{9}\dfrac{Ze^2}{R^7}, \end{cases} \quad (16.16)$$

where R is again the distance of each charge $-Ze$ from the centre of the cube.

For a regular tetrahedron whose vertices are 4 of the 8 apexes of the cube, the even part of the potential is half of that of the cube and

$$A_{4,6} \text{ (tetrahedron)} = \tfrac{1}{2} A_{4,6} \text{ (cube)}. \quad (16.17)$$

Since the tetrahedron (unlike the cube or octahedron) does not possess a centre of inversion symmetry, its potential includes also terms of odd degree involving spherical harmonics Y_k^q of odd k. These have no matrix elements within a manifold of electrons of a given value of l (see the remarks following eqn (16.4)).

More detailed evaluations of these expansions can be found in Hutchings (1964).

16.2. Equivalent operators

Having written down the expansions (16.1) or (16.10) for the crystal field, the next problem is to calculate the matrix elements $(\Psi' |\, V\, | \Psi'')$. A simple pedestrian approach would involve expanding Ψ' and Ψ'' into Slater determinants, thus reducing $(\Psi'|\, V\, |\Psi'')$ to a sum of one-electron matrix elements of the type $(l, m_l |\, V\, | l, m_l') \, \delta(m_s, m_s')$.

It is very much preferable to use for Ψ' and Ψ'' eigenstates of L, S, M_L, M_S or L, S, J, M_J and apply the Wigner–Eckart theorem as has been explained in § 13.5. The L, S, M_L, M_S representation is suitable for crystal fields of intermediate strength (iron group) and the J, M_J representation for weak crystal fields (rare-earth group).

For matrix elements within a manifold L or J such as

$$(L, M_L |\, V\, | L, M_L') \text{ or } (J, M_J |\, V\, | J, M_J')$$

the best procedure is to use the equivalent operators mentioned in § 13.5. With the components L_x, L_y, L_z of $\mathbf{L}$ or J_x, J_y, J_z of $\mathbf{J}$, tensor operators O_k^q are constructed which have the same transformation

ELEMENTARY THEORY OF THE CRYSTAL FIELD

properties as the polynomials P_k^q defined in Table 15. Thus, within each manifold L we shall be able to write

$$\langle L, M_L| \sum_i P_k^q(\mathbf{r}_i) |L, M_L'\rangle = a_k \langle r^k \rangle \langle L, M_L| O_k^q(\mathbf{L}) |L, M_L'\rangle \quad (16.18)$$

where the summation $\sum_i$ is over all the electrons.

For the $|J, M_J\rangle$ representation, L is to be replaced throughout by J. The value of the constant a_k depends on the structure of the level L or J under consideration and has to be determined in each case by a direct calculation. The current and unfortunate practice is to write α, β, γ instead of a_2, a_4, a_6, and we shall write $\langle l\| \alpha \|l\rangle$, $\langle L\| \alpha \|L\rangle$, $\langle J\| \alpha \|J\rangle$, etc. to make it clear whether we are dealing with single electrons or representations in terms of L or J.

The construction of the polynomials O_k^q is not trivial because the various components J_x, J_y, J_z do not commute with each other. As a result, whenever an expression $x^\lambda y^\mu z^\nu$ appears in P_k^q it should not be replaced by $J_x^\lambda J_y^\mu J_z^\nu$ in O_k^q but rather by the symmetrized product, average of all the products where J_x, J_y, J_z appear respectively λ, μ, ν times. This average can then be simplified using the commutation rules of J_x, J_y, J_z. A list of the O_k^q is given in Table 16, where we use the notation $\{A, B\}_S = \frac{1}{2}(AB+BA)$. The non vanishing matrix elements of these operators O_k^q are given in Tables 17 for values of J varying from $\frac{1}{2}$ to $\frac{15}{2}$ and from 0 to 8.

To show how α, β, γ are calculated we start first with a single electron with orbital angular momentum l, for which the parameters we require can be written as $\langle l\| \alpha \|l\rangle$, etc. We take as an example the operator P_4^0. In the state l, $m_l = l$ the expectation value of P_4^0 can be found by direct calculation:

$$\langle l; l| P_4^0 |l, l\rangle = \langle r^4\rangle \{(2l+1)/4\pi\} \int (Y_l^l)^2 (P_4^0/r^4) \, d\Omega$$

$$= \langle r^4\rangle \frac{12l(l-1)}{(2l+3)(2l+5)}. \quad (16.19)$$

On the other hand, the value of $\langle l, l| O_4^0 |l, l\rangle$ is found from Table 16 by writing $J_z = J = l$ to be

$$\langle l, l| O_4^0 |l, l\rangle = 2l(2l-1)(l-1)(2l-3), \quad (16.20)$$

which incidentally shows that the matrix element of this operator vanishes for all quantum numbers less than 2. Hence we have, since $\langle l, l| P_4^0 |l, l\rangle \equiv \langle r^4\rangle\langle l\| \beta \|l\rangle\langle l, l| O_4^0 |l, l\rangle$,

$$\langle l\| \beta \|l\rangle = \frac{6}{(2l-1)(2l-3)(2l+3)(2l+5)}. \quad (16.21)$$

Similar calculations can be carried out for α, γ and the results are tabulated in Table 18, together with the numerical values for p-, d-, and f-electrons ($l = 1, 2, 3$).

To proceed to the next step we must describe the structure of the spectral terms (L, S) and multiplets (L, S, J) of the ground levels of the free ion, encountered in the transition group; we begin with the ions of the iron group listed in Table 19.

We notice in Table 19 that the values of L and S given by Hund's rule are symmetrical with respect to the middle of the $3d$ shell and the values of α and β (there is no γ for $l = 2$) antisymmetrical. This is a consequence of a general rule that we derive as follows.

Consider two complementary configurations $\mathscr{C}$ and $\mathscr{C}'$ containing respectively x and $2(2l+1)-x$ electrons outside closed shells. It is possible to establish a one-to-one correspondence between a function Ψ of $\mathscr{C}$ and Ψ'' of $\mathscr{C}'$ such that

$$\begin{cases} \langle \Psi | L_z | \Psi \rangle = \langle \Psi'' | L_z | \Psi'' \rangle, \\ \langle \Psi | S_z | \Psi \rangle = \langle \Psi'' | S_z | \Psi'' \rangle. \end{cases} \quad (16.22)$$

More generally, given a tensor operator $\mathscr{T}_k = \sum_i \mathscr{T}_k(\mathbf{l}_i, \mathbf{s}_i)$ with $k > 0$,

$$\langle \Psi | \mathscr{T}_k | \Psi \rangle = \pm \langle \Psi'' | \mathscr{T}_k | \Psi'' \rangle, \quad (16.23)$$

where the $+$ sign applies when $\mathscr{T}_k$ is time-odd and the negative sign when it is time-even; that is, odd or even through inversion of all $\mathbf{l}_i$ and $\mathbf{s}_i$.

The correspondence between Ψ and Ψ'' is as follows. To each determinant of $\mathscr{C}$, $\Phi = (m_l^1, m_s^1; \ldots m_l^x, m_s^x)$ we associate the determinant of $\mathscr{C}'$,

$$\Phi' = (-m_l'^1, -m_s'^1; \ldots; -m_l'^{2(2l+1)-x}, -m_s'^{2(2l+1)-x}),$$

where $m_l'^1, m_s'^1 \ldots$, etc. are the one-electron states *unoccupied* in Φ. The proof of (16.23) then goes as follows.

We introduce also the Slater determinant

$$\Phi'' = (m_l'^1, m_s'^1; \ldots; m_l'^{2(2l+1)-x}, m_s'^{2(2l+1)-x})$$

which (apart from possible phase factors) is the time-reversed of Φ'. The sum $\langle \Phi | \mathscr{T}_k | \Phi \rangle + \langle \Phi'' | \mathscr{T}_k | \Phi'' \rangle$ vanishes, since, from the rules given in § 11.6 for the calculation of diagonal matrix elements of Slater determinants this sum is a trace over a closed shell of one-electron tensor operators

$$\sum_{\substack{m_l, m_s \\ \text{closed shell}}} (m_l, m_s | \mathscr{T}_k | m_l, m_s).$$

Hence $(\Phi|\mathcal{T}_k|\Phi) = -(\Phi''|\mathcal{T}_k|\Phi'') = \mp(\Phi'|\mathcal{T}_k|\Phi')$ where the negative sign applies for time-even operators. One says sometimes that the configuration $\mathscr{C}'$ consists of holes. According to (16.22), (16.23) the holes must be treated as positive charges for time-even operators such as the crystal field or the spin-orbit coupling, but as negative charges for the coupling with a magnetic field either applied or produced by the nuclear magnetic moment.

As a consequence, the order in energy of the various terms (L, S) of the free ion, which is determined by the electrostatic repulsion between the particles, whether electrons or holes, is the same for $\mathscr{C}$ and $\mathscr{C}'$ (we have not really *proved* this since our reasoning applied only to one-electron operators, but it turns out to be true). On the other hand, the spin-orbit coupling constant λ in $\lambda(\mathbf{L}\cdot\mathbf{S})$, and the constants α, β, have opposite signs and the order of sub-levels in the splitting of a term (L, S) is reversed.

We now extend our previous calculation of $\langle l\|\alpha\|l\rangle$, $\langle l\|\beta\|l\rangle$ to find the equivalent factors for a term (L, S), and consider first $\langle L\|\alpha\|L\rangle$, for which an explicit formula can be obtained rather easily. Essentially we wish to replace the operator $\langle l\|\alpha\|l\rangle \sum \{3l_z^2 - l(l+1)\}$ summed over all electrons, by the operator

$$\langle L\|\alpha\|L\rangle\{3L_z^2 - L(L+1)\}.$$

To find the value of $\langle L\|\alpha\|L\rangle$ we consider a term (L, S) originating from a configuration with r electrons of orbital momentum l, and obeying Hund's rule so that the state $L_z = L$, $S_z = S$ is a single Slater determinant. Assume first that we are in the first half of the shell, so that $r \leq 2l+1$ and $r = 2S$. According to Hund's rule,

$$L = l + (l-1) + \ldots + (l-r+1) = \tfrac{1}{2}r(2l-r+1) = S(2l+1-2S) \tag{16.24}$$

and

$$\sum \{3l_z^2 - l(l+1)\} = 3\{l^2 + (l-1)^2 + \ldots + (l-r+1)^2\} - 2Sl(l+1)$$
$$= S(2l+1-r)(2l+1-2r) = L(2l+1-4S)$$

while $\{3L_z^2 - L(L+1)\}_{L_z=L} = L(2L-1)$. Hence, remembering that the sign of $\langle L\|\alpha\|L\rangle$ must be reversed for a shell that is more than half-filled, we have

$$\langle L\|\alpha\|L\rangle = \pm\frac{(2l+1-4S)}{2L-1}\langle l\|\alpha\|l\rangle$$
$$= \mp\frac{2(2l+1-4S)}{(2l-1)(2l+3)(2L-1)}, \tag{16.25}$$

where the upper sign applies to a shell less than half-full, and the lower sign to a shell more than half-full.

Similar methods are used to find the values of $\langle L\| \beta \|L\rangle$ and $\langle L\| \gamma \|L\rangle$, but in these cases it is often as simple to carry out the calculation numerically for each ion as to use a general formula. As an example we calculate the value of β for the term ^4F of the configuration $3d^3$. The state $L = 3$, $L_z = 3$, $S = \frac{3}{2}$, $S_z = \frac{3}{2}$ of this term is a single Slater determinant $|\xi\rangle = (2^+, 1^+, 0^+)$, for which

$$\langle l\| \beta \|l\rangle \{\langle 2, 2| O_4^0(1) |2, 2\rangle + \langle 2, 1| O_4^0(1) |2, 1\rangle + \langle 2, 0| O_4^0(1) |2, 0\rangle\}$$
$$= \langle L\| \beta \|L\rangle \langle 3, 3| O_4^0(\mathbf{L}) |3, 3\rangle,$$

whence, using Tables 17 and 18, we have

$$\langle L\| \beta \|L\rangle = \tfrac{1}{5}\langle l\| \beta \|l\rangle = \tfrac{2}{315}$$

for this particular ion.

The numerical values of $\langle L\| \alpha \|L\rangle$ and $\langle L\| \beta \|L\rangle$ for the ground states of the iron-group ions are listed in Table 19, together with general formulae for these quantities.

We pass on now to the rare-earth group where the various terms are given together with the values of the constants α, β, γ in Table 20. The calculation of α, β, γ is more complicated than in the iron group because not only do we couple the individual l and s to a total **L** and **S** but we then couple **L** and **S** to a total **J**. However the problem is much simpler in the second half of the rare-earth group, where the ground multiplet is a state not only of maximum S and L but also of maximum $J = L+S$. It follows that the state $J_z = J$ is a single Slater determinant and we can use the same method as for the iron group: first we calculate α, β, γ for a single f-electron by the same procedure as before (the results are listed in Table 18) and find, for example, that

$$\langle l\| \alpha \|l\rangle = -\frac{2}{45}.$$

Consider then, for instance, $Tm^{3+}(4f^{12}, {}^3H_6)$. The state $J = 6$, $J_z = 6$, or $|6, 6\rangle$ is also a state $L = L_z = 5$; $S = S_z = 1$ and therefore a single Slater determinant of holes $(3^+, 2^+)$. Hence

$$\langle J\| \alpha \|J\rangle (4f^{12}, {}^3H_6)\langle 6, 6| O_2^0(\mathbf{J}) |6, 6\rangle$$
$$= -\langle J\| \alpha \|J\rangle (4f^2, {}^3H_6)\langle 6, 6| O_2^0(\mathbf{J}) |6, 6\rangle$$
$$= -\langle l\| \alpha \|l\rangle \{\langle 3, 3| O_2^0(1) |3, 3\rangle + \langle 3, 2| O_2^0(1) |3, 2\rangle\} \quad (16.26)$$

and, using Tables 17, we obtain for $(4f^{12}, {}^3H_6)$

$$\langle J\| \alpha \|J\rangle = \tfrac{1}{99}.$$

ELEMENTARY THEORY OF THE CRYSTAL FIELD

The first half of the shell is more difficult to handle for in the ground multiplet (J, L, S), J is not $L+S$ but $|L-S|$ and no substate of this multiplet is a single Slater determinant. We can, however, expand any substate of this multiplet into eigenstates of $|L, M_L\rangle$ and $|S, M_S\rangle$ using Clebsch–Gordan coefficients,

$$|L, S, J, M_J\rangle = \sum_{\substack{M_L+M_S \\ =M_J}} \langle L, M_L; S, M_S \mid L, S; J, M_J\rangle |L, M_L\rangle |S, M_S\rangle.$$

(16.27)

Let us assume we wish to calculate $\langle J\| \alpha \|J\rangle$. We find first the expectation value of $\sum_i P_2^0(\mathbf{r}_i)$ in the state (16.27). In the state $|L, S, J, M_J\rangle$ it is equal to

$$\langle J\| \alpha \|J\rangle \langle J, M_J| O_2^0(\mathbf{J}) |J, M_J\rangle,$$

and in a state $|L, M_L\rangle |S, M_S\rangle$ to

$$\langle L\| \alpha \|L\rangle \langle L, M_L| O_2^0(\mathbf{L}) |L, M_L\rangle.$$

Equation (16.27) then yields

$$\begin{cases} \langle J\| \alpha \|J\rangle \langle J, M_J| O_2^0(\mathbf{J}) |J, M_J\rangle \\ \qquad = \langle L\| \alpha \|L\rangle \sum_{\substack{M_L+M_S \\ =M_J}} |\langle L, M_L; S, M_S \mid L, S; J, M_J\rangle|^2 \times \\ \qquad\qquad\qquad\qquad \times \langle L, M_L| O_2^0(\mathbf{L}) |L, M_L\rangle, \end{cases}$$
(16.28)

where the $\langle L\| \alpha \|L\rangle$ can be calculated by the same method as in the iron group.

Actually, as we explained in § 13.5 dealing with the coupling of angular momenta, the use of equivalent operators is a paraphrase of the fundamental formula (13.38) which expresses the Wigner–Eckart theorem, and the Tables 17 of the matrix elements of the O_k^q are nothing but tables of Clebsch–Gordan coefficients apart from normalization constants. The quantities $\langle J\| \alpha \|J\rangle$, $\langle J\| \beta \|J\rangle$, $\langle J\| \gamma \|J\rangle$ that we have tabulated are proportional to the diagonal values of reduced matrix elements in formula (13.38),

$$\frac{1}{\sqrt{(2J+1)}}(J\| T_k \|J).$$

Similarly $\langle L\| \alpha \|L\rangle$, $\langle L\| \beta \|L\rangle$, $\langle L\| \gamma \|L\rangle$ are proportional to reduced matrix elements $(1/\sqrt{(2L+1)})(L\| T_k \|L)$ and a relation between them can be obtained in a closed form by means of the vector recoupling 6j-symbol of Racah.

Indeed the passage from $(L\| T_k \|L)$ to $(J\| T_k \|J)$ involves two

vector coupling schemes: on one hand,
$$L+S = J, \quad J+k = J;$$
and on the other hand,
$$L+k = L, \quad L+S = J,$$
which, from the vector coupling properties of the $6j$-symbol described in § 13.5 leads to the formula

$$(J\| \, T_k \, \|J) = (-1)^{L+S+J+k}(2J+1)\begin{pmatrix} J & k & J \\ L & S & L \end{pmatrix}(SL\| \, T_k \, \|SL). \quad (16.29)$$

This enables us to rewrite formula (16.28) in a closed form,

$$(J\| \, \alpha, \beta, \gamma \, \|J) = (L, S\| \, \alpha, \beta, \gamma \, \|L, S)(-1)^{L+S+J+k} \times$$

$$\times \frac{\sqrt{(2J+1)}}{\sqrt{(2L+1)}}\begin{pmatrix} J & k & J \\ L & S & L \end{pmatrix}. \quad (16.30)$$

16.3. Off-diagonal matrix elements of the crystal field

The approximation of the intermediate crystal field implies that its magnitude is small compared to the distance between two terms (L, S) and (L', S) between which it has matrix elements. A few calculations have been made in the iron group taking into account off-diagonal effects of the crystal field. Even more important is the coupling through the crystal field between two multiplets (L, S, J) and (L, S, J') in the rare earths. The method of equivalent operators becomes inadequate in that case and one must go back to the Wigner–Eckart formula (13.38). One of the advantages of this more powerful approach is that the calculation of a reduced off-diagonal matrix element $(J\| \, T_k \, \|J')$, knowing $(L, S\| \, T_k \, \|L, S)$ or $(L, S\| \, T_k \, \|L', S)$, is no more difficult than that of the diagonal elements $(J\| \, T_k \, \|J)$ which are proportional to the $(J\| \, \alpha, \beta, \gamma \, \|J)$.

The formula generalizing (16.29) is

$$(J\| \, T_k \, \|J') = (-1)^{L+S+J'+k}\sqrt{(2J+1)}\sqrt{(2J'+1)}\begin{pmatrix} J & k & J' \\ L' & S & L \end{pmatrix} \times$$

$$\times (SL\| \, T_k \, \|SL'). \quad (16.31)$$

This formula (16.31) combined with the fundamental formula (13.38) enables one to calculate all the matrix elements

$$\langle L, S, J, M_J |\, T_k^q \,| L, S, J', M_J' \rangle \quad (16.32)$$

ELEMENTARY THEORY OF THE CRYSTAL FIELD

in closed form once the constants $(SL\| T_k \|SL)$ have been computed.

It is in a way unfortunate that the pioneer work of Stevens (1952a) has not been expressed in the more rational formalism of Racah which culminates in the powerful formulae (13.38) and (16.31). However, so much of the published work in paramagnetic resonance has been couched in the Stevens language that more would be lost than gained by attempting to translate it all into Racah's language. Therefore even for off-diagonal elements we shall keep the normalization (or lack of it) of Stevens and give the following formulae useful in the theory of magnetic resonance for rare-earth ions.

$$\langle J+1, J_z| \sum_i P_2^0(\mathbf{r}_i) |J, J_z\rangle = \langle r^2\rangle\langle J+1\| \alpha \|J\rangle J_z \sqrt{\{(J+1)^2 - J_z^2\}} \quad (16.33)$$

$$\langle J+1, J_z| \sum_i P_4^0(\mathbf{r}_i) |J, J_z\rangle$$
$$= \langle r^4\rangle\langle J+1\| \beta \|J\rangle J_z(7J_z^2 - 3J^2 - 6J + 2)\sqrt{\{(J+1)^2 - J_z^2\}} \quad (16.34)$$

$$\langle J+1, J_z| \sum_i P_6^0(\mathbf{r}_i) |J, J_z\rangle = \langle r^6\rangle\langle J+1\| \gamma \|J\rangle \times$$
$$\times J_z\{33J_z^4 - 5J_z^2(6J^2 + 12J - 15) + 5J^4 + 20J^3 - 5J^2 - 50J + 12\} \quad (16.35)$$

$$\langle J+1, J_z \pm 6| \sum_i P_6^6(\mathbf{r}_i) |J, J_z\rangle$$
$$= \pm \frac{\langle r^6\rangle}{14}\langle J+1\| \gamma \|J\rangle \sqrt{\frac{(J\pm J_z+7)!\,(J\mp J_z)!}{(J\mp J_z-5)!\,(J\pm J_z)!}}. \quad (16.36)$$

The coefficients $\langle J+1\| \alpha \|J\rangle$, $\langle J+1\| \beta \|J\rangle$, $\langle J+1\| \gamma \|J\rangle$ are listed for the rare earths in Table 20.

16.4. The electronic Zeeman interaction

We continue the list of electronic interactions by including first the equivalent operators for the electronic Zeeman interaction. The magnetic moment operator is $\boldsymbol{\mu} = -\beta(\mathbf{L}+2\mathbf{S})$ and inside a manifold (J, L, S) we can write

$$\mathbf{L} + 2\mathbf{S} = \langle J\| \Lambda \|J\rangle \mathbf{J}, \quad (16.37)$$

where $\langle J\| \Lambda \|J\rangle$ is just the Landé factor. It is given by eqn (11.12b) in the approximation we have made of taking $g_s = 2$, and numerical values for the ground states of the rare-earth ions are listed in Table 20.

There are also off-diagonal elements $\langle JM| \mathbf{L}+2\mathbf{S} |J'M'\rangle$. Here it is useful to note that the matrix elements of $\mathbf{J} = \mathbf{L}+\mathbf{S}$, being diagonal in J, are zero between two manifolds for which $J \neq J'$. Hence

$$\langle JM| \mathbf{S} |J'M'\rangle = -\langle JM| \mathbf{L} |J'M'\rangle = \langle JM| \mathbf{L}+2\mathbf{S} |J'M'\rangle \quad (16.38)$$

and these matrix elements vanish except for $|J-J'| = 1$. The matrix elements we require are

$$\langle J+1, J_z| L_z+2S_z |J, J_z\rangle = \langle J+1\| \Lambda \|J\rangle\{(J+1)^2-J_z^2\}^{\frac{1}{2}} \quad (16.39)$$

$$\langle J+1, J_z\pm1| L_x+2S_x |J, J_z\rangle$$
$$= \mp\langle J+1\| \Lambda \|J\rangle\tfrac{1}{2}\{(J\pm J_z+1)(J\pm J_z+2)\}^{\frac{1}{2}}, \quad (16.40)$$

where the reduced matrix element is given by

$$\langle J+1\| \Lambda \|J\rangle$$
$$= \left\{\frac{(J+L+S+2)(-J+S+L)(J+S-L+1)(L+J-S+1)}{4(J+1)^2(2J+1)(2J+3)}\right\}^{\frac{1}{2}}, \quad (16.41)$$

whose values are also listed in Table 20.

16.5. Electron spin-spin interactions

In addition to the magnetic interaction between the electrons and the external magnetic field, there are also spin-spin interactions between electrons which may contribute to the fine structure of the resonance line.

As shown by Breit (1929), the magnetic interaction between two electrons can be written:

$$\mathscr{H}_{\text{SS}} = 4\beta^2\left\{\frac{\mathbf{s}_1 \cdot \mathbf{s}_2}{r_{12}^3} - \frac{3(\mathbf{s}_1 \cdot \mathbf{r}_{12})(\mathbf{s}_2 \cdot \mathbf{r}_{12})}{r_{12}^5}\right\} - \frac{32\pi\beta^2}{3}(\mathbf{s}_1 \cdot \mathbf{s}_2)\,\delta(\mathbf{r}_{12}). \quad (16.42)$$

The last term of (16.42) is invariant under rotations of both spin and orbital coordinates. It commutes therefore with $\mathbf{L}^2$ and $\mathbf{S}^2$, and within a term (L, S) behaves as an additive constant. This constant vanishes for a term obeying Hund's rule which, being completely symmetrical with respect to spin coordinates, is completely antisymmetrical with respect to orbital coordinates. The first term of (16.42), by arguments (see § 17.4) similar to those that lead to eqn (17.45), can be rewritten within a term (L, S) as

$$W_{\text{SS}} = -\rho \sum_{p,q}\{\tfrac{1}{2}(L_pL_q+L_qL_p)-\tfrac{1}{3}L(L+1)\,\delta_{pq}\}S_pS_q$$
$$= -\rho\{(\mathbf{L} \cdot \mathbf{S})^2+\tfrac{1}{2}(\mathbf{L} \cdot \mathbf{S})-\tfrac{1}{3}L(L+1)S(S+1)\}. \quad (16.43)$$

The constant ρ can be evaluated by the usual methods of operator equivalents. For an (L, S) term that obeys Hund's rule and is built from d-electrons, ρ is given by (Pryce 1950)

$$\rho = \frac{-4}{7(2L-1)}\left\{\left(\frac{5}{S}-4\right)p+\frac{1}{7}\left(62-\frac{100}{S}\right)q\right\}\beta^2, \quad (16.44)$$

where

$$p = \int_0^\infty \frac{1}{r} R^2(r)\,\mathrm{d}r \int_0^r r'^2 R^2(r')\,\mathrm{d}r',$$

$$q = \int_0^\infty \frac{1}{r^3} R^2(r)\,\mathrm{d}r \int_0^r r'^4 R^2(r')\,\mathrm{d}r',$$

in which $R(r)$ is the radial wave-function of the d-electrons. For free ions, ρ can be estimated from the departure from the Landé interval rule, leading to values of ρ of the order of 0·5 cm^{-1} in the iron group (Pryce 1950), but, as pointed out by Trees (1951), this includes contributions from second-order effects of the spin-orbit coupling. Later calculations by Watson and Blume (1965) give rather smaller values, which are listed in Table 7.6 for ions of the $3d$ group. Just what value should be used in interpreting a paramagnetic resonance experiment is rather complicated. If an interaction of the form W_{SS} is present, its magnitude should include contributions from all sources, including second-order effects of the spin-orbit coupling. However the latter may not be the same as for the free ion, just as the spin-orbit coupling itself is modified through bonding effects.

17

HYPERFINE STRUCTURE

HYPERFINE structure in paramagnetic resonance lines was discovered by Penrose (1949). It is due to the coupling of the electrons with the electric and magnetic moments of the nucleus. In this chapter we shall derive the form of the interactions responsible for these couplings. We begin with the electrostatic interactions which, although less important in practice, are conceptually simpler.

17.1. Electrostatic hyperfine interactions

Atomic energy levels are usually calculated under the simplifying assumption of a point nucleus with a charge Ze. If the finite nuclear dimensions are taken into account, the changes in the atomic energy levels are exceedingly small and, for nuclear spins different from zero, are best observed through the removal of the degeneracy of the atomic levels associated with the various orientations of the nuclear spin. The Hamiltonian responsible for the splittings between those levels can be determined as follows, using the correspondence principle as a starting point. If we describe the nucleus and the electron cloud as two classical charge distributions $\rho_n(r_n)$ and $\rho_e(r_e)$, their mutual electrostatic energy is

$$W_E = \iint \frac{\rho_e(r_e)\rho_n(r_n)\,dr_e\,dr_n}{|\mathbf{r}_n - \mathbf{r}_e|}, \tag{17.1}$$

which can be expanded through the classical formula

$$\frac{1}{|\mathbf{r}_n - \mathbf{r}_e|} = \sum_{k=0}^{\infty} \sum_{q=-k}^{k} \frac{r_<^k}{r_>^{k+1}} C_k^q(\theta_n, \varphi_n) C_k^{q*}(\theta_e, \varphi_e), \tag{17.2}$$

where the functions $C_k^q(\theta, \varphi)$ are normalized spherical harmonics: $C_k^q(\theta, \varphi) = \sqrt{\{4\pi/(2k+1)\}} Y_k^q(\theta, \varphi)$ and where the symbols $r_<$ and $r_>$ mean that the larger of the two numbers r_e and r_n is in the denominator, and the smaller in the numerator. If the small penetration of the electron inside the nucleus is neglected we may assume $r_e > r_n$ and write

$$W_E = \sum_{k,q} A_k^q B_k^{q*}, \tag{17.3}$$

where

$$\begin{cases} A_k^q = \int \rho_n(r_n) r_n^k C_k^q(\theta_n, \varphi_n) \, dr_n, \\ B_k^q = \int \rho_e(r_e) r_e^{-(k+1)} C_k^q(\theta_e, \varphi_e) \, dr_e. \end{cases} \quad (17.4)$$

If the state of the nucleus is described by a wave-function $\Psi_n(R_1, \ldots, R_A)$ of the coordinates of its A nucleons, the nuclear charge density can be written as the expectation value of the operator density of charge at the point r_n:

$$\rho_n(r_n) = \langle \Psi_n | \sum_{i=1}^{A} e_i \, \delta(r_n - R_i) | \Psi_n \rangle, \quad (17.5)$$

where $e_i = e$ for a proton and zero for a neutron. From (17.3) and (17.4) A_k^q can be written as an expectation value $A_k^q = \langle \mathscr{A}_k^q \rangle$ where the nuclear operator $\mathscr{A}_k^q$ is defined by

$$\mathscr{A}_k^q = \sum_i e_i R_i^k C_k^q(\Theta_i, \Phi_i), \quad (17.6)$$

R_i, Θ_i, Φ_i being the polar coordinates of the A nucleons. Similarly, B_k^q is the expectation value of the electron operator $\mathscr{B}_k^q$:

$$\mathscr{B}_k^q = -e \sum_{i=1}^{N} r_i^{-(k+1)} C_k^q(\theta_i, \varphi_i), \quad (17.7)$$

where r_i, θ_i, φ_i are the coordinates of the electrons and the minus sign occurs because of the negative charge ($-e$) of the electron. The energy of electrostatic interaction between the electrons and the nucleus is then the expectation value of a Hamiltonian

$$\mathscr{H}_E = \sum_{k,q} \mathscr{A}_k^q \mathscr{B}_k^{q*}. \quad (17.8)$$

From the definitions (17.6) and (17.7) it is clear that the operators $\mathscr{A}_k^q$ and $\mathscr{B}_k^q$ are tensor operators of order k. The tensor operator $\mathscr{A}_k$ with $2k+1$ components $\mathscr{A}_k^q$ is called the multipole moment of order k of the nucleus. Although for the calculation of the splittings of the atomic energy levels, only the expectation value $A_k^q = \langle \mathscr{A}_k^q \rangle$ of the nuclear multipole operator, taken over the wave-function of a nuclear eigenstate, is required, this operator does in fact have off-diagonal matrix elements between various nuclear states of different energy. The study of these matrix elements, which are responsible for γ-ray transitions between those states, is outside the scope of this book.

For the diagonal matrix elements $A_k^q = \langle \mathscr{A}_k^q \rangle$ odd values of k are forbidden if we assume, as seems well established experimentally, that stationary nuclear states have well-defined parities. In particular, nuclei should have no permanent electric dipole moments ($k = 1$) in agreement with experimental evidence. Invariance with respect to time reversal (cf. the discussion in § 15.10) also forbids the existence of nuclear electric dipole moments. Off-diagonal matrix elements of the electric dipole nuclear moment between nuclear states of different parity may of course exist. Further information and limitations on the values of the matrix elements of the nuclear multipole operators result from their tensor character and are based on the Wigner–Eckart theorem. It follows from this theorem that for a nucleus of spin I, $A_k^q = \langle \mathscr{A}_k^q \rangle$ will only be different from zero if $k \leqslant 2I$. Thus nuclei of spin $I \geqslant 1$ may have quadrupole moments, nuclei with $I \geqslant 2$ may have moments of order 4, etc, ... The term $k = 0$ of the electrostatic interaction between electron and nucleus clearly corresponds to the coupling with a point charge Ze. Since the nuclear radius R is much smaller than the electronic radius a, the various terms of W_E as given by eqns (17.3), (17.4) decrease rapidly, roughly as $(R/a)^k$. This explains why there is little experimental evidence of electrostatic interactions with $k > 2$.

Only quadrupole interactions will be considered from now on. The components of the nuclear quadrupole moment operator can be written in the form

$$\mathscr{A}_2^0 = \tfrac{1}{2} \sum_i e_i(3z_i^2 - r_i^2),$$

$$\mathscr{A}_2^{\pm 1} = \mp \sqrt{(\tfrac{3}{2})} \sum_i e_i z_i (x_i \pm iy_i), \qquad (17.9)$$

$$\mathscr{A}_2^{\pm 2} = \sqrt{(\tfrac{3}{8})} \sum_i e_i (x_i \pm iy_i)^2.$$

According to the Wigner–Eckart theorem the $\mathscr{A}_2^q$ have, for a nuclear spin I, within the manifold of the $(2I+1)$ substates $I_z = m$, the same matrix elements as the Hermitian tensor operator formed from the components of the vector I,

$$Q_2^0 = \langle I \| \alpha \| I \rangle (\tfrac{1}{2}) \{ 3I_z^2 - I(I+1) \},$$

$$Q_2^{\pm 1} = \mp \langle I \| \alpha \| I \rangle \sqrt{(\tfrac{3}{2})} \, \tfrac{1}{2}(I_z I_\pm + I_\pm I_z),$$

$$Q_2^{\pm 2} = \langle I \| \alpha \| I \rangle \sqrt{(\tfrac{3}{8})} \, I_\pm^2.$$

The constant $\langle I \| \alpha \| I \rangle$ is determined, for instance, by the condition that Q_2^0 and $\mathscr{A}_2^0$ have the same expectation value in the substate $I_z = I$ denoted by $|II\rangle$. The usual convention is to represent by the

symbol eQ the quantity
$$eQ = \langle II| \sum_{i=1}^{A} e_i(3z_i^2 - r_i^2) |II\rangle.$$
Then writing
$$eQ = 2\langle II| Q_2^0 |II\rangle = 2\langle II| \mathscr{A}_2^0 |II\rangle$$
$$= \langle I\| \alpha \|I\rangle\langle II| 3I_z^2 - I(I+1) |II\rangle,$$
we find that $\langle I\| \alpha \| I\rangle = eQ/\{I(2I-1)\}$, whence

$$Q_2^0 = \frac{eQ}{I(2I-1)} \tfrac{1}{2}\{3I_z^2 - I(I+1)\},$$

$$Q_2^{\pm 1} = \mp \frac{eQ}{I(2I-1)} \sqrt{(\tfrac{3}{2})} \tfrac{1}{2}(I_z I_\pm + I_\pm I_z), \qquad (17.10)$$

$$Q_2^{\pm 2} = \frac{eQ}{I(2I-1)} \sqrt{(\tfrac{3}{8})} (I_\pm)^2.$$

The components $\mathscr{B}_2^q$ of the electronic tensor (17.7) can be rewritten in the same way as the $\mathscr{A}_2^q$ in eqn (17.9), giving

$$\mathscr{B}_2^0 = -\tfrac{1}{2}\sum_{i=1}^{N} \frac{e_i(3z_i^2 - r_i^2)}{r_i^5},$$

$$\mathscr{B}_2^{\pm 1} = \pm\sqrt{(\tfrac{3}{2})}\sum_{i=1}^{N} \frac{e_i z_i(x_i \pm iy_i)}{r_i^5}, \qquad (17.11)$$

$$\mathscr{B}_2^{\pm 2} = -\sqrt{(\tfrac{3}{8})}\sum_{i=1}^{N} \frac{e_i(x_i \pm iy_i)^2}{r_i^5},$$

where the sum is over all electrons. The electron operators can be replaced by equivalent operators (as described in the last chapter), involving the orbital angular momentum vector of each electron. This is then summed over all electrons, giving

$$\mathscr{B}_2^0 = -e \sum_i \langle l\| \alpha \|l\rangle \langle r_q^{-3}\rangle_i \tfrac{1}{2}\{3l_z^2 - l(l+1)\},$$

$$\mathscr{B}_2^{\pm 1} = \pm e \sum_i \langle l\| \alpha \|l\rangle \langle r_q^{-3}\rangle_i \sqrt{(\tfrac{3}{2})} \tfrac{1}{2}(l_z l_\pm + l_\pm l_z), \qquad (17.12)$$

$$\mathscr{B}_2^{\pm 2} = -e \sum_i \langle l\| \alpha \|l\rangle \langle r_q^{-3}\rangle_i \sqrt{(\tfrac{3}{8})} l_\pm^2.$$

For a term in LS- coupling the sums can be replaced by operators involving the components of the total orbital angular momentum just as in the case of the crystal field. This is straightforward if the value of $\langle r_q^{-3}\rangle_i$ is the same for every electron involved but even if not all the values are the same it can be shown that in the non-relativistic limit only one constant is needed in the quadrupole interaction (Sandars and Beck 1965). Hence we can write

$$\mathscr{B}_2^0 \equiv -e\langle r_q^{-3}\rangle\langle L\| \alpha \|L\rangle\tfrac{1}{2}\{3L_z^2-L(L+1)\},$$
$$\mathscr{B}_2^{\pm 1} \equiv \pm e\langle r_q^{-3}\rangle\langle L\| \alpha \|L\rangle\sqrt{(\tfrac{3}{2})}\,\tfrac{1}{2}(L_zL_\pm+L_\pm L_z), \quad (17.13)$$
$$\mathscr{B}_2^{\pm 2} \equiv -e\langle r_q^{-3}\rangle\langle L\| \alpha \|L\rangle\sqrt{(\tfrac{3}{8})}\,(L_\pm)^2,$$

where $\langle r_q^{-3}\rangle$ is the mean inverse third power of the electron distance from the nucleus, averaged over the electronic wave-functions.

For an atom or ion with LS- coupling the form of the Hamiltonian for the electric quadrupole interaction is thus

$$\mathscr{H} = -\frac{e^2Q\langle r_q^{-3}\rangle}{I(2I-1)}\langle L\| \alpha \|L\rangle \times$$
$$\times \begin{bmatrix} \tfrac{1}{4}\{3L_z^2-L(L+1)\}\{3I_z^2-I(I+1)\}+ \\ +\tfrac{3}{8}\{(L_zL_++L_+L_z)(I_zI_-+I_-I_z)+(L_zL_-+L_-L_z)(I_zI_++I_+I_z)\}+ \\ +\tfrac{3}{8}(L_+^2I_-^2+L_-^2I_+^2). \end{bmatrix}$$

(17.14)

For the ground terms of the iron group the values of $\langle L\| \alpha \|L\rangle$ are listed in Table 19. The angular momentum operator expression in square brackets in (17.14) is in a form that is generally the most useful in calculations for weakly-bonded ions of the $3d$ group; it is equivalent to the more usual form

$$\tfrac{1}{2}\{3(\mathbf{L}\cdot\mathbf{I})^2+\tfrac{3}{2}(\mathbf{L}\cdot\mathbf{I})-L(L+1)I(I+1)\} \quad (17.15)$$

and in the notation of Abragam and Pryce (1951a,b,c)

$$-\langle L\| \alpha \|L\rangle = +\eta_L. \quad (17.16)$$

When L and S couple to a resultant J, similar formulae for the quadrupole interaction can be obtained in which L is replaced everywhere by J and $\langle L\| \alpha \|L\rangle$ by $\langle J\| \alpha \|J\rangle$. This representation is useful when considering ions of the rare-earth group, and is similar to that used in atomic beam measurements with a constant B which is given by the relation

$$B = -J(2J-1)e^2Q\langle r_q^{-3}\rangle\langle J\| \alpha \|J\rangle. \quad (17.17)$$

We can of course retain a nomenclature similar to that of Abragam and Pryce by introducing a symbol η_J such that (cf. eqn (17.16))

$$-\langle J\| \alpha \|J\rangle = +\eta_J. \quad (17.18)$$

For the state $J = L+S$ which is the ground state of the second half of the rare-earth group we have the simple relationship

$$\frac{\langle J\|\alpha\|J\rangle}{\langle L\|\alpha\|L\rangle} = \frac{\eta_J}{\eta_L} = \frac{\langle JJ|\,3L_z^2-L(L+1)\,|JJ\rangle}{\langle JJ|\,3J_z^2-J(J+1)\,|JJ\rangle} = \frac{L(2L-1)}{J(2J-1)}. \quad (17.19)$$

For any other value of J we find from eqn (16.30)

$$\eta_J = \eta_L(-1)^{L+S+J}\left(\frac{2J+1}{2L+1}\right)^{\frac{1}{2}}\begin{Bmatrix}J & 2 & J \\ L & S & L\end{Bmatrix}. \quad (17.20)$$

We shall defer until later the discussion of the influence of the nuclear electric quadrupole coupling on the paramagnetic resonance spectrum in bulk matter. However we notice that electron operators such as (17.11) are time-even and will therefore have the same expectation value for the two time-conjugate states of a Kramers doublet and will behave as constants as far as paramagnetic resonance is concerned. In cubic symmetry we meet groups of substates whose degeneracy is greater than twofold, and since for cubic symmetry a symmetrical Cartesian form of the quadrupole operator is often convenient, we make here some remarks about spatial symmetry of the operator in connection with the transformation of (17.14) to a symmetrical Cartesian form.

The electronic and nuclear tensor components (eqns (17.11) and (17.9)) contain two types of terms, one of which transforms like Γ_3, that is like $(3z^2-r^2)$, (x^2-y^2), and the other like Γ_5, that is like xy, yz, zx, which are of course just like the orbital wave-functions for a d-electron with $l = 2$. Since the nuclear electric quadrupole operator as a whole must remain unchanged under rotations of *both* electric and nuclear variables belonging to the cubic group it follows that nuclear components of Γ_5 symmetry can only be multiplied by electronic components of Γ_5 symmetry, and nuclear Γ_3 components by electronic Γ_3 components.

It can be seen that these distinctions are preserved in the following transformations of the nuclear electric quadrupole operator components to and from simple Cartesian forms:

$$(m/6)\begin{bmatrix}\{3L_x^2-L(L+1)\}\{3I_x^2-I(I+1)\}+ \\ +\{3L_y^2-L(L+1)\}\{3I_y^2-I(I+1)\}+ \\ +\{3L_z^2-L(L+1)\}\{3I_z^2-I(I+1)\}\end{bmatrix}$$

$$= (m/4)\{3L_z^2-L(L+1)\}\{3I_z^2-I(I+1)\}+$$
$$+(3m/4)(L_x^2-L_y^2)(I_x^2-I_y^2) \quad (17.21)$$
$$= (m/4)\{3L_z^2-L(L+1)\}\{3I_z^2-I(I+1)\}+$$
$$+(3m/16)(L_+^2+L_-^2)(I_+^2+I_-^2) \quad (17.22)$$

and

$$(3n/4)\begin{bmatrix} (L_xL_y+L_yL_x)(I_xI_y+I_yI_x)+ \\ +(L_yL_z+L_zL_y)(I_yI_z+I_zI_y)+ \\ +(L_zL_x+L_xL_z)(I_zI_x+I_xI_z) \end{bmatrix}$$
$$= (3n/8)\begin{bmatrix} (L_+L_z+L_zL_+)(I_-I_z+I_zI_-)+ \\ +(L_-L_z+L_zL_-)(I_+I_z+I_zI_+) \end{bmatrix}$$
$$-(3n/16)(L_+^2-L_-^2)(I_+^2-I_-^2). \quad (17.23)$$

These terms contain products $L_+^2I_+^2$, $L_-^2I_-^2$ which are absent in (17.14). However, if we take $m = n = 1$ and sum (17.22) and (17.23) such products vanish and other products occur in just the ratios appropriate to (17.14). Thus we have the symmetrical Cartesian form, equivalent to (17.14),

$$-\frac{e^2Q\langle r_q^{-3}\rangle}{I(2I-1)}\langle L\| \alpha \|L\rangle \begin{bmatrix} \tfrac{1}{6}\{3L_x^2-L(L+1)\}\{3I_x^2-I(I+1)\}+ \\ +\tfrac{1}{6}\{3L_y^2-L(L+1)\}\{3I_y^2-I(I+1)\}+ \\ +\tfrac{1}{6}\{3L_z^2-L(L+1)\}\{3I_z^2-I(I+1)\}+ \\ +\tfrac{3}{4}(L_xL_y+L_yL_x)(I_xI_y+I_yI_x)+ \\ +\tfrac{3}{4}(L_yL_z+L_zL_y)(I_yI_z+I_zI_y)+ \\ +\tfrac{3}{4}(L_zL_x+L_xL_z)(I_zI_x+I_xI_z) \end{bmatrix}.$$
(17.24)

This is appropriate to a free atom or ion with full rotational symmetry (which imposes a relation between the coefficients m and n), but with cubic symmetry m and n are not necessarily equal (see § 18.4) in an equivalent operator where the components of L are replaced by those of an effective spin $\tilde{S}$.

In electron paramagnetic resonance experiments we are faced with the difficulty, which exists also for free atoms, of deciding what value to use for the quantity $\langle r_q^{-3}\rangle$. The assumption often made is that it has the same value as the parameter $\langle r^{-3}\rangle$ that appears in the magnetic hyperfine structure and that it can be determined from the latter if the value of the nuclear magnetic moment is known independently. However, distortion of the charge cloud of the inner closed shells of electrons produces an appreciable change in the electric field gradient at the nucleus due to the magnetic electrons, in contrast to the analogous effect ('diamagnetic shielding') in the magnetic case, which is quite small and generally negligible compared to other effects that produce uncertainties in the value of $\langle r^{-3}\rangle$, particularly in solids. We may formally represent the effect of this distortion by writing $\langle r_q^{-3} \rangle = (1-R_q)\langle r^{-3} \rangle$, where the contributions to R_q from the various closed shells may be either positive (shielding) or negative (antishielding).

Hitherto we have been concerned primarily with the 'internal' electric field gradient due to the asymmetric charge distribution of the electrons 'on' the magnetic ion. There may also be an 'external' electric field gradient arising from the ligand ions and the lattice generally. If the paramagnetic ion is at a site of local cubic symmetry in the lattice, the 'external' electric field gradient vanishes; otherwise, it is related to the terms of second degree V_2^q in the crystal potential, eqn (16.1). The relation is not however a simple one as there are shielding effects, both for the magnetic electrons and for the nucleus, which may be quite different in size from one another as well as from the effect discussed in the previous paragraph. For the nucleus, the external electric field gradient is multiplied by a factor $(1-\gamma_\infty)$, where γ_∞ is estimated to be large and negative, with values ranging in some cases up to -100. We cannot simply use this factor to multiply the values of V_2^q obtained empirically by fitting the electronic spectra, since any shielding effects for the magnetic electrons are already included in the latter. However, it is clear that a large value of $|\gamma_\infty|$ will greatly enhance the effect of the external electric field gradient at the nucleus, making it comparable with the internal electric field gradient in some cases, and almost certainly the larger of the two for ions in S-states such as Mn^{2+}, $3d^5$ on non-cubic lattice sites. The general question of shielding and antishielding, based on the well-known work of Sternheimer, is discussed by Watson and Freeman (1967a), where further references can be found, and also briefly in § 17.7.

17.2. Magnetic hyperfine interactions

A logical procedure would be to develop the theory of the magnetic interactions between the electron and the nucleus along the same lines as for the electrostatic interactions, that is, to assign electric-current densities to the electrons and to the nucleus (rather than charge densities as in the previous section) and to calculate their interactions according to the laws of classical electromagnetism. One would thus define magnetic multipole operators for the nucleus which, like the electric ones, would be tensor operators of integral order k.

If one recalls the opposite parity properties of the electric field (a polar vector) and of the magnetic field (an axial vector) it is understandable that even, rather than odd, values of k are forbidden for permanent magnetic multipoles, on the assumption of a well-defined parity for the nuclear energy states. The first non-vanishing nuclear magnetic multipole is thus a magnetic dipole, the next a magnetic octopole, etc. Here again the fact that the magnetic field is a time-odd vector leads to the same conclusion because of invariance through time reversal.

Although the existence of magnetic octopoles may be established by atomic-beam methods, they have never been observed by means of magnetic resonance in bulk matter. Furthermore, the description of the magnetic properties of a nucleus as those of a system of currents is more complicated and at the same time, in our present state of knowledge, much less satisfying than the description of its electrostatic properties as those of a system of charges. We shall therefore be content to describe the magnetic properties of the nucleus as those of a magnetic dipole $\boldsymbol{\mu}_I = \gamma_n \hbar \mathbf{I}$ (see however § 17.7). The reason why the magnetic dipole is collinear with the spin vector $\mathbf{I}$ is again that, within the manifold of the substates of a given nuclear state of spin $\mathbf{I}$, all tensor operators of given k (vectors in the present case) have the same matrix elements. Magnetic fields of impossibly high values, of the order of 10^{16} G or more, would have to be applied to the nucleus before its magnetic energy $-\boldsymbol{\mu}_I \cdot \mathbf{H}$ became comparable to the interval between two different nuclear energy states, invalidating the approximation $\boldsymbol{\mu}_I = \gamma_n \hbar \mathbf{I}$.

The interaction of the nuclear dipole $\boldsymbol{\mu}_I$ with the electronic shell is small even compared with atomic-energy splittings (let alone the nuclear ones) and may be computed by a perturbation method.

The behaviour of an electron in a magnetic field $\mathbf{H}$ is obtained by replacing the momentum $\mathbf{p}$ by $\mathbf{p}+(e/c)\mathbf{A}$ in its Hamiltonian, where $\mathbf{A}$ is the magnetic vector potential defined by

$$\text{div } \mathbf{A} = 0, \quad \text{curl } \mathbf{A} = \mathbf{H}.$$

According to classical electromagnetic theory a magnetic dipole $\boldsymbol{\mu}$ produces a magnetic field, at a point removed from it by a vector $\mathbf{r}$, deriving from a vector potential

$$\mathbf{A} = \frac{\boldsymbol{\mu} \wedge \mathbf{r}}{r^3} = \text{curl}\left(\frac{\boldsymbol{\mu}}{r}\right). \tag{17.25}$$

Near the dipole the vector potential $\mathbf{A}$ has a singularity of order r^{-2} and $\mathbf{H} = \text{curl } \mathbf{A}$ a singularity of order r^{-3}, so that some care must be exercised in the calculation of its interaction with an electron. In the

HYPERFINE STRUCTURE

non-relativistic Pauli description of the electron the Hamiltonian in the presence of $\mathbf{A}$ is

$$\mathscr{H} = \frac{1}{2m}\left(\mathbf{p}+\frac{e}{c}\mathbf{A}\right)^2 + g_s\beta(\mathbf{s}\cdot\operatorname{curl}\mathbf{A}), \tag{17.26}$$

where β is the Bohr magneton and $\mathbf{s}$ the electron spin. In a first-order perturbation calculation the only terms of (17.26) to be retained are those linear in $\mathbf{A}$:

$$\mathscr{H}_1 = \frac{e}{2mc}(\mathbf{p}\cdot\mathbf{A}+\mathbf{A}\cdot\mathbf{p}) + g_s\beta(\mathbf{s}\cdot\operatorname{curl}\mathbf{A}). \tag{17.26a}$$

This can be written, using (17.25), as

$$\mathscr{H}_1 = 2\beta\frac{(\mathbf{l}\cdot\boldsymbol{\mu})}{r^3} + g_s\beta\left\{\mathbf{s}\cdot\operatorname{curl}\operatorname{curl}\left(\frac{\boldsymbol{\mu}}{r}\right)\right\}, \tag{17.27}$$

where $\hbar\mathbf{l} = \mathbf{r}\wedge\mathbf{p}$ is the orbital momentum of the electron.

The spin-dependent part of (17.27) gives

$$\mathscr{H}_1^s = g_s\beta\mathbf{s}\cdot\left\{\nabla\wedge\left(\nabla\wedge\frac{\boldsymbol{\mu}}{r}\right)\right\} = g_s\beta\{(\mathbf{s}\cdot\nabla)(\boldsymbol{\mu}\cdot\nabla)-(\mathbf{s}\cdot\boldsymbol{\mu})\nabla^2\}\frac{1}{r}, \tag{17.28}$$

which for reasons to appear presently we rewrite as

$$\mathscr{H}_1^s = g_s\beta\{(\mathbf{s}\cdot\nabla)(\boldsymbol{\mu}\cdot\nabla)-\tfrac{1}{3}(\mathbf{s}\cdot\boldsymbol{\mu})\nabla^2\}\left(\frac{1}{r}\right) - \frac{2\beta}{3}g_s(\mathbf{s}\cdot\boldsymbol{\mu})\nabla^2\left(\frac{1}{r}\right). \tag{17.29}$$

The magnetic interaction of the nuclear moment with the electron spin $W_m^s = (\Psi_e|\,\mathscr{H}_1^s\,|\Psi_e)$ is obtained by multiplying (17.29) by the electronic density $\rho = \Psi_e^*\Psi_e$ and integrating over the electron coordinates. For $r \neq 0$, $\mathscr{H}_1^s$, as given by (17.29), is a regular function where the first term is equal to $g_s\beta\{3(\mathbf{s}\cdot\mathbf{r})(\boldsymbol{\mu}\cdot\mathbf{r})/r^5 - (\mathbf{s}\cdot\boldsymbol{\mu})/r^3\}$, which is the usual dipole-dipole interaction, and the second term vanishes because of Laplace's equation. When r tends toward zero we may remark that the first term $\mathscr{H}_1^{s'}$ of (17.29) behaves as a spherical harmonic of order 2 under a rotation of the coordinate system. Hence if Ψ_e is expanded in a sum of spherical harmonics, $\Psi_e = \sum_l a_l \Psi_l$, the only non-vanishing contributions to $(\Psi_e|\,\mathscr{H}_1^{s'}\,|\Psi_e)$ will come from terms $(\Psi_l|\,\mathscr{H}_1^{s'}\,|\Psi_{l'})$ such that $l+l' \geqslant 2$. It is well known that a wave-function Ψ_l is of order r^l near the origin so that in the matrix element

$$(\Psi_l|\,\mathscr{H}_1^{s'}\,|\Psi_{l'}) = \int \Psi_l^* \mathscr{H}_1^{s'} \Psi_{l'} r^2\, dr\, d\Omega$$

the integrand varies as $r^{(l+l'+2-3)}$ and the corresponding integral always remains finite since $l+l' \geqslant 2$. According to the theory of the Coulomb potential the second term of (17.29) is equal to $(8g_s/3)\pi\beta(\mathbf{S} \cdot \boldsymbol{\mu}) \delta(\mathbf{r})$ and by integration gives

$$\frac{8g_s}{3}\pi\beta(\mathbf{S} \cdot \boldsymbol{\mu}) |\Psi_e(0)|^2,$$

which is finite for s electrons and zero for the others. The Hamiltonian for the magnetic interaction of the electron with the nucleus can then be written without ambiguity as (writing $g_s = 2$)

$$\mathcal{H}_1 = 2\beta\gamma_n\hbar\mathbf{I} \cdot \left\{\frac{1}{r^3} - \frac{\mathbf{S}}{r^3} + 3\frac{\mathbf{r}(\mathbf{S} \cdot \mathbf{r})}{r^5} + \frac{8}{3}\pi\mathbf{S}\,\delta(\mathbf{r})\right\}. \qquad (17.30)$$

If several electrons surround the nucleus, the interaction Hamiltonian is the sum of the contributions of the individual electrons. Although the expression (17.30) has been derived for the purpose of calculating its expectation value $(\Psi_e|\,\mathcal{H}_1\,|\Psi_e)$, it is clear that it also gives unambiguous results for off-diagonal matrix elements $(\Psi_e|\,\mathcal{H}_1\,|\Phi_e)$ between, say, the ground state and an excited state of the electronic system. Use will be made of this to calculate some effects of $\mathcal{H}_1$, using second-order perturbation theory.

We may write the expression in curly brackets in (17.30) simply as a vector $\mathbf{N}$ so that (17.30) becomes

$$\mathcal{H}_1 = 2\beta\gamma_n\hbar(\mathbf{N} \cdot \mathbf{I}). \qquad (17.31)$$

This is equivalent to an interaction

$$-\gamma_n\hbar(\mathbf{H}_e \cdot \mathbf{I})$$

where

$$\mathbf{H}_e = -2\beta\langle r^{-3}\rangle\mathbf{N} \qquad (17.32)$$

is the electronic magnetic field at the nucleus.

17.3. Alternative derivation of the magnetic hyperfine interaction

There is a weak point in our derivation of (17.30), namely, the statement that the first part of (17.29) $g_s\beta\{(\mathbf{S} \cdot \nabla)(\boldsymbol{\mu} \cdot \nabla) - \tfrac{1}{3}(\mathbf{S} \cdot \boldsymbol{\mu})\nabla^2\}(1/r)$ gives a vanishing result when the second term $-(2g_s\beta/3)(\mathbf{S} \cdot \boldsymbol{\mu})\nabla^2(1/r)$ does not; that is, for an s-electron. This is based on an angular integration, performed before a radial integration. If we were to interchange these operations we would find an infinite result instead. An alternative derivation based on the relativistic equation of Dirac does not have this drawback.

HYPERFINE STRUCTURE

In a magnetic field Dirac's equation can be written

$$\{c\boldsymbol{\alpha} \cdot (\mathbf{p}+e\mathbf{A}/c)+\beta mc^2-eV\}\Psi = (W+mc^2)\Psi. \quad (17.33)$$

In this equation $\boldsymbol{\alpha}$ stands for the three Dirac 4 by 4 matrices $\alpha_1, \alpha_2, \alpha_3$; β is the fourth Dirac matrix (not to be confused with the Bohr magneton); W is the kinetic energy of the electron and Ψ the four-component spinor which is the electron wave-function in Dirac theory.

The spinor Ψ can be written as $\begin{pmatrix}\varphi\\\chi\end{pmatrix}$ where φ and χ are 2 two-component spinors describing what is known as the large and small components of the Dirac wave-function. If we neglect $e\boldsymbol{\alpha} \cdot \mathbf{A}$, which is a small perturbation, it is known from the structure of the matrices α_i and β that (17.33) can be rewritten:

$$\begin{aligned} c(\boldsymbol{\sigma} \cdot \mathbf{p})\chi &= (W+eV)\varphi, \\ c(\boldsymbol{\sigma} \cdot \mathbf{p})\varphi &= (W+2mc^2+eV)\chi, \end{aligned} \quad (17.34)$$

where $\boldsymbol{\sigma}$ is the set of the three Pauli matrices. Neglecting $W+eV$ in comparison with $2mc^2$ we get

$$\chi \cong \frac{1}{2mc}(\boldsymbol{\sigma} \cdot \mathbf{p})\varphi. \quad (17.35)$$

We may treat the coupling $e(\boldsymbol{\alpha} \cdot \mathbf{A})$ with the magnetic field as a perturbation and proceed to calculate its expectation value:

$$\langle\Psi|\, e(\boldsymbol{\alpha} \cdot \mathbf{A})\,|\Psi\rangle = \langle\varphi|\, e(\boldsymbol{\sigma} \cdot \mathbf{A})\,|\chi\rangle + \langle\chi|\, e(\boldsymbol{\sigma} \cdot \mathbf{A})\,|\varphi\rangle$$
$$\approx \frac{e}{2mc}\langle\varphi|\,(\boldsymbol{\sigma} \cdot \mathbf{A})(\boldsymbol{\sigma} \cdot \mathbf{p})+(\boldsymbol{\sigma} \cdot \mathbf{p})(\boldsymbol{\sigma} \cdot \mathbf{A})\,|\varphi\rangle. \quad (17.36)$$

Using the commutation relations of the components of $\boldsymbol{\sigma}$ we find that (17.36) becomes

$$\frac{e}{2mc}\langle\varphi|\,(\mathbf{p} \cdot \mathbf{A})+(\mathbf{A} \cdot \mathbf{p})+i\boldsymbol{\sigma} \cdot (\mathbf{A} \wedge \mathbf{p}+\mathbf{p} \wedge \mathbf{A})\,|\varphi\rangle \quad (17.37)$$

and from the relation $\mathbf{p} = (\hbar/i)\nabla$ we get

$$i(\mathbf{A} \wedge \mathbf{p}+\mathbf{p} \wedge \mathbf{A}) = \hbar\, \mathbf{curl} \cdot \mathbf{A}$$

whence

$$\langle\Psi|\, e(\boldsymbol{\alpha} \cdot \mathbf{A})\,|\Psi\rangle = \langle\varphi|\,\frac{e}{2mc}\{(\mathbf{p} \cdot \mathbf{A})+(\mathbf{A} \cdot \mathbf{p})\}+\beta\boldsymbol{\sigma} \cdot \mathbf{curl}\,\mathbf{A}\,|\varphi\rangle, \quad (17.38)$$

which was precisely our starting point (eqn (17.26a)) for the non-relativistic derivation above, provided that we take $g_s = 2$.

The operation $\langle\varphi|\ \ |\varphi\rangle$ corresponds to an integration over orbital and spin variables. In the Pauli approximation $|\varphi\rangle$ is the product of the Schrödinger orbital wave-function Ψ_e times a spin function (or a sum of such products). Leaving out the integration over spin variables, the second part of (17.37) is a spin-operator

$$i\frac{\beta}{\hbar}\boldsymbol{\sigma}\cdot\langle\Psi_e|\ \mathbf{A}\wedge\mathbf{p}+\mathbf{p}\wedge\mathbf{A}\ |\Psi_e\rangle, \tag{17.39}$$

which, since $\mathbf{p}$ is a Hermitian operator, can be rewritten in terms of the density $\rho = \varphi_e^*\varphi_e$ as

$$\beta\boldsymbol{\sigma}\cdot\int(\nabla\rho\wedge\mathbf{A})\,d\tau = \beta\int\text{curl}\{(\rho\boldsymbol{\sigma})\cdot\mathbf{A}\}\,d\tau. \tag{17.40}$$

If one assigns to the electron a spin current density

$$\mathbf{j}_s = \beta\,\text{curl}\,(\rho\boldsymbol{\sigma}), \tag{17.41}$$

we can write (17.40) in the form

$$\int(\mathbf{j}_s\cdot\mathbf{A})\,d\tau. \tag{17.42}$$

The expression (17.40) is equivalent to our starting point (17.38) but has the advantage that if we replace $\mathbf{A}$ by $\text{curl}\,(\boldsymbol{\mu}/r)$ the integral converges absolutely; the singularity of the integrand is of order $1/r^2$ rather than $1/r^3$ and for an s-electron it leads unambiguously to the last term of (17.30).

17.4. Equivalent operators for the magnetic hyperfine structure

The dipole-dipole coupling $\{3(\mathbf{I}\cdot\mathbf{r})(\mathbf{s}\cdot\mathbf{r})-r^2(\mathbf{I}\cdot\mathbf{s})\}/r^5$ is a product of two second-rank tensors and can be rewritten as

$$\sum_{j,k}\{I_j s_k - \tfrac{1}{3}\delta_{jk}(\mathbf{I}\cdot\mathbf{s})\}(3x_j x_k - r^2\delta_{jk})r^{-5}. \tag{17.43}$$

By a process similar to that used in § 17.1 in deriving the nuclear electric quadrupole interaction, this is found for a single electron l to be equivalent to

$$\langle r^{-3}\rangle\frac{2}{(2l+3)(2l-1)}\{l(l+1)(\mathbf{I}\cdot\mathbf{s})-\tfrac{3}{2}(\mathbf{l}\cdot\mathbf{I})(\mathbf{l}\cdot\mathbf{s})-\tfrac{3}{2}(\mathbf{l}\cdot\mathbf{s})(\mathbf{l}\cdot\mathbf{I})\}. \tag{17.44}$$

Similarly for a term (L, S) obeying Hund's rule we obtain

$$\sum_i\frac{3(\mathbf{I}\cdot\mathbf{r}_i)(\mathbf{s}_i\cdot\mathbf{r}_i)-r_i^2(\mathbf{I}\cdot\mathbf{s}_i)}{r_i^5}$$
$$= \xi\langle r^{-3}\rangle\{L(L+1)(\mathbf{I}\cdot\mathbf{S})-\tfrac{3}{2}(\mathbf{L}\cdot\mathbf{I})(\mathbf{L}\cdot\mathbf{S})-\tfrac{3}{2}(\mathbf{L}\cdot\mathbf{S})(\mathbf{L}\cdot\mathbf{I})\}, \tag{17.45}$$

in which
$$\xi = \frac{2l+1-4S}{S(2l-1)(2l+3)(2L-1)} = \pm\frac{\eta_L}{2S} = \mp\frac{1}{2S}\langle L\| \alpha \|L\rangle, \quad (17.46)$$

where η_L is given by (17.16) and $\langle L\| \alpha \|L\rangle$ by (16.25).

Since the operator (17.43) is time-odd with respect to electron variables, ξ has the same sign for two complementary configurations of x and $2(2l+1)-x$ electrons, in contrast to η_L or $\langle L\| \alpha \|L\rangle$ which were introduced in dealing with interactions that are time-even with respect to the electron variables.

Within a ground term (L, S) of electrons for which $l \neq 0$ the magnetic hyperfine coupling can then be written

$$2\beta\gamma_n\hbar\langle r^{-3}\rangle\{(\mathbf{L}\cdot\mathbf{I})+\xi L(L+1)(\mathbf{I}\cdot\mathbf{S})-\tfrac{3}{2}\xi(\mathbf{L}\cdot\mathbf{I})(\mathbf{L}\cdot\mathbf{S})-\tfrac{3}{2}\xi(\mathbf{L}\cdot\mathbf{S})(\mathbf{L}\cdot\mathbf{I})\}.$$
(17.47)

We now construct an equivalent operator for the hfs interaction within the manifold (J, L, S). We define the vector $\mathbf{N}$ by

$$\mathbf{N} = \sum_i \left\{\mathbf{l}_i - \mathbf{s}_i + 3\frac{(\mathbf{r}_i\cdot\mathbf{s}_i)}{r_i^2}\mathbf{r}_i\right\}, \quad (17.48)$$

which is equivalent to (17.31) except that we have assumed the term $\delta(r)$ in (17.30) to be absent. Inside the manifold (L, S) this is equivalent to, using (17.47),

$$\langle LS\| \mathbf{N} \|LS\rangle = \mathbf{L}+\xi L(L+1)\mathbf{S}-\tfrac{3}{2}\xi\{\mathbf{L}(\mathbf{L}\cdot\mathbf{S})+(\mathbf{L}\cdot\mathbf{S})\mathbf{L}\}. \quad (17.49)$$

Within the manifold (J, L, S) we can replace $\mathbf{N}$ by a vector collinear with $\mathbf{J}$

$$\frac{(\mathbf{N}\cdot\mathbf{J})\mathbf{J}}{J(J+1)} = \langle J\| N \|J\rangle\mathbf{J}, \quad (17.50)$$

where

$$\langle J\| N \|J\rangle = \frac{1}{J(J+1)}\{(\mathbf{L}\cdot\mathbf{J})+\xi L(L+1)(\mathbf{S}\cdot\mathbf{J})-3\xi(\mathbf{L}\cdot\mathbf{J})(\mathbf{L}\cdot\mathbf{S})\},$$
(17.51)

in which

$$(\mathbf{L}\cdot\mathbf{J}) = \tfrac{1}{2}\{J(J+1)+L(L+1)-S(S+1)\},$$
$$(\mathbf{S}\cdot\mathbf{J}) = \tfrac{1}{2}\{J(J+1)+S(S+1)-L(L+1)\}, \quad (17.52)$$
$$(\mathbf{L}\cdot\mathbf{S}) = \tfrac{1}{2}\{J(J+1)-L(L+1)-S(S+1)\}.$$

Here it may be useful to note that

$$(\mathbf{L}\cdot\mathbf{J})/J(J+1) = 2 - \langle J\| \Lambda \|J\rangle,$$
$$(\mathbf{S}\cdot\mathbf{J})/J(J+1) = \langle J\| \Lambda \|J\rangle - 1, \qquad (17.53)$$

and that the form of the magnetic hyperfine interaction is now simply

$$2\beta\gamma_n\hbar\langle r^{-3}\rangle\langle J\| N \|J\rangle(\mathbf{J}\cdot\mathbf{I}). \qquad (17.54)$$

The values of $\langle J\| N \|J\rangle$ are tabulated for the rare-earth group in Table 20.

For a single electron with $j = l \pm \tfrac{1}{2}$ a direct derivation leads to a much simpler result for (17.51) since

$$(\mathbf{j}\cdot\mathbf{N}) = \mathbf{j}\cdot\left\{1 - \mathbf{s} + 3\frac{(\mathbf{r}\cdot\mathbf{s})}{r^2}\mathbf{r}\right\} = \mathbf{l}\cdot\left\{1 - \mathbf{s} + 3\frac{(\mathbf{r}\cdot\mathbf{s})}{r^2}\mathbf{r}\right\} + \mathbf{s}\cdot\left\{1 - \mathbf{s} + 3\frac{(\mathbf{r}\cdot\mathbf{s})}{r^2}\mathbf{r}\right\}$$

$$= l(l+1) - s(s+1) + 3\frac{(\mathbf{r}\cdot\mathbf{s})^2}{r^2} = l(l+1),$$

where we have used the fact that $(\mathbf{r}\cdot\mathbf{l})$ is proportional to $\mathbf{r}\cdot(\mathbf{r}\wedge\mathbf{p}) = 0$. Hence

$$\langle j\| N \|j\rangle = \frac{l(l+1)}{j(j+1)}. \qquad (17.55)$$

To find the off-diagonal elements $\langle J'M'| \mathbf{N} |JM\rangle$ of the vector $\mathbf{N}$ between two manifolds J', J belonging to the same (L, S) multiplet we make use of the fact that, from eqn (17.49), the matrix elements of $\mathbf{N}$ are a linear combination of those of $\mathbf{L}$ and $\mathbf{S}$. It then follows from (16.38) that the off-diagonal elements of $\mathbf{N}$ are proportional to those of $\mathbf{L}+2\mathbf{S}$, and vanish except for $|J'-J| = 1$. Introducing a symbol $\langle J+1\| N \|J\rangle$, we can therefore write

$$\langle J+1, M'| \mathbf{N} |J, M\rangle = \frac{\langle J+1\| N \|J\rangle}{\langle J+1\| \Lambda \|J\rangle}\langle J+1, M'| \mathbf{L}+2\mathbf{S} |J, M\rangle \qquad (17.56)$$

and from (17.49) and (16.38) we have

$$\frac{\langle J+1\| N \|J\rangle}{\langle J+1\| \Lambda \|J\rangle} = -1 + \xi L(L+1) + \tfrac{3}{2}\xi\{\langle J|(\mathbf{L}\cdot\mathbf{S})|J\rangle + \langle J+1|(\mathbf{L}\cdot\mathbf{S})|J+1\rangle\}, \qquad (17.57)$$

where ξ (which is equal to $-\nu$ in the notation of Elliott and Stevens (1953a)) is given by eqn (17.46) and $\langle J|(\mathbf{L}\cdot\mathbf{S})|J\rangle$ by eqn (17.52).

Values of ξ and $\langle J+1 \| N \| J \rangle$, as well as of $\langle J+1 \| \Lambda \| J \rangle$ (eqn (16.41)), are listed for the rare-earth group in Table 20. The off-diagonal elements of $\mathbf{L} + 2\mathbf{S}$ are given by eqns (16.39), (16.40), whence we have

$$\langle J+1, J_z | N_z | J, J_z \rangle = \langle J+1 \| N \| J \rangle \{(J+1)^2 - J_z^2\}^{\frac{1}{2}}, \tag{17.58}$$

$$\langle J+1, J_z \pm 1 | N_x | J, J_z \rangle = \mp \langle J+1 \| N \| J \rangle \tfrac{1}{2}\{(J \pm J_z+1)(J \pm J_z+2)\}^{\frac{1}{2}}.$$
$$\tag{17.59}$$

17.5. The effect of s-electrons: configuration interaction

We have already commented on the fact that, to calculate the coupling between an atom (or an ion) and a homogeneous applied field, a detailed knowledge of the atomic wave-function is not required (§ 11.3); in particular, for a term (L, S) it is not necessary to know in what configuration this term originated. The position is quite different for the coupling with the very inhomogeneous magnetic field produced by a nuclear moment; a formula such as (17.45) rests on the explicit assumption that the term (L, S) is built from a specific configuration of i electrons each with angular momentum l.

In our calculations of the effects of the crystal field we have assumed that the configuration is a good quantum number and on the basis of this assumption we would have expected eqn (17.47) to give reasonable agreement with experiment. In fact in the important case of the iron group the agreement is miserable. The explanation resides in the fact that s-electrons, especially those of low principal quantum number n, have hyperfine couplings very much stronger than those of other electrons with $l \neq 0$. It follows that if a term (L, S) belonging to an ion of the transition group, whose ground configuration has no unpaired s-electrons, is 'contaminated' even slightly with excited configurations containing unpaired s-electrons (for brevity s-configurations), there may be an appreciable change in the hyperfine coupling although all other properties of the term remain practically unaffected. If there is an admixture from s-configurations the contribution of the so-called contact term, the last term of (17.30), will not be zero; within the manifold of the term (L, S), this contribution can be represented quite generally as $A_s \mathbf{I} \cdot \mathbf{S}$, the value of the constant A_s depending on the nature and amount of s-configuration present. It is customary (though perhaps unfortunate) to define a dimensionless constant κ through the relation

$$A_s = -2\gamma_n \beta \hbar \langle r^{-3} \rangle \kappa. \tag{17.60}$$

To sum up, within a term (L, S) the total hyperfine structure, electric and magnetic, can be written

$$W_n = 2\beta\gamma_n\hbar\langle r^{-3}\rangle\{(\mathbf{L}\cdot\mathbf{I})+\{\xi L(L+1)-\kappa\}(\mathbf{S}\cdot\mathbf{I})\}-$$
$$-\tfrac{3}{2}\xi\{(\mathbf{L}\cdot\mathbf{S})(\mathbf{L}\cdot\mathbf{I})+(\mathbf{L}\cdot\mathbf{I})(\mathbf{L}\cdot\mathbf{S})\}-$$
$$-\langle L\|\alpha\|L\rangle\frac{e^2Q}{2I(2I-1)}\langle r^{-3}\rangle\{3(\mathbf{L}\cdot\mathbf{I})^2+\tfrac{3}{2}(\mathbf{L}\cdot\mathbf{I})-L(L+1)I(I+1)\}.$$

(17.61)

We examine below and in § 17.6 methods by which theoretical estimates can be made of the magnitude of the constant κ.

The original idea that configuration interaction is responsible for anomalies in atomic magnetic hyperfine structure is due to Fermi and Segré (1933a, b). Among the examples they studied, the most relevant for our purpose, as will appear shortly, is the $6s^26p$, ^2P ground term of atomic thallium perturbed by the excited ^2P term $6s6p7s$. Fermi and Segré used a perturbation method to estimate the small admixture α from the excited term into the ground manifold; the unperturbed one-electron wave-functions in each configuration were eigenstates of an approximate central-field Hamiltonian, the matrix elements of the Coulomb repulsion $V = \sum_{i<k} e^2/r_{ik}$ between the two terms were computed numerically, and the energy separation Δ_{10} between the terms was taken from spectroscopic data.

Let Ψ_0 and Ψ_1 be atomic states, belonging respectively to the ground and to the excited configuration, and linked by the electronic repulsion V, which in first order changes Ψ_0 into $\Psi = \Psi_0 + \alpha\Psi_1$, where

$$\alpha \approx -\frac{\langle\Psi_0|V|\Psi_1\rangle}{\Delta_{10}}.$$ (17.62)

Since the coupling Hamiltonian V is invariant through spatial rotation and does not contain the spin variables it is clear that the states Ψ_0 and Ψ_1 must have the same quantum numbers L, S, M_L, M_S.

For our purpose the essential feature of the excited configuration $6s6p7s$ is that it is obtained from the ground configuration $6s^26p$ by promotion of one electron from an s-orbit φ_{6s} to another s-orbit φ_{7s}. Therefore the contact part $\mathscr{H}_c$ of the hyperfine Hamiltonian does have a non-vanishing matrix element between Ψ_0 and Ψ_1, and according to the rules given in § 11.6 this is proportional to a one-electron matrix element $(\varphi_{6s}|\delta(\mathbf{r})|\varphi_{7s})$, that is, to the product $\varphi_6(0)\cdot\varphi_7(0)$. The expectation value $\langle\Psi|\mathscr{H}_c|\Psi\rangle$ of the contact hyperfine interaction in the state Ψ given by (17.62) contains therefore a term proportional to $2\alpha\varphi_{6s}(0)\varphi_{7s}(0)$. This linear dependence on α of $\langle\Psi|\,\mathscr{H}_c\,|\Psi\rangle$ has two far-reaching consequences: first, as α is small, the correction to the hyperfine structure is very much larger than the second-order term $\alpha^2\langle\Psi_1|\,\mathscr{H}_c\,|\Psi_1\rangle$, which is proportional to $\alpha^2\varphi_{7s}^2(0)$; secondly, this correction can have either sign, again in contrast to the second-order term.

This feature proves to be of paramount importance in the explanation, that now follows, of anomalous hyperfine structure in the paramagnetic resonance spectrum of ions belonging to the iron group (Abragam 1950; Abragam and Pryce 1951a, b, c; Abragam, Horowitz, and Pryce 1955). The conventional configuration assigned to the ions of the iron group in their ground state is

$$\mathscr{C}_0 = 1s^2 2s^2 2p^6 3s^2 3p^6 3d^x. \tag{17.63}$$

In the ground term, constructed from this configuration according to Hund's rule, the total spin S is either $x/2$ or $5-(x/2)$ depending on whether $x \leqslant 5$ or $x \geqslant 5$. This configuration does not allow for a term such as $A_s\mathbf{I}\cdot\mathbf{S} = -2\gamma_n\beta\hbar\langle r^{-3}\rangle\kappa\mathbf{I}\cdot\mathbf{S}$ (eqn (17.60)) in the hyperfine Hamiltonian (17.61); however such a term is found to be necessary in order to fit the experimental data in the iron group (see Chapter 7). The sign of the coefficient κ is found to be positive, that is such as to make the electronic hyperfine field $\mathbf{H}_e$ at the nucleus parallel to the total ionic spin $\mathbf{S}$. In contrast the field $\mathbf{H}_e$ produced at the nucleus by a single s-electron is antiparallel to its spin.

It is convenient to introduce a quantity χ characteristic of the density of unpaired spins at the nucleus:

$$\chi = \frac{4\pi}{S}\left\langle \sum_k \delta(\mathbf{r}_k) s_{kz} \right\rangle_{S_z=S}. \tag{17.64}$$

The expectation value in (17.64) is to be taken over the state $S_z = S$ of the ground term. According to (17.30) and (17.60) χ is related to κ by

$$\kappa\langle r^{-3}\rangle = -(\tfrac{2}{3})\chi \tag{17.65}$$

so that χ must be negative if κ is positive.

For each ion of the iron group χ is found to depend little on the surroundings of the ion, which makes it plausible that it is a feature of the free ion. χ is also found experimentally to vary little (by no more than 25 per cent) inside the iron group, the mean value being of the order of -3 in atomic units, a fact for which there is no simple physical explanation. The proportionality coefficient between χ, expressed in atomic units, and the electronic field H_e at the nucleus, expressed in gauss, is given by $H_e = -2S \times 4 \cdot 21 \times 10^4 \chi$.

In contrast with the hyperfine structure the other features of the spectrum such as the g-factor and fine structure agree well with the ground configuration (17.63). It is therefore reasonable to look for small admixtures into this configuration that contain unpaired s-electrons. In order for such admixtures to be able to modify appreciably the hyperfine structure, while leaving practically unaffected the other features of the spectrum, and also to produce hyperfine fields $\mathbf{H_e}$ of the right sign, they must differ from the ground configuration through promotion of one electron from an s-orbit into another s-orbit, as explained in connection with the work of Fermi and Segré. Let us denote by $\mathscr{C}_3$ one such configuration, in which a $3s$ electron has been promoted,

$$\mathscr{C}_3 = 1s^2 2s^2 2p^6 3s Rs 3p^6 3d^x = 3s Rs(\mathscr{C}_0 - 3s^2). \qquad (17.66)$$

In (17.66) Rs means an s-orbital that is orthogonal to all the s-orbitals of $\mathscr{C}_0$, $1s$, $2s$, $3s$ (it is automatically orthogonal to p- and d-orbitals), and at this point our treatment departs from that of Fermi and Segré. In the latter, the excited s-orbital $7s$ is a physical orbital of the thallium atom; states belonging to the excited configuration $6s6p7s$ have actually been observed and their energies have been measured by the methods of optical spectroscopy. In a configuration such as $\mathscr{C}_3$, given by (17.66), it is not clear what meaning one should attach to the s-orbit Rs. In order to elucidate this point let us return to the ground configuration $\mathscr{C}_0$ and recall how the orbits $1s$ to $3d$ of which it is composed are obtained in the variational Hartree–Fock method. As explained in § 11.4 we build from these orbits, each of which has the usual form

$$\varphi_{n,l,m_l,m_s} = P_{nl}(r) Y_l^{m_l}(\theta, \varphi) \chi(m_s), \qquad (17.67)$$

a many-electron wave-function Ψ_0 which has the required symmetry and quantum numbers L, S, M_L, M_S. Terms obeying Hund's rule have a simpler structure than the others: the magnetic substate of such a term with $M_L = L$, $M_S = S$ (sometimes called the 'stretched' state) can be written as a single Slater determinant. Then we write down the expectation value in the state Ψ_0 of an electronic Hamiltonian $\mathscr{H}$ which is the sum of the kinetic energies of the electrons and of their potential energy due to their attraction by the nucleus and to their mutual repulsion (smaller magnetic interactions are neglected at this stage). The expression $\langle \Psi_0 | \mathscr{H} | \Psi_0 \rangle$ is expanded into a sum of one-electron and two-electron matrix elements by the rules of § 11.6, and minimized with respect to the trial radial functions $P_{nl}(r)$ which are considered as unknown. The generalization whereby the configuration $\mathscr{C}_3$ of (17.63) is introduced and the orbit Rs as defined is now straightforward: as a trial many-electron wave-function we use the linear combination

$$\Psi = \cos \alpha\, \Psi_0 + \sin \alpha\, \Psi_3, \qquad (17.68)$$

where Ψ_3 is a wave-function belonging to $\mathscr{C}_3$ and having the same quantum numbers L, S, M_L, M_S as Ψ_0. The latter, being the 'stretched' substate of a Hund's term, is a single Slater determinant. (We have tacitly assumed that the configuration $\mathscr{C}_3$ contains only one state Ψ_3 with the right quantum numbers; we shall return to this point later.) We formulate again the expression $\langle \Psi | \mathscr{H} | \Psi \rangle$ and minimize it with respect to the functions $P_{nl}(r)$ and also with respect to the new trial function Rs, which appears in the configuration $\mathscr{C}_3$, and with respect to the normalizing variable α. Since the set of trial functions Ψ is wider than that of the functions Ψ_0, which it contains, it must be better than or at least as good as the latter for the calculation of every physical parameter. The parameter of interest to us is the hyperfine structure and this is naturally the reason why we have augmented the set of the trial functions by adding to it components to which the hyperfine structure is most sensitive.

We now write out the wave-function Ψ_3 which has the same quantum numbers as Ψ_0. Since $\mathscr{C}_3$ results from $\mathscr{C}_0$ through promotion of an electron from one s-orbit to another it is clear that Ψ_3 will have the same M_L and L as Ψ_0, provided each electron has the same m_l in Ψ_3 and Ψ_0. Let us write the configuration $\mathscr{C}_0$ as P^2Q and the stretched state Ψ_0 as

$$\Psi_0 = P^+P^-Q^S, \qquad (17.69)$$

where P is the orbital $3s$, Q stands for $3d^x$, and Q^S is the Slater determinant formed from the $3d^x$ electrons with all their spins up, thus adding to a total spin $S_z = S$; the orbitals belonging to the other closed shells of $\mathscr{C}_0$ are omitted in this shorthand notation. Ψ_0 itself is a Slater determinant so that

$$P^+P^-Q^S = -P^-P^+Q^S = (-1)^x P^+Q^S P^-. \qquad (17.70)$$

We can similarly define

$$Q^{S-1} = \frac{1}{\sqrt{(2S)}}(S_x - iS_y)Q^S \qquad (17.71)$$

as the state of spin S with $M_S = S-1$.

The configuration $\mathscr{C}_3$ can be written as PRQ and two states with total spin S and $M_S = S$ can be formed from it: Ψ_{3a} where the spins of the electrons Ps and Rs couple to zero and which is obviously of the form

$$\Psi_{3a} = \frac{1}{\sqrt{2}}(P^+R^- - P^-R^+)Q^S \qquad (17.72)$$

and the state where the spins of P and R couple to a spin $S' = 1$ which in turn couples to the spin S of Q to give a spin S again. This state can be written

$$\Psi_{3b} = \frac{\sqrt{(2S)}}{2\sqrt{(S+1)}}(P^+R^-Q^S + P^-R^+Q^S - \frac{2}{\sqrt{(2S)}}P^+R^+Q^{S-1}), \qquad (17.73)$$

which is justified by the fact that, from (17.72), (17.73),

$$\langle \Psi_{3b} | \Psi_{3a} \rangle = 0; \qquad S_z\Psi_{3b} = S\Psi_{3b}; \qquad S_+\Psi_{3b} = 0. \qquad (17.74)$$

We can now write the expression to be minimized as

$$\langle \Psi | \mathscr{H} | \Psi \rangle = \cos^2\alpha \langle \Psi_0 | \mathscr{H} | \Psi_0 \rangle +$$
$$+ \sin 2\alpha \langle \Psi_0 | \mathscr{H} | \Psi_3 \rangle + \sin^2\alpha \langle \Psi_3 | \mathscr{H} | \Psi_3 \rangle, \qquad (17.75)$$

where Ψ_0 and $\Psi_3 = \Psi_{3b}$ are given in shorthand notation by (17.69) and (17.73). (We shall show shortly that the state Ψ_{3a} given by (17.72) is not coupled to Ψ_0.) Minimizing (17.75) with respect to α, we find

$$\tan 2\alpha = -2\frac{\langle \Psi_0 | \mathscr{H} | \Psi_3 \rangle}{\langle \Psi_3 | \mathscr{H} | \Psi_3 \rangle - \langle \Psi_0 | \mathscr{H} | \Psi_0 \rangle}. \qquad (17.76)$$

If we use our assumption that the admixture from the excited configuration is small, (17.76) becomes, in a good approximation,

$$\alpha = -\frac{\langle \Psi_0 | \mathscr{H} | \Psi_3 \rangle}{\langle \Psi_3 | \mathscr{H} | \Psi_3 \rangle - \langle \Psi_0 | \mathscr{H} | \Psi_0 \rangle}, \qquad (17.77)$$

which resembles the perturbation formula (17.62). It remains to vary (17.75) with respect to all the radial functions P_{nl} of $\mathscr{C}_0$ and with respect to Rs. Since the admixture α is very small, it is a good approximation to take for the P_{nl} the unperturbed functions of the configuration $\mathscr{C}_0$, that is the functions that minimized $\langle \Psi_0 | \mathscr{H} | \Psi_0 \rangle$. This enables us to prove the statement that the state Ψ_{3a} given by (17.72) is not coupled to Ψ_0 in first order in α; using (17.69) and (17.72) we can rewrite $\Psi_0 + \alpha \Psi_{3a}$, where α is an arbitrary small quantity, as

$$\Psi = \Psi_0 + \alpha\Psi_{3a} = \left(P^+ + \frac{\alpha}{\sqrt{2}}R^+, P^- + \frac{\alpha}{\sqrt{2}}R^-\right)Q^S + O(\alpha^2). \qquad (17.78)$$

HYPERFINE STRUCTURE 701

This equation shows that as far as terms $O(\alpha^2)$, Ψ results from Ψ_0 by giving a first-order increase $\alpha R/\sqrt{2}$ to P. Since, however, the functions P have been determined precisely by the condition that they should make $\langle \Psi_0| \mathscr{H} |\Psi_0\rangle$ a minimum, it follows that with Ψ given by (17.78)

$$\langle \Psi| \mathscr{H} |\Psi\rangle = \langle \Psi_0| \mathscr{H} |\Psi_0\rangle + O(\alpha^2). \qquad (17.79)$$

Since we also have from (17.78)

$$\langle \Psi| \mathscr{H} |\Psi\rangle = \langle \Psi_0| \mathscr{H} |\Psi_0\rangle + 2\alpha \langle \Psi_0| \mathscr{H} |\Psi_{3a}\rangle + O(\alpha^2), \qquad (17.80)$$

we must have

$$\langle \Psi_0| \mathscr{H} |\Psi_{3a}\rangle = 0 \qquad (17.81)$$

and, according to (17.77), the value of α that minimizes $\langle \Psi| \mathscr{H} |\Psi\rangle$ and gives the amplitude of Ψ_{3a} in Ψ vanishes. If we now minimize (17.75) with respect to Rs we get for this function an integro-differential equation that must be integrated numerically. We shall not write out its expression which is lengthy (see Abragam, Horowitz, and Pryce (1955) where slightly different notations are used). Once this is done and Rs is known we can calculate χ using eqns (17.64), (17.68), and (17.73) and we find, neglecting terms of order $O(\alpha^2)$,

$$\chi = \frac{4\pi}{S} \langle \Psi| \sum_k \delta(\mathbf{r}_k) s_{zk} |\Psi\rangle \approx -2\alpha \frac{4\pi}{S} \langle \Psi_0| \sum_k \delta(\mathbf{r}_k) s_{zk} |\Psi_{3b}\rangle$$

$$= 4\pi\alpha \sqrt{\left\{\frac{2}{S(S+1)}\right\}} \varphi_{3s}(0) \varphi_{Rs}(0), \qquad (17.82)$$

or if we write

$$\varphi_{3s} = \frac{P_{3s}(r)}{r} \frac{1}{\sqrt{(4\pi)}}, \; \varphi_{Rs} = \frac{Rs(r)}{r} \frac{1}{\sqrt{(4\pi)}}, \qquad (17.83)$$

$$\chi = -\alpha \sqrt{\left\{\frac{2}{S(S+1)}\right\}} P'_{1s}(0) R's(0). \qquad (17.84)$$

It was hoped originally (Abragam and Pryce 1951a) that the bulk of the anomalous hyperfine structure in the iron group was due to the configuration $\mathscr{C}_3$ and a numerical calculation of χ as given by (17.84) was attempted for Mn^{2+} (Abragam, Horowitz, and Pryce 1955) leading to a value of $\chi = -0\cdot3\,.\,10^{-3}$ in atomic units (instead of the experimental value $\chi \approx -3$). This dismal failure is due to three facts.

(i) The unperturbed wave-functions P_{nl} for the configuration $\mathscr{C}_0$ of Mn^{2+} were not available at the time of the calculation and those of the cuprous ion Cu^+ for the configuration $1s^2 2s^2 2p^6 3s^2 3p^6 3d^{10}$ were used instead. It turns out that in the equation satisfied by Rs there occur overlap integrals of the type $\langle 3s, 3d|\ 1/r_{12}\ |3d, Rs\rangle$ with large positive and negative parts which nearly cancel each other, making the integrals very sensitive to small changes in the wave-function. In consequence, large errors may result through starting from incorrect unperturbed wave-functions.

(ii) The assumption that the configuration $\mathscr{C}_3$ is the only one to contribute appreciably to the anomalous hyperfine structure is unwarranted, as will appear shortly. Instead of the set of functions Ψ given by (17.63) a wider set should be used in the expression $\langle\Psi|\ \mathscr{H}\ |\Psi\rangle$ to be minimized, namely,

$$\Psi = \nu_0\Psi_0 + \sum_{i=1}^{3}\nu_i\Psi_i. \qquad (17.85)$$

In this equation Ψ_i belongs to a configuration $\mathscr{C}_i$ where an s-electron from the shell i is promoted. Such a calculation was not attempted through lack of adequate computing facilities.

(iii) Last but not least, it is not impossible that a numerical error was made in the calculation by those in charge of the numerical computation (W. Marshall, private communication). All further calculations of anomalous hyperfine structure were made using the so-called core polarization method to be described now.

17.6. The effect of s-electrons: core polarization

The idea of magnetic polarization of an atomic or ionic core of closed shells by an unfilled external shell with a total spin S goes back to the pioneer work of Sternheimer (1951). This idea is physically much easier to grasp than the configuration interaction method outlined in the previous section: let us consider a paramagnetic ion, say Mn^{2+}, in the state $S_z = S = +\frac{5}{2}$ where the spins of its five $3d$ electrons are all 'up'. The electrostatic repulsion between an s-electron of an inner shell, with spin up, say $1s^+$, and these $3d$ electrons will not be the same as that for an electron in the state $1s^-$ (for brevity, in the following, spin up or $+$ will mean spin parallel to the total spin of the ion). This is a consequence of the exclusion principle that prevents two electrons with parallel spins from occupying the same position in space. Mathematically this principle is built into the formalism by using wave-functions that are antisymmetric with respect to interchange of both spin and orbital coordinates of any two electrons. For wave-functions approximated by Slater determinants this results in the vanishing of the exchange matrix elements of the electrostatic repulsion between electrons of opposite spins, such as $\langle 3s^-(1), 3d^+(2)| e^2/r_{12} |3d^+(1), 3s^-(2)\rangle$ (eqn (11.25)). It is therefore not unreasonable to think that for the electron with spin up, the density at the origin, $|\varphi_{1s}^+(0)|^2$, will be different from $|\varphi_{1s}^-(0)|^2$. Can one predict the sign of the difference? Hand-waving arguments can be produced for either sign. One could argue that the inner $1s^+$ electron with spin up is kept apart from the outer $3d^+$ electrons by the exclusion principle and thus has a greater probability to be pushed near the nucleus. One would then have

$$|\varphi_{1s}^+(0)|^2 - |\varphi_{1s}^-(0)|^2 > 0 \qquad (17.86)$$

for the spin density at the origin. On the other hand, one can also argue that for electrons with parallel spins the fact of being kept apart by the exclusion principle reduces the expectation value of their electrostatic repulsion by so much that on the whole they are less repelled by each other than by electrons with antiparallel spins. The sign of (17.86) would then be reversed. Whatever the conclusion reached for the sign of the spin density at the origin for innermost s-electrons it would be reversed for s-electrons which are farther removed from the nucleus than the magnetic $3d$ electrons, such as the two $4s$ electrons of a free atom of the iron group.

In order to settle this question, and more generally to calculate χ, eqn (17.64) is rewritten as

$$\chi = \frac{4\pi}{2S} \sum_i \{|\varphi_{is}^+(0)|^2 - |\varphi_{is}^-(0)|^2\}. \qquad (17.87)$$

For the $3d$ group, i goes from 1 to 3 for ions and from 1 to 4 for atoms, and we again use a variational method. In contrast to the configuration interaction approach, the wave-function Ψ for the stretched state of the ground term ($S_z = S$, $L_z = L$, L and S maximum) is now written as a single Slater determinant constructed solely from orbitals belonging to the unperturbed configuration $\mathscr{C}_0$ given by (17.63). The difference from the conventional function Ψ_0 is in the choice of the trial one-electron functions which, in the conventional Hartree–Fock method, are given by (17.67). The trial wave-functions $P_{nl}(r)$ of (17.67) are assumed to be independent of the orientation of the spin, a restriction made for the sake of simplicity. On the other hand, we have explained that the radial functions of electrons with spin up and down are expected to be different because of their different coupling with the magnetic d-electrons. This is why it is reasonable to use trial one-electron functions where $P^+_{nl}(r)$ is not required to be equal to $P^-_{nl}(r)$. This method is called the unrestricted Hartree–Fock (UHF) method. The new ionic wave-functions Ψ form a wider set than the functions Ψ_0 of the conventional Hartree–Fock method, sometimes called restricted Hartree–Fock method (RHF) because there the restriction is imposed that the orbitals must be independent of the orientation of the spin. The functions Ψ are therefore at least as good as the functions Ψ_0; the foregoing argument suggests that they can be expected to be considerably better for the description of the spin-density near the nucleus, that is of the anomalous magnetic hyperfine structure. We shall not describe the procedure used for writing down and solving the equations obeyed by the spin-polarized orbitals in the UHF method (any more than we did for the conventional RHF method). Some details and an abundant bibliography can be found in the review articles of Freeman and Watson (1965), Watson and Freeman (1967a).

The calculated values of χ are in surprisingly good agreement with experiment considering the uncertainties involved in the calculation. First, the spin densities, that is the differences between the electronic densities $|\varphi^+_{ns}(0)|^2$ and $|\varphi^-_{ns}(0)|^2$ are, for each shell, very small fractions of the electronic densities themselves. Secondly, the contributions of different shells have opposite signs and cancel each other to a large extent. The calculation shows that the electronic densities behave as if electrons with parallel spins did attract each other, the second of our hand-waving conclusions. This is why, for the inner shells $1s$, $2s$ where the electrons with spin up are attracted away from the nucleus by the magnetic electrons $3d$, the spin density is negative, i.e. opposite to the orientation of the ionic spin. On the other hand, the spin density is positive for $3s$ electrons and, in free atoms, for $4s$ electrons also. Table 17.1 (Freeman and Watson 1965; Watson and Freeman 1967a)

TABLE 17.1

Ion	$Mn^{2+}(3d^5)$	$Fe^{2+}(3d^6)$	$Ni^{2+}(3d^8)$	$Mn^0(3d^54s^2)$
χ (atomic units)	−3·34	−3·29	−3·94	−0·54
$1s$ shell contribution to χ	−0·16	−0·21	−0·27	−0·03
$2s$ shell contribution to χ	−6·73	−7·80	−9·62	−6·63
$3s$ shell contribution to χ	+3·55	+4·72	+5·95	+3·23
$4s$ shell contribution to χ				+2·89

illustrates these statements. The ratio

$$\rho_n = \frac{|\varphi^+_{ns}|^2 - |\varphi^-_{ns}|^2}{|\varphi^+_{ns}|^2 + |\varphi^-_{ns}|^2}$$

HYPERFINE STRUCTURE 705

is always a very small number; for example, for Mn^{2+}: $\rho_1 \approx 6 \times 10^{-6}$, $\rho_2 \approx 3 \times 10^{-3}$, $\rho_3 \approx 1 \cdot 1 \times 10^{-2}$.

The foregoing discussion was centered on the hyperfine structure in the iron group. For the $4f$ group of the rare earths where the orbital momentum is largely unquenched the effects of core polarization are only a small correction and cannot be separated from orbital hyperfine structure except for ions with the half-closed shell configuration $4f^7$ (see § 5.4 for a discussion).

Comparison between core polarization and configuration interaction

The explanations provided for the origin of the anomalous magnetic hyperfine interaction by these two approaches appear to be qualitatively different. It is true that quantitatively both are variational methods that go beyond the conventional Hartree–Fock calculations by enlarging the set of the many-electron trial functions Ψ from which are chosen the solutions that minimize $\langle \Psi | \mathscr{H} | \Psi \rangle$. However the ways in which these sets are enlarged in either method appear to differ considerably. In the configuration-interaction method the one-electron orbitals, from which are constructed the many-electron Slater determinants, are the conventional functions (17.67) with a radial part independent of the orientation of the individual electronic spins, but more than one configuration $\mathscr{C}$ is required to represent adequately the ionic functions Ψ. In contrast, a single configuration is used in the core-polarization method but the one-electron orbitals from which it is built are spin-dependent as explained earlier. We now show that this distinction is more apparent than real and in the process we shall point out the main theoretical weakness of the core-polarization method, however attractive its other features. Let us write again the ionic wave-function Ψ,

$$\Psi = \Psi_0 + \alpha \Psi_{3b} + O(\alpha^2), \quad (17.88)$$

used in the configuration-interaction method, with Ψ_0 given by (17.69) and Ψ_{3b} by (17.73). Equation (17.88) can be rewritten

$$\Psi = \left(P^+ - \frac{\alpha\sqrt{(2S)}}{2\sqrt{(S+1)}} R^+, P^- + \frac{\alpha\sqrt{(2S)}}{2\sqrt{(S+1)}} R^- \right) Q^S +$$
$$+ O(\alpha^2) - \frac{\alpha}{\sqrt{(S+1)}} P^+ R^+ Q^{S-1}. \quad (17.89)$$

If we disregard the last term of (17.89), which is there to make Ψ an eigenstate of S, we see that the wave-function Ψ is nothing but a spin-polarized UHF function where the two radial trial functions with spin

46

up and spin down used in the core-polarization method are, respectively,

$$P - \frac{\alpha\sqrt{(2S)}}{2\sqrt{(S+1)}}R \text{ and } P + \frac{\alpha\sqrt{(2S)}}{2\sqrt{(S+1)}}R. \tag{17.90}$$

It remains to explain the last term of (17.89). Its absence in the core polarization method is the main weakness of the method, in that the Slater determinants it uses are no longer eigenstates of S. In principle this is a serious matter, since the true ionic state *is* an eigenstate of S, and the errors committed through disregard of this fact are very difficult to evaluate theoretically (see, for a discussion and references, Freeman and Watson 1965; Watson and Freeman 1967a; Moser 1967). The justification for the neglect of a proper symmetrization of the wave-function Ψ is twofold: first, beyond the lightest atoms the calculations become unmanageable; secondly, even in its imperfect form the agreement of the UHF method with experiment is satisfactory in most cases.

To conclude this discussion one can say that the origin of the anomalous magnetic hyperfine structure in the iron group is well understood qualitatively. It is due to the existence of a finite spin-density at the nucleus for which the method of core polarization provides both a satisfactory physical model and calculated values in reasonable agreement with experiment.

17.7. Finer effects in the theory of hyperfine structure

Since the discovery of Endor (see Chapter 4) the accuracy in the measurements of the various parameters of hyperfine structure has increased considerably. Many small effects that hitherto could be observed only by the very precise method of atomic beams have become accessible to paramagnetic resonance measurements. Some of these effects are described below.

Hyperfine structure anomaly

This effect (not to be confused with the anomalous hyperfine structure discussed in the last section) is related to the finite size of the atomic nucleus. When an s-electron penetrates inside the nucleus the magnetic properties of the latter are no longer those of a point-like magnetic moment. Instead the electron probes the detailed distribution of the magnetization inside the nucleus which, like the electronic magnetization of an atom, has a part due to the magnetic spin moments of the individual nucleons and a part related to their orbital motion.

For two different isotopes the shapes of these distributions may be quite different and the ratio of their contributions to the magnetic hyperfine structure has no reason to be equal to the ratio of the nuclear magnetic moments. Another effect related to the former, but not identical with it, is the fact that although two isotopes have the same charge Ze, it may be distributed differently inside the two nuclei. This modifies the electronic wave-function inside the nucleus, which is another reason why the ratio of the hyperfine structures may differ from that of the nuclear moments.

Let $A_S \mathbf{I} \cdot \mathbf{S}$ and $A'_S \mathbf{I}' \cdot \mathbf{S}$ be the scalar parts of the hyperfine interactions for two isotopes with spins I and I' and magnetic moments $\mu_I = \gamma_n \hbar \mathbf{I}$, $\mu'_I = \gamma'_n \hbar \mathbf{I}'$. The hyperfine anomaly can be defined as the parameter Δ in the following relation

$$\frac{A_S}{A'_S} = \frac{\gamma_n}{\gamma'_n}(1+\Delta). \tag{17.91}$$

The ratios A_S/A'_S and γ_n/γ'_n can be obtained in principle from an analysis of Endor measurements (or for γ_n/γ'_n from NMR measurements performed on diamagnetic compounds) and Δ, extracted from (17.91), can be compared with theoretical values derived from nuclear models, thus providing a test of their validity. For light nuclei Δ is very small, of the order of a few parts in a million, but for heavy nuclei where, because of the large Z, the electron spends an appreciable part of its time inside the nucleus, Δ can exceed 1 per cent. In the analysis of Endor measurements leading to Δ one should beware of spurious effects (see, for example, § 4.8) which can simulate a hyperfine anomaly. These are the second-order magnetic hyperfine structure and the pseudo-nuclear Zeeman effects to be described in § 18.1; also, very light nuclei have a zero-point motion that modifies the value of the wave-function at the nucleus, making the ratio A_S/A'_S different from γ_n/γ'_n even though the electronic penetration inside the nucleus is negligible. The best example of this situation is provided by the nuclei of ^{6}Li and ^{7}Li in silicon where an apparent Δ of the order of 1 per cent was observed (Feher 1959).

The Sternheimer antishielding factor

We have already mentioned at the end of § 17.1 the fact that in contrast to magnetic hyperfine structure, quadrupole interactions in ions do not require the existence of an unfilled shell of d or f electrons; if a diamagnetic ion such as say F$^-$, with only closed shells, is placed in a non-cubic environment, there will be an electric-field gradient at the nucleus. This gradient can be written as

$$q = q_c(1-\gamma_\infty), \qquad (17.92)$$

where q_c is the gradient produced by the charges outside the ion and γ_∞ is a coefficient, introduced and calculated by Sternheimer, which represents the effect of the polarization of the closed shells by the external charges. It is large and negative, reaching values of the order of 100 for the heaviest ions. In paramagnetic ions the electrons belonging to the unfilled shells d and f produce at the nucleus electric field gradients proportional to $\langle r^{-3} \rangle$, which are usually larger than those given by (17.92) for diamagnetic ions even when the anti-shielding amplification is included. We must also add to the electric-field gradient at the nucleus the effect of the polarization of the closed shells of the paramagnetic ions by the electrons of the unfilled shell. This problem has also been considered by Sternheimer and the polarization is taken into account by writing

$$\langle r_q^{-3} \rangle = (1-R_q)\langle r^{-3} \rangle. \qquad (17.93)$$

The remarkable fact is that in contrast to $|\gamma_\infty|$, which is large, $|R_q|$ is a small number of the order 0·1–0·2.

Without going into details of the heavy calculations of γ_∞ and R_q it is interesting to explain qualitatively the disparity between their magnitudes. Let us consider a spherical electronic shell polarized by a point charge e with polar coordinates R and Θ. In the absence of this charge the ground state $|0\rangle$ of the electronic shell is spherically symmetrical and the expectation value of the quadrupole gradient $\langle V_{zz} \rangle = e \langle 0| P_2(\cos \theta)/r^3 |0\rangle$ vanishes. The point charge itself produces at the nucleus a gradient $q_c = e(P_2(\cos \Theta)/R^3)$. Its electrostatic interaction with the closed shell is $e^2/|\mathbf{r}-\mathbf{R}|$, which admixes excited states $|n\rangle$ into the ground state $|0\rangle$ so that the expectation value of $P_2(\cos \theta)/r^3$ has now a finite value that can be written symbolically

$$\frac{e}{\Delta_{on}} \langle 0| \frac{P_2(\cos \theta)}{r^3} |n\rangle\langle n| \frac{e^2}{|\mathbf{r}-\mathbf{R}|} |0\rangle. \qquad (17.94)$$

Let us assume first that the charge density of the closed shell vanishes for $r \geqslant R$ (external polarizing charge). A typical term in the expansion of $e^2/|\mathbf{r}-\mathbf{R}|$ will be

$$e^2 P_2(\cos \theta) P_2(\cos \Theta) r^2/R^3, \qquad (17.95)$$

HYPERFINE STRUCTURE 709

and (17.94) can be rewritten

$$\frac{1}{\Delta_{on}} \langle O| \frac{P_2(\cos \theta)}{r^3} |n\rangle\langle n| e^2 r^2 P_2(\cos \theta) |O\rangle \left\{\frac{eP_2(\cos \Theta)}{R^3}\right\} \quad (17.96)$$

where the expression multiplying $eP_2(\cos \Theta)/R^3$ appears as an anti-shielding factor.

Suppose instead that the charge density of the closed shell vanishes for $r \leqslant R$ (internal polarizing charge). In the expansion of $e^2/|\mathbf{r}-\mathbf{R}|$ we must now replace (17.95) by $e^2 P_2(\cos \theta) P_2(\cos \Theta) R^2/r^3$ and (17.96) is replaced by

$$\frac{1}{\Delta_{on}} \langle O| \frac{P_2(\cos \theta)}{r^3} |n\rangle\langle n| e^2 r^2 P_2(\cos \theta) \left(\frac{R}{r}\right)^5 |O\rangle \left\{\frac{eP_2(\cos \Theta)}{R^3}\right\}. \quad (17.97)$$

A comparison of (17.96) and (17.97) shows that because of the factor $(R/r)^5$ in (17.97), where $R/r < 1$, the anti-shielding factor is very much larger when the polarizing charge is outside the closed shell than when it is inside. In applying these arguments to the electrons of the unfilled shells d or f which play the role of the polarizing charges, we have the problem that they are neither inside nor outside the closed shells but are spread all over them. However the main contribution to their quadrupole interactions comes from regions near the nucleus, where they play the part of inside charges with respect to the closed shells and thus should have a small anti-shielding factor. The foregoing argument (Watson and Freeman 1967a) is admittedly very crude but perhaps sufficient to give an inkling as to the causes of the smallness of $|R_q|$ relative to $|\gamma_\infty|$.

The different values of $\langle r^{-3} \rangle$

Equation (17.61), which is the operator expression of the hyperfine structure for the manifold of states belonging to a term (L, S), contains (besides the universal constants β and $\hbar$) three types of parameters; first, the purely nuclear constants γ_n and Q; secondly, what one might call angular parameters, namely ξ and $\langle L\| \alpha \|L\rangle$, which are determined by the algebra of vector coupling of angular momenta; and thirdly, the so-called radial constants, $\langle r^{-3} \rangle$ and $\kappa \langle r^{-3} \rangle$, which have to be computed numerically from a detailed knowledge of the electronic wave-functions. We have explained in some detail in §§ 17.5 and 17.6 the origin of the scalar interaction proportional to $\kappa \langle r^{-3} \rangle$. Whatever the language used to express it, core polarization or configuration interaction, this interaction is due to the presence at the nucleus of a finite density of unpaired s-electrons and it is a large effect, responsible in particular for the bulk of magnetic hyperfine structure in S-state ions. Apart from this 'anomalous' scalar interaction, the remainder of (17.61) or what we might call its 'normal' part, results from the following assumptions: the term (L, S) is constructed from a single configuration $\mathscr{C}$, which contains a single unfilled shell; the one-electron orbitals are of the restricted Hartree–Fock type given by (17.67); the values of L and S are such that only one (L, S) term can be constructed from the configuration $\mathscr{C}$. The last condition is always fulfilled if the term (L, S) obeys Hund's rule; in that case ξ and $\langle L\| \alpha \|L\rangle$ are given by (17.46). A direct consequence of these assumptions is that a single 'radial' parameter $\langle r^{-3}\rangle$ multiplies all the 'normal' terms of (17.61). We have already shown in the preceding section that this assumption is invalid for the quadrupole interaction, because of the polarization of the inner shells by the electrons of the unfilled shell.

On the other hand it can be shown, using rotational invariance arguments, that (17.61) is the most general expression that can be written *a priori* for magnetic dipole and electric quadrupole hyperfine structure within a term (L, S) if the constants ξ and $\langle L\| \alpha \|L\rangle$ can be considered as adjustable parameters rather than well-defined algebraic functions such as those given by (17.46). This is a direct consequence of the fact that the hyperfine interaction is a sum of one-electron operators and that individual electrons each have a spin $\frac{1}{2}$ (Sandars and Beck 1965). However the common practice is to retain for ξ and $\langle L\| \alpha \|L\rangle$ their calculated values and to assume that for the three terms of (17.61), proportional respectively to $\langle r^{-3}\rangle$, $\xi\langle r^{-3}\rangle$, and $\langle L\| \alpha \|L\rangle\langle r^{-3}\rangle$, it is the constant $\langle r^{-3}\rangle$, which can have three different values, that we write as $\langle r_l^{-3}\rangle$, $\langle r_{sc}^{-3}\rangle$, and $\langle r_q^{-3}\rangle$. We had already done so for the quadrupole interactions in eqn (17.93). We can even extend it to the 'anomalous' scalar term by taking $-\kappa = 1$ and introducing a fourth value $\langle r_s^{-3}\rangle$. We can then rewrite the magnetic hyperfine structure as

$$\frac{W_n^m}{\beta\gamma_n\hbar} = 2\langle r_l^{-3}\rangle(\mathbf{L}\cdot\mathbf{I}) + g_s\langle r_s^{-3}\rangle(\mathbf{S}\cdot\mathbf{I}) +$$
$$+ g_s\xi\langle r_{sc}^{-3}\rangle\{L(L+I)(\mathbf{S}\cdot\mathbf{I}) - \tfrac{3}{2}(\mathbf{L}\cdot\mathbf{S})(\mathbf{L}\cdot\mathbf{I}) - \tfrac{3}{2}(\mathbf{L}\cdot\mathbf{I})(\mathbf{L}\cdot\mathbf{S})\}, \quad (17.98)$$

where we have used for the spin gyromagnetic factor the correct value g_s rather than the approximate value 2.

In contrast to the comparatively large effects connected with the 'anomalous' scalar coupling the difference between $\langle r_l^{-3}\rangle$ and $\langle r_{sc}^{-3}\rangle$ is rather small. Experimentally a difference of 10 per cent has been observed in oxygen and fluorine atoms (Harvey 1965) and 1 per cent for Sm (Woodgate 1966). There is no evidence so far for such a difference from paramagnetic resonance experiments. The origin of this difference must again be sought in the insufficient accuracy of restricted Hartree–Fock functions. The arguments we used in § 17.6 to convince ourselves that radial functions with spin up and spin down should be slightly different is by no means restricted to s-electrons but applies equally well to all the electrons of the ion whether they belong to closed shells or to the unfilled shell, provided there *is* an unfilled shell. Furthermore there is no reason to restrict the one-electron orbits with different values of the magnetic quantum number m_l to have the same radial part: as long as there is an imbalance between the values of m_l in the unfilled shell, the variational equations to be satisfied by two trial radial functions P_{nlm_l} and $P_{nlm_{l'}}$ will be different, and in particular P_{nlm_l} will differ from $P_{nl,-m_l}$. It is clear that under these conditions closed shells will bring a finite contribution to the orbital and to the dipolar hyperfine couplings and that the corresponding values of $\langle r_l^{-3} \rangle$ and $\langle r_{sc}^{-3} \rangle$ have no reason to coincide. The same will naturally also be true for the contributions of the unfilled shell.

For quadrupole interactions the fact that $P_{nlm_l} \neq P_{nl,-m_l}$ will lead to a finite quadrupole interaction for S-state ions with half-filled shells such as $3d^5$ or $4f^7$ since it is proportional to

$$\sum_{m_l=-l}^{l} \{3m_l^2 - l(l+1)\} \langle r_q^{-3} \rangle_{m_l}. \tag{17.99}$$

Such a small but finite quadrupole interaction has been observed in paramagnetic resonance for the $\text{Eu}^{2+}(4f^7)$ ion in cubic symmetry in CaF_2 (Baker and Williams 1962) and also in the free atoms Mn and Eu (Evans, Sandars, and Woodgate 1965).

In the foregoing we had implicitly assumed that the non-relativistic expression (17.30) for the one-electron Hamiltonian was correct. However, as Sandars and Beck (1965) have shown, the magnetic part of the hyperfine interaction can still be represented correctly by the expression (17.98) when relativistic corrections, neglected in the derivation of (17.30), are taken into account. It is therefore very difficult to separate relativistic corrections from those due to the use of unrestricted Hartree–Fock functions (UHF). Nevertheless this can still be achieved under favorable circumstances for the scalar term $g_s \langle r_s^{-3} \rangle (\mathbf{S} \cdot \mathbf{I})$ of (17.98). If this term originates in core polarization it implies an unpaired spin-density at the nucleus and thus makes possible the existence of a hyperfine structure anomaly observable by comparing the spectra of two isotopes. A scalar term of relativistic origin is not necessarily associated with a finite spin-density at the nucleus and in this case does not lead to a hyperfine structure anomaly. The hyperfine structure measurements on the europium atom ($4f^7$, ^8S) show a small isotropic magnetic hyperfine structure but no hyperfine anomaly for the isotopes (151,153). Apart from contributions arising from the breakdown of LS-coupling, the hyperfine structure is therefore attributed mainly to relativistic effects (Evans, Sandars, and Woodgate 1965) (see § 4.7).

18

IONS IN A WEAK
CRYSTAL FIELD (f ELECTRONS)

WE are now in a position to apply the mathematical techniques described in the previous chapters to the determination of the wave-functions and energy intervals of the low-lying levels of paramagnetic ions embedded in a crystal field. In this chapter we shall examine the weak-field case that applies to the $4f$ shell of the rare earths, delaying the intermediate field case applicable to the $3d$ shell of the iron group until Chapter 19.

18.1. Kramers ions in a weak crystal field

For experimental reasons the observability of paramagnetic resonance in an ion is dependent on the existence of at least two energy levels separated by a frequency gap $\Delta E/h$ which, using the technique of microwave spectrometers rather than infra-red spectrometers, rarely exceeds, say, five wave numbers (2 mm wavelength).

This condition singles out the so-called Kramers ions with an odd number of electrons which, in the absence of a magnetic field, will always possess at least a pair of degenerate levels. These levels, through the application of a magnetic field, can be separated by the desired amount. Kramers ions also have other features such as relatively long relaxation times and relatively narrow lines that make them particularly suitable for magnetic resonance work. Among them S-state ions, which are much less sensitive to the crystal field than any other ions, also occupy a special position.

If we exclude the cubic multiplet Γ_8 which requires a special treatment, the description of the resonance properties of all Kramers ions in the rare-earth group follows a uniform pattern. The ground multiplet (J, L, S) of the free ion will be split by the crystal field into $J+\frac{1}{2}$ doublets, and resonance, with a few exceptions, will in general be observed only between the two substates of the lowest doublet, split by the magnetic field. To calculate this splitting, that is the resonance frequency, we must know the wave-functions that span the ground level. This knowledge will enable us to calculate the magnetic 'tensor' $g_{p\alpha}$ and, for an ion with a nuclear spin, the hyperfine 'tensor' $a_{p\alpha}$ and the nuclear electric quadrupole coupling constant.

In the course of the determination of the wave-functions of the ground doublet the positions and the wave-functions of the excited doublets will also become known. These data are of interest, first because they can be confronted with the results of optical spectroscopy and susceptibility measurements, secondly because, for temperatures that are not too low, resonance can be observed between substates of some excited doublets as well, and thirdly because, as discussed in Chapter 10, down to the lowest temperatures obtainable, the relaxation properties of the ground doublet are determined by the positions and the nature of the excited levels.

The calculation proceeds in a series of steps:

(1) In first approximation one assumes that J is a good quantum number. From the assumed symmetry of the environment an expansion of the type (16.1) or (16.10) is written for the crystal potential, and the P_k^q are then replaced by the equivalent operators O_k^q. Secular equations are set up and solved for the wave-functions of the various doublets. In much paramagnetic work, the coefficients $A_k^q \langle r^k \rangle$ in the expansion (16.10) are largely unknown and are adjusted to give the best fit with the various experimental quantities listed above.

(2) The first-order approximation will often be inadequate and the crystal potential will drag in admixtures from excited multiplets. These can be handled mathematically by means of the off-diagonal matrix elements $(J+1\| \alpha, \beta, \gamma \|J)$ listed in Table 20. It may sometimes (e.g with one electron or one hole) be simpler to go back to the (L, S) representation.

(3) Further theoretical refinements can be introduced if warranted by the accuracy of the experimental data. For example, the spin-orbit coupling $\sum \zeta_i(\mathbf{l}_i \cdot \mathbf{s}_i)$ results in a small departure from LS-coupling, represented by an admixture into the ground state (J, L, S) of contributions from excited states (J, L', S'). Thus for the free ion we have a state of the form

$$(1+\alpha'^2+\alpha''^2+\ldots)^{-\frac{1}{2}}(|J, L, S\rangle+\alpha' |J, L', S'\rangle+\alpha'' |J, L'', S''\rangle+\ldots).$$
(18.1)

Here all terms have the same value of J, and otherwise follow the selection rules $\Delta L = 0, \pm 1$; $\Delta S = 0, \pm 1$; but $\Delta L = \Delta S = 0$ is excluded because the ground term, which obeys Hund's rules, is unique.

For the magnetic moment $\boldsymbol{\mu} = -\beta(\mathbf{L}+2\mathbf{S})$ of the ion the matrix elements $\langle LS, \ldots | \boldsymbol{\mu} | L'S', \ldots \rangle$ vanish unless $L = L'$ and $S = S'$, and it therefore follows that changes in the magnetic moment due to the admixtures will be of order α^2.

For the magnetic hyperfine operator (17.30), summed over all the electrons, the orbital term obeys the selection rule $\Delta L = 0$, $\Delta S = 0$ and therefore has no matrix elements linear in α. However the spin dipolar term, which obeys the rule $\Delta L = 0, \pm 1, \pm 2; \Delta S = 0, \pm 1$ may well have matrix elements between $|J, L, S\rangle$ and $|J, L', S'\rangle$ leading to a first-order correction due to intermediate coupling.

Fortunately for electron paramagnetic resonance experiments in solids, intermediate coupling plays a less important role in the ground (L, S) multiplet than in excited multiplets. Nevertheless, its effects are quite appreciable in the $5f$ group (see Chapter 6).

For ions with a large hyperfine structure or low-lying excited states or both, other second-order effects may be observable. Cross terms between the electron Zeeman coupling and the hyperfine coupling give a term, linear in the nuclear spin and in the applied field, which adds to the nuclear Zeeman coupling (we shall call it the pseudo-nuclear Zeeman coupling); and in second order the magnetic hyperfine coupling can give a pseudo-quadrupole interaction.

The pseudo-nuclear Zeeman coupling is given by the formula (where c.c. stands for the complex conjugate expression)

$$\sum_n{}' \frac{\langle 0| -\mathbf{\mu} \cdot \mathbf{H} |n\rangle\langle n| -\gamma_n \hbar \mathbf{H}^e \cdot \mathbf{I} |0\rangle}{W_0 - W_n} + \text{c.c.}, \quad (18.2)$$

where $-\gamma_n \hbar \mathbf{H}^e \cdot \mathbf{I}$ is the operator for the magnetic hyperfine structure. $|0\rangle$ represents the ground doublet where the two substates are left unspecified. (18.2) can be written as

$$-\gamma_n \hbar \mathbf{H} \cdot \mathbf{\alpha} \cdot \mathbf{I} \quad (18.3)$$

where $\mathbf{\alpha}$ is a tensor with components

$$\alpha_{pq} = -\sum_n{}' \frac{\langle 0| \mu_p |n\rangle\langle n| H_q^e |0\rangle + \langle 0| H_q^e |n\rangle\langle n| \mu_p |0\rangle}{W_0 - W_n}. \quad (18.4)$$

The restricted summation in (18.4) is over all the excited doublets where the two Kramers-conjugate substates of each excited doublet give the same contribution. (18.4) can be written as the expectation value $\langle 0| \mathscr{A}_{pq} |0\rangle$ in the ground doublet of the operator

$$\mathscr{A}_{pq} = -\left\{\mu_p \left(\sum_n{}' \frac{|n\rangle\langle n|}{W_0 - W_n}\right) H_q^e + H_q^e \left(\sum_n{}' \frac{|n\rangle\langle n|}{W_0 - W_n}\right) \mu_p\right\}. \quad (18.5)$$

Here the Hermitian operator $\mathscr{A}_{pq}$ is the symmetrized product of three operators: μ_p, $C = \sum_n' \{|n\rangle\langle n|/(W_0 - W_n)\}$, H_q^e. The first and the last are time-odd, the middle one time-even. The symmetrized product $\mathscr{A}_{pq}$ is time-even and therefore has the same expectation value for the two substates of the ground doublet $|O\rangle$. We can therefore treat α_{pq} as a constant. We shall see later that the situation is different if the ground state is a quadruplet Γ_8.

If J is a good quantum number we may write $\mathbf{\mu}/\beta = -\Lambda \mathbf{J}$, $\mathbf{H}^e/\beta = -2N\mathbf{J}\langle r^{-3}\rangle$ and (18.4) becomes

$$\alpha_{pq} = -2\beta^2 \langle r^{-3}\rangle \Lambda N \sum_n' \frac{\langle O| J_p |n\rangle\langle n| J_q |O\rangle + \langle O| J_q |n\rangle\langle n| J_p |O\rangle}{W_0 - W_n},$$
(18.6)

where for brevity we have written Λ for $(J\| \Lambda \|J)$ and N for $(J\| N \|J)$.

The relative importance of the pseudo-nuclear Zeeman coupling is given by the dimensionless expressions (18.4) or (18.6) for α_{pq}. Roughly speaking, its magnitude is of the order of a Zeeman electron energy βH^e divided by the splitting between two Kramers doublets, arising from the crystal field. Electronic hyperfine fields H^e of the order of a million gauss are not uncommon in the rare-earth group and βH^e can be as large as a 100 cm^{-1}, which is comparable to the crystal field splittings. In many cases the pseudo-nuclear Zeeman coupling will be as large or even larger than the direct nuclear Zeeman coupling. One may notice that the tensor $\boldsymbol{\alpha}$ is not fundamentally different from the chemical shift tensor $\boldsymbol{\sigma}$, well known in nuclear resonance. The difference in order of magnitude between $\boldsymbol{\alpha}$ and $\boldsymbol{\sigma}$ is due to the proximity of excited levels in the rare-earth group (10 to 100 cm^{-1} as contrasted with, say, 10 eV or 10^5 cm^{-1} for an excited electronic state of a molecule). Also, compared with nuclear magnetic resonance measurements on light atoms we have a much larger value of $\langle r^{-3}\rangle$ due to the high atomic numbers Z in the rare-earth group.

The second-order term from the magnetic hyperfine structure can be written

$$\gamma_n^2 \hbar^2 \sum_n' \frac{\langle O| \mathbf{H}^e \cdot \mathbf{I} |n\rangle\langle n| \mathbf{H}^e \cdot \mathbf{I} |O\rangle}{W_0 - W_n} = \gamma_n^2 \hbar^2 \sum_{p,q} \langle O| H_p^e C H_q^e |O\rangle I_p I_q. \quad (18.7)$$

Some caution should be exercised here in using the concept of time-even operators, as pointed out in § 15.4: the operator $H_p^e C H_q^e$ with $p \neq q$ is not time-even, for its transform by time reversal, $\theta H_p^e C H_q^e \theta^{-1}$

IONS IN A WEAK CRYSTAL FIELD (f ELECTRONS)

is equal to itself, $H^e_p CH^e_q$, rather than to its Hermitian conjugate $(H^e_p CH^e_q)^+ = H^e_q CH^e_p$. Eqn (18.7) can be rewritten as

$$\tfrac{1}{2}\gamma_n^2 \hbar^2 \Big\{ \tfrac{1}{2} \sum_{p,q} \langle 0| H^e_p CH^e_q + H^e_q CH^e_p |0\rangle (I_p I_q + I_q I_p) +$$
$$+ \tfrac{1}{2} \sum_{p,q} \langle 0| H^e_p CH^e_q - H^e_q CH^e_p |0\rangle (I_p I_q - I_q I_p) \Big\}. \quad (18.8)$$

Here $\tfrac{1}{2}\gamma_n^2\hbar^2(H^e_p CH^e_q + H^e_q CH^e_p)$ is a Hermitian time-even operator and therefore within the manifold of a Kramers doublet, a component b_{pq} of a symmetric tensor, which is a real number. On the other hand $\tfrac{1}{2}\gamma_n^2\hbar^2\{H^e_p CH^e_q - H^e_q CH^e_p\}$ is a time-odd anti-Hermitian operator, which within the same manifold can be written $i\sum_\alpha d^{pq}_\alpha \sigma_\alpha$ where $d^{pq}_\alpha = -d^{qp}_\alpha$ is real. The commutator $(I_p I_q - I_q I_p)$ can be written $i\sum_r \varepsilon_{pqr} I_r$, where ε_{pqr} is the usual antisymmetrization symbol and (18.8) can be rewritten

$$\sum_{p,q} b_{pq}(I_p I_q + I_q I_p) - \sum_{r,\alpha} u_{r\alpha} I_r \sigma_\alpha, \quad (18.9)$$

where $u_{r\alpha} = \sum_{p,q} \varepsilon_{pqr} d^{pq}_\alpha$, or

$$u_{1,\alpha} = d^{23}_\alpha, \quad u_{2,\alpha} = d^{31}_\alpha, \quad u_{3,\alpha} = d^{12}_\alpha.$$

The first term in (18.9) is the pseudo-quadrupole hyperfine interaction and the second a contribution that adds to the magnetic hyperfine interaction. Both are smaller than the first-order magnetic hyperfine interaction by a factor of the order $|\hbar\gamma_n H^e|/\Delta W$, a ratio of the order of 10^{-3}. The pseudo-quadrupole interaction may be a very much larger fraction of the true quadrupole interaction, but since an accurate theoretical estimate of the latter is very difficult one might well question the interest of calculating these terms. A point to remember in calculating the second term of (18.9) is that it makes the ratio of the hyperfine structure coefficients for two isotopes different from that of their γ_n. This effect may be a nuisance in so far as it interferes with the measurement of the hfs anomaly, due to the penetration of the electron inside the nucleus.

The experimental results for ions of the rare-earth group have been discussed in Chapter 5, together with theoretical interpretations that embody most of the features outlined above. The extra hyperfine terms given above are of course most important in Endor measurements, some specific examples being given in Chapter 4. (Further small terms are mentioned in § 5.10.) It is thus unnecessary to give here an extensive discussion of the application of the theory to Kramers ions. We may however remark that for such ions in crystal fields possessing an axis of threefold or sixfold symmetry, there may not be an allowed resonance transition within the ground Kramers doublet. The selection rule $\Delta J_z = \pm 1$ requires that the wave-function for one member of the doublet must contain at least one value of $J_z = M$ that differs by one unit from one of the values in the conjugate wave-function for the other member of the doublet. This produces the following result.

If there is an axis of q-fold symmetry, the doublet wave-functions will be generally of the form

$$a_M |J, M\rangle + a_{M \pm q} |J, M \pm q\rangle + a_{M \pm 2q} |J, M \pm 2q\rangle + \ldots$$
$$a'_M |J, -M\rangle + a'_{M \pm q} |J, -(M \pm q)\rangle + a'_{M \pm 2q} |J, -(M \pm 2q)\rangle + \ldots \quad (18.10)$$

For resonance to be observable we need $-(M \pm nq) = M \pm 1$, or

$$|nq| = 2M \pm 1 \quad (18.11)$$

where $n = 0, 1, 2, \ldots$. For sixfold symmetry ($q = 6$) this means that resonance is observable in the doublets spanned by

$$M = \pm \tfrac{1}{2};\ (\pm \tfrac{7}{2}, \mp \tfrac{5}{2});\ (\pm \tfrac{13}{2}, \pm \tfrac{1}{2}, \mp \tfrac{11}{2})$$

but not in doublets spanned by

$$M = \pm \tfrac{3}{2};\ (\pm \tfrac{9}{2}, \mp \tfrac{3}{2});\ (\pm \tfrac{15}{2}, \pm \tfrac{3}{2}, \mp \tfrac{9}{2}).$$

For threefold symmetry the wave-functions are more complicated in that they contain all values of M differing by 3 instead of just by 6, but the doublets in which resonance is allowed are otherwise unaltered. For fourfold symmetry ($q = 4$) the position is different; all possible doublets are of the form

$$(\pm \tfrac{15}{2},\ \pm \tfrac{7}{2},\ \mp \tfrac{1}{2},\ \mp \tfrac{9}{2})$$

or

$$(\pm \tfrac{13}{2},\ \pm \tfrac{5}{2},\ \mp \tfrac{3}{2},\ \mp \tfrac{11}{2}),$$

where the higher values of M do not of course occur for lower values of J. Inspection shows that resonance will be allowed in all such doublets for values of $J \geqslant \tfrac{5}{2}$, the lowest half-integral value to occur for the ground state of a $4f$ or $5f$ ion. For twofold symmetry the rules are clearly the same (with the addition that resonance is also possible in either doublet of $J = \tfrac{3}{2}$). We see therefore that we should always expect a resonance in the ground state of a Kramers rare-earth ion in a substance such as $CaWO_4$, even if the fourfold symmetry of the Ca^{2+} site is not reduced by local charge compensation when a tripositive rare-earth ion replaces it. On the other hand resonance cannot be observed, for example, in the ground doublet $\pm\frac{3}{2}$ of $4f^{13}$, Yb^{3+}, $^2F_{\frac{7}{2}}$ in lanthanide ethylsulphate or trichloride, for which $q = 6$.

18.2. Rare-earth ions in cubic symmetry

Following a suggestion by Bleaney (1959a,b), that rare-earth Kramers ions in cubic environment should provide suitable maser materials if their ground states were the quadruplet Γ_8, there have been many studies both theoretical and experimental of the properties of rare-earth ions in cubic surroundings. Trivalent rare-earth ions M can be introduced substitutionally to replace divalent positive ions in such cubic crystals as, say, CaO, CaF_2, MgO. Charge compensation will then demand the presence of a negative charge in the neighbourhood of M^{3+}. If the compensating charge is, with respect to M, in a direction $\langle 100 \rangle$ the symmetry of the distorted crystal field 'seen' by M will be tetragonal; it will be trigonal for a direction $\langle 111 \rangle$; for any other relative orientation of the compensating charge the symmetry will be lower. The compensation may also be long range, the negative charge being sufficiently removed from M to leave it in a symmetry indistinguishable from cubic. It is this situation that we propose to examine in this section. Often the cubic resonance spectrum will be observed simultaneously with spectra of lower symmetry, coming from ions M, less fortunate in their charge-compensating environment.

It has also proved possible to create in these environments divalent states of rare-earth ions. If in their normal trivalent state these ions, such as holmium or thulium, have an even number of electrons they will now in the divalent state be Kramers ions. No charge compensation is needed in this case and the symmetry is cubic.

Theoretically the problem of a cubic crystal potential is attractive because for f-electrons this potential depends on two constants only as in eqn (16.13),

$$V = A_4 P_4 + A_6 P_6,$$

the polynomials P_4 and P_6 being defined in (16.11) and (16.12). The wave-functions of the various crystal levels and the ratios of the intervals between the different levels depend on a single constant, the ratio of the fourth to sixth order potential; this makes it possible to tabulate them for all the values of J occurring in the ions of the rare-earth group at least in the framework of first-order theory, where J is taken to be a good quantum number. Such tables (see Lea, Leask, and Wolf (1962)) are organized as follows.

For each value of J it is convenient for practical reasons to write the cubic potential as

$$\mathscr{H} = B_4 F(4)\frac{O_4}{F(4)} + B_6 F(6)\frac{O_6}{F(6)}. \quad (18.12)$$

Here O_4 and O_6 are the equivalent operators

$$O_4 = (O_4^0 + 5O_4^4); \quad O_6 = (O_6^0 - 21 O_6^4), \quad (18.13)$$

where the O_k^q are defined and tabulated in Table 16 and

$$B_4 = (J\| \beta \|J)\langle r^4 \rangle A_4; \quad B_6 = (J\| \gamma \|J)\langle r^6 \rangle A_6.$$

The values of $(J\| \beta \|J)$, $(J\| \gamma \|J)$ are tabulated in Table 20. The quantities $F(4)$ and $F(6)$ are certain positive factors common to all the matrix elements of the O_k^q for a given k separated out to keep all the eigenvalues in the same numerical range. (In Lea, Leask and Wolf (1962) they are not always the same as the values of F given in the Tables 17.)

Two constants W and x are then defined by

$$B_4 F(4) = Wx; \quad B_6 F(6), = W(1-|x|); \quad (18.14\text{a})$$

or

$$\frac{B_4}{B_6} = \frac{x}{1-|x|}\frac{F(6)}{F(4)}. \quad (18.14\text{b})$$

With these definitions, $B_4/B_6 = 0$ for $x = 0$ and $\pm\infty$ for $x = \pm 1$ so that all possible ratios of B_4/B_6 are included in the range $-1 \leqslant x \leqslant 1$. The Hamiltonian is then rewritten as

$$\mathscr{H} = W\left\{x\frac{O_4}{F(4)} + (1-|x|)\frac{O_6}{F(6)}\right\}. \quad (18.15)$$

For each value of x all the matrix elements of the operator in the curly brackets are known and its eigenvalues and eigenstates are then found by numerical calculations and tabulated, W being the single energy parameter.

We know from Tables 3 and 7 the decomposition of each manifold J into the irreducible representations Γ_i of the cubic group. If the representation occurs only once, the kets that span it are determined from symmetry alone and are independent of x. Their expressions are given in a closed form in Tables 4 and 9 and the eigenvalues are linear functions of x. We shall call such representations 'isolated'.

From (18.14a,b) we see that the sign of W is that of B_6 and the sign of x

IONS IN A WEAK CRYSTAL FIELD (f ELECTRONS) 721

is determined by B_4/B_6. Since the β and γ are tabulated, to know the signs of B_4 and B_6 it is necessary to know only those of A_4 and A_6. If we accept the qualitative validity of the point-charge model the latter are given in the various coordinations by eqns (16.15)–(16.17).

Resonance in Γ_6 and Γ_7

The pattern of resonance is very similar to that observed for lower symmetry since Γ_6 and Γ_7 are Kramers doublets. The only difference is that the magnetic 'tensor' g and the hyperfine 'tensor' a are isotropic and the nuclear electric quadrupole interaction vanishes.

For 'isolated' Γ_6 or Γ_7 doublets the values of $g' = g/\Lambda$ are independent of x; they are given in Table 22.

A number of experimental results are given in Tables 5.14 and 5.15. They are close to the simple theoretical values given in Table 22, but comparison with the free ions shows that even when intermediate coupling is taken into account there are small discrepancies that are presumably to be attributed to a certain amount of covalent bonding, or to effects of the zero-point vibrations of the lattice (see § 5.8), or to both.

18.3. The quadruplet Γ_8

We have been faced several times with the problem of finding the matrix elements of a vector operator such as say the magnetic moment $\boldsymbol{\mu}$ or the electronic hyperfine field $\mathbf{H}_e$, within a certain manifold. We saw in particular that thanks to the Wigner–Eckart theorem, a vector $\mathbf{V}$ could be replaced within a manifold J by $a\mathbf{J}$ where a was a constant depending on the vector $\mathbf{V}$. Similarly in cubic symmetry within a manifold Γ_4 or Γ_5 a vector could be taken proportional to a fictitious angular momentum $\tilde{\mathbf{l}}$ defined by eqn (14.5). We have emphasized the point that this representation was possible because the reduction of the direct products $\mathscr{D}^J \times \mathscr{D}^J$ in the rotation group, or of $\Gamma_4 \times \Gamma_4$ and $\Gamma_5 \times \Gamma_5$ in the cubic group, contained only once the representation of the group spanned by the components of a vector, namely $\mathscr{D}^1$ for the rotation group and Γ_4 for the cubic group.

On the other hand, we have shown in § 15.6 that within a Kramers doublet we could represent a time-odd vector $\mathbf{V}$ as $\mathbf{V} = T \cdot \mathbf{s}$, where $\mathbf{s} = \boldsymbol{\sigma}/2$ was a fictitious spin and T a so-called 'tensor'. This result had *nothing* to do with spatial symmetry but resulted solely from invariance with respect to time reversal.

We require now to find the matrix elements of a vector within a

722   IONS IN A WEAK CRYSTAL FIELD (f ELECTRONS)

quadruplet Γ_8. This quadruplet is spanned by four kets $|\tilde{m}\rangle$ which we have represented in § 14.4 as $|\pm\tilde{\tfrac{3}{2}}\rangle$ and $|\pm\tilde{\tfrac{1}{2}}\rangle$, meaning thereby that under the rotations of the cubic group they transform in the same way as the eigenstates $|J, m\rangle = |\tfrac{3}{2}, \pm\tfrac{3}{2}\rangle$ and $|\tfrac{3}{2}, \pm\tfrac{1}{2}\rangle$ of an angular momentum $J = \tfrac{3}{2}$. When the quadruplet Γ_8 results from the reduction of a given representation $\mathscr{D}^J$, the states $|\tilde{m}\rangle$ can be expanded as a linear combination $\sum_{J,m} C_{J,m} |J, m\rangle$ as shown in Table 9, but an arbitrary Γ_8 can naturally be spanned by kets with arbitrary values of J.

We can define in the manifold Γ_8 three operators $\tilde{S}_x$, $\tilde{S}_y$, $\tilde{S}_z$, components of a fictitious spin $\tilde{S} = \tfrac{3}{2}$, by the condition that they have the same matrix elements in the basis $|\tilde{m}\rangle$ as the components $\mathscr{J}_x$, $\mathscr{J}_y$, $\mathscr{J}_z$ of an angular momentum $\mathscr{J} = \tfrac{3}{2}$ have with respect to the states $|\tfrac{3}{2}, m\rangle$. $\tilde{S}_x$, $\tilde{S}_y$, $\tilde{S}_z$ have naturally the usual commutation relations $[\tilde{S}_x, \tilde{S}_y] = i\tilde{S}_z$, etc.

We see from Table 8 that the reduction of the direct product $\Gamma_8^* \times \Gamma_8$ (which is equivalent to that of $\Gamma_8 \times \Gamma_8$, since all the characters of Γ_8 are real) contains Γ_4 *twice*. It is not too difficult to show by an adaptation of the proof of the Wigner–Eckart theorem that the matrix elements of a vector within a manifold Γ_8 can be specified by two constants. The detailed argument can be found, for instance, in Koster and Statz (1959). We also see from Table 8 that it is the symmetrical direct product $[\Gamma_8 \times \Gamma_8]_S$ which contains Γ_4 twice, and therefore according to the rules given in § 15.9 only time-odd vectors have non-vanishing matrix elements inside the manifold Γ_8.

This result and also the form of the operator equivalent to a vector **V** within Γ_8 can be obtained by an elementary method (Ayant, Beloritzky, and Rosset 1962). In the representation $|\tilde{m}\rangle$ the component V_z has no off-diagonal matrix elements $\langle \tilde{m}| V_z |\tilde{m}'\rangle$, since a rotation of $\pi/2$ around the z-axis, which must leave V_z invariant, multiplies such matrix elements by $e^{\tfrac{1}{2}i\pi(\tilde{m}-\tilde{m}')}$. A rotation by π around the axis Oy changes $\langle \tilde{m}| V_z |\tilde{m}\rangle$ into

$$\langle \tilde{m}| e^{-i\pi\tilde{S}_y} V_z e^{i\pi\tilde{S}_y} |\tilde{m}\rangle = -\langle \tilde{m}| V_z |\tilde{m}\rangle$$
$$= (-1)^{2(j-\tilde{m})} \langle -\tilde{m}| V_z |-\tilde{m}\rangle = \langle -\tilde{m}| V_z |-\tilde{m}\rangle.$$

Hence we see that the non-vanishing matrix elements of V_z are of the form

$$\begin{aligned}
\langle \tilde{\tfrac{3}{2}}| V_z |\tilde{\tfrac{3}{2}}\rangle &= -\langle -\tilde{\tfrac{3}{2}}| V_z |-\tilde{\tfrac{3}{2}}\rangle = P, \\
\langle \tilde{\tfrac{1}{2}}| V_z |\tilde{\tfrac{1}{2}}\rangle &= -\langle -\tilde{\tfrac{1}{2}}| V_z |-\tilde{\tfrac{1}{2}}\rangle = Q.
\end{aligned} \qquad (18.16)$$

As a consequence, V_z can be written as

$$V_z = a\tilde{S}_z + b\tilde{S}_z^3 \qquad (18.17)$$

with

$$P = \frac{3a}{2} + \frac{27b}{8}, \qquad Q = \frac{a}{2} + \frac{b}{8};$$

or

$$a = -\frac{P}{12} + \frac{9Q}{4}, \qquad b = \frac{P}{3} - Q. \qquad (18.18)$$

It is obvious from the nature of cubic symmetry that we must also have

$$V_x = a\tilde{S}_x + b\tilde{S}_x^3,$$
$$V_y = a\tilde{S}_y + b\tilde{S}_y^3. \qquad (18.19)$$

It may be more convenient to use instead of (18.17) the form

$$V_z = a'\tilde{S}_z + b\{\tilde{S}_z^3 - \tfrac{3}{5}\tilde{S}(\tilde{S}+1)\tilde{S}_z + \tfrac{1}{5}\tilde{S}_z\} \qquad (18.20)$$

with $a' = a + \tfrac{1}{5}b\{3\tilde{S}(\tilde{S}+1) - 1\}$. This has the advantage that the curly bracket in (18.20) is the operator O_3^0 which vanishes for $\tilde{S} < \tfrac{3}{2}$. Using (18.20) we can write the Zeeman and magnetic hyperfine Hamiltonians in the form

$$g\beta\tilde{\mathbf{S}}\cdot\mathbf{H} + A\tilde{\mathbf{S}}\cdot\mathbf{I} - g_n\beta_n\mathbf{H}\cdot\mathbf{I} +$$
$$+ u\beta[\tilde{S}_x^3 H_x + \tilde{S}_y^3 H_y + \tilde{S}_z^3 H_z - \tfrac{1}{5}(\tilde{\mathbf{S}}\cdot\mathbf{H})\{3\tilde{S}(\tilde{S}+1) - 1\}] +$$
$$+ U[\tilde{S}_x^3 I_x + \tilde{S}_y^3 I_y + \tilde{S}_z^3 I_z - \tfrac{1}{5}(\tilde{\mathbf{S}}\cdot\mathbf{I})\{3\tilde{S}(\tilde{S}+1) - 1\}]. \qquad (18.21)$$

Diagonalization of the spin Hamiltonian for Γ_8 quadruplet

Complete diagonalization of the spin Hamiltonian (18.21) is obviously rather complicated and we shall confine our treatment to two special cases: (A) the remaining terms are all small compared with the first term $g\beta(\tilde{\mathbf{S}}\cdot\mathbf{H})$, which we therefore diagonalize first and then treat the remaining terms by first-order perturbation theory; (B) the nuclear spin is zero, and the complete electronic Zeeman interaction is diagonalized exactly for an arbitrary direction of $\mathbf{H}$. In the former of these two cases we are assuming that the terms of third degree involving the parameter U are also small compared with $A(\tilde{\mathbf{S}}\cdot\mathbf{I})$; this situation is likely to occur in the 3d group rather than the 4f group, except for the half-filled shell $4f^7$, ^8S, where small terms of the third and even higher degree may be required (see § 5.9).

(A) *Perturbation treatment*

Suppose that the external field $\mathbf{H}$ is directed along an axis OZ whose direction cosines are (n_1, n_2, n_3) with respect to the fourfold axes

(x, y, z) of the cube. Since g is isotropic the first term in (18.21) is diagonalized simply by changing to a coordinate system (X, Y, Z) of which OZ is the Z-axis. We introduce the component

$$\tilde{S}_Z = n_1\tilde{S}_x + n_2\tilde{S}_y + n_3\tilde{S}_z \tag{18.22}$$

and two other components $\tilde{S}_X$, $\tilde{S}_Y$ along two orthogonal axes that we need not specify. We can write

$$\tilde{S}_x = n_1\tilde{S}_Z + \alpha_1\tilde{S}_X + \beta_1\tilde{S}_Y \tag{18.23}$$

together with two similar equations for $\tilde{S}_y$, $\tilde{S}_z$.

On our assumption that all the terms except $g\beta(\tilde{\mathbf{S}} \cdot \mathbf{H}) = g\beta\tilde{S}_ZH$ are small, first-order perturbation theory requires that we calculate the expectation values $\langle \tilde{S}_x^3 \rangle$, $\langle \tilde{S}_y^3 \rangle$, $\langle \tilde{S}_z^3 \rangle$ in a state where $\tilde{S}_Z = M$. From (18.23), we get

$$\langle \tilde{S}_x^3 \rangle = n_1^3\langle \tilde{S}_Z^3 \rangle + n_1\langle (\alpha_1\tilde{S}_X + \beta_1\tilde{S}_Y)^2\tilde{S}_Z + \tilde{S}_Z(\alpha_1\tilde{S}_X + \beta_1\tilde{S}_Y)^2 +$$
$$+ (\alpha_1\tilde{S}_X + \beta_1\tilde{S}_Y)\tilde{S}_Z(\alpha_1\tilde{S}_X + \beta_1\tilde{S}_Y) \rangle \tag{18.24}$$

where we have omitted terms of odd powers in $\tilde{S}_X$ and $\tilde{S}_Y$ which clearly have a vanishing expectation value in the state $\tilde{S}_Z = M$. The first term of (18.24) is $n_1^3 M^3$ and the second and the third have the same expectation value:

$$n_1 \frac{(\alpha_1^2 + \beta_1^2)}{2} \{\tilde{S}(\tilde{S}+1) - M^2\}M = n_1 \frac{(1-n_1^2)}{2} \{\tilde{S}(\tilde{S}+1) - M^2\}M. \tag{18.25}$$

The fourth term is more complicated:

$$n_1\langle (\alpha_1\tilde{S}_X + \beta_1\tilde{S}_Y)\tilde{S}_Z(\alpha_1\tilde{S}_X + \beta_1\tilde{S}_Y)\rangle$$
$$= n_1\langle \tilde{S}_Z(\alpha_1\tilde{S}_X + \beta_1\tilde{S}_Y)^2 + [\alpha_1\tilde{S}_X + \beta_1\tilde{S}_Y, \tilde{S}_Z](\alpha_1\tilde{S}_X + \beta_1\tilde{S}_Y)\rangle$$
$$= n_1 \frac{1-n_1^2}{2} \{\tilde{S}(\tilde{S}+1) - M^2\}M + n_1\langle [\alpha_1\tilde{S}_X + \beta_1\tilde{S}_Y, \tilde{S}_Z](\alpha_1\tilde{S}_X + \beta_1\tilde{S}_Y)\rangle. \tag{18.26}$$

The last term of (18.26) can be written

$$n_1\langle (-i\alpha_1\tilde{S}_Y + i\beta_1\tilde{S}_X)(\alpha_1\tilde{S}_X + \beta_1\tilde{S}_Y)\rangle$$
$$= n_1\langle -i\alpha_1^2\tilde{S}_Y\tilde{S}_X + i\beta_1^2\tilde{S}_X\tilde{S}_Y + i\beta_1\alpha_1(\tilde{S}_X^2 - \tilde{S}_Y^2)\rangle$$
$$= n_1\left\langle i\left(\frac{\beta_1^2 + \alpha_1^2}{2}\right)[\tilde{S}_X, \tilde{S}_Y] + \right.$$
$$\left. + i\left(\frac{\beta_1^2 - \alpha_1^2}{2}\right)(\tilde{S}_X\tilde{S}_Y + \tilde{S}_Y\tilde{S}_X) + i\beta_1\alpha_1(\tilde{S}_X^2 - \tilde{S}_Y^2)\right\rangle$$
$$= n_1\left\langle -\left(\frac{\beta_1^2 + \alpha_1^2}{2}\right)\tilde{S}_Z + \frac{\beta_1^2 - \alpha_1^2}{4}(\tilde{S}_+^2 - \tilde{S}_-^2) + i\frac{\beta_1\alpha_1}{2}(\tilde{S}_+^2 - \tilde{S}_-^2)\right\rangle. \tag{18.27}$$

IONS IN A WEAK CRYSTAL FIELD (f ELECTRONS)   725

On summing over the similar terms for $\langle \tilde{S}_y^3 \rangle$, $\langle \tilde{S}_z^3 \rangle$ we find that in the approximation in which we are working the eigenvalues of (18.21) are given by

$$W_{M,m} = g\beta HM + AMm - g_n\beta_n Hm + (u\beta H + Um) \times$$
$$\times \{M^3 - \tfrac{1}{5}(3S(S+1) - 1)M\}\{1 - 5(n_1^2 n_2^2 + n_2^2 n_3^2 + n_3^2 n_1^2)\}, \quad (18.28)$$

where the angular variation of the last term is just that of the quantity $p = 1 - 5(n_1^2 n_2^2 + n_2^2 n_3^2 + n_3^2 n_1^2) = \tfrac{5}{2}(n_1^4 + n_2^4 + n_3^4 - \tfrac{3}{5})$ (cf. § 3.4).

(B) *Exact diagonalization of the electronic Zeeman interaction*

This involves finding the eigenvalues of the Zeeman Hamiltonian $Z = -\boldsymbol{\mu} \cdot \mathbf{H}$ in the general case when the cubic terms are not small (Ayant, Beloritzky, and Rosset 1962).

In order to deal with dimensionless quantities we take for $V_z = -\mu_z/\beta$ the matrix form (18.16). Then, taking for $V_x = -\mu_x/\beta$, $V_y = -\mu_y/\beta$ the expressions (18.19), where a and b are related to P and Q by (18.18), and calculating from the known matrices for $\tilde{S}_x$ and $\tilde{S}_y$ those for $\tilde{S}_x^3$ and $\tilde{S}_y^3$, we find the matrix expressions for V_x and V_y. In fact it is sufficient to give the non-vanishing matrix elements of $V_+ = V_x + iV_y$ that determine those of V_x and V_y:

$$(\tfrac{\tilde{3}}{2}|\,V_+\,|\tfrac{\tilde{1}}{2}) = (-\tfrac{\tilde{1}}{2}|\,V_+\,|-\tfrac{\tilde{3}}{2}) = \frac{\sqrt{3}}{2}(P+Q),$$
$$(\tfrac{\tilde{1}}{2}|\,V_+\,|-\tfrac{\tilde{1}}{2}) = \frac{3P-Q}{2}, \qquad (-\tfrac{\tilde{3}}{2}|\,V_+\,|\tfrac{\tilde{3}}{2}) = \frac{P-3Q}{2}. \quad (18.29)$$

For $P = 3Q$ the cubic coefficient b in (18.19) vanishes and the matrix elements (18.29) take their normal values for an ordinary multiplet $\mathscr{J} = \tfrac{3}{2}$. The secular equation for the eigenvalues $(W/\beta H) = y$ of $Z = -\boldsymbol{\mu} \cdot \mathbf{H}$ is obtained from (18.16) and (18.29). It is given by

$$y^4 - (P^2 + Q^2)y^2 + P^2Q^2 +$$
$$+ \tfrac{3}{16}(P-3Q)(3P-Q)(P+Q)^2(n_1^2 n_2^2 + n_2^2 n_3^2 + n_3^2 n_1^2) = 0, \quad (18.30)$$

where n_1, n_2, n_3 are the cosines of the magnetic field with respect to the fourfold axes.

The absence of odd powers in (18.30) is a straightforward consequence of invariance with respect to time reversal. The energy levels are linear with H but they are neither equidistant nor independent of the direction of the field, unlike an ordinary quadruplet $S = \tfrac{3}{2}$. For an arbitrary orientation of the magnetic field with respect to the cubic axes, transitions can be induced between any of the four levels.

If **H** is along the z-axis the roots of (18.30) are $\pm P$, $\pm Q$, as follows immediately from (18.16). The selection rules and relative intensities obtained from (18.29) are, in this case, with the oscillatory field normal to Oz, as shown in Table 18.1.

TABLE 18.1

Transition	Energy	Relative intensity
$\frac{3}{2} \to -\frac{3}{2}$	$2P$	$(P-3Q)^2$
$\frac{1}{2} \to -\frac{1}{2}$	$2Q$	$(3P-Q)^2$
$\frac{3}{2} \to \frac{1}{2} = -\frac{1}{2} \to -\frac{3}{2}$	$P-Q$	$3(P+Q)^2$

The transitions $|\Delta \tilde{S}_z| = 2$ are forbidden for this particular orientation. It may be noted that if a rotating field H_1 is used, some of the transitions may require right-handed and others left-handed circular polarization (Bleaney 1959a,b).

It may be convenient, especially for a study of trigonal distortions of the cubic field to take as axis of quantization OZ a body diagonal (see below). Three new operators $\tilde{S}_X$, $\tilde{S}_Y$, $\tilde{S}_Z$ are defined through

$$\tilde{S}_Z = \frac{\tilde{S}_x + \tilde{S}_y + \tilde{S}_z}{\sqrt{3}},$$

$$\tilde{S}_Y = \frac{\tilde{S}_x - \tilde{S}_y}{\sqrt{2}}, \qquad (18.31)$$

$$\tilde{S}_X = -\frac{\tilde{S}_x + \tilde{S}_y - 2\tilde{S}_z}{\sqrt{6}};$$

$$\tilde{S}_z = \frac{2\tilde{S}_X}{\sqrt{6}} + \frac{\tilde{S}_Z}{\sqrt{3}},$$

$$\tilde{S}_y = -\frac{\tilde{S}_X}{\sqrt{6}} - \frac{\tilde{S}_Y}{\sqrt{2}} + \frac{\tilde{S}_Z}{\sqrt{3}}, \qquad (18.32)$$

$$\tilde{S}_x = -\frac{\tilde{S}_X}{\sqrt{6}} + \frac{\tilde{S}_Y}{\sqrt{2}} + \frac{\tilde{S}_Z}{\sqrt{3}},$$

together with operators V_X, V_Y, V_Z given by

$$V_Z = a\tilde{S}_Z + \frac{b}{\sqrt{3}}(\tilde{S}_x^3 + \tilde{S}_y^3 + \tilde{S}_z^3),$$

$$V_Y = a\tilde{S}_Y + \frac{b}{\sqrt{2}}(\tilde{S}_x^3 - \tilde{S}_y^3), \qquad (18.33)$$

$$V_X = a\tilde{S}_X - \frac{b}{\sqrt{6}}(\tilde{S}_x^3 + \tilde{S}_y^3 - 2\tilde{S}_z^3).$$

IONS IN A WEAK CRYSTAL FIELD (f ELECTRONS)

In the representation where $\tilde{S}_Z$ is diagonal and $\tilde{S}_X$ and $\tilde{S}_Y$ have the usual matrix form we find from (18.33) the non-vanishing matrix elements of V_X, V_Y, V_Z using (18.32):

$$(\tfrac{\tilde{3}}{2}|\ V_Z\ |\tfrac{\tilde{3}}{2}) = -(-\tfrac{\tilde{3}}{2}|\ V_Z\ |-\tfrac{\tilde{3}}{2}) = \frac{5P+3Q}{6},$$

$$(\tfrac{\tilde{1}}{2}|\ V_Z\ |\tfrac{\tilde{1}}{2}) = -(-\tfrac{\tilde{1}}{2}|\ V_Z\ |-\tfrac{\tilde{1}}{2}) = \frac{P-Q}{2}, \qquad (18.34)$$

$$(\tfrac{\tilde{3}}{2}|\ V_Z\ |-\tfrac{\tilde{3}}{2}) = (-\tfrac{\tilde{3}}{2}|\ V_Z\ |\tfrac{\tilde{3}}{2}) = \frac{1}{\sqrt{2}}\left(\frac{P}{3}-Q\right).$$

Apropos of the last term in (18.34) it is worth pointing out that when a body diagonal is chosen as axis of quantization OZ, a matrix element such as $\langle\tfrac{\tilde{3}}{2}|\ V_Z\ |-\tfrac{\tilde{3}}{2}\rangle$ need not vanish since it is left unchanged by a rotation $\pm 2\pi/3$ around OZ. Also,

$$\langle\tfrac{\tilde{3}}{2}|\ V_X+\mathrm{i}V_Y\ |\tfrac{\tilde{1}}{2}\rangle = \langle -\tfrac{\tilde{1}}{2}|\ V_X+\mathrm{i}V_Y\ |-\tfrac{\tilde{3}}{2}\rangle = \frac{2P}{\sqrt{3}},$$

$$\langle\tfrac{\tilde{1}}{2}|\ V_X+\mathrm{i}V_Y\ |-\tfrac{\tilde{1}}{2}\rangle = P+Q, \qquad (18.35)$$

$$\langle -\tfrac{\tilde{1}}{2}|\ V_X+\mathrm{i}V_Y\ |\tfrac{\tilde{3}}{2}\rangle = -\langle -\tfrac{\tilde{3}}{2}|\ V_X+\mathrm{i}V_Y\ |\tfrac{\tilde{3}}{2}\rangle = \frac{\sqrt{6}}{2}\left(\frac{P}{3}-Q\right).$$

From (18.34) we find that when the field is along OZ, two of the eigenstates are $|\pm\tfrac{\tilde{1}}{2}\rangle$ and the other two a mixture of the states $|\pm\tfrac{\tilde{3}}{2}\rangle$. The eigenvalues are

$$y_{1,2} = \pm\left(\frac{P-Q}{2}\right),$$

$$y_{3,4} = \pm\tfrac{1}{2}\sqrt{\{3(P^2+Q^2)+2PQ\}}. \qquad (18.36)$$

Obviously (18.36) could have been obtained directly by solving the secular equation (18.30). It is clear from (18.35) that the transitions $\Delta M = 0$, ± 3 between the levels y_3 and y_4 are forbidden for an oscillatory magnetic field perpendicular to the trigonal axis, and it can be shown that they are also forbidden for an oscillatory field along this axis.

Comparison with experiment

A good example for which we can compare the foregoing theory with experiment is the spectrum of Er^{3+}, $4f^{11}$, $^4I_{\frac{15}{2}}$ in MgO. In a cubic field the $J = \tfrac{15}{2}$ manifold splits into five levels, of which three are Γ_8 quadruplets; resonance has been observed in the lowest of these, which is the ground state of the ion in MgO. Assuming a value of $\langle J\|\ \Lambda\ \|J\rangle = \tfrac{6}{5}$

(the simple Landé value) Descamps and Merle d'Aubigné (1964) found that the resonance results, both for **H** along a fourfold and along a threefold axis, corresponded to

$$P' = \langle \tilde{\tfrac{3}{2}}| J_z |\tilde{\tfrac{3}{2}}\rangle = 4 \cdot 925,$$
$$Q' = \langle \tilde{\tfrac{1}{2}}| J_z |\tilde{\tfrac{1}{2}}\rangle = 1 \cdot 925.$$

In the expansions

$$|\tilde{\tfrac{3}{2}}\rangle = \sum_{M_J} C_{M_J} |J, M_J\rangle,$$
$$|\tilde{\tfrac{1}{2}}\rangle = \sum_{M_J} C'_{M_J} |J, M_J\rangle,$$
(18.37)

the C_{M_J} and C'_{M_J}, tabulated by Lea, Leask, and Wolf (1962), are functions of the parameter x defined in eqn (18.14). This parameter is overdetermined by the two relations,

$$P' = \sum_{M_J} M_J |C_{M_J}(x)|^2,$$
$$Q' = \sum_{M_J} M_J |C'_{M_J}(x)|^2,$$
(18.38)

and, using the tables of Lea, Leask and Wolf (1962) slightly different values of x are obtained from the two eqns (18.38), respectively:

$$x = 0 \cdot 765 \pm 0 \cdot 005 \text{ from } P',$$

and

$$x = 0 \cdot 715 \pm 0 \cdot 005 \text{ from } Q'.$$

However a unique value of x can be obtained if in the relation $\mu = -\beta \Lambda \mathbf{J}$, Λ is given a value smaller than $\tfrac{6}{5}$ by a factor $0 \cdot 983$. Such a reduction could perhaps be caused by a spread of the wave-functions of the magnetic electrons on the neighbouring atoms, a phenomenon which as we shall see in the chapter on covalent bonding leads to a reduction of the orbital g-factor below its value of one. This reduction is similar to those found necessary in several Γ_6, Γ_7 doublets, and in this case it leads to corrected values $P'' = 5 \cdot 01$, $Q'' = 1 \cdot 960$ which correspond to a unique value of $x = 0 \cdot 72$.

A number of other results in reasonable if not perfect agreement with the theory of the Γ_8 quadruplet are mentioned in § 5.8. Here we merely note that if Γ_8 occurs only once in the decomposition of D^J, the values of P', Q' are independent of x and can be calculated from eqns (18.37) and (18.38) using the values of C_{M_J}, C'_{M_J} given in Table 9, yielding the results:

$$J = \tfrac{5}{2};\ P' = -\tfrac{11}{6},\ Q' = +\tfrac{1}{2},$$
$$J = \tfrac{7}{2};\ P' = +\tfrac{1}{2},\ Q =' -\tfrac{11}{6}.$$
(18.39)

IONS IN A WEAK CRYSTAL FIELD (f ELECTRONS)

In general there are two possible ways of choosing which pair of states to call $|\pm\tilde{\tfrac{3}{2}}\rangle$ and which $|\pm\tilde{\tfrac{1}{2}}\rangle$. A natural choice, which has been followed above, is to denote by $|\tilde{M}\rangle$ that state which under a rotation through an angle $\omega = \pi/2$ or π about the Oz axis simply multiplies the state by $e^{-i\omega M}$. Then reference to Table 9 shows that in any Γ_8 quadruplet for J up to $\tfrac{7}{2}$ the $|J, M_J\rangle$ states always contain one pair in which $M_J = \pm\tfrac{1}{2}$ occurs, and on the above choice these become $|\tilde{M}\rangle = |\pm\tilde{\tfrac{1}{2}}\rangle$, while the companion pair contains $M_J = \pm\tfrac{3}{2}$, which similarly become $|\pm\tilde{\tfrac{3}{2}}\rangle$. The same result holds for higher values of J.

The effect of crystalline distortions on a Γ_8 quadruplet

There is an important qualitative difference between the transitions $+\tfrac{1}{2}\leftrightarrow -\tfrac{1}{2}$, $+\tfrac{3}{2}\leftrightarrow -\tfrac{3}{2}$ on the one hand, and the transitions $+\tfrac{3}{2}\leftrightarrow +\tfrac{1}{2}$, $-\tfrac{1}{2}\leftrightarrow -\tfrac{3}{2}$ on the other: the former occur within two pairs of Kramers-conjugate states. It follows that in first order they are insensitive to small distortions of the crystal potential. Also, when higher-order Zeeman effects are considered, the change in resonant frequency involves terms proportional to H^3. On the other hand, the frequencies of the transitions $+\tfrac{3}{2}\leftrightarrow +\tfrac{1}{2}$, $-\tfrac{1}{2}\leftrightarrow -\tfrac{3}{2}$ (and obviously also $\pm\tfrac{3}{2}\leftrightarrow \mp\tfrac{1}{2}$, if allowed) are affected by a crystalline distortion in first order, and by a higher-order Zeeman effect in second order. In large magnetic fields along a fourfold axis, for example, random crystalline distortions are likely to result in two relatively broad transitions of energy $(P-Q)\pm\alpha H^2$, while the transitions at $2P$ and $2Q$ (modified by terms in H^3) are much sharper.

For a *tetragonal* distortion, whose effect is small compared with the energy separation to other cubic levels, we may add a term $D\{\tilde{S}_z^2 - \tfrac{1}{3}\tilde{S}(\tilde{S}+1)\}$ to the Zeeman Hamiltonian $-(\boldsymbol{\mu}\cdot\mathbf{H})$, the components of $-\boldsymbol{\mu}/\beta$ being given by eqns (18.17), (18.19) or the matrix elements (18.16), (18.20).

For a *trigonal* distortion a similar term can be added, the OZ axis in this case being the threefold axis. The components of $-\boldsymbol{\mu}/\beta$ are then given by the matrix elements (18.34), (18.35).

If the effect of the distortion, expressed by the constant D, is much larger than the Zeeman energy, we can regard the quadruplet Γ_8 as split into two independent Kramers doublets, each of which has a g-'tensor' (neglecting off-diagonal effects of the Zeeman interaction between the two doublets) given by the following formulae.

(a) *Tetragonal distortion*

The two doublets are clearly $|\pm\tilde{\tfrac{3}{2}}\rangle$ and $|\pm\tilde{\tfrac{1}{2}}\rangle$. For the former, it follows from the last matrix element in (18.29) that we must identify one pair of states as $|+\rangle = |-\tilde{\tfrac{3}{2}}\rangle$ and $|-\rangle = |+\tilde{\tfrac{3}{2}}\rangle$. Then, using also eqn (18.16), we have

$$g_\| = 2\langle +|-\mu_z/\beta|+\rangle = -2P,$$
$$|g_\perp| = 2|\langle -|-\mu_x/\beta|+\rangle| = \tfrac{1}{2}|(P-3Q)|; \tag{18.40}$$

and

$$g^2 = (2P\cos\theta)^2 + \{\tfrac{1}{2}(P-3Q)\sin\theta\}^2, \tag{18.41}$$

where θ is the angle that **H** makes with the axis of tetragonal distortion. Here we note that $g_\perp$ is not zero as it would be if our Γ_8 were an ordinary $D^{\tfrac{3}{2}}$ quadruplet, for which $(P-3Q) = 0$.

For the other doublet $|\pm\rangle = |\pm\tilde{\tfrac{1}{2}}\rangle$, we have

$$g_\| = 2Q, \quad |g_\perp| = \tfrac{1}{2}|(3P-Q)|; \tag{18.42}$$

and

$$g^2 = (2Q\cos\theta)^2 + \{\tfrac{1}{2}(3P-Q)\sin\theta\}^2. \tag{18.43}$$

(b) *Trigonal distortion*

For the doublet $|\pm\tilde{\tfrac{3}{2}}\rangle$ we may take $|+\tilde{\tfrac{3}{2}}\rangle$ and $|-\tilde{\tfrac{3}{2}}\rangle$ as basic states and we find from (18.34) that the 'tensor' $g_{q\alpha}$ defined in (15.31) has off-diagonal components

$$|g_{Z1}| = \sqrt{(2)}\left|\tfrac{P}{3}-Q\right|, \quad g_{Z2} = 0, \quad g_{Z3} = \frac{5P+3Q}{3}, \tag{18.44}$$

whereas $g_{X\alpha} = g_{Y\alpha} = 0$.

The tensor G_{pq} defined by (15.38) has one non-vanishing component

$$G_{ZZ} = g_{Z1}^2 + g_{Z2}^2 + g_{Z3}^2 = 3(P^2+Q^2) + 2PQ, \tag{18.45}$$

and

$$g^2 = G_{ZZ}\cos^2\theta = \{3(P^2+Q^2) + 2PQ\}\cos^2\theta. \tag{18.46}$$

No resonance is observable in principle in this doublet.

The doublet $|\pm\tilde{\tfrac{1}{2}}\rangle$ behaves quite normally and from (18.34), (18.35) we find

$$g_\| = P-Q,$$
$$g_\perp = P+Q, \tag{18.47}$$

with

$$g^2 = (P-Q)^2\cos^2\theta + (P+Q)^2\sin^2\theta. \tag{18.48}$$

18.4. Representation of an irreducible tensor within the quadruplet Γ_8—quadrupole coupling

To find how many constants are necessary to define an irreducible tensor T_2^q whose components provide a representation D^2 of the rotation group, we turn again to group theory. A matrix element such as $\langle \tilde{m} | T_2^q | \tilde{m}' \rangle$, where $|\tilde{m}\rangle$ and $|\tilde{m}'\rangle$ are the kets that span Γ_8, transforms under the rotations of the cubic group like

$$\Gamma_8 \times D^2 \times \Gamma_8 = \Gamma_8 \times (\Gamma_3 + \Gamma_5) \times \Gamma_8.$$

The product $\Gamma_8 \times \Gamma_8$ contains Γ_3 once and Γ_5 twice. *A priori* three constants are thus necessary to define a tensor T_2 within a multiplet Γ_8. However, as we saw in § 15.9, more detailed predictions can be made if the tensor T_2 has a definite time-reversal parity. This is the case for the quadrupole tensor which is time-even. Therefore according to the rules given in § 15.9 it is the antisymmetric product $[\Gamma_8 \times \Gamma_8]_A$ which must contain Γ_3 and Γ_5, and we see in Table 8 that it contains each of them once. Two constants are thus necessary to specify the quadrupole moment tensor of the ion inside a multiplet Γ_8.

We can obtain the same result by writing out the combinations of various products of the components $\tilde{S}_x$, $\tilde{S}_y$, $\tilde{S}_z$ of the fictitious spin $\tfrac{3}{2}$, which transform according to a given representation of the cubic group.

We have already seen that $\tilde{S}_x$, $\tilde{S}_y$, $\tilde{S}_z$ and $\tilde{S}_x^3$, $\tilde{S}_y^3$, $\tilde{S}_z^3$ each provide a representation of Γ_4. $\tilde{S}_x^2 - \tilde{S}_y^2$ and $3\tilde{S}_z^2 - \tilde{S}(\tilde{S}+1)$ transform like Γ_3 and $(\tilde{S}_x \tilde{S}_y + \tilde{S}_y \tilde{S}_x)$, etc. like Γ_5. A second representation of Γ_5 is provided by $\tilde{S}_z(\tilde{S}_x^2 - \tilde{S}_y^2) + (\tilde{S}_x^2 - \tilde{S}_y^2)\tilde{S}_z$, etc., but it is time-odd and cannot represent an electric quadrupole moment. It is contained in the symmetrical product $[\Gamma_8 \times \Gamma_8]_S$.

Within Γ_8 we have the following relations:

$$\sum_i \frac{3z_i^2 - r_i^2}{r_i^5} = a\{3\tilde{S}_z^2 - \tilde{S}(\tilde{S}+1)\},$$

$$\sum_i \frac{x_i y_i}{r_i^5} = b(\tilde{S}_x \tilde{S}_y + \tilde{S}_y \tilde{S}_x). \tag{18.49}$$

The quadrupole interaction within Γ_8 can thus be written (cf. § 17.1) as

$$\frac{-e^2 Q \langle r_q^{-3} \rangle}{I(2I-1)} \langle L \| \alpha \| L \rangle \left[\frac{m}{6} \sum_{q=1}^{3} \{3I_q^2 - I(I+1)\}\{3\tilde{S}_q^2 - \tilde{S}(\tilde{S}+1)\} + \right. \\ \left. + \frac{3n}{4} \sum_{p \neq q} (I_p I_q + I_q I_p)(\tilde{S}_p \tilde{S}_q + \tilde{S}_q \tilde{S}_p) \right], \tag{18.50}$$

where m and n are two constants that depend on the particular manifold Γ_8 under consideration. They are equal if Γ_8 is an ordinary quadruplet, $J = \frac{3}{2}$.

If the multiplet Γ_8 results from the decomposition of a single J,

and
$$m = \frac{\langle J\| \alpha \|J\rangle}{\langle L\| \alpha \|L\rangle} \frac{\langle \tilde{\tfrac{3}{2}}| \, 3J_z^2 - J(J+1)\, |\tilde{\tfrac{3}{2}}\rangle}{3}$$

$$n = \frac{\langle J\| \alpha \|J\rangle}{\langle L\| \alpha \|L\rangle} \frac{\langle \tilde{\tfrac{3}{2}}| \, J_z J_+ + J_+ J_z \, |\tilde{\tfrac{1}{2}}\rangle}{2\sqrt{3}}.$$
(18.51)

The constants m and n can be calculated from the expansions (18.37) of the kets $|\tilde{m}\rangle$ that span Γ_8.

The pseudo-nuclear Zeeman coupling given in formula (18.4) will have the following general form quite different from that of a Kramers doublet:

$$\frac{Z_n^*}{-\beta_n g_n} = p(\mathbf{H} \cdot \mathbf{I}) + q[H_x I_x\{3\tilde{S}_x^2 - \tilde{S}(\tilde{S}+1)\} + H_y I_y\{3\tilde{S}_y^2 - \tilde{S}(\tilde{S}+1)\} +$$
$$+ H_z I_z\{3\tilde{S}_z^2 - \tilde{S}(\tilde{S}+1)\}] + r\{(H_x I_y + H_y I_x)(\tilde{S}_x \tilde{S}_y + \tilde{S}_y \tilde{S}_x) +$$
$$+ (H_y I_z + H_z I_y)(\tilde{S}_y \tilde{S}_z + \tilde{S}_z \tilde{S}_y) + (H_z I_x + H_x I_z)(\tilde{S}_z \tilde{S}_x + \tilde{S}_x \tilde{S}_z)\}. \quad (18.52)$$

18.5. Non-Kramers ions in the rare-earth group

These ions have been much less studied by resonance methods than the Kramers ions, for several reasons.

(a) There may be no degeneracy left even in the absence of an applied magnetic field and the distance between two neighbouring levels may be outside the range of microwave spectrometers. This requirement is all the more restrictive because, as we saw in Chapter 10, the relaxation times of non-Kramers ions are so short in many cases that only at helium temperatures can the resonance be observed at all. The two sublevels between which the resonance can be observed must then necessarily be the lowest or among the lowest of the bound ion, in order to have non-negligible populations.

(b) When the symmetry of the surroundings is such that there is still some degeneracy left in the absence of a magnetic field, this degeneracy can be lifted by slight local departures from the general symmetry due to crystal imperfections. This will result in an inhomogeneous broadening of the line.

We now consider briefly the main features of magnetic resonance for non-Kramers ions of the rare-earth group.

IONS IN A WEAK CRYSTAL FIELD (f ELECTRONS) 733

If the crystal potential had purely axial symmetry, a manifold J would be split into a singlet $J_z = 0$ together with J doublets $J_z = \pm M$. However in practice we find axes with 6-, 3-, 4-, or 2-fold symmetry.

(a) Sixfold symmetry effectively exists in the ethylsulphates and anhydrous trichlorides: with C_{3h} symmetry there are matrix elements $|\Delta M| = 6$, which couple together the states $M = \pm 3$, and also couple states $M = \pm 6$ with the state $M = 0$ and hence indirectly with one another. The behaviour of states $|\xi\rangle = \sum C_M |M\rangle$ with $M = 3p$ is thus different from those with $M = 3p \pm 1$.

(b) Threefold symmetry, typical of the double nitrates with nearly C_{3v} symmetry, gives additional matrix elements $|\Delta M| = 3$ which couple more states together but do not further lift the degeneracy found with sixfold symmetry.

(c) Fourfold symmetry, for which $CaWO_4$ is a typical host lattice, provided the full local symmetry at the Ca^{2+} site is not disturbed by charge compensation when a tripositive lanthanide ion is introduced, produces matrix elements $|\Delta M| = 4$. These couple together the states $M = \pm 2$, and hence indirectly those with $M = \pm 6$; also states $M = \pm 4$ (and hence indirectly $M = \pm 8$) are coupled to $M = 0$. Thus in this case states with even values of M behave differently from those with odd values of M.

(d) Twofold symmetry produces matrix elements $|\Delta M| = 2$, which couple all states of even M to one another, and all states of odd M to one another.

As an example of time-conjugate states that are not coupled with one another we may take the lowest doublet of Pr^{3+}, $4f^2$, $^3H_4$ in the double nitrate, for which the states given by eqn (5.67) are

$$|\xi_0\rangle = \alpha |+4\rangle + \beta |+1\rangle + \gamma |-2\rangle,$$
$$|\bar{\xi}_0\rangle = \alpha |-4\rangle - \beta |-1\rangle + \gamma |+2\rangle. \quad (18.53)$$

The Zeeman Hamiltonian $Z = \Lambda\beta(\mathbf{J} \cdot \mathbf{H})$ has opposite expectation values in the two states $|\xi_0\rangle$ and $|\bar{\xi}_0\rangle$, separating them in energy by

$$2\langle \xi_0| Z |\xi_0\rangle = 2\Lambda\beta H_z(4\alpha^2 + \beta^2 - 2\gamma^2) = g_\parallel \beta H_z. \quad (18.54)$$

There are no off-diagonal elements between the two states (18.53), making it impossible to induce resonance transitions between them. This is a direct consequence of the discussion in § 15.4: for an even number of electrons, $\theta^2 = +1$, and a time-odd operator such as Z has opposite expectation values in two states conjugated by time reversal but with no matrix elements between them. Note that although values of M differing by unity appear in $|\xi_0\rangle$ and $|\bar{\xi}_0\rangle$, the change in sign of the coefficient of the $|\pm 1\rangle$ states (which follows from the rules for the time-reversal operator given in § 15.5) makes the matrix element $\langle \bar{\xi}_0 | J_x | \xi_0 \rangle = 0$.

A local distortion of crystal symmetry would result in a change $\Delta V(\mathbf{r})$ in the crystal potential seen by the ion. $\Delta V(\mathbf{r})$ is a time-even operator and, as shown in § 15.4, it will have the same expectation value in $|\xi_0\rangle$ and $|\bar{\xi}_0\rangle$, which displaces both levels by the same amount and will not affect the resonance frequency. However there is nothing to prevent $\Delta V(\mathbf{r})$ from having off-diagonal matrix elements between $|\xi_0\rangle$ and $|\bar{\xi}_0\rangle$ of the general form

$$(\xi_0 | \Delta V(\mathbf{r}) | \bar{\xi}_0) = \langle \bar{\xi}_0 | \Delta V(\mathbf{r}) | \xi_0 \rangle^* = \frac{\Delta_x + i\Delta_y}{2}.$$

Inside the doublet spanned by $|\xi_0\rangle$ and $|\bar{\xi}_0\rangle$ we may introduce a fictitious spin $\tilde{S} = \tfrac{1}{2}$ with

$$\begin{aligned}\langle \xi_0 | \tilde{S}_z | \xi_0 \rangle &= -\langle \bar{\xi}_0 | \tilde{S}_z | \bar{\xi}_0 \rangle = \tfrac{1}{2},\\ \langle \xi_0 | \tilde{S}_x | \bar{\xi}_0 \rangle &= i\langle \xi_0 | \tilde{S}_y | \bar{\xi}_0 \rangle = \tfrac{1}{2}.\end{aligned} \quad (18.55)$$

The spin Hamiltonian for this doublet can then be written as

$$\mathscr{H} = g_\| \beta H_z \tilde{S}_z + \Delta_x \tilde{S}_x + \Delta_y \tilde{S}_y, \quad (18.56)$$

with

$$g_\| = 2\langle \xi_0 | \tilde{S}_z | \xi_0 \rangle = 2\langle J \| \Lambda \| J \rangle \langle \xi_0 | J_z | \xi_0 \rangle, \quad (18.57)$$

where the last expression for $g_\|$ is appropriate to a rare-earth ion. The eigenvalues of (18.56) are given by

$$W = \pm \tfrac{1}{2}(\hbar\omega) = \pm \tfrac{1}{2}\{(g_\| \beta H_z)^2 + \Delta^2\}^{\tfrac{1}{2}} \quad (18.58)$$

where $\Delta^2 = \Delta_x^2 + \Delta_y^2$. The eigenstates are no longer just $|\xi_0\rangle$, $|\bar{\xi}_0\rangle$ but become

$$\begin{aligned}|\xi\rangle &= a|\xi_0\rangle + b|\bar{\xi}_0\rangle,\\ |\eta\rangle &= b^*|\xi_0\rangle - a^*|\bar{\xi}_0\rangle.\end{aligned} \quad (18.59)$$

These are no longer conjugate states under time reversal, and the operator $\tilde{S}_z$ has matrix elements between them. This makes it possible for an oscillatory field, *parallel* to the z-axis and at the frequency corresponding to $(\hbar\omega)$, to induce transitions between them. The transition probability is given by the square of the matrix element, $|\mu_z \mu_z^*|$, whose value is given by

$$|\mu_z \mu_z^*| = \tfrac{1}{4} g_\|^2 \beta^2 \frac{\Delta^2}{(\hbar\omega)^2}. \quad (18.60)$$

IONS IN A WEAK CRYSTAL FIELD (f ELECTRONS)

If $\Delta^2 \ll (\hbar\omega)^2$, this will of course be considerably smaller than that of an ordinary allowed transition.

The existence of distortions represented by the terms in Δ_x, Δ_y is to be expected on the basis of the Jahn–Teller theorem, to be discussed in Chapter 21, which predicts that a state with electronic non-Kramers degeneracy may be unstable against small distortions of the surroundings, which would lower the symmetry of the crystal potential and lift the degeneracy. One may suppose that Δ_x and Δ_y vary from one site to another, independently of each other, and in a random way. If we assume a Gaussian distribution for Δ_x and Δ_y, Δ^2 will have an exponential distribution of the form

$$P(\Delta^2)\,\mathrm{d}(\Delta^2) = \exp\left(\frac{-\Delta^2}{\Delta_0^2}\right)\mathrm{d}\left(\frac{\Delta^2}{\Delta_0^2}\right) \tag{18.61}$$

where Δ_0^2 is the average value of Δ^2.

Since the transition probability according to (18.60) is proportional to Δ^2 the resonance line will have an unusual shape which (assuming all other causes of broadening negligible) is given by

$$I(\Delta^2)\,\mathrm{d}(\Delta^2) = \frac{\Delta^2}{\Delta_0^2}\exp\left(\frac{-\Delta^2}{\Delta_0^2}\right)\mathrm{d}\left(\frac{\Delta^2}{\Delta_0^2}\right). \tag{18.62}$$

The resonance is usually observed by keeping the microwave frequency ($\omega/2\pi$) constant and sweeping the field, which we assume from now on to be parallel to the crystal axis: $H_z = H$. We define the symbols

$$H_0 = \frac{\hbar\omega}{g_\parallel \beta}, \quad h = H_0 - H, \tag{18.63}$$

in terms of which eqn (18.58) can be rewritten

$$(\hbar\omega)^2 = \Delta^2 + g_\parallel^2 \beta^2 (H_0 - h)^2,$$

giving

$$h \approx \frac{\Delta^2}{2g_\parallel^2 \beta^2 H_0}. \tag{18.64}$$

For values of $h \ll H_0$, we further introduce the symbol

$$h_0 = \frac{\Delta_0^2}{2g_\parallel^2 \beta^2 H_0}, \tag{18.65}$$

and the line shape obtained by sweeping the field will have the form

$$I(h)\,\mathrm{d}h = \frac{h}{h_0}\exp\left(-\frac{h}{h_0}\right)\mathrm{d}\left(\frac{h}{h_0}\right). \tag{18.66}$$

The line will have a cut-off for $h = 0$ which from (18.63) is at the high-field end, a maximum for $h = h_0$, or $H = H_0 - h_0$ and a rather long tail towards low fields. Increasing the microwave frequency should make the line narrower since h_0 is proportional to ω^{-1}, but also less intense since the transition probability varies as ω^{-2}. The existence of such a cut-off is independent of the particular form (18.61) assumed for the distribution $P(\Delta^2)$ and follows from (18.58) and (18.60). It permits a more accurate measurement of $g_\parallel$ than would be expected from the overall line width.

If a hyperfine structure is present, it is necessarily of the form AI_zS_z and the only change is, for each hyperfine component, the replacement in (18.58) of $g_\parallel \beta H$ by $(g_\parallel \beta H + Am)$, where $m = I_z$. The hyperfine lines are identical and separated from each other by an interval A. If A is smaller than $g_\parallel \beta h_0$, they will overlap. A quadrupole term of the form $P_\parallel I_z^2$ will not affect the resonant frequencies since all transitions are of the type $\Delta m = 0$.

Para-electric transitions

It must be pointed out that the fictitious spin $\tilde{S} = \frac{1}{2}$ introduced above for non-Kramers doublets differs in an important respect from that discussed in § 15.6 for Kramers doublets. There the three components of $\tilde{S}$ were time-odd operators, whereas here $\tilde{S}_z$ is time-odd but $\tilde{S}_x, \tilde{S}_y$ are necessarily time-even operators. This is a serious drawback in the spin $\tilde{S} = \frac{1}{2}$ formalism for a non-Kramers doublet, and Mueller (1968) has suggested alternative formalisms in which this drawback does not occur. A simple possibility is to take an $\tilde{S} = 1$ manifold in which the $\tilde{S}_z = 0$ state is suppressed, or removed by a term such as $D\tilde{S}_z^2$ away from the two $\tilde{S}_z = \pm 1$ states, which are left as a non-Kramers doublet, well isolated if $D \gg (\hbar\omega)$; (cf. § 3.14, where we led up to the subject of non-Kramers doublets by discussing the case of $S = 2$ with a splitting of the form $D\tilde{S}_z^2$).

An important point made independently by Williams (1967) and by Culvahouse, and Schinke and Foster (1967) is that for a non-Kramers ion at a site lacking in full inversion symmetry, there may be terms linear in an electric field for a non-Kramers doublet. For example, if the crystal field has odd parity for reflection in a plane normal to the x-axis, the system may have a permanent *electric* dipole along this axis whose interaction with an applied *electric* field may be represented (in the $\tilde{S} = \frac{1}{2}$ formalism) by a term of the form $g_x^{(E)}\beta \tilde{S}_x E_x$. In C_{3h} symmetry, where the only plane of even reflection symmetry is normal to the z-axis, there will be an equivalent term for the y-axis, giving altogether

$$g_\perp^{(E)}\beta(\tilde{S}_x E_x + \tilde{S}_y E_y). \tag{18.67}$$

Thus if random crystal distortions are also present, the Hamiltonian becomes (with $\tilde{S} = \tfrac{1}{2}$)

$$\mathscr{H} = g_\parallel \beta H_z \tilde{S}_z + (\Delta_x + g_\perp^{(E)}\beta E_x)\tilde{S}_x + (\Delta_y + g_\perp^{(E)}\beta E_y)\tilde{S}_y, \tag{18.68}$$

which gives again eigenstates of the form (18.59), with suitably modified values of the coefficients a, b. The change in resonance frequency resulting from an applied static electric field follows from (18.58) if we replace Δ^2 by

$$\Delta'^2 = (\Delta_x + g_\perp^{(E)}\beta E_x)^2 + (\Delta_y + g_\perp^{(E)}\beta E_y)^2, \tag{18.69}$$

but with experimentally feasible values of the electric field it may not be easy to detect. More important is the fact that transitions between the two components of the doublet may be induced by a resonant microwave *electric* field *perpendicular* to the z-axis. The matrix elements for this and the transition rate are given by eqns (3.109)–(3.111); a significant feature is that the line shape is different (see Fig. 3.26). Transitions are now allowed for $\Delta_x = \Delta_y = 0$, because the perpendicular electric field has matrix elements between the states $|\xi_0\rangle$, $|\tilde{\xi}_0\rangle$, in contrast with the magnetic field along the z-axis, for which matrix elements between the states exist only for $|\xi\rangle$, $|\eta\rangle$ and hence depend on the presence of a crystalline distortion. Experimental verification that the transitions in some praseodymium compounds are primarily due to the microwave electric field is discussed in § 3.14 (see particularly Fig. 3.27).

Non-Kramers doublets with 'allowed' magnetic transitions

It was shown in § 3.14 that for an ion with $S = 2$ and a large splitting of the form DS_z^2, the $S_z = \pm 2$ states are slightly split if a term $E(S_x^2 - S_y^2) = V_2^2$ is present. The latter has matrix elements between the $S_z = 0$ state and the $S_z = \pm 2$ states, giving a splitting of the latter states, with an allowed transition between them when an oscillatory magnetic field at the resonance frequency is applied along the z-axis. This differs from the situation considered above because the V_2^2 term may have a unique value determined by the crystal potential in a perfect crystal instead of a range of values associated with crystal imperfections or Jahn–Teller distortions. The transition $+2 \leftrightarrow -2$ then has a sharp energy as far as the crystal potential is concerned, giving a line of normal shape whose width is determined by other interactions such as spin-spin and spin-lattice relaxation.

A similar situation can arise for non-Kramers ions of the rare-earth group on sites of 2-, 3-, 4-, or 6-fold symmetry. Since it is easier to discuss a concrete example, we consider the case of Tb^{3+}, $4f^8$, $^7F_6$ in the ethylsulphate (Baker and Bleaney 1958), where the ground doublet is $|J_z\rangle = |\pm 6\rangle$ with a small admixture of the state $|J_z\rangle = |0\rangle$. The component μ_z of the magnetic moment parallel to the applied field does have matrix elements between the two states spanning the ground doublet even in the absence of local distortions of the crystal potential and the lines are relatively narrow. The experimental value $g_\parallel = 17\cdot 72$ is very nearly the value $12\Lambda = 18$ that one would expect for a doublet $|\pm 6\rangle$, but it is precisely the difference $g_\parallel - 18$, together with a splitting $\Delta = 0\cdot 387$ cm^{-1}, that we have to explain.

In the absence of an applied field it is simplest to start from the states

$$|6^s\rangle = \frac{1}{\sqrt{2}}\{|6\rangle + |-6\rangle\},$$
$$|6^a\rangle = \frac{1}{\sqrt{2}}\{|6\rangle - |-6\rangle\}. \tag{18.70}$$

The term V_6^6 of the crystal potential admixes into $|6^s\rangle$ a certain amplitude q of the higher state $|0\rangle$ and pushes it down by a certain amount Δ; but $|6^a\rangle$ is not coupled to $|0\rangle$. In first-order perturbation theory

$$q = -\frac{c\sqrt{2}}{d}, \quad \Delta = \frac{2c^2}{d},$$

where

$$c = \langle 0|\ V_6^6\ |6\rangle,\ d = \langle 0|\ V\ |0\rangle - \langle 6|\ V\ |6\rangle. \tag{18.71}$$

The use of perturbation theory is justified *a posteriori* by the fact that experimentally it turns out that $|c| \ll d$.

The basic states are then, with $p^2 + q^2 = 1$,

$$|\xi'\rangle = p\ |6^s\rangle + q\ |0\rangle,$$
$$|\eta'\rangle = |6^a\rangle. \tag{18.72}$$

In the presence of a magnetic field it is convenient to replace $|\xi'\rangle$ and $|\eta'\rangle$ by the linear combinations

$$|\xi\rangle = \frac{|\xi'\rangle + |\eta'\rangle}{\sqrt{2}} = \frac{1+p}{2}|6\rangle - \frac{1-p}{2}|-6\rangle + \frac{q}{\sqrt{2}}|0\rangle,$$
$$|\eta\rangle = \frac{|\xi'\rangle - |\eta'\rangle}{\sqrt{2}} = -\frac{1-p}{2}|6\rangle + \frac{1+p}{2}|-6\rangle + \frac{q}{\sqrt{2}}|0\rangle. \tag{18.73}$$

IONS IN A WEAK CRYSTAL FIELD (f ELECTRONS) 739

The Zeeman coupling $-\boldsymbol{\mu}\cdot\mathbf{H} = \Lambda\beta(\mathbf{J}\cdot\mathbf{H})$ has no matrix elements between $|\xi\rangle$ and $|\eta\rangle$ and opposite expectation values in either state:

$$\langle\xi|\,Z\,|\xi\rangle = -\langle\eta|\,Z\,|\eta\rangle = 6\Lambda\beta p H_z = 6\Lambda\beta(1-q^2)^{\frac{1}{2}}H_z.$$
(18.74)

With a fictitious spin $\tilde{S} = \frac{1}{2}$ the spin Hamiltonian becomes

$$\mathscr{H} = g_{\|}\beta H_z \tilde{S}_z + \Delta \tilde{S}_x,$$
(18.75)

where

$$g_{\|} = 12\Lambda(1-q^2)^{\frac{1}{2}} \approx 12\Lambda\left(1-\frac{c^2}{d^2}\right); \quad \Delta = \frac{2c^2}{d}.$$

The experimental results for terbium in the ethylsulphate are discussed in § 5.6. It turns out that the major part of the difference between the measured value $g_{\|} = 17\cdot72$ and the simple theoretical value of 18 is due to the influence of intermediate coupling, which leads to a value $\Lambda = 1\cdot491$ rather than $\frac{3}{2}$. Also there are appreciable corrections to Δ through matrix elements of the crystal field with excited states such as $J = 5$.

18.6. Non-Kramers rare-earth ions in cubic surroundings

There are few experimental resonance results on non-Kramers rare-earth ions in cubic surroundings and we shall limit ourselves to a brief group theoretical discussion of the form to be expected for the spin Hamiltonian.

The ground level will belong to one of the five non-specific representations of the cubic groups Γ_1 to Γ_5. Γ_1 and Γ_2 are non-degenerate singlet states and as shown in § 15.4 have no permanent moment. There is (in first order) no Zeeman or magnetic hyperfine energy in those states. The quadrupole interactions vanish also because the spatial symmetry is cubic; they would not vanish for a non-cubic singlet.

For the representations Γ_4 and Γ_5, the direct products have the same decomposition:

$$\Gamma_4 \times \Gamma_4 = \Gamma_5 \times \Gamma_5 = \Gamma_1 + \Gamma_3 + \Gamma_4 + \Gamma_5.$$
(18.76)

Since the representation Γ_4 appears only *once* on the right-hand side of (18.76) it follows that, as already explained at the beginning of § 18.3, all vectors $\mathbf{V}$ have within either Γ_4 or Γ_5 the same representation, apart from a proportionality constant. Furthermore, as shown in § 15.9, Γ_4 is contained in $[\Gamma_4 \times \Gamma_4]_A$ and only time-odd vectors $\mathbf{V}$ have non-vanishing matrix elements inside Γ_4 or Γ_5. We can thus write the Zeeman coupling, as $g\beta(\mathbf{H}\cdot\tilde{\mathbf{S}})$, and the hyperfine coupling as $A(\mathbf{I}\cdot\tilde{\mathbf{S}})$, where $\tilde{\mathbf{S}}$ is a fictitious spin $\tilde{S} = 1$.

On the other hand, just as for Γ_8, *two* constants are necessary to describe the nuclear quadrupole coupling. This coupling is represented by the same formula (18.50) as for Γ_8; the only change is that the fictitious spin is $\tilde{S} = 1$ rather than $\tfrac{3}{2}$. The formulae (18.51) giving m and n should be replaced by the following:

$$m = \frac{\langle J\| \alpha \|J\rangle}{\langle L\| \alpha \|L\rangle} \langle \tilde{1}|\, 3J_z^2 - J(J+1)\, |\tilde{1}\rangle,$$

$$n = \frac{\langle J\| \alpha \|J\rangle}{\langle L\| \alpha \|L\rangle} \frac{\langle \tilde{1}|\, J_z J_+ + J_+ J_z\, |\tilde{0}\rangle}{\sqrt{2}}.$$
(18.77)

For instance in the Γ_5 triplet for $J = 3$, we find from the kets in Table 4 that

$$\langle \tilde{1}|\, 3J_z^2 - J(J+1)\, |\tilde{1}\rangle = 0,$$

$$\frac{1}{\sqrt{2}} \langle \tilde{1}|\, J_z J_+ + J_+ J_z\, |\tilde{0}\rangle = -\tfrac{15}{2}.$$

The doublet Γ_3

This has some rather peculiar features. The direct product $\Gamma_3 \times \Gamma_3 = \Gamma_1 + \Gamma_2 + \Gamma_3$ and the matrix elements of any vector vanish within the manifold Γ_3, which has neither Zeeman nor magnetic hyperfine energy. Since $\Gamma_3 \times \Gamma_3$ (more precisely $[\Gamma_3 \times \Gamma_3]_S$) does contain Γ_3, this doublet does however have a non-vanishing quadrupole coupling.

The components $-e \sum_i (3z_i^2 - r_i^2)$ and $-e \sum_i (x_i^2 - y_i^2)$ of the electronic quadrupole moment, which transform according to Γ_3, will have the same representation, apart from a proportionality factor, in *all* doublets Γ_3 and we can therefore without loss of generality choose for the states spanning Γ_3 the eigenstates of $J = 2$,

$$|a'\rangle = \frac{|2, 2\rangle + |2, -2\rangle}{\sqrt{2}},$$

$$|b'\rangle = |2, 0\rangle;$$
(18.78)

and use as components of the electronic quadrupole moment, $3J_z^2 - J(J+1)$ and $(J_x^2 - J_y^2)$.

The matrix representation of the quadrupole interaction which is proportional to (see eqn (17.21))

$$\tfrac{1}{3}\{3J_z^2 - J(J+1)\}\{3I_z^2 - I(I+1)\} + (J_x^2 - J_y^2)(I_x^2 - I_y^2) \qquad (18.79)$$

becomes in the basis $|a'\rangle$, $|b'\rangle$

$$2\begin{vmatrix} 3I_z^2-I(I+1) & \sqrt{3}(I_x^2-I_y^2) \\ \sqrt{3}(I_x^2-I_y^2) & -\{3I_z^2-I(I+1)\} \end{vmatrix} \quad (18.80)$$

which, by introducing a fictitious spin $\tilde{s} = \frac{1}{2}$ with

$$\langle a'|\,\tilde{s}_3\,|a'\rangle = -\langle b'|\,\tilde{s}_3\,|b'\rangle = \tfrac{1}{2},$$

corresponds to a spin Hamiltonian

$$\tilde{s}_3\{3I_z^2-I(I+1)\}+\sqrt{3}\,\tilde{s}_1(I_x^2-I_y^2). \quad (18.81)$$

We can always define two new states $|a\rangle$ and $|b\rangle$ through a unitary unimodular substitution of the basis states

$$\begin{aligned}|a'\rangle &= p\,|a\rangle+q\,|b\rangle,\\ |b'\rangle &= -q^*\,|a\rangle+p\,|b\rangle, \\ pp^*+qq^* &= 1.\end{aligned} \quad (18.82)$$

This is equivalent to a 'rotation' of the components of $\tilde{s}$ (which has *nothing* to do with a spatial rotation of the coordinate axes). If we choose for the 'rotation'

$$\begin{aligned}\tilde{s}_1 &= \frac{-\tilde{s}_x+\tilde{s}_y}{\sqrt{2}}, \\ \tilde{s}_2 &= \frac{\tilde{s}_x+\tilde{s}_y+\tilde{s}_z}{\sqrt{3}}, \\ \tilde{s}_3 &= \frac{\tilde{s}_x+\tilde{s}_y-2\tilde{s}_z}{\sqrt{6}},\end{aligned} \quad (18.83)$$

where $\tilde{s}_x$, $\tilde{s}_y$, $\tilde{s}_z$ are the new components of the fictitious spin, the Hamiltonian (18.81) takes the symmetrical form

$$-\frac{2}{\sqrt{6}}[\tilde{s}_x\{3I_x^2-I(I+1)\}+\tilde{s}_y\{3I_y^2-I(I+1)\}+\tilde{s}_z\{3I_z^2-I(I+1)\}]. \quad (18.84)$$

We have explained in § 13.2 how coefficients such as p and q of the unimodular transformation (18.82) are related to those of the three-dimensional 'rotation' (18.83).

19
INTERMEDIATE CRYSTAL FIELDS
(THE IRON GROUP)

19.1. Effect of the cubic crystal potential

WE have stated earlier that in the iron group, as contrasted with the rare-earth group, the crystal potential is as a rule much stronger than the spin-orbit coupling and that in a first step of the calculation the latter can be disregarded altogether. At this stage the distinction between Kramers and non-Kramers ions disappears, for the relevant quantum number is L rather than J and it is an integer for both types of ions.

A second difference from the rare-earth group arises from the important role played by the cubic crystal potential. In most compounds where resonance of iron-group elements has been observed, even if the local symmetry 'seen' by the paramagnetic ion is lower than cubic, the cubic part of the crystal potential is much larger than components of lower symmetry and it is very instructive for a first orientation to examine the behaviour of the various ions in a field of cubic symmetry.

An inspection of Table 19 shows that the ground terms in the iron group are S, D, F. The first is an orbital singlet unaffected by the crystal potential in the absence of spin-dependent forces. Table 3 shows that D terms are split into a doublet Γ_3 and a triplet Γ_5, while F terms split into a singlet Γ_2 and two triplets Γ_4 and Γ_5. The cubic potential energy may be written (eqn (16.13)) as $V_{\text{cubic}} = A_4 P_4$ (the sixth-order term gives a vanishing contribution for d-electrons) where P_4 is defined in (16.11). Within a (L, S) term, P_4 can be replaced by the equivalent operator $\beta A_4 \langle r^4 \rangle O_4(\mathbf{L})$, the values of $\beta = \langle L\| \beta \|L\rangle$ being given in Table 19 while $O_4 = O_4^0 + 5O_4^4$ is defined in Table 16. Using the matrix elements of the O_k^q operators given in Table 17 and the expansions of the wave-functions that span the various levels Γ_i, given in Table 4, the energy levels of the paramagnetic ions in the cubic field are easily computed. For D terms, with $L = 2$, we have

$$W(\Gamma_3) = 72\beta A_4 \langle r^4 \rangle,$$

$$W(\Gamma_5) = -48\beta A_4 \langle r^4 \rangle, \qquad (19.1)$$

$$W(\Gamma_3) - W(\Gamma_5) = 120\beta A_4 \langle r^4 \rangle,$$

INTERMEDIATE CRYSTAL FIELDS (THE IRON GROUP) 743

and for F terms, with $L = 3$, we have

$$W(\Gamma_2) = -720\beta A_4 \langle r^4 \rangle,$$
$$W(\Gamma_5) = -120\beta A_4 \langle r^4 \rangle, \qquad (19.2)$$
$$W(\Gamma_4) = 360\beta A_4 \langle r^4 \rangle,$$

whence

$$\frac{W(\Gamma_2) - W(\Gamma_5)}{W(\Gamma_5) - W(\Gamma_4)} = \frac{5}{4}.$$

The triplet Γ_5 thus always lies between the singlet Γ_2 and the triplet Γ_4 and nearer to the latter by a factor $\frac{4}{5}$. With $B_4 = \beta A_4 \langle r^4 \rangle$, these results correspond to the splittings shown in Figs. 7.3–7.6.

If we assume that the point-charge model gives correctly at least the sign if not the magnitude of the cubic potential, we see from eqn (16.15) that A_4 is positive for octahedral coordination but negative for cubic or tetrahedral coordination, eqns (16.16), (16.17). Since, for a single d-electron, $\langle l \| \beta \| l \rangle = \frac{2}{63}$ is positive, (see Table 18), it follows from (19.1) that for such an electron in octahedral coordination the triplet Γ_5 is lower than the doublet Γ_3 by

$$120 \langle l \| \beta \| l \rangle A_4 \langle r^4 \rangle = \tfrac{80}{21} A_4 \langle r^4 \rangle.$$

In the point-charge model this becomes $\tfrac{5}{3}(e^2/R)(\langle r^4 \rangle / R^4)$. More generally from the signs of $\langle l \| \beta \| l \rangle$ in Table 18 and the formulae (19.1), (19.2), the nature of the ground multiplet within the iron group in octahedral coordination will be

$$\begin{cases} d^1 & d^2 & d^3 & d^4 & d^5 & d^6 & d^7 & d^8 & d^9 \\ \Gamma_5 & \Gamma_4 & \Gamma_2 & \Gamma_3 & \Gamma_1 & \Gamma_5 & \Gamma_4 & \Gamma_2 & \Gamma_3. \end{cases} \qquad (19.3)$$

In cubic or tetrahedral coordination the ground multiplets will be

$$\Gamma_3 \quad \Gamma_2 \quad \Gamma_4 \quad \Gamma_5 \quad \Gamma_1 \quad \Gamma_3 \quad \Gamma_2 \quad \Gamma_4 \quad \Gamma_5. \qquad (19.4)$$

The degeneracy and the spin of the ground level of an iron-group ion in a field of cubic symmetry can be predicted by a simple qualitative argument that makes it easy to memorize the important results (19.3), (19.4). Let us choose, for instance, octahedral coordination; assuming the validity of the point-charge approximation, let us consider for a single d-electron the two wave-functions $x^2 - y^2$ and xy which belong respectively to Γ_3 and Γ_5 and are obtained the one from the other by a $\pi/4$ rotation around the z-axis. The function xy vanishes along the x-axis and y-axis where four of the six negative charges surrounding the paramagnetic ion are situated. The electrostatic repulsion between a d-electron and these charges will thus be smaller for the xy function than for the x^2-y^2 function (which has maxima along these axes) and the triplet Γ_5 will be lower than the doublet Γ_3. The $3d$ wave-functions that span Γ_5 are usually called $d\varepsilon$ in the literature, while the two functions of Γ_3 are called $d\gamma$ (an alternative nomenclature is respectively t_{2g} and e_g).

With more than one d-electron, we obviously lower the energy of the ion by placing further electrons in the orbitals $d\varepsilon$ at lower energy.

On the other hand, because of Hund's rule we also reduce the energy of electrostatic repulsion between the electrons by giving to the total spin S the maximum possible value. When there is a conflict between the two rules, as will be the case for more than three d-electrons, we assume that Hund's rule should be the determining factor. This corresponds to the assumption of a crystal potential too weak to break the LS-coupling (we shall return to this question in Chapter 20 when discussing covalent bonding). The ground states are then as shown in Fig. 7.7. With one or two electrons in the $d\varepsilon$ shell, there are three possible arrangements with parallel spin, giving a triplet orbital ground state. However, three electrons with parallel spin give a half-filled $d\varepsilon$ shell, with only one possible arrangement; thus d^3 has a singlet orbital ground state. At d^4 one electron must enter the $d\gamma$ shell in order to have parallel spin, giving a twofold orbital degeneracy, while at d^5 both the $d\varepsilon$ and the $d\gamma$ shells are half-filled, giving an orbital singlet and the maximum spin $S = \frac{5}{2}$ for the half-filled d-shell. At d^6 the extra electron enters the $d\varepsilon$ shell with antiparallel spin, giving a lower total spin; as for d^1, the antiparallel electron has three orbital possibilities, giving an orbital triplet ground state. Further electrons can only be added with anti-parallel spin, giving a progressive reduction in the value of S; the orbital ground state of d^7, like that of d^2, is a triplet, while for d^8, like d^3, we have an orbital singlet, in this case because the $d\varepsilon$ shell is full and the $d\gamma$ shell half-full. Finally at d^9 we have just one hole in the $d\gamma$ shell, with a twofold orbital degeneracy; at d^{10} the whole shell is closed.

We have assumed above octahedral symmetry, which gives the $d\varepsilon$ orbitals a lower energy than the $d\gamma$ orbitals. In cubic (tetrahedral) symmetry we can proceed similarly, but the $d\gamma$ shell now lies below the $d\varepsilon$ shell in energy. It is readily seen that the required results are simply equivalent to working backwards in Fig. 7.7, interchanging the configurations d^n and d^{10-n}, or just by considering the 'holes' rather than the electrons.

The results for the orbital ground state are summed up in Table 7.2.

This gives a useful classification in preparation for the deduction of the appropriate spin Hamiltonian to describe the magnetic resonance behaviour. In fact the singlet and doublet ground states can often be treated together, for the following reason. As we have seen in § 14.2, the doublet Γ_3 is 'non-magnetic', and within its manifold there are no non-vanishing matrix elements of the spin-orbit coupling $\lambda(\mathbf{L}\cdot\mathbf{S})$. If a crystal field of lower symmetry is superimposed on the cubic field, the doublet Γ_3 splits into two orbitals $|O\rangle$ and $|O'\rangle$ (except in the case of a trigonal field which does *not* split Γ_3 and requires a special treatment). Since $\lambda(\mathbf{L}\cdot\mathbf{S})$ does not couple the ground state $|O\rangle$ to $|O'\rangle$ but only to states $|n\rangle$ lying above $|O\rangle$ by energies of the order of the cubic splitting, which is much larger than $|\lambda|$, it will prove possible to use the same perturbation methods for treating the spin-orbit coupling for the doublet states as for the singlet orbital ground states. (See however § 21.3).

For simplicity we shall call these ions with singlet or doublet ground orbital states 'ions of Type A', while those with triplet orbital ground states will be called 'ions of Type B'.

19.2. 'Singlet' orbital ground states (ions of type A)

In this section we are concerned with ions that have an orbital ground singlet state well-separated from all excited states. By this we mean that any excited states $|n\rangle$ that have matrix elements through the spin-orbit coupling with the ground state $|O\rangle$, lie at energies W_n such that $W_n - W_0 \gg |\lambda|$. We have written 'singlet' in inverted commas in the title of this section to signify that, when appropriate, split Γ_3 doublet states are also included, as discussed in the preceding paragraph. Our aim is to discover how the ground manifold, which has no orbital degeneracy left but which still possesses a spin degeneracy of $(2S+1)$, is split by the interactions we have not so far included. These interactions are as follows.

Spin orbit coupling:
$$W_{LS} = W''_{LS} = \lambda(\mathbf{L}\cdot\mathbf{S}). \tag{19.5}$$

Spin-spin interaction (see eqn (16.43)):
$$W_{SS} = W'_{SS} = -\rho\sum_{p,q}\left\{\frac{L_pL_q+L_qL_p}{2}-\tfrac{1}{3}L(L+1)\,\delta_{pq}\right\}S_pS_q. \tag{19.6}$$

Zeeman interaction:
$$Z = Z'+Z'' = \beta g_s(\mathbf{H}\cdot\mathbf{S})+\beta(\mathbf{H}\cdot\mathbf{L}). \tag{19.7}$$

Magnetic hyperfine structure (see eqn (17.61)):

$$W_\mu = W'_\mu + W''_\mu,$$

where

$$W'_\mu = \mathscr{P}[\{\xi L(L+1)-\kappa\}(\mathbf{S}\cdot\mathbf{I}) - \tfrac{3}{2}\xi(\mathbf{L}\cdot\mathbf{S})(\mathbf{L}\cdot\mathbf{I}) - \tfrac{3}{2}\xi(\mathbf{L}\cdot\mathbf{I})(\mathbf{L}\cdot\mathbf{S})], \quad (19.8)$$

$$W''_\mu = \mathscr{P}(\mathbf{L}\cdot\mathbf{I}), \quad (19.9)$$

$$\mathscr{P} = 2\beta\gamma_n\hbar\langle r^{-3}\rangle. \quad (19.10)$$

Quadrupole hyperfine structure (eqn (17.24)):

$$W_q = W'_q = \sum_{p,q} q'\{\tfrac{1}{2}(L_pL_q+L_qL_p) - \tfrac{1}{3}\delta_{pq}L(L+1)\}\{\tfrac{1}{2}(I_pI_q+I_qI_p) - \tfrac{1}{3}\delta_{pq}I(I+1)\}, \quad (19.11)$$

where

$$q' = \frac{-3e^2Q}{2I(2I-1)}\langle r^{-3}\rangle\langle L\|\alpha\|L\rangle. \quad (19.12)$$

Nuclear Zeeman interaction:

$$Z_n = Z'_n = -\gamma_n\hbar(\mathbf{H}\cdot\mathbf{I}). \quad (19.13)$$

We have put a single prime on the interactions that are even with respect to components of the orbital momentum **L** and have non-vanishing expectation values in the ground manifold, and a double prime on those that are linear with respect to components of **L** and whose expectation value, according to Van Vleck's theorem, vanishes in the ground state. Thus in first-order perturbation theory, only the single primed interactions contribute to the splitting of the ground manifold.

The states spanning this manifold can be written $|O\rangle|i\rangle$ where $|O\rangle$ represents the non-degenerate orbital wave-function, and $|i\rangle$ specifies the various $(2S+1)$ spin-substates.

In taking the expectation values of the single primed interactions it is convenient to perform the integration over the orbital variables only, leaving out the spin-dependent parts in the operator form. We obtain thus a first contribution to what is called the spin-Hamiltonian

$$-\rho l_{pq}S_pS_q + g_s\beta\mathbf{H}\cdot\mathbf{S} - \mathscr{P}(\kappa\delta_{pq}+3\xi l_{pq})S_pI_q + q'l_{pq}I_pI_q - \gamma_n\hbar\mathbf{H}\cdot\mathbf{I}, \quad (19.14)$$

where

$$l_{pq} = \tfrac{1}{2}\langle O|\,L_pL_q+L_qL_p\,|O\rangle - \tfrac{1}{3}L(L+1)\delta_{pq}. \quad (19.15)$$

It would, however, be grossly incorrect to content ourselves with first-order perturbation theory because the spin–orbit coupling is very much greater than the other interactions listed in (19.6)–(19.13), and second-order contributions involving the spin-orbit coupling, either quadratically or as cross terms with other interactions, may well be comparable to the first-order terms in (19.14). We can use for the excited levels of the bound ion, which are eigenstates of the crystal potential V, the same description as for the ground manifold: $|n\rangle |j\rangle$ where $|n\rangle$ represents an orbital state with an eigenvalue W_n and $|j\rangle$ the spin substates. If we remain inside the term (L, S) of the free ion (which is a necessary condition for eqns (19.5)–(19.12) for the various interactions to be valid), there are as many spin substates, $(2S+1)$, as for the ground level. In second-order perturbation theory the relevant operator is

$$C = \sum_{\substack{n \neq 0 \\ j}} \frac{|n\rangle |j\rangle\langle j| \langle n|}{W_0 - W_n}, \qquad (19.16)$$

or, since $\sum_j |j\rangle\langle j| = 1$, we can write

$$C = -\sum_{n \neq 0} \frac{|n\rangle\langle n|}{W_n - W_0} \qquad (19.17)$$

which is a purely orbital operator that is clearly time-even.

Using the same procedure as for the first-order terms, namely, performing the integration over orbital variables only, we find that the double-primed operators bring in second order the following contributions to the splitting of the ground state:

$$\langle 0| \lambda(\mathbf{L}\cdot\mathbf{S})C\lambda(\mathbf{L}\cdot\mathbf{S}) |0\rangle = -\lambda^2 \sum_{p,q} \Lambda_{pq} S_p S_q, \qquad (19.18)$$

where

$$\Lambda_{pq} = \sum_n{}' \frac{\langle 0| L_p |n\rangle\langle n| L_q |0\rangle}{W_n - W_0} \qquad (19.19)$$

for the term quadratic in the spin-orbit coupling. From the definition (19.19), we see immediately that

$$\Lambda_{pq} = \Lambda_{qp}^*. \qquad (19.20)$$

Actually it can be shown from time-reversal symmetry that Λ_{pq} is symmetrical, i.e. $\Lambda_{pq} = \Lambda_{qp}$ and is therefore real. For, since the state $|0\rangle$ is a non-degenerate eigenstate of the crystal field Hamiltonian, it must coincide within a phase factor with its time-reversed state $\theta |0\rangle$ and hence

$$-\Lambda_{pq} = \langle 0| L_p C L_q |0\rangle = (0, L_p C L_q 0) = (\theta L_p C L_q 0, \theta 0)$$
$$= (\theta L_p C L_q \theta^{-1} \theta 0, \theta 0) = (L_p C L_q \theta 0, \theta 0) = (\theta 0, (L_p C L_q)^\dagger \theta 0)$$
$$= (\theta 0, L_q C L_p \theta 0) = \langle 0| L_q C L_p |0\rangle = -\Lambda_{qp}. \qquad (19.21)$$

The cross term between $\lambda(\mathbf{L}\cdot\mathbf{S})$ and $Z'' = \beta(\mathbf{L}\cdot\mathbf{H})$ gives

$$\lambda\beta\langle 0|\,(\mathbf{L}\cdot\mathbf{S})C(\mathbf{L}\cdot\mathbf{H})+(\mathbf{L}\cdot\mathbf{H})C(\mathbf{L}\cdot\mathbf{S})\,|0\rangle = -2\beta\lambda\sum_{p,q}\Lambda_{pq}H_pS_q \quad (19.22)$$

where Λ_{pq} is defined in (19.19).

The cross term between $\lambda(\mathbf{L}\cdot\mathbf{S})$ and W''_μ gives similarly:

$$-\lambda\mathscr{P}\langle 0|\,(\mathbf{L}\cdot\mathbf{S})C(\mathbf{L}\cdot\mathbf{I})+(\mathbf{L}\cdot\mathbf{I})C(\mathbf{L}\cdot\mathbf{S})\,|0\rangle = -2\lambda\mathscr{P}\sum_{p,q}\Lambda_{pq}S_pI_q. \quad (19.23)$$

We may also consider cross terms between $\lambda(\mathbf{L}\cdot\mathbf{S})$ and single primed terms of (19.6)–(19.13). First, the spin Zeeman interaction

$$Z' = \beta g_s(\mathbf{H}\cdot\mathbf{S})$$

has no off-diagonal matrix elements between $|0\rangle$ and $|n\rangle$ and there is thus no second-order contribution from that term. There is, however, a cross term with the spin-spin interaction which is

$$-\rho\lambda\langle 0|\,(\mathbf{L}\cdot\mathbf{S})C\left(\frac{L_pL_q+L_qL_p}{2}\right)S_pS_q + \left(\frac{L_pL_q+L_qL_p}{2}\right)S_pS_qC(\mathbf{L}\cdot\mathbf{S})\,|0\rangle$$

$$= \frac{\rho\lambda}{2}\sum_{p,q,r}(\Lambda_{r,pq}S_rS_pS_q+\Lambda_{pq,r}S_pS_qS_r), \quad (19.24)$$

where

$$\Lambda_{r,pq} = \sum_n{}' \frac{\langle 0|L_r|n\rangle\langle n|L_pL_q+L_qL_p|0\rangle}{W_n-W_0},$$

$$\Lambda_{pq,r} = \sum_n{}' \frac{\langle 0|L_pL_q+L_qL_p|n\rangle\langle n|L_r|0\rangle}{W_n-W_0}. \quad (19.25)$$

A time-reversal argument very similar to that used to demonstrate the symmetry of Λ_{pq} shows that $\Lambda_{r,pq} = -\Lambda_{pq,r}$. Since from Hermitian properties $\Lambda_{r,pq} = \Lambda^*_{pq,r}$, it follows that $\Lambda_{r,pq}$ is purely imaginary. Eqn (19.24) can then be rewritten

$$\frac{\lambda\rho}{2}\sum_{pq,r}\Lambda_{r,pq}[S_r,S_pS_q] \quad (19.26)$$

or, using the commutation relations $[S_r, S_p] = i\sum_j \varepsilon_{rpj}S_j$, as

$$+\lambda\rho(\Lambda'_{pq}+\Lambda'_{qp})S_pS_q, \quad (19.27)$$

where

$$\Lambda'_{pq} = -\frac{i}{2}\sum_{t,r}\varepsilon_{ptr}\Lambda_{r,qt}$$

$$= -\frac{i}{2}\sum_{t,r}\varepsilon_{ptr}\sum_n{}' \frac{\langle 0|L_r|n\rangle\langle n|L_qL_t+L_tL_q|0\rangle}{W_n-W_0}. \quad (19.28)$$

From the symmetry of the $\Lambda_{r,qt}$ it follows that the coefficient Λ'_{pq} is real but not necessarily symmetrical. In practice (19.27) is smaller than the first-order effect of spin-spin interaction $-\rho l_{pq}S_p S_q$ given in (19.14) and indistinguishable from it.

The cross term between $\lambda(\mathbf{L} \cdot \mathbf{S})$ and W'_μ will give similarly:

$$3\xi\lambda\,\mathscr{P}\Lambda'_{pq}S_p I_q. \tag{19.29}$$

Finally, to be consistent we should add but will not write out the small cross term between $\lambda(\mathbf{L} \cdot \mathbf{S})$ and the quadrupole interaction W'_q.

Another second-order term which, although very small is by no means unobservable because it competes with a first-order term also small, is the nuclear pseudo-Zeeman coupling defined by eqn (18.2). This is a cross term between the magnetic hyperfine structure W''_μ and the electronic Zeeman coupling $Z'' = \beta(\mathbf{L} \cdot \mathbf{H})$. The calculation is clearly identical to that which has given the second-order hyperfine terms (19.23), provided $\lambda(\mathbf{L} \cdot \mathbf{S})$ is replaced by $\beta(\mathbf{L} \cdot \mathbf{H})$, and yields

$$-2\mathscr{P}\beta\Lambda_{pq}H_p I_q = -\gamma_n\hbar(\mathbf{H} \cdot \boldsymbol{\alpha} \cdot \mathbf{I}), \tag{19.30}$$

with

$$\alpha_{pq} = 4\beta^2\langle r^{-3}\rangle\Lambda_{pq}. \tag{19.31}$$

Gathering together all the above terms, we obtain the so-called spin Hamiltonian

$$\mathscr{H}_\mathrm{S} = \{-\lambda^2\Lambda_{pq}-\rho l_{pq}+\lambda\rho(\Lambda'_{pq}+\Lambda'_{qp})\}S_p S_q+$$
$$+\beta(g_s\delta_{pq}-2\lambda\Lambda_{pq})H_p S_q - \mathscr{P}(\kappa\delta_{pq}+3\xi l_{pq}+2\lambda\Lambda_{pq}-3\xi\lambda\Lambda'_{pq})S_p I_q+$$
$$+q'l_{pq}I_p I_q-\gamma_n\hbar H_p(\delta_{pq}+\alpha_{pq})I_q, \tag{19.32}$$

which we rewrite as

$$\mathscr{H}_\mathrm{S} = D_{pq}S_p S_q+\beta g_{pq}H_p S_q+A_{pq}S_p I_q+$$
$$+ P_{pq}I_p I_q-\gamma_n\hbar\{(\mathbf{H} \cdot \mathbf{I})+(\mathbf{H} \cdot \boldsymbol{\alpha} \cdot \mathbf{I})\}. \tag{19.33}$$

We have already met various spin-Hamiltonians in previous chapters; for instance in the description of Kramers doublets (eqn (15.36)) or of the cubic multiplet Γ_8 (eqn (18.21)). There are, however, important differences between those expressions and the spin-Hamiltonian (19.33). The former spin-Hamiltonians were written *a priori* as the most general expressions in operator form capable of describing the splitting of, say, a Kramers doublet or a Γ_8 quadruplet. On the contrary (19.33) does not claim such generality; it is an expression valid to the second order of perturbation theory only but to that order of accuracy explicit theoretical expressions are given for the various coefficients that appear in it, which may at least in principle be compared with the numerical values extracted from the study of the resonance spectrum.

An even more important difference between the previous spin Hamiltonians and (19.33) resides in the fact that the former were constructed from components of fictitious spins that bear little relation to the actual spin of the ion. This was particularly striking in the description of non-Kramers ions in the rare-earth group where, as pointed out in § 18.5, different components of the fictitious spin had different time-reversal properties.

In contrast the spin S in (19.33) is almost exactly the real spin of the ion. We say almost exactly for the following reason: the coupling of states of the ground manifold $|O\rangle |i\rangle$ with excited states $|n\rangle |j\rangle$ changes the ground manifold into perturbed states

$$|Oi\rangle = T |O\rangle |i\rangle = (1-\alpha_i) |O\rangle |i\rangle + \sum_{n,j}' \beta_{nj} |n\rangle |j\rangle, \quad (19.34)$$

where T is a unitary operator. The β_{nj} are of first order in the ratio $\lambda/(W_n - W_0)$ while α_i is of second order. A matrix element $\langle Oi| S_q |Oi'\rangle$ will thus differ from the usual spin matrix element $\langle i| S_q |i'\rangle$ by terms of the order of α_i and $\beta_{nj} \beta_{nj'}^*$, that is in second order only.

It follows that to the same approximation the coefficients D_{pq}, g_{pq}, A_{pq} in (19.33) are indeed tensors (P_{pq} and α_{pq} are always tensors since I and H are always real vectors). The tensors g_{pq} and D_{pq} are always symmetrical but A_{pq} is not necessarily so because of the terms Λ_{pq}'. However for symmetry no lower than rhombic, that is if the environment possesses three orthogonal binary symmetry axes, the off-diagonal elements D_{pq}, g_{pq}, A_{pq} vanish and all tensors admit these symmetry axes as principal axes. We shall demonstrate it for Λ_{pq}' which is the least obvious. According to eqn (19.28)

$$\Lambda_{12}' = i\langle O| L_3 C L_2^2 - L_2 C \left(\frac{L_2 L_3 + L_3 L_2}{2}\right) |O\rangle = i\langle O| A |O\rangle. \quad (19.35)$$

Since $|O\rangle$ is a non-degenerate eigenstate of the system, a rotation by an angle π: $R = e^{-i\pi L_1}$ around the axis Ox_1 changes $|O\rangle$ into itself (within a phase factor) whereas it changes the sign of the operator inside the symbol $\langle O| \quad |O\rangle$. Hence

$$\Lambda_{12}' = i\langle O| A |O\rangle = i\langle O| R^+ R A R^{-1} R |O\rangle = i\langle O| R A R^{-1} |O\rangle$$
$$= -i\langle O| A |O\rangle = -\Lambda_{12}' = 0. \quad (19.36)$$

For lower symmetry it is conceivable in principle to have, in the magnetic hyperfine structure, terms of the form

$$(\Lambda'_{12}-\Lambda'_{21})(S_1 I_2 - S_2 I_1) = (\Lambda'_{12}-\Lambda'_{21})(S \wedge I)_3,$$

although no such terms appear to have been observed in practice. Experimentally at least, the tensors D, g, A, P, and α have so far always been found to have the same principal axes.

With respect to these axes the spin Hamiltonian (19.33) is usually rewritten as (see Chapter 3)

$$\mathcal{H}_S = D\{S_z^2 - \tfrac{1}{3}S(S+1)\} + E(S_x^2 - S_y^2) + \beta(g_z H_z S_z + g_x H_x S_x + g_y H_y S_y) +$$

$$+ A_z S_z I_z + A_x S_x I_x + A_y S_y I_y + P_\parallel \left\{ I_z^2 - \tfrac{1}{3}I(I+1) + \tfrac{1}{3}\eta(I_x^2 - I_y^2) \right\} -$$

$$- \gamma_n \hbar \{(1+\alpha_x) H_x I_x + (1+\alpha_y) H_y I_y + (1+\alpha_z) H_z I_z\}. \quad (19.37)$$

For tetragonal or trigonal symmetry

$$E = 0, \quad g_x = g_y = g_\perp, \quad g_z = g_\parallel,$$
$$A_x = A_y = A_\perp, \quad A_z = A_\parallel, \quad \eta = 0, \quad (19.38)$$
$$\alpha_x = \alpha_y = \alpha_\perp, \quad \alpha_z = \alpha_\parallel.$$

We wish to emphasize once more that (19.33) is *not* the most general expression for the spin-Hamiltonian permitted by symmetry considerations. For instance in cubic symmetry in order to describe correctly the observed spectrum, terms of the form $S_x^4 + S_y^4 + S_z^4$ have to be added, and sometimes small Zeeman and hyperfine couplings involving S_x^3, S_y^3, S_z^3 as given in eqn (18.21). These terms can be obtained by methods similar to those used in deriving (19.33) by going to higher orders in perturbation theory.

A spin Hamiltonian such as (19.33) (augmented if necessary by higher-order spin polynomials) is the meeting point between the theoretician who attempts to calculate the various coefficients in (19.33) from what is known or can be surmised about the crystal field and the wave-functions of the free ion, and the experimentalist who, using the methods of computation described in Chapter 3, extracts from the observed spectrum numerical values for the same coefficients.

19.3. Triplet orbital ground state (ions of type B)

We consider now those ions which in a cubic field have as their ground level an orbital triplet Γ_4 or Γ_5.

Cubic symmetry

Assume first that the symmetry is exactly cubic. The degeneracy of the ground level is then $3(2S+1)$ including that due to spin. As we pointed out earlier in § 14.2, within this manifold every orbital vector can be replaced by an equivalent vector operator $\alpha \tilde{\mathbf{l}}$, where $\tilde{\mathbf{l}}$ is a fictitious angular momentum of magnitude 1. In particular, the spin-orbit coupling can be written within this manifold as $\alpha\lambda(\tilde{\mathbf{l}} \cdot \mathbf{S})$, the values of α being tabulated in Table 4. If we neglect the admixtures into the ground manifold Γ_i, where i is 4 or 5, from excited orbital multiplets Γ_j, through spin-orbit coupling, we can couple $\tilde{\mathbf{l}}$ and $\mathbf{S}$ according to the usual rules of angular momentum coupling and obtain $\tilde{J}$-multiplets where $\tilde{J}$ is a fictitious total angular momentum that takes the values $S+1$, S, $|S-1|$. (The tilde emphasizes the fictitious character of $\tilde{J}$.) In order to have a concrete example, we consider the ion Co^{2+} with a ground term ^4F, split in an octahedral environment into a ground orbital triplet Γ_4, an excited triplet Γ_5, and an excited singlet Γ_2. The spin-orbit coupling will lift the twelvefold degeneracy of the ground manifold into three pseudo-$\tilde{J}$ multiplets $\tilde{J} = \frac{1}{2}$, $\tilde{J} = \frac{3}{2}$, $\tilde{J} = \frac{5}{2}$ separated respectively by

$$W(\Gamma_4, \tfrac{5}{2}) - W(\Gamma_4, \tfrac{3}{2}) = \tfrac{5}{2}\lambda\alpha,$$
$$W(\Gamma_4, \tfrac{3}{2}) - W(\Gamma_4, \tfrac{1}{2}) = \tfrac{3}{2}\lambda\alpha. \tag{19.39}$$

Since for Co^{2+}, which is in the second half of the $3d$ shell, λ is negative and α, from Table 4, is $-\tfrac{3}{2}$, the doublet $\tilde{J} = \tfrac{1}{2}$ will be lowest. Its g-factor can be calculated by an adaptation of Landé's formula where the orbital gyromagnetic factor α or, more generally, $\tilde{g}_l$, is $-\tfrac{3}{2}$ rather than unity:

$$g(\Gamma_4, \tilde{J}) = \frac{1}{\tilde{J}(\tilde{J}+1)}\{\tilde{g}_l(\tilde{\mathbf{l}} \cdot \tilde{\mathbf{J}}) + g_s(\mathbf{S} \cdot \tilde{\mathbf{J}})\} \tag{19.40}$$

which for $\tilde{J} = \tfrac{1}{2}$ gives

$$\frac{5g_s - 2\tilde{g}_l}{3} \tag{19.40a}$$

or for $g_s = 2$, $\tilde{g}_l = -\tfrac{3}{2}$, $g_J = \tfrac{13}{3} = 4\cdot 33$, to be compared with experimental value obtained for Co^{2+} in MgO: $g = 4\cdot 28$. (Reasons for the slight discrepancy will be discussed in Chapters 20 and 21.)

If we tackle the same problem using orthodox group theory we find that the twelvefold degenerate ground manifold provides a representation $\Gamma_4 \times D^{\frac{3}{2}} = \Gamma_4 \times \Gamma_8$ of the cubic group which, according to Table 8, is split into:

$$\Gamma_6 + \Gamma_7 + 2\Gamma_8. \tag{19.41}$$

We recognize Γ_6 as being the level $\tilde{J} = \tfrac{1}{2}$ and one of the representations Γ_8 as being $\tilde{J} = \tfrac{3}{2}$. The other representation Γ_8, and Γ_7, originate in the multiplet $\tilde{J} = \frac{5}{2}$ which in cubic symmetry, according to Table 7, contains the representations Γ_7 and Γ_8. However, whereas the abstract group-theory gives us no indication about the distances between the four multiplets (19.41), our perturbation approach is much more explicit. In first approximation the distances between the various levels are given by (19.39), whereas the splitting of $\tilde{J} = \frac{5}{2}$ into Γ_7 and Γ_8 is very much smaller and due to a very much higher-order perturbation mechanism involving matrix elements of the spin-orbit coupling (for which now the correct form $\lambda(\mathbf{L}\cdot\mathbf{S})$ rather than the reduced form $\alpha\lambda(\tilde{\mathbf{l}}\cdot\mathbf{S})$ must be used) between the ground orbital triplet Γ_4 and the excited orbital multiplets Γ_2 and Γ_5. Phenomenologically the splitting of $\tilde{J} = \frac{5}{2}$ into Γ_7 and Γ_8 can be described by an equivalent operator $(a/6)(\tilde{J}_x^4+\tilde{J}_y^4+\tilde{J}_z^4)$ added to the spin-Hamiltonian of $J = \frac{3}{2}$. The constant a would involve the spin-orbit constant λ in fourth order, or λ in second order and the spin-spin constant ρ in first order, or ρ in second order. Similarly the Zeeman and hyperfine structure would involve small terms in $\tilde{J}_x^3$ and $\tilde{J}_x^5$ for $\tilde{J} = \frac{5}{2}$ and terms $\tilde{J}_x^3$ for $\tilde{J} = \frac{3}{2}$ with coefficients that would again be polynomials in λ and ρ: (λ^4, $\lambda^2\rho$, ρ^2 for the $\tilde{J}_x^5$ terms and λ^2, ρ for the $\tilde{J}_x^3$ terms.)

Coming back to earth, that is to first-order perturbation theory, we are in a situation very similar to that prevailing in the rare-earth group. The main part of the magnetic hyperfine coupling will be of the form $A_J(\mathbf{I}\cdot\tilde{\mathbf{J}})$ where the constant A_J will be obtained by calculating in, say the state $|\tilde{J}, \tilde{J}_z = \tilde{J}\rangle$ the expectation value of the magnetic hyperfine structure as defined in eqn (17.61). Thus

$$A_J = \frac{\mathscr{P}}{\tilde{J}} \langle \mid L_z + \{\xi L(L+1)-\kappa\}S_z - \tfrac{3}{2}\xi\{L_z(\mathbf{L}\cdot\mathbf{S})+(\mathbf{L}\cdot\mathbf{S})L_z\} \mid \rangle. \tag{19.42}$$

For instance, the state $\tilde{J} = \frac{5}{2}$, $\tilde{J}_z = \frac{5}{2}$ is the state $\tilde{l}_z = 1$, $S_z = \frac{3}{2}$. The state $\tilde{l}_z = 1$ of a multiplet Γ_4 originating in the reduction of an F-term is found from Table 4 to be: $|\tilde{1}\rangle = \sqrt{(\tfrac{5}{8})}\mid-3\rangle+\sqrt{(\tfrac{3}{8})}\mid1\rangle$ where $\mid-3\rangle$ and $\mid1\rangle$ refer to the value of L_z, the true orbital momentum. On replacing L by 3 in (19.42) and $\mid\ \rangle$ by $|\tilde{1}\rangle\,|\tfrac{3}{2}\rangle$ where $|\tfrac{3}{2}\rangle$ is an eigenstate of S_z, it is easily found that

$$A = \frac{2\mathscr{P}}{5}\left\{\frac{-3\kappa}{2}-\frac{3}{2}-9\xi\right\}. \tag{19.43}$$

The first term in (19.43) is the contact hyperfine interaction, the second the orbital hyperfine interaction and the third the dipolar hyperfine interaction. For $l=2$, $L=3$, $S=\frac{3}{2}$, ξ, given by (17.46), has the value $\xi = -2/(9\times 7\times 5)$, so that

$$A_{\frac{5}{2}} = -\frac{3\mathscr{P}}{5}\left(1+\kappa-\frac{4}{105}\right). \tag{19.44}$$

For $\tilde{J} = \frac{1}{2}$ the calculation is complicated by the fact that the state $\tilde{J} = \frac{1}{2}, \tilde{J}_z = \frac{1}{2}$ is

$$|\,\rangle = p\,|\tilde{1}\rangle\,|-\tfrac{1}{2}\rangle + q\,|\tilde{0}\rangle\,|\tfrac{1}{2}\rangle + r\,|-\tilde{1}\rangle\,|\tfrac{3}{2}\rangle,$$

where $p = 1/\sqrt{6}$, $q = -1/\sqrt{3}$, $r = 1/\sqrt{2}$ (cf. Table 7.18) are Clebsch–Gordan coefficients; $|\tilde{1}\rangle$, $|\tilde{0}\rangle$, $|-\tilde{1}\rangle$ the three fictitious $\tilde{l}_z$ eigenstates given in Table 4; and $|-\tfrac{1}{2}\rangle, |\tfrac{1}{2}\rangle, |\tfrac{3}{2}\rangle$ eigenstates of S_z. The result is

$$A_{\frac{1}{2}} = \mathscr{P}\left(1-\frac{5\kappa}{3}-\frac{5}{2}\xi\right) = \mathscr{P}\left(1-\frac{5\kappa}{3}-\frac{1}{63}\right). \tag{19.45}$$

It will be seen that the dipolar hyperfine contribution, that is, the last term of (19.44) and (19.45) is very small, indicating that the distribution of spin magnetization is not far from spherical.

The nuclear electric quadrupole interaction will obviously vanish for the state $\tilde{J} = \frac{1}{2}$, since it is a Kramers doublet Γ_6 of cubic symmetry similar to those discussed in § 18.2. For higher values of $\tilde{J}$ it can be derived as follows.

Within an orbital triplet state $\tilde{l} = 1$ we can use for the nuclear electric quadrupole Hamiltonian a form that is an obvious modification of eqn (17.24) in the light of the discussion preceding that equation (cf. also § 18.6):

$$\frac{-e^2Q\langle r_q^{-3}\rangle}{I(2I-1)}\langle L\|\,\alpha\,\|L\rangle\Big[\frac{m}{6}\sum_{x,y,z}\{3\tilde{l}_x^2-\tilde{l}(\tilde{l}+1)\}\{3I_x^2-I(I+1)\}+$$

$$+\frac{3n}{4}\sum_{x,y,z}(\tilde{l}_x\tilde{l}_y+\tilde{l}_y\tilde{l}_x)(I_xI_y+I_yI_x)\Big], \tag{19.46}$$

in which the coefficients m, n are found from the equivalences

$$\begin{aligned}\{3L_z^2-L(L+1)\} &= m\{3\tilde{l}_z^2-\tilde{l}(\tilde{l}+1)\},\\ (L_xL_y+L_yL_x) &= n(\tilde{l}_x\tilde{l}_y+\tilde{l}_y\tilde{l}_x).\end{aligned} \tag{19.47}$$

The values of m, n for $L=2$ and $L=3$ are given in Table 21.

Within a manifold $(\Gamma_i, \tilde{J})$ we can then replace the operators that occur in (19.46) in terms of $\tilde{l}$ by similar operators in terms of $\tilde{J}$ giving a

Hamiltonian of the form

$$\frac{-e^2Q\langle r_q^{-3}\rangle}{I(2I-1)}\langle \tilde{J}\| \alpha \|\tilde{J}\rangle\Big[\frac{m}{6}\sum_{x,y,z}\{3\tilde{J}_x^2-\tilde{J}(\tilde{J}+1)\}\{3I_x^2-I(I+1)\}+$$

$$+\frac{3n}{4}\sum_{x,y,z}(\tilde{J}_x\tilde{J}_y+\tilde{J}_y\tilde{J}_x)(I_xI_y+I_yI_x)\Big], \quad (19.48)$$

where the value of the reduced matrix element $\langle \tilde{J}\| \alpha \|\tilde{J}\rangle$ is found from the relation

$$\frac{\langle \tilde{J}\| \alpha \|\tilde{J}\rangle}{\langle L\| \alpha \|L\rangle} = \frac{\langle \tilde{J}, J_z = \tilde{J}|\, 3\tilde{l}_z^2-2\,|\tilde{J}, J_z = \tilde{J}\rangle}{\tilde{J}(2\tilde{J}-1)}. \quad (19.49)$$

The quadrupole interaction (19.48) has a form similar to that given by eqn (18.50) for a Γ_8 quartet, with $\tilde{J}$ replacing $\tilde{S}$.

Finally, we compute the pseudo-nuclear Zeeman effect, arising from the cross term between the electron Zeeman interaction $-(\mathbf{\mu}\cdot\mathbf{H})$ and the hyperfine interaction $-\gamma_n\hbar(\mathbf{I}\cdot\mathbf{H}_e)$. This is

$$Z_n^* = -\gamma_n\hbar\sum_{\tilde{J}'\neq\tilde{J}}\frac{\langle\Gamma_i,\tilde{J}|-\mathbf{\mu}\cdot\mathbf{H}|\Gamma_i,\tilde{J}'\rangle\langle\Gamma_i,\tilde{J}'|\mathbf{I}\cdot\mathbf{H}_e|\Gamma_i,\tilde{J}\rangle}{W(\Gamma_i,\tilde{J})-W(\Gamma_i,\tilde{J}')}+\text{c.c.} \quad (19.50)$$

Here $|\Gamma_i,\tilde{J}\rangle$ is the ground $\tilde{J}$-multiplet and

$$W(\Gamma_i,\tilde{J})-W(\Gamma_i,\tilde{J}') = \frac{\lambda\alpha}{2}\{\tilde{J}(\tilde{J}+1)-\tilde{J}'(\tilde{J}'+1)\}. \quad (19.51)$$

The excited cubic levels Γ_j with $j\neq i$ bring to (19.50) a much smaller contribution because of much larger energy denominators. For $\tilde{J}>\tfrac{1}{2}$ Z_n^* has the same form as that given in (18.52) for Γ_8, $\tilde{S}$ being replaced by $\tilde{J}$.

19.4. Departures from cubic symmetry

If we superimpose on the cubic field K another field T of lower symmetry the situation does not change markedly for the ions of type 'A' considered in § 19.2. However for ions of type 'B', the situation may become a good deal more complicated and the general predictions of the magnetic behaviour of the ion much less detailed. A tetragonal or trigonal field will split an orbital triplet Γ_4 or Γ_5 into a singlet and a doublet. A field of lower symmetry will split it into three orbital singlets.

(a) If the lowest level is a singlet and the distance to the nearest excited orbital level (doublet or singlet) is very much larger than the spin-orbit coupling, that is if $T \gg \lambda$, we are back to the situation for which we have derived a spin Hamiltonian in § 19.2.

(b) If T is comparable to λ, and hence very much smaller than the cubic field K, we can in first approximation neglect admixtures from the higher cubic multiplets Γ_j. Inside the manifold of degeneracy $3(2S+1)$ we can represent the non-cubic field T by the equivalent operator $T = \sum_{i \leqslant j} \Delta_{ij}(\tilde{l}_i \tilde{l}_j + \tilde{l}_j \tilde{l}_i)$, which by a suitable rotation of coordinate axes can be brought to the form

$$T = \Delta_1 \tilde{l}_x^2 + \Delta_2 \tilde{l}_y^2 + \Delta_3 \tilde{l}_z^2 \qquad (19.52)$$

with

$$\Delta_1 + \Delta_2 + \Delta_3 = 0$$

and the spin-orbit coupling is again represented by $\alpha\lambda(\tilde{\mathbf{l}} \cdot \mathbf{S})$.

The secular equation for the diagonalization of $T + \alpha\lambda(\tilde{\mathbf{l}} \cdot \mathbf{S})$ will factorize into equations of order no higher than cubic. Once the eigenstates are found as linear combinations of states $|\tilde{l}_z\rangle |S_z\rangle$, the expectation values of $Z = \beta \mathbf{H} \cdot (g_s \mathbf{S} + \alpha \tilde{\mathbf{l}})$ can be calculated immediately.

In particular, if T has tetragonal or trigonal symmetry around the z-axis, $\Delta_1 = \Delta_2 = -\Delta_3/2 = -\Delta/3$ and T can be written

$$T = \Delta(\tilde{l}_z^2 - \tfrac{2}{3}) \qquad (19.53)$$

and $\tilde{J}_z = \tilde{l}_z + S_z$ is a good quantum number (while $\tilde{J}$ is no longer so), which simplifies the classification of the eigenstates according to the values of $\tilde{J}_z$. However, for the calculation of the magnetic and quadrupole hyperfine structure as expectation values of the operators in eqn (17.61) one must use the expansions of the $|\tilde{l}_z\rangle$ as sums of states $|L_z\rangle$ given in Tables 4 or 5.

Finally, if the field T is not very much smaller than K it may bring into the lower orbital cubic multiplet Γ_i admixtures from excited cubic levels Γ_j. Thus for an F-term a tetragonal or a trigonal field will mix into the triplet Γ_4, formed from an F-term, wave-functions from Γ_5 (see Figs. 7.5, 7.6). Using Table 4 it is easy to show that the three lower orbital eigenstates of $K+T$ will be of the form

$$\xi'_{+1} = p\xi_{+1} + q\eta_{-1}, \quad \xi'_0 = \xi_0, \quad \xi'_{-1} = p\xi_{-1} + q\eta_{+1}, \qquad (19.54)$$

where ξ_m and η_m are the states $|\tilde{m}\rangle$ respectively for Γ_4 and Γ_5 for tetragonal symmetry; p and q depend on the field T and $p^2 + q^2 = 1$. Slightly different formulae are needed for trigonal symmetry, which also admixes Γ_2 into $|\Gamma_4, \tilde{0}\rangle$ (see Fig. 7.6). Similarly, it is worth noticing that Γ_5 formed from a D-term is coupled to the doublet Γ_3 by a trigonal but not by a tetragonal field; see Figs. 7.3, 7.4. We can still define a fictitious momentum $\tilde{1}$ by taking ξ'_{+1}, ξ'_0, ξ'_{-1} of (19.54) as eigenstates of $\tilde{l}_z = 1, 0, -1$. Within the manifold of these states T keeps the form $\Delta(\tilde{l}_z^2-\tfrac{2}{3})$ but the equivalence

$$\mathbf{L} = \alpha\tilde{\mathbf{l}} \tag{19.55}$$

with a single constant is replaced by

$$L_x = \alpha'\tilde{l}_x,\ L_y = \alpha'\tilde{l}_y,\ L_z = \alpha\tilde{l}_z. \tag{19.56}$$

In the new Hamiltonian for the $3(2S+1)$ manifold

$$\Delta(\tilde{l}_z^2-\tfrac{2}{3})+\lambda\{\alpha\tilde{l}_zS_z+\alpha'(\tilde{l}_xS_x+\tilde{l}_yS_y)\}, \tag{19.57}$$

$\tilde{J}_z = \tilde{l}_z+S_z$ is still a good quantum number. The expectation value of

$$Z = \beta g_\text{s}(\mathbf{H}\cdot\mathbf{S})+\beta\{\alpha H_z\tilde{l}_z+\alpha'(H_x\tilde{l}_x+H_y\tilde{l}_y)\} \tag{19.58}$$

is calculated as before. For the evaluation of the expectation values of the hyperfine interaction, care should be taken to use for the states $|\tilde{1}\rangle$, $|\tilde{0}\rangle$, $|-\tilde{1}\rangle$ the correct expansion (either (19.54) or its equivalent for trigonal symmetry) and Tables 4 or 5.

To proceed further one must consider separately the cases of Kramers and non-Kramers ions.

(i) *Kramers ions*

For these, $\tilde{J}_z$ is a good quantum number and there will be a certain number of Kramers doublets $|\tilde{J}_z| = m$. For instance for Co^{2+}, ^4F there will be one doublet $|\tilde{J}_z| = \tfrac{5}{2}$, two doublets $|\tilde{J}_z| = \tfrac{3}{2}$ and three doublets $|\tilde{J}_z| = \tfrac{1}{2}$. The secular equation factors into three equations of the third, second, and first order. Resonance will be observable in the doublets $|\tilde{J}_z| = \tfrac{1}{2}$ only. A doublet $|\tilde{J}_z| = \tfrac{1}{2}$ happens to be the lowest in the case of Co^{2+} in tutton salts where the field T is approximately tetragonal and in the fluosilicate where it is trigonal. If we add to T a component of lower symmetry it will naturally not be able to lift the Kramers degeneracy, but as $\tilde{J}_z$ ceases to be a good quantum number, the resonance may become observable in all doublets.

(ii) *Non-Kramers ions*

In tetragonal or trigonal symmetry there will be a singlet $\tilde{J}_z = 0$ and doublets $|\tilde{J}_z| = m$.

Higher-order perturbations may lift the degeneracy of some of the

19.5. The influence of excited terms

In the foregoing we have consistently assumed that the wave-function of the ground level of the bound ion was constructed from eigenstates of the ground term (L, S) of the free ion. We have used this assumption in several places: in the expressions (19.5)–(19.12) for the spin-orbit coupling, spin-spin coupling, and hyperfine structure couplings, in the expansion given in Tables 4 or 5 for the kets spanning the various multiplets Γ_i, etc. ... This procedure rests on the assumption that off-diagonal matrix elements of the crystal potential between the ground term (L, S) of the free ion and excited terms (L', S) are small compared to the energy differences $W(L, S) - W(L', S)$. We shall defer to a later chapter the study of situations where this assumption is completely unwarranted, as in covalent complexes, and describe briefly the changes introduced in our description of the paramagnetic ion by relatively small admixtures from excited terms brought by the crystal potential (admixtures caused by the spin-orbit coupling are very much smaller and will be disregarded).

Let us consider first the case of ions of type B and to be specific the example of Co^{2+}, $3d^7$, where to the cubic multiplet Γ_4 of the ground term ^4F the cubic field admixes the excited term ^4P, the amplitude of the admixture being of the order of $\tau \approx 0.2$. The lifting of the twelvefold degeneracy of the ground manifold can still be described for tetragonal or trigonal symmetry by a reduced equivalent Hamiltonian of the same form as (19.57)

$$\Delta(\tilde{l}_z^2 - \tfrac{2}{3}) + \lambda\{\alpha \tilde{l}_z S_z + \alpha'(\tilde{l}_x S_x + \tilde{l}_y S_y)\},$$

but the eigenstates $|\tilde{1}\rangle$, $|\tilde{0}\rangle$, $|-\tilde{1}\rangle$ of $\tilde{l}_z$ will involve admixtures from states belonging to the ^4P level.

To obtain the constants $\lambda\alpha$, $\lambda\alpha'$ of the spin-orbit coupling in (19.57) it is necessary in principle to start from the general expression $\sum_i \zeta(\mathbf{l}_i \cdot \mathbf{S}_i)$ for the spin-orbit coupling. However, because of the selection rule $|\Delta L| = 0, 1$, the operator $\sum_i \zeta(\mathbf{l}_i \cdot \mathbf{S}_i)$ has no matrix elements between the terms ^4F and ^4P and the representation of the spin-orbit coupling by $\lambda(^4F)(\mathbf{L} \cdot \mathbf{S})$ leads to an error only of the order of τ^2. Similarly, for the Zeeman coupling $\beta\mathbf{H} \cdot (g_s\mathbf{S} + \mathbf{L})$, which has no cross terms between

^4P and ^4F, the admixture from the excited term ^4P changes its expectation value by a quantity of order τ^2.

For instance, for Co^{2+} in octahedral environment, for the lower triplet Γ_4, the admixture of ^4P changes $\tilde{g}_l$ from $-\frac{3}{2}$ into $-\frac{3}{2}+5\tau^2/2$, causing according to (19.40a) a change δg of $-5\tau^2/3$. To explain the discrepancy $-0\cdot 05$ from the observed value it is sufficient to take $\tau^2 = 0\cdot 03$, or $|\tau| = 0\cdot 17$ which is of the right order of magnitude.

It is in the calculation of hyperfine structure that the changes are most important. The expressions (19.8)–(19.11) for the hyperfine structure can be used for diagonal terms $\langle LS|$ $|LS\rangle$; for instance, for Co^{2+}, this yields

$$(1-\tau^2)\langle 3, \tfrac{3}{2}|\, W_n\, |3, \tfrac{3}{2}\rangle + \tau^2 \langle 1, \tfrac{3}{2}|\, W_n\, |1, \tfrac{3}{2}\rangle, \tag{19.59}$$

the first term being the contribution of the ^4F and the second of the ^4P term. The constant ξ in (19.8) has a different value for the terms ^4F and ^4P and for the latter *cannot* be computed by eqn (17.46) since ^4P is a term that does not obey Hund's rule. Off-diagonal terms $(LS|\, W_n\, |L'S)$ or, for Co^{2+}, $(^4F|\, W_n\, |^4P)$ must be computed using for the hyperfine interactions the more general expressions (17.12) for the quadrupole interaction, and (17.30) or (17.42) for the magnetic interaction, summed over all the electrons. The actual calculations although straightforward are too lengthy to be outlined in a general survey.

Passing on to ions of type A for which the spin Hamiltonian (19.33) has been derived by a perturbation method, the following remarks can be made. The assumption that the ground state has no orbital degeneracy and that it is an eigenstate of S (but not necessarily of L) is sufficient for the derivation of a spin Hamiltonian of the form (19.33). The only other important assumption is that of the smallness of the spin-orbit coupling relative to the crystal field, that permits the treatment of the operator C in (19.16) as a purely orbital operator.

The assumption that the ground state and the excited states are eigenstates of L and, for the excited states, the assumption that they are eigenstates of the same value of S as the ground state, are not essential. On using for the various interactions, the more general expressions $\zeta \sum (\mathbf{l}_i \cdot \mathbf{s}_i)$ for the spin-orbit coupling, (17.12), (17.30) for the hyperfine interaction, and (16.42) for the spin-spin interaction, we end up in the ground state with an operator expression linear or quadratic with respect to the components of the spins of the individual electrons, which by the Wigner–Eckart theorem can be cast into the form (19.33). The various symmetry properties of the coefficients of

(19.33) based on time-reversal and spatial symmetry will be valid as well. On the other hand, the various analytical expressions (19.32) giving the coefficients of (19.33) will no longer be valid.

The main advantage of these expressions is their relative simplicity. For the more complicated situations where L is not a good quantum number general formulae become unwieldy and each special case is best dealt with separately.

20
THE EFFECTS OF COVALENT BONDING

20.1. Summary of the foregoing theory

IN the theory of paramagnetic ions described in the previous chapters we have assumed throughout that the magnetic electrons were strictly localized on the paramagnetic ion itself.

The neglect of configuration interaction (with one exception, see § 17.5) resulted in the even more restrictive assumption that, apart from closed shells, the one-electron wave-functions, from which the wave-function of the ion was built, were f or d Hartree–Fock wave-functions with the same radial function as in the free ion. Many-electron wave-functions, eigenstates of (L, S, L_z, S_z) or (L, S, J, J_z), can be built as linear combinations of Slater determinants constructed using these one-electron wave-functions. Finally the low-lying states of the bound ion were represented as linear combinations of states $|L, S, L_z, S_z\rangle$ or $|L, S, J, J_z\rangle$, the former representation being more convenient for the iron group and the latter for the rare-earth group. In first approximation a single set of values was required for L, S in the iron group and for L, S, J in the rare-earth group. The amplitudes of the component states $|L, S, L_z, S_z\rangle$ or $|L, S, J, J_z\rangle$ were obtained by solving a secular equation whose coefficients were matrix elements

$$\langle L, S, M_L, M_S| V |L', S', M'_L, M'_S\rangle \text{ or } \langle L, S, J, M_J| V |L', S', J', M_J\rangle$$

of a certain crystal potential $V = \sum_i V(\mathbf{r}_i)$ which was intended to represent the influence of the surroundings of the paramagnetic ion. The number of independent constants in V was severely limited by symmetry considerations and even more so by the neglect of configuration interaction.

The fact that a wealth of magnetic resonance data could be made to agree with calculated values, simply by adjusting these few constants to 'reasonable' values, might be considered as a justification of the crystal field model. The fact that a single set of (L, S) values was generally sufficient in the iron group and a single set of (L, S, J) values in the rare-earth group pointed further to the approximate correctness of the hypothesis of 'intermediate crystal field' and 'weak crystal field' respectively. There are, however, serious reasons for believing that for $3d$ electrons (and even more so for $4d$ and $5d$) the above picture is incomplete. Among the evidence pointing in this direction we can list the following facts.

(a) Some paramagnetic compounds such as cyanides have anomalous values of spin and magnetic moment which contradict Hund's rule. Thus, for instance, $K_3Fe(CN)_6$ and $K_3Co(CN)_6$, where in an ionic model the iron should be Fe^{3+} and the cobalt Co^{3+}, would be expected through Hund's rule to have in their ground states respectively spins $S = \frac{5}{2}$ and $S = 2$, but turn out to have an effective spin $\frac{1}{2}$ for $K_3Fe(CN)_6$ and 0 for $K_3Co(CN)_6$.

(b) In compounds with 'normal' values of the spin the orbital contribution to the magnetic moment turns out to be systematically smaller than the theoretical value calculated by using the crystal field model.

(c) While crude calculations of the magnitude of the crystal field using the point-charge model can give values in qualitative agreement with those extracted from the analysis of resonance and spectroscopic data, attempts to improve the model by taking into account the spatial extension of neighbouring ions and the overlap of their wave-functions with those of the central ion, destroy the agreement, leading to values for the crystal field wrong in magnitude or in sign or in both.

(d) The most cogent argument against the purely ionic model is the observation in the resonance spectrum of hyperfine structure with neighbouring nuclei, demonstrating unambiguously the presence of unpaired electrons on these ligand ions.

While some of the discrepancies above can be explained away by suitable modifications of the purely ionic crystal field picture, the molecular orbital model appears to provide a consistent explanation for them all and will be discussed next.

20.2. The molecular orbitals model for covalent bonding

We shall describe this model using the specific example of an octahedral complex XY_6 where the central atom X is surrounded by six identical atoms Y, called ligands, each at the apex of a regular octahedron. In the purely ionic picture X is a paramagnetic ion and the ligands are six diamagnetic ions with closed shells. The complex NiF_6 constitutes a good example. The central ion Ni^{++} has $3d^8$ electrons outside closed shells and each surrounding ion F^- has the closed-shell configuration $1s^2 2s^2 2p^6$. In the ionic picture we study the $3d$ electrons of the central ion under the influence of the crystal potential created by the neighbouring fluorines (and also by ions further removed). The next step is to treat the XY_6 complex as a single isolated molecule.

In principle we could conceive of a generalized self-consistent field method where one-electron orbitals are determined in a self-consistent field that has the octahedral symmetry of the molecule rather than the full rotational symmetry of the isolated atom. In practice this is neither possible nor indeed necessary. Instead we assume first that the wave-functions of the inner electrons, $1s^2$ through $3p^6$ of the central atom X, and of the inner electrons $1s^2$ of the surrounding atoms Y, do not participate in the bonding and are correctly described by the Hartree–Fock functions of the corresponding atoms. For the external or valence orbitals of X and Y we make the drastic approximation that they are linear combinations of the atomic Hartree–Fock orbitals of these atoms. The number and nature of these combinations is predetermined to a large extent by symmetry considerations. Once these combinations have been written down, the remaining independent amplitudes can be determined either from experiment by adjusting them to fit the observed data, or from theory by a variational method: the molecular orbitals, consisting of linear combinations of atomic Hartree–Fock orbitals are considered as trial functions and the amplitudes are determined from the usual condition that the energy be stationary. The two sets of values can be compared to provide a test of the validity of the model.

We begin by examining the symmetry limitations on the linear combinations of atomic orbitals to be used as molecular orbitals. The three C_4 axes of the octahedron are chosen as coordinate axes Ox, Oy, Oz. We label respectively 1 and 4 the ligands on the positive and negative side of Ox, 2 and 5 those on Oy, 3 and 6 those on Oz. Each ligand has a $2s$ orbital which we label $\sigma_{1,s}$ through $\sigma_{6,s}$. It also has 3 p-orbitals. The p-orbitals with zero orbital momentum along the direction towards the central atom, we label $\sigma_{1,p}$ through $\sigma_{6,p}$. Since each function $\sigma_{i,p}$ has a positive and a negative lobe we adopt the convention that all the positive (we could choose all the negative) lobes of the $\sigma_{i,p}$ point towards the central ion. There are twelve other p-orbitals on the ligands, the so-called π-orbitals. We call Y_1 and Z_1 the two π-orbitals of ligand 1, whose angular momentum components are zero along axes parallel to Oy and Oz respectively and passing through the nucleus of ligand 1. The other orbitals will be $Y_4, Z_4; X_2, Z_2; X_5, Z_5; X_3, Y_3; X_6, Y_6$. Their relative signs can be defined by the convention that the symbols X, Y, Z in X_i, Y_i, Z_i transform under the operation of the full cubic group O_h (that includes the inversion I through the centre of the octahedron) in the same way as the coordinates x, y, z, the transformation of the indices i being obvious. Thus for instance a rotation of 90° around Oz will turn Y_4 into $-X_5$ while the inversion operation I turns Y_4 into $-Y_1$. In either case,

$$(Y_1-Y_4+X_2-X_5) \to -(Y_1-Y_4+X_2-X_5).$$

We could have labelled the σ_p functions: X_1, X_4; Y_2, Y_5; Z_3, Z_6; however, if we applied to their transformation the rules outlined for the π-functions, we would contradict the earlier convention that all positive lobes point towards the centre.

Different sign conventions may be adopted, and are sometimes used in the literature. Since we are going to use definite linear combinations of the σ_s, σ_p, and π-orbitals, the signs of the coefficients in these combinations depend on the sign conventions made for the σ_s, σ_p, π and the latter must therefore be clearly specified.

Returning for a moment to the central atom we recall that its d-orbitals provide a basis for the irreducible representations Γ_3 and Γ_5 of the cubic group. Since the d-functions are even under inversion these representations are customarily written as e_g and t_{2g} (g stands for 'gerade' or 'even' in German). Similarly the orbital $4s$ provides the even representation Γ_1 or a_{1g} and $4p$ the odd representation Γ_4 or t_{1u} (u for 'ungerade' which means 'odd'). Similarly the orbitals of the ligands, which transform into each other under operations of O_h, provide representations of this group which are reducible and can be reduced by the usual methods using the character tables of O_h. In these transformations ($\sigma_{i,s}$, $\sigma_{i,p}$) and π-orbitals clearly do not mix and the representations they span can be reduced separately. It is also clear that $\sigma_{i,s}$ and $\sigma_{i,p}$ transform in the same way (thanks to our sign convention for the $\sigma_{i,p}$) and will give rise to the same irreducible representations. It is easy to verify that the reduction of the representations spanned by the six functions σ_s (or σ_p) yields $a_{1g}+e_g+t_{1u}$, whereas the reduction of that spanned by the π-functions yields $t_{1u}+t_{2u}+t_{1g}+t_{2g}$. The corresponding proper linear combinations can be calculated by the method outlined in § 14.4 for the reduction of the rotation group.

The results are given in Table 23 for the representations a_{1g}, e_g, t_{1u}, t_{2g}, which are the only ones that can be coupled by a Hamiltonian of octahedral symmetry to orbitals $3d$, $4s$, $4p$ of the central ion. The representation t_{2u} would couple to f-electrons on the central atoms.

20.3. Bonding and anti-bonding orbitals, overlap, and covalency

We consider for the sake of definiteness an orbital of the complex XY_6, belonging to the representation t_{2g} and denoted by

$$\psi_\zeta = \alpha\varphi_\zeta + \beta\chi_\zeta, \qquad (20.1)$$

THE EFFECTS OF COVALENT BONDING 765

where φ_ζ is the usual d-function of the central ion

$$\varphi_\zeta = d_{xy} = \frac{1}{i\sqrt{2}}\{|2\rangle - |-2\rangle\} \tag{20.2}$$

and χ_ζ is the linear combination, given in Table 23, of p_π functions belonging to the ligands 1, 2, 4, 5,

$$\chi_\zeta = \tfrac{1}{2}(Y_1 - Y_4 + X_2 - X_5). \tag{20.3}$$

We use (20.1) as a trial function where the 'best' values must be selected for the amplitudes α and β. It is important to notice that since the functions φ_ζ and χ_ζ which belong to different atoms are not orthogonal, the normalization condition for (20.1) is not $\alpha^2 + \beta^2 = 1$ but rather

$$\alpha^2 + \beta^2 + 2\alpha\beta S = 1, \tag{20.4}$$

where $S = \langle \varphi_\zeta | \chi_\zeta \rangle$ is an overlap integral. If h is an approximate one-electron Hamiltonian that contains both the kinetic energy of the electron and its electrostatic interactions (including exchange terms) with the electrons on the central ion and on the ligands, the energy W of the orbital ψ_ζ is

$$W = (\psi_\zeta | h | \psi_\zeta) = \alpha^2(\varphi_\zeta | h | \varphi_\zeta) + \beta^2(\chi_\zeta | h | \chi_\zeta) + 2\alpha\beta(\varphi_\zeta | h | \chi_\zeta). \tag{20.5}$$

By varying α and β in order to make (20.5) stationary, under the subsidiary normalization condition (20.4) we find two sets of values (α_a, β_a), (α_b, β_b) for (α, β) and two values W_a and W_b for the energy W. Writing $W_1 = (\varphi_\zeta | h | \varphi_\zeta)$, $W_2 = (\chi_\zeta | h | \chi_\zeta)$, $W_{12} = (\varphi_\zeta | h | \chi_\zeta)$, the equations for α and β can be written

$$\begin{aligned}\alpha W_1 + \beta W_{12} &= W(\alpha + \beta S),\\ \alpha W_{12} + \beta W_2 &= W(\alpha S + \beta).\end{aligned} \tag{20.6}$$

The secular equation $(W_1 - W)(W_2 - W) - (W_{12} - WS)^2 = 0$ has clearly two roots, W_a and W_b, with $W_a > W_1, W_2$; $W_b < W_1, W_2$. Let α_a, β_a and α_b, β_b be the amplitudes of φ_ζ and χ_ζ in the orbitals whose energies are W_a and W_b. If we assume $W_1 > W_2$ we can extract from (20.6) the following expressions for W_a and W_b:

$$\begin{aligned}W_a &= W_1 + \frac{(\beta_a/\alpha_a)^2}{1-(\beta_a/\alpha_a)^2}(W_1 - W_2),\\ W_b &= W_2 - \frac{(\alpha_b/\beta_b)^2}{1-(\alpha_b/\beta_b)^2}(W_1 - W_2).\end{aligned} \tag{20.7}$$

It is clear from (20.7) that since $W_a > W_1 > W_2 > W_b$ we must have

$(\alpha_a/\beta_a)^2 > 1$, $(\alpha_b/\beta_b)^2 < 1$. The two orbits with energies W_a and $W_b < W_a$ are called respectively anti-bonding and bonding. The anti-bonding orbital contains a greater admixture α_a^2 of the orbital of higher energy φ_ζ, and the bonding orbital a greater admixture β_b^2 of the orbital of lower energy, χ_ζ.

It is convenient to rewrite eqn (20.1) for the anti-bonding and bonding orbitals in the form

$$\psi_\zeta^a = N_t^{-\frac{1}{2}}(\varphi_\zeta - \lambda\chi_\zeta),$$
$$\psi_\zeta^b = N_t'^{-\frac{1}{2}}(\chi_\zeta + \gamma\varphi_\zeta),$$
(20.8)

where $\lambda = -\beta_a/\alpha_a$, $\gamma = \alpha_b/\beta_b$, and the normalization conditions are

$$N_t = 1 - 2\lambda S + \lambda^2,$$
$$N_t' = 1 + 2\gamma S + \gamma^2.$$
(20.9)

So far no assumption has been made about the magnitudes of λ and γ. For instance $\lambda = \gamma = 1$ corresponds to a complete sharing of the electron between the two atomic orbitals φ and χ. However a complex XY_6 such as, say, NiF_6 is still mainly ionic and we expect the electrons to stay mainly either on the central ion or on the ligands. This implies that the overlap $S = (\varphi \mid \chi)$ is small compared to unity and that the coupling matrix element $W_{12} = (\varphi \mid h \mid \chi)$ is small compared to the difference $(W_1 - W_2) = (\varphi \mid h \mid \varphi) - (\chi \mid h \mid \chi)$. If we treat S, $W_{12}/(W_1 - W_2)$, λ, and γ as small quantities of first order and neglect terms of higher order the formulae become much simpler. From the orthogonality of ψ_ζ^a and ψ_ζ^b as given in (20.8) we get

$$\lambda = S + \gamma. \qquad (20.10)$$

The value of $\gamma = \alpha_b/\beta_b$, extracted from (20.6), takes the simple form

$$\gamma = -\frac{(\varphi \mid h \mid \chi) - S(\chi \mid h \mid \chi)}{(\varphi \mid h \mid \varphi) - (\chi \mid h \mid \chi)}. \qquad (20.11)$$

The expression (20.11) resembles the usual first-order perturbation formula except for the term $S(\chi \mid h \mid \chi)$ which takes into account the lack of orthogonality of φ and χ. Similarly,

$$\lambda = -\frac{(\varphi \mid h \mid \chi) - S(\varphi \mid h \mid \varphi)}{(\varphi \mid h \mid \varphi) - (\chi \mid h \mid \chi)}. \qquad (20.12)$$

It is reasonable to assume that the unperturbed orbitals χ of the ligands constructed from $2s$ and $2p$ electrons lie much deeper than the $3d$ orbitals (and *a fortiori* deeper than $4s$ and $4p$) of the central atom, and to assign the bonding orbitals ψ_b to the ligands and the anti-bonding orbitals to the central ion.

Since the overlap $S = (\varphi \mid \chi)$ is not zero, the definition of covalency needs to be sharpened. We shall agree to call the complex purely ionic if the surrounding ions Y carry no charge transferred from the central ion, that is if in the bonding orbitals (20.8), $\gamma = 0$. But from (20.10) it follows that $\lambda = \gamma + S = S$ is *not* zero for the anti-bonding orbitals.

From a knowledge of the Hartree–Fock wave-functions of the atoms X and Y it is possible to calculate from first principles, using (20.7), (20.10), and (20.11), the admixture coefficients λ and γ, the overlaps S, and the energies W_a of the various anti-bonding orbitals. The description of such a calculation is outside of the scope of this book. Performed for the complex NiF_6 by Sugano and Shulman (1963) it has led, for the admixtures λ and γ and the anti-bonding energies W_a, to values in excellent agreement with most of the observed data, contrasting with earlier calculations based on the ionic model (even when generalized to take into account the overlap between orbits of the central atoms and of the ligands). Some of the results of this calculation will be mentioned in a later discussion (§ 20.6).

We have so far considered the case when an orbit φ of the central ion was coupled by the Hamiltonian to a single orbit χ of the ligands. The generalization is straightforward. With the functions $\varphi_e = d\gamma$ belonging to the representation e_g we have an example of a coupling with two orbitals $\chi^e_{\sigma,s}$ and $\chi^e_{\sigma,p}$ of the ligands. We now have three orbitals, one which is mainly $\chi_{\sigma,s}$ with admixtures from φ_e, a second $\chi_{\sigma,p}$ which is mainly $\chi_{\sigma,p}$ and a third anti-bonding orbital

$$\psi_e = N_e^{-\frac{1}{2}}(\varphi_e - \lambda_s \chi^e_{\sigma,s} - \lambda_p \chi^e_{\sigma,p}) \tag{20.13}$$

where

$$N_e = 1 + \lambda_s^2 + \lambda_p^2 - 2\lambda_s S_s - 2\lambda_p S_p, \tag{20.14}$$

which is mainly $\varphi_e = d\gamma$ and is primarily localized on the central ion. Finally, on the ligands there are non-bonding orbitals which transform according to t_{1g} and t_{2u} and which have no counterpart on the central ion.

20.4. The ground states in weakly covalent compounds

We now pass on to the description of the magnetic behaviour of the central ion by means of anti-bonding molecular orbitals, adopting therefore a 'strong field' representation. Let $3d^n$ be the configuration of the central ion in the ionic picture. In this picture, since we have assumed that all the orbits $2s$ and $2p$ of the ligands lie much lower in energy than the orbits $3d$ (and *a fortiori* lower than $4s$ and $4p$) of the central ion, the number of bonding and non-bonding orbitals will be exactly equal to the number of ligand orbitals $2s$ and $2p$, i.e. $6 \times (1+3) = 24$. The corresponding $6 \times 8 = 48$ electrons, which fill these $2s$ and $2p$ orbitals in the ionic picture, will be housed in the bonding and non-bonding orbitals. The n remaining electrons that are purely $3d$ in the ionic picture, will be housed in the anti-bonding orbitals whose wave-functions are listed in Table 24.

The model we use to construct the ground state of a many-electron ion is the following. The effects of covalency are taken into account by using the orbitals (20.13) as basis one-electron wave-functions and by assigning a different energy to the orbits t_{2g} and e_g. The splitting $W(e_g) - W(t_{2g})$ is not here considered to arise from a classical crystal field but now represents the difference

$$\Delta = (\psi^a_{e_g}|\, h\, |\psi^a_{e_g}) - (\psi^a_{t_{2g}}|\, h\, |\psi^a_{t_{2g}}), \qquad (20.15)$$

computed by means of the formalism outlined in the last section, a straightforward calculation in principle but very heavy in practice. In the case of NiF_6 where such a calculation has actually been carried out completely (Sugano and Shulman 1963) it turns out that probably most of the splitting Δ is due to terms proportional to the covalency coefficients γ, thus explaining why the best calculations based on the ionic model have led to values of Δ which are an order of magnitude too small. (For further discussion see Owen and Thornley (1966).)

The eigenstates of the ion are obtained by filling the various orbitals t_2 and e with electrons and constructing from these orbitals many-electron wave-functions that transform according to irreducible representations of O_h. If at the first stage we neglect spin-orbit coupling, these functions must be eigenstates of the total spin S and belong to one of the five single-valued representations of O: Γ_1, Γ_2, Γ_3, Γ_4, Γ_5, also called A_1, A_2, E, T_1, T_2. We may by analogy with free ions call these states cubic terms. Very many cubic terms can be built in this fashion which are needed for the interpretation of optical absorption spectra of the transition elements. In magnetic resonance we may take a narrower view of the problem by concentrating on the cubic term of lowest energy, and possibly one or two low-lying excited cubic terms.

The assumption of weak covalency means that the nature of the cubic term is the same as in the ionic picture, given in § 19.1. To illustrate this point consider, for instance, the configuration $3d^5$ which in the ionic model gives the sextet term ^6S, which is also the cubic term $(t_2^3 e^2)^6 A_1$. By placing five electrons in five t_2 orbitals we are able to form the doublet cubic term $(t_2^5)^2 T_1$. With little or no covalency this doublet is well above the sextet in energy, but with increasing covalency the separation $\Delta = W(e) - W(t_2)$ may increase to such an extent that it becomes energetically more advantageous to place five electrons in t_2 orbitals giving a doublet spin state, rather than to decrease their electrostatic repulsion by having their spins parallel. If we were to plot the energy of the ion as a function of Δ, the two levels would cross for a certain value of Δ and the doublet would come below the sextet (cf. Fig. 7.13). This is the situation that prevails in the $4d$, $5d$ groups and in the complex cyanides of the $3d$ group (see Chapter 8).

We return now to the case of weak covalency for which the ground cubic terms for the various ions are listed in Table 25. The weak covalency assumption implies that they are the same as in the purely ionic picture as given in § 19.1, but their expression in terms of one-electron orbitals has a somewhat different form.

We see from Table 25 that the cubic term $^3T_1$ which is the ground state for the configuration $3d^2$ can be built from either one of the two cubic configurations t_2^2 or $t_2 e$. Similarly the cubic term $^4T_1$ which is the ground state for seven electrons $3d^7$ (equivalent to three holes $3d^3$) can be built from either one of the two 3-hole configurations $t_2 e^2$ or $t_2^2 e$. This shows that the simple-minded argument that we used in § 19.1 to predict the structure of the ground states in cubic symmetry had only approximate validity. For instance, for $3d^2$ we had argued that by putting two electrons with parallel spins into two orbits t_2 of lower energy we minimized both the cubic field energy and the electrostatic electron-electron repulsion. Actually, by allowing one electron to spend some time in an upper orbit e, that is by using a combination of the two configurations t_2^2 and $t_2 e$, it is possible further to lower the overall energy of the cubic term $^3T_1$. The same is true for the cubic term $^4T_1$ with three holes. In fact $^3T_1(3d^2)$ and $^4T_1(3d^3)$ are the only cubic terms of maximum spin for which the state $S_z = S$ is not a single Slater determinant. This is related to the fact that in the weak-field representation they are respectively mixtures of $(^3F, {}^3P)$ and $(^4F, {}^4P)$. As an illustration we discuss in some detail the form of the wave-function of $^4T_1(3d^7 \equiv 3d^3)$ which applies to Co^{2+} in octahedral environment. We use the formalism of the fictitious orbital momentum introduced in § 14.2 where the one-electron functions η_x, η_y, η_z that span the triplet t_2 are related to the eigenstates $|\tilde{m}\rangle$ of $\tilde{l}_z$ by the formulae (14.5) which we rewrite here for convenience:

$$|\pm\tilde{1}\rangle = \mp\frac{\eta_x\pm i\eta_y}{\sqrt{2}}, \quad |\tilde{0}\rangle = \eta_z. \tag{20.16}$$

If η_x, η_y, η_z are purely d-functions (which they are not if covalent bonding is allowed) the states $|\tilde{m}\rangle$ are related to the states $l_z = m$ of the *true* orbital-momentum by the formulae given in Table 4,

$$|\tilde{1}\rangle = |2, -1\rangle;\ |-\tilde{1}\rangle = -|2, 1\rangle;\ |\tilde{0}\rangle = \frac{1}{\sqrt{2}}\{|2, 2\rangle - |2, -2\rangle\}, \tag{20.16a}$$

where the symbol $|2, -1\rangle$ represents the state $l = 2$, $l_z = -1$.

The functions θ and ε that span $\Gamma_3 = E$ transform like $3z^2 - r^2$ and $\sqrt{(3)}(x^2 - y^2)$. Their transformation law under permutation of the coordinates is given by eqns (14.4).

For the cubic term T_1 we use the symbols

$$|\overline{\pm 1}\rangle = \mp\frac{\bar{X}\pm i\bar{Y}}{\sqrt{2}}, \quad |\bar{0}\rangle = \bar{Z}, \tag{20.17}$$

to represent again eigenstates of the fictitious orbital momentum. Equations (20.16) and (20.17) differ in two respects: the former represent one-electron wave-functions belonging to the cubic triplet t_2; the latter deal with many-electron wave-functions spanning a cubic triplet term T_1.

If now we single out the state $S = \frac{3}{2}$, all electrons have spin $+\frac{1}{2}$. The three orbital states of $^4T_1(t_2e^2)$ are then simply

$$\begin{cases} |^4T_1(t_2e^2), \bar{1}\rangle = \{\tilde{1}, \theta, \varepsilon\}, \\ |^4T_1(t_2e^2), \bar{0}\rangle = \{\tilde{0}, \theta, \varepsilon\}, \\ |^4T_1(t_2e^2), -\bar{1}\rangle = \{-\tilde{1}, \theta, \varepsilon\}. \end{cases} \tag{20.18}$$

The cubic term $^4T_1(t_2^2e)$ is less simple since from t_2^2e we can also construct a term $^4T_2(t_2^2e)$. We give the result first and then explain how to obtain it simply.

$$\begin{cases} |^4T_1(t_2^2e), \bar{1}\rangle = -\tfrac{1}{2}\{\tilde{1}, \tilde{0}, \theta\} + \frac{\sqrt{3}}{2}\{-\tilde{1}, \tilde{0}, \varepsilon\}, \\ |^4T_1(t_2^2e), \bar{0}\rangle = \{\tilde{1}, -\tilde{1}, \theta\}, \\ |^4T_1(t_2^2e), -\bar{1}\rangle = \tfrac{1}{2}\{-\tilde{1}, \tilde{0}, \theta\} - \frac{\sqrt{3}}{2}\{\tilde{1}, \tilde{0}, \varepsilon\}. \end{cases} \tag{20.19}$$

The expression $\{\tilde{1}, -\tilde{1}, \theta\} = i\{\eta_x, \eta_y, \theta\}$ for $|^4T_1(t_2^2e), \bar{0}\rangle$ is obvious;

THE EFFECTS OF COVALENT BONDING 771

it is the only Slater determinant unchanged by a rotation of $\pm \pi/2$ around Oz. To find the expressions for $|\bar{1}\rangle$ and $|-\bar{1}\rangle$ in (20.19) we notice that $|\bar{X}\rangle = -(|\bar{1}\rangle - |-\bar{1}\rangle)/\sqrt{2}$ and $\bar{Y} = \mathrm{i}(|\bar{1}\rangle + |-\bar{1}\rangle)/\sqrt{2}$ are deduced from $\bar{Z} = |\bar{0}\rangle$ by the cyclic permutations R and R^2 of $\{x, y, z\}$. We can then write

$$\begin{cases} |\bar{X}\rangle = R\,|\bar{Z}\rangle = \mathrm{i}R\{\eta_x, \eta_y, \theta\} = \mathrm{i}\{\eta_y, \eta_z, R\theta\} \\ |\bar{Y}\rangle = R^2\,|\bar{Z}\rangle = \mathrm{i}\{\eta_z, \eta_x, R^2\theta\} \end{cases} \quad (20.20)$$

and on replacing $R\theta$ and $R^2\theta$ by the expressions (14.4) we get $|\bar{X}\rangle$ and $|\bar{Y}\rangle$ and then $|\bar{1}\rangle$ and $|-\bar{1}\rangle$ in (20.19).

In general we can write the cubic term of the ground state as

$$\cos \alpha \; {}^4T_1(t_2 e^2) + \sin \alpha \; {}^4T_1(t_2^2 e). \quad (20.21)$$

For Co^{2+} where the hole orbitals e are below the hole orbitals t_2 we expect the configuration to be mainly $t_2 e^2$, and α to be small. To determine the amplitudes $\cos \alpha$ and $\sin \alpha$ we must know the matrix elements of the electrostatic energy between the two states ${}^4T_1(t_2 e^2)$ and ${}^4T_1(t_2^2 e)$. These can be found by going back to the weak-field representation and expressing the states ${}^4T_1(t_2^2 e)$ and ${}^4T_1(t_2 e^2)$, where we assume that the orbitals t_2 and e are pure d-orbitals, as linear combinations of the free ion states, ${}^4P$ and ${}^4F$, using (20.18) and (20.19). A simple calculation gives

$$\begin{cases} {}^4T_1(t_2^2 e) = \dfrac{2}{\sqrt{5}}|{}^4P\rangle - \dfrac{1}{\sqrt{5}}|{}^4F\rangle, \\ {}^4T_1(t_2 e^2) = \dfrac{1}{\sqrt{5}}|{}^4P\rangle + \dfrac{2}{\sqrt{5}}|{}^4F\rangle. \end{cases} \quad (20.22)$$

Since the electrostatic interaction has no off-diagonal matrix elements between ${}^4F$ and ${}^4P$ the secular matrix can be written as

$$\begin{pmatrix} \dfrac{4\delta}{5}+\Delta, & \dfrac{2\delta}{5} \\ \dfrac{2\delta}{5}, & \dfrac{\delta}{5} \end{pmatrix} \quad (20.23)$$

where we have taken the electrostatic energy as zero in ${}^4F$ and δ in ${}^4P$ and denote by Δ the cubic splitting $W(t_2) - W(e)$ (which is positive for holes). The parameter δ is related to a certain radial integral B, known in the literature as a Racah parameter, by $\delta = 15B$. Then from (20.23) we get

$$\tan 2\alpha = \frac{12B}{9B+\Delta}. \tag{20.24}$$

When B/Δ goes to infinity we are back to the purely ionic case where the admixture from ^4P is negligible and our state $^4T_1$ is pure ^4F. In this case $\tan 2\alpha = -\frac{4}{3}$, $\cos \alpha = 2/\sqrt{5}$, $\sin \alpha = -1/\sqrt{5}$. In the opposite case when $B/\Delta \to 0$ we find $\alpha = 0$ and the configuration is pure (t_2e^2).

For the sake of completeness we also consider the other cubic term of Table 25 which has components belonging to two different strong field configurations, namely $^3T_1(t_2^2, t_2e)$ where the orbitals t_2 and e are either electrons or holes, depending on whether there are two or eight electrons outside of closed shells. We use a notation similar to that for the term $^4T_1(t_2e^2, t_2^2e)$ in (20.18) and (20.19) and obtain the formulae

$$\begin{cases} |^3T_1(t_2^2), \bar{1}\rangle = (\tilde{1}, \tilde{0}), \\ |^3T_1(t_2^2), \bar{0}\rangle = (1, -1), \\ |^3T_1(t_2^2), -\bar{1}\rangle = (\tilde{0}, -\tilde{1}); \end{cases} \tag{20.18a}$$

$$\begin{cases} |^3T_1(t_2e), \bar{1}\rangle = \frac{\sqrt{3}}{2}(-\tilde{1}, \theta) - \frac{1}{2}(\tilde{1}, \varepsilon), \\ |^3T_1(t_2e), \bar{0}\rangle = (\tilde{0}, \varepsilon), \\ |^3T_1(t_2e), -\bar{1}\rangle = \frac{\sqrt{3}}{2}(\tilde{1}, \theta) - \frac{1}{2}(-\tilde{1}, \varepsilon). \end{cases} \tag{20.19a}$$

The wave-functions (20.18a) can be written down at once: they are the same as those of a ^3P term constructed from two p-electrons. The relations (20.19a) are obtained in the same manner as (20.19): we start from the expression $|\tilde{0}, \tilde{\varepsilon}\rangle$ for $|^3T_1(t_2e), \bar{0}\rangle$ as the only Slater determinant unchanged by a rotation $\pm \pi/2$ around Oz, and then use a cyclic permutation to generate the two other states of (20.19a), as explained for (20.19). A calculation similar to that which gave (20.22) yields

$$\begin{cases} ^3T_1(t_2e) = \frac{2}{\sqrt{5}}|^3P\rangle - \frac{1}{\sqrt{5}}|^3F\rangle, \\ ^3T_1(t_2^2) = \frac{1}{\sqrt{5}}|^3P\rangle + \frac{2}{\sqrt{5}}|^3F\rangle. \end{cases} \tag{20.22a}$$

The signs in (20.22a), as in (20.22), are not completely determined since there is a certain freedom in the choice of the phases of the different states. We have made the choice that gives the same coefficients and thus the same secular equation as (20.22) (cf. § 7.3).

20.5. Orbital momentum and spin-orbit coupling in the presence of covalent bonding

In a uniform external magnetic field the fact that the electron spin may no longer be localized on the central ion does not affect its contribution to the magnetic moment. The situation is different for the orbital momentum and we now show that covalent bonding leads to a reduction of the matrix elements of the orbital momentum.

The matrix elements of the orbital momentum taken between the anti-bonding orbitals of Table 24 are somewhat different from those taken between pure d-functions. Using (20.16) we rewrite the formulae of Table 24 in the self-evident form

$$|+\tilde{1}\rangle = N_t^{-\frac{1}{2}}\{|\tilde{1}_d\rangle - \lambda_t |\tilde{1}_\pi\rangle\}, \text{ etc.,}$$
$$|\theta\rangle = N_\sigma^{-\frac{1}{2}}\{|\theta_d\rangle - \lambda_{\sigma s} |\theta_s\rangle - \lambda_{\sigma p} |\theta_p\rangle\}, \quad (20.25)$$
$$|\varepsilon\rangle = N_\sigma^{-\frac{1}{2}}\{|\varepsilon_d\rangle - \lambda_{\sigma s} |\varepsilon_s\rangle - \lambda_{\sigma p} |\varepsilon_p\rangle\}.$$

There are two types of matrix elements of **L**: those inside the manifold t_2 and those between t_2 and e. (We have shown in § 14.2 that they all vanish within e.)

If the λ in (20.25) are all zero, the matrix elements of **L** within t_2 are those of a fictitious angular momentum $\tilde{\mathbf{l}} = \alpha \mathbf{L}$ with $\alpha = -1$. Between e and t_2 a typical matrix element is

$$\langle \tilde{0}_d| L_z |\varepsilon_d\rangle = \langle \eta_{zd}| L_z |\varepsilon_d\rangle = 2. \quad (20.26)$$

Use of the augmented orbitals (20.25) results in multiplication of the $\langle t_{2d}| \mathbf{L} |t_{2d}\rangle$ matrix elements by a certain numerical factor, written in the literature as $k_{\pi\pi}$, and the $\langle t_{2d}| \mathbf{L} |e_d\rangle$ matrix elements by another factor $k_{\pi\sigma}$.

The calculation of $k_{\pi\pi}$ is straightforward:

$$k_{\pi\pi} = \frac{\langle 1| L_z |\tilde{1}\rangle}{\langle \tilde{1}_d| L_z |\tilde{1}_d\rangle} = -\langle 1| L_z |\tilde{1}\rangle$$
$$= -N_t^{-1}\{\langle \tilde{1}_d| L_z |\tilde{1}_d\rangle - 2\lambda_t \langle \tilde{1}_d| L_z |\tilde{1}_\pi\rangle + \lambda_t^2 \langle \tilde{1}_\pi| L_z |\tilde{1}_\pi\rangle\}. \quad (20.27)$$

The first two terms in the curly bracket give $-1 + 2\lambda_t S_t$, where $S_t = \langle \tilde{1}_d | \tilde{1}_\pi \rangle = (\varphi_t | \chi_t)$ is the overlap integral for π-bonding. To calculate the last term, we get from Table 24,

$$|\tilde{1}_\pi\rangle = \frac{1}{2}\left\{\frac{Z_1 - Z_4 - i(Z_2 - Z_5)}{\sqrt{2}} + \frac{(X_3 - iY_3)}{\sqrt{2}} - \frac{(X_6 - iY_6)}{\sqrt{2}}\right\}. \quad (20.28)$$

We introduce the vector operators,

$$\mathbf{L}(3) = \mathbf{L} - \frac{1}{\hbar}(\mathbf{a}_3 \wedge \mathbf{p}),$$
$$\mathbf{L}(6) = \mathbf{L} - \frac{1}{\hbar}(\mathbf{a}_6 \wedge \mathbf{p}),$$
(20.29)

where $\mathbf{L}(i)$ is the orbital momentum operator with respect to the position $\mathbf{a}_i$ of the ligand i. Clearly $L_z(3) = L_z(6) = L_z$; $(X_3 - iY_3)/\sqrt{2}$ and $(X_6 - iY_6)/\sqrt{2}$ are normalized eigenstates of respectively $L_z(3)$ and $L_z(6)$ with eigenvalue -1. Then in the expansion of $\langle +\tilde{1}_\pi | L_z | +\tilde{1}_\pi \rangle$ all terms can be seen to vanish except the expectation values of $L_z(3)$ and $L_z(6)$ in these two states, whence,

$$\langle \tilde{1}_\pi | L_z | \tilde{1}_\pi \rangle = -\tfrac{1}{2}$$
(20.30)

and

$$k_{\pi\pi} = N_t^{-1}\left\{1 - 2\lambda_t S_t + \frac{\lambda_t^2}{2}\right\} = \frac{\left(1 - 2\lambda_t S_t + \frac{\lambda_t^2}{2}\right)}{1 - 2\lambda_t S_t + \lambda_t^2} = 1 - \frac{\lambda_t^2}{2} N_t^{-1} \approx 1 - \frac{\lambda_t^2}{2}.$$
(20.31)

The calculation of

$$k_{\pi\sigma} = \frac{\langle \tilde{0} | L_z | \varepsilon \rangle}{\langle \tilde{0}_d | L_z | \varepsilon_d \rangle}$$
(20.32)

is more involved and its details are outlined below. From (20.26) we have

$$k_{\pi\sigma} = \tfrac{1}{2}\langle \tilde{0} | L_z | \varepsilon \rangle$$
(20.33)

or, using (20.25),

$$k_{\pi\sigma} = \frac{N_t^{-\frac{1}{2}} N_\sigma^{-\frac{1}{2}}}{2} \langle \tilde{0}_d - \lambda_t \tilde{0}_\pi | L_z | \varepsilon_d - \lambda_{\sigma s}\varepsilon_s - \lambda_{\sigma p}\varepsilon_p \rangle$$

$$= N_t^{-\frac{1}{2}} N_\sigma^{-\frac{1}{2}} \bigg(1 - \lambda_t S_t - \lambda_{\sigma s} S_s - \lambda_{\sigma p} S_p + $$

$$+ \frac{\lambda_t \lambda_{\sigma s}}{2} \langle \tilde{0}_\pi | L_z | \varepsilon_s \rangle + \frac{\lambda_t \lambda_{\sigma p}}{2} \langle \tilde{0}_\pi | L_z | \varepsilon_p \rangle \bigg).$$
(20.34)

To compute the last two terms of (20.34) we replace ε_s, ε_p and $|\tilde{0}_\pi\rangle$ by their expressions from Table 24. We now encounter terms such as $\langle \chi_{\pi j} | L_z | \chi_{\sigma i} \rangle$ where $\chi_{\pi j}$ and $\chi_{\sigma i}$ are orbitals located on ligands i and j. Replacing L_z by $L_z(i) + (\mathbf{a}_i \wedge \mathbf{p})_z/\hbar$ (cf. eqn (20.29)), we get

$$\langle \chi_{\pi j} | L_z | \chi_{\sigma i} \rangle = \delta_{ij}\left\{ \langle \chi_{\pi i} | L_z(i) | \chi_{\sigma i} \rangle + \langle \chi_{\pi i} | \frac{(\mathbf{a}_i \wedge \mathbf{p})_z}{\hbar} | \chi_{\sigma i} \rangle \right\}.$$
(20.35)

THE EFFECTS OF COVALENT BONDING 775

The first term on the right-hand side of (20.35) vanishes if $\chi_{\sigma i}$ is a σ_s orbital that carries zero orbital momentum, and the second vanishes if $\chi_{\sigma i}$ is a σ_p orbital because of parity conservation.

We thus find

$$\frac{\langle \tilde{0}_\pi | L_z | \varepsilon_s \rangle}{2} = -\frac{a}{2}\langle p_y | \frac{\partial}{\partial_y} | s \rangle, \quad (20.36)$$

where p_y and s are wave-functions $2p$ and $2s$ of a ligand and a its distance from the origin. Also

$$\frac{\langle \tilde{0}_\pi | L_z | \varepsilon_p \rangle}{2} = -\tfrac{1}{2}, \quad (20.37)$$

whence

$$k_{\pi\sigma} = N_t^{-\frac{1}{2}} N_\sigma^{-\frac{1}{2}} \left(1 - \lambda_t S_t - \lambda_{\sigma s} S_s - \lambda_{\sigma p} S_p - \tfrac{1}{2} \lambda_t \lambda_{\sigma p} - \frac{\lambda_t \lambda_{\sigma s}}{2} a \langle p_y | \frac{\partial}{\partial_y} | s \rangle \right). \quad (20.38)$$

From the equations for N_σ and N_t in Table 24 we have approximately,

$$(N_t N_\sigma)^{-\frac{1}{2}} = 1 + \lambda_t S_t + \lambda_{\sigma s} S_s + \lambda_{\sigma p} S_p - \tfrac{1}{2}(\lambda_t^2 + \lambda_{\sigma s}^2 + \lambda_{\sigma p}^2) \quad (20.39)$$

and in the same approximation we have, from (20.38),

$$k_{\pi\sigma} = 1 - \tfrac{1}{2}\left(\lambda_t^2 + \lambda_{\sigma s}^2 + \lambda_{\sigma p}^2 + \lambda_t \lambda_{\sigma p} + \lambda_t \lambda_{\sigma s} a \langle p_y | \frac{\partial}{\partial_y} | s \rangle \right). \quad (20.40)$$

The dimensionless quantity $a\langle p_y | \partial/\partial_y | s \rangle$ in the last term of (20.40) must be computed numerically.

For instance, for $Ni^{2+}F_6^-$, Owen and Thornley (1966) find it has the value -1.6, using analytical Hartree–Fock functions for the $2s$ and $2p$ orbitals of F^-, and taking $a = 0.21$ nm. For $Ni^{2+}F_6^-$ (20.40) can then be rewritten as

$$k_{\pi\sigma} = 1 - \tfrac{1}{2}\{\lambda_t^2 + \lambda_{\sigma s}^2 + \lambda_{\sigma p}^2 + \lambda_t(\lambda_{\sigma p} - 1.6\lambda_{\sigma s})\}. \quad (20.41)$$

Spin-orbit coupling

It is sometimes said that the spin-orbit constant ζ is reduced by covalent bonding. In this form the statement is misleading and needs to be clarified.

The introduction of covalent bonding modifies the spin-orbit coupling in a way which is related to, but not identical with, the reduction of the orbital momentum through the coefficients $k_{\pi\pi}$ and $k_{\pi\sigma}$ of (20.31) and (20.40). The covalent bonding does not affect the form of the operator

L in the x-representation and the change in its matrix elements is solely due to the change in the one-electron wave-functions that are the augmented orbitals of Table 24 rather than pure d-functions. On the other hand the spin-orbit coupling operator is already changed in the x-representation. We recall that the spin-orbit coupling is represented in the Pauli approximation by the operator

$$\frac{1}{2}\left(\frac{\hbar}{mc}\right)^2\left(\frac{\nabla V \wedge \mathbf{p}}{\hbar}\right) \cdot \mathbf{s}, \qquad (20.42)$$

where V is the potential energy of the electron moving in a self-consistent field. It could only be cast into the familiar form (11.19) of $\zeta \mathbf{l} \cdot \mathbf{s}$ with $\zeta = \frac{1}{2}(\hbar/mc)^2(1/r)(dV/dr)$ because V was assumed to have spherical symmetry. We can write quite generally the spin-orbit coupling operator in the form $\mathbf{U} \cdot \mathbf{s}$. The vector $\mathbf{U}$, like all vectors, has matrix elements within t_2 proportional to those of $\mathbf{L}$, and others between t_2 and e also proportional to those of $\mathbf{L}$ but with a different proportionality constant. We can then write

$$\langle t_2| \mathbf{U} |t_2\rangle = \zeta_{\pi\pi}\langle t_2| \mathbf{L} |t_2\rangle = \zeta_{\pi\pi}k_{\pi\pi}\langle t_{2d}| \mathbf{L} |t_{2d}\rangle \qquad (20.43)$$

and

$$\langle t_2| \mathbf{U} |e\rangle = \zeta_{\pi\sigma}\langle t_2| \mathbf{L} |e\rangle = \zeta_{\pi\sigma}k_{\pi\sigma}\langle t_{2d}| \mathbf{L} |e_d\rangle, \qquad (20.44)$$

where $k_{\pi\pi}$ and $k_{\pi\sigma}$ are given by (20.31) and (20.40).

The actual matrix elements of the spin-orbit coupling between t_2 and e, responsible for the orbital contribution $g-2$ of the orbital momentum for ions of type A (see § 19.2), are thus reduced in the ratio $\zeta_{\pi\sigma}k_{\pi\sigma}/\zeta_0$ where ζ_0 is the spin-orbit coupling constant for the free ion. On the other hand, the matrix elements of the spin-orbit coupling inside t_2 that occur for ions of type B such as, say, Co^{2+} considered in § 19.3, are reduced in the ratio $\zeta_{\pi\pi}k_{\pi\pi}/\zeta_0$.

It is sometimes argued that the main contribution to the matrix elements of the spin-orbit coupling comes from the region near the nucleus of the central ion where (a) the potential energy $V(r)$ is practically equal to that $V_0(r)$ of the free ion and (b) the values of the wave-functions from the ligands, admixed into the augmented orbitals of Table 24, are small. In that case one should have

$$\langle t_2| \mathbf{U} |t_2\rangle = N_t^{-1}\langle t_{2d}| \mathbf{U}_0 |t_{2d}\rangle \qquad (20.45)$$

and

$$\langle t_2| \mathbf{U} |e\rangle = N_t^{-\frac{1}{2}}N_\sigma^{-\frac{1}{2}}\langle t_{2d}| \mathbf{U}_0 |e_d\rangle, \qquad (20.46)$$

where $\mathbf{U}_0 = \zeta_0 \mathbf{L}$. In practice there is considerable uncertainty about the actual value of the matrix elements of the spin-orbit coupling and it is preferable to consider $\zeta_{\pi\pi}/\zeta_0$ and $\zeta_{\pi\sigma}/\zeta_0$ as adjustable parameters. (For a discussion of these points see Owen and Thornley (1966).)

20.6. Ligand hyperfine structure for ions of type A

As we have already stated, the presence in the resonance spectrum of hyperfine structure due to the coupling of the electronic spins with the nuclei of the ligands is the most clear-cut evidence for the transfer of spin density onto the ligands.

If the ground state of the central ion has a spin S (real or fictitious), in cubic symmetry we shall expect the hyperfine structure of say ligand 3 placed on the positive side of the z-axis to have a hyperfine structure of the form

$$A_\parallel I_z S_z + A_\perp (I_x S_x + I_y S_y). \tag{20.47}$$

Here we use $A_\parallel$ for the value along the bond axis, that is the line joining the nucleus of the ligand ion to the nucleus of the central ion. Thus for the ligand 1, for instance, (20.47) becomes obviously $A_\parallel I_x S_x + A_\perp (I_z S_z + I_y S_y)$.

Here our problem is clearly that of calculating the coefficients $A_\parallel$ and $A_\perp$. As we shall see, these coefficients are directly correlated to the admixture coefficients λ of Table 24 and their measurement affords the most direct information on the amount of covalent bonding. In this and following sections the more detailed formulae (in particular those where specific reference is made to $2s$-, $2p$-orbitals) assume that the ligand ions are fluorines.

We consider first the case when the central ion is of type A, that is effectively without orbital degeneracy in its ground state, as in Cr^{3+}, $V^{2+}(d^3)$; Cr^+, Mn^{2+}, $Fe^{3+}(d^5)$; $Ni^{2+}(d^8)$. If in a first approximation we neglect admixtures from higher cubic terms through the spin-orbit coupling, the substate of the ground state with $S_z = S$ is a single Slater determinant of molecular orbitals, and the expectation value of the operator $\mathscr{H}(3)$, describing the hyperfine structure coupling with the nucleus of ligand 3, will be a sum of expectation values of $\mathscr{H}(3)$ taken over the orbitals composing the ground term. Let ψ be one of the orbitals of Table 24. It can be written as

$$\psi = \alpha \varphi_d + \sum_{\lambda, i} \beta_{\lambda, i} \chi_\lambda(i), \tag{20.48}$$

where the index i refers to the ligand i, the coefficient α is of the order of unity, and the $\beta_{\lambda,i}$ are small. The hyperfine structure operator $\mathscr{H}$ (3)

decreases rapidly with the distance from the nucleus of ligand 3 and it is reasonable to keep in the expectation value $(\psi|\ \mathscr{H}(3)\ |\psi)$, only the terms

$$\sum_{\lambda,\mu} \beta^*_{\lambda 3}\beta_{\mu 3}\{(\chi_\lambda(3)|\ \mathscr{H}(3)\ |\chi_\mu(3))+|\alpha|^2(\varphi_d|\ \mathscr{H}(3)\ |\varphi_d)\}. \quad (20.49)$$

We have retained the second term because $|\alpha|^2$ is of order unity, but even so it is often small compared to the first and will be disregarded for the time being. We rewrite (20.48) as $\psi = F(3)+\psi'$, where $F(3)$ is the part of ψ constructed from orbitals of ligand 3, and we wish to calculate $(F(3)|\ \mathscr{H}(3)\ |F(3))$.

From Table 24 we extract the $F(3)$ that differ from zero; these are

$$F_{\eta_x}(3) = -\frac{\lambda_t}{2} N_t^{-\frac{1}{2}} Y_3,$$

$$F_{\eta_y}(3) = -\frac{\lambda_t}{2} N_t^{-\frac{1}{2}} X_3, \quad (20.50)$$

$$F_\theta(3) = -\frac{\lambda_{\sigma s}}{\sqrt{3}} N_\sigma^{-\frac{1}{2}} \sigma_{s3} + \frac{\lambda_{\sigma p}}{\sqrt{3}} N_\sigma^{-\frac{1}{2}} Z_3.$$

Note that in $F_\theta(3)$ our sign convention for the σ_p leads to $\sigma_{p3} = -Z_3$. Also it is customary in the literature to introduce the so-called spin-densities

$$f_t = \frac{\lambda_t^2}{4} N_t^{-1}, \quad f_{\sigma s} = \frac{\lambda_{\sigma s}^2}{3} N_\sigma^{-1}, \quad f_{\sigma p} = \frac{\lambda_{\sigma p}^2}{3} N_\sigma^{-1}, \quad (20.51)$$

which are the squares of the amplitudes in (20.50).

For p-ligand orbitals we take as magnetic hyperfine operator the form (17.45) with $L = 1$, $S = \frac{1}{2}$, and ξ given by (17.46) = $\frac{2}{5}$. For s-ligand orbitals we take $(16\pi/3)\beta\gamma_n\hbar\delta(r)(\mathbf{I}\cdot\mathbf{s})$. Omitting as usual the electron-spin operator from the expectation value we obtain

$$\langle Z_3|\ \mathscr{H}(3)\ |Z_3\rangle = a_p\{2I_zS_z-(I_xS_x+I_yS_y)\}, \quad (20.52)$$

with

$$a_p = \tfrac{4}{5}\beta\gamma_n\hbar\langle r^{-3}\rangle_{2p}. \quad (20.53)$$

Clearly $\langle X_3|\ \mathscr{H}(3)\ |X_3\rangle$ and $\langle Y_3|\ \mathscr{H}(3)\ |Y_3\rangle$ are deduced from (20.52) by cyclic permutation of x, y, z.

Similarly

$$\left\langle \sigma_{s3}\left|\frac{16\pi}{3}\beta\gamma_n\hbar\delta(r)\right|\sigma_{s3}\right\rangle(\mathbf{I}\cdot\mathbf{S}) = a_s(\mathbf{I}\cdot\mathbf{S}) \quad (20.54)$$

with

$$a_s = \frac{16\pi}{3}\beta\gamma_n\hbar\,|\psi_{2s}(0)|^2. \quad (20.55)$$

Disregarding the second term in (20.49), we find

$$\langle \eta_x | \mathcal{H}(3) | \eta_x \rangle \approx \langle F_{\eta_x}(3) | \mathcal{H}(3) | F_{\eta_x}(3) \rangle = f_t a_p (2I_y S_y - I_x S_x - I_z S_z),$$
$$\langle \eta_y | \mathcal{H}(3) | \eta_y \rangle \approx \langle F_{\eta_y}(3) | \mathcal{H}(3) | F_{\eta_y}(3) \rangle = f_t a_p (2I_x S_x - I_y S_y - I_z S_z),$$
$$\langle \theta | \mathcal{H}(3) | \theta \rangle \approx \langle F_\theta(3) | \mathcal{H}(3) | F_\theta(3) \rangle$$
$$= f_{\sigma s} a_s (\mathbf{I} \cdot \mathbf{S}) + f_{\sigma p} a_p (2I_z S_z - I_x S_x - I_y S_y). \quad (20.56)$$

Let us apply (20.56) to our three ions of type A which have singlet ground states.

(1) V^{2+}, Cr^{3+} (d^3)

The ground cubic term is $^4A_2(t_2^3)$, and the substate $S_z = S = \tfrac{3}{2}$ is $\Psi = (\eta_x^+, \eta_y^+, \eta_z^+)$ for which

$$\langle \Psi | \mathcal{H}(3) | \Psi \rangle = \langle \eta_x | \mathcal{H}(3) | \eta_x \rangle + \langle \eta_y | \mathcal{H}(3) | \eta_y \rangle$$
$$= (f_t a_p (2I_y S_y^{(1)} - I_x S_x^{(1)} - I_z S_z^{(1)}) + f_t a_p (2I_x S_x^{(2)} - I_y S_y^{(2)} - I_z S_z^{(2)}),) \quad (20.57)$$

where $S^{(1)}$, $S^{(2)}$ are the spins of the two electrons in the orbits $|\eta_x\rangle$ and $|\eta_y\rangle$.

Within the manifold $S = \tfrac{3}{2}$ of $^4A_2(t_2^3)$, the expectation value of a component of the spin of each electron is a fraction $(1/2S)$ (that is for $S = \tfrac{3}{2}$, one-third), of that of the total spin $\mathbf{S}$. Hence in (20.47) we find from (20.56), (20.57) expressions for $A_\parallel$ and $A_\perp$:

$$A_\parallel = -2A_\perp = -\frac{2f_t}{2S} a_p = -2A_\pi \quad (20.58)$$

where

$$A_\pi = \frac{f_t a_p}{2S}. \quad (20.59)$$

(2) Cr^+, Mn^{2+}, Fe^{3+} (d^5)

The ground cubic term is $^6A_1(t_2^3 e^2)$ and the substate $S_z = S = \tfrac{5}{2}$ is $(\eta_x^+, \eta_y^+, \eta_z^+, \theta^+, \varepsilon^+)$. We find by the same method as before

$$A_\parallel = \frac{2}{2S}(f_{\sigma p} - f_t)a_p + f_{\sigma s}\frac{a_s}{2S} = 2(A_\sigma - A_\pi) + A_s, \quad (20.60)$$

$$A_\perp = \frac{-1}{2S}(f_{\sigma p} - f_t)a_p + f_{\sigma s}\frac{a_s}{2S} = -(A_\sigma - A_\pi) + A_s,$$

where

$$A_\sigma = \frac{f_{\sigma p} a_p}{2S}; \qquad A_s = \frac{f_{\sigma s} a_s}{2S}. \quad (20.61)$$

(3) Ni^{2+} (d^8)

The ground cubic term expressed by means of holes is $^3A_2(e^2)$ and the ground substate $S_z = S = 1$ is $(\theta^+\varepsilon^+)$. We find

$$A_\parallel = \frac{2}{2S}f_{\sigma p}a_p + \frac{1}{2S}f_{\sigma s}a_s = 2A_\sigma + A_s,$$
$$A_\perp = \frac{-1}{2S}f_{\sigma p}a_p + \frac{1}{2S}f_{\sigma s}a_s = -A_\sigma + A_s.$$
(20.62)

From a measurement of $A_\parallel$ and $A_\perp$ and from a Hartree–Fock calculation of a_{2p} and a_{2s} for the fluorine ligand ions we can obtain the spin densities $f_t, f_{\sigma p}, f_{\sigma s}$.

It is reasonable to assume that since the orbits $2s$ of the ligands lie well below the orbits $2p$, they must have a smaller amplitude λ_s in the anti-bonding orbitals and the corresponding spin density $f_{\sigma s} = (\lambda_{\sigma s}^2/3)N_\sigma^{-1}$ must be a good deal smaller than $f_{\sigma p}$ and f_t. (This is not to say that $A_s = f_{\sigma s}a_{2s}/2S$ should be much smaller than A_σ or A_π for the smallness of $f_{\sigma s}$ is more than compensated by the fact that $a_{2s} \gg a_{2p}$.) On the other hand, it had generally been assumed that the σ_p orbitals of the ligands, directed towards the central ion, were much more strongly bound to it than π-orbitals and that $f_{\sigma p}$ should correspondingly be much larger than f_t. It came therefore as a great surprise that in the first experiment on $Mn^{2+}(d^5)$ in $KMnF_3$ by Shulman and Knox (1960) the value of $f_{\sigma p} - f_t$ (attributed for the larger part to $f_{\sigma p}$) turned out to be ~ 0.3 per cent, whereas $f_{\sigma s}$ was of the order of 0.5 per cent.

One possible explanation was that actually $f_{\sigma p}$ and f_t were comparable and that the small value observed for $f_{\sigma p} - f_t$ resulted from a near compensation between two larger quantities. Striking confirmation of this explanation and of the general correctness of the model was provided by the study of d^3 compounds, for which $f_{\sigma p} = 0$, and d^8 compounds, for which $f_t = 0$. For $Cr^{3+}(d^3)$ in K_2NaCrF_6 it was found that $f_t = 4.90$ per cent whereas values of $f_{\sigma p} = 4.95$ per cent and $f_{\sigma s} = 0.5$ per cent were found for $Ni^{2+}(d^8)$ in $KNiF_3$ establishing that $f_{\sigma p} \approx f_t \gg f_{\sigma s}$.

Finally, further evidence that the bulk of the isotropic hyperfine structure on the ligands does indeed result from covalent σ_s bonds and not from some other mechanism, such as fluorine core-polarization, has been provided by the small value of the isotropic spin density $f_{\sigma s}$ in d^3 compounds where σ_s bonds cannot be formed. Values of $f_{\sigma s} \sim 0.02$ per cent, which is some twenty times smaller than for d^5 and d^8 compounds, have been observed for $V^{2+}(d^3)$ and $Cr^{3+}(d^3)$. The reader is referred to the review article of Owen and Thornley (1966) for further discussion of the experimental evidence.

20.7. Orbital singlets: correction terms for the ligand hyperfine structure

Direct coupling with the central ion

We have so far neglected the interaction of the nuclear moment $\mathbf{\mu}_I$ of the ligand with the distribution of magnetization carried by the d-part of the anti-bonding orbitals of the central ion, which is described by the second term of (20.49). If the distance R from the ligand nucleus to the nucleus of the central ion is large compared to the mean radius of d-orbitals, we can argue that the field $\mathbf{H}_\mu$ created by $\mathbf{\mu}_I$ is uniform all over the ion and that its coupling with the central ion is given by $\mathbf{H}_\mu \cdot \mathbf{g} \cdot \beta \mathbf{S}$, where $-\beta \mathbf{g} \cdot \mathbf{S}$ is the magnetic dipole moment of the central ion. Even if R is not very large compared to the ionic radius we can still treat the central ion as a point dipole if its charge distribution has spherical symmetry as for d^5 compounds (Cr^+, Mn^{2+}, Fe^{3+}). Under these conditions for, say, the ligand 3, the corresponding coupling can be written

$$A_d(2I_z S_z - I_x S_x - I_y S_y), \qquad (20.63)$$

with

$$A_d = \frac{2\beta\hbar\gamma_n}{R^3}. \qquad (20.64)$$

For $R = 0.2$ nm, a typical distance in fluorine compounds, we find $A_d \approx 3 \times 10^{-4}$ cm^{-1}. This is to be compared with values of A_s of the order of 15×10^{-4} cm^{-1} observed in d^5 fluorine compounds.

The comparison of A_d with values of $A_\sigma - A_\pi$ observed for d^5 compounds is not very instructive since, as we mentioned earlier, the fact that $(A_\sigma - A_\pi)$ is experimentally found to be small does not necessarily imply that A_σ, A_π taken separately are small. For d^3 fluorine compounds, values of A_π of the order of 7×10^{-4} cm^{-1}, and for d^8 compounds values of A_σ of the order of 10×10^{-4} cm^{-1}, have been observed.

Multipole corrections to the coupling with the central ion

For d^3 and d^8 compounds small corrections arising from higher multipole moments of the distribution of magnetization of the central ion have to be introduced. These corrections have different forms for the distributions of magnetization due to spin and to unquenched orbital momentum. We quote only the results (Marshall (1963)).

For the *spin magnetization* the spin constant $A_\mathrm{d} = 2\beta\gamma_\mathrm{n}\hbar/R^3$ has to be multiplied by the following correction factors in which $\langle r^4 \rangle$ is the mean fourth power of a $3d$ wave-function:

$$3d^8(^3F): \quad \left(1 + \frac{5}{2}\frac{\langle r^4 \rangle}{R^4}\right), \tag{20.65}$$

$$3d^3(^4F): \quad \left(1 - \frac{5}{2}\frac{\langle r^4 \rangle}{R^4}\right). \tag{20.66}$$

For the unquenched *orbital magnetization* the constant

$$\frac{(g-2)\beta\gamma_\mathrm{n}\hbar}{R^3} = A_\mathrm{d}'' \tag{20.67}$$

must be multiplied by a correction factor (which has the same form for d^3 and d^8)

$$\left(1 - \frac{5}{28}\frac{\langle r^2 \rangle}{R^2}\right). \tag{20.68}$$

A second correction is a small scalar coupling

$$A_\mathrm{d}'''(\mathbf{I}\cdot\mathbf{S}) = -\tfrac{1}{2}A_\mathrm{d}''(\langle r^2\rangle/R^2)(\mathbf{I}\cdot\mathbf{S}), \tag{20.69}$$

which must be added to the overall dipolar coupling.

Such multipole corrections are usually quite small.

Effects of the spin-orbit coupling on the ligand hyperfine structure

We saw in § 19.2 how in second order the spin-orbit coupling modifies the hyperfine structure on the central nucleus by adding to the hyperfine structure operator (17.30), where $\mathbf{n}_i$ is a unit vector along $\mathbf{r}_i$,

$$\mathscr{H} = \sum_i \frac{2\beta\gamma_\mathrm{n}\hbar}{r^3}\mathbf{I}\cdot\{\mathbf{l}_i - \mathbf{s}_i + 3\mathbf{n}_i(\mathbf{s}_i\cdot\mathbf{n}_i)\}, \tag{20.70}$$

the operator

$$\mathscr{H}' = \mathscr{H}C(\lambda\mathbf{L}\cdot\mathbf{S}) + (\lambda\mathbf{L}\cdot\mathbf{S})C\cdot\mathscr{H} \tag{20.71}$$

with

$$C = \sum_n{}' \frac{|n\rangle\langle n|}{W_0 - W_n}. \tag{20.72}$$

The same procedure yields correction terms to the hyperfine structure of the ligands (Marshall 1963). We quote the results only for $3d^3(^4F)$ and $3d^8(^3F)$.

The orbital part $(2\beta\gamma_\mathrm{n}\hbar/r^3)(\mathbf{l}\cdot\mathbf{I})$ of (20.70) adds through (20.71) a contribution of the same form for $(3d^3, {}^4F)$ and $(3d^8, {}^3F)$, namely

$$\frac{\beta\hbar\gamma_n(g-2)}{12}\langle r^{-3}\rangle\lambda_{\sigma p}\lambda_t\{3(\mathbf{S}\cdot\mathbf{n})(\mathbf{I}\cdot\mathbf{n})-(\mathbf{I}\cdot\mathbf{S})-2(\mathbf{I}\cdot\mathbf{S})\}. \quad (20.73)$$

This represents a contribution to $(A_\sigma - A_\pi)$ equal to

$$\delta_l(A_\sigma - A_\pi) = \frac{\beta\hbar\gamma_n(g-2)}{12}\langle r^{-3}\rangle\lambda_{\sigma p}\lambda_t \approx \frac{\beta\hbar\gamma_n}{2\sqrt{3}}\langle r^{-3}\rangle(f_{\sigma p}f_t)^{\frac{1}{2}}(g-2) \quad (20.74)$$

and a contribution to A_s equal to

$$\delta_l A_s = -\frac{\beta\hbar\gamma_n}{\sqrt{3}}\langle r^{-3}\rangle(f_{\sigma p}f_t)^{\frac{1}{2}}(g-2). \quad (20.75)$$

Similarly the spin part of (20.70) adds to $(A_\sigma - A_\pi)$ the following corrections:

$(3d^8, {}^3F)$ $\delta_s(A_\sigma - A_\pi) = \tfrac{3}{80}\beta\hbar\gamma_n(g-2)\langle r^{-3}\rangle\lambda_{\sigma p}\lambda_t$

$$\approx \frac{3\sqrt{3}}{40}\beta\hbar\gamma_n\langle r^{-3}\rangle(f_{\sigma p}f_t)^{\frac{1}{2}}(g-2); \quad (20.76)$$

$(3d^3, {}^4F)$ $\delta_s(A_\sigma - A_\pi) = \tfrac{1}{40}\beta\hbar\gamma_n(g-2)\langle r^{-3}\rangle\lambda_{\sigma p}\lambda_t$

$$\approx \frac{\sqrt{3}}{20}\beta\hbar\gamma_n\langle r^{-3}\rangle(f_{\sigma p}f_t)^{\frac{1}{2}}(g-2). \quad (20.77)$$

Effects of the fluorine 1s wave-functions on the ligand hyperfine structure (Marshall and Stuart 1961)

In our model of covalent bonding we had assumed that the augmented anti-bonding orbitals did not contain any admixtures from the 1s orbitals of the ligands. Although such admixtures are undoubtedly small their contribution to the ligand hyperfine structure may not be negligible, for $a_{1s} = (16\pi/3)\beta\gamma_n\hbar\,|\varphi_{1s}(0)|^2$ is very much larger than a_{2s}. If instead of a single admixture coefficient λ_s we introduce two, namely λ_{1s} and λ_{2s}, into the anti-bonding σ-orbitals we find that the isotropic hyperfine structure constant A_s instead of being proportional to $\lambda_{2s}^2\,|\varphi_{2s}(0)|^2$ is now proportional to

$$\lambda_{2s}^2\varphi_{2s}^2(0)\left\{1+\frac{\lambda_{1s}\varphi_{1s}(0)}{\lambda_{2s}\varphi_{2s}(0)}\right\}^2. \quad (20.78)$$

The magnitude of the second term in the curly bracket is difficult to estimate: $\varphi_{1s}(0)/\varphi_{2s}(0)$ in F^- is of the order of $-4\cdot 5$. For λ_{2s}, assuming that the effect of φ_{1s} is a small correction, we can take the value extracted from the measurement of $A_s \approx (\lambda_{2s}^2/3)(a_{2s}/2S)$.

For λ_{1s} we could tentatively take the overlap integral

$$S_{1s} = (3d_\sigma \mid \chi_{\sigma,1s}),$$

which is equivalent to taking the covalent bonding constant

$$\gamma_{1s} = \lambda_{1s} - S_{1s} = 0.$$

Unfortunately this integral depends very critically on the shape of the $3d$ wave-function near the fluorine ion, which may be quite different from that of the $3d$ wave-function of the free ion. For instance, with $S_{1s} = \lambda_{1s} = 0.0045$, calculated using $3d$ functions of the free ion Mn^{2+}, and with $\lambda_{2s} \simeq 0.12$ compatible with $A_s = 16 \times 10^{-4}$ cm^{-1} in $KMnF_3$, we get $\{1+(S_{1s}\varphi_{1s}(0)/\lambda_{2s}\varphi_{2s}(0))\}^2 \approx 0.70$, which is a sizeable reduction. The correction is negative because φ_{2s} has a node whereas φ_{1s} has not. If we choose φ_{1s} and φ_{2s} to have the same sign in the overlap region with $3d$, they will have opposite signs at the fluorine nucleus, in accordance with the sign convention we have chosen. However in view of the uncertainty in the magnitude of the correction some authors prefer to omit it altogether.

20.8. Ligand hyperfine structure for ions of type B

When the ground cubic term of the central ion is degenerate the calculation of the hyperfine interaction for the ligands is much more complicated than for orbital singlets. The spin-orbit coupling, which now acts in first order, scrambles orbital and spin states in a way that makes each substate of the ground level a sum of many Slater determinants, and the expectation value of the hyperfine coupling with the nuclear moment of a ligand is a sum of many terms. To illustrate the principle of the calculation we shall take the specific example of Co^{2+} (Thornley, Windsor, and Owen 1965).

To start with, the ground cubic term $^4T_1$ of Co^{2+}, composed of three holes, must be described in the strong-field formalism, which is suitable for the study of covalent bonding, using the formulae (20.18), (20.19), (20.21). On introducing the spin-orbit coupling, its effect in first approximation is to split the ground level which has a degeneracy of $3(2S+1) = 12$ into three $\tilde{J}$ multiplets, $\tilde{J} = \frac{5}{2}, \frac{3}{2}, \frac{1}{2}$, resulting from the vector coupling of the fictitious orbital momentum $\tilde{l} = 1$ with the spin $S = \frac{3}{2}$. It also mixes into the ground cubic term $^4T_1$ the excited cubic term $^4T_2$ but we shall disregard this smaller effect to keep our treatment as simple as possible. The ground level is $\tilde{J} = \frac{1}{2}$, for which the states

THE EFFECTS OF COVALENT BONDING

$\tilde{J}_z = \frac{1}{2}$ and $\tilde{J}_z = -\frac{1}{2}$ are given by the vector coupling formulae

$$|\pm\rangle = \frac{1}{\sqrt{6}}|\pm\tilde{1}\rangle|\mp\tfrac{1}{2}\rangle - \frac{1}{\sqrt{3}}|\tilde{0}\rangle|\pm\tfrac{1}{2}\rangle + \frac{1}{\sqrt{2}}|\mp\tilde{1}\rangle|\pm\tfrac{3}{2}\rangle. \quad (20.79)$$

The notation in this equation requires some clarification. Let $|A\rangle$ be an orbital state which is a single Slater determinant $\{a, b, c\}$. The combined orbit-spin state $|A\rangle|\tfrac{3}{2}\rangle$ will be the Slater determinant $\{a^+b^+c^+\}$ and can be written symbolically

$$|A\rangle\{+, +, +\} = \{a, b, c\}[+, +, +].$$

The state $|A\rangle|\tfrac{1}{2}\rangle$ can similarly be written symbolically $\{a, b, c\}$ $[+, +, -]$ where $[+, +, -]$ is a symmetrized spin Slater determinant $(1/\sqrt{3})[\{+, +, -\}+\{+, -, +\}+\{-, +, +\}]$. Thus $(a, b, c)[+, +, -]$ means $(1/\sqrt{3})\{(a^+b^+c^-)+(a^+b^-c^+)+(a^-b^+c^+)\}$.

This factorization of orbit and spin variables is a special feature of Hund's states with maximum spin. The state $|+\rangle$ of (20.79) can now be rewritten in detail using (20.18), (20.19), and (20.21),

$$|+\rangle = \frac{1}{\sqrt{6}}\Big\{C(\tilde{1}, \theta, \varepsilon) - \frac{S}{2}(\tilde{1}, \tilde{0}, \theta) + \frac{S\sqrt{3}}{2}(-\tilde{1}, \tilde{0}, \varepsilon)\Big\}[-, -, +] -$$

$$-\frac{1}{\sqrt{3}}\Big\{C(\tilde{0}, \theta, \varepsilon) + S(\tilde{1}, -\tilde{1}, \theta)\Big\}[+, +, -] +$$

$$+\frac{1}{\sqrt{2}}\Big\{C(-\tilde{1}, \theta, \varepsilon) + \frac{S}{2}(-\tilde{1}, \tilde{0}, \theta) - \frac{S\sqrt{3}}{2}(\tilde{1}, \tilde{0}, \varepsilon)\Big\}[+, +, +], \quad (20.80)$$

with a similar formula for $|-\rangle$. In this formula C and S are the amplitudes $\cos \alpha$ and $\sin \alpha$ (see eqn (20.21)) of the cubic configurations (t_2e^2) and (t_2^2e) in the ground term $^4T_1$. The one-electron orbital states $|\tilde{1}\rangle, |\tilde{0}\rangle, |-\tilde{1}\rangle$ (not to be confused with the many-electron states $|\tilde{1}\rangle, |\tilde{0}\rangle, |-\tilde{1}\rangle$) are defined in (20.16) where η_x, η_y, η_z and also θ and ε are the augmented antibonding orbitals of Table 24.

The calculation of the hyperfine structure with, say, the nucleus of ligand 3 is now straightforward in principle. The result can again be written in the form $A_{\parallel}I_z\tilde{S}_z + A_{\perp}(I_x\tilde{S}_x + I_y\tilde{S}_y)$ where the fictitious spin $\tilde{S}$ is the fictitious total angular momentum $\tilde{J} = \tfrac{1}{2}$. The values of $A_{\parallel}$ and $A_{\perp}$ are given by

$$\tfrac{1}{2}I_zA_{\parallel} = \langle +| \mathscr{H}(3) |+\rangle,$$

$$\tfrac{1}{2}I_xA_{\perp} = \langle +| \mathscr{H}(3) |-\rangle, \quad (20.81)$$

where $\mathscr{H}(3)$ is the Hamiltonian for the hyperfine coupling with the nucleus of ligand 3. In order to evaluate (20.81) we must replace the orbitals $|\tilde{1}\rangle$, $|\tilde{0}\rangle$, $|-\tilde{1}\rangle$, $|\theta\rangle$, $|\varepsilon\rangle$ in (20.80) by their expansions given in Table 24, using (20.16). The calculation is lengthy but offers no special difficulties. To illustrate it we consider in $\mathscr{H}(3)$ the contact interaction part

$$\mathscr{H}_c(3) = \frac{16\pi}{3}\beta\gamma_n\hbar\delta(r_3)\,(\mathbf{I}_3\cdot\mathbf{s}). \tag{20.82}$$

The only orbital in (20.80) that gives a non-vanishing contribution to $\langle +|\,\mathscr{H}_c(3)\,|+\rangle$ and $\langle +|\,\mathscr{H}_c(3)\,|-\rangle$ is the orbital $|\theta\rangle$, which contains the wave-function $\varphi_{2s}(3)$ with an amplitude $(-\lambda_s/\sqrt{3})(N_\sigma^{-\frac{1}{2}}) = -f_s^{\frac{1}{2}}$. From (20.79) we obtain

$$\tfrac{1}{2}I_z A_{\|}^s = \langle +|\,\mathscr{H}_c(3)\,|+\rangle = \tfrac{1}{6}\langle\bar{1},-\tfrac{1}{2}|\,\mathscr{H}_c(3)\,|\bar{1},-\tfrac{1}{2}\rangle +$$
$$+\tfrac{1}{3}\langle\bar{0},\tfrac{1}{2}|\,\mathscr{H}_c(3)\,|\bar{0},\tfrac{1}{2}\rangle + \tfrac{1}{2}\langle -\bar{1},\tfrac{3}{2}|\,\mathscr{H}_c(3)\,|-\bar{1},\tfrac{3}{2}\rangle \tag{20.83}$$

which, using (20.80), becomes

$$A_{\|}^s = \frac{16\pi}{3}\beta\gamma_n\hbar f_{\sigma s}|\varphi_{2s}(0)|^2\left\{-\frac{1}{6}\left(C^2+\frac{S^2}{4}\right)\frac{1}{3}+\frac{1}{3}(C^2+S^2)\frac{1}{3}+\frac{1}{2}\left(C^2+\frac{S^2}{4}\right)\right\}$$
$$= a_{2s}f_{\sigma s}\frac{5C^2+2S^2}{9}. \tag{20.84}$$

Similarly

$$\tfrac{1}{2}I_x A_{\perp}^s = \langle +|\,\mathscr{H}_c(3)\,|-\rangle$$
$$= \frac{1}{\sqrt{12}}\langle\bar{1},-\tfrac{1}{2}|\,\mathscr{H}_c|\bar{1},-\tfrac{3}{2}\rangle + \frac{1}{3}\langle\bar{0},\tfrac{1}{2}|\,\mathscr{H}_c|\bar{0},-\tfrac{1}{2}\rangle +$$
$$+\frac{1}{\sqrt{12}}\langle -\bar{1},\tfrac{3}{2}|\,\mathscr{H}_c|-\bar{1},\tfrac{1}{2}\rangle, \tag{20.85}$$

whence

$$A_{\perp}^s = a_{2s}f_{\sigma s}\left\{\frac{5C^2}{9}+\frac{11S^2}{36}\right\}. \tag{20.86}$$

A feature of this result, which we have not met before, is that the contact interaction is not perfectly isotropic. The anisotropy, as measured by $(A_{\|}^s - A_{\perp}^s)/A_{\|}^s = (-\tfrac{3}{20})(\tan^2\alpha)/(1+\tfrac{2}{5}\tan^2\alpha)$ which is quite small. For a vanishing splitting $\Delta = W(t_{2g}) - W(e)$ the ground cubic term is pure ^4F and $\tan^2\alpha = \tfrac{1}{4}$ (see after eqn (20.24)) giving an anisotropy of about -3 per cent. It is actually much less because experimentally $\tan^2\alpha$ turns out to be of the order of 0·08.

The calculation of $A_{\|}^p$ and $A_{\perp}^p$ follows the same principle but is more involved, in particular because the p-hyperfine structure Hamiltonian

THE EFFECTS OF COVALENT BONDING 787

$\mathcal{H}_p$, in contrast with $\mathcal{H}_c$, can have off-diagonal matrix elements between two different antibonding orbitals.

Since α is small, it is not a bad approximation and a great simplification to take in (20.80) $C = 1$, $S = 0$. Then the part of (20.80) that contains p-orbitals of ligand 3, which we need to calculate the expectation value of the p-hyperfine structure Hamiltonian $\mathcal{H}_p(3)$, can be written

$$|+\rangle = \frac{1}{\sqrt{6}}\{-f_t^{\frac{1}{2}}(-1_p, \theta, \varepsilon) - f_{\sigma p}^{\frac{1}{2}}(0_p, \tilde{1}, \varepsilon)\}[-, -, +] +$$

$$+ \frac{1}{\sqrt{3}} f_{\sigma p}^{\frac{1}{2}}(0_p, \tilde{0}, \varepsilon)[+, +, -] +$$

$$+ \frac{1}{\sqrt{2}} \{f_t^{\frac{1}{2}}(1_p, \theta, \varepsilon) - f_{\sigma p}^{\frac{1}{2}}(0_p, -\tilde{1}, \varepsilon)\}[+, +, +], \qquad (20.87)$$

where $|-1_p\rangle$, $|0_p\rangle$, $|1_p\rangle$ are p-orbitals of ligand 3, eigenstates of $l_z(3)$ with eigenvalues $-1, 0, 1$. A similar formula can be written for the Kramers conjugate state $|-\rangle$.

The one-electron p-hyperfine structure Hamiltonian $\mathcal{H}_p(3)$ can be written

$$\begin{cases} \mathscr{P}(\mathbf{I} \cdot \mathbf{N}), \text{ where} \\ \mathscr{P} = 2\beta\hbar\gamma_n \langle r^{-3} \rangle \text{ and} \\ \mathbf{N} = \mathbf{l} + \tfrac{4}{5}\mathbf{s} - \tfrac{3}{5}\{\mathbf{l}(\mathbf{l} \cdot \mathbf{s}) + (\mathbf{l} \cdot \mathbf{s})\mathbf{l}\} - \kappa\mathbf{s}. \end{cases} \qquad (20.88)$$

Here the last term in **N** allows phenomenologically for possible core polarization on the fluorine ion.

The hyperfine structure coefficients $A_\parallel^p$ and $A_\perp^p$ are given by

$$A_\parallel^p = 2\mathscr{P}\langle +| N_z |+\rangle, \; A_\perp^p = 2\mathscr{P}\langle +| N_x |-\rangle$$

and, using (20.87), (20.88), we obtain

$$\begin{cases} A_\parallel = \mathscr{P}\left\{f_{\sigma p}\left(\dfrac{4}{9} - \dfrac{5\kappa}{9}\right) + f_t\left(\dfrac{22}{45} - \dfrac{4\kappa}{9}\right)\right\}, \\ A_\perp = \mathscr{P}\left\{f_{\sigma p}\left(-\dfrac{2}{9} - \dfrac{5\kappa}{9}\right) + f_t\left(\dfrac{1}{15} - \dfrac{\kappa}{3}\right)\right\}. \end{cases} \qquad (20.89)$$

The complete expressions for $A_\parallel$ and $A_\perp$ for $\alpha \neq 0$ can be found in Thornley et al. (1965), where finer effects such as spin-orbit mixing of $^4T_1$ and $^4T_2$ are also considered.

We have outlined this calculation of ligand hyperfine structure in some detail because it is the first example of a situation where we could not use the Wigner–Eckart theorem generalized for the cubic group and have had to write out the Slater determinants explicitly. The reason for this state of affairs is clear: the hyperfine structure with *one given* ligand is *not* invariant under cubic symmetry.

20.9. Ligand quadrupole hyperfine structure

If the ligand nuclei have spins $I \geq 1$ there exists, besides the hyperfine magnetic coupling, an electric quadrupole coupling between the nuclear quadrupole moments and the electronic field gradients at the nuclear sites. Since most studies of transferred hyperfine couplings have been performed on compounds where the ligands are fluorine ions with nuclear spin $I = \frac{1}{2}$, the experimental evidence on quadrupole effects is scarce. Furthermore, the relationship between these couplings and covalency is somewhat more complicated than in the case of magnetic hyperfine coupling. One obvious difference between magnetic and quadrupole couplings with the nuclei of the ligands is that s-orbitals, so important for the former, do not contribute to the latter. Another essential difference is the fact that in contrast to magnetic hyperfine structure, bonding orbitals do contribute to the quadrupole couplings. In order to minimize the total energy of the complex, each bonding orbital is filled with two electrons with opposite spins whose contributions to the magnetic hyperfine structure of a ligand nucleus cancel each other. On the other hand, their contributions to the quadrupole coupling of a ligand are spin independent and *add*.

Let us again consider ligand 3, and the contributions from its orbitals to the anti-bonding orbitals of the complex given by (20.50). The quadrupole hyperfine interaction will have axial symmetry around the z-axis. If $\mathscr{H}_Q$ is the quadrupole Hamiltonian we know that for a p-electron

$$\langle Z_3| \mathscr{H}_Q |Z_3\rangle = -2\langle X_3| \mathscr{H}_Q |X_3\rangle = -2\langle Y_3| \mathscr{H}_Q |Y_3\rangle$$
$$= -\frac{12}{5} \frac{e^2 Q \langle r^{-3}\rangle}{4I(2I-1)} \{I_z^2 - \tfrac{1}{3}I(I+1)\}, \quad (20.90)$$

where $\langle r^{-3}\rangle$ is of course the value for a p-orbital on the ligand ion. Using (20.50) and (20.51) we find the following contributions to the quadrupole couplings from the antibonding orbitals of ions in singlet states:

$$3d^3 \quad \langle \mathscr{H}_Q \rangle = -f_t \langle Z_3| \mathscr{H}_Q |Z_3\rangle,$$
$$3d^5 \quad \langle \mathscr{H}_Q \rangle = (f_{\sigma p} - f_t)\langle Z_3| \mathscr{H}_Q |Z_3\rangle, \quad (20.91)$$
$$3d^8 \quad \langle \mathscr{H}_Q \rangle = (f_{\sigma p} - 2f_t)\langle Z_3| \mathscr{H}_Q |Z_3\rangle.$$

To these we must add the contribution of the filled bonding orbitals. If these orbitals were exactly p-orbitals of the ligand their contribution to the quadrupole interaction would be that of a closed shell and therefore vanish. Actually each bonding p-orbital contains admixtures from $3d$ orbitals of the central ion (and possibly admixtures from other orbitals) which behave like holes in the p-shell. This remark makes possible an estimate of the quadrupole interaction of the ligands if the following simplifying assumptions are made (Owen and Thornley 1966): bonding ligand p-orbitals contain admixtures from central $3d$-orbitals which have the *same* magnitude as the corresponding admixtures from ligand p-orbitals in the central antibonding $3d$-orbitals (this assumption is equivalent to taking the overlap S as much smaller than the covalency coefficient γ in eqn (20.10)).

With these assumptions the contribution of an electron in a bonding p-orbital to the quadrupole interaction is equal and *opposite* to that of an electron in the corresponding anti-bonding orbital. The contribution of the bonding orbitals to the quadrupole interaction is thus, using (20.50), (20.51),

$$-2\langle Z_3| \mathscr{H}_Q |Z_3\rangle \{f_{\sigma p}-f_t\}. \tag{20.92}$$

Adding (20.91) and (20.92) we find the following approximate expressions for the quadrupole interaction:

$$\begin{aligned} 3d^3 &\quad \langle Z_3| \mathscr{H}_Q |Z_3\rangle (f_t-2f_{\sigma p}), \\ 3d^5 &\quad \langle Z_3| \mathscr{H}_Q |Z_3\rangle (f_t-f_{\sigma p}), \\ 3d^8 &\quad -\langle Z_3| \mathscr{H}_Q |Z_3\rangle f_{\sigma p}. \end{aligned} \tag{20.93}$$

These expressions should be compared to the eqns (20.58), (20.60), and (20.62), which give the magnetic hyperfine structure due to p-electrons. It will be noticed that while for $3d^5$ and $3d^8$ the magnetic hyperfine structure contains the same admixture coefficients, namely $f_t-f_{\sigma p}$, and $f_{\sigma p}$, the situation is different for $3d^3$ where the magnetic hyperfine structure is proportional to f_t.

21

THE JAHN–TELLER EFFECT
IN PARAMAGNETIC RESONANCE

21.1. Introduction

PARAMAGNETIC resonance of transition elements is studied most conveniently in dilute samples. Given a crystal containing a regular array of diamagnetic ions such as, say, Zn^{2+}, a few of these are replaced by paramagnetic ions such as, say, Cu^{2+} or Mn^{2+}, whose paramagnetic resonance is then observed. To interpret the results of the observations, the following simplifying assumptions are usually made.

(a) the influence of the environment on the electronic wave-function of the ion can be described by means of a *static* crystal field.

(b) if the paramagnetic ion has the same charge state as the diamagnetic ion which it has replaced, so that no charge compensation occurs, the symmetry (if not the magnitude) of the crystal field it 'sees', is unchanged.

There are, however, cases when these assumptions are invalid. Let us begin by neglecting spin-orbit coupling, a usual procedure in the iron group, and assume that the symmetry of the crystal potential is sufficiently high to leave some orbital degeneracy in the electronic ground state of the paramagnetic ion. In a cubic field this state will be a Γ_3 doublet or a Γ_5 triplet for a D-state of the free ion, and a Γ_4 triplet for an F-state. There exists a general theorem due to Jahn and Teller (1937) which predicts that such a degenerate state is unstable with respect to small displacements of the neighbouring atoms which lower the symmetry of the crystal field 'seen' by the ion. As a consequence of this theorem, whose precise meaning will be discussed in the following sections, either assumption (a) or (b) may be violated.

(a) The environment of the paramagnetic impurity ion may keep the same symmetry that it had for the diamagnetic ion but the description of this environment by the device of a static crystal field is invalid. This is the so-called dynamic Jahn–Teller effect.

(b) The approximation of using a static crystal field is still valid but this field has a lower symmetry than in the original crystal. This is the so-called static Jahn–Teller effect.

In either case, the consequences for the features of the paramagnetic resonance spectrum are far-reaching. In the following sections we shall outline the general theoretical treatment of the Jahn–Teller effect, which will then be applied to situations that actually occur in paramagnetic ions.

Since the original work of Jahn and Teller (1937) and the pioneer work of Van Vleck (1939) a large number of articles, some of them of a highly sophisticated nature, have been published on the theory of static and dynamic Jahn–Teller effects. The reader will find in the detailed bibliography of the review article by Ham (1969) the references to the important work of Öpik and Pryce, Liehr, Moffitt and Thorson, Longuet-Higgins, Child, O'Brien, Bersuker, and many others, whose contributions have helped to clarify the various aspects of the theory. In spite of this theoretical effort and in contrast to it, experimental evidence of the existence of the Jahn–Teller effect had until recently been remarkably scarce, with the exception of the pioneer work of Bleaney and his colleagues on Cu^{2+}. Neither the reasons for the failure of the Jahn–Teller effect to materialize in the form in which it was expected nor its dependence on the temperature, when observed, were fully understood, in spite of the early work of Abragam and Pryce (1950), which has been considerably expanded, along different lines, by Bersuker (1962, 1963) and O'Brien (1964).

A major advance was achieved with the theoretical work of Ham (1965, 1968, 1969) who was the first to push to its logical consequences the idea, which somehow had escaped general attention, that the Jahn–Teller coupling between the nuclear and the electronic motion had necessarily the same high symmetry as the electronic Hamiltonian responsible for the degeneracy of the paramagnetic ion, and thus could not possibly lift that degeneracy. Ham pointed out that the Jahn–Teller effect whose influence had seemed to be mysteriously absent from certain paramagnetic spectra had been there all the time as large as life if one knew where to look for it. A second major advance has been the discovery by Coffman (1965) and Höchli (1966) of a new type of Jahn–Teller spectrum previously unobserved, which fitted beautifully into the general theory of the dynamic Jahn–Teller effect.

The treatment of this chapter owes much to the ideas and even the presentation of Ham in his two major papers (1965, 1968) and also to his review article (1969) which through his kindness was made available to the authors prior to its publication. This treatment, where the ion and its nearest neighbours are considered as an isolated molecule interacting only through relaxation mechanisms with the rest of the crystal in which it is embedded, is undoubtedly a gross oversimplification. Its justification is twofold: first a rigorous theory of the Jahn–Teller coupling of the paramagnetic ion to the rest of the lattice simply does not exist at present; secondly, as elsewhere in this book, the most important features of the phenomenon are already determined by the symmetry of the problem, which is built into the approximate treatment to be given below.

We limit the discussion throughout to cubic symmetry. The changes needed when a field of lesser symmetry still leaves an orbital degeneracy, such as a trigonal field applied to a cubic Γ_3 doublet, are straightforward. Furthermore, we shall see that if these fields are much weaker than the cubic field, their effect is often appreciably quenched by the dynamic Jahn–Teller effect itself.

21.2. The Born-Oppenheimer approximation and the Jahn-Teller theorem

Let us make the simplifying assumption that for the electronic state of the paramagnetic ion under consideration we may disregard the influence of the positions (and motions) of all but its nearest neighbours in the crystal. We can then treat the complex XY_n formed by the paramagnetic ion X and its n nearest neighbours Y ($n = 6$ for octahedral and 4 for tetrahedral coordination) as a single molecule and we start by recalling the general method for finding its equilibrium configuration. The Hamiltonian $\mathscr{H}$ of the molecule can be written

$$\mathscr{H} = T+V = T_\mathrm{e}+T_\mathrm{N}+V_\mathrm{ee}+V_\mathrm{eN}+V_\mathrm{NN}, \qquad (21.1)$$

where the various terms of (21.1) are respectively the kinetic energy of the electrons and of the nuclei, together with the Coulomb interactions between the electrons, between electrons and nuclei, and between nuclei (smaller magnetic interactions are disregarded at this stage of approximation). Let the symbol q represent the $3n$ coordinates of the n electrons of the molecule and Q the $3N$ coordinates of its N nuclei. Equation (21.1) can then be rewritten as

$$\mathscr{H} = T+V = T_\mathrm{e}(p)+T_\mathrm{N}(P)+V_\mathrm{ee}(q)+V_\mathrm{eN}(q,Q)+V_\mathrm{NN}(Q) \qquad (21.1\mathrm{a})$$

where p and P are momenta conjugate to q and Q. An eigenfunction $\Phi(q, Q)$ of (21.1a) satisfies the Schrödinger equation

$$\mathscr{H}\Phi(q, Q) = W\Phi(q, Q). \qquad (21.1\mathrm{b})$$

These eigenfunctions are found by the following approximate

procedure due to Born and Oppenheimer (1927). The nuclear coordinates Q are given some fixed numerical values Q' (the nuclei are 'nailed down'), and the electronic Hamiltonian

$$\mathscr{H}_e(q, Q') = \mathscr{H} - T_N(P) = T_e(p) + V(q, Q'), \qquad (21.2)$$

where Q' is a set of *numerical* parameters, is then diagonalized. Let $\varphi_{Q'}(q)$ be an eigenstate and $U(Q')$ the corresponding eigenvalue of (21.2) so that

$$\mathscr{H}_e(q, Q')\varphi_{Q'}(q) = U(Q')\varphi_{Q'}(q). \qquad (21.3)$$

This task is repeated for all values of Q' in order to determine the function $U(Q')$. The complete wave-function of the molecule, the solution of (21.1b), is then sought in the so-called Born–Oppenheimer form

$$\Phi(q, Q) = \varphi_Q(q)\Psi(Q), \qquad (21.4)$$

where $\Psi(Q)$ is the nuclear part of the molecular wave-function. Introducing (21.4) into (21.1b) and making use of (21.2) and (21.3), one obtains the following equation for $\Psi(Q)$:

$$\{T_N(P) + U(Q)\}\Psi(Q) \approx W\Psi(Q). \qquad (21.5)$$

Equation (21.5) is only approximately correct: it is obtained by making the assumption

$$T_N\{\varphi_Q(q)\Psi(Q)\} \approx \varphi_Q(q)T_N\Psi(Q)$$

and implies that the electronic wave-function $\varphi_Q(q)$ is relatively insensitive to small nuclear displacements. A general discussion of the validity of this assumption, which relies on the smallness of the ratio of the electronic to nuclear masses, is outside the scope of this book. It is clear that it will in general become invalid in cases of degeneracy or near degeneracy when two eigenvalues of (21.3), $U_i(Q')$ and $U_j(Q')$, corresponding to two eigenstates $\varphi_{Q'}^i(q)$ and $\varphi_{Q'}^j(q)$ come very close to each other, because the kinetic energy terms neglected in (21.5) may appreciably mix these two states. This constitutes a failure of the Born–Oppenheimer approximation.

The Born–Oppenheimer procedure can now be summarized as follows. With the nuclei nailed down at positions Q', the electronic eigenfunction $\varphi_{Q'}(q)$ and the electronic energy $U(Q')$ are found for all values of Q'. One then forgets completely about the electrons and solves a Schrödinger equation of motion for the nuclei, the inter-nuclear potential energy being precisely the electronic energy $U(Q)$ obtained previously by solving (21.3).

To find the equilibrium shape of the molecule (or of the paramagnetic complex XY_n, embedded in the crystal) one then seeks the minimum of the potential energy $U(Q)$, which requires a knowledge of the derivatives $\partial U/\partial Q_j$. Their calculation is facilitated by a remark due to Feynman (1939). From the relation

$$U(Q) = \langle \varphi_Q(q) | \mathcal{H}^e_Q(q) | \varphi_Q(q) \rangle, \qquad (21.6)$$

we find

$$\frac{\partial U}{\partial Q_j} = \langle \varphi_Q(q) | \frac{\partial \mathcal{H}^e_Q(q)}{\partial Q_j} | \varphi_Q(q) \rangle + U(Q) \frac{\partial}{\partial Q_j} \langle \varphi_Q | \varphi_Q \rangle. \qquad (21.6a)$$

The last term vanishes because $\varphi_Q(q)$ is normalized, so that

$$\frac{\partial U}{\partial Q_j} = \left\langle \frac{\partial \mathcal{H}^e_Q(q)}{\partial Q_j} \right\rangle = \left\langle \frac{\partial V}{\partial Q_j} \right\rangle, \qquad (21.7)$$

the expectation value being taken over an eigenstate $\varphi_Q(q)$ of (21.3). What the relation (21.7) means is this: suppose that for a set Q_0 of values of Q the electronic wave-function $\varphi_{Q_0}(q)$ is either known or can be guessed from, say, symmetry considerations. In order to determine whether this set Q_0 corresponds to a minimum of $U(Q)$ it is *not* necessary to calculate $\partial U/\partial Q_j$ at Q_0, that is $U(Q)$ in the neighbourhood of Q_0. Rather, it is enough to calculate the expectation value, over $\varphi_{Q_0}(q)$, of the derivative $\partial V/\partial Q_j$ of the Coulomb energy whose expression, $V_{ee}(q) + V_{eN}(q, Q) + V_{NN}(Q)$, is naturally known. Formula (21.7) also follows directly from the well-known result of first-order perturbation theory whereby the first-order change in the energy is equal to the expectation value of the change in the potential, taken over the unperturbed wave-function.

Once the set of values Q_0 has been found which minimizes $U(Q)$, the latter is expanded as a quadratic form of the differences $Q_i - Q_{0i}$ and the vibrational motion of the nuclei around the equilibrium position Q_0 can be considered as harmonic. If one is not interested in nuclear vibrations one may solve the electronic equation (21.3) by giving to the nuclear coordinates Q' the experimental values as found, for instance, from X-ray measurements of internuclear distances. This, however, is only an approximation, as will appear later in more detail.

If the central ion of the complex is a diamagnetic ion already non-degenerate when free, the minimum of $U(Q)$ will often occur for a set of values Q_0 such that the complex has a high symmetry, such as cubic. If the nucleus of the diamagnetic ion X' has a spin $I > \frac{1}{2}$ this symmetry can be verified, for example, from the observation of its nuclear resonance and the absence of any quadrupole splitting. What happens then if we replace the diamagnetic ion X' by a paramagnetic ion X which, when free, possesses an orbital degeneracy in its ground state? Let us nail down the nuclei at positions Q_0 in the paramagnetic complex XY_n such that it has the same *symmetry* as the equilibrium position of the diamagnetic complex $X'Y_n$ (the overall *scale* may be different). The ground level Γ of the ion X in the environment created by its neighbours Y may then still have some degeneracy left (Γ_3, Γ_4, or Γ_5). What the Jahn–Teller theorem says is that it is always possible, by shifting some of the nuclei, to lower the energy of at least one of the states that span the ground level.

In order to put the discussion on a quantitative basis some definitions are required. Let us consider as an example the octahedral complex XY_6 where the coordinates of the various atoms ($i = 1, \ldots, 6$) have in the cubic symmetry position Q_0 the following values x_α^{0i}: $(a, 0, 0)$, $(0, a, 0)$, $(0, 0, a)$, $(-a, 0, 0)$, $(0, -a, 0)$, $(0, 0, -a)$. Any distortion or displacement of the complex is defined by the values of the 18 displacements $X_\alpha^i = x_\alpha^i - x_\alpha^{0i}$. Under symmetry operations of the cubic group the quantities X_α^i transform into themselves, providing a representation of the cubic group which is reducible. It is more convenient to introduce the so-called normal coordinates Q_j, linear combinations of the X_α^i which transform according to *irreducible* representations of the cubic group. These combinations are obtained by the standard methods of group theory. Three linear combinations of the X_α^i correspond to rotations without distortion of the octahedron XY_6 as a whole. They are of no interest to us in so far as we disregard interactions between the complex and the crystal wherein it is embedded. The fifteen remaining combinations transform according to a representation $\bar{\Gamma}$ which can be reduced as follows:

$$\bar{\Gamma} = \Gamma_1 + \Gamma_{3g} + \Gamma_{5g} + \Gamma_{5u} + \Gamma_{4u}^a + \Gamma_{4u}^b \qquad (21.8)$$

where the suffixes g and u mean even and odd. The indices a and b mean that Γ_{4u} is contained twice in $\bar{\Gamma}$. As explained in § 14.6, only the even displacements are expected to change the potential energy of the complex in first order.

The explicit expressions of the Q_i as functions of the X_α^i are given for the even representations in Table 26. Similar expressions can be written for Q_7 to Q_{15} but we shall not require them. From now on when we write the Coulomb energy as $V(q, Q)$ and the potential energy as

$U(Q)$, where Q stands for the set of normal coordinates defined above (suitably modified for groups other than O_h).

We now return to the ground manifold of the paramagnetic ion X within the complex XY_n, a manifold spanned by the wave-functions $\varphi_Q^p(q)$ which obey the relations

$$\langle \varphi_Q^p(q)| \mathscr{H}_Q^e(q) |\varphi_Q^r(q)\rangle = \delta_p^r U_p(Q). \tag{21.9}$$

We have made the assumption that this manifold is degenerate when the complex has a symmetrical shape invariant through a group $\mathscr{G}$. This occurs when all normal coordinates Q_i are set equal to $Q_i^0 = 0$. Equation (21.9) then becomes

$$\langle \varphi_{Q_0}^p(q)| \mathscr{H}_{Q_0}^e(q) |\varphi_{Q_0}^r(q)\rangle = \langle p| \mathscr{H}_0^e(q) |r\rangle = \delta_r^p U(0) \tag{21.10}$$

where we write for brevity $|\varphi_{Q_0}^r(q)\rangle = |r\rangle$ and $\mathscr{H}_{Q_0}^e(q) = \mathscr{H}_0^e$.

The functions $|r\rangle$ provide a basis for an irreducible representation Γ of the group $\mathscr{G}$. If the normal coordinates are given small values Q^j, the potential $V(q, Q)$ will change in first approximation by an amount

$$\delta V \approx \sum_j V_j(q, 0) Q_j. \tag{21.11}$$

Since δV is clearly left unchanged by an operation of the group $\mathscr{G}$ acting simultaneously on the nuclear coordinates Q *and* on the electronic coordinates q, the electronic operators $V_j(q, 0)$ must provide the same irreducible representations Γ' of $\mathscr{G}$ as the Q_j. According to the rules of first-order perturbation theory, the changes in the energy $U(0)$ of the degenerate manifold caused by the term (21.11) are given by the eigenvalues of the matrix

$$\langle p| \delta V |r\rangle = \sum_j \langle p| V_j |r\rangle Q_j.$$

As shown in § 12.6 the degeneracy of the manifold Γ will be lifted by the perturbation V_j belonging to a representation Γ' of $\mathscr{G}$ (different from the unit representation) if the direct product $\Gamma^* \times \Gamma \times \Gamma'$ contains the unit representation. The case where Γ' is the unit representation corresponds to the totally symmetric (or breathing) mode, which simply changes the scale of the complex without altering its shape. The degenerate energy level of the complex is then shifted without being split.

The proof of the Jahn–Teller theorem consists in a systematic check, for all point-groups $\mathscr{G}$, that there always exists at least one representation Γ', different from the unit representation, such that $\Gamma^* \times \Gamma \times \Gamma'$ contains the unit representation (excluding the case of a linear complex and that of Kramers degeneracy). Γ is an irreducible representation of $\mathscr{G}$ with more than one dimension, and Γ' is one of the representations spanned by the normal coordinates of the complex which is left invariant through operations of $\mathscr{G}$.

The sum of the energy displacements induced by V_j is equal to the trace $\sum_p \langle p| V_j |p \rangle$ which has been shown to vanish (eqn (12.40b)), (unless V_j belongs to the unit representation Γ_1). It follows that for $|Q| \neq 0$ and sufficiently small there will always be at least one electronic energy level *lower* than the degenerate level for $Q = 0$.

We shall concern ourselves almost exclusively with cubic symmetry. The situation is very different depending on whether the degenerate multiplet is the doublet Γ_3 or the triplet Γ_4 or Γ_5 and we shall deal with them successively, starting with the doublet.

21.3. The magnetic properties of a ^2E level

It has already been stated (§ 14.2) that the doublet Γ_3 is non-magnetic. This statement, which means that inside the manifold Γ_3 a vector operator **V** has vanishing matrix elements, is a consequence of the reduction formula of the direct product (Table 2)

$$\Gamma_3 \times \Gamma_3 = \Gamma_3 + \Gamma_1 + \Gamma_2. \tag{21.12}$$

Let θ and ε be the two wave-functions that span the doublet Γ_3 and transform respectively like $(3z^2 - r^2)$ and $\sqrt{(3)}(x^2 - y^2)$ under the operation of the cubic group. Under a rotation R of $2\pi/3$ or R^2 of $4\pi/3$ around the body diagonal [111] of the cube, these functions transform according to the formulae (14.4) which we reproduce here for convenience:

$$R\theta = -\frac{\theta}{2} + \frac{\sqrt{3}}{2}\varepsilon, \qquad R^2\theta = -\frac{\theta}{2} - \frac{\sqrt{3}}{2}\varepsilon;$$

$$R\varepsilon = -\frac{\sqrt{3}}{2}\theta - \frac{\varepsilon}{2}, \qquad R^2\varepsilon = \frac{\sqrt{3}}{2}\theta - \frac{\varepsilon}{2}. \tag{21.13}$$

The normal coordinates Q_3 and Q_2 of Table 26 transform under the operation of the cubic group in the same way as θ and ε, and for this reason are often written as Q_θ and Q_ε.

Let $X = X_1 - X_4$, $Y = Y_2 - Y_5$, $Z = Z_3 - Z_6$ be the axial distortions of an octahedral complex (XY_6). If the totally symmetrical (or 'breathing') coordinate $Q_1 = X + Y + Z$ is given the value zero, the quantities X, Y, and Z are expressed by the following relations extracted from Table 26:

$$X = Q_2 - \frac{Q_3}{\sqrt{3}}, \qquad Y = -Q_2 - \frac{Q_3}{\sqrt{3}}, \qquad Z = \frac{2Q_3}{\sqrt{3}}. \tag{21.14}$$

If we define two new variables φ and ρ by the formulae

$$Q_2 \equiv Q_\varepsilon = \rho \sin \varphi, \qquad Q_3 \equiv Q_\theta = \rho \cos \varphi, \tag{21.15}$$

eqn (21.14) can be rewritten

$$Z = \frac{2\rho \cos \varphi}{\sqrt{3}}, \qquad X = \frac{2\rho}{\sqrt{3}} \cos\left(\varphi - \frac{2\pi}{3}\right), \qquad Y = \frac{2\rho}{\sqrt{3}} \cos\left(\varphi + \frac{2\pi}{3}\right). \tag{21.16}$$

We see that a cyclical permutation $X \to Y \to Z \to X$ can be described by successive changes $2\pi/3$ in the angle φ, which can thus be given the geometrical interpretation of a rotation through the angle $2\pi/3$ around the body diagonal [111]. For a given value of ρ, $\varphi = 0$ corresponds, according to (21.16), to an elongation of the octahedron along the z-axis, $\varphi = 2\pi/3$ along the x-axis, and $\varphi = -2\pi/3$ along the y-axis. Values of $\varphi = \pi$, $\pi+2\pi/3$, and $\pi-2\pi/3$ correspond to compressions of the octahedron along the same axes.

If two representations Γ_3 are spanned respectively by the two pairs (θ, ε) and (θ', ε') the direct product $\Gamma_3 \times \Gamma_3$ is spanned by the four products $\theta\theta'$, $\varepsilon\varepsilon'$, $\theta\varepsilon'$, $\varepsilon\theta'$. The linear combinations of these which span the irreducible representations, Γ_1, Γ_2, Γ_3 of the right hand of (21.12) are

$$(E)\Gamma_3: \begin{cases} \theta'' = \varepsilon\varepsilon' - \theta\theta', \\ \varepsilon'' = \varepsilon\theta' + \theta\varepsilon', \end{cases}$$

$$(A_1)\Gamma_1: \qquad \theta\theta' + \varepsilon\varepsilon', \tag{21.17}$$

$$(A_2)\Gamma_2: \qquad \theta\varepsilon' - \varepsilon\theta'.$$

The formulae (21.17) are easily obtained by standard methods of group theory or, in view of their simplicity, by direct inspection. It is clear from (21.15) and (21.17) that $\rho^2 = Q_\theta^2 + Q_\varepsilon^2$ is an invariant and that $\cos \varphi$ and $\sin \varphi$ transform like θ and ε; $\cos 2\varphi = \cos^2\varphi - \sin^2\varphi$ and $\sin 2\varphi = 2 \cos \varphi \sin \varphi$ transform like $-\theta$ and ε. The quantity

$$\cos 3\varphi = \cos 2\varphi \cos \varphi - \sin 2\varphi \sin \varphi$$

transforms like $-(\theta\theta' + \varepsilon\varepsilon')$ and is thus an invariant belonging to the unit representation Γ_1, whereas $\sin 3\varphi = \sin 2\varphi \cos \varphi + \sin \varphi \cos 2\varphi$ transforms like $\varepsilon'\theta - \theta'\varepsilon$ and belongs to the unidimensional representation $\Gamma_2(A_2)$.

Inside the manifold Γ_3 spanned by the functions $|\theta\rangle$ and $|\varepsilon\rangle$ all Hermitian operators can be expressed as a linear combination of the four Pauli matrices

$$\sigma_1 = \begin{pmatrix} 0 & 1 \\ 1 & 0 \end{pmatrix} = |\theta\rangle\langle\varepsilon| + |\varepsilon\rangle\langle\theta|, \quad \mathscr{I} = \begin{pmatrix} 1 & 0 \\ 0 & 1 \end{pmatrix} = |\theta\rangle\langle\theta| + |\varepsilon\rangle\langle\varepsilon|,$$

$$\sigma_2 = \begin{pmatrix} 0 & -i \\ i & 0 \end{pmatrix} = i\{|\varepsilon\rangle\langle\theta| - |\theta\rangle\langle\varepsilon|\}, \quad \sigma_3 = \begin{pmatrix} 1 & 0 \\ 0 & -1 \end{pmatrix} = |\theta\rangle\langle\theta| - |\varepsilon\rangle\langle\varepsilon|.$$

(21.18)

A comparison of (21.18) with (21.17) shows that σ_1 and σ_3 transform respectively as ε and $-\theta$ and for that reason, when they operate on the substates $|\theta\rangle$ and $|\varepsilon\rangle$ of a doublet Γ_3, they are usually written in the literature as: $\sigma_1 = U_\varepsilon$ and $\sigma_3 = -U_\theta$. Similarly σ_2, according to (21.17) and (21.18), belongs to the unidimensional representation Γ_2 and is usually written as $\sigma_2 = A_2$.

An alternative proof of the fact that σ_3, σ_1, σ_2 transform like $-\theta$, ε, and A_2, of which we shall make use very shortly, is as follows.

Consider the following Hermitian tensor operators

$$U'_\theta = \tfrac{1}{6}\{3L_z^2 - L(L+1)\}, \quad U'_\varepsilon = \tfrac{1}{6}\sqrt{(3)}(L_x^2 - L_y^2),$$

$$A'_2 = -\frac{1}{\sqrt{3}}P(L_x L_y L_z),$$

(21.19)

where L_x, L_y, L_z are the components of an orbital momentum in a D state, $L = 2$, and $P(L_x L_y L_z)$ is the symmetrized product

$$\tfrac{1}{6}(L_x L_y L_z + L_y L_x L_z + \ldots).$$

From their expressions (21.19) it is clear that the operators U'_θ, U'_ε transform respectively like θ and ε of Γ_3 and that A'_2 belongs to the representation Γ_2.

Let us consider the particular doublet Γ_3, which is obtained from the splitting of a D term and, according to Table 4, is spanned by the two functions

$$\theta = |0\rangle, \quad \varepsilon = \frac{|2\rangle + |-2\rangle}{\sqrt{2}}.$$

(21.20)

An elementary calculation of the matrix elements of the operators (21.19) shows that inside this manifold the operators U'_θ, U'_ε, and A'_2 are represented by the matrices $-\sigma_3$, σ_1, and σ_2, which shows that the latter do indeed transform like θ, ε, and A_2. Being true for a particular manifold Γ_3 spanned by the functions (21.20), this result is clearly true whatever the nature of the manifold.

Although the existence of a Γ_3 doublet in a static crystal field is not allowed by the Jahn–Teller theorem, it is interesting, and also very useful as we shall see shortly, to derive the spin Hamiltonian and the paramagnetic spectrum to be expected from a $^2\Gamma_3$ ground level, originating in the ^2D term of a paramagnetic ion, split by a cubic field. It turns out that our former treatment of the ions of type A with a non-degenerate ground state $|O\rangle$ (in § 19.2) can be adapted with very little change to the present situation provided it is understood that the symbol $|O\rangle$ now means the twofold degenerate orbital ground manifold Γ_3. An expectation value such as say $\langle O|\,L_p L_q + L_q L_p\,|O\rangle$ now becomes an operator within this manifold represented by a 2×2 matrix, that is by linear combination of the Pauli matrices $-U_\theta$, U_ε, A_2, and $\mathscr{I}$.

Let us consider, for instance, with this new interpretation where $|O\rangle$ is the ground manifold Γ_3, the expression given by formula (19.19),

$$\Lambda_{pq} = \sum_n{}' \frac{\langle O|\,L_p\,|n\rangle\langle n|\,L_q\,|O\rangle}{W_n - W_0}$$

or, since $\Lambda_{pq} = \Lambda_{qp}$,

$$\Lambda_{pq} = -\tfrac{1}{2}\langle O|\,L_p C L_q + L_q C L_p\,|O\rangle, \tag{21.21}$$

where C(eqn (19.17)) is the operator $-\sum_n{}' |n\rangle\langle n|/(W_n - W_0)$. The energy difference $W_n - W_0$ has the single value $\Delta = 10Dq$ equal to the distance between the ground doublet Γ_3 and the excited triplet Γ_5. We can thus write $C = -(1/\Delta)\{1 - |O\rangle\langle O|\}$ where **1** is the unit operator (**1** is the unit operator inside the whole manifold D rather than inside the more restricted manifold Γ_3 where it is represented by the unit 2×2 matrix $\mathscr{I}$). Carrying this expression of C into (21.21) and taking into account that $\langle O|\,L_p\,|O\rangle = 0$ for all p (because the Γ_3 doublet is non-magnetic) we obtain

$$\Lambda_{pq} = \frac{1}{2\Delta}\langle O|\,L_p L_q + L_q L_p\,|O\rangle \tag{21.22}$$

or, since it is more convenient to deal with traceless tensors,

$$\left.\begin{aligned}\Lambda_{pq} &= \tilde{\Lambda}_{pq} + \delta_{pq}\Lambda_0 \text{ where } \Lambda_0 = \frac{L(L+1)\mathscr{I}}{3\Delta} = \frac{2\mathscr{I}}{\Delta},\\ \tilde{\Lambda}_{pq} &= \frac{1}{3\Delta}\langle O|\,\tfrac{3}{2}(L_p L_q + L_q L_p) - L(L+1)\,|O\rangle.\end{aligned}\right\} \tag{21.23}$$

Because of the reduction formula (21.12) the only components of the tensor operator $\tilde{\Lambda}$ that do not vanish are

$$\tilde{\Lambda}_\theta = \frac{1}{6\Delta}\langle O|\, 3L_z^2 - L(L+1)\,|O\rangle,$$
$$\tilde{\Lambda}_\varepsilon = \frac{1}{6\Delta}\sqrt{(3)}\langle O|\, L_x^2 - L_y^2\,|O\rangle, \qquad (21.24)$$

which we can identify, using (21.19), with U_θ/Δ and U_ε/Δ and also $\Lambda_0 = 2\mathscr{I}/\Delta$. The Zeeman coupling given, according to (19.22) by the formula $Z = \beta H_p S_q(g_s \delta_{pq} - 2\lambda \Lambda_{pq})$, can now be written as

$$Z = \beta(\mathbf{H}\cdot\mathbf{S})(g_s\mathscr{I} - 2\lambda\Lambda_0) -$$
$$-2\lambda\beta\{(3H_z S_z - \mathbf{H}\cdot\mathbf{S})\tilde{\Lambda}_\theta + (H_x S_x - H_y S_y)\sqrt{(3)}\tilde{\Lambda}_\varepsilon\}, \quad (21.25)$$

which we can rewrite as

$$Z = g_1\beta(\mathbf{H}\cdot\mathbf{S}) + \frac{g_2\beta}{2}\{(3H_z S_z - \mathbf{H}\cdot\mathbf{S})U_\theta + \sqrt{(3)}(H_x S_x - H_y S_y)U_\varepsilon\} \quad (21.26)$$

with, according to (21.24) and (21.25),

$$g_1 = g_s - \frac{4\lambda}{\Delta},$$
$$g_2 = -\frac{4\lambda}{\Delta}. \qquad (21.27)$$

It should be noted that formula (21.26) is more general than (21.27). Indeed, once we had established through eqns (21.17) and (21.18) the transformation laws of the operators U_θ and U_ε, eqn (21.26) could have been written *a priori* as the only invariant Zeeman spin Hamiltonian for any $^2\Gamma_3$ level. On the other hand, eqn (21.27) assumes explicitly that the level $^2\Gamma_0$ originates in the splitting by a cubic field of a ^2D term.

Taking (21.26) as a starting point, the values (21.27) for g_1 and g_2 could have been derived directly by correlating them to the values of $g_\|(\theta)$ and $g_\|(\varepsilon)$ which are the values of $g_\|$ in tetragonal symmetry with either θ or ε as the non-degenerate ground level. It is indeed sufficient to take the expectation value of (21.26) either in the state θ or ε to see that

$$g_\|(\theta) = g_1 - g_2, \qquad g_\|(\varepsilon) = g_1 + g_2; \qquad (21.28)$$

$g_\|(\theta)$ and $g_\|(\varepsilon)$ are given directly by the formulae (19.19) and (19.32) for ions of type A.

This remark will be helpful in writing down the hyperfine coupling with the nuclear spin. From invariance arguments this coupling is necessarily of the form

$$A_1(\mathbf{I}\cdot\mathbf{S})+\frac{A_2}{2}\{(3I_zS_z-\mathbf{I}\cdot\mathbf{S})U_\theta+\sqrt{(3)}(I_xS_x-I_yS_y)U_\varepsilon\} \quad (21.29)$$

where, as before, A_1 and A_2 can be correlated to $A_\parallel(\theta)$ and $A_\parallel(\varepsilon)$ obtained in tetragonal symmetry,

$$A_\parallel(\theta) = A_1-A_2, \quad A_\parallel(\varepsilon) = A_1+A_2. \quad (21.30)$$

Using the formula (19.32) for the hyperfine coupling where ξ is given by (17.46) we find, writing $\mathscr{P}$ for $2\gamma_n\hbar\beta\langle r^{-3}\rangle = 2g_n\beta\beta_n\langle r^{-3}\rangle$ (see eqn (19.10)),

$$A_1 = -\mathscr{P}\left(\kappa+\frac{4\lambda}{\Delta}\right),$$

$$A_2 = -\mathscr{P}\left\{6\xi+\frac{4\lambda}{\Delta}+\left(\frac{9\lambda}{\Delta}\right)\xi\right\}. \quad (21.31)$$

For a ^2D term $\xi = \frac{2}{21}$.

The next problem is to find the resonance spectrum associated with the Zeeman Hamiltonian (21.26). Actually this problem has already been solved earlier in connection with the Zeeman effect of the Γ_8 quadruplet. When a spin $\frac{1}{2}$ is included the doublet Γ_3 becomes the fourfold level $D^{\frac{1}{2}}\times\Gamma_3 = \Gamma_6\times\Gamma_3 = \Gamma_8$. The four states $|\theta, +\rangle$, $|\theta, -\rangle$, $|\varepsilon, +\rangle$, $|\varepsilon, -\rangle$ that span $\Gamma_3\times\Gamma_6$ are linear combinations of the states $|\tilde{m}\rangle = |\tilde{\tfrac{3}{2}}\rangle, |\tilde{\tfrac{1}{2}}\rangle, |-\tilde{\tfrac{1}{2}}\rangle, |-\tilde{\tfrac{3}{2}}\rangle$ that span Γ_8 and were introduced in § 14.4. Actually the correspondence turns out to be very simple; it is

$$|\tilde{\tfrac{3}{2}}\rangle = |\varepsilon, -\rangle; \; |\tilde{\tfrac{1}{2}}\rangle = |\theta, +\rangle; \; |-\tilde{\tfrac{1}{2}}\rangle = -|\theta, -\rangle; \; |-\tilde{\tfrac{3}{2}}\rangle = -|\varepsilon, +\rangle. \quad (21.32)$$

To check the correspondence (21.32) we take again for θ and ε the kets $|0\rangle$ and $(1/\sqrt{2})\{|2\rangle+|-2\rangle\}$ used previously. Under a rotation of $\pi/2$ around the axis Oz the ket $|\tilde{\tfrac{3}{2}}\rangle$ is multiplied by definition by $e^{3i\pi/4}$. The ket $\varepsilon = (1/\sqrt{2})\{|2\rangle+|-2\rangle\}$ is multiplied by $e^{i\pi}$ and the ket $|\varepsilon, -\rangle$ by $e^{3i\pi/4}$. It is easy to check that it is the only such one among the four kets $|\theta, \pm\rangle, |\varepsilon, \pm\rangle$. The first equality (21.32) is thus established and similarly for the others. The eigenvalues of the spin Hamiltonian (21.26) (divided by βH) are then given by the secular equation (18.30) where from (21.26) and (21.32) the coefficients P and Q are the expectation values,

$$P = \langle\varepsilon, -|\, g_1S_z\mathscr{I}+g_2S_zU_\theta\,|\varepsilon, -\rangle = -\frac{(g_1+g_2)}{2},$$

$$Q = \langle\theta, +|\, g_1S_z\mathscr{I}+g_2S_zU_\theta\,|\theta, +\rangle = \frac{g_1-g_2}{2}. \quad (21.33)$$

In practice we shall be concerned only with the case when $|g_2| \ll |g_1|$, that is $P \approx -Q$ or $|P+Q| \ll |P, Q|$. The roots of the secular equation (18.30) then take an approximate form that is rather simple. We shall not write them down, however, for in the case when $|g_2| \ll |g_1|$ it is more illuminating to write directly an approximate solution of (21.26).

Let ξ, η, ζ be the cosines of the direction OZ of the magnetic field. The first term of (21.26) is $g_1 \beta H S_Z$ and the second term becomes, in first-order perturbation theory,

$$\frac{g_2 \beta H S_Z}{2}\{(3\zeta^2-1)U_\theta + \sqrt{(3)}(\xi^2-\eta^2)U_\varepsilon\}.$$

The eigenvalues of (21.26) are then clearly (with $M = \pm\frac{1}{2}$)

$$g_1 \beta H M \pm g_2 \beta \frac{HM}{2}\{(3\zeta^2-1)^2 + 3(\xi^2-\eta^2)^2\}^{\frac{1}{2}}$$
$$= \beta H M [g_1 \pm g_2\{1-3(\xi^2\eta^2+\eta^2\zeta^2+\zeta^2\xi^2)\}^{\frac{1}{2}}]. \quad (21.34)$$

There are two resonance frequencies,

$$h\nu = \beta H[g_1 \pm g_2\{1-3(\xi^2\eta^2+\eta^2\zeta^2+\zeta^2\xi^2)\}^{\frac{1}{2}}], \quad (21.34a)$$

and the four eigenstates are

$$|\pm\rangle\left\{\theta \cos\frac{\omega}{2} - \varepsilon \sin\frac{\omega}{2}\right\}, \; |\pm\rangle\left\{\theta \sin\frac{\omega}{2} + \varepsilon \cos\frac{\omega}{2}\right\}, \quad (21.35)$$

where ω is defined by

$$\cos\omega = \frac{(3\zeta^2-1)}{2}\{1-3(\xi^2\eta^2+\eta^2\zeta^2+\zeta^2\xi^2)\}^{-\frac{1}{2}},$$
$$\sin\omega = \frac{\sqrt{3}}{2}(\xi^2-\eta^2)\{1-3(\xi^2\eta^2+\eta^2\zeta^2+\zeta^2\xi^2)\}^{-\frac{1}{2}}. \quad (21.36)$$

The spectrum is clearly anisotropic, and the two frequencies ν_1 and ν_2 become equal when the field is along a body diagonal of the cube where the cubic invariant in curly brackets vanishes.

Whereas the assumption $|g_1| \gg |g_2|$ is legitimate when g_2 is of the order (λ/Δ), a similar assumption $|A_1| \gg |A_2|$ is not warranted unless the coefficient κ, which represents the core polarization in (21.31), happens to be large. When $|A_1| \gg |A_2|$ the eigenvalues of the spin Hamiltonian, sum of (21.26) and (21.29), are given by

$$W_{M,m} = M[(g_1\beta H + A_1 m) \pm (g_2 \beta H + A_2 m)\{1-3(\xi^2\eta^2+\eta^2\zeta^2+\zeta^2\xi^2)\}^{\frac{1}{2}}], \quad (21.37)$$

where $m = I_Z$ is the quantized projection of the nuclear spin along the applied field H. The resonance frequencies ν_m, observed in an electronic spin transition $|\Delta M| = 1$ are given by

$$h\nu_m = (g_1\beta H + A_1 m) \pm (g_2\beta H + A_2 m)\{1 - 3(\xi^2\eta^2 + \eta^2\zeta^2 + \zeta^2\xi^2)\}^{\frac{1}{2}}. \quad (21.37a)$$

The reader may well ask what is the point of calculating the shape of spectra such as (21.34a) or (21.37a) if the assumptions of cubic symmetry under which they are derived are made invalid by the Jahn–Teller theorem. The remarkable fact is that anisotropic spectra of this form *have* actually been observed. Furthermore, as will be shown later, their existence is by no means forbidden by the Jahn–Teller theorem but rather is a striking manifestation of a dynamic Jahn–Teller effect.

21.4. The static Jahn-Teller effect in a ^2E state

Using Table 26, which describes the various irreducible representations Γ' spanned by the normal coordinates, and the reduction formula (21.12), we find that only the $\Gamma_3(E)$ mode of distortion $(Q_\theta, Q_\varepsilon)$ can lift the orbital degeneracy of a Γ_3 doublet. From what has been said about the transformation properties of the Pauli matrices $\sigma_1 = U_\varepsilon$ and $\sigma_3 = -U_\theta$ within the manifold Γ_3 it is clear that in first order the electron-nuclear coupling given by (21.11) is necessarily of the form

$$\delta V = V(Q_\theta U_\theta + Q_\varepsilon U_\varepsilon), \quad (21.38)$$

where V is a constant. To it we must add the potential energy $\kappa(Q_\theta^2 + Q_\varepsilon^2)$ associated with the vibrations of the normal coordinates $(Q_\theta, Q_\varepsilon)$. We can write $\kappa = \mu\omega^2/2$ where ω is the harmonic frequency associated with the vibrational mode E and where in first approximation we can equate the mass coefficient μ to the mass M of each of the nuclei Y of the complex XY_6.

The resultant potential energy operator

$$U = V(Q_\theta U_\theta + Q_\varepsilon U_\varepsilon) + \frac{M\omega^2}{2}(Q_\theta^2 + Q_\varepsilon^2)\mathscr{I} \quad (21.39)$$

can be rewritten as a function of the variables ρ and φ introduced in (21.15):

$$U = V\rho(U_\theta \cos \varphi + U_\varepsilon \sin \varphi) + \frac{M\omega^2\rho^2}{2}. \quad (21.39a)$$

The two eigenfunctions of (21.39a) are

$$\psi_- = \theta \cos\frac{\varphi}{2} - \varepsilon \sin\frac{\varphi}{2}, \quad \psi_+ = \theta \sin\frac{\varphi}{2} + \varepsilon \cos\frac{\varphi}{2}, \quad (21.40)$$

and the two eigenvalues are

$$U = \pm V\rho + \frac{M\omega^2\rho^2}{2} \qquad (21.41)$$

with the lower sign for ψ_-. It is clear that $\rho = 0$ is not a position of stable equilibrium for the complex and that a continuum of positions of minimum energy is obtained for $\rho_0 = |V|/M\omega^2$. The value of the minimum energy is

$$W_{\mathrm{JT}} = \frac{|V|\,\rho_0}{2} = \frac{V^2}{2M\omega^2} = \frac{M\omega^2\rho_0^2}{2}. \qquad (21.42)$$

The expression (21.41) for the potential energy suffers from a serious drawback: its symmetry far exceeds that of the cubic group since it is independent of φ whereas, as we showed earlier, it need only be invariant when φ is changed by $2\pi/3$. The picture can be made more realistic by what one calls warping of the potential energy surface, that is by adding higher-order terms either to the quasi-elastic second-order term $M\omega^2\rho^2/2$ or to the first-order Jahn–Teller coupling (21.39). The first term with cubic symmetry to be added to the potential energy will be of the form $V_3\rho^3\cos 3\varphi$ since, as we showed in § 21.3, $\cos 3\varphi$ is a cubic invariant. We can also add to the Jahn–Teller coupling a second-order term which, in order to be cubically invariant, must necessarily be of the form

$$V_2\rho^2(-U_\theta\cos 2\varphi + U_\varepsilon \sin 2\varphi) \qquad (21.43)$$

since, as we also showed in § 21.3, $\cos 2\varphi$ transforms like $-\theta$, and $\sin 2\varphi$ like ε. The electronic Hamiltonian becomes then

$$V\rho\left\{U_\theta\left(\cos\varphi - \frac{V_2\rho}{V}\cos 2\varphi\right) + U_\varepsilon\left(\sin\varphi + \frac{V_2\rho}{V}\sin 2\varphi\right)\right\} +$$
$$+ \left(\frac{M\omega^2\rho^2}{2} + V_3\rho^3 \cos 3\varphi\right). \qquad (21.44)$$

The eigenvalues of this Hamiltonian are

$$\pm V\rho\left\{\left(\cos\varphi - \frac{V_2\rho}{V}\cos 2\varphi\right)^2 + \left(\sin\varphi + \frac{V_2\rho}{V}\sin 2\varphi\right)^2\right\}^{\frac{1}{2}} +$$
$$+ \left(\frac{M\omega^2\rho^2}{2} + V_3\rho^3 \cos 3\varphi\right)$$
$$= \pm V\rho\left(1 + \frac{V_2^2\rho^2}{V^2} - \frac{2V_2\rho}{V}\cos 3\varphi\right)^{\frac{1}{2}} + \frac{M\omega^2\rho^2}{2} + V_3\rho^3\cos 3\varphi, \qquad (21.45)$$

which exhibit cubic symmetry by their dependence on φ being solely as $\cos 3\varphi$. The expression (21.45) shows that the minima of the potential energy no longer form a continuum but rather are obtained for discrete values of φ. These values are either $\varphi = 0, 2\pi/3, 4\pi/3$ or $\pi/3, \pi, 5\pi/3$, depending on the relative signs and magnitudes of the coefficients V_2 and V_3. For simplicity we shall from now on, unless stated otherwise, take $V_2 = 0$. This has in particular the advantage of preserving for the eigenfunctions of (21.44) the simple form (21.40) whilst leaving intact all the qualitative features of the problem. If $V_3 < 0$ the minima occur for $\varphi = 0, 2\pi/3, 4\pi/3$ and correspond to an elongation of the octahedron along one of its three fourfold axes. If $V_3 > 0$ the minima are obtained for a compression along the same axes.

Consider in particular a distortion along the z-axis, that is $\varphi = 0$ or π. The ground state is given by Table 21.1. The electronic wave-functions that correspond to distortions along Ox or Oy can be obtained

TABLE 21.1

V	V_3	φ	Ground state
+	+	π	$\psi_-(\pi) = -\varepsilon$
+	−	0	$\psi_-(0) = \theta$
−	+	π	$\psi_+(\pi) = \theta$
−	−	0	$\psi_+(0) = \varepsilon$

from these by the formulae (21.13), which are equivalent to a cyclic permutation of the coordinates x, y, z in the wave-functions θ and ε. In a simple-minded picture one is led to predict that since the three axes are equivalent, one-third of the octahedra present in the crystal will be distorted along each axis, and the observed paramagnetic spectrum should be a superposition of three normal tetragonal spectra in equal proportions. The parameters $g_\parallel$ and $g_\perp$ of these spectra are obtained by taking the expectation value of the coefficient of S_z or S_x in eqn (21.26) over the electronic wave-function of the appropriate ground state given in Table 21.1. For the parameters $A_\parallel$ and $A_\perp$ one takes similarly the expectation value of the coefficients of $I_z S_z$ or $I_x S_x$ in eqn (21.29). The results are given below (cf. Table 7.22):

$$g_\parallel^\pm = g_1 \pm g_2, \qquad g_\perp^\pm = g_1 \mp \frac{g_2}{2}; \qquad (21.46)$$

$$A_\parallel^\pm = A_1 \pm A_2, \qquad A_\perp^\pm = A_1 \mp \frac{A_2}{2}; \qquad (21.47)$$

where the sign $\pm$ in $g^\pm$ or $A^\pm$ is, according to Table 21.1, that of the product VV_3, and g_1, g_2, A_1, A_2 are given by (21.27) and (21.31). Such a pattern is indeed observed in some copper salts at 20°K as discussed in detail in § 7.16. Experimentally it turns out that for Cu^{2+}, $3d^9$ it is the upper sign that is needed in (21.47).

21.5. Dynamic features of the static Jahn-Teller effect

In the foregoing we have implicitly assumed that at each minimum the nuclear parameters ρ and φ had fixed values, neglecting completely the zero-point motion of the nuclei; this is certainly incorrect. Actually for a distortion along Oz, assuming for the sake of definiteness that $V > 0$, $V_3 < 0$, the wave-function of the complex XY_6 should be written, if the Born–Oppenheimer approximation is valid, as

$$\Phi_Z = \psi_- \Psi_Z(\rho, \varphi), \qquad (21.48)$$

where the nuclear wave-function $\Psi_Z(\rho, \varphi)$ is localized in the neighbourhood of $\rho = \rho_0$, $\varphi = 0$. For the Born–Oppenheimer approximation to be valid in the first place, we require that the distance between the two electronic energy levels that correspond to the wave-functions ψ_- and ψ_+, which is of the order of $2|V|\rho_0 = 4W_{JT}$, be much larger than the nuclear zero-point energy $\hbar\omega/2$. This is what is meant by strong Jahn–Teller coupling. Furthermore, for the nuclear wave-function Ψ_Z to be strongly localized in the trough of the minimum corresponding to $\varphi = 0$, we require that the height $V_\varphi = |2V_3\rho^3|$ of the barrier that separates this minimum from the other two at $\varphi = \pm 2\pi/3$ be much greater than the zero-point energy $\hbar\omega'/2$ that corresponds to small oscillations of the coordinate φ.

The frequency ω' is of the order of

$$\omega' \approx \left(\frac{9V_\varphi}{2M\rho_0^2}\right)^{\frac{1}{2}} \approx \frac{3\omega}{2}\left(\frac{V_\varphi}{W_{JT}}\right)^{\frac{1}{2}}. \qquad (21.49)$$

To obtain an order of magnitude for Cu^{2+} surrounded by six water-molecules, we take an admittedly crude estimate of Öpik and Pryce (1957) that leads to

$$W_{JT} \approx 3000 \text{ cm}^{-1}, \qquad V_\varphi \sim 600 \text{ cm}^{-1},$$

$$\omega \sim 350 \text{ cm}^{-1}, \quad \text{and from (21.49)}, \ \omega' \approx \frac{2\omega}{3}.$$

Our calculation of the parameters (21.46), (21.47) of the paramagnetic spectra, using the wave-functions of Table 21.1, was tantamount to assuming that the nuclear wave-function $\Psi_Z(\rho, \varphi)$ of (21.48) had the form of a δ-function $\delta(\varphi)$. A more realistic calculation should take into account the finite extension of $\Psi_Z(\varphi)$. Making use of the relations

$$(\psi_\pm|\ U_\theta\ |\psi_\pm) = \pm\cos\varphi, \qquad (\psi_\pm|\ U_\varepsilon\ |\psi_\pm) = \pm\sin\varphi, \qquad (21.50)$$

we find for the parameters g and A,

$$g_\parallel^\pm = g_1 \pm ug_2, \qquad g_\perp^\pm = g_1 \mp \frac{ug_2}{2}$$

$$A_\parallel^\pm = A_1 \pm uA_2, \qquad A_\perp^\pm = A_1 \mp \frac{uA_2}{2} \qquad (21.51)$$

where

$$u = \langle\cos\varphi\rangle = \iint \cos\varphi\ |\Psi_Z(Q)|^2\ \mathrm{d}Q_\theta\ \mathrm{d}Q_\varepsilon.$$

The point is that while at the precise position of the minimum, say $\varphi = 0$, the electronic wave-function ψ_- is exactly θ, because of the finite spatial extent of $\Psi_Z(\rho, \varphi)$ it is also partly ε, which gives a change in the values of g and A (O'Brien 1964). As discussed in § 7.16, it may also explain some features of the hyperfine structure on the ligands in the spectrum of Ni^+ in NaF.

The foregoing discussion has some very strong limitations. First, it applies only to the ground state of the complex, whereas there are many excited states connected with nuclear vibrations which may be appreciably populated at the temperature of the experiment. Second, by considering the complex XY_6 as an isolated system and neglecting the coupling with its surroundings in the crystal, one disregards the possibility of fast relaxation transitions from one configuration to another, which may drastically alter the observed patterns. Last, and this is really the main point, the dynamic features of the problem, caused by the existence of the kinetic nuclear energy, have far-reaching consequencies of which the finite extension of the nuclear wave-function in each trough, leading to the corrected formulae (21.51), is only one minor aspect.

We pass now to the study of the dynamic Jahn–Teller effect, at the end of which it will become clearer under what conditions a static effect such as the one described in the present section can be observed.

21.6. The dynamic Jahn–Teller effect in a ^2E state
General

In the last section we said that for a strong Jahn–Teller coupling, that is when the distance, $4W_{JT}$, between the two electronic energy states ψ_+ and ψ_- was much greater than the nuclear zero-point energy

$\hbar\omega$, the Born–Oppenheimer approximation was valid and the wave-function of the system could be written as a single product of an electronic wave-function, say θ, and a nuclear wave-function, say Ψ_Z. Actually a closer look shows that even in this case things are more complicated and electronic and nuclear motions are closely intertwined. There are actually three potential minima with three wave-functions $\Phi_Z = \theta\Psi_Z$, $\Phi_X = R\theta\Psi_X$, $\Phi_Y = R^2\theta\Psi_Y$ and the total energy of the complex XY_6 is the same in all of these. This degeneracy, as will be discussed later in more detail, is lifted at least partially because of the finite overlap between the nuclear functions Ψ_X, Ψ_Y, Ψ_Z. The 'good' zero order wave-functions will then be suitable linear combinations of the three Φ. States described by such linear combinations, where electronic and nuclear motions cannot be separated, are called vibronic states.

We take up now a study of these states in the general case, that is for an arbitrary strength of the Jahn–Teller coupling. The total vibronic Hamiltonian $\mathscr{H}_v$ is obtained by adding to the expression (21.44) the nuclear kinetic energy $(1/2M)(P_\theta^2 + P_\varepsilon^2)$. This Hamiltonian operates on the two nuclear variables Q_θ and Q_ε and on two electronic variables that are the amplitudes C_θ and C_ε of the two electronic wave-functions θ and ε which span the electronic doublet Γ_3. Under operations of the cubic group, Q_θ and Q_ε on one hand, C_θ and C_ε on the other, transform according to Γ_3 and the Hamiltonian $\mathscr{H}_v$ is invariant under a *simultaneous* transformation

$$\begin{pmatrix} Q'_\theta \\ Q'_\varepsilon \end{pmatrix} = D^{\Gamma_3} \begin{pmatrix} Q_\theta \\ Q_\varepsilon \end{pmatrix}; \qquad \begin{pmatrix} C'_\theta \\ C'_\varepsilon \end{pmatrix} = D^{\Gamma_3} \begin{pmatrix} C_\theta \\ C_\varepsilon \end{pmatrix}$$

where D^{Γ_3} is one of the 2×2 transformation matrices of Γ_3. $\mathscr{H}_v$ is thus invariant through a transformation of its *four* variables $Q_\theta, Q_\varepsilon, C_\theta, C_\varepsilon$ by $\Gamma_3 \times \Gamma_3 = \Gamma_3 + \Gamma_2 + \Gamma_1$.

The very important consequence of this fact is that all the vibronic eigenstates of $\mathscr{H}_v$ are either doublets of symmetry Γ_3 or singlets of symmetry Γ_1 or Γ_2. This result is quite general and would still be true if even higher-order terms, either for the Jahn–Teller coupling or the potential energy, were added to (21.44).

It is actually misleading to say that the Jahn–Teller effect lifts a degeneracy due to a high symmetry, since the Jahn–Teller coupling term itself has the same symmetry. What the Jahn–Teller coupling does is to replace a purely electronic degeneracy by a vibronic degeneracy of exactly the same symmetry. Consider in particular the ground vibronic level of the complex XY_6. In the absence of Jahn–Teller coupling, the nuclear motion is that of a two-dimensional harmonic oscillator and its ground state $\Psi_0(Q)$ is *non-degenerate*. The overall degeneracy is then that of an electronic doublet Γ_3 spanned by the two Born–Oppenheimer functions $\Psi_0(Q)\theta$ and $\Psi_0(Q)\varepsilon$. A Jahn–Teller coupling will change the ground level into a *vibronic* Γ_3 doublet but a doublet it will remain.

We must now examine successively what must be the nature of the paramagnetic spectrum if the vibronic state is (a) a doublet, (b) a singlet, assuming in each case that the vibronic level under consideration is sufficiently removed from other vibronic levels for their mixing by the Zeeman Hamiltonian to be negligible. Finally, we shall have to explain the existence of a static Jahn–Teller effect whose occurrence in the form of three spectra of lower symmetry does not seem to fit at all into our picture of vibronic singlets and doublets.

The vibronic doublet

A vibronic doublet is spanned by two wave-functions that we shall call Θ and $\mathscr{E}$. They contain both electronic and nuclear variables and transform like θ and ε, when *both* electronic and nuclear variables undergo a transformation of the cubic group. Following Ham (1968), we define the operators $U_{g\theta}$ and $U_{g\varepsilon}$, which are represented by the matrices $-\sigma_3$ and σ_1 when Θ and $\mathscr{E}$ are chosen as basis states and therefore can be written as

$$U_{g\theta} = |\mathscr{E}\rangle\langle\mathscr{E}| - |\Theta\rangle\langle\Theta|,$$
$$U_{g\varepsilon} = |\mathscr{E}\rangle\langle\Theta| + |\Theta\rangle\langle\mathscr{E}|. \qquad (21.52)$$

Operators such as U_θ and U_ε, which were represented by the Pauli matrices $-\sigma_3$ and σ_1 when the purely electronic states θ and ε were chosen as basic states, will have the following matrix elements inside the manifold of the vibronic doublet:

$$\langle\mathscr{E}|U_\theta|\mathscr{E}\rangle = -\langle\Theta|U_\theta|\Theta\rangle = \langle\Theta|U_\varepsilon|\mathscr{E}\rangle = \langle\mathscr{E}|U_\varepsilon|\Theta\rangle = q, \qquad (21.53)$$

where q is a number smaller than unity, whereas all the other matrix elements of U_θ and U_ε vanish. This is a consequence of the Wigner–Eckart theorem since $(U_\theta, U_\varepsilon)$ have the same transformation properties as $(U_{g\theta}, U_{g\varepsilon})$. The unit operator $\mathscr{I}$ has naturally the same matrix elements within the vibronic doublet as within a purely electronic doublet. An electronic operator such as the Zeeman coupling (21.26), which in the absence of the Jahn–Teller effect could be written as

$$G_0\mathscr{I} + G_\varepsilon U_\varepsilon + G_\theta U_\theta, \qquad (21.54)$$

can now be written

$$G_0\mathscr{I}+q(G_\varepsilon U_{g\varepsilon}+G_\theta U_{g\theta}), \qquad (21.54\text{a})$$

where we repeat that, with our choice of Θ and $\mathscr{E}$ as the basic vibronic states, $U_{g\theta}$ and $U_{g\varepsilon}$ are simply the Pauli matrices $-\sigma_3$ and σ_1. Actually (21.54a) is not the most general operator acting inside the vibronic doublet and should be complemented by a term $G_2 A_2$ proportional to the third Pauli matrix $A_2 = \sigma_2$. We define a coefficient $p = i\langle\Theta|A_2|\mathscr{E}\rangle$ and an operator $A_{g_2} = i(|\mathscr{E}\rangle\langle\Theta|-|\Theta\rangle\langle\mathscr{E}|)$, and (21.54a) then becomes

$$G_0\mathscr{I}+q(G_\varepsilon U_{g\varepsilon}+G_\theta U_{g\theta})+pG_2 A_{g_2}. \qquad (21.55)$$

It is worth noting that p, the proportionality coefficient for the operator A_2, is different from that for U_ε and U_θ, because A_2 belongs to a different representation of the cubic group. Since the Zeeman coupling (21.26) does not contain the operator A_2, the resonance spectrum in the presence of the Jahn–Teller coupling must have exactly the same form as that calculated in § 21.3, provided that in the formulae (21.26) and (21.29) and also in (21.34) and (21.37) the splitting factor g_2 and the hyperfine constant A_2 (not to be confused with the operator A_2 proportional to σ_2) are replaced by qg_2 and qA_2.

One of the crucial problems of the theory is naturally the calculation of the coefficient q for each vibronic doublet. It is a difficult problem that requires the knowledge of the eigenfunctions of a fairly complicated Hamiltonian $\mathscr{H}_\text{v}$.

If the Jahn–Teller coupling is weak (meaning thereby $W_\text{JT}/\hbar\omega \ll 1$) it is legitimate to simplify the problem by neglecting the warping terms, which can be expected to be small for values of $\rho \sim |V|/M\omega^2$. One of the undesirable features of this so-called linear approximation is the introduction of a spurious degeneracy. When the warping terms are neglected all states are doubly degenerate; not only the doublets Γ_3 but also the singlets belonging to the representations Γ_1 and Γ_2 which coincide in pairs. This can be seen as follows. The linear vibronic Hamiltonian $\mathscr{H}_\text{v}^\text{L}$, which has the form

$$\mathscr{H}_\text{v}^\text{L}(Q) = \frac{1}{2M}(P_\theta^2+P_\varepsilon^2)+\frac{M\omega^2}{2}(Q_\theta^2+Q_\varepsilon^2)+V(Q_\theta U_\theta+Q_\varepsilon U_\varepsilon), \qquad (21.56)$$

is invariant under the transformation

$$\mathscr{H}_\text{v}^\text{L}(Q_\theta, Q_\varepsilon) = U_\theta^{-1}\mathscr{H}_\text{v}^\text{L}(Q_\theta, -Q_\varepsilon)U_\theta \qquad (21.56\text{a})$$

which is a consequence of the commutation relations of U_θ and U_ε. It follows that if $\Phi(Q_\theta, Q_\varepsilon) = \theta\Psi_\theta(Q_\theta, Q_\varepsilon) + \varepsilon\Psi_\varepsilon(Q_\theta, Q_\varepsilon)$ is an eigenstate of (21.56), so is $\tilde{\Phi}(Q_\theta, Q_\varepsilon) = U_\theta \Phi(Q, -Q_\varepsilon)$. It will shortly be shown explicitly (eqn (21.58a)) that $\tilde{\Phi} \neq \Phi$, thus demonstrating the twofold degeneracy. When the warping terms are introduced as a perturbation the singlets A_1 and A_2 separate, whereas the doublets Γ_3 naturally remain degenerate.

If we are interested in the ground doublet only, and if the first excited singlet is sufficiently removed from it, this spurious degeneracy is not very serious and the linear approximation is useful. To obtain the form of the wave-functions that span the ground doublet we take a trial function of the form $\Psi_\pm(Q)\psi_\pm$, where the electronic functions $\psi_\pm$ are given by (21.40). The vibronic Hamiltonian $\mathscr{H}_v^L$ acting on this function gives

$$\mathscr{H}_v^L\{\Psi_\pm(Q)\psi_\pm\} = \left(T + \frac{M\omega^2\rho^2}{2} \pm V\rho\right)\Psi_\pm(\rho, \varphi)\psi_\pm, \quad (21.57)$$

where T is the kinetic energy operator

$$T = -\frac{\hbar^2}{2M}\left\{\frac{1}{\rho}\frac{\partial}{\partial\rho}\left(\rho\frac{\partial}{\partial\rho}\right) + \frac{1}{\rho^2}\frac{\partial^2}{\partial\varphi^2}\right\}.$$

Since eqn (21.57) does not contain explicitly the variable φ it is separable and we may seek a vibronic eigenfunction of the following form:

$$e^{in\varphi}\{f_n(\rho)\psi_- + ig_n(\rho)\psi_+\}. \quad (21.58)$$

Here the two unknown functions $f_n(\rho)$ and $g_n(\rho)$ obey the following system of coupled equations, derived from (21.57),

$$-\frac{\hbar^2}{2M}\left\{\frac{1}{\rho}\frac{\partial}{\partial\rho}\left(\rho\frac{\partial}{\partial\rho}\right) - \frac{n^2+\frac{1}{4}}{\rho^2}\right\}\binom{f_n}{g_n} +$$
$$+ (\tfrac{1}{2}M\omega^2\rho^2 \mp V\rho - W_n)\binom{f_n}{g_n} = -\frac{\hbar^2 n}{2M\rho^2}\binom{g_n}{f_n}. \quad (21.59)$$

Since (21.59) has real coefficients the complex conjugate of (21.58)

$$e^{-in\varphi}\{f_n(\rho)\psi_- - ig_n(\rho)\psi_+\} \quad (21.58a)$$

is also a solution with the same energy. This corresponds to the twofold degeneracy of all eigenstates of $\mathscr{H}_v^L$, alluded to earlier. Instead of the two degenerate solutions (21.58) and (21.58a) it is often more convenient to use their real and imaginary parts. Since the electronic wave-functions $\psi_\pm$ of (21.40) are functions of $\varphi/2$, in order for the vibronic functions (21.58) and (21.58a) to be single-valued, the constant n must be half-integer.

For $n = \frac{1}{2}+3m$ or $n = \frac{5}{2}+3m$ the doublets are bona fide Γ_3 doublets and remain such in the presence of warping, whereas for $n = \frac{3}{2}+3m$ we have two states belonging to the unidimensional representations A_1 and A_2 which are split by the warping. In particular, the wave-functions $\Theta_{\frac{1}{2}}$ and $\mathscr{E}_{\frac{1}{2}}$, which span the ground vibronic doublet and are proportional to the real and imaginary part of (21.58), are given by

$$\sqrt{(2)}\Theta_{\frac{1}{2}} = (f_{\frac{1}{2}}+g_{\frac{1}{2}})(-\theta \cos \varphi + \varepsilon \sin \varphi) + (g_{\frac{1}{2}}-f_{\frac{1}{2}})\theta,$$
$$\sqrt{(2)}\mathscr{E}_{\frac{1}{2}} = (f_{\frac{1}{2}}+g_{\frac{1}{2}})(\theta \sin \varphi + \varepsilon \cos \varphi) + (g_{\frac{1}{2}}-f_{\frac{1}{2}})\varepsilon.$$
(21.60)

Here we have written them in a way that displays explicitly their transformation character as θ and ε in accordance with eqn (21.17).

If the Jahn–Teller energy $W_{\rm JT}$ is small compared to $\hbar\omega$, the first excited state for $n = \frac{3}{2}$ will be approximately separated from the ground doublet by one vibrational quantum $\hbar\omega$ which is much greater than the Zeeman energy, and at helium or hydrogen temperatures it will not be populated. The coefficients p and q which are unity in the absence of Jahn–Teller coupling will be only slightly smaller than unity for a weak coupling. Within the linear approximation they have been computed numerically (Child and Longet-Higgins 1962) and their variation is represented in Fig. 21.1. Figure 21.2 gives the distance from the ground doublet to the first excited state.

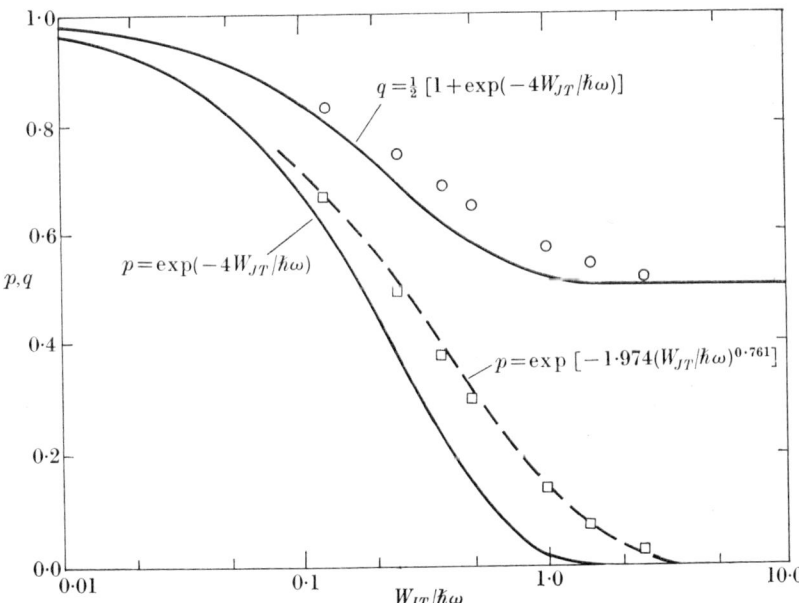

FIG. 21.1. Plot of the coefficients p and q against $W_{\rm JT}/\hbar\omega$, together with approximate analytical expressions for these variations.

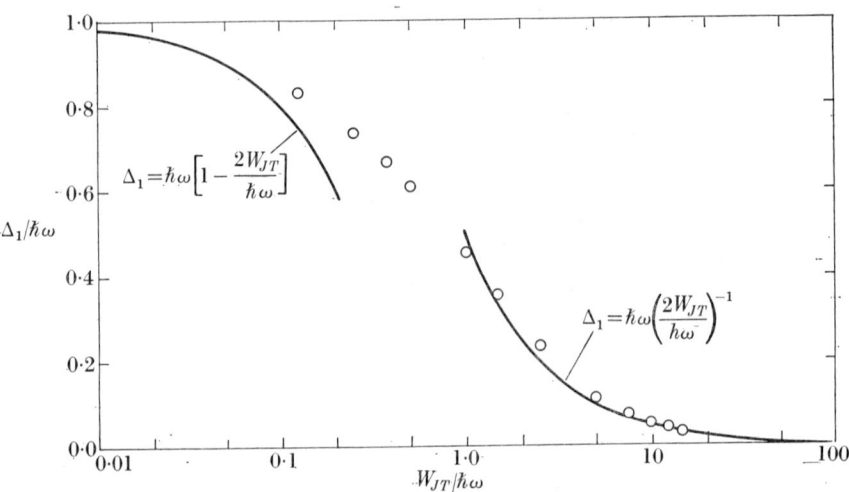

Fig. 21.2. Plot of the singlet-doublet splitting Δ_1 against $W_{JT}/\hbar\omega$.

When the Jahn–Teller effect becomes large the Born–Oppenheimer approximation becomes valid, one of the components of the wavefunction (21.58), say g_n, vanishes and we may disregard the right-hand side of eqn (21.59). The energy gap Δ_1 between the ground doublet and the first excited state then becomes approximately

$$\Delta_1 \approx \frac{\hbar^2}{2M}\left\{(\tfrac{3}{2})^2\langle f_{\frac{3}{2}}|\frac{1}{\rho^2}|f_{\frac{3}{2}}\rangle - (\tfrac{1}{2})^2\langle f_{\frac{1}{2}}|\frac{1}{\rho^2}|f_{\frac{1}{2}}\rangle\right\}.$$

Since ρ is localized in the neighbourhood of the energy minimum $\rho_0 \sim |V|/M\omega^2$, we have $\langle f_{\frac{3}{2}}|\,1/\rho^2\,|f_{\frac{3}{2}}\rangle \approx \langle f_{\frac{1}{2}}|\,1/\rho^2\,|f_{\frac{1}{2}}\rangle \approx M^2\omega^4/V^2$ and

$$\Delta_1 \approx 2\frac{\hbar^2}{2M}\frac{M^2\omega^4}{V^2} \approx \hbar\omega\frac{\hbar\omega}{2W_{JT}}. \tag{21.60a}$$

If we take the values suggested by Öpik and Pryce (1957) for Cu^{2+} we find $\Delta_1 \approx 20$ cm^{-1}. Unfortunately the linear approximation loses much of its significance when the Jahn–Teller effect becomes large. In particular, the energy of the first excited state depends very critically on the warping. A different approach, more suitable when the Jahn–Teller coupling *and* the warping terms are large, will be outlined shortly.

In view of the difficulty of the problem in the general case it is interesting to obtain the maximum of information from symmetry arguments only. Elementary group-theoretical considerations based on the relations (21.12) and (21.17) show that the most general expression

for two wave-functions that span a vibronic doublet Γ_3 is
$$\Theta = -b_\theta\theta + b_\varepsilon\varepsilon + a_1\theta + a_2\varepsilon,$$
$$\mathscr{E} = b_\varepsilon\theta + b_\theta\varepsilon + a_1\varepsilon - a_2\theta. \tag{21.61}$$

Here θ and ε are two electronic wave-functions, whose coefficients a b are functions of the nuclear variables Q and transform as follows: $b_\theta(Q)$ and $b_\varepsilon(Q)$ belong to Γ_3 and transform like θ and ε; $a_1(Q)$ belongs to Γ_1 and $a_2(Q)$ to Γ_2. As a consequence the following orthogonality relations are evident:

$$\langle b_\theta^2 \rangle = \langle b_\varepsilon^2 \rangle = \langle b^2 \rangle; \ \langle b_\theta \cdot b_\varepsilon \rangle = \langle b_\theta \cdot a_{1,2} \rangle = \langle b_\varepsilon \cdot a_{1,2} \rangle = \langle a_1 \cdot a_2 \rangle = 0, \tag{21.62}$$

where the symbol $\langle \ \rangle$ means integration over the variables Q. We see that the wave-functions (21.60) obtained in the linear approximation for the ground doublet are of this form with $a_2(Q) = 0$.

From the normalization and orthogonality of Θ and $\mathscr{E}$ in (21.61) and making use of (21.62) we find

$$\langle a_1^2 \rangle + \langle a_2^2 \rangle + 2\langle b^2 \rangle = 1. \tag{21.63}$$

For the coefficients p and q we get

$$p = i\langle \Theta | A_2 | \mathscr{E} \rangle = \langle a_1^2 \rangle + \langle a_2^2 \rangle - 2\langle b^2 \rangle,$$
$$q = \langle \Theta | U_\varepsilon | \mathscr{E} \rangle = \langle a_1^2 \rangle - \langle a_2^2 \rangle. \tag{21.64}$$

The relations (21.64), which are based on symmetry considerations only, are quite general and independent of the strength and the form of both the Jahn–Teller coupling and the warping terms. From (21.63) and (21.64) we get

$$q = \frac{1+p}{2} - 2\langle a_2^2 \rangle. \tag{21.65}$$

For the ground doublet and in the linear approximation, since $a_2 = 0$, we have $q = \frac{1}{2}(1+p)$, a result due to Ham (1968).

If the Jahn–Teller coupling is very strong the two electronic wave-functions $\psi_\pm$ which make it diagonal (and which have the form (21.40) when this coupling is linear) differ in energy by a large amount, the Born–Oppenheimer approximation is valid and for the ground doublet only one of these, say ψ_-, multiplies both components of the doublet. This means that in (21.61) the coefficients of θ and ε are in the same ratio for Θ and $\mathscr{E}$. Hence

$$\frac{a_1 - b_\theta}{a_2 + b_\varepsilon} = \frac{b_\varepsilon - a_2}{a_1 + b_\theta}$$

or
$$\langle a_1^2 \rangle + \langle a_2^2 \rangle = 2\langle b^2 \rangle = \tfrac{1}{2}. \tag{21.66}$$

When (21.66) is valid we get from (21.64) and (21.65)
$$p = 0; \qquad q = \tfrac{1}{2} - 2\langle a_2^2 \rangle \leqslant \tfrac{1}{2}. \tag{21.67}$$
The results (21.67) for a strong Jahn–Teller effect with warping have been previously obtained by Ham (1968) for the ground doublet and with the assumption that the Jahn–Teller coupling contains first-order terms only.

The vibronic singlets

These states can be dealt with very quickly. For singlet states, which belong to representations A_1 or A_2, it is clear that the expectation values of U_θ and U_ε, which transform like θ and ε of Γ_3, vanish. The Zeeman spin-Hamiltonian and the hyperfine coupling then have necessarily the fully isotropic form
$$g_1\beta(\mathbf{H}\cdot\mathbf{S}) + A_1(\mathbf{I}\cdot\mathbf{S}). \tag{21.68}$$
Spectra corresponding to such a Hamiltonian have actually been observed. Indeed it was the first observation by Bleaney and Ingram in 1950 of an isotropic spectrum for Cu^{2+} in copper fluosilicate, inexplicable in the framework of static crystal-field theory, which led Abragam and Pryce (1950) to postulate a dynamic Jahn–Teller effect as a possible explanation. However, as we shall see in a later section, fast relaxation may lead to a spectrum consistent with (21.68) even when the state of the system is a vibronic doublet and also when conditions are such that a static Jahn–Teller effect would be observable in the absence of relaxation. Thus an isotropic spectrum does not necessarily mean that the system is in a singlet state.

The effects of strain

The paramount importance of the effects of strain in connection with the Jahn–Teller effect has been emphasized by Ham (1965, 1968). A uniform strain is best defined by its six components belonging respectively to
$$\begin{aligned}&\Gamma_1\ (e_{xx}+e_{yy}+e_{zz}=e_0),\\ &\Gamma_5\ (e_{xy},\ e_{yz},\ e_{zx}),\\ &\Gamma_3\left\{e_\theta = \frac{3e_{zz}-e_0}{2};\quad e_\varepsilon = \frac{\sqrt{(3)}}{2}(e_{xx}-e_{yy})\right\}.\end{aligned} \tag{21.69}$$
Inside a cubic manifold Γ_3 the components e_θ and e_ε are the only ones that split the energy level of the ion in first order. A trigonal strain e_{xy} can affect it in second order only, via the spin-orbit coupling (Ham 1968).

The first-order change in energy can be rewritten in operator form as

$$V_{ES}(e_\theta U_\theta + e_\varepsilon U_\varepsilon),\qquad(21.69\text{a})$$

where V_{ES} is a constant (the index E stands for $E = \Gamma_3$ and S for strain). Consider a distortion of the complex XY_n described by the normal coordinates Q_i and caused by a strain e. In order of magnitude we shall have $Q_i \sim eR$ where R is the nearest neighbour distance. We are thus led to expect a relationship between the stress constant V_{ES} and the Jahn–Teller coupling constant V of the form $V \sim V_{ES}/R$. An elementary calculation based on the expressions of the Q_i given in Table 26 leads to the following formulae (Ham 1968):

$$\left.\begin{array}{ll}\text{sixfold octahedral coordination,} & V = \dfrac{\sqrt{3}}{2}(V_{ES}/R); \\[4pt] \text{eightfold cubic coordination,} & V = \tfrac{3}{4}(V_{ES}/R); \\[4pt] \text{fourfold tetrahedral coordination,} & V = \dfrac{3}{2\sqrt{2}}(V_{ES}/R).\end{array}\right\}\quad(21.69\text{b})$$

We see that ions with the largest Jahn–Teller effect will be the most sensitive to strain. In the numerical example of Öpik and Pryce for the Cu^{2+} ion,

$$W_{JT} = 3000 \text{ cm}^{-1} = \frac{V\rho_0}{2} = V \times 0\cdot 3 \times 10^{-8},$$

and if we take $R = 2 \times 10^{-8}$ cm for the distance to the nearest neighbour we find

$$V_{ES} = \frac{2VR}{\sqrt{3}} = \frac{V\rho_0}{2}\left(\frac{R}{\rho_0}\right)\frac{4}{\sqrt{3}} \approx 50000 \text{ cm}^{-1}.$$

A strain as small as 10^{-5} would in this particular case split the orbital doublet Γ_3 by an amount of the order of a wave-number, and even in the best crystals local random strains can be expected to influence the paramagnetic spectrum considerably.

Inside a vibronic doublet, the strain-induced energy change (21.69a) must naturally be multiplied by the same reduction factor q defined in (21.53) as the anisotropy constants g_2 and A_2 of the Zeeman and hyperfine Hamiltonian (21.26) and (21.29). The total Hamiltonian, Zeeman plus hyperfine, plus strain, can then be written

$$\begin{aligned}\mathscr{H} = {} & g_1\beta(\mathbf{S}\cdot\mathbf{H}) + A_1(\mathbf{I}\cdot\mathbf{S}) + \\ & + qU_\theta\left\{V_{ES}e_\theta + \frac{g_2\beta}{2}(3H_zS_z - \mathbf{S}\cdot\mathbf{H}) + \frac{A_2}{2}(3I_zS_z - \mathbf{I}\cdot\mathbf{S})\right\} + \\ & + qU_\varepsilon\left\{V_{ES}e_\varepsilon + \frac{g_2\beta\sqrt{3}}{2}(H_xS_x - H_yS_y) + \frac{A_2\sqrt{3}}{2}(I_xS_x - I_yS_y)\right\}.\quad(21.70)\end{aligned}$$

To keep the problem simple we maintain the assumptions $|g_2| \ll |g_1|$, $|A_2| \ll |A_1|$, which enables us to rewrite (21.70) as

$$\mathcal{H} = g_1\beta HM + A_1 Mm +$$
$$+ qU_\theta \left\{ V_{ES} e_\theta + \frac{g_2\beta HM}{2}(3\zeta^2-1) + \frac{A_2 Mm}{2}(3\zeta^2-1) \right\} +$$
$$+ qU_\varepsilon \left\{ V_{ES} e_\varepsilon + \frac{g_2\beta\sqrt{(3)}HM}{2}(\xi^2-\eta^2) + \frac{\sqrt{(3)}A_2 Mm}{2}(\xi^2-\eta^2) \right\}. \quad (21.71)$$

With respect to the magnitude of the strain we consider two extreme cases

(a) *Very small strains*

These are defined by the inequality

$$V_{ES}(e_\theta^2 + e_\varepsilon^2)^{\frac{1}{2}} \ll \tfrac{1}{2} |g_2\beta H + A_2 m| \{1 - 3(\xi^2\eta^2 + \eta^2\zeta^2 + \zeta^2\xi^2)\}^{\frac{1}{2}}. \quad (21.72)$$

In this case the correct eigenfunctions of the Hamiltonian (21.71) are still given by formulae (21.35) and (21.36), provided the vibronic functions Θ and $\mathscr{E}$ replace the electronic functions θ and ε; the random strain affects the resonance frequencies and broadens the lines in second order in the strain by an amount of the order of

$$V_{ES}^2(e_\theta^2 + e_\varepsilon^2)/|g_2\beta H + A_2 m|.$$

The condition (21.72) may conceivably be fulfilled for certain hyperfine lines and certain orientations but not for others (and never with the field along a body diagonal). In view of the smallness of $g_2\beta H$ (of the order of 0·1 cm^{-1} in Cu^{2+} for a field of 10 kG) an exceptionally strain-free crystal may be needed for (21.72) to be satisfied.

(b) *Large strains*

With the inequality (21.72) reversed, the correct wave-functions for the Hamiltonian (21.71), different from those given by eqn (21.35), are now

$$\Phi_a^M = |M\rangle \left(\Theta \cos\frac{\alpha}{2} - \mathscr{E} \sin\frac{\alpha}{2} \right),$$
$$\Phi_b^M = |M\rangle \left(\Theta \sin\frac{\alpha}{2} + \mathscr{E} \cos\frac{\alpha}{2} \right), \quad (21.73)$$

where

$$\cos\alpha = e_\theta (e_\theta^2 + e_\varepsilon^2)^{-\frac{1}{2}},$$
$$\sin\alpha = e_\varepsilon (e_\theta^2 + e_\varepsilon^2)^{-\frac{1}{2}}. \quad (21.73\text{a})$$

The replacement in (21.71) of U_θ and U_ε by their expectation values within the states $\Phi_{a,b}$, respectively $\pm\cos\alpha$ and $\pm\sin\alpha$, leads to the following formulae for the spectrum corresponding to (21.71):

$$h\nu_{a,b}(m) = (g_1\beta H + A_1 m) \pm$$

$$\pm q(g_2\beta H + A_2 m)\{1-3(\xi^2\eta^2+\eta^2\zeta^2+\zeta^2\xi^2)\}^{\frac{1}{2}}\cos(\omega-\alpha), \quad (21.74)$$

where $\cos\omega$ and $\sin\omega$ have been defined in eqn (21.36). Since the angle α related to the random strain by (21.73a) can be assumed to take all values between 0 and 2π it would seem that each line of the spectrum (21.74) should be broadened beyond recognition over a region

$$2q(g_2\beta H + A_2 m)\{1-3(\xi^2\eta^2+\eta^2\zeta^2+\zeta^2\xi^2)\}^{\frac{1}{2}}.$$

However, as pointed out by Ham (1969), each line of (21.74) so broadened has a singularity in its spectral density for

$$|\cos(\omega-\alpha)| = 1.$$

If all values of the random variable α are equally probable the spectral density is proportional to

$$d\nu \Big/ \left|\frac{d\nu}{d\alpha}\right| \propto \frac{1}{|\sin(\omega-\alpha)|} \propto \frac{1}{\sqrt{|\nu-\nu_0|}}, \quad (21.75)$$

where ν_0 is the sharp frequency that would obtain in the absence of strain broadening. The quasi-continuous spectrum (21.74) exhibits sharp peaks at the same frequencies as in the absence of strain, namely

$$h\nu_{a,b}(m) = g_1\beta H + A_1 m \pm q(g_2\beta_2 H + A_2 m)\{1-3(\xi^2\eta^2+\eta^2\zeta^2+\zeta^2\xi^2)\}^{\frac{1}{2}}.$$

$$(21.76)$$

We have already stated, when establishing formula (21.37), that the assumption $|A_2| \ll A_1$ was often unwarranted. If we give up this assumption while keeping $|g_2| \ll |g_1|$ the formula (21.74) has to be replaced by the following,

$$h\nu_{a,b} = g_1\beta H \pm qg_2\beta H(1-3\bar{u})^{\frac{1}{2}}\cos(\omega-\alpha)+m\mathscr{A}, \quad (21.74a)$$

where we write for brevity $\xi^2\eta^2+\eta^2\zeta^2+\zeta^2\xi^2 = \bar{u}$, and

$$\mathscr{A}^2 = A_1^2 \pm 2qA_1A_2(1-3\bar{u})^{\frac{1}{2}}\cos(\omega-\alpha) + \frac{q^2 A_2^2}{2}\{1+(1-3\bar{u})^{\frac{1}{2}}\cos(\omega+2\alpha)\}.$$

Here m is the projection of $\mathbf{I}$ along a direction whose cosines are given by

$$\mathscr{A}\xi' = \xi\left\{A_1 \pm \frac{qA_2}{2}(-\cos\alpha + \sqrt{(3)}\sin\alpha)\right\},$$

$$\mathscr{A}\eta' = \eta\left\{A_1 \pm \frac{qA_2}{2}(\cos\alpha + \sqrt{(3)}\sin\alpha)\right\}, \quad (21.74b)$$

$$\mathscr{A}\zeta' = \zeta(A_1 \pm qA_2\cos\alpha).$$

These formulae are very easy to establish if one notices that for strong strain, U_θ and U_ε can be replaced in (21.70) by their expectation values over the functions (21.73), namely $\pm\cos\alpha$ and $\pm\sin\alpha$. Strain also plays an important role in the passage from the static to the dynamic Jahn–Teller effect, a problem to be examined presently.

The tunnelling model and the transition from the dynamic to the static Jahn–Teller effect

In the foregoing studies of the effects of perturbations such as Zeeman and hyperfine couplings, and also strain, on the vibronic ground doublet, we have consistently assumed that the relevant Hamiltonians are very much smaller than the doublet-singlet distance and that their off-diagonal doublet-singlet matrix elements could therefore be disregarded. We saw, however, that even in the absence of warping, an increasingly strong Jahn–Teller coupling had a tendency to bring together the ground doublet and the first two excited (spuriously degenerate) singlet levels, the separation Δ_1 or 3Γ as we shall call it henceforth, being, see (21.60a), approximately equal to $\hbar\omega(\hbar\omega/2W_{\mathrm{JT}})$. A numerical study (O'Brien 1964) has shown that the introduction of warping has two effects: first, it separates the singlets A_1 and A_2; second, it brings the lower of the two very much nearer to the ground doublet than predicted by the linear approximation.

We saw earlier (eqn (21.54a)) that any perturbation, such as caused by magnetic coupling or strain, whose action on a purely electronic Γ_3 doublet was expressed by an operator

$$G_0\mathscr{I} + G_\varepsilon U_\varepsilon + G_\theta U_\theta, \quad (21.54)$$

could be written within a vibronic doublet spanned by two vibronic functions Θ and $\mathscr{E}$ as $G_0\mathscr{I} + q(G_\varepsilon U_{g\varepsilon} + G_\theta U_{g\theta})$ and within a vibronic singlet A_1 or A_2 as $G_0\mathscr{I}$. If the distance 3Γ between the singlet and the doublet is comparable to the interaction (21.54) we need the matrix elements of U_θ and U_ε between the singlet and the doublet. They obey the following relations, easily deduced using eqn (21.13):

$$\langle A_1| U_\theta |\Theta\rangle = \langle A_1| U_\varepsilon |\mathscr{E}\rangle = r,$$
$$\langle A_2| U_\theta |\mathscr{E}\rangle = -\langle A_2| U_\varepsilon |\Theta\rangle = r', \qquad (21.77)$$

all other singlet-doublet matrix elements of U_θ and U_ε being zero.

We now set up the secular matrix relative to the perturbation (21.54), which has a slightly different form depending on whether the singlet is A_1 or A_2. For A_1 it is given by (21.78),

$$\begin{array}{c c} & \begin{array}{c c c} A_1 & \Theta & \mathscr{E} \end{array} \\ \begin{array}{c} A_1 \\ \Theta \\ \mathscr{E} \end{array} & \left(\begin{array}{c c c} 3\Gamma+G_0 & rG_\theta & rG_\varepsilon \\ rG_\theta & -qG_\theta+G_0 & qG_\varepsilon \\ rG_\varepsilon & qG_\varepsilon & qG_\theta+G_0 \end{array}\right) \end{array} \qquad (21.78)$$

but if the excited singlet is A_2, rG_θ and rG_ε are replaced in (21.78) respectively by $-r'G_\varepsilon$ and $r'G_\theta$. Since the quantities G_θ, G_ε are usually smaller than one wave-number, in order for 3Γ to be of the same order of magnitude both the Jahn–Teller coupling and the warping must be strong, an assumption we had made in the study of the static Jahn–Teller effect. This provides a basis for the so-called tunnelling model (Bersuker 1962).

Let us return to the static Jahn–Teller effect, assuming for the sake of definiteness that $V > 0, V_3 < 0$. We then found a potential minimum for $\varphi = 0$ and we described the complete wave-function of the system near that minimum by the Born–Oppenheimer product

$$\Phi_Z = \psi_z(\varphi)\Psi_Z(\rho, \varphi). \qquad (21.79)$$

The nuclear wave-function $\Psi_Z(\rho, \varphi)$ is strongly localized in the neighbourhood of $\rho = \rho_0 \approx |V|/M\omega^2$. A reasonable approximation for $\Psi_Z(\rho, \varphi)$ would be a Gaussian function, the ground state of a two-dimensional harmonic oscillator,

$$\Psi_Z(\rho, \varphi) = \left(\frac{\alpha\alpha'}{\pi}\right)^{\frac{1}{2}} \exp\,-\tfrac{1}{2}\{\alpha^2(\rho-\rho_0)^2 + \alpha'^2\rho_0^2\varphi^2\}, \qquad (21.80)$$

where $\alpha = (\hbar/M\omega)^{\frac{1}{2}}$, $\alpha' = (\hbar/M\omega')^{\frac{1}{2}}$. Here ω is the frequency for radial oscillation, which appears in the quadratic potential energy $\tfrac{1}{2}M\omega^2\rho^2$ and ω', given by eqn (21.49), is the frequency for azimuthal oscillation inside the potential trough at $\varphi = 0$, produced by the warping. Because of this strong localization of the nuclear wave-function one may in a first approximation replace the electronic wave-function $\psi_z(\varphi)$ by

$\psi_z(0)$. If one wishes to take into account the finite spread of the nuclear wave-function (21.80), one can replace $\psi_z(\varphi)$ in the neighbourhood of $\varphi = 0$ by

$$\psi_z(\varphi) = \theta \cos\frac{\varphi}{2} - \varepsilon \sin\frac{\varphi}{2}, \qquad (21.81)$$

which was used to obtain the corrected formulae (21.51) for $g_\parallel$ and $g_\perp$.

From the wave-function

$$\Phi_Z = \psi_z \Psi_Z \approx \theta \Psi_Z \propto (3z^2 - r^2)\Psi_Z \qquad (21.82)$$

we can deduce two other wave-functions, Φ_X and Φ_Y, relative to the minima $\varphi = \pm 2\pi/3$, by the condition that Φ_X, Φ_Y, Φ_Z derive from each other by the cyclical permutation, $x \to y \to z \to x$,

$$\begin{aligned}\Phi_X &\approx (3x^2 - r^2)\Psi_X(\rho, \varphi) = (3x^2 - r^2)\Psi_Z\left(\rho, \varphi - \frac{2\pi}{3}\right), \\ \Phi_Y &\approx (3y^2 - r^2)\Psi_Y(\rho, \varphi) = (3y^2 - r^2)\Psi_Z\left(\rho, \varphi + \frac{2\pi}{3}\right).\end{aligned} \qquad (21.82a)$$

A more accurate representation of Φ_X in the neighbourhood of $\varphi = 2\pi/3$, following (21.81), is

$$\left\{(3x^2 - r^2)\cos\tfrac{1}{2}\left(\varphi - \frac{2\pi}{3}\right) - \sqrt{(3)}(y^2 - z^2)\sin\tfrac{1}{2}\left(\varphi - \frac{2\pi}{3}\right)\right\}\Psi_Z\left(\rho, \varphi - \frac{2\pi}{3}\right), \qquad (21.82b)$$

with a similar expression for Φ_Y.

The three wave-functions Φ_X, Φ_Y, Φ_Z, which correspond to the same energy, provide a reducible representation of the cubic group. The combination

$$A_1 = \frac{1}{\sqrt{3}}\{\Phi_X + \Phi_Y + \Phi_Z\}, \qquad (21.83)$$

which is normalized if the small overlap between the functions is neglected, is clearly invariant under the operations of the cubic group and belongs to the representation A_1. The two combinations

$$\begin{aligned}\Theta &= \frac{1}{\sqrt{6}}(2\Phi_Z - \Phi_X - \Phi_Y), \\ \mathscr{E} &= \frac{1}{\sqrt{2}}(\Phi_X - \Phi_Y),\end{aligned} \qquad (21.83a)$$

belong obviously to Γ_3 and transform like θ and ε. If we introduce the overlap $\delta = \langle \Phi_X | \Phi_Z \rangle$ between two functions Φ, eqns (21.83),(21.83a) are replaced by

$$A_1 = \frac{1}{\sqrt{(3+6\delta)}}(\Phi_X+\Phi_Y+\Phi_Z),$$
$$\Theta = \frac{1}{\sqrt{(6-6\delta)}}(2\Phi_Z-\Phi_X-\Phi_Y). \tag{21.84}$$

It is sometimes stated (Bersuker 1962, Höchli 1967) that the overlap $\delta = \langle \Phi_X | \Phi_Z \rangle$ can be represented as the product of the overlap $\gamma = \langle \Psi_X | \Psi_Z \rangle$ of the nuclear wave-functions and the overlap $\langle 3x^2-r^2 | 3z^2-r^2 \rangle = -\frac{1}{2}$ (an immediate consequence of the relations (21.13)) of the electronic wave-functions. It should be remembered, however, that for ψ_x, for instance, the expression $\psi_x(2\pi/3) = 3x^2-r^2$, and even the more accurate expression (21.82b), are only valid in the neighbourhood of $\varphi = 2\pi/3$, and the same is true for the expression (21.81) for $\psi_z(\varphi)$ in the neighbourhood of $\varphi = 0$. (Expressions such as (21.81) could not possibly be valid for all values of φ since Φ_Z as given by (21.79),(21.80) would not be a single-valued function of Q_θ, Q_ε.) The main contribution to the overlap between the two functions Φ_X and Φ_Y comes from regions of φ distant from the minima. This means that the electronic scalar product $\langle \psi_x(\varphi) | \psi_z(\varphi) \rangle$ is an unknown function of φ and a formula such as $\delta = -\gamma/2$ is unwarranted.

If we now assume that the product VV_3 of Table 21.1 is positive, the three wave-functions Φ_X, Φ_Y, Φ_Z become

$$\Phi_Z \approx \sqrt{(3)}(x^2-y^2)\Psi_Z,$$
$$\Phi_X \approx \sqrt{(3)}(y^2-z^2)\Psi_X, \tag{21.85}$$
$$\Phi_Y \approx \sqrt{(3)}(z^2-x^2)\Psi_Y,$$

where the nuclear wave-functions Ψ are centered around $\varphi = \pi$ and $\varphi = \pi \pm 2\pi/3$. In contrast to eqn (21.83), the combination

$$A_2 = \frac{1}{\sqrt{(3)}}(\Phi_X+\Phi_Y+\Phi_Z) \tag{21.86}$$

belongs this time to the unidimensional representation A_2 rather than to A_1. This can be checked by noticing that a rotation through $\pi/2$ around, say, Oz changes A_2 as given by (21.86) into $-A_2$. The two combinations,

$$\Theta' = (\Phi_Y-\Phi_X)/\sqrt{2},$$
$$\mathscr{E}' = (2\Phi_Z-\Phi_X-\Phi_Y)/\sqrt{6}, \tag{21.86a}$$

belong again to Γ_3 and transform like θ and ε (we have neglected in (21.86) and (21.86a) the overlap δ).

We have now, with eqns (21.83), (21.83a) or (21.86), (21.86a), constructed approximate expressions for the vibronic wave-functions that span the ground doublet and the first excited singlet, the latter being A_1 or A_2 depending on the sign of VV_3. The three wave-functions Φ_X, Φ_Y, Φ_Z are three degenerate eigenfunctions of an approximate Hamiltonian, the approximation being essentially that of an infinite barrier height between two neighbouring potential minima. We can take into account as a perturbation $\mathscr{H}_1$ the possibility of tunnelling through the barriers between the various minima. We need not concern ourselves with the actual form of this perturbation: the simple fact that it has cubic symmetry insures automatically that A, Θ, $\mathscr{E}$ given by (21.83) etc. or (21.86) etc. are the 'good' zero-order wave-functions and that the triply degenerate level spanned by Φ_X, Φ_Y, Φ_Z will be split into a singlet A and a doublet (Θ, $\mathscr{E}$). This approximation should be good only when the barrier is very high and the overlap 3Γ very small. We expect the splitting to be positive by continuity, starting from no Jahn–Teller coupling when we know that the doublet is lower. We shall not discuss the approximate methods for calculating Γ, by the WKB approximation or otherwise (Bersuker 1962, Sturge 1967). This calculation, like that of the overlap δ, is sensitive to the rather uncertain shape of the tails of the wave-functions Φ_X, Φ_Y, Φ_Z in between the minima.

Since in (21.83), (21.83a) or (21.86), (21.86a) we have explicit expressions for the vibronic functions A, Θ, $\mathscr{E}$, we can calculate the coefficients q, r, and r' (we have demonstrated in eqn (21.67) that $p = 0$ for strong Jahn–Teller coupling). We can obtain q for instance from the relation $q = \langle \mathscr{E} | U_\theta | \mathscr{E} \rangle$ where $\mathscr{E}$ is given by (21.83a). Taking into account the expression (21.82b) for Φ_X (and a similar one for Φ_Y) we obtain $q = \tfrac{1}{2}u$, where

$$u = \langle \cos \varphi \rangle = \int \Psi_Z'^2(\rho, \varphi) \cos \varphi \, d\varphi \qquad (21.87)$$

is the average value of $\cos \varphi$ in the trough $\varphi = 0$. In this calculation we have neglected the overlap between Ψ_X and Ψ_Y. If the approximate form (21.82a) is used for Φ_X and Φ_Y in the calculation of the overlap this leads to a small correction to q, equal to $3\gamma/4$ (Höchli 1967). Actually, as we mentioned earlier, we do not know the form of the electronic wave-functions ψ_x and ψ_y in the regions where overlap is appreciable. While the correction to q due to overlap must indeed be of the order of γ, there is no particular reason to believe it to be $3\gamma/4$. Disregarding this correction it is satisfactory to notice that the value of $\tfrac{1}{2}u$ for q is in accordance with the general inequality, $q \leqslant \tfrac{1}{2}$, derived in eqn (21.67). The reader may check that the same value is obtained for q when the functions Φ are of the type (21.85) rather than (21.82), (21.82a).

A similar calculation of the matrix elements $\langle A_1|\,U_\varepsilon\,|\mathscr{E}\rangle$ or $\langle A_2|\,U_\varepsilon\,|\Theta\rangle$ yields

$$r = -\frac{u}{\sqrt{2}} = -q\sqrt{2}; \qquad r' = \frac{u}{\sqrt{2}} = q\sqrt{2}. \qquad (21.88)$$

In order to see what happens when the singlet-doublet splitting 3Γ is smaller than the Hamiltonian (21.54), it is illuminating to transcribe the matrix (21.78) for the representation Φ_X, Φ_Y, Φ_Z inverting the equations (21.82), (21.82a) or (21.86), (21.86a). We obtain

$$\begin{array}{c} \\ \Phi_Z \\ \Phi_X \\ \Phi_Y \end{array} \begin{pmatrix} \Phi_Z & \Phi_X & \Phi_Y \\ \Gamma+G_0\mp 2qG_\theta & \Gamma & \Gamma \\ \Gamma & \Gamma+G_0\mp q(-G_\theta+\sqrt{(3)}G_\varepsilon) & \Gamma \\ \Gamma & \Gamma & \Gamma+G_0\mp q(-G_\theta-\sqrt{(3)}G_\varepsilon) \end{pmatrix}$$
$$(21.89)$$

where the upper sign corresponds to the situation (21.82), (21.82a) and the lower to (21.85), and where we have used the relations (21.88).

Let us temporarily assume that G_θ and G_ε are much larger than Γ. It is then apparent in (21.89) that we may disregard the small off-diagonal matrix elements Γ and that Φ_X, Φ_Y, Φ_Z become the good zero-order functions. They are precisely those we used for the static Jahn–Teller effect. If G_0, G_θ, and G_ε result from the Zeeman coupling (21.26) with

$$G_0 = g_1\beta(\mathbf{H}\cdot\mathbf{S}),$$

$$G_\theta = \frac{g_2\beta}{2}(3H_zS_z-\mathbf{H}\cdot\mathbf{S}),$$

$$G_\varepsilon = \frac{g_2\beta\sqrt{3}}{2}(H_xS_x-H_yS_y),$$

we find, as expected for the static Jahn–Teller effect, in the state Φ_Z,

$$\langle\Phi_Z|\,Z\,|\Phi_Z\rangle = g_\|\beta H_zS_z+g_\perp\beta(H_xS_x+H_yS_y) \qquad (21.90)$$

and similar expressions for Φ_X and Φ_Y where, in accordance with (21.51)

$$g_\| = g_1\mp 2qg_2 \approx g_1\mp ug_2,$$
$$g_\perp = g_1\pm qg_2 \approx g_1\pm\frac{ug_2}{2}, \qquad (21.90a)$$

the upper sign corresponding to the case when Φ_Z is defined by (21.82).

We have already alluded to the smallness of $g_2\beta H$ ($\sim$0.1 cm^{-1} for Cu^{2+} in a field of 10 kG) which makes it unlikely that it will be larger than the tunnelling splitting 3Γ. There is, however, a second origin for G_θ and G_ε, namely a strain-induced change in energy of the form (21.69a) $V_{ES}(e_\theta U_\theta + e_\varepsilon U_\varepsilon)$, whence $G_\theta = V_{ES}e_\theta$, $G_\varepsilon = V_{ES}e_\varepsilon$. From (21.89) and the definitions (21.69) of e_θ and e_ε, this gives for the state Φ_Z a change in energy $\mp qV_{ES}(2e_{zz}-e_{xx}-e_{yy})$ and similar expressions for Φ_X, Φ_Y.

These strain-induced energy changes may be much larger than $g_2\beta H$ and larger than Γ. We realize now the true origin of the static Jahn–Teller effect, as explained by Ham (1968). When, because of a strong Jahn–Teller coupling and a high warping barrier, the frequency $3\Gamma/\hbar$ for tunnelling between the three minima (which in the language of the dynamic Jahn–Teller effect is the distance between the ground doublet and the first excited singlet) becomes smaller than the strain energy, the latter 'locks' the system in one of the three potential minima corresponding to a static tetragonal distortion (with the possibility of small zero-point motion near each minimum). If the strain splitting is smaller than kT all three states Φ_X, Φ_Y, Φ_Z will be equally populated but, even if kT is very small, there will be an equal number of ions in the crystal in each state because of the random character of the local strain.

The intermediate situation where Γ is neither large nor small compared to $g_2\beta H$ has been discussed in the literature by means of eqn (21.89) (Bersuker 1963, O'Brien 1964) before the importance of strain had been fully realized. We do not reproduce these discussions here since, especially for ions with strong Jahn–Teller coupling, the effect of strain is likely to be predominant.

21.7. Motional narrowing of the Jahn–Teller spectrum

A survey of the various types of spectra

In the foregoing we saw that three types of spectra could be expected from a ^2E level undergoing a Jahn–Teller effect:

(a) a superposition of three anisotropic spectra of tetragonal symmetry with parameters given by (21.46), (21.47) or better (21.51), characteristic of a static Jahn–Teller effect;

(b) a purely isotropic spectrum with isotropic g-factor and hyperfine structure constant, equal respectively to g_1 and A_1, characteristic of a vibronic singlet;

(c) an anisotropic spectrum described by parameters g_1, qg_2, A_1, qA_2, characteristic of a vibronic doublet. Actually this last spectrum can be subdivided into two types:

(c1) where the average local strain is larger than the anisotropic part of the magnetic interactions

$$V_{ES}(e_\theta^2+e_\varepsilon^2)^{\frac{1}{2}} \gg |g_2\beta H+A_2 m| \qquad (21.91)$$

and where the resonance frequencies are given by eqn (21.74), and

(c2) where the inequality (21.91) is reversed and the spectrum is given again by (21.74) but with the factor $\cos(\omega-\alpha)$ missing in the anisotropic part.

The occurrence of these various spectra is determined by the relative magnitude of the following three quantities: the doublet-singlet splitting 3Γ, the average strain $V_{ES}(e_\theta^2+e_\varepsilon^2)^{\frac{1}{2}}$ henceforth called for brevity Δ_s, and the anisotropic magnetic coupling $(g_2\beta H+A_2 m)$, called for brevity Δ_m.

If $3\Gamma \gg \Delta_s, \Delta_m$, which will always occur for a weak or moderately strong Jahn–Teller coupling, the vibronic doublet, well separated from the first excited singlet, should exhibit either a spectrum (c1) or (c2) depending on the relative magnitude of Δ_s and Δ_m. The isotropic spectrum (b) of the vibronic singlet will be observable if $kT/3\Gamma$ is sufficiently large for the singlet to be appreciably populated.

If the Jahn–Teller coupling and the warping are very strong, 3Γ becomes much smaller, while the strain splitting Δ_s is expected to be relatively large as suggested by eqn (21.69b). It is likely that in this case $\Delta_s \gg \Delta_m$, and the only question is the relative size of 3Γ and Δ_s.

If $3\Gamma \gg \Delta_s$, a singlet spectrum (b), and a doublet spectrum (c1) should be observed.

If $3\Gamma \ll \Delta_s$, a static Jahn–Teller spectrum of type (a) should be observed as explained in the last section.

Motional narrowing

In our discussion we have hitherto considered the complex XY_6 as an isolated system, disregarding its coupling with the other degrees of freedom of the crystal in which the complex is embedded, such as lattice vibrations. If this coupling is switched on, fast transitions may occur between various states of the complex, with emission or absorption of one or more phonons resulting in an appreciable change in the observed pattern. The main change to be expected if these transitions are sufficiently fast is the transformation of the anisotropic spectra (a) and (c) into an isotropic singlet-like spectrum (b). Thus the origin of the appearance of such a spectrum when the temperature is raised is not unambiguous: it may be due either to an increased population of an excited singlet or to an increase in relaxation rate that turns an anisotropic spectrum (a) or (c) into an isotropic spectrum (b) in ways now to be described.

Let us start with a static spectrum (a) and call $1/\tau$ the rate of change, due to thermal relaxation, of the population of anyone of the three distorted configurations described by the wave-functions Φ_X, Φ_Y, Φ_Z. The following relation is easily established from the rate equations: $1/\tau = 3P_{XY}$, where on the right-hand side P_{XY} represents the thermally induced transition probability from a configuration Φ_X to another configuration Φ_Y. The total spread $\Delta \nu$ in the paramagnetic spectrum due to anisotropy is given by

$$h\Delta\nu = |(g_\parallel - g_\perp)\beta H + (A_\parallel - A_\perp)m| = \tfrac{3}{2}u\,|g_2\beta H + A_2 m|, \quad (21.92)$$

where we made use of (21.51). If τ is sufficiently short to satisfy the inequality $\Delta\nu \cdot \tau \ll 1$, then, according to the well-known theory of motional narrowing, only the average frequency of the three spectra will be observed, given by

$$h\nu(m) = (\tfrac{1}{3}g_\parallel + \tfrac{2}{3}g_\perp)\beta H + (\tfrac{1}{3}A_\parallel + \tfrac{2}{3}A_\perp)m = g_1\beta H + A_1 m, \quad (21.93)$$

which is identical with the spectrum of the singlet.

The reorientation from one configuration to another can occur through several processes among which we single out for a brief description, as in the theory of spin-lattice relaxation, the Orbach process, the direct process, and the Raman process, outlined in Chapter 10.

In the Orbach process a real transition to an excited vibronic state lying at a distance W_0 above the ground state will have a probability of the order of

$$\tau^{-1} \simeq \nu_0 \exp\left(-\frac{W_0}{kT_0}\right) \quad (21.94)$$

where, as explained in Chapter 10, ν_0 is essentially the inverse lifetime of the excited state and $\exp(-W_0/kT_0)$ the relative probability of finding in the crystal an energetic phonon of frequency W_0/h at the temperature of the experiment.

At the temperatures too low for the Orbach process to be appreciable the direct process corresponds to absorption or emission of a single phonon of energy equal to the energy difference Δ_{XY} induced by stress between two distorted configurations Φ_X and Φ_Y. We now consider this process.

As explained in § 21.6, the observability of a spectrum of type (a) characteristic of a static Jahn–Teller effect implies the assumption $|\Gamma| \ll |\Delta_{XY}, \Delta_{YZ}, \Delta_{ZX}|$. The calculation of the corresponding relaxation time, treating phonons as time-dependent strains, is standard except for one feature: only those phonons that have the symmetry properties of the strain components e_θ and e_ε have non-vanishing matrix elements between various states of our system (unless higher-order terms involving spin-orbit coupling are brought in). On the other hand, the strain Hamiltonian $V_{ES}(e_\theta U_\theta + e_\varepsilon U_\varepsilon)$ does *not* have any off-diagonal matrix elements between the states Φ_X, as can be seen from eqn (21.89). Such matrix elements can only appear because the states Φ_X are not exact eigenstates of the system, and because the off-diagonal matrix elements Γ in (21.89) bring into, say Φ_Z, admixtures of order Γ/Δ of Φ_X and Φ_Y. It follows that the transition probability P_{XY}, instead of being proportional (as is usually the case) to the square of the energy difference Δ_{XY} (see chapter 10), is reduced by the square of the admixture coefficient $(\Gamma/\Delta)^2$ and thus is independent of Δ.

Apart from these changes the transition rate due to the direct process has the usual form (Ham 1969)

$$P_{XY} = \frac{9\Gamma^2(qV_{ES})^2 kT_0}{5\pi\rho\hbar^4 v_t^5}\left\{1 + \frac{2}{3}\left(\frac{v_t}{v_l}\right)^5\right\}, \tag{21.95}$$

where ρ is the density of the crystal and V_t and V_l the transverse and longitudinal velocities of sound.

The Raman process, which has an unusual temperature-dependence with a transition rate proportional to T^3, is given by (Pirc, Zeks, and Gosar 1966)

$$P_{XY} = \frac{27\Gamma^2(qV_{ES})^4(kT_0)^3}{50\pi\hbar^7\rho^2 v_t^{10}}\left\{1 + \frac{2}{3}\left(\frac{v_t}{v_l}\right)^5\right\}^2. \tag{21.96}$$

We consider also the transition from a spectrum of type (c1), a vibronic doublet with stress energy greater than the magnetic anisotropy but much smaller than the singlet-doublet splitting 3Γ, to an isotropic singlet-like spectrum (b). The orbital eigenstates, given by eqns (21.73), (21.73a), are

$$\Psi_a = \Theta\cos\frac{\alpha}{2} - \mathscr{E}\sin\frac{\alpha}{2}, \quad \cos\alpha = \frac{e_\theta}{\sqrt{(e_\theta^2 + e_\varepsilon^2)}};$$
$$\Psi_b = \Theta\sin\frac{\alpha}{2} + \mathscr{E}\cos\frac{\alpha}{2}, \quad \sin\alpha = \frac{e_\varepsilon}{\sqrt{(e_\theta^2 + e_\varepsilon^2)}}; \tag{21.73a}$$

and are separated by an energy

$$\Delta = W_b - W_a = 2qV_{ES}\sqrt{(e_\theta^2 + e_\varepsilon^2)}.$$

The spread in frequency due to anisotropy is

$$h\Delta\nu(m) = 2q(g_2\beta H + A_2 m)\{1 - 3(\xi^2\eta^2 + \eta^2\zeta^2 + \zeta^2\xi^2)\}^{\frac{1}{2}}. \quad (21.97)$$

In order to be wiped out by fast transitions between Ψ_a and Ψ_b, the relaxation rate $1/\tau$ must again be such that $\Delta\nu \cdot \tau \ll 1$.

The calculation is standard and yields for the direct process:

$$\frac{1}{\tau} = \frac{3|\Delta|^3 (qV_{ES})^2}{20\pi\hbar^4 \rho v_t^5} \left\{1 + \frac{2}{3}\left(\frac{v_t}{v_l}\right)^5\right\} \coth\left(\frac{\Delta}{2kT_0}\right) \quad (21.98)$$

or, if $kT_0 \gg |\Delta|$,

$$\frac{1}{\tau} = \frac{3}{10} \frac{\Delta^2 (qV_{ES})^2}{\pi\hbar^4 \rho v_t^5} \left\{1 + \frac{2}{3}\left(\frac{v_t}{v_l}\right)^5\right\} kT_0. \quad (21.98')$$

For the Raman process a temperature-dependence of the rate on the fifth power is obtained (Ham 1968):

$$\frac{1}{\tau} = \frac{6\pi(qV_{ES})^4(kT_0)^5}{125\hbar^7 \rho^2 v_t^{10}} \left\{1 + \frac{2}{3}\left(\frac{v_t}{v_l}\right)^5\right\}^2. \quad (21.99)$$

All the transitions described so far are purely orbital with no spin flip involved. Spin flips can occur because the orbital parts of states of opposite spin belonging to different configurations are not strictly orthogonal. This lack of orthogonality is connected with paramagnetic anisotropy and leads to spin-lattice relaxation rates, $1/\tau_1$, smaller than the previously calculated purely orbital transition rates by reduction factors of the order of $(g_2/g_1)^2$ or $(A_2/g_1\beta H)^2$ as the case may be.

Even so, these rates may be much faster than those of ions in less symmetrical environments which are not subject to the Jahn–Teller effects. A detailed discussion of these and other spin-lattice relaxation processes characteristic of Jahn–Teller ^2E ions can be found elsewhere (Ham 1969, Williams et al. 1969).

21.8. Comparison with experiment

The anisotropic spectra of type (a) which correspond to a static Jahn–Teller effect, exhibited by the ion Cu^{2+}, $3d^9$ in cubic or nearly cubic environment, have already been discussed in § 7.16 and need not be considered again here. The review article of Ham (1969), which has provided the inspiration and the information for the writing of the present chapter, contains a thorough discussion of the experimental evidence on the observation of the Jahn–Teller effect in EPR spectra of ^2E levels, to which the reader is referred. We shall be content to examine briefly, as an illustration of the theory, the measurements of Coffman on Cu^{2+}, $3d^9$ in MgO (Coffman 1965, 1966, 1968) and Höchli and Estle on Sc^{2+} in CaF_2 and SrF_2 (Höchli and Estle 1967, Höchli 1967).

An anisotropic spectrum of type (c), characteristic of a vibronic doublet, was observed for Cu^{2+} in MgO at $1 \cdot 2°K$ with the parameters

$$g_1 = 2 \cdot 195, \qquad qg_2 = 0 \cdot 108, \qquad q = \frac{qg_2}{g_1 - g_s} \sim 0 \cdot 5.$$

This value of q indicates a strong or moderately strong Jahn–Teller effect. It is not quite clear whether this spectrum is the no-strain spectrum (c2) or the strain broadened spectrum (c1) given by eqn (21.74). Using for the residual strain at the Cu^{2+} site the same average value $\sim 10^{-4}$ as found elsewhere for Fe^{2+} in MgO, and for the strain coefficient V_{ES} the same value as that estimated for Ni^{3+} in Al_2O_3, Ham finds a value of the order of 1 cm^{-1} for the strain splitting $\Delta_s \sim 2qV_{ES}(e_\theta^2 + e_\varepsilon^2)^{\frac{1}{2}}$, which is much greater than the anisotropy energy Δ_m. Unless this estimate turns out to be grossly incorrect the spectrum of Cu^{2+} in MgO observed by Coffman at $1 \cdot 2°K$ should thus be of the strain-broadened type (c1). At $77°K$ this spectrum is replaced by an isotropic singlet spectrum (b), which is interpreted as a vibronic doublet spectrum (c1) narrowed by fast-relaxation transitions between the orbital states (21.73).

The resonance spectrum of Sc^{2+}, $3d^1$ in CaF_2 and SrF_2 provides an interesting example of a vibronic doublet spectrum (c) that corresponds to a rather weak Jahn–Teller effect. It should be noted that although Sc^{2+} has one unpaired electron, rather than a hole as in Cu^{2+}, $3d^9$, the electronic ground multiplet is still a doublet Γ_3 because in CaF_2 the coordination is eightfold instead of sixfold as in MgO. Since the spin-orbit coupling is now positive, g_1 should be smaller than g_s and g_2 should be negative. From the values, $g_1 = 1 \cdot 973$, $qg_2 = -0 \cdot 022$, one extracts for q the value $q = qg_2/(g_1 - g_s) = 0 \cdot 75$; according to the inequality (21.67) this precludes a strong Jahn–Teller effect and leads for $W_{JT}/\hbar\omega$ to an estimate (Ham 1968) of the order of $0 \cdot 25$. (In SrF_2 the value of q is $0 \cdot 71$ and the estimate for $W_{JT}/\hbar\omega$ is $0 \cdot 34$.) For such weak couplings the linear theory based on the eqns (21.59) should be adequate. The assumption that here again the spectrum is of the strain-broadened type is supported by the fact that a stress of up to 250 kg/cm^2, applied by Höchli to his sample, did not induce any observable change in the spectrum. This is to be expected if large random local strains exist already in the crystal.

When the temperature is raised to a few degrees K the (c) spectrum is replaced by an isotropic singlet-like spectrum. This spectrum was interpreted by Höchli as due to the increased population of the first excited singlet, and the singlet-doublet splitting was estimated from the temperature dependence of the intensity as 10 cm^{-1} for Sc^{2+} in CaF$_2$ and 8 cm^{-1} in SrF$_2$. This is clearly incompatible with the assumption, made necessary by a coefficient $q \sim 0.75$, of a weak Jahn–Teller effect since the singlet-doublet distance is then of the order of $\hbar\omega \sim 200$ or 300 cm^{-1}. Ham therefore interprets the isotropic spectrum of Sc^{2+} as due to the same motional narrowing of the low temperature spectrum as for Cu^{2+} in MgO.

21.9. The Jahn–Teller effect in a triplet state

The problem of the Jahn–Teller effect in orbital triplets Γ_4 or Γ_5 differs from that of the orbital doublet in several respects that are all in some way connected with the reduction formula, which replaces eqn (21.12) applicable to Γ_3,

$$\Gamma_4 \times \Gamma_4 = \Gamma_5 \times \Gamma_5 = \Gamma_4 + \Gamma_5 + \Gamma_3 + \Gamma_1. \tag{21.100}$$

Already in the absence of Jahn–Teller coupling this relation has important consequences for the ions with a degenerate triplet as ground state, classified as ions of type B in § 19.3. For convenience, we summarize here some of the results.

The first far-reaching consequence of eqn (21.100) arises from the presence on its right-hand side of Γ_4, a tridimensional representation whose components transform like those of a vector. This means that, in contrast to Γ_3, the components of an orbital momentum will have non-vanishing components inside the triplets Γ_4 or Γ_5. We have shown in § 14.2 that it is convenient, using the Wigner–Eckart theorem, to introduce a pseudo-angular momentum whose components inside the manifold Γ_4 or Γ_5 have the same matrix elements as those of an orbital momentum $L = 1$ within a p-state. This is summed up by the relation

$$\langle \mathbf{L} \rangle = \alpha \tilde{\mathbf{l}}, \tag{21.101}$$

where the coefficient α depends on the detailed structure of the orbital triplet Γ_4 or Γ_5 and the symbol $\langle \ \rangle$ means projection inside the triplet manifold. In particular, α has the value -1 for a Γ_5 triplet originating in a D-term, $\alpha = +\frac{1}{2}$ if Γ_5 comes from an F-term, and $\alpha = -\frac{3}{2}$ for the triplet Γ_4 originating in an F-term. As basis states spanning Γ_4 or Γ_5 one can use either the eigenstates $|\pm\tilde{1}\rangle$, $|\tilde{0}\rangle$ of $\tilde{l}_z = \pm 1$, 0 or real wave-functions ψ_X, ψ_Y, ψ_Z related to the former by

$$|\pm\tilde{1}\rangle = \mp\frac{\psi_X \pm i\psi_Y}{\sqrt{2}}\ ;\ |\tilde{0}\rangle = \psi_Z. \quad (21.102)$$

The existence of a finite orbital momentum $\langle \mathbf{L} \rangle = \alpha \mathbf{l}$ entails also a finite spin-orbit coupling

$$\lambda \langle \mathbf{L} \rangle . \mathbf{S} = \alpha\lambda(\tilde{\mathbf{l}} . \mathbf{S}). \quad (21.103)$$

If (in contrast to our treatment of Γ_3) we neglect second-order effects arising from the matrix elements of $\lambda(\mathbf{L} . \mathbf{S})$ to the excited cubic multiplets, the total spin $\mathbf{S}$ and the pseudo-momentum $\tilde{\mathbf{l}}$ combine to yield the multiplets $\tilde{\mathscr{J}}$, where $\tilde{\mathscr{J}}$ takes all the values from $|S-1|$ to $S+1$. The spectroscopic splitting factor of the ion in a $\tilde{\mathscr{J}}$ multiplet is given by the Landé formula

$$g(\tilde{\mathscr{J}}) = \frac{1}{\tilde{\mathscr{J}}(\tilde{\mathscr{J}}+1)} \{\tilde{g}_l(\tilde{\mathbf{l}} . \tilde{\mathscr{J}}) + g_s(\mathbf{S} . \tilde{\mathscr{J}})\}$$

$$= \frac{1}{2\tilde{\mathscr{J}}(\tilde{\mathscr{J}}+1)}[\tilde{g}_l\{\tilde{\mathscr{J}}(\tilde{\mathscr{J}}+1)+2-S(S+1)\}+$$

$$+g_s\{\tilde{\mathscr{J}}(\tilde{\mathscr{J}}+1)+S(S+1)-2\}]. \quad (21.104)$$

In (21.104) $\tilde{g}_l = \alpha g_L$, where g_L is the *true* orbital gyromagnetic factor that may be smaller than unity because of covalency as explained in § 20.5.

The presence of Γ_5 and Γ_3 on the right-hand side of (21.100) means that the two even vibrational modes of Table 26 can both be coupled to the electronic states of the degenerate triplet with two coupling constants which we denote by V_E and V_T (E and T_2 being alternative names for Γ_3 and Γ_5).

Our problem is to find, inside the degenerate triplet manifold, electronic operators transforming respectively like components of Γ_3 and Γ_5, thus taking the part played by U_θ and U_ε for the degenerate doublet. These operators can be easily formed using the components of the pseudo-momentum $\tilde{\mathbf{l}}$. For Γ_3 they are

$$\mathscr{E}_\theta = \tfrac{1}{2}\{3\tilde{l}_z^2 - l(l+1)\} = \tfrac{1}{2}(3\tilde{l}_z^2 - 2),$$

$$\mathscr{E}_\varepsilon = \frac{\sqrt{3}}{2}(\tilde{l}_x^2 - \tilde{l}_y^2), \quad (21.105)$$

and for Γ_5

$$T_{2X} = (\tilde{l}_y \tilde{l}_z + \tilde{l}_z \tilde{l}_y), \quad T_{2Y} = (\tilde{l}_z \tilde{l}_x + \tilde{l}_x \tilde{l}_z), \quad T_{2Z} = (\tilde{l}_x \tilde{l}_y + \tilde{l}_y \tilde{l}_x).$$
(21.105a)

It is easily checked, by writing them in the representation ψ_X, ψ_Y, ψ_Z, where they are diagonal (rather than in the representation $|\pm 1\rangle$, $|0\rangle$), that the operators $\mathscr{E}_\theta$ and $\mathscr{E}_\varepsilon$ commute with each other. (Naturally this is *not* a general property of tensor operators of the form (21.105) as we know very well from the example of the doublet where the non-commuting operators U_θ and U_ε were, by eqn (21.19), of the form (21.105), with $L = 2$ rather than 1.) This makes the study of the Jahn–Teller coupling with the mode Γ_3 much simpler than that with the mode Γ_5 whose operators (21.105a) do not commute with each other. In the representation ψ_X, ψ_Y, ψ_Z these operators are given by the following matrices:

$$\mathscr{E}_\theta = \begin{pmatrix} \tfrac{1}{2} & & \\ & \tfrac{1}{2} & \\ & & -1 \end{pmatrix}, \quad \mathscr{E}_\varepsilon = \begin{pmatrix} -\sqrt{(3)}/2 & & \\ & \sqrt{(3)}/2 & \\ & & 0 \end{pmatrix},$$

$$T_{2X} = \begin{pmatrix} 0 & 0 & 0 \\ 0 & 0 & -1 \\ 0 & -1 & 0 \end{pmatrix}$$
(21.106)

with obvious changes for T_{2Y} and T_{2Z}.

The vibronic Hamiltonian can be written as $\mathscr{H}_v = \mathscr{H}_v^E + \mathscr{H}_v^T$, where

$$\mathscr{H}_v^E = \frac{1}{2\mu}(P_\theta^2 + P_\varepsilon^2) + \frac{\mu\omega^2}{2}(Q_\theta^2 + Q_\varepsilon^2) + V_E(Q_\theta \mathscr{E}_\theta + Q_\varepsilon \mathscr{E}_\varepsilon), \quad (21.107)$$

$$\mathscr{H}_v^T = \frac{1}{2\mu'}(P_4^2 + P_5^2 + P_6^2) + \frac{\mu'\omega'^2}{2}(Q_4^2 + Q_5^2 + Q_6^2) + $$
$$+ V_T(Q_4 T_{2X} + Q_5 T_{2Y} + Q_6 T_{2Z}). \quad (21.107a)$$

We shall see in the next section that if the Γ_3 mode only is coupled to the triplet ($V_T = 0$) the problem has an exact and rather simple solution. On the other hand, if the coupling is to the mode Γ_5 or to both, only qualitative statements can be made.

Before we discuss these two couplings we must make some assumptions about the magnitude of the spin-orbit coupling. As usual we shall consider the two cases when this coupling is either much stronger or much weaker than the Jahn–Teller coupling, shirking the complicated intermediate situation. It has been shown (Öpik and Pryce 1957, Van Vleck 1960) that a spin-orbit coupling appreciably stronger than the Jahn–Teller coupling may quench the latter to a large extent. This can be seen as follows: consider a level $\tilde{\mathscr{J}}$ resulting from the spin-orbit coupling, represented as $\alpha\lambda(\tilde{\mathbf{l}}\cdot\mathbf{S})$. If $|\alpha\lambda| \gg V_E$, V_T we may disregard off-diagonal matrix elements of the Jahn–Teller coupling between two different $\tilde{\mathscr{J}}$ multiplets and keep only those inside the multiplet $\tilde{\mathscr{J}}$ under consideration. Inside this multiplet the five operators (21.105), (21.105a) can be expressed by the Wigner–Eckart theorem as similar combinations of $\tilde{\mathscr{J}}_x$, $\tilde{\mathscr{J}}_y$, $\tilde{\mathscr{J}}_z$ multiplied by a constant σ which is the same for all five tensor components:

$$\langle 3\tilde{l}_z^2 - \tilde{l}(\tilde{l}+1)\rangle_{\tilde{\mathscr{J}}} = \sigma\{3\tilde{\mathscr{J}}_z^2 - \tilde{\mathscr{J}}(\tilde{\mathscr{J}}+1)\}, \text{ etc.} \qquad (21.108)$$

Here the constant σ can be computed either by means of the Racah $6j$- symbol (eqn (16.29)) or by a direct calculation. If the ground $\tilde{\mathscr{J}}$ multiplet is $\tilde{\mathscr{J}} = \frac{1}{2}$, the five operators (21.105; 21.105a) obviously vanish. This is to be expected, since otherwise the Jahn–Teller coupling would have been able to lift a Kramers-degeneracy which it cannot possibly do, being a time-even operator.

For values of $\tilde{\mathscr{J}}$ higher than $\frac{1}{2}$ the coefficient σ often turns out to be small. Thus, for instance, for Fe^{2+}, $3d^6$ in MgO, where the ground orbital triplet is $\Gamma_5({}^5D)$, the ground $\tilde{\mathscr{J}}$ multiplet is also a triplet $\tilde{\mathscr{J}} = 1$. A direct calculation of the expectation value of $(\frac{1}{2})\{3\tilde{l}_z^2 - \tilde{l}(\tilde{l}+1)\}$ in the state $\tilde{\mathscr{J}} = \tilde{\mathscr{J}}_z = 1$ yields $\frac{1}{20}$, whence $\sigma = \frac{1}{10}$. The effective Jahn–Teller coupling is reduced in the ratio $\frac{1}{10}$ and the Jahn–Teller energy in the ratio $1/100$.

In the following sections we shall make the opposite assumption of a weak spin-orbit coupling, introduced as a perturbation once the Jahn–Teller coupling has been included, which implies that the spin-orbit coupling is also smaller than the vibrational quantum $\hbar\omega$.

21.10. The Jahn–Teller effect in an orbital triplet with Γ_3 coupling

The vibronic wave-functions

The eigenfunctions of the vibronic Hamiltonian $\mathscr{H}_v^E$ (eqn (21.107)) are Born–Oppenheimer products of the form

$$\Phi_X = \psi_X \Psi_X, \qquad \Phi_Y = \psi_Y \Psi_Y, \qquad \Phi_Z = \psi_Z \Psi_Z,$$

where ψ_X, ψ_Y, ψ_Z are the electronic wave-functions that span the electronic triplet and Ψ_X, Ψ_Y, Ψ_Z three nuclear functions of Q_θ and Q_ε. Applying $\mathscr{H}_v^E$ to $\psi_Z \Psi_Z$, we find for Ψ_Z the equation

$$\left\{\frac{1}{2\mu}(P_\theta^2+P_\varepsilon^2)+\frac{\mu\omega^2}{2}(Q_\theta^2+Q_\varepsilon^2)-V_E Q_\theta\right\}\Psi_Z = W\Psi_Z, \qquad (21.109)$$

where use has been made of the form (21.106) of $\mathscr{E}_\theta$ and $\mathscr{E}_\varepsilon$; (21.109) is the equation of a two-dimensional harmonic oscillator in the plane $(Q_\theta, Q_\varepsilon)$ centred on the point

$$Q_\varepsilon^Z = 0, \qquad Q_\theta^Z = \frac{V_E}{\mu\omega^2}, \qquad (21.110)$$

or in polar coordinates $\rho_0 = V_E/\mu\omega^2$, $\varphi = 0$, the point at which the potential energy in (21.109) is a minimum and has the Jahn–Teller value, $W_{JT} = V_E^2/2\mu\omega^2$. Because of cubic symmetry the two other functions Ψ_X and Ψ_Y must be similar oscillator functions but centred around the points $\rho = \rho_0$, $\varphi = \pm 2\pi/3$, which are the points

$$Q_\theta^X = -\frac{1}{2}\frac{V_E}{\mu\omega^2}, \qquad Q_\varepsilon^X = \frac{\sqrt{3}}{2}\frac{V_E}{\mu\omega^2}, \qquad (21.110a)$$
$$Q_\theta^Y = -\frac{1}{2}\frac{V_E}{\mu\omega^2}, \qquad Q_\varepsilon^Y = -\frac{\sqrt{3}}{2}\frac{V_E}{\mu\omega^2},$$

in accordance with the form of $\mathscr{E}_\theta$ and $\mathscr{E}_\varepsilon$ given by (21.106). If, as is customary in the theory of the static Jahn–Teller effect, one neglects the nuclear motion, the three points in Q space

$$\{Q_\varepsilon^X, Q_\theta^X; Q_\varepsilon^Y, Q_\theta^Y; Q_\varepsilon^Z, Q_\theta^Z\}$$

represent the stable equilibrium positions of the distorted complex with electronic functions ψ_X, ψ_Y, ψ_Z. At each of these points, say $(Q_\theta^Z, Q_\varepsilon^Z)$, the energy of the two other states ψ_X, ψ_Y is greater by the amount $3W_{JT} = 3V_E^2/2\mu\omega^2$.

The potential energy surfaces relative to the three functions ψ_X, ψ_Y, ψ_Z are three paraboloids shown in Fig. (21.3), which result from each other by a rotation of $2\pi/3$. This situation resembles to some extent the tunnelling model described in § 21.6 with the three potential minima depressed by the Jahn–Teller energy W_{JT} and the three wave-functions (21.82, 21.82a) or (21.85). There is, however, an important difference: those three functions Φ were not orthogonal and their degeneracy was only approximate, due to the closeness of a singlet with a wave-function A to the doublet E spanned by two wave-functions Θ and $\mathscr{E}$. A, Θ, and $\mathscr{E}$ were the 'good' wave-functions, linear combinations

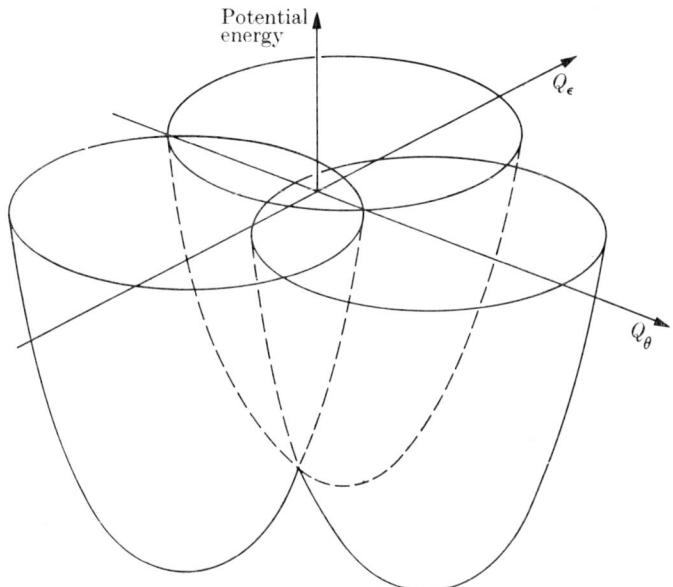

Fig. 21.3. Potential energy surfaces for a triplet with a Γ_3 Jahn–Teller coupling.

of Φ_X, Φ_Y, Φ_Z, and the doublet and the singlet were split apart by an amount 3Γ when proper account was taken of the lack of orthogonality of Φ_X, Φ_Y, Φ_Z.

In the present situation Φ_X, Φ_Y, Φ_Z are exactly orthogonal because their electronic parts ψ_X, ψ_Y, ψ_Z are exactly orthogonal, and they span a bona fide triplet. It was already a degenerate triplet Γ_4 or Γ_5 before the Jahn–Teller coupling was switched on (the ground vibrational function is then the non-degenerate ground state of a two-dimensional harmonic oscillator and the overall degeneracy is that of the electronic triplet), and must stay a triplet of the same nature, Γ_4 or Γ_5, when the Jahn–Teller coupling is included.

The eigenstates of $\mathcal{H}_v^E$ (eqn (21.107)) can be written

$$\Phi_{i, n_\theta, n_\varepsilon}(q, Q) = \psi_i(q) F_{n_\theta}(Q_\theta - Q_\theta^i) F_{n_\varepsilon}(Q_\varepsilon - Q_\varepsilon^i). \tag{21.111}$$

i stands for X, Y, Z in this formula; Q_θ^i, Q_ε^i are the equilibrium positions of the harmonic oscillators given by (21.110, 21.110a), n_θ, n_ε are integer quantum numbers, and $F_{n_\theta}(Q_\theta)$, $F_{n_\varepsilon}(Q_\varepsilon)$ the usual quantum states of one-dimensional harmonic oscillators. The energy of a state (21.111) is

$$W_{i, n_\theta, n_\varepsilon} = W_0 - \frac{V_E^2}{2\mu\omega^2} + (n_\theta + n_\varepsilon + 1)\hbar\omega. \tag{21.112}$$

Here excited levels have a high and partly spurious degeneracy due to the linear character of the Hamiltonian $\mathscr{H}_v^E$ (21.107). We did not find it necessary to introduce into $\mathscr{H}_v^E$ higher-order warping terms, as we were forced to in the study of the electronic doublet, whose effect among other things was to split the spurious degeneracy of the singlets A_1, A_2. In the present case, from the relation (see Table 2)

$$\Gamma_3 \times \Gamma_4 = \Gamma_3 \times \Gamma_5 = \Gamma_4 + \Gamma_5,$$

using the same reasoning as in § 21.6, we see that the energy levels of the most general vibronic Hamiltonian, describing a Γ_3 Jahn–Teller coupling with an electronic triplet Γ_4 or Γ_5, are triplets Γ_4 *and* Γ_5, the ground level having the same symmetry as in the absence of Jahn Teller coupling.

First-order effects

Armed with the wave-functions (21.111) we can compute the matrix elements of the various electronic operators of interest for paramagnetic resonance, first and foremost being those of the orbital momentum $\tilde{\mathbf{l}}$. From the form $\Phi_{in} = \psi_i \Psi_{in}$, where n stands for $(n_\theta, n_\varepsilon)$, of the vibronic functions we see that for any electronic operator O_d diagonal in the representation ψ_i we have

$$\langle \Phi_{in} | O_\mathrm{d} | \Phi_{jn'} \rangle = \delta_{nn'} \langle \psi_i | O_\mathrm{d} | \psi_j \rangle. \tag{21.113}$$

Such operators are not affected by Jahn–Teller coupling. On the other hand, for an operator O_od with off-diagonal matrix elements, such as the orbital momentum, we get

$$\langle \Phi_{in} | O_\mathrm{od} | \Phi_{jn'} \rangle = \langle \psi_i | O_\mathrm{od} | \psi_j \rangle \langle \Psi_{in} | \Psi_{jn'} \rangle, \tag{21.114}$$

in which

$$\langle \Psi_{in} | \Psi_{jn'} \rangle = \langle F_{in_\theta} | F_{jn'_\theta} \rangle \langle F_{in_\varepsilon} | F_{jn'_\varepsilon} \rangle$$
$$= \langle F_{n_\theta}(Q_\theta - Q_\theta^i) | F_{n'_\theta}(Q_\theta - Q_\theta^j) \rangle \langle F_{n_\varepsilon}(Q_\varepsilon - Q_\varepsilon^i) | F_{n'_\varepsilon}(Q_\varepsilon - Q_\varepsilon^j) \rangle, \tag{21.115}$$

where the scalar products $\langle F_{in_\theta} | F_{jn'_\theta} \rangle$ can be computed exactly from the known wave-functions of the harmonic oscillator (Ham 1965). In the ground triplet the relation (21.114) takes the form

$$\langle \Phi_{ioo} | O_\mathrm{od} | \Phi_{joo} \rangle$$
$$= \langle \psi_i | O_\mathrm{od} | \psi_j \rangle \langle F_o^i | F_o^j \rangle^2 = \langle \psi_i | O_\mathrm{od} | \psi_j \rangle \exp(-3W_\mathrm{JT}/2\hbar\omega). \tag{21.116}$$

This relation (21.116) means that in the ground triplet the orbital momentum, which in the representation ψ_i has off-diagonal matrix elements only, is quenched by a factor $\exp(-3W_{\rm JT}/2\hbar\omega)$.

In the same way as we introduced for the vibronic doublet vibronic operators $U_{g\theta}$, $U_{g\varepsilon}$ (eqn (21.52)), we introduce here an operator $\mathbf{1}_g$ using the condition $\langle \Phi_{i0}| \mathbf{1}_g |\Phi_{j0}\rangle = \langle \psi_i| \tilde{\mathbf{l}} |\psi_j\rangle$. Within the ground triplet we can replace $\tilde{\mathbf{l}}$ by

$$\mathbf{1}_g \exp\left(-\frac{3W_{\rm JT}}{2\hbar\omega}\right).$$

It is easy to see that all electronic tensor components, with off-diagonal matrix elements only, which transform like Γ_4, that is as $\tilde{\mathbf{l}}_x, \tilde{\mathbf{l}}_y, \tilde{\mathbf{l}}_z$; or like Γ_5, that is as T_{2X}, T_{2Y}, T_{2Z}; will have the same reduction factors. On the other hand, tensor components that transform like $\mathscr{E}_\theta$ and $\mathscr{E}_\varepsilon$ are diagonal and unaffected by Jahn–Teller coupling. Following Ham (1969), we transcribe these results as

$$\kappa(T_1) = \kappa(T_2) = \exp\left(-\frac{3W_{\rm JT}}{\hbar\omega}\right), \qquad \kappa(E) = 1. \qquad (21.117)$$

Equation (21.117) shows in particular that a small static distortion of the cubic field will be partially quenched if it is trigonal (transforming as T_2), but unaffected if it is tetragonal (transforming as E).

The same reduction factor $\kappa(T_1) = \kappa_1$ applies naturally to the spin-orbit coupling $\lambda(\mathbf{L} \cdot \mathbf{S}) = \alpha\lambda(\tilde{\mathbf{l}} \cdot \mathbf{S})$, which now becomes in first order $\kappa_1\alpha\lambda(\mathbf{1}_g \cdot \mathbf{S})$. We can still apply the method developed for ions of type B in the absence of a Jahn–Teller effect and couple the vibronic operator $\mathbf{1}_g$ to $\mathbf{S}$ to obtain the multiplets $\tilde{\mathscr{J}}$, but the separation $\alpha\lambda\tilde{\mathscr{J}}$ between two multiplets $\tilde{\mathscr{J}}$ and $\tilde{\mathscr{J}}-1$ is also reduced by the factor κ_1.

Second-order terms and the transition to the static effect

Obviously this procedure can only be valid as long as the reduced spin-orbit coupling $\kappa_1\alpha\lambda(\mathbf{1}_g \cdot \mathbf{S})$ remains much larger than either the magnetic or strain perturbations whose expectation values we wish to calculate inside the multiplet $\tilde{\mathscr{J}}$. There is another limitation, which is the following: $\alpha\kappa_1\lambda(\mathbf{1}_g \cdot \mathbf{S})$ represents only the first-order effect of the spin-orbit coupling inside the vibronic ground triplet. Second-order terms due to this coupling are small in a static crystal field because of the large energy denominators which are the cubic splittings. In the presence of Jahn–Teller coupling, however, matrix elements to excited vibronic states, such as $\langle \Phi_{i0}| \alpha\lambda(\tilde{\mathbf{l}} \cdot \mathbf{S}) |\Phi_{jn}\rangle$, which is equal to

$\langle \psi_i | \, \alpha\lambda(\tilde{\mathbf{l}} \cdot \mathbf{S}) \, | \psi_j \rangle \langle \Psi_{i0} | \Psi_{jn} \rangle$, do not vanish for $i \neq j$, and because the energy denominators $n\hbar\omega$ are much smaller than the cubic splittings, second-order terms may become appreciable especially in view of the quenching of the first-order terms. From symmetry arguments these second-order terms can be expressed as follows:

$$\mathcal{H}_{\text{SO}}^{(2)} = -\frac{\lambda^2 \alpha^2}{\hbar\omega} \{k_1(\mathbf{1}_g \cdot \mathbf{S})^2 + k_2(l_{gx}^2 S_x^2 + l_{gy}^2 S_y^2 + l_{gz}^2 S_z^2)\}. \quad (21.118)$$

The coefficients k_1 and k_2 can be computed exactly (Ham 1965).

If we write $x = 3W_{\text{JT}}/\hbar\omega$, the coefficients $k_1(x)$ and $k_2(x)$ are given by

$$k_1(x) = e^{-x} G(\tfrac{1}{2}x), \qquad k_2(x) = e^{-x}\{G(x) - G(\tfrac{1}{2}x)\}, \quad (21.119)$$

where

$$G(x) = \int_0^x \frac{1}{u}(e^u - 1)\, du \underset{x\to\infty}{\sim} \frac{e^x}{x}\left(1 + \frac{1}{x} + \dots\right). \quad (21.119a)$$

The coefficients k_1 and k_2 must naturally vanish in the absence of Jahn–Teller coupling, in accordance with (21.119). For x relatively small, when the second-order spin-orbit term $\mathcal{H}_{\text{SO}}^{(2)}$ is much smaller than the first-order term $\mathcal{H}_{\text{SO}}^{(1)} = \alpha\lambda e^{-x/2}(\mathbf{1}_g \cdot \mathbf{S})$, its only effect is to violate the Landé interval rule between the multiplets $\tilde{\mathscr{J}}$ and to split those for which $\tilde{\mathscr{J}} \geq 2$.

For very strong Jahn–Teller coupling $W_{\text{JT}}/\hbar\omega \to \infty$ the orbital momentum and the spin-orbit coupling are quenched completely in first order by the reduction factor $\exp -(3W_{\text{JT}}/2\hbar\omega)$. The second-order term (21.118) has the asymptotic form

$$-\frac{\lambda^2 \alpha^2}{(\hbar\omega)x}(l_{gx}^2 S_x^2 + l_{gy}^2 S_y^2 + l_{gz}^2 S_z^2) = -\frac{\lambda^2 \alpha^2}{3W_{\text{JT}}}\{\dots\}. \quad (21.120)$$

In the ground state Φ_{ZO}, (21.120) has the expectation value

$$-\frac{\lambda^2 \alpha^2}{3W_{\text{JT}}}(S_x^2 + S_y^2) = D\{S_z^2 - S(S+1)\} \text{ with } D = \frac{\lambda^2 \alpha^2}{3W_{\text{JT}}}. \quad (21.121)$$

This is the value predicted by static crystal-field theory for an ion of type A in the non-degenerate electronic state ψ_Z at the bottom $(Q_\theta^Z, Q_\varepsilon^Z)$ of the paraboloid Z, separated at that point by $3W_{\text{JT}}$ from the paraboloids X and Y where the electronic wave-functions are ψ_X and ψ_Y. This is exactly what one expects from a static Jahn–Teller effect.

The Zeeman energy Z is expressed by the operator

$$Z = \beta g_L(\mathbf{L} \cdot \mathbf{H}) + g_s \beta(\mathbf{S} \cdot \mathbf{H}) = \beta \tilde{g}_l(\tilde{\mathbf{l}} \cdot \mathbf{H}) + g_s \beta(\mathbf{S} \cdot \mathbf{H}). \quad (21.122)$$

IN PARAMAGNETIC RESONANCE 841

If second-order terms are negligible Z can be rewritten as

$$Z^{(1)} = \beta\kappa_1\tilde{g}_l(\mathbf{l}_g \cdot \mathbf{H}) + g_S\beta(\mathbf{S} \cdot \mathbf{H}). \qquad (21.122\text{a})$$

The formula (21.104) can still be used for the gyromagnetic factor $g(\tilde{\mathscr{J}})$ provided $g_l = \alpha g_L$ is replaced by $\kappa_1 g_l = \alpha\kappa_1 g_L$, where $\kappa_1 = \kappa(T_1)$ in eqn (21.117). In second order, cross terms between the Zeeman Hamiltonian and the spin-orbit coupling yield an expression $Z^{(2)}$ very similar to (21.118),

$$Z^{(2)} = g_1\beta\{(\mathbf{l}_g \cdot \mathbf{S})(\mathbf{l}_g \cdot \mathbf{H}) + (\mathbf{l}_g \cdot \mathbf{H})(\mathbf{l}_g \cdot \mathbf{S})\} +$$
$$+ g_2\beta(l_{g_x}^2 S_x H_x + l_{g_y}^2 S_y H_y + l_{g_z}^2 S_z H_z), \qquad (21.123)$$

where

$$g_1 = -(\lambda\alpha^2/\hbar\omega)g_L k_1(x), \qquad g_2 = -(2\lambda\alpha^2/\hbar\omega)g_L k_2(x).$$

For a very strong Jahn–Teller effect when $x \to \infty$

$$g_1 \to 0, \qquad g_2 \to -\frac{2\lambda\alpha^2}{3W_{\text{JT}}}g_L.$$

In the state $\Phi_{ZO} = \psi_Z\Psi_{ZO}$, $Z^{(2)}$ has a limiting expectation value

$$\langle\Phi_{ZO}| Z^{(2)} |\Phi_{ZO}\rangle = -\frac{2\lambda\alpha^2}{3W_{\text{JT}}}g_L(S_x H_x + S_y H_y). \qquad (21.124)$$

Again, this is what the theory of the static Jahn–Teller effect would predict.

This study of the orbital triplet coupled to the Γ_3 vibration mode can be summed up as follows. For moderate Jahn–Teller coupling, formally the system behaves very much as if the Jahn–Teller effect did not exist except for a reduction in some parameters such as the spin-orbit coupling, the orbital splitting factor, the trigonal distortion, and an increase in second-order terms such as (21.118). For a very strong Jahn–Teller coupling, the naïve theory of the static Jahn– Teller effect is valid, with orbital Zeeman energy and spin-orbit energy completely quenched in first order, and given in second order by the usual formulae for non-degenerate ions A, the excitation energy being $3W_{\text{JT}}$, which at a point of stable equilibrium is the distance to the upper sheets of the potential energy surfaces.

21.11. The Jahn-Teller effect in an orbital triplet with Γ_5 coupling

We have already mentioned the fact that this problem is made difficult because the coupling operators T_{2X}, T_{2Y}, T_{2Z} in the vibronic

Hamiltonian $\mathscr{H}_v^T$ (eqn (21.107a)) do not commute. It is easiest to approach from the limit of a very strong, nearly static, Jahn–Teller effect, when the tunnelling model can be used. We begin therefore by neglecting the nuclear kinetic energy and look for the points in Q space where the potential energy is a minimum, using the method of Öpik and Pryce (1957). With the small change in notation, $Q_4 \to Q_X$, $Q_5 \to Q_Y$, $Q_6 \to Q_Z$, (and similarly for P_4, P_5, P_6), we can rewrite the vibronic Hamiltonian (21.107a) as

$$\mathscr{H}_v^T = T(P) + V + U_0(Q), \qquad (21.125)$$

where the nuclear kinetic energy is

$$T(P) = \frac{1}{2\mu}(P_X^2 + P_Y^2 + P_Z^2); \qquad (21.125\text{a})$$

the Jahn–Teller coupling operator is

$$V(Q) = V_T(Q_X T_{2X} + Q_Y T_{2Y} + Q_Z T_{2Z}); \qquad (21.125\text{b})$$

and the quasi-elastic potential energy is

$$U_0(Q) = \frac{\mu'\omega'^2}{2}(Q_X^2 + Q_Y^2 + Q_Z^2). \qquad (21.125\text{c})$$

Let $\psi = a_X\psi_X + a_Y\psi_Y + a_Z\psi_Z$ be an eigenstate of V with the eigenvalue $\varepsilon'(Q)$, such that

$$V|a\rangle = \varepsilon'(Q)|a\rangle; \quad \varepsilon'(Q) = \langle a|V|a\rangle; \quad \langle a|a\rangle = 1, \qquad (21.126)$$

where $|a\rangle$ is the set of amplitudes (a_X, a_Y, a_Z). The points in Q space where the potential energy $U(Q) = U_0(Q) + \varepsilon'(Q)$ is a minimum are given by

$$\frac{\partial U_0}{\partial Q_i} + \frac{\partial \varepsilon'}{\partial Q_i} = \mu'\omega'^2 Q_i + \langle a|\frac{\partial V}{\partial Q_i}|a\rangle = 0. \qquad (21.127)$$

If we replace T_{2X}, T_{2Y}, T_{2Z} by their matrix expression (21.106), eqns (21.126) become

$$\left.\begin{array}{l}V_T(Q_Y a_Z + Q_Z a_Y) = -\varepsilon' a_X, \\ V_T(Q_Z a_X + Q_X a_Z) = -\varepsilon' a_Y, \\ V_T(Q_X a_Y + Q_Y a_X) = -\varepsilon' a_Z,\end{array}\right\} \qquad (21.126\text{a})$$

in which

$$\varepsilon' = -2V_T(Q_X a_Y a_Z + Q_Y a_Z a_X + Q_Z a_X a_Y) \qquad (21.126\text{b})$$

and

$$a_X^2 + a_Y^2 + a_Z^2 = 1. \qquad (21.126\text{c})$$

Similarly eqns (21.127) become

$$Q_X = \frac{2V_T}{\mu'\omega'^2}a_Y a_Z, \quad Q_Y = \frac{2V_T}{\mu'\omega'^2}a_Z a_X, \quad Q_Z = \frac{2V_T}{\mu'\omega'^2}a_X a_Y.$$
(21.127a)

Carrying (21.127a) into (21.126a) and making use of (21.126b) and (21.126c) we find the following results:

$$a_X^2 = a_Y^2 = a_Z^2 = \tfrac{1}{3},$$

$$\varepsilon' = -\frac{4V_T^2}{3\mu'\omega'^2}, \quad U_0 = \frac{2V_T^2}{3\mu'\omega'^2},$$

$$U = \varepsilon' + U_0 = -\frac{2V_T^2}{3\mu'\omega'^2}, \quad Q_X = \frac{2V_T}{\mu'\omega'^2}a_Y a_Z.$$
(21.128)

The outcome of eqns (21.127) is that there are four equivalent equilibrium positions in Q space, that is four equivalent distortions of the paramagnetic complex. The electronic wave-functions at these four points are given by

$$\psi_\nu = \frac{m_X \psi_X + m_Y \psi_Y + m_Z \psi_Z}{\sqrt{3}},$$
(21.128a)

where m_X, m_Y, m_Z are ± 1 and $\nu = 1, 2, 3, 4$. (A change of sign of all the m naturally does not change the electronic state.) The equilibrium positions in Q space are given by

$$(Q_X)_\nu = \frac{2V_T}{3\mu'\omega'^2}(m_Y m_Z)_\nu.$$
(21.129)

The Jahn–Teller energy at each point, namely the depression with respect to the energy of the undistorted complex, is

$$W_{\mathrm{JT}} = -U(Q_0) = 2V_T^2/3\mu'\omega'^2.$$
(21.130)

We have tacitly assumed that the condition (21.127) that is necessary for the positions (21.129) to be minima of energy in the Q space was also a sufficient condition. This has been checked by Öpik and Pryce (1957) who have also shown that when Γ_3 and Γ_5 couplings are *both* present, the only relative minima are *still* those obtained when the two couplings are considered separately, a point which is by no means trivial. The position of the absolute minimum is determined by comparing the two Jahn–Teller energies,

$$(W_{\mathrm{JT}})_{\Gamma_3} = \frac{V_E^2}{2\mu\omega^2}, \quad (W_{\mathrm{JT}})_{\Gamma_5} = \frac{2V_T^2}{3\mu'\omega'^2}.$$

The naive description of the static Jahn–Teller effect leads to the assumption that the system would be locked preferentially in the distortion that corresponds to the lowest minimum or minima, ignoring the higher ones. We know that because of the dynamic effects tunnelling will occur and the system will not sit in the lowest minima, unless helped by some external cause such as strain. The combined effects of Γ_3 and Γ_5 coupling simply cannot be predicted quantitatively on the basis of the existing theories.

Returning to Γ_5 coupling, we associate with each electronic function $\psi_\nu(m_X, m_Y, m_Z)$ a nuclear function $\Psi_\nu(Q-Q_\nu)$ localized around the minimum Q_ν and define four vibronic functions, $\Phi_\nu = \psi_\nu \Psi_\nu$. The index ν is associated with the values m_X, m_Y, m_Z as follows:

$$\nu = 1, 2, 3, 4 \rightarrow (1, 1, 1), (-1, -1, 1), (1, -1, -1), (-1, 1, -1).$$

Having started with an orbital triplet in cubic symmetry we are faced here with an unexpected fourfold degeneracy. The situation is very similar to that described in § 21.6 where, starting with an orbital doublet, we had a threefold degeneracy. The answer is similar. Our present quadruplet, spanned by the functions Φ_ν, is not a true quadruplet but results from the merging of a singlet and a triplet, spanned by wave-functions which are 'good' linear combinations of the Φ_ν. The wave-function for the singlet is obviously

$$\Phi_S = (2\Lambda)^{-1}(\Phi_1 + \Phi_2 + \Phi_3 + \Phi_4), \tag{21.131}$$

where Λ is a number of the order of unity that takes into account the small overlap of the Φ_ν. For the triplet the proper wave-functions are

$$\begin{aligned}\Phi_X &= (2\Lambda')^{-1}(\Phi_1 - \Phi_2 + \Phi_3 - \Phi_4),\\ \Phi_Y &= (2\Lambda')^{-1}(\Phi_1 - \Phi_2 - \Phi_3 + \Phi_4),\\ \Phi_Z &= (2\Lambda')^{-1}(\Phi_1 + \Phi_2 - \Phi_3 - \Phi_4).\end{aligned} \tag{21.132}$$

We can use the explicit expressions (21.132) to calculate the limiting values of the quenching factors $\kappa(T_1) = \kappa(\tilde{\mathbf{l}})$, $\kappa(T_2)$ and $\kappa(E)$. Let us consider first the purely imaginary matrix element

$$\langle \Phi_X | \tilde{l}_z | \Phi_Y \rangle = -i\kappa(T_1).$$

If we use the expressions (21.132) for Φ_X and Φ_Y we find

$$\langle \Phi_X | \tilde{l}_z | \Phi_Y \rangle = 2(\Lambda')^{-2} \langle \Phi_3 | \tilde{l}_z | \Phi_1 \rangle. \tag{21.133}$$

Notice that the diagonal terms $\langle \Phi_\nu | \tilde{l}_z | \Phi_\nu \rangle$ vanish since the Φ_ν are real functions and $\tilde{l}_z$ is Hermitian and purely imaginary. As the strength of the Jahn–Teller coupling increases, the overlap between the localized functions Φ_ν goes to zero and so does $\kappa(T_1)$. In the same way we find

$$\langle \Phi_X | \mathscr{E}_\varepsilon | \Phi_X \rangle = 2(\Lambda')^{-2} \langle \Phi_3 | \mathscr{E}_\varepsilon | \Phi_1 \rangle, \qquad (21.134)$$

which also goes to zero with increasing Jahn–Teller coupling in contrast to what happened when this coupling was of the Γ_3 type and $\kappa(E)$ was equal to unity. Finally,

$$\langle \Phi_X | T_{2Z} | \Phi_Y \rangle = (\Lambda')^{-2} \{ \langle \Phi_1 | T_{2Z} | \Phi_1 \rangle - \langle \Phi_1 | T_{2Z} | \Phi_2 \rangle \}. \qquad (21.135)$$

As the Jahn–Teller coupling increases, the limiting value of $\kappa(T_2)$ is $\frac{2}{3}$. This difference of behaviour between the $\mathscr{E}$ and T_2 tensors can be understood in simple terms (Ham 1965). A strain along the [100] axis, which is of the $\mathscr{E}$ type, cannot distinguish between the four distorted configurations that are equivalent with respect to that direction and causes no splitting, whilst a [111] strain of the T_2 type does distinguish between the four configurations.

The three quenching factors have been computed numerically by Caner and Englman (1966). Their variation with the strength of the Jahn–Teller coupling is shown in Fig. (21.4), together with the approximate expressions proposed by Ham (1965) as an extrapolation of a

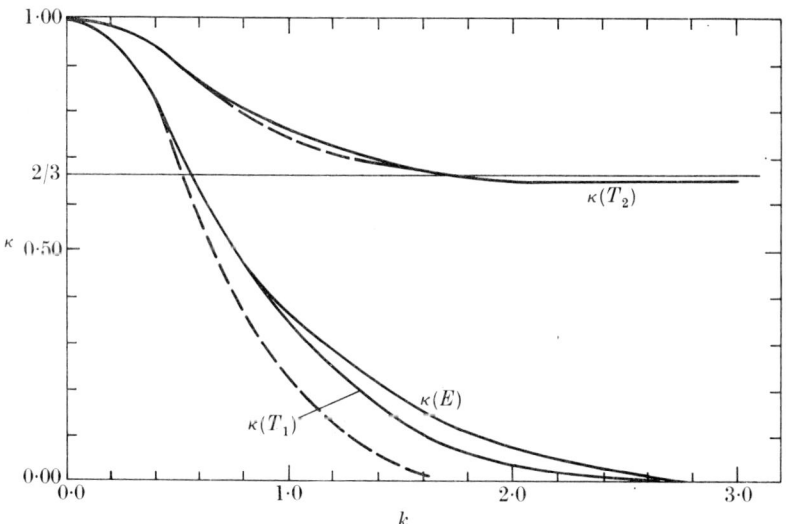

Fig. 21.4. Variation of the reduction factors $\kappa(T_1)$, $\kappa(T_2)$, $\kappa(E)$ against k, where $k^2 = \frac{3}{2}(W_{JT}/\hbar\omega)$. The broken curves are the approximate expressions (21.136) (Ham 1965).

perturbation calculation of Moffitt and Thorson (1957). These are

$$\kappa(E) \approx \kappa(T_1) \approx \exp\{-\tfrac{9}{4}(W_{JT}/\hbar\omega)\}, \\ \kappa(T_2) \approx \tfrac{1}{3}[2+\exp\{-\tfrac{9}{4}(W_{JT}/\hbar\omega)\}].$$ (21.136)

21.12. Comparison with experiment

The reader is again referred for a more detailed discussion of the experimental evidence to the review article by Ham (1969). There are several instances of anisotropic spectra in environments of apparently cubic symmetry which could be interpreted as due to a static Jahn–Teller distortion. A rather convincing example is that of V^{2+} in CaF_2 (Höchli 1966; Zaripov, Kropotov, Lipanova, and Stepanov 1967). Since V^{2+} has the same charge as Ca^{2+}, no charge compensation is expected and, in the absence of Jahn–Teller effect, the symmetry of the crystal potential should be cubic. Spectra having trigonal symmetry are observed, pointing, it would seem, to a predominant Γ_5 coupling.

Several remarkable examples of a dynamic Jahn–Teller effect are provided by the spectra of impurities belonging to the iron group, which occupy interstitial positions in silicon or substitutional positions in MgO and CaO (Ham 1965). They are summarized in Table 21.2 (Ham 1965).

For the ions $3d^6$, $^5T_2$ in the absence of orbital quenching the splitting factor g should be given by (21.104) where $\tilde{\mathscr{J}} = 1, S = 2$, and $\alpha = -1$. This gives $g_{\tilde{\mathscr{J}}} = g_l = 1 \cdot 5 g_s + 0 \cdot 5 g_L$, where g_L is the true orbital gyromagnetic factor, possibly slightly reduced by covalency. A value $g_l \approx 3 \cdot 5$ is thus expected. It is clear from the values given in the table for Mn^+ and Cr^0 that there is a drastic reduction of the orbital momentum. On the other hand, in Fe^{2+}: MgO there is only a small quenching of the orbital momentum of the order of $0 \cdot 8$ which can be attributed to covalent bonding. This absence of Jahn–Teller quenching

TABLE 21.2

	Ion	g
$3d^6$, $^5T_2$	Fe^{2+}:MgO	3·428
	Fe^{2+}:CaO	3·30
	Mn^+:Si	3·01
	Cr^0:Si	2·97
$3d^7$, $^4T_1$	Co^{2+}:MgO	4·278
	Fe^+:MgO	4·15
	Mn^0:Si	3·362
	Fe^+:Si	3·524

can arise from the stabilizing effect of the spin–orbit coupling as explained in § 21.9. In Fe^{2+}:CaO, however, the experimental value $g = 3\cdot30$ requires a reduction of the orbital momentum by a factor $0\cdot60$. One has the choice between greater covalency or stronger Jahn–Teller coupling than in Fe^{2+}:MgO. The former is unlikely since Mössbauer measurements of the isomer shift in the compounds would indicate *less* covalency for Fe^{2+} in CaO than in MgO (Chappert, Frankel, and Blum 1967). The greater orbital quenching would thus imply a stronger Jahn–Teller coupling in CaO than in MgO.

For the $3d^7$, $^4T_1$ ions, eqn (21.104) with $S = \tfrac{3}{2}$, $\tilde{\mathscr{J}} = \tfrac{1}{2}$, $\alpha = -\tfrac{3}{2}$ predicts for a static field, $g_{\frac{1}{2}} = (\tfrac{5}{3})g_s + g_L \approx \tfrac{13}{3} \approx 4\cdot33$ in the absence of all quenching. In fact the quenching factor is almost zero for Mn^0:Si and $0\cdot2$ for Fe^+:Si. It would appear that in these ions the spin–orbit coupling is probably smaller than the Jahn–Teller coupling and thus unable to stabilize it. The opposite situation must clearly prevail for Co^{2+}:MgO, where the g-factor is almost exactly that of the crystal-field theory, and to a lesser extent for Fe^+:MgO.

We refer the reader to the article by Ham (1969) for the discussion of a recent explanation by the Jahn–Teller effect of the puzzling spectrum of Ti^{3+} in Al_2O_3 (see also the end of § 7.8).

APPENDIX A

THERMAL AND MAGNETIC PROPERTIES OF A PARAMAGNETIC SUBSTANCE

ALTHOUGH we have not been concerned in this book with the bulk properties (thermal and magnetic) of paramagnetic substances, it may be convenient to the reader to have a summary of the relations between these quantities and the parameters of the spin Hamiltonian. We give first the fundamental relations, from statistical mechanics and thermodynamics, from which the entropy and magnetization can be calculated generally, and then proceed to a number of useful formulae that are valid in a high-temperature approximation.

For a system with a set of energy levels W_i, the magnetic and thermal properties are calculated from the partition function

$$Z = \sum_i \exp(-W_i/kT) \tag{A.1}$$

by means of the thermodynamic potential

$$G = U - TS - HM \tag{A.2}$$

which is connected with the partition function Z by the relation

$$G = -NkT \ln Z = -RT \ln Z \tag{A.3}$$

in which we have assumed that we are dealing with a gram mole of substance. From the second law of thermodynamics,

$$T\, dS = dQ = dU - H\, dM, \tag{A.4}$$

we have

$$dG = -S\, dT - M\, dH \tag{A.5}$$

whence

$$(\partial G/\partial T)_H = -S \tag{A.6}$$

and

$$(\partial G/\partial H)_T = -M. \tag{A.7}$$

The entropy and magnetization are therefore related to the partition function by

$$S = R\frac{\partial}{\partial T_H}(T \ln Z) \tag{A.8}$$

and

$$M = RT\frac{\partial}{\partial H_T}(\ln Z), \tag{A.9}$$

APPENDIX A—THERMAL AND MAGNETIC PROPERTIES 849

which of course satisfy the thermodynamic relation

$$(\partial S/\partial H)_T = (\partial M/\partial T)_H \qquad (A.10)$$

that follows from the fact that dG is a perfect differential.

If the energy levels W_i are known, the thermodynamic and magnetic properties can be calculated from these formulae. For example, we can find how the internal energy U varies with the field H, since we have from the second law of thermodynamics (A.4) and eqn (A.10),

$$\left(\frac{\partial U}{\partial H}\right)_T = T\left(\frac{\partial S}{\partial H}\right)_T + H\left(\frac{\partial M}{\partial H}\right)_T = T\left(\frac{\partial M}{\partial T}\right)_H + H\left(\frac{\partial M}{\partial H}\right)_T. \qquad (A.11)$$

For an ideal paramagnet, which we define as one in which the only interaction energy is the Zeeman energy,

$$W = -\boldsymbol{\mu}\cdot\mathbf{H} = g\beta(\mathbf{H}\cdot\mathbf{S}), \qquad (A.12)$$

the energy levels vary directly as H and the partition function is just a function of (H/T). So also are quantities such as S, M, and if we write $M = f(H/T)$ we find in this case that

$$(\partial U/\partial H)_T = T(-Hf'/T^2) + H(f'/T) = 0, \qquad (A.13)$$

showing that the internal energy U is a function only of T and not of H. It follows that the same is true of the specific heat at constant magnetization, $C_M = (\partial U/\partial T)_M$, for our 'ideal' paramagnetic. For many purposes we need to know the specific heat at constant field,

$$C_H = T(\partial S/\partial T)_H, \qquad (A.14)$$

which is related to C_M, from eqn (A.4), by

$$C_H = C_M - H(\partial M/\partial T)_H. \qquad (A.15)$$

At high temperatures our ideal paramagnet obeys Curie's law

$$M/H = \frac{Ng^2\beta^2 S(S+1)}{3kT} \qquad (A.16)$$

from which we find

$$C_H = C_M + \frac{Rg^2\beta^2 S(S+1)H^2}{3(kT)^2}. \qquad (A.17)$$

We shall find that a similar relation holds in the high-temperature approximation for paramagnetic substances that are not 'ideal' in the sense in which we have defined it.

High-temperature approximation

When we are mainly concerned with a region of temperature in which the differences in the energy levels W_i are small compared with kT, we may use a power series in inverse powers of T. For the partition function Z, expansion of the exponentials gives

$$Z/m = 1 - \langle W \rangle / kT + \langle W^2 \rangle / 2(kT)^2 - \ldots$$
$$= 1 + \sum_{p=1}^{\infty} b_p T^{-p}, \qquad (A.18)$$

in which m is the total number of levels W_i and the coefficient

$$b_p = \frac{(-1)^p \langle (W/k)^p \rangle}{p!} \qquad (A.19)$$

where $\langle (W/k)^p \rangle$ is the average value of the pth powers of the levels measured in °K. The series expansion for $\ln Z$ is then

$$\ln Z = \ln m - \sum_{n=1}^{\infty} a_n T^{-n} \qquad (A.20)$$

where

$$a_n = b_n - \tfrac{1}{2} \sum_{p=1}^{n-1} b_{n-p} b_p + \tfrac{1}{3} \sum_{p=1}^{n-2} \sum_{q=1}^{n-2} b_{n-p-q} b_p b_q + \text{etc.} \qquad (A.21)$$

From our thermodynamic formulae we now obtain

$$S/R = \ln m - \sum_{n=2}^{\infty} (n-1) a_n T^{-n}, \qquad (A.22)$$

$$C/R = \sum_{n=2}^{\infty} n(n-1) a_n T^{-n}, \qquad (A.23)$$

and

$$M/R = \sum_{n=1}^{\infty} (\partial a_n / \partial H)_T T^{-(n-1)}. \qquad (A.24)$$

Considerable simplification of these formulae is obtained if the origin of energy is chosen to make $\langle W \rangle = 0$, in which case the leading terms in the expansions for C and M become

$$C/R = \langle (W/k)^2 \rangle T^{-2} - \langle (W/k)^3 \rangle T^{-3} + \ldots \qquad (A.25)$$

$$M/N = \frac{1}{2kT} \left(\frac{\partial \langle W^2 \rangle}{\partial H} \right)_T - \frac{1}{6(kT)^2} \left(\frac{\partial \langle W^3 \rangle}{\partial H} \right)_T + \ldots \qquad (A.26)$$

In these formulae C is the specific heat at constant field C_H since we assume H to be constant in evaluating quantities such as $\langle W^n \rangle$.

System of non-interacting ions

We now relate the susceptibility and specific heat to the spin Hamiltonian for a system of non-interacting ions, by which we mean ions subjected to ligand fields but which do not interact with other paramagnetic ions, i.e. we neglect spin-spin interaction. The effect of the ligand field on the magnetic properties and on hyperfine structure is expressed in the spin Hamiltonian

$$\mathcal{H} = \beta(g_x H_x S_x + g_y H_y S_y + g_z H_z S_z) +$$
$$+ D_x S_x^2 + D_y S_y^2 + D_z S_z^2 +$$
$$+ A_x S_x I_x + A_y S_y I_y + A_z S_z I_z +$$
$$+ P_x I_x^2 + P_y I_y^2 + P_z I_z^2 -$$
$$- g_n \beta_n (\mathbf{H} \cdot \mathbf{I}). \tag{A.27}$$

This is not the most general type of spin Hamiltonian but is the one encountered in most cases of interest. The principal axes of the various 'tensor' interactions are all the same, and coincide with the (x, y, z) axes. Electronic multipole terms of higher than the second degree have been omitted; so has any 'pseudo-nuclear' Zeeman interaction. To satisfy the condition $\langle W \rangle = 0$, we impose the conditions

$$D_x + D_y + D_z = 0; \quad P_x + P_y + P_z = 0. \tag{A.28}$$

For the other terms, $\langle W \rangle = 0$ because components of S, I occur equally often with positive and negative signs.

The advantage of the high-temperature expansions (A.25) and (A.26) is that we do not have to evaluate the energy levels but can use the theorem of the invariance of the spur of a matrix. Thus to find the value of $\langle W^2 \rangle$ we square the Hamiltonian and pick out only the diagonal terms; on summing these over all possible values, many of the sums vanish because positive and negative values occur with equal probability. For example $\sum_{-S}^{+S} S_z = 0$. This makes the algebra quite manageable for the lowest powers such as $\langle W^2 \rangle$ and $\langle W^3 \rangle$. Again, in evaluating the susceptibility we can obviously neglect all terms that do not involve H.

We begin by neglecting the hyperfine terms, and assuming that H is in an arbitrary direction, whose direction cosines are (l, m, n) with respect to the (x, y, z) axes. The first term in the susceptibility involves $\langle W^2 \rangle$, which is found to be

$$\langle W^2 \rangle = \tfrac{1}{3} S(S+1)\beta^2 H^2(l^2 g_x^2 + m^2 g_y^2 + n^2 g_z^2) + \tag{A.29}$$
$$+ \tfrac{1}{30}(D_x^2 + D_y^2 + D_z^2) S(S+1)(2S-1)(2S+3). \tag{A.30}$$

Hence on writing
$$g^2 = l^2 g_x^2 + m^2 g_y^2 + n^2 g_z^2 \quad (A.31)$$
we find for the first term in the susceptibility
$$\chi = \frac{M}{H} = \frac{Ng^2 S(S+1)\beta^2}{3kT}, \quad (A.32)$$
which is just the usual expression for Curie's law. The additional terms in D make no contribution in this approximation because they are independent of H. A similar argument applies to the hyperfine terms except for the nuclear Zeeman interaction, so that their presence only adds to the T^{-1} term in the susceptibility an additional amount
$$\chi_n = \frac{Ng_n^2 I(I+1)\beta_n^2}{3kT} \quad (A.33)$$
which can almost invariably be neglected in comparison with (A.32), being smaller by a factor of order $(\beta_n/\beta)^2 \approx 10^{-6}$.

On continuing the calculation to find the term in T^{-2} in the susceptibility we obtain as the only contribution
$$-\frac{N\beta^2}{30(kT)^2}(l^2 g_x^2 D_x + m^2 g_y^2 D_y + n^2 g_z^2 D_z)S(S+1)(2S-1)(2S+3) \quad (A.34)$$
apart from a similar term of order $P\beta_n^2$ which we can neglect for most purposes. Thus the effect of the hyperfine terms vanishes for practical purposes both in the T^{-1} and T^{-2} terms in the susceptibility.

Since $D_x + D_y + D_z = 0$, the term in T^{-2} given in (A.34) will change sign with direction. A further result is that if g is isotropic, i.e. if $g_x = g_y = g_z$, the susceptibility of a powder sample will contain no term in T^{-2} because the mean value of $(l^2 g_x^2 D_x + m^2 g_y^2 D_y + n^2 g_z^2 D_z)$ becomes just $\tfrac{1}{3}g^2(D_x + D_y + D_z) = 0$.

In giving the formulae for the specific heat we shall restrict ourselves to terms in T^{-2}, which for the Hamiltonian (A.27) are
$$\frac{CT^2}{R} = \frac{1}{3k^2}g^2\beta^2 H^2 S(S+1) + \quad (A.35)$$
$$+ \frac{1}{30k^2}(D_x^2 + D_y^2 + D_z^2)S(S+1)(2S-1)(2S+3) + \quad (A.36)$$
$$+ \frac{1}{9k^2}(A_x^2 + A_y^2 + A_z^2)S(S+1)I(I+1) + \quad (A.37)$$
$$+ \frac{1}{30k^2}(P_x^2 + P_y^2 + P_z^2)I(I+1)(2I-1)(2I+3) + \quad (A.38)$$
$$+ \frac{1}{3k^2}g_n^2\beta_n^2 H^2 I(I+1). \quad (A.39)$$

This specific heat is that measured at constant field C_H, and we can find the specific heat at constant magnetization C_M using eqn (A.15). From eqns (A.32), (A.33) we have

$$-H\left(\frac{\partial M}{\partial T}\right)_H = \frac{RH^2}{3(kT)^2}\{g^2\beta^2 S(S+1)+g_n^2\beta_n^2 I(I+1)\}, \quad (A.40)$$

which just corresponds to the first and last terms (A.35), (A.39). It follows that C_M is just equal to the sum of the three middle terms (A.36), (A.37), (A.38). Thus in this approximation the substance behaves like an 'ideal' paramagnet, since by (A.32), (A.33) M is just a function of (H/T), and the internal energy U and its derivative $C_M = (\partial U/\partial T)_M$ are independent of H, just as in eqn (A.13). This is not necessarily true of terms of higher order (T^{-3}, etc.) in the specific heat, since it is obvious already for eqn (A.34) that the magnetization is no longer simply a function of (H/T).

Effect of spin-spin interaction

We assume that the spin-spin interaction is of the form $(\mathbf{S}_i \cdot \mathcal{J}_{ij} \cdot \mathbf{S}_j)$ and that the principal axes of the tensor $\mathcal{J}_{ij}$ coincide with those of the g-tensor. If an external field is applied along the z-axis (one of the principal axes), the Hamiltonian for a set of interacting spins is similar to eqn (9.46), viz.

$$\mathscr{H} = g_z\beta H_z \sum_i S_{iz} + \sum_{j>i}(\mathbf{S}_i \cdot \mathcal{J}_{ij} \cdot \mathbf{S}_j). \quad (A.41)$$

To find the leading term in the effect of the spin-spin interaction on the susceptibility we need the contributions to $\langle W^2 \rangle$, $\langle W^3 \rangle$ which contain H^2. From (A.41) we find

$$\langle W^2 \rangle = \tfrac{1}{3}g_z^2\beta^2 H_z^2 S(S+1) + \tfrac{1}{2}(\tfrac{1}{3}S(S+1))^2 \sum_{j>i}(\mathcal{J}_{xx}^2 + \mathcal{J}_{yy}^2 + \mathcal{J}_{zz}^2)_{ij}$$
$$(A.42)$$

in which the first term is just that given by (A.29) taking H along the z-axis. The second term does not contain H^2, so that the term in T^{-1} in the susceptibility is not altered by the spin-spin interaction. The relevant term in H^2 contained in $\langle W^3 \rangle$ gives a term in T^{-2} for the susceptibility along the z-axis amounting to

$$-\frac{Ng_z^2\beta^2 S^2(S+1)^2}{9(kT)^2}\sum_{j>i}(\mathcal{J}_{zz})_{ij} \quad (A.43)$$

and analogous terms along x, y are found by a simple change of subscripts. We make the following comments on this result.

(a) As far as terms in T^{-2} are concerned, the result is the same as that given by the molecular field approximation. For, if we write

$$M = \frac{c}{T}(H+\lambda M),$$

we have

$$\chi = \frac{M}{H} = \frac{c}{T\{1-(\lambda c/T)\}} = \frac{c}{T} + \frac{\lambda c^2}{T^2} + \ldots, \quad (A.44)$$

so that the term in T^{-2} in the susceptibility along the z-axis is

$$\frac{\lambda c^2}{T^2} = \lambda \left\{ \frac{Ng_z^2\beta^2 S(S+1)}{3kT} \right\}^2. \quad (A.45)$$

In the molecular field approximation we are replacing $\sum_{j>i}(\mathscr{I}_{zz})_{ij}S_{iz}S_{jz}$ by

$$S_{iz}\langle S_{jz}\rangle \sum_{j>i}(\mathscr{I}_{zz})_{ij} = S_{iz}(-M_z/Ng_z\beta)\sum_{j>i}(\mathscr{I}_{zz})_{ij}.$$

This is equivalent to a term $g_{iz}\beta S_{iz}(\lambda M_z)$, so that

$$\lambda = -\frac{1}{Ng_z^2\beta^2}\sum_{j>i}(\mathscr{I}_{zz})_{ij}$$

and

$$\frac{\lambda c^2}{T^2} = -\frac{Ng_z^2\beta^2 S^2(S+1)^2}{9(kT)^2}\sum_{j>i}(\mathscr{I}_{zz})_{ij},$$

which is just our previous result (A.43).

(b) The summation $\sum(\mathscr{I}_{zz})_{ij}$ is not necessarily convergent; for dipolar interaction the shape of the sample is involved, since the 'demagnetizing field' is included.

(c) If g is isotropic (and not otherwise), then for a powder only the isotropic part of the spin-spin interaction contributes to the term in T^{-2}, since (cf. eqns (9.46)–(9.48)) we have

$$\tfrac{1}{3}(\mathscr{I}_{xx}+\mathscr{I}_{yy}+\mathscr{I}_{zz})_{ij} = \mathscr{I}_{ij} + \tfrac{1}{3}(\mathscr{I}'_{xx}+\mathscr{I}'_{yy}+\mathscr{I}'_{zz})_{ij} = \mathscr{I}_{ij}.$$

(d) The term in T^{-2} from spin-spin interaction is similar in form to that due to an electronic quadrupole splitting $D_xS_x^2+D_yS_y^2+D_zS_z^2$, which along the z-axis gives, from eqn (A.34),

$$-\frac{Ng_z^2\beta^2}{30(kT)^2}D_z S(S+1)(2S-1)(2S+3).$$

The two terms are additive, so that for $S \geq 1$ we cannot use the T^{-2} term to determine uniquely either contribution except in special circumstances such as cubic symmetry, for which D vanishes, or if g is isotropic, when for a powder only the isotropic spin-spin parameter $\mathscr{J}_{ij}$ contributes to the term in T^{-2}.

The contribution to the specific heat of order T^{-2} comes immediately from $\langle W^2 \rangle$, and we find the extra term to be

$$\frac{CT^2}{R} = \frac{1}{2}\left\{\frac{S(S+1)}{3k}\right\}^2 \sum_{j>i}(\mathscr{J}_{xx}^2+\mathscr{J}_{yy}^2+\mathscr{J}_{zz}^2)_{ij} \qquad (A.46)$$

which is additive to the other terms in (A.35)–(A.39). We can separate the contributions from isotropic and anisotropic exchange, since

$$\mathscr{J}_{xx}^2+\mathscr{J}_{yy}^2+\mathscr{J}_{zz}^2 = 3\mathscr{J}^2+(\mathscr{J}_{xx}'^2+\mathscr{J}_{yy}'^2+\mathscr{J}_{zz}'^2)$$

but we cannot separate the two contributions from anisotropic exchange and dipolar interaction, since they are similar in form.

The specific heat contribution from spin-spin interaction can be assigned to a mean square internal field $\langle H_i^2 \rangle$ by writing it in the form

$$\frac{CT^2}{R} = \frac{1}{2}\frac{g^2\beta^2\langle H_i^2\rangle S(S+1)}{3k^2},$$

which is similar to (A.35). Clearly

$$g^2\beta^2\langle H_i^2\rangle = \tfrac{1}{3}S(S+1)\sum_{j>i}\{\mathscr{J}_{xx}^2+\mathscr{J}_{yy}^2+\mathscr{J}_{zz}^2\}_{ij} \qquad (A.47)$$

and, using the Curie constant $c = Ng^2\beta^2 S(S+1)/3k$, we have

$$C_H = \frac{c}{T^2}(H^2+\tfrac{1}{2}\langle H_i^2\rangle). \qquad (A.48)$$

The factor $\tfrac{1}{2}$ which appears in the second term of (A.42) and subsequent terms in the specific heat arises because spin-spin interaction represents mutual energy, which must not be counted twice in the summations.

More elaborate calculations and further formulae can be found in Van Vleck (1937), with a correction in Joseph and Van Vleck (1960), and Daniels (1953).

APPENDIX B

Some General Tables

Table 1

Character table for the twenty-four proper rotations of the cubic group O

Representation		Classes			
	E	$(3)C_2$	$(6)C_4$	$(6)C'_2$	$(6)C_3$
A_1, Γ_1	1	1	1	1	1
A_2, Γ_2	1	1	-1	-1	1
E, Γ_3	2	2	0	0	-1
T_1, Γ_4	3	-1	1	-1	0
T_2, Γ_5	3	-1	-1	1	0

Table 2

Decomposition into irreducible representations of the double direct products $\Gamma_i \times \Gamma_j$ of the cubic group O. For the diagonal products $\Gamma_i \times \Gamma_i$ the superscript S or A indicates whether a representation belongs to the symmetrical or antisymmetrical product

	Γ_1	Γ_2	Γ_3	Γ_4	Γ_5
Γ_1	Γ_1^S	Γ_2	Γ_3	Γ_4	Γ_5
Γ_2	Γ_2	Γ_1^S	Γ_3	Γ_5	Γ_4
Γ_3	Γ_3	Γ_3	$\Gamma_1^S+\Gamma_2^A+\Gamma_3^S$	$\Gamma_4+\Gamma_5$	$\Gamma_4+\Gamma_5$
Γ_4	Γ_4	Γ_5	$\Gamma_4+\Gamma_5$	$\Gamma_1^S+\Gamma_3^S+\Gamma_4^A+\Gamma_5^S$	$\Gamma_2+\Gamma_3+\Gamma_4+\Gamma_5$
Γ_5	Γ_5	Γ_4	$\Gamma_4+\Gamma_5$	$\Gamma_2+\Gamma_3+\Gamma_4+\Gamma_5$	$\Gamma_1^S+\Gamma_3^S+\Gamma_4^A+\Gamma_5^S$

Table 3

Irreducible representations of the cubic group O for integral values of J from 0 to 10

J	Irreducible representations
0	Γ_1
1	Γ_4
2	$\Gamma_3+\Gamma_5$
3	$\Gamma_2+\Gamma_4+\Gamma_5$
4	$\Gamma_1+\Gamma_3+\Gamma_4+\Gamma_5$
5	$\Gamma_3+2\Gamma_4+\Gamma_5$
6	$\Gamma_1+\Gamma_2+\Gamma_3+\Gamma_4+2\Gamma_5$
7	$\Gamma_2+\Gamma_3+2\Gamma_4+2\Gamma_5$
8	$\Gamma_1+2\Gamma_3+2\Gamma_4+2\Gamma_5$
9	$\Gamma_1+\Gamma_2+\Gamma_3+3\Gamma_4+2\Gamma_5$
10	$\Gamma_1+\Gamma_2+2\Gamma_3+2\Gamma_4+3\Gamma_5$

APPENDIX B—SOME GENERAL TABLES

TABLE 4

States corresponding to the decomposition of the cubic representation D^J for $J = 1$ to 4 inclusive. The constant α is the proportionality coefficient between $\mathbf{J}$ and the fictitious momentum $\tilde{\mathbf{I}}$, i.e. $\mathbf{J} = \alpha\tilde{\mathbf{I}}$

$$J = 1 \quad \begin{array}{c} \Gamma_4 \\ \alpha = 1 \end{array} \begin{cases} |\tilde{1}\rangle = |1\rangle \\ |\tilde{0}\rangle = |0\rangle \\ |-\tilde{1}\rangle = |-1\rangle \end{cases}$$

$$J = 2 \begin{cases} \begin{array}{c} \Gamma_5 \\ \alpha = -1 \end{array} \begin{cases} |\tilde{1}\rangle = |-1\rangle \\ |\tilde{0}\rangle = \dfrac{1}{\sqrt{2}}\{|2\rangle - |-2\rangle\} \\ |-\tilde{1}\rangle = -|+1\rangle \end{cases} \\ \Gamma_3 \begin{cases} \theta = |0\rangle \\ \varepsilon = \dfrac{1}{\sqrt{2}}\{|2\rangle + |-2\rangle\} \end{cases} \end{cases}$$

$$J = 3 \begin{cases} \Gamma_2 \quad \left\{ \dfrac{1}{\sqrt{2}}\{|2\rangle - |-2\rangle\} \right\} \\ \begin{array}{c} \Gamma_4 \\ \alpha = -\tfrac{3}{2} \end{array} \begin{cases} |\tilde{1}\rangle = \sqrt{(\tfrac{5}{8})}\,|-3\rangle + \sqrt{(\tfrac{3}{8})}\,|1\rangle \\ |\tilde{0}\rangle = -|0\rangle \\ |-\tilde{1}\rangle = \sqrt{(\tfrac{5}{8})}\,|3\rangle + \sqrt{(\tfrac{3}{8})}\,|-1\rangle \end{cases} \\ \begin{array}{c} \Gamma_5 \\ \alpha = +\tfrac{1}{2} \end{array} \begin{cases} |\tilde{1}\rangle = \sqrt{(\tfrac{5}{8})}\,|-1\rangle - \sqrt{(\tfrac{3}{8})}\,|3\rangle \\ |\tilde{0}\rangle = \dfrac{1}{\sqrt{2}}\{|2\rangle + |-2\rangle\} \\ |-\tilde{1}\rangle = \sqrt{(\tfrac{5}{8})}\,|1\rangle - \sqrt{(\tfrac{3}{8})}\,|-3\rangle \end{cases} \end{cases}$$

$$J = 4 \begin{cases} \Gamma_1 \quad \left\{ \dfrac{1}{\sqrt{24}}\{\sqrt{14}\,|0\rangle + \sqrt{5}\,|4\rangle + \sqrt{5}\,|-4\rangle\} \right\} \\ \Gamma_3 \begin{cases} \theta = \dfrac{1}{\sqrt{24}}\{-\sqrt{10}\,|0\rangle + \sqrt{7}\,|4\rangle + \sqrt{7}\,|-4\rangle\} \\ \varepsilon = \dfrac{1}{\sqrt{2}}\{|2\rangle + |-2\rangle\} \end{cases} \\ \begin{array}{c} \Gamma_4 \\ \alpha = \tfrac{1}{2} \end{array} \begin{cases} |\tilde{1}\rangle = -\dfrac{1}{\sqrt{8}}\{|-3\rangle + \sqrt{7}\,|1\rangle\} \\ |\tilde{0}\rangle = \dfrac{1}{\sqrt{2}}\{|4\rangle - |-4\rangle\} \\ |-\tilde{1}\rangle = \dfrac{1}{\sqrt{8}}\{|3\rangle + \sqrt{7}\,|-1\rangle\} \end{cases} \\ \begin{array}{c} \Gamma_5 \\ \alpha = \tfrac{5}{2} \end{array} \begin{cases} |\tilde{1}\rangle = \dfrac{1}{\sqrt{8}}\{\sqrt{7}\,|3\rangle - |-1\rangle\} \\ |\tilde{0}\rangle = \dfrac{1}{\sqrt{2}}\{|2\rangle - |-2\rangle\} \\ |-\tilde{1}\rangle = -\dfrac{1}{\sqrt{8}}\{\sqrt{7}\,|-3\rangle - |1\rangle\} \end{cases} \end{cases}$$

Table 5

States for the cubic representations Γ_4 and Γ_5 contained in $J = 2$ and $J = 3$, taking a threefold axis of the cube as axis of quantization. They differ from those given in Figs. 7.4 and 7.6, being equivalent to those derived from a spin Hamiltonian $O_4^0 - 20\sqrt{(2)}O_4^3$ instead of $O_4^0 + 20\sqrt{(2)}O_4^3$: either representation is valid (though care must be exercised in taking the correct set of states for each spin Hamiltonian), since they differ only by a rotation of the x, y axes about the z-axis through an angle π

$J = 2$, Γ_5, $\alpha = -1$:
$$|\tilde{1}\rangle_T = -\frac{1}{\sqrt{3}}\{\sqrt{(2)}|-2\rangle + |1\rangle\}$$
$$|\tilde{0}\rangle_T = |0\rangle$$
$$|-\tilde{1}\rangle_T = \frac{1}{\sqrt{3}}\{\sqrt{(2)}|2\rangle - |-1\rangle\}$$

$J = 3$, Γ_4, $\alpha = -\frac{3}{2}$:
$$|\tilde{1}\rangle_T = \frac{1}{\sqrt{6}}\{-\sqrt{(5)}|-2\rangle + |1\rangle\}$$
$$|\tilde{0}\rangle_T = -\tfrac{2}{3}|0\rangle - \tfrac{1}{3}\sqrt{(\tfrac{5}{2})}\{|3\rangle - |-3\rangle\}$$
$$|-\tilde{1}\rangle_T = \frac{1}{\sqrt{6}}\{\sqrt{(5)}|2\rangle + |-1\rangle\}$$

$J = 3$, Γ_5, $\alpha = +\frac{1}{2}$:
$$|\tilde{1}\rangle_T = \frac{1}{\sqrt{6}}\{|-2\rangle + \sqrt{(5)}|1\rangle\}$$
$$|\tilde{0}\rangle_T = \frac{|3\rangle + |-3\rangle}{\sqrt{2}}$$
$$|-\tilde{1}\rangle_T = \frac{1}{\sqrt{6}}\{|2\rangle - \sqrt{(5)}|-1\rangle\}$$

Table 6

Character table for the extra representations of the double cubic group O^+. The characters for the representations Γ_1 to Γ_5 are the same as those given in Table I

Representation	Classes							
	E	R	C_2	C_2'	C_3	RC_3	C_4	RC_4
Γ_6, E'	2	-2	0	0	1	-1	$\sqrt{2}$	$-\sqrt{2}$
Γ_7, E''	2	-2	0	0	1	-1	$-\sqrt{2}$	$\sqrt{2}$
Γ_8, U'	4	-4	0	0	-1	1	0	0

Table 7

Irreducible representations of the double cubic group O^+ for half-integral values of J from $\frac{1}{2}$ to $\frac{15}{2}$

J		J	
$\frac{1}{2}$	Γ_6	$\frac{9}{2}$	$\Gamma_6 + 2\Gamma_8$
$\frac{3}{2}$	Γ_8	$\frac{11}{2}$	$\Gamma_6 + \Gamma_7 + 2\Gamma_8$
$\frac{5}{2}$	$\Gamma_7 + \Gamma_8$	$\frac{13}{2}$	$\Gamma_6 + 2\Gamma_7 + 2\Gamma_8$
$\frac{7}{2}$	$\Gamma_6 + \Gamma_7 + \Gamma_8$	$\frac{15}{2}$	$\Gamma_6 + \Gamma_7 + 3\Gamma_8$

TABLE 8

Decomposition into irreducible representations of the direct products for the double cubic group O^+. The superscripts S and A refer to symmetrical and antisymmetrical products

	Γ_1	Γ_2	Γ_3	Γ_4	Γ_5	Γ_6	Γ_7	Γ_8
Γ_1	Γ_1^S	Γ_2	Γ_3	Γ_4	Γ_5	Γ_6	Γ_7	Γ_8
Γ_2		Γ_1^S	Γ_3	Γ_5	Γ_4	Γ_7	Γ_6	Γ_8
Γ_3			$\Gamma_1^S+\Gamma_2^A+\Gamma_3^S$	$\Gamma_4+\Gamma_5$	$\Gamma_4+\Gamma_5$	Γ_8	Γ_8	$\Gamma_6+\Gamma_7+\Gamma_8$
Γ_4				$\Gamma_1^S+\Gamma_3^S+\Gamma_4^A+\Gamma_5^S$	$\Gamma_2+\Gamma_3+\Gamma_4+\Gamma_5$	$\Gamma_6+\Gamma_8$	$\Gamma_7+\Gamma_8$	$\Gamma_6+\Gamma_7+2\Gamma_8$
Γ_5					$\Gamma_1^S+\Gamma_3^S+\Gamma_4^A+\Gamma_5^S$	$\Gamma_7+\Gamma_8$	$\Gamma_6+\Gamma_8$	$\Gamma_6+\Gamma_7+2\Gamma_8$
Γ_6						$\Gamma_1^A+\Gamma_4^S$	$\Gamma_2+\Gamma_5$	$\Gamma_3+\Gamma_4+\Gamma_5$
Γ_7							$\Gamma_1^A+\Gamma_4^S$	$\Gamma_3+\Gamma_4+\Gamma_5$
Γ_8								$\Gamma_1^A+\Gamma_2^S+\Gamma_3^A+2\Gamma_4^S+2\Gamma_5^S+\Gamma_5^A$

TABLE 9

Zero-order wave-functions or states corresponding to the irreducible representations of the double cubic group O^+, for the half-integral values of J that occur as the ground states of paramagnetic ions

$J = \frac{1}{2}$ $\quad \Gamma_6 \quad |\widetilde{\pm\tfrac{1}{2}}\rangle = |\pm\tfrac{1}{2}\rangle$

$J = \frac{3}{2}$ $\quad \Gamma_8 \begin{bmatrix} |\widetilde{\pm\tfrac{3}{2}}\rangle = |\pm\tfrac{3}{2}\rangle \\ |\widetilde{\pm\tfrac{1}{2}}\rangle = |\pm\tfrac{1}{2}\rangle \end{bmatrix}$

$J = \frac{5}{2}$ $\begin{bmatrix} \Gamma_7 \begin{bmatrix} \alpha = \dfrac{1}{\sqrt{6}}\{|\tfrac{5}{2}\rangle - \sqrt{(5)}\,|-\tfrac{3}{2}\rangle\} \\[4pt] \beta = \dfrac{1}{\sqrt{6}}\{|-\tfrac{5}{2}\rangle - \sqrt{(5)}\,|\tfrac{3}{2}\rangle\} \end{bmatrix} \\[30pt] \Gamma_8 \begin{bmatrix} |\widetilde{\tfrac{3}{2}}\rangle = -\dfrac{1}{\sqrt{6}}\{|\tfrac{3}{2}\rangle + \sqrt{(5)}\,|-\tfrac{5}{2}\rangle\} \\ |\widetilde{\tfrac{1}{2}}\rangle = |\tfrac{1}{2}\rangle \\ |-\widetilde{\tfrac{1}{2}}\rangle = -|-\tfrac{1}{2}\rangle \\ |-\widetilde{\tfrac{3}{2}}\rangle = \dfrac{1}{\sqrt{6}}\{|-\tfrac{3}{2}\rangle + \sqrt{(5)}\,|\tfrac{5}{2}\rangle\} \end{bmatrix} \end{bmatrix}$

$J = \frac{7}{2}$ $\begin{bmatrix} \Gamma_6 \begin{bmatrix} |\widetilde{\tfrac{1}{2}}\rangle = \dfrac{1}{\sqrt{12}}\{\sqrt{(7)}\,|\tfrac{1}{2}\rangle + \sqrt{(5)}\,|-\tfrac{7}{2}\rangle\} \\ |-\widetilde{\tfrac{1}{2}}\rangle = -\dfrac{1}{\sqrt{12}}\{\sqrt{(7)}\,|-\tfrac{1}{2}\rangle + \sqrt{(5)}\,|\tfrac{7}{2}\rangle\} \end{bmatrix} \\[30pt] \Gamma_7 \begin{bmatrix} \alpha = \sqrt{(\tfrac{3}{4})}\,|\tfrac{5}{2}\rangle - \tfrac{1}{2}|-\tfrac{3}{2}\rangle \\ \beta = -\sqrt{(\tfrac{3}{4})}\,|-\tfrac{5}{2}\rangle + \tfrac{1}{2}|\tfrac{3}{2}\rangle \end{bmatrix} \\[20pt] \Gamma_8 \begin{bmatrix} |\widetilde{\tfrac{3}{2}}\rangle = \tfrac{1}{2}\{|-\tfrac{5}{2}\rangle + \sqrt{(3)}\,|\tfrac{3}{2}\rangle\} \\ |\widetilde{\tfrac{1}{2}}\rangle = \dfrac{1}{\sqrt{12}}\{\sqrt{(7)}\,|-\tfrac{7}{2}\rangle - \sqrt{(5)}\,|\tfrac{1}{2}\rangle\} \\ |-\widetilde{\tfrac{1}{2}}\rangle = \dfrac{1}{\sqrt{12}}\{\sqrt{(7)}\,|\tfrac{7}{2}\rangle - \sqrt{(5)}\,|-\tfrac{1}{2}\rangle\} \\ |-\widetilde{\tfrac{3}{2}}\rangle = \tfrac{1}{2}\{|\tfrac{5}{2}\rangle + \sqrt{(3)}\,|-\tfrac{3}{2}\rangle\} \end{bmatrix} \end{bmatrix}$

$J = \frac{9}{2}$ $\quad \Gamma_6 \begin{bmatrix} |\widetilde{\tfrac{1}{2}}\rangle = \dfrac{1}{\sqrt{24}}\{3\,|\tfrac{9}{2}\rangle + \sqrt{(14)}\,|\tfrac{1}{2}\rangle + |-\tfrac{7}{2}\rangle\} \\[6pt] |-\widetilde{\tfrac{1}{2}}\rangle = \dfrac{1}{\sqrt{24}}\{3\,|-\tfrac{9}{2}\rangle + \sqrt{(14)}\,|-\tfrac{1}{2}\rangle + |\tfrac{7}{2}\rangle\} \end{bmatrix}$

$J = \frac{15}{2}$ $\begin{bmatrix} \Gamma_6 \begin{bmatrix} |\widetilde{\tfrac{1}{2}}\rangle = \tfrac{1}{8}\{\sqrt{(\tfrac{7}{3})}\,|\tfrac{9}{2}\rangle + \sqrt{(33)}\,|\tfrac{1}{2}\rangle + \sqrt{(7)}\,|-\tfrac{7}{2}\rangle + \sqrt{(\tfrac{6.5}{3})}\,|-\tfrac{15}{2}\rangle\} \\ |-\widetilde{\tfrac{1}{2}}\rangle = -\tfrac{1}{8}\{\sqrt{(\tfrac{7}{3})}\,|-\tfrac{9}{2}\rangle + \sqrt{(33)}\,|-\tfrac{1}{2}\rangle + \sqrt{(7)}\,|\tfrac{7}{2}\rangle + \sqrt{(\tfrac{6.5}{3})}\,|\tfrac{15}{2}\rangle\} \end{bmatrix} \\[25pt] \Gamma_7 \begin{bmatrix} \alpha = \dfrac{1}{8\sqrt{3}}\{\sqrt{(77)}\,|\tfrac{13}{2}\rangle + \sqrt{(65)}\,|\tfrac{5}{2}\rangle - \sqrt{(39)}\,|-\tfrac{3}{2}\rangle - \sqrt{(11)}\,|-\tfrac{11}{2}\rangle\} \\ \beta = -\dfrac{1}{8\sqrt{3}}\{\sqrt{(77)}\,|-\tfrac{13}{2}\rangle + \sqrt{(65)}\,|-\tfrac{5}{2}\rangle - \sqrt{(39)}\,|\tfrac{3}{2}\rangle - \sqrt{(11)}\,|\tfrac{11}{2}\rangle\} \end{bmatrix} \end{bmatrix}$

APPENDIX B—SOME GENERAL TABLES

TABLE 10

Character table for the representations Γ_k^t of the tetragonal group

Representation	E	C_2	$(2)C_4$	$(2)C_2'$	$(2)C_2''$
Γ_1^t	1	1	1	1	1
Γ_2^t	1	1	1	-1	-1
Γ_3^t	1	1	-1	1	-1
Γ_4^t	1	1	-1	-1	1
Γ_5^t	2	-2	0	0	0

TABLE 11

Character table for the representations Γ_k^T of the trigonal group

	E	C_3	C_2'
Γ_1^T	1	1	1
Γ_2^T	1	1	-1
Γ_3^T	2	-1	0

TABLE 12

Irreducible representations of the trigonal group Γ_k^T for integral values of J from 0 to 6

J	D^J
0	Γ_1^T
1	$\Gamma_2^T + \Gamma_3^T$
2	$\Gamma_1^T + 2\Gamma_3^T$
3	$\Gamma_1^T + 2\Gamma_2^T + 2\Gamma_3^T$
4	$2\Gamma_1^T + \Gamma_2^T + 3\Gamma_3^T$
5	$\Gamma_1^T + 2\Gamma_2^T + 4\Gamma_3^T$
6	$3\Gamma_1^T + 2\Gamma_2^T + 4\Gamma_3^T$

TABLE 13

Character table for the extra representations Γ_k^T of the double trigonal group

	E	R	C_3	RC_3	C_2'	RC_2'
Γ_4^T	1	-1	-1	1	i	$-i$
Γ_5^T	1	-1	-1	1	$-i$	i
Γ_6^T	2	-2	1	-1	0	0

Table 14

Irreducible representations Γ_k^T of the double trigonal group for half-integral values of J for $J = \tfrac{1}{2}$ to $\tfrac{15}{2}$. Note that for values of $J \geqslant \tfrac{9}{2}$ we have: $D^J = D^{J-3} + \Gamma_4^T + \Gamma_5^T + 2\Gamma_6^T$

$J = \tfrac{1}{2}$	Γ_6^T
$J = \tfrac{3}{2}$	$\Gamma_4^T + \Gamma_5^T + \Gamma_6^T$
$J = \tfrac{5}{2}$	$\Gamma_4^T + \Gamma_5^T + 2\Gamma_6^T$
$J = \tfrac{7}{2}$	$\Gamma_4^T + \Gamma_5^T + 3\Gamma_6^T$
$J = \tfrac{9}{2}$	$2\Gamma_4^T + 2\Gamma_5^T + 3\Gamma_6^T$
$J = \tfrac{11}{2}$	$2\Gamma_4^T + 2\Gamma_5^T + 4\Gamma_6^T$
$J = \tfrac{13}{2}$	$2\Gamma_4^T + 2\Gamma_5^T + 5\Gamma_6^T$
$J = \tfrac{15}{2}$	$3\Gamma_4^T + 3\Gamma_5^T + 5\Gamma_6^T$

Table 15

Some homogeneous polynomials P_k^q of even degree which occur frequently in electron paramagnetic resonance, together with the relations between the coefficients A_k^q in eqn (16.10) and the coefficients B_k^q in eqn (16.1)

$P_2^0 = 3z^2 - r^2$

$P_4^0 = 35z^4 - 30r^2z^2 + 3r^4$

$P_4^3 = xz(x^2 - 3y^2)$

$P_6^0 = 231z^6 - 315r^2z^4 + 105r^4z^2 - 5r^6$

$P_6^4 = (11z^2 - r^2)(x^4 - 6x^2y^2 + y^4)$

$P_2^2 = x^2 - y^2$

$P_4^2 = (7z^2 - r^2)(x^2 - y^2)$

$P_4^4 = x^4 - 6x^2y^2 + y^4$

$P_6^3 = (11z^2 - 3r^2)(x^2 - 3y^2)xz$

$P_6^6 = x^6 - 15x^4y^2 + 15x^2y^4 - y^6$

$A_2^0 = \dfrac{1}{\sqrt{(2\pi)}} \sqrt{(\tfrac{5}{8})} B_2^0$

$A_4^0 = \dfrac{1}{\sqrt{(2\pi)}} \dfrac{3\sqrt{2}}{16} B_4^0$

$A_4^3 = \dfrac{1}{\sqrt{(2\pi)}} \dfrac{3\sqrt{(70)}}{4} B_4^3$

$A_6^0 = \dfrac{1}{\sqrt{(2\pi)}} \dfrac{\sqrt{(26)}}{32} B_6^0$

$A_6^4 = \dfrac{1}{\sqrt{(2\pi)}} \dfrac{3\sqrt{(13 \times 28)}}{32} B_6^4$

$A_2^2 = \dfrac{1}{\sqrt{(2\pi)}} \sqrt{(\tfrac{15}{4})} B_2^2$

$A_4^2 = \dfrac{1}{\sqrt{(2\pi)}} \dfrac{3\sqrt{5}}{4} B_4^2$

$A_4^4 = \dfrac{1}{\sqrt{(2\pi)}} \dfrac{3\sqrt{(35)}}{8} B_4^4$

$A_6^3 = \dfrac{1}{\sqrt{(2\pi)}} \dfrac{\sqrt{(26 \times 105)}}{16} B_6^3$

$A_6^6 = \dfrac{1}{\sqrt{(2\pi)}} \dfrac{\sqrt{(13 \times 21 \times 22)}}{32} B_6^6$

Table 16

Equivalent operators for the polynomials in Table 15. The notation $\{A, B\}_S$ is used as a shorthand for $(\frac{1}{2})(AB+BA)$. Thus $O_4^3 = \frac{1}{2}\{J_z(J_+^3+J_-^3)\}$ denotes $O_4^3 = \frac{1}{4}\{J_z(J_+^3+J_-^3)+(J_+^3+J_-^3)J_z\}$

$k = 2$
$$\begin{aligned}O_2^0 &= 3J_z^2 - J(J+1) \\ O_2^2 &= \tfrac{1}{2}(J_+^2+J_-^2)\end{aligned}$$

$k = 4$
$$\begin{aligned}O_4^0 &= 35J_z^4 - 30J(J+1)J_z^2 + 25J_z^2 - 6J(J+1) + 3J^2(J+1)^2 \\ O_4^2 &= \tfrac{1}{2}\{(7J_z^2 - J(J+1) - 5)(J_+^2+J_-^2)\}_S \\ O_4^3 &= \tfrac{1}{2}\{J_z(J_+^3+J_-^3)\}_S \\ O_4^4 &= \tfrac{1}{2}(J_+^4+J_-^4)\end{aligned}$$

$k = 6$
$$\begin{aligned}O_6^0 &= 231J_z^6 - 315J(J+1)J_z^4 + 735J_z^4 + 105J^2(J+1)^2J_z^2 - 525J(J+1)J_z^2 + \\ &\quad + 294J_z^2 - 5J^3(J+1)^3 + 40J^2(J+1)^2 - 60J(J+1) \\ O_6^3 &= \tfrac{1}{2}\{(11J_z^3 - 3J(J+1)J_z - 59J_z)(J_+^3+J_-^3)\}_S \\ O_6^4 &= \tfrac{1}{2}\{(11J_z^2 - J(J+1) - 38)(J_+^4+J_-^4)\}_S \\ O_6^6 &= \tfrac{1}{2}(J_+^6+J_-^6)\end{aligned}$$

TABLE 17

Matrix elements of the operators O_k^q in Table 16. The numbers in column F are multiplying factors common to all elements in a row. Rows where all matrix elements are zero have been omitted. Further matrix elements for $O_2^{\pm 1}$, $O_4^{\pm 1}$, $O_6^{\pm 1}$, and $O_6^{\pm 5}$ are listed by Buckmaster (1962)

Matrix elements of $O_2^0 = 3J_z^2 - J(J+1)$

$J_z =$	J	F	$\pm\tfrac{1}{2}$	$\pm\tfrac{3}{2}$	$\pm\tfrac{5}{2}$	$\pm\tfrac{7}{2}$	$\pm\tfrac{9}{2}$	$\pm\tfrac{11}{2}$	$\pm\tfrac{13}{2}$	$\pm\tfrac{15}{2}$
	$\tfrac{3}{2}$	3	−1	1						
	$\tfrac{5}{2}$	2	−4	−1	5					
	$\tfrac{7}{2}$	3	−5	−3	1	7				
	$\tfrac{9}{2}$	6	−4	−3	−1	2	6			
	$\tfrac{11}{2}$	1	−35	−29	−17	1	25	55		
	$\tfrac{13}{2}$	6	−8	−7	−5	−2	2	7	13	
	$\tfrac{15}{2}$	3	−21	−19	−15	−9	−1	9	21	35

$J_z =$	J	F	± 1	± 2	± 3	± 4	± 5	± 6	± 7	± 8
	1	1	1							
	2	3	−1	2						
	3	3	−3	0	5					
	4	1	−17	−8	7	28				
	5	3	−9	−6	−1	6	15			
	6	3	−13	−10	−5	2	11	22		
	7	1	−53	−44	−29	−8	19	52	91	
	8	3	−23	−20	−15	−8	1	12	25	40

APPENDIX B—SOME GENERAL TABLES

Matrix elements of
$$O_4^0 = 35J_z^4 - 30J(J+1)J_z^2 + 25J_z^2 - 6J(J+1) + 3J^2(J+1)^2$$

$J_z =$		$\pm\frac{1}{2}$	$\pm\frac{3}{2}$	$\pm\frac{5}{2}$	$\pm\frac{7}{2}$	$\pm\frac{9}{2}$	$\pm\frac{11}{2}$	$\pm\frac{13}{2}$	$\pm\frac{15}{2}$
J	F								
$\frac{5}{2}$	60	2	−3	1					
$\frac{7}{2}$	60	9	−3	−13	7				
$\frac{9}{2}$	84	18	3	−17	−22	18			
$\frac{11}{2}$	120	28	12	−13	−33	−27	33		
$\frac{13}{2}$	60	108	63	−13	−92	−132	−77	143	
$\frac{15}{2}$	60	189	129	23	−101	−201	−221	−91	273

$J_z =$		± 0	± 1	± 2	± 3	± 4	± 5	± 6	± 7	± 8
J	F									
2	12	0	−4	1						
3	60	6	1	−7	3					
4	60	6	9	−11	−21	14				
5	420	18	4	−1	−6	−6	6			
6	60	6	64	11	−54	−96	−66	99		
7	12	84	621	251	−249	−704	−869	−429	1001	
8	420	756	31	17	−3	−24	−39	−39	−13	52
		36								

TABLE 17 (*continued*)

Matrix elements of

$$O_6^0 = 231J_z^6 - 315J(J+1)J_z^4 + 735J_z^4 + 105J^2(J+1)^2J_z^2 - 525J(J+1)J_z^2 + \\ + 294J_z^2 - 5J^3(J+1)^3 + 40J^2(J+1)^2 - 60J(J+1)$$

J	F	$J_z = \pm\frac{1}{2}$	$\pm\frac{3}{2}$	$\pm\frac{5}{2}$	$\pm\frac{7}{2}$	$\pm\frac{9}{2}$	$\pm\frac{11}{2}$	$\pm\frac{13}{2}$	$\pm\frac{15}{2}$
$\frac{7}{2}$	1260	−5	9	−5	1				
$\frac{9}{2}$	5040	−8	6	10	−11	3			
$\frac{11}{2}$	7560	−20	4	25	11	−31	11		
$\frac{13}{2}$	2160	−200	−25	185	227	−11	−319	143	
$\frac{15}{2}$	13860	−75	−25	45	87	59	−39	−117	65

J	F	$J_z = 0$	± 1	± 2	± 3	± 4	± 5	± 6	± 7	± 8
3	180	−20	15	−6	1					
4	1260	−20	1	22	−17	4				
5	2520	−40	−12	36	29	−48	15			
6	7560	−40	−20	22	43	8	−55	22		
7	3780	−200	−125	50	197	176	−55	−286	143	
8	13860	−120	−85	2	93	128	65	−78	−169	104

APPENDIX B—SOME GENERAL TABLES

Matrix elements of
$$O_2^0 = \tfrac{1}{2}(J_+^2 + J_-^2)$$

J	$\langle \pm 1 \| \mp 1 \rangle$	$\langle \pm 2 \| 0 \rangle$	$\langle \pm 3 \| \mp 1 \rangle$	$\langle \pm 4 \| \pm 2 \rangle$	$\langle \pm 5 \| \pm 3 \rangle$	$\langle \pm 6 \| \pm 4 \rangle$	$\langle \mp 7 \| \pm 5 \rangle$	$\langle \mp 8 \| \pm 6 \rangle$
1	1							
2	3	$\sqrt{6}$						
3	6	$\sqrt{30}$	$\sqrt{15}$					
4	10	$3\sqrt{10}$	$3\sqrt{7}$	$2\sqrt{7}$				
5	15	$\sqrt{210}$	$2\sqrt{42}$	$6\sqrt{3}$	$3\sqrt{5}$			
6	21	$2\sqrt{105}$	$6\sqrt{10}$	$3\sqrt{30}$	$\sqrt{165}$	$\sqrt{66}$		
7	28	$6\sqrt{21}$	$15\sqrt{3}$	$5\sqrt{22}$	$6\sqrt{11}$	$3\sqrt{26}$	$\sqrt{91}$	
8	36	$6\sqrt{35}$	$\sqrt{1155}$	$3\sqrt{110}$	$2\sqrt{195}$	$\sqrt{546}$	$3\sqrt{35}$	$2\sqrt{30}$

J	$\langle \pm\tfrac{3}{2} \| \mp\tfrac{1}{2} \rangle$	$\langle \pm\tfrac{5}{2} \| \pm\tfrac{1}{2} \rangle$	$\langle \mp\tfrac{7}{2} \| \pm\tfrac{3}{2} \rangle$	$\langle \mp\tfrac{9}{2} \| \pm\tfrac{5}{2} \rangle$	$\langle \mp\tfrac{11}{2} \| \pm\tfrac{7}{2} \rangle$	$\langle \mp\tfrac{13}{2} \| \pm\tfrac{9}{2} \rangle$	$\langle \mp\tfrac{15}{2} \| \pm\tfrac{11}{2} \rangle$	
$\tfrac{3}{2}$	$\sqrt{3}$							
$\tfrac{5}{2}$	$3\sqrt{2}$	$\sqrt{10}$						
$\tfrac{7}{2}$	$2\sqrt{15}$	$3\sqrt{5}$	$\sqrt{21}$					
$\tfrac{9}{2}$	$5\sqrt{6}$	$3\sqrt{14}$	$2\sqrt{21}$	6				
$\tfrac{11}{2}$	$3\sqrt{35}$	$2\sqrt{70}$	$6\sqrt{6}$	$3\sqrt{15}$	$\sqrt{55}$			
$\tfrac{13}{2}$	$14\sqrt{3}$	$6\sqrt{15}$	$15\sqrt{2}$	$\sqrt{330}$	$3\sqrt{22}$	$\sqrt{78}$		
$\tfrac{15}{2}$	$12\sqrt{7}$	$3\sqrt{105}$	$5\sqrt{33}$	$2\sqrt{165}$	$6\sqrt{13}$	$\sqrt{273}$	$\sqrt{105}$	

Matrix elements of
$$O_4^2 = \tfrac{1}{4}[\{7J_z^2 - J(J+1) - 5\}(J_+^2 + J_-^2) + (J_+^2 + J_-^2)\{7J_z^2 - J(J+1) - 5\}]$$

J	F	$\langle \pm 1 \| \mp 1 \rangle$	$\langle \pm 2 \| 0 \rangle$	$\langle \pm 3 \| \mp 1 \rangle$	$\langle \mp 4 \| \pm 2 \rangle$	$\langle \pm 5 \| \pm 3 \rangle$	$\langle \mp 6 \| \pm 4 \rangle$	$\langle \mp 7 \| \pm 5 \rangle$	$\langle +8 \| \pm 6 \rangle$
2	3	-4	$\sqrt{6}$						
3	3	-20	$-\sqrt{30}$	$6\sqrt{15}$					
4	3	-60	$-11\sqrt{10}$	$10\sqrt{7}$	$30\sqrt{7}$				
5	21	-20	$-\sqrt{210}$	0	$10\sqrt{3}$	$12\sqrt{5}$			
6	3	-280	$-22\sqrt{105}$	$-24\sqrt{10}$	$23\sqrt{30}$	$24\sqrt{165}$	$45\sqrt{66}$		
7	3	-504	$-94\sqrt{21}$	$-130\sqrt{3}$	$15\sqrt{22}$	$116\sqrt{11}$	$121\sqrt{26}$	$66\sqrt{91}$	
8	7	-360	$-54\sqrt{35}$	$-6\sqrt{1155}$	$-3\sqrt{110}$	$12\sqrt{195}$	$15\sqrt{546}$	$78\sqrt{35}$	$78\sqrt{30}$

J	F	$\langle \pm\tfrac{3}{2} \| =\tfrac{1}{2} \rangle$	$\langle \mp\tfrac{5}{2} \| \pm\tfrac{1}{2} \rangle$	$\langle \mp\tfrac{7}{2} \| \pm\tfrac{3}{2} \rangle$	$\langle \mp\tfrac{9}{2} \| \pm\tfrac{5}{2} \rangle$	$\langle \mp\tfrac{11}{2} \| \pm\tfrac{3}{2} \rangle$	$\langle \mp\tfrac{13}{2} \| \pm\tfrac{9}{2} \rangle$	$\langle \mp\tfrac{15}{2} \| \pm\tfrac{11}{2} \rangle$	
$\tfrac{5}{2}$	3	$-5\sqrt{2}$	$3\sqrt{10}$	$5\sqrt{21}$					
$\tfrac{7}{2}$	6	$-4\sqrt{15}$	$\sqrt{5}$	$2\sqrt{21}$	18				
$\tfrac{9}{2}$	21	$-5\sqrt{6}$	$-\sqrt{14}$	$5\sqrt{6}$	$13\sqrt{15}$	$9\sqrt{55}$			
$\tfrac{11}{2}$	12	$-8\sqrt{35}$	$-3\sqrt{70}$	$-15\sqrt{2}$	$13\sqrt{330}$	$95\sqrt{22}$	$55\sqrt{78}$		
$\tfrac{13}{2}$	3	$-210\sqrt{3}$	$-62\sqrt{15}$	$-15\sqrt{33}$	$8\sqrt{165}$	$80\sqrt{13}$	$25\sqrt{273}$	$39\sqrt{105}$	
$\tfrac{15}{2}$	6	$-120\sqrt{7}$	$-23\sqrt{105}$						

TABLE 17 (continued)

Matrix elements of

$$O_6^2 = \tfrac{1}{4}\{33J_z^4 - 18J_z^2J(J+1) - 123J_z^2 + J^2(J+1)^2 + 10J(J+1) + 102\}(J_+^2 + J_-^2)$$
$$+ (J_+^2 + J_-^2)\{33J_z^4 - 18J_z^2J(J+1) - 123J_z^2 + J^2(J+1)^2 + 10J(J+1) + 102\}]$$

J	F	$\langle \pm 1\|\|\mp 1\rangle$	$\langle \pm 2\|\|0\rangle$	$\langle \pm 3\|\|\pm 1\rangle$	$\langle \pm 4\|\|\pm 2\rangle$	$\langle \pm 5\|\|\pm 3\rangle$	$\langle \pm 6\|\|\pm 4\rangle$	$\langle \pm 7\|\|\pm 5\rangle$	$\langle \pm 8\|\|\mp 6\rangle$
3	24	15	$-2\sqrt{30}$	$\sqrt{15}$					
4	30	84	0	$-36\sqrt{7}$	$24\sqrt{7}$				
5	120	84	$2\sqrt{210}$	$-11\sqrt{42}$	$-42\sqrt{3}$	$42\sqrt{5}$			
6	72	420	$22\sqrt{105}$	$-63\sqrt{10}$	$-84\sqrt{30}$	$-14\sqrt{165}$	$70\sqrt{66}$		
7	1080	70	$10\sqrt{21}$	$-7\sqrt{3}$	$-14\sqrt{22}$	$-21\sqrt{11}$	0	$11\sqrt{91}$	
8	264	630	$78\sqrt{35}$	$\sqrt{1155}$	$-42\sqrt{110}$	$-49\sqrt{195}$	$-20\sqrt{546}$	$39\sqrt{35}$	$182\sqrt{30}$

J	F	$\langle \pm\tfrac{3}{2}\|\|\mp\tfrac{1}{2}\rangle$	$\langle \pm\tfrac{5}{2}\|\|\mp\tfrac{1}{2}\rangle$	$\langle \pm\tfrac{7}{2}\|\|\mp\tfrac{3}{2}\rangle$	$\langle \pm\tfrac{9}{2}\|\|\pm\tfrac{5}{2}\rangle$	$\langle \pm\tfrac{11}{2}\|\|\pm\tfrac{7}{2}\rangle$	$\langle \pm\tfrac{13}{2}\|\|\pm\tfrac{9}{2}\rangle$	$\langle \pm\tfrac{15}{2}\|\|\pm\tfrac{11}{2}\rangle$	
$\tfrac{7}{2}$	24	$7\sqrt{15}$	$-21\sqrt{5}$	$5\sqrt{21}$	21				
$\tfrac{9}{2}$	240	$7\sqrt{6}$	$-3\sqrt{14}$	$-5\sqrt{21}$	$-14\sqrt{15}$	$14\sqrt{55}$			
$\tfrac{11}{2}$	216	$12\sqrt{35}$	$-\sqrt{70}$	$-35\sqrt{6}$	$-4\sqrt{330}$	$-3\sqrt{22}$	$11\sqrt{78}$		
$\tfrac{13}{2}$	720	$35\sqrt{3}$	$3\sqrt{15}$	$-36\sqrt{2}$	$-7\sqrt{165}$	$-21\sqrt{13}$	$\sqrt{273}$	$13\sqrt{105}$	
$\tfrac{15}{2}$	1320	$30\sqrt{7}$	$3\sqrt{105}$	$-7\sqrt{33}$					

APPENDIX B—SOME GENERAL TABLES

Matrix elements of
$$O_4^3 = \tfrac{1}{4}\{J_z(J_+^3+J_-^3)+(J_+^3+J_-^3)J_z\}$$

J	F	$\langle 2\|\|-1\rangle$	$\langle 3\|\|0\rangle$	$\langle 4\|\|1\rangle$	$\langle 5\|\|2\rangle$	$\langle 6\|\|3\rangle$	$\langle 7\|\|4\rangle$	$\langle 8\|\|5\rangle$
2	3	1						
3	3	$\sqrt{10}$	$3\sqrt{5}$					
4	3	$5\sqrt{2}$	$3\sqrt{35}$	$5\sqrt{14}$				
5	3	$5\sqrt{7}$	$6\sqrt{35}$	$10\sqrt{21}$	$7\sqrt{30}$			
6	3	$7\sqrt{10}$	$6\sqrt{105}$	$50\sqrt{3}$	$7\sqrt{165}$	$9\sqrt{55}$		
7	3	$14\sqrt{6}$	$15\sqrt{42}$	$25\sqrt{33}$	$35\sqrt{22}$	$9\sqrt{286}$	$11\sqrt{91}$	
8	3	$6\sqrt{70}$	$3\sqrt{2310}$	$25\sqrt{77}$	$7\sqrt{1430}$	$9\sqrt{910}$	$11\sqrt{455}$	$26\sqrt{35}$

J	F	$\langle \tfrac{5}{2}\|\|-\tfrac{1}{2}\rangle$	$\langle \tfrac{7}{2}\|\|\tfrac{1}{2}\rangle$	$\langle \tfrac{9}{2}\|\|\tfrac{3}{2}\rangle$	$\langle \tfrac{11}{2}\|\|\tfrac{5}{2}\rangle$	$\langle \tfrac{13}{2}\|\|\tfrac{7}{2}\rangle$	$\langle \tfrac{15}{2}\|\|\tfrac{9}{2}\rangle$	
$\tfrac{5}{2}$	3	$\sqrt{10}$						
$\tfrac{7}{2}$	3	$4\sqrt{5}$	$2\sqrt{35}$					
$\tfrac{9}{2}$	3	$5\sqrt{14}$	$8\sqrt{14}$	$6\sqrt{21}$				
$\tfrac{11}{2}$	$12\sqrt{5}$	$\sqrt{14}$	$\sqrt{42}$	$3\sqrt{6}$	$\sqrt{33}$			
$\tfrac{13}{2}$	3	$14\sqrt{15}$	$40\sqrt{6}$	$15\sqrt{66}$	$16\sqrt{55}$	$5\sqrt{286}$		
$\tfrac{15}{2}$	6	$4\sqrt{105}$	$5\sqrt{231}$	$30\sqrt{11}$	$4\sqrt{715}$	$10\sqrt{91}$	$3\sqrt{455}$	

Note that for these matrix elements $\langle -M\,|\,-M'\rangle = -\langle M\,|\,M'\rangle$.

TABLE 17 (continued)

Matrix elements of

$$O_6^3 = \tfrac{1}{4}[(11J_z^3 - 3J_z J(J+1) - 59J_z)(J_+^3 + J_-^3) + (J_+^3 + J_-^3)(11J_z^3 - 3J_z J(J+1) - 59J_z)]$$

J	F	$\langle 2\|-1\rangle$	$\langle 3\|0\rangle$	$\langle 4\|1\rangle$	$\langle 5\|2\rangle$	$\langle 6\|3\rangle$	$\langle 7\|4\rangle$	$\langle 8\|5\rangle$
3	18	$-3\sqrt{10}$	$2\sqrt{5}$					
4	90	$-7\sqrt{2}$	$-2\sqrt{35}$	$4\sqrt{14}$				
5	180	$-6\sqrt{7}$	$-5\sqrt{35}$	$-\sqrt{21}$	$7\sqrt{30}$			
6	36	$-63\sqrt{10}$	$-43\sqrt{105}$	$-175\sqrt{3}$	$14\sqrt{165}$	$84\sqrt{55}$		
7	90	$-70\sqrt{6}$	$-64\sqrt{42}$	$-70\sqrt{33}$	$-21\sqrt{22}$	$21\sqrt{286}$	$66\sqrt{91}$	
8	198	$-18\sqrt{70}$	$-8\sqrt{2310}$	$-50\sqrt{77}$	$-7\sqrt{1430}$	$3\sqrt{910}$	$22\sqrt{455}$	$104\sqrt{35}$

J	F	$\langle \tfrac{5}{2}\|-\tfrac{1}{2}\rangle$	$\langle \tfrac{7}{2}\|\tfrac{1}{2}\rangle$	$\langle \tfrac{9}{2}\|\tfrac{3}{2}\rangle$	$\langle \tfrac{11}{2}\|\tfrac{5}{2}\rangle$	$\langle \tfrac{13}{2}\|\tfrac{7}{2}\rangle$	$\langle \tfrac{15}{2}\|\tfrac{9}{2}\rangle$	
$\tfrac{7}{2}$	36	$-7\sqrt{5}$	$2\sqrt{35}$					
$\tfrac{9}{2}$	360	$-2\sqrt{14}$	$-\sqrt{14}$	$2\sqrt{21}$				
$\tfrac{11}{2}$	$36\sqrt{5}$	$-27\sqrt{14}$	$-16\sqrt{42}$	$7\sqrt{6}$	$28\sqrt{33}$			
$\tfrac{13}{2}$	360	$-14\sqrt{15}$	$-29\sqrt{6}$	$-4\sqrt{66}$	$6\sqrt{55}$	$6\sqrt{286}$		
$\tfrac{15}{2}$	1980	$-2\sqrt{105}$	$-2\sqrt{231}$	$-7\sqrt{11}$	0	$3\sqrt{91}$	$2\sqrt{455}$	

Note that for these matrix elements $\langle -M \mid -M' \rangle = -\langle M \mid M' \rangle$.

APPENDIX B—SOME GENERAL TABLES

Matrix elements of
$$O_4^4 = \tfrac{1}{2}(J_+^4 + J_-^4)$$

J	F	$\langle 2\|-2\rangle$	$\langle \pm 3\|\mp 1\rangle$	$\langle \mp 4\|0\rangle$	$\langle \pm 5\|\mp 1\rangle$	$\langle \pm 6\|\mp 2\rangle$	$\langle \mp 7\|\pm 3\rangle$	$\langle \mp 8\|\pm 4\rangle$
2	12	1						
3	12	5	$\sqrt{15}$					
4	12	15	$5\sqrt{7}$	$\sqrt{70}$				
5	12	35	$5\sqrt{42}$	$3\sqrt{70}$	$\sqrt{210}$			
6	12	70	$21\sqrt{10}$	$15\sqrt{14}$	$5\sqrt{66}$	$3\sqrt{55}$		
7	12	126	$70\sqrt{3}$	$5\sqrt{462}$	$15\sqrt{33}$	$5\sqrt{143}$	$\sqrt{1001}$	
8	12	210	$6\sqrt{1155}$	$15\sqrt{154}$	$5\sqrt{1001}$	$\sqrt{15015}$	$5\sqrt{273}$	$2\sqrt{455}$

J	F	$\langle \pm\tfrac{5}{2}\|\mp\tfrac{3}{2}\rangle$	$\langle \pm\tfrac{7}{2}\|\mp\tfrac{1}{2}\rangle$	$\langle \mp\tfrac{9}{2}\|\pm\tfrac{1}{2}\rangle$	$\langle \mp\tfrac{11}{2}\|\pm\tfrac{3}{2}\rangle$	$\langle \pm\tfrac{13}{2}\|\mp\tfrac{5}{2}\rangle$	$\langle \pm\tfrac{15}{2}\|\mp\tfrac{7}{2}\rangle$
$\tfrac{5}{2}$	12	$\sqrt{5}$					
$\tfrac{7}{2}$	12	$5\sqrt{3}$	$\sqrt{35}$				
$\tfrac{9}{2}$	$12\sqrt{7}$	$5\sqrt{3}$	$5\sqrt{2}$	$3\sqrt{2}$			
$\tfrac{11}{2}$	$12\sqrt{2}$	35	$3\sqrt{105}$	$5\sqrt{21}$	$\sqrt{165}$		
$\tfrac{13}{2}$	12	$42\sqrt{5}$	$35\sqrt{6}$	$15\sqrt{22}$	$15\sqrt{11}$	$\sqrt{715}$	
$\tfrac{15}{2}$	12	$42\sqrt{15}$	$10\sqrt{231}$	$15\sqrt{77}$	$5\sqrt{429}$	$\sqrt{5005}$	$\sqrt{1365}$

TABLE 17 (*continued*)

Matrix elements of
$$O_6^4 = \tfrac{1}{4}\{(11J_z^2-J(J+1)-38)(J_+^4+J_-^4)+(J_+^4+J_-^4)(11J_z^2-J(J+1)-38)\}$$

J	F	$\langle 2\|\|-2\rangle$	$\langle \pm 3\|\|\mp 1\rangle$	$\langle \pm 4\|\|0\rangle$	$\langle \pm 5\|\|\pm 1\rangle$	$\langle \pm 6\|\|\pm 2\rangle$	$\langle \pm 7\|\|\pm 3\rangle$	$\langle \mp 8\|\|\pm 4\rangle$
3	60	-6	$\sqrt{15}$	$2\sqrt{70}$				
4	180	-14	$-\sqrt{7}$	$12\sqrt{70}$	$15\sqrt{210}$			
5	60	-168	$-13\sqrt{42}$	$8\sqrt{14}$	$21\sqrt{66}$	$28\sqrt{55}$		
6	180	-168	$-35\sqrt{10}$	$-6\sqrt{462}$	$147\sqrt{33}$	$126\sqrt{143}$	$45\sqrt{1001}$	
7	60	-1260	$-546\sqrt{3}$	$-6\sqrt{154}$	$3\sqrt{1001}$	$2\sqrt{15015}$	$19\sqrt{273}$	$12\sqrt{455}$
8	660	-252	$-6\sqrt{1155}$					

J	F	$\langle \pm\tfrac{5}{2}\|\|\mp\tfrac{3}{2}\rangle$	$\langle \pm\tfrac{7}{2}\|\|\mp\tfrac{1}{2}\rangle$	$\langle \pm\tfrac{9}{2}\|\|\pm\tfrac{1}{2}\rangle$	$\langle \pm\tfrac{11}{2}\|\|\pm\tfrac{3}{2}\rangle$	$\langle \pm\tfrac{13}{2}\|\|\pm\tfrac{5}{2}\rangle$	$\langle \pm\tfrac{15}{2}\|\|\pm\tfrac{7}{2}\rangle$	
$\tfrac{7}{2}$	60	$-7\sqrt{3}$	$3\sqrt{35}$	$30\sqrt{2}$				
$\tfrac{9}{2}$	$60\sqrt{7}$	$-16\sqrt{3}$	$6\sqrt{2}$	$13\sqrt{21}$	$7\sqrt{165}$			
$\tfrac{11}{2}$	$180\sqrt{2}$	-63	$-\sqrt{105}$	$13\sqrt{22}$	$46\sqrt{11}$	$6\sqrt{715}$		
$\tfrac{13}{2}$	360	$-56\sqrt{5}$	$-21\sqrt{6}$	$3\sqrt{77}$	$7\sqrt{429}$	$3\sqrt{5005}$	$5\sqrt{1365}$	
$\tfrac{15}{2}$	660	$-42\sqrt{15}$	$-6\sqrt{231}$					

Matrix elements of
$$O_6^6 = \tfrac{1}{2}(J_+^6+J_-^6)$$

J	F	$\langle 3\|\|-3\rangle$	$\langle \pm 4\|\|\mp 2\rangle$	$\langle \pm 5\|\|\mp 1\rangle$	$\langle \pm 6\|\|0\rangle$	$\langle \mp 7\|\|\pm 1\rangle$	$\langle \mp 8\|\|\pm 2\rangle$	
3	360	1						
4	360	7	$2\sqrt{7}$					
5	360	28	$14\sqrt{3}$	$\sqrt{210}$				
6	360	84	$14\sqrt{30}$	$7\sqrt{66}$	$2\sqrt{231}$			
7	360	210	$42\sqrt{22}$	$28\sqrt{33}$	$2\sqrt{(21\times 143)}$	$\sqrt{(21\times 143)}$		
8	360	462	$42\sqrt{110}$	$12\sqrt{(13\times 77)}$	$14\sqrt{(3\times 143)}$	$7\sqrt{(13\times 55)}$	$2\sqrt{(14\times 143)}$	

J	F	$\langle \pm\tfrac{7}{2}\|\|\mp\tfrac{5}{2}\rangle$	$\langle \pm\tfrac{9}{2}\|\|\mp\tfrac{3}{2}\rangle$	$\langle \pm\tfrac{11}{2}\|\|\mp\tfrac{1}{2}\rangle$	$\langle \pm\tfrac{13}{2}\|\|\pm\tfrac{1}{2}\rangle$	$\langle \pm\tfrac{15}{2}\|\|\pm\tfrac{3}{2}\rangle$		
$\tfrac{7}{2}$	360	$\sqrt{7}$						
$\tfrac{9}{2}$	360	14	$2\sqrt{21}$					
$\tfrac{11}{2}$	360	$28\sqrt{3}$	$7\sqrt{30}$	$\sqrt{462}$				
$\tfrac{13}{2}$	720	$21\sqrt{10}$	$7\sqrt{66}$	$7\sqrt{33}$	$\sqrt{429}$			
$\tfrac{15}{2}$	$360\sqrt{11}$	$42\sqrt{5}$	84	$4\sqrt{273}$	$7\sqrt{39}$	$\sqrt{(35\times 13)}$		

APPENDIX B—SOME GENERAL TABLES

TABLE 18

Values of α, β, γ for a single electron

$$\langle l \| \alpha \| l \rangle = -\frac{2}{(2l-1)(2l+3)}$$

$$\langle l \| \beta \| l \rangle = +\frac{6}{(2l-1)(2l-3)(2l+3)(2l+5)}$$

$$\langle l \| \gamma \| l \rangle = -\frac{20}{(2l-1)(2l-3)(2l-5)(2l+3)(2l+5)(2l+7)}$$

	$l = 1$	$l = 2$	$l = 3$
$\langle l \| \alpha \| l \rangle$	$-\frac{2}{5}$	$-\frac{2}{21}$	$-\frac{2}{45}$
$\langle l \| \beta \| l \rangle$	—	$+\frac{2}{63}$	$+\frac{2}{11 \cdot 45}$
$\langle l \| \gamma \| l \rangle$	—	—	$-\frac{4}{11 \cdot 13 \cdot 27}$

TABLE 19

Values of $\langle L \| \alpha \| L \rangle$, $\langle L \| \beta \| L \rangle$ for the ground states of the 3d group. General formulae for a ground state obeying Hund's rules are (the upper sign is required for a shell less than half full, and the lower sign for a shell more than half full)

$$\langle L \| \alpha \| L \rangle = \mp \frac{2(2l+1-4S)}{(2l-1)(2l+3)(2L-1)}$$

$$\langle L \| \beta \| L \rangle = \langle L \| \alpha \| L \rangle \frac{3\{3(l-1)(l+2) - 7(l-2S)(l+1-2S)\}}{2(2l-3)(2l+5)(L-1)(2L-3)}$$

		$\langle L \| \alpha \| L \rangle$	$\langle L \| \beta \| L \rangle$
$3d^1(Ti^{3+})$	^2D	$-\frac{2}{21}$	$+\frac{2}{63}$
$3d^2(V^{3+})$	^3F	$-\frac{2}{105}$	$-\frac{2}{315}$
$3d^3(V^{2+}, Cr^{3+})$	^4F	$+\frac{2}{105}$	$+\frac{2}{315}$
$3d^4(Cr^{2+}, Mn^{3+})$	^5D	$+\frac{2}{21}$	$-\frac{2}{63}$
$3d^5(Mn^{2+}, Fe^{3+})$	^6S		
$3d^6(Fe^{2+}, Co^{3+})$	^5D	$-\frac{2}{21}$	$+\frac{2}{63}$
$3d^7(Co^{2+})$	^4F	$-\frac{2}{105}$	$-\frac{2}{315}$
$3d^8(Ni^{2+})$	^3F	$+\frac{2}{105}$	$+\frac{2}{315}$
$3d^9(Cu^{2+})$	^2D	$+\frac{2}{21}$	$-\frac{2}{63}$

TABLE 20
Multiplicative factors for 4f ions

	Ce^{3+} $4f^1\,^2F_{\frac{5}{2}}$	Pr^{3+} $4f^2\,^3H_4$	Nd^{3+} $4f^3\,^4I_{\frac{9}{2}}$	Pm^{3+} $4f^4\,^5I_4$	Sm^{3+} $4f^5\,^6H_{\frac{5}{2}}$	Tb^{3+} $4f^8\,^7F_6$
$\langle J\|\Lambda\|J\rangle$	$\frac{6}{7}$	$\frac{4}{5}$	$\frac{8}{11}$	$\frac{3}{5}$	$\frac{2}{7}$	$\frac{3}{2}$
$\langle J\|\Lambda\|J+1\rangle$	$\frac{1}{7}$	$\frac{\sqrt{66}}{5\cdot 11}$	$\frac{\sqrt{14}}{2\cdot 11}$	$\frac{1}{5}\sqrt{\frac{14}{11}}$	$\frac{\sqrt{30}}{2\cdot 7}$	$\frac{1}{2\sqrt{11}}$
$v=-\xi$	$\frac{-2}{3^2\cdot 5}$	$\frac{-1}{3^3\cdot 5}$	$\frac{-2}{3^3\cdot 5\cdot 11}$	$\frac{1}{2\cdot 3^2\cdot 5\cdot 11}$	$\frac{2}{3^3\cdot 5^2}$	$\frac{1}{3^3\cdot 5}$
$\langle J\|N\|J\rangle$	$\frac{2^4\cdot 3}{5\cdot 7}$	$\frac{2^3\cdot 37}{3^3\cdot 5^2}$	$\frac{2^2\cdot 7\cdot 17}{3\cdot 11^2}$	$\frac{2^5\cdot 7}{3\cdot 5\cdot 11}$	$\frac{2^3\cdot 61}{3^2\cdot 5\cdot 7}$	$\frac{5}{9}$
$\langle J\|N\|J+1\rangle$	$\frac{-1}{14}$	$\frac{-7\sqrt{66}}{2\cdot 3^2\cdot 5^2}$	$\frac{-193\sqrt{14}}{2^2\cdot 3^2\cdot 11^2}$	$\frac{-133}{5\cdot 2^2\cdot 3\cdot 11}\sqrt{\frac{14}{11}}$	$\frac{-19\cdot 23\cdot\sqrt{30}}{4\cdot 3^2\cdot 5^2\cdot 7}$	$\frac{-\sqrt{11}}{2\cdot 3^2}$
$\langle J\|\alpha\|J\rangle$	$\frac{-2}{5\cdot 7}$	$\frac{-2^2\cdot 13}{3^2\cdot 5^2\cdot 11}$	$\frac{-7}{3^2\cdot 11^2}$	$\frac{2\cdot 7}{3\cdot 5\cdot 11^2}$	$\frac{13}{3^2\cdot 5\cdot 7}$	$\frac{-1}{3^2\cdot 11}$
$\langle J\|\beta\|J\rangle$	$\frac{2}{3^2\cdot 5\cdot 7}$	$\frac{-2^2}{3^2\cdot 5\cdot 11^2}$	$\frac{-2^3\cdot 17}{3^3\cdot 11^3\cdot 13}$	$\frac{2^3\cdot 7\cdot 17}{3^3\cdot 5\cdot 11^3\cdot 13}$	$\frac{2\cdot 13}{3^3\cdot 5\cdot 7\cdot 11}$	$\frac{2}{3^3\cdot 5\cdot 11^2}$
$\langle J\|\gamma\|J\rangle$	0	$\frac{2^4\cdot 17}{3^4\cdot 5\cdot 7\cdot 11^2\cdot 13}$	$\frac{-5\cdot 17\cdot 19}{3^3\cdot 7\cdot 11^3\cdot 13^2}$	$\frac{2^3\cdot 17\cdot 19}{3^3\cdot 7\cdot 11^3\cdot 13^2}$	0	$\frac{-1}{3^4\cdot 7\cdot 11^2\cdot 13}$
$\langle J\|\alpha\|J+1\rangle$	$\frac{2^2}{3\cdot 5\cdot 7}$	$\frac{13\sqrt{66}}{3^3\cdot 5^2\cdot 11}$	$\frac{2\sqrt{14}}{11^2\cdot 13}$	$\frac{-1}{3\cdot 5\cdot 11}\sqrt{\frac{14}{13}}$	$\frac{-2^2\cdot 13}{3^3\cdot 7\cdot\sqrt{30}}$	$\frac{1}{3\cdot 5\cdot\sqrt{11}}$
$\langle J\|\beta\|J+1\rangle$	$\frac{-2^3}{3^2\cdot 7\cdot 11}$	$\frac{2^2\cdot\sqrt{66}}{3^3\cdot 7\cdot 11^2}$	$\frac{2^5\cdot 17\cdot\sqrt{14}}{3^3\cdot 7\cdot 11^3\cdot 13}$	$\frac{-2^5\cdot 17}{3^3\cdot 11^2\cdot 13\cdot\sqrt{154}}$	$\frac{-2^4\cdot 5\cdot 13}{3^3\cdot 7\cdot 11^2\cdot\sqrt{30}}$	$\frac{-1}{3^3\cdot 11\cdot\sqrt{11}}$
$\langle J\|\gamma\|J+1\rangle$	$\frac{2^3}{3^2\cdot 11\cdot 13}$	$\frac{-17\sqrt{66}}{3^4\cdot 5\cdot 11^2\cdot 13}$	$\frac{2\cdot 5\cdot 19\cdot\sqrt{14}}{3\cdot 7\cdot 11^3\cdot 13^2}$	$\frac{-2\cdot 17\cdot 19}{3^2\cdot 11^2\cdot 13^2\cdot\sqrt{154}}$	$\frac{2^4\cdot 5\cdot 17}{3^4\cdot 11^2\cdot 13\cdot\sqrt{30}}$	$\frac{1}{3^4\cdot 11\cdot 13\cdot\sqrt{11}}$

APPENDIX B—SOME GENERAL TABLES   875

	Dy³⁺ $4f^9\,{}^6H_{15/2}$	Ho³⁺ $4f^{10}\,{}^5I_8$	Er³⁺ $4f^{11}\,{}^4I_{15/2}$	Tm³⁺ $4f^{12}\,{}^3H_6$	Yb³⁺ $4f^{13}\,{}^2F_{7/2}$
$\langle J\|\Lambda\|J\rangle$	$\frac{4}{3}$	$\frac{5}{4}$	$\frac{6}{5}$	$\frac{7}{6}$	$\frac{8}{7}$
$\langle J\|\Lambda\|J+1\rangle$	$\dfrac{1}{3\sqrt{7}}$	$\dfrac{1}{4\sqrt{5}}$	$\dfrac{2}{5\sqrt{14}}$	$\dfrac{5}{6\sqrt{55}}$	$\dfrac{1}{7}$
$v=-\xi$	2	1	-2	$\dfrac{-1}{3^3\cdot 5}$	$\dfrac{-2}{3^2\cdot 5}$
$\langle J\|N\|J\rangle$	$\dfrac{3^3\cdot 5^2}{2^5}$	$\dfrac{1}{2\cdot 3^2\cdot 5\cdot 11}$	$\dfrac{5\sqrt{14}}{3^3\cdot 5\cdot 11}$	$\dfrac{7}{3^2}$	$\dfrac{2^4}{3\cdot 7}$
$\langle J\|N\|J+1\rangle$	$\dfrac{2^5}{3^2\cdot 5}$	$\dfrac{23}{2\cdot 3\cdot 5}$	$\dfrac{2^4\cdot 11}{3^2\cdot 5^2}$	$\dfrac{-11}{2\cdot 3^2\cdot \sqrt{55}}$	$\dfrac{-1}{14}$
$\langle J\|\alpha\|J\rangle$	$\dfrac{-\sqrt{7}}{2\cdot 3^2}$	$\dfrac{-4}{3\cdot 5\cdot \sqrt{5}}$	$\dfrac{-83}{3^2\cdot 5^2\cdot \sqrt{14}}$	$\dfrac{1}{3^2\cdot 11}$	$\dfrac{2}{3^2\cdot 7}$
$\langle J\|\beta\|J\rangle$	$\dfrac{-2}{3^2\cdot 5\cdot 7}$	$\dfrac{-1}{2\cdot 3^2\cdot 5^2}$	$\dfrac{2^2}{3^2\cdot 5^2\cdot 7}$	$\dfrac{2^3}{3^4\cdot 5\cdot 11^2}$	$\dfrac{-2}{3\cdot 5\cdot 7\cdot 11}$
$\langle J\|\gamma\|J\rangle$	$\dfrac{-2^3}{3^3\cdot 5\cdot 7\cdot 11\cdot 13}$	$\dfrac{-1}{2\cdot 3\cdot 5\cdot 7\cdot 11\cdot 13}$	$\dfrac{2}{3^2\cdot 5\cdot 7\cdot 11\cdot 13}$	$\dfrac{-5}{3^4\cdot 7\cdot 11^2\cdot 13}$	$\dfrac{2^2}{3^3\cdot 7\cdot 11\cdot 13}$
$\langle J\|\alpha\|J+1\rangle$	$\dfrac{2^2\sqrt{7}}{3\cdot 5\cdot 7\cdot 13}$	$\dfrac{-5}{3^3\cdot 7\cdot 11^2\cdot 13^2}$	$\dfrac{-2^2\sqrt{14}}{3\cdot 5^2\cdot 7\cdot 13}$	$\dfrac{-1}{3\cdot 5\cdot \sqrt{55}}$	$\dfrac{-2^2}{3\cdot 5\cdot 7}$
$\langle J\|\beta\|J+1\rangle$	$\dfrac{2^5\sqrt{7}}{3^3\cdot 7\cdot 11^2\cdot 13}$	$\dfrac{1}{2\cdot 3\cdot 5\cdot 7\cdot 11\cdot 13}$	$\dfrac{-2^2\sqrt{14}}{3^2\cdot 7\cdot 11^2\cdot 13}$	$\dfrac{2^2}{3^4\cdot 11\cdot \sqrt{55}}$	$\dfrac{2^3}{3^2\cdot 7\cdot 11}$
$\langle J\|\gamma\|J+1\rangle$	$\dfrac{-2^3\sqrt{7}}{3^4\cdot 11^2\cdot 13^2}$	$\dfrac{5\sqrt{5}}{3^3\cdot 11^2\cdot 13^2}$	$\dfrac{-2^3\sqrt{14}}{3^4\cdot 11^2\cdot 13^2}$	$\dfrac{\sqrt{55}}{3^4\cdot 11^2\cdot 13}$	$\dfrac{-2^3}{3^2\cdot 11\cdot 13}$

TABLE 21

Values of the coefficients m, n defined by eqns (19.47) for orbital triplet states, calculated from the wave-functions of Table 4

Orbital triplet	m	n
$L=2, \Gamma_5$	-3	$+3$
$L=3, \Gamma_4$	$+6$	$-\frac{3}{2}$
$L=3, \Gamma_5$	0	$-\frac{15}{2}$

TABLE 22

Values of g for 'isolated' Γ_6 and Γ_7 Kramers doublets for various rare-earth ions, in cubic symmetry

J	Γ	$g' = g/\Lambda$	Ion	Λ	g
$\frac{5}{2}$	Γ_6	$-\frac{5}{3}$	Ce^{3+}	$\frac{6}{7}$	$-\frac{10}{7}$
$\frac{7}{2}$	Γ_6	$-\frac{7}{3}$	$Yb^{3+}Tm^{2+}$	$\frac{8}{7}$	$-\frac{8}{3}$
	Γ_7	3	$Yb^{3+}Tm^{2+}$	$\frac{8}{7}$	$\frac{24}{7}$
$\frac{9}{2}$	Γ_6	$\frac{11}{3}$	Nd^{3+}	$\frac{8}{11}$	$\frac{8}{3}$
$\frac{15}{2}$	Γ_6	-5	$Er^{3+}Ho^{2+}$	$\frac{6}{5}$	-6
			Dy^{3+}	$\frac{4}{3}$	$-\frac{20}{3}$
	E_7	$\frac{17}{3}$	$Er^{3+}Ho^{2+}$	$\frac{6}{5}$	$\frac{34}{5}$
			Dy^{3+}	$\frac{4}{3}$	$\frac{68}{9}$

APPENDIX B—SOME GENERAL TABLES

TABLE 23

Combinations of central ion orbitals and ligand orbitals for the XY_6 complex. Geometrical illustrations are given in Figs. 7.17 and 7.18

Representations	Central orbitals	Ligand σ-orbitals (s or p)	Ligand π-orbitals
a_{1g}	$4s$	$\frac{1}{\sqrt{6}}(\sigma_1+\ldots+\sigma_6)$	
e_g	$3d\begin{cases}3z^2-r^2 \\ \sqrt{(3)}(x^2-y^2)\end{cases}$	$\frac{1}{\sqrt{12}}(2\sigma_3+2\sigma_6-\sigma_1-\sigma_4-\sigma_2-\sigma_5)$ $\frac{1}{2}(\sigma_1+\sigma_4-\sigma_2-\sigma_5)$	
t_{2g}	$3d\begin{cases}xy \\ yz \\ zx\end{cases}$		$\frac{1}{2}(Y_1-Y_4+X_2-X_5)$ $\frac{1}{2}(Z_2-Z_5+Y_3-Y_6)$ $\frac{1}{2}(X_3-X_6+Z_1-Z_4)$
t_{1u}	$4p\begin{cases}x \\ y \\ z\end{cases}$	$\frac{1}{\sqrt{2}}(\sigma_1-\sigma_4)$ $\frac{1}{\sqrt{2}}(\sigma_2-\sigma_5)$ $\frac{1}{\sqrt{2}}(\sigma_3-\sigma_6)$	$\frac{1}{2}(X_2+X_3+X_5+X_6)$ $\frac{1}{2}(Y_1+Y_3+Y_4+Y_6)$ $\frac{1}{2}(Z_1+Z_2+Z_4+Z_5)$

TABLE 24

Molecular orbital combinations of symmetry t_{2g} and e_g for the XY_6 complex, where X is an ion of a d-electron transition group

$$t_{2g}\begin{cases}\eta_x = \Psi_\xi = iN_t^{-\frac{1}{2}}\left\{d_\xi - \frac{\lambda_t}{2}(Z_2-Z_5+Y_3-Y_6)\right\} \\ \eta_y = \Psi_\eta = iN_t^{-\frac{1}{2}}\left\{d_\eta - \frac{\lambda_t}{2}(X_3-X_6+Z_1-Z_4)\right\} \\ \eta_z = \Psi_\zeta = iN_t^{-\frac{1}{2}}\left\{d_\zeta - \frac{\lambda_t}{2}(Y_1-Y_4+X_2-X_5)\right\}\end{cases}$$

$$e_g\begin{cases}\theta = \Psi_\theta = N_\sigma^{-\frac{1}{2}}\Big[d_\theta - \frac{\lambda_{\sigma s}}{\sqrt{12}}\{2\sigma_{s3}+2\sigma_{s6}-(\sigma_{s1}+\sigma_{s2}+\sigma_{s4}+\sigma_{s5})\} \\ \qquad\qquad - \frac{\lambda_{\sigma p}}{\sqrt{12}}\{2\sigma_{p3}+2\sigma_{p6}-(\sigma_{p1}+\sigma_{p2}+\sigma_{p4}+\sigma_{p5})\}\Big] \\ \varepsilon = \Psi_\varepsilon = N_\sigma^{-\frac{1}{2}}\Big\{d_\varepsilon - \frac{\lambda_{\sigma s}}{2}(\sigma_{s1}+\sigma_{s4}-\sigma_{s2}-\sigma_{s5}) \\ \qquad\qquad - \frac{\lambda_{\sigma p}}{2}(\sigma_{p1}+\sigma_{p4}-\sigma_{p2}-\sigma_{p5})\Big\}\end{cases}$$

where: $d_{\xi,\eta,\zeta}$ transform like yz, zx, xy;
$d_{\theta,\varepsilon}$ transform like $3z^2-r^2$, $\sqrt{(3)}(x^2-y^2)$;
and $N_t = 1 - 2\lambda_t S_t + \lambda_t^2$
$N_\sigma = 1 - 2\lambda_{\sigma s}S_s - 2\lambda_{\sigma p}S_p + \lambda_{\sigma s}^2 + \lambda_{\sigma p}^2$

Table 25

Lowest terms for the iron group.
For $3d^n$ electrons with $n > 5$ the free-ion terms are the same as for $3d^{10-n}$; the sequence in energy of the cubic terms is reversed (for instance for $3d^7$ the ground cubic term is $^4T_1$)

Free ion configuration	Lowest free ion terms of maximum spin	Lowest cubic terms of maximum spin	Strong field configurations
$3d^1$	^2D	$^2T_2$	t_2
$3d^2$	$\begin{bmatrix}^3F\\^3P\end{bmatrix}$	$\begin{bmatrix}^3T_1\\^3T_2\end{bmatrix}$	$(t_2)^2, t_2e$ t_2e
$3d^3$	$\begin{bmatrix}^4F\\^4P\end{bmatrix}$	$^4A_2$ $^4T_2$ $^4T_1$	$(t_2)^3$ $(t_2)^2e$ $(t_2)^2e, t_2e^2$
$3d^4$	^5D	^5E $^5T_2$	$(t_2)^3e$ $(t_2)^2e^2$
$3d^5$	^6S	$^6A_1$	$(t_2)^3e^2$

Table 26

Normal coordinates in the octahedral complex XY_6 which transform according to even representations of O_h

Γ_{1g}	$\{Q_1 = \{(X_1-X_4)+(Y_2-Y_5)+(Z_3-Z_6)\}/\sqrt{6}$
Γ_{3g}	$\begin{cases}Q_2 = \{(X_1-X_4)-(Y_2-Y_5)\}/2\\Q_3 = \{2(Z_3-Z_6)-(X_1-X_4)-(Y_2-Y_5)\}/\sqrt{12}\end{cases}$
Γ_{5g}	$\begin{cases}Q_4 = \{(Z_2-Z_5)+(Y_3-Y_6)\}/2\\Q_5 = \{(X_3-X_6)+(Z_1-Z_4)\}/2\\Q_6 = \{(Y_1-Y_4)+(X_2-X_5)\}/2\end{cases}$

BIBLIOGRAPHY

ABE, H. and SHIMADA, J. (1953) *Phys. Rev.* **90**, 316.
—— —— (1957) *J. phys. Soc. Japan* **12**, 1255.
ABRAGAM, A. (1950) *Phys. Rev.* **79**, 534.
—— (1955) *Phys. Rev.* **98**, 1729.
—— (1961) *The principles of nuclear magnetism.* Clarendon Press, Oxford.
—— and BORGHINI, M. (1964) *Prog. in Low Temp. Phys.* **4**, 384.
—— HOROWITZ, J. and PRYCE, M. H. L. (1955) *Proc. R. Soc.* A **230**, 169.
—— and PROCTOR, W. G. (1958) *C.r. hebd. Séanc. Acad. Sci., Paris* **246**, 2253
—— and PRYCE, M. H. L. (1949) *Nature, Lond.* **163**, 992.
—— —— (1950) *Proc. phys. Soc.* A **63**, 409.
—— —— (1951a) *Proc. R. Soc.* A **205**, 135.
—— —— (1951b) *Proc. R. Soc.* A **206**, 164.
—— —— (1951c) *Proc. R. Soc.* A **206**, 173.
ABRAHAM, M. M., FINCH, C. B., and CLARK, G. W. (1968) *Phys. Rev.* **168**, 933.
—— JUDD, B. R. and WICKMAN, H. H. (1963) *Phys. Rev.* **130**, 611.
—— KEDZIE, R. W. and JEFFRIES, C. D. (1960) *Phys. Rev.* **117**, 1070.
—— WEEKS, R. A., CLARK, G. W., and FINCH, C. B. (1965) *Phys. Rev.* **137**, A 138.
—— —— —— —— (1966) *Phys. Rev.* **148**, 350.
ALBERTSON, W. (1937) *Phys. Rev.* **52**, 644.
ALLEN, S. J. (1968) *Phys. Rev.* **166**, 530.
ANDERSON, P. W. (1951) *Phys. Rev.* **82**, 342.
—— (1959) *Phys. Rev.* **115**, 2.
—— (1963) *Adv. solid St. Phys.* **14**, 99.
—— and WEISS, P. R. (1953) *Rev. mod. Phys.* **25**, 269.
ARMSTRONG, L. and MARRUS, R. (1966) *Phys. Rev.* **144**, 994.
AXE, J. D. and BURNS, G. (1966) *Phys. Rev.* **152**, 331.
—— and DIEKE, G. H. (1962) *J. chem. Phys.* **37**, 2364.
—— STAPLETON, H. J. and JEFFRIES, C. D. (1961) *Phys. Rev.* **121**, 1630.
AYANT, Y. and BELORIZKY, É. (1964) *C.r. hebd. Séanc. Acad. Sci., Paris* **259**, 3748.
—— —— and ROSSET, J. (1962) *J. Phys. Radium, Paris* **23**, 201.
AZARBAYEJANI, G. H. and MERLO, A. L. (1965) *Phys. Rev.* **137**, A 489.

BAGGULEY, D. M. S. (1955) *Proc. R. Soc.* A **228**, 549.
—— and GRIFFITHS, J. H. E. (1948) *Nature, Lond.* **162**, 538.
—— —— (1950) *Proc. R. Soc.* A **201**, 366.
—— and VELLA-COLLEIRO, G. (1969) *J. Phys.* C (*Proc. phys. Soc.*) **2**, 2310.
BAKER, J. M. (1964a) *Phys. Rev.* **136**, A 1341.
—— (1964b) *Phys. Rev.* **136**, A 1633.
—— BIRGENEAU, R. J., HUTCHINGS, M. T., and RILEY, J. D. (1968) *Phys. Rev. Lett.* **21**, 620.
—— BLAKE, W. B. J. and COPLAND, G. M. (1969) *Proc. R. Soc.* A **309**, 119.
—— and BLEANEY, B. (1958) *Proc. R. Soc.* A **245**, 156.
—— —— and BOWERS, K. D. (1956) *Proc. phys. Soc.* B **69**, 1205.

BAKER, J. M., CHADWICK, J. R., GARTON, G., and HURRELL, J. P. (1965) *Proc. R. Soc.* A **286**, 352.
—— COPLAND, G. M. and WANKLYN, B. M. (1969) *J. Phys.* C (*Proc. phys. Soc.*) **2**, 862.
—— and FORD, N. C. (1964) *Phys. Rev.* **136**, A 1692.
—— HAYES, W. and JONES, D. A. (1959) *Proc. phys. Soc.* **73**, 942.
—— —— and O'BRIEN, M. C. M. (1960) *Proc. R. Soc.* A **254**, 273.
—— and HURRELL, J. P. (1963) *Proc. phys. Soc.* **82**, 742.
—— and MAU, A. E. (1967) *Can. J. Phys.* **45**, 403.
—— and RUBINS, R. S. (1961) *Proc. phys. Soc.* **78**, 1353.
—— and WILLIAMS, F. I. B. (1961) *Proc. phys. Soc.* **78**, 1340.
—— —— (1962) *Proc. R. Soc.* A **267**, 283.
BARNES, R. G., MÖSSBAUER, R. L., KANKELEIT, E., and POINDEXTER, J. M. (1964) *Phys. Rev.* **136**, A 175.
BELORIZKY, E., AYANT, Y., DESCAMPS, D., and MERLE D'AUBIGNÉ, Y. (1966) *J. Phys. Radium, Paris* **27**, 313.
BENZIE, R. J. and COOKE, A. H. (1950) *Proc. phys. Soc.* A **63**, 201.
—— —— and WHITLEY, S. (1955) *Proc. R. Soc.* A **232**, 277.
BERSUKER, I. B. (1962) *Zh. éksp. teor. Fiz.* **43**, 1315; *Soviet Phys. JETP* **16**, 933.
—— (1963) *Zh. éksp. teor. Fiz.* **44**, 1239; *Soviet Phys. JETP* **17**, 836.
BESSENT, R. G. and HAYES, W. (1965) *Proc. R. Soc.* A **285**, 430.
—— —— and HODBY, J. W. (1967) *Proc. R. Soc.* A **297**, 376.
BETHE, H. (1929) *Annln. Phys.* **3**, 133.
BIERIG, R. W. and WEBER, M. J. (1963) *Phys. Rev.* **132**, 164.
—— —— and WARSHAW, S. I. (1964) *Phys. Rev.* **134**, A 1504.
BIRGENEAU, R. J. (1967a) *Phys. Rev. Lett.* **19**, 160.
—— (1967b) *Can. J. Phys.* **45**, 3761.
—— HUTCHINGS, M. T., BAKER, J. M., and RILEY, J. D. (1969) *J. Appl. Phys.* **40**, 1070.
BLEANEY, B. (1950) *Proc. R. Soc.* A **204**, 203.
—— (1951a) *Proc. phys. Soc.* A **64**, 315.
—— (1951b) *Phil. Mag.* **42**, 441.
—— (1955) *Discuss. Faraday Soc.* No. 19, p. 112.
—— (1959a) *Proc. phys. Soc.* **73**, 937.
—— (1959b) *Proc. phys. Soc.* **73**, 939.
—— (1960) *Proc. phys. Soc.* **75**, 621.
—— (1961) *Proc. phys. Soc.* **77**, 113.
—— (1964a) *Proc. R. Soc.* A **277**, 289.
—— (1964b) *Proceedings of The Third Quantum Electronics Conference.* Dunod, Paris.
—— (1967) 'La structure hyperfine magnétique des atomes et des molécules', p. 13. Colloques Internationaux du C.N.R.S., Paris.
—— BOGLE, G. S., COOKE, A. H., DUFFUS, H. J., O'BRIEN, M. C. M., and STEVENS, K. W. H. (1955) *Proc. phys. Soc.* A **68**, 57.
—— and BOWERS, K. D. (1952a) *Phil. Mag.* **43**, 372.
—— —— (1952b) *Proc. R. Soc.* A **214**, 451.
—— —— and INGRAM, D. J. E. (1955) *Proc. R. Soc.* A **228**, 147.
—— —— and PRYCE, M. H. L. (1955) *Proc. R. Soc.* A **228**, 166.
—— —— and TRENAM, R. S. (1955) *Proc. R. Soc.* A **228**, 157.
—— DANIELS, J. M., GRACE, M. A., HALBAN, H., KURTI, N., ROBINSON, F. N. H., and SIMON, F. E. (1954) *Proc. R. Soc.* A **221**, 170.

BIBLIOGRAPHY

BLEANEY, B., ELLIOTT, R. J., and SCOVIL, H. E. D. (1951) *Proc. phys. Soc.* A **64**, 933.
—— HAYES, W. and LLEWELLYN, P. M. (1957) *Nature, Lond.* **179**, 140.
—— HUTCHISON, C. A., LLEWELLYN, P. M., and POPE, D. F. D. (1956) *Proc. phys. Soc.* B **69**, 1167.
—— and INGRAM, D. J. E. (1950) *Proc. phys. Soc.* A **63**, 408.
—— —— (1951a) *Proc. R. Soc.* A **205**, 336.
—— —— (1951b) *Proc. R. Soc.* A **208**, 143.
—— LLEWELLYN, P. M. and JONES, D. A. (1956) *Proc. phys. Soc.* B **69**, 858.
—— —— PRYCE, M. H. L., and HALL, G. R. (1954a) *Phil. Mag.* **45**, 991.
—— —— —— —— (1954b) *Phil. Mag.* **45**, 992.
—— and O'BRIEN, M. C. M. (1956) *Proc. phys. Soc.* B **69**, 1216.
—— and OWEN, J. (1965) *Physical properties of diamond* (Ed. R. Berman), p. 274. Clarendon Press, Oxford.
—— PENROSE, R. P. and PLUMPTON, B. I. (1949) *Proc. R. Soc.* A **198**, 406.
—— and RUBINS, R. S. (1961) *Proc. phys. Soc.* **77**, 103.
—— and SCOVIL, H. E. D. (1950) *Proc. phys. Soc.* A **63**, 1369.
—— —— and TRENAM, R. S. (1954) *Proc. R. Soc.* A **223**, 15.
—— and TRENAM, R. S. (1954) *Proc. R. Soc.* A **223**, 1.
BLOCH, F. (1946) *Phys. Rev.* **70**, 460.
BLOEMBERGEN, N. (1956) *Phys. Rev.* **104**, 324.
—— PURCELL, E. M. and POUND, R. V. (1948) *Phys. Rev.* **73**, 679.
BLOK, J. and SHIRLEY, D. A. (1966) *Phys. Rev.* **143**, 278.
BLUMBERG, W. E., EISINGER, J., and GESCHWIND, S. (1963) *Phys. Rev.* **130** 900.
BLUME, M., FREEMAN, A. J., and WATSON, R. E. (1964) *Phys. Rev.* **134**, A 320.
—— and ORBACH, R. (1961) *Phys. Rev.* **127**, 1587.
—— and WATSON, R. E. (1962) *Proc. R. Soc.* A **270**, 127.
—— —— (1963) *Proc. R. Soc.* A **271**, 565.
BORDARIER, Y., JUDD, B. R., and KLAPISCH, M. (1966) *Proc. R. Soc.* A **289**, 81.
BORN, M. and OPPENHEIMER, R. (1927) *Annln. Phys.* **84**, 457.
BOWERS, K. D. and OWEN, J. (1955) *Rep. Prog. Phys.* **18**, 305.
BRANDON, R. W., GERKIN, R. E., and HUTCHISON, C. A. (1962) *J. chem. Phys.* **37**, 447.
BREEN, D. P., KRUPKA, D. C., and WILLIAMS, F. I. B. (1969) *Phys. Rev.* **179**, 241.
BREIT, G. (1929) *Phys. Rev.* **34**, 553.
BROCHARD, J. and HELLWEGE, K. H. (1953) *Z. Phys.* **135**, 620.
BROWER, K. L. and STAPLETON, H. J. (1967) *J. chem. Phys.* **46**, 888.
BRYA, W. J. and WAGNER, P. E. (1965) *Phys. Rev. Lett.* **14**, 431.
—— —— (1967) *Phys. Rev.* **157**, 400.
BUCKMASTER, H. A. (1962) *Can. J. Phys.* **40**, 1670.
BURNS, G. (1962) *Phys. Rev.* **128**, 2121.

CANER, M. and ENGLMAN, R. (1966) *J. chem. Phys.* **44**, 4054.
CARNALL, W. T. and WYBOURNE, B. G. (1964) *J. chem. Phys.* **40**, 3428.
CASTNER, T., NEWELL, G. S., HOLTON, W. C., and SLICHTER, C. P. (1960) *J. chem. Phys.* **32**, 668.
CAVENETT, B. C. (1964) *Proc. phys. Soc.* **84**, 1.
CHAKRAVARTY, A. S. (1959) *Proc. phys. Soc.* **74**, 711.
CHANG, TE-TSE (1964) *Phys. Rev.* **136**, A 1413.
CHAPPERT, J., FRANKEL, R. B., and BLUM, N. A. (1967) *Physics Lett.* **25 A**, 149.
CHILD, M. S. and LONGUET-HIGGINS, H. C. (1962) *Phil. Trans. R. Soc.* A **254**, 259.

COFFMAN, R. E. (1965) *Physics Lett.* **19**, 475.
—— (1966) *Physics Lett.* **21**, 381.
—— (1968) *J. chem. Phys.* **48**, 609.
—— LYLE, D. L., and MATTISON, D. R. (1968) *J. phys. Chem., Ithaca* **72**, 1392.
CONDON, E. U. and SHORTLEY, G. H. (1935) *Theory of atomic spectra.* Cambridge University Press.
CONWAY, J. G. (1964) *J. chem. Phys.* **40**, 2504.
—— and WYBOURNE, B. G. (1963) *Phys. Rev.* **130**, 2325.
COOK, R. J. and WHIFFEN, D. H. (1964) *Proc. phys. Soc.* **84**, 845.
—— —— (1965) *J. chem. Phys.* **43**, 2908.
COOKE, A. H. (1950) *Rep. Prog. Phys.* **13**, 276.
—— and DUFFUS, H. J. (1955) *Proc. phys. Soc.* **68**, 32.
—— EDMONDS, D. T., McKIM, F. R., and WOLF, W. P. (1959) *Proc. R. Soc.* A **252**, 246.
—— LAZENBY, R., McKIM, F. R., OWEN, J., and WOLF, W. P. (1959) *Proc. R. Soc.* A **250**, 97.
CROSSWHITE, H. M. and DIEKE, G. H. (1961) *J. chem. Phys.* **35**, 1535.
CULVAHOUSE, J. W. (1962) *J. chem. Phys.* **36**, 2720.
—— SCHINKE, D. P. and FOSTER, D. L. (1967) *Phys. Rev. Lett.* **18**, 117.

DABBS, J. W. T., ROBERTS, L. D., and PARKER, G. W. (1958) *Proceedings of the Kamerlingh Onnes Low Temperature Conference, Physica, 's Grav.*, S 69.
DANIELS, J. M. (1953) *Proc. phys. Soc.* A **66**, 673.
—— GRACE, M. A. and ROBINSON, F. N. H. (1951) *Nature, Lond.*, **168**, 780.
DAVIES, E. R. and HURRELL, J. P. (1968) *J. Sci. Inst. (J. Phys. E)*, **1**, 847.
DAVIES, J. J., SMITH, S. R. P., and WERTZ, J. E. (1969) *Phys. Rev.* **178**, 608.
DAWSON, J. K. (1952) *Nucleonics* **10**, Sept. 39.
DELBECQ, C. J., HAYES, W., O'BRIEN, M. C. M., and YUSTER, P. H. (1963) *Proc. R. Soc.* A **271**, 243.
DESCAMPS, D. and MERLE D'AUBIGNÉ, Y. (1964) *Physics Lett.* **8**, 5.
DEVINE, S. D. (1967) *J. chem. Phys.* **47**, 1844.
DE WIT, M. and REINBERG, A. R. (1967) *Phys. Rev.* **163**, 261.
DIEKE, G. H. and HEROUX, L. (1956) *Phys. Rev.* **103**, 1227.
—— and PANDEY, B. (1964) *J. chem. Phys.* **41**, 1952.
DIETZ, R. E., KAMIMURA, H., STURGE, M. D., and YARIV, A. (1963) *Phys. Rev.* **132**, 1559.
DIRAC, P. A. M. (1930) *The principles of quantum mechanics.* Oxford University Press.
DORAIN, P. B., HUTCHISON, C. A., and WONG, E. (1957) *Phys. Rev.* **105**, 1307.
—— and WHEELER, R. G. (1966) *J. chem. Phys.* **45**, 1172.
DREISCH, T. and TROMMER, W. (1937) *Z. phys. Chem.* B **37**, 37.
DUNN, T. M. (1961) *Trans. Faraday Soc.* **57**, 1441.
DYER, G. L. and WOONTON, G. A. (1967) *Can. J. Phys.* **45**, 2975.
DZYALOSHINSKY, I. (1958) *Physics Chem. Solids* **4**, 241.

EDELSTEIN, N. and EASLEY, W. (1968) *J. chem. Phys.* **48**, 2110.
EDMONDS, A. R. (1957) *Angular momentum in quantum mechanics.* Princeton University Press, Princeton, N.J.
EDMONDS, D. T. (1963) *Phys. Rev. Lett.* **10**, 129.
EISENSTEIN, J. C. (1963a) *J. chem. Phys.* **39**, 2128.
—— (1963b) *J. chem. Phys.* **39**, 2134.

EISENSTEIN, J. C. and PRYCE, M. H. L. (1955) *Proc. R. Soc.* A **229**, 20.
—— —— (1956) *Proc. R. Soc.* A **238**, 31.
—— —— (1960) *Proc. R. Soc.* A **255**, 181.
EISINGER, J. and FEHER, G. (1958) *Phys. Rev.* **109**, 1172.
ELLIOTT, R. J. (1953) *Phys. Rev.* **89**, 659.
—— (1957) *Proc. phys. Soc.* B **70**, 119.
—— and STEVENS, K. W. H. (1952) *Proc. R. Soc.* A **215**, 437.
—— —— (1953a) *Proc. R. Soc.* A **218**, 553.
—— —— (1953b) *Proc. R. Soc.* A **219**, 387.
—— and THORPE, M. F. (1968) *J. appl. Phys.* **39**, 802.
ELLIS, M. M. and NEWMAN, D. J. (1966) *Physics Lett.* **21**, 508.
ERATH, E. E. (1961) *J. chem. Phys.* **34**, 1985.
ERB, E., MOTCHANE, J. L., and UEBERSFELD, J. (1958) *C.r. hebd. Séanc. Acad. Sci., Paris* **246**, 2121, 3050.
ESTLE, T. L., WALTERS, G. K., and DE WIT, M. (1963) *Symposium on paramagnetic resonance* vol. **1**, p. 144. Academic Press, New York.
EVANS, L., SANDARS, P. G. H., and WOODGATE, G. K. (1965) *Proc. R. Soc.* A **289**, 114.
EVANS, R. C. (1964) *An introduction to crystal chemistry*. Cambridge University Press.

FANO, U. and RACAH, G. (1958) *Irreducible tensorial sets*, Academic Press, N.Y.
FAUGHNAN, B. W. and STRANDBERG, M. W. P. (1961) *Physics. Chem. Solids* **19**, 155.
FAUST, J., MARRUS, R., and NIERENBERG, W. A. (1965) *Physics Lett.* **16**, 71.
FEHER, E. R. (1964) *Phys. Rev.* **136**, A 145.
FEHER, G. (1956a) *Phys. Rev.* **103**, 500.
—— (1956b) *Phys. Rev.* **103**, 834.
—— (1959) *Phys. Rev.* **114**, 1219.
—— and GERE, E. A. (1956) *Phys. Rev.* **103**, 501.
—— —— (1959) *Phys. Rev.* **114**, 1245.
FELDMAN, D. W., CASTLE, J. G., and WAGNER, G. R. (1966) *Phys. Rev.* **145**, 237.
FERMI, E. and SEGRÉ, E. (1933a) *Z. Phys.* **82**, 729.
—— —— (1933b) *Memorie R. Accad. Ital.* **4**, 131.
FEYNMAN, R. P. (1939) *Phys. Rev.* **56**, 340.
FIDONE, I. and STEVENS, K. W. H. (1959) *Proc. phys. Soc.* **73**, 116.
FINN, C. B. P., ORBACH, R., and WOLF, W. P. (1961) *Proc. phys. Soc.* **77**, 261.
FITZWATER, D. R. and RUNDLE, R. E. (1959) *Z. Kristallogr. Kristallgeom.* **112**, 362.
FLETCHER, J. R. and STEVENS, K. W. H. (1969) *J. Phys.* C (*Proc. phys. Soc.*) **2**, 444.
FOEX, G., KARANTASSIS, T., and PERAKIS, N. (1953) *C.r. hebd. Séanc. Acad. Sci., Paris* **237**, 982.
FREEMAN, A. J. and WATSON, R. E. (1962) *Phys. Rev.* **127**, 2058.
—— —— (1964) *Phys. Rev.* **133**, A 1571.
—— —— (1965) *Magnetism*, Vol. II A (Eds. G. T. Rado and H. Suhl), p. 167. Academic Press.
FRIEDERICH, A., HELLWEGE, K. H., and LÄMMERMAN, H. (1960) *Z. Phys.* **159**, 524.
FRY, D. J. I. and LLEWELLYN, P. M. (1962) *Proc. R. Soc.* A **266**, 84.
FULLER, G. H. and COHEN, V. W. (1965) *Nuclear moments* (Appendix 1 to *Nuclear Data Sheets*).

GABRIEL, J. R., JOHNSTON, D. F., and POWELL, M. J. D. (1961) *Proc. R. Soc.* A **264**, 503.
GESCHWIND, S. (1959) *Phys. Rev. Lett.* **3**, 207.
—— (1967) *Hyperfine interactions*, p. 225. Academic Press.
—— and REMEIKA, J. P. (1961) *Phys. Rev.* **122**, 757.
—— —— (1962) *J. appl. Phys.* Suppl. to vol. 33, 370.
GHATIKAR, M. N., RAYCHAUDHURI, A. K., and RAY, D. K. (1965) *Proc. phys. Soc.* **86**, 1239.
GILL, J. C. (1963) *Proc. phys. Soc.* **82**, 1066.
—— and ELLIOTT, R. J. (1961) *Advances in quantum electronics*, (Ed. J. R. Singer), p. 399. Columbia University Press.
GIORDMAINE J. A. and NASH, F. R. (1965) *Phys. Rev.* **138**, A 1510.
GORTER, C. J. (1947) *Paramagnetic relaxation*. Elsevier.
—— (1948) *Physica, 's Grav.* **14**, 504.
—— and VAN VLECK, J. H. (1947) *Phys. Rev.* **72**, 1128.
GRAMBERG, G. (1960) *Z. Phys.* **159**, 125.
GREENSLADE, D. J. and STEVENS, K. W. H. (1967) *Proc. phys. Soc.* **91**, 627.
GRIFFITH, J. S. (1961) *The theory of transition-metal ions*. Cambridge University Press.
GRIFFITHS, J. H. E., O'BRIEN, M. C. M., OWEN, J., and WARD, I. M. (1955) Quoted by Bowers, K. D. and Owen, J., *Rep. Prog. Phys.* **18**, 304.
—— and OWEN, J. (1952a) *Proc. R. Soc.* A **213**, 459.
—— —— (1952b) *Proc. phys. Soc.* A **65**, 951.
—— —— (1954) *Proc. R. Soc.* A **226**, 96.
—— —— PARK, J. G., and PARTRIDGE, M. F. (1959) *Proc. R. Soc.* A **250**, 84.
—— —— and WARD, I. M. (1953) *Proc. R. Soc.* A **219**, 526.
GRUBER, J. B. (1963) *J. chem. Phys.* **38**, 946.
—— COCHRAN, W. R., CONWAY, J. G. and NICOL, A. T. (1966) *J. chem. Phys.* **45**, 1423.
—— and SATTEN, R. A. (1963) *J. chem. Phys.* **39**, 1455.

HAHN, E. L. (1950) *Phys. Rev.* **80**, 580.
HALFORD, D. (1962) *Phys. Rev.* **127**, 1940.
HALL, J. L. and SCHUMACHER, R. T. (1962) *Phys. Rev.* **127**, 1892.
HALL, T. P. P., HAYES, W., STEVENSON, R. W. H., and WILKENS, J. (1963) *J. chem. Phys.* **38**, 1977.
HAM, F. S. (1961) *Phys. Rev. Lett.* **7**, 242.
—— (1965) *Phys. Rev.* A **138**, 1727.
—— (1967) *Phys. Rev.* **160**, 328.
—— (1968) *Phys. Rev.* **166**, 307.
—— (1969) *Jahn–Teller effects in electron paramagnetic resonance spectra, Electron Paramagnetic Resonance*. Plenum Publishing Corp., New York.
—— LUDWIG, G. W., WATKINS, G. D., and WOODBURY, H. H. (1960) *Phys. Rev. Lett.* **5**, 468.
—— SCHWARZ, W. M. and O'BRIEN, M. C. M. (1969) *Phys. Rev.* **185**, 548.
HANAUER, S. H., DABBS, J. W. T., ROBERTS, L. D., and PARKER, G. W. (1961) *Phys. Rev.* **124**, 1512.
HARGREAVES, W. A. (1967) *Phys. Rev.* **156**, 331.
HARRIS, E. A. and OWEN, J. (1963) *Phys. Rev. Lett.* **11**, 9.
—— —— (1965) *Proc. R. Soc.* A **289**, 122.
—— and YNGVESSON, K. S. (1966) *Physics Lett.* **21**, 252.

BIBLIOGRAPHY

HARRIS, E. A. and YNGVESSON, K. S. (1968) *J. Phys. C (Proc. phys. Soc.)* **1**, 990, 1011.
HARVEY, J. S. M. (1965) *Proc. R. Soc.* A **285**, 581.
HAUSMANN, A. and SCHREIBER, P. (1969) *Solid St. Communs.* **7**, 631.
HAYES, W. and TWIDELL, J. W. (1961) *J. chem. Phys.* **35**, 1521.
—— —— (1963) *Proc. phys. Soc.* **82**, 330.
—— and WILKENS, J. (1964) *Proc. R. Soc.* A **281**, 340.
HEITLER, W. and TELLER, E. (1936) *Proc. R. Soc.* A **155**, 629.
HELLWEGE, A. M. and HELLWEGE, K. H. (1951) *Z. Phys.* **130**, 549.
—— —— (1953) *Z. Phys.* **135**, 92.
HELLWEGE, K. H., ORLICH, E., and SCHAACK, G. (1965) *Phys. Kondens. Materie*, **4**, 196.
HENDERSON, A. J. and ROGERS, R. N. (1966) *Phys. Rev.* **152**, 218.
HILL, J. C. and WHEELER, R. G. (1966) *Phys. Rev.* **152**, 482.
HÖCHLI, U. T. (1966) *Bull. Am. phys. Soc.* **11**, 203.
—— (1967) *Phys. Rev.* **162**, 262.
—— and ESTLE, T. L. (1967) *Phys. Rev. Lett.* **18**, 128.
—— MÜLLER, K. A., and WYSLING, P. (1965) *Physics Lett.* **15**, 5.
HODGES, J. A., SERWAY, R. A., and MARSHALL, S. A. (1966) *Phys. Rev.* **151**, 196.
HOLTON, W. C., SCHNEIDER, J., and ESTLE, T. L. (1964) *Phys. Rev.* **133**, A 1638.
HORAK, J. B. and NOLLE, A. W. (1967) *Phys. Rev.* **153**, 372.
HOSKINS, R. H. and SOFFER, B. H. (1964) *Phys. Rev.* **133** A, 490.
HUANG, CHAO-YUAN (1965) *Phys. Rev.* **139**, A 241.
HÜFNER, S. (1962) *Z. Phys.* **169**, 417.
HURRELL, J. P. (1965) *Br. J. appl. Phys.* **16**, 755.
HUTCHINGS, M. T. (1964) *Solid St. Phys.* **16**, 227.
—— and RAY, D. K. (1963) *Proc. phys. Soc.* **81**, 663.
HUTCHISON, C. A. and CANDELA, G. A. (1957) *J. chem. Phys.* **27**, 707.
—— JUDD, B. R. and POPE, D. F. D. (1957) *Proc. phys. Soc.* B **70**, 514.
—— and LEWIS, W. B. (1954) *Phys. Rev.* **95**, 1096.
—— LLEWELLYN, P. M., WONG, E., and DORAIN, P. (1956) *Phys. Rev.* **102**, 292.
—— and MANGUM, B. W. (1958) *J. chem. Phys.* **29**, 952.
—— TSANG, T. and WEINSTOCK, B. (1962) *J. chem. Phys.* **37**, 555.
—— and WEINSTOCK, B. (1960) *J. chem. Phys.* **32**, 56.
—— and WONG, E. (1958) *J. chem. Phys.* **29**, 754.

IBERS, J. A. and SWALEN, J. D. (1962) *Phys. Rev.* **127**, 1914.
INGRAM, D. J. E. (1949) *Proc. phys. Soc.* A **62**, 664.
INOUE, M. (1963) *Phys. Rev. Lett.* **11**, 196.
ISHIGURO, E., KAMBE, K., and USUI, T. (1951) *Physica, 's Grav.* **17**, 310.

JACOBSEN, E. H. and STEVENS, K. W. H. (1963) *Phys. Rev.* **129**, 2036.
JAHN, H. A. and TELLER, E. (1937) *Proc. R. Soc.* A **161**, 220.
JEFFRIES, C. D. (1957) *Phys. Rev.* **106**, 164.
—— (1960) *Phys. Rev.* **117**, 1056.
—— (1963) *Dynamic nuclear orientation*. Wiley, New York.
JOHNSTON, D. R., SATTEN, R. A., SCHREIBER, C. L., and WONG, E. Y. (1966) *J. chem. Phys.* **44**, 3141.
JONES, D. A., BAKER, J. M., and POPE, D. F. D. (1959) *Proc. phys. Soc.* **74**, 249.
JORGENSEN, C. K. (1962) *Absorption spectra and chemical bonding in complexes.* Pergamon Press, Oxford.
JOSEPH, R. I. and VAN VLECK, J. H. (1960) *J. chem. Phys.* **32**, 1573.

JOYCE, R. R. and RICHARDS, P. L. (1969) *Phys. Rev.* **179**, 375.
JUDD, B. R. (1955) *Proc. R. Soc.* A **232**, 458.
—— (1957a) *Proc. R. Soc.* A **241**, 122.
—— (1957b) *Proc. R. Soc.* A **241**, 414.
—— (1959a) *Proc. R. Soc.* A **251**, 134.
—— (1959b) *Molec. Phys.* **2**, 407.
—— (1959c) *Proc. R. Soc.* A **250**, 110.
—— (1963a) *Proc. phys. Soc.* **82**, 874.
—— (1963b) *Operator techniques in atomic spectroscopy*. McGraw-Hill, New York.
—— and LINDGREN, I. (1961) *Phys. Rev.* **122**, 1802.
—— LOVEJOY, C. A. and SHIRLEY, D. A. (1962) *Phys. Rev.* **128**, 1733.
—— and WONG, E. (1958) *J. chem. Phys.* **28**, 1097.

KAMBE, K. and USUI, T. (1952) *Prog. theor. Phys., Osaka* **8**, 302.
KAMIMURA, H. (1956) *J. phys. Soc. Japan* **11**, 1171.
—— KOIDE, S., SEKIYAMA, H. and SUGANO, S. (1960) *J. phys. Soc. Japan* **15**, 1264.
KARPLUS, R. and SCHWINGER, J. (1948) *Phys. Rev.* **73**, 1020.
KASK, N. E., KORNIENKO, L. S., MANDEL'SHTAM, T. S., and PROKHOROV, A. M. (1964) *Fizika tverd. Tela* **5**, 2306; *Soviet Phys. solid St.* **5**, 1677.
KETELAAR, J. A. A. (1937) *Physica, 's Grav.* **4**, 619.
KIEL, A. and MIMS, W. B. (1967) *Phys. Rev.* **161**, 386.
KINGSLEY, J. D. and AVEN, M. (1967) *Phys. Rev.* **155**, 235.
KISS, Z. J. (1962) *Phys. Rev.* **127**, 718.
—— (1965) *Phys. Rev.* **137**, A 1749.
—— ANDERSON, C. H. and ORBACH, R. (1965) *Phys. Rev.* **137**, A 1761.
KITTEL, C. and ABRAHAMS, E. (1953) *Phys. Rev.* **90**, 238.
KOKOSZKA, G. F., ALLEN, H. C., and GORDON, G. (1965) *J. chem. Phys.* **42**, 3693.
KONYUKHOV, V. K., PASHININ, P. P., and PROKHOROV, A. M. (1962) *Fizika tverd. Tela* **4**, 246; *Soviet Phys. solid St.* **4**, 175.
KOSTER, G. F. and STATZ, H. (1959) *Phys. Rev.* **113**, 445.
KOTANI, M. (1949) *J. phys. Soc. Japan* **4**, 293.
—— (1960) *Prog. theor. Phys., Osaka* Suppl. No. 14, p. 1.
KRONIG, R. DE L. (1939) *Physica, 's Grav.* **6**, 33.
—— and BOUWKAMP, C. J. (1939) *Physica, 's Grav.* **6**, 290.
KYI, RU-TAO (1962) *Phys. Rev.* **128**, 151.

LACROIX, R. (1957) *Helv. phys. Acta* **30**, 374.
—— HÖCHLI, U. and MÜLLER, K. A. (1964) *Helv. phys. Acta* **37**, 627.
LAMBE, J., LAURANCE, N., MCIRVINE, E. C., and TERHUNE, R. W. (1961) *Phys. Rev.* **122**, 1161.
LARSON, G. H. and JEFFRIES, C. D. (1966a) *Phys. Rev.* **141**, 461.
—— —— (1966b) *Phys. Rev.* **145**, 311.
LAURANCE, N. and LAMBE, J. (1963) *Phys. Rev.* **132**, 1029.
LEA, K. R., LEASK, M. J. M., and WOLF, W. P. (1962) *Physics Chem. Solids* **23**, 1381.
LEASK, M. J. M., ORBACH, R., POWELL, M. J. D., and WOLF, W. P. (1963) *Proc. R. Soc.* A **272**, 371.
LEVY, P. M. (1964) *Phys. Rev.* **135**, A 155.
—— (1968) *Phys. Rev. Lett.* **20**, 1366.
LEWIS, H. R. and SABISKY, E. S. (1963) *Phys. Rev.* **130**, 1370.

LINDGREN, I. (1962) *Nucl. Phys.* **32**, 151.
LINES, M. E. (1963) *Proc. R. Soc.* A **271**, 105.
LOCHER, P. R. and GESCHWIND, S. (1963) *Phys. Rev. Lett.* **11**, 333.
—— —— (1965) *Phys. Rev.* **139**, A 991.
LOW, W. (1956) *Phys. Rev.* **101**, 1827.
—— (1958) *Phys. Rev.* **109**, 265.
—— (1960) *Paramagnetic resonance in solids.* Academic Press, New York.
—— (1967) Physics Lett. **24A**, 46.
—— and LLEWELLYN, P. M. (1958) *Phys. Rev.* **110**, 842.
—— and ROSENBERGER, U. (1959) *Phys. Rev.* **116**, 621.
—— and RUBINS, R. S. (1963a) *Symposium on paramagnetic resonance*, Vol. **1**, p. 79. Academic Press, New York.
—— —— (1963b) *Phys. Rev.* **131**, 2527.
—— and SHALTIEL (1958) *Physics Chem. Solids* **6**, 315.
—— and SUSS, J. T. (1963) *Physics Lett.* **7**, 310.
—— —— (1965) *Phys. Rev. Lett.* **15**, 519.
—— and WEGER, M. (1960) *Phys. Rev.* **118**, 1119.
LUDWIG, G. W. and HAM, F. S. (1962) *Phys. Rev. Lett.* **8**, 210.
—— and WOODBURY, H. H. (1961) *Phys. Rev. Lett.* **7**, 240.

MACFARLANE, R. M., WONG, J. Y., and STURGE, M. D. (1968) *Phys. Rev.* **166**, 250.
MCCLURE, D. S. (1959) *Solid St. Phys.*, **9**, 399.
—— (1962) *J. chem. Phys.* **36**, 2757.
MCCONNELL, H. M. (1956) *J. chem. Phys.* **25**, 709.
MCDONALD, P. F. (1969) *Phys. Rev.* **177**, 447.
MCGARVEY, B. R. (1956) *J. phys. Chem., Ithaca* **60**, 71.
MCLAUGHLAN, S. D. (1966) *Phys. Rev.* **150**, 119.
MCLELLAN, A. G. (1961) *J. chem. Phys.* **34**, 1350.
MCMAHON, D. H. (1964) *Phys. Rev.* **134**, A 128.
—— and SILSBEE, R. H. (1964) *Phys. Rev.* **135**, A 91.
MCMILLAN, M. and OPECHOWSKI, W. (1960) *Can. J. Phys.* **38**, 1168.
—— —— (1961) *Can. J. Phys.* **39**, 1369.
MANENKOV, A. A. and PROKHOROV, A. M. (1955) *Zh. éksp. teor. Fiz.* **28**, 762; *Soviet Phys. JETP* **1**, 611.
—— —— (1962) *Zh. éksp. teor. Fiz.* **42**, 1371; *Soviet Phys. JETP* **15**, 951.
MARGOLIS, J. S. (1961) *J. Chem. Phys.* **35**, 1367.
MARSHALL, F. G. and RAMPTON, V. W. (1968) *J. Phys.* C (*Proc. phys. Soc.*) **1**, 594.
MARSHALL, S. A., KIKUCHI, T. T., and REINBERG, A. R. (1962) *Phys. Rev.* **125**, 453.
MARSHALL, W. (1963) *Symposium on paramagnetic resonance*, Vol. **1**, p. 347. Academic Press, New York.
—— and STUART, R. (1961) *Phys. Rev.* **123**, 2048.
MEHRA, A. and VENKATESWARLU, P. (1967) *Phys. Rev. Lett.* **19**, 145.
MERGERIAN, D., HARROP, I. H., STOMBLER, M. P., and KRIKORIAN, K. C. (1967) *Phys. Rev.* **153**, 349.
—— STOMBLER, M. P. and HARROP, I. H. (1965) *Bull. Am. phys. Soc.* **10**, 1109.
MERRITT, F. R., GUGGENHEIM, H., and GARRETT, C. G. B. (1966) *Phys. Rev.* **145**, 188.
MESSIAH, A. (1961) *Quantum mechanics.* North-Holland, Amsterdam.
MIKKELSON, R. C. and STAPLETON, H. J. (1965) *Phys. Rev.* **140**, A 1968.

MIMS, W. B., DEVLIN, G. E., GESCHWIND, S., and JACCARINO, V. (1967) *Physics Lett.* **24 A**, 481.
MIYATA, N. and ARGYLE, B. E. (1967) *Phys. Rev.* **157**, 448.
MOFFITT, W. and THORSON, W. (1957) *Phys. Rev.* **108**, 1251.
MORIGAKI, K. (1964a) *J. phys. Soc. Japan* **19**, 187.
—— (1964b) *J. phys. Soc. Japan* **19**, 2064.
MORIYA, T. (1960) *Phys. Rev.* **120**, 91.
MOSER, C. M. (1967) *Hyperfine interactions.* p. 95. Academic Press.
MUELLER, K. A. (1968) *Phys. Rev.* **171**, 350.
MÜLLER, W., STEUDEL, A., and WALTHER, H. (1965) *Z. Phys.* **183**, 303.
MURPHY, J. (1966) *Phys. Rev.* **145**, 241.

NAIMAN, C. S. (1961) *J. chem. Phys.* **35**, 323.
NELSON, E. D., WONG, J. Y., and SCHAWLOW, A. L. (1967) *Phys. Rev.* **156**, 298.

O'BRIEN, M. C. M. (1964) *Proc. R. Soc.* A **281**, 323.
O'CONNOR, J. R. and CHEN, J. H. (1964) *Applied Phys. Lett.* **5**, 100.
OFELT, G. S. (1963) *J. chem. Phys.* **38**, 2171.
OLSCHEWSKI, L. and OTTEN, E. W. (1967) *Z. Phys.* **200**, 224.
ONO, K., KOIDE, S., SEKIYAMA, H., and ABE, H. (1954) *Phys. Rev.* **96**, 38.
—— and OHTSUKA, M. (1958) *J. phys. Soc. Japan* **13**, 206.
OPECHOWSKI, W. (1940) *Physica, 's Grav.* **7**, 552.
ÖPIK, U. and PRYCE, M. H. L. (1957) *Proc. R. Soc.* A **238**, 425.
ORBACH, R. (1961a,b) *Proc. R. Soc.* A **264**, 458, 485.
—— and TACHIKI, M. (1967) *Phys. Rev.* **158**, 524.
ORTON, J. W. (1959) *Rep. Prog. Phys.* **22**, 204.
—— AUZINS, P., GRIFFITHS, J. H. E., and WERTZ, J. E. (1961) *Proc. phys. Soc.* **78**, 554.
—— —— and WERTZ, J. E. (1960a) *Phys. Rev. Lett.* **4**, 128.
—— —— —— (1960b) *Phys. Rev.* **119**, 1691.
OVERHAUSER, A. W. (1953) *Phys. Rev.* **89**, 689; **92**, 411.
OVERMEYER, J. and GAMBINO, R. J. (1964) *Physics Lett.* **9**, 108.
OWEN, J. (1955) *Proc. R. Soc.* A **227**, 183.
—— (1961) *J. appl. Phys.* Suppl. **32**, 213 S.
—— and STEVENS, K. W. H. (1953) *Nature, Lond.* **171**, 836.
—— and THORNLEY, J. H. M. (1966) *Rep. Prog. Phys.* **29**, 675.
—— and WARD, I. M. (1956) *Phys. Rev.* **102**, 591.

PAKE, G. E. and PURCELL, E. M. (1948) *Phys. Rev.* **74**, 1184.
PENROSE, R. P. (1949) *Nature, Lond.* **163**, 992.
—— and STEVENS, K. W. H. (1950) *Proc. phys. Soc.* A **63**, 29.
PETERSON, R. L. (1965) *Phys. Rev.* **137**, A 1444.
—— (1967) *Phys. Rev.* **159**, 227.
PICHANICK, F. M., SANDARS, P. G. H., and WOODGATE, G. K. (1960) *Proc. R. Soc.* A **257**, 277.
PIKSIS, A. H., DIEKE, G. H., and CROSSWHITE, H. M. (1967) *J. chem. Phys.* **47**, 5083.
PIPKIN, F. M. and CULVAHOUSE, J. W. (1957) *Phys. Rev.* **106**, 1102.
PIRC, R., ZEKS, B., and GOSAR, P. (1966) *Physics Chem. Solids* **27**, 1219.
PORTIS, A. M. (1953) *Phys. Rev.* **91**, 1071.
POUND, R. V. (1949) *Phys. Rev.* **76**, 1410.
POWELL, M. J. D., GABRIEL, J. R., and JOHNSTON, D. F. (1960) *Phys. Rev. Lett.* **5**, 145.

POWELL, M. J. D. and ORBACH, R. (1961) *Proc. phys. Soc.* **78**, 753.
PRYCE, M. H. L. (1948) *Nature, Lond.* **162**, 539.
—— (1949) *Nature, Lond.* **164**, 117.
—— (1950) *Phys. Rev.* **80**, 1107.
—— (1959) *Phys. Rev. Lett.* **3**, 375.
—— and RUNCIMAN, W. A. (1958) *Discuss. Faraday. Soc.* **26**, 34.
—— and STEVENS, K. W. H. (1950) *Proc. phys. Soc.* A **63**, 36.

RABI, I. I. (1937) *Phys. Rev.* **51**, 652.
—— RAMSEY, N. F. and SCHWINGER, J. (1954) *Rev. mod. Phys.* **26**, 167.
RAHN, R. O. and DORAIN, P. B. (1964) *J. chem. Phys.* **41**, 3249.
RAJNAK, K. and KRUPKE, W. F. (1967) *J. chem. Phys.* **46**, 3532.
—— —— (1968) *J. chem. Phys.* **48**, 3343.
RANON, U. and HYDE, J. S. (1966) *Phys. Rev.* **141**, 259.
—— and LOW, W. (1963) *Phys. Rev.* **132**, 1609.
RAUBENHEIMER, L. J., BOESMAN, E., and STAPLETON, H. J. (1965) *Phys. Rev.* **137**, A 1449.
RAY, TUHINA (1964) *Proc. R. Soc.* A **277**, 76.
RIMMER, D. E. and JOHNSTON, D. F. (1966) *Proc. phys. Soc.* **89**, 943, 953.
RODBELL, D. S., JACOBS, I. S., OWEN, J., and HARRIS, E. A. (1963) *Phys. Rev. Lett.* **11**, 10.
ROGERS, R. N., CARBONI, F., and RICHARDS, P. M. (1967) *Phys. Rev. Lett.* **19**, 1016.
ROSE, M. E. (1949) *Phys. Rev.* **75**, 213.
ROTENBERG, M., BIVINS, R., METROPOLIS, N., and WOOTEN, J. K. (1959) *The 3j- and 6j-symbols*, Technology Press, Cambridge, Mass.
RUBY, R. H., BENOIT, H., and JEFFRIES, C. D. (1962) *Phys. Rev.* **127**, 51.

SABISKY, E. S. (1964) *J. chem. Phys.* **41**, 892.
—— and ANDERSON, C. H. (1966) *Phys. Rev.* **148**, 194.
SANDS, R. H. (1955) *Phys. Rev.* **99**, 1222.
SANDARS, P. G. H. and BECK, J. (1966) *Proc. R. Soc.* A **289**, 97.
—— and WOODGATE, G. K. (1960) *Proc. R. Soc.* A **257**, 269.
SATTEN, R. A., SCHREIBER, C. L., and WONG, E. Y. (1965) *J. chem. Phys.* **42**, 162.
SAUZADE, M., PONTNAU, J., LESAS, P., and SILHOUETTE, D. (1966) *Physics Lett.* **19**, 617.
SCHMIDT, J. and SOLOMON, I. (1966) *J. appl. Phys.* **37**, 3719.
SCHNEIDER, J., DISCHLER, B., and RÄUBER, A. (1967) *Solid St. Communs.* **5**, 603.
—— and RÄUBER, A. (1966) *Physics Lett.* **21**, 380.
SCOTT P. L. and JEFFRIES, C. D. (1962) *Phys. Rev.* **127**, 32.
SEABORG, G. T. (1949) *Nucleonics* **5**, Nov. 16.
SEARL, J. W., SMITH, R. C., and WYARD, S. J. (1961) *Proc. phys. Soc.* A **78**, 1174.
SHALTIEL, D. and LOW, W. (1961) *Phys. Rev.* **124**, 1062.
DE SHAZER, L. G. and DIEKE, G. H. (1963) *J. chem. Phys.* **38**, 2190.
SHIREN, N. S. (1962) *Phys. Rev.* **128**, 2103.
—— (1966) *Phys. Rev. Lett.* **17**, 958.
—— (1967) *Proc. Int. Conf. on Magnetic Resonance and Relaxation* (Ljubljana, 1966) North-Holland, Amsterdam, p. 213.
SHULMAN, R. G. and JACCARINO, V. (1956) *Phys. Rev.* **103**, 1126.
—— and KNOX, K. (1960) *Phys. Rev.* **119**, 94.
—— and SUGANO, S. (1963) *Phys. Rev.* **130**, 506.
SHUSKUS, A. J. (1962) *Phys. Rev.* **127**, 2022.

SIEGERT, A. (1937) *Physica, 's Grav.* **4**, 138.
SIERRO, J. (1967) *Physics Chem. Solids* **28**, 417
ŠIMÁNEK, E., ŠROUBEK, Z., ŽDÁNSKÝ, K., KACZÉR, J., and NOVÁK, L. (1966) *Solid St. Phys.* **14**, 333.
SIMON, F. E. (1939) *Comptes Rendus du Congrès sur le Magnétisme Strasbourg*, Vol. 3, p. 1.
SLACK, G. A., HAM, F. S., and CHRENKO, R. M. (1966) *Phys. Rev.* **152**, 376.
—— ROBERTS, S., and HAM, F. S. (1967) *Phys. Rev.* **155**, 170.
SMITH, D. and THORNLEY, J. H. M. (1966) *Proc. phys. Soc.* **89**, 779.
SMITH, K. F. and UNSWORTH, P. J. (1965) *Proc. phys. Soc.* **86**, 1249.
SMITH, S. R. P., DRAVNIEKS, F. and WERTZ, J. E. (1969) *Phys. Rev.* **178**, 471.
SPALDING, I. J. (1963) *Proc. phys. Soc.* **81**, 156.
ŠROUBEK, Z. and ŽDÁNSKY, K. (1966) *J. chem. Phys.* **44**, 3078.
STAPLETON, H. J., JEFFRIES, C. D., and SHIRLEY, D. A. (1961) *Phys. Rev.* **124**, 1455.
STERNHEIMER, R. M. (1951) *Phys. Rev.* **84**, 244.
—— (1966) *Phys. Rev.* **146**, 140.
—— BLUME, M. and PEIERLS, R. F. (1968) *Phys. Rev.* **173**, 376.
STEVENS, K. W. H. (1952a) *Proc. phys. Soc.* **65**, 209.
—— (1952b) *Proc. R. Soc.* A **214**, 237.
—— (1953a) *Proc. R. Soc.* A **219**, 542.
—— (1953b) *Rev. mod. Phys.* **25**, 166.
—— (1962) *Proc. phys. Soc.* A **65**, 1952.
—— (1967) *Rep. Prog. Phys.* **30**, 189.
STONEHAM, A. M. (1966) *Proc. phys. Soc.* **89**, 909.
STURGE, M. D. (1967) *Solid St. Phys.* **20**, 91.
SUGANO, S. (1960) *Prog. theor. Phys., Osaka* Suppl. No. 14, p. 66.
—— and PETER, M. (1961) *Phys. Rev.* **122**, 381.
—— and SHULMAN, R. G. (1963) *Phys. Rev.* **130**, 517.
—— and TANABE, Y. (1958) *J. phys. Soc. Japan* **13**, 880.
SUGIHARA, K. (1959) *J. phys. Soc. Japan* **14**, 1231.
SVARE, I. and SEIDEL, G. (1964) *Phys. Rev.* **134**, A 172.

TANABE, Y. (1960) *Prog. theor. Phys., Osaka* Suppl. No. 14, p. 17.
—— and SUGANO, S. (1954) *J. phys. Soc. Japan* **9**, 753, 766.
THOMAS, K. S., SINGH, S., and DIEKE, G. H. (1963) *J. chem. Phys.* **38**, 2180.
THORNLEY, J. H. M. (1963) *Phys. Rev.* **132**, 1492.
—— (1968) *J. Phys.* C (*Proc. phys. Soc.*) **1**, 1024.
—— WINDSOR, C. G. and OWEN, J. (1965) *Proc. R. Soc.* A **284**, 252.
TINKHAM, M. (1956a,b) *Proc. R. Soc.* A **236**, 535, 549.
—— (1964) *Group theory and quantum mechanics*, McGraw-Hill.
TINSLEY, B. M. (1963) *J. chem. Phys.* **39**, 3503.
TITLE, R. S., SOROKIN, P. P., STEVENSON, M. J., PETTIT, G. D., SCARDEFIELD, J. E., and LANKARD, J. R. (1962) *Phys. Rev.* **128**, 62.
TRAMMELL, G. T., ZELDES, H., and LIVINGSTONE, R. (1958) *Phys. Rev.* **110**, 630.
TREES, R. E. (1951) *Phys. Rev.* **82**, 683.
TRENAM, R. S. (1953) *Proc. phys. Soc.* A **66**, 118.
TUCKER, E. B. (1966) *Phys. Rev.* **143**, 264.

VAN DEN HANDEL, J. and SIEGERT, A. (1937) *Physica, 's Grav.* **4**, 871.
VAN NIEKERK, J. N. and SCHOENING, F. R. L. (1953) *Acta crystallogr.* **6**, 227.

VAN VLECK, J. H. (1932) *Electric and magnetic susceptibilities.* Oxford University Press.
—— (1935) *J. chem. Phys.* **3**, 807.
—— (1937) *J. chem. Phys.* **5**, 320.
—— (1939) *J. chem. Phys.* **7**, 72.
—— (1940) *Phys. Rev.* **57**, 426.
—— (1941a) *Phys. Rev.* **59**, 724.
—— (1941b) *Phys. Rev.* **59**, 730.
—— (1948) *Phys. Rev.* **74**, 1168.
—— (1957) *Nuovo Cim.* Suppl. No. 3, **6**, 993.
—— (1960) *Physica, 's Grav.* **26**, 544.
—— and WEISSKOPF, V. F. (1945) *Rev. mod. Phys.* **17**, 227.
VAN WIERINGEN, J. S. (1955) *Discuss Faraday Soc.* **19**, 118.

WALKER, M. B. (1968) *Can. J. Phys.* **46**, 1347.
WALLER, I. (1932) *Z. Phys.* **79**, 370.
WALSH, W. M. (1958) *Bull. Am. phys. Soc.* **3**, 178.
—— and BLOEMBERGEN, N. (1957) *Phys. Rev.* **107**, 904.
WATANABE, H. (1957) *Prog. theor. Phys., Osaka* **18**, 405.
—— (1964) *Physics Chem. Solids*, **25**, 1471.
WATSON, R. E. and BLUME, M. (1965) *Phys. Rev.* **139**, A 1209.
—— and FREEMAN, A. J. (1967a) *Hyperfine interactions*, p. 53, Academic Press.
—— —— (1967b) *Phys. Rev.* **156**, 251.
WEAKLIEM, H. A. and KISS, Z. J. (1967) *Phys. Rev.* **157**, 277.
WEIL, J. A. and ANDERSON, J. H. (1961) *J. chem. Phys.* **35**, 1410.
WERTZ, J. E. and AUZINS, P. (1957) *Phys. Rev.* **106**, 484.
WETSEL, G. C. and DONOHO, P. L. (1965) *Phys. Rev.* **139**, A 334.
WHIFFIN, P. A. C. and ORTON, J. W. (1965) *Br. J. appl. Phys.* **16**, 567.
WICKERSHEIM, K. A. and WHITE, R. L. (1962) *Phys. Rev. Lett.* **8**, 483
WICKMAN, H. H., TROZZOLO, A. M., WILLIAMS, H. J., HULL, G. W., and MERRITT, F. R. (1967) *Phys. Rev.* **155**, 563.
WIELINGA, R. F., BLÖTE, H. W. J., ROEST, J. A., and HUISKAMP, W. J. (1967) *Physica, 's Grav.* **34**, 223.
WIGNER, E. P. (1959) *Group theory.* Academic Press, New York.
WILKENS, J., DE GRAAG, D. P., and HELLE, J. N. (1965) *Physics Lett.* **19**, 178.
WILLIAMS, F. I. B. (1967) *Proc. phys. Soc.* **91**, 111.
—— KRUPKA, D. C. and BREEN, D. P. (1969) *Phys. Rev.* **179**, 255.
WOLF, W. P. and BIRGENEAU, R. J. (1968) *Phys. Rev.* **166**, 376.
WONG, J. Y. (1968) *Phys. Rev.* **168**, 337.
—— BERGGREN, M. J. and SCHAWLOW, A. L. (1968) *J. chem. Phys.* **49**, 835.
WONG, E. Y. and RICHMAN, I. (1961) *J. chem. Phys.* **34**, 1182.
WOOD D. L. and KAISER, W. (1962) *Phys. Rev.* **126**, 2079.
WOODBURY, H. H. and LUDWIG, G. W. (1961) *Phys. Rev.* **124**, 1083.
WOODGATE, G. K. (1966) *Proc. R. Soc.* A **293**, 117.
WOONTON, G. A. and DYER, G. L. (1967) *Can. J. Phys.* **45**, 2265.
WU, C. S., AMBLER, E., HAYWARD, R. W., HOPPES, D. D., and HUDSON, R. P. (1957) *Phys. Rev.* **105**, 1413.
WYBOURNE, B. G. (1964) *J. chem. Phys.* **40**, 1456.
—— (1965) *Spectroscopic properties of rare earths.* Interscience Publishers (John Wiley), New York.
—— (1966) *Phys. Rev.* **148**, 317.

YARIV, A. (1962) *Phys. Rev.* **128**, 1588.

ZALKIN, A., FORRESTER, J. D., and TEMPLETON, D. H. (1963) *J. chem. Phys.* **39**, 2881.
—— and TEMPLETON, D. H. (1964) *J. chem. Phys.* **40**, 501.
ZARIPOV, M. M., KROPOTOV, V. S., LIVANOVA, L. D., and STEPANOV, V. G. (1967) *Fizika tverd. Tela* **9**, 209; *Soviet Phys. solid St.* **9**, 992.
ZAVOISKY, E. (1945) *Fiz. Zh.* **9**, 211, 245.
ZVEREV, G. M., KORNIENKO, L. S., PROKHOROV, A. M., and SMIRNOV, A. I. (1962) *Fizika tverd. Tela* **4**, 392; *Soviet Phys. solid St.* **4**, 284.
—— and PROKHOROV, A. M. (1960) *Zh. éksp. teor. Fiz.* **38**, 449; *Soviet Phys. JETP* **11**, 330.
—— —— (1961) *Zh. éksp. teor. Fiz.* **40**, 1016; *Soviet Phys. JETP* **13**, 714.

AUTHOR INDEX

Abe, H., 434, 506–7
Abragam, A., 80, 84, 87, 248, 427, 446, 457–8, 461, 465, 528, 684, 697, 701, 791, 816
Abraham, M. M., 82, 331, 333, 349, 352
Abrahams, E., 531
Albertson, W., 283
Allen, H. C., 506, 508
Allen, S. J., 502
Ambler, E., 77
Anderson, C. H., 329, 333–4
Anderson, J. H., 198
Anderson, P. W., 495, 528, 531
Argyle, B. E., 337
Armstrong, L., 354
Auzins, P., 208, 432, 453–5, 466, 487
Aven, M., 331–2
Axe, J. D., 305, 328, 334, 347, 354
Ayant, Y., 331–3, 722, 725
Azarbayejani, G. H., 489–90

Bagguley, D. M. S., 58, 98, 305, 312, 514
Baker, J. M., 140, 199, 228, 252–3, 260, 263, 269, 271, 279, 313, 317, 319, 333–8, 340–1, 343–5, 479–80, 482, 501–2, 514, 529, 560, 711, 738
Barnes, R. G., 303
Beck, J., 253, 684, 710–1
Belorizky, E., 331–2, 722, 725
Benoit, H., 322, 569, 579
Benzie, R. J., 57
Berggren, M. J., 425
Bersuker, I. B., 791, 821, 823–4, 826
Bessent, R. G., 40, 228, 262–3, 269, 271, 333–4
Bethe, H., 595, 638
Bierig, R. W., 328–9, 572
Birgeneau, R. J., 304, 312, 501–2, 628
Bivins, R., 623
Blake, W. B. J., 333–4, 344–5
Bleaney, B., 30, 40, 56–7, 75, 77, 140, 158, 163, 169, 172, 176, 186–7, 190–1, 201, 204–5, 218, 258–9, 279, 298–300, 305, 313, 317, 319, 327, 330, 334, 339, 346, 348–50, 360, 363, 424, 437, 455, 457–9, 476, 479–82, 485, 501, 506–9, 518, 523, 525, 535–6, 719, 726, 738, 791, 816
Bloch, F., 109–10, 115
Bloembergen, N., 73, 119, 451
Blok, J., 303
Blöte, H. W. J., 470

Blum, N. A., 847
Blumberg, W. E., 391, 451–3
Blume, M., 283–4, 305, 391, 399, 474–5, 560, 565, 593, 679
Boesman, E., 354, 358
Bogle, G. S., 424
Bohr, N., 346
Bordarier, Y., 254
Borghini, M., 87
Born, M., 793
Bouwkamp, C. J., 144, 146, 437
Bowers, K. D., 186, 279, 322, 455, 457–9, 479–80, 482, 506–9
Brandon, R. W., 153
Breen, D. P., 459, 463, 465, 830
Breit, G., 678
Brochard, J., 324
Brower, K. L., 318
Brya, W. J., 70, 580–1
Buckmaster, H. A., 628, 864
Burns, G., 304–5, 328, 334, 399

Candela, G. A., 355
Caner, M., 845
Carboni, F., 528
Carnall, W. T., 347, 350
Castle, J. G., 573
Castner, T., 203
Cavenett, B. C., 191
Chadwick, J. R., 335–6, 340–1
Chakravarty, A. S., 427–8
Chang, Tse-Tse, 479
Chappert, J., 847
Chen, J. H., 489
Child, M. S., 791, 813
Chrenko, R. M., 468, 470
Clark, G. W., 331, 333, 352
Cochran, W. R., 347, 350
Coffman, R. E., 466, 791, 831
Cohen, V. W., 416
Condon, E. U., 593
Conway, J. G., 282–3, 347–8, 350
Cook, R. J., 226
Cooke, A. H., 57, 317, 424, 480, 532, 545, 566
Copland, G. M., 333–4, 340, 344–5
Crosswhite, H. M., 305
Culvahouse, J. W., 84, 451, 736

Dabbs, J. W. T., 360
Daniels, J. M., 76–7, 855

AUTHOR INDEX

Davies, E. R., 228
Davies, J. J., 432
Dawson, J. K., 346
de Graag, D. P., 488
de Haas, W. J., 16
Delbecq, C. J., 40, 488
Descamps, D., 331–2, 728
de Shazer, L. G., 305
Devine, S. D., 279
Devlin, G. E., 441
de Wit, M., 470–1
Dieke, G. H., 305, 324
Dietz, R. E., 468, 471
Dirac, P. A. M., 495, 584, 587
Dischler, B., 469
Donoho, P. L., 351
Dorain, P. B., 349, 480
Dravnieks, F., 208
Dreisch, T., 457
Duffus, H. J., 424, 480
Dunn, T. M., 399, 474–5
Dyer, G. L., 434, 441
Dzyaloshinsky, I., 495

Easley, W., 351–3
Edelstein, N., 351–3
Edmonds, A. R., 615, 617
Edmonds, D. T., 259, 302, 316–7
Eisenstein, J. C., 305, 348, 355–60, 362–3
Eisinger, J., 239, 391, 451–3
Elliott, R. J., 287–8, 290, 303, 310, 316, 360, 501–2, 525, 535–6, 694
Ellis, M. M., 305
Englman, R., 845
Erath, E. E., 304
Erb, E., 84
Estle, T. L., 469–70, 831
Evans, L., 252–4, 711–2
Evans, R. C., 276, 347

Fano, U., 615, 658
Faughnan, B. W., 577
Faust, J., 364
Feher, E. R., 208
Feher, G., 40, 83, 87, 90, 218, 221, 223, 234, 236, 238–242, 707
Feldman, D. W., 573
Fermi, E., 696, 698
Feynman, R. P., 794
Fidone, I., 441
Finch, C. B., 331, 333, 352
Finn, C. B. P., 68, 323, 560
Fitzwater, D. R., 277
Fletcher, J. R., 436
Fock, V., 590
Foex, G., 507
Ford, N. C., 560
Forrester, J. D., 278–80

Foster, D. L., 736
Frankel, R. B., 847
Freeman, A. J., 276, 283–4, 299, 304–5, 328, 399, 411, 413, 415, 449, 453, 474–5, 478, 480, 687, 704, 706, 709
Friederich, A., 321, 324
Fry, D. J. I., 270, 447
Fuller, G. H., 416

Gabriel, J. R., 441–2
Gambino, R. J., 337
Garrett, C. G. B., 328
Garton, G., 335–6, 340–1
Gere, E. A., 83, 239–42
Gerkin, R. E., 153
Geschwind, S., 162, 265–9, 338–9, 391, 413, 434, 441, 451–4, 484, 487
Ghatikar, M. N., 303
Gill, J. C., 318
Giordmaine, J. A., 575
Gordon, G., 506, 508
Gorter, C. J., 75, 77, 124, 527, 545, 566, 573
Gosar, P., 829
Grace, M. A., 76–7
Gramberg, G., 317
Greenslade, D. J., 481
Griffith, J. S., 392, 658
Griffiths, J. H. E., 50, 58, 365, 451, 453–5, 466, 482, 484–7, 498, 514, 531–2
Gruber, J. B., 304, 347, 350
Guggenheim, H., 328

Hahn, E. L., 114
Halban, H., 77
Halford, D., 226, 255–9, 314–5
Hall, G. R., 348, 360, 363
Hall, J. L., 40
Hall, T. P. P., 445
Ham, F. S. 445–6, 468, 470, 664, 791, 810, 815–7, 819, 826, 829, 830–2, 838–40, 845–7
Hanauer, S. H., 360
Hargreaves, W. A., 351
Harris, E. A., 532, 535, 552, 561
Harrop, I. H., 329–30
Hartree, D. R., 590
Harvey, J. S. M., 164, 411, 711
Hausmann, A., 471
Hayes, W., 40, 199, 228, 262–3, 269, 271, 279, 333–4, 445, 464–5, 488–90
Hayward, R. W., 77
Heisenberg, W., 495
Heitler, W., 557
Helle, J. N., 488
Hellwege, A. M., 324
Hellwege, K. H., 305, 310–1, 321, 324
Henderson, A. J., 528
Heroux, L., 324

AUTHOR INDEX

Hill, J. C., 304
Höchli, U. T., 469, 487, 791, 823–4, 831–2, 846
Hodby, J. W., 40
Hodges, J. A., 487
Holton, W. C., 203, 469
Hoppes, D. D., 77
Horak, J. B., 572
Horowitz, J., 697, 701
Hoskins, R. H., 429
Huang, Chao-Yuan, 572–3
Hudson, R. P., 77
Hüfner, S., 304, 317
Huiskamp, W. J., 470
Hull, G. W., 389
Hurrell, J. P., 228, 260, 263, 271, 335–7, 340–1
Hutchings, M. T., 305, 501–2, 670
Hutchison, C. A., 15, 153, 277, 287, 307, 313, 315, 339, 347–9, 355, 358, 364
Hyde, J. S., 262–3

Ibers, J. A., 201
Ingram, D. J. E., 158, 163, 172, 176, 187, 458, 518, 816
Inoue, M., 312, 330, 334
Ishiguro, E., 529

Jaccarino, V., 365, 441
Jacobs, I. S., 535
Jacobsen, E. H., 575
Jahn, H. A., 371, 403, 790–1
Jeffries, C. D., 67, 69, 81–2, 87, 248, 304, 315, 322, 324–5, 347, 354, 558, 560, 567, 569–71, 579
Johnston, D. F., 434, 441–2
Johnston, D. R., 347
Jones, D. A., 338, 350
Jorgensen, C. K., 476
Joseph, R. L., 855
Joyce, R. R., 425–6, 429
Judd, B. R., 254, 256, 279, 282–4, 299–300, 304–5, 314–6, 320–5, 335, 339, 349, 498, 532, 615

Kaczér, J., 487
Kaiser, W., 283
Kambe, K., 529
Kamimura, H., 468, 471, 476, 478–9, 481
Kankeleit, E., 303
Karantissis, T., 507
Karplus, R., 124
Kask, N. E., 425
Kedzie, R. W., 82
Ketelaar, J. A. A., 277
Kiel, A., 115, 565, 568
Kikuchi, T. T., 452
Kingsley, J. D., 331–2

Kiss, Z. J., 328–9, 332, 334
Kittel, C., 531
Klapisch, M., 254
Knox, K., 780
Koide, S., 434, 476, 478–9, 481
Kokoszka, G. F., 506, 508
Konyukhov, V. K., 333
Kornienko, L. S., 331, 425
Koster, G. F., 343, 722
Kotani, M., 476
Krikorian, K. C., 330
Kronig, R. de L., 65, 144, 146, 437, 557
Kropotov, V. S., 846
Krupka, D. C., 459, 463, 465, 830
Krupke, W. F., 305
Kurti, N., 77
Kyi, Ru-Tao, 479

Lacroix, R., 149–50, 337, 487
Lambe, J., 273, 432
Lämmerman, H., 321, 324
Lankard, J. R., 351
Larson, G. H., 67, 304, 325, 560, 567, 570–1
Laurance, N., 273, 432
Lazenby, R., 532
Lea, K. R., 327–9, 332, 351, 719–20, 728
Leask, M. J. M., 321, 323, 327–9, 332, 351, 719–20, 728
Lesas, P., 429
Levy, P. M., 502
Lewis, H. R., 167, 228, 331–2
Lewis, W. B., 348, 364
Liehr, A. D., 791
Lindgren, I., 282, 284, 299–300, 335
Lines, M. E., 532
Livanova, L. D., 846
Livingstone, R., 199
Llewellyn, P. M., 270, 279, 348–50, 360, 363, 447, 479
Locher, P. R., 265–9, 441, 452–4
Longuet-Higgens, H. C., 791, 813
Lovejoy, C. A., 316
Low, W., 149–50, 207, 329, 331–6, 432, 444, 453, 479, 487
Ludwig, G. W., 198, 470, 664
Lyle, D. L., 466

MacFarlane, R. M., 426
Mandel'shtam, T. S., 425
Manenkov, A. A., 560
Mangum, B. W., 39
Margolis, J. S., 305
Marrus, R., 354, 364
Marshall, F. G., 436
Marshall, S. A., 452, 487
Marshall, W. C., 395, 702, 781–3
Mattison, D. R., 466
Mau, A. E., 502

McClure, D. S., 378, 389, 390–1
McConnell. H. M., 466, 536
McDonald, P. F., 351
McGarvey, B. R., 466, 536
McIrvine, E. C., 273
McKim, F. R., 317, 532
McLaughlan, S. D., 351
McLellan, A. G., 323
McMahon, D. H., 208, 502
McMillan, M., 529, 540
Mehra, A., 442
Mergerian, D., 329–30
Merle D'Aubigné, Y., 331–2, 728
Merlo, A. L., 489–90
Merritt, F. R., 328, 389
Messiah, A., 615
Metropolis, N., 623
Mikkelson, R. C., 305, 551
Mims, W. B., 115, 441, 565, 568
Miyata, N., 337
Moffitt, W., 791, 846
Morigaki, K., 470
Moriya, T., 495
Moser, C. M., 706
Mössbauer, R. L., 303, 445
Motchane, J. L., 84
Müller, K. A., 487, 736
Müller, W., 254
Murphy, J., 573

Naiman, C. S., 476
Nash, F. R., 575
Nelson, E. D., 425
Newell, G. S., 203
Newman, D. J., 305
Nicol, A. T., 347, 350
Nierenberg, W. A., 364
Nolle, A. W., 572
Novák, L., 487

O'Brien, M. C. M., 40, 169, 199, 424, 446, 460, 464, 476, 481–2, 485, 488, 791, 808, 820, 826
O'Connor, J. R., 489
Ofelt, G. S., 283
Ohtsuka, M., 514
Olschewski, L., 344–5
Ono, K., 434, 514
Opechowski, W., 529, 540, 634
Öpik, U., 462, 791, 807, 814, 817, 835, 842–3
Oppenheimer, R., 793
Orbach, R., 65–8, 304, 318, 321, 323, 329, 445, 502, 557, 560, 565
Orlich, E., 305, 310–1
Orton, J. W., 208, 322, 453–5, 466, 487
Otten, E. W., 344–5
Overhauser, A. W., 87
Overmeyer, J., 337

Owen, J., 40, 49–50, 322, 365, 392, 394–5, 400, 451, 476, 479, 482, 484–6, 498, 531–2, 534–5, 768, 775, 777, 781, 784, 787–9

Pake, G. E., 271
Pandey, B., 305
Park, J. G., 498, 531
Parker, G. W., 360
Partridge, M. F., 498, 531
Pashinin, P. P., 333
Pauli, W., 586
Peierls, R. F., 305
Penrose, R. P., 56–7, 365, 451, 457, 680
Perakis, N., 507
Peter, M., 395, 433
Peterson, R. L., 583
Pettit, G. D., 351
Pichanick, F. M., 253
Piksis, A. H., 305
Pipkin, F. M., 84
Pirc, R., 829
Plumpton, B. I., 56–7
Poindexter, J. M., 303
Pontnau, J., 429
Pope, D. F. D., 338–9
Portis, A. M., 123
Pound, R. V., 75, 78, 119
Powell, M. J. D., 304, 318, 321, 323, 441–2
Proctor, W. G., 84
Prokhorov, A. M., 331, 333, 425, 429–30, 560
Pryce, M. H. L., 170, 348, 355–60, 362–3, 391, 427, 430, 446–7, 455, 457–8, 461–2, 465, 514, 529, 678–9, 684, 697, 701, 791, 807, 814, 816–7, 835, 842–3
Purcell, E. M., 119, 271

Rabi, I. I., 98, 100–1
Racah, G., 383, 615, 623, 658, 675, 677
Rahn, R. O., 480
Rajnak, K., 305
Rampton, V. W., 436
Ramsey, N. F., 101
Ranon, U., 262–3, 331
Raubenheimer, L. J., 354, 359
Räuber, A., 469
Ray, D. K., 303, 305
Ray, Tuhina, 343
Raychaudhuri, A. K., 303
Reinberg, A. R., 452, 471
Remeika, J. P., 338–9, 484, 487
Richards, P. L., 425–6, 429
Richards, P. M., 528
Richman, I., 304
Riley, J. D., 501–2
Rimmer, D. E., 434
Roberts, L. D., 360
Roberts, S., 470

Robinson, F. N. H., 76-7
Rodbell, D. S., 535
Roest, J. A., 470
Rogers, R. N., 528
Rose, M. E., 75, 77
Rosenberger, U., 335
Rosset, J., 722, 725
Rotenberg, M., 623
Rubins, R. S., 187, 190-1, 204-5, 279, 329, 332-5
Ruby, R. H., 322, 569, 579
Runciman, W. A., 391, 430
Rundle, R. E., 277

Sabisky, E. S., 167, 228, 329, 331-4
Sandars, P. G. H., 252-4, 684, 710-2
Sands, R. H., 201-2
Satten, R. A., 304, 347, 355
Sauzade, M., 429
Scardefield, J. E., 351
Schaak, G., 305, 310-1
Schawlow, A. L., 425
Schinke, D. P., 736
Schmidt, J., 130
Schneider, J., 469
Schoening, F. R. L., 506
Schreiber, C. L., 347, 355
Schreiber, P., 471
Schumacher, R. T., 40
Schwarz, W. M., 446
Schwinger, J., 101, 124
Scott, P. L., 69, 324, 558
Scovil, H. E. D., 30, 140, 218, 339, 525, 535-6
Seaborg, G. T., 346
Searl, J. W., 201
Segré, E., 696, 698
Seidel, G. S., 538-40
Sekiyama, H., 434, 476, 478-9, 481
Serway, R. A., 487
Shaltiel, D., 207, 335
Shimada, J., 506-7
Shiren, N. S., 71, 551, 582
Shirley, D. A., 303, 315-6
Shortley, G. H., 593
Shulman, R. G., 365, 397, 767-8, 780
Shuskus, A. J., 335-6
Siegert, A., 427-8
Sierro, J., 488, 490
Silhouette, D., 429
Silsbee, R. H., 502
Šimánek, E., 487
Simon, F. E., 74, 77
Singh, S., 305
Slack, G. A., 468, 470
Slichter, C. P., 203
Smirnov, A. L., 331
Smith, D., 628

Smith, K. F., 258-9
Smith, R. C., 201
Smith, S. R. P., 208, 432
Soffer, B. H., 429
Solomon, I., 130
Sorokin, P. P., 351
Spalding, I. J., 259
Šroubek, Z., 487, 574
Stapleton, H. J., 305, 315, 318, 322, 347, 354, 359, 551
Statz, H., 343, 722
Stepanov, V. G., 846
Sternheimer, R. M., 302, 305, 687, 702, 707-8
Steudel, A., 254
Stevens, K. W. H., 50, 287-8, 290, 303, 310, 424, 436, 441, 451, 481, 484, 495, 501, 529, 541, 575, 628, 677, 694
Stevenson, M. J., 351
Stevenson, R. W. H., 445
Stombler, M. P., 329-30
Stoneham, A. M., 208
Strandberg, M. W. P., 577
Stuart, R., 395, 783
Sturge, M. D., 426, 468, 471, 824
Sugano, S., 383-90, 394-5, 397, 432-3, 476, 478-9, 481, 767-8
Sugihara, K., 501
Suss, J. T., 453, 487
Svare, I., 538-40
Swalen, J. D., 201

Tachiki, M., 502
Tanabe, V., 383-90, 394, 432, 476
Teller, E., 371, 403, 557, 790-1
Templeton, D. H., 278-80
Terhune, R. W., 273
Thomas, K. S., 305
Thornley, J. H. M., 49, 323, 392, 394-5, 400, 476, 483, 486, 628, 768, 775, 777, 781, 784, 787, 789
Thorpe, M. F., 502
Thorson, W., 791, 846
Tinkham, M., 212, 365, 615
Tinsley, B. M., 321, 324
Title, R. S., 351
Trammell, G. T., 199
Trees, R. E., 679
Trenam, R. S., 140, 186, 218, 279, 339, 437, 458-9
Trommer, W., 457
Trozzolo, A. M., 389
Tsang, T., 358
Tucker, E. B., 399
Twidell, J. W., 489-90

Uebersfeld, J., 84
Unsworth, P. J., 258-9
Usui, T., 529

Van den Handel, J., 427
van Niekerk, J. N., 506
van Vleck, J. H., 1, 65, 69, 124, 134, 271, 285, 376, 460, 475, 516, 519, 523, 527, 557, 559, 564, 578, 595–6, 648–9, 746, 791, 835, 855
van Wieringen, J. S., 413
Vella-Colleiro, G., 305
Venkateswarlu, P., 442

Wagner, G. R., 573
Wagner, P. E., 70, 580–1
Walker, M. B., 565
Waller, I., 63–4, 515, 552–3, 556, 562
Walsh, W. M., 451
Walters, G. K., 470
Walther, H., 254
Wanklyn, B. M., 340
Ward, I. M., 479, 482
Warshaw, S. I., 572
Watanabe, H., 441–2
Watkins, G. D., 470
Watson, R. E., 276, 283–4, 299, 304–5, 328, 391, 399, 411, 413, 415, 449, 453, 474–5, 478, 480, 593, 679, 687, 704, 706, 709
Weakliem, H. A., 332
Weber, M. J., 328–9, 572
Weeks, R. A., 331, 333
Weger, M., 444
Weil, J. A., 198
Weinstock, B., 15, 347, 355, 358
Weiss, P. R., 528
Weisskopf, V. F., 124
Wertz, J. E., 208, 432, 453–5, 466, 487
Wetsel, G. C., 351
Wheeler, R. G., 304, 480
Whiffen, D. H., 226
Whiffin, P. A. C., 487

White, R. L., 502
Whitley, S., 57
Wickersheim, K. A., 502
Wickman, H. H., 349, 389
Wielinga, R. F., 470
Wigner, E. P., 615, 644, 649
Wilkens, J., 445, 464–5, 488
Williams, F. I. B., 140, 215–6, 252–3, 269, 279, 313, 335, 338, 340, 343, 459, 463, 465, 711, 736, 830
Williams, H. J., 389
Windsor, C. G., 784, 787
Wolf, W. P., 68, 317, 321, 323, 327–9, 332, 351, 502, 532, 560, 719–20, 728
Wong, E. Y., 277, 304, 307, 313, 315, 322, 325, 347, 349, 355
Wong, J. Y., 425–6, 446
Wood, D. L., 283
Woodbury, H. H., 198, 470, 664
Woodgate, G. K., 252–4, 411, 711–2
Woonton, G. A., 434, 441
Wooten, J. K., 623
Wu, C. S., 77
Wyard, S. I., 201
Wybourne, B. G., 277, 282–3, 323, 341, 347–8, 350
Wysling, P., 487

Yariv, A., 351, 468, 471
Yngvesson, K. S., 552, 561
Yuster, P. H., 40, 488

Zalkin, A., 278–80
Zaripov, M. M., 846
Zavoisky, E., 365
Ždánský, K., 487, 574
Zeks, B., 829
Zeldes, H., 199
Zverev, G. M., 331, 429–30

SUBJECT INDEX

A-tensor, 26, 166, 168, 170, 750
abelian group, 601
acoustic paramagnetic resonance, 436
actinide group, 346–64
— ions, 347
— — in CaF_2, 350
— — octahedral symmetry, 354
— series, 347
adiabatic condition, 103–4, 106
— demagnetization, 365
— rapid passage, 73, 82–3, 90, 104–8, 112–4, 220–1, 237–8, 240–2, 578
Ag in KCl, 40
Ag^{2+} in KCl, 488
— in LiCl, 490
— in NaCl, 490
Al in quartz, 40
Al_2O_3, 389, 475
allowed transitions, 28
Am^{2+} in CaF_2, 351–3
Am^{2+}, g_J, 354
angular momentum, 615–7
anhydrous rare earth chlorides, 305
anisotropic exchange, 59, 495–9
— — interaction, 501, 508, 855
— — modulated by phonons, 552
— g-factor, 135–9, 402, 420, 455
— spectrum, vibronic doublet, 827
anisotropy, 41
anti-bonding orbitals, 764–8, 773, 780, 787–8
antilinear antiunitary operator, 644
anti-shielding effects, 55, 491, 501, 687, 707–9
antisymmetric direct products, 657, 661
antiunitary operator, 645
atomic beam triple resonance, 252–4, 258, 299
available power, 127
axial field, 419
— symmetry, 18, 157, 178, 200
α, β, γ for a single electron, 873

Back–Goudsmit region, 6–7
BaF_2, 281, 325, 467
BeO, 467
biquadratic exchange, 534–5
Bloch equations, 115
Bohr magneton, 585
bolometer detection, 130–2
— sensitivity, 131–2

bonding effects, 392–8, 679
— — orbital reduction factor, 424–6
— orbitals, 764–8, 788
Born–Oppenheimer approximation, 792, 807, 809, 814–5
— — validity of, 793
— functions, 810
'brute force' method, 74, 78, 84
'burning a hole in a line', 72, 236–7

C_{3h} symmetry, 285–95, 733, 736
C_{3v} symmetry, 733
$CaCO_3$, 439
CaF_2, 281, 325, 467–8, 490, 719
— dipolar interactions, 524
CaO, 281, 325, 719
cavity coupling, 130
— design, 129
— detuning, 126
— geometry, 125
— optimum coupling, 127, 129
— resonator, 125
$CaWO_4$, 280, 467–8, 475, 718, 733
CdF_2, 281, 331
CdO, 467
CdS, 467–8
CdSe, 467
CdTe, 467–8
Ce^{3+}, 33, 308
— in C_{3h} symmetry, 308
— in C_{3v} symmetry, 323
— in $La_2Mg_3(NO_3)_{12}$, $24H_2O$, 323, 579
$CeCl_3$, 310–1
central field approximation, 42
CeO_2, 282, 331
cerium ethylsulphate, 310, 501
— magnesium nitrate, 68, 279–80, 581
$[(CH_3)_4N]_2PtCl_6$, 347
$[(CH_3)_4N]_2UCl_6$, 355
characters, 604–5, 607, 620, 630, 657
chemical shift, 716
chlorides, lanthanide, 303–25
circular polarization, 15, 120–2, 132, 178, 726
circularly polarized nuclear field, 225–6, 232
— — radiation, 9, 120, 139, 146, 290, 355
classical equations of motion, 102–3
Clebsch–Gordan coefficients, 620–3, 626, 675, 754

SUBJECT INDEX

Cm^{3+} in CaF_2, 351–3
— in $La(C_2H_5SO_4)_3$, $9H_2O$, 350
— in $LaCl_3$, 349–50
Cm^{3+}, g_J, 354
$^{244}Cm^{3+}$ in CeO_2, ThO_2, 352
$Co(NH_3)_6Cl_3$, 475
$Co(NH_4)_2(SO_4)_2$, $6H_2O$, 124
Co^+ in CaO, 453
— in MgO, 453
Co^{2+} in $CdS(\alpha)$, 470
— in $CdTe$, 470
— in $CoCs_3Br_5$, 470
— in $CoCs_3Cl_5$, 470
— in cubic $ZnS(\beta)$, 470
— in MgO, 752, 846–7
— octahedral symmetry, 497, 752, 769–72
Co^{2+}, $3d^7$, 758
$^{59}Co^{2+}$ in MgO, 270, 447
cobalt ammonium sulphate, 77, 176, 518
cobaltous salts, high anisotropy, 449
$CoCs_3Br_5$, 467
$CoCs_3Cl_5$, 467–8
commutation relations, 722, 748
— rules, 616
complex conjugation, 644–6
— cyanides, 380–1, 762, 769
— susceptibility, 116–7, 120–4, 126
— XY_6, 392, 804
concentration dependence of τ_1, 553
conduction electrons, 39, 132
configuration interaction, 49, 591, 675–705, 709, 761
configurations, 42, 590–1
contact interaction, 47, 256, 695, 753–4, 786
copper ammonium selenate, 56
— — sulphate, 57
— caesium sulphate, 57
core polarization, 46–7, 251, 254, 258, 267, 297–300, 411, 413, 432, 434, 452, 458, 480, 484–5, 702–6, 709, 711, 780, 787, 803
Coulomb energy, 282, 382–3, 795
— field, 42
— interaction, 379, 384, 394, 491, 499, 502, 792
— repulsion, 363, 696
coupled spin–phonon system, 575
covalency, 764–7, 833
covalent bonding, 476, 480, 598, 661, 721, 728, 761–89, 846
— — orbital reduction, 773–5
— — spin–orbit coupling, 775–7
— complexes, 758
$[Cr(H_2O)_6]^{3+}$, 431
$CrSO_4$, $5H_2O$, 434–5
Cr^0 in Si, 846
Cr^+, Mn^{2+}, $Fe^{3+}(d^5)$, 779, 781
Cr^+, Mn^{2+}, Fe^{3+} in octahedral field, 436–42

Cr^{2+} in $CdS(\alpha)$, 470
— in MgO, 436
— in octahedral field, 434–6
Cr^{3+}, 43
— in Al_2O_3, 273, 432–3, 451
— in complex cyanides, 479
— in MgO, 431
$Cr^{3+}(d^3)$, 779–80
— in K_2NaCrF_6, 780
Cr^{3+}, $3d^3$ in octahedral field, 412
$^{53}Cr^{3+}$ in Al_2O_3, 432–4
— in MgO, 434
Cr^{4+} in α-Al_2O_3, 429
cross relaxation, 71, 265, 269–73
crystal electric potential, 12, 45, 285, 287, 557, 758–9, 761
— field, 12, 595–9, 673
— — configuration, 598
— — energy, 379
— — parameters, 304–5, 321
— — terms, 598
— — theory, C_{3h} symmetry, 285
— imperfections, 205–8
— orientation using forbidden transitions, 227
$CsCr(Sr_4)_2$, $12H_2O$, 523
$CsTi(SO_4)_2$, $12H_2O$, 417
Cs_2ZrCl_6, 347
$Cu(CH_3COO)_2$, H_2O, 506–9
$[Cu(H_2O)_6]^{2+}$, 456, 462
$(Cu, Mg)_3Bi_2(NO_3)_{12}$, $24H_2O$, 460
$(Cu, Mg)O$, 460
$(Cu, Mg)SiF_6$, $6H_2O$, 460
$Cu(NH_3)_4SO_4$, H_2O, 528
$CuSO_4$, $5H_2O$, 58, 435, 513–4, 529
$(Cu, Zn)(BrO_3)_2$, $6H_2O$, 460
Cu^{2+}, 807, 814, 817–8, 826
— hyperfine structure, 457–9
— in BeO, 471
— in CaO, 466
— in $CdWO_4$, 574
— in copper fluosilicate, 816
— in glass, 202
— in $La_2Mg_3(NO_3)_{12}$, $24D_2O$, 458–9
— in $La_2Mg_3(NO_3)_{12}$, $24H_2O$, 458, 463, 465
— in MgO, 466, 832
— in ZnO, 468, 471
Cu^{2+}, $3d^9$, 807
— in cubic symmetry, 830–2
— in MgO, 831
Cu^{3+} in Al_2O_3, 451–4, 476, 487
cubic coordination, 325, 374, 743
— field, 372–6
— — operator, 373–4
— — parameter (Dq), 377–8, 391, 468
— — splitting, 44, 142–51, 325–38, 341–5, 351–9, 372–96, 629–42, 719–32, 739–60

cubic field states, 375
— group, 667, 720–2, 731, 795, 811, 822
— — O, 629, 661–4
— — — character table, 856
— — — direct products, 856
— — — irreducible representations, 630, 856
— — O_h, 641, 763
— potential, 367, 667–70, 719
— — octahedron, 669
— — tetrahedron, 670
— representation D_J, states of, 857
— splitting for $4f^7$, 335
— symmetry, 14, 24–5, 367, 372, 649, 667, 685–6, 742, 751, 794, 797, 805
— — departures from, 405–6
— — nuclear electric quadrupole interaction, 303
— terms, 768–9, 878
— zinc blende, 467
Curie's law, 40, 545, 849, 852
cyanides, 762

D-tensor, 151–2, 170, 750
$d\epsilon$ triplet, 379, 744
$d\gamma$ doublet, 379, 744
$(d\epsilon)^6$ sub-shell, 477
d^1 eightfold coordination, 490
— octahedral field, 404–5
— strong octahedral field, 478
— trigonal distortion, 417–26
d^2 octahedral field, 405
— strong octahedral field, 381–4, 479, 769, 772–3
— trigonal distortion, 426–30
d^3 core polarization, 434
— covalency effects, 433
— strong octahedral field, 383–5, 479–80
— tetrahedral symmetry, 405
— trigonal distortion, 432
d^4 orthorhombic distortion, 434
— strong octahedral field, 385–6, 480–1
— tetragonal distortion, 434
— tetrahedral symmetry, 404
d^5 fine structure, 436–42
— hyperfine structure, 436
— — strong octahedral field, 481–3, 485
— strong octahedral field, 385–7, 481–6, 768–9
d^6 octahedral field, 404
— strong octahedral field, 388, 486
d^7 octahedral field, 383, 405
— strong octahedral field, 389, 486–7, 769
d^8 strong octahedral field, 390, 487, 772–3
— tetrahedral symmetry, 405
d^9 octahedral field, 455
— — tetragonal distortion, 456–8
— — trigonal distortion, 458–66

— strong octahedral field, 487–90
— tetrahedral symmetry, 404
$3d$ group, 703
— cubic terms for, 878
— cyanides, 476
— (Dq), ζ, 379, 391
— λ, 399–400
— nuclear properties, 416
— $\langle r^{-3}\rangle$, $\langle r^2\rangle$, $\langle r^4\rangle$, ρ, 399
— Racah parameters, B, C, 378, 391, 395
— tetrahedral symmetry, 467–71
$3d^1$, tetrahedral symmetry, 469
$3d^2$, tetrahedral symmetry, 469
$3d^3$, tetrahedral symmetry, 469
$3d^3(^4F)$, 782–3
$3d^4$, tetrahedral symmetry, 469–70
$3d^5$, tetrahedral symmetry, 440, 470
$3d^5$, ^6S ions, representative values, 440
$3d^6$ octahedral field, 443
— tetrahedral symmetry, 470
$3d^7$ octahedral field, 446–9, 769
— tetrahedral symmetry, 470
$3d^8$ octahedral field, 449–55
— tetrahedral symmetry, 470
$3d^8(^3F)$, 782–3
$3d^9$ octahedral field, 455
— tetrahedral symmetry, 470–1
$4d$ group, 472–89, 769
— — core configuration, 472
— — nuclear properties, 473
— — spin–orbit coupling, 474
— — $\langle r^{-3}\rangle$, $\langle r^2\rangle$, $\langle r^4\rangle$, 474
— hyperfine structure, 478
— ions, 474
$5d$ group, 472–89, 769
— core configuration, 472
— nuclear properties, 473
— hyperfine structure, 478
— ions, 474
— — spin–orbit coupling, 475
defects, 206
degeneracy, 601
demagnetizing field, 854
$d\epsilon$, $d\gamma$ states, 379
diamagnetic shielding, 686
— susceptibility, 585
— term, 585–6
diamagnetism, 1
diamond, 55
dipolar broadening, 516
— hyperfine interaction, 753–4
— interaction, 396, 523, 536, 854–5
dipole–dipole interaction, 689
Dirac equation, 586, 691
direct products, 612–3, 625, 630, 721, 796, 798
— relaxation process, 63–7, 240, 553–5, 558–60, 563–74, 582, 828–30

displacement in time D_t, 643
distant Endor, 273
donors in silicon, 221, 234–42
Doppler effect, 52
double cubic group O^+, 634–7
—— character table, 858
—— direct products, 859
—— irreducible representations, 858
—— states of, 860
— nitrates, 278, 320–5, 365
— quantum line, 208, 469
— tetragonal group, 639
— trigonal group, 640, 649
—— character table, 861
—— irreducible representations, 862
doublet E, 770
— orbital ground state, 402–3
Dy^{2+} in BaF_2, 328
— in CaF_2, 327–30, 494
— in SrF_2, 328
Dy^{3+} in CaF_2, 328
— in CaO, 329
— in C_{3h} symmetry, 317
Dy^{3+}, Er^{3+} in C_{3v} symmetry, 325
dynamic Jahn–Teller effect, 401, 403, 417, 426, 446, 459, 462, 469, 490, 790–2, 804, 846
—— ^2E state, 808–32
— nuclear orientation, 74–87
—— polarization, 78, 80, 273–4
— orbit–lattice interaction, 557
— saturation, 118, 130, 220, 240–1
$\Delta M = \pm 2$ transition, 21, 155, 159, 208, 454
$\Delta m = 0$ transition, 181
$\Delta m = \pm 1, \pm 2$ transition, 182

^2E level, 797–804
— resonance spectrum, 802
e (doublet), 379
$\frac{10}{3}$ effect, 528
effective spin, 10–16, 133
— Hamiltonian, 35, 407
effet solide, 84
Einstein coefficients of absorption and emission, 542
electric dipole moment, 215, 295, 659–664, 736
—— transitions, 214–6
— dipole–dipole interaction, 499
— field effect, 659–64
—— gradient, 345, 415
— quadrupole tensor, 628
— quadrupole–quadrupole interaction, 491, 500–2
electromagnetic radiation bath, 542
— density, 543
— field, heat capacity, 547
electron paramagnetic resonance, advantages and limitations of, 93–4

——phonon interactions, 334
— spin density, 408
electronic and nuclear magnetic dipole moments, 1–5
— magnetic octupole moment, 342
— multipole terms, 139–63, 851
— oscillatory spin Hamiltonian, 136–9, 230
— quadrupole splitting, 151–63, 854
—— tensor, 683
— spin Hamiltonian, 398–406
— Zeeman interaction, 7, 13, 34–6, 398, 400, 677–8
electrostatic interaction, 55, 371
— repulsion, 594
— shielding, 415
Endor, 87–93, 299, 354, 656
— accuracy, 91–2, 217
— crystal orientation, 261
— enhancement effects, 222, 228–33
— level populations, 245
— line widths, 264–71, 454
— on donors in silicon, 234–42
—— relaxation effects, 239–42
— relaxation effects in, 243–51
— sign determination from, 223–7, 233
— transition, 223, 225–7
—— indirect observation, 272
entropy, 848
equivalent noise temperature, 128
— operator, 628, 670–9, 683, 692, 863
— representation, 602
Er^{3+} in CaF_2, CdF_2, ThO_2, CeO_2, 330–2
— in CaO, 332
— in C_{3h} symmetry, 320
— in C_{3v} symmetry, 325
— in cubic field, 330–2
— in MgO, 331–2, 727
— in ZnSe, 331–2
Er^{3+}, Ho^{2+} in CaF_2, 330
ethylsulphates, 303–25
Eu^{2+} in BaO, 337
— in CaF_2, 252, 263, 269, 271, 335, 337, 340, 342–3, 572–3, 711
Eu^{2+} in CaO, 335–7
— in $SrCl_2$, 335
— in SrO, 337
Eu^{3+} in C_{3h} symmetry, 316
Euler angles, 604, 617
EuO, 281
EuS, 281
EuSe, 281
EuTe, 281
exchange broadening, 59, 529
— interactions, 41, 57, 59, 410–1, 491–2, 494–9, 502, 594
—— line shape, 527
—— modulated by phonons, 552
— narrowing, 57–9, 527–9

excited configurations, 667
— terms, 377, 381, 758–60
— vibronic state, 828
exclusion principle, 702–3

$4f$ group, 276–345, 705
— crystal field parameters, 304–5, 321, 335
— hyperfine constants, 286, 298
— ionic size, 276
— nuclear data, 286
— $\langle r^2 \rangle$, $\langle r^4 \rangle$, $\langle r^6 \rangle$, 276
— $\langle r^{-3} \rangle$, 300
— resonance data, 306–7, 322
— ζ, 284
$4f$ ions, multiplicative factors, 874–5
$4f^7$, hyperfine constants, 340
— — interactions, 341
— in C_{3h} symmetry, 338
— in C_{3v} symmetry, 338
— in cubic symmetry, 336, 343
— ions, 335
$4f^{13}$, Yb^{3+}, 719
$5f$ group, 344–64, 715
— ionic size, 347
— ζ, 347
$5f^1$, octahedral symmetry, 354–9
$5f^2$, U^{4+} in CaF$_2$, 350
$5f^3$, U^{3+} in CaF$_2$, 350
$5f^4$, U^{2+} in CaF$_2$, 350
$5f^7$, $^8S_{\frac{7}{2}}$, Γ_6 doublet, 352
Faraday rotation, 566
Fe$^+$ in MgO, 846–7
— in Si, 846–7
Fe^{2+}(CN$^-$)$_6$, 476
Fe^{2+} in CaO, 446, 846–7
— in CdTe, 470
— in MgAl$_2$O$_4$, 470
— in MgO, 208, 444, 446, 501–2, 582, 831, 835, 846–7
— in NaF, 445
— in ZnF$_2$, 212
— in ZnS(β), 470
Fe^{2+}(3d^6), 704
— in MgO, 835
Fe^{3+}, 162, 779
— in complex cyanides, 482
— in glass, 203
— in K$_3$Co(CN)$_6$, 482–3
— in MgO, 441
Fe^{3+}(CN$^-$)$_6$, 476
fictitious angular momentum, 373, 403–4, 427, 632, 659, 752, 857
— orbital momentum, 769–70
— rotation R, 653
— spin, 651, 722, 734, 739, 750
fine structure, 16–25, 108, 139–63, 172
— axial symmetry, 157

— constant, 3
— in cubic fields, 142–51
— in strong magnetic fields, 156–63
— spin operators, 140
— splittings, 402
first moment, 538
flip-flop transitions, 78
fluorine hyperfine structure, 260, 262–4, 269, 397
— ligands, 365
forbidden hyperfine lines, 204–5
— hyperfine transitions, 186–91, 207–8
— transitions, 560, 727
forced precession, 98, 105
— resonance, 106, 116
Fortran notation, 628
fourth moment, 516, 522–3
free radicals, 39

g-factor, 135–9, 759
— ellipse, 422
— negative, 144
— sign of, 15, 138–9, 422–3
g, isotropic in S-state, 151, 440
$\mathbf{g}$-tensor, 13, 15, 23, 134, 167, 169–70, 729, 750
— effect of electric field, 663–4
$g_\parallel$, $g_\perp$ for Γ_5 orbital triplet, d^1, 421
garnets, 468
gauge invariance, 585
Gaussian distribution, 55, 213, 735
— line shape, 55, 522–4, 528
Gd^{3+} in Al$_2$O$_3$, 338–9
— in bismuth magnesium nitrate, 339
— in CaF$_2$, 335, 572
— in CaO, 335
— in CeO$_2$, 340
— in LaCl$_3$, 339, 341
— in lanthanum ethylsulphate, 339
— in neodymium ethylsulphate, 535–6
— in SrCl$_2$, 335
— in ThO$_2$, 207, 335, 340, 343
ground configurations, $4f$ ions, 282
— multiplet of $4f^6$, ^7F, 283
Γ_1, Γ_2 singlets, 325–6
Γ_1 singlet, 632, 739, 809, 811
Γ_2 singlet, 430–1, 449–50, 469, 632, 739, 742, 752–3, 809, 811
Γ_3 doublets, 325–7, 329, 342, 375, 417, 436, 455, 459, 469, 494, 501, 631, 640, 740–4, 757, 770, 790, 792, 797, 799, 804, 809–11, 813, 816–7, 820, 831
— non-magnetic, 633, 745
$^2\Gamma_3$ level, Zeeman spin Hamiltonian, 801
Γ_4, Γ_5 triplets, 325–7
— states for cubic representations, 858
Γ_4 in trigonal axes, 633

Γ_4 triplet, 329, 383, 427, 430, 446, 469, 497–8, 501, 631, 639, 661, 663, 721, 739–40, 742–3, 752–3, 755, 790, 832–47
Γ_5 triplet, 329, 373, 375, 417, 419, 421, 423, 430, 443, 451, 455, 469, 501, 631, 661, 663, 721, 739–40, 742–4, 752–3, 755, 757, 790, 832–47
— d^1 in octahedral field, 418
— in trigonal axes, 633
Γ_6, Γ_7 doublets, 325–6, 337
Γ_6 doublet, 148–50, 329–31, 334, 345, 357, 471, 636, 650, 721, 752–4
— in cubic field, anisotropy of, 353
Γ_6, Γ_7 Kramers doublets, values of g, 876
Γ_7 doublet, 143–4, 148–50, 330–2, 334–5, 344–5, 357, 636, 650, 721, 752–3
Γ_8 quartet, 143–6, 149–50, 165, 325–6, 328, 331–2, 337, 342, 345, 351, 353–4, 357, 463, 470, 480, 501, 636, 716, 721–32, 752–3, 755, 802
— circular polarization, 145
— crystal distortions, 729–30
— nuclear electric quadrupole interaction, 731–2

H in CaF_2, 40
H_1, optimum orientation of, 137–8
half-field line, 454, 518
Hartree–Fock calculations, 705
— — $4d$ group, 475
— method, 590–2, 698, 703
— wave-functions, 595, 599, 667–8, 703, 710, 761, 763, 767, 775
high-order terms in the spin Hamiltonian, 341
— Zeeman effects, 35
high-temperature approximation, 850
Ho^{2+} in CaF_2, 330–2
Ho^{2+}, Er^{3+} in a cubic field, 331
Ho^{3+} in C_{3h} symmetry, 318
hole configurations, 673, 744
holmium ethylsulphate, 319
homogeneous broadening, 53, 72, 92, 110, 112, 522, 535
Hund's rule, 42, 282, 379, 479, 592, 672–3, 678, 692, 698, 710, 714, 744, 759, 762, 873
hyperfine anomaly, 239, 251, 253, 452, 706–7, 712, 717
— coupling, 801
— energy levels, 8, 27
— fields for $3d^8$ ions, 453
— interaction, 50
— — sign, 29
— line widths, 465
— multiplet, 5
— operators, 48
— structure, 5–8, 172, 589, 759

— — contact interaction, 410
— — contact term, 695–6
— — covalent reduction, 414
— — cupric acetate, 509
— — diamagnetic effects, 408
— — energy levels, 183
— — equivalent operators, 409
— — europium, 251
— — free atoms, 163
— — Γ_3 doublet, 412–3, 415
— — orbital singlet, 412–3
— — — triplet, 414
— — relative intensity, 168
— — relativistic corrections, 711–2
— — — effects, 297, 408, 411
— — s-electrons, 410
— — S-state ions, 710–1
— — sign determination, 185
— — tensor, 134, 651, 655, 713
— — transition, 183
— — zero-field transitions, 179–80
— terms, 852

icosahedral crystal field, 325
— symmetry, 321
icosahedron, 278
'ideal' paramagnet, 853
improper rotations, 640–2, 661
induced absorption, 60, 558
— emission, 60, 558, 560
inhomogeneity in D and g, 454
inhomogeneous broadening, 52–3, 72, 92, 110–1, 114, 221, 265, 267, 521, 535, 732
intensity, 119, 162
— distribution, 202
intermediate coupling, 335, 593, 714–5, 739
— — $5d$ group, 480
— crystal field approach, 372–7, 380, 597–8, 676, 761
— ligand field, 44, 47, 371, 398, 407–8, 414
internal energy, 544, 848–9, 853
invariance, 601
inversion, 660
— operation, 641, 764
— symmetry, 670, 736
inverted population, 73
ionic bonding, 661
— complex, 767
— diffusion, 536
— size, $4d$ and $5d$ ions, 473
— —, $4f$ ions, 276
— —, $5f$ ions, 347
ions of type A, 745, 777–81
— B, 751
$[IrBr_6]^{2-}$, 484
$[IrCl_6]^{2-}$, 484
iron group, 559, 598, 670, 676, 684, 695, 697, 703, 705, 761

irreducible representation, 603–9, 612, 614, 617, 620, 625, 764, 795–6, 804
— — of O, 630
— — of O^+, 634–7
— — of O_h, 768
— tensor operators, 624–8
$(IrX_6)^{2+}$ complexes, 486
Ir^{4+} in Na_2PtCl_6, $6H_2O$, 483–4
— in K_2PtCl_6, 531–2
— in $(NH_4)_2PtCl_6$, 483–4, 486, 532
isolated representations, 720
isomer shift, 847
isotropic exchange interaction, 57, 495–9, 498, 508, 510–14, 516, 527–8, 855
— singlet spectrum, 826–7, 829, 831–2

$3j$-symbols, 621
$6j$-symbols, 623, 675–6, 835
Jahn–Teller coupling, 791, 807, 826, 835, 837, 845, 847
— distortions, 490
— — for a d^9 ion, 460
— effect, 51, 406, 436, 449, 460, 464–5, 486, 790–847
— — orbital triplet, 832–47
— — Γ_3 coupling, 835–41
— — Γ_5 coupling, 841–6
— — tunnelling, 820–6, 842–4
— energy, 461, 805, 843
— spectrum, motional narrowing, 826–30
— theorem, 735, 792, 795–6, 804

$KMgF_3$, 366, 396
$KMnF_3$, 784
K_2CuCl_4, $2H_2O$, 528
$K_2(Ir, Pt)Cl_6$, 531–2
$K_2[Ir, PtBr_6]$, 485
K_2OsCl_6, 481
K_2PtCl_6, 475
K_2RuCl_6, 481
$K_3Co(CN)_6$, 475, 762
$K_3Fe(CN)_6$, 482, 762
$K_4Fe(CN)_6$, $3H_2O$, 475
Kramers degeneracy, 289, 647–56, 797, 835
— doublets, 12–13, 342, 406, 414, 420, 501, 558–9, 564–5, 649–64, 685, 729, 736
— ion, 563, 566, 757
— — weak crystal field, 713
— theorem, 640
Kramers conjugate state, 647–8, 729
Kramers–Kronig relations, 123

$\langle L\| \alpha \|L \rangle$, $\langle L\| \beta \|L \rangle$, values for $3d$ group, 873
$La(C_2H_5SO_4)$, $9H_2O$, 346, 349
$LaCl_3$, 346, 348
— phonon velocity, 551
La^{2+} in CaF_2, 489–90

Landé factor, 587–8, 627, 677
— formula, 4, 752, 833
— interval rule, 6, 282, 400, 533–4, 593, 679, 840
lanthanide chlorides, 277, 287, 303–20, 733
— compounds, 277
— double nitrates, 277, 287, 320–5, 733
— ethylsulphates, 277–8, 303–20, 733
— ions in cubic symmetry, 325
— nitrides, 501
— series, 277
lanthanum ethylsulphate, 30
— hexa-antipyrene iodide, 279
Laplace equation, 631
Larmor velocity, 115
lattice strains, 436, 548, 552
level populations, 272
ligand field, 11, 39, 41, 43
— — energy, 371
— — modulation, 64
— hyperfine interaction, 108, 264
— — structure, 50, 93, 192–200, 250, 271, 365, 396, 485, 762, 777–89
— — — Co^{2+}, 784–8
— — — correction terms, 781–4
— — — effects of $1s$ wave-functions, 783
— — — Endor measurements of, 259
— — — ions of type A, 777–84
— — — ions of type B, 784–8
— ions, 392
— orbitals, 392
— quadrupole hyperfine structure, 788–9
line breadth parameter, 522
— broadening by τ_1, 574
— — spin–spin interaction, 514–29
— shape due to dipolar interaction, 521–7
— — method of moments, 515–20
— shift at low temperatures, 537–40
— width, 54, 119
linear operator, 601–2, 606
— polarization, 121–2, 124–5, 132, 136
— representation, 602, 608
— unitary operator, 644
— vibronic Hamiltonian, 811
local fields, 109
longitudinal relaxation time, 109
Lorentzian function, 110
— line shape, 58, 522, 528, 574
— — dilute salts, 531
low temperatures, change of intensity, 162
LS-coupling, 3, 12, 42, 282–5, 348, 371, 407–10, 469, 587–9, 593, 597–8, 683–4, 712, 714
— breakdown, 253–4, 284–5

magnetic cooling, 40, 78, 365
— dilution, 41, 54, 111, 530–1
— dipole, 165, 286, 688

magnetic dipole interaction, 52, 57–8, 499, 521, 529
— dipole–dipole interaction, 491–5
— hyperfine constant A_J, values of, 286, 298
— — Hamiltonians, 296, 723
— — interaction, 6–7, 36, 78, 227–8, 326, 364, 406–14, 445, 560, 570–2, 687–95
— — — second-order correction to, 257
— — — thermal modulation, 248
— — operator, 47, 778
— — structure, 25–31, 169, 171, 174–5, 196–200, 296–301, 480, 746, 753, 756, 789
— — — $3d$-group, 406–14
— — — nuclear electric quadrupole interaction for $3d^8$, 452–4
— — — orbital contribution, 407–8
— — — second-order effects, 173, 176
— — — sign determination, 177
— — — spin contribution, 408–9
— ion–lattice interaction, 558
— moment, 663
— multipole operators, 687
— octopole, 688
— resonance intensity, 92
— spin–spin interaction, 562
— vector potential, 584, 688
magnetization, 102, 848
magnetogyric ratio, 2, 3
— tensor, 651
magnetostatic modes, 539
manganese ammonium sulphate, 172
— fluosilicate, 158
maser amplifier, 73–4
matrices of spin–orbit coupling, 448
matrix elements of operator equivalents, 864–72
— transformation, 662
— unitary, 617
measurements in strong fields, 23
$MgAl_2O_4$, 468
$(Mg, Cu)_3La_2(NO_3)_{12}, 24D_2O$, 186
MgO, 55, 281, 325, 475, 719
— phonon velocity, 551
minimum detectable susceptibility, 128
— — signal, 128
Mn^+ in Si, 846
Mn^0 in Si, 846–7
$Mn^0(3d^54s^2)$, 704
Mn^{2+}, 701–2, 705
— in BaF_2, 572
— in CaO, 441
— in $CaCo_3$, 439
— in complex cyanides, 482
— in $K_4Fe(CN)_6, 3H_2O$, 482–3
— in MgO, 208, 441–2
— in powder, 203–4
— in $RbMnF_3$, 442

— in SrF_2, 572
— in ZnS, 441
— pairs, in MgO, 535
$Mn^{2+}(d^5)$ in $KMnF_3$, 780
$Mn^{2+}(3d^5)$, 704, 779
Mn^{3+} in $K_3Mn(CN)_6$, 480
Mn^{4+} in MgO, 431
$^{55}Mn^{4+}$ in Al_2O_3, 434
Mo^{5+} in $CaWO_4$, 489–90
— in K_2SnCl_6, 479
— in $K_3(InCl_6), 3H_2O$, 478
— in TiO_2, 479
molecular field approximation, 854
— orbitals, 12, 49, 762–89
mosaic structure, 205
Mössbauer measurements, 847
motional narrowing, 465, 527, 536, 831
— in liquids, 536
multiplet ground state, 565
multipole
— interactions, 499–502
— moment, 681
mutual spin flips, 522

N in diamond, 40
$Na_2[Ir, PtBr_6]6H_2O$, 485
$Na_2(Ir, Pt)Cl_6, 6H_2O$, 50, 485
Nd^{3+}, 30, 67, 216
— in C_{3h} symmetry, 313
— in C_{3v} symmetry, 324
— in $LaCl_3$, 255–9, 349
— in $La_2Mg_3(NO_3)_{12}, 24H_2O$, 324, 568–71
— in yttrium ethylsulphate, 67, 216, 568–71
^{143}Nd in lanthanum ethylsulphate, 179
negative temperature, 70, 73, 578
neodymium ethylsulphate, 515, 527, 529, 535–6, 538
— — dipolar interaction, 524–5
$(NH_4)_2(Ir, Pt)Cl_6$, 485, 532
$(NH_4)_2PtCl_6$, 475
$[NiF_6]^{2+}$, 762, 766, 768, 775
$[Ni(H_2O)_6]^{2+}$, 451
$NiSiF_6, 6H_2O$, 451
^{61}Ni in Al_2O_3, 265–9
Ni^+ in MgO, 466
— in LiF, NaF, 464, 808
Ni^{2+} in Al_2O_3, 266, 451–4, 476, 487
— in CaO, 453
— in $KNiF_3$, 476, 780
— in MgO, 208, 453, 582
— in $Zn_3La_2(NO_3)_{12}, 24H_2O$, 451
$Ni^{2+}(d^8)$, 780
$Ni^{2+}(3d^8)$, 704
Ni^{3+} in Al_2O_3, 487, 831
— in CaO, 487
— in MgO, 487
noise power, 73

non-bonding orbitals, 768
non-Hermitian operators, 647
non-interacting ions, 851
non-Kramers doublets, 24, 32, 213–6, 313, 315, 318, 324, 342, 363, 454, 470, 501, 565, 736
—— hyperfine interaction, 295
—— line shape, 214–6, 295
—— spin Hamiltonian, 294
— ions, 209–16, 292, 317, 558–9, 563, 566–7, 732–41, 750, 757
non-linear shielding, 304
normal coordinates, 795, 797, 804
— modes of vibration, 460
NpF_6, 346–7, 362
— in UF_6, 355–9
NpO_2, 501
$(NpO_2)^{2+}$, 347–8
— in $(UO_2)Rb(NO_3)_3$, 360–363
nuclear alignment, 40, 76, 365
— electric quadrupole
—— effects, 184
—— interaction, 8, 31–3, 36–7, 163–71, 178, 181, 225–8, 254, 260, 286, 301–3, 326, 342, 345, 363–4, 414, 434, 445, 458, 478, 680–7, 740, 746, 754–5
——— lattice contribution, 259, 302
——— ligand field contribution, 415
——— thermal modulation of, 248
—— moments, 3d group, 416
———, 4d group, 473
———, 5d group, 473
———, 4f group, 286, 340
———, definition, 683
—— operator, 46–7, 680–7,
— induction, 106
— magnetic resonance in K_2OsF_6, 480
— moments, $4f$, 286; $3d$, 416; $4d$, $5d$, 473
— multipole, 682
— polarization, 74, 77, 79, 80–1, 83, 85–6
— quadrupolarization, 75–7, 79–81, 83, 86
— transition probability, 232
—— enhancement of, 221–2, 228, 250
— transitions, 31, 89
— Zeeman effect, 6, 196–200
—— interaction, 7, 27, 74, 224, 228–9, 414–6, 746, 852

octahedral complex, 49, 797
— coordination, 367, 374, 743
— crystal field splitting Δ, 476
— symmetry, 325
one-dimensional harmonic oscillators, 837
one-electron operators, 594–5, 666, 673
operator equivalents, 626
—— matrix elements, 864–72
optical spectra, 276

optimum sensitivity, 129
Orbach–Blume process, 560
Orbach relaxation process, 66–8, 323, 445, 560–6, 568–9, 572, 579, 582, 828
orbital contributions, g-values, 464
— degeneracy, 596
— g-factor, 377, 383, 404, 728, 833
—— reduction of, 395, 400
— hyperfine interaction, 753–4
— magnetic moment, 50, 585
— magnetization, 782
— moment reduction, 424–6, 432, 443, 482, 773–5, 847
— momentum, 401, 585, 838, 840
—— quenching, 51, 596, 648–9, 846
——— by Jahn–Teller effect, 839
— splitting factor, reduction in, 841
— triplet, coefficients m, n, 876
orthogonality relation, 603, 609
Os^{4+}, 480
oscillatory electronic Zeeman interaction, 136–9, 230–1
— nuclear g-factor g_1, 230
—— Hamiltonian, 233
—— Zeeman interaction, 230
overlap integral, 764–5

P-tensor, 166, 170, 750
p-ligand orbitals, 778
Pa^{4+} in Cs_2ZrCl_6, 354–5, 359
packet-shifting mechanism, 237
pair spectra, 502–14, 524–35
— anisotropic exchange, 505–6
— dissimilar ions, 509–14
— isotropic exchange, 504–5
— similar ions, 502–9
palladium group, 598
para-electric transitions, 736–7
paramagnetic anti-shielding, 223
— relaxation, 124, 573
— resonance data, ethylsulphates, 306
——— trichlorides, 307
—— double nitrates, 322
— shielding, 167
— susceptibility, 585
parity, 642, 660, 687
partition function, 848
Pauli matrices, 617, 651–2, 691, 799–800, 804, 810–1
— principle, 590
Pd^{3+} in Al_2O_3, 487
permutation group S_4, 641
phase angle, 116–7
— coherence, 110–2
phonon avalanche, 70, 574–83
— -bath relaxation, 582
—— time, 575
— bottle neck, 69, 566, 574–83

phonon effects on g-values, 312, 334
— energy density, 548, 554–5, 558, 580
— heat capacity, 574, 580
— -induced transitions, 465
— interactions, 51
— line width, 575
— mean free path, 574–5
— occupation number, 548, 577
— -phonon coupling, 582
— radiation bath, 63, 547–51
— — — energy density, 547
— — field, 491, 547
— scattering, 574
— system, specific heat, 549–51
— temperature, 548, 576
— wavelength, 551
platinum group, 598
plutonyl sodium acetate, 364
Pm^{3+} in C_{3h} symmetry, 315
point charge approximation, 665, 743
— — model, 762
— defects, 39
— dipolar interaction, 260, 263–4
polarization effects, 665
— rules, 30
polarized nuclear targets, 84
population inversion, 70, 83, 221
potential energy, 795, 837
powder spectrum, 200–5
power absorbed, 120, 125
— saturation, 62, 72, 80, 88–90
Pr^{2+} in CaF_2, 328
Pr^{3+} in C_{3h} symmetry, 312
— in C_{3v} symmetry, 324
— in CaF_2, 328, 351
— in double nitrate, 733
— in $LaCl_3$, 499
— in yttrium ethylsulphate, 216
precessing magnetization, 104–5, 116, 118, 122
precession, 9, 97
— frequency, 10
— magnetic dipole, 2
— velocity, 3, 95
pseudo-magnetic hyperfine interaction, 38, 257–8, 345, 717
pseudo-nuclear electric quadrupole interaction, 37–9, 257–9, 316, 349, 417, 715, 717
— Zeeman effect, 38–9, 164, 257–8, 267, 316, 354, 707, 715–6, 732, 749, 755
— — interaction, 31, 91, 167, 228, 254, 416, 851
Pt^{3+} in Al_2O_3, 487
— in $BaTiO_3$, 487
— in yttrium aluminium garnet, 478
PuF_6, 346
pulse 90 degree, 113

— 180 degree, 113–4
pulses, 113
$(PuO_2)^{2+}$, 347–8
— in $(UO_2)Rb(NO_3)_3$, 363
π-bonding, 379, 424
— orbitals, 392, 477, 763

quadrupole–quadrupole interaction, 501–2
quantum-mechanical tunnelling effect, 463
quenching factors, 840, 844, 847
— orbital momentum, 51, 403, 840, 846
— spin–orbit coupling, 840

$\langle r^{-3} \rangle$ values, 300, 710–1
Racah operators, 628
— parameters B, 386–7, 392, 395, 476–7, 479, 772
— — $(3B + C)$, 478
Racah parameters C, 386–7, 392, 394–5, 479
— — reduction of, 395
— 6j-symbols, 622
Raman relaxation process, 63–7, 554–7, 560, 562–6, 568, 571–2, 582, 649, 828–30
— — at high temperatures, 556–7, 572
rare-earth ethylsulphates, 287–8, 303–20
— group, 276–345, 559, 597, 670, 674, 676–7, 684, 694, 750, 761
— — cubic symmetry, 719–32
$Rb(UO_2)(NO_3)_3$, 75, 348, 360–4
Re^{4+} in K_2PtCl_6, 480
reducible representation, 603, 605, 607, 612–3, 641
regular octahedron, 369–70
relative intensities, 24, 159, 726
— function of temperature, 141, 151, 160, 162–3, 429, 505–7, 534, 538
relativistic contributions, 47, 253–4, 267, 299–300
relaxation effects, 108
— paths, 79, 81–2, 85, 88–9, 219, 243, 246–7, 272
— processes, 248
— rates, 247
restricted Hartree–Fock functions, 704, 711
Rh^0 in $AgCl$, 488
Rh^{2+} in $ZnWO_4$, 487
rhombic group, 639, 654
— symmetry, 750
rotating coordinate system, 95, 103, 105, 107, 112–4
— magnetic field, 97, 120
rotation group, 600, 604, 607, 616–27, 629, 634, 651, 721, 764
— of axes, 137

SUBJECT INDEX

rotational invariance, 587
— symmetry, 589, 596
Ru^{3+} in Al_2O_3, 484
— in $Co(NH_3)_6Cl_3$, 482–4
rubidium aluminium sulphate, 162
Russell–Saunders coupling, 587

$S = \frac{3}{2}$, rhombic symmetry, 155–6
$S = \frac{5}{2}$, axial symmetry, 160
— cubic field, 142–8
— zero magnetic field, 438
$S = \frac{5}{2}, \frac{7}{2}$, 142–51
$S = \frac{7}{2}$, cubic field, 147–51
$S = 1$, 152–5
— rhombic field, 153–5
$\tilde{S} = 1$, 19–22
— in strong field, 22
$S = 2$, weak-field Zeeman effect, 209–13
s-configurations, 695
^6S, cubic field splitting, 441–2
^6S, g-value, 441
S-state ions, 573
satellite transitions, 198
saturation, 79, 81, 123–4, 220, 240–1, 244, 270
— effect on Endor line width, 249
— parameter, 118–9
Sc^{2+} in CaF_2, 469, 831–2
— in SrF_2, 469, 831–2
second moment, 55, 516, 518–20, 522–3, 526, 528, 530
— — dipolar interaction, 521
— — dissimilar ions, 520
— -order effects, 29
— — Zeeman effect, 210, 291
selection rules, 99, 660–1, 726
— time reversal, 656
self-consistent field, 590
semiconductors, 39
shape function, 100, 118, 120
shielding effects, 48, 55, 259, 304, 491, 501, 597, 687
sign of D, 23, 161
— g, 15, 138–9
— hyperfine parameters, 225–8, 233, 262
similarity transformation, 602, 608
singlet orbital ground state, 401, 745
— -triplet transitions, 503
six-j symbol, 623, 675–6
sixfold symmetry, 718
Slater determinant, 590–3, 666, 670, 672–5, 698–9, 702–3, 705–6, 761, 769, 771–2, 777, 784–5, 788
Sm^{2+} in SrF_2, 283
Sm^{3+} in C_{3h} symmetry, 315
— in C_{3v} symmetry, 324
solid effect, 84
— state maser, 73

spatial rotation S, 652–3, 655, 696, 741
— symmetry, 760
specific heat, 852, 855
— — constant field, 849–50, 853
— — — magnetization, 849, 853
— representations, 635–6, 640, 661
spectrometer sensitivity, 93, 125–132
spin-bath relaxation, 69
spin coupling, energy, 379
— densities, 397, 408, 410, 703–4, 706, 712, 778, 780
— -dependent forces, 587–9
— diffusion, 71, 265, 273
spin echoes, 113–5
— flips, 71, 111–3, 199, 520, 522, 830
spin-Hamiltonians, 93, 133–216, 340, 372, 650–6, 746, 749–51, 756, 848, 851
— — ^2E state, 817
— — signs of parameters, 223
spin-lattice interaction, 60–74, 123
— — relaxation, 41, 61, 69, 73, 78, 92–4, 828, 830
— — — concentration dependence, 553
— — — ligand field modulation, 557–74
— — — thermodynamic approach, 545
— — — time, 61, 109, 117, 129, 465, 535–6, 544, 732
— — — Waller process, 551–7
— magnetic moment, 586
— magnetization, 782
— operators, 140–1, 287–8, 371–2, 377
spin-orbit coupling, 4, 42, 50, 283–4, 347, 371, 377–8, 383, 398–400, 402–6, 419, 423, 428, 431, 433–5, 448, 455, 491, 495–9, 508, 557–8, 592–3, 597–8, 648, 673, 679, 745–7, 753, 756, 758–9, 775–7, 782, 784, 816, 829, 831, 833–5, 839–41, 847
— — reduction in, 395–6, 400, 432, 477, 775–7, 835, 839–41
spin-orbit coupling parameters, $3d$, 378, 391, 399–400
— — — $4d$, 474
— — — $5d$, 475
— — — $4f$, 284
— — — $5f$, 347
spin packet, 54, 56, 72, 92, 112, 130, 236–7, 244, 249, 264, 269
— quadruplet, 430
— quintuplet, 434
spin-spin interactions, 41–2, 52–60, 93, 109, 123, 271, 398–402, 405–6, 414, 430, 433–5, 442, 444, 546, 593–4, 648, 678–9, 745, 748–9, 753, 759, 851, 853–5
— — axial symmetry, 525–7
— — modulated by phonons, 552–7
— — parameter ρ, 399, 678–9
— — relaxation, 71
— — time, 264, 522, 535

SUBJECT INDEX

spin-spin interaction, temperature dependent effects, 535–40
spin system internal energy, 544
— — specific heat, 549–51
— temperature, 61, 71, 117–8, 123, 273, 545, 576
— wave theory, 532
spinels, 468
spontaneous emission, 60, 70, 544, 560
$SrCl_2$, 281
SrF_2, 281, 325, 467
$SrWO_4$, 280, 467
Stark splitting, 12
static crystal field, 839–40
— Jahn–Teller distortion, 846
— — effect, 459, 462, 470, 490, 790, 804–8, 821, 825–7, 829, 836, 840–1, 844
— — — dynamic features, 807–8
— magnetic saturation, 129, 537
steady-state conditions, 115
— methods, 566
— solution, 105
Sternheimer shielding, 302
stimulated absorption, 580
— emission, 70, 74, 580
strain-broadened spectrum, 831
— effects, 429, 816–20
— Hamiltonians, 829
— -induced energy, 826
strains, large, 818–20
— random, 213, 817
— very small, 818
stress constant V_{ES}, 817
strong crystal field approach, 377–92, 598
— field Zeeman effect, 146–7, 156–63
— Jahn–Teller coupling, 807
— ligand field, 45, 48, 371, 407–8
— — octahedral complex, 476–8
strongly allowed transitions, 32
susceptibility, 851–2, 854
symmetric direct product, 657, 722
— tensor, 653, 750
σ-bonding orbitals, 392, 763
σ-bonds, 359, 379

$(t_2)^6$ sub-shell, 477
Tb^{3+} in C_{3h} symmetry, 317
— in ethylsulphate, 738–9
— in yttrium ethylsulphate, 567
Tb^{4+} in ThO_2, 335–7, 340, 343
Tc^{4+}, 480
— in K_2PtCl_6, 479
temperature fluctuations, 131
— -independent susceptibility, 559
tensor components, 839
— G_{p_q}, 652
— $g_{q\alpha}$, 651–5, 713
— operators, 670, 672, 681, 688, 799–801

tetragonal distortion, 367, 369, 372–5, 729–30, 755, 826, 839
— group, 638–9, 641–2
— — character table, 861
— symmetry, 751, 756, 801–2
tetrahedral coordination, 369, 374, 743
— symmetry, 661–4
— — group T_d, 641–2
thermal analysis, classical, 583
— bath, 574
— equilibrium, 61–2, 106–7, 113–4, 120, 541–7, 583
— magnetic properties, 848–55
— noise power, 131
— relaxation, 828
— reservoirs, 583
thermodynamic potential, 848
third-order Zeeman effect, 35, 291–2, 425
ThO_2, 282, 325, 331
threefold symmetry, 654–6, 718
Ti^{2+} in ZnS, 469
Ti^{3+} in Al_2O_3, 417, 425, 847
Ti^{3+}, V^{4+} octahedral field, 417
time-conjugate states, 563
— -dependent perturbation, 99
— -even operator, 647–8, 716–7, 734, 747, 835
— -odd operators, 342–3, 647–8, 651, 717, 733
— reversal, 12, 141, 313, 558–9, 643–64, 682, 688, 716, 721, 725, 731, 733–4, 747–8, 750, 760
— operator, 646–50, 657
Tm^{2+} in CaF_2, 263, 271, 327–8, 332–4, 337
— in cubic symmetry, 333
— in SrF_2, 328
$^{169}Tm^{2+}$ in CaF_2, 269
Tm^{3+} in C_{3h} symmetry, 320
$Tm^{3+}(4f^{12}, {}^3H_6)$, 674
transient methods, 566
transition groups, 40
— probability, 100, 102, 122, 136, 564, 734–5
— rate, 60, 119, 249, 542
transitions, satellite, 198–9
transmission spectrometer, 127
transverse relaxation time, 109, 114
triad spectra, 532
trigonal distortion, 369–70, 372, 374, 726, 729–30, 755, 839
— — reduction, 841
— field parameter v, 391
— fields, 374, 376, 745, 755
— group character table, 861
— — D_3, 639–40
— — irreducible representations, 861
— strain, 816
— symmetry C_3, 654–6, 751, 756, 846

triplet orbital ground state, 403–6, 751
— t_2, 379, 769–70
— T_1, T_2, 770
tripositive ions in Al_2O_3, 389–91
truncated Hamiltonian, 517
tunnelling, 820–6, 844
two-dimensional harmonic oscillator, 836–7
— -electron operators, 595, 666
— -quantum transition, 454
— -valued representation, 619
τ_1 angular dependence, 570–1
— measurement, 566
τ_b, spin–bath relaxation time, 576–83

U^{3+} in $LaCl_3$, 349
U^{2+}, U^{3+}, U^{4+} in CaF_2, 350–1
UF_6, 346
ultimate sensitivity, 126
unimodular group, 634
unitary matrix, 617
— operator, 617
— transformation, 644
— unimodular group, 618
unnormalized spherical harmonics, 628
unrestricted Hartree–Fock method, 704, 711
UO_2, 501
$(UO_2)^{2+}$, 347–8, 359
— in $(UO_2)Rb(NO_3)_2$, 360
$(UO_2)Na(C_2H_3O_2)_3$, 348
$(UO_2)Rb(NO_3)_3$, 348, 360

V^{2+} in CaF_2, 846
— in complex cyanides, 479
— in MgO, 431
V^{2+}, $Cr^{3+}(d^3)$, 779–80
V^{2+}, Cr^3, Mn^4 in octahedral field, 430–4
$^{51}V^{2+}$ in Al_2O_3, 434
— in cubic ZnS, 469
V^{3+}, 383
— in α-Al_2O_3, 429
— in ZnS, 469
V^{3+}, Cr^4 in octahedral field, 426–36
V^{4+} in Al_2O_3, 417, 425–6
Van Vleck cancellation, 564, 649
— paramagnetism, 1
vanadium fluosilicate, 191
vector coupling, 622
vibronic doublet, 827, 829, 831, 839
— functions, 818, 824
— Γ_3 doublet, 810–6, 815
— Hamiltonian, 812, 835, 838, 841–2
— operators, 839
— singlets, 810, 816, 826
— states, 809–42
virtual phonon coupling, 60
— — interaction, 502
$V(NH_4)(SO_4)_2$, $12H_2O$, 426–8

W_{JT}, Jahn–Teller energy, 805
Waller process, 552–7
warping, 813, 816, 820
— terms, 811, 814–5, 838
weak crystal field, 597, 713–9, 761
— field Zeeman interaction, 165
— — — effect, 143–6, 209
— ligand field, 43, 45, 407
weakly allowed transitions, 30
— covalent compounds, 767
Wigner coefficients, 620–3, 625
Wigner–Eckart theorem, 586, 593, 624, 653, 658–9, 670, 675–6, 682, 721–2, 759, 788, 810, 832, 835
WKB approximation, 824
wurtzite, 467

XY_6 complex, 804
— — e_g orbitals, 392–4
— — ligand orbitals, 877
— — molecular orbital combinations, 877
— — t_2 orbitals, 392–4
— octahedral complex, normal coordinates, 878

Y^{2+} in CaF_2, 489–90
Yb^{3+} in CaF_2, 263, 332–5, 344–5
— in CaO, 333–4
— in CdF_2, ThO_2, CeO_2, 333
— in C_{3h} symmetry, 320
— in cubic symmetry, 333–5
yttrium ethylsulphate, 67

Zeeman effect, in free atom, 586–9
— energy, 849
— Hamiltonians, 723, 801
— interaction, 11, 402, 406, 586, 677, 745, 748, 758, 840
zero field measurements, 161, 163
— — spectrum, 153
— magnetic field for $S = \frac{5}{2}$, 438
— -point energy, 807
— — motion, Jahn–Teller effect, 464–5
— — vibrations, 541, 721
ZnO, 467–8
ZnS, 467–8
ZnSe, 467
ZnTe, 467

QC
762
A27

OCT 2 9 1971